STUDENT'S SOLUTIONS MANUAL

MATH MADE VISIBLE, LLC

ELEMENTARY AND INTERMEDIATE ALGEBRA: GRAPHS AND MODELS

FOURTH EDITION

Marvin L. Bittinger

Indiana University Purdue University Indianapolis

David J. Ellenbogen

Community College of Vermont

Barbara L. Johnson

Indiana University Purdue University Indianapolis

PEARSON

Boston San Francisco New York
London Toronto Sydney Tokyo Singapore Madrid
Mexico City Munich Paris Cape Town Hong Kong Montreal

Copyright © 2012, 2008, 2004 Pearson Education, Inc.
Publishing as Pearson, 75 Arlington Street, Boston, MA 02116.

ISBN-13: 978-0-321-72660-5
ISBN-10: 0-321-72660-X

2 3 4 5 6 EBM 16 15 14 13

www.pearsonhighered.com

PEARSON

Table of Contents

Chapter 1

Introduction to Algebraic Expressions

1. Expression

3. Equation

5. Equation

7. Expression

9. Equation

11. Expression

13. Substitute 9 for a and multiply.
$3a = 3 \cdot 9 = 27$

15. Substitute 2 for t and add.
$t + 6 = 2 + 6 = 8$

17. Substitute 2 for x, 14 for y. Add and then divide the sum by 4.
$\frac{x+y}{4} = \frac{2+14}{4} = \frac{16}{4} = 4$

19. Substitute 20 for m, 6 for n, subtract and then divide the difference by 2.
$\frac{m-n}{2} = \frac{20-6}{2}$
$= \frac{14}{2} = 7$

21. Substitute 6 for m and 18 for q.
$\frac{9m}{q} = \frac{9 \cdot 6}{18} = \frac{54}{18} = 3$

23. Enter the expression in the graphing calculator, replacing a with 136 and b with 13. We see that $27a - 18b = 3438$ for $a = 136$ and $b = 13$.

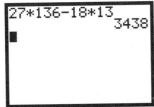

25. $A = bh$
$= (6\text{ft})(3\frac{1}{2}\text{ft})$
$= (6)(3\frac{1}{2})(\text{ft})(\text{ft})$
$= 21\text{ft}^2$, or 21 square feet

27. $A = \frac{1}{2} \cdot bh$
$= \frac{1}{2}(5\text{cm})(6\text{cm})$
$= \frac{1}{2}(5)(6)(\text{cm})(\text{cm})$
$= 15\text{cm}^2$, or 15 square centimeters

29. $A = bh$
$= (67\,\text{ft})(12\,\text{ft})$
$= (67)(12)(\text{ft})(\text{ft})$
$= 804\,\text{ft}^2$, or 804 square feet

31. Let r represent Ron's age. 5 more than Ron's age, or Ron's age plus 5 can be expressed as $5 + r$, or $r + 5$

33. Product is the result of multiplication, so we have $a \cdot 4$, or $4a$

35. 9 less than c; $c - 9$

37. 6 increased by q; to increase is to add; $6 + q$, or $q + 6$

39. The difference of m and n; $m - n$

41. x less than y, or decrease y by x; $y - x$

43. x divided by w (Note: x is the dividend, or numerator, and w is the divisor, or denominator); $\frac{x}{w}$, or $x \div w$

45. Let l and h represent the box's length and height, respectively. We have the sum $l + h$ or $h + l$.

47. Let p represent Panya's speed and let w represent the speed of the wind. $p - 2w$

49. Let y represent "some number";
$\frac{1}{4}y - 12$, or $\frac{y}{4} - 12$

51. Let a and b represent the numbers; $8(a - b)$

53. $64\% = 0.64$, so 64% of w, where $w =$ the number of women, is $0.64w$. (Note: replace "of" with multiplication)

55. $\underline{x + 17 = 32}$ Writing the equation

$15 + 17 \,?\, 32$ Substituting 15 for x
$\quad 32 \mid 32$ $32 = 32$ is TRUE
Since the left-hand and right-hand sides are the same, 15 is a solution.

57. $\underline{a - 28 = 75}$ Writing the equation

$93 - 28 \,?\, 75$ Substituting 93 for a
$\quad 65 \mid 75$ $65 = 75$ is FALSE
Since the left-hand and right-hand sides are not the same, 93 is not a solution.

59. $\frac{t}{7} = 9$

$\frac{63}{7} \,?\, 9$
$\quad 9 \mid 9$ $9 = 9$ is TRUE
Since the left-hand and right-hand sides are the same, 63 is a solution.

61. $\frac{108}{x} = 36$

$\frac{108}{3} \,?\, 36$
$\quad 36 \mid 36$ $36 = 36$ is TRUE
Since the left-hand and right-hand sides are the same, 3 is a solution.

63. Let x represent the number.
$\qquad$ 7 times $\underline{\text{what number}}$ is 1596?
$\qquad\qquad \downarrow \quad \downarrow \qquad \downarrow \qquad \downarrow \;\; \downarrow$
Translating: 7 $\quad \cdot \qquad x \qquad = 1596$
$7x = 1596$

65. Let x represent the number.
$\quad$ Rewording: 42 times $\underline{\text{what number}}$ is 2352?
$\qquad\qquad\qquad \downarrow \;\; \downarrow \qquad\quad \downarrow \qquad \downarrow \;\; \downarrow$
Translating: 42 $\quad \cdot \qquad\quad x \qquad = 2352$
$42x = 2352$

67. Let s represent the number of unoccupied squares.

Rewording: $\underbrace{\text{Number of occupied}}$ plus $\underbrace{\text{number of unoccupied}}$ is 64.
$\qquad\qquad\quad \downarrow \qquad\quad \downarrow \qquad\quad \downarrow \qquad \downarrow \;\; \downarrow$
Translating: $\quad 19 \qquad\quad + \qquad\quad s \qquad = \; 64$
$19 + s = 64$

69. Let w represent the total amount of waste generated, in millions of tons.

Rewording: 33.4% of $\underbrace{\begin{array}{c}\text{the total amount of waste}\end{array}}$ is $\begin{array}{c}85\text{ million tons.}\end{array}$
$\qquad\qquad\quad \downarrow \;\; \downarrow \qquad \downarrow \qquad \downarrow \qquad \downarrow$
Translating: $33.4\% \;\cdot \qquad w \qquad = \qquad 85$
$33.4\% \, w = 85$, or $0.334 \, w = 85$

71. f

73. d

75. g

77. e

79. Look for a pattern in the data. Observe that the daily recommended number of grams of dietary fiber for each age is 5 more than that age. We reword and translate.

$\underline{\text{Grams of fiber}}$ is 5 $\underline{\text{more than}}$ the age
$\qquad \downarrow \qquad\quad \downarrow \; \downarrow \qquad \downarrow \qquad\quad \downarrow$
$\qquad f \qquad\quad = \; 5 \qquad + \qquad\quad a$

We have $f = 5 + a$. This equation could also be written as $f = a + 5$.

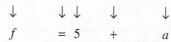

81. Look for a pattern in the data. Observe that it costs $2.21 more to process nonmachinable packages than to process machinable packages. We reword and translate.

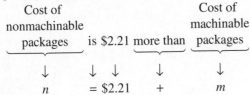

Cost of nonmachinable packages	is $2.21	more than	Cost of machinable packages
↓	↓ ↓	↓	↓
n	= $2.21	+	m

We have $n = \$2.21 + m$. This equation could also be written as $n = m + 2.21$.

83. Observe that the number of vehicle miles traveled is 10,000 times the number of drivers. Reword and translate.

Number of Vehicle miles	is	10,000	times	Number of drivers
↓	↓	↓	↓	↓
v	=	10,000	·	d

We have $v = 10,000d$.

85. *Thinking and Writing Exercise.*

87. *Thinking and Writing Exercise.*

89. The sign is a triangle whose area is
$$\frac{1}{2} \cdot b \cdot h, \text{ or } \frac{1}{2}(3\,\text{ft})(2.5\,\text{ft}) = 3.75\,\text{ft}^2$$
The cost is $120 for each square foot.
Thus, $C = \$120(3.75)$
$$= \$450$$

91. When y is twice x and $x = 6$,
$$y = 2 \cdot 6 = 12$$
$$\frac{x+y}{2} = \frac{6+12}{2}$$
$$= \frac{18}{2} = 9$$

93. The next whole number is one more than $w + 3$:
$$w + 3 + 1 = w + 4$$

95. Adding the length of the sides, we have:
$$l + w + l + w, \text{ or } 2l + 2w$$

97. Ella's time is 5 more than Kyle's time, or $E = 5 + K$. Kyle's time is 3 more than Ava's time (t), or $K = 3 + t$.
Substituting $3 + t$ for K into $E = 5 + K$
We have: $5 + (3 + t)$, or $8 + t$,
or $t + 8$.
Also: Kyle is 3 more than Ava, or $t + 3$ and Ella is 5 more than Kyle, or $(t + 3) + 5 = t + 8$

99. *Thinking and Writing Exercise.*

Exercise Set 1.2

1. commutative

3. associative

5. distributive

7. associative

9. commutative

For exercises 11–18, change the order of the *addends*.

11. $x + 7$

13. $3y + x$

15. $c + ab$

17. $5(1 + a)$

For exercises 19–26, change the order of the *factors*.

19. $a \cdot 2$

21. ts

23. $5 + ba$

25. $(a + 1)5$

For exercises 27–32, regroup the *addends*.

27. $a + (5 + b)$

29. $(r + t) + 7$

31. $ab + (c + d)$

For exercises 33–38, regroup the *factors*.

33. $7(mn)$

35. $(2a)b$

37. $(3 \cdot 2)(a + b)$

39. $2 + (t + 6) = 2 + (6 + t)$ Commutative law of addition

          $= (2 + 6) + t$ Associative law of addition

          $= 8 + t$ Simplification

Answers may vary.

41. $(3a) \cdot 7 = (a \cdot 3) \cdot 7$ Commutative law of multiplication

          $= a(3 \cdot 7)$ Associative law of multiplication

          $= a \cdot 21$ or $21a$ Simplification

Answers may vary.

43. $(5 + x) + 2$

     $= (x + 5) + 2$ Commutative law

     $= x + (5 + 2)$ Associative law

     $= x + 7$ Simplifying

45. $(m \cdot 3)7 = m(3 \cdot 7)$ Associative law

     $= (3 \cdot 7)m$ Commutative law

     $= 21m$ Simplification

47. $4(a + 3) = 4 \cdot a + 4 \cdot 3$

           $= 4a + 12$

49. $6(1 + x) = 6 \cdot 1 + 6 \cdot x$

           $= 6 + 6x$

51. $(n + 5)2 = n \cdot 2 + 5 \cdot 2$

         $= 2n + 10$

53. $8(3x + 5y) = 8 \cdot 3x + 8 \cdot 5y$

             $= 24x + 40y$

55. $9(2x + 6) = 9 \cdot 2x + 9 \cdot 6$

           $= 18x + 54$

57. $5(r + 2 + 3t) = 5 \cdot r + 5 \cdot 2 + 5 \cdot 3t$

             $= 5r + 10 + 15t$

59. $(a + b)2 = a \cdot 2 + b \cdot 2$

          $= 2a + 2b$

61. $(x + y + 2)5 = x \cdot 5 + y \cdot 5 + 2 \cdot 5$

             $= 5x + 5y + 10$

For exercises 63–66, the terms are separated by the addition/plus signs.

63. x, xyz, 19

65. $2a$, $\dfrac{a}{b}$, $5b$

67. $2a + 2b = 2 \cdot a + 2 \cdot b$ The common factor is 2

         $= 2(a + b)$ Using the distributive law

Check:

$2(a + b) = 2 \cdot a + 2 \cdot b = 2a + 2b$

69. $7 + 7y = 7 \cdot 1 + 7 \cdot y$ The common factor is 7

         $= 7(1 + y)$ Using the distributive law

Check:

$7(1 + y) = 7 \cdot 1 + 7y = 7 + 7y$

71. $18x + 3 = 3 \cdot 6x + 3 \cdot 1$ The common factor is 3.

         $= 3(6x + 1)$ Using the distributive law

Check:

$3(6x + 1) = 3 \cdot 6x + 3 \cdot 1 = 18x + 3$

73. $5x + 10 + 15y = 5 \cdot x + 5 \cdot 2 + 5 \cdot 3y$

               $= 5(x + 2 + 3y)$

Check:

$5(x + 2 + 3y) = 5 \cdot x + 5 \cdot 2 + 5 \cdot 3y$

             $= 5x + 10 + 15y$

75. $12x + 9 = 3 \cdot 4x + 3 \cdot 3$
$$= 3(4x + 3)$$
Check:
$$3(4x + 3) = 3 \cdot 4x + 3 \cdot 3$$
$$= 12x + 9$$

77. $3a + 9b = 3 \cdot a + 3 \cdot 3b$
$$= 3(a + 3b)$$
Check:
$$3(a + 3b) = 3a + 3 \cdot 3b$$
$$= 3a + 9b$$

79. $44x + 88y + 66z = 22 \cdot 2x + 22 \cdot 4y + 22 \cdot 3z$
$$= 22(2x + 4y + 3z)$$
Check:
$$22(2x + 4y + 3z) = 22 \cdot 2x + 22 \cdot 4y + 22 \cdot 3z$$
$$= 44x + 88y + 66z$$

81. s and t

83. 3 and $(x + y)$

85. 7, a, and b

87. $(a - b)$ and $(x - y)$

89. Terms are separated by addition and subtraction signs. A single term may include plus or minus signs only if they are inside parentheses, brackets, or other grouping symbols.
Factors are numbers, variables, or expressions that are multiplied together.

91. Let k represent Kylie's salary. Then we have $\frac{1}{2}k$, or $\frac{k}{2}$.

93. *Thinking and Writing Exercise.*

95. The expressions are equivalent by the distributive law.
$$8 + 4(a + b) = 8 + 4 \cdot a + 4 \cdot b$$
$$= 4 \cdot 2 + 4 \cdot a + 4 \cdot b$$
$$= 4(2 + a + b)$$

97. The expressions are not equal. For example, let $m = 1$. Then we have
$$7 \div 3 \cdot 1 = \frac{7}{3} \cdot 1 = \frac{7}{3}, \text{ but}$$
$$3 \cdot 1 \div 7 = 3 \div 7 = \frac{3}{7}.$$

99. The expressions are not equivalent.
$$5[2(x + 3y)] = 5 \cdot 2(x + 3y)$$
$$= 10(x + 3y)$$
$$= 10 \cdot x + 10 \cdot 3y$$
$$= 10x + 30y$$
$$= 30y + 10x$$

101. *Thinking and Writing Exercise.*

Exercise Set 1.3

1. b

3. d

5. The paired factors of 50 are:
$1 \cdot 50$, $2 \cdot 25$, and $5 \cdot 10$
The factors of 50 are:
1, 2, 5, 10, 25, 50.

7. The paired factors of 42 are:
$1 \cdot 42$, $2 \cdot 21$, $3 \cdot 14$, and $6 \cdot 7$
The factors of 42 are:
1, 2, 3, 6, 7, 14, 21, 42

9. Factors of 21 are 1, 3, 7, 21. Composite.

11. The factors of 31 are: 1 and 31. Prime.

13. The factors of 25 are: 1, 5, 25. Composite.

15. The factors of 2 are: 1 and 2. Prime.

17. 0 is neither prime nor composite.
(Note: The number must be a counting number *greater* than 1).

19. The factors of 40 are: 1, 2, 4, 5, 8, 10, 20, 40. Composite.

21. $26 = 2 \cdot 13$

23. $30 = 2 \cdot 15 = 2 \cdot 3 \cdot 5$

25. $27 = 3 \cdot 9 = 3 \cdot 3 \cdot 3$

27. $40 = 2 \cdot 20 = 2 \cdot 2 \cdot 10 = 2 \cdot 2 \cdot 2 \cdot 5$

29. 43; Prime

31. $210 = 2 \cdot 105 = 2 \cdot 3 \cdot 35 = 2 \cdot 3 \cdot 5 \cdot 7$

33. $115 = 5 \cdot 23$

35. $\dfrac{14}{21} = \dfrac{2 \cdot 7}{3 \cdot 7}$ Factoring numerator and denominator

$\quad = \dfrac{2}{3} \cdot \dfrac{7}{7}$ Rewriting as a product of two fractions

$\quad = \dfrac{2}{3} \cdot 1$ $\dfrac{7}{7} = 1$

$\quad = \dfrac{2}{3}$ Using the identity property of 1

37. $\dfrac{16}{56} = \dfrac{2 \cdot 2 \cdot 2 \cdot 2}{2 \cdot 2 \cdot 2 \cdot 7}$

$\quad = \dfrac{2}{2} \cdot \dfrac{2}{2} \cdot \dfrac{2}{2} \cdot \dfrac{2}{7} = \dfrac{2}{7}$

$\quad = 1 \cdot 1 \cdot 1 \cdot \dfrac{2}{7} = \dfrac{2}{7}$

$\quad = \dfrac{2}{7}$

39. $\dfrac{6}{48} = \dfrac{\cancel{6}}{\cancel{6} \cdot 8} = \dfrac{1}{8}$

41. $\dfrac{52}{13} = \dfrac{\cancel{13} \cdot 4}{\cancel{13} \cdot 1} = 4$

43. $\dfrac{19}{76} = \dfrac{\cancel{19}}{4 \cdot \cancel{19}} = \dfrac{1}{4}$

45. $\dfrac{150}{25} = \dfrac{6 \cdot \cancel{25}}{\cancel{25}} = 6$

47. $\dfrac{42}{50} = \dfrac{\cancel{2} \cdot 21}{\cancel{2} \cdot 25} = \dfrac{21}{25}$

49. $\dfrac{120}{82} = \dfrac{\cancel{2} \cdot 60}{\cancel{2} \cdot 41} = \dfrac{60}{41}$

51. $\dfrac{210}{98} = \dfrac{\cancel{2} \cdot 3 \cdot 5 \cdot \cancel{7}}{\cancel{2} \cdot 7 \cdot \cancel{7}} = \dfrac{15}{7}$

53. $\dfrac{1}{2} \cdot \dfrac{3}{7} = \dfrac{1 \cdot 3}{2 \cdot 7} = \dfrac{3}{14}$

55. $\dfrac{12}{5} \cdot \dfrac{10}{9} = \dfrac{12 \cdot 10}{5 \cdot 9} = \dfrac{\cancel{3} \cdot 4 \cdot \cancel{5} \cdot 2}{\cancel{5} \cdot \cancel{3} \cdot 3} = \dfrac{4 \cdot 2}{3} = \dfrac{8}{3}$

57. $\dfrac{1}{8} + \dfrac{3}{8} = \dfrac{1+3}{8} = \dfrac{4}{8} = \dfrac{\cancel{4}}{\cancel{4} \cdot 2} = \dfrac{1}{2}$

59. $\dfrac{4}{9} + \dfrac{13}{18} = \dfrac{2}{2} \cdot \dfrac{4}{9} + \dfrac{13}{18}$ Use 18 as the common denominator

$\quad = \dfrac{8}{18} + \dfrac{13}{18} = \dfrac{21}{18} = \dfrac{\cancel{3} \cdot 7}{\cancel{3} \cdot 6}$

$\quad = \dfrac{7}{6}$

61. $\dfrac{3}{a} \cdot \dfrac{b}{7} = \dfrac{3 \cdot b}{a \cdot 7} = \dfrac{3b}{7a}$

63. $\dfrac{4}{a} + \dfrac{3}{a} = \dfrac{4+3}{a} = \dfrac{7}{a}$

65. $\dfrac{3}{10} + \dfrac{8}{15} = \dfrac{3}{3} \cdot \dfrac{3}{10} + \dfrac{2}{2} \cdot \dfrac{8}{15}$ Use 30 as the common denominator

$\quad = \dfrac{9}{30} + \dfrac{16}{30} = \dfrac{9+16}{30} = \dfrac{25}{30}$

$\quad = \dfrac{\cancel{5} \cdot 5}{\cancel{5} \cdot 6} = \dfrac{5}{6}$

67. $\dfrac{9}{7} - \dfrac{2}{7} = \dfrac{9-2}{7} = \dfrac{7}{7} = 1$

69. $\dfrac{13}{18} - \dfrac{4}{9} = \dfrac{13}{18} - \dfrac{2}{2} \cdot \dfrac{4}{9}$ Use 18 as the common denominator

$\quad = \dfrac{13}{18} - \dfrac{8}{18}$

$\quad = \dfrac{13-8}{18} = \dfrac{5}{18}$

71. $\dfrac{20}{30} - \dfrac{2}{3} = \dfrac{2 \cdot \cancel{10}}{3 \cdot \cancel{10}} - \dfrac{2}{3}$

$\quad = \dfrac{2}{3} - \dfrac{2}{3} = 0$

73. $\dfrac{7}{6} \div \dfrac{3}{5} = \dfrac{7}{6} \cdot \dfrac{5}{3}$ Multiply by the reciprocal
of the divisor

$= \dfrac{7 \cdot 5}{6 \cdot 3}$

$= \dfrac{35}{18}$

75. $\dfrac{8}{9} \div \dfrac{4}{15} = \dfrac{8}{9} \cdot \dfrac{15}{4}$

$= \dfrac{8 \cdot 15}{9 \cdot 4}$

$= \dfrac{2 \cdot \cancel{4} \cdot \cancel{3} \cdot 5}{\cancel{3} \cdot 3 \cdot \cancel{4}}$

$= \dfrac{10}{3}$

77. $12 \div \dfrac{3}{7} = 12 \cdot \dfrac{7}{3}$

$= \dfrac{\cancel{3} \cdot 4 \cdot 7}{\cancel{3}} = 28$

79. $\dfrac{7}{13} \div \dfrac{7}{13} = \dfrac{7}{13} \cdot \dfrac{13}{7} = \dfrac{\cancel{7} \cdot \cancel{13}}{\cancel{13} \cdot \cancel{7}} = 1$

81. $\dfrac{\frac{2}{7}}{\frac{5}{3}} = \dfrac{2}{7} \div \dfrac{5}{3} = \dfrac{2}{7} \cdot \dfrac{3}{5} = \dfrac{6}{35}$

83. $\dfrac{9}{\frac{1}{2}} = 9 \div \dfrac{1}{2} = 9 \cdot 2 = 18$

85. *Thinking and Writing Exercise.*

87. $5(x+3) = (x+3)5$
Commutative Law of Multiplication
$5(x+3) = 5(3+x)$
Commutative Law of Addition

89. *Thinking and Writing Exercise.*

91.

Product	56	63	36	72	140	96	168
Factor	7	7	2	36	14	8	8
Factor	8	9	18	2	10	12	21
Sum	15	16	20	38	24	20	29

93. $\dfrac{16 \cdot 9 \cdot 4}{15 \cdot 8 \cdot 12} = \dfrac{\cancel{4} \cdot \cancel{4} \cdot \cancel{3} \cdot \cancel{3} \cdot \cancel{2} \cdot 2}{\cancel{3} \cdot 5 \cdot \cancel{2} \cdot \cancel{4} \cdot \cancel{3} \cdot \cancel{4}} = \dfrac{2}{5}$

95. $\dfrac{27pqrs}{9prst} = \dfrac{3 \cdot \cancel{9} \cdot \cancel{p} \cdot q \cdot \cancel{r} \cdot \cancel{s}}{\cancel{9} \cdot \cancel{p} \cdot \cancel{r} \cdot \cancel{s} \cdot t} = \dfrac{3q}{t}$

97. $\dfrac{15 \cdot 4xy \cdot 9}{6 \cdot 25x \cdot 15y} = \dfrac{\cancel{15} \cdot \cancel{2} \cdot 2 \cdot \cancel{x} \cdot \cancel{y} \cdot \cancel{3} \cdot 3}{\cancel{2} \cdot \cancel{3} \cdot 25 \cdot \cancel{x} \cdot \cancel{15} \cdot \cancel{y}} = \dfrac{6}{25}$

99. $\dfrac{\frac{27ab}{15mn}}{\frac{18bc}{25np}} = \dfrac{27ab}{15mn} \div \dfrac{18bc}{25np} = \dfrac{27ab}{15mn} \cdot \dfrac{25np}{18bc} = \dfrac{27ab \cdot 25np}{15mn \cdot 18bc}$

$= \dfrac{\cancel{3} \cdot \cancel{9} \cdot a \cdot \cancel{b} \cdot \cancel{5} \cdot 5 \cdot \cancel{n} \cdot p}{\cancel{3} \cdot \cancel{5} \cdot m \cdot \cancel{n} \cdot 2 \cdot \cancel{9} \cdot \cancel{b} \cdot c} = \dfrac{5ap}{2cm}$

101. $\dfrac{5\frac{3}{4}rs}{4\frac{1}{2}st} = 5\dfrac{3}{4}rs \div 4\dfrac{1}{2}st$

$= \dfrac{23rs}{4} \div \dfrac{9st}{2} = \dfrac{23rs}{4} \cdot \dfrac{2}{9st}$

$= \dfrac{23r\cancel{s} \cdot \cancel{2}}{2 \cdot 2 \cdot 9\cancel{s} \cdot t} = \dfrac{23r}{18t}$

Note: $5\dfrac{3}{4} = 5 + \dfrac{3}{4} = \dfrac{4}{4} \cdot \dfrac{5}{1} + \dfrac{3}{4}$

$= \dfrac{20+3}{4} = \dfrac{23}{4}$

Similarly: $4\dfrac{1}{2} = \dfrac{2}{2} \cdot \dfrac{4}{1} + \dfrac{1}{2} = \dfrac{8+1}{2} = \dfrac{9}{2}$

103. $A = lw = \left(\dfrac{4}{5}\text{m}\right)\left(\dfrac{7}{9}\text{m}\right)$

$= \left(\dfrac{4}{5}\right)\left(\dfrac{7}{9}\right)(\text{m})(\text{m})$

$= \dfrac{28}{45}\text{m}^2$, or $\dfrac{28}{45}$ square meters

105. $P = 4s = 4\left(3\dfrac{5}{9}\text{m}\right)$

$= 4 \cdot \dfrac{32}{9}\text{m}$

$= \dfrac{128}{9}\text{m}$, or $14\dfrac{2}{9}\text{m}$

107. The cube has 12 edges, all with length of
$2\frac{3}{10}$ cm.

$$TL = 12 \cdot 2\frac{3}{10}$$

$$= 12 \cdot \frac{23}{10} = \frac{\cancel{2} \cdot 6 \cdot 23}{\cancel{2} \cdot 5}$$

$$= \frac{138}{5} = 27\frac{3}{5} \text{ cm}$$

Exercise Set 1.4

1. repeating

3. integer

5. rational number

7. natural number

9. 100 corresponds to the highest temperature recorded in Alaska. −80 corresponds to the lowest.

11. −777.68 corresponds to falling 777.68 points, and 936.42 corresponds to gaining 936.42 points.

13. −12,500 corresponds to taking on a debt of $12,500, and 5000 corresponds to receiving $5000 as a scholarship.

15. 8 corresponds to the 8 yard gain, and −5 corresponds to the 5 yard loss.

17. −10 corresponds to 10 seconds before liftoff, and 235 corresponds to 235 seconds after liftoff.

19. ![number line with point at −2]

21. ![number line with point at −4.3]

23. ![number line with point at 10/3]

25. $\frac{7}{8}$ means $7 \div 8$, so we divide.

$$\begin{array}{r} 0.875 \\ 8\overline{)7.000} \\ 64 \\ \hline 60 \\ 56 \\ \hline 40 \\ 40 \\ \hline 0 \end{array}$$

$\frac{7}{8} = 0.875.$

27. We first find decimal notation for $\frac{3}{4}$. Since $\frac{3}{4}$ means $3 \div 4$, we divide.

$$\begin{array}{r} 0.75 \\ 4\overline{)3.00} \\ 2\,8 \\ \hline 20 \\ 20 \\ \hline 0 \end{array}$$

$-\frac{3}{4} = -0.75.$

29. $-\frac{7}{6}$ means $-7 \div 6$, so we divide.

$$\begin{array}{r} 1.166 \\ 6\overline{)7.000} \\ 6 \\ \hline 10 \\ 6 \\ \hline 40 \\ 36 \\ \hline 40 \\ 36 \\ \hline 4 \end{array}$$

$-\frac{7}{6} = -1.1\overline{6}.$

31. $\frac{2}{3}$ means $2 \div 3$, so we divide.

$$\begin{array}{r} 0.666\ldots \\ 3\overline{)2.000} \\ 1\,8 \\ \hline 20 \\ 18 \\ \hline 20 \\ 18 \\ \hline 2 \end{array}$$

$\frac{2}{3} = 0.\overline{6}.$

33. $-\frac{1}{2}$ means $-(1 \div 2)$, so we divide

$$
\begin{array}{r}
0.5 \\
2\overline{)1.0} \\
1\ 0 \\
\hline
0
\end{array}
$$

$\frac{1}{2} = 0.5$, so $-\frac{1}{2} = -0.5$.

35. Since the denominator is 100, we know that $\frac{13}{100} = 0.13$. We could also divide 13 by 100 to find this result.

37.

39.

41. $7 > 0$, since 7 is to the right of 0.

43. $-6 < 6$, since -6 is to the left of 6.

45. $-8 < -5$, since -8 is to the left of -5.

47. $-5 > -11$, since -5 is to the right of -11.

49. $-12.5 < -9.4$, since -12.5 is to the left of -9.4.

51. $-\frac{5}{12} > -\frac{11}{25}$, since $-\frac{5}{12}$ is to the right of $-\frac{11}{25}$.

53. $-7 > x$ has the same meaning as $x < -7$.

55. $-10 \le y$ has the same meaning as $y \ge -10$.

57. $-3 \ge -11$ is true.

59. $0 \ge 8$ is false.

61. $-8 \le -8$ is true.

63. $|-58| = 58$

65. $|5.6| = 5.6$

67. $\left|\sqrt{2}\right| = \sqrt{2}$

69. $\left|\frac{-9}{7}\right| = \frac{9}{7}$

71. $|0| = 0$

73. $|x| = |-8| = 8$

75. $18, -4.7, 0, -\frac{5}{9}, 2.\overline{16}, -37$

77. $18, 0, -37$

79. All of them:

$18, -4.7, 0, -\frac{5}{9}, \pi, \sqrt{17}, 2.\overline{16}, -37$

81. *Thinking and Writing Exercise.*

83. $3xy = 3 \cdot 2 \cdot 7 = 42$

85. *Thinking and Writing Exercise.*

87. *Thinking and Writing Exercise.*

89. List the numbers as they occur on the number line, from left to right: $-23, -17, 0, 4$

91. Converting to decimal notation, we can write $\frac{4}{5}, \frac{4}{3}, \frac{4}{8}, \frac{4}{6}, \frac{4}{9}, \frac{4}{2}, -\frac{4}{3}$ as $0.8, 1.3\overline{3}, 0.5, 0.6\overline{6}, 0.4\overline{4}, 2, -1.3\overline{3}$, respectively. List the numbers (in fraction form) as they occur on the number line, from left to right: $-\frac{4}{3}, \frac{4}{9}, \frac{4}{8}, \frac{4}{6}, \frac{4}{5}, \frac{4}{3}, \frac{4}{2}$

93. $|4| = 4$ and $|-7| = 7$, so $|4| < |-7|$.

95. $|23| = 23$ and $|-23| = 23$, so $|23| = |-23|$.

97. $|x| < 3$

x represents an integer whose distance from 0 is less than 3 units. Thus, $x = -2, -1, 0, 1, 2$.

99. $0.1\overline{1} = \frac{0.\overline{3}}{3} = \frac{\frac{1}{3}}{3} = \frac{1}{3} \cdot \frac{1}{3} = \frac{1}{9}$

101. $5.5\overline{5} = 50\left(0.\overline{1}\right) = 50 \cdot \frac{1}{9} = \frac{50}{9}$

(See Exercise 99).

103. $a < 0$

105. $|x| \le 10$

107. *Thinking and Writing Exercise.*

Mid-Chapter Review

Guided Solutions

1. $\dfrac{x-y}{3} = \dfrac{22-10}{3} = \dfrac{12}{3} = 4$

2. $40x + 8 = 8 \cdot 5x + 8 \cdot 1 = 8(5x+1)$

Mixed Review

1. $x + y = 3 + 12 = 15$

2. $\dfrac{2a}{5} = \dfrac{2 \cdot 10}{5} = \dfrac{20}{5} = 4$

3. $d - 10$

4. Let h be the number of hours worked. Then eight times the number of hours work is: $8h$.

5. Let n be the number of students that originally enrolled in Janine's class. Then reword "five fewer than the number that originally enrolled" as "the number that originally enrolled minus five" and write the equation:

Twenty-seven	is	The number that originally enrolled	minus	five
↓	↓	↓	↓	↓
27	=	n	−	5

$$27 = n - 5$$

6. $13 \cdot 8 = 104 \neq 94$, so $t = 8$ is not a solution to $13t = 94$.

7. $7 + 10x = 10x + 7$

8. $3(ab) = (3a)b$

9. $4(2x + 8) = 4 \cdot 2x + 4 \cdot 8 = 8x + 32$

10. $3(2m + 5n + 10) = 3 \cdot 2m + 3 \cdot 5n + 3 \cdot 10$
$= 6m + 15n + 30$

11. $18x + 9 = 9 \cdot 2x + 9 \cdot 1 = 9(2x + 1)$

12. $8a + 24y + 20 = 4 \cdot 2a + 4 \cdot 6y + 4 \cdot 5$
$= 4(2a + 6y + 5)$

13. $84 = 2 \cdot 42 = 2 \cdot 2 \cdot 21 = 2 \cdot 2 \cdot 3 \cdot 7$

14. $\dfrac{48}{40} = \dfrac{\cancel{8} \cdot 6}{\cancel{8} \cdot 5} = \dfrac{6}{5}$

15. $135 = 3 \cdot 45 = 3 \cdot 3 \cdot 15 = 3 \cdot 3 \cdot 3 \cdot 5$
$315 = 3 \cdot 105 = 3 \cdot 3 \cdot 35 = 3 \cdot 3 \cdot 5 \cdot 7$
$\dfrac{135}{315} = \dfrac{\cancel{3} \cdot \cancel{3} \cdot 3 \cdot \cancel{5}}{\cancel{3} \cdot \cancel{3} \cdot \cancel{5} \cdot 7} = \dfrac{3}{7}$

16. $12 = 2 \cdot 6 = 2 \cdot 2 \cdot 3$ and $8 = 2 \cdot 2 \cdot 2$, so the LCM of 12 and 8 is $2 \cdot 2 \cdot 2 \cdot 3 = 24$.
$\dfrac{11}{12} = \dfrac{11}{12} \cdot \dfrac{2}{2} = \dfrac{22}{24}$ and $\dfrac{3}{8} = \dfrac{3}{8} \cdot \dfrac{3}{3} = \dfrac{9}{24}$, so
$\dfrac{11}{12} - \dfrac{3}{8} = \dfrac{22}{24} - \dfrac{9}{24} = \dfrac{22-9}{24} = \dfrac{13}{24}$

17. $\dfrac{8}{15} \div \dfrac{6}{11} = \dfrac{8}{15} \cdot \dfrac{11}{6} = \dfrac{2 \cdot 2 \cdot \cancel{2}}{3 \cdot 5} \cdot \dfrac{11}{\cancel{2} \cdot 3} = \dfrac{44}{45}$

18. -2.5 is 2.5 units to the left of zero on the number line:

19. $-\dfrac{3}{20} = -(3 \div 20)$ and

$$3 \div 20 = 20\overline{)3.00}$$
$$\begin{array}{r} 0.15 \\ \hline 3.00 \\ \underline{20} \\ 10 \\ \underline{0} \\ 100 \\ \underline{100} \\ 0 \end{array}$$

So, $-\dfrac{3}{20} = -0.15$

20. $-16 \boxed{>} -24$, because -16 is to the right of -24 on the number line.

21. $22 = 2 \cdot 11$ and $15 = 3 \cdot 5$, so the LCM of 22 and 15 is $2 \cdot 3 \cdot 5 \cdot 11 = 330$.

$$-\frac{3}{22} = \frac{-3}{22} \cdot \frac{15}{15} = \frac{-45}{330}$$

$$-\frac{2}{15} = \frac{-2}{15} \cdot \frac{22}{22} = \frac{-44}{330}$$

Therefore $-\dfrac{3}{22} = \dfrac{-45}{330} \boxed{<} \dfrac{-44}{330} = \dfrac{-2}{15}$

since -45 is to the left of -44 on the number line.

22. $x \geq 9$ is equivalent to $9 \leq x$.

23. -6 is to the left of -5 on the number line, so the statement $-6 \leq -5$ is true.

24. $|-5.6| = 5.6$

25. $|0| = 0$

Exercise Set 1.5

1. f

3. e

5. b

7. Start at 5. Move 8 units to the left.

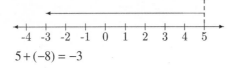

$5 + (-8) = -3$

9. Start at –5. Move 9 units to the right.

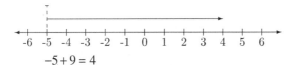

$-5 + 9 = 4$

11. Start at -4. Move 0 units.

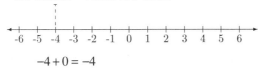

$-4 + 0 = -4$

13. Start at –3. Move 5 units to the left.

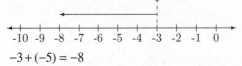

$-3 + (-5) = -8$

15. $-6 + (-5)$. Two negatives. Add their absolute values . Make the answer negative.
$-6 + (-5) = -(6 + 5) = -11$

17. $10 + (-15)$. The absolute values are 10 and 15. Their difference is $15 - 10 = 5$. The negative number has the larger absolute value, so the answer is negative.
$10 + (-15) = -(15 - 10) = -5$

19. $12 + (-12)$. The numbers have the same absolute value and their difference is 0.
$12 + (-12) = 0$

21. $-24 + (-17)$. Two negatives. Add their absolute values. Make the answer negative.
$-24 + (-17) = -(24 + 17) = -41$

23. $-13 + 13$. The numbers have the same absolute value and their difference is 0.
$-13 + 13 = 0$

25. $18 + (-11)$. The absolute values are 18 and 11. Their difference is $18 - 11 = 7$. The positive number has the greater absolute value, so the answer is positive.
$18 + (-11) = 18 - 11 = 7$

27. $-36 + 0$. 0 is the additive identity.
$-36 + 0 = -36$

29. $-3 + 14$. The absolute values are 3 and 14. Their difference is $14 - 3 = 11$. The positive number has the larger absolute value, so the answer is positive.
$-3 + 14 = 14 - 3 = 11$

31. $-14 + (-19)$. Two negatives. Add their absolute values. Make the answer negative.
$-14 + (-19) = -(14 + 19) = -33$

33. $19 + (-19)$. The numbers have the same absolute value and their difference is 0.
$$19 + (-19) = 0$$

35. $23 + (-5)$. The absolute values are 23 and 5. Their difference is $23 - 5$ or 18. The positive number has the larger absolute value, so the answer is positive.
$$23 + (-5) = 23 - 5 = 18$$

37. $-31 + (-14)$. Two negatives. Add their absolute values. Make the answer negative.
$$-31 + (-14) = -(31 + 14) = -45$$

39. $40 + (-40)$. The numbers have the same absolute value and their difference is 0.
$$40 + (-40) = 0$$

41. $85 + (-65)$. The absolute values are 85 and 65. Their difference is $85 - 65 = 20$. The positive number has the larger absolute value, so the answer is positive.
$$85 + (-65) = 85 - 65 = 20$$

43. $-3.6 + 1.9$. The absolute values are 3.6 and 1.9. Their difference is $3.6 - 1.9 = 1.7$. The negative number has the larger absolute value, so the answer is negative.
$$-3.6 + 1.9 = -(3.6 - 1.9) = -1.7$$

45. $-5.4 + (-3.7)$. Two negatives. Add their absolute values. Make the answer negative.
$$-5.4 + (-3.7) = -(5.4 + 3.7) = -9.1$$

47. $\frac{-3}{5} + \frac{4}{5}$. The absolute values are $\frac{3}{5}$ and $\frac{4}{5}$. Their difference is $\frac{4}{5} - \frac{3}{5} = \frac{1}{5}$. The positive number has the larger absolute value, so the answer is positive.
$$\frac{-3}{5} + \frac{4}{5} = \frac{4}{5} - \frac{3}{5} = \frac{1}{5}$$

49. $\frac{-4}{7} + \frac{-2}{7}$. Two negatives. Add their absolute values. Make the answer negative.
$$\frac{-4}{7} + \frac{-2}{7} = -\left(\frac{4}{7} + \frac{2}{7}\right) = -\frac{6}{7}$$

51. $-\frac{2}{5} + \frac{1}{3}$. The absolute values are $\frac{2}{5}$ and $\frac{1}{3}$. Their difference is $\frac{6}{15} - \frac{5}{15} = \frac{1}{15}$. The negative number has the larger absolute value, so the answer is negative.
$$\frac{-2}{5} + \frac{1}{3} = \frac{-6}{15} + \frac{5}{15} = -\left(\frac{6}{15} - \frac{5}{15}\right) = -\frac{1}{15}$$

53. $\frac{-4}{9} + \frac{2}{3}$. The absolute values are $\frac{4}{9}$ and $\frac{2}{3}$. Their difference is $\frac{6}{9} - \frac{4}{9} = \frac{2}{9}$. The positive number has the larger absolute value, so the answer is positive.
$$\frac{-4}{9} + \frac{2}{3} = \frac{-4}{9} + \frac{6}{9} = \frac{6}{9} - \frac{4}{9} = \frac{2}{9}$$

55. $35 + (-14) + (-19) + (-5)$
$$= 35 + [(-14) + (-19) + (-5)] \quad \text{Using the associative law of addition}$$
$$= 35 + (-38) \quad \text{Adding the negatives}$$
$$= -3 \quad \text{Adding a positive and a negative}$$

57. $-4.9 + 8.5 + 4.9 + (-8.5)$ Note that we have two pairs of numbers with different signs and the same absolute value: -4.9 and 4.9, 8.5 and -8.5. The sum of each pair is 0, so the result is $0 + 0$, or 0.

59. Rewording:

First change	plus	second change	plus
↓	↓	↓	↓

Translating: -0.05 $+$ (-0.03) $+$

third change	is	total change
↓	↓	↓
0.07	$=$	total change

Since $-0.05 + (-0.03) + 0.07 = -0.01$, the price dropped 1¢, or the cost changed $-\$0.01$.

61. Rewording: First change plus second change plus

$$\left(-\tfrac{1}{2}\right) \quad + \quad \left(\tfrac{6}{5}\right) \quad +$$

third change plus fourth change is level change

$$\left(\tfrac{3}{4}\right) \quad + \quad \left(-\tfrac{3}{2}\right) \quad = \quad \text{total change}$$

Since $\left(-\dfrac{1}{2}\right)+\left(\dfrac{6}{5}\right)+\left(\dfrac{3}{4}\right)+\left(-\dfrac{3}{2}\right)$

$=\left(-\dfrac{10}{20}\right)+\left(\dfrac{24}{20}\right)+\left(\dfrac{15}{20}\right)+\left(-\dfrac{30}{20}\right)$

$=\left(\dfrac{-10+24+15-30}{20}\right)=\left(-\dfrac{1}{20}\right)$

the lake level dropped by $\left(\dfrac{1}{20}\right)$ of a foot.

63. Rewording: First try plus second try plus third try is Total gain

Translating: $13 \quad + \quad (-12) \quad + \quad 21 = \text{Total gain}$

Since $13+(-12)+21=22,$ the total gain was 22 yd.

65. Rewording: Original balance plus change from check plus

Translating: $82 \quad + \quad (-50) \quad +$

Change from August charges is New balance

$63 \quad = \quad \text{New balance}$

Since $82+(-50)+63=95,$ his new balance is \$95.

67. Rewording: base plus rise is elevation

Translating: $(-19{,}684) \quad + \quad 33{,}480 = \text{elevation}$

Since $(-19{,}684)+33{,}480=13{,}796,$ the elevation of the peak is 13,796 ft above sea level.

69. $5a+(-8a)=-(8-5)\,a=-3a$

71. $-3x+12x=(12-3)x=9x$

73. $-5a+-2a=-(5+2)\,a=-7a$

75. $-3+8x+4+(-10x)$

$\quad = -3+4+8x+(-10x)$ Using the commutative law of addition

$\quad = (-3+4)+[8+(-10)]x$ Using the distributive law

$\quad = 1-2x$ Adding

77. $6m+9n+(-9n)+(-10m)$

$\quad = -(10-6)m+(9-9)n=-4m$

79. $-4x+6.3+(-x)+(-10.2)$

$\quad = [(-4)+(-1)]x-(10.2-6.3)$

$\quad = -5x-3.9$

81. Perimeter $=8+5x+9+7x$

$\quad = 8+9+5x+7x$

$\quad = (8+9)+(5+7)x$

$\quad = 12x+17$

83. Perimeter $=3t+3r+7+5t+9+4r$

$\quad = (3t+5t)+(3r+4r)+(7+9)$

$\quad = (3+5)t+(3+4)r+(7+9)$

$\quad = 8t+7r+16$

85. Perimeter $=9+6n+7+8n+4n$

$\quad = 9+7+6n+8n+4n$

$\quad = (9+7)+(6+8+4)n$

$\quad = 18n+16$

87. *Thinking and Writing Exercise.*

89. $7(3z+y+2)=7\cdot 3z+7\cdot y+7\cdot 2$

$\quad\quad\quad\quad = 21z+7y+14$

91. *Thinking and Writing Exercise.*

93. We're looking for the difference between the amount in Travis's account after the deposit and his eventual overdrawn amount.

Rewording:

Original amount   plus   deposit   minus

$\downarrow$ $\downarrow$ $\downarrow$ $\downarrow$

Translating: 257.33   +   (152)   −

overdrawn amount   is   check amount

$\downarrow$ $\downarrow$ $\downarrow$

−42.37   =   check amount

Since $257.33 + 152 - (-42.37) = 451.70$, the check's amount $451.70.

95. $4x + \underline{} + (-9x) + (-2y) = -5x - 7y$

Simplify the left side of the equation.

$4x + \underline{} + (-9x) + (-2y)$

$= [4x + (-9x)] + \underline{} + (-2y)$

$= [4 + (-9)]x + \underline{} + (-2y)$

$= -5x + \underline{} + (-2y)$

We now have:

$-5x + \underline{} + (-2y) = -5x - 7y$

$-5x = -5x,$ so $\underline{} + (-2y) = -7y$

The missing term is $-5y$, since

$-5y + -2y = -7y$

97. $3m + 2n + \underline{} + (-2m) = 2n + (-6m)$

Simplify the left side of the equation.

$3m + 2n + \underline{} + (-2m)$

$= [3m + (-2m)] + 2n + \underline{}$

$= [3 + (-2)]m + 2n + \underline{}$

$= m + 2n + \underline{}$

$= 2n + m + \underline{}$

We now have:

$2n + m + \underline{} = 2n + (-6m)$

$2n = 2n,$ so $m + \underline{} = -6m$

The missing term is $-7m$, since

$m + (-7m) = -6m$

99. Note that, in order for the sum to be 0, the two missing terms must be the opposites of the given terms. Thus, the missing terms are $-7t$ and -23.

101. $-3 + (-3) + 2 + (-2) + 1 = -5$

Since the total is 5 under par after the five rounds and

$-5 = -1 + (-1) + (-1) + (-1) + (-1),$ the golfer was 1 under par on average.

Exercise Set 1.6

1. d

3. f

5. a

7. b

9. four minus ten

11. two minus negative nine

13. the negative/opposite of x minus y

15. negative three minus the negative/opposite of n

17. The additive inverse of 39 is –39 because $39 + (-39) = 0$

19. The additive inverse of $-\frac{11}{2}$ is $\frac{11}{2}$, because

$$-\frac{11}{2} + \frac{11}{2} = 0$$

21. The additive inverse of -3.14 is 3.14 because $-3.14 + 3.14 = 0$

23. If $x = -45$, then $-x = -(-45) = 45$.

25. If $x = -\frac{14}{3}$, then $-x = -\left(-\frac{14}{3}\right) = \frac{14}{3}$.

27. If $x = 0.101$, then $-x = -0.101$.

29. If $x = 72$, then $-(-x) = x = 72$.

31. If $x = -\frac{2}{5}$, then $-(-x) = x = -\frac{2}{5}$.

33. $-(-1) = 1$

35. $-(7) = -7$

37. $6 - 8 = -(8 - 6) = -2$

39. $0-5 = 0+(-5) = -5$

41. $-4-3 = -(4+3) = -7$

43. $-9-(-3) = -9+3 = -6$

45. Note that we are subtracting a number from itself. The result is 0. We could also do this exercise as follows:
$-8-(-8) = -8+8 = 0$

47. $30-40 = 30+(-40) = -10$

49. $-7-(-9) = -7+9 = 2$

51. $-9-(-9) = -9+9 = 0$

53. $5-5 = 5+(-5) = 0$

55. $4-(-4) = 4+4 = 8$

57. $-7-4 = -7+(-4) = -11$

59. $6-(-10) = 6+10 = 16$

61. $-6-(-5) = -6+5 = -1$

63. $5-(-12) = 5+12 = 17$

65. $0-(-10) = 0+10 = 10$

67. $-5-(-2) = -5+2 = -3$

69. $-7-14 = -7+(-14) = -21$

71. $-8-0 = -8+0 = -8$

73. $0-11 = 0+(-11) = -11$

75. $2-25 = 2+(-25) = -23$

77. $-4.2-3.1 = -(4.2+3.1) = -7.3$

79. $-1.8-(-2.4) = -1.8+2.4 = 2.4-1.8 = 0.6$

81. $3.2-8.7 = 3.2+(-8.7) = -5.5$

83. $0.072-1 = 0.072+(-1) = -0.928$

85. $\dfrac{2}{11}-\dfrac{9}{11} = -\left(\dfrac{9}{11}-\dfrac{2}{11}\right) = -\dfrac{7}{11}$

87. $\dfrac{-1}{5}-\dfrac{3}{5} = -\left(\dfrac{1}{5}+\dfrac{3}{5}\right) = -\dfrac{4}{5}$

89. $\dfrac{-4}{17}-\left(-\dfrac{9}{17}\right) = -\dfrac{4}{17}+\dfrac{9}{17} = \dfrac{5}{17}$

91. $-21-37 = -21+(-37) = -58$

93. $9-(-25) = 9+25 = 34$

95. We subtract (difference) the lesser number from the greater.
$3.8-(-5.2) = 3.8+5.2 = 9$

97. We subtract the lesser number from the greater.
$114-(-79) = 114+79 = 193$

99. $25-(-12)-7-(-2)+9$
$= 25+12+(-7)+2+9 = 41$

101. $-31+(-28)-(-14)-17$
$= (-31)+(-28)+14+(-17) = -62$

103. $-34-28+(-33)-44$
$= (-34)+(-28)+(-33)+(-44) = -139$

105. $-93+(-84)-(-93)-(-84)$
Note that we are subtracting -93 from -93 and -84 from -84. Thus, the result will be 0.

107. $-7x-4y = -7x+(-4y)$, so the terms are $-7x$ and $-4y$

109. $9-5t-3st = 9+(-5t)+(-3st)$, so the terms are $9, -5t,$ and $-3st$

111. $4x-7x = 4x+(-7x)$ Adding the opposite
$= \big(4+(-7)\big)x$ Using the distributive law
$= -3x$

113. $7a - 12a + 4$

$= 7a + (-12a) + 4$ Adding the opposite

$= (7 + (-12))a + 4$ Using the distributive law

$= -5a + 4$

115. $-8n - 9 + 7n$

$= (-8 + 7)n - 9$ Using the distributive law

$= -n - 9$

117. $2 - 6t - 9 + t = (-6 + 1)t + (2 - 9)$

$= -5t + (-7)$

$= -5t - 7$

119. $5y + (-3x) - 9x + 1 - 2y + 8$

$= 5y + (-3x) + (-9x) + 1 + (-2y) + 8$

$= 5y + (-2y) + (-3x) + (-9x) + 1 + 8$

$= 3y - 12x + 9$

or $-12x + 3y + 9$

121. $13x - (-2x) + 45 - (-21) - 7x$

$= 13x + 2x + 45 + 21 + (-7x)$

$= 13x + 2x + (-7x) + 45 + 21$

$= 8x + 66$

123. We subtract the lower elevation from the higher elevation:

$102,880 \text{ ft} - (-35,797 \text{ ft})$

$= 102,880 \text{ ft} + 35,797 \text{ ft}$

$= 138,677 \text{ ft}$

The difference in elevations is 138,677 ft.

125. We subtract the lower temperature from the higher temperature:

$15 - (-32) = 15 + 32 = 47$

The temperature rose 47°F.

127. We subtract the lesser differential from the greater:

$8.5 - (-0.4) = 8.5 + 0.4 = 8.9$

The Cavaliers improved 8.9 points.

129. *Thinking and Writing Exercise.*

131. Area $= lw = (36\text{ft})(12\text{ft}) = 432\,\text{ft}^2$

133. *Thinking and Writing Exercise.*

135. Rewrite: Actual   Clock   Time power
time minus   time is   was out

Translate: 3:00 pm   –   8:00 am = Time power was out

3:00 pm (our 15:00 hrs) – 8:00 am = 7 hrs. Since the power was restored 7 hrs after it was lost, 4:00 pm + 7 hrs = 11:00 pm. The power was restored at 11:00 pm on August 14.

137. False. For example, let $m = -3$ and $n = -5$. Then $-3 > -5$, but $-3 + (-5) = -8 \not> 0$.

139. True. For example, for $m = 4$ and $n = -4, 4 = -(-4)$ and $4 + (-4) = 0$; for $m = -3$ and $n = 3, -(3) = -3$ and $-3 + 3 = 0$.

141. $\boxed{(-)}\boxed{9}\boxed{-}\boxed{(-)}\boxed{7}\boxed{Enter}$

Exercise Set 1.7

1. 1

3. 0

5. 0

7. 1

9. 1

11. $-3 \cdot 8 = -24$ Multiply the absolute values, $3 \cdot 8$; the product is negative.

13. $-8 \cdot 7 = -56$ Multiply the absolute values, $8 \cdot 7$; the product is negative.

15. $8 \cdot (-3) = -(8 \cdot 3) = -24$

17. $-6 \cdot (-7) = 42$ Multiply the absolute values, $6 \cdot 7$; the product of two negative numbers is positive.

19. $19 \cdot (-10) = -190$

21. $-12 \cdot 12 = -144$

23. $-25 \cdot (-48) = 1200$

25. $4.5 \cdot (-28) = -126$

27. $-5 \cdot (-2.3) = 11.5$

29. $-25 \cdot 0 = 0$

31. $\frac{2}{3} \cdot \left(-\frac{3}{5}\right) = -\left(\frac{2}{3} \cdot \frac{3}{5}\right) = -\left(\frac{2 \cdot \cancel{3}}{\cancel{3} \cdot 5}\right) = -\frac{2}{5}$

33. $-\frac{3}{8} \cdot \left(-\frac{2}{9}\right) = \frac{3}{8} \cdot \frac{2}{9} = \frac{3 \cdot 2}{8 \cdot 9} = \frac{\cancel{3} \cdot \cancel{2}}{\cancel{2} \cdot 4 \cdot \cancel{3} \cdot 3} = \frac{1}{12}$

35. $(-5.3)(2.1) = -11.13$

37. $-\frac{5}{9} \cdot \frac{3}{4} = -\left(\frac{5}{9} \cdot \frac{3}{4}\right) = -\left(\frac{5 \cdot \cancel{3}}{\cancel{3} \cdot 3 \cdot 4}\right) = -\frac{5}{12}$

39. $3 \cdot (-7) \cdot (-2) \cdot 6$
    $= -21 \cdot -12$ Multiplying the first two numbers and the last two numbers
    $= 252$

41. $\frac{-1}{3} \cdot \frac{1}{4} \cdot \left(-\frac{3}{7}\right) = \frac{-1}{12} \cdot \left(\frac{-3}{7}\right)$
    $= \frac{1 \cdot \cancel{3}}{\cancel{3} \cdot 4 \cdot 7} = \frac{1}{28}$

43. $-2 \cdot (-5) \cdot (-3) \cdot (-5) = 10 \cdot 15 = 150$

45. $(-31) \cdot (-27) \cdot 0 = 0.$

47. $(-8)(-9)(-10) = 72 \cdot (-10)$
    $= -720$

49. $(-6)(-7)(-8)(-9)(-10) = 42 \cdot 72 \cdot (-10)$
    $= 3024 \cdot (-10)$
    $= -30,240$

51. $14 \div (-2) = -7$ Check: $-7 \cdot (-2) = 14$

53. $-26 \div (-13) = 2$ Check: $2 \cdot (-13) = -26$

55. $-50 \div 5 = -10$ Check: $-10 \cdot 5 = -50$

57. $-10.2 \div (-2) = 5.1$ Check: $5.1 \cdot (-2) = -10.2$

59. $-100 \div (-11) = \frac{100}{11}$

61. $\frac{400}{-50} = -\frac{\cancel{50} \cdot 8}{\cancel{50}} = -8$

63. $\frac{28}{0}$ is undefined

65. $-4.8 \div 1.2 = -4$ Check: $-4 \cdot 1.2 = -4.8$

67. $\frac{0}{-9} = 0$ Check: $0 \cdot (-9) = 0$

69. $\frac{9.7(-2.8)0}{4.3} = \frac{0}{4.3} = 0$ Check: $0 \cdot 4.3 = 0$

71. $\frac{-8}{3} = \frac{8}{-3}$ and $\frac{-8}{3} = -\frac{8}{3}$

73. $\frac{29}{-35} = -\frac{29}{35}$ and $\frac{29}{-35} = \frac{-29}{35}$

75. $-\frac{7}{3} = \frac{-7}{3}$ and $-\frac{7}{3} = \frac{7}{-3}$

77. $\frac{-x}{2} = -\frac{x}{2}$ and $\frac{-x}{2} = \frac{x}{-2}$

79. $-\frac{5}{4}$,   since $\left(-\frac{5}{4}\right) \cdot \left(-\frac{4}{5}\right) = 1$

81. $-\frac{10}{51}$;   since $-\frac{10}{51} \cdot \left(\frac{51}{-10}\right) = 1$

83. $-\frac{1}{10}$,   since $-\frac{1}{10} \cdot (-10) = 1$

85. $\frac{1}{4.3}$,   since $\frac{1}{4.3} \cdot 4.3 = 1.$
    This can also be written as $\frac{1}{4.3} \cdot \frac{10}{10} = \frac{10}{43}$

87. -4,   since $-4 \cdot \left(\frac{-1}{4}\right) = \frac{-4 \cdot (-1)}{4} = \frac{4}{4} = 1$

89. Does not exist

91. $\left(\dfrac{-7}{4}\right)\left(-\dfrac{3}{5}\right) = \dfrac{7 \cdot 3}{4 \cdot 5} = \dfrac{21}{20}$

93. $\dfrac{-3}{8} + \dfrac{-5}{8} = \dfrac{-3 + (-5)}{8} = \dfrac{-8}{8} = -1$

95. $\left(\dfrac{-9}{5}\right)\left(\dfrac{5}{-9}\right) = 1$

Note: This is the product of reciprocals.

97. $\left(-\dfrac{3}{11}\right) - \left(-\dfrac{6}{11}\right) = \dfrac{-3 - (-6)}{11} = \dfrac{-3 + 6}{11} = \dfrac{3}{11}$

99. $\dfrac{7}{8} \div \left(-\dfrac{1}{2}\right) = \dfrac{7}{8} \cdot \left(\dfrac{-2}{1}\right) = -\left(\dfrac{7 \cdot 2}{8 \cdot 1}\right)$

$= -\left(\dfrac{7 \cdot \cancel{2}}{\cancel{2} \cdot 4}\right) = -\dfrac{7}{4}$

101. $\dfrac{9}{5} \cdot \dfrac{-20}{3} = -\left(\dfrac{9 \cdot 20}{5 \cdot 3}\right) = -\left(\dfrac{\cancel{3} \cdot 3 \cdot 4 \cdot 5}{\cancel{5} \cdot \cancel{3}}\right) = -12$

103. $\left(-\dfrac{18}{7}\right) + \left(-\dfrac{3}{7}\right) = \dfrac{-18 + (-3)}{7} = \dfrac{-21}{7} = -3$

105. $-\dfrac{5}{9} \div \left(-\dfrac{5}{9}\right) = 1$

Any non-zero number divided by itself equals 1.

107. $\dfrac{5}{9} - \dfrac{7}{9} = \dfrac{5 - 7}{9} = \dfrac{-2}{9}, \text{ or } -\dfrac{2}{9}$

109. $\dfrac{-3}{10} + \dfrac{2}{5} = \dfrac{-3}{10} + \dfrac{4}{10} = \dfrac{-3 + 4}{10} = \dfrac{1}{10}$

111. $\dfrac{7}{10} \div \left(\dfrac{-3}{5}\right) = -\left(\dfrac{7}{10} \cdot \dfrac{5}{3}\right) = -\left(\dfrac{7 \cdot 5}{10 \cdot 3}\right)$

$= -\left(\dfrac{7 \cdot \cancel{5}}{\cancel{5} \cdot 2 \cdot 3}\right) = -\dfrac{7}{6}$

113. $\dfrac{14}{-9} \div \dfrac{0}{3} = -\dfrac{14}{9} \cdot \dfrac{3}{0} \text{ is undefined.}$

115. $\dfrac{-4}{15} + \dfrac{2}{-3} = -\dfrac{4}{15} + \left(-\dfrac{10}{15}\right) = \dfrac{-4 + (-10)}{15} = -\dfrac{14}{15}$

117. *Thinking and Writing Exercise.*

119. $\dfrac{264}{468} = \dfrac{\cancel{2} \cdot \cancel{2} \cdot 2 \cdot \cancel{3} \cdot 11}{\cancel{2} \cdot \cancel{2} \cdot \cancel{3} \cdot 3 \cdot 13} = \dfrac{22}{39}$

121. *Thinking and Writing Exercise.*

123. $\dfrac{1}{a + b}$ Answers may vary.

125. $-(a + b)$ Answers may vary.

127. $x = -x$

129. Consider the sum $2 + 3$. Its reciprocal is $\dfrac{1}{2+3}$, or $\dfrac{1}{5}$, but $\dfrac{1}{2} + \dfrac{1}{3} = \dfrac{5}{6}$. Answers may vary.

131. When n is negative, $-n$ is positive, so $\dfrac{m}{-n}$ is the quotient of a negative number and a positive number, which is negative.

133. When n is negative, $-n$ is positive, so $\dfrac{-n}{m}$ is the quotient of a positive and a negative number and, thus, is negative. When m is negative, $-m$ is positive, so $-m \cdot \left(\dfrac{-n}{m}\right)$ is the product of a positive and a negative number and, thus, is negative.

135. $m + n$ is the sum of two negative numbers, so it is negative; $\dfrac{m}{n}$ is the quotient of two negative numbers, so it is positive. Then $(m + n) \cdot \dfrac{m}{n}$ is the product of a negative and a positive number and, thus, is negative.

137. a) m and n have different signs;
b) either m or n is zero;
c) m and n have the same sign.

139. $a(-b) + ab = a[-b + b]$ Distributive law
$\quad\quad\quad\quad\quad = a(0)$ Law of opposites
$\quad\quad\quad\quad\quad = 0$ Multiplicative property of 0

Therefore $a(-b)$ is the opposite of ab by the law of opposites.

Exercise Set 1.8

1. a) Division

   b) Subtraction

   c) Addition

   d) Multiplication

   e) Subtraction

   f) Multiplication

3. $\underbrace{x \cdot x \cdot x \cdot x \cdot x \cdot x \cdot x}_{7 \, \text{factors}} = x^7$

5. $(-5)(-5)(-5) = (-5)^3$

7. $\underbrace{3t \cdot 3t \cdot 3t \cdot 3t \cdot 3t}_{5 \, \text{factors}} = (3t)^5$

9. $2 \cdot n \cdot n \cdot n \cdot n = 2n^4$

11. $3^2 = 3 \cdot 3 = 9$

13. $(-4)^2 = -4 \cdot (-4) = 16$

15. $-4^2 = -4 \cdot 4 = -16$

17. $4^3 = 4 \cdot 4 \cdot 4 = 64$

19. $(-5)^4 = -5 \cdot (-5) \cdot (-5) \cdot (-5) = 625$

21. $7^1 = 7$ (1 factor of 7)

23. $(-2)^5 = (-2)(-2)(-2)(-2)(-2) = -32$

25. $(3t)^4 = 3t \cdot 3t \cdot 3t \cdot 3t$
    $= 3 \cdot 3 \cdot 3 \cdot 3 \cdot t \cdot t \cdot t \cdot t$
    $= 81t^4$

27. $(-7x)^3 = -7x \cdot (-7x) \cdot (-7x)$
    $= -7 \cdot (-7) \cdot (-7) \cdot x \cdot x \cdot x$
    $= -343x^3$

Exercises 29–55. Review the "Rules for Order of Operations" on Page 65.

29. $5 + 3 \cdot 7 = 5 + 21 = 26$

31. $8 \cdot 7 + 6 \cdot 5 = 56 + 30 = 86$

33. $9 \div 3 + 16 \div 8 = 3 + 2 = 5$

35. $14 \cdot 19 \div (19 \cdot 14) = 1$ Note: we are dividing a number by itself.

37. $3(-10)^2 - 8 \div 2^2$
    $= 3 \cdot 100 - 8 \div 4$ Simplifying the exponential expressions
    $= 300 - 2$ Multiplying and dividing from left to right
    $= 298$ Subtracting

39. $8 - (2 \cdot 3 - 9)$
    $= 8 - (6 - 9)$ Multiplying inside the parentheses
    $= 8 - (-3)$ Subtracting inside the parentheses
    $= 11$ Subtracting

41. $(8 - 2)(3 - 9) = 6 \cdot (-6) = -36$

43. $5 \cdot 3^2 - 4^2 \cdot 2 = 5 \cdot 9 - 16 \cdot 2$
    $= 45 - 32 = 13$

45. $5 + 3(2 - 9)^2 = 5 + 3(-7)^2 = 5 + 3 \cdot 49$
    $= 5 + 147 = 152$

47. $[2 \cdot (5 - 8)]^2 - 12 = [2 \cdot (-3)]^2 - 12$
    $= (-6)^2 - 12 = 36 - 12 = 24$

49. $\dfrac{7 + 2}{5^2 - 4^2} = \dfrac{9}{25 - 16} = \dfrac{9}{9} = 1$

51. $8(-7) + |6(-5)| = -56 + |-30|$
    $= -56 + 30 = -26$

53. $\dfrac{(-2)^3 + 4^2}{3 - 5^2 + 3 \cdot 6} = \dfrac{-8 + 16}{3 - 25 + 3 \cdot 6} = \dfrac{8}{3 - 25 + 18}$

$\qquad\qquad = \dfrac{8}{-22 + 18} = \dfrac{8}{-4} = -2$

55. $\dfrac{-3^3 - 2 \cdot 3^2}{8 \div 2^2 - (6 - |2 - 15|)} = \dfrac{-27 - 2 \cdot 9}{8 \div 4 - (6 - |-13|)}$

$\qquad = \dfrac{-27 - 18}{2 - (6 - 13)} = \dfrac{-45}{2 - (-7)} = \dfrac{-45}{2 + 7} = \dfrac{-45}{9} = -5$

57. This expression is equivalent to expression (a).

$\dfrac{5(3 - 7) + 4^3}{(-2 - 3)^2} = \dfrac{5(-4) + 4^3}{(-5)^2}$

$\qquad\qquad = \dfrac{5(-4) + 64}{25}$

$\qquad\qquad = \dfrac{-20 + 64}{25}$

$\qquad\qquad = \dfrac{44}{25}$

59. This expression is equivalent to expression (d).

$5(3 - 7) + 4^3 \div (-2 - 3)^2 = 5(-4) + 4^3 \div (-5)^2$

$\qquad\qquad\qquad = 5(-4) + 64 \div 25$

$\qquad\qquad\qquad = -20 + 2.56$

$\qquad\qquad\qquad = -17.44$

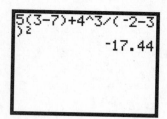

61.

Rounded to the nearest thousandth, the answer is 21.563.

63. (13.4−5abs(1.2+4
.6))/(9.3−5.4)²
        −1.025641026

Rounded to the nearest thousandth, the answer is −1.026.

65. $9 - 4x = 9 - 4 \cdot 5$ Substituting 5 for x

$\qquad = 9 - 20$ Multiplying

$\qquad = -11$ Subtracting

67. $24 \div t^3 = 24 \div (-2)^3$ Substituting -2 for t

$\qquad = 24 \div (-8)$ Simplifying the exponential expression

$\qquad = -3$ Dividing

69. $45 \div 3 \cdot a = 45 \div 3 \cdot (-1)$ Substituting -1 for a

$\qquad = 15 \cdot (-1)$ Dividing

$\qquad = -15$ Multiplying

71. $5x \div 15x^2$

$\qquad = 5 \cdot 3 \div 15(3)^2$ Substituting 3 for x

$\qquad = 5 \cdot 3 \div 15 \cdot 9$ Simplifying the exponential expression

$\qquad = 15 \div 15 \cdot 9$ Dividing and multiplying in order from left to right

$\qquad = 1 \cdot 9$

$\qquad = 9$

73. $45 \div 3^2 x(x-1)$

    $= 45 \div 3^2 \cdot 3(3-1)$ Substituting 3 for x

    $= 45 \div 3^2 \cdot 3 \cdot 2$ Simplifying inside the parentheses

    $= 45 \div 9 \cdot 3 \cdot 2$ Simplifying the exponential expression

    $= 5 \cdot 3 \cdot 2$ Dividing and multiplying

    $= 15 \cdot 2$ in order from

    $= 30$ left to right.

75. $-x^2 - 5x = -(-3)^2 - 5(-3)$

          $= -9 - 5(-3)$

          $= -9 + 15 = 6$

77. $\dfrac{3a - 4a^2}{a^2 - 20} = \dfrac{3 \cdot 5 - 4(5)^2}{(5)^2 - 20} = \dfrac{3 \cdot 5 - 4 \cdot 25}{25 - 20}$

          $= \dfrac{15 - 100}{5} = \dfrac{-85}{5} = -17$

79.
```
13-(6-4)^3+10
            15
```

81.
```
3(1.6+2*5.9)/1.6
          25.125
```

83.
```
(1/2)(141+5/.2)²
          13778
```

85. $-(9x+1) = -9x - 1$ Removing parenthesis and changing the sign of each term.

87. $-[5 - 6x] = -5 + 6x$ Removing parenthesis and changing the sign of each term

89. $-(4a - 3b + 7c) = -4a + 3b - 7c$

91. $-(3x^2 + 5x - 1) = -3x^2 - 5x + 1$

93. $8x - (6x + 7) = 8x - 6x - 7 = 2x - 7$

95. $2x - 7x - (4x - 6) = 2x - 7x - 4x + 6$

                $= -9x + 6$

97. $9t - 5r + 2(3r + 6t) = 9t - 5r + 6r + 12t$

                    $= 21t + r$

99. $15x - y - 5(3x - 2y + 5z)$

    $= 15x - y - 15x + 10y - 25z$ Multiplying each term in parentheses by -5

    $= 9y - 25z$

101. $3x^2 + 7 - (2x^2 + 5) = 3x^2 + 7 - 2x^2 - 5$

                    $= x^2 + 2$

103. $5t^3 + t + 3(t - 2t^3) = 5t^3 + t + 3t - 6t^3$

                    $= -t^3 + 4t$

105. $12a^2 - 3ab + 5b^2 - 5(-5a^2 + 4ab - 6b^2)$

    $= 12a^2 - 3ab + 5b^2 + 25a^2 - 20ab + 30b^2$

    $= 37a^2 - 23ab + 35b^2$

107. $-7t^3 - t^2 - 3(5t^3 - 3t) = -7t^3 - t^2 - 15t^3 + 9t$

                    $= -22t^3 - t^2 + 9t$

109. $5(2x - 7) - [4(2x - 3) + 2]$

    $= 5(2x - 7) - [8x - 12 + 2]$

    $= 5(2x - 7) - [8x - 10]$

    $= 10x - 35 - 8x + 10$

    $= 2x - 25$

111. *Thinking and Writing Exercise.*

113. Let n represent "a number." Then we have $2n + 9$.

115. *Thinking and Writing Exercise.*

117. $5t - \{7t - [4r - 3(t - 7)] + 6r\} - 4r$

$= 5t - \{7t - [4r - 3t + 21] + 6r\} - 4r$

$= 5t - \{7t - 4r + 3t - 21 + 6r\} - 4r$

$= 5t - \{10t + 2r - 21\} - 4r$

$= 5t - 10t - 2r + 21 - 4r$

$= -6r - 5t + 21$

119. $\{x - [f - (f - x)] + [x - f]\} - 3x$

$= \{x - [f - f + x] + [x - f]\} - 3x$

$= \{x - [x] + [x - f]\} - 3x$

$= \{x - x + x - f\} - 3x$

$= x - f - 3x$

$= -2x - f$

121. *Thinking and Writing Exercise.*

123. True; $m - n = -n + m = -(n - m)$

125. False; let $m = 2$ and $n = 1$. Then
$-2(1 - 2) = -2(-1) = 2$, but
$-(2 \cdot 1 + 2^2) = -(2 + 4) = -6$.

127. $[x + 3(2 - 5x) \div 7 + x](x - 3)$. When $x = 3$, the factor $x - 3$ is 0, so the product is 0.

129. $\dfrac{x^2 + 2^x}{x^2 - 2^x} = \dfrac{3^2 + 2^3}{3^2 - 2^3}$ for $x = 3$

$= \dfrac{9 + 8}{9 - 8} = \dfrac{17}{1} = 17$

131. $4 \cdot 20^3 + 17 \cdot 20^2 + 10 \cdot 20 + 0 \cdot 20^0$

$= 4 \cdot 8000 + 17 \cdot 400 + 10 \cdot 20 + 0 \cdot 1$

$= 32{,}000 + 6800 + 200 + 0$

$= 39{,}000$

133. There are 6 tiers | rows of blocks, each have dimension of $x \cdot x \cdot x = x^3$. The topmost tier has 5; the second tier has $5 + 1$, or 6; the third tier has $6 + 1$, or 7; the fourth tier has $7 + 1$, or 8; the bottom two tiers both have $8 + 1$ or 9. $5 + 6 + 7 + 8 + 9 + 9 = 44$. The volume is $44x^3$.

Chapter 1 Study Summary

1. $3 + 5c - d = 3 + 5 \cdot 3 - 10 = 3 + 15 - 10 = 8$

2. $A = lw = (8 \text{ ft})(\frac{1}{2} \text{ ft}) = 4 \text{ ft}^2$

3. Let $n =$ the unknown number.
$78 = n - 92$

4. $29 + t = 29 + 13 = 42 \neq 43$, so $t = 13$ is NOT a solution to the equation $29 + t = 43$.

5. $6 + 10n = 10n + 6$

6. $3(ab) = (3a)b$

7. $10(5m + 9n + 1) = 10 \cdot 5m + 10 \cdot 9n + 10 \cdot 1$
$= 50m + 90n + 10$

8. $26x + 13 = 13 \cdot 2x + 13 \cdot 1 = 13(2x + 1)$

9. Whole numbers are numbers in the set: $\{0, 1, 2, 3, \ldots\}$. Only 7 and 0 are in this set.

10. $15 = 3 \cdot 5$. It is a composite number.

11. $84 = 2 \cdot 42 = 2 \cdot 2 \cdot 21 = 2 \cdot 2 \cdot 3 \cdot 7$

12. If $t \neq 0$, then $\dfrac{t}{t} = 1$. $\dfrac{t}{t}$ is undefined at $t = 0$.

13. $\dfrac{9}{10} \cdot \dfrac{13}{13} = \dfrac{9}{10} \cdot 1 = \dfrac{9}{10}$

14. $\dfrac{2}{3} + \dfrac{5}{6} = \dfrac{2}{3} \cdot \dfrac{2}{2} + \dfrac{5}{6} = \dfrac{4}{6} + \dfrac{5}{6}$
$= \dfrac{4 + 5}{6} = \dfrac{9}{6} = \dfrac{3 \cdot 3}{2 \cdot 3} = \dfrac{3}{2}$

15. $\dfrac{3}{4} - \dfrac{3}{10} = \dfrac{3}{4} \cdot \dfrac{5}{5} - \dfrac{3}{10} \cdot \dfrac{2}{2}$
$= \dfrac{15}{20} - \dfrac{6}{20} = \dfrac{15 - 6}{20} = \dfrac{9}{20}$

16. $\dfrac{15}{14} \cdot \dfrac{35}{9} = \dfrac{3 \cdot 5}{2 \cdot 7} \cdot \dfrac{5 \cdot 7}{3 \cdot 3} = \dfrac{5 \cdot 5}{2 \cdot 3} = \dfrac{25}{6}$

17. $15 \div \frac{3}{5} = \frac{15}{1} \cdot \frac{5}{3} = \frac{\cancel{3} \cdot 5}{1} \cdot \frac{5}{\cancel{3}} = 25$

18. Integers are numbers in the set:
$\{...,-3,-2,-1,0,1,2,3,...\}$.
$0, -15,$ and $\frac{30}{3} = \frac{\cancel{3} \cdot 10}{\cancel{3}} = 10$ are in this set.

19. $\frac{10}{9} = 9\overline{)10.000}$, so $-\frac{10}{9} = -1.\overline{1}$
$\phantom{\frac{10}{9} = 9)}\underline{9}$
$\phantom{\frac{10}{9} = 9)}10$

20. -15 is to the right of -16 on the number line, so $-15 > -16$. The statement is FALSE.

21. $|-1.5| = 1.5$

22. $-15 + (-10) + 20$
$= -(15 + 10) + 20$
$= -25 + 20 = -(25 - 20) = -5$

23. $-2.9 + 0 = -2.9$

24. $-(-x) = x = -12$

25. $6 - (-9) = 6 + 9 = 15$

26. $3c + d - 10c - 2 + 8d$
$= (3 - 10)c + (1 + 8)d - 2$
$= -(10 - 3)c + 9d - 2$
$= -7c + 9d - 2$

27. The product of two negative numbers is positive: $-3(-7) = 21$

28. $10 \div (-2.5) = \frac{10}{-2.5} = -\frac{10}{2.5} \cdot \frac{10}{10} = -\frac{100}{25} = -4$

29. $-10^2 = -10 \cdot 10 = -100$

30. $120 \div (-10) \cdot 2 - 3(4 - 5)$
$= [120 \div (-10)] \cdot 2 - 3(-1)$
$= -12 \cdot 2 - (-3)$
$= -24 + 3 = -(24 - 3) = -21$

31. $-(-a + 2b - 3c)$
$= -1 \cdot (-a) + (-1) \cdot (2b) - (-1) \cdot (3c)$
$= a + (-2b) - (-3c) = a - 2b + 3c$

32. $2m + n - 3(5 - m - 2n) - 12$
$= 2m + n + (-3)(5 - m - 2n) - 12$
$= 2m + n + (-3) \cdot 5 - (-3) \cdot m - (-3) \cdot 2n - 12$
$= 2m + n + (-15) - (-3m) - (-6n) - 12$
$= 2m + n - 15 + 3m + 6n - 12$
$= (2 + 3)m + (1 + 6)n - 15 - 12$
$= 5m + 7n - 27$

Chapter 1 Review Exercises

1. True

2. True

3. False

4. True

5. False

6. False

7. True

8. False

9. False

10. True

11. $5t$; substitute 3 for t; $5 \cdot 3 = 15$

12. $9 - y^2 = 9 - (-5)^2$
$= 9 - (25)$
$= -(25 - 9) = -16$

13. $-10 + a^2 \div (b + 1) = -10 + (5)^2 \div (-6 + 1)$
$= -10 + 25 \div (-5) = -10 + (-5) = -15$

14. 7 less than $z \rightarrow z - 7$

15. Ten more than the product of x and z is:
$xz + 10$, or $10 + xz$.

16. Let b be Brent's speed and w be the wind's speed. Then fifteen times the difference of Brent's speed and the wind's speed is: $15(b-w)$.

17. Substitute 35 for x into the equation

$$\frac{x}{5} = 8$$

$$\frac{35}{5} \overset{?}{=} 8$$

$$7 \neq 8$$

No; 35 is not a solution.

18. Let b represent the number of calories burned per hour backpacking and h represent the number of calories burned per hour housecleaning. Then saying, "Backpacking burns twice as many calories per hour as cleaning" is equivalent to saying: $b = 2h$. Thus if Katie burns 237 calories cleaning, the number of calories she would burn per hour backpacking could be represented by the equation: $b = 2 \cdot 237$.

19. Note that each value of c is 200 times the corresponding value of t,

$$
\begin{array}{ccccc}
 & c & \text{is} & 200 & \text{times} & t \\
 & \downarrow & \downarrow & \downarrow & \downarrow & \downarrow \\
\text{or} & c & = & 200 & \cdot & t
\end{array}
$$

20. $3t + 5 = t \cdot 3 + 5$

21. $(2x + y) + z = 2x + (y + z)$

22. Answers may vary.
$$4(xy) = (4x)y$$
$$4(xy) = 4(yx)$$
$$4(xy) = 4(yx) = (4y)x$$

23. $6(3x + 5y) = 6 \cdot 3x + 6 \cdot 5y$
$$= 18x + 30y$$

24. $8(5x + 3y + 2) = 8 \cdot 5x + 8 \cdot 3y + 8 \cdot 2$
$$= 40x + 24y + 16$$

25. $21x + 15y = 3 \cdot 7x + 3 \cdot 5y$
$$= 3(7x + 5y)$$

26. $35x + 77y + 7 = 7 \cdot 5x + 7 \cdot 11y + 7 \cdot 1$
$$= 7(5x + 11y + 1)$$

27. $52 = 2 \cdot 26 = 2 \cdot 2 \cdot 13$

28. $\dfrac{20}{48} = \dfrac{\cancel{2} \cdot \cancel{2} \cdot 5}{\cancel{2} \cdot \cancel{2} \cdot 2 \cdot 2 \cdot 3} = \dfrac{5}{12}$

29. $\dfrac{18}{8} = \dfrac{\cancel{2} \cdot 3 \cdot 3}{\cancel{2} \cdot 2 \cdot 2} = \dfrac{9}{4}$

30. $\dfrac{5}{12} + \dfrac{4}{9}$ Use 36 as the common denominator

$$= \frac{3}{3} \cdot \frac{5}{12} + \frac{4}{4} \cdot \frac{4}{9}$$

$$= \frac{15}{36} + \frac{16}{36}$$

$$= \frac{15 + 16}{36} = \frac{31}{36}$$

31. $\dfrac{9}{16} \div 3 = \dfrac{9}{16} \cdot \dfrac{1}{3}$ Multiply by the reciprocal of the divisor

$$= \frac{\cancel{3} \cdot 3 \cdot 1}{16 \cdot \cancel{3}}$$

$$= \frac{3}{16}$$

32. $\dfrac{2}{3} - \dfrac{1}{15} = \dfrac{5}{5} \cdot \dfrac{2}{3} - \dfrac{1}{15}$ Use 15 as the common denominator

$$= \frac{5 \cdot 2}{15} - \frac{1}{15}$$

$$= \frac{10 - 1}{15} = \frac{9}{15}$$

$$= \frac{\cancel{3} \cdot 3}{\cancel{3} \cdot 5} = \frac{3}{5}$$

33. $\dfrac{9}{10} \cdot \dfrac{16}{5} = \dfrac{9 \cdot 16}{10 \cdot 5} = \dfrac{9 \cdot \cancel{2} \cdot 8}{\cancel{2} \cdot 5 \cdot 5} = \dfrac{72}{25}$

34. 172 corresponds to the highest dive. -820 corresponds to the deepest free dive.

35. $-\dfrac{1}{3}$ is $\dfrac{1}{3}$ of a unit to the left of zero:

36. $-3 < x$ has the same meaning as $x > -3$.

37. $0 \le -1$. False; 0 is *not* left of -1.

38. $-\dfrac{7}{8} = -\left(\dfrac{7}{8}\right) = -(7 \div 8)$, so we divide.

$$\begin{array}{r} 0.875 \\ 8\overline{)7.000} \\ \underline{6\,4} \\ 60 \\ \underline{56} \\ 40 \\ \underline{40} \\ 0 \end{array}$$

$\dfrac{7}{8} = 0.875$, so $-\dfrac{7}{8} = -0.875$

39. $|-1| = 1$, since -1 is 1 unit from 0.

40. $-(-x) = x = -9$

41. $-3 + (-7) = -10$

42. $-\dfrac{2}{3} + \dfrac{1}{12} = \dfrac{-8}{12} + \dfrac{1}{12}$ The absolute values are $\frac{8}{12}$ and $\frac{1}{12}$. The difference is $\frac{8}{12} - \frac{1}{12}$, or $\frac{8-1}{12} = \frac{7}{12}$. The negative number has the greater absolute value, so the answer is negative. $-\frac{2}{3} + \frac{1}{12} = -\frac{7}{12}$

43. $10 + (-9) + (-8) + 7 = (10+7) + \big[(-9) + (-8)\big]$
    Using the commutative and associative laws of addition.
    $= 17 + (-17) = 0$. The numbers have the same absolute value.

44. $-3.8 + 5.1 + (-12) + (-4.3) + 10$
    $= (5.1 + 10) + \big[(-3.8) + (-12) + (-4.3)\big]$
    $= 15.1 + [-20.1]$ Adding positive nos. and adding negative nos.
    $= -5$ Adding a positive and a negative nos.

45. $-2 - (-7) = -2 + 7 = 5$

46. $\dfrac{1}{2} - \dfrac{9}{10} = \dfrac{1}{2} \cdot \dfrac{5}{5} - \dfrac{9}{10}$

$= \dfrac{5}{10} - \dfrac{9}{10} = \dfrac{5-9}{10}$

$= \dfrac{-4}{10} = -\dfrac{4}{10} = -\dfrac{2}{5}$

47. $-3.8 - 4.1 = -3.8 + (-4.1) = -7.9$

48. $-9 \cdot (-6) = 9 \cdot 6 = 54$

49. $-2.7(3.4) = -(2.7 \cdot 3.4) = -9.18$

50. $\dfrac{2}{3} \cdot \left(-\dfrac{3}{7}\right) = -\left(\dfrac{2}{3} \cdot \dfrac{3}{7}\right) = -\left(\dfrac{2 \cdot \cancel{3}}{\cancel{3} \cdot 7}\right) = -\dfrac{2}{7}$

51. $2 \cdot (-7) \cdot (-2) \cdot (-5) = -14 \cdot 10 = -140$

52. $35 \div (-5) = -7$

53. $-5.1 \div 1.7 = -3$

54. $-\dfrac{3}{5} \div \left(-\dfrac{4}{5}\right) = \dfrac{3}{5} \cdot \dfrac{5}{4} = \dfrac{3 \cdot \cancel{5}}{\cancel{5} \cdot 4} = \dfrac{3}{4}$

55. $|-3 \cdot 4 - 12 \cdot 2| - 8(-7) = |-12 - 24| - (-56)$
    $= |-36| + 56$
    $= 36 + 56$
    $= 92$

56. $16 \div (-2)^3 - 5[3 - 1 + 2(4 - 7)]$
    $= 16 \div (-8) - 5[2 + 2(-3)]$
    $= -2 - 5[2 + (-6)] = -2 - 5(-4)$
    $= -2 - (-20) = -2 + 20 = 18$

57. $120 - 6^2 \div 4 \cdot 8 = 120 - 36 \div 4 \cdot 8$
    $= 120 - 9 \cdot 8$
    $= 120 - 72$
    $= 48$

58. $(120 - 6^2) \div 4 \cdot 8 = (120 - 36) \div 4 \cdot 8$
    $= 84 \div 4 \cdot 8$
    $= 21 \cdot 8$
    $= 168$

59. $\left(120-6^2\right)\div\left(4\cdot8\right)=\left(120-36\right)\div\left(4\cdot8\right)$

$\qquad\qquad = 84\div32$

$\qquad\qquad = \dfrac{84}{32}=\dfrac{\cancel{4}\cdot21}{\cancel{4}\cdot8}=\dfrac{21}{8}$

60. $\dfrac{4\left(18-8\right)+7\cdot9}{9^2-8^2}=\dfrac{4\left(18-8\right)+7\cdot9}{81-64}$

$\qquad\qquad = \dfrac{4\left(10\right)+7\cdot9}{81-64}$

$\qquad\qquad = \dfrac{40+63}{81-64}$

$\qquad\qquad = \dfrac{103}{17}$

61. $11a+2b+\left(-4a\right)+\left(-5b\right)$

$\qquad = 11a+\left(-4a\right)+2b+\left(-5b\right)$

$\qquad = \left[11+\left(-4\right)\right]a=\left[2+\left(-5\right)\right]b$

$\qquad = 7a-3b$

62. $7x-3y-9x+8y$

$\qquad = 7x+\left(-3y\right)+\left(-9x\right)+8y$

$\qquad = 7x+\left(-9x\right)+\left(-3y\right)+8y$

$\qquad = \left[7+\left(-9\right)\right]x+\left(-3+8\right)y$

$\qquad = -2x+5y$

63. The opposite of -7 is 7, since $7+\left(-7\right)=0$

64. The reciprocal of -7 is $-\frac{1}{7}$, since $-\frac{1}{7}\cdot-7=1$

65. $\underbrace{2x\cdot2x\cdot2x\cdot2x}_{\text{4 Factors}}=\left(2x\right)^4$

66. $\left(-5x\right)^3=-5x\cdot\left(-5x\right)\cdot\left(-5x\right)$

$\qquad = -5\cdot\left(-5\right)\cdot\left(-5\right)\cdot x\cdot x\cdot x$

$\qquad = -125x^3$

67. $2a-\left(5a-9\right)=2a+\left(-5a\right)+9$

$\qquad\qquad = \left[2+\left(-5\right)\right]a+9=-3a+9$

68. $11x^4+2x+8(x-x^4)=11x^4+2x+8x-8x^4$

$\qquad = (11-8)x^4+(2+8)x=3x^4+10x$

69. $2n^2-5(-3n^2+m^2-4mn)+6m^2$

$\qquad = 2n^2-5\cdot(-3n^2)-5\cdot m^2-5\cdot(-4mn)+6m^2$

$\qquad = 2n^2+15n^2-5m^2+20mn+6m^2$

$\qquad = (2+15)n^2+(-5+6)m^2+20mn$

$\qquad = 17n^2+m^2+20mn$

70. $8\left(x+4\right)-6-\left[3\left(x-2\right)+4\right]$

$\qquad = 8x+32-6-[3x-6+4]$

$\qquad = 8x+26-[3x-2]$

$\qquad = 8x+26-3x+2$

$\qquad = (8-3)x+28=5x+28$

71. *Thinking and Writing Exercise.* The value of a constant never varies. A variable can represent a variety of numbers.

72. *Thinking and Writing Exercise.* A term is one of the parts of an expression that is separated from the other parts by plus signs. A factor is part of a product, or term.

73. *Thinking and Writing Exercise.* The distributive law is used in factoring algebraic expressions, multiplying algebraic expressions, combining like terms, finding the opposite of a sum, and subtracting algebraic expressions.

74. *Thinking and Writing Exercise.* A negative number raised to an even exponent is positive; a negative number raised to an odd exponent is negative.

75. Substitute 1 for a, 2 for b, and evaluate:

$a^{50}-20a^{25}b^4+100b^8$

$\qquad = 1^{50}-20\cdot1^{25}2^4+100\cdot2^8$

$\qquad = 1-20\cdot1\cdot16+100\cdot256$

$\qquad = 1-320+25{,}600$

$\qquad = 25{,}281$

76. a.   Since $0.090909\cdots+0.181818\cdots=$

      we have $\dfrac{1}{11}+\dfrac{2}{11}=\dfrac{3}{11}$; $0.272727\cdots=\dfrac{3}{11}$

   b.   Since $10\cdot0.090909\cdots=0.909090\cdots$,

      we have $10\cdot\dfrac{1}{11}=\dfrac{10}{11}$; $0.909090\cdots=\dfrac{10}{11}$

77. $-\left|\dfrac{7}{8}-\left(-\dfrac{1}{2}\right)-\dfrac{3}{4}\right|$ Use 8: the common denominator

$-\left|\dfrac{7}{8}-\left(-\dfrac{4}{8}\right)-\dfrac{6}{8}\right| = -\left|\dfrac{7}{8}+\dfrac{4}{8}+\dfrac{-6}{8}\right|$

$\qquad\qquad\qquad = -\left|\dfrac{5}{8}\right| = -\dfrac{5}{8}$

78. $\left(|2.7-3|+3^2-|-3|\right)\div(-3)$

$= \left(|2.7-3|+9-|-3|\right)\div(-3)$

$= \left(|-0.3|+9-|-3|\right)\div(-3)$

$= (0.3+9-3)\div(-3)$

$= 6.3\div(-3)$

$= -2.1$

79. i

80. j

81. a

82. h

83. k

84. b

85. c

86. e

87. d

88. f

89. g

Chapter 1 Test

1. Substitute 10 for x and 5 for y.

   $\dfrac{2x}{y} = \dfrac{2\cdot 10}{5} = \dfrac{2\cdot 2\cdot 5}{5} = 4$

2. Let x and y represent the numbers: $xy-9$

3. $A = \dfrac{1}{2}\cdot b\cdot h$

$= \dfrac{1}{2}\cdot(16\text{ ft})\cdot(30\text{ ft})$

$= \dfrac{1}{2}\cdot 16\cdot 30\cdot\text{ft}\cdot\text{ft}$

$= 8\cdot 30\text{ ft}\cdot\text{ft}$

$= 240\text{ ft}^2$, or 240 square feet

4. $3p+q = q+3p$

5. $x\cdot(4\cdot y) = (x\cdot 4)\cdot y$

6. Substitute 7 for x into the equation.
   $65-x=69$

   $65-7\overset{?}{=}69$

   $58\neq 69$

   No; 7 is not a solution.

7. Let $x =$ the number of tamarins that live in zoos. Then translate, "The number of tamarins that live in the wild equals the number of tamarins that live in zoos plus 1050."
   $1500 = x+1050$

8. $7(5+x) = 7\cdot 5+7\cdot x = 35+7x$

9. $-5(y-2) = -5\cdot y-5(-2) = -5y+10$

10. $11+44x = 11\cdot 1+11\cdot 4x = 11(1+4x)$

11. $7x+7+14y = 7\cdot x+7\cdot 1+7\cdot 2y$
    $\qquad\qquad\qquad = 7(x+1+2y)$

12. $300 = 2\cdot 150 = 2\cdot 2\cdot 75$
    $\qquad = 2\cdot 2\cdot 3\cdot 25 = 2\cdot 2\cdot 3\cdot 5\cdot 5$

13. $\dfrac{10}{35} = \dfrac{2\cdot 5}{5\cdot 7} = \dfrac{2}{7}$

14. $-4 < 0$, since -4 is left of 0.

15. $-3 > -8$, since -3 is right of -8.

16. $\left|\dfrac{9}{4}\right| = \dfrac{9}{4}$, since $\frac{9}{4}$ is $\frac{9}{4}$ units from 0.

17. $|-3.8| = 3.8$

18. $\frac{2}{3}$, since $\frac{2}{3} + \left(-\frac{2}{3}\right) = 0$

19. $-\frac{7}{4}$, since $\left(-\frac{7}{4}\right) \cdot \left(-\frac{4}{7}\right) = 1$

20. Let $-x = -(-10) = 10$

21. $x \le -5$ has the same meaning as $-5 \ge x$.

22. $3.1 - (-4.7) = 3.1 + 4.7 = 7.8$

23. $-8 + 4 + (-7) + 3 = \left[-8 + (-7)\right] + \left[4 + 3\right]$
$\quad\quad\quad\quad\quad\quad = -15 + 7 = -8$

24. $3.2 - 5.7 = 3.2 + (-5.7) = -2.5$

25. $-\frac{1}{8} - \frac{3}{4} = -\frac{1}{8} - \frac{3}{4} \cdot \frac{2}{2} = -\frac{1}{8} - \frac{6}{8}$
$\quad\quad\quad = \frac{-1 + (-6)}{8} = -\frac{7}{8}$

26. $4 \cdot (-12) = -(4 \cdot 12) = -48$

27. $-\frac{1}{2} \cdot \left(-\frac{4}{9}\right) = \frac{1}{2} \cdot \frac{4}{9} = \frac{1 \cdot 4}{2 \cdot 9} = \frac{1 \cdot \cancel{2} \cdot 2}{\cancel{2} \cdot 9} = \frac{2}{9}$

28. $-66 \div 11 = -6$

29. $-\frac{3}{5} \div \left(-\frac{4}{5}\right) = \frac{3}{5} \cdot \frac{5}{4} = \frac{3 \cdot \cancel{5}}{\cancel{5} \cdot 4} = \frac{3}{4}$

30. $4.864 \div (-0.5) = -9.728$

31. $10 - 2(-16) \div 4^2 + |2 - 10|$
$\quad = 10 - 2(-16) \div 16 + |-8|$
$\quad = 10 + 32 \div 16 + 8$
$\quad = 10 + 2 + 8 = 20$

32. $256 \div (-16) \cdot 4 = -256 \div 16 \cdot 4 = -16 \cdot 4 = -64$

33. $2^3 - 10\left[4 - (-2 + 18)3\right]$
$\quad = 8 - 10\left[4 - (-2 + 18)3\right]$
$\quad = 8 - 10\left[4 - 16 \cdot 3\right]$
$\quad = 8 - 10\left[4 - 48\right]$
$\quad = 8 - 10\left[4 + (-48)\right]$
$\quad = 8 - 10\left[-44\right]$
$\quad = 8 - (-440) = 8 + 440 = 448$

34. $-18y + 30a - 9a + 4y$
$\quad = (30 - 9)a + (-18 + 4)y$
$\quad = 21a + (-14)y$
$\quad = 21a - 14y$

35. $(-2x)^4 = -2x \cdot (-2x) \cdot (-2x) \cdot (-2x)$
$\quad\quad\quad = -2 \cdot (-2) \cdot (-2) \cdot (-2) \cdot x \cdot x \cdot x \cdot x$
$\quad\quad\quad = 16x^4$

36. $4x - (3x - 7) = 4x - 3x + 7$
$\quad\quad\quad\quad\quad = (4 - 3)x + 7 = x + 7$

37. $4(2a - 3b) + a - 7 = 8a - 12b + a - 7$
$\quad\quad\quad\quad\quad\quad = (8a + a) - 12b - 7$
$\quad\quad\quad\quad\quad\quad = (8 + 1)a - 12b - 7$
$\quad\quad\quad\quad\quad\quad = 9a - 12b - 7$

38. $3\left[5(y - 3) + 9\right] - 2(8y - 1)$
$\quad = 3[5y - 15 + 9] - 16y + 2$
$\quad = 3[5y - 6] - 16y + 2$
$\quad = 15y - 18 - 16y + 2$
$\quad = (15 - 16)y - 18 + 2$
$\quad = (-1)y - 16 = -y - 16$

39. y is 4 less than half of x

Rewording: $\underbrace{\text{half of } x}$ $\underbrace{\text{less } 4}$ is y

  Translating: $\frac{1}{2}x$ $\quad -4$ $= y$

Since $x = 20$, $\frac{1}{2}(20) - 4 = y$

$$10 - 4 = y$$
$$y = 6$$

Substitute 20 for x and 6 for y.

$$\frac{5y - x}{2} = \frac{5 \cdot 6 - 20}{2}$$
$$= \frac{30 - 20}{2} = \frac{10}{2} = 5$$

40. $9 - (3 - 4) + 5 = 15$

41. $|-27 - 3(4)| - |-36| + |-12|$

$= |-27 - 12| - |-36| + |-12|$

$= |-27 + (-12)| - |-36| + |-12|$

$= |-39| - |-36| + |-12|$

$= 39 - 36 + 12$

$= 15$

42. $a - \left\{ 3a - \left[4a - (2a - 4a) \right] \right\}$

$= a - \left\{ 3a - \left[4a - (-2a) \right] \right\}$

$= a - \left\{ 3a - \left[4a + 2a \right] \right\}$

$= a - \left\{ 3a - 6a \right\}$

$= a - \left\{ -3a \right\}$

$= a + 3a = 4a$

43. Let $a = 2$, $b = 3$, $c = 4$.

$a|b - c| = |ab| - |ac|$

$2|3 - 4| \overset{?}{=} |2 \cdot 3| - |2 \cdot 4|$

$2|-1| \overset{?}{=} |6| - |8|$

$2 \cdot 1 \overset{?}{=} 6 - 8$

$2 \neq -2$

False.

Chapter 2

Equations, Inequalities, and Problem Solving

1. c

3. f

5. d

7. d

9. c

11. $6 - x = -2$

$$\frac{6 - 4 \mid -2}{2 \mid -2} \quad \text{FALSE}$$

No, 4 is not a solution.

13. $\frac{2}{3}t = 12$

$$\frac{\frac{2}{3}(18) \mid 12}{12 \mid 12} \quad \text{TRUE}$$

Yes, 18 is a solution.

15. $x + 7 = 3 - x$

$$\frac{-2 + 7 \mid 3 - (-2)}{5 \mid 5} \quad \text{TRUE}$$

Yes, -2 is a solution.

17. $4 - \frac{1}{5}n = 8$

$$\frac{4 - \frac{1}{5}(-20) \mid 8}{4 - (-4) \mid 8}$$
$$8 \mid 8 \quad \text{TRUE}$$

Yes, -20 is a solution.

19. $x + 6 = 23$

$x + 6 - 6 = 23 - 6$ Subtracting 6 from both sides

$x = 17$ Simplifying

Check: $x + 6 = 23$

$$\frac{17 + 6 \mid 23}{23 \mid 23} \quad \text{TRUE}$$

The solution is 17.

21. $y + 7 = -4$

$y + 7 - 7 = -4 - 7$ Subtracting 7 from both sides

$y = -11$ Simplifying

Check: $y + 7 = -4$

$$\frac{-11 + 7 \mid -4}{-4 \mid -4} \quad \text{TRUE}$$

The solution is -11.

23. $-6 = y + 25$

$-6 - 25 = y + 25 - 25$ Subtracting 25 from both sides

$-31 = y$ Simplifying

Check: $-6 = y + 25$

$$\frac{-6 \mid -31 + 25}{-6 \mid -6} \quad \text{TRUE}$$

The solution is -31.

25. $x - 8 = 5$

$x - 8 + 8 = 5 + 8$ Adding 8 to both sides

$x = 13$ Simplifying

Check: $x - 8 = 5$

$$\frac{13 - 8 \mid 5}{5 \mid 5} \quad \text{TRUE}$$

The solution is 13.

27. $12 = -7 + y$

$7 + 12 = 7 + (-7) + y$ Adding 7 to both sides

$19 = y$ Simplifying

Check: $12 = -7 + y$

$$\frac{12 \mid -7 + 19}{12 \mid 12} \quad \text{TRUE}$$

The solution is 19.

29.
$$-5 + t = -9$$
$$5 + (-5) + t = 5 + (-9) \qquad \text{Adding 5 to both sides}$$
$$t = -4 \qquad \text{Simplifying}$$

Check:
$$\begin{array}{c|c} -5 + t = -9 & \\ \hline -5 + (-4) & -9 \\ -9 & -9 \quad \text{TRUE} \end{array}$$

The solution is –4.

31.
$$r + \tfrac{1}{3} = \tfrac{8}{3}$$
$$r + \tfrac{1}{3} + \left(-\tfrac{1}{3}\right) = \tfrac{8}{3} + \left(-\tfrac{1}{3}\right) \qquad \text{Subtracting } \tfrac{1}{3} \text{ from both sides}$$
$$r = \tfrac{7}{3} \qquad \text{Simplifying}$$

Check:
$$\begin{array}{c|c} r + \tfrac{1}{3} = \tfrac{8}{3} & \\ \hline \tfrac{7}{3} + \tfrac{1}{3} & \tfrac{8}{3} \\ \tfrac{8}{3} & \tfrac{8}{3} \quad \text{TRUE} \end{array}$$

The solution is $\tfrac{7}{3}$.

33.
$$x - \tfrac{3}{5} = -\tfrac{7}{10}$$
$$x - \tfrac{3}{5} + \left(\tfrac{3}{5}\right) = -\tfrac{7}{10} + \left(\tfrac{3}{5}\right) \qquad \text{Adding } \tfrac{3}{5} \text{ to both sides}$$
$$x = -\tfrac{7}{10} + \left(\tfrac{6}{10}\right) \qquad \text{Simplifying}$$
$$x = -\tfrac{1}{10} \qquad \text{Simplifying}$$

Check:
$$\begin{array}{c|c} x - \tfrac{3}{5} = -\tfrac{7}{10} & \\ \hline -\tfrac{1}{10} - \tfrac{3}{5} & -\tfrac{7}{10} \\ -\tfrac{1}{10} - \tfrac{6}{10} & -\tfrac{7}{10} \\ -\tfrac{7}{10} & -\tfrac{7}{10} \quad \text{TRUE} \end{array}$$

The solution is $-\tfrac{1}{10}$.

35.
$$-\tfrac{1}{5} + z = -\tfrac{1}{4}$$
$$\tfrac{1}{5} + \left(-\tfrac{1}{5}\right) + z = \tfrac{1}{5} + \left(-\tfrac{1}{4}\right) \qquad \text{Adding } \tfrac{1}{5} \text{ to both sides}$$
$$z = \tfrac{4}{20} + \left(-\tfrac{5}{20}\right) \qquad \text{Simplifying}$$
$$z = -\tfrac{1}{20} \qquad \text{Simplifying}$$

Check:
$$\begin{array}{c|c} -\tfrac{1}{5} + z = -\tfrac{1}{4} & \\ \hline -\tfrac{1}{5} + \left(-\tfrac{1}{20}\right) & -\tfrac{1}{4} \\ -\tfrac{4}{20} + \left(-\tfrac{1}{20}\right) & -\tfrac{1}{4} \\ -\tfrac{5}{20} & -\tfrac{1}{4} \\ -\tfrac{1}{4} & -\tfrac{1}{4} \quad \text{TRUE} \end{array}$$

The solution is $-\tfrac{1}{20}$.

37.
$$m + 3.9 = 5.4$$
$$m + 3.9 - 3.9 = 5.4 - 3.9 \qquad \text{Subtracting 3.9 from both sides}$$
$$m = 1.5 \qquad \text{Simplifying}$$

Check:
$$\begin{array}{c|c} m + 3.9 = 5.4 & \\ \hline 1.5 + 3.9 & 5.4 \\ 5.4 & 5.4 \quad \text{TRUE} \end{array}$$

The solution is 1.5.

39.
$$-9.7 = -4.7 + y$$
$$4.7 + -9.7 = 4.7 + (-4.7) + y \qquad \text{Adding 4.7 to both sides}$$
$$-5 = y \qquad \text{Simplifying}$$

Check:
$$\begin{array}{c|c} -9.7 = -4.7 + y & \\ \hline -9.7 & -4.7 + (-5) \\ -9.7 & -9.7 \quad \text{TRUE} \end{array}$$

The solution is –5.

41.
$$5x = 70$$
$$\frac{5x}{5} = \frac{70}{5} \qquad \text{Dividing both sides by 5}$$
$$1 \cdot x = 14 \qquad \text{Simplifying}$$
$$x = 14 \qquad \text{Identity property of 1}$$

Check:
$$\begin{array}{c|c} 5x = 70 & \\ \hline 5 \cdot 14 & 70 \\ 70 & 70 \quad \text{TRUE} \end{array}$$

The solution is 14.

43.
$$84 = 7n$$
$$\frac{84}{7} = \frac{7n}{7} \qquad \text{Dividing both sides by 7}$$
$$12 = 1 \cdot n \qquad \text{Simplifying}$$
$$12 = n \qquad \text{Identity property of 1}$$

Check:
$$\begin{array}{c|c} 84 = 7x & \\ \hline 84 & 7 \cdot 12 \\ 84 & 84 \quad \text{TRUE} \end{array}$$

The solution is 12.

45. $\quad -x = 23$

$\quad (-1)(-x) = (-1)23$ Multiply both sides by -1

$\quad\quad x = -23$ Simplifying

Check: $\quad -x = 23$

$$\begin{array}{c|c} -(-23) & 23 \\ \hline 23 & 23 \end{array} \quad \text{TRUE}$$

The solution is -23.

47. $\quad -t = -8$

$\quad (-1)(-t) = (-1)(-8)$ Multiply both sides by -1

$\quad\quad t = 8$ Simplifying

Check: $\quad -t = -8$

$$\begin{array}{c|c} -(8) & -8 \\ \hline -8 & -8 \end{array} \quad \text{TRUE}$$

The solution is 8.

49. $\quad 2x = -5$

$\quad \dfrac{2x}{2} = \dfrac{-5}{2}$ Dividing both sides by 2

$\quad 1 \cdot x = -\dfrac{5}{2}$ Simplifying

$\quad\quad x = -\dfrac{5}{2}$ Identity property of 1

Check: $\quad 2x = -5$

$$\begin{array}{c|c} 2\left(-\frac{5}{2}\right) & -5 \\ \hline -5 & -5 \end{array} \quad \text{TRUE}$$

The solution is $-\frac{5}{2}$.

51. $\quad -1.3a = -10.4$

$\quad \dfrac{-1.3a}{-1.3} = \dfrac{-10.4}{-1.3}$ Dividing both sides by -1.3

$\quad 1 \cdot a = 8$ Simplifying

$\quad\quad a = 8$ Identity property of 1

Check: $\quad -1.3a = -10.4$

$$\begin{array}{c|c} -1.3(8) & -10.4 \\ \hline -10.4 & -10.4 \end{array} \quad \text{TRUE}$$

The solution is 8.

53. $\quad \dfrac{y}{-8} = 11$

$\quad \left(-\dfrac{1}{8}\right) \cdot y = 11$ Rewriting

$\quad (-8)\left(-\dfrac{1}{8}\right) \cdot y = (-8)11$ Multiplying both sides by 8

$\quad\quad 1 \cdot y = -88$ Simplifying

$\quad\quad y = -88$ Identity property of 1

Check: $\quad \dfrac{y}{-8} = 11$

$$\begin{array}{c|c} \frac{-88}{-8} & 11 \\ \hline 11 & 11 \end{array} \quad \text{TRUE}$$

The solution is -88.

55. $\quad \dfrac{4}{5}x = 16$

$\quad \dfrac{5}{4} \cdot \left(\dfrac{4}{5}\right) \cdot x = \dfrac{5}{4} \cdot 16$ Multiplying both sides by $\frac{5}{4}$

$\quad\quad 1 \cdot x = 20$ Simplifying

$\quad\quad x = 20$ Identity property of 1

Check: $\quad \dfrac{4}{5}x = 16$

$$\begin{array}{c|c} \frac{4}{5} \cdot 20 & 16 \\ \hline 16 & 16 \end{array} \quad \text{TRUE}$$

The solution is 20.

57. $\quad \dfrac{-x}{6} = 9$

$\quad 6 \cdot \left(\dfrac{1}{6}\right) \cdot (-x) = 6 \cdot 9$ Rewriting and multiplying both sides by 6

$\quad\quad -x = 54$ Simplifying

$\quad\quad x = -54$ Multiplying both sides by -1

Check: $\quad \dfrac{-x}{6} = 9$

$$\begin{array}{c|c} \frac{-(-54)}{6} & 9 \\ \frac{54}{6} & 9 \\ \hline 9 & 9 \end{array} \quad \text{TRUE}$$

The solution is -54.

59. $\dfrac{1}{9} = \dfrac{z}{5}$

$5 \cdot \left(\dfrac{1}{9}\right) = 5 \cdot \left(\dfrac{1}{5}\right) \cdot z$ Rewriting and multiplying both sides by 5

$\dfrac{5}{9} = z$ Simplifying

Check: $\dfrac{1}{9} = \dfrac{z}{5}$

$\begin{array}{c|c} \frac{1}{9} & \left(\frac{5}{9}\right)/5 \\ \hline \frac{1}{9} & \frac{1}{9} \end{array}$ TRUE

The solution is $\dfrac{5}{9}$.

61. $-\dfrac{3}{5}r = -\dfrac{3}{5}$

Note: The solution of this equation is the number multiplied by $-\frac{3}{5}$ to get $-\frac{3}{5}$. That number is 1. We could also do this exercise as follows:

$-\dfrac{3}{5}r = -\dfrac{3}{5}$

$\left(-\dfrac{5}{3}\right)\left(-\dfrac{3}{5}\right)r = \left(-\dfrac{5}{3}\right)\left(-\dfrac{3}{5}\right)$ Multiplying both sides by $-\frac{5}{3}$

$r = 1$ Simplifying

Check: $-\dfrac{3}{5}r = -\dfrac{3}{5}$

$\begin{array}{c|c} -\frac{3}{5} \cdot 1 & -\frac{3}{5} \\ \hline -\frac{3}{5} & -\frac{3}{5} \end{array}$ TRUE

The solution is 1.

63. $\dfrac{-3r}{2} = -\dfrac{27}{4}$

$\left(-\dfrac{3}{2}\right)r = -\dfrac{27}{4}$ Rewriting

$\left(-\dfrac{2}{3}\right)\left(-\dfrac{3}{2}\right)r = \left(-\dfrac{2}{3}\right)\left(-\dfrac{27}{4}\right)$ Multiplying both sides by $-\frac{2}{3}$

$r = \dfrac{9}{2}$ Simplifying

Check: $\dfrac{-3r}{2} = -\dfrac{27}{4}$

$\begin{array}{c|c} \left[(-3)\left(\frac{9}{2}\right)\right]/2 & -\frac{27}{4} \\ \hline \left(-\frac{27}{2}\right)/2 & -\frac{27}{4} \\ -\frac{27}{4} & -\frac{27}{4} \end{array}$ TRUE

The solution is $\dfrac{9}{2}$.

65. $4.5 + t = -3.1$

$4.5 + t - 4.5 = -3.1 - 4.5$

$t = -7.6$

The solution is -7.6.

67. $-8.2x = 20.5$

$\dfrac{-8.2}{-8.2}x = \dfrac{20.5}{-8.2}$

$x = -2.5$

The solution is -2.5.

69. $x - 4 = -19$

$x - 4 + 4 = -19 + 4$

$x = -15$

The solution is -15.

71. $-12x = 72$

$\dfrac{-12}{-12}x = \dfrac{72}{-12}$

$x = -6$

The solution is -6.

73. $48 = -\dfrac{3}{8}y$

$\left(-\dfrac{8}{3}\right) \cdot 48 = \left(-\dfrac{8}{3}\right)\left(-\dfrac{3}{8}\right)y$

$-128 = y$

The solution is -128.

75. $a - \dfrac{1}{6} = -\dfrac{2}{3}$

$a - \dfrac{1}{6} + \dfrac{1}{6} = -\dfrac{2}{3} + \dfrac{1}{6}$

$a = -\dfrac{4}{6} + \dfrac{1}{6}$

$a = -\dfrac{3}{6}$

$a = -\dfrac{1}{2}$

The solution is $-\dfrac{1}{2}$.

77.
$$-24 = \frac{8x}{5}$$
$$\left(\frac{5}{8}\right)(-24) = \left(\frac{5}{8}\right)\left(\frac{8}{5}\right)x$$
$$-15 = x$$
The solution is –15.

79.
$$-\frac{4}{3}t = -16$$
$$\left(-\frac{3}{4}\right)\left(-\frac{4}{3}\right)t = \left(-\frac{3}{4}\right)(-16)$$
$$t = 12$$
The solution is 12.

81.
$$-483.297 = -794.053 + t$$
$$794.053 - 483.297 = 794.053 - 794.053 + t$$
$$310.756 = t$$
The solution is 310.756.

83. *Thinking and Writing Exercise.*

85. $3 \cdot 4 - 18$
$$= 12 - 18 \qquad \text{Multiplying}$$
$$= -6 \qquad \text{Subtracting}$$

87. $16 \div (2 - 3 \cdot 2) + 5$
$$= 16 \div (2 - 6) + 5 \qquad \text{Simplifying inside}$$
$$= 16 \div (-4) + 5 \qquad \text{the parentheses}$$
$$= -4 + 5 \qquad \text{Dividing}$$
$$= 1 \qquad \text{Adding}$$

89. *Thinking and Writing Exercise.*

91. $mx = 9.4m$
$$\frac{mx}{m} = \frac{9.4m}{m}$$
$$x = 9.4$$
The solution is 9.4.

93. $cx + 5c = 7c$
$$-5c + cx + 5c = -5c + 7c$$
$$cx = 2c$$
$$\frac{cx}{c} = \frac{2c}{c}$$
$$x = 2$$
The solution is 2.

95.
$$7 + |x| = 20$$
$$-7 + 7 + |x| = -7 + 20$$
$$|x| = 13$$
x represents a number whose distance from 0 is 13. Thus, $x = -13$ or $x = 13$.

97.
$$t - 3590 = 1820$$
$$t - 3590 + 7180 = \qquad \text{Adding 7180}$$
$$1820 + 7180 \qquad \text{to both sides}$$
$$t + 3590 = 9000 \qquad \text{Simplifying}$$
It follows that $t + 3590$ equals 9000.

99. To "undo" the last step, divide 22.5 by 0.3.
$22.5 \div 0.3 = 75$
Now divide 75 by 0.3.
$75 \div 0.3 = 250$
The answer should be 250 not 22.5.

Exercise Set 2.2

1. c

3. a

5. b

7.
$$2x + 9 = 25$$
$$2x + 9 - 9 = 25 - 9 \qquad \text{Subtracting 9 from both sides}$$
$$2x = 16 \qquad \text{Simplifying}$$
$$\frac{2x}{2} = \frac{16}{2} \qquad \text{Dividing both sides by 2}$$
$$x = 8 \qquad \text{Simplifying}$$
Check: $\quad 2x + 9 = 25$

$2 \cdot 8 + 9$	25
$16 + 9$	25
	25 $\mid$ 25 TRUE

The solution is 8.

9.
$$6z + 4 = -20$$
$$6z + 4 - 4 = -20 - 4 \qquad \text{Subtracting 4 from both sides}$$
$$6z = -24 \qquad \text{Simplifying}$$
$$\frac{6z}{6} = \frac{-24}{6} \qquad \text{Dividing both sides by 6}$$
$$z = -4 \qquad \text{Simplifying}$$

Check:　$6z + 4 = -20$

$$
\begin{array}{c|c}
6\cdot(-4)+4 & -20 \\
\hline
-24+4 & -20 \\
-20 & -20 \ \text{TRUE}
\end{array}
$$

The solution is –4.

11.　$7t - 8 = 27$

$7t - 8 + 8 = 27 + 8$　Adding 8 to both sides

$7t = 35$　Simplifying

$\dfrac{7t}{7} = \dfrac{35}{7}$　Dividing both sides by 7

$t = 5$　Simplifying

Check:　$7t - 8 = 27$

$$
\begin{array}{c|c}
7\cdot 5 - 8 & 27 \\
\hline
35 - 8 & 27 \\
27 & 27 \ \text{TRUE}
\end{array}
$$

The solution is 5.

13.　$3x - 9 = 33$

$3x - 9 + 9 = 33 + 9$　Adding 9 to both sides

$3x = 42$　Simplifying

$\dfrac{3x}{3} = \dfrac{42}{3}$　Dividing both sides by 3

$x = 14$　Simplifying

Check:　$3x - 9 = 33$

$$
\begin{array}{c|c}
3\cdot 14 - 9 & 33 \\
\hline
42 - 9 & 33 \\
33 & 33 \ \text{TRUE}
\end{array}
$$

The solution is 14.

15.　$-91 = 9t + 8$

$-91 - 8 = 9t + 8 - 8$　Subtracting 8 from both sides

$-99 = 9t$　Simplifying

$\dfrac{-99}{9} = \dfrac{9t}{9}$　Dividing both sides by 9

$t = -11$　Simplifying

Check:　$-91 = 9t + 8$

$$
\begin{array}{c|c}
-91 & 9\cdot(-11)+8 \\
\hline
-91 & -99 + 8 \\
-91 & -91 \quad\quad \text{TRUE}
\end{array}
$$

The solution is –11.

17.　$12 - t = 16$

$-12 + 12 - t = -12 + 16$　Subtracting 12 from both sides

$-t = 4$　Simplifying

$\dfrac{-t}{-1} = \dfrac{4}{-1}$　Dividing both sides by -1

$t = -4$　Simplifying

Check:　$12 - t = 16$

$$
\begin{array}{c|c}
12 - (-4) & 16 \\
\hline
12 + 4 & 16 \\
16 & 16 \ \text{TRUE}
\end{array}
$$

The solution is –4.

19.　$-6z - 18 = -132$

$-6z - 18 + 18 = -132 + 18$　Adding 18 to both sides

$-6z = -114$　Simplifying

$\dfrac{-6z}{-6} = \dfrac{-114}{-6}$　Dividing both sides by -6

$z = 19$　Simplifying

Check:　$-6z - 18 = -132$

$$
\begin{array}{c|c}
-6(19) - 18 & -132 \\
\hline
-114 - 18 & -132 \\
-132 & -132 \ \text{TRUE}
\end{array}
$$

The solution is 19.

21.　$5.3 + 1.2n = 1.94$

$-5.3 + 5.3 + 1.2n = -5.3 + 1.94$　Subtracting 5.3 from both sides

$1.2n = -3.36$　Simplifying

$\dfrac{1.2n}{1.2} = \dfrac{-3.36}{1.2}$　Dividing both sides by 1.2

$x = -2.8$　Simplifying

Check:　$5.3 + 1.2n = 1.94$

$$
\begin{array}{c|c}
5.3 + 1.2(-2.8) & 1.94 \\
\hline
5.3 - 3.36 & 1.94 \\
1.94 & 1.94 \ \text{TRUE}
\end{array}
$$

The solution is –2.8.

23. $4x + 5x = 10$

$9x = 10$ Simplifying

$\dfrac{9x}{9} = \dfrac{10}{9}$ Dividing both sides by 9

$x = \dfrac{10}{9}$ Simplifying

Check: $4x + 5x = 10$

$$\begin{array}{c|c} 4\cdot\frac{10}{9} + 5\cdot\frac{10}{9} & 10 \\ \frac{40}{9} + \frac{50}{9} & 10 \\ \frac{90}{9} & 10 \\ 10 & 10 \quad \text{TRUE} \end{array}$$

The solution is $\dfrac{10}{9}$.

25. $32 - 7x = 11$

$-32 + 32 - 7x = -32 + 11$ Subtracting 32 from both sides

$-7x = -21$ Simplifying

$\dfrac{-7x}{-7} = \dfrac{-21}{-7}$ Dividing both sides by -7

$x = 3$ Simplifying

Check: $32 - 7x = 11$

$$\begin{array}{c|c} 32 - 7\cdot 3 & 11 \\ 32 - 21 & 11 \\ 11 & 11 \quad \text{TRUE} \end{array}$$

The solution is 3.

27. $\frac{3}{5}t - 1 = 8$

$\frac{3}{5}t - 1 + 1 = 8 + 1$ Adding 1 to both sides

$\frac{3}{5}t = 9$ Simplifying

$\left(\frac{5}{3}\right)\left(\frac{3}{5}\right)t = \left(\frac{5}{3}\right)\cdot 9$ Multiplying both sides by $\frac{5}{3}$

$t = 15$ Simplifying

Check: $\frac{3}{5}t - 1 = 8$

$$\begin{array}{c|c} \frac{3}{5}\cdot(15) - 1 & 8 \\ 9 - 1 & 8 \\ 8 & 8 \quad \text{TRUE} \end{array}$$

The solution is 15.

29. $4 + \frac{7}{2}x = -10$

$-4 + 4 + \frac{7}{2}x = -4 - 10$ Subtracting 4 from both sides

$\frac{7}{2}x = -14$ Simplifying

$\left(\frac{2}{7}\right)\left(\frac{7}{2}\right)x = \left(\frac{2}{7}\right)(-14)$ Multiplying both sides by $\frac{2}{7}$

$x = -4$ Simplifying

Check: $4 + \frac{7}{2}x = -10$

$$\begin{array}{c|c} 4 + \frac{7}{2}(-4) & -10 \\ 4 + (-14) & -10 \\ -10 & -10 \quad \text{TRUE} \end{array}$$

The solution is –4.

31. $-\dfrac{3a}{4} - 5 = 2$

$-\dfrac{3}{4}a - 5 = 2$ Rewriting

$-\dfrac{3}{4}a - 5 + 5 = 2 + 5$ Adding 5 to both sides

$-\dfrac{3}{4}a = 7$ Simplifying

$\left(-\dfrac{4}{3}\right)\left(-\dfrac{3}{4}\right)a = \left(-\dfrac{4}{3}\right)\cdot 7$ Multiplying both sides by $-\frac{4}{3}$

$a = -\dfrac{28}{3}$ Simplifying

Check: $-\dfrac{3a}{4} - 5 = 2$

$$\begin{array}{c|c} \left(-\frac{3}{4}\right)\left(-\frac{28}{3}\right) - 5 & 2 \\ 7 - 5 & 2 \\ 2 & 2 \quad \text{TRUE} \end{array}$$

The solution is $-\dfrac{28}{3}$.

33. $2x = x + x$

This equation is true regardless of the replacement for x, so all real numbers are solutions. The equation is an identity.

35. $4x - 6 = 6x$

$4x - 6 - 4x = 6x - 4x$ Adding $-4x$ to both sides

$-6 = 2x$ Simplifying

$\dfrac{-6}{2} = \dfrac{2x}{2}$ Dividing both sides by 2

$-3 = x$ Simplifying

Check: $4x - 6 = 6x$

$4 \cdot (-3) - 6$	$6 \cdot (-3)$	
$-12 - 6$	-18	
-18	-18	TRUE

The solution is –3.

37. $2 - 5y = 26 - y$

$2 - 5y + y = 26 - y + y$ Adding y to both sides

$2 - 4y = 26$ Simplifying

$-2 + 2 - 4y = -2 + 26$ Subtracting 2 from both sides

$-4y = 24$ Simplifying

$\dfrac{-4y}{-4} = \dfrac{24}{-4}$ Dividing both sides by -4

$y = -6$ Simplifying

Check: $2 - 5y = 26 - y$

$2 - 5(-6)$	$26 - (-6)$	
$2 + 30$	$26 + 6$	
32	32	TRUE

The solution is –6.

39. $7(2a - 1) = 21$

$14a - 7 = 21$ Distributive law

$14a - 7 + 7 = 21 + 7$ Adding 7 to both sides

$14a = 28$ Simplifying

$\dfrac{14a}{14} = \dfrac{28}{14}$ Dividing both sides by 14

$a = 2$ Simplifying

Check: $7(2a - 1) = 21$

$7(2 \cdot 2 - 1)$	21	
$7(4 - 1)$	21	
$7(3)$	21	
21	21	TRUE

The solution is 2.

41. $8 = 8(x + 1)$

$8 = 8x + 8$ Distributive law

$8 - 8 = 8x + 8 - 8$ Subtracting 8 from both sides

$0 = 8x$ Simplifying

$\dfrac{0}{8} = \dfrac{8x}{8}$ Dividing both sides by 8

$0 = x$ Simplifying

Check: $8 = 8(x + 1)$

8	$8(0 + 1)$	
8	$8(1)$	
8	8	TRUE

The solution is 0.

43. $2(3 + 4m) - 6 = 48$

$6 + 8m - 6 = 48$ Distributive law

$8m = 48$ Simplifying

$\dfrac{8m}{8} = \dfrac{48}{8}$ Dividing both sides by 8

$m = 6$ Simplifying

Check: $2(3 + 4m) - 6 = 48$

$2(3 + 4 \cdot 6) - 6$	48	
$2(3 + 24) - 6$	48	
$2(27) - 6$	48	
$54 - 6$	48	
48	48	TRUE

The solution is 6.

45. $3(y + 4) = 3(y - 1)$

$3y + 12 = 3y - 3$ Distributive law

$3y + 12 - 3y = 3y - 3 - 3y$ Subtracting $3y$ from both sides

$12 = -3$ Simplifying

This is a false statement. The equation is a contradiction and has no solution.

47.
$$2r + 8 = 6r + 10$$

$-6r + 2r + 8 = -6r + 6r + 10$ Subtracting $6r$ from both sides

$-4r + 8 = 10$ Simplifying

$-4r + 8 - 8 = 10 - 8$ Subtracting 8 from both sides

$-4r = 2$ Simplifying

$\dfrac{-4r}{-4} = \dfrac{2}{-4}$ Dividing both sides by -4

$r = -\dfrac{1}{2}$ Simplifing

Check: $2r + 8 \quad = 6r + 10$

$2 \cdot \left(-\dfrac{1}{2}\right) + 8$	$6 \cdot \left(-\dfrac{1}{2}\right) + 10$
$-\dfrac{2}{2} + 8$	$-\dfrac{6}{2} + 10$
$-1 + 8$	$-3 + 10$
7	7 TRUE

The solution is $-\dfrac{1}{2}$.

49. $5 - 2x = 3x - 7x + 25$

$5 - 2x = -4x + 25$ Simplifying

$5 - 2x + 4x = -4x + 25 + 4x$ Adding $4x$ to both sides

$5 + 2x = 25$ Simplifying

$-5 + 5 + 2x = -5 + 25$ Adding -5 to both sides

$2x = 20$ Simplifying

$\dfrac{2x}{2} = \dfrac{20}{2}$ Dividing both sides by 2

$x = 10$ Simplifying

Check: $5 - 2x = 3x - 7x + 25$

$5 - 2 \cdot 10$	$3 \cdot 10 - 7 \cdot 10 + 25$
$5 - 20$	$30 - 70 + 25$
-15	$-40 + 25$
-15	-15 TRUE

The solution is 10.

51. $7 + 3x - 6 = 3x + 5 - x$

$1 + 3x = 2x + 5$ Simplifying

$1 + 3x - 2x = 2x + 5 - 2x$ Subtracting $2x$ from both sides

$1 + x = 5$ Simplifying

$-1 + 1 + x = -1 + 5$ Subtracting 1 from both sides

$x = 4$ Simplifying

Check: $7 + 3x - 6 = 3x + 5 - x$

$7 + 3 \cdot 4 - 6$	$3 \cdot 4 + 5 - 4$
$7 + 12 - 6$	$12 + 5 - 4$
$19 - 6$	$17 - 4$
13	13 TRUE

The solution is 4.

53. $4y - 4 + y + 24 = 6y + 20 - 4y$

$5y + 20 = 2y + 20$ Simplifying

$5y + 20 - 2y = 2y + 20 - 2y$ Subtracting $2y$ from both sides

$3y + 20 = 20$ Simplifying

$3y + 20 - 20 = 20 - 20$ Subtracting 20 from both sides

$3y = 0$ Simplifying

$\dfrac{3y}{3} = \dfrac{0}{3}$ Dividing both sides by 3

$y = 0$ Simplifying

Check:

$4y - 4 + y + 24 = 6y + 20 - 4y$

$4 \cdot 0 - 4 + 0 + 24$	$6 \cdot 0 + 20 - 4 \cdot 0$
$0 - 4 + 0 + 24$	$0 + 20 - 0$
20	20 TRUE

The solution is 0.

55. $4 + 7a = 7(a - 1)$

$4 + 7a = 7a - 7$ Distributive law

$4 + 7a - 7a = 7a - 7a - 7$ Subtracting $7a$ from both sides

$4 = -7$ Simplifying

This is a false statement. The equation is a contradiction and has no solution.

57. $13-3(2x-1)=4$

$\qquad 13-6x+3=4 \qquad$ Distributive law

$\qquad 16-6x=4 \qquad$ Simplifying

$\qquad 16-6x-16=4-16 \qquad$ Subtracting 16 from both sides

$\qquad -6x=-12 \qquad$ Simplifying

$\qquad \dfrac{-6x}{-6}=\dfrac{-12}{-6} \qquad$ Dividing both sides by -6

$\qquad x=2 \qquad$ Simplifying

Check:

$$\dfrac{13-3(2x-1)=4}{\begin{array}{c|c} 13-3(2\cdot2-1) & 4 \\ 13-3(4-1) & 4 \\ 13-3(3) & 4 \\ 13-9 & 4 \\ 4 & 4 \ \text{TRUE} \end{array}}$$

The solution is 2.

59. $\qquad 7(5x-2)=6(6x-1)$

$\qquad 35x-14=36x-6 \qquad$ Distributive law

$\qquad 35x-14-35x= \qquad$ Subtracting $35x$ from both sides

$\qquad\qquad 36x-6-35x$

$\qquad -14=x-6 \qquad$ Simplifying

$\qquad -14+6=x-6+6 \qquad$ Adding 6 to both sides

$\qquad -8=x \qquad$ Simplifing

Check:

$$7(5x-2)=6(6x-1)$$

$$\begin{array}{c|c} 7[5(-8)-2] & 6[6(-8)-1] \\ 7[-40-2] & 6[-48-1] \\ 7[-42] & 6[-49] \\ -294 & -294 \ \text{TRUE} \end{array}$$

The solution is –8.

61. $\qquad 2(7-x)-20=7x-3(2+3x)$

$\qquad 14-2x-20=7x-6-9x \qquad$ Distributive law

$\qquad -2x-6=-2x-6 \qquad$ Simplifying

$\qquad -2x-6+6=-2x-6+6 \qquad$ Adding 6 to both sides

$\qquad -2x=-2x \qquad$ Simplifying

This equation is true regardless of the replacement for x, so all real numbers are solutions. The equation is an identity.

63. $\qquad 19-(2x+3)=2(x+3)+x$

$\qquad 19-2x-3=2x+6+x \qquad$ Distributive law

$\qquad 16-2x=3x+6 \qquad$ Simplifying

$\qquad 16-2x+2x= \qquad$ Adding $2x$ to both sides

$\qquad\qquad 3x+6+2x$

$\qquad 16=5x+6 \qquad$ Simplifying

$\qquad 16-6=5x+6-6 \qquad$ Subtracting 6 from both sides

$\qquad 10=5x \qquad$ Simplifying

$\qquad \dfrac{10}{5}=\dfrac{5x}{5} \qquad$ Dividing both sides by 5

$\qquad 2=x \qquad$ Simplifying

Check:

$$19-(2x+3)=2(x+3)+x$$

$$\begin{array}{c|c} 19-(2\cdot2+3) & 2(2+3)+2 \\ 19-(4+3) & 2(5)+2 \\ 19-(7) & 10+2 \\ 12 & 12 \ \text{TRUE} \end{array}$$

The solution is 2.

65. $\frac{2}{3}+\frac{1}{4}t=2$

The number 12 is the least common denominator, so we multiply by 12 on both sides.

$12\left(\frac{2}{3}+\frac{1}{4}t\right)=12\cdot2$

$\qquad 8+3t=24$

$\qquad -8+8+3t=-8+24 \qquad$ Subtracting 8 from both sides

$\qquad 3t=16 \qquad$ Simplifying

$\qquad \dfrac{3t}{3}=\dfrac{16}{3} \qquad$ Dividing both sides by 3

$\qquad t=\dfrac{16}{3} \qquad$ Simplifying

Check:

$$\frac{2}{3}+\frac{1}{4}t=2$$

$$\begin{array}{c|c} \frac{2}{3}+\frac{1}{4}\left(\frac{16}{3}\right) & 2 \\ \frac{2}{3}+\frac{16}{12} & 2 \\ \frac{2}{3}+\frac{4}{3} & 2 \\ \frac{6}{3} & 2 \\ 2 & 2 \ \text{TRUE} \end{array}$$

The solution is $\dfrac{16}{3}$.

67. $\dfrac{2}{3} + 4t = 6t - \dfrac{2}{15}$

The number 15 is the least common denominator, so we multiply by 15 on both sides.

$$15\left(\dfrac{2}{3} + 4t\right) = 15\left(6t - \dfrac{2}{15}\right)$$

$$10 + 60t = 90t - 2$$

$10 + 60t - 60t =$	Subtracting $60t$ from both sides
$\quad 90t - 2 - 60t$	
$10 = 30t - 2$	Simplifying
$10 + 2 = 30t - 2 + 2$	Adding 2 to both sides
$12 = 30t$	Simplifying
$\dfrac{12}{30} = \dfrac{30t}{30}$	Dividing both sides by 30
$\dfrac{2}{5} = t$	Simplifying

Check:

$$\dfrac{2}{3} + 4t = 6t - \dfrac{2}{15}$$

$\dfrac{2}{3} + 4\left(\dfrac{2}{5}\right)$	$6\left(\dfrac{2}{5}\right) - \dfrac{2}{15}$
$\dfrac{2}{3} + \dfrac{8}{5}$	$\dfrac{12}{5} - \dfrac{2}{15}$
$\dfrac{10}{15} + \dfrac{24}{15}$	$\dfrac{36}{15} - \dfrac{2}{15}$
$\dfrac{34}{15}$	$\dfrac{34}{15}$ TRUE

The solution is $\dfrac{2}{5}$.

69. $\dfrac{1}{3}x + \dfrac{2}{5} = \dfrac{4}{15} + \dfrac{3}{5}x - \dfrac{2}{3}$

The number 15 is the least common denominator, so we multiply by 15 on both sides.

$$15\left(\dfrac{1}{3}x + \dfrac{2}{5}\right) = 15\left(\dfrac{4}{15} + \dfrac{3}{5}x - \dfrac{2}{3}\right)$$

$$5x + 6 = 4 + 9x - 10$$

$5x + 6 = 9x - 6$	Simplifying
$5x + 6 - 5x = 9x - 6 - 5x$	Subtracting $5x$ from both sides
$6 = 4x - 6$	Simplifying
$6 + 6 = 4x - 6 + 6$	Adding 6 to both sides
$12 = 4x$	Simplifying
$\dfrac{12}{4} = \dfrac{4x}{4}$	Dividing both sides by 4
$3 = x$	Simplifying

Check:

$$\dfrac{1}{3}x + \dfrac{2}{5} = \dfrac{4}{15} + \dfrac{3}{5}x - \dfrac{2}{3}$$

$\dfrac{1}{3}(3) + \dfrac{2}{5}$	$\dfrac{4}{15} + \dfrac{3}{5}(3) - \dfrac{2}{3}$
$1 + \dfrac{2}{5}$	$\dfrac{4}{15} + \dfrac{9}{5} - \dfrac{2}{3}$
$\dfrac{5}{5} + \dfrac{2}{5}$	$\dfrac{4}{15} + \dfrac{27}{15} - \dfrac{10}{15}$
$\dfrac{7}{5}$	$\dfrac{21}{15}$
$\dfrac{7}{5}$	$\dfrac{7}{5}$ TRUE

The solution is 3.

71. $2.1x + 45.2 = 3.2 - 8.4x$

We can clear decimals by multiplying both sides by 10.

$$10(2.1x + 45.2) = 10(3.2 - 8.4x)$$

$$21x + 452 = 32 - 84x$$

$21x + 452 + 84x =$	Adding $84x$ to both sides
$\quad 32 - 84x + 84x$	
$105x + 452 = 32$	Simplifying
$105x + 452 - 452 =$	Subtracting 452 from both sides
$\quad 32 - 452$	
$105x = -420$	Simplifying
$\dfrac{105x}{105} = \dfrac{-420}{105}$	Dividing both sides by 105
$x = -4$	Simplifing

Check:

$$2.1x + 45.2 = 3.2 - 8.4x$$

$2.1(-4) + 45.2$	$3.2 - 8.4(-4)$
$-8.4 + 45.2$	$3.2 + 33.6$
36.8	36.8 TRUE

The solution is -4.

73. $0.76 + 0.21t = 0.96t - 0.49$

We can clear decimals by multiplying both sides by 100.

$100(0.76 + 0.21t) = 100(0.96t - 0.49)$

$76 + 21t = 96t - 49$

$76 + 21t - 76 =$	Subtracting 76
$\qquad 96t - 49 - 76$	from both sides
$21t = 96t - 125$	Simplifying
$21t - 96t =$	Subtracting $96t$
$\qquad 96t - 125 - 96t$	from both sides
$-75t = -125$	Simplifying
$\dfrac{-75t}{-75} = \dfrac{-125}{-75}$	Dividing both sides by -75
$t = 1.\overline{6}$	Simplifing

Check:
$$0.76 + 0.21t = 0.96t - 0.49$$

$0.76 + 0.21\left(1.\overline{6}\right)$	$0.96\left(1.\overline{6}\right) - 0.49$
$0.76 + 0.35$	$1.6 - 0.49$
1.11	1.11 TRUE

The solution is $1.\overline{6}$.

75. $\frac{2}{5}x - \frac{3}{2}x = \frac{3}{4}x + 2$

The number 20 is the least common denominator, so we multiply by 20 on both sides.

$20\left(\frac{2}{5}x - \frac{3}{2}x\right) = 20\left(\frac{3}{4}x + 2\right)$

$8x - 30x = 15x + 40$

$-22x = 15x + 40$	Simplifying
$-22x - 15x =$	Subtract $15x$ from
$\qquad 15x + 40 - 15x$	both sides
$-37x = 40$	Simplifying
$\dfrac{-37x}{-37} = \dfrac{40}{-37}$	Divide both sides by -37
$x = -\dfrac{40}{37}$	Simplifying

Check:
$$\tfrac{2}{5}x - \tfrac{3}{2}x = \tfrac{3}{4}x + 2$$

$\frac{2}{5}\left(-\frac{40}{37}\right) - \frac{3}{2}\left(-\frac{40}{37}\right)$	$\frac{3}{4}\left(-\frac{40}{37}\right) + 2$
$-\frac{16}{37} + \frac{60}{37}$	$-\frac{30}{37} + \frac{74}{37}$
$\frac{44}{37}$	$\frac{44}{37}$ TRUE

The solution is $-\frac{40}{37}$.

77. $\frac{1}{3}(2x - 1) = 7$

The number 3 is the least common denominator, so we multiply by 3 on both sides.

$3\left[\frac{1}{3}(2x - 1)\right] = 3 \cdot 7$

$2x - 1 = 21$

$2x - 1 + 1 = 21 + 1$	Add 1 to both sides
$2x = 22$	Simplifying
$\dfrac{2x}{2} = \dfrac{22}{2}$	Divide both sides by 2
$x = 11$	Simplifying

Check:
$$\tfrac{1}{3}(2x - 1) = 7$$

$\frac{1}{3}(2 \cdot 11 - 1)$	7
$\frac{1}{3}(22 - 1)$	7
$\frac{1}{3}(21)$	7
7	7 TRUE

The solution is 11 .

79. $\frac{3}{4}(3t - 6) = 9$

$\frac{4}{3}\left[\frac{3}{4}(3t - 6)\right] = \frac{4}{3}(9)$	Multiply both sides by $\frac{4}{3}$
$3t - 6 = 12$	Simplifying
$3t - 6 + 6 = 12 + 6$	Add 6 to both sides
$3t = 18$	Simplifying
$\dfrac{3t}{3} = \dfrac{18}{3}$	Divide both sides by 3
$t = 6$	Simplifying

Check:
$$\tfrac{3}{4}(3t - 6) = 9$$

$\frac{3}{4}(3 \cdot 6 - 6)$	9
$\frac{3}{4}(18 - 6)$	9
$\frac{3}{4}(12)$	9
9	9 TRUE

The solution is 6.

81. $\frac{1}{6}\left(\frac{3}{4}x-2\right)=-\frac{1}{5}$

$30\left[\frac{1}{6}\left(\frac{3}{4}x-2\right)\right]=30\left(-\frac{1}{5}\right)$ Multiply both sides by 30

$5\left(\frac{3}{4}x-2\right)=-6$ Simplifying

$\frac{15}{4}x-10=-6$ Distributive Law

$\frac{15}{4}x-10+10=-6+10$ Add 10 to both sides

$\frac{15}{4}x=4$ Simplifying

$\left(\frac{4}{15}\right)\left(\frac{15}{4}\right)x=\left(\frac{4}{15}\right)(4)$ Multiply both sides by $\frac{4}{15}$

$x=\frac{16}{15}$ Simplifying

Check:

$\frac{1}{6}\left(\frac{3}{4}x-2\right)=-\frac{1}{5}$

$\begin{array}{c|c} \frac{1}{6}\left[\frac{3}{4}\left(\frac{16}{15}\right)-2\right] & -\frac{1}{5} \\ \frac{1}{6}\left(\frac{4}{5}-2\right) & -\frac{1}{5} \\ \frac{1}{6}\left(\frac{4}{5}-\frac{10}{5}\right) & -\frac{1}{5} \\ \frac{1}{6}\left(-\frac{6}{5}\right) & -\frac{1}{5} \\ -\frac{1}{5} & \frac{1}{5} \end{array}$ TRUE

The solution is $\frac{16}{15}$.

83. $0.7(3x+6)=1.1-(x+2)$

We can clear decimals by multiplying both sides by 10.

$10\left[0.7(3x+6)\right]=10\left[1.1-(x+2)\right]$

$7(3x+6)=11-10(x+2)$

$21x+42=11-10x-20$ Distributive Law

$21x+42=-10x-9$ Simplifying

$21x+42+10x=$ Adding $10x$ to both sides
$-10x-9+10x$

$31x+42=-9$ Simplifying

$31x+42-42=-9-42$ Subtracting 42 from both sides

$31x=-51$ Simplifying

$\frac{31x}{31}=\frac{-51}{31}$ Dividing both sides by 31

$x=-\frac{51}{31}$ Simplifying

Check:

$0.7(3x+6)=1.1-(x+2)$

$\begin{array}{c|c} \frac{7}{10}\left[3\left(-\frac{51}{31}\right)+6\right] & \frac{11}{10}-\left(-\frac{51}{31}+2\right) \\ \frac{7}{10}\left(-\frac{153}{31}+\frac{186}{31}\right) & \frac{11}{10}-\left(-\frac{51}{31}+\frac{62}{31}\right) \\ \frac{7}{10}\left(\frac{33}{31}\right) & \frac{11}{10}-\left(\frac{11}{31}\right) \\ \frac{231}{310} & \frac{341}{310}-\frac{110}{310} \\ \frac{231}{110} & \frac{231}{110} \end{array}$ TRUE

The solution is $-\frac{51}{31}$.

85. $a+(a-3)=(a+2)-(a+1)$

$a+a-3=a+2-a-1$ Simplifying Distributive Law

$2a-3=1$ Simplifying

$2a-3+3=1+3$ Adding 3 to both sides

$2a=4$ Simplifying

$\frac{2a}{2}=\frac{4}{2}$ Dividing both sides by 2

$a=2$ Simplifying

Check:

$a+(a-3)=(a+2)-(a+1)$

$\begin{array}{c|c} 2+(2-3) & (2+2)-(2+1) \\ 2+(-1) & 4-3 \\ 1 & 1 \end{array}$ TRUE

The solution is 2.

87. *Thinking and Writing Exercise.*

89. $3-5a$

We substitute 2 for a and evaluate.

$3-5a=3-5\cdot 2$

$=3-10$

$=-7$

91. $7x-2x$

We substitute -3 for x and evaluate.

$7x-2x=7(-3)-2(-3)$

$=-21+6$

$=-15$

93. *Thinking and Writing Exercise*

95. $8.43x - 2.5(3.2 - 0.7x) = -3.455x + 9.04$

We can clear decimals by multiplying both sides by 1000.

$$1000\big[8.43 - 2.5(3.2 - 0.7x)\big] = 1000\big[-3.455x + 9.04\big]$$

$8430x - 2500(3.2 - 0.7x) = -3455x + 9040$	
$8430x - 8000 + 1750x = -3455x + 9040$	Distributive Law
$10,180x - 8000 = -3455x + 9040$	Simplifying
$10,180x - 8000 + 3455x = -3455x + 9040 + 3455x$	Adding $3455x$ to both sides
$13,635x - 8000 = 9040$	Simplifying
$13,635x - 8000 + 8000 = 9040 + 8000$	Adding 8000 to both sides
$13,635x = 17,040$	Simplifying
$\dfrac{13,635x}{13,635} = \dfrac{17,040}{13,635}$	Dividing both sides by 13,635
$x = \dfrac{1136}{909}$	Simplifing

The solution is , $\frac{1136}{909}$, or $1.\overline{2497}$.

97. $-2\big[3(x - 2) + 4\big] = 4(5 - x) - 2x$

$-2\big[3x - 6 + 4\big] = 20 - 4x - 2x$	Distributive Law
$-6x + 12 - 8 = 20 - 4x - 2x$	Distributive Law
$-6x + 4 = 20 - 6x$	Simplifying
$-6x + 4 + 6x = 20 - 6x + 6x$	Add $6x$ to both sides
$4 = 20$	Simplifying

This is a false statement. The equation is a contradiction and has no solution.

99. $2x(x + 5) - 3(x^2 + 2x - 1) = 9 - 5x - x^2$

$$2x^2 + 10x - 3x^2 - 6x + 3 = 9 - 5x - x^2$$
$$-x^2 + 4x + 3 = 9 - 5x - x^2$$
$$9x = 6$$
$$x = \frac{2}{3}$$

The solution is $\frac{2}{3}$.

101. $9 - 3x = 2(5 - 2x) - (1 - 5x)$

$$9 - 3x = 10 - 4x - 1 + 5x$$
$$0 = -9 + 3x + 10 - 4x - 1 + 5x$$
$$0 = 4x$$
$$0 = x$$

The solution is 0.

103. $$\frac{x}{14} - \frac{5x + 2}{49} = \frac{3x - 4}{7}$$

$$98\left(\frac{x}{14} - \frac{5x + 2}{49}\right) = 98\left(\frac{3x - 4}{7}\right)$$
$$7x - 10x - 4 = 14(3x - 4)$$
$$7x - 10x - 4 = 42x - 56$$
$$-3x - 4 = 42x - 56$$
$$-4 + 56 = 42x + 3x$$
$$52 = 45x$$
$$\frac{52}{45} = x$$

The solution is $\frac{52}{45}$.

105. $2\{9-3[-2x-4]\} = 12x+42$

$2\{9+6x+12\} = 12x+42$

$2\{21+6x\} = 12x+42$

$42+12x = 12x+42$

$42 = 42$

Since this equation is true for all real numbers, this equation is an identity.

107. $3|x|-2 = 10$

$3|x| = 12$

$|x| = 4$

$x = \pm 4$

The solution is -4 and 4.

Exercise Set 2.3

1. We substitute 21,345 for n and calculate f.

$f = \dfrac{n}{15} = \dfrac{21{,}345}{15} = 1423$

The number of students is 1423.

3. We substitute 30 for I and 115 for V and calculate P.

$P = I \cdot V = 30 \cdot 115 = 3450$

The power consumed is 3450 watts.

5. We substitute 0.025 for I and 0.044 for U and solve for f.

$f = 8.5 + 1.4(I - U)$

$= 8.5 + 1.4(0.025 - 0.044)$

$= 8.5 + 1.4(-0.019)$

$= 8.5 - 0.0266$

$= 8.4734$

The federal funds rate should be 8.4734.

7. We substitute 1 for t and solve for n.

$n = 0.5t^4 + 3.45t^3 - 96.65t^2 + 347.7t$

$= 0.5(1)^4 + 3.45(1)^3 - 96.65(1)^2 + 347.7(1)$

$= 0.5 + 3.45 - 96.65 + 347.7$

$= 255$

After 1 hour, 255 mg of ibuprofen remains in the bloodstream.

9. $A = bh$

$\dfrac{A}{h} = \dfrac{bh}{h}$ Dividing both sides by h

$\dfrac{A}{h} = b$

11. $I = Prt$

$\dfrac{I}{rt} = \dfrac{Prt}{rt}$ Dividing both sides by rt

$\dfrac{I}{rt} = P$

13. $H = 65 - m$

$H + m = 65$ Adding m to both sides

$m = 65 - H$ Subtracting H from both sides

15. $P = 2l + 2w$

$P - 2w = 2l + 2w - 2w$ Subtracting $2w$ from both sides

$P - 2w = 2l$

$\dfrac{P - 2w}{2} = \dfrac{2l}{2}$ Dividing both sides by 2

$\dfrac{P - 2w}{2} = l$, or

$\dfrac{P}{2} - w = l$

17. $A = \pi r^2$

$\dfrac{A}{r^2} = \dfrac{\pi r^2}{r^2}$ Divide both sides by r^2

$\dfrac{A}{r^2} = \pi$

19. $A = \frac{1}{2}bh$

$2A = bh$ Multiplying both sides by 2

$\dfrac{2A}{b} = \dfrac{bh}{b}$ Dividing both sides by b

$\dfrac{2A}{b} = h$

21. $E = mc^2$

$\dfrac{E}{c^2} = \dfrac{mc^2}{c^2}$ Divide both sides by c^2

$\dfrac{E}{c^2} = m$

23. $Q = \dfrac{c+d}{2}$

$2Q = c + d$ — Multipling both sides by 2

$2Q - c = c + d - c$ — Subtracting c from both sides

$2Q - c = d$

25. $p - q + r = 2$

$p - q + r + q = 2 + q$ — Adding q to both sides

$p + r - 2 = 2 + q - 2$ — Subtracting 2 from both sides

$q = p + r - 2$

27. $w = \dfrac{r}{f}$

$wf = r$ — Multiplying both sides by f

29. $H = \dfrac{TV}{550}$

$550H = TV$ — Multiplying both sides by 550

$\dfrac{550H}{V} = T$ — Dividing both sides by V

31. $F = \dfrac{9}{5}C + 32$

$F - 32 = \dfrac{9}{5}C + 32 - 32$ — Subtracting 32 from both sides

$F - 32 = \dfrac{9}{5}C$

$\dfrac{5}{9}(F - 32) = \dfrac{5}{9}\left(\dfrac{9}{5}C\right)$ — Multiplying both sides by $\dfrac{5}{9}$

$\dfrac{5}{9}(F - 32) = C$

33. $2x - y = 1$

$2x - y + y = 1 + y$ — Adding y to both sides

$2x - 1 = 1 - 1 + y$ — Subtracting 1 from both sides

$2x - 1 = y$

35. $2x + 5y = 10$

$-2x + 2x + 5y = -2x + 10$ — Subtracting $2x$ from both sides

$\dfrac{5y}{5} = \dfrac{-2x + 10}{5}$ — Dividing both sides by 5

$y = -\dfrac{2}{5}x + 2$

37. $4x - 3y = 6$

$-4x + 4x - 3y = -4x + 6$ — Subtracting $4x$ from both sides

$\dfrac{-3y}{-3} = \dfrac{-4x + 6}{-3}$ — Dividing both sides by -3

$y = \dfrac{4}{3}x - 2$

39. $9x + 8y = 4$

$-9x + 9x + 8y = -9x + 4$ — Subtracting $9x$ from both sides

$\dfrac{8y}{8} = \dfrac{-9x + 4}{8}$ — Dividing both sides by 8

$y = -\dfrac{9}{8}x + \dfrac{1}{2}$

41. $3x - 5y = 8$

$-3x + 3x - 5y = -3x + 8$ — Subtracting $3x$ from both sides

$\dfrac{-5y}{-5} = \dfrac{-3x + 8}{-5}$ — Dividing both sides by -5

$y = \dfrac{3}{5}x - \dfrac{8}{5}$

43. $A = at + bt$

$A = (a + b)t$ — Factoring

$\dfrac{A}{a+b} = \dfrac{(a+b)t}{a+b}$ — Dividing both sides by $a + b$

$\dfrac{A}{a+b} = t$

45. $A = \frac{1}{2}ah + \frac{1}{2}bh$

$2A = 2\left(\frac{1}{2}ah + \frac{1}{2}bh\right)$ Multiplying both sides by 2

$2A = ah + bh$

$2A = (a+b)h$ Factoring

$\dfrac{2A}{a+b} = \dfrac{(a+b)h}{a+b}$ Dividing both sides by $a+b$

$\dfrac{2A}{a+b} = h$

47. $z = 13 + 2(x+y)$

$z = 13 + 2x + 2y$ Distributive law

$z - 13 = 2x + 2y$ Subtracting both sides by 13

$z - 13 - 2y = 2x$ Subtracting both sides by $2y$

$\dfrac{z - 13 - 2y}{2} = \dfrac{2x}{2}$ Dividing both sides by 2

$\dfrac{z-13}{2} - y = x$

49. $t = 27 - \dfrac{1}{4}(w - l)$

$t = 27 - \dfrac{1}{4}w + \dfrac{1}{4}l$ Distributive law

$t - 27 = -\dfrac{1}{4}w + \dfrac{1}{4}l$ Subtracting both sides by 27

$t - 27 + \dfrac{1}{4}w = \dfrac{1}{4}l$ Adding $-\dfrac{1}{4}w$ to both sides

$4(t - 27) + w = l$ Multiplying both sides by 4

51. $R = r + \dfrac{400(W - L)}{N}$

$R - r = r + \dfrac{400(W - L)}{N} - r$ Subtract r from both sides

$R - r = \dfrac{400(W - L)}{N}$

$N(R - r) = N\left[\dfrac{400(W - L)}{N}\right]$ Multiplying both sides by N

$NR - Nr = 400(W - L)$

$NR - Nr = 400W - 400L$ Distributive Law

$\begin{array}{l} NR - Nr - 400W = \\ \quad 400W - 400L - 400W \end{array}$ Subtracting $400W$ from both sides

$NR - Nr - 400W = -400L$

$\dfrac{NR - Nr - 400W}{-400} = \dfrac{-400L}{-400}$ Dividing both sides by -400

$\dfrac{N(R - r)}{-400} + W = L$, or

$W - \dfrac{N(R - r)}{400} = L$, or

$\dfrac{400W - NR + Nr}{400} = L$

53. *Thinking and Writing Exercise.*

55. $-2 + 5 - (-4) - 17$

$= 3 - (-4) - 17$

$= 7 - 17$

$= -10$

57. $4.2(-11.75)(0) = -49.35(0) = 0$

59. $20 \div (-4) \cdot 2 - 3 = -5 \cdot 2 - 3$

$= -10 - 3$

$= -13$

61. *Thinking and Writing Exercise.*

63. We substitute 80 for w, 190 for h, and 2852 for K and solve for a.

$$K = 21.235w + 7.75h - 10.54a + 102.3$$
$$2852 = 21.235(80) + 7.75(190) - 10.54a + 102.3$$
$$2852 = 1698.8 + 1472.5 - 10.54a + 102.3$$
$$2852 = 3273.6 - 10.54a$$
$$2852 - 3273.6 = 3273.6 - 10.54a - 3273.6$$
$$-421.6 = -10.54a$$
$$40 = a$$

Janos is 40 years old.

65. We substitute 54 for A and solve for s.

$$A = 6s^2$$
$$54 = 6s^2$$
$$\frac{54}{6} = \frac{6s^2}{6}$$
$$9 = s^2$$

So, $s = -3$ and $s = 3$. Since length cannot be negative, we disregard -3.

The volume of a cube is given by $V = s^3$, so we substitute 3 for s and solve for V.

$$V = s^3$$
$$= 3^3$$
$$= 27$$

The volume of the cube is 27 in^3.

67. $\dfrac{y}{z} \div \dfrac{z}{t} = 1$

$\dfrac{y}{z} \times \dfrac{t}{z} = 1$ Rewriting

$\dfrac{yt}{z^2} = 1$

$z^2 \left(\dfrac{yt}{z^2} \right) = z^2 \cdot 1$ Multiplying both sides by z^2

$yt = z^2$

$\dfrac{yt}{t} = \dfrac{z^2}{t}$ Dividing both sides by t

$y = \dfrac{z^2}{t}$

69. $qt = r(s + t)$

$qt = rs + rt$

$qt - rt = rs + rt - rt$ Subtracting rt from both sides

$(q - r)t = rs$

$\dfrac{(q - r)t}{q - r} = \dfrac{rs}{q - r}$ Dividing both sides by $q - r$

$t = \dfrac{rs}{q - r}$

71. We subtract the minimum output for a well-insulated house with a square feet from the minimum output for a poorly-insulated house with a square feet. Let S represent the number of Btu's saved.

$$S = 50a - 30a = 20a$$

Mid-Chapter Review

Guided Solutions

1. $2x + 3 - 3 = 10 - 3$

$2x = 7$

$\frac{1}{2} \cdot 2x = \frac{1}{2} \cdot 7$

$x = \frac{7}{2}$

2. $6 \cdot \frac{1}{2}(x - 3) = 6 \cdot \frac{1}{3}(x - 4)$

$3(x - 3) = 2(x - 4)$

$3x - 9 = 2x - 8$

$3x - 9 + 9 = 2x - 8 + 9$

$3x - 2x = 2x + 1 - 2x$

$x = 1$

Mixed Review

1. $x - 2 = -1$

   $x - 2 + 2 = -1 + 2$

   $x = 1$

2. $2 - x = -1$

   $2 - 2 - x = -1 - 2$

   $-x = -3$

   $-1 \cdot (-x) = -1 \cdot (-3)$

   $x = 3$

3. $3t = 5$

   $\dfrac{3}{3} t = \dfrac{5}{3}$

   $t = \dfrac{5}{3}$

4. $-\frac{3}{2} x = 12$

   $\left(-\frac{2}{3}\right) \cdot -\frac{3}{2} x = \left(-\frac{2}{3}\right) 12$

   $x = -8$

5. $\dfrac{y}{8} = 6$

   $\left(\dfrac{8}{1}\right) \cdot \dfrac{y}{8} = \left(\dfrac{8}{1}\right) \cdot 6$

   $y = 48$

6. $0.06x = 0.03$

   $\dfrac{0.06x}{0.06} = \dfrac{0.03}{0.06}$

   $t = \dfrac{1}{2}$ or 0.5

7. $3x - 7x = 20$

   $-4x = 20$

   $\dfrac{-4x}{-4} = \dfrac{20}{-4}$

   $x = -5$

8. $9x - 7 = 17$

   $9x - 7 + 7 = 17 + 7$

   $9x = 24$

   $\dfrac{9x}{9} = \dfrac{24}{9}$

   $x = \dfrac{8}{3}$

9. $4(t - 3) - t = 6$

   $4t - 12 - t = 6$

   $3t - 12 = 6$

   $3t - 12 + 12 = 6 + 12$

   $3t = 18$

   $\dfrac{3t}{3} = \dfrac{18}{3}$

   $t = 6$

10. $3(y + 5) = 8y$

    $3y + 15 = 8y$

    $3y + 15 - 3y = 8y - 3y$

    $15 = 5y$

    $\dfrac{15}{5} = \dfrac{5y}{5}$

    $3 = y$

11. $8n - (3n - 5) = 5 - n$

    $8n - 3n + 5 = 5 - n$

    $5n + 5 = 5 - n$

    $5n + 5 + n = 5 - n + n$

    $6n + 5 = 5$

    $6n + 5 - 5 = 5 - 5$

    $6n = 0$

    $n = 0$

12. $\frac{9}{10} y - \frac{7}{10} = \frac{21}{5}$

    $10 \left(\frac{9}{10} y - \frac{7}{10} \right) = 10 \cdot \left(\frac{21}{5} \right)$

    $9y - 7 = 42$

    $9y - 7 + 7 = 42 + 7$

    $9y = 49$

    $y = \frac{49}{9}$

13. $2(t-5)-3(2t-7)=12-5(3t+1)$
$2t-10-6t+21=12-15t-5$
$-4t+11=-15t+7$
$-4t+11+15t=-15t+7+15t$
$11t+11=7$
$11t+11-11=7-11$
$11t=-4$
$t=-\frac{4}{11}$

14. $\frac{2}{3}(x-2)-1=-\frac{1}{2}(x-3)$
$6\left(\frac{2}{3}(x-2)-1\right)=6\left(-\frac{1}{2}(x-3)\right)$
$4(x-2)-6=-3(x-3)$
$4x-8-6=-3x+9$
$4x-14=-3x+9$
$4x-14+3x=-3x+9+3x$
$7x-14=9$
$7x-14+14=9+14$
$7x=23$
$x=\frac{23}{7}$

15. $E=wA$
$\dfrac{E}{w}=\dfrac{wA}{w}$
$\dfrac{E}{w}=A$

16. $V=lwh$
$\dfrac{V}{lh}=\dfrac{lwh}{lh}$
$\dfrac{V}{lh}=w$

17. $Ax+By=C$
$Ax+By-Ax=C-Ax$
$By=C-Ax$
$\dfrac{By}{B}=\dfrac{C-Ax}{B}$
$y=\dfrac{C-Ax}{B}$

18. $at+ap=m$
$a(t+p)=m$
$\dfrac{a(t+p)}{(t+p)}=\dfrac{m}{(t+p)}$
$a=\dfrac{m}{(t+p)}$

19. $m=\dfrac{F}{a}$
$m\cdot a=\dfrac{F}{a}\cdot a$
$ma=F$
$\dfrac{ma}{m}=\dfrac{F}{m}$
$a=\dfrac{F}{m}$

20. $v=\dfrac{d_2-d_1}{t}$
$v\cdot t=\dfrac{d_2-d_1}{t}\cdot t$
$vt=d_2-d_1$
$vt-d_2=d_2-d_1-d_2$
$vt-d_2=-d_1$
$-1(vt-d_2)=-1\cdot(-d_1)$
$d_2-vt=d_1$

Exercise Set 2.4

1. d

3. e

5. c

7. f

9. b

11. $67\%=67\times0.01$ Replacing % by $\times0.01$
$=0.67$

13. $2\%=2\times0.01$ Replacing % by $\times0.01$
$=0.02$

15. $3.5\% = 3.5 \times 0.01$ Replacing % by $\times 0.01$

$= 0.035$

17. $40\% = 40 \times 0.01$ Replacing % by $\times 0.01$

$= 0.40$

19. 62.58%

$= 62.58 \times 0.01$ Replacing % by $\times 0.01$

$= 0.6258$

21. 0.7%

$= 0.7 \times 0.01$ Replacing % by $\times 0.01$

$= 0.007$

23. 125%

$= 125 \times 0.01$ Replacing % by $\times 0.01$

$= 1.25$

25. 0.13
First move the decimal point two places to the right; then write a % symbol:

$0.13.$

13%

27. 0.014
First move the decimal point two places to the right; then write a % symbol:

$0.01.4$

1.4%

29. 0.326
First move the decimal point two places to the right; then write a % symbol:

$0.326.$

32.6%

31. 0.9
First move the decimal point two places to the right; then write a % symbol:

$0.247.$

24.7%

33. 0.0049
First move the decimal point two places to the right; then write a % symbol:

$0.1.9$

10%

35. 1.08
First move the decimal point two places to the right; then write a % symbol:

$1.08.$

108%

37. 2.3
First move the decimal point two places to the right; then write a % symbol:

$2.30.$

230%

39. $\dfrac{4}{5}$ $\left(\text{Note: } \dfrac{4}{5} = 0.8\right)$

First move the decimal point two places to the right; then write a % symbol:

$0.80.$

80%

41. $\dfrac{8}{25}$ $\left(\text{Note: } \dfrac{8}{25} = 0.32\right)$

First move the decimal point two places to the right; then write a % symbol:

$0.32.$

32%

43. **Translate.**

What percent of 68 is 17?

y · 68 = 17

We solve the equation and then convert to percent notation.

$y \cdot 68 = 17$

$y = \dfrac{17}{68}$

$y = 0.25 = 25\%$

The answer is 25%.

45. **Translate.**

What percent of 125 is 30?

y · 125 = 30

We solve the equation and then convert to percent notation.

$y \cdot 125 = 30$

$y = \dfrac{30}{125}$

$y = 0.24 = 24\%$

The answer is 24%.

47. **Translate.**

14 is 30% of what number?

$\downarrow\downarrow\downarrow \quad \downarrow \qquad \downarrow$
$14 = 30\% \quad \cdot \qquad y$

We convert to percent notation, then solve the equation.

$14 = 0.3y \quad (30\% = 0.3)$

$\dfrac{14}{0.3} = y$

$46.\overline{6} = y$

The answer is $46.\overline{6}$, or $46\dfrac{2}{3}$, or $\dfrac{140}{3}$.

49. **Translate.**

0.3 is 12% of what number?

$\downarrow\downarrow\downarrow \quad \downarrow \qquad \downarrow$
$0.3 = 12\% \quad \cdot \qquad y$

We convert to percent notation, then solve the equation.

$0.3 = 0.12y \quad (12\% = 0.12)$

$\dfrac{0.3}{0.12} = y$

$2.5 = y$

The answer is 2.5.

51. **Translate.**

What number is 35% of 240?

$\quad\downarrow \qquad \downarrow\downarrow \quad \downarrow \ \downarrow$
$\quad y \qquad = 35\% \ \cdot \ 240$

We convert to percent notation, then solve the equation.

$y = 0.35 \cdot 240 \quad (35\% = 0.35)$

$y = 84 \qquad \text{Multiplying}$

The answer is 84.

53. **Translate.**

What percent of 60 is 75?

$\quad\downarrow \qquad \downarrow\downarrow\downarrow\downarrow$
$\quad y \qquad \cdot 60 = 75$

We solve the equation and then convert to percent notation.

$y \cdot 60 = 75$

$y = \dfrac{75}{60}$

$y = 1.25 = 125\%$

The answer is 125%.

55. **Translate.**

What is 2% of 40?

$\downarrow \ \downarrow\downarrow \ \ \downarrow\downarrow$
$y \ = 2\% \ \cdot \ 40$

We convert to percent notation, then solve the equation.

$y = 0.02 \cdot 40 \quad (2\% = 0.02)$

$y = 0.8 \qquad \text{Multiplying}$

The answer is 0.8.

57. Observe that 25 is half of 50. Thus, the answer is 0.5, or 50%. We could also do this exercise by translating to an equation.

Translate.

25 is what percent of 50?

$\downarrow\downarrow \qquad \downarrow \qquad \downarrow\downarrow$
$25 = \qquad y \qquad \cdot 50$

We solve the equation and then convert to percent notation.

$25 = y \cdot 50$

$\dfrac{25}{50} = y$

$0.5 = y$, or $50\% = y$

The answer is 50%.

59. **Translate.**

What percent of 69 is 23?

$\qquad\quad\downarrow \qquad\quad \downarrow\downarrow\downarrow\downarrow$
$\qquad\quad p \qquad \cdot \ 69 = 23$

$\dfrac{69p}{69} \quad \dfrac{23}{69}$

$p \quad 0.3\overline{3}$

The answer is $33.\overline{3}\%$, or $33\frac{1}{3}\%$.

61. Let v = the cost of surgical vet visits. Then we have:

v is 32% of $1425
$\downarrow\downarrow\downarrow \quad \downarrow \qquad \downarrow$
$v = 0.32 \ \cdot \ \$1425$
$v = \$456$

The cost of surgical vet visits is $456.

63. Let t = the cost of treats. Then we have:

t is 5% of $1425
$\downarrow\downarrow\downarrow \quad \downarrow \qquad \downarrow$
$t = 0.05 \ \cdot \ \$1425$
$t = \$71.25$

The cost of treats is $71.25.

65. Let c = the number of credit hours
Clayton completed. Then we have:
c is 60% of 125?
↓ ↓ ↓ ↓ ↓
$c = 0.6 \quad \cdot \quad 125$
$c = 75$

Clayton has completed 75 credit hours.

67. Let x = the number of at-bats. Then we have:
35.5% of x is 213.
↓ ↓ ↓ ↓ ↓
$0.355 \cdot x = 213$

$x = \dfrac{213}{0.355}$

$x = 600$
Ichiro Suzuki had 600 at-bats.

69. a) Let x = the percent of the cost of the meal
representing Shane's tip. Then we have:
$x \cdot \$25$ is $4.
↓ ↓ ↓ ↓ ↓
$x \cdot \$25 = \$4$

$x = \dfrac{4}{25}$

$x = 0.16$, or 16%
The tip was 16% of the cost of the meal.

b) The total cost of the meal, including the
tip was $25 + $4, or $29.

71. Let x = the percent of crude oil imports from
Canada and Mexixo. Then we have:
$x \cdot 8.9$ is 3.1.
↓ ↓ ↓ ↓
$x \cdot 8.9 = 3.1$

$x = \dfrac{3.1}{8.9}$

$x = 0.34831461$
The percent of crude oil imports from Canada
and Mexixo was about 34.8%. The percent
that came from the rest of the world was
about $100 - 34.8 = 65.2\%$.

73. Let I = the amount of interest Irena will pay.
Then we have:
I is 6% of $3500?
↓ ↓ ↓ ↓   ↓
$I = 0.06 \quad \cdot \quad \$3500$
$I = \$210$

Irena will pay $210 in interest.

75. Let x = the number of women who had babies
in good or excellent health. Then we have:
x is 95% of 300?
↓ ↓ ↓ ↓   ↓
$x = 0.95 \quad \cdot \quad 300$
$x = 285$

285 women had babies in good or excellent
health.

77. Let x = the amount of income Sara would
need to earn on her own to be comparable to
her hourly rate at Village Copy. Then we
have:
x is 120% of $16?
↓ ↓ ↓ ↓   ↓
$x = 1.2 \quad \cdot \quad \$16$
$x = 19.2$

Sara would need to earn $19.20/ hr if she
worked on her own.

79. Let x = the percentage by which the number
of minutes users spent on Facebook
increased. Since the number of minutes grew
from 1.7 million to 13.9 million, the increase
was 12.2 million minutes. Then we have:
$x \cdot 1.7$ is 12.2.
↓ ↓ ↓ ↓
$x \cdot 1.7 = 12.2$

$x = \dfrac{12.2}{1.7}$

$x \approx 7.17647059$, or 718%
The percentage by which the number of
minutes users spent on Facebook increased is
about 718%.

81. Let c = the cost of the merchandise. Then we
have:
Cost plus tax is $37.80.
↓   ↓ ↓ ↓   ↓
$c \quad + \quad 0.05c = \$37.80$

$1.05c = \$37.80$

$c = \dfrac{\$37.80}{1.05}$

$c = \$36$
The cost of the merchandise is $36.

83. Let c = the cost of only the software. Then we have:

Cost plus tax is $157.41.
$$\downarrow \quad \downarrow \quad \downarrow \quad \downarrow \quad \downarrow$$
$$c \;\; + \;\; 0.06c = \$157.41$$
$$1.06c = \$157.41$$
$$c = \frac{\$157.41}{1.06}$$
$$c = \$148.50$$

The cost of the software alone would have been $148.50.

85. Let p = the number of pounds of body weight in fat. Then we have:

16.5% of 191 pounds is weight of fat.
$$\downarrow \quad \downarrow \quad \downarrow \quad \downarrow \quad \downarrow$$
$$0.165 \; \cdot \quad\;\; 191 \quad = \quad\;\; p$$
$$0.165 \cdot 191 = p$$
$$31.515 = p$$

The part, in pounds, of the author's body weight in fat is about 31.5 pounds.

87. Let x = the number of a sale or a response from a customer. Then we have:

Number of sales or response is 2.15% of 114 billion.
$$\downarrow \qquad\qquad \downarrow \quad \downarrow \quad \downarrow \;\; \downarrow$$
$$x \qquad\qquad = 0.0215 \; \cdot \; 114$$
$$x = 0.0215 \cdot 114$$
$$x = 2.451$$

The number of pieces of mail that led to a sale or a response from a customer was about 2.45 billion pieces of mail.

89. Let c = the number of calories in a 1-oz serving of Lay's® Classic potato chips. Then we have:

Baked Lay's calories is 20% less calories than Lay's Classic
$$\downarrow \qquad\qquad \downarrow \qquad\qquad \downarrow$$
$$120 \qquad\quad = \qquad c - 0.20c$$
$$120 = c - 0.20c$$
$$120 = 0.80c$$
$$150 = c$$

The number of calories in a serving of Lay's Classic Potato Chips is 150 calories.

91. *Thinking and Writing Exercise.*

93. Let l represent the length and w represent the width; $2l + 2w$.

95. Let p represent the number of points Tino scored. 5 fewer means subtract 5 from p, or $p - 5$.

97. The product of 10 and half of a, or $\frac{1}{2}a$, $10 \cdot \left(\frac{1}{2}a\right)$.

99. Let l represent the length and w represent the width; $w = l - 2$.

101. a) Together Monday and Friday make up $\frac{2}{5}$ or 40% of the work week. So citing the survey indicates that Monday and Friday do not appear to involve excessive sick leave.
b) Memorial Day is the last Monday of May and Labor Day is the first Monday in September. This means there are 3 months and a few days between these two dates, and $\frac{3}{12} = \frac{1}{4}$ which is about 25% of the entire year. So, if 26% of home burglaries occur between Memorial Day and Labor Day that is not more than would be expected.

103. Let p = the population of Bardville. Then we have:

1332 is 15% of 48% of the population.
$$\downarrow \downarrow \downarrow \downarrow \downarrow \quad \downarrow \qquad\quad \downarrow$$
$$1332 = 0.15 \; \cdot \; 0.48 \; \cdot \qquad p$$
$$\frac{1332}{0.15(0.48)} = p$$
$$18,500 = p$$

The population of Bardville is 18,500.

105. Let h = Dana's final adult height. Dana's height of 4 ft 8 in is equivalent to 56 in (4 ft × 12 in/ft + 8 in). Then we have:

56 in is 84.4% of adult height.
$$\downarrow \downarrow \downarrow \quad \downarrow \qquad\quad \downarrow$$
$$56 \; = 0.844 \; \cdot \qquad h$$
$$56 = 0.844h$$
$$\frac{56}{0.844} = h$$
$$66.3507109 = h$$

Dana's adult height would be about 5 ft 6 in.

107. Between 2005 and 2007, the high school dropout rate in the US decreased from 94 to 87 per thousand, or by 94 − 87 = 7. Let p = the percent by which the dropout rate decreased. Then we have:

7 is what percent of 94?

$$7 = p \cdot 94$$

$$\frac{7}{94} = p$$

$$0.07446809 \approx p$$

The dropout rate between 2005 and 2007 decreased by about 7.4%, which is about $7.4 \div 2$ years, or 3.7% per year. Assuming the dropout rate continues to decrease at the same rate, in 2008 it will be $87 - (0.037 \times 87) \approx 83.8$, or 84 per thousand, and in in 2009 it will be $83.8 - (0.037 \times 83.8) \approx 80.7$, or 81 per thousand.

109. *Thinking and Writing Exercise.*

Exercise Set 2.5

1. *Familiarize.* Let x = the number. Then "two fewer than ten times a number" translates to $10x - 2$.
Translate.

Two fewer than ten times a number is 78.

$$10x - 2 = 78$$

Carry out. We solve the equation.
$$10x - 2 = 78$$
$$10x = 80 \quad \text{Adding 2}$$
$$x = 8 \quad \text{Dividing by 10}$$
Check. Ten times 8 is 80. Two less than 80 is 78. The answer checks.
State. The number is 8.

3. *Familiarize.* Let x = the number. Then "five times the sum of 3 and some number" translates to $5(3 + x)$.
Translate.

Five times the sum of three and some number is 70.

$$5(3 + x) = 70$$

Carry out. We solve the equation.

$$5(3 + x) = 70$$
$$15 + 5x = 70 \quad \text{Distributive Law}$$
$$5x = 55 \quad \text{Subtracting 15}$$
$$x = 11 \quad \text{Dividing by 5}$$
Check. Three plus 11 is 14. Five times 14 is 70. The answer checks.
State. The number is 11.

5. *Familiarize.* Let p = the price of the 8-GB iPod Nano. At 20% more, the 16-GB iPod Nano is 120% of the 8-GB iPod Nano.
Translate.
$180 is 120% of the 8-GB iPod Nano.

$$180 = 1.20 \cdot p$$

Carry out. We solve the equation.
$$180 = 1.20 \cdot p$$

$$\frac{180}{1.20} = p \quad \text{Dividing by 1.20}$$

$$150 = p$$

Check. 120% of $150, or 1.20($150), is $180. The answer checks.
State. The price of the 8-GB iPod Nano was $150.

7. *Familiarize.* Let p = the cost of the Nike running shoes. The sales tax is 7%.
Translate.

Price of Nike shoes plus sales tax is $90.95.

$$p + 0.07p = 90.95$$

Carry out. We solve the equation.
$$p + 0.07p = 90.95$$
$$1.07p = 90.95 \quad \text{Combining like terms}$$

$$p = \frac{90.95}{1.07} \quad \text{Dividing by 1.07}$$

$$p = 85$$

Check. The Nike running shoes cost $85 and the sales tax is $0.07 \cdot \$85 = \$5.95$. Because $85 + $5.95 = $ 90.95, the answer checks.
State. The cost of the running shoes is $85.

9. **Familiarize.** Let d = Kouros' distance, in miles, from the start after 16 hr. Then the distance from the finish line is $2d$.

Translate.

$$\underbrace{\text{Distance from start}} \underset{\text{plus}}{} \underbrace{\text{distance from finish}} \text{ is 433 km.}$$

$$\begin{array}{ccccc} \downarrow & \downarrow & \downarrow & \downarrow & \downarrow \\ d & + & 2d & = & 433 \end{array}$$

Carry out. We solve the equation.

$$d + 2d = 433$$
$$3d = 433 \qquad \text{Combining like terms}$$
$$d = \frac{433}{3} \qquad \text{Dividing by 3}$$
$$d = 144\frac{1}{3}$$

Check. If Kouros is $\frac{433}{3}$ mi from the start, then he is $2 \cdot \frac{433}{3}$, or $\frac{866}{3}$ mi from the finish.

Since $\frac{433}{3} + \frac{866}{3} = \frac{1299}{3} = 433$, the total distance run. The answer checks.

State. Kouros had run approximately $144\frac{1}{3}$ mi.

11. **Familiarize.** Let d = the distance the driver had traveled when he was 20 mi closer to the finish line than the start. At that point, his distance to the finish line is $d - 20$.

Translate.

$$\underbrace{\text{Distance from start}} \underset{\text{plus}}{} \underbrace{\text{distance from finish}} \text{ is 300 mi.}$$

$$\begin{array}{ccccc} \downarrow & \downarrow & \downarrow & \downarrow & \downarrow \\ d & + & d-20 & = & 300 \end{array}$$

Carry out. We solve the equation.

$$d + (d - 20) = 300$$
$$2d - 20 = 300 \qquad \text{Combining like terms}$$
$$2d = 320 \qquad \text{Adding 20}$$
$$d = 160 \qquad \text{Dividing by 2}$$

Check. When the driver is 160 mi from the start, he has $300 - 160$, or 140 mi left to go. 140 mi is 20 mi less than 160 mi. The answer checks.

State. The driver had traveled 160 mi.

13. **Familiarize.** Let n = Lara's apartment

number. Then her next-door neighbor's apartment number is $n + 1$.

Translate.

$$\underbrace{\text{Lara's number}} \underset{\text{plus}}{} \underbrace{\text{next-door's number}} \text{ is 2409.}$$

$$\begin{array}{ccccc} \downarrow & \downarrow & \downarrow & \downarrow & \downarrow \\ n & + & n+1 & = & 2409 \end{array}$$

Carry out. We solve the equation.

$$n + (n + 1) = 2409$$
$$2n + 1 = 2409 \qquad \text{Combining like terms}$$
$$2n = 2408 \qquad \text{Subtracting 1}$$
$$n = 1204 \qquad \text{Dividing by 2}$$

Check. If = Lara's apartment number is 1204, then her next-door neighbor's apartment number is 1205. The sum of these two numbers is 2409. The answer checks.

State. The apartment numbers are 1204 and 1205.

15. **Familiarize.** Let n = Chrissy's house number. Then Bryan's house number is $n + 2$.

Translate.

$$\underbrace{\text{Chrissy's number}} \underset{\text{plus}}{} \underbrace{\text{Bryan's number}} \text{ is 794.}$$

$$\begin{array}{ccccc} \downarrow & \downarrow & \downarrow & \downarrow & \downarrow \\ n & + & n+2 & = & 794 \end{array}$$

Carry out. We solve the equation.

$$n + (n + 2) = 794$$
$$2n + 2 = 794 \qquad \text{Combining like terms}$$
$$2n = 792 \qquad \text{Subtracting 2}$$
$$n = 396 \qquad \text{Dividing by 2}$$

Check. If Chrissy's house number is 396, then Bryan's house number is 398. The sum of these number is 794. The answer checks.

State. The house numbers are 396 and 398.

17. **Familiarize.** Let x = the first page number. Then $x+1$ = the second page number, and $x+2$ = the third page number.
Translate.

$\underbrace{\text{The sum of three consecutive page numbers}}$ is 60.

$$x+(x+1)+(x+2) \;=\; 60$$

Carry out. We solve the equation.
$x+(x+1)+(x+2)=60$

$3x+3=60$ Combining like terms

$3x=57$ Subtracting 3

$x=19$ Dividing by 3

If $x=19$, then $x+1=20$, and $x+2=21$.
Check 19, 20, and 21 are consecutive integers, and $19+20+21=60$. The result checks.
State. The page numbers are 19, 20, and 21.

19. **Familiarize.** Let x = the age of the groom. Then $x+19$ = the age of the bride.
Translate.

$\underbrace{\text{The sum of their ages}}$ is 185.

$$x+(x+19) \;=\; 185$$

Carry out. We solve the equation.
$x+(x+19)=185$

$2x+19=185$ Combining like terms

$2x=166$ Subtracting 19

$x=83$ Dividing by 2

If the groom was 83, then the bride was 83 + 19, or 102.
Check The sum of the ages of the groom and the bride is 83 + 102, or 185. The result checks.
State. The groom was 83, and the bride was 102.

21. **Familiarize.** Let x = the number of nonspam messages. Then $5x$ = the number of spam messages.
Translate.

$\underbrace{\text{The sum of spam and nonspam messages}}$ is 210 billion e-mails.

$$x+5x \;=\; 210$$

Carry out. We solve the equation.
$x+5x=210$

$6x=210$ Combining like terms

$x=35$ Dividing by 6

If the number of nonspam messages was 35 billion, then number of spam messages was $5\cdot35$, or 175 billion.
Check The sum of the nonspam and spam messages was 35 + 175 = 210 billion messages. The results check.
State. If the number of nonspam messages was 35 billion and the number of spam messages was 175 billion.

23. **Familiarize.** Let x = the page number of the left page and $x+1$ = the page number of the right page.
Translate.

$\underbrace{\text{The sum of the pages}}$ is 281.

$$x+(x+1) \;=\; 281$$

Carry out. We solve the equation.
$x+(x+1)=281$

$2x+1=281$ Combining like terms

$2x=280$ Subtracting 1

$x=140$ Dividing by 2

Check The pages are numbered 140 and 141, so they are consecutive. The sum of the pages is $140+141=281$. The results check
State. The facing pages are numbered 140 and 141.

25. **_Familiarize._** Let x = the width of the rectangular top, in feet. Then $x+60$ = the length of the rectangular top, in feet. Recall that the perimeter is given by the formula:
$P = 2w + 2l$

Translate.

$\underbrace{\text{The perimeter}}$ is 520 ft.

$\downarrow \qquad \downarrow \downarrow$

$2x + 2(x + 60) = 520$

Carry out. We solve the equation.

$2x + 2(x + 60) = 520$

$2x + 2x + 120 = 520$ Distributive Law

$4x + 120 = 520$ Combining like terms

$4x = 400$ Subtracting 120

$x = 100$ Dividing by 4

If the width is 100 ft, then the length is 100 + 60, or 160 ft.

Check The length, 160 ft, is 60 ft more than the width, 100 ft. The perimeter is $2 \cdot 100 + 2 \cdot 160 = 520$. These results check

State. The width of the rectangular top of the building is 100 ft, and the length of the top is 160 ft. The area of the top is length times width, or $100 \text{ ft} \cdot 160 \text{ ft}$, or $16{,}000 \text{ ft}^2$.

27. **_Familiarize._** Let x = the width of the basketball court, in feet. Then $x+34$ is the length of the court, in feet. Recall that the perimeter is given by the formula:
$P = 2w + 2l$

Translate.

$\underbrace{\text{The perimeter}}$ is 268 ft.

$\downarrow \qquad \downarrow \downarrow$

$2x + 2(x + 34) = 268$

Carry out. We solve the equation.

$2x + 2(x + 34) = 268$

$2x + 2x + 68 = 268$ Distributive Law

$4x + 68 = 268$ Combining like terms

$4x = 200$ Subtracting 68

$x = 50$ Dividing by 4

If the width of the court is 50 ft, then the length is 50 ft + 34 ft, or 84 ft.

Check The length, 84 ft, is 34 ft more than the width, 50 ft. The perimeter is

$2 \cdot 50 \text{ ft} + 2 \cdot 84 \text{ ft} = 268 \text{ ft}$. These results check.

State. The width of the basketball court is 50 ft, and the length is 84 ft.

29. **_Familiarize._** Let x = the width of the cross section and $x+2$ = the length of the cross section. Recall that the perimeter is given by the formula: $P = 2w + 2l$

Translate.

$\underbrace{\text{The perimeter}}$ is 10 in.

$\downarrow \qquad \downarrow \downarrow$

$2x + 2(x + 2) \ = \ 10$

Carry out. We solve the equation.

$2x + 2(x + 2) = 10$

$2x + 2x + 4 = 10$

$4x + 4 = 10$ Combining like terms

$4x + 4 - 4 = 10 - 4$ Subtracting 4 from both sides

$\dfrac{4x}{4} = \dfrac{6}{4}$ Dividing by 4

$x = 1\tfrac{1}{2}$

If the width of the cross section is $1\tfrac{1}{2}$ in., then the length is $3\tfrac{1}{2}$ in. $\left(1\tfrac{1}{2} + 2 = 3\tfrac{1}{2}\right)$.

Check The length of the cross section, $3\tfrac{1}{2}$ in. is two more than the width, $1\tfrac{1}{2}$ in. The perimeter is $2 \cdot \left(1\tfrac{1}{2}\right) + 2 \cdot \left(3\tfrac{1}{2}\right) = 10$ in. These results check.

State. The width of the cross section is $1\tfrac{1}{2}$ in., and the length is $3\tfrac{1}{2}$ in.

31. **_Familiarize._** Let x = the measure of the first angle, in degrees. Then the measure of the second angle is $3x$ degrees, and the third angle measures $x+30$ degrees. Recall that the sum of the measures of the angles of any triangle is 180°.

Translate.

$\underbrace{\text{The sum of the measures}\atop\text{of the three angles}}$ is 180 degrees.

$\downarrow \qquad\qquad \downarrow \downarrow$

$x + 3x + (x + 30) \qquad = 180$

Carry out.   We solve the equation.

$x + 3x + (x + 30) = 180$

$5x + 30 = 180$ Combining like terms

$5x = 150$ Subtracting 30

$x = 30$ Dividing by 5

If the first angle measures 30°, then the second angle measures $3 \cdot 30°$, or 90°, and the third angle measures $30° + 30°$, or 60°.

Check   The second angle has a measure of 90°, which is three times the measure of the first angle, 30°. The third angle measures 60°, which is 30° more than the first angle. The sum of the measures of the three angles is $30° + 90° + 60°$, or 180°. These results check.

State. The angles measure 30°, 90°, and 60°, respectively.

33. ***Familiarize.*** Let x = the measure of the first angle, in degrees. Then the measure of the second angle is $3x$ degrees, and the third angle measures $x + 3x + 10$ degrees. Recall that the sum of the measures of the angles of any triangle is 180°.

Translate.

The sum of the measures of the three angles is 180 degrees.

$$x + 3x + (x + 3x + 10) = 180$$

Carry out.   We solve the equation.

$x + 3x + (x + 3x + 10) = 180$

$8x + 10 = 180$ Combining like terms

$8x = 170$ Subtracting 10

$x = 21.25$ Dividing by 8

If the first angle measures 21.25°, then the second angle measures $3 \cdot 21.25$, or 63.75°, and the third angle measures $21.25° + 63.75° + 10°$, or 95°.

Check   The measure of the second angle, 63.75°, is three times the measure of the first angle, 21.25°. The third angle has a measure of 95°, which is 10° more than the sum of the first and second angles, $21.25° + 63.75°$. The sum of the measures of the three angles is $21.25° + 63.75° + 95° = 180°$. These results check.

State. The angles measure 21.25°, 63.75°,

and 95°, respectively.

35. ***Familiarize.*** Let x = the length of bottom section, in ft. Then the top section is $\dfrac{x}{6}$ ft, and the middle section is $\dfrac{x}{2}$ ft.

Translate.

The sum of the lengths is 240 ft.

$$x + \frac{x}{6} + \frac{x}{2} = 240$$

Carry out.   We solve the equation.

$x + \dfrac{x}{6} + \dfrac{x}{2} = 240$

$6x + x + 3x = 1440$ Multiply by LCD = 6

$10x = 1440$ Combining like terms

$x = 144$ Dividing by 10

If the bottom section is 144 ft, the top section is 144 ft / 6, or 24 ft, and the middle section is 144 ft / 2, or 72 ft.

Check   The top section is 24 ft, which is one sixth of the bottom section, and the middle section is 72 ft, which is half the length of the bottom section. The sum of the lengths is 24 ft + 72 ft + 144 ft = 240 ft. The result checks.

State. The top section of the rocket is 24 ft, the middle section is 72 ft, and the bottom section is 144 ft.

37. ***Familiarize.*** Let x = the number of miles Debbie can travel. Therefore, the ride will cost $3.25 + $1.80x$.

Translate.

The total cost of the taxi ride is $19.

$$3.25 + 1.80x = 19$$

Carry out.   We solve the equation.

$3.25 + 1.80x = 19$

$1.80x = 15.75$ Subtract 3.25

$x = 8.75$ Dividing by 1.8

Check   The total cost of the ride is $3.25 + $1.80/mi × 8.75 mi, or $19. This results checks.

State. On a $19 budget, Debbie can travel $8\frac{3}{4}$, or 8.75 mi.

39. **Familiarize.** Let x = the number of miles Concert Productions can travel. The total of their rental will be $49.95 + $0.39x.
Translate.
The total cost of the rental is $100.

$$\underbrace{49.95 + 0.39x}_{\downarrow} \quad \underset{\downarrow}{\underset{=}{}} \quad \underset{\downarrow}{100}$$

Carry out. We solve the equation.
$$49.95 + 0.39x = 100$$
$$0.39x = 50.05 \quad \text{Subtract 49.95}$$
$$x = 128\tfrac{1}{3} \quad \text{Dividing by 0.39}$$
Check The total amount for the rental will be $49.95 + $0.39/mi $\times$ $128\tfrac{1}{3}$ mi, or $100. This results checks.
State. Concert Productions can travel a total of $128\tfrac{1}{3}$ mi on their $100 budget.

41. **Familiarize.** Let x = the measure of the first angle, in degrees. Then the complement of the first angle is $90° - x$.
Translate.

The measure of an angle	is 15°	more than	twice the measure of its complement.

$$\underset{x}{\downarrow} \quad \underset{= 15}{\downarrow\downarrow} \quad \underset{+}{\downarrow} \quad \underset{2(90-x)}{\downarrow}$$

Carry out.   We solve the equation.
$$x = 15 + 2(90 - x)$$
$$x = 15 + 180 - 2x \quad \text{Distributive Law}$$
$$x = 195 - 2x \quad \text{Combining like terms}$$
$$3x = 195 \quad \text{Adding } 2x$$
$$x = 65 \quad \text{Dividing by 3}$$
Check If the angle measures 65°, then its complement measures 90° – 65°, or 25°. Twice the measure of the complement is $2 \cdot 25°$, or 50°. The original angle's measure of 65° is 15° more than 50°. These results check.
State. The angle measures 65°, and its complement measures 25°.

43. **Familiarize.** Let l = the length of the paper, in cm. Then the width is $l - 6.3$ cm. Recall that the perimeter of a rectangle is calculated using the formula $P = 2w + 2l$.

Translate.
The perimeter of the paper is 99 cm.

$$\underbrace{2(l - 6.3) + 2l}_{\downarrow} \quad \underset{\downarrow\downarrow}{\underset{= 99}{}}$$

Carry out. We solve the equation.
$$2(l - 6.3) + 2l = 99$$
$$2l - 12.6 + 2l = 99 \quad \text{Distributive Law}$$
$$4l - 12.6 = 99 \quad \text{Combining like terms}$$
$$4l = 111.6 \quad \text{Adding 12.6}$$
$$l = 27.9 \quad \text{Dividing by 4}$$
Check If the length of the paper is 27.9 cm, then the width is 27.9 cm – 6.3 cm, or 21.6 cm. Therefore the perimeter is $2 \cdot 21.6$ cm + $2 \cdot 27.9$ cm , or 99 cm.
State. The width is the paper is 21.6 cm, and the length is 27.9 cm.

45. **Familiarize.** Let x = the amount Amber invested in her savings account, in dollars. Her investment increased by 6%, or by $0.06x$.
Translate.

Investment	plus	6% of the investment	is $6996.

$$\underset{x}{\downarrow} \quad \underset{+}{\downarrow} \quad \underset{0.06x}{\downarrow} \quad \underset{= 6996}{\downarrow\downarrow}$$

Carry out.   We solve the equation.
$$x + 0.06x = 6996$$
$$1.06x = 6996 \quad \text{Combining like terms}$$
$$x = 6600 \quad \text{Dividing by 1.06}$$
Check If Amber invested $6600, then her investment would have grown by 6%($6600), or $396. Thus her investment would have grown to $6600 + $396, or $6996. These results check.
State. Amber 's original investment would have been $6600.

47. **Familiarize.** Let x = the losing score. If the margin of victory was 323 points, then the winning score would have been $x + 323$ points.

Translate.

$\underbrace{\text{Sum of the winning and losing scores}}$ is 1127 points.

$$x + (x + 323) \quad = 1127$$

Carry out. We solve the equation.

$x + (x + 323) = 1127$

$\quad 2x + 323 = 1127$ Combining like terms

$\qquad\quad 2x = 804$ Subtracting 323

$\qquad\qquad x = 402$ Dividing by 2

Check If the losing score was 402, then the winning score, assuming a margin of victory of 323 points, would have been 402 + 323, or 725 points. The sum of the winning and losing scores is 725 + 402, or 1127 points. These results check.

State. The winning score was 725 points.

49. *Familiarize.* Let x = the final bid. Let $0.08x$ = the seller's premium on the final bid.

Translate.

$\underbrace{\text{Final bid minus seller's premium of 8\%}}$ is $\underbrace{\text{the amount left}}$

$$x - 0.08x \qquad = \qquad 1150$$

Carry out. We solve the equation.

$x - 0.08x = 1150$

$\quad 0.92x = 1150$ Combine like terms

$\qquad x = 1250$ Dividing by 0.92

Check If the final bid was $1250, the 8% of $1250 is $0.08(1250) = \$100$. So, $1250 – $100 = $1150. This result checks.

State. The final bid is $1250.

51. *Familiarize.* Let N = the number of times the cricket chirps per minute. Let T = the corresponding Fahrenheit temperature.

Translate.

We are already given the equation relating Fahrenheit temperature T, in degrees, and the number of times a cricket chirps per minute N. We are to determine the number of chips corresponding to 80°F.

$\underbrace{\text{Temperature}}$ is $\underbrace{\text{40 more than one fourth the number of chirps.}}$

$$T \qquad = \qquad \tfrac{1}{4}N + 40$$

Carry out. We substitute 80 for T and solve the equation for N.

$T = \tfrac{1}{4}N + 40$

$80 = \tfrac{1}{4}N + 40$ Substituting 80 for T

$40 = \tfrac{1}{4}N$ Subtracting 40

$160 = N$ Multiplying by 4

Check If a cricket chirps 160 times per minute, then the Fahrenheit temperature would be $\tfrac{1}{4}(160) + 40$, or 40 + 40, or 80°F. This result checks.

State. The number of chirps per minute corresponding to 80°F is 160.

53. *Familiarize.* We examine the table to determine the size of aquarium and the recommended stocking density. We notice that:

$100 = 20 \times 5$

$120 = 24 \times 5$

$200 = 40 \times 5$

$250 = 50 \times 5$

This leads us to conclude that size of aquarium is 5 times the stocking density. We let x = the stocking density and we let S = the size of the aquarium.

Translate.

$\underbrace{\text{Size of Aquarium}}$ is 5 times $\underbrace{\text{the stocking density.}}$

$$S \quad = 5 \cdot \qquad x$$

Carry out. We substitute 30 in. for x and solve for S.

$S = 5x$

$S = 5(30)$ Substituting 30 for x

$S = 150$ Simplify

Check. If the aquarium is 150 gallons, then the recommended stocking density $150 \div 5$, or 30 inches. This result checks.

State. The aquarium size for 30 in. of fish is 150 gallons.

55. *Familiarize.* We examine the table to determine the relationship between the day in August and weight, in pounds, of the pumpkin. Observe that the pumpkin's weight increases by 30 pounds each day:

From Aug. 1 to Aug. 2: 410 – 380 = 30 lb

From Aug. 2 to Aug. 3: 440 – 410 = 30 lb

From Aug. 3 to Aug. 4: 470 – 440 = 30 lb

From Aug. 4 to Aug. 11 (11 – 4, or 7 days): 680 – 470 = 210 = 7·30 lb.

From Aug. 11 to Aug. 25 (25 – 11, or 14

days): $1100 - 680 = 420 = 14 \cdot 30$ lb.

Let d = the number of days after August 1 on which the pumpkin weighed 920 pounds.

Translate.

Weight on August 1	plus 30 times	number of days after August 1	is	920 pounds.
↓	↓ ↓ ↓	↓	↓	↓
380	+ 30 ·	d	=	920

Carry out. We solve the equation.

$$380 + 30 \cdot d = 920$$
$$30d = 920 - 380 \quad \text{Subtracting 380}$$
$$30d = 540 \quad \text{Collecting like terms}$$
$$d = 18 \quad \text{Dividing by 30}$$

Check. We use a table to check. Enter $y = 380 + 30x$ in a graphing calculator and set up a table in Ask mode. Enter the x-values as the number of days after August 1: 0, 1, 2, 3, 10, 24, and 18. We see that we get the values given in the statement of the problem for the first 6 x-values and the x-value 18 gives a weight of 920 pounds. The answer checks.

X	Y₁	
0	380	
1	410	
2	440	
3	470	
10	680	
24	1100	
18	920	

X=18

State. The pumpkin weighed 920 pounds 18 days after August 1, or on August 19.

57. **Familiarize.** Let w = the walking speed, in feet per minute. Then $w + 250$ = the running speed, in feet per minute. Since $d = rt$, the walking distance is $w \cdot 10$, or $10w$, and the running distance is $(w + 250) \cdot 20$, or $20(w + 250)$.

Translate.

We are given that Samantha ran and walked a total of 15,500 feet.

running distance	plus	walking distance	is 15,500.	
↓	↓	↓	↓	↓
$20(w + 250)$	+	$10w$	=	15,500

Carry out. We solve the equation for w.

$$20(w + 250) + 10w = 15,500$$
$$20w + 5000 + 10w = 15,500 \quad \text{Distributive Law}$$
$$30w + 5000 = 15,500 \quad \text{Combine like terms}$$
$$30w = 10,500 \quad \text{Subtracting by 5000}$$
$$w = 350 \quad \text{Dividing by 30}$$

Check If the walking speed is 350 feet per minute, the running speed is $350 + 250 = 600$ feet per minute. The total distance is: $20 \cdot 600 + 10 \cdot 350 = 12,000 + 3500 = 15,500$. This result checks.

State. Samantha ran at 600 feet per minute.

59. **Familiarize.** Let c = the driving speed in clear weather, in mph. Let $\frac{1}{2}c$ = the driving speed in the snowstorm, in mph. Since $d = rt$, the clear weather driving distance is $c \cdot 5$, or $5c$, and the snowstorm driving distance is $\frac{1}{2}c \cdot 2$, or c.

Translate.

We are given the Anthony drove 240 more miles in clear weather than the snowstorm.

distance in clear weather	is	distance in snowstorm	plus 240	
↓	↓	↓	↓	↓
$5c$	=	c	+	240

Carry out. We solve the equation for c.

$$5c = c + 240$$
$$4c = 240 \quad \text{Subtracting by } c$$
$$c = 60 \quad \text{Dividing by 4}$$

Check If the driving speed in clear weather was 60 mph, the driving speed in the snowstorm was $\frac{1}{2} \cdot 60 = 30$ mph. The distance equation is: $5 \cdot 60 = 60 + 240$. This result checks.

State. Anthony drove 30 mph in the snow.

61. **Familiarize.** Let t = the time spend bicycling at a slower rate of 10 mph *in hours*. Then $t + 30$ = the time *in minutes* spent bicycling at 15 mph. Converting this to time into hours gives $t + \frac{1}{2}$. Since $d = rt$, the distance at a slower rate is $10t$, and the distance at a faster rate is $15\left(t + \frac{1}{2}\right)$.

Translate.

We are given the total distance bicycled of 25 miles.

$$\underbrace{\text{distance at faster rate}} \text{ plus } \underbrace{\text{distance at slower rate}} \text{ is } 25.$$

$$15\left(t+\tfrac{1}{2}\right) \quad + \quad 10t \quad = 25$$

Carry out. We solve the equation for w.

$$15\left(t+\tfrac{1}{2}\right)+10t = 25$$
$$15t+7.5+10t = 25 \quad \text{Distributive Law}$$
$$25t+7.5 = 25 \quad \text{Combine like terms}$$
$$25t = 17.5 \quad \text{Subtracting by } 7.5$$
$$t = 0.7 \quad \text{Dividing by } 25$$

Check The time spend bicycling at a slower rate is 0.7 hours, or $0.7 \cdot 60 = 42$ minutes. So the time spent bicycling at a faster rate is 42 + 30 = 72 minutes, or 0.7 + 0.5 = 1.2 hours. The total distance is:
$15 \cdot 1.2 + 10 \cdot 0.7 = 18 + 7 = 25$. This result checks.

State. Justin rode for 1.2 hours, or 72 minutes, at a faster speed.

63. *Thinking and Writing Exercise.*

65. Since –9 is to the left of 5 on the number line, we have $-9 < 5$.

67. Since –4 is to the left of 7 on the number line, we have $-4 < 7$.

69. Simply exchange the numbers and change the inequality symbol; $-4 \leq x$.

71. Simply exchange the numbers and change the inequality symbol; $y < 5$.

73. *Thinking and Writing Exercise.*

75. **Familiarize.** Let c = the amount the meal originally cost. The 15% tip is calculated on the original cost of the meal, so the tip is $0.15c$.

Translate.

$$\underbrace{\text{Original cost}} \text{ plus tip } \text{ less } \$10 \text{ is } \$32.55.$$

$$c \quad + 0.15c - \quad 10 = 32.55$$

Carry out. We solve the equation.

$$c+0.15c-10 = 32.55$$
$$1.15c-10 = 32.55$$
$$1.15c = 42.55$$
$$c = 37$$

Check. If the meal originally cost \$37, the tip was 15% of \$37, or 0.15(\$37), or \$5.55. Since \$37 + \$5.55 – \$10 = \$32.55, the answer checks.

State. The meal originally cost \$37.

77. **Familiarize.** Let x = a score. Lincoln referred to "four score and seven,", which can be expressed as $4x+7$. The Gettysburg Address was given in 1863, and Lincoln refers to 1776, a total of 87 years prior.

Translate.

$$\underbrace{\text{Four score and seven years}} \text{ is } \underbrace{87 \text{ years}}.$$

$$4x+7 \quad = \quad 87$$

Carry out. We solve the equation.

$$4x+7 = 87$$
$$4x = 80$$
$$x = 20$$

Check. If a score is 20 years, then four score and seven is $4 \cdot 20 + 7$, or 87. If 87 is added to the year 1776, we have $1776 + 87$, or 1863. This results checks.

State. A score is 20 years.

79. **Familiarize.** Let x = the number of half-dollars. Then $2x$ would represent the number of quarters, $2(2x)$, or $4x$ would represent the number of dimes, and $3(4x)$, or $12x$ would represent the number of nickels. Furthermore, the value of the half-dollars, in cents, is represented by $50x$, the value of the quarters, in cents, is represented by $25(2x)$, or $50x$, the value of the dimes is represented by $10(4x)$, or $40x$, and the value of the nickels, in cents is represented by $5(12x)$, or $60x$.

Translate.

$$\underbrace{\text{Total value of the change, in cents,}} \text{ is } 1000.$$

$$50x+50x+40x+60x \quad = 1000$$

Carry out. We solve the equation.

$$50x+50x+40x+60x = 1000$$
$$200x = 1000$$
$$x = 5$$

If $x = 5$, then there are 5 half-dollars; there are 10 quarters, there are 20 dimes, and there are 60 nickels.

Check. The total value of the change is:
5(50¢) + 10(25¢) + 20(10¢) + 60(5¢), or 250¢ + 250¢ + 200¢ + 300¢ = 1000¢, or $10.00. These results check.

State. There are 5 half-dollars, 10 quarters, 20 dimes, and 60 nickels.

81. **Familiarize.** Let x = the original price of the camera before the two discounts. Julio's credit account agreement gives him a 10% discount; so, he would pay 90% of the original price, or $0.9x$. His coupon give him an additional 10% discount off of the reduced price; so, he would pay 90% of $0.9x$, or $0.9(0.9x)$, or $0.81x$.

Translate.

$$\underbrace{\text{Final discounted price of the camera}}\ \text{is \$77.75.}$$

$$\underset{0.81x}{\downarrow}\qquad\underset{= 77.75}{\downarrow\quad\downarrow}$$

Carry out. We solve the equation.
$$0.81x = 77.75$$
$$x = 95.99 \quad \text{(Rounded)}$$

Check. If the original price of the camera was $95.99, then after the credit account discount is applied, Julio will owe 90% of $95.99, or $86.39. After applying the coupon discount of 10%, he will owe 90% of the first discounted price, or 90% of $86.39, or $77.75 (rounded). These results check.

State. The original price of the camera was $95.99.

83. **Familiarize.** Let x = the number of additional games they must play. Then the total number of games they will win is $\frac{1}{2}x + 15$, and the total number of games they will play, including the additional games, is $x + 20$.

Translate.

$$\underbrace{\text{The total number of games they win}}\ \text{is 60\% of }\underbrace{\text{the number of games played.}}$$

$$\underset{\frac{1}{2}x+15}{\downarrow}\quad\underset{= 0.6}{\downarrow\quad\downarrow}\quad\underset{\cdot}{\downarrow}\quad\underset{(x+20)}{\downarrow}$$

Carry out. We solve the equation.
$$\frac{1}{2}x + 15 = 0.6(x + 20)$$
$$0.5x + 15 = 0.6x + 12$$
$$3 = 0.1x$$
$$30 = x$$

Check. If they play 30 additional games, then they will have played 20 + 30, or 50 games. If they win 60% of all the games they play, they will win 60% of 50, or 0.6(50), or 30 games. They have already won 15 games, and if they win half of the additional ones, or $\frac{1}{2}(30) = 15$, they will have won 15 + 15, or 30 games. These results check.

State. They must play a additional 30 games.

85. **Familiarize.** Let x = the score on the third test. If the average on the first two exams was 85, then it follows that the total of her exam percentages on the first two exams was 170 $(170 \div 2 = 85)$.

Translate.

$$\underbrace{\text{Average of the three exams}}\ \text{is 82.}$$

$$\underset{\dfrac{170+x}{3}}{\downarrow}\qquad\underset{= 82}{\downarrow\quad\downarrow}$$

Carry out. We solve the equation.
$$\frac{170 + x}{3} = 82$$
$$3\left(\frac{170 + x}{3}\right) = 3 \cdot 82$$
$$170 + x = 246$$
$$x = 76$$

Check. If Elsa got an average of 85 on her first two exams, then she must have received a total of 170 points on the two exams since $170 \div 2 = 85$. If she received a score of 76 on her third exam, then her total exam points would be 170 + 76, or 246. The average on all three exams would be $246 \div 3 = 82$. This result checks.

State. The score on the third exam must have been 76.

87. **Thinking and Writing Exercise.**

89. **Familiarize.** Let x = the width of the rectangle, in cm. Then the length is $x + 4.25$ cm. Recall that the perimeter is $P = 2w + 2l$.
Translate.

$\underbrace{\text{The perimeter of the rectangle}}$ is 101.74.

$$\downarrow \qquad\qquad \downarrow \quad \downarrow$$
$$2x + 2(x + 4.25) \qquad = 101.74$$

Carry out. We solve the equation.
$$2x + 2(x + 4.25) = 101.74$$
$$2x + 2x + 8.5 = 101.74$$
$$4x + 8.5 = 101.74$$
$$4x = 93.24$$
$$x = 23.31$$

Check. If the width is 23.32 cm, then the length is $23.31 + 4.25$, or 27.56 cm. The perimeter is 2(23.31 cm) + 2(27.56 cm), or 101.74 cm. These results check.
State. The width is 23.31 cm and the length is 27.56 cm.

Exercise Set 2.6

1. $\geq$

3. $<$

5. Equivalent

7. Equivalent

9. $x > -2$
   a) Since $5 > -2$ is true, 5 is a solution.
   b) Since $0 > -2$ is true, 0 is a solution.
   c) Since $-3 > -2$ is false, -3 is not a solution.

11. $y \leq 19$
   a) Since $18.99 \leq 19$ is true, 18.99 is a solution.
   b) Since $19.01 \leq 19$ is false, 19.01 is not a solution.
   c) Since $19 \leq 19$ is true, 19 is a solution.

13. $a \geq -6$
   a) Since $-6 \geq -6$ is true, -6 is a solution.
   b) Since $-6.1 \geq -6$ is false, -6.1 is not a solution.
   c) Since $-5.9 \geq -6$ is true, -5.9 is a solution.

15. The solutions of $x \leq 7$ are shown by using a bracket at the point 7 and shading all points to the left of 7. The bracket indicates that 7 is part of the graph.

17. The solutions of $t > -2$ are shown by using a parenthesis at the point -2 and shading all points to the right of -2. The parenthesis indicates that -2 is not part of the graph.

19. The solutions of $1 \leq m$ are shown by using a bracket at the point 1 and shading all points to the right of 1. The bracket indicates that 1 is part of the graph.

21. In order to be the solution of the inequality $-3 < x \leq 5$, a number must be a solution of both $-3 < x$ and $x \leq 5$. The solution set is graphed as follows:

23. In order to be the solution of the inequality $0 < x < 3$, a number must be a solution of both $0 < x$ and $x < 3$. The solution set is graphed as follows:

25. The solutions of $y < 6$ are shown by using a parenthesis at the point 6 and shading all points to the left of 6. The parenthesis indicates that 6 is not part of the graph. The solution set is $\{y \mid y < 6\}$, or $(-\infty, 6)$.,

27. The solutions of $x \geq -4$ are shown by using a bracket at the point -4 and shading all points to the right of -4. The bracket indicates that -4 is part of the graph. The solution set is $\{x \mid x \geq -4\}$, or $[-4, \infty)$.

29. The solutions of $t > -3$ are shown by using parenthesis at the point -3 and shading all points to the right of -3. The parenthesis indicates that -3 is not part of the graph. The solution set is $\{t \mid t > -3\}$, or $(-3, \infty)$.

31. The solutions of $x \le -7$ are shown by using a bracket at the point -7 and shading all points to the left of -7. The bracket indicates that -7 is part of the graph. The solution set is $\{x \mid t \le -7\}$, or $(-\infty, -7]$.

33. All the points to the right of -4 are shaded. The parenthesis at -4 indicates that -4 is not part of the graph. We have $\{x \mid x > -4\}$, or $(-4, \infty)$.

35. All the points to the left of 2 are shaded. The bracket at 2 indicates that 2 is part of the graph. We have $\{x \mid x \le 2\}$, or $(-\infty, 2]$.

37. All the points to the left of -1 are shaded. The parenthesis at -1 indicates that -1 is not part of the graph. We have $\{x \mid x < -1\}$, or $(-\infty, -1)$.

39. All the points to the right of 0 are shaded. The bracket at 0 indicates that 0 is part of the graph. We have $\{x \mid x \ge 0\}$, or $[0, \infty)$.

41.
$$y + 2 > 9$$
$$y + 2 - 2 > 9 - 2 \quad \text{Adding } -2 \text{ to both sides}$$
$$y > 7 \qquad \text{Simplifying}$$
The solution set is $\{y \mid y > 7\}$, or $(7, \infty)$.
The graph is as follows:

43.
$$x - 8 \le -10$$
$$x - 8 + 8 \le -10 + 8 \quad \text{Adding 8 to both sides}$$
$$x \le -2 \qquad \text{Simplifying}$$
The solution set is $\{x \mid x \le -2\}$, or $(-\infty, -2]$. The graph is as follows:

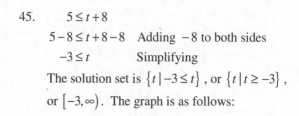

45.
$$5 \le t + 8$$
$$5 - 8 \le t + 8 - 8 \quad \text{Adding } -8 \text{ to both sides}$$
$$-3 \le t \qquad \text{Simplifying}$$
The solution set is $\{t \mid -3 \le t\}$, or $\{t \mid t \ge -3\}$, or $[-3, \infty)$. The graph is as follows:

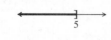

47.
$$2x + 4 \le x + 9$$
$$2x + 4 - 4 \le x + 9 - 4 \quad \text{Adding } -4 \text{ to both sides}$$
$$2x \le x + 5 \qquad \text{Simplifying}$$
$$2x - x \le x + 5 - x \quad \text{Adding } -x$$
$$x \le 5 \qquad \text{Simplifying}$$
The solution set is $\{x \mid x \le 5\}$, or $(-\infty, 5]$.

The graph is as follows:

49.
$$y + \frac{1}{3} \le \frac{5}{6}$$
$$y + \frac{1}{3} - \frac{1}{3} \le \frac{5}{6} - \frac{1}{3}$$
$$y \le \frac{5}{6} - \frac{2}{6}$$
$$y \le \frac{3}{6}$$
$$y \le \frac{1}{2}$$

The solution set is $\left\{ y \,\middle|\, y \le \frac{1}{2} \right\}$, or $\left(-\infty, \frac{1}{2} \right]$.

51.
$$t - \frac{1}{8} > \frac{1}{2}$$
$$t - \frac{1}{8} + \frac{1}{8} > \frac{1}{2} + \frac{1}{8}$$
$$t > \frac{4}{8} + \frac{1}{8}$$
$$t > \frac{5}{8}$$

The solution set is $\left\{ t \middle| t > \frac{5}{8} \right\}$, or $\left(\frac{5}{8}, \infty \right)$.

53.
$$-9x + 17 > 17 - 8x$$
$$-9x + 17 - 17 > 17 - 8x - 17$$
$$-9x > -8x$$
$$-9x + 9x > -8x + 9x$$
$$0 > x$$

The solution set is $\{ x \mid 0 > x \}$, or $\{ x \mid x < 0 \}$, or $(-\infty, 0)$.

55. $-23 < -t$

The inequality states that the opposite of 23 is less than the opposite of t. Thus, t must be less than 23, so the solution set is $\{ t \mid t < 23 \}$.

To solve this inequality using the addition principle, we would proceed as follows:
$$-23 < -t$$
$$-23 + t < -t + t$$
$$-23 + t < 0$$
$$-23 + t + 23 < 0 + 23$$
$$t < 23$$

The solution set is $\{ t \mid t < 23 \}$, or $(-\infty, 23)$.

57. $5x < 35$
$$\frac{1}{5} \cdot 5x < \frac{1}{5} \cdot 35 \qquad \text{Multiplying by } \frac{1}{5}$$
$$x < 7$$

The solution set is $\{ x \mid x < 7 \}$, or $(-\infty, 7)$.
The graph is as follows:

59. $-24 > 8t$
$$\frac{1}{8} \cdot (-24) > \frac{1}{8} \cdot 8t \qquad \text{Multiplying by } \frac{1}{8}$$
$$-3 > t \qquad \text{Simplifying}$$

The solution set is $\{ t \mid -3 > t \}$, or $\{ t \mid t < -3 \}$, or $(-\infty, -3)$. The graph is as follows:

61. $1.8 \geq -1.2n$
$$\frac{1.8}{-1.2} \leq \frac{-1.2n}{-1.2} \qquad \text{Dividing by } -1.2$$
$$\qquad\qquad\qquad \text{The symbol has to be reversed}$$
$$-1.5 \leq n$$

The solution set is $\{ n \mid n \geq -1.5 \}$, or $[-1.5, \infty)$.

63.
$$-2y \leq \frac{1}{5}$$
$$-\frac{1}{2} \cdot (-2y) \geq -\frac{1}{2} \cdot \frac{1}{5} \qquad \text{Multiplying by } -\frac{1}{2}$$
$$\qquad\qquad\qquad \text{The symbol has to be reversed}$$
$$y \geq -\frac{1}{10}$$

The solution set is $\left\{ y \middle| y \geq -\frac{1}{10} \right\}$, or $\left[-\frac{1}{10}, \infty \right)$.

65.
$$-\frac{8}{5} > -2x$$
$$-\frac{1}{2} \cdot \left(-\frac{8}{5} \right) < -\frac{1}{2} \cdot (-2x) \qquad \text{Multiplying by } -\frac{1}{2}$$
$$\qquad\qquad\qquad \text{The symbol has to be reversed.}$$
$$\frac{8}{10} < x$$
$$\frac{4}{5} < x, \text{ or } x > \frac{4}{5}$$

The solution set is $\left\{ x \middle| \frac{4}{5} < x \right\}$, or $\left\{ x \middle| x > \frac{4}{5} \right\}$, or $\left(\frac{4}{5}, \infty \right)$.

67. $7 + 3x < 34$

$7 + 3x - 7 < 34 - 7$ Adding -7

$3x < 27$ Simplifying

$x < 9$ Multiplying both sides by $\dfrac{1}{3}$

The solution set is $\{x \mid x < 9\}$, or $(-\infty, 9)$.

69. $4t - 5 \leq 23$

$4t - 5 + 5 \leq 23 + 5$ Adding 5

$4t \leq 28$ Simplifying

$t \leq 7$ Multiplying both sides by $\dfrac{1}{4}$

The solution set is $\{t \mid t \leq 7\}$, or $(-\infty, 7]$.

71. $16 < 4 - a$

$16 - 4 < 4 - a - 4$ Adding -4

$12 < -a$ Simplifying

$-12 > a$ Multiplying both sides by -1 and reversing the inequality symbol

The solution set is $\{a \mid -12 > a\}$, or $\{a \mid a < -12\}$, or $(-\infty, -12)$.

73. $5 - 7y \geq 5$

$5 - 7y - 5 \geq 5 - 5$ Subtracting 5

$-7y \geq 0$ Simplifying

$y \leq 0$ Multiplying both sides by $-\dfrac{1}{7}$ and reversing the inequality symbol

The solution set is $\{y \mid y \leq 0\}$, or $(-\infty, 0]$.

75. $-3 < 8x + 7 - 7x$

$-3 < x + 7$ Simplifying

$-3 - 7 < x + 7 - 7$ Adding -7

$-10 < x$ Simplifying

The solution set is $\{x \mid -10 < x\}$, or $\{x \mid x > -10\}$, or $(-10, \infty)$.

77. $6 - 4y > 4 - 3y$

$6 - 4y - 4 > 4 - 3y - 4$ Adding -4

$-4y + 2 > -3y$ Simplifying

$-4y + 2 + 4y > -3y + 4y$ Adding $4y$

$2 > y$ Simplifying

The solution set is $\{y \mid 2 > y\}$, or $\{y \mid y < 2\}$, or $(-\infty, 2)$.

79. $7 - 9y \leq 4 - 8y$

$7 - 9y - 4 \leq 4 - 8y - 4$ Adding -4

$3 - 9y \leq -8y$ Simplifying

$3 - 9y + 9y \leq -8y + 9y$ Adding $9y$

$3 \leq y$ Simplifying

The solution set is $\{y \mid 3 \leq y\}$, or $\{y \mid y \geq 3\}$, or $[3, \infty)$.

81. $2.1x + 43.2 > 1.2 - 8.4x$

$2.1x + 43.2 - 43.2 > 1.2 - 8.4x - 43.2$ Adding -43.2

$2.1x > -8.4x - 42$ Simplifying

$2.1x + 8.4x > -8.4x - 42 + 8.4x$ Adding $8.4x$

$10.5x > -42$ Simplifying

$x > -4$ Multiplying by $\dfrac{1}{10.5}$

The solution set is $\{x \mid x > -4\}$, or $(-4, \infty)$.

83. $0.7n - 15 + n \geq 2n - 8 - 0.4n$

$1.7n - 15 \geq 1.6n - 8$ Simplifying

$1.7n - 15 - 1.6n \geq 1.6n - 8 - 1.6n$ Adding $-1.6n$

$0.1n - 15 \geq -8$ Simplifying

$0.1n - 15 + 15 \geq -8 + 15$ Adding 15

$0.1n \geq 7$ Simplifying

$n \geq 70$ Multiplying by $\dfrac{1}{0.1}$

The solution set is $\{n \mid n \geq 70\}$, or $[70, \infty)$.

85. $\dfrac{x}{3} - 4 \le 1$

$\dfrac{x}{3} - 4 + 4 \le 1 + 4$ Adding 4

$\dfrac{x}{3} \le 5$ Simplifying

$3\left(\dfrac{x}{3}\right) \le 3 \cdot 5$ Multiplying by 3

$x \le 15$ Simplifying

The solution set is $\{x \mid x \le 15\}$, or $(-\infty, 15]$.

87. $3 < 5 - \dfrac{t}{7}$

$7 \cdot 3 < 7\left(5 - \dfrac{t}{7}\right)$ Multiplying by 7

$21 < 35 - t$ Simplifying

$21 - 21 < 35 - t - 21$ Adding -21

$0 < 14 - t$ Simplifying

$0 + t < 14 - t + t$ Adding t

$t < 14$ Simplifying

The solution set is $\{t \mid t < 14\}$ or $(-\infty, 14)$.

89. $4(2y - 3) \le -44$

$8y - 12 \le -44$ Distributive Law

$8y - 12 + 12 \le -44 + 12$ Adding 12

$8y \le -32$ Simplifying

$y \le -4$ Multiplying by $\dfrac{1}{8}$

The solution set is $\{y \mid y \le -4\}$, or $(-\infty, -4]$.

91. $3(t - 2) \ge 9(t + 2)$

$3t - 6 \ge 9t + 18$ Distributive Law

$3t - 6 + 6 \ge 9t + 18 + 6$ Adding 6

$3t \ge 9t + 24$ Simplifying

$3t - 9t \ge 9t + 24 - 9t$ Adding $-9t$

$-6t \ge 24$ Simplifying

$t \le -4$ Multiplying by $-\dfrac{1}{6}$ and reversing the inequality symbol

The solution set is $\{t \mid t \le -4\}$, or $(-\infty, -4]$.

93. $3(r - 6) + 2 < 4(r + 2) - 21$

$3r - 18 + 2 < 4r + 8 - 21$ Distributive Law

$3r - 16 < 4r - 13$ Simplifying

$3r - 16 + 13 < 4r - 13 + 13$ Adding 13

$3r - 3 < 4r$ Simplifying

$3r - 3 - 3r < 4r - 3r$ Adding $-3r$

$-3 < r$ Simplifying

The solution set is $\{r \mid -3 < r\}$, or $\{r \mid r > -3\}$, or $(-3, \infty)$.

95. $\dfrac{2}{3}(2x - 1) \ge 10$

$\dfrac{3}{2}\left[\dfrac{2}{3}(2x - 1)\right] \ge \dfrac{3}{2} \cdot 10$ Multiplying by $\dfrac{3}{2}$

$2x - 1 \ge 15$ Simplifying

$2x - 1 + 1 \ge 15 + 1$ Adding 1

$2x \ge 16$ Simplifying

$x \ge 8$ Multiplying by $\dfrac{1}{2}$

The solution set is $\{x \mid x \ge 8\}$, or $[8, \infty)$.

97. $\dfrac{3}{4}\left(3x - \dfrac{1}{2}\right) - \dfrac{2}{3} < \dfrac{1}{3}$

$12\left[\dfrac{3}{4}\left(3x - \dfrac{1}{2}\right) - \dfrac{2}{3}\right] < 12 \cdot \dfrac{1}{3}$ Multiplying by 12

$9\left(3x - \dfrac{1}{2}\right) - 8 < 4$ Simplifying

$27x - 4.5 - 8 < 4$ Distributive Law

$27x - 12.5 < 4$ Simplifying

$27x - 12.5 + 12.5 < 4 + 12.5$ Adding 12.5

$27x < 16.5$ Simplifying

$270x < 165$ Multiplying by 10

$x < \dfrac{165}{270}$ Multiplying by $\dfrac{1}{270}$

$x < \dfrac{11}{18}$ Simplifying

The solution set is $\left\{x \mid x < \dfrac{11}{18}\right\}$, or $\left(-\infty, \dfrac{11}{18}\right)$.

99. *Thinking and Writing Exercise.*

101. $4 - x = 8 - 5x$

$4 - x - 4 = 8 - 5x - 4$ Subtracting 4

$-x = 4 - 5x$ Simplifying

$-x + 5x = 4 - 5x + 5x$ Adding $5x$

$4x = 4$ Simplifying

$x = 1$ Multiplying by $\dfrac{1}{4}$

The solution is $x = 1$.

103. $2(5-x) = \frac{1}{2}(x+1)$

$$2\left[2(5-x)\right] = 2\left[\frac{1}{2}(x+1)\right] \quad \text{Multiplying by 2}$$
$$4(5-x) = x+1 \quad \text{Simplifying}$$
$$20-4x = x+1 \quad \text{Distributive Law}$$
$$20-4x-1 = x+1-1 \quad \text{Adding } -1$$
$$19-4x = x \quad \text{Simplifying}$$
$$19-4x+4x = x+4x \quad \text{Adding } 4x$$
$$19 = 5x \quad \text{Simplifying}$$
$$\frac{19}{5} = x \quad \text{Multiplying by } \frac{1}{5}$$

The solution is $x = \dfrac{19}{5}$.

109. $27 - 4\left[2(4x-3)+7\right] \geq 2\left[4-2(3-x)\right]-3$

$$27 - 4[8x-6+7] \geq 2[4-6+2x]-3 \quad \text{Distributive Law}$$
$$27 - 4(8x+1) \geq 2(-2+2x)-3 \quad \text{Simplifying}$$
$$27 - 32x-4 \geq -4+4x-3 \quad \text{Distributive Law}$$
$$23 - 32x \geq -7+4x \quad \text{Simplifying}$$
$$23 - 32x-23 \geq -7+4x-23 \quad \text{Adding } -23$$
$$-32x \geq -30+4x \quad \text{Simplifying}$$
$$-32x-4x \geq -30+4x-4x \quad \text{Adding } -4x$$
$$-36x \geq -30 \quad \text{Simplifying}$$
$$x \leq \frac{30}{36} \quad \text{Multiplying by } -\frac{1}{36} \text{ and reversing the inequality symbol}$$
$$x \leq \frac{5}{6} \quad \text{Simplifying}$$

The solution set is $\left\{x \mid x \leq \dfrac{5}{6}\right\}$, or $\left(-\infty, \dfrac{5}{6}\right)$.

111. $\frac{1}{2}(2x+2b) > \frac{1}{3}(21+3b)$

$$x+b > 7+b \quad \text{Distributive Law}$$
$$x+b-b > 7+b-b \quad \text{Subtracting } b$$
$$x > 7 \quad \text{Simplifying}$$

The solution set is $\left\{x \mid x > 7\right\}$, or $(7, \infty)$.

113. $\quad y < ax+b$

$$y-b < ax+b-b \quad \text{Adding } -b$$
$$y-b < ax \quad \text{Simplifying}$$
$$\frac{y-b}{a} > x \quad \text{Multiplying by } \frac{1}{a}$$

107. One more than some number x, or $x+1$ is always going to be larger than x for any x, so the solution set is $\left\{x \mid x \text{ is a real number}\right\}$, or $(-\infty, \infty)$. We could have also solved this inequality as follows:

$$x < x+1$$
$$x-x < x+1-x \quad \text{Adding } -x$$
$$0 < 1 \quad \text{Simplifying}$$

Since $0 < 1$ is a true statement, x can be any real number.

The solution set is $\left\{x \mid \dfrac{y-b}{a} > x\right\}$, or

$\left\{x \mid x < \dfrac{y-b}{a}\right\}$, or $\left(-\infty, \dfrac{y-b}{a}\right)$.

115. $|x| > -3$

The absolute value of any real number is greater than or equal to 0, so the solution set is $\left\{x \mid x \text{ is a real number}\right\}$, or $(-\infty, \infty)$.

Exercise Set 2.7

1. $b \le a$

3. $a \le b$

5. $b \le a$

7. $b < a$

9. Let n represent the number. Then we have $n < 10$.

11. Let t represent the temperature. Then we have $t \le -3$.

13. Let a represent the age of the Mayan altar. Then we have $a > 1200$.

15. Let d represent the distance to Normandale Community College. Then we have $d \le 15$.

17. Let d represent the number of years of driving experience. Then we have $d \ge 5$.

19. Let c represent the cost of production. Then we have $c \le 12,500$.

21. **Familiarize.** Let h = the number of hours RJ worked. The cost of an emergency call is \$55 plus \$40 times the number of hours, or \$40h.
Translate.

Emergency fee	plus	hourly fee	is more than	\$100.
↓	↓	↓	↓	↓
55	+	40h	>	100

Carry out. We solve the inequality.
$$55 + 40h > 100$$
$$40h > 45$$
$$h > \tfrac{9}{8} \text{ or } 1.125$$

Check. As a partial check, we can determine how much the bill would have been if the plumber had worked 1 hour. The bill would have been $\$55 + \$40(1) = \$95$. From this calculation, it would appear that $h > 1.125$ is correct.
State. The plumber worked more than 1.125 hours.

23. **Familiarize.** Let n = Chloe's grade point average. An unconditional acceptance is given to students whose GMAT score plus 200 times the undergraduate grade point average is at least 950.
Translate.

GMAT score	plus	200 times the undergrad GPA		at least 950
↓	↓	↓	↓	↓
500	+	200n	≥	950

Carry out. We solve the inequality.
$$500 + 200n \ge 950$$
$$200n \ge 450$$
$$n \ge 2.25$$

Check. As a partial check, we can determine the score for a grade point average of 2. $500 + 200(2) = 500 + 400 = 900$. 900 is less than the 950 score required, so it appears that $n \ge 2.25$ is correct.
State. Chloe must earn at least a 2.25 grade points average to an unconditional acceptance into the Master of Business Administration (MBA) program at Arkansas State University.

25. **Familiarize.** The average of the five scores is their sum divided by the number of quizzes, 5. We let s represent Rod's score on the last quiz.
Translate.
The average of the four scores is given by
$$\frac{73 + 75 + 89 + 91 + s}{5}.$$
Since this average must be at least 85, this means that it must be greater than or equal to 85. Thus, we can translate the problem to the inequality
$$\frac{73 + 75 + 89 + 91 + s}{5} \ge 85.$$
Carry out. We first multiply by 5 to clear the fraction.
$$5\left(\frac{73 + 75 + 89 + 91 + s}{5}\right) \ge 5 \cdot 85$$
$$73 + 75 + 89 + 91 + s \ge 425$$
$$328 + s \ge 425$$
$$s \ge 97$$
Check. As a partial check, we show that Rod can get a score of 97 on the fourth test and have an average of at least 85:

$$\frac{73+75+89+91+97}{5} = \frac{425}{5} = 85$$

State. Scores of 97 and higher will earn Rod at least an 85.

27. **Familiarize.** The average of the credits for the four quarters is their sum divided by the number of quarters, 4. We let c represent the number of credits for the fourth quarter.
Translate.
The average of the credits for the four quarters is given by

$$\frac{5+7+8+c}{4}.$$

Since this average must be at least 7, this means that it must be greater than or equal to 7. Thus, we can translate the problem to the inequality

$$\frac{5+7+8+c}{4} \ge 7.$$

Carry out. We first multiply by 4 to clear the fraction.

$$4\left(\frac{5+7+8+c}{4}\right) \ge 4\cdot 7$$
$$5+7+8+c \ge 28$$
$$20+c \ge 28$$
$$c \ge 8$$

Check. As a partial check, we show that Millie can complete 8 credits during the fourth quarter and have an average of at least 7 credits:

$$\frac{5+7+8+8}{4} = \frac{28}{4} = 7$$

State. Millie can average 7 credits per quarter per year if she takes 8 credits or more in the fourth quarter.

29. **Familiarize.** Let b represent the plate appearances in the tenth game. The average plate appearances per game must be at least 3.1 to qualify for a batting title.
Translate.

$$\underbrace{\text{Average plate appearances}}\ \underbrace{\text{is at least}}\ 3.1.$$
$$\downarrow \qquad\qquad \downarrow\ \ \downarrow$$
$$\frac{5+1+4+2+3+4+4+3+2+b}{10} \ge\ \ 3.1$$

Carry out. We solve the inequality.

$$\frac{5+1+4+2+3+4+4+3+2+b}{10} \ge 3.1$$
$$10\left(\frac{5+1+4+2+3+4+4+3+2+b}{10}\right) \ge 3.1$$
$$5+1+4+2+3+4+4+3+2+b \ge 31$$
$$28+b \ge 31$$
$$b \ge 3$$

Check. As a partial check, we show that if 3 plate appearances occur in the tenth game,

$$\frac{5+1+4+2+3+4+4+3+2+3}{10} = \frac{31}{10} = 3.1$$

and the answer checks.
State. The player must have at least 3 plate appearances in the tenth game in order to qualify for a batting title.

31. We first make a drawing. Let b = the length of the base, in cm. The one side is $b-2$ and the other side is $b+3$.

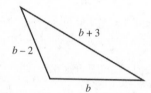

Translate.
$$\underbrace{\text{The perimeter}}\quad \underbrace{\text{is greater than}}\ 19\ \text{cm}.$$
$$\downarrow \qquad\qquad\quad \downarrow \qquad\quad \downarrow$$
$$b+(b-2)+(b+3) \qquad > \qquad 19$$
Carry out.
$$b+(b-2)+(b+3) > 19$$
$$3b+1 > 19$$
$$3b > 18$$
$$b > 6$$

Check. We check to see if the solution seems reasonable. If $b=5$ cm, the perimeter is $5+(5-2)+(5+3)$, or 16 cm. If $b=6$ cm, the perimeter is $6+(6-2)+(6+3)$, or 19 cm. If $b=7$ cm, the perimeter is $7+(7-2)+(7+3)$, or 22 cm.

State. If the base is greater than 6 cm, then the perimeter of the triangle will be greater than 19 cm.

33. **Familiarize.** Let d = the depth of the well. Under the "pay-as-you-go" plan, the charge is $500 + \$8d$. Under the "guaranteed-water" plan, the charge is $4000.
Translate.

Pay-as-you-go plan	is less than	the guaranteed-water plan.
$\downarrow$	$\downarrow$	$\downarrow$
$500 + 8d$	$<$	4000

Carry out. We solve the inequality.
$$500 + 8d < 4000$$
$$8d < 3500$$
$$d < 437.50$$

Check. We compute the cost of the well under the "pay-as-you-go" plan for various depths. If $d = 437$ ft, the cost is $\$500 + \$8 \cdot 437 = \$3996$. If $d = 438$ ft, the cost is $\$500 + \$8 \cdot 438 = \$4004$.
State. The "pay-as-you-go" plan is cheaper if the depth is less than 437.5 ft.

35. **Familiarize.** Let b = the blue-book value of Michelle's car. The car was repaired rather than being replaced.
Translate.

Cost of the repair	did not exceed	80% of the blue-book value.
$\downarrow$	$\downarrow$	$\downarrow$
8500	$\leq$	$0.8b$

Carry out. We solve the inequality.
$$8500 \leq 0.8b$$
$$10,625 \leq b$$

Check. If the blue-book value of the vehicle is $10,625, then 80% of that amount would be $\$10,625 \cdot 0.8 = \$8500$. Since the repairs, $8500, did not exceed this amount, the car would have been repaired rather than being replaced.
State. The blue-book value of the vehicle was at least $10,625.

37. **Familiarize.** Let L = the length of the envelope.
Translate.
The area of the envelop is $A = L \cdot W$.

Area of the envelope	must be at least	$17\frac{1}{2}$ in².
$\downarrow$	$\downarrow$	$\downarrow$
$L \cdot \left(3\frac{1}{2}\right)$	$\geq$	$17\frac{1}{2}$

Carry out. We solve the inequality.
$$L \cdot \left(3\tfrac{1}{2}\right) \geq 17\tfrac{1}{2}$$
$$\frac{7}{2}L \geq \frac{35}{2}$$
$$7L \geq 35$$
$$L \geq 5$$

Check. We can do a partial check by calculating the are when the length is 5 in. The area would be
$$3\tfrac{1}{2} \cdot 5 = \frac{7}{2} \cdot 5 = \frac{35}{2} = 17\tfrac{1}{2} \text{ in}^2.$$
State. The envelopes used must have lengths greater than or equal to 5 in.

39. **Familiarize.** We let C = the body temperature of the person, in degrees Celsius and let F = the body temperature of the person, in degrees Fahrenheit. We will use the formula $F = \dfrac{9}{5}C + 32$.
Translate.

Fahrenheit temperature	is above	98.6°.
$\downarrow$	$\downarrow$	$\downarrow$
F	$>$	98.6

Substituting $\dfrac{9}{5}C + 32$ for F, we have
$$\frac{9}{5}C + 32 > 98.6.$$

Carry out. We solve the inequality.
$$\frac{9}{5}C + 32 > 98.6$$
$$\frac{9}{5}C > 66.6$$
$$C > \frac{333}{9}$$
$$C > 37$$

Check. We check to see if the solution seems reasonable.

When $C = 36$, $\dfrac{9}{5} \cdot 36 + 32 = 96.8$.

When $C = 37$, $\dfrac{9}{5} \cdot 37 + 32 = 98.6$.

When $C = 38$, $\dfrac{9}{5} \cdot 38 + 32 = 100.4$.

It would appear that the solution is correct, considering that rounding occurred.
State. The human body is feverish for Celsius temperatures greater than 37°C.

41. **Familiarize.** Let h = the height of the triangular flag, in ft. The base of the flag is $1\frac{1}{2}$ ft. We will use the formula for the area of a triangle, $A = \frac{1}{2}bh$, where b = the base and h = the height.
Translate.

$\underbrace{\text{Area of the flag}}$ $\underbrace{\text{is at least}}$ $3\ \text{ft}^2.$
$\quad\quad\downarrow\quad\quad\quad\quad\downarrow\quad\quad\downarrow$
$\quad\frac{1}{2}\cdot\left(1\frac{1}{2}\right)\cdot h\quad\quad\geq\quad\quad 3$

Carry out. We solve the inequality.
$$\frac{1}{2}\cdot\left(1\frac{1}{2}\right)\cdot h \geq 3$$
$$\frac{1}{2}\left(\frac{3}{2}\right)h \geq 3$$
$$\frac{3}{4}h \geq 3$$
$$h \geq 4$$

Check. As a partial check we compute the area of the flag with a height of 4 ft to be $\frac{1}{2}\left(1\frac{1}{2}\right)(4) = \frac{1}{2}\left(\frac{3}{2}\right)(4) = 3\ \text{ft}^2$. Any increase in the height would result in an area greater than $3\ \text{ft}^2$.
State. The height of the flag must be greater than or equal to 4 ft.

43. **Familiarize.** Let r = the number of grams of fat in a serving of regular Oreo® cookies. Reduced fat Oreo® cookies have 4.5 g of fat per serving. If reduced fat Oreo® cookies contains at least 25% less fat than regular Oreo® cookies , then reduced fat Oreo® cookies contains at most 75% as much fat as the regular Oreo® cookies .
Translate.

$\underbrace{\text{4.5 g of fat}}$ $\underbrace{\substack{\text{is at} \\ \text{most}}}$ 75% of $\underbrace{\substack{\text{the amount} \\ \text{of fat in regular} \\ \text{Oreo® cookies}}}$
$\quad\downarrow\quad\quad\quad\downarrow\quad\ \downarrow\ \downarrow\quad\quad\quad\downarrow$
$\quad 4.5\quad\quad\leq\quad 0.75\ \cdot\quad\quad\quad r$

Carry out. We solve the inequality.
$$4.5 \leq 0.75r$$
$$6 \leq r$$

Check. As a partial check, we show that 4.5 g of fat does not exceed 75% of 6 g of fat:
$$0.75(6) = 4.5$$
State. Regular Oreo® cookies contain at least 6 g of fat per serving.

45. **Familiarize.** Let d = the number of days after September 5. Then d days after September 5, Charlotte's pumpkin will weigh $532 + 26d$.

Translate.

$\underbrace{\substack{\text{The weight of} \\ \text{Charlotte's pumpkin}}}$ $\underbrace{\text{will exceed}}$ 818 lb.
$\quad\quad\downarrow\quad\quad\quad\quad\quad\downarrow\quad\quad\downarrow$
$\quad 532 + 26d\quad\quad\quad\ >\quad\quad 818$

Carry out. We solve the inequality.
$$532 + 26d > 818$$
$$26d > 286$$
$$d > 11$$

Check. As a partial check, we calculate the weight of Charlotte's pumpkin on different days.
If $d = 10$, the weight is $532 + 26\cdot 10 = 792$.
If $d = 11$, the weight is $532 + 26\cdot 11 = 818$.
If $d = 12$, the weight is $532 + 26\cdot 12 = 844$.
State. Charlotte's pumpkin will weigh more than 818 lb more than 11 days after September 5, or for dates after September 16.

47. **Familiarize.** Let n = the number of text messages sent or received. The monthly fee of $39.95 plus taxes of $6.65 is $46.60. The cost per text message is $0.1n$, and Braden's budget is $60 per month.
Translate.

$\underbrace{\substack{\text{Cost of monthly} \\ \text{fees and text messages}}}$ $\underbrace{\substack{\text{cannot} \\ \text{exceed}}}$ $60.
$\quad\quad\downarrow\quad\quad\quad\quad\quad\quad\downarrow\quad\ \downarrow$
$\quad 46.60 + 0.1n\quad\quad\quad\leq\quad 60$

Carry out. We solve the inequality.
$$46.60 + 0.1n \leq 60$$
$$0.1n \leq 13.40$$
$$n \leq 134$$
Check. As a partial check, if 314 text messages are sent or received, the monthly bill will be
$$\$46.60 + \$0.1\cdot 134 = \$46.60 + \$13.40 = \$60.$$
State. Braden can send or receive no more than 134 text messages.

49. **Familiarize.** Let $t =$ the number of years after 1900 for which the world record will be less than 3.6 min. We use the equation $R = -0.0065t + 4.3259$ where $R =$ the world record, in minutes.

Translate.

$$\underbrace{\text{The world record}}_{\downarrow \atop R} \quad \underbrace{\text{is less than}}_{\downarrow \atop <} \quad \underbrace{3.6 \text{ min.}}_{\downarrow \atop 3.6}$$

We substitute $-0.0065t + 4.3259$ for R to get the inequality $-0.0065t + 4.3259 < 3.6$.

Carry out. We solve the inequality.
$$-0.0065t + 4.3259 < 3.6$$
$$-0.0065t < -0.7259$$
$$t > 111.67692308$$

Check. As a partial check, we calculate the record time for different values of t. If $t = 112$, the time would be $-0.0065(112) + 4.3259$, or 3.5979 min. If $t = 111$, the time would be $-0.0065(111) + 4.3259$, or 3.6044.

State. The record time will be less than 3.6 min for years greater than 112 years after 1900, or years after 2012.

51. **Familiarize.** Let $x =$ the number of miles driven and let $y =$ the cost of driving on the toll road, in dollars. We will use the equation $y = 0.06x + 0.50$.

Translate.

$$\underbrace{\text{The cost of driving on the toll road}}_{\downarrow \atop y} \quad \underbrace{\text{is at most}}_{\downarrow \atop \le} \quad \underbrace{\$14.}_{\downarrow \atop 14}$$

Carry out. We substitute $0.06x + 0.50$ for y to obtain the inequality $0.06x + 0.50 \le 14$. We then solve this inequality.
$$0.06x + 0.50 \le 14$$
$$0.06x \le 13.50$$
$$x \le 225$$

Check. As a partial check we determine the toll cost if 225 miles are driven to be $0.06(226) + 0.50 = 13.50 + 0.50$, or $14.

State. If the toll is at most $14, the number of miles driven must be less than or equal to 225 miles.

53. **Familiarize.** Let $n =$ the number of checks

written. For the Anywhere plan, the total cost for checks would be $0.20n$, and for the Acu-checking plan, the total cost for checks would be $2 + 0.12n$.

Translate.

$$\underbrace{\text{Cost of the Acu-checking plan}}_{\downarrow \atop 2 + 0.12n} \quad \underbrace{\text{is less than}}_{\downarrow \atop <} \quad \underbrace{\text{the cost of the Anywhere checking plan.}}_{\downarrow \atop 0.20n}$$

Carry out. We solve the inequality.
$$2 + 0.12n < 0.20n$$
$$2 < 0.08n$$
$$25 < n$$

Check. As a partial check, we can calculate the costs of each plan for different numbers of checks.

No.	Acu-checking	Any-where
24	$2 + $0.12(24) = $4.88	$0.20(24) = $4.80
25	$2 + $0.12(25) = $5	$0.20(25) = $5
26	$2+$0.12(26) = $5.12	$0.20(26) = $5.20

State. The Acu-checking plan costs less if more than 25 checks are written.

55. **Familiarize.** We list the given information in a table. Let $s =$ gross sales.

Plan A: Monthly Income	Plan B: Monthly Income
$400 salary	$610
8% of sales	5% of sales
Total:	Total:
400 + 8% of sales	610 + 5% of sales

Translate.

$$\underbrace{\text{Income from plan A}}_{\downarrow \atop 400 + 0.08s} \quad \underbrace{\text{is greater than}}_{\downarrow \atop >} \quad \underbrace{\text{income from plan B.}}_{\downarrow \atop 610 + 0.05s}$$

Carry out. We solve the inequality.
$$400 + 0.08s > 610 + 0.05s$$
$$400 + 0.03s > 610$$
$$0.03s > 210$$
$$s > 7000$$

Check. For $s = \$7000$, the income from plan A is $\$400 + 0.08(\$7000)$, or $\$960$, and the income from plan B is $\$610 + 0.05(\$7000)$, or $\$960$. This shows that for sales of $\$7000$ Toni's income is the same from each plan. For $s = \$6990$, the income from plan A is $\$400 + 0.08(\$6990)$, or $\$959.20$, and the income from plan B is $\$610 + 0.05(\$6990)$, or $\$959.50$. For $s = \$7010$, the income from plan A is $\$400 + 0.08(\$7010)$, or $\$960.80$, and the income from plan B is $\$610 + 0.05(\$7010)$, or $\$960.50$.

State. For gross sales greater than $\$7000$, plan A provides Toni with the greater income.

57. **Familiarize.** Let $g =$ the number of gallons of gasoline used. If Abriana chooses the first option, paying for an entire tank of gasoline, she would pay $\$3.099(14)$, or $\$43.386$. If she chooses the second option, paying for only the gasoline required to fill the tank, she would pay $\$6.34g$.

    Translate.

    $$\underbrace{\text{Paying for only the gallons used}} \quad \underbrace{\text{is less than}} \quad \underbrace{\text{paying for an entire tank.}}$$
    $$\downarrow \qquad\qquad \downarrow \qquad\qquad \downarrow$$
    $$6.34g \qquad\qquad < \qquad\qquad 43.386$$

    Carry out. We solve the inequality.
    $$6.34g < 43.386$$
    $$g < 6.843217666$$

    Check. As a partial check, we can calculate the cost under both options if 6.8 gallons of gas was used. Paying for only the gas used, Abriana would owe $\$6.34(6.8)$, or about $\$43.11$.

    State. Abriana should use the second plan, paying for only the gas used, if she uses about 6.8 gallons of gasoline or less.

59. *Thinking and Writing Exercise.*

61. $-2 + (-5) - 7 = -7 - 7 = -14$

63. $3 \cdot (-10) \cdot (-1) \cdot (-2) = -30 \cdot (-1) \cdot (-2)$
    $$= 30 \cdot (-2)$$
    $$= -60$$

65. $(3 - 7) - (4 - 8) = -4 - (-4) = -4 + 4 = 0$

67. $\dfrac{-2 - (-6)}{8 - 10} = \dfrac{4}{-2} = -2$

69. *Thinking and Writing Exercise.*

71. **Familiarize.** Let $n =$ the number of wedding guests. For plan A, the cost for the guests would be $\$30n$. For plan B, the cost for the guests would be $\$1300 + \$20(n - 25)$, assuming that more than 25 guests attend.

    Translate.

    $$\underbrace{\substack{\text{The cost} \\ \text{for plan B}}} \quad \underbrace{\text{is less than}} \quad \underbrace{\substack{\text{the cost} \\ \text{for plan A.}}}$$
    $$\downarrow \qquad\qquad \downarrow \qquad\qquad \downarrow$$
    $$1300 + 20(n - 25) \quad < \quad 30n$$

    Carry out. We solve the inequality.
    $$1300 + 20(n - 25) < 30n$$
    $$1300 + 20n - 500 < 30n$$
    $$800 < 10n$$
    $$80 < n$$

    Check. As a partial check, we calculate the cost for the guests under both plans for different numbers of guests.

No	Plan A	Plan B
79	$30(79)$ $= \$2370$	$1300 + \$20(79 - 25)$ $= \$2380$
80	$30(80)$ $= \$2400$	$1300 + \$20(80 - 25)$ $= \$2400$
81	$30(81)$ $= \$2430$	$1300 + \$20(81 - 25)$ $= \$2420$

    State. Plan B is cheaper if there are more than 80 guests attending the wedding.

73. **Familiarize.** Let $h =$ the number of hours the car was parked. The cost to park the car can be expressed as $\$4 + \$2.50(h - 1)$, assuming the car was parked at least one hour.

    Translate.

    $$\underbrace{\text{The charge to park the car}} \quad \text{exceeds } \$16.50.$$
    $$\downarrow \qquad\qquad \downarrow \quad \downarrow$$
    $$4 + 2.5(h - 1) \qquad > \quad 16.50$$

    Carry out. We solve the inequality.
    $$4 + 2.5(h - 1) > 16.50$$
    $$4 + 2.5h - 2.5 > 16.50$$
    $$2.5h + 1.5 > 16.50$$
    $$2.5h > 15$$
    $$h > 6$$

Check. As a partial check, we can calculate the cost to park the car for 6 hours. The cost would be $4 + $2.5(6 − 1), or $4 + $2.5(5), or $16.50.

State. The car must have been parked more than 6 hours.

75. **Familiarize.** Let s = the length of the side of the square, in cm. The area of the square would be s^2 cm^2.

Translate.

The area of a square | can be no more than | 64 cm^2.

$$s^2 \qquad \le \qquad 64$$

Carry out. We solve the inequality.

$$s^2 \le 64$$
$$s \le 8$$

Check. As a partial check, we find the area of a square having length of side equal to 8. The area is $8^2 = 64$.

State. The length can be less than or equal to 8 cm.

77. **Familiarize.** Let x = the amount of fat in a serving of nacho cheese tortilla chips, in grams. If reduced fat Tortilla Pops contain 60% less fat than regular nacho cheese tortilla chips, then they must contain 40% of the fat in regular nacho cheese tortilla chips, or $0.4x$ g.

Translate.

Reduced fat Tortilla Pops | contain at least | 3 g of fat.

$$0.4x \qquad \ge \qquad 3$$

Carry out. We solve the inequality.

$$0.4x \ge 3$$
$$x \ge 7.5$$

Check. As a partial check, if a serving of nacho cheese tortilla chips have 7.5 g of fat, then a serving of reduced fat Tortilla Pops must contain 60% less fat, or 40% of the fat in nacho cheese tortilla chips, or 0.4(7.5), or 3 g of fat. In order to be labeled "lowfat," the reduced fat Tortilla Pops would have to contain less than 3 of fat.

State. A serving of nacho cheese tortilla chips contain at least 7.5 g of fat.

79. **Familiarize.** Let p = the price of Neoma's tenth book. If the average price of each of the first 9 books is $12, then the total price of the 9 books is $9 \cdot \$12$, or $108. The average price of the first 10 books will be $\dfrac{\$108 + p}{10}$.

Translate.

The average price of 10 books | is at least | $15.

$$\dfrac{108 + p}{10} \qquad \ge \qquad 15$$

Carry out. We solve the inequality.

$$\dfrac{108 + p}{10} \ge 15$$
$$108 + p \ge 150$$
$$p \ge 42$$

Check. As a partial check, we show that the average price of the 10 books is $15 when the price of the tenth book is $42.

$$\dfrac{\$108 + \$42}{10} = \dfrac{150}{10} = \$15$$

State. Neoma's tenth book should cost at least $42 if she wants to select a $15 book for her free book.

81. *Thinking and Writing Exercise.*

Chapter 2 Study Summary

1. $\quad x - 8 = -3$

   $\quad x - 8 + 8 = -3 + 8 \quad$ Adding 8

   $\qquad\qquad x = 5 \qquad\quad$ Simpifying

   The solution is 5.

2. $\quad \dfrac{1}{4}x = 1.2$

   $\quad 4\left(\dfrac{1}{4}x\right) = 4(1.2) \quad$ Multiplying by 4

   $\qquad\qquad x = 4.8 \qquad$ Simplifying

   The solution is 4.8.

3. $$4 - 3x = 7$$

$$4 - 3x + (-4) = 7 + (-4) \quad \text{Adding } -4$$

$$-3x = 3 \quad \text{Simplifying}$$

$$\frac{-3x}{-3} = \frac{3}{-3} \quad \text{Dividing by } -3$$

$$x = -1 \quad \text{Simplifying}$$

The solution is –1.

4. $$\frac{1}{6}t - \frac{3}{4} = t - \frac{2}{3}$$

$$12\left(\frac{1}{6}t - \frac{3}{4}\right) = 12\left(t - \frac{2}{3}\right) \quad \text{Multiplying by 12}$$

$$12 \cdot \frac{1}{6}t - 12 \cdot \frac{3}{4} = 12 \cdot t - 12 \cdot \frac{2}{3} \quad \text{Distributive Law}$$

$$2t - 9 = 12t - 8 \quad \text{Simplifying}$$

$$2t - 9 + 9 = 12t - 8 + 9 \quad \text{Adding 9}$$

$$2t = 12t + 1 \quad \text{Simplifying}$$

$$2t - 12t = 12t + 1 - 12t \quad \text{Adding } -12t$$

$$-10t = 1 \quad \text{Simplifying}$$

$$\frac{-10t}{-10} = \frac{1}{-10} \quad \text{Dividing by } -10$$

$$t = -\frac{1}{10} \quad \text{Simplifying}$$

The solution is $-\dfrac{1}{10}$.

5. $$ac - bc = d$$

$$c(a - b) = d \quad \text{Factoring out } c$$

$$\frac{c(a-b)}{(a-b)} = \frac{d}{(a-b)} \quad \text{Dividing by } a - b$$

$$c = \frac{d}{(a-b)} \quad \text{Simplifying}$$

6. 12 is 15% of <u>what number</u>?

↓↓↓   ↓   ↓
$$12 = 0.15 \cdot \qquad y$$

We solve the equation.
$$12 = 0.15 \cdot y$$

$$\frac{12}{0.15} = y$$

$$80 = y$$

The answer is 80.

7. ***Familiarize.*** Let x = the length of the shorter bicycle tour, in miles. Then $x + 25$ is the length of the longer tour, in miles.
Translate.

Length of two
<u>bicycle tours</u>   was a total of 120 miles.

↓   ↓   ↓
$$x \ + \ (x + 25) \ = \qquad 120$$

Carry out. We solve the equation.
$$x + (x + 25) = 120$$

$$2x + 25 = 120$$

$$2x = 95$$

$$x = 47\tfrac{1}{2}$$

Check. If the short tour was $47\tfrac{1}{2}$, then the total length is:
$$47\tfrac{1}{2} + \left(47\tfrac{1}{2} + 25\right) = 47\tfrac{1}{2} + 72\tfrac{1}{2} = 120 \text{ miles.}$$
This checks.
State. The length of tours were $47\tfrac{1}{2}$ miles and $72\tfrac{1}{2}$ miles.

8. $\{x \mid x \leq 0\}$ in interval notation is $(-\infty, 0]$

9. $$x - 11 > -4$$

$$x - 11 + 11 > -4 + 11 \quad \text{Adding 11}$$

$$x > 7 \quad \text{Simplifying}$$

The solution set is $\{x \mid x > 7\}$, or $(7, \infty)$.

10. $$-8x \leq 2$$

$$\frac{-8x}{-8} \geq \frac{2}{-8} \quad \text{Dividing by } -8$$

$$x \geq -\frac{1}{4} \quad \text{Simplifying}$$

The solution set is $\left\{x \mid x \geq -\tfrac{1}{4}\right\}$, or $\left[-\tfrac{1}{4}, \infty\right)$.

11. Let d represent the distance Luke runs, in miles. If Luke runs no less than 3 mi per day, then $d \geq 3$.

Chapter 2 Review Exercises

1. True

2. False

3. True

4. True

5. True

6. False

7. True

8. True

9. $x + 9 = -16$

$x + 9 - 9 = -16 - 9$ Adding -9

$x = -25$ Simplifying

The solution is -25.

10. $-8x = -56$

$\left(-\dfrac{1}{8}\right)(-8x) = \left(-\dfrac{1}{8}\right)(-56)$ Multiplying by $-\dfrac{1}{8}$

$x = 7$ Simplifying

The solution is 7.

11. $-\dfrac{x}{5} = 13$

$-5\left(-\dfrac{x}{5}\right) = -5(13)$ Multiplying by -5

$x = -65$ Simplifying

The solution is -65.

12. $-8 = n - 11$

$-8 + 11 = n - 11 + 11$ Adding 11

$3 = n$ Simpifying

The solution is 3.

13. $\dfrac{2}{5}t = -8$

$\dfrac{5}{2}\left(\dfrac{2}{5}t\right) = \dfrac{5}{2}(-8)$ Multiplying by $\dfrac{5}{2}$

$t = -20$ Simplifying

The solution is -20.

14. $x - 0.1 = 1.01$

$x - 0.1 + 0.1 = 1.01 + 0.1$ Adding 0.1

$x = 1.11$ Simplifying

The solution is 1.11.

15. $-\dfrac{2}{3} + x = -\dfrac{1}{6}$

$6\left(-\dfrac{2}{3} + x\right) = 6\left(-\dfrac{1}{6}\right)$ Multiplying by 6

$-4 + 6x = -1$ Simplifying

$-4 + 6x + 4 = -1 + 4$ Adding 4

$6x = 3$ Simplifying

$x = \dfrac{1}{2}$ Multiplying by $\dfrac{1}{6}$

The solution is $\dfrac{1}{2}$.

16. $5z + 3 = 41$

$5z + 3 - 3 = 41 - 3$ Adding -3

$5z = 38$ Simplifying

$z = \dfrac{38}{5}$ Multiplying by $\dfrac{1}{5}$

The solution is $\dfrac{38}{5}$.

17. $5 - x = 13$

$5 - x - 5 = 13 - 5$ Adding -5

$-x = 8$ Simplifying

$x = -8$ Multiplying by -1

The solution is -8.

18. $5t + 9 = 3t - 1$

$5t + 9 - 9 = 3t - 1 - 9$ Adding -9

$5t = 3t - 10$ Simplifying

$5t - 3t = 3t - 10 - 3t$ Adding $-3t$

$2t = -10$ Simplifying

$t = -5$ Multiplying by $\dfrac{1}{2}$

The solution is -5.

19. $7x - 6 = 25x$

$7x - 6 - 7x = 25x - 7x$ Adding $-7x$

$-6 = 18x$ Simplifying

$-\dfrac{1}{3} = x$ Multiplying by $\dfrac{1}{18}$

The solution is $-\dfrac{1}{3}$.

20. $\dfrac{1}{4}a - \dfrac{5}{8} = \dfrac{3}{8}$

$8\left(\dfrac{1}{4}a - \dfrac{5}{8}\right) = 8\left(\dfrac{3}{8}\right)$ Multiplying by 8

$2a - 5 = 3$ Simplifying

$2a - 5 + 5 = 3 + 5$ Adding 5

$2a = 8$ Simplifying

$\dfrac{2a}{2} = \dfrac{8}{2}$ Dividing by 2

$a = 4$ Simplifying

The solution is 4.

21.
$$14y = 23y - 17 - 9y$$
$$14y = 14y - 17 \qquad \text{Simplifying}$$
$$14y - 14y = 14y - 17 - 14y \qquad \text{Adding } -14y$$
$$0 = -17 \qquad \text{Simplifying}$$

This is a false statement. The equation is a contradiction, and has no solution.

22.
$$0.22y - 0.6 = 0.12y + 3 - 0.8y$$
$$0.22y - 0.6 = -0.68y + 3 \qquad \text{Simplifying}$$
$$0.22y - 0.6 + 0.68y = -0.68y + 3 + 0.68y \qquad \text{Adding } 0.68y$$
$$0.9y - 0.6 = 3 \qquad \text{Simplifying}$$
$$0.9y - 0.6 + 0.6 = 3 + 0.6 \qquad \text{Adding } 0.6$$
$$0.9y = 3.6 \qquad \text{Simplifying}$$
$$y = 4 \qquad \text{Multiplying by } \frac{1}{0.9}$$

The solution is 4.

23.
$$\tfrac{1}{4}x - \tfrac{1}{8}x = 3 - \tfrac{1}{16}x$$
$$16\left(\tfrac{1}{4}x - \tfrac{1}{8}x\right) = 16\left(3 - \tfrac{1}{16}x\right) \qquad \text{Multiplying by 16}$$
$$4x - 2x = 48 - x \qquad \text{Distributive Law}$$
$$2x = 48 - x \qquad \text{Simplifying}$$
$$2x + x = 48 - x + x \qquad \text{Adding } x$$
$$3x = 48 \qquad \text{Simplifying}$$
$$x = 16 \qquad \text{Multiplying by } \tfrac{1}{3}$$

The solution is 16.

24
$$3(5 - n) = 36$$
$$15 - 3n = 36 \qquad \text{Distributive Law}$$
$$15 - 3n - 15 = 36 - 15 \qquad \text{Adding } -15$$
$$-3n = 21 \qquad \text{Simplifying}$$
$$x = -7 \qquad \text{Multiplying by } -\tfrac{1}{3}$$

The solution is –7.

25.
$$4(5x - 7) = -56$$
$$20x - 28 = -56 \qquad \text{Distributive Law}$$
$$20x - 28 + 28 = -56 + 28 \qquad \text{Adding 28}$$
$$20x = -28 \qquad \text{Simplifying}$$
$$x = -\frac{28}{20} \qquad \text{Multiplying by } \tfrac{1}{20}$$
$$x = -\frac{7}{5} \qquad \text{Simplifying}$$

The solution is $-\dfrac{7}{5}$.

26.
$$8(x - 2) = 5(x + 4)$$
$$8x - 16 = 5x + 20 \qquad \text{Distributive Law}$$
$$8x - 16 + 16 = 5x + 20 + 16 \qquad \text{Adding 16}$$
$$8x = 5x + 36 \qquad \text{Simplifying}$$
$$8x - 5x = 5x + 36 - 5x \qquad \text{Adding } -5x$$
$$3x = 36 \qquad \text{Simplifying}$$
$$x = 12 \qquad \text{Multiplying by } \tfrac{1}{3}$$

The solution is 12.

27.
$$3(x - 4) + 2 = x + 2(x - 5)$$
$$3x - 12 + 2 = x + 2x - 10 \qquad \text{Distributive Law}$$
$$3x - 10 = 3x - 10 \qquad \text{Simplifying}$$
$$3x - 10 - 3x = 3x - 10 - 3x \qquad \text{Subtracting } 3x$$
$$-10 = -10 \qquad \text{Simplifying}$$

Because $-10 = -10$ is a true statement, this equation is an identity and the solution is all real numbers.

28.
$$C = \pi d$$
$$C\left(\frac{1}{\pi}\right) = \pi d\left(\frac{1}{\pi}\right) \qquad \text{Multipling by } \frac{1}{\pi}$$
$$\frac{C}{\pi} = d \qquad \text{Simplifying}$$

29.
$$V = \frac{1}{3}Bh$$
$$3 \cdot V = 3\left(\frac{1}{3}Bh\right) \qquad \text{Multiplying by 3}$$
$$3V = Bh \qquad \text{Simplifying}$$
$$\frac{1}{h}(3V) = \frac{1}{h}(Bh) \qquad \text{Multiplying by } \frac{1}{h}$$
$$\frac{3V}{h} = B \qquad \text{Simplifying}$$

30.
$$5x - 2y = 10$$
$$5x - 2y - 5x = 10 - 5x \qquad \text{Subtracting } 5x$$
$$-2y = 10 - 5x \qquad \text{Simplifying}$$
$$\frac{-2y}{-2} = \frac{10 - 5x}{-2} \qquad \text{Dividing by } -2$$
$$y = \frac{5}{2}x - 5 \qquad \text{Simplifying}$$

31.
$$tx = ax + b$$

$$tx - ax = ax + b - ax \quad \text{Subtracting } ax$$

$$tx - ax = b \quad \text{Simplifying}$$

$$x(t-a) = b \quad \text{Factor out } x$$

$$\frac{x(t-a)}{t-a} = \frac{b}{t-a} \quad \text{Dividing by } t-a$$

$$x = \frac{b}{t-a} \quad \text{Simplifying}$$

32. $0.9\% = 0.9 \times 0.01 \quad \text{Replacing \% by } \times 0.01$

$\quad\quad = 0.009$

33. $\frac{11}{25} = \frac{4}{4} \cdot \frac{11}{25} = \frac{44}{100} = 0.44$

First, move the decimal point two places to the right; then write a % symbol:
The answer is 44%.

0.44.

44%

34. What percent of 60 is 42?

$\downarrow \quad\quad \downarrow\downarrow\downarrow$

$y \quad\quad\quad \cdot\ 60 = 42$

We solve the equation and then convert to percent notation.

$$y \cdot 60 = 42$$

$$y = \frac{42}{60}$$

$$y = 0.70 = 70\%$$

The answer is 70%.

35. 42 is 30% of what number?

$\downarrow\downarrow\downarrow \quad \downarrow \quad\quad \downarrow$

$42 = 0.30 \ \cdot \quad\quad y$

We solve the equation.

$$42 = 0.30 \cdot y$$

$$\frac{42}{0.30} = y$$

$$140 = y$$

The answer is 140.

36. $x \le -5$

We substitute -3 for x giving $-3 \le -5$, which is a false statement since -3 is to the right of -5 on the number line. So -3 is not a solution of the inequality $x \le -5$.

37. $x \le -5$

We substitute -7 for x giving $-7 \le -5$, which is a true statement since -7 is to the left of -5 on the number line. So -7 is a solution of the inequality $x \le -5$.

38. $x \le -5$

We substitute 0 for x giving $0 \le -5$, which is a false statement since 0 is to the right of -5 on the number line. So 0 is not a solution of the inequality $x \le -5$.

39.
$$5x - 6 < 2x + 3$$

$$5x - 6 + 6 < 2x + 3 + 6 \quad \text{Adding 6}$$

$$5x < 2x + 9 \quad \text{Simplifying}$$

$$5x - 2x < 2x + 9 - 2x \quad \text{Adding } -2x$$

$$3x < 9 \quad \text{Simplifying}$$

$$x < 3 \quad \text{Multiplying by } \tfrac{1}{3}$$

The solution set is $\{x \mid x < 3\}$, or $(-\infty, 3)$.

The graph is as follows:

$5x - 6 < 2x + 3$

$\xleftarrow{\quad} \text{+++++++}) \text{++} \xrightarrow{\quad}$
$\quad -4\ -2\ \ 0\ \ 2\ \ 4$

40. $-2 < x \le 5$

The solution set is $\{x \mid -2 < x \le 5\}$, or $(-2, 5]$. The graph is as follows:

$-2 < x \le 5$

$\xleftarrow{\quad} \text{++} (\text{+++++++}] \xrightarrow{\quad}$
$\quad -4\ -2\ \ 0\ \ 2\ \ 4\ \ 6$

41. $t > 0$

The solution set is $\{t \mid t > 0\}$, or $(0, \infty)$. The graph is as follows:

$t > 0$

$\xleftarrow{\quad} \text{++++++} (\text{++++++} \xrightarrow{\quad}$
$-5\ -4\ -3\ -2\ -1\ \ 0\ \ 1\ \ 2\ \ 3\ \ 4\ \ 5$

42.
$$t + \tfrac{2}{3} \ge \tfrac{1}{6}$$

$$6\left(t + \tfrac{2}{3}\right) \ge 6\left(\tfrac{1}{6}\right) \quad \text{Multiplying by 6}$$

$$6t + 4 \ge 1 \quad \text{Simplifying}$$

$$6t + 4 - 4 \ge 1 - 4 \quad \text{Adding } -4$$

$$6t \ge -3 \quad \text{Simplifying}$$

$$\tfrac{1}{6}(6t) \ge \tfrac{1}{6}(-3) \quad \text{Multiplying by } \tfrac{1}{6}$$

$$t \ge -\tfrac{1}{2} \quad \text{Simplifying}$$

The solution set is $\left\{t \mid t \ge -\tfrac{1}{2}\right\}$, or $\left[-\tfrac{1}{2}, \infty\right)$.

43.

$$9x \geq 63$$
$$\tfrac{1}{9}(9x) \geq \tfrac{1}{9} \cdot 63 \quad \text{Multiplying by } \tfrac{1}{9}$$
$$x \geq 7 \qquad \text{Simplifying}$$

The solution set is $\{x \mid x \geq 7\}$, or $[7, \infty)$.

44.

$$2 + 6y > 20$$
$$2 + 6y - 2 > 20 - 2 \quad \text{Adding } -2$$
$$6y > 18 \qquad \text{Simplifying}$$
$$\tfrac{1}{6}(6y) > \tfrac{1}{6} \cdot 18 \quad \text{Multiplying by } \tfrac{1}{6}$$
$$y > 3 \qquad \text{Simplifying}$$

The solution set is $\{y \mid y > 3\}$, or $(3, \infty)$.

45.

$$7 - 3y \geq 27 + 2y$$
$$7 - 3y - 7 \geq 27 + 2y - 7 \quad \text{Adding } -7$$
$$-3y \geq 20 + 2y \qquad \text{Simplifying}$$
$$-3y - 2y \geq 20 + 2y - 2y \quad \text{Adding } -2y$$
$$-5y \geq 20 \qquad \text{Simplifying}$$
$$y \leq -4 \qquad \begin{array}{l}\text{Multiplying by} \\ -\tfrac{1}{5} \text{ and reversing} \\ \text{the inequality} \\ \text{symbol}\end{array}$$

The solution set is $\{y \mid y \leq -4\}$, or $(-\infty, -4]$.

46.

$$3x + 5 < 2x - 6$$
$$3x + 5 - 2x < 2x - 6 - 2x \quad \text{Adding } -2x$$
$$x + 5 < -6 \qquad \text{Simplifying}$$
$$x + 5 - 5 < -6 - 5 \qquad \text{Adding } -5$$
$$x < -11 \qquad \text{Simplifying}$$

The solution set is $\{x \mid x < -11\}$, or $(-\infty, -11)$.

47.

$$-4y < 28$$
$$-\tfrac{1}{4}(-4y) > -\tfrac{1}{4} \cdot 28 \quad \begin{array}{l}\text{Multiplying by } -\tfrac{1}{4} \\ \text{and reversing the} \\ \text{inequality symbol}\end{array}$$
$$y > -7 \qquad \text{Simplifying}$$

The solution set is $\{y \mid y > -7\}$, or $(-7, \infty)$.

48.

$$3 - 4x < 27$$
$$3 - 4x - 3 < 27 - 3 \quad \text{Adding } -3$$
$$-4x < 24 \qquad \text{Simplifying}$$
$$-\tfrac{1}{4}(-4x) > -\tfrac{1}{4} \cdot 24 \quad \begin{array}{l}\text{Multiplying by } -\tfrac{1}{4} \\ \text{and reversing the} \\ \text{inequality symbol}\end{array}$$
$$x > -6 \qquad \text{Simplifying}$$

The solution set is $\{x \mid x > -6\}$, or $(-6, \infty)$.

49.

$$4 - 8x < 13 + 3x$$
$$4 - 8x - 4 < 13 + 3x - 4 \quad \text{Adding } -4$$
$$-8x < 9 + 3x \qquad \text{Simplifying}$$
$$-8x - 3x < 9 + 3x - 3x \quad \text{Adding } -3x$$
$$-11x < 9 \qquad \text{Simplifying}$$
$$-\tfrac{1}{11}(-11x) > -\tfrac{1}{11} \cdot 9 \quad \text{Multiplying by } -\tfrac{1}{11}$$
$$x > -\tfrac{9}{11} \qquad \text{Simplifying}$$

The solution set is $\left\{x \mid x > -\tfrac{9}{11}\right\}$, or $\left(-\tfrac{9}{11}, \infty\right)$.

50.

$$13 \leq -\tfrac{2}{3}t + 5$$
$$13 + \tfrac{2}{3}t \leq -\tfrac{2}{3}t + 5 + \tfrac{2}{3}t \quad \text{Adding } \tfrac{2}{3}t$$
$$13 + \tfrac{2}{3}t \leq 5 \qquad \text{Simplifying}$$
$$13 + \tfrac{2}{3}t - 13 \leq 5 - 13 \qquad \text{Adding } -13$$
$$\tfrac{2}{3}t \leq -8 \qquad \text{Simplifying}$$
$$\tfrac{3}{2}\left(\tfrac{2}{3}t\right) \leq \tfrac{3}{2}(-8) \qquad \text{Multipying by } \tfrac{3}{2}$$
$$t \leq -12 \qquad \text{Simplifying}$$

The solution set is $\{t \mid t \leq -12\}$, or $(-\infty, -12]$.

51.

$$7 \leq 1 - \tfrac{3}{4}x$$
$$7 - 1 \leq 1 - \tfrac{3}{4}x - 1 \quad \text{Adding } -1$$
$$6 \leq -\tfrac{3}{4}x \qquad \text{Simplifying}$$
$$-\tfrac{4}{3} \cdot 6 \geq -\tfrac{4}{3}\left(-\tfrac{3}{4}x\right) \quad \text{Multiplying by } -\tfrac{4}{3}$$
$$-8 \geq x \qquad \text{Simplifying}$$

The solution set is $\{x \mid -8 \geq x\}$, or $\{x \mid x \leq -8\}$, or $(-\infty, -8]$.

52. **Familiarize.** Let x = the amount given to charities in 2008, in dollars.

Translate.

35% of the amount given to charities was $106.9 billion.

$\downarrow$ $\downarrow$ $\downarrow$ $\downarrow$ $\downarrow$
0.35 · x = 106.9

Carry out. We solve the equation.
$$0.35x = 106.9$$
$$x = \frac{106.9}{0.35}$$
$$x \approx 305.4286$$

Check. If about $305.4 billion was given to charities, then 35% of that amount was given to religious organizations. So 35%·$305.4 billion , or $106.9.billion was given to religious organizations. This amount checks.

State. The amount given to charities in 2008 was about $305.4 billion.

53. **Familiarize.** Let x = the length of the first piece, in ft. Since the second piece is 2 ft longer than the first piece, it must be $x+2$ ft.

Translate.

The sum of the lengths of the two pieces is 32 ft.

$\downarrow$ $\downarrow$ $\downarrow$
$x+(x+2)$ = 32

Carry out. We solve the equation.
$$x+(x+2) = 32$$
$$2x+2 = 32$$
$$2x = 30$$
$$x = 15$$

Check. If the first piece is 15 ft long, then the second piece must be 15 + 2, or 17 ft long. The sum of the lengths of the two pieces is 15 ft + 17 ft, or 32 ft. The answer checks.

State. The lengths of the two pieces are 15 ft and 17 ft.

54. **Familiarize.** Let x = the first odd integer and let $x+2$ = the next consecutive odd integer.

Translate.

The sum of the two consecutive odd integers is 116.

$\downarrow$ $\downarrow$ $\downarrow$
$x+(x+2)$ = 116

Carry out. We solve the equation.

$$x+(x+2) = 116$$
$$2x+2 = 116$$
$$2x = 114$$
$$x = 57$$

Check. If the first odd integer is 57, then the next consecutive odd integer would be 57 + 2, or 59. The sum of these two integers is 57 + 59, or 116. This result checks.

State. The integers are 57 and 59.

55. **Familiarize.** Let x = the length of the rectangle, in cm. The width of the rectangle is $x-6$ cm. The perimeter of a rectangle is given by $P = 2l+2w$, where l is the length and w is the width.

Translate.

The perimeter of the rectangle is 56 cm.

$\downarrow$ $\downarrow$ $\downarrow$
$2x+2(x-6)$ = 56

Carry out. We solve the equation.

$$2x+2(x-6) = 56$$
$$2x+2x-12 = 56$$
$$4x-12 = 56$$
$$4x = 68$$
$$x = 17$$

Check. If the length is 17 cm, then the width is 17 cm − 6 cm, or 11 cm. The perimeter is 2·17 cm + 2·11 cm ,or 34 cm + 22 cm, or 56 cm. These results check.

State. The length is 17 cm and the width is 11 cm.

56. **Familiarize.** Let x = the regular price of the picnic table. Since the picnic table was reduced by 25%, it actually sold for 75% of its original price.

Translate.

75% of the original price is $120.

$\downarrow$ $\downarrow$ $\downarrow$ $\downarrow$ $\downarrow$
0.75 · x = 120

Carry out. We solve the equation.
$$0.75x = 120$$
$$x = \frac{120}{0.75}$$
$$x = 160$$

Check. If the original price was $160 with a 25% discount, then the purchaser would have

paid 75% of $160, or $0.75 \cdot \$160$, or $120.
This result checks.
State. The original price was $160.

57. **Familiarize.** Let $x =$ the amount of sleep that
infants need. 12 hours is 25% less than x, or
75% of x.
Translate.

75% of   the hours   is 12 hours.
          infants sleep

$\downarrow$ $\downarrow$ $\downarrow$ $\downarrow$ $\downarrow$
0.75   · x =   12

Carry out. We solve the equation.
$0.75x = 12$

$$x = \frac{12}{0.75}$$

$$x = 16$$

Check. If infants sleep 16 hours, then 75% of
16 hours is 12 hours. This result checks.
State. Infants need 16 hours of sleep per day.

58. **Familiarize.** Let $x =$ the measure of the first
angle. The measure of the second angle is $x +$
$50°$, and the measure of the third angle is $2x -$
$10°$. The sum of the measures of the angles
of a triangle is $180°$.
Translate.

The sum of the measures   is $180°$.
   of the angles

$\downarrow$ $\downarrow$ $\downarrow$
$x + (x + 50) + (2x - 10)$ = 180

Carry out. We solve the equation.
$x + (x + 50) + (2x - 10) = 180$

$$4x + 40 = 180$$

$$4x = 140$$

$$x = 35$$

Check. If the measure of the first angle is
$35°$, then the measure of the second angle is
$35° + 50°$, or $85°$, and the measure of the third
angle is $2 \cdot 35° - 10°$, or $60°$. The sum of the
measures of the first, second, and third angles
is $35° + 85° + 60°$, or $180°$. These results
check.
State. The measures of the angles are $35°$,
$85°$, and $60°$.

59. **Familiarize.** We examine the values in the
table and determine the relationship between
the number of gift subscriptions and the total
cost of the subscriptions.

Number of Gift Subscriptions	Total Cost of Subscriptions
1	$1 \cdot 10 + 3 = 10 + 3 = 13$
2	$2 \cdot 10 + 3 = 20 + 3 = 23$
4	$4 \cdot 10 + 3 = 40 + 3 = 43$
10	$10 \cdot 10 + 3 = 100 + 3 = 103$

We note that $10 times the number of gift
subscriptions plus $3 gives the total cost of
the subscriptions. We let x represent the
number of gift subscriptions.

Translate.
The total cost of gift subscriptions is $73.

$\downarrow$ $\downarrow$ $\downarrow$
$10x + 3$ = 73

Carry out. We solve the equation.
$10x + 3 = 73$

$$10x = 70$$

$$x = 7$$

Check. If the number of gift subscriptions is
7, then the total cost of the subscription is
$\$10 \cdot 7 + \$3$, or $73. This results checks.
State. Tonya purchased 7 gift subscriptions.

60. **Familiarize.** Let $x =$ the amounts Caroline
can spend during the sixth month.
Translate.

The average of the   does not $95.
amounts spent   exceed
for entertainment

$\downarrow$ $\downarrow$ $\downarrow$
$\dfrac{98 + 89 + 110 + 85 + 83 + x}{6}$ $\leq$ 95

Carry out. We solve the inequality.

$$\frac{98 + 89 + 110 + 85 + 83 + x}{6} \leq 95$$

$$6\left(\frac{98 + 89 + 110 + 85 + 83 + x}{6}\right) \leq 6 \cdot 95$$

$$98 + 89 + 110 + 85 + 83 + x \leq 570$$

$$465 + x \leq 570$$

$$465 + x - 465 \leq 570 - 465$$

$$x \leq 105$$

Check. As a partial check we calculate the
average amount spent on entertainment if
Caroline spends $105 during the sixth
month. The average is

$$\frac{\$98 + \$89 + \$110 + \$85 + \$83 + \$105}{6}$$

$$= \frac{\$570}{6} = \$95.$$

If Caroline spent any less than \$105 during the sixth month, her 6-month average would be less than \$95. These results check.
State. Caroline should spend \$105 or less during the sixth month.

61. **Familiarize.** Let x = the widths of the rectangle. The perimeter of a rectangle is given by $P = 2l + 2w$.
Translate.

The perimeter of the rectangle $\underbrace{\qquad}$ is greater than $\underbrace{\qquad}$ 120 cm.

$$\downarrow \qquad\qquad \downarrow \qquad \downarrow$$
$$2 \cdot 43 + 2x \qquad > \qquad 120$$

Carry out. We solve the inequality.

$$2 \cdot 43 + 2x > 120$$
$$86 + 2x > 120$$
$$86 + 2x - 86 > 120 - 86$$
$$2x > 34$$
$$x > 17$$

Check. As a partial check, we calculate the perimeter when the width is 17 cm. The perimeter is $2 \cdot 43$ cm $+ 2 \cdot 17$ cm, or 86 cm $+$ 34 cm, or 120 cm. If the width exceeded 17 cm, the perimeter would exceed 120 cm. These results check.
State. The width must be greater than 17 cm.

62. *Thinking and Writing Exercise.* Multiplying both sides of an equation by *any* nonzero number results in an equivalent equation. When multiplying on both sides of an inequality, the sign of the number being multiplied by must be considered. If the number is positive, the direction of the inequality symbol remains unchanged; if the number is negative, the direction of the inequality symbol must be reversed to produce an equivalent inequality.

63. *Thinking and Writing Exercise.* The solutions of an equation can usually each be checked. The solutions of an inequality are normally too numerous to check. Checking a few numbers from the solution set found cannot guarantee that the answer is correct, although

if any number does not check, the answer found is incorrect.

64. **Familiarize.** Let x = the amount of time that sixth- and seventh-graders spend reading or doing homework each day. 108% more than this is 208%x. Note that 3 hr 20 min is equivalent to 200 min.
Translate.

208% of the time spent by children on reading or homework $\underbrace{\qquad}$ is 200 min.

$$\downarrow \quad \downarrow \qquad\qquad \downarrow \qquad\qquad \downarrow \quad \downarrow$$
$$2.08 \quad \cdot \qquad\qquad x \qquad\qquad = \quad 200$$

Carry out. We solve the equation.

$$2.08x = 200$$
$$x = \frac{200}{2.08}$$
$$x \approx 96.153846$$

Check. If children spend 96.153846 minutes reading or doing homework, they spend $(2.08)(96.153846)$ minutes watching TV or playing video games, or about 200 minutes. The result checks.
State. Children spend about 96 minutes, or 1 hour 36 minutes, reading or doing homework each day.

65. **Familiarize.** Let x = the length of the Nile River, in miles. Let $x + 65$ represent the length of the Amazon River, in miles.
Translate.

The combined length of both rivers $\underbrace{\qquad}$ is 8385 miles.

$$\downarrow \qquad\qquad \downarrow \quad \downarrow$$
$$x + (x + 65) \qquad = 8385$$

Carry out. We solve the equation.

$$x + (x + 65) = 8385$$
$$2x + 65 = 8385$$
$$2x = 8320$$
$$x = 4160$$

Check. If the Nile River is 4160 miles long, then the Amazon River is $4160 + 65$, or 4225 miles. The combined length of both rivers is then $4160 + 4225$, or 8385 miles. These results check.
State. The Nile River is 4160 miles long, and the Amazon River is 4225 miles long.

66. **Familiarize.** Let x = the sticker price of the car. The sticker price minus 20% of the sticker price would be $x - 0.20x$.

Translate.

The sticker price minus 20% of the sticker price   plus $200 is $15,080.

$$\underbrace{x - 0.20x} \quad + \quad 200 \quad = \quad 15{,}080$$

Carry out. We solve the equation.

$$x - 0.20x + 200 = 15080$$
$$0.80x + 200 = 15{,}080$$
$$0.80x = 14{,}880$$
$$x = \frac{14{,}880}{0.80}$$
$$x = 18{,}600$$

Check. If the sticker price is $18,600, then 20% of the sticker price is $0.20 \cdot \$18{,}600$, or $3720. The sticker price minus 20% of the sticker price is $18,600 – $3720, or $14,880. Adding $200, we get $14,880 + $200, or $15,080, which is the purchase price.

State. The sticker price of the car is $18,600.

67. $2|n| + 4 = 50$

$$2|n| = 46$$
$$|n| = 23$$

The distance from some number n and the origin is 23 units. The solution is $n = 23$, or $n = -23$.

68. $|3n| = 60$

The distance from some number, $3n$, to the origin is 60 units. So we have:

$$3n = -60 \qquad 3n = 60$$
$$n = -20 \qquad n = 20$$

The solution is –20 and 20.

69. $y = 2a - ab + 3$

$$y = a(2 - b) + 3$$
$$y - 3 = a(2 - b)$$
$$\frac{y - 3}{2 - b} = a$$

The solution is $a = \dfrac{y - 3}{2 - b}$.

Chapter 2 Test

1. $t + 7 = 16$

$$t + 7 - 7 = 16 - 7 \quad \text{Adding } -7$$
$$t = 9 \qquad \text{Simplifying}$$

The solution is 9.

2. $t - 3 = 12$

$$t - 3 + 3 = 12 + 3 \quad \text{Adding } 3$$
$$t = 15 \qquad \text{Simplifying}$$

The solution is 15.

3. $6x = -18$

$$\tfrac{1}{6}(6x) = \tfrac{1}{6}(-18) \quad \text{Multiplying by } \tfrac{1}{6}$$
$$x = -3 \qquad \text{Simplifying}$$

The solution is –3.

4. $-\tfrac{4}{7}x = -28$

$$-\tfrac{7}{4}\left(-\tfrac{4}{7}x\right) = -\tfrac{7}{4}(-28) \quad \text{Multiplying by } -\tfrac{7}{4}$$
$$x = 49 \qquad \text{Simplifying}$$

The solution is 49.

5. $3t + 7 = 2t - 5$

$$3t + 7 - 7 = 2t - 5 - 7 \qquad \text{Adding } -7$$
$$3t = 2t - 12 \qquad \text{Simplifying}$$
$$3t - 2t = 2t - 12 - 2t \qquad \text{Adding } -2t$$
$$t = -12 \qquad \text{Simplifying}$$

The solution is –12.

6. $\tfrac{1}{2}x - \tfrac{3}{5} = \tfrac{2}{5}$

$$\tfrac{1}{2}x - \tfrac{3}{5} + \tfrac{3}{5} = \tfrac{2}{5} + \tfrac{3}{5} \quad \text{Adding } \tfrac{3}{5}$$
$$\tfrac{1}{2}x = \tfrac{5}{5} \qquad \text{Simplifying}$$
$$\tfrac{1}{2}x = 1 \qquad \text{Simplifying}$$
$$2\left(\tfrac{1}{2}x\right) = 2 \cdot 1 \quad \text{Multiplying by } 2$$
$$x = 2 \qquad \text{Simplifying}$$

The solution is 2.

7. $8 - y = 16$

$$8 - y - 8 = 16 - 8 \quad \text{Adding } -8$$
$$-y = 8 \qquad \text{Simplifying}$$
$$y = -8 \qquad \text{Multiply by } -1$$

The solution is –8.

8. $4.2x + 3.5 = 1.2 - 2.5x$

$4.2x + 3.5 - 3.5 = 1.2 - 2.5x - 3.5$ Subtracting 3.5

$4.2x = -2.3 - 2.5x$ Simplifying

$4.2x + 2.5x = -2.3 - 2.5x + 2.5x$ Adding $2.5x$

$6.7x = -2.3$ Simplifying

$\dfrac{6.7x}{6.7} = \dfrac{-2.3}{6.7}$ Dividing by 6.7

$x = \dfrac{-23}{67}$ Simplifying

The solution is $-\frac{23}{67}$.

9. $4(x+2) = 36$

$4x + 8 = 36$ Distributive Law

$4x + 8 - 8 = 36 - 8$ Adding -8

$4x = 28$ Simplifying

$\frac{1}{4}(4x) = \frac{1}{4}(28)$ Multiplying by $\frac{1}{4}$

$x = 7$ Simplifying

The solution is 7.

10. $9 - 3x = 6(x+4)$

$9 - 3x = 6x + 24$ Distributive Law

$9 - 3x - 24 = 6x + 24 - 24$ Adding -24

$-3x - 15 = 6x$ Simplifying

$-3x - 15 + 3x = 6x + 3x$ Adding $3x$

$-15 = 9x$ Simplifying

$\frac{1}{9}(-15) = \frac{1}{9}(9x)$ Multiplying by $\frac{1}{9}$

$-\frac{5}{3} = x$ Simplifying

The solution is $-\frac{5}{3}$.

11. $\frac{5}{6}(3x+1) = 20$

$\frac{6}{5}\left[\frac{5}{6}(3x+1)\right] = \frac{6}{5} \cdot 20$ Multiplying by $\frac{6}{5}$

$3x + 1 = 24$ Simplifying

$3x + 1 - 1 = 24 - 1$ Adding -1

$3x = 23$ Simplifying

$\frac{1}{3}(3x) = \frac{1}{3}(23)$ Multiplying by $\frac{1}{3}$

$x = \frac{23}{3}$ Simplifying

The solution is $\frac{23}{3}$.

12. $3(2x-8) = 6(x-4)$

$6x - 24 = 6x - 24$ Distributive Law

Since this is true for all x, the equation is an identity.

13. $x + 6 > 1$

$x + 6 - 6 > 1 - 6$ Adding -6

$x > -5$ Simplifying

The solution set is $\{x \mid x > -5\}$, or $(-5, \infty)$.

14. $14x + 9 > 13x - 4$

$14x + 9 - 9 > 13x - 4 - 9$ Adding -9

$14x > 13x - 13$ Simplifying

$14x - 13x > 13x - 13 - 13x$ Adding $-13x$

$x > -13$ Simplifing

The solution set is $\{x \mid x > -13\}$, or $(-13, \infty)$.

15. $-2y \le 26$

$-\frac{1}{2}(-2y) \ge -\frac{1}{2}(26)$ Multiplying by $-\frac{1}{2}$ and reversing the inequality symbol

$y \ge -13$ Simplifying

The solution set is $\{y \mid y \ge -13\}$, or $[-13, \infty)$.

16. $4y \le -32$

$\frac{1}{4}(4y) \le \frac{1}{4}(-32)$ Multiplying by $\frac{1}{4}$

$y \le -8$ Simplifying

The solution set is $\{y \mid y \le -8\}$, or $(-\infty, -8]$.

17. $4n + 3 < -17$

$4n + 3 - 3 < -17$ Subtracting 3

$4n < -20$ Simplifying

$\dfrac{4n}{4} < \dfrac{-20}{4}$ Dividing by 4

$n < -5$

The solution set is $\{n \mid n < -5\}$, or $(-\infty, -5)$.

18. $\frac{1}{2}t - \frac{1}{4} \le \frac{3}{4}t$

$4\left(\frac{1}{2}t - \frac{1}{4}\right) \le 4 \cdot \left(\frac{3}{4}t\right)$ Multiplying by 4

$2t - 1 \le 3t$ Simplifying

$2t - 1 + 1 \le 3t + 1$ Adding 1

$2t \le 3t + 1$ Simplifying

$2t - 3t \le 3t + 1 - 3t$ Subtracting $3t$

$-t \le 1$ Simplifying

$t \ge -1$ Multiplying by -1

The solution set is $\{t \mid t \ge -1\}$, or $[-1, \infty)$.

19. $5 - 9x \ge 19 + 5x$

$5 - 9x - 5 \ge 19 + 5x - 5$ Adding -5

$-9x \ge 14 + 5x$ Simplifying

$-9x - 5x \ge 14 + 5x - 5x$ Adding $-5x$

$-14x \ge 14$ Simplifying

$-\frac{1}{14}(-14x) \le -\frac{1}{14}(14)$ Multiplying by $-\frac{1}{14}$ and reversing the inequality symbol

$x \le -1$ Simplifying

The solution set is $\{x \mid x \le -1\}$, or $(-\infty, -1]$.

20. $A = 2\pi rh$

$\frac{1}{2\pi h} \cdot A = \frac{1}{2\pi h}(2\pi rh)$ Multiplying by $\frac{1}{2\pi h}$

$\frac{A}{2\pi h} = r$ Simplifying

The solution is $r = \frac{A}{2\pi h}$.

21. $w = \frac{P + l}{2}$

$2 \cdot w = 2\left(\frac{P + l}{2}\right)$ Multiplying by 2

$2w = P + l$ Simplifying

$2w - P = P + l - P$ Adding $-P$

$2w - P = l$ Simplifying

The solution is $l = 2w - P$.

22. $230\% = 230 \times 0.01$ Replacing % by $\times 0.01$

$= 2.3$

$0.05.4$

$\underset{\llcorner\lrcorner}{}$

5.4%

23. 0.054

First move the decimal point two places to the right; then write a % symbol. The answer is 5.4%.

24. **Translate.**

$\underset{\downarrow}{\underline{\text{What number}}}$ is 32% of 50?

$x \quad = 0.32 \cdot 50$

We solve the equation.

$x = 0.32 \cdot 50$

$x = 16$

The solution is 16.

25. **Translate.**

$\underset{\downarrow}{\underline{\text{What percent}}}$ of 75 is 33?

$y \qquad \cdot 75 = 33$

We solve the equation and then convert to percent notation.

$y \cdot 75 = 33$

$y = \frac{33}{75}$

$y = 0.44 = 44\%$

The solution is 44%.

26. $y < 4$

27. $-2 \le x \le 2$

28. **Familiarize.** Let w = the width of the calculator, in cm. Then the length is $w + 4$, in cm. The perimeter of a rectangle is given by $P = 2l + 2w$.

Translate.

$\underline{\text{The perimeter of the rectangle}}$ is 36.

$2(w + 4) + 2w \qquad = 36$

Carry out. We solve the equation.

$2(w + 4) + 2w = 36$

$2w + 8 + 2w = 36$

$4w + 8 = 36$

$4w = 28$

$w = 7$

Check. If the width is 7 cm, then the length is 7 + 4, or 11 cm. The perimeter is then

$2 \cdot 11 + 2 \cdot 7$, or $22 + 14$, or 36 cm. The results checks.

State. The width is 7 cm and the length is 11 cm.

29. *Familiarize.* Let x = the number of miles Kari has already ridden. Then she has $3x$ miles to go.
Translate.

$\underbrace{\text{The total length of Kari's bicycle trip}}$ is 240 miles.

$$x + 3x \qquad = 240$$

Carry out. We solve the equation.
$$x + 3x = 240$$
$$4x = 240$$
$$x = 60$$

Check. If Kari has already ridden 60 mi, then she has yet to ride $3 \cdot 60$ mi, or 180 mi. The total length of her ride is 60 mi + 180 mi, or 240 mi. This result checks.
State. Kari has already ridden 60 miles.

30. *Familiarize.* Let x = the length of the first side, in mm. Then the length of the second side is $x + 2$ mm, and the length of the third side is $x + 4$ mm. The perimeter of a triangle is the sum of the lengths of the three sides.
Translate.

$\underbrace{\text{The perimeter of the triangle}}$ is 249 mm.

$$x + (x + 2) + (x + 4) \qquad = 249$$

Carry out. We solve the equation.
$$x + (x + 2) + (x + 4) = 249$$
$$3x + 6 = 249$$
$$3x = 243$$
$$x = 81$$

Check. If the length of the first side is 81 mm, then the length of the second side is 81 + 2, or 83 mm, and the length of the third side is 81 + 4, or 85 mm. The perimeter of the triangle is 81 + 83 + 85, or 249 mm. These results check.
State. The lengths of the sides are 81 mm, 83 mm, and 85 mm.

31. *Familiarize.* Let x = the electric bill before the temperature of the water heater was lowered. If the bill dropped by 7%, then the Kelly's paid 93% of their original bill.
Translate.

93% of $\underbrace{\text{the original bill}}$ is $60.45.

$$0.93 \quad \cdot \qquad x \qquad = 60.45$$

Carry out. We solve the equation.
$$0.93x = 60.45$$
$$x = \frac{60.45}{0.93}$$
$$x = 65$$

Check. If the original bill was $65, and the bill was reduced by 7%, or $0.07 \cdot \$65$, or $4.55, the new bill would be $65 – $4.55, or $60.45. This result checks.
State. The original bill was $65.

32. *Familiarize.* Let x = the number of miles that will allow the business to stay within budget. The cost for the rental would be $14.95 + 0.59x$.
Translate.

$\underbrace{\text{The cost of the rental}}$ $\underbrace{\text{must not exceed}}$ $250.

$$14.95 + 0.59x \qquad \leq \qquad 250$$

Carry out. We solve the inequality.
$$14.95 + 0.59x \leq 250$$
$$0.59x \leq 235.05$$
$$x \leq \frac{235.05}{0.59}$$
$$x \leq 398.3898$$

Check. As a partial check, we let $x = 398.39$ mi and determine the cost of the rental. The rental would be $14.95 + $0.59(398.39), or $250 (rounded to the nearest dollar). If the mileage were less, the rental would be under $250, so the result checks. If 398.4 miles is used, the result will be slightly more than $250, so we round down to 398.3 miles to the nearest tenth of a mile.
State. The mileage is less than or equal to 398.3 mi, rounded to the nearest tenth.

33.
$$c = \frac{2cd}{a-d}$$

$$(a-d)c = (a-d)\left(\frac{2cd}{a-d}\right) \quad \text{Multiplying by } a-d$$

$$ac - dc = 2cd \quad \text{Simplifying}$$

$$ac - dc + dc = 2cd + dc \quad \text{Adding } dc$$

$$ac = 3cd \quad \text{Simplifying}$$

$$\frac{1}{3c}(ac) = \frac{1}{3c}(3cd) \quad \text{Multiplying by } \frac{1}{3c}$$

$$\frac{ac}{3c} = d \quad \text{Simplifying}$$

$$\frac{a}{3} = d \quad \text{Simplifying}$$

The solution is $d = \dfrac{a}{3}$.

34.
$$3|w| - 8 = 37$$

$$3|w| - 8 + 8 = 37 + 8 \quad \text{Adding 8}$$

$$3|w| = 45 \quad \text{Simplifying}$$

$$\tfrac{1}{3}(3|w|) = \tfrac{1}{3} \cdot 45 \quad \text{Multiplying by } \tfrac{1}{3}$$

$$|w| = 15 \quad \text{Simplifying}$$

This tells us that the number w is 15 units from the origin. The solutions are $w = -15$ and $w = 15$.

35. Let $h =$ the number of hours of sun each day. Then at least 4 hr but no more than 6 hr of sun is $4 \le h \le 6$.

36. **Familiarize.** Let $x =$ the number of tickets given away. The following shows the distribution of the tickets:

First person received $\frac{1}{3}x$ tickets.

Second person received $\frac{1}{4}x$ tickets.

Third person received $\frac{1}{5}x$ tickets.

Fourth person received 8 tickets.

Fifth person received 5 tickets.

Translate.

The number of tickets the five people received	is	the total number of tickets.
↓	↓	↓
$\frac{1}{3}x + \frac{1}{4}x + \frac{1}{5}x + 8 + 5$	$=$	x

Carry out. We solve the equation.

$$\tfrac{1}{3}x + \tfrac{1}{4}x + \tfrac{1}{5}x + 8 + 5 = x$$

$$60\left(\tfrac{1}{3}x + \tfrac{1}{4}x + \tfrac{1}{5}x + 8 + 5\right) = 60x$$

$$60\left(\tfrac{1}{3}x + \tfrac{1}{4}x + \tfrac{1}{5}x + 13\right) = 60x$$

$$20x + 15x + 12x + 780 = 60x$$

$$47x + 780 = 60x$$

$$780 = 13x$$

$$60 = x$$

Check. If the total number of tickets given away was 60, then the first person received $\frac{1}{3}(60)$, or 20 tickets; the second person received $\frac{1}{4}(60)$, or 15 tickets; the third person received $\frac{1}{5}(60)$, or 12 tickets. We are told that the fourth person received 8 tickets, and the fifth person received 5 tickets. The sum of the tickets distributed is 20 + 15 + 12 + 8 + 5, or 60 tickets. These results check.

State. A total of 60 tickets were given away.

Chapter 3

Introduction to Graphing and Functions

Exercise Set 3.1

1. a

3. b

5. We go to the top of the bar that is above the body weight 100 lb. Then we move horizontally from the top of the bar to the vertical scale listing numbers of drinks. It appears that consuming approximately 2 drinks in one hour will give a 100 lb person a blood-alcohol level of 0.08%.

7. From 3 on the vertical scale we move horizontally until we reach a bar whose top is above the horizontal line on which we are moving. The first such bar corresponds to a body weight of 160 lb. This means that for body weights represented by bars to the left of this one, consuming 3 drinks will yield a blood-alcohol level of 0.08%. The bar immediately to the left of the 160-pound bar represents 140 pounds. Thus, we can conclude an individual weighs more than 140 lb if 3 drinks are consumed in one hour without reaching a blood-alcohol level of 0.08%.

9. **Familiarize.** We are told the total amount of student aid distributed. In order to find the average federal loan per full-time equivalent student, we must first calculate the total of all federal loans using information given in the circle graph, and then divide by the number of full-time equivalent students. Let x = the average federal loan per full-time equivalent student in 2008.
    Translate. From the circle graph we see that 47% of student aid were federal loans. The total amount distributed was $143.4 billion, or $143,400,000,000, so we have
    federal loans $= 0.47 \cdot 143,400,000,000$
    $$= 67,398,000,000$$
    We reword the problem.

The average federal loan	is	the amount of all federal loans	divided by	the number of students.
↓	↓	↓	↓	↓
x	=	67,398,000,000	÷	13,803,201

Carry out.
   $x = 67,398,000,000 \div 13,803,201 \approx 4883$
Check. If each student received $4883, the total amount of aid distributed would be
   $4883 \cdot 13,803,201 \approx 67,401,030,480$. Since this is approximately 47% of the total student aid for 2008, our answer checks.
State. In 2008, the average federal loan per full-time equivalent student in the United States was about $4883.

11. **Familiarize.** From the circle graph, we see that 6% of student aid in 2008 came from state grants. Therefore the portion of the 143.4 billion dollars in student aid coming from state grants is 6% of 143.4 billion dollars. Of this amount, 70% is need-based. Let g = the amount, in dollars, of student aid from state grants that is need-based.
    Translate. We reword and translate the problem.

What is 70% of (6% of 143.4 billion)?
   ↓ ↓ ↓ ↓ ↓ ↓   ↓
   g = 70% · (6% · 143,400,000,000)

Carry out. We solve the equation.
   $g = 0.70 \cdot (0.06 \cdot 143,400,000,000)$
   $$= 6,022,800,000$$
Check. If the amount of need-based state aid distributed were $6,022,800,000, then
   $\dfrac{6,022,800,000}{0.06 \cdot 0.70} = 143,400,000,000$, which is
equal to the total amount, in dollars, of student aid.
State. About $6,022,800,000 in student aid was provided from need-based state grants.

13. **Familiarize.** From the pie chart we see that 5.3% of solid waste is glass. Let x = the amount of glass, in millions of tons, in the waste generated in 2007.
    Translate. We reword the problem.

What is 5.3% of 254 million?

$$\downarrow \quad \downarrow \quad \downarrow \quad \downarrow \quad \downarrow$$

$$x \quad = 5.3\% \cdot 254$$

Carry out.

$$x = 0.053 \cdot 254 \approx 13.5$$

Check. If 13.5 million tons of glass were generated , the total amount of solid waste, in millions of tons, would be $\dfrac{13.5}{0.053} \approx 254.72$, which is approximately equal to the total amount of solid waste generated in 2007.

State. In 2007, about 13.5 million tons of solid waste consisted of glass.

15. **Familiarize.** From the pie chart we see that 5.3% of solid waste is glass. Therefore the portion of the 254 million tons of solid waste generated in 2007 that is glass is 5.3% of 254 million tons. Of this amount, 23.7% is recycled. Let g = the amount, in millions of tons, of recycled glass in 2007.

Translate. We reword the problem.

What is 23.7% of (5.3% of 254)?

$$\downarrow \quad \downarrow \quad \downarrow \quad \downarrow \quad \downarrow \quad \downarrow \quad \downarrow$$

$$g \quad = 23.7\% \quad \cdot \quad (5.3\% \quad \cdot \quad 254)$$

Carry out.

$$g = 0.237 \cdot (0.053 \cdot 254) \approx 3.2$$

Check. If the amount of recycled glass was 3.2 million tons, then $\dfrac{3.2}{0.053 \cdot 0.237} \approx 254.8$, which is approximately equal to the total amount of solid waste in millions of tons.

State. In 2007, Americans recycled about 3.2 million tons of glass.

17. Locate 2005 on the horizontal scale and then move up to the point that represents the number of wireless-only households in that year. Now move to the vertical axis and read that about 9,000,000 U.S. households had only wireless telephones in 2005.

19. Locate 15 million on the vertical scale. Move right until we reach a point representing the number of wireless-only households that is above the horizontal line on which we are moving.

The first such point corresponds to the year 2007. The points to the left of this one represent 2005 and 2006. Thus, we can conclude that in 2005 and 2006, the number

of U.S. households that had only wireless telephoneswere fewer than 15 million.

21. Starting at the origin:

$(1, 2)$ is 1 unit right and 2 units up.

$(-2, 3)$ is two units left and 3 units up.

$(4, -1)$ is 4 units right and 1 unit down.

$(-5, -3)$ is 5 units left and 3 units down.

$(4, 0)$ is 4 units right and 0 units up or down.

$(0, -2)$ is 0 units left or right and 2 units down.

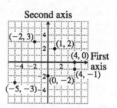

23. Starting at the origin:

$(4, 4)$ is 4 units right and 4 units up.

$(-2, 4)$ is 2 units left and 4 units up.

$(5, -3)$ is 5 units right and 3 units down.

$(-5, -5)$ is 5 units left and 5 units down.

$(0, 4)$ is 0 units left or right and 4 units up.

$(0, -4)$ is 0 units left or right and 4 units down.

$(3, 0)$ is 3 units right and 0 units up or down.

$(-4, 0)$ is 4 units left and 0 units up or down.

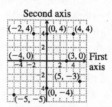

25.

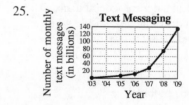

27. Point A is 4 units left and 5 units up. The coordinates of A are $(-4,5)$.

Point B is 3 units left and 3 units down. The coordinates of B are $(-3,-3)$.

Point C is 0 units left or right and 4 units up. The coordinates of C are $(0,4)$.

Point D is 3 units right and 4 units up. The coordinates of D are $(3,4)$.

The point E is 3 units right and 4 units down. The coordinates of E are $(3,-4)$.

29. The point A is 4 units right and 1 unit up. The coordinates of A are $(4,1)$.

The point B is 0 units left or right and 5 units down. The coordinates of B are $(0,-5)$.

The point C is 4 units left and 0 units up or down. The coordinates of C are $(-4,0)$.

The point D is 3 units left and 2 units down. The coordinates of D are $(-3,-2)$.

The point E is 3 units right and 0 units up or down. The coordinates of E are $(3,0)$.

31.

33.

35.

37.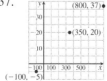

39.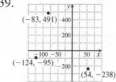

41. Since the first coordinate is positive and the second coordinate is negative, the point $(7,-2)$ is in quadrant IV.

43. Since the first coordinate is negative and the second coordinate is negative, the point $(-4,-3)$ is in quadrant III.

45. Since the first coordinate is 0, the point $(0,-3)$ must lie on the y-axis.

47. Since the first coordinate is negative and the second coordinate is positive, the point $(-4.9, 8.3)$ is in quadrant II.

49. Since the second coordinate is 0, the point $\left(-\dfrac{5}{2}, 0\right)$ must lie on the x-axis.

51. Since the first coordinate is positive and the second coordinate is positive, the point $(160, 2)$ is in quadrant I.

53. The first coordinates are positive in quadrants I and IV.

55. Both coordinates have the same sign in quadrants I and III.

57.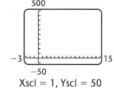

Xscl $= 1$, Yscl $= 50$

59.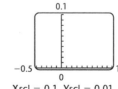

Xscl $= 0.1$, Yscl $= 0.01$

61. *Thinking and Writing Exercise.*

63. $5y = 2x$

   $\frac{1}{5}(5y) = \frac{1}{5}(2x)$

   $y = \frac{2}{5}x$

65. $x - y = 8$

   $x - y - x = 8 - x$

   $-y = 8 - x$

   $-1(-y) = -1(8 - x)$

   $y = -8 + x$, or $y = x - 8$

67. $2x + 3y = 5$

   $2x + 3y - 2x = 5 - 2x$

   $3y = 5 - 2x$

   $\frac{1}{3}(3y) = \frac{1}{3}(5 - 2x)$

   $y = \frac{5}{3} - \frac{2}{3}x$, or $y = -\frac{2}{3}x + \frac{5}{3}$

69. *Thinking and Writing Exercise.*

71. The coordinates can be opposites of one another only in quadrants II and IV.

73.

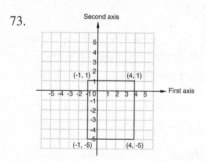

75. Answers may vary. We select eight points such that the sum of the coordinates for each point is 7.

   $(0, 7)$ $0 + 7 = 7$

   $(1, 6)$ $1 + 6 = 7$

   $(2, 5)$ $2 + 5 = 7$

   $(3, 4)$ $3 + 4 = 7$

   $(4, 3)$ $4 + 3 = 7$

   $(5, 2)$ $5 + 2 = 7$

   $(6, 1)$ $6 + 1 = 7$

   $(7, 0)$ $7 + 0 = 7$

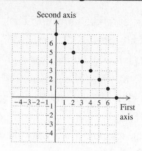

77. Plot the three given points.

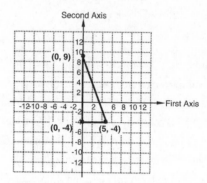

The base of the triangle is 5 units and the height is 13 units.

$A = \frac{1}{2}bh = \frac{1}{2} \cdot 5 \cdot 13 = \frac{65}{2}$ sq units, or $32\frac{1}{2}$ sq units.

79. Latitude 27° North; longitude 81° West

Exercise Set 3.2

1. The graph of the set of solutions of a linear equation is a line of points, so there will be infinitely many solutions.

3. The graph of $y = 3x - 7$ is the line that contains all ordered pairs that are solutions to the equation. The statement is therefore true.

5. By selecting a value for x and substituting it into the equation $y = 3x - 7$, the uniqe value of y that solves the equation will be calculated. The statement is therefore true.

7. We substitute 2 for x and 1 for y (alphabetical order of variables).

$$y = 4x - 7$$

$$
\begin{array}{c|c}
1 & 4 \cdot 2 - 7 \\
  & 8 - 7 \\
1 & 1 \qquad \text{TRUE}
\end{array}
$$

Since $1 = 1$ is true, the pair $(2,1)$ is a solution.

9. We substitute 4 for x and 2 for y (alphabetical order of variables).

$$3y + 2x = 12$$

$$
\begin{array}{c|c}
3 \cdot 2 + 2 \cdot 4 & 12 \\
6 + 8 & \\
14 & 12 \quad \text{FALSE}
\end{array}
$$

Since $14 = 12$ is false, the pair $(4,2)$ is not a solution.

11. We substitute 3 for m and -1 for n (alphabetical order of variables).

$$4m - 5n = 7$$

$$
\begin{array}{c|c}
4 \cdot 3 - 5(-1) & 7 \\
12 + 5 & \\
17 & 7 \quad \text{FALSE}
\end{array}
$$

Since $17 = 7$ is false, the pair $(3,-1)$ is not a solution.

13. To show that a pair is a solution, we substitute, replacing x with the first coordinate and y with the second coordinate in each pair.

$$
\begin{array}{c|c} \qquad
y = x + 3 & \\
\hline
2 & -1 + 3 \\
  & 2 \\
2 & 2 \qquad \text{TRUE}
\end{array}
\qquad
\begin{array}{c|c}
y = x + 3 & \\
\hline
7 & 4 + 3 \\
  & 7 \\
7 & 7 \qquad \text{TRUE}
\end{array}
$$

In each case, the substitution results in a true equation. Thus, $(-1,2)$ and $(4,7)$ are both solutions of $y = x + 3$. We graph these points and sketch the line passing through them.

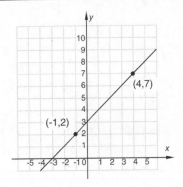

The line appears to pass through $(0,3)$ also. We check to determine if $(0,3)$ is a solution of $y = x + 3$.

$$y = x + 3$$

$$
\begin{array}{c|c}
3 & 0 + 3 \\
  & 2 \\
3 & 3 \qquad \text{TRUE}
\end{array}
$$

Thus, $(0,3)$ is another solution. There are other correct answers, including $(-5,-2)$, $(-4,-1)$, $(-3,0)$, $(-2,1)$, $(1,4)$, $(2,5)$, and $(3,6)$.

15. To show that a pair is a solution, we substitute, replacing x with the first coordinate and y with the second coordinate in each pair.

$$
\begin{array}{c|c}
y = \frac{1}{2}x + 3 & \\
\hline
5 & \frac{1}{2} \cdot 4 + 3 \\
  & 2 + 3 \\
5 & 5 \qquad \text{TRUE}
\end{array}
\qquad
\begin{array}{c|c}
y = \frac{1}{2}x + 3 & \\
\hline
2 & \frac{1}{2}(-2) + 3 \\
  & -1 + 3 \\
2 & 2 \qquad \text{TRUE}
\end{array}
$$

In each case, the substitution results in a true equation. Thus, $(4,5)$ and $(-2,2)$ are both solutions of $y = \frac{1}{2}x + 3$. We graph these points and sketch the line passing through them.

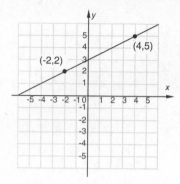

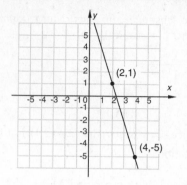

The line appears to pass through $(0,3)$ also. We check to determine if $(0,3)$ is a solution of $y = \frac{1}{2}x + 3$.

$$
\begin{array}{c|c}
y = \frac{1}{2}x + 3 & \\
\hline
3 & \frac{1}{2} \cdot 0 + 3 \\
& 3 \\
3 & 3 \qquad \text{TRUE}
\end{array}
$$

Thus, $(0,3)$ is another solution. There are other correct answers, including $(-6,0)$, $(-4,1)$, $(2,4)$, and $(6,6)$.

17. To show that a pair is a solution, we substitute, replacing x with the first coordinate and y with the second coordinate in each pair.

$$
\begin{array}{c|c}
y + 3x = 7 & \\
\hline
1 + 3 \cdot 2 & 7 \\
1 + 6 & 7 \\
7 & 7 \qquad \text{TRUE}
\end{array}
\qquad
\begin{array}{c|c}
y + 3x = 7 & \\
\hline
-5 + 3 \cdot 4 & 7 \\
-5 + 12 & 7 \\
7 & 7 \qquad \text{TRUE}
\end{array}
$$

In each case, the substitution results in a true equation. Thus, $(2,1)$ and $(4,-5)$ are both solutions of $y + 3x = 7$. We graph these points and sketch the line passing through them.

The line appears to pass through $(1,4)$ also. We check to determine if $(1,4)$ is a solution of $y + 3x = 7$.

$$
\begin{array}{c|c}
y + 3x = 7 & \\
\hline
4 + 3 \cdot 1 & 7 \\
4 + 3 & 7 \\
7 & 7 \qquad \text{TRUE}
\end{array}
$$

Thus, $(1,4)$ is another solution. There are other correct answers, including $(3,-2)$.

19. To show that a pair is a solution, we substitute, replacing x with the first coordinate and y with the second coordinate in each pair.

$$
\begin{array}{c|c}
4x - 2y = 10 & \\
\hline
4 \cdot 0 - 2 \cdot (-5) & 10 \\
10 & 10 \\
10 & 10 \qquad \text{TRUE}
\end{array}
$$

$$
\begin{array}{c|c}
4x - 2y = 10 & \\
\hline
4 \cdot 4 - 2 \cdot 3 & 10 \\
16 - 6 & 10 \\
10 & 10 \qquad \text{TRUE}
\end{array}
$$

In each case, the substitution results in a true equation. Thus, $(0,-5)$ and $(4,3)$ are both solutions of $4x - 2y = 10$. We graph these points and sketch the line passing through them.

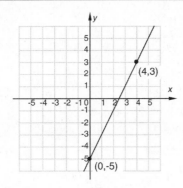

The line appears to pass through $(2,-1)$ also.
We check to determine if $(2,-1)$ is a
solution of $4x-2y=10$.

$$4x-2y=10$$

$$
\begin{array}{c|c}
4(2)-2(-1) & 10 \\
8+2 & 10 \\
10 & 10 \text{ TRUE}
\end{array}
$$

Thus, $(2,-1)$ is another solution. There are
other correct answers, including $(1,-3)$,
$(3,1)$, and $(5,5)$.

21. $y = x+1$
When $x = 0$, $y = 0+1 = 1$.
When $x = 2$, $y = 2+1 = 3$.
When $x = -3$, $y = -3+1 = -2$.

x	y
0	1
2	3
-3	-2

Plot these points, draw the line they
determine, and label the graph $y = x+1$.

23. $y = -x$
When $x = 0$, $y = -0 = 0$.
When $x = 3$, $y = -3$.
When $x = -5$, $y = -(-5) = 5$.

x	y
0	0
3	-3
-5	5

Plot these points, draw the line they
determine, and label the graph $y = -x$.

25. $y = 2x$
When $x = 0$, $y = 2 \cdot 0 = 0$.
When $x = -2$, $y = 2(-2) = -4$.
When $x = 3$, $y = 2 \cdot 3 = 6$.

x	y
0	0
-2	-4
3	6

Plot these points, draw the line they
determine, and label the graph $y = 2x$.

27. $y = 2x+2$
When $x = 0$, $y = 2 \cdot 0 + 2 = 0 + 2 = 2$.
When $x = 1$, $y = 2 \cdot 1 + 2 = 2 + 2 = 4$.
When $x = -3$, $y = 2(-3) + 2 = -6 + 2 = -4$.

x	y
0	2
1	4
-3	-4

Plot these points, draw the line they
determine, and label the graph $y = 2x+2$.

29. $y = -\frac{1}{2}x$

When $x = 0$, $y = -\frac{1}{2} \cdot 0 = 0$.

When $x = -4$, $y = -\frac{1}{2}(-4) = 2$.

When $x = 2$, $y = -\frac{1}{2} \cdot 2 = -1$.

x	y
0	0
-4	2
2	-1

Plot these points, draw the line they determine, and label the graph $y = -\frac{1}{2}x$.

31. $y = \frac{1}{3}x - 4$

When $x = 0$, $y = \frac{1}{3} \cdot 0 - 4 = 0 - 4 = -4$.

When $x = 3$, $y = \frac{1}{3} \cdot 3 - 4 = 1 - 4 = -3$.

When $x = -3$, $y = \frac{1}{3}(-3) - 4 = -1 - 4 = -5$.

x	y
0	-4
3	-3
-3	-5

Plot these points, draw the line they determine, and label the graph $y = \frac{1}{3}x - 4$.

33. $x + y = 4$

When $x = 0$, we have

$0 + y = 4$

$y = 4$.

When $y = 0$, we have

$x + 0 = 4$

$x = 4$.

When $x = 2$, we have

$2 + y = 4$

$y = 2$.

x	y
0	4
4	0
2	2

Plot these points, draw the line they determine, and label the graph $x + y = 4$.

35. $x - y = -2$

When $x = 0$, we have

$0 - y = -2$

$-y = -2$

$y = 2$.

When $y = 0$, we have

$x - 0 = -2$

$x = -2$.

When $x = -1$, we have

$-1 - y = -2$

$y = 1$.

x	y
0	2
-2	0
-1	1

Plot these points, draw the line they determine, and label the graph $x - y = -2$.

37. $x + 2y = -6$

When $x = 0$, we have
$0 + 2y = -6$
$2y = -6$
$y = -3$.

When $y = 0$, we have
$x + 2 \cdot 0 = -6$
$x + 0 = -6$
$x = -6$.

When $x = -8$, we have
$-8 + 2y = -6$
$2y = 2$
$y = 1$.

x	y
0	-3
-6	0
-8	1

Plot these points, draw the line they determine, and label the graph $x + 2y = -6$.

39. $y = -\frac{2}{3}x + 4$

When $x = 0$, $y = -\frac{2}{3} \cdot 0 + 4 = 0 + 4 = 4$.

When $x = 6$, $y = -\frac{2}{3} \cdot 6 + 4 = -4 + 4 = 0$.

When $x = 3$, $y = -\frac{2}{3} \cdot 3 + 4 = -2 + 4 = 2$.

x	y
0	4
6	0
3	2

Plot these points, draw the line they determine, and label the graph $y = -\frac{2}{3}x + 4$.

41. $4x = 3y$

When $x = 0$, we have
$4 \cdot 0 = 3y$
$0 = 3y$
$y = 0$.

When $x = 3$, we have
$4 \cdot 3 = 3y$
$12 = 3y$
$y = 4$.

When $x = -3$, we have
$4(-3) = 3y$
$-12 = 3y$
$y = -4$.

x	y
0	0
3	4
-3	-4

Plot these points, draw the line they determine, and label the graph $4x = 3y$.

43. $8x - 4y = 12$

When $x = 0$, we have
$8 \cdot 0 - 4y = 12$
$0 - 4y = 12$
$y = -3$.

When $x = 2$, we have
$8 \cdot 2 - 4y = 12$
$16 - 4y = 12$
$-4y = -4$
$y = 1$.

When $x = 4$, we have
$8 \cdot 4 - 4y = 12$
$32 - 4y = 12$
$-4y = -20$
$y = 5$.

x	y
0	−3
2	1
4	5

Plot these points, draw the line they determine, and label the graph $8x - 4y = 12$.

45. $6y + 2x = 8$

When $x = 4$, we have
$$6y + 2 \cdot 4 = 8$$
$$6y + 8 = 8$$
$$6y = 0$$
$$y = 0.$$

When $x = 1$, we have
$$6y + 2 \cdot 1 = 8$$
$$6y + 2 = 8$$
$$6y = 6$$
$$y = 1.$$

When $x = -5$, we have
$$6y + 2(-5) = 8$$
$$6y - 10 = 8$$
$$6y = 18$$
$$y = 3.$$

x	y
4	0
1	1
−5	3

Plot these points, draw the line they determine, and label the graph $6x + 2y = 8$.

47. $y = x^2 + 1$

When $x = -2$, $y = (-2)^2 + 1 = 4 + 1 = 5$.

When $x = -1$, $y = (-1)^2 + 1 = 1 + 1 = 2$.

When $x = 0$, $y = 0^2 + 1 = 0 + 1 = 1$.

When $x = 1$, $y = 1^2 + 1 = 1 + 1 = 2$.

When $x = 2$, $y = 2^2 + 1 = 4 + 1 = 5$.

x	y
−2	5
−1	2
0	1
1	2
2	5

Plot these points, connect them with a smooth curve, and label the graph $y = x^2 + 1$.

49. $y = 2 - x^2$

When $x = -2$, $y = 2 - (-2)^2 = 2 - 4 = -2$.

When $x = -1$, $y = 2 - (-1)^2 = 2 - 1 = 1$.

When $x = 0$, $y = 2 - 0^2 = 2 - 0 = 2$.

When $x = 1$, $y = 2 - 1^2 = 2 - 1 = 1$.

When $x = 2$, $y = 2 - 2^2 = 2 - 4 = -2$.

x	y
−2	−2
−1	1
0	2
1	1
2	−2

Plot these points, connect them with a smooth curve, and label the graph $y = 2 - x^2$.

51. a) $x + 1 = 4$

$x = 4 - 1$

$x = 3$

We graph the solution of the equation, 3, on the number line.

<+++++++++◆++>
 −4 −2 0 2 4

b)   We graph $x + 1 = y$.

When $x = -3$, $y = -3 + 1 = -2$.

When $x = 0$, $y = 0 + 1 = 1$.

When $x = 2$, $y = 2 + 1 = 3$.

x	y
−3	−2
0	1
2	3

Plot the points, draw the line they determine, and label the graph $x + 1 = y$.

53. a)   We graph $y = 2x - 3$.

When $x = -1$, $y = 2(-1) - 3 = -2 - 3 = -5$.

When $x = 0$, $y = 2 \cdot 0 - 3 = 0 - 3 = -3$.

When $x = 3$, $y = 2 \cdot 3 - 3 = 6 - 3 = 3$.

x	y
−1	−5
0	−3
3	3

Plot the points, draw the line they determine, and label the graph $y = 2x - 3$.

b) $-7 = 2x - 3$

$-4 = 2x$

$-2 = x$

We graph the solution of the equation, −2, on the number line.

<+++◆++++++++>
 −4 −2 0 2 4

55. $y = (-3/2)\,x + 1$

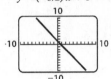

57. $4y - 3x = 1$, or

$y = (3x + 1)/4$

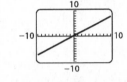

59. $y = -2$

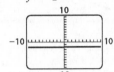

61. $y = -x^2$

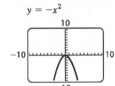

63. $x^2 - y = 3$, or

$y = x^2 - 3$

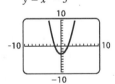

65. $y = x \wedge 3$

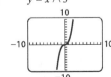

67.   Only window (b) shows where the graph crosses the x- and y-axes.

69. Only window (a) shows where the graph crosses the x- and y-axes.

71. Only window (b) shows where the graph crosses the x- and y-axes.

73. We graph $a = 0.08t + 2.5$. Since the number of years since 1994 cannot be negative in this application, we select only nonnegative values for t.

If $t = 0$, $n = 0.08 \cdot 0 + 2.5 = 0 + 2.5 = 2.5$.
If $t = 4$, $n = 0.08 \cdot 4 + 2.5 = 0.32 + 2.5 = 2.82$.
If $t = 8$, $n = 0.08 \cdot 8 + 2.5 = 0.64 + 2.5 = 3.14$.

t	n
0	2.50
4	2.82
8	3.14

We plot the points and draw the graph.

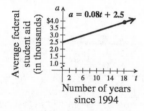

To estimate the average amount of federal student aid, in thousands of dollars, per student in 2012, we find the second coordinate associated with 18 (2012 is 18 years after 1994). Locate the point on the line that is above 18 and then find the value on the vertical axis that corresponds to that point. That value is approximately 4, so we estimate that in 2012, the average amount of federal student aid per student will be about 4 thousand dollars.

75. We graph $c = 3.1w + 29.07$. Since the weight above 1 lb cannot be negative in this application, we select only nonnegative values for w.

If $w = 0$, then
$c = 3.1 \cdot 0 + 29.07 = 0 + 29.07 = 29.07$.
If $w = 5$, then
$c = 3.1 \cdot 5 + 29.07 = 15.5 + 29.07 = 44.57$.
If $w = 10$, then
$c = 3.1 \cdot 10 + 29.07 = 31 + 29.07 = 60.07$.

w	c
0	29.07
5	44.57
10	60.07

We plot these points and draw the graph.

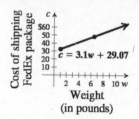

To estimate the cost of shipping a $6\frac{1}{2}$-lb package, we find the second coordinate associated with $6\frac{1}{2}$. Locate the point on the line that is above $6\frac{1}{2}$ and then find the value on the vertical axis that corresponds to that point. That value is about 49, so we estimate that the cost of shipping a $6\frac{1}{2}$-lb package is $49.

77. We graph $p = 3.5n + 9$. Since the number of pages in the bound scrapbook cannot be negative, we select only nonnegative numbers.

If $n = 0$, $p = 3.5 \cdot 0 + 9 = 0 + 9 = 9$.
If $n = 10$, $p = 3.5 \cdot 10 + 9 = 35 + 9 = 44$.
If $n = 30$, $p = 3.5 \cdot 30 + 9 = 105 + 9 = 114$.

n	p
0	9
10	44
30	114

We plot the points and draw the graph.

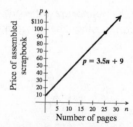

To estimate the cost of printing a 25-page scrapbook, we find the second coordinate associated with 25. Locate the point on the line that is above 25 and then find the value on the vertical axis that corresponds to that point. That value is about 96, so we estimate that it will cost $96 to print the scrapbook.

79. We graph $T = \frac{5}{4}c + 2$. Since credits cannot be negative in this application, we select only nonnegative values for c.

If $c = 0$, $T = \frac{5}{4} \cdot 0 + 2 = 0 + 2 = 2$.

If $c = 10$, $T = \frac{5}{4} \cdot 10 + 2 = 12.5 + 2 = 14.5$.

If $c = 20$, $T = \frac{5}{4} \cdot 20 + 2 = 25 + 2 = 27$.

c	T
0	2
10	14.5
20	27

We plot these points and draw the graph.

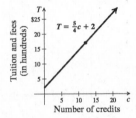

To estimate the cost of tuition and fees when a student registers for 4 three-credit courses, we find the second coordinate associated with 12 (4 three-credit courses equals 12 credits). Locate the point on the line that is above 12 and then find the value on the vertical axis that corresponds to that point. That value is 17, so we estimate that it will cost $1700 in tuition and fees to register for 4 three-credit courses.

81. We graph $T = -2m + 54$. Since time cannot be negative in this application, we select only nonnegative values for m.

If $m = 0$, $T = -2 \cdot 0 + 54 = 0 + 54 = 54$.

If $m = 10$, $T = -2 \cdot 10 + 54 = -20 + 54 = 34$.

If $m = 20$, $T = -2 \cdot 20 + 54 = -40 + 54 = 14$.

n	T
0	54
10	34
20	14

We plot the points and draw the graph.

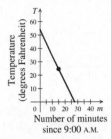

To estimate the Fahrenheit temperature in Spearfish at 9:15 A.M., we find the second coordinate associated with 15 (9:15 A.M. is 15 minutes past 9:00 A.M). Locate the point on the line that is above 15 and then find the value on the vertical axis that corresponds to that point. That value is 24, so we estimate that the temperature at 9:15 A.M. was 24° F.

83. *Thinking and Writing Exercise.*

85. $5x + 3 \cdot 0 = 12$

$5x + 0 = 12$

$5x = 12$

$x = \dfrac{12}{5}$

Check: $5x + 3 \cdot 0 = 12$

$$5 \cdot \frac{12}{5} + 3 \cdot 0 \;\bigg|\; 12$$

$$12 + 0 \;\bigg|$$

$$12 \;\bigg|\; 12 \quad \text{TRUE}$$

The solution is $\dfrac{12}{5}$.

87. $5x + 3(2 - x) = 12$

$5x + 6 - 3x = 12$

$2x + 6 - 6 = 12 - 6$

$2x = 6$

$x = 3$

Check: $5x + 3(2 - x) = 12$

$$5 \cdot 3 + 3(2 - 3) \;\bigg|\; 12$$

$$15 + 3(-1) \;\bigg|$$

$$12 \;\bigg|\; 12 \quad \text{TRUE}$$

The solution is 3.

89. $pt + p = w$

     $p(t+1) = w$ Distributive Law

     $p = \dfrac{w}{t+1}$ Dividing by $t+1$

91. $A = \dfrac{T+Q}{2}$

     $2A = T + Q$ Multiplying by 2

     $2A - T = Q$ Adding $-T$

93.   *Thinking and Writing Exercise.*

95.   Let s represent the gear that Lauren uses on the southbound portion of her ride and n represent the gear she uses on the northbound portion. Then we have $s + n = 18$. We graph this equation, using only positive integer values for s and n.

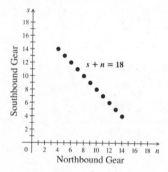

97.   Note that the sum of the coordinates of each point on the graph is 2. Thus, we have $x + y = 2$, or $y = -x + 2$.

99.   Note that when $x = 0$, $y = -5$ and when $y = 0$, $x = 3$. An equation that fits this situation is $5x - 3y = 15$, or $y = \dfrac{5}{3}x - 5$.

101. The equation is $25d + 5l = 225$.
Since the number of dinners cannot be negative, we choose only nonnegative values of d when graphing the equation. The graph stops at the horizontal axis since the number of lunches cannot be negative.

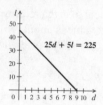

We see that three points on the graph are $(1, 40)$, $(5, 20)$, and $(8, 5)$. Thus, three combinations of dinners and lunches that total \$225 are

    1 dinner, 40 lunches
    5 dinners, 20 lunches
    8 dinners, 5 lunches

103. $y = -|x|$

x	y
-3	-3
-2	-2
-1	-1
0	0
1	-1
2	-2
3	-3

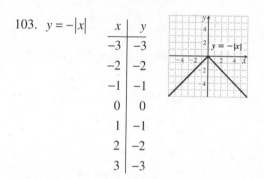

105. $y = -|x| + 2$

x	y
-3	-1
-2	0
-1	1
0	2
1	1
2	0
3	-1

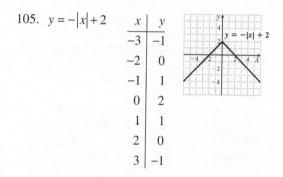

107. First find the fuel efficiency for a truck moving at 55 mph. Locating on the graph the point on the line that is above 55, and then finding the value on the t-axis that corresponds to this point, the fuel efficiency may be found to be about 7.6 mpg. Similarly, the fuel efficiency at 70 mph is about 6.1 mpg. To find the amount of gallons of fuel saved, divide 500 miles by each rate, then subtract the smaller number of gallons from the larger.

$$\frac{500}{6.1} - \frac{500}{7.6} = x$$

$$16.2 \approx x$$

So about 16.2 gallons are saved by the driver traveling at 55 mph. To find the savings in terms of dollars, multiply the gallons saved by the cost per gallon of fuel.
$$16.2 \cdot 3.50 = 56.70,$$
or about \$56.70 in savings.

Exercise Set 3.3

1. f

3. d

5. b

7. $5x - 3y = 15$
    The equation is already in the form $Ax + By = C$, so $5x - 3y = 15$ is a linear equation.

9. $7y = x - 5$
    We attempt to write this equation in the form $Ax + By = C$. We have
    $$7y = x - 5$$
    $$-x + 7y = -5$$
    Thus, $7y = x - 5$ is a linear equation.

11. $xy = 7$
    The x and y are multiplied. This equation cannot be written in the form $Ax + By = C$, so it is not a linear equation.

13. $16 + 4y = 0$
    Although there is no x term, we can rewrite this equation in the form $Ax + By = C$. We have
    $$16 + 4y = 0$$
    $$4y = -16$$
    $$0 \cdot x + 4y = -16$$
    Thus, $16 + 4y = 0$ is a linear equation.

15. $2y - \dfrac{3}{x} = 5$
    We attempt to write this equation in the form $Ax + By = C$. Multiplying both sides of the equation by x, we have
    $$2y - \dfrac{3}{x} = 5$$
    $$2xy - 3 = 5x$$

This is not a linear equation since we have a product of x and y.

17. The y-intercepts are those points whose first coordinate is 0. The y-intercept is $(0,5)$.
    The x-intercepts are those points whose second coordinate is 0. The x-intercept is $(2,0)$.

19. The y-intercepts are those points whose first coordinate is 0. The y-intercept is $(0,-4)$.
    The x-intercepts are those points whose second coordinate is 0. The x-intercept is $(3,0)$.

21. The y-intercepts are those points whose first coordinate is 0. The y-intercept is $(0,-2)$.
    The x-intercepts are those points whose second coordinate is 0. The x-intercepts are $(-3,0)$ and $(3,0)$.

23. The y-intercepts are those points whose first coordinate is 0. The y-intercept is $(0,4)$.
    The x-intercepts are those points whose second coordinate is 0. The x-intercepts are $(-3,0)$, $(3,0)$, and $(5,0)$.

25. $5x + 3y = 15$
    To find the y-intercept, we let $x = 0$ and solve for y:
    $$5 \cdot 0 + 3y = 15$$
    $$3y = 15$$
    $$y = 5$$
    Thus, the y-intercept is $(0,5)$.
    To find the x-intercept, we let $y = 0$ and solve for x:
    $$5x + 3 \cdot 0 = 15$$
    $$5x = 15$$
    $$x = 3$$
    Thus, the x-intercept is $(3,0)$.

27. $7x - 2y = 28$
    To find the y-intercept, we let $x = 0$ and solve for y:

$7 \cdot 0 - 2y = 28$

$-2y = 28$

$y = -14$

Thus, the y-intercept is $(0, -14)$.

To find the x-intercept, we let $y = 0$ and solve for x:

$7x - 2 \cdot 0 = 28$

$7x = 28$

$x = 4$

Thus, the x-intercept is $(4, 0)$.

29. $-4x + 3y = 150$

To find the y-intercept, we let $x = 0$ and solve for y:

$-4 \cdot 0 + 3y = 150$

$3y = 150$

$y = 50$

Thus, the y-intercept is $(0, 50)$.

To find the x-intercept, we let $y = 0$ and solve for x:

$-4x + 3 \cdot 0 = 150$

$-4x = 150$

$x = -\frac{75}{2}$

Thus, the x-intercept is $\left(-\frac{75}{2}, 0\right)$.

31. $y = 9$

y is equal to 9 for any x, so the y-intercept is $(0, 9)$.

The graph of this equation is a horizontal line passing through $(0, 9)$, so there is no x-intercept.

33. $x = -7$

x is equal to -7 for any y, so the x-intercept is $(-7, 0)$.

The graph of this equation is a vertical line passing through $(-7, 0)$, so there is no y-intercept.

35. $3x + 2y = 12$

To find the y-intercept, we let $x = 0$ and solve for y:

$3 \cdot 0 + 2y = 12$

$2y = 12$

$y = 6$

Thus, the y-intercept is $(0, 6)$.

To find the x-intercept, we let $y = 0$ and solve for x:

$3x + 2 \cdot 0 = 12$

$3x = 12$

$x = 4$

Thus, the x-intercept is $(4, 0)$.

Before drawing the line, we plot a third point as a check. We substitute any convenient value for x and solve for y. If we let $x = 2$, then

$3 \cdot 2 + 2y = 12$

$6 + 2y = 12$

$2y = 6$

$y = 3$

The point $(2, 3)$ appears to line up with the intercepts. To finish, we draw and label the line.

37. $x + 3y = 6$

To find the y-intercept, we let $x = 0$ and solve for y:

$0 + 3y = 6$

$3y = 6$

$y = 2$

Thus, the y-intercept is $(0, 2)$.

To find the x-intercept, we let $y = 0$ and solve for x:

$x + 3 \cdot 0 = 6$

$x = 6$

Thus, the x-intercept is $(6, 0)$.

Before drawing the line, we plot a third point as a check. We substitute any convenient value for x and solve for y. If we let $x = 3$, then

$3 + 3y = 6$

$3y = 3$

$y = 1$

The point $(3, 1)$ appears to line up with the intercepts. To finish, we draw and label the line.

39. $-x + 2y = 8$

To find the y-intercept, we let $x = 0$ and solve for y:
$$-0 + 2y = 8$$
$$2y = 8$$
$$y = 4$$

Thus, the y-intercept is $(0, 4)$.

To find the x-intercept, we let $y = 0$ and solve for x:
$$-x + 2 \cdot 0 = 8$$
$$-x = 8$$
$$x = -8$$

Thus, the x-intercept is $(-8, 0)$.

Before drawing the line, we plot a third point as a check. We substitute any convenient value for x and solve for y. If we let $x = -4$, then
$$-(-4) + 2y = 8$$
$$4 + 2y = 8$$
$$2y = 4$$
$$y = 2$$

The point $(-4, 2)$ appears to line up with the intercepts. To finish, we draw and label the line.

41. $3x + y = 9$

To find the y-intercept, we let $x = 0$ and solve for y:
$$3 \cdot 0 + y = 9$$
$$y = 9$$

Thus, the y-intercept is $(0, 9)$.

To find the x-intercept, we let $y = 0$ and solve for x:

$$3x + 0 = 9$$
$$3x = 9$$
$$x = 3$$

Thus, the x-intercept is $(3, 0)$.

Before drawing the line, we plot a third point as a check. We substitute any convenient value for x and solve for y. If we let $x = 2$, then
$$3 \cdot 2 + y = 9$$
$$6 + y = 9$$
$$y = 3$$

The point $(2, 3)$ appears to line up with the intercepts. To finish, we draw and label the line.

43. $y = 2x - 6$

To find the y-intercept, we let $x = 0$ and solve for y:
$$y = 2 \cdot 0 - 6$$
$$y = -6$$

Thus, the y-intercept is $(0, -6)$.

To find the x-intercept, we let $y = 0$ and solve for x:
$$0 = 2x - 6$$
$$6 = 2x$$
$$3 = x$$

Thus, the x-intercept is $(3, 0)$.

Before drawing the line, we plot a third point as a check. We substitute any convenient value for x and solve for y. If we let $x = 1$, then
$$y = 2 \cdot 1 - 6$$
$$y = 2 - 6$$
$$y = -4$$

The point $(1, -4)$ appears to line up with the intercepts. To finish, we draw and label the line.

45. $3x - 9 = 3y$

To find the y-intercept, we let $x = 0$ and solve for y:

$3 \cdot 0 - 9 = 3y$

$-9 = 3y$

$-3 = y$

Thus, the y-intercept is $(0, -3)$.

To find the x-intercept, we let $y = 0$ and solve for x:

$3x - 9 = 3 \cdot 0$

$3x - 9 = 0$

$3x = 9$

$x = 3$

Thus, the x-intercept is $(3, 0)$.

Before drawing the line, we plot a third point as a check. We substitute any convenient value for x and solve for y. If we let $x = 1$, then

$3 \cdot 1 - 9 = 3y$

$3 - 9 = 3y$

$-6 = 3y$

$-2 = y$

The point $(1, -2)$ appears to line up with the intercepts. To finish, we draw and label the line.

47. $2x - 3y = 6$

To find the y-intercept, we let $x = 0$ and solve for y:

$2 \cdot 0 - 3y = 6$

$-3y = 6$

$y = -2$

Thus, the y-intercept is $(0, -2)$.

To find the x-intercept, we let $y = 0$ and solve for x:

$2x - 3 \cdot 0 = 6$

$2x = 6$

$x = 3$

Thus, the x-intercept is $(3, 0)$.

Before drawing the line, we plot a third point as a check. We substitute any convenient value for x and solve for y. If we let $x = 6$, then

$2 \cdot 6 - 3y = 6$

$12 - 3y = 6$

$-3y = -6$

$y = 2$

The point $(6, 2)$ appears to line up with the intercepts. To finish, we draw and label the line.

49. $4x + 5y = 20$

To find the y-intercept, we let $x = 0$ and solve for y:

$4 \cdot 0 + 5y = 20$

$5y = 20$

$y = 4$

Thus, the y-intercept is $(0, 4)$.

To find the x-intercept, we let $y = 0$ and solve for x:

$4x + 5 \cdot 0 = 20$

$4x = 20$

$x = 5$

Thus, the x-intercept is $(5, 0)$.

Before drawing the line, we plot a third point as a check. We substitute any convenient value for x and solve for y. If we let $x = 2$, then

$4 \cdot 2 + 5y = 20$

$8 + 5y = 20$

$5y = 12$

$y = \dfrac{12}{5}$

The point $\left(2, \frac{12}{5}\right)$ appears to line up with the intercepts. To finish, we draw and label the line.

51. $3x + 2y = 8$

To find the y-intercept, we let $x = 0$ and solve for y:
$$3 \cdot 0 + 2y = 8$$
$$2y = 8$$
$$y = 4$$

Thus, the y-intercept is $(0, 4)$.

To find the x-intercept, we let $y = 0$ and solve for x:
$$3x + 2 \cdot 0 = 8$$
$$3x = 8$$
$$x = \frac{8}{3}$$

Thus, the x-intercept is $\left(\frac{8}{3}, 0\right)$.

Before drawing the line, we plot a third point as a check. We substitute any convenient value for x and solve for y. If we let $x = 4$, then
$$3 \cdot 4 + 2y = 8$$
$$12 + 2y = 8$$
$$2y = -4$$
$$y = -2$$

The point $(4, -2)$ appears to line up with the intercepts. To finish, we draw and label the line.

53. $2x + 4y = 6$

To find the y-intercept, we let $x = 0$ and solve for y:

$$2 \cdot 0 + 4y = 6$$
$$4y = 6$$
$$y = \frac{3}{2}$$

Thus, the y-intercept is $\left(0, \frac{3}{2}\right)$.

To find the x-intercept, we let $y = 0$ and solve for x:
$$2x + 4 \cdot 0 = 6$$
$$2x = 6$$
$$x = 3$$

Thus, the x-intercept is $(3, 0)$.

Before drawing the line, we plot a third point as a check. We substitute any convenient value for x and solve for y. If we let $x = -3$, then
$$2(-3) + 4y = 6$$
$$-6 + 4y = 6$$
$$4y = 12$$
$$y = 3$$

The point $(-3, 3)$ appears to line up with the intercepts. To finish, we draw and label the line.

55. $5x + 3y = 180$

To find the y-intercept, we let $x = 0$ and solve for y:
$$5 \cdot 0 + 3y = 180$$
$$3y = 180$$
$$y = 60$$

Thus, the y-intercept is $(0, 60)$.

To find the x-intercept, we let $y = 0$ and solve for x:
$$5x + 3 \cdot 0 = 180$$
$$5x = 180$$
$$x = 36$$

Thus, the x-intercept is $(36, 0)$.

Before drawing the line, we plot a third point as a check. We substitute any convenient value for x and solve for y. If we let $x = 20$, then

$$5 \cdot 20 + 3y = 180$$
$$100 + 3y = 180$$
$$3y = 80$$
$$y = \frac{80}{3}$$

The point $\left(20, \frac{80}{3}\right)$ appears to line up with the intercepts. To finish, we draw and label the line.

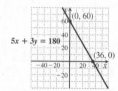

57. $y = -30 + 3x$

To find the y-intercept, we let $x = 0$ and solve for y:
$$y = -30 + 3 \cdot 0$$
$$y = -30$$

Thus, the y-intercept is $(0, -30)$.

To find the x-intercept, we let $y = 0$ and solve for x:
$$0 = -30 + 3x$$
$$-3x = -30$$
$$x = 10$$

Thus, the x-intercept is $(10, 0)$.

Before drawing the line, we plot a third point as a check. We substitute any convenient value for x and solve for y. If we let $x = 5$, then
$$y = -30 + 3 \cdot 5$$
$$y = -30 + 15$$
$$y = -15$$

The point $(5, -15)$ appears to line up with the intercepts. To finish, we draw and label the line.

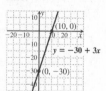

59. $-4x = 20y + 80$

To find the y-intercept, we let $x = 0$ and solve for y:

$$-4 \cdot 0 = 20y + 80$$
$$-20y = 80$$
$$y = -4$$

Thus, the y-intercept is $(0, -4)$.

To find the x-intercept, we let $y = 0$ and solve for x:
$$-4x = 20 \cdot 0 + 80$$
$$-4x = 80$$
$$x = -20$$

Thus, the x-intercept is $(-20, 0)$.

Before drawing the line, we plot a third point as a check. We substitute any convenient value for x and solve for y. If we let $x = -30$, then

$$-4(-30) = 20y + 80$$
$$120 = 20y + 80$$
$$40 = 20y$$
$$2 = y$$

The point $(-30, 2)$ appears to line up with the intercepts. To finish, we draw and label the line.

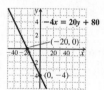

61. $y - 3x = 0$

To find the y-intercept, we let $x = 0$ and solve for y:
$$y - 3 \cdot 0 = 0$$
$$y = 0$$

Thus, the y-intercept is $(0, 0)$.

To find the x-intercept, we let $y = 0$ and solve for x:
$$0 - 3x = 0$$
$$-3x = 0$$
$$x = 0$$

Thus, the x-intercept is $(0, 0)$.

Before drawing the line, we need a second point. We substitute any convenient value for x and solve for y. If we let $x = 1$, then

$$y - 3 \cdot 1 = 0$$
$$y - 3 = 0$$
$$y = 3$$

The point $(1, 3)$ is our second point. A third point is $(2, 6)$, which lines up with the other two points. To finish, we draw and label the line.

63. $y = 5$

We can write this equation as $0 \cdot x + y = 5$.

No matter what number we choose for x, we find that y must be 5. Consider the following table.

x	y	(x, y)
-4	5	$(-4, 5)$
0	5	$(0, 5)$
4	5	$(4, 5)$

We plot the ordered pairs and connect the points.

65. $x = -1$

We can write this equation as $x + 0 \cdot y = -1$.

No matter what number we choose for y, we find that x must be -1. Consider the following table.

x	y	(x, y)
-1	-4	$(-1, -4)$
-1	0	$(-1, 0)$
-1	4	$(-1, 4)$

We plot the ordered pairs and connect the points.

67. $y = -15$

We can write this equation as $0 \cdot x + y = -15$.

No matter what number we choose for x, we find that y must be -15. Consider the following table.

x	y	(x, y)
-4	-15	$(-4, -15)$
0	-15	$(0, -15)$
4	-15	$(4, -15)$

We plot the ordered pairs and connect the points.

69. $y = 0$

We can write this equation as $0 \cdot x + y = 0$.

No matter what number we choose for x, we find that y must be 0. Consider the following table.

x	y	(x, y)
-4	0	$(-4, 0)$
0	0	$(0, 0)$
4	0	$(4, 0)$

We plot the ordered pairs and connect the points.

71. $x = -\frac{5}{2}$

We can write this equation as $x + 0 \cdot y = -\frac{5}{2}$.

No matter what number we choose for y, we find that x must be $-\frac{5}{2}$. Consider the following table.

x	y	(x, y)
$-\frac{5}{2}$	-4	$\left(-\frac{5}{2}, -4\right)$
$-\frac{5}{2}$	0	$\left(-\frac{5}{2}, 0\right)$
$-\frac{5}{2}$	4	$\left(-\frac{5}{2}, 4\right)$

We plot the ordered pairs and connect the points.

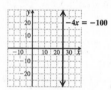

73. $-4x = -100$

We can write this equation as $x = 25$, or $x + 0 \cdot y = 25$. No matter what number we choose for y, we find that x must be 25. Consider the following table.

x	y	(x, y)
25	-4	$(25, -4)$
25	0	$(25, 0)$
25	4	$(25, 4)$

We plot the ordered pairs and connect the points.

75. $35 + 7y = 0$

We can write this equation as $7y = -35$, or $y = -5$, or $0 \cdot x + y = -5$. No matter what number we choose for x, we find that y must be -5. Consider the following table.

x	y	(x, y)
-4	-5	$(-4, -5)$
0	-5	$(0, -5)$
4	-5	$(4, -5)$

We plot the ordered pairs and connect the points.

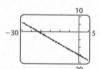

77. This graph is a horizontal line with a y-intercept of $(0, -1)$. The equation for this graph is $y = -1$.

79. This is a vertical line with x-intercept of $(4, 0)$. The equation for this graph is $x = 4$.

81. This is a horizontal line with a y-intercept of $(0, 0)$. The equation for this graph is $y = 0$.

83. $y = 20 - 4x$

To find the y-intercept, we let $x = 0$ and solve for y:
$$y = 20 - 4 \cdot 0$$
$$y = 20$$

Thus, the y-intercept is $(0, 20)$.

To find the x-intercept, we let $y = 0$ and solve for x:
$$0 = 20 - 4x$$
$$4x = 20$$
$$x = 5$$

Thus, the x-intercept is $(5, 0)$.

Viewing window $[-10, 10, -10, 30]$ will show both intercepts. This is choice (c).

85. $y = -35x + 7000$

To find the y-intercept, we let $x = 0$ and solve for y:
$$y = -35 \cdot 0 + 7000$$
$$y = 7000$$

Thus, the y-intercept is $(0, 7000)$.

To find the x-intercept, we let $y = 0$ and solve for x:
$$0 = -35x + 7000$$
$$35x = 7000$$
$$x = 200$$

Thus, the x-intercept is $(200, 0)$.

Viewing window $[0, 500, 0, 10,000]$ will show both intercepts. This is choice (d).

87. $y = -0.72x - 15$

Xscl = 5, Yscl = 5

89. $5x + 6y = 84$, or $y = (84 - 5x)/6$

Xscl = 2, Yscl = 2

91. $19x - 17y = 200$, or $y = (19x - 200)/17$

Xscl = 5, Yscl = 5

93. *Thinking and Writing Exercise.*

95. 7 less than d
     $d - 7$

97. The sum of 7 and four times a number.
     Let n represent the number. We have $7 + 4n$,
     or $4n + 7$.

99. Twice the sum of two numbers.
     Let x and y represent the numbers. We have
     $2(x + y)$.

101. *Thinking and Writing Exercise.*

103. The x-axis would correspond to a horizontal
     line, and all horizontal lines are of the form
     $y = b$, where the y-intercept is $(0, b)$. The y-
     intercept of the x-axis would be the point
     $(0, 0)$, so $b = 0$. Thus, the equation for the x-
     axis would be $y = 0$.

105. A line parallel to the y-axis through the point
     $(-2, 7)$ would be a vertical line having x-
     intercept $(-2, 0)$. So, the equation would be
     $x = -2$.

107. The graph of the equation $y = x$ is a line
     running through the origin and containing the
     points (c, c) (e.g., $(-3, -3)$, $(-1, -1)$, $(5, 5)$,
     etc.). The graph of the equation $x = -3$ is a
     vertical line with x-intercept of $(-3, 0)$.
     These lines would intersect at $(-3, -3)$.

109. Note that when $x = 0$, $y = 5$, and when
     $y = 0$, $x = -3$. An equation that fits this
     situation is $-5x + 3y = 15$, or $y = \frac{5}{3}x + 5$.

111. Given the equation $4x = C - 3y$, we
     determine the y-intercept by letting $x = 0$ and
     solving for y.
     $4 \cdot 0 = C - 3y$

     $0 = C - 3y$

     $3y = C$

     $y = \dfrac{C}{3}$

     If the y-intercept is $(0, -8)$, then $\dfrac{C}{3} = -8$, or

     $C = -24$.

Exercise Set 3.4

---

1. The units of rate for 100 miles in 5 hours are
    miles per hour, or miles/hour.

3. The units of rate for \$300 for 150 miles
    are dollars per mile, or dollars/mile.

5. The units of rate for 40 minutes used to run 8
    errands are minutes per errand, or
    minutes/errand.

7. a) Rate, in miles per gallon
        $= \dfrac{14,014 \text{ mi} - 13,741 \text{ mi}}{13 \text{ gal}}$

        $= \dfrac{273 \text{ mi}}{13 \text{ gal}}$

        $= 21 \text{ mi/gal}$
        The rate is 21 mpg.

    b) From June 5 to June 8 is $8 - 5 = 3$ days.
        Average cost, in dollars per day
        $= \dfrac{\$118}{3 \text{ days}} = \$39.3\overline{3}/\text{day}$
        The average cost per day is
        approximately \$39.33/day.

    c) The rate of travel, in miles per day
        $= \dfrac{273 \text{ mi}}{3 \text{ days}} = 91 \text{ mi/day}$
        The rate of travel is 91 mi/day.

    d) Total rental was \$118, or $118 \times 100 =$
        11800¢. The total miles driven was 273
        miles.
        The rental rate, in cents per mile
        $= \dfrac{11800¢}{273 \text{ mi}} = 43.22344322¢/\text{mi}$
        The rental rate is approximately 43¢/mi.

9. a) Perry rented the bike from 2:00 to 5:00, for a total of 3 hours. His average speed, in miles per hour

$$= \frac{18 \text{ mi}}{3 \text{ hr}} = 6 \text{ mi/hr}$$

Perry's average speed was 6 mph.

b) The rental rate, in dollars per hour

$$= \frac{\$12}{3 \text{ hr}} = \$4/\text{hr}$$

The rental rate was \$4/hr.

c) The rental rate, in dollars per mile

$$= \frac{\$12}{18 \text{ mi}} = \$0.6\overline{6}/\text{mi}$$

The rental rate was \$0.67/mi.

11. a) The proofreader worked 5 hours (from 9 a.m. until 2 p.m. is 5 hours). The rate of pay, in dollars per hour

$$= \frac{\$110}{5 \text{ hr}} = \$22/\text{hr}$$

The rate of pay is \$22/hr.

b) The proofreader read 103 pages (from the top of page 93 to the end of page 195, which is to say the top of page 196, or $196 - 93 = 103$). The rate, in number of pages per hour

$$= \frac{103 \text{ pages}}{5 \text{ hr}} = 20.6 \text{ pages/hr}$$

The rate is 20.6 pages/hr.

c) The rate of pay, in dollars per page

$$= \frac{\$110}{103 \text{ pages}} \approx \$1.07/\text{page}$$

The rate of pay is approximately \$1.07/page.

13. The rate at which the price of an LCD television decreased

$$= \frac{\$460 - \$700}{18 \text{ months}} = \frac{-\$240}{18 \text{ months}}$$

$$= -\$13.33/\text{month}$$

The price decreased at a rate of \$13.33/month.

15. Peter spent a total of 2 min (from 2:38 until 2:40 is 2 minutes). He descended 29 floors (34th floor to the 5th floor is 29 floors).

a) The average rate of travel, in floors per minute

$$= \frac{29 \text{ floors}}{2 \text{ min}} = 14.5 \text{ floors/min}$$

The average rate of travel is 14.5 floors/min.

b) Note that 2 minutes = 120 seconds. The average rate of travel, in seconds per floor

$$= \frac{120 \text{ sec}}{29 \text{ floors}} = 4.137931034 \text{ sec/floor}$$

The average rate of travel is approximately 4.14 sec/floor.

17. a) The total of time of ascent was 8 hours, 10 minutes, or $8 \times 60 + 10 = 490$ minutes. The average rate of ascent

$$= \frac{29{,}035 \text{ ft} - 17{,}552 \text{ ft}}{490 \text{ min}}$$

$$= \frac{11{,}483 \text{ ft}}{490 \text{ min}} \approx 23.4347 \text{ ft/min}$$

The average rate of ascent was approximately 23.43 ft/min.

b) The average rate of ascent, in minutes per foot

$$= \frac{490 \text{ min}}{11{,}483 \text{ ft}} \approx 0.042672 \text{ min/ft}$$

The average rate of ascent was approximately 0.04 min/ft.

19. The rate is given in pounds per person per year, so we list the amount of paper, in pounds per person, on the vertical axis and the year on the horizontal axis. If we count by tens of pounds per person on the vertical axis, we can easily reach 340 pounds per person without needing a very large graph. A jagged line at the base of the axis indicates that we are not showing amounts smaller than 300. We plot the point $(2003, 340)$ and then move to a point that represents 5 lb per person more 1 year later. The coordinates of this point are $(2003 + 1, 340 + 5)$, or $(2004, 345)$. Finally, we draw a line through the points.

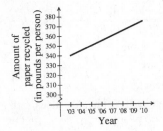

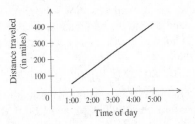

21. The rate is given in crimes per year, so we list number of crimes, in millions, on the vertical axis and years on the horizontal axis. If we count by 0.2's of millions on the vertical axis, we can easily reach 10 million without needing a very large graph. A jagged line at the base of the axis indicates that we are not showing amounts smaller than 9 million. We plot the point $(2006, 10 \text{ million})$. Then to display the rate of growth, we move to a point that represents 0.1 million fewer crimes 1 year later. The coordinates of this point are $(2006+1, 10 \text{ million} - 0.1 \text{ million})$, or $(2007, 9.9 \text{ million})$. Finally, we draw a line through these two points.

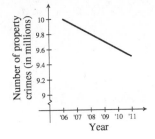

23. The rate is given in miles per hour, so we list the number of miles traveled on the vertical axis and the time of day on the horizontal axis. If we count by 100's of miles on the vertical axis we can easily reach 230 without needing a very large graph. We plot the point $(3:00, 230)$. Then to display the rate of travel, we move from that point to a point that represents 90 more miles traveled 1 hour later. The coordinates of this point are $(3:00+1, 230+90)$, or $(4:00, 320)$. Finally, we draw a line through the two points.

25. The rate is given in dollars per hour so we list money earned on the vertical axis and time of day on the horizontal axis. We can count by 20's on the vertical axis and reach \$50 without needing a terribly large graph. Next we plot the point $(2:00, 50)$. To display the rate we move from that point to a point that represents \$15 more 1 hour later. The coordinates of this point are $(2:00+1, 50+15)$, or $(3:00, 65)$. Finally, we draw a line through the two points.

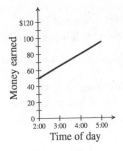

27. The rate is given in cost per minute so we list the amount of the telephone bill on the vertical axis and the number of additional minutes on the horizontal axis. We begin with \$7.50 on the vertical axis and count by \$0.50. A jagged line at the base of the axis indicates that we are not showing amounts smaller than \$7.50. We begin with 0 additional minutes on the horizontal axis and plot the point $(0, 7.50)$. We move from there to a point that represents \$0.10 more 1 minute later. The coordinates of this point are $(0+1, 7.50+0.10)$, or $(1, 7.60)$. Then we draw a line through the two points.

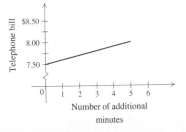

29. Because the vertical axis shows the number of calls handled and the horizontal axis lists the time in hour-long increments, we can find the rate, *number of calls handled per hour*. The points $(10:00 \text{ A.M.}, 30)$ and $(1:00 \text{ P.M.}, 90)$ are both on the graph. This tells us that in the 3 hours between 10:00 A.M. and 1:00 P.M., there were $90 - 30 = 60$ calls handled. Thus the rate is

$$\frac{90 \text{ calls} - 30 \text{ calls}}{1:00 \text{ P.M.} - 10:00 \text{ A.M.}} = \frac{60 \text{ calls}}{3 \text{ hours}}$$
$$= 20 \text{ calls/hr}$$

The calls are handled at a rate of 20 calls/hr.

31. Because the vertical axis shows the distance from Chicago, in miles and the horizontal axis lists the time in hour-long increments, we can find the rate, *miles per hour*. The points $(12 \text{ noon}, 100)$ and $(2:00 \text{ P.M.}, 250)$ are both on the graph. This tells us that in the 2 hours between 12:00 Noon and 2:00 P.M., there were $250 - 100 = 150$ miles traveled. Thus the rate is

$$\frac{250 \text{ miles} - 100 \text{ miles}}{2:00 \text{ P.M.} - 12:00 \text{ Noon}} = \frac{150 \text{ miles}}{2 \text{ hours}}$$
$$= 75 \text{ miles/hr}$$

The train traveled at a rate of 75 mi/hr.

33. Because the vertical axis shows the cost of phone calls, in cents, and the horizontal axis lists the length of phone calls, in minutes, we can find the rate, *cents per minute*. The points $(5, 60)$ and $(10, 120)$ are both on the graph. This tells us that when phone calls increase from 5 min to 10 min, the cost increases from 60¢ to 120¢. Thus the rate is

$$\frac{120¢ - 60¢}{10 \text{ min} - 5 \text{ min}} = \frac{60¢}{5 \text{ min}} = 12¢ / \text{min}$$

The customer was billed 12¢/min.

35. Because the vertical axis shows the population of Youngstown, in thousands of people, and the horizontal axis lists time, in given years, we can find the rate, *thousands of people per year*. The points $(1970, 140,000)$ and $(2000, 80,000)$ are both on the graph. This tells us that during the years from 1970 to 2000, the population of Youngstown decreased from 140 thousand people to 80 thousand people. Thus the rate is

$$\frac{80,000 \text{ people} - 140,000 \text{ people}}{2000 - 1970}$$
$$= \frac{-60,000 \text{ people}}{30 \text{ years}}$$
$$= -2000 \text{ people/yr}$$

The population of Youngstown changed at a rate of −2000 people/yr, or decreased by 2000 people/yr.

37. Because the vertical axis shows the gasoline consumed, in gallons, and the horizontal axis shows number of miles driven, we can find the rate, *gallons per mile*. The points $(80, 2)$ and $(240, 6)$ are both on the graph.

This tells us that as the miles driven increased from 80 miles to 240 miles, the number of gallons consumed increased from 2 gallons to 6 gallons. Thus the rate is

$$\frac{6 \text{ gallons} - 2 \text{ gallons}}{240 \text{ miles} - 80 \text{ miles}} = \frac{4 \text{ gallons}}{160 \text{ miles}}$$
$$= 0.025 \text{ gal/mi}$$

The vehicle was consuming gas at a rate of 0.025 gal/mi.

39. e

41. d

43. b

45. *Thinking and Writing Exercise.*

47. $-2 - (-7) = -2 + 7 = 5$

49. $\dfrac{5 - (-4)}{-2 - 7} = \dfrac{5 + 4}{-2 - 7} = \dfrac{9}{-9} = -1$

51. $\dfrac{-4 - 8}{7 - (-2)} = \dfrac{-4 - 8}{7 + 2} = \dfrac{-12}{9} = -\dfrac{4}{3}$

53. $\dfrac{-6 - (-6)}{-2 - 7} = \dfrac{-6 + 6}{-2 - 7} = \dfrac{0}{-9} = 0$

55. *Thinking and Writing Exercise.*

57.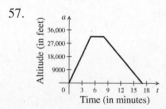

Locate the point on the horizontal axis corresponding to $2700. We locate the point on the graph that is directly above this point and move horizontally to the vertical scale and determine that approximately $540 in wages are paid to this salesperson.

59.

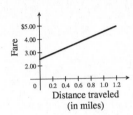

61. First determine Penny's rate with respect to the tugboat. The rate is

$$= \frac{24 \text{ ft}}{3 \text{ sec}} = 8 \text{ ft/sec}.$$

The tugboat is moving at a rate of 5 ft/sec. So, Penny's rate with respect to land is the sum of these two rates, or

$$\text{Rate}_{\text{land}} = 8 \text{ ft/sec} + 5 \text{ ft/sec} = 13 \text{ ft/sec}.$$

63. Zoe ran a total distance of 3 km (distance from the 4-km mark to the 7-km mark is 7 km − 4 km, or 3 km). She ran this distance in 15.5 min. We can compute Zoe's rate, in mi/min:

$$\frac{7 \text{ km} - 4 \text{ km}}{15.5 \text{ min}} = \frac{3 \text{ km}}{15.5 \text{ min}} \times \frac{0.62 \text{ mi}}{1 \text{ km}}$$
$$= 0.12 \text{ mi/min}$$

To determine how long it would take Zoe to run a 5-mi race, we note that

$$\text{Time} = \frac{\text{Distance}}{\text{Rate}}.$$

So

$$\frac{5 \text{ mi}}{0.12 \text{ mi/min}} = 41.6\overline{6} \text{ min}.$$

It would take Zoe approximately 41.7 min to run a 5-mile race.

65. First determine the rate at which Camden and Natalie can make candles, in candles per minute.

$$\text{rate} = \frac{100 \text{ candles} - 46 \text{ candles}}{5:00 - 3:00}$$
$$= \frac{54 \text{ candles}}{2 \text{ hr}} \times \frac{1 \text{ hr}}{60 \text{ min}}$$
$$= 0.45 \text{ candles/min}$$

Let t = the number of minutes past 3:00 P.M. Then the number of candles made, N, can be represented by

$N = 0.45t + 46$.

Let $N = 82$ and solve for t.

$$82 = 0.45t + 46$$
$$36 = 0.45t$$
$$\frac{36}{0.45} = t$$
$$80 = t$$

So, 80 minutes after 3:00 P.M., or 1 hr and 20 minutes after 3:00, or at 4:20, Camden and Natalie had made their 82nd candle.

Exercise Set 3.5

1. Positive

3. Negative

5. Positive

7. Zero

9. Negative

11. The rate can be found using the coordinates of any two points on the line. We use $(2006, 298)$ and $(2010, 310)$.

$$\text{Rate} = \frac{\text{change in the U.S. population}}{\text{corresponding change in time}}$$
$$= \frac{310 \text{ million} - 298 \text{ million}}{2010 - 2006}$$
$$= \frac{12 \text{ million}}{4 \text{ years}} = 3 \text{ million people/yr}$$

13. The rate can be found using the coordinates of any two points on the line. We use $(1940, 4.1)$ and $(1980, 3.8)$.

$$\text{Rate} = \frac{\text{change in running times}}{\text{corresponding change in years}}$$
$$= \frac{3.8 \text{ minutes} - 4.1 \text{ minutes}}{1980 - 1940}$$
$$= \frac{-0.3 \text{ minutes}}{40 \text{ years}}$$
$$= -0.0075 \text{ min/yr}$$
$$= -0.0075 \frac{\text{min}}{\text{yr}} \times \frac{60 \text{ sec}}{1 \text{ min}} = -0.45 \text{ sec/yr}$$

15. The rate can be found using the two coordinates on the line. We use $(60{,}000,\ 500)$ and $(100{,}000,\ 530)$.

Rate

$= \dfrac{\text{change in SAT math scores}}{\text{corresponding change in family income}}$

$= \dfrac{530\ \text{points} - 500\ \text{points}}{\$100{,}000 - \$60{,}000}$

$= \dfrac{30\ \text{points}}{\$40{,}000} = \tfrac{3}{4}\ \text{point/\$1000}$

17. The rate can be found using the two coordinates on the line. We use $(0, 54)$ and $(27, -4)$.

Rate

$= \dfrac{\text{change in degrees Fahrenheit}}{\text{corresponding change in time}}$

$= \dfrac{(-4)^{\circ}\,\text{F} - 54^{\circ}\,\text{F}}{27 - 0\ \text{min}} = \dfrac{-58^{\circ}\,\text{F}}{27\ \text{min}}$

$= -2.\overline{148}^{\circ}\,\text{F/min} \approx -2.1^{\circ}\ \text{F/min}$

19. We can use any two points on the line such as $(-4, -2)$ and $(4, 4)$.

$m = \dfrac{\text{change in } y}{\text{change in } x}$

$= \dfrac{-2 - 4}{-4 - 4} = \dfrac{-6}{-8} = \dfrac{3}{4}$

21. We can use any two points on the line such as $(-3, -4)$ and $(3, -2)$.

$m = \dfrac{\text{change in } y}{\text{change in } x}$

$= \dfrac{-4 - (-2)}{-3 - 3} = \dfrac{-4 + 2}{-6} = \dfrac{-2}{-6} = \dfrac{1}{3}$

23. We can use any two points on the line such as $(-3, 5)$ and $(4, -2)$.

$m = \dfrac{\text{change in } y}{\text{change in } x}$

$= \dfrac{5 - (-2)}{-3 - 4} = \dfrac{5 + 2}{-7} = \dfrac{7}{-7} = -1$

25. We can use any two points on the line such as $(-5, 4)$ and $(3, 4)$.

$m = \dfrac{\text{change in } y}{\text{change in } x}$

$= \dfrac{4 - 4}{-5 - 3} = \dfrac{0}{-8} = 0$

27. We can use any two points on the line such as $(2, 4)$ and $(5, -2)$.

$m = \dfrac{\text{change in } y}{\text{change in } x}$

$= \dfrac{4 - (-2)}{2 - 5} = \dfrac{4 + 2}{-3} = \dfrac{6}{-3} = -2$

29. We can use any two points on the line such as $(-3, 4)$ and $(-3, -4)$.

$m = \dfrac{\text{change in } y}{\text{change in } x}$

$= \dfrac{4 - (-4)}{-3 - (-3)} = \dfrac{4 + 4}{-3 + 3} = \dfrac{8}{0}$

Since division by 0 is undefined, the slope of this line is undefined.

31. We can use any two points on the line such as $(-2, 3)$ and $(2, 2)$.

$m = \dfrac{\text{change in } y}{\text{change in } x}$

$= \dfrac{3 - 2}{-2 - 2} = \dfrac{1}{-4} = -\dfrac{1}{4}$

33. We can use any two points on the line such as $(2, -4)$ and $(3, 1)$.

$m = \dfrac{\text{change in } y}{\text{change in } x}$

$= \dfrac{-4 - 1}{2 - 3} = \dfrac{-5}{-1} = 5$

35. We can use any two points on the line such as $(-5, -1)$ and $(5, 3)$.

$m = \dfrac{\text{change in } y}{\text{change in } x}$

$= \dfrac{3 - (-1)}{5 - (-5)} = \dfrac{3 + 1}{5 + 5} = \dfrac{4}{10} = \dfrac{2}{5}$

37. $(1, 2)$ and $(5, 8)$

$m = \dfrac{8 - 2}{5 - 1} = \dfrac{6}{4} = \dfrac{3}{2}$

39. $(-2,4)$ and $(3,0)$

$$m = \frac{0-4}{3-(-2)} = \frac{-4}{3+2} = \frac{-4}{5} = -\frac{4}{5}$$

41. $(-4,0)$ and $(2,3)$

$$m = \frac{3-0}{2-(-4)} = \frac{3-0}{2+4} = \frac{3}{6} = \frac{1}{2}$$

43. $(0,7)$ and $(-3,10)$

$$m = \frac{10-7}{-3-0} = \frac{3}{-3} = -1$$

45. $(-2,3)$ and $(-6,5)$

$$m = \frac{5-3}{-6-(-2)} = \frac{2}{-6+2} = \frac{2}{-4} = -\frac{1}{2}$$

47. $\left(-2,\frac{1}{2}\right)$ and $\left(-5,\frac{1}{2}\right)$

Note that the y-coordinates of both points are the same, so the points lie on a horizontal line which has slope 0. We could have solved for the slope as follows:

$$m = \frac{\frac{1}{2}-\frac{1}{2}}{-5-(-2)} = \frac{0}{-5+2} = \frac{0}{-3} = 0$$

49. $(5,-4)$ and $(2,-7)$

$$m = \frac{-7-(-4)}{2-5} = \frac{-7+4}{2-5} = \frac{-3}{-3} = 1$$

51. $(6,-4)$ and $(6,5)$

$$m = \frac{5-(-4)}{6-6} = \frac{5+4}{0} = \frac{9}{0}$$

Since division by zero is undefined, the slope of this line is undefined.

53. The line $y = 3$ is a horizontal line. The slope is 0.

55. The line $x = -1$ is a vertical line. The slope is undefined.

57. The line $x = 9$ is a vertical line. The slope is undefined.

59. The line $y = -9$ is a horizontal line. The slope is 0.

61. The grade is expressed as a percent.

$$m = \frac{106 \text{ m}}{1325 \text{ m}} = 0.08 = 8\%$$

63. The rise is 1 ft and the run is 20 ft.

$$m = \frac{1 \text{ ft}}{20 \text{ ft}} = 0.05 = 5\%$$

65. The rise is 2 ft 5 in, or $2 \times 12 + 5 = 29$ in., and the run is 8 ft 2 in., or $8 \times 12 + 2 = 98$ in.

$$m = \frac{29 \text{ in.}}{98 \text{ in.}} = \frac{29}{98} \approx 30\%$$

67. The rise is 5400 ft – 3500 ft, or 1900 ft, and the run is 37,000 ft.

$$m = \frac{1900 \text{ ft}}{37,000 \text{ ft}} \approx 0.05135 \approx 5.1\%$$

A rated climb for the Tour de France must be at least 4%. Because the Dooley Mountain ascent has a grade of 5.1%, it would be a rated climb if it were a part of the Tour de France.

69. a) Graph II indicates that 200 ml of fluid was dripped in the first 3 hr, a rate of $\frac{200}{3}$ ml/hr. It also indicates that 400 ml of fluid was dripped in the next 3 hours, a rate of $\frac{400}{3}$ ml/hr, and that this rate continues until the end of the time period shown. Since the rate of $\frac{400}{3}$ ml/hr is double the rate of $\frac{200}{3}$ ml/hr, this graph is appropriate for the given situation.

   b) Graph IV indicates that 300 ml of fluid was dripped in the first 2 hr, a rate of $\frac{300}{2}$, or 150 ml/hr. In the next 2 hr, 200 ml was dripped. This is a rate of $\frac{200}{2}$, or 100 ml/hr. Then 100 ml was dripped in the next 3 hours, a rate of $\frac{100}{3}$, or $33\frac{1}{3}$ ml/hr. Finally, in the remaining 2 hr, 0 ml of fluid was dripped, a rate of $\dfrac{0}{2}$, or 0 ml/hr. Since the rate at which the fluid was given decreased as time progressed and eventually became 0, this graph is appropriate for the given situation.

c) Graph I is the only graph that shows a constant rate for 5 hours, in this case from 3 PM to 8 PM. Thus, it is appropriate for the given situation.

d) Graph III indicates that 100 ml of fluid was dripped in the first 4 hr, a rate of $\dfrac{100}{4}$, or 25 ml/hr. In the next 4 hr, 300 ml was dripped. This is a rate of $\dfrac{300}{4}$, or 75 ml/hr. In the last hour 200 ml was dripped, a rate of 200 ml/hr. Since the rate at which the fluid was given gradually increased, this graph is appropriate for the given situation.

71. *Thinking and Writing Exercise.*

73. $ax + by = c$

$by = c - ax$ Adding $-ax$ to both sides

$y = \dfrac{c - ax}{b}$ Dividing both sides by b

75. $ax - by = c$

$ax = c + by$ Adding by to both sides

$ax - c = by$ Adding $-c$ to both sides

$\dfrac{ax - c}{b} = y$ Dividing both sides by b

77. $8x + 6y = 24$

To find the y-intercept, we let $x = 0$ and solve for y:

$8 \cdot 0 + 6y = 24$

$6y = 24$

$y = 4$

Thus, the y-intercept is $(0, 4)$.

To find the x-intercept, we let $y = 0$ and solve for x:

$8x + 6 \cdot 0 = 24$

$8x = 24$

$x = 3$

Thus, the x-intercept is $(3, 0)$.

Before drawing the line, we plot a third point as a check. We substitute any convenient value for x and solve for y. If we let $x = 1\frac{1}{2}$, then

$8 \cdot \left(1\frac{1}{2}\right) + 6y = 24$

$12 + 6y = 24$

$6y = 12$

$y = 2$

The point $\left(1\frac{1}{2}, 2\right)$ appears to line up with the intercepts. To finish, we draw and label the line.

79. *Thinking and Writing Exercise.*

81. If the line passes through $(4, -7)$ and never enters the first quadrant, then it slants down from left to right or is horizontal. This means that its slope is not positive $(m \le 0)$. The line will slant most steeply if it passes through $(0, 0)$. In this case, $m = \dfrac{-7 - 0}{4 - 0} = -\dfrac{7}{4}$.

Thus, the numbers the line could have for its slope are $\left\{ m \left| -\dfrac{7}{4} \le m \le 0 \right. \right\}$.

83. $x + y = 18$

$y = 18 - x$

The slope is $\dfrac{y}{x}$, or $\dfrac{18 - x}{x}$.

85.   Let $t =$ the number of units each tick mark on the vertical axis represents. Note that the graph drops 1 unit for every 6 units of horizontal change. Then we have:

$\dfrac{-1}{6t} = -\dfrac{2}{3}$

$-1 = -4t$ Multiplying by $6t$

$\dfrac{1}{4} = t$ Dividing by -4

Each tick mark on the vertical axis represents $\dfrac{1}{4}$ unit.

87.   The rise is 61 cm and the run is 167.6 cm.

$m = \dfrac{61 \text{ cm}}{167.6 \text{ cm}} = 0.363961813$

≈ 0.364, or 36.4%

Mid-Chapter Review

Guided Solutions

1.   For the line $y - 3x = 6$, the intercepts are

$y\text{-}intercept$: $y - 3 \cdot 0 = 6$

$y = 6$

The y-intercept $(0, 6)$.

$x\text{-}intercept$: $0 - 3x = 6$

$-3x = 6$

$x = -2$

The x-intercept $(-2, 0)$.

2.   For the line containing $(1, 5)$ and $(3, -1)$, the slope is

$m = \dfrac{y_2 - y_1}{x_2 - x_1} = \dfrac{-1 - 5}{3 - 1} = \dfrac{-6}{2} = -3$

Mixed Review

1.   Starting at the origin, $(0, -3)$ is 0 units right or left and 3 units down.

2.   Since the first coordinate is positive and the second coordinate is negative, the point $(4, -15)$ is in quadrant IV.

3.   We substitute -2 for x and -3 for y (alphabetical order of variables).

$y = 5 - x$

$\begin{array}{c|c} -3 & 5 - (-2) \\ & 5 + 2 \\ \hline -3 & 7 \qquad \text{FALSE} \end{array}$

Since $-3 = 7$ is false, the pair $(-2, -3)$ is not a solution.

4. $y = x - 3$

When $x = 0$, $y = 0 - 3 = -3$.

When $x = 3$, $y = 3 - 3 = 0$.

When $x = 5$, $y = 5 - 3 = 2$.

x	y
0	-3
3	0
5	2

Plot these points, draw the line they determine, and label the graph $y = x - 3$.

5. $y = -3x$

When $x = -1$, $y = -3(-1) = 3$.

When $x = 0$, $y = -3 \cdot 0 = 0$.

When $x = 2$, $y = -3 \cdot 2 = -6$.

x	y
-1	3
0	0
2	-6

Plot these points, draw the line they determine, and label the graph $y = -3x$.

6. $3x - y = 2$

$3x - y - 2 = 2 - 2$

$3x - y + y - 2 = 0 + y$

$3x - 2 = y$

When $x = -1$, $y = 3(-1) - 2 = -3 - 2 = -5$.

When $x = 0$, $y = 3 \cdot 0 - 2 = 0 - 2 = -2$.

When $x = 2$, $y = 3 \cdot 2 - 2 = 6 - 2 = 4$.

x	y
-1	-5
0	-2
2	4

Plot these points, draw the line they determine, and label the graph $3x - y = 2$.

7. $4x - 5y = 20$

When $x = 0$, $y = \dfrac{20}{-5} = -4$.

When $y = 0$, $x = \dfrac{20}{4} = 5$.

When $x = 2\frac{1}{2}$,

$$y = \frac{20 - 4 \cdot \left(2\frac{1}{2}\right)}{-5} = \frac{20 - 10}{-5} = \frac{10}{-5} = -2.$$

x	y
0	-4
$2\frac{1}{2}$	-2
5	0

Plot these points, draw the line they determine, and label the graph $4x - 5y = 20$.

8. $y = -2$

We can write this equation as $0 \cdot x + y = -2$. No matter what number we choose for x, we find that $y = -2$. Consider the following table.

x	y	(x, y)
-4	-2	$(-4, -2)$
0	-2	$(0, -2)$
4	-2	$(4, -2)$

We plot the ordered pairs and connect the points.

9. $x = 1$

We can write this equation as $x + 0 \cdot y = 1$.

No matter what number we choose for y, we find that $x = 1$. Consider the following table.

x	y	(x, y)
1	-4	$(1, -4)$
1	0	$(1, 0)$
1	4	$(1, 4)$

We plot the ordered pairs and connect the points.

10.

$$y = 2x^2 + x$$

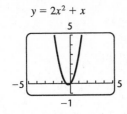

11. $4x - 5y = 20$

We can write this equation in the form $Ax + By = C$, as $4x + (-5)y = 20$, so it is a linear equation.

12. $y = 2x^2 + x$

Since the x is squared, this is not a linear equation.

13. Construction Builders had worked 2 months (from the end of June to the end of August, or all of July and August). In that time, they winterized a total of $38 - 10 = 28$ homes. The rate of winterization, in homes per month,

$$= \frac{28 \text{ homes}}{2 \text{ months}} = 14 \text{ homes/month}$$

The rate was 14 homes/month.

14. The rise is 14,255 ft – 9600 ft, or 4655 ft, and the run is 15,840 ft.

$$m = \frac{4655 \text{ ft}}{15,840 \text{ ft}} \approx 0.293876 \approx 29\%$$

15. $(-5, -2)$ and $(1, 8)$

$$m = \frac{8 - (-2)}{1 - (-5)} = \frac{8 + 2}{1 + 5} = \frac{10}{6} = \frac{5}{3}$$

16. $(1, 2)$ and $(4, -7)$

$$m = \frac{-7 - 2}{4 - 1} = \frac{-9}{3} = -3$$

17. $(0, 0)$ and $(0, -2)$

$$m = \frac{0 - (-2)}{0 - 0} = \frac{2}{0}$$

Since division by zero is undefined, the slope of this line is undefined.

18. The line $y = 4$ is a horizontal line. The slope is 0.

19. The line $x = -7$ is a vertical line. The slope is undefined.

20. $2y - 3x = 12$

To find the x-intercept, we let $y = 0$ and solve for x:
$$2 \cdot 0 - 3x = 12$$
$$-3x = 12$$
$$x = -4$$

Thus, the x-intercept is $(-4, 0)$.

To find the y-intercept, we let $x = 0$ and solve for y:
$$2y - 3 \cdot 0 = 12$$
$$2y = 12$$
$$y = 6$$

Thus, the y-intercept is $(0, 6)$.

Exercise Set 3.6

1. f

3. d

5. e

7. Slope $\frac{2}{5}$; y-intercept $(0, 1)$

We plot $(0, 1)$ and from there move up 2 units and right 5 units. This locates the point $(5, 3)$. We plot $(5, 3)$ and draw a line passing through $(0, 1)$ and $(5, 3)$.

9.   Slope $\frac{5}{3}$; y-intercept $(0,-2)$

We plot $(0,-2)$ and from there move up 5 units and right 3 units. This locates the point $(3,3)$. We plot $(3,3)$ and draw a line passing through $(0,-2)$ and $(3,3)$.

11.   Slope $-\frac{1}{3}$; y-intercept $(0,5)$

We can think of the slope as $\frac{-1}{3}$, so from $(0,5)$ we move down 1 unit and right 3 units. This locates the point $(3,4)$. We plot $(3,4)$ and draw a line passing through $(0,5)$ and $(3,4)$.

13.   Slope 2; y-intercept $(0,0)$

We can think of the slope as $\frac{2}{1}$, so from $(0,0)$ we move up 2 units and right 1 unit. This locates the point $(1,2)$. We plot $(1,2)$ and draw a line passing through $(0,0)$ and $(1,2)$.

15.   Slope -3; y-intercept $(0,2)$

We can think of the slope as $\frac{-3}{1}$, so from $(0,2)$ we move down 3 units and right 1

unit. This locates the point $(1,-1)$. We plot $(1,-1)$ and draw a line passing through $(0,2)$ and $(1,-1)$.

17.   Slope 0; y-intercept $(0,-5)$

We can think of the slope as $\frac{0}{x}$, so from $(0,-5)$ we move up or down 0 units for all units of x left and right. Among the points on this line is $(3,-5)$. We plot $(3,-5)$ and draw a line passing through $(0,-5)$ and $(3,-5)$.

19.   We read the slope and y-intercept from the equation.

$$y = -\frac{2}{7}x + 5$$

The slope is $-\frac{2}{7}$.   The y-intercept is $(0,5)$.

21.   We read the slope and y-intercept from the equation.

$$y = \frac{1}{3}x + 7$$

The slope is $\frac{1}{3}$.   The y-intercept is $(0,7)$.

23.   We read the slope and y-intercept from the equation.

$$y = \frac{9}{5}x - 4$$

The slope is $\frac{9}{5}$.   The y-intercept is $(0,-4)$.

25. $-3x + y = 7$

We solve for y to rewrite the equation in the form $y = mx + b$.
$$-3x + y = 7$$
$$y = 3x + 7$$
Slope is 3, and the y-intercept is $(0, 7)$.

27. $4x + 2y = 8$

We solve for y to rewrite the equation in the form $y = mx + b$.
$$4x + 2y = 8$$
$$2y = -4x + 8$$
$$y = -2x + 4$$
Slope is -2, and the y-intercept is $(0, 4)$.

29. $y = 4$

Observe that this is the equation of a horizontal line that lies 4 units above the x-axis. Thus, the slope is 0, and the y-intercept is $(0, 4)$. We could also write the equation in slope intercept form.
$$y = 4$$
$$y = 0x + 4$$
The slope is 0, and the y-intercept is $(0, 4)$.

31. $2x - 5y = -8$

We solve for y to rewrite the equation in the form $y = mx + b$.
$$2x - 5y = -8$$
$$-5y = -2x - 8$$
$$y = \frac{2}{5}x + \frac{8}{5}$$
The slope is $\frac{2}{5}$, and the y-intercept is $\left(0, \frac{8}{5}\right)$.

33. $9x - 8y = 0$

We solve for y to rewrite the equation in the form $y = mx + b$.
$$9x - 8y = 0$$
$$-8y = -9x$$
$$y = \frac{-9}{-8}x$$
$$y = \frac{9}{8}x + 0$$

The slope is $\frac{9}{8}$, and the y-intercept is $(0, 0)$.

35. a) $y = 3x - 5$

The slope is 3, and the y-intercept is $(0, -5)$. This matches with graph II.

b) $y = 0.7x + 1$

The slope is 0.7, and the y-intercept is $(0, 1)$. This matches with graph IV.

c) $y = -0.25x - 3$

The slope is -0.25, and the y-intercept is $(0, -3)$. This matches with graph III.

d) $y = -4x + 2$

The slope is -4, and the y-intercept is $(0, 2)$. This matches with graph I.

37. Slope: 3; y-intercept $(0, 7)$

Use the slope intercept form, $y = mx + b$, and substitute 3 for m and 7 for b. This gives $y = 3x + 7$.

39. Slope $\frac{7}{8}$; y-intercept $(0, -1)$

Use the slope intercept form, $y = mx + b$, and substitute $\frac{7}{8}$ for m and -1 for b. This gives $y = \frac{7}{8}x - 1$.

41. Slope $-\frac{5}{3}$; y-intercept $(0, -8)$

Use the slope intercept form, $y = mx + b$, and substitute $-\frac{5}{3}$ for m and -8 for b. This gives $y = -\frac{5}{3}x - 8$.

43. Slope 0; y-intercept $\left(0, \frac{1}{3}\right)$

Since the slope is 0, we know that the line is horizontal. Since its y-intercept is $\left(0, \frac{1}{3}\right)$, the equation must be $y = \frac{1}{3}$. We could also use the slope intercept form, $y = mx + b$, and substitute 0 for m and $\frac{1}{3}$ for b. This gives $y = 0x + \frac{1}{3}$, or $y = \frac{1}{3}$.

45. From the graph, we note that the y-intercept is $(0,70)$. Also, taking any two points on the line, we can compute the slope. We choose $(0,70)$ and $(10,100)$. The slope is

$m = \dfrac{100-70}{10-0} = \dfrac{30}{10} = 3$. Using the slope

intercept form, $y = mx+b$, we substitute 3 for m and 70 for b. The equation of the line is $y = 3x+70$, where y is the number of jobs, in thousands, and x is the number of years since 2006.

47. From the graph, we note that the y-intercept is $(0,2.2)$. Also, taking any two points on the line, we can compute the slope. We choose $(0,2.2)$ and $(8,2.5)$. The slope is

$m = \dfrac{2.5-2.2}{8-0} = \dfrac{0.3}{8} = 0.0375$. Using the

slope intercept form, $y = mx+b$, we substitute 0.0375 for m and 2.2 for b. The equation of the line is $y = 0.0375x+2.2$, where y is the number of registered nurses, in millions, and x is the number of years since 2000.

49. $y = \frac{3}{5}x+2$

Slope: $\dfrac{3}{5}$; y-intercept: $(0,2)$

First we plot the y-intercept $(0,2)$. We can

start at the y-intercept and use the slope, $\dfrac{3}{5}$,

to find another point. We move up 3 units and right 5 units to get a new point $(5,5)$.

Thinking of the slope as $\dfrac{-3}{-5}$ we can start at

$(0,2)$ and move down 3 units and left 5 units

to get another point $(-5,-1)$.

51. $y = -\frac{3}{5}x+1$

Slope: $-\dfrac{3}{5}$; y-intercept: $(0,1)$

First we plot the y-intercept $(0,1)$. We can start at the y-intercept and thinking of the

slope as, $\dfrac{-3}{5}$, find another point. We move

down 3 units and right 5 units to get a new

point $(5,-2)$. Thinking of the slope as $\dfrac{3}{-5}$

we can start at $(0,1)$ and move up 3 units and left 5 units to get another point $(-5,4)$.

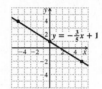

53. $y = \frac{5}{3}x+3$

Slope: $\dfrac{5}{3}$; y-intercept: $(0,3)$

First we plot the y-intercept $(0,3)$. We can

start at the y-intercept and use the slope, $\dfrac{5}{3}$,

to find another point. We move up 5 units and right 3 units to get a new point $(3,8)$.

Thinking of the slope as $\dfrac{-5}{-3}$ we can start at

$(0,3)$ and move down 5 units and left 3

units to get another point $(-3,-2)$.

55. $y = -\frac{3}{2}x-2$

Slope: $-\dfrac{3}{2}$; y-intercept: $(0,-2)$

First we plot the y-intercept $(0,-2)$. We can start at the y-intercept and thinking of the

slope as, $\dfrac{-3}{2}$, find another point. We move

down 3 units and right 2 units to get a new

point $(2,-5)$. Thinking of the slope as $\dfrac{3}{-2}$ we can start at $(0,-2)$ and move up 3 units and left 2 units to get another point $(-2,1)$.

57. We first rewrite the equation in slope-intercept form.
$$2x + y = 1$$
$$y = -2x + 1$$

Slope: -2; y-intercept: $(0,1)$

First we plot the y-intercept $(0,1)$. We can start at the y-intercept and, thinking of the slope as $\dfrac{-2}{1}$ slope, find another point by moving down 2 units and right 1 unit to the point $(1,-1)$. In a similar manner, we can move from the point $(1,-1)$ to find a third point $(2,-3)$.

59. We first rewrite the equation in slope-intercept form.
$$3x + y = 0$$
$$y = -3x + 0$$

Slope: -3; y-intercept: $(0,0)$

First we plot the y-intercept $(0,0)$. We can start at the y-intercept and, thinking of the slope as $\dfrac{-3}{1}$ slope, find another point by moving down 3 units and right 1 unit to the point $(1,-3)$. In a similar manner, we can move from the point $(1,-3)$ to find a third point $(2,-6)$.

61. We first rewrite the equation in slope-intercept form.
$$2x + 3y = 9$$
$$3y = -2x + 9$$
$$y = \frac{1}{3}(-2x + 9)$$
$$y = -\frac{2}{3}x + 3$$

Slope: $-\dfrac{2}{3}$; y-intercept: $(0,3)$

First we plot the y-intercept $(0,3)$. We can start at the y-intercept and thinking of the slope as, $\dfrac{-2}{3}$, find another point. We move down 2 units and right 3 units to get a new point $(3,1)$. Thinking of the slope as $\dfrac{2}{-3}$ we can start at $(0,3)$ and move up 2 units and left 3 units to get another point $(-3,5)$.

63. We first rewrite the equation in slope-intercept form.
$$x - 4y = 12$$
$$-4y = -x + 12$$
$$y = -\frac{1}{4}(-x + 12)$$
$$y = \frac{1}{4}x - 3$$

Slope: $\dfrac{1}{4}$; y-intercept: $(0,-3)$

First we plot the y-intercept $(0,-3)$. We can start at the y-intercept and use the slope $\dfrac{1}{4}$ to find another point. We move up 1 unit and right 4 units to get a new point $(4,-2)$.

Thinking of the slope as $\dfrac{-1}{-4}$ we can start at $(0,-3)$ and move down 1 unit and left 4 units to get another point $(-4,-4)$.

65. $y=\frac{2}{3}x+7$: The slope is $\dfrac{2}{3}$, and the y-intercept is $(0,7)$.

$y=\frac{2}{3}x-5$: The slope is $\dfrac{2}{3}$, and the y-intercept is $(0,-5)$.

Since both lines have slope $\dfrac{2}{3}$ but different y-intercepts, their graphs are parallel.

67. $y=2x-5$: The slope is 2, and the y-intercept is $(0,-5)$.

We rewrite the second equation in slope-intercept form:
$4x+2y=9$

$2y=-4x+9$

$y=\dfrac{1}{2}(-4x+9)$

$y=-2x+\dfrac{9}{2}$

The slope is –2, and the y-intercept is $\left(0,-\dfrac{9}{2}\right)$.

Since the lines have different slopes, the graphs of their equations are not parallel.

69. We rewrite the first equation in slope-intercept form:
$3x+4y=8$

$4y=-3x+8$

$y=\dfrac{1}{4}(-3x+8)$

$y=-\dfrac{3}{4}x+2$

The slope is $-\dfrac{3}{4}$, and the y-intercept is $(0,2)$.

We rewrite the second equation in slope-intercept form:
$7-12y=9x$

$-12y=9x-7$

$y=-\dfrac{1}{12}(9x-7)$

$y=-\dfrac{9}{12}x+\dfrac{7}{12}$

$y=-\dfrac{3}{4}x+\dfrac{7}{12}$

The slope is $-\dfrac{3}{4}$, and the y-intercept is $\left(0,\dfrac{7}{12}\right)$.

Since both lines have slope $-\dfrac{3}{4}$ but different y-intercepts, the graphs of their equations are parallel.

71. $y=4x-5$
$4y=8-x$

The first equation is in slope-intercept form. It represents a line with slope 4. Now we rewrite the second equation in slope-intercept form.
$4y=8-x$

$4y=-x+8$

$y=\dfrac{1}{4}(-x+8)$

$y=-\dfrac{1}{4}x+2$

The slope of the line is $-\dfrac{1}{4}$.

Since $4\left(-\dfrac{1}{4}\right)=-1$, the equations represent perpendicular lines.

73. $x-2y=5$
$2x+4y=8$

We write both equations in slope-intercept form:
$x-2y=5$

$-2y=-x+5$

$y=-\dfrac{1}{2}(-x+5)$

$y=\dfrac{1}{2}x-\dfrac{5}{2}$

The slope of the line is $\dfrac{1}{2}$.

$$2x + 4y = 8$$
$$4y = -2x + 8$$
$$y = \frac{1}{4}(-2x + 8)$$
$$y = \frac{-2}{4}x + \frac{8}{4}$$
$$y = -\frac{1}{2}x + 2$$

The slope of the line is $-\dfrac{1}{2}$.

Since $\dfrac{1}{2}\left(-\dfrac{1}{2}\right) = -\dfrac{1}{4} \neq -1$, the equations do not represent perpendicular lines.

75. $2x + 3y = 1$
$3x - 2y = 1$
We rewrite the equations in slope-intercept form:
$$2x + 3y = 1$$
$$3y = -2x + 1$$
$$y = \frac{1}{3}(-2x + 1)$$
$$y = -\frac{2}{3}x + \frac{1}{3}$$
The slope is $-\dfrac{2}{3}$.
$$3x - 2y = 1$$
$$-2y = -3x + 1$$
$$y = -\frac{1}{2}(-3x + 1)$$
$$y = \frac{3}{2}x - \frac{1}{2}$$
The slope is $\dfrac{3}{2}$.

Since $-\dfrac{2}{3}\left(\dfrac{3}{2}\right) = -1$, the equations represent perpendicular lines.

77. $y = \frac{7}{8}x - 3$

a)   The slope of this line is $\dfrac{7}{8}$, so the slope of a parallel line is also $\dfrac{7}{8}$.

b)   The reciprocal of this slope is $\dfrac{8}{7}$, and the opposite of this number is $-\dfrac{8}{7}$, so the slope of a perpendicular line is $-\dfrac{8}{7}$.

79. $y = -\frac{1}{4}x - \frac{5}{8}$

a)   The slope of this line is $-\dfrac{1}{4}$, so the slope of a parallel line is also $-\dfrac{1}{4}$.

b)   The reciprocal of this slope is -4, and the opposite of this number is 4, so the slope of a perpendicular line is 4.

81. $20x - y = 12$
We rewrite the second equation in slope-intercept form:
$$20x - y = 12$$
$$y = 20x - 12$$

a)   The slope of this line is 20, so the slope of a parallel line is also 20.

b)   The reciprocal of this slope is $\dfrac{1}{20}$, and the opposite of this number is $-\dfrac{1}{20}$, so the slope of a perpendicular line is $-\dfrac{1}{20}$.

83. $x + y = 4$
We rewrite the second equation in slope-intercept form:
$$x + y = 4$$
$$y = -x + 4$$

a)   The slope of this line is -1, so the slope of a parallel line is also -1.

b)   The reciprocal of this slope is $-\dfrac{1}{1} = -1$, and the opposite of this number is 1, so the slope of a perpendicular line is 1.

85. The slope of the line represented by $y = 5x - 7$ is 5. Then a line parallel to the graph of $y = 5x - 7$ has slope 5 also. Since the y-intercept is $(0, 11)$, the desired equation is $y = 5x + 11$.

87. First find the slope of the line represented by $2x + y = 0$.

$2x + y = 0$

$y = -2x$

The slope is –2. Then the slope of a line perpendicular to the graph of $2x + y = 0$ is the negative reciprocal of –2, or $\frac{1}{2}$. Since the y-intercept is $(0,0)$, the desired equation is

$y = \frac{1}{2}x + 0$, or $y = \frac{1}{2}x$.

89. The slope of the line represented by $y = x$ is 1. Then a line parallel to this line also has slope 1. Since the y-intercept is $(0,3)$, the desired equation is $y = 1 \cdot x + 3$, or $y = x + 3$.

91. First find the slope of the line represented by $x + y = 3$.

$x + y = 3$

$y = -x + 3$

The slope is –1. Then the slope of a line perpendicular to this line is the negative reciprocal of –1, or 1. Since the y-intercept is $(0,-4)$, the desired equation is $y = 1 \cdot x - 4$, or $y = x - 4$.

93. *Thinking and Writing Exercise.*

95.
$y - k = m(x - h)$
$y - k + k = m(x - h) + k$
$y = m(x - h) + k$

97. $-10 - (-3) = -10 + 3 = -(10 - 3) = -7$

99. $-4 - 5 = -(4 + 5) = -9$

101. *Thinking and Writing Exercise.*

103. Given $y = mx + b$, we first substitute 0 and 1 for x to determine the coordinate of two points.

If $x = 0$, then $y = m \cdot 0 + b = b$, giving the point $(0,b)$.

If $x = 1$, then $y = m \cdot 1 + b = m + b$, giving the point $(1, m + b)$.

We then determine the slope of the line containing these two points:

$m = \frac{\text{change in } y}{\text{change in } x} = \frac{(m + b) - b}{1 - 0} = \frac{m}{1} = m$

105. Let $x =$ the number of residents in excess of 2, and let $y =$ the size of the refrigerator in ft³. If $x = 0$, then the size of the refrigerator should be 16 ft³. If $x = 1$, then the size of the refrigerator should be $16 + 1.5$, or 17.5 ft³. Hence, we have the coordinates of two points, $(0,16)$ and $(1,17.5)$. The slope of the line containing these two points is

$m = \frac{17.5 - 16}{1 - 0} = \frac{1.5}{1} = 1.5$.

The y-intercept is $(0,16)$, so the equation is $y = 1.5x + 16$.

107. $rx + py = s$

Rewrite the equation in slope-intercept form.

$rx + py = s$
$py = -rx + s$
$y = \frac{1}{p}(-rx + s)$
$y = -\frac{r}{p}x + \frac{s}{p}$

The slope is $-\frac{r}{p}$, and the y-intercept is $\left(0, \frac{s}{p}\right)$.

109. *Thinking and Writing Exercise.*

Exercise Set 3.7

1. g

3. e

5. b

7. f

9. Examining the graph, we can see that the line goes down by $1\frac{1}{2}$ units, for every 1 unit to the right. Therefore the slope of the line is $-\frac{3}{2}$. The coordinates of the point are $(1,-4)$. This matches with the point-slope form of the equation choice c.

11. Examining the graph, we can see that the line goes up by $1\frac{1}{2}$ units, for every 1 unit to the right. Therefore the slope of the line is $\frac{3}{2}$. The coordinates of the point are $(1,-4)$. This matches with the point-slope form of the equation choice d.

13. $m = 5; (6,2)$

The point-slope equation is
$y - y_1 = m(x - x_1)$.
We substitute 5 for m, 6 for x_1, and 2 for y_1 giving $y - 2 = 5(x - 6)$.

15. $m = -4; (3,1)$

The point-slope equation is
$y - y_1 = m(x - x_1)$.
We substitute -4 for m, 3 for x_1, and 1 for y_1 giving $y - 1 = -4(x - 3)$.

17. $m = \frac{3}{2}; (5,-4)$

The point-slope equation is
$y - y_1 = m(x - x_1)$.
We substitute $\frac{3}{2}$ for m, 5 for x_1, and -4 for y_1 giving $y - (-4) = \frac{3}{2}(x - 5)$, or
$y + 4 = \frac{3}{2}(x - 5)$.

19. $m = -\frac{5}{4}; (-2,6)$

The point-slope equation is
$y - y_1 = m(x - x_1)$.
We substitute $-\frac{5}{4}$ for m, -2 for x_1, and 6 for y_1 giving $y - 6 = -\frac{5}{4}(x - (-2))$, or
$y - 6 = -\frac{5}{4}(x + 2)$.

21. $m = -2; (-4,-1)$

The point-slope equation is
$y - y_1 = m(x - x_1)$.
We substitute -2 for m, -4 for x_1, and -1 for y_1 giving $y - (-1) = -2(x - (-4))$.

23. $m = 1; (-2,8)$

The point-slope equation is
$y - y_1 = m(x - x_1)$.
We substitute 1 for m, -2 for x_1, and 8 for y_1 giving $y - 8 = 1(x - (-2))$.

25. $y - 9 = \frac{2}{7}(x - 8)$

Slope: $\dfrac{2}{7}$; coordinates of a point $(8,9)$

27. $y + 2 = -5(x - 7)$

Rewriting the equation in point-slope form:
$y - (-2) = -5(x - 7)$
Slope: -5; coordinates of a point $(7,-2)$

29. $y - 4 = -\frac{5}{3}(x + 2)$

Rewriting the equation in point-slope form:
$y - 4 = -\frac{5}{3}(x - (-2))$
Slope: $-\frac{5}{3}$; coordinates of a point $(-2,4)$

31. $y = \frac{4}{7}x$

Rewriting the equation in point-slope form:
$y - 0 = \frac{4}{7}(x - 0)$
Slope: $\frac{4}{7}$; coordinates of a point $(0,0)$
Alternatively, the equation is of the form $y = mx$, so we know that its graph is a line through the origin with slope m. Thus, the slope is $\frac{4}{7}$ and a point on the graph is $(0,0)$.

33. $m = 2; (5,7)$

$y - y_1 = m(x - x_1)$ Point-slope equation

$y - 7 = 2(x - 5)$ Substituting 2 for m, 5 for x_1, and 7 for y_1

$y - 7 = 2x - 10$ Simplifying

$y = 2x - 3$ Adding 7 to both sides

35. $m = \frac{7}{4}; \ (4, -2)$

$y - y_1 = m(x - x_1)$ Point-slope equation

$y - (-2) = \frac{7}{4}(x - 4)$ Substituting $\frac{7}{4}$ for m, 4 for x_1, and -2 for y_1

$y + 2 = \frac{7}{4}x - 7$ Simplifying

$y = \frac{7}{4}x - 9$ Adding -2 to both sides

37. $m = -3; \ (-1, 6)$

$y - y_1 = m(x - x_1)$ Point-slope equation

$y - 6 = -3(x - (-1))$ Substituting -3 for m, -1 for x_1, and 6 for y_1

$y - 6 = -3(x + 1)$ Simplifying

$y - 6 = -3x - 3$ Simplifying

$y = -3x + 3$ Adding 6 to both sides

39. $m = -4; \ (-2, -1)$

$y - y_1 = m(x - x_1)$ Point-slope equation

$y - (-1) = -4(x - (-2))$ Substituting -4 for m, -2 for x_1, and -1 for y_1

$y + 1 = -4(x + 2)$ Simplifying

$y + 1 = -4x - 8$ Simplifying

$y = -4x - 9$ Adding -1 to both sides

41. $m = -\frac{5}{6}; \ (0, 4)$

$y - y_1 = m(x - x_1)$ Point-slope equation

$y - 4 = -\frac{5}{6}(x - 0)$ Substituting $-\frac{5}{6}$ for m, 0 for x_1, and 4 for y_1

$y - 4 = -\frac{5}{6}x$ Simplifying

$y = -\frac{5}{6}x + 4$ Adding 4 to both sides

43. $(4, 7), \ x + 2y = 6$

Find the slope of $x + 2y = 6$.

$x + 2y = 6$

$2y = -x + 6$

$y = -\frac{1}{2}x + 3$

The slope of $x + 2y = 6$ is $-\frac{1}{2}$. This is also the slope of a line parallel to $x + 2y = 6$.

$y - 7 = -\frac{1}{2}(x - 4)$

$y - 7 = -\frac{1}{2}x + 2$

$y = -\frac{1}{2}x + 9$

45. $(0, -7), \ y = 2x + 1$

Find the slope of $y = 2x + 1$.

Since $y = 2x + 1$ is already in slope-intercept form, the slope is 2. This is also the slope of a line parallel to $y = 2x + 1$.

$y - (-7) = 2(x - 0)$

$y + 7 = 2x$

$y = 2x - 7$

47. $(2, -6), \ 5x - 3y = 8$

Find the slope of $5x - 3y = 8$.

$5x - 3y = 8$

$-3y = -5x + 8$

$y = \frac{5}{3}x - \frac{8}{3}$

The slope of $5x - 3y = 8$ is $\frac{5}{3}$. This is also the slope of a line parallel to $5x - 3y = 8$.

$y - (-6) = \frac{5}{3}(x - 2)$

$y + 6 = \frac{5}{3}(x - 2)$

$y + 6 = \frac{5}{3}x - \frac{10}{3}$

$y = \frac{5}{3}x - 6 - \frac{10}{3}$

$y = \frac{5}{3}x - \frac{18}{3} - \frac{10}{3}$

$y = \frac{5}{3}x - \frac{28}{3}$

49. $(5, -4), \ x = 2$

The slope of $x = 2$ is undefined, as it is a vertical line. A line parallel to $x = 2$ must also be vertical. Therefore, the vertical line that passes through the point $(5, -4)$ is $x = 5$.

51. $(3, -2), \ 3x - 6y = 5$

Find the slope of $3x - 6y = 5$.

$3x - 6y = 5$

$-6y = -3x + 5$

$y = \frac{-3}{-6}x + \frac{5}{-6}$

$y = \frac{1}{2}x - \frac{5}{6}$

The slope of $3x - 6y = 5$ is $\frac{1}{2}$. So, the slope of a line perpendicular to this line is -2.

$$y-(-2)=-2(x-3)$$
$$y+2=-2x+6$$
$$y=-2x+4$$

53. $(-4,2)$, $x+y=6$

Find the slope of $x+y=6$.

$$x+y=6$$
$$y=-x+6$$

The slope of $x+y=6$ is -1. So, the slope of a line perpendicular to this line is 1.

$$y-2=1(x-(-4))$$
$$y-2=x+4$$
$$y=x+6$$

55. $(0,6)$, $2x-5=y$

We can write $2x-5=y$ as $y=2x-5$. The slope of this line is 2 so the slope of a line perpendicular to this line is $-\frac{1}{2}$.

$$y-6=-\frac{1}{2}(x-0)$$
$$y-6=-\frac{1}{2}x$$
$$y=-\frac{1}{2}x+6$$

57. $(-3,7)$, $y=5$

Find the slope of $y=5$.

$$y=5$$
$$y=0\cdot x+5$$

The slope of $y=5$ is 0. So, the slope of the line perpendicular to this line is undefined. The vertical line that passes through the point $(-3,7)$ is $x=-3$.

59. $(1,5)$ and $(4,2)$

First find the slope of the line:

$$m=\frac{2-5}{4-1}=\frac{-3}{3}=-1$$

Use the point-slope equation with $m=-1$ and $(x_1,y_1)=(1,5)$. (We could let $(x_1,y_1)=(4,2)$ instead and obtain an equivalent equation.)

$$y-5=-1(x-1)$$
$$y-5=-x+1$$
$$y=-x+6$$

61. $(-3,1)$ and $(3,5)$

First find the slope of the line:

$$m=\frac{5-1}{3-(-3)}=\frac{4}{3+3}=\frac{4}{6}=\frac{2}{3}$$

Use the point-slope equation with $m=\frac{2}{3}$ and $(x_1,y_1)=(-3,1)$. (We could let $(x_1,y_1)=(3,5)$ instead and obtain an equivalent equation.)

$$y-1=\frac{2}{3}(x-(-3))$$
$$y-1=\frac{2}{3}(x+3)$$
$$y-1=\frac{2}{3}x+2$$
$$y=\frac{2}{3}x+3$$

63. $(5,0)$ and $(0,-2)$

First find the slope of the line:

$$m=\frac{-2-0}{0-5}=\frac{-2}{-5}=\frac{2}{5}$$

Use the point-slope equation with $m=\frac{2}{5}$ and $(x_1,y_1)=(5,0)$. (We could let $(x_1,y_1)=(0,-2)$ instead and obtain an equivalent equation.)

$$y-0=\frac{2}{5}(x-5)$$
$$y=\frac{2}{5}x-2$$

65. $(-2,-4)$ and $(2,-1)$

First find the slope of the line:

$$m=\frac{-1-(-4)}{2-(-2)}=\frac{-1+4}{2+2}=\frac{3}{4}$$

Use the point-slope equation with $m=\frac{3}{4}$ and $(x_1,y_1)=(-2,-4)$. (We could let $(x_1,y_1)=(2,-1)$ instead and obtain an equivalent equation.)

$$y-(-4)=\frac{3}{4}(x-(-2))$$
$$y+4=\frac{3}{4}(x+2)$$
$$y+4=\frac{3}{4}x+\frac{6}{4}$$
$$y=\frac{3}{4}x-4+\frac{6}{4}$$
$$y=\frac{3}{4}x-\frac{16}{4}+\frac{6}{4}$$
$$y=\frac{3}{4}x-\frac{10}{4}$$
$$y=\frac{3}{4}x-\frac{5}{2}$$

67. Slope $\frac{4}{3}$; passing through the point $(1,2)$

We plot $(1,2)$ and then find a second point by moving up 4 units and to the right 3 units giving the point $(4,6)$. Thinking of the slope as $\frac{-4}{-3}$, we could also find a third point by moving down 4 units and left 3 units giving the point $(-2,-2)$. The line can then be drawn as shown below:

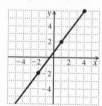

69. Slope $-\frac{3}{4}$; passing through the point $(2,5)$

We plot $(2,5)$ and then thinking of the slope as $\frac{-3}{4}$ find a second point by moving down 3 units and to the right 4 units giving the point $(6,2)$. Thinking of the slope as $\frac{3}{-4}$, we could also find a third point by moving up 3 units and left 4 units giving the point $(-2,8)$.

The line can then be drawn as shown below:

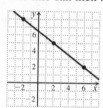

71. $y - 2 = \frac{1}{2}(x - 1)$

The equation is in point-slope form. We see that the line passes through $(1,2)$ and has slope $\frac{1}{2}$. Plot $(1,2)$ and move up 1 unit and right 2 units to $(3,3)$ or, thinking of the slope as $\frac{-1}{-2}$, move down 1 unit and left 2 units to $(-1,1)$ and draw the line.

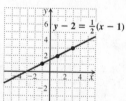

73. $y - 1 = -\frac{1}{2}(x - 3)$

The equation is in point-slope form. We see that the line passes through $(3,1)$ and has slope $-\frac{1}{2}$. Plot $(3,1)$ and thinking of the slope as $\frac{-1}{2}$ move down 1 unit and right 2 units to $(5,0)$ or, thinking of the slope as $\frac{1}{-2}$, move up 1 unit and left 2 units to $(1,2)$ and draw the line.

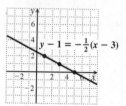

75. $y + 4 = 3(x + 1)$

Rewrite the equation in point-slope form.

$y - (-4) = 3(x - (-1))$

We see that the line passes through $(-1,-4)$ and has slope 3. Plot $(-1,-4)$ and thinking of the slope as $\frac{3}{1}$ move up 3 units and right 1 unit to $(0,-1)$ or, thinking of the slope as $\frac{6}{2}$, move up 6 units and right 2 units to $(1,2)$ and draw the line.

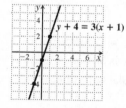

77. $y+3=-(x+2)$

Rewrite the equation in point-slope form.

$y-(-3)=-1(x-(-2))$

We see that the line passes through $(-2,-3)$ and has slope -1. Plot $(-2,-3)$ and thinking of the slope as $\frac{-1}{1}$ move down 1 unit and right 1 unit to $(-1,-4)$ or, thinking of the slope as $\frac{1}{-1}$, move up 1 unit and left 1 unit to $(-3,-2)$ and draw the line.

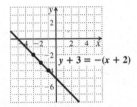

79. $y+1=-\frac{3}{5}(x+2)$

Rewrite the equation in point-slope form.

$y-(-1)=-\frac{3}{5}(x-(-2))$

We see that the line passes through $(-2,-1)$ and has slope $-\frac{3}{5}$. Plot $(-2,-1)$ and thinking of the slope as $\frac{-3}{5}$ move down 3 units and right 5 units to $(3,-4)$ or, thinking of the slope as $\frac{3}{-5}$, move up 3 units and left 5 units to $(-7,2)$ and draw the line.

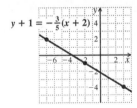

81. a) We form pairs of type (x,y), where x is the number of years since 2005 and y is the number of college students, in millions, attending public meetings. We have two pairs: $(1,392.5)$ and $(3,468.2)$. These two points are on the graph of the linear function we are seeking. We use the point-slope form to write an equation relating y and x:

$m=\dfrac{468.2-392.5}{3-1}=\dfrac{75.7}{2}=37.85$

$y-392.5=37.85(x-1)$

$y-392.5=37.85x-37.85$

$y=37.85x-37.85+392.5$

$y=37.85x+354.65$

b) To calculate the number of college students attending public meetings in 2007, we note that 2007 is 2 years after 2005, so we substitute 2 for x and solve for y.

$y=37.85x+354.65$

$y=37.85(2)+354.65$

$y=75.7+354.65$

$y=430.35$

The number of college students attending public meetings in 2007 was 430.35 million.

c) To predict the number of college students that will attend public meetings in 2012, we note that 2012 is 7 years after 2005, so we substitute 7 for x and solve for y.

$y=37.85x+354.65$

$y=37.85(7)+354.65$

$y=264.95+354.65$

$y=619.6$

The predicted number of college students attending public meetings in 2012 is 619.6 million.

83. a) We form pairs of type (t,A), where t is the number of years since 2000 and A is the number of acres, in millions, in the National Park system. We have two pairs: $(0,78.2)$ and $(7,78.8)$. These two points are on the graph of the linear function we are seeking. We use the point-slope form to write an equation relating t and A:

$m=\dfrac{78.8-78.2}{7-0}=\dfrac{0.6}{7}\approx 0.0857$

$A-78.2=0.0857(t-0)$

$A-78.2=0.0857t$

$A=0.0857t+78.2$

b) To estimate the number of acres in 2003, we note that 2003 is 3 years after 2000, so we substitute 3 for t and solve for A.

$A = 0.0857t + 78.2$

$A = 0.0857(3) + 78.2$

$A = 0.2571 + 78.2$

$A = 78.4571 \approx 78.5$

The estimated number of acres for 2003 is about 78.5 million acres.

c) To predict the number of acres in 2010, we note that 2010 is 10 years after 2000, so we substitute 10 for t and solve for A.

$A = 0.0857t + 78.2$

$A = 0.0857(10) + 78.2$

$A = 0.857 + 78.2$

$A = 79.057 \approx 79.1$

The predicted number of acres for 2010 is about 79.1 million acres.

85. a) We form pairs of type (t, C), where t is the number of years since 2006 and C is the percentage of Americans who are familiar with the term "carbon footprint." We have two pairs: $(1, 38)$ and $(3, 57)$.

These two points are on the graph of the linear function we are seeking. We use the point-slope form to write an equation relating t and C:

$m = \dfrac{57 - 38}{3 - 1} = \dfrac{19}{2} = 9.5$

$C - 38 = 9.5(t - 1)$

$C - 38 = 9.5t - 9.5$

$C = 9.5t - 9.5 + 38$

$C = 9.5t + 28.5$

b) To predict the percentage of Americans who will be familiar with the term "carbon footprint" in 2012, we note that 2012 is 6 years after 2006. We substitute 6 for t and solve for C.

$C = 9.5t + 28.5$

$C = 9.5(6) + 28.5$

$C = 57 + 28.5$

$C = 85.5$

In 2012, the percentage of Americans who will be familiar with the term "carbon footprint" will be 85.5%.

c) To predict when 100% of Americans will be familiar with the term "carbon footprint," substitute 100 for C and solve for t.

$C = 9.5t + 28.5$

$100 = 9.5t + 28.5$

$100 - 28.5 = 9.5t$

$71.5 = 9.5t$

$\dfrac{71.5}{9.5} = t$

$t \approx 7.5$

The predicted year when all Americans will be familiar with the term "carbon footprint" will be about 7.5 years after 2006, or about 2014.

87. a) We form pairs of type (t, E), where t is the number of years since 1990 and E is life expectancy at birth of females. We have two pairs: $(3, 78.8)$ and $(16, 80.2)$. These two points are on the graph of the linear function we are seeking. We use the point-slope form to write an equation relating t and E:

$m = \dfrac{80.2 - 78.8}{16 - 3} = \dfrac{1.4}{13} \approx 0.1077$

$E - 78.8 = 0.1077(t - 3)$

$E - 78.8 = 0.1077t - 0.3231$

$E = 0.1077t - 0.3231 + 78.8$

$E = 0.1077t + 78.4769$

b) To predict the life expectancy of females in 2010, we note that 2010 is 2010 − 1990 = 20 years after 1990. We substitute 20 for t and solve for E.

$E = 0.1077t + 78.4769$

$E = 0.1077(20) + 78.4769$

$E = 2.154 + 78.4769$

$E = 80.639 \approx 80.6$

By 2010, the life expectancy for females will be about 80.6 years.

89. a) We form pairs of type (t, N), where t is the number of years since 1996 and N is the amount of PAC contributions, in millions. We have two pairs: $(0, 217.8)$ and $(10, 372.1)$. These two points are on the graph of the linear function we are seeking. We use the point-slope form to write an equation relating t and N:

$$m = \frac{372.1 - 217.8}{10 - 0} = \frac{154.3}{10} = 15.43$$

$$N - 217.8 = 15.43(t - 0)$$

$$N - 217.8 = 15.43t$$

$$N = 15.43t + 217.8$$

b) To predict the amount of PAC contributions in 2010, we note that 2010 is 2010 − 1996 = 14 years after 1996. We substitute 14 for t and solve for N.

$$N = 15.43t + 217.8$$

$$N = 15.43(14) + 217.8$$

$$N = 216.02 + 217.8$$

$$N = 433.82$$

In 2010, the PAC contributions are expected to be $433.82 million.

91. The points lie in an approximately straight line, so the graph of this data is linear.

93. The points do not lie on a straight line, so the data are not linear.

95. The points lie in an approximately straight line, so the graph of this data is linear.

97. a) Let x = the number of years since 1900 and let W = the life expectancy of a woman in years. We enter the data with the number of years since 1900 as L_1 and the life expectancy, in years, as L_2. Using the LinReg feature of the graphing calculator, we find:

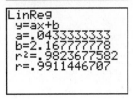

The second screen indicates that the equation is
$W = 0.1611x + 63.6983$.

b) To predict the life expectancy of a woman in 2010, we note that 2010 is

2010 − 1900 = 110 years after 1900. We substitute 110 for x and solve for W. We can do this using the TABLE feature and determine that the predicted life expectancy is approximately 81.4 years. This estimate is 81.4 − 80.6 = 0.8 year higher than the result we obtained in Exercise 87.

99. a) Let x = the number of years since 2000 and let N = the number of registered nurses, in millions, employed in the United States. We enter the data with the number of years since 2000 as L_1 and the number of registered nurses, in millions, as L_2. Using the LinReg feature of the graphing calculator, we find:

The second screen indicates that the equation is
$N = 0.0433t + 2.1678$.

b) To predict the number of registered nurses employed in the United States in 2012, we note that 2012 is 2012 − 2000 = 12 years after 2000. We substitute 12 for t and solve for N. We can do this using the VALUE option of the CALC menu and determine that the number of registered nurses employed in the United States in 2012 will be approximately 2.69 million.

101. *Thinking and Writing Exercise.*

103. $3 - 4x$, for $x = 5$
$$3 - 4(5) = 3 - 20 = -17$$

105. $n^2 - 5n$, for $n = -1$

$(-1)^2 - 5(-1) = 1 + 5 = 6$

107. $\dfrac{x-6}{2x+8}$, for $x = 4$

$\dfrac{(4)-6}{2(4)+8} = \dfrac{-2}{8+8} = \dfrac{-2}{16} = -\dfrac{1}{8}$

109. *Thinking and Writing Exercise.*

111. $y - 3 = 0(x - 52)$

Since the slope of this line is 0, it is a horizontal line. One point on this line is $(52, 3)$, and every point on this line would be of the form $(x, 3)$.

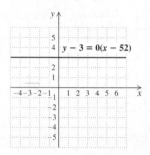

113. We have two points with coordinates $(4, -1)$ and $(3, -3)$. We first determine the slope of the line containing these two points.

$m = \dfrac{-1-(-3)}{4-3} = \dfrac{-1+3}{4-3} = \dfrac{2}{1} = 2$

Use the point-slope equation with $m = 2$ and $(x_1, y_1) = (4, -1)$ to obtain the equation. (We could let $(x_1, y_1) = (3, -3)$ instead and obtain an equivalent equation.)

$y - (-1) = 2(x - 4)$

$y + 1 = 2x - 8$

$y = 2x - 9$

115. Rewrite $x - 3y = 6$ in slope-intercept form.

$x - 3y = 6$

$-3y = -x + 6$

$y = \tfrac{1}{3}x - 2$

This line has y-intercept $(0, -2)$.

Next determine the equation of a line containing the points $(0, -2)$ and $(5, -1)$.

We first determine the slope of the line containing these two points.

$m = \dfrac{-1-(-2)}{5-0} = \dfrac{-1+2}{5} = \dfrac{1}{5}$

Use the point-slope equation with $m = \tfrac{1}{5}$ and $(x_1, y_1) = (0, -2)$ to obtain the equation. (We could let $(x_1, y_1) = (5, -1)$ instead and obtain an equivalent equation.)

$y - (-2) = \tfrac{1}{5}(x - 0)$

$y + 2 = \tfrac{1}{5}x$

$y = \tfrac{1}{5}x - 2$

117. First determine the slope of the line $4x - 8y = 12$ by rewriting it in slope-intercept form.

$4x - 8y = 12$

$-8y = -4x + 12$

$y = \tfrac{-4}{-8}x + \tfrac{12}{-8}$

$y = \tfrac{1}{2}x - \tfrac{3}{2}$

This line has slope $\tfrac{1}{2}$.

Use the point-slope equation with $m = \tfrac{1}{2}$ and $(x_1, y_1) = (-2, 0)$ to obtain the equation.

$y - 0 = \tfrac{1}{2}(x - (-2))$

$y = \tfrac{1}{2}(x + 2)$

$y = \tfrac{1}{2}x + 1$

119. We form pairs of the type (x, y) where $x =$ the Fahrenheit temperature, in degrees, and y is the Celsius temperature, in degrees. We have two points $(32, 0)$ and $(212, 100)$. We compute the slope of the line containing these points.

$m = \dfrac{100-0}{212-32} = \dfrac{100}{180} = \dfrac{5}{9}$

Use the point-slope equation with $m = \dfrac{5}{9}$ and $(x_1, y_1) = (32, 0)$ to obtain the equation. (We could let $(x_1, y_1) = (212, 100)$ instead and obtain an equivalent equation.)

$y - 0 = \tfrac{5}{9}(x - 32)$

$y = \tfrac{5}{9}x - \tfrac{160}{9}$

To find the Celsius temperature corresponding to 70°F, we substitute 70 for x and solve for y.

$y = \frac{5}{9}x - \frac{160}{9}$

$y = \frac{5}{9}(70) - \frac{160}{9}$

$y = \frac{350}{9} - \frac{160}{9}$

$y = \frac{190}{9}$

$y = 21.\overline{1}$

The Celsius temperature corresponding to 70°F is approximately 21.1°C.

121. We form pairs of the type (x, y) where $x =$ the number of months in operation, and y is the cost of operation. We have two points $(4, 7500)$ and $(7, 9250)$. We compute the slope of the line containing these points.

$m = \dfrac{9250 - 7500}{7 - 4} = \dfrac{1750}{3}$

Use the point-slope equation with $m = \frac{1750}{3}$ and $(x_1, y_1) = (4, 7500)$ to obtain the equation. (We could let $(x_1, y_1) = (7, 9250)$ instead and obtain an equivalent equation.)

$y - 7500 = \frac{1750}{3}(x - 4)$

$y - 7500 = \frac{1750}{3}x - \frac{7000}{3}$

$y = \frac{1750}{3}x - \frac{7000}{3} + 7500$

$y = \frac{1750}{3}x - \frac{7000}{3} + \frac{22,500}{3}$

$y = \frac{1750}{3}x + \frac{15,500}{3}$

To predict the cost after 10 months, we substitute 10 for x and solve for y.

$y = \frac{1750}{3}x + \frac{15,500}{3}$

$y = \frac{1750}{3}(10) + \frac{15,500}{3}$

$y = \frac{17,500}{3} + \frac{15,500}{3}$

$y = \frac{33,000}{3}$

$y = 11,000$

The predicted cost after 10 months is $11,000.

Exercise Set 3.8

---

1. correspondence

3. domain

5. horizontal

7. "f of 3," "f at 3," or "the value of f at 3"

9. The correspondence is a function because each member of the domain corresponds to just one member of the range.

11. The correspondence is a function because each member of the domain corresponds to just one member of the range.

13. The correspondence is not a function because two members of the domain, 2008 and 2009, each correspond to more than one member of the range.

15. The correspondence is a function because each member of the domain corresponds to just one member of the range.

17. This corresponds to a function because each USB flash drive can have only one storage capacity.

19. The correspondence is a function because each member of the team has a unique number on his or her uniform.

21. a) To determine $f(1)$, locate 1 on the x-axis, then move down to the point on the graph whose x-coordinate is 1. Then move horizontally to the vertical axis and determine the y-coordinate. It appears that $f(1) = -2$.

    b) The domain is the set $\{x \mid -2 \le x \le 5\}$, or $[-2, 5]$.

    c) To locate any x-values for which $f(x) = 2$, locate 2 on the y-axis, then move horizontally to any points on the graph having second coordinate equal to 2. For any such points move vertically to the x-axis and read the x-coordinate. It appears that $f(x) = 2$ for $x = 4$.

    d) The range is the set $\{y \mid -3 \le y \le 4\}$, or $[-3, 4]$.

23. a) To determine $f(1)$, locate 1 on the x-axis, then move up to the point on the graph whose x-coordinate is 1. Then

move horizontally to the vertical axis and determine the y-coordinate. It appears that $f(1) = 3$.

b) The domain is the set $\{x \mid -1 \le x \le 4\}$, or $[-1, 4]$.

c) To locate any x-values for which $f(x) = 2$, locate 2 on the y-axis, then move horizontally to any points on the graph having second coordinate equal to 2. For any such points move vertically to the x-axis and read the x-coordinate. It appears that $f(x) = 2$ for $x = 3$.

d) The range is the set $\{y \mid 1 \le y \le 4\}$, or $[1, 4]$.

25. a) To determine $f(1)$, locate 1 on the x-axis, then move down to the point on the graph whose x-coordinate is 1. Then move horizontally to the vertical axis and determine the y-coordinate. It appears that $f(1) = -2$.

b) The domain is the set $\{x \mid -4 \le x \le 2\}$, or $[-4, 2]$.

c) To locate any x-values for which $f(x) = 2$, locate 2 on the y-axis, then move horizontally to any points on the graph having second coordinate equal to 2. For any such points move vertically to the x-axis and read the x-coordinate. It appears that $f(x) = 2$ for $x = -2$.

d) The range is the set $\{y \mid -3 \le y \le 3\}$, or $[-3, 3]$.

27. a) To determine $f(1)$, locate 1 on the x-axis, then move up to the point on the graph whose x-coordinate is 1. Then move horizontally to the vertical axis and determine the y-coordinate. It appears that $f(1) = 3$.

b) The domain is the set $\{x \mid -4 \le x \le 3\}$, or $[-4, 3]$.

c) To locate any x-values for which $f(x) = 2$, locate 2 on the y-axis, then move horizontally to any points on the graph having second coordinate equal to 2. For any such points move vertically to the x-axis and read the x-coordinate. It appears that $f(x) = 2$ for $x = -3$.

d) The range is the set $\{y \mid -2 \le y \le 5\}$, or $[-2, 5]$.

29. a) To determine $f(1)$, locate 1 on the x-axis, then move up to the point on the graph whose x-coordinate is 1. Then move horizontally to the vertical axis and determine the y-coordinate. It appears that $f(1) = 1$.

b) The domain is the set $\{-3, -1, 1, 3, 5\}$.

c) To locate any x-values for which $f(x) = 2$, locate 2 on the y-axis, then move horizontally to any points on the graph having second coordinate equal to 2. For any such points move vertically to the x-axis and read the x-coordinate. It appears that $f(x) = 2$ for $x = 3$.

d) The range is the set $\{-1, 0, 1, 2, 3\}$.

31. a) To determine $f(1)$, locate 1 on the x-axis, then move up to the point on the graph whose x-coordinate is 1. Then move horizontally to the vertical axis and determine the y-coordinate. It appears that $f(1) = 4$.

b) The domain is the set $\{x \mid -3 \le x \le 4\}$, or $[-3, 4]$.

c) To locate any x-values for which $f(x) = 2$, locate 2 on the y-axis, then move horizontally to any points on the graph having second coordinate equal to 2. For any such points move vertically to the x-axis and read the x-coordinate. It appears that $f(x) = 2$ for $x = -1$ and $x = 3$.

d) The range is the set $\{y \mid -4 \le y \le 5\}$, or $[-4, 5]$.

33. a) To determine $f(1)$, locate 1 on the x-axis, then move up to the point on the graph whose x-coordinate is 1. Then move horizontally to the vertical axis and determine the y-coordinate. It appears that $f(1) = 1$.

b) The domain is the set $\{x \mid -4 < x \le 5\}$, or $(-4, 5]$.

c) To locate any x-values for which $f(x) = 2$, locate 2 on the y-axis, then move horizontally to any points on the graph having second coordinate equal to 2. For any such points move vertically to the x-axis and read the x-coordinate. It appears that $f(x) = 2$ for $\{x \mid 2 < x \le 5\}$, or $(2, 5]$.

d) The range is the set $\{-1, 1, 2\}$.

35. Domain: $\mathbb{R}$; range: $\mathbb{R}$

37. Domain: $\mathbb{R}$; range: $\{4\}$

39. Domain: $\mathbb{R}$; range: $\{y \mid y \ge 1\}$, or $[1, \infty)$

41. Domain: $\{x \mid x \text{ is a real number } and \ x \ne -2\}$; range: $\{y \mid y \text{ is a real number } and \ y \ne -4\}$

43. Domain: $\{x \mid x \ge 0\}$, or $[0, \infty)$; range: $\{y \mid y \ge 0\}$, or $[0, \infty)$

45. This graph passes the vertical line test, so this is a function.

47. This graph passes the vertical line test, so this is a function.

49. This graph does not pass the vertical line test, so this is not a function.

51. This graph does not pass the vertical line test, so this is not a function.

53. $g(x) = 2x + 3$

a) $g(0) = 2 \cdot 0 + 3 = 0 + 3 = 3$

b) $g(-4) = 2(-4) + 3 = -8 + 3 = -5$

c) $g(-7) = 2(-7) + 3 = -14 + 3 = -11$

d) $g(8) = 2 \cdot 8 + 3 = 16 + 3 = 19$

e) $g(a+2) = 2(a+2) + 3$
$= 2a + 4 + 3 = 2a + 7$

f) $g(a) + 2 = 2 \cdot a + 3 + 2 = 2a + 5$

55. $f(n) = 5n^2 + 4n$

a) $f(0) = 5 \cdot 0^2 + 4 \cdot 0 = 0$

b) $f(-1) = 5(-1)^2 + 4(-1)$
$= 5(1) - 4 = 5 - 4 = 1$

c) $f(3) = 5(3)^2 + 4(3) = 5(9) + 12$
$= 45 + 12 = 57$

d) $f(t) = 5t^2 + 4t$

e) $f(2a) = 5(2a)^2 + 4(2a) = 5(4a^2) + 8a$
$= 20a^2 + 8a$

f) $2 \cdot f(a) = 2(5a^2 + 4a) = 10a^2 + 8a$

57. $f(x) = \dfrac{x-3}{2x-5}$

a) $f(0) = \dfrac{0-3}{2 \cdot 0 - 5} = \dfrac{-3}{0-5} = \dfrac{-3}{-5} = \dfrac{3}{5}$

b) $f(4) = \dfrac{4-3}{2 \cdot 4 - 5} = \dfrac{1}{8-5} = \dfrac{1}{3}$

c) $f(-1) = \dfrac{-1-3}{2(-1)-5} = \dfrac{-4}{-2-5} = \dfrac{-4}{-7} = \dfrac{4}{7}$

d) $f(3) = \dfrac{3-3}{2 \cdot 3 - 5} = \dfrac{0}{6-5} = \dfrac{0}{1} = 0$

e) $f(x+2) = \dfrac{(x+2)-3}{2(x+2)-5}$
$= \dfrac{x-1}{2x+4-5} = \dfrac{x-1}{2x-1}$

59. $f(a) = a^2 + a - 1$

a) $f(-6) = (-6)^2 + (-6) - 1$
$= 36 - 6 - 1$
$= 29$

b) $f(1.7) = (1.7)^2 + (1.7) - 1$
$\quad = 2.89 + 1.7 - 1$
$\quad = 3.59$

61. $h(n) = 8 - n - \dfrac{1}{n}$

a) $h(0.2) = 8 - 0.2 - \dfrac{1}{0.2}$
$\quad = 8 - 0.2 - 5$
$\quad = 2.8$

b) $h\left(-\tfrac{1}{4}\right) = 8 - \left(-\tfrac{1}{4}\right) - \dfrac{1}{\left(-\tfrac{1}{4}\right)}$
$\quad = 8 + \tfrac{1}{4} + 4$
$\quad = 12\tfrac{1}{4} = 12.25$

63. $A(s) = s^2 \dfrac{\sqrt{3}}{4}$

$A(4) = 4^2 \cdot \dfrac{\sqrt{3}}{4} = 4\sqrt{3} \approx 6.9282032$

The area is $4\sqrt{3}$ cm^2, or approximately 6.93 cm^2.

65. $V(r) = 4\pi r^2$

$V(3) = 4\pi \cdot 3^2 = 4\pi \cdot 9 = 36\pi$

The surface area is 36π in^2, or approximately 113.10 in^2.

67. $P(d) = 1 + \left(\dfrac{d}{33}\right)$

$P(20) = 1 + \left(\dfrac{20}{33}\right) = 1\tfrac{20}{33}$ atm

$P(30) = 1 + \left(\dfrac{30}{33}\right) = 1\tfrac{10}{11}$ atm

$P(100) = 1 + \left(\dfrac{100}{33}\right) = 1 + 3\tfrac{1}{33} = 4\tfrac{1}{33}$ atm

69. $f(x) = 2x - 5$

$f(8) = 2 \cdot 8 - 5 = 16 - 5 = 11$

71. $f(x) = 2x - 5$
$-5 = 2x - 5$
$\quad 0 = 2x$
$\quad 0 = x$

73. $f(x) = \tfrac{1}{3}x + 4$
$\tfrac{1}{2} = \tfrac{1}{3}x + 4$
$\tfrac{1}{2} - 4 = \tfrac{1}{3}x$
$\tfrac{1}{2} - \tfrac{8}{2} = \tfrac{1}{3}x$
$-\tfrac{7}{2} = \tfrac{1}{3}x$
$-\tfrac{21}{2} = x$

75. $f(x) = \tfrac{1}{3}x + 4$
$f\left(\tfrac{1}{2}\right) = \tfrac{1}{3}\left(\tfrac{1}{2}\right) + 4 = \tfrac{1}{6} + \tfrac{24}{6} = \tfrac{25}{6}$

77. $f(x) = 4 - x$
If $f(x) = 7$, we must find x.
$7 = 4 - x$
$x = 4 - 7$
$x = -3$

79. $f(x) = 0.1x - 0.5$
If $f(x) = -3$, we must find x.
$-3 = 0.1x - 0.5$
$-3 + 0.5 = 0.1x$
$-2.5 = 0.1x$
$-25 = x$

81. $f(x) = \dfrac{5}{x - 3}$

The domain is any real number that does not make the denominator zero. If $x = 3$, then $x - 3 = 0$. So the domain is $\{x \mid x \text{ is any real number } and \ x \neq 3\}$.

83. $f(x) = \dfrac{x}{2x - 1}$

The domain is any real number that does not make the denominator zero. If $x = \tfrac{1}{2}$, then $2x - 1 = 0$. So the domain is $\{x \mid x \text{ is any real number } and \ x \neq \tfrac{1}{2}\}$.

85. $f(x) = 2x + 1$

The domain of this function is any real number, $\mathbb{R}$.

87. $f(x) = |5 - x|$

The domain of this function is any real number, $\mathbb{R}$.

89. $f(x) = \dfrac{5}{x-9}$

The domain is any real number that does not make the denominator zero. The denominator will be zero if $x = 9$. So the domain is $\{x \mid x$ is any real number *and* $x \neq 9\}$.

91. $f(x) = x^2 - 9$

The domain of this function is any real number, $\mathbb{R}$.

93. $f(x) = \dfrac{2x-7}{5}$

The domain of this function is any real number, $\mathbb{R}$.

95. $f(x) = \begin{cases} x, & \text{if } x < 0, \\ 2x+1, & \text{if } x \geq 0 \end{cases}$

a) $f(-5)$

Since $-5 < 0$, $f(x) = x$. Thus $f(-5) = -5$.

b) $f(0)$

Since $0 \geq 0$, $f(x) = 2x+1$. Thus,

$f(0) = 2 \cdot 0 + 1 = 1$.

c) $f(10)$

Since $10 \geq 0$, $f(x) = 2x+1$. Thus,

$f(10) = 2 \cdot 10 + 1 = 20 + 1 = 21$.

97. $G(x) = \begin{cases} x-5, & \text{if } x < -1, \\ x, & \text{if } -1 \leq x \leq 2 \\ x+2, & \text{if } x > 2 \end{cases}$

a) $G(0)$

Since $-1 \leq 0 \leq 2$, $G(x) = x$. Thus,

$G(0) = 0$.

b) $G(2)$

Since $-1 \leq 2 \leq 2$, $G(x) = x$. Thus,

$G(2) = 2$.

c) $G(5)$

Since $5 > 2$, $G(x) = x+2$. Thus,

$G(5) = 5 + 2 = 7$

99. $f(x) = \begin{cases} x^2 - 10 & \text{if } x < -10 \\ x^2, & \text{if } -10 \leq x \leq 10 \\ x^2 + 10, & \text{if } x > 10 \end{cases}$

a) $f(-10)$

Since $-10 \leq -10 \leq 10$, $f(x) = x^2$.

Thus, $f(-10) = (-10)^2 = 100$.

b) $f(10)$

Since $-10 \leq 10 \leq 10$, $f(x) = x^2$. Thus,

$f(10) = 10^2 = 100$.

c) $f(11)$

Since $11 > 10$, $f(x) = x^2 + 10$. Thus,

$f(11) = 11^2 + 10 = 121 + 10 = 131$.

101. $C(x) = 25x + 75$

25 signifies that the cost per person is $25. 75 signifies that the setup cost for the party is $75.

103. $D(t) = \frac{2}{3}t + \frac{10}{3}$

$\frac{2}{3}$ signifies that the consumption of renewable energy increases $\frac{2}{3}$ quadrillion Btu's per year, for years after 1960. $\frac{10}{3}$ signifies that the consumption of renewable energy was $\frac{10}{3}$ quadrillion Btu's in 1960.

105. $G(t) = \frac{1}{8}t + 2$

$\frac{1}{8}$ signifies that the grass grows $\frac{1}{8}$ in. per day. 2 signifies that the grass is 2 in. long when cut.

107. $P(t) = 0.21t + 5.43$

0.21 signifies that the price increases $0.21 per year, for years since 2000. 5.43 signifies that the average cost of a movie ticket in 2000 was $5.43.

109. $F(t) = -5000t + 90,000$

a) -5000 signifies that the depreciation is $5000 per year. $90,000$ signifies that the original value of the truck was $90,000.

b) We need to find the value of t for which $F(t) = 0$. We substitute 0 for $F(t)$ and solve for t.

$$0 = -5000t + 90,000$$
$$5000t = 90,000$$
$$t = 18$$

After 18 years, the truck will have fully depreciated.

c) The domain of F is the set $\{t \mid 0 \le t \le 18\}$, or $[0, 18]$.

After 7.5 years, the truck will have fully depreciated.

c) The domain of V is the set $\{t \mid 0 \le t \le 7.5\}$, or $[0, 7.5]$.

111. $R(t) = 46.8 - 0.075t$

a) 46.8 signifies that the 400-m record time in 1930 was 46.8 sec. -0.075 signifies that the record time decreases by 0.075 sec per year for years after 1930.

b) We need to find the value of t for which $R(t) = 38.7$. We substitute 38.7 for $R(t)$ and solve for t.

$$38.7 = 46.8 - 0.075t$$
$$0.075t = 46.8 - 38.7$$
$$0.075t = 8.1$$
$$t = \frac{8.1}{0.075} = 108$$

The record time for the 400-m run will occur 108 years after 1930, or in 2038.

c) To determine the domain, we must determine what value of t will result in $R(t) = 0$. We replace $R(t)$ by 0 and solve for t.

$$0 = 46.8 - 0.075t$$
$$0.075t = 46.8$$
$$t = \frac{46.8}{0.075} = 624$$

The domain of R is the set $\{t \mid 0 \le t < 624\}$, or $[0, 624)$, with the assumption that a time of 0 sec will not actually be reached.

113. *Thinking and Writing Exercise.*

115. $(-5)^3 = (-5) \cdot (-5) \cdot (-5)$
$$= -(5) \cdot (5 \cdot 5)$$
$$= -5 \cdot 25$$
$$= -125$$

117. $-2^6 = -(2 \cdot 2 \cdot 2 \cdot 2 \cdot 2 \cdot 2)$
$$= -(2 \cdot 2) \cdot (2 \cdot 2) \cdot (2 \cdot 2)$$
$$= -(4 \cdot 4 \cdot 4)$$
$$= -(4 \cdot 16)$$
$$= -64$$

119. $2 - (3 - 2^2) + 10 \div 2 \cdot 5 = 2 - (3 - 4) +$
$$(10 \div 2) \cdot 5$$
$$= 2 - (-1) + 5 \cdot 5$$
$$= 2 + 1 + 25$$
$$= 28$$

121. *Thinking and Writing Exercise.*

123. $f(x) = 3x^2 - 1$
$g(x) = 2x + 5$
$f(g(-4)) = f(2(-4) + 5)$
$$= f(-8 + 5) = f(-3) = 3(-3)^2 - 1$$
$$= 3 \cdot 9 - 1 = 27 - 1 = 26$$
$g(f(-4)) = g(3(-4)^2 - 1)$
$$= g(3 \cdot 16 - 1) = g(48 - 1)$$
$$= g(47) = 2 \cdot 47 + 5$$
$$= 94 + 5 = 99$$

125. $f\big(f\big(f\big(f\,(tiger)\big)\big)\big) =$
$$= f\big(f\big(f\,(dog)\big)\big)$$
$$= f\big(f\,(cat)\big)$$
$$= f\,(fish)$$
$$= worm$$

127. To find the time during the test when the largest contraction occurred, locate the highest point on the graph, and find the corresponding time on the *x*-axis, which is the first coordinate of the point. The time of the largest contraction was approximately 2 min 50 sec into the test.

129. The two largest contractions occurred at about 2 minutes, 50 seconds and 5 minute, 40 seconds. The difference in these times is 2 minutes 50 seconds, so the frequency is about 1 every 3 minutes.

131. We know that $(-1,-7)$ and $(3,8)$ are both solutions of $g(x) = mx + b$. Substituting, we have

$$-7 = m(-1) + b, \text{ or } -7 = -m + b,$$

and $8 = m(3) + b$, or $8 = 3m + b$.

Solve the first equation for b and substitute that expression into the second equation.

$-7 = -m + b$	First equation
$m - 7 = b$	Solving for b
$8 = 3m + b$	Second equation
$8 = 3m + (m - 7)$	Substituting
$8 = 3m + m - 7$	
$8 = 4m - 7$	
$15 = 4m$	
$\dfrac{15}{4} = m$	

We know that $m - 7 = b$, so $\dfrac{15}{4} - 7 = b$, or

$-\dfrac{13}{4} = b$. We have $m = \dfrac{15}{4}$ and $b = -\dfrac{13}{4}$, so

$$g(x) = \dfrac{15}{4}x - \dfrac{13}{4}.$$

133. $f(x) = mx + b$

$$f(cd) = m(cd) + b = mcd + b$$
$$f(c)f(d) = (mc + b)(md + b)$$
$$= m^2cd + mbd + mbc + b^2$$

Since $f(cd) \neq f(c)f(d)$, the statement is false.

135. $f(x) = mx + b$

$$f(c - d) = m(c - d) + b = mc - md + b$$
$$f(c) - f(d) = mc + b - (md + b)$$
$$= mc + b - md - b = mc - md$$

Since $f(c - d) \neq f(c) - f(d)$, the statement is false.

Chapter 3 Study Summary

1.

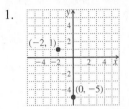

2. Since the first coordinate is negative and the second coordinate is negative, the point $(-10, -20)$ is in quadrant III.

3. $y = 2x + 1$

When $x = -2$, $y = 2(-2) + 1 = -4 + 1 = -3$.
When $x = 0$, $y = 2 \cdot 0 + 1 = 0 + 1 = 1$.
When $x = 2$, $y = 2 \cdot 2 + 1 = 4 + 1 = 5$.

x	y
-2	-3
0	1
2	5

Plot these points, draw the line they determine, and label the graph $y = 2x + 1$.

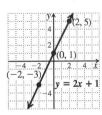

4. $10x - y = 10$

To find the *x*-intercept, we let $y = 0$ and solve for *x*:

$$10x - 0 = 10$$
$$10x = 10$$
$$x = 1$$

Thus, the *x*-intercept is $(1, 0)$.

To find the *y*-intercept, we let $x = 0$ and solve for *y*:

$$10 \cdot 0 - y = 10$$
$$-y = 10$$
$$y = -10$$

Thus, the y-intercept is $(0, -10)$.

5. $y = -2$

We can write this equation as $0 \cdot x + y = -2$.
No matter what number we choose for x, we find that y must be –2.

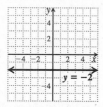

6. $x = 3$

We can write this equation as $x + 0 \cdot y = 3$.
No matter what number we choose for y, we find that x must be 3.

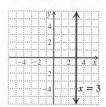

7.   The time interval for the breakfast was 45 min (from 8:30 until 9:15 is 45 minutes). During this time, the number of people served was $67 - 47 = 20$. The serving rate, in meals per minute, was

$$= \frac{20 \text{ meals}}{45 \text{ min}} = \frac{4}{9} \text{ meals/min}$$

The serving rate is $\frac{4}{9}$ meals/min.

8. $(1, 4)$ and $(-9, 3)$

$$m = \frac{3 - 4}{-9 - 1} = \frac{-1}{-10} = \frac{1}{10}$$

9.   The line $y = 10$ is a horizontal line. The slope is 0.

10. $y = -4x + \frac{2}{5}$

Slope is –4, and the y-intercept is $\left(0, \frac{2}{5}\right)$.

11. $y = \frac{1}{2}x + 2$

Slope: $\frac{1}{2}$; y-intercept: $(0, 2)$

First we plot the y-intercept $(0, 2)$. We can start at the y-intercept and use the slope, $\frac{1}{2}$, to find another point. We move up 1 unit and right 2 units to get a new point $(2, 3)$.

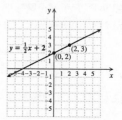

12. $y = 4x - 12$: The slope is 4, and the y-intercept is $(0, -12)$.

We rewrite the second equation in slope-intercept form:
$$4y = x - 9$$
$$y = \frac{1}{4}x - \frac{9}{4}$$

The slope is $\frac{1}{4}$, and the y-intercept is $\left(0, -\frac{9}{4}\right)$. Since the lines have different slopes, the graphs of their equations are not parallel.

13. $y = x - 7$
$$x + y = 3$$

The first equation is in slope-intercept form. It represents a line with slope 1. Now we rewrite the second equation in slope-intercept form.
$$x + y = 3$$
$$y = -x + 3$$

The slope is –1.
Since $1(-1) = -1$, the equations represent perpendicular lines.

14. $m = \frac{1}{4}; \left(-1, 6\right)$

The point-slope equation is
$y - y_1 = m\left(x - x_1\right)$.

We substitute $\frac{1}{4}$ for m, -1 for x_1, and 6 for
y_1 giving $y - 6 = \frac{1}{4}\left(x - \left(-1\right)\right)$.

15. $f\left(x\right) = 2 - 3x$
$$f\left(-1\right) = 2 - 3\left(-1\right)$$
$$= 2 + 3 = 5$$

16. This graph passes the vertical line test, so this is a function.

17. Domain: $\mathbb{R}$; range: $\left\{y \mid y \geq -2\right\}$, or $\left[-2, \infty\right)$

18. $f\left(x\right) = \frac{1}{4}x - 5$

The domain of this function is any real number, $\mathbb{R}$.

Chapter 3 Review Exercises

1. False

2. True

3. False

4. False

5. True

6. True

7. True

8. False

9. True

10. True

11. The total amount of online searches for August 2009 was 10.8 billion. Of that total, 16.0% were searches at Yahoo. Let $x =$ the number of searches for Yahoo alone. We reword and translate the problem.

What is 16.0% of 10.8 billion?

↓ ↓↓   ↓   ↓

$x \quad = 16.0\% \quad \cdot \quad 10.8$ billion

$x = 0.16 \cdot 10.8 = 1.728 \approx 1.7$

To check this answer, we note that if there were 1.7 billion searches at Yahoo, the total number of searches would be $\frac{1.7}{0.16} = 10.625$.

Since this is approximately the total number of online searches for August 2009, our answer checks.

In August 2009, the number of on-line searches that were for Yahoo alone was about 1.7 billion.

12. From the circle graph, we see that 64.6% of all on-line searches in August 2009 were at Google. Therefore the portion of the 4200 searches performed at Advanced Graphics was 64.6% of 4200 billion. Of this amount, 55% were for image searches. Let $x =$ the number of image searches on Google performed in August 2009 at Advanced Graphics. We reword and translate the problem.

What is 55% of $\left(64.6\% \text{ of } 4200\right)$?

↓   ↓↓   ↓   ↓   ↓   ↓

$x \quad = 55\% \quad \cdot \quad \left(64.6\% \quad \cdot \quad 4200\right)$

$x = 0.55 \cdot \left(0.646 \cdot 4200\right) \approx 1492$

To check the answer, we note that if the amount of image searches at Google in August 2009 was about 1492, then

$\dfrac{1492}{0.646 \cdot 0.55} \approx 4199$, which is nearly equal to

the total number of searches performed during that time at Adcanced Graphics.

About 1492 image searches were performed using Google by Advanced Graphics in August 2009.

13 – 15. Starting at the origin:

$\left(5, -1\right)$ is 5 units right and 1 unit down.

$\left(2, 3\right)$ is two units right and 3 units up.

$\left(-4, 0\right)$ is 4 units left and 0 units up or down.

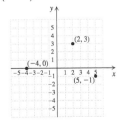

16. Since the first coordinate is positive and the second coordinate is negative, the point $(2,-6)$ is in quadrant IV.

17. Since the first coordinate is negative and the second coordinate is negative, the point $(-0.5,-12)$ is in quadrant III.

18. Since the second coordinate is 0, the point $(-8,0)$ must lie on the x-axis.

19. Point A is 5 units left and 1 unit down. The coordinates of A are $(-5,-1)$.

20. Point B is 2 units left and 5 units up. The coordinates of B are $(-2,5)$.

21. Point C is 3 units right and 0 units up or down. The coordinates of C are $(3,0)$.

22.

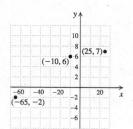

23. $y = 2x - 5$

   a) Test point $(3,1)$:

$$y = 2x - 5$$

$$\begin{array}{c|c} 1 & 2 \cdot 3 - 5 \\ 1 & 6 - 5 \\ 1 & 1 \qquad \text{TRUE} \end{array}$$

Since $1 = 1$ is a true statement, the ordered pair $(3,1)$ is a solution.

   b) Test point $(-3,1)$:

$$y = 2x - 5$$

$$\begin{array}{c|c} 1 & 2 \cdot (-3) - 5 \\ 1 & -6 - 5 \\ 1 & -11 \qquad \text{FALSE} \end{array}$$

Since $1 = -11$ is a false statement, the ordered pair $(-3,1)$ is not a solution.

24. First test the points $(0,-3)$ and $(2,1)$.

$$2x - y = 3$$

$$\begin{array}{c|c} 2 \cdot 0 - (-3) & 3 \\ 0 + 3 & 3 \\ 3 & 3 \quad \text{TRUE} \end{array}$$

$$2x - y = 3$$

$$\begin{array}{c|c} 2 \cdot 2 - 1 & 3 \\ 4 - 1 & 3 \\ 3 & 3 \quad \text{TRUE} \end{array}$$

Both points are solutions of the equation $2x - y = 3$.

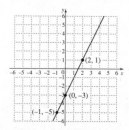

We can select several other points that also lie on this line, for example, $(-1,-5)$.

25. $y = x - 5$

Slope: 1; y-intercept: $(0,-5)$

We begin by plotting the y-intercept $(0,-5)$ and thinking of the slope as $\frac{1}{1}$, we move up 1 unit and right 1 unit to the point $(1,-4)$. Thinking of the slope as $\frac{3}{3}$ we can begin at the y-intercept $(0,-5)$ and move up 3 units and right 3 units to the point $(3,-2)$. We finish by connecting these points to form a line.

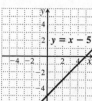

26. $y = -\frac{1}{4}x$

Rewriting this equation in slope-intercept form, we have $y = -\frac{1}{4}x + 0$.

Slope: $-\frac{1}{4}$; y-intercept: $(0,0)$

We begin by plotting the y-intercept $(0,0)$ and thinking of the slope as $\frac{-1}{4}$ we obtain a second point by moving down 1 unit and right 4 units to the point $(4,-1)$. Thinking of the slope as $\frac{1}{-4}$ we can begin at the y-intercept $(0,0)$ and move up 1 unit and left 4 units to the point $(-4,1)$. We finish by connecting these points to form a line.

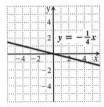

27. $y = -x + 4$

Slope: -1; y-intercept: $(0,4)$

We begin by plotting the y-intercept $(0,4)$ and thinking of the slope as $\frac{-1}{1}$, we move down 1 unit and right 1 unit to the point $(1,3)$. Thinking of the slope as $\frac{1}{-1}$ we can begin at the y-intercept $(0,4)$ and move up 1 units and left 1 unit to the point $(-1,5)$. We finish by connecting these points to form a line.

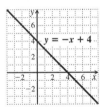

28. $4x + y = 3$

Rewriting this equation in slope-intercept form, we have:
$$4x + y = 3$$
$$y = -4x + 3$$

Slope: -4; y-intercept: $(0,3)$

We begin by plotting the y-intercept $(0,3)$ and thinking of the slope as $\frac{-4}{1}$, we move down 4 units and right 1 unit to the point

$(1,-1)$. Thinking of the slope as $\frac{4}{-1}$ we can begin at the y-intercept $(0,3)$ and move up 4 units and left 1 unit to the point $(-1,7)$. We finish by connecting these points to form a line.

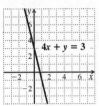

29. $4x + 5 = 3$

Rewriting this equation, we have:
$$4x + 5 = 3$$
$$4x = -2$$
$$x = -\frac{1}{2}$$

This is the equation of a vertical line with x-intercept $\left(-\frac{1}{2},0\right)$.

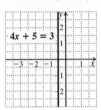

30. $5x - 2y = 10$

Rewriting this equation in slope-intercept form, we have:
$$5x - 2y = 10$$
$$-2y = -5x + 10$$
$$2y = 5x - 10$$
$$y = \frac{5}{2}x - 5$$

Slope: $\frac{5}{2}$; y-intercept: $(0,-5)$

We begin by plotting the y-intercept $(0,-5)$ and using the slope $\frac{5}{2}$, we move up 5 units and right 2 units to the point $(0,2)$. We finish by connecting these points to form a line.

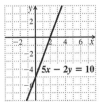

31. $y = x^2 + 1$

32. $2y - x = 8$

First solve the equation for y.

$2y - x = 8$

$2y = x + 8$

$y = \frac{1}{2}x + 4$

33.   Let the vertical axis represent g, the number of U.S. households, in millions, that use only natural garden products. Let the horizontal axis represent t, the number of years since 2004.

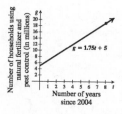

To estimate the number of households that use natural garden products in 2012, we note that 2012 is 8 years after 2004. Locate the point on the line above 8 on the horizontal axis. Then move left from that point to the vertical axis. It appears that the nuber of household using only natural garden products in 2012 will be about 19 million.

34.   The equation $y = x^2 - 2$ is not linear because the x-variable is squared.

35.   The equation $y = x - 2$ is in the form $y = mx + b$; therefore, it is a linear equation.

36.   a)   The driving rate in miles per minute is given by

$$\text{rate} = \frac{\text{change in distance}}{\text{change in time}}$$

$$= \frac{23 \text{ miles} - 17 \text{ miles}}{4\text{:}45 - 4\text{:}00}$$

$$= \frac{6 \text{ miles}}{45 \text{ min}} = \frac{2}{15} \text{ mi/min}$$

b)   The driving rate in minutes per mile is

$$\frac{15 \text{ min}}{2 \text{ miles}} = 7.5 \text{ min/mile} .$$

37.   We have two points on the graph, $(60, 5)$ and $(120, 10)$. The rate at which the vehicle was consuming gas is

$$\text{rate} = \frac{\text{change in amount of gas consumed}}{\text{change in number of miles driven}}$$

$$= \frac{10 \text{ gal} - 5 \text{ gal}}{120 \text{ mi} - 60 \text{ mi}}$$

$$= \frac{5 \text{ gal}}{60 \text{ mi}} = \frac{1}{12} \text{ gal/mi}$$

38.   a) y-intercept: $(0, -2)$

b) x-intercept: None

c)   Slope: 0

39.   a) y-intercept: $(0, 2)$

b) x-intercept: $(4, 0)$

c)   We can use the coordinates of the intercepts to determine the slope:

$$\text{slope} = \frac{0 - 2}{4 - 0} = \frac{-2}{4} = -\frac{1}{2}$$

40.   a) y-intercept: $(0, -3)$

b) x-intercept: $(2, 0)$

c)   We can use the coordinates of the intercepts to determine the slope:

$$\text{slope} = \frac{0 - (-3)}{2 - 0} = \frac{0 + 3}{2} = \frac{3}{2}$$

41. $(-2, 5)$ and $(3, -1)$

$$m = \frac{-1 - 5}{3 - (-2)} = \frac{-6}{3 + 2} = -\frac{6}{5}$$

42. $(5, 1)$ and $(-1, 1)$

$$m = \frac{1 - 1}{-1 - 5} = \frac{0}{-6} = 0$$

43. $(-3,0)$ and $(-3,5)$

$$m = \frac{5-0}{-3-(-3)} = \frac{5}{-3+3} = \frac{5}{0}$$

The slope of the line containing these points is undefined.

44. $(-8.3, 4.6)$ and $(-9.9, 1.4)$

$$m = \frac{1.4-4.6}{-9.9-(-8.3)} = \frac{1.4-4.6}{-9.9+8.3} = \frac{-3.2}{-1.6} = 2$$

45. The rise is 1 ft and the run is 12 ft.

$$\text{grade} = \frac{1 \text{ ft}}{12 \text{ ft}} = 0.083\overline{3} \approx 8.3\%$$

46. $y = -\frac{7}{10}x - 3$

The equation is already in slope-intercept form. The slope is $-\frac{7}{10}$.

47. $y = 5$

We can rewrite the equation in slope-intercept form as $y = 0 \cdot x + 5$. The slope is 0.

48. $x = -\frac{1}{3}$

This is the equation of a vertical line. The slope is undefined.

49. $3x - 2y = 6$

Rewrite the equation in slope-intercept form.
$$3x - 2y = 6$$
$$-2y = -3x + 6$$
$$2y = 3x - 6$$
$$y = \frac{3}{2}x - 3$$

The slope is $\frac{3}{2}$.

50. $5x - y = 30$

We can find the x-intercept by letting $y = 0$ and solving for x.
$$5x - y = 30$$
$$5x + 0 = 30$$
$$5x = 30$$
$$x = 6$$

The x-intercept is $(6,0)$.

We can find the y-intercept by letting $x = 0$ and solving or y.

$$5x - y = 30$$
$$5 \cdot 0 - y = 30$$
$$0 - y = 30$$
$$-y = 30$$
$$y = -30$$

The y-intercept is $(0,-30)$.

51. $2x + 4y = 20$

Rewrite the equation in slope-intercept form.
$$2x + 4y = 20$$
$$4y = -2x + 20$$
$$y = -\frac{1}{2}x + 5$$

The slope is $-\frac{1}{2}$ and the y-intercept is $(0,5)$.

52. $y + 5 = -x$
$$x - y = 2$$

Write both equations in slope-intercept form.
$$y + 5 = -x \quad \text{and} \quad x - y = 2$$
$$y = -x - 5 \qquad -y = -x + 2$$
$$\qquad\qquad\qquad y = x - 2$$

The slope of the first line is -1 and the slope of the second line is 1. The product of these two slopes is -1. Thus, the lines are perpendicular.

53. $3x - 5 = 7y$
$$7y - 3x = 7$$

Write both equations in slope-intercept form.
$$3x - 5 = 7y \quad \text{and} \quad 7y - 3x = 7$$
$$7y = 3x - 5 \qquad 7y = 3x + 7$$
$$y = \frac{3}{7}x - \frac{5}{7} \qquad y = \frac{3}{7}x + 1$$

Both lines have the same slope but different y-intercepts so the lines are parallel.

54. Slope: $-\frac{3}{4}$; y-intercept: $(0,6)$

The slope-intercept equation is $y - \frac{3}{4}x + 6$.

55. Slope: $-\frac{1}{2}$; point: $(3,6)$

The point-slope equation is
$$y - 6 = -\frac{1}{2}(x - 3).$$

56. $(1,-2)$ and $(-3,-7)$

First find the slope of the line containing these two points.

$$m = \frac{-7-(-2)}{-3-1} = \frac{-7+2}{-3-1} = \frac{-5}{-4} = \frac{5}{4}$$

We substitute either point into the point-slope equation and let $m = \frac{5}{4}$. We choose the point $(1,-2)$.

$$y-(-2) = \tfrac{5}{4}(x-1)$$
$$y+2 = \tfrac{5}{4}x - \tfrac{5}{4}$$
$$y = \tfrac{5}{4}x - 2 - \tfrac{5}{4}$$
$$y = \tfrac{5}{4}x - \tfrac{8}{4} - \tfrac{5}{4}$$
$$y = \tfrac{5}{4}x - \tfrac{13}{4}$$

57. Rewrite the equation $3x-5y=9$ in slope-intercept form.
$$3x-5y=9$$
$$-5y = -3x+9$$
$$y = \tfrac{3}{5}x - \tfrac{9}{5}$$

The slope of this line is $\frac{3}{5}$. The slope of a line perpendicular to this line is $-\frac{5}{3}$.

Substituting $-\frac{5}{3}$ for m and using the point $(2,-5)$ in the point-slope equation, we have

$$y-(-5) = -\tfrac{5}{3}(x-2)$$
$$y+5 = -\tfrac{5}{3}x + \tfrac{10}{3}$$
$$y = -\tfrac{5}{3}x - 5 + \tfrac{10}{3}$$
$$y = -\tfrac{5}{3}x - \tfrac{15}{3} + \tfrac{10}{3}$$
$$y = -\tfrac{5}{3}x - \tfrac{5}{3}$$

58. a) We have two points of the form (t,a):
$(3, 11.4)$ and $(6, 12)$. First determine the slope of the line containing these points.
$$m = \frac{12-11.4}{6-3} = \frac{0.6}{3} = 0.2$$
Using either point and this slope, we can determine the equation of the line. We choose the first point.
$$a-11.4 = 0.2(t-3)$$
$$a = 0.2t - 0.6 + 11.4$$
$$a = 0.2t + 10.8$$

b) To calculate the attendance in 2004, we note that 2004 is 4 years after 2000, and substitute 4 for t and solve for a.
$$a = 0.2t + 10.8$$
$$a = 0.2(4) + 10.8$$
$$a = 0.8 + 10.8$$
$$a = 11.6$$
In 2004, the attendance for U.S. arts performances was 11.6 million.

c) To predict the attendance in 2012, we note that 2012 is 12 years after 2000, and substitute 12 for t and solve for a.
$$a = 0.2t + 10.8$$
$$a = 0.2(12) + 10.8$$
$$a = 2.4 + 10.8$$
$$a = 13.2$$
In 2012, the attendance for U.S. arts performances is predicted to be 13.2 million.

59. $y = \tfrac{2}{3}x - 5$

Slope: $\frac{2}{3}$; y-intercept: $(0,-5)$

We begin by plotting the y-intercept $(0,-5)$ and using the slope $\frac{2}{3}$ we obtain a second point by moving up 2 units and right 3 units to the point $(3,-3)$. Thinking of the slope as $\frac{4}{6}$ we can begin at the y-intercept $(0,-5)$ and move up 4 units and right 6 units to the point $(6,-1)$. We finish by connecting these points to form a line.

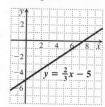

60. $2x + y = 4$

Rewriting the equation in slope-intercept form we have:
$$2x + y = 4$$
$$y = -2x + 4$$

Slope: -2; y-intercept: $(0,4)$

We begin by plotting the y-intercept $(0,4)$ and thinking of the slope as $\frac{-2}{1}$, we obtain a second point by moving down 2 units and right 1 unit to the point $(1,2)$. Thinking of the slope as $\frac{2}{-1}$ we can begin at the y-intercept $(0,4)$ and move up 2 units and left 1 unit to the point $(-1,6)$. We finish by connecting these points to form a line.

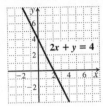

61. $y = 6$

This is the equation of a horizontal line with y-intercept $(0,6)$.

62. $x = -2$

This is the equation of a vertical line with x-intercept $(-2,0)$.

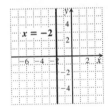

63. $y + 2 = -\frac{1}{2}(x-3)$

We rewrite the equation in point-slope form.
$y - (-2) = -\frac{1}{2}(x-3)$
The line has slope $-\frac{1}{2}$ and contains the point $(3,-2)$. We plot the point $(3,-2)$ and thinking of the slope as $\frac{1}{-2}$, we move up 1 unit and left 2 units to the point $(1,-1)$. We finish by connecting the points to form a line.

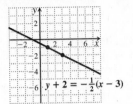

64.   Let list L_1 represent the number of years after 2000 and let list L_2 represent the number of revenue in billions and graph the points. The data appear to be linear.

65.   Let t = the number of years since 2000 and let F = the revenue, in billions of dollars, of U.S fitness and recreation centers. We enter the data with the number of years since 2000 as L_1 and the revenue, in bllions of dollars, as L_2. Using the LinReg feature of the graphing calculator, we find:

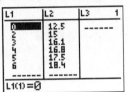

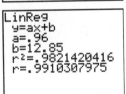

The second screen indicates that the equation is $F = 0.96t + 12.85$.

66.   To estimate the revenue in 2012, we note that 2012 is $2012 - 2000 = 12$ years after year 2000. We substitute 12 for t and solve for F.
$F = 0.96t + 12.85$
$F = 0.96(12) + 12.85$
$F = 11.52 + 12.85$
$F = 24.37$
The estimated revenue in 2012 will be about $24.4 billion.

67. a) To determine $f(2)$, locate 2 on the horizontal axis, then move up to the point on the graph having first coordinate 2. Then move horizontally to the vertical axis and read the y coordinate. It appears that $f(2) = 3$.

    b) The domain is the set $\{x \mid -2 \le x \le 4\}$, or $[-2, 4]$.

    c) To locate the x value for which $f(x) = 2$, locate 2 on the vertical axis, then move horizontally to the point on the graph having second coordinate 2. Then move down to the horizontal axis and read the x-coordinate. It appears that $f(x) = 2$ when $x = -1$.

    d) The range is the set $\{y \mid 1 \le y \le 5\}$ or $[1, 5]$.

68. $g(x) = x^2 - 10x$

    To find $g(-3)$, substitute -3 for x and simplify.
    $$g(-3) = (-3)^2 - 10(-3)$$
    $$= 9 + 30$$
    $$= 39$$

69. $f(x) = x^2 + 2x - 3$

    To find $f(2a)$, substitute $2a$ for x and simplify.
    $$f(2a) = (2a)^2 + 2(2a) - 3 = 4a^2 + 4a - 3$$

70. $A(t) = 0.11t + 7.9$

    a) To find $A(20)$, substitute 20 for t and simplify.
    $$A(20) = 0.11(20) + 7.9$$
    $$= 2.2 + 7.9$$
    $$= 10.1$$
    In 2015, the median age of the cars in the United States will be 10.1 years.

    b) 0.11 signifies that the median age of cars increases 0.11 yr per year; 7.9 signifies that the median age of cars was 7.9 years in 1995.

71. a) This graph passes the vertical line test so it represents a function.

    b) The domain is the set of all real numbers, $\mathbb{R}$, and the range is the set $\{y \mid y \ge 0\}$, or $[0, \infty)$.

72. This graph does not pass the vertical line test. It does not represent a function.

73. This graph does not pass the vertical line test. It does not represent a function.

74. a) This graph passes the vertical line test so it represents a function.

    b) The domain is the set of all real numbers, $\mathbb{R}$, and the range is the set $\{y \mid y = -2\}$, or $\{-2\}$.

75. $f(x) = 3x^2 - 7$

    The domain is the set of all real numbers, $\mathbb{R}$.

76. $g(x) = \dfrac{x^2}{x-1}$

    The domain is all real numbers, except any that make the denominator zero. The denominator is equal to 0 when $x = 1$. Therefore, the domain is the set $\{x \mid x \text{ is a real number } and\ x \ne 1\}$.

77. $f(x) = \begin{cases} 2-x, & \text{for } x \le -2 \\ x^2, & \text{for } -2 < x \le 5 \\ x+10 & \text{for } x > 5 \end{cases}$

    a) $f(-3)$

       Since $-3 \le -2$, $f(x) = 2 - x$; so $f(-3) = 2 - (-3) = 2 + 3 = 5$.

    b) $f(-2)$

       Since $-2 \le -2$, $f(x) = 2 - x$; so $f(-2) = 2 - (-2) = 2 + 2 = 4$.

    c) $f(4)$

       Since $-2 < 4 \le 5$, $f(x) = x^2$; so $f(4) = 4^2 = 16$.

d) $f(25)$

Since $25 > 5$, $f(x) = x + 10$; so

$f(25) = 25 + 10 = 35$.

b) $f(-2)$

Since $-2 \le -2$, $f(x) = 2 - x$; so

$f(-2) = 2 - (-2) = 2 + 2 = 4$.

c) $f(4)$

Since $-2 < 4 \le 5$, $f(x) = x^2$; so

$f(4) = 4^2 = 16$.

d) $f(25)$

Since $25 > 5$, $f(x) = x + 10$; so

$f(25) = 25 + 10 = 35$.

78. *Thinking and Writing Exercise.* Two perpendicular lines share the same y-intercept if their point of intersection is on the y-axis.

79. *Thinking and Writing Exercise.* Two functions that have the same domain and range are not necessarily identical. For example, the functions $f : \{(-2,1),(-3,2)\}$ and $g : \{(-2,2),(-3,1)\}$ have the same domain and range but are different functions.

80. If the point whose coordinates are $(-2,5)$ lies on the line having equation $y = mx + 3$, then it must satisfy the equation of that line. Therefore, substituting -2 for x and 5 for y, we have

$5 = m(-2) + 3$

$5 = -2m + 3$

$2 = -2m$

$-1 = m$

81. If the point whose coordinates are $(3,4)$ lies on the line having equation $y = -5x + b$, then it must satisfy the equation of that line. Therefore, substituting 3 for x and 4 for y, we have

$4 = -5(3) + b$

$4 = -15 + b$

$4 + 15 = b$

$19 = b$

82. We begin by making a sketch and plotting the points $(-2,2)$, $(7,2)$, and $(7,-3)$.

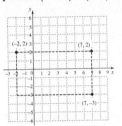

The rectangle is $2 - (-3) = 2 + 3 = 5$ units wide and $7 - (-2) = 7 + 2 = 9$ units long.

Therefore the area of the rectangle is $5 \cdot 9 = 45$ sq units, and the perimeter is 2×5 units $+ 2 \times 9$ units $= 10 + 18$ $= 28$ units.

83. $y = 4 - |x|$

Answers may vary. The domain includes all real numbers, $\mathbb{R}$. So, we can select any numbers we like to substitute for x. We choose 0, 1, and -1.

If $x = 0$, then $y = 4 - |0| = 4 - 0 = 4$.

If $x = 1$, then $y = 4 - |1| = 4 - 1 = 3$.

If $x = -1$, the $y = 4 - |-1| = 4 - 1 = 3$.

Thus, the possible solutions are $(0,4)$, $(1,3)$, and $(-1,3)$.

84. The domain is the set $\{x \mid x \ge -4 \text{ and } x \ne 2\}$.

The range is the set $\{y \mid y \ge 0 \text{ and } y \ne 3\}$.

Chapter 3 Test

1. The portion of the 1200 students at Rolling Hills College that volunteer is 25%. From the circle graph, we see that 27.7% of all volunteers between the ages of 16 and 24 volunteer in education or youth services. Let $x =$ the number of students at Rolling Hills College that volunteer for education or youth services. We reword and translate the problem.

What is 27.7% of $(25\%$ of $1200)$?

$\downarrow \quad \downarrow \quad \downarrow \quad \downarrow \quad \downarrow \quad \downarrow \quad \downarrow$

$x \;\; = 27.7\% \;\; \cdot \;\; (25\% \;\; \cdot \;\; 1200)$

$x = 0.277 \cdot (0.25 \cdot 1200) \approx 83$

To check this answer, we note that if there were 83 students at Rolling Hills College who volunteered for education or youth servies, the total number of students would be $\dfrac{83}{0.25 \cdot 0.277} \approx 1199.$ Since this is approximately the total number of students at Rolling Hills College, our answer checks.

The number of students at Rolling Hills College who volunteer for education or youth services is about 83.

2. The portion of the 3900 students at Valley Universiey that volunteer is $\frac{1}{3}$. From the circle graph, we see that 8.8% of all volunteers between the ages of 16 and 24 volunteer in hospital or health care. Let $x =$ the number of students at Valley University that volunteer for hospital or health care. We reword and translate the problem.

What is 8.8% of $\left(\frac{1}{3}$ of $3900\right)$?

$\downarrow \quad \downarrow \downarrow \quad \downarrow \quad \downarrow \downarrow \quad \downarrow$

$x \;\; = 8.8\% \;\; \cdot \;\; \left(\frac{1}{3} \;\; \cdot \;\; 3900\right)$

$x = 0.088 \cdot \left(\dfrac{3900}{3}\right) \approx 114$

To check this answer, we note that if there were 114 students at Valley University who volunteered for hospital or health care, the total number of students would be $\dfrac{114}{\frac{1}{3} \cdot 0.088} \approx 3886.$ Since this is nearly equal to the total number of students at Valley University, our answer checks.

The number of students at Valley University who volunteer for hospital or health care is about 114.

3. Since the first coordinate is 0, the point $(0,6)$ must lie on the y-axis.

4. Since the first coordinate is negative and the second coordinate is positive, the point $(-1.6, 2.3)$ is in quadrant II.

5. Point A is 3 units right and 4 units up. The coordinates of A are $(3,4)$.

6. Point B is 0 units left or right and 4 units down. The coordinates of B are $(0,-4)$.

7. Point C is 5 units left and 2 units up. The coordinates of C are $(-5,2)$.

8. $y = 2x - 1$

Slope: 2; y-intercept: $(0,-1)$.

First we plot the y-intercept $(0,-1)$. We can start at the y-intercept and thinking of the slope as $\dfrac{2}{1}$ we find another point. We move up 2 units and right 1 unit to get a new point $(1,1)$. Thinking of the slope as $\dfrac{4}{2}$ we can start at $(0,-1)$ and move up 4 units and right 2 units to get another point $(2,3)$. To finish, we draw and label the line.

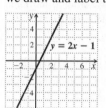

9. $2x - 4y = -8$

We rewrite the equation in slope-intercept form.

$2x - 4y = -8$

$\qquad -4y = -2x - 8$

$\qquad 4y = 2x + 8$

$\qquad y = \tfrac{1}{2}x + 2$

Slope: $\frac{1}{2}$; y-intercept: $(0,2)$

First we plot the y-intercept $(0,2)$. We can start at the y-intercept and use the slope $\dfrac{1}{2}$ to find another point. We move up 1 unit and right 2 units to get a new point $(2,3)$.

Thinking of the slope as $\dfrac{-1}{-2}$ we can start at $(0,2)$ and move down 1 unit and left 2 units to get another point $(-2,1)$. To finish, we draw and label the line.

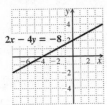

10. $y + 4 = -\frac{1}{2}(x - 3)$

Rewriting this equation in point-slope form, we have:

$y - (-4) = -\frac{1}{2}(x - 3)$

We have slope of $-\frac{1}{2}$ and a point having coordinates $(3, -4)$. Thinking of the slope as $\frac{-1}{2}$, we start at $(3, -4)$ and move down 1 unit and right 2 units to get a new point $(5, -5)$.

Thinking of the slope as $\frac{1}{-2}$, we move up 1 unit and left 2 units to get the point $(1, -3)$. To finish, we draw and label the line.

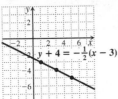

11. $y = \frac{3}{4}x$

We rewrite this equation in slope-intercept form.

$y = \frac{3}{4}x + 0$

Slope: $\frac{3}{4}$; y-intercept: $(0, 0)$

First we plot the y-intercept $(0, 0)$. We can start at the y-intercept and using the slope as $\frac{3}{4}$ we find another point. We move up 3 units and right 4 units to get a new point $(4, 3)$. Thinking of the slope as $\frac{-3}{-4}$ we can start at $(0, 0)$ and move down 3 units and left 4 units to get another point $(-4, -3)$. To finish, we draw and label the line.

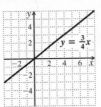

12. $2x - y = 3$

We rewrite this equation in slope-intercept form.

$2x - y = 3$

$-y = -2x + 3$

$y = 2x - 3$

Slope: 2; y-intercept: $(0, -3)$

First we plot the y-intercept $(0, -3)$. We can start at the y-intercept and thinking of the slope as $\frac{4}{2}$ we move up 4 units and right 2 units to get a new point $(2, 1)$. To finish, we draw and label the line.

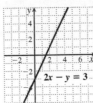

13. $x = -1$

This is a vertical line with x-intercept $(-1, 0)$.

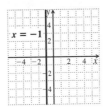

14. $1.2x - y = 5$

$-y = -1.2x + 5$

$y = 1.2x - 5$

15. $x - 2y = 16$

To find the x-intercept, we let $y = 0$ and solve for x.
$$x - 2 \cdot 0 = 16$$
$$x - 0 = 16$$
$$x = 16$$

The x-intercept is $(16, 0)$.

To find the y-intercept, we let $x = 0$ and solve for y.
$$0 - 2y = 16$$
$$-2y = 16$$
$$y = -8$$

The y-intercept is $(0, -8)$.

16. $(4, -1)$ and $(6, 8)$
$$m = \frac{8 - (-1)}{6 - 4} = \frac{8 + 1}{6 - 4} = \frac{9}{2}$$

17.   rate $= \dfrac{\text{change in distance}}{\text{change in time}}$

$\phantom{\text{rate }} = \dfrac{6 \text{ km} - 3 \text{ km}}{2:24 \; - 2:15}$

$\phantom{\text{rate }} = \dfrac{3 \text{ km}}{9 \text{ min}} = \dfrac{1}{3} \text{ km/min}$

18.   The rise is 63 ft and the run is 200 ft.

grade $= \dfrac{63 \text{ ft}}{200 \text{ ft}} = 0.315 = 31.5\%$

19. $y - 8x = 10$

Rewrite the equation in slope-intercept form.
$$y - 8x = 10$$
$$y = 8x + 10$$

The line has slope 8 and y-intercept $(0, 10)$.

20. $4y + 2 = 3x$

$-3x + 4y = -12$

Rewrite the equations in slope-intercept form.
$$4y + 2 = 3x$$
$$4y = 3x - 2$$
$$y = \tfrac{3}{4}x - \tfrac{1}{2}$$

Slope: $\tfrac{3}{4}$; y-intercept: $\left(0, -\tfrac{1}{2}\right)$

$-3x + 4y = -12$
$$4y = 3x - 12$$
$$y = \tfrac{3}{4}x - 3$$

Slope $\tfrac{3}{4}$; y-intercept: $(0, -3)$

Both lines have the same slope, but different y-intercepts so they are parallel.

21. $y = -2x + 5$

$2y - x = 6$

Rewrite the second equation in slope-intercept form.
$$2y - x = 6$$
$$2y = x + 6$$
$$y = \tfrac{1}{2}x + 3$$

The slope of the first line is -2 and the slope of the second line is $\tfrac{1}{2}$. The product of these slopes is $-2\left(\tfrac{1}{2}\right) = -1$. Therefore, the lines are perpendicular.

22.   Slope: -3; y-intercept: $(6, 8)$

Substituting -3 for m and 6 for x_1 and 8 for y_1, in the point slope formula,
$$y - y_1 = m(x - x_1).$$
$$y - 8 = -3(x - 6)$$

23.   First find the slope of the equation, $x - y = 3$, by rewriting it in slope-intercept form.
$$x - y = 3$$
$$-y = -x + 3$$
$$y = x - 3$$

The slope of this line is 1, so the slope of a line perpendicular to this line would have slope -1.

We need to determine the equation of a line having slope -1 and containing the point $(3, 7)$. Using the point-slope form of the equation of a line, we have:
$$y - 7 = -1(x - 3)$$
$$y - 7 = -x + 3$$
$$y = -x + 10$$

24. a) Let c = the number of hours the average commuter sits in traffic, and let t = the number of years after 1982. We have two points with coordinates $(0,16)$ and $(25,41)$. (Note: 2007 is 2007 – 1982 = 25 years after 1982.) We first determine the slope of the line containing these two points.

$$m = \frac{41-16}{25-0} = \frac{25}{25} = 1$$

Use the point-slope equation with $m = 1$ and $(t_1, c_1) = (0,16)$ to obtain the equation. (We could let $(t_1, c_1) = (25,41)$ instead and obtain an equivalent equation.)

$$c - 16 = 1(t - 0)$$
$$c - 16 = t$$
$$c = t + 16$$

b) To calculate the number of hours the average commuter was sitting in traffic in 2000, we note that 2000 is 2000 – 1982 = 18 years after 1982. We substitute 18 for t and solve for c.

$$c = t + 16$$
$$c = 18 + 16$$
$$c = 34$$

The average commuter spent 34 hours sitting in traffic in 2000.

c) To predict the number of hours the average commuter will sit in traffic in 2012, we note that 2012 is 2012 – 1982 = 30 years after 1982. We substitute 30 for t and solve for c.

$$c = t + 16$$
$$c = 30 + 16$$
$$c = 46$$

By the year 2012, the average commuter will spend 46 hours sitting in traffic.

25. Let t = the number of years since 1980 and let B = the number of twin births, in thousands. We enter the data with the number of years since 1980 as L_1 and the number of twin births, in thousands, as L_2. Using the LinReg feature of the graphing calculator, we find:

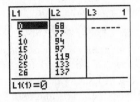

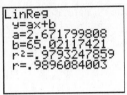

The second screen indicates that the equation is
$B = 2.6718t + 65.0212$.

26. To predict the number of twin births in 2012, we note that 2012 is 2012 – 1980 = 32 years after 1980. We substitute 32 for t and solve for B.

$$B = 2.6718t + 65.0212$$
$$B = 2.6718(32) + 65.0212$$
$$B = 85.4976 + 65.0212$$
$$F = 150.5188$$

The predicted number of twin births in 2012 will be about 151 thousand.

27. a) This graph passes the vertical line test so it represents a function.

b) The domain is the set $\{x \mid -4 \le x \le 5\}$, or $[-4,5]$. The range is the set $\{y \mid -2 \le y \le 4\}$, or $[-2,4]$.

28. a) This graph passes the vertical line test so it represents a function.

b) The domain is the set of all real numbers, $\mathbb{R}$, and the range is the set $\{y \mid y \ge 1\}$, or $[1,\infty)$.

29. This graph does not pass the vertical line test, so it is not a function.

30. a) To determine $f(3)$, locate 3 on the x-axis, then move up to the point on the graph whose x-coordinate is 3. Then move horizontally to the vertical axis and determine the y-coordinate. It appears that $f(3) = 0$.

b)   To locate any x-values for which $f(x) = -2$, locate –2 on the y-axis, then move horizontally to any points on the graph having a second coordinate equal to –2. For any such points move vertically to the x-axis and read the x-coordinate. It appears that $f(x) = -2$ for $x = -4$.

31. $g(x) = \dfrac{4}{2x+1}$

   a)   To find $g(1)$, substitute 1 for x in the expression and evaluate.

$$g(1) = \frac{4}{2 \cdot 1 + 1} = \frac{4}{2+1} = \frac{4}{3}$$

   b)   The domain is all real numbers except any that make $2x+1$ equal 0. Note that $2x+1 = 0$ only when $x = -\frac{1}{2}$. Therefore, the domain is the set $\left\{ x \mid x \text{ is any real number } and \ x \neq -\frac{1}{2} \right\}$.

32. $f(x) = \begin{cases} x^2, & \text{for } x < 0 \\ 3x - 5, & \text{for } 0 \le x \le 2 \\ x + 7 & \text{for } x > 2 \end{cases}$

   a) $f(0)$

       Since $0 \le 0 \le 2$, $f(x) = 3x - 5$; so $f(0) = 3 \cdot 0 - 5 = -5$.

   b) $f(3)$

       Since $3 > 2$, $f(x) = x + 7$; so $f(3) = 3 + 7 = 10$.

33.   We first find the slope of the line having equation $2x - 5y = 6$ by writing it in slope-intercept form.

$$2x - 5y = 6$$
$$-5y = -2x + 6$$
$$5y = 2x - 6$$
$$y = \tfrac{2}{5}x - \tfrac{6}{5}$$

This line has slope $\frac{2}{5}$ and any line parallel to this line would also have slope $\frac{2}{5}$.

Second, we must find the y-intercept of the line $3x + y = 9$ by writing it in slope-intercept form.

$$3x + y = 9$$
$$y = -3x + 9$$

The y-intercept of this line is $(0, 9)$.

We use the point-slope form of the line to determine the equation of a line having slope $\frac{2}{5}$ and containing the point $(0, 9)$.

$$y - 9 = \tfrac{2}{5}(x - 0)$$
$$y - 9 = \tfrac{2}{5}x$$
$$y = \tfrac{2}{5}x + 9$$

34.   First make a sketch.

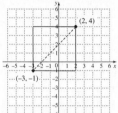

The height of the square is $4 - (-1) = 4 + 1 = 5$ and the width is $2 - (-3) = 2 + 3 = 5$. Therefore, the area of the square is $5 \times 5 = 25$ sq units and the perimeter is $4 \times 5 = 20$ units.

Chapters 1 – 3

Cumulative Review

1. $\dfrac{x}{5y}$

   We substitute 70 for x and 2 for y and evaluate.

   $$\dfrac{x}{5y} = \dfrac{70}{5(2)} = \dfrac{70}{10} = 7$$

2. $6(2a - b + 3) = 12a - 6b + 18$

3. $8x - 4y + 4 = 4(2x - y + 1)$

4. $54 = 2 \cdot 3 \cdot 3 \cdot 3 = 2 \cdot 3^3$

5. $-\dfrac{3}{20} = -\dfrac{15}{100} = -0.15$

6. $|-37| = 37$

7. $-\left(-\dfrac{1}{10}\right) = \dfrac{1}{10}$

8. $\dfrac{1}{-\frac{1}{10}} = -10$

9. $36.7\% = 0.367$

10. $\dfrac{3}{5} - \dfrac{5}{12} = \dfrac{36}{60} - \dfrac{25}{60} = \dfrac{11}{60}$

11. $(-2)(-1.4)(2.6) = 7.28$

12. $\dfrac{3}{8} \div \left(-\dfrac{9}{10}\right) = \dfrac{\overset{1}{\cancel{3}}}{\cancel{8}_{4}} \times \left(-\dfrac{\overset{5}{\cancel{10}}}{\cancel{9}_{3}}\right) = -\dfrac{5}{12}$

13. $1 + 6 \cdot 10 \div (-1) \cdot 2^2 = 1 + 6 \cdot 10 \div (-1) \cdot 4$
    $= 1 + 60 \div (-1) \cdot 4$
    $= 1 - 60 \cdot 4$
    $= 1 - 240$
    $= -239$

14. $1 - \left[32 \div \left(4 + 2^2\right)\right] = 1 - \left[32 \div (4 + 4)\right]$
    $= 1 - [32 \div 8]$
    $= 1 - 4$
    $= -3$

15. $-5 + 16 \div 2 \cdot 4 = -5 + 8 \cdot 4 = -5 + 32 = 27$

16. $y - (3y + 7) = y - 3y - 7 = -2y - 7$

17. $3(x - 1) - 2\left[x - (2x + 7)\right]$
    $= 3(x - 1) - 2[x - 2x - 7]$
    $= 3(x - 1) - 2(-x - 7)$
    $= 3x - 3 + 2x + 14$
    $= 5x + 11$

18. $2.7 = 5.3 + x$
    $2.7 - 5.3 = x$
    $-2.6 = x$

19. $\dfrac{5}{3}x = -45$
    $x = \dfrac{3}{5}(-45)$
    $x = -27$

20. $3x - 7 = 41$
    $3x = 48$
    $x = 16$

21. $\dfrac{3}{4} = \dfrac{-n}{8}$
    $8\left(\dfrac{3}{4}\right) = 8\left(\dfrac{-n}{8}\right)$
    $6 = -n$
    $-6 = n$

22. $14 - 5x = 2x$
    $14 = 2x + 5x$
    $14 = 7x$
    $2 = x$

23. $3(5-x)=2(3x+4)$
$15-3x=6x+8$
$-3x=6x+8-15$
$-3x-6x=6x-6x-7$
$-9x=-7$
$x=\dfrac{-7}{-9}=\dfrac{7}{9}$

24. $\frac{1}{4}x-\frac{2}{3}=\frac{3}{4}+\frac{1}{3}x$
$12\left(\frac{1}{4}x-\frac{2}{3}\right)=12\left(\frac{3}{4}+\frac{1}{3}x\right)$
$3x-8=9+4x$
$3x=9+4x+8$
$3x=4x+17$
$3x-4x=17$
$-x=17$
$x=-17$

25. $y+5-3y=5y-9$
$-2y+5=5y-9$
$-2y=5y-9-5$
$-2y=5y-14$
$-2y-5y=-14$
$-7y=-14$
$y=2$

26. $x-28<20-2x$
$x<20-2x+28$
$x<48-2x$
$x+2x<48$
$3x<48$
$x<16$
The solution set is $\{x\mid x<16\}$, or
$(-\infty,16)$.

27. $2(x+2)\ge 5(2x+3)$
$2x+4\ge 10x+15$
$2x\ge 10x+15-4$
$2x\ge 10x+11$
$2x-10x\ge 11$
$-8x\ge 11$
$x\le -\dfrac{11}{8}$

The solution set is $\left\{x\mid x\le -\frac{11}{8}\right\}$, or
$\left(-\infty,-\frac{11}{8}\right]$.

28. $A=2\pi rh+\pi r^2$ Solve for h.
$A-\pi r^2=2\pi rh$
$\dfrac{A-\pi r^2}{2\pi r}=h$

29. Since the first coordinate is positive and the second coordinate is negative, the point $(3,-1)$ is in quadrant IV.

30. $-1<x\le 2$
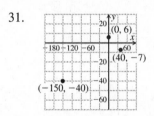

31.

32. $x=3$
This is the equation of a vertical line through the x-intercept $(3,0)$.

33. $2x-5y=10$
We can determine the x- and y-intercepts. We determine the y-intercept by letting $x=0$ and solving for y.
$2\cdot 0-5y=10$
$0-5y=10$
$-5y=10$
$y=-2$
The y-intercept is $(0,-2)$.

We can determine the x-intercept by letting $y=0$ and solving for x.

$$2x - 5 \cdot 0 = 10$$

$$2x = 10$$

$$x = 5$$

The x-intercept is $(5,0)$. We plot these two points and finish by drawing a line through them.

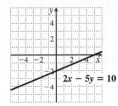

34. We graph $y = -2x + 1$.

When $x = -2$, $y = -2(-2) + 1 = 4 + 1 = 5$.

When $x = 0$, $y = -2(0) + 1 = 0 + 1 = 1$.

When $x = 2$, $y = -2(2) + 1 = -4 + 1 = -3$.

x	y
-2	5
0	1
2	-3

Plot the points, draw the line they determine, and label the graph $y = -2x + 1$.

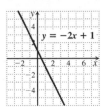

35. We graph $y = \frac{2}{3}x$.

When $x = -3$, $y = \frac{2}{3}(-3) = -2$.

When $x = 0$, $y = \frac{2}{3}(0) = 0$.

When $x = 3$, $y = \frac{2}{3}(3) = 2$.

x	y
-3	-2
0	0
3	2

Plot the points, draw the line they determine, and label the graph $y = \frac{2}{3}x$.

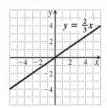

36. $y = -\frac{3}{4}x + 2$

Slope $-\frac{3}{4}$; y-intercept $(0, 2)$

We can think of the slope as $\frac{-3}{4}$, so from $(0, 2)$ we move down 3 units and right 4 units. This locates the point $(4, -1)$. We plot $(4, -1)$ and draw a line passing through $(0, 2)$ and $(4, -1)$.

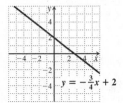

37. $2y - 5 = 3$

First, we rewrite the equation in slope-intercept form.

$$2y = 3 + 5$$

$$2y = 8$$

$$y = 4$$

This is the equation of a horizontal line through the y-intercept $(0, 4)$.

38. $2x - 7y = 21$

To find the x-intercept, we substitute 0 for y and solve for x.

$$2x - 7(0) = 21$$

$$2x = 21$$

$$x = \frac{21}{2}$$

The x-intercept is $\left(\frac{21}{2}, 0\right)$.

To find the y-intercept, we substitute 0 for x and solve for y.

$$2(0) - 7y = 21$$
$$-7y = 21$$
$$y = -3$$

The y-intercept is $(0, -3)$.

39. $y = 4x + 5$

To find the x-intercept, we substitute 0 for y and solve for x.

$$y = 4x + 5$$
$$0 = 4x + 5$$
$$-4x = 5$$
$$x = -\frac{5}{4}$$

The x-intercept is $\left(-\frac{5}{4}, 0\right)$.

To find the y-intercept, we substitute 0 for x and solve for y.

$$y = 4x + 5$$
$$y = 4(0) + 5$$
$$y = 5$$

The y-intercept is $(0, 5)$.

40. $3x - y = 2$

Rewrite the equation in the form slope-intercept form, $y = mx + b$.

$$3x - y = 2$$
$$-y = -3x + 2$$
$$y = 3x - 2$$

We can see that slope is 3 and the y-intercept is $(0, -2)$.

41. $(-4, 1)$ and $(2, -1)$

$$m = \frac{-1 - 1}{2 - (-4)} = \frac{-2}{2 + 4} = \frac{-2}{6} = -\frac{1}{3}$$

42.   Slope $\frac{2}{7}$ and y-intercept $(0, -4)$

Using the slope-intercept form, $y = mx + b$, substitute $\frac{2}{7}$ for m and -4 for b, giving

$$y = \frac{2}{7}x - 4.$$

43. $m = -\frac{3}{8}; (-6, 4)$

The point-slope equation is

$$y - y_1 = m(x - x_1).$$

We substitute $-\frac{3}{8}$ for m, -6 for x_1, and 4 for y_1 giving $y - 4 = -\frac{3}{8}(x - (-6))$.

44. $y - 4 = -\frac{3}{8}(x - (-6))$

Rewrite the equation in the form slope-intercept form, $y = mx + b$.

$$y - 4 = -\frac{3}{8}(x + 6)$$
$$y - 4 = -\frac{3}{8}x - \frac{18}{8}$$
$$y = -\frac{3}{8}x - \frac{18}{8} + 4$$
$$y = -\frac{3}{8}x - \frac{18}{8} + \frac{32}{8}$$
$$y = -\frac{3}{8}x + \frac{14}{8} = -\frac{3}{8}x + \frac{7}{4}$$

45.   From the graph, we note that the y-intercept is $(0, 1)$. Also, taking any two points on the line, we can compute the slope. We choose $(0, 1)$ and $(2, 5)$. The slope is

$$m = \frac{5 - 1}{2 - 0} = \frac{4}{2} = 2.$$ Using the slope intercept form, $y = mx + b$, we substitute 2 for m and 1 for b. The equation of the line is $y = 2x + 1$.

46.   Domain: $\{-5, -3, -1, 1, 3\}$

Range: $\{-3, -2, 1, 4, 5\}$

$$f(-3) = -2$$

The x-value for which $f(x) = 5$ is 3.

47. $f(x) = \dfrac{7}{2x - 1}$

a) $f(0) = \dfrac{7}{2(0) - 1} = \dfrac{7}{0 - 1} = \dfrac{7}{-1} = -7$

b)   The domain is any real number that does not make the denominator zero. The only real number that causes the denominator, $2x - 1$, to equal 0 is $\frac{1}{2}$. So the domain is the set $\left\{x \mid x \text{ is a real number } and \ x \neq \frac{1}{2}\right\}$.

48. a) Let r = the rate at which the person is cycling and c = the number of calories burned per hour. We have two points of the form (r,c). One point is $(6, 240)$ and one is $(12, 410)$. First determine the slope:

$$m = \frac{410 - 240}{12 - 6} = \frac{170}{6} = \frac{85}{3}$$

Next we write the point-slope form and solve for c.

$$c - 240 = \tfrac{85}{3}(r - 6)$$

$$c - 240 = \tfrac{85}{3}r - 170$$

$$c = \tfrac{85}{3}r + 70$$

Using the two given points, the line can then be drawn as shown below:

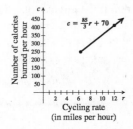

b) Let $r = 10$ mph. Substitute 10 for r in the equation $c = \tfrac{85}{3}r + 70$ and solve for c.

$$c = \tfrac{85}{3}(10) + 70$$

$$= \frac{850}{3} + \frac{210}{3} = \frac{1060}{3} = 353\tfrac{1}{3}$$

The number of calories burned will be approximately 353 per hour.

49. a) Let w = the weight of the bicyclist in pounds, and let $c =$ = the number of calories burned per hour at a speed of 12 mph. We enter the data with weight as L_1 and number of calories as L_2. Using the LinReg feature of the graphing calculator, we find:

L1	L2	L3	1
100	270	------	
150	410		
200	534		
------	------		
L1(1)=100			

```
LinReg
y=ax+b
a=2.64
b=8.666666667
r²=.9987771324
r=.9993883791
```

The second screen indicates that the equation is $c = 2.64w + 8.6\overline{6}$.

b) A 135-pound person would be expected to burn

$$c(135) = 2.64(135) + 8.6\overline{6}$$

$$= 356.4 + 8.6\overline{6}$$

$$\approx 365 \text{ calories per hour}$$

50. Let x = the mean earnings of an individual with a bachelor's degree in 2007.

$$\underbrace{54.7\% \text{ of the earnings with a bachelor's degree}}_{\downarrow} \quad \underbrace{\text{is}}_{\downarrow} \quad \underbrace{\$31,286.}_{\downarrow}$$

$$0.547 \cdot x \qquad = 31,286$$

$$0.547 \cdot x = 31,286$$

$$x \approx 57,196$$

To check this answer, we note that if an individual with a bachelor's degree earned $57,196, then $0.547 \cdot 57,196 = 31,286.21$. Since this is approximately the amount an individual with a high-school diploma would earn, our answer checks.

The mean earnings of an individual with a bachelors degree is about $57,196.

51. Let x = the number of Americans with O-negative blood, in millions. Then $x + 115$ represents the number of Americans, in millions, with O-positive blood.

$$\underbrace{\text{Number of Americans with O-positive or O-negative blood}}_{\downarrow} \quad \underbrace{\text{is}}_{\downarrow} \quad \underbrace{\text{136 million.}}_{\downarrow}$$

$$x + (x + 115) \qquad = 136$$

$$x + (x + 115) = 136$$

$$2x + 115 = 136$$

$$2x = 21$$

$$x = 10.5$$

If the number of people with O-negative blood is 10.5 million, then the number of people with O-positive blood is 10.5 + 115, or 125.5 million. Thus the total number of people with O-negative and O-positive blood is 10.5 + 125.5, or 136 million. These results check.

The number of Americans with O-negative blood is 10.5 million.

52. Let x = the cost of just the drill.

$$\underbrace{\text{Cost of the drill}}_{\downarrow} \text{ plus } \underbrace{\text{sales tax}}_{} \text{ is } \$126.$$

$$x \quad + \quad 0.05x \quad = \quad 126$$

$$x + 0.05x = 126$$
$$1.05x = 126$$
$$x = 120$$

If the original price of the drill alone was $120, then sales tax on the drill would have been 5% of $120, or $6. Thus, the total amount charged would have been $120 + $6, or $126, which checks.

The original price of the drill was $120.

53. Let x = the length of the first piece of wire, in meters. Then $x + 3$ = the length of the second piece, and $\frac{4}{5}x$ = the length of the third piece.

$$\underbrace{\text{The sum of the lengths of the wire}}_{\downarrow} \text{ is } 143 \text{ m.}$$

$$x + (x+3) + \frac{4}{5}x \qquad = 143$$

$$x + (x+3) + \frac{4}{5}x = 143$$
$$x + x + \frac{4}{5}x + 3 = 143$$
$$\frac{5}{5}x + \frac{5}{5}x + \frac{4}{5}x + 3 = 143$$
$$\frac{14}{5}x + 3 = 143$$
$$\frac{14}{5}x = 140$$
$$x = \frac{5}{14}(140)$$
$$x = 50$$

If the first length is 50 m, the second length is 50 + 3, or 53 m, and the third length is $\frac{4}{5}(50) = 40$ m. The sum of these lengths is 50 + 53 + 40, or 143 m, which checks.

The lengths of the three pieces are 50 m, 53 m, and 40 m.

54. Let x = the number of hours Clint must work on the fifth day (Friday) to qualify for availability pay.

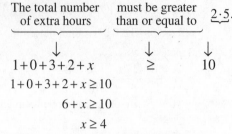

$$\underbrace{\text{The total number of extra hours}}_{\downarrow} \quad \underbrace{\text{must be greater than or equal to}}_{\downarrow} \quad \underbrace{2 \cdot 5}_{\downarrow}.$$

$$1+0+3+2+x \qquad \geq \qquad 10$$

$$1+0+3+2+x \geq 10$$
$$6+x \geq 10$$
$$x \geq 4$$

If Clint worked 4 hours on Friday, he has worked a total of 1 + 0 + 3 + 2 + 4, or 10 hours. If he worked 5 hours on Friday, he would have worked 1 + 0 + 3 + 2 + 5, or 11 hours. Because the answer is an inequality, we cannot check all possible values for x, but it is obvious that if the number of hours worked on Friday is 4 or greater, the inequality will be satisfied.

Clint must work at least 4 extra hours on the Friday.

55. Let x = Anya's salary at the beginning of the year.
If Anya received a 4% increase in February, then her new salary would have been $1.04x$. Then in June, when she received an additional cost-of-living adjustment, her new salary would have been $1.03(1.04x)$, which we are told equaled $26,780. So, we solve for x.

$$1.03(1.04x) = 26,780$$

$$x = \frac{26,780}{(1.03)(1.04)} = 25,000$$

Anya's salary at the beginning of the year was $25,000.

56. $4|x| - 13 = 3$
$$4|x| = 3 + 13$$
$$4|x| = 16$$
$$|x| = 4$$
$$x = -4 \text{ and } x = 4$$

57. $\dfrac{2+5x}{4} = \dfrac{11}{28} + \dfrac{8x+3}{7}$

$28\left(\dfrac{2+5x}{4}\right) = 28\left(\dfrac{11}{28} + \dfrac{8x+3}{7}\right)$

$7(2+5x) = 11 + 4(8x+3)$

$14+35x = 11+32x+12$

$14+35x = 32x+23$

$35x-32x = 23-14$

$3x = 9$

$x = 3$

58. $5(7+x) = (x+6)5$

$5(7+x) = 5(x+6)$

$35+5x = 5x+30$

$35 = 30$

Since this is a false statement, this equation has no solution. This is a contradiction.

59. $p = \dfrac{2}{m+Q}$ Solve for Q.

$(m+Q)p = (m+Q)\left(\dfrac{2}{m+Q}\right)$

$mp + Qp = 2$

$Qp = 2 - mp$

$Q = \dfrac{2-mp}{p}$, or $\dfrac{2-pm}{p}$

60. We begin by making a sketch and plotting the points $(-3,0)$, $(0,7)$, $(3,0)$, and $(0,-7)$.

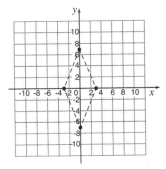

Find the equation of the line containing the points $(0,7)$ and $(3,0)$.

$m = \dfrac{0-7}{3-0} = \dfrac{-7}{3} = -\dfrac{7}{3}$

The y-intercept is $(0,7)$.

The equation is $y = -\dfrac{7}{3}x + 7$.

Find the equation of the line containing the points $(-3,0)$ and $(0,-7)$.

$m = \dfrac{-7-0}{0-(-3)} = \dfrac{-7}{3} = -\dfrac{7}{3}$

The y-intercept is $(0,-7)$.

The equation is $y = -\dfrac{7}{3}x - 7$.

Find the equation of the line containing the points $(3,0)$ and $(0,-7)$.

$m = \dfrac{0-(-7)}{3-0} = \dfrac{7}{3}$

The y-intercept is $(0,-7)$.

The equation is $y = \dfrac{7}{3}x - 7$.

Find the equation of the line containing the points $(-3,0)$ and $(0,7)$.

$m = \dfrac{7-0}{0-(-3)} = \dfrac{7}{3}$

The y-intercept is $(0,7)$.

The equation is $y = \dfrac{7}{3}x + 7$.

Chapter 4

Systems of Equations in Two Variables

1. True

3. True

5. True

7. False

9. We use alphabetical order for the variables.
Substitute 1 for x and 2 for y.

$4x - y = 2$

$$\begin{array}{c|c} 4(1) - (2) & 2 \\ \hline 4 - 2 & 2 \\ 2 & 2 \quad \text{TRUE} \end{array}$$

$10x - 3y = 4$

$$\begin{array}{c|c} 10(1) - 3(2) & 4 \\ \hline 10 - 6 & 4 \\ 4 & 4 \quad \text{TRUE} \end{array}$$

The ordered pair $(1, 2)$ is a solution of the system of equations.

11. We use alphabetical order for the variables.
Substitute -5 for x and 1 for y.

$x + 5y = 0$

$$\begin{array}{c|c} (-5) + 5(1) & 0 \\ \hline -5 + 5 & 0 \\ 0 & 0 \quad \text{TRUE} \end{array}$$

$y = 2x + 9$

$$\begin{array}{c|c} (1) & 2(-5) + 9 \\ \hline 1 & -10 + 9 \\ 1 & -1 \quad \text{FALSE} \end{array}$$

The ordered pair $(-5, 1)$ is not a solution of the system of equations.

13. We use alphabetical order for the variables.
Substitute 0 for x and -5 for y.

$x - y = 5$

$$\begin{array}{c|c} (0) - (-5) & 5 \\ \hline 0 - (-5) & 5 \\ 5 & 5 \quad \text{TRUE} \end{array}$$

$y = 3x - 5$

$$\begin{array}{c|c} (-5) & 3(0) - 5 \\ \hline -5 & 0 - 5 \\ -5 & -5 \quad \text{TRUE} \end{array}$$

The ordered pair $(0, -5)$ is a solution of the system of equations.

15. Observe that if we multiply both sides of the first equation by 2, we get the second equation. Thus, if we find that the given point makes the one equation true, we will know that it makes the other equation true also. Substitute 3 for x and 1 for y.

$3x + 4y = 13$

$$\begin{array}{c|c} 3(3) + 4(1) & 13 \\ \hline 9 + 4 & 13 \\ 13 & 13 \quad \text{TRUE} \end{array}$$

The ordered pair $(3, 1)$ is a solution of the system of equations.

17. Graph both equations.

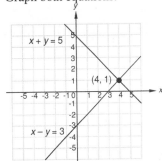

The solution (point of intersection) is apparently $(4, 1)$.

Check:

$x - y = 3$

$$\begin{array}{c|c} 4 - 1 & 3 \\ \hline 3 & 3 \quad \text{TRUE} \end{array}$$

$x + y = 5$

$$\begin{array}{c|c} 4 + 1 & 5 \\ \hline 5 & 5 \quad \text{TRUE} \end{array}$$

The solution is $(4, 1)$.

19. Graph the equations.

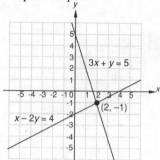

The solution (point of intersection) is apparently $(2,-1)$.

Check:

$$\begin{array}{c|c}
\multicolumn{2}{l}{3x+y=5} \\
\hline
3\cdot 2+(-1) & 5 \\
6-1 & 5 \\
5 & 5 \quad \text{TRUE}
\end{array}
\qquad
\begin{array}{c|c}
\multicolumn{2}{l}{x-2y=4} \\
\hline
2-2(-1) & 4 \\
2+2 & \\
4 & 4 \quad \text{TRUE}
\end{array}$$

The solution is $(2,-1)$.

21. Graph both equations.

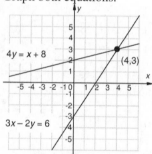

The solution point of intersection) is apparently $(4,3)$.

Check:

$$\begin{array}{c|c}
\multicolumn{2}{l}{4y=x+8} \\
\hline
4\cdot 3 & 4+8 \\
12 & 12 \quad \text{TRUE}
\end{array}
\qquad
\begin{array}{c|c}
\multicolumn{2}{l}{3x-2y=6} \\
\hline
3\cdot 4-2\cdot 3 & 6 \\
12-6 & \\
6 & 6 \quad \text{TRUE}
\end{array}$$

The solution is $(4,3)$.

23. Graph both equations.

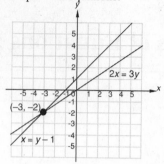

The solution (point of intersection) is apparently $(-3,-2)$.

Check:

$$\begin{array}{c|c}
\multicolumn{2}{l}{x=y-1} \\
\hline
-3 & -2-1 \\
-3 & -3 \quad \text{TRUE}
\end{array}
\qquad
\begin{array}{c|c}
\multicolumn{2}{l}{2x=3y} \\
\hline
2\cdot(-3) & 3\cdot(-2) \\
-6 & -6 \quad \text{TRUE}
\end{array}$$

The solution is $(-3,-2)$.

25. Graph both equations.

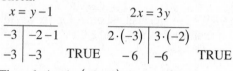

The solution (point of intersection) is apparently $(-3,2)$.

The ordered pair $(-3,2)$ checks in both equations. It is the solution.

27. Graph both equations.

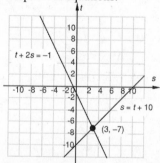

The solution (point of intersection) is apparently $(3,-7)$.

Check:

$$\begin{array}{c|c} t+2s=-1 \\ \hline -7+2\cdot 3 & -1 \\ -7+6 & \\ \hline & -1 \, | \, -1 \;\; \text{TRUE} \end{array} \qquad \begin{array}{c|c} s=t+10 \\ \hline 3 & -7+10 \\ 3 & 3 \quad \text{TRUE} \end{array}$$

The solution is $(3,-7)$.

29. Graph both equations.

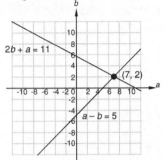

The solution (point of intersection) is apparently $(7,2)$.

Check:

$$\begin{array}{c|c} 2b+a=11 \\ \hline 2\cdot 2+7 & 11 \\ 4+7 & \\ \hline 11 & 11 \;\; \text{TRUE} \end{array} \qquad \begin{array}{c|c} a-b=5 \\ \hline 7-2 & 5 \\ & 5 \, | \, 5 \;\; \text{TRUE} \end{array}$$

The solution is $(7,2)$.

31. Graph both equations.

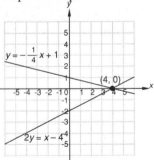

The solution (point of intersection) is apparently $(4,0)$.

Check:

$$\begin{array}{c|c} y=-\dfrac{1}{4}x+1 \\ \hline 0 & -\dfrac{1}{4}\cdot(4)+1 \\ & -1+1 \\ \hline 0 & 0 \qquad \text{TRUE} \end{array} \qquad \begin{array}{c|c} 2y=x-4 \\ \hline 2\cdot 0 & 4-4 \\ 0 & 0 \quad \text{TRUE} \end{array}$$

The solution is $(4,0)$.

33. Graph both equations.

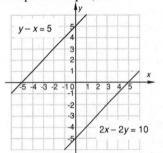

The lines are parallel. The system has no solution.

35. Graph both equations.

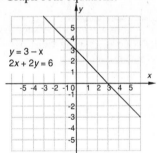

The graphs are the same. Any solution of one equation is a solution of the other. Each equation has infinitely many solutions. The solution set is the set of all pairs (x,y) for which $y=3-x$, or $\{(x,y)\,|\,y=3-x\}$. (In place of $y=3-x$, we could have used $2x+2y=6$ since the two equations are equivalent.)

37. Enter $y_1 = -5.43x + 10.89$ and

$y_2 = 6.29x - 7.04$ on a graphing calculator and use the INTERSECT feature.

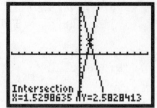

Intersection
X=1.5298635 Y=2.5828413

The solution is about $(1.53, 2.58)$.

39. Solve each equation for y. We get

$y = \dfrac{-2.6x + 4}{-1.1}$ and $y = \dfrac{3.12x - 5.04}{1.32}$. Graph

these equations on a graphing calculator, using a window that shows both graphs clearly. One choice is [–1, 1, –5, 0]. The graphs appear to be parallel.

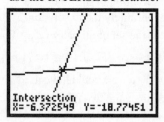

This is confirmed by the error message "NO SIGN CHNG" that is returned when we use the INTERSECT feature. The system of equations has no solution.

41. Solve each equation for y. We get

$y = 0.2x - 17.5$ and $y = \dfrac{10.6x + 30}{2}$. Graph

these equations on a graphing calculator and use the INTERSECT feature.

Intersection
X=-6.372549 Y=-18.77451

The solution is about $(-6.37, -18.77)$.

43. A system of equations is consistent if it has at least one solution. Of the systems under consideration, only the ones in Exercises 33 and 39 have no solution. Therefore, all except the systems in Exercise 33 and 39 are consistent.

45. A system of two equations in two variables is dependent if it has infinitely many solutions. Only the system in Exercise 35 is dependent.

47. (a) Let x represent the number of years since 1980 and y represent the number of full-time faculty, in thousands, for the first regression line, and the number of part-time faculty, in thousands, for the second regression line. To determine the regression equation for each type of employment, enter the following data in lists L1, L2, and L3:

L1	L2	L3
0	450	236
5	459	256
11	536	291
15	551	381
19	591	437
25	676	615

Press **STAT** and move the cursor to **CALC**. Under **CALC** select **LinReg(a*x* + b)** and select lists L1 and L2 to display the equation: $y \approx 9.0524x + 430.6778$.

(b) Press **STAT** again and move the cursor to **CALC**. Select **LinReg(a*x* + b)** and select lists L1 and L3 to display the equation: $y \approx 14.7175x + 185.3643$.

(c) Graph these equations on your calculator and use the INTERSECT feature. According to these equations, the number of full-time and part-time faculty will be the same in 2023 $(x \approx 43.30)$.

49. (a) Let x represent the number of years since 2004 and y represent the independent financial advisers, in thousands, for the first regression line, and the number of financial advisors with national firms, in thousands, for the second regression line. To determine the regression equation for each type of adviser, enter the following data in lists L1, L2, and L3:

L1	L2	L3
0	21	60
1	23	62
2	25	59
3	28	57
4	32	55

Press **STAT** and move the cursor to **CALC.**
Under **CALC** select **LinReg(ax + b)** and
select lists L1 and L2 to display the equation:
$y = 2.7x + 20.4$.

(b) Press **STAT** again and move the cursor
to **CALC.** Select **LinReg(ax + b)** and select
lists L1 and L3 to display the equation:
$y = -1.5x + 61.6$.

(c) Graph these equations on your calculator
and use the INTERSECT feature. According
to these equations, the number of independent
advisers will equal the number of financial
advisors with national firms in about 2014
$(x \approx 9.8)$.

51. *Thinking and Writing Exercise.*

53. $2(4x - 3) - 7x = 9$

$\quad\quad 8x - 6 - 7x = 9$ Removing parentheses

$\quad\quad\quad x - 6 = 9$ Collecting like terms

$\quad\quad\quad\quad x = 15$ Adding 6 to both sides

55. $4x - 5x = 8x - 9 + 11x$

$\quad -x = 19x - 9$ Collecting like terms

$\quad -20x = -9$ Adding $-19x$ to both sides

$\quad x = \dfrac{9}{20}$ Mult. both sides by $-\dfrac{1}{20}$

57. $3x + 4y = 7$

$\quad 4y = -3x + 7$ Add $-3x$ both sides

$\quad y = \dfrac{1}{4}(-3x + 7)$ Mult. both sides by $\dfrac{1}{4}$

$\quad y = -\dfrac{3}{4}x + \dfrac{7}{4}$

59. *Thinking and Writing Exercise.*

61. (a) There are many correct answers. One
can be found by expressing the sum and
difference of the two numbers:

$x + y = 6$

$x - y = 4$

(b) There are many correct answers. For
example, write an equation in two variables.
Then write a second equation by multiplying
the left side of the first equation by one
nonzero constant and multiplying the right
side by another nonzero constant.

$x + y = 1$

$2x + 2y = 3$

(c) There are many correct answers. One
can be found by writing an equation in two
variables and then writing a nonzero constant
multiple of that equation:

$x + y = 1$

$2x + 2y = 2$

63. Substitute 4 for x and -5 for y in the first
equation:

$A(4) - 6(-5) = 13$

$\quad 4A + 30 = 13$

$\quad 4A = -17$

$\quad A = -\dfrac{17}{4}$

Substitute 4 for x and -5 for y in the second
equation:

$4 - B(-5) = -8$

$\quad 4 + 5B = -8$

$\quad 5B = -12$

$\quad B = -\dfrac{12}{5}$

We have $A = -\dfrac{17}{4}$ and $B = -\dfrac{12}{5}$.

65.

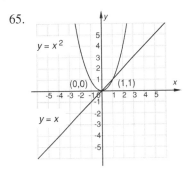

The solutions are apparently $(0,0)$ and $(1,1)$. Both pairs check.

67. The equations have the same slope and the same y-intercept. Thus their graphs are the same. Graph (c) matches this system.

69. The equations have the same slope and different y-intercepts. Thus their graphs are parallel lines. Graph (b) matches this system.

71. (a). *Familiarize.* The number of notebook PCs shipped in 2004 was 46 million and growing at a rate of $22\frac{2}{3} = \frac{68}{3}$ million per year. The number of desktop PCs shipped in 2004 was 140 million and growing at a rate of 4 million per year. Let $t =$ the number of years since 2004, and $n =$ the number of PCs shipped, in millions.

Translate. For a linear equation, the rate of growth represents the slope of the line, and the value at $t = 0$, in this case, the amount shipped in 2004 is the y-intercept. Thus for notebooks, the amount in millions shipped worldwide satisfies the equation:

$$n = \frac{68}{3}t + 46.$$

And for desktops the amount in millions is:
$$n = 4t + 140$$

(b) Graph $y_1 = \frac{68}{3}t + 46$ and $y_2 = 4t + 140$ on a graphing calculator and use the INTERSECT feature. The x value (or t value) is approximately 5. According to these equations, approximately 5 years after 2004, in 2009, the numbers of notebook PCs shipped worldwide will equal the number of desktop PCs shipped worldwide.

Exercise Set 4.2

---

1. False

3. True

5. $x + y = 5$ (1)

    $y = x + 3$ (2)

Substitute $x + 3$ for y in the first equation and solve for x.

$$x + (x + 3) = 5$$
$$x + x + 3 = 5$$
$$2x + 3 = 5$$
$$2x = 2$$
$$x = 1$$

Next substitute 1 for x in either equation of the original system and solve for y.

$$y = (1) + 3 = 4$$

Check the ordered pair (1, 4).

$x + y = 5$		$y = x + 3$	
$(1) + (4)$	5	(4)	$(1) + 3$
$1 + 4$	5	4	$1 + 3$
5	5 TRUE	4	4 TRUE

The ordered pair (1, 4) is the solution.

7. $x = y + 1$ (1)

   $x + 2y = 4$ (2)

Substitute $y + 1$ for x in the second equation and solve for y.

$$(y + 1) + 2y = 4$$
$$y + 1 + 2y = 4$$
$$3y + 1 = 4$$
$$3y = 3$$
$$y = 1$$

Next substitute 1 for y in either equation of the original system and solve for x.

$$x = y + 1 = 1 + 1 = 2$$

Check the ordered pair (2, 1).

$x = y + 1$		$x + 2y = 4$	
(2)	$(1) + 1$	$(2) + 2(1)$	4
2	$1 + 1$	$2 + 2$	4
2	2 TRUE	4	4 TRUE

The ordered pair (2,1) is the solution.

9. $y = 2x - 5$ (1)

  $3y - x = 5$ (2)

Substitute $2x - 5$ for y into the second equation and solve for x.

$$3y - x = 5 \quad (2)$$
$$3(2x - 5) - x = 5$$
$$6x - 15 - x = 5$$
$$5x - 15 = 5$$
$$5x = 20$$
$$x = 4$$

Next substitute 4 for x in either equation of the original system and solve for y.

$y = 2x - 5$ (1)

$y = 2 \cdot 4 - 5$

$y = 8 - 5$

$y = 3$

Check the ordered pair $(4, 3)$.

$$\begin{array}{c|c} y = 2x-5 & 3y-x=5 \\ \hline (3) & 2(4)-5 \\ 3 & 8-5 \\ 3 & 3 \text{ TRUE} \end{array} \qquad \begin{array}{c|c} 3y-x=5 \\ \hline 3(3)-4 & 5 \\ 9-4 & 5 \\ 5 & 5 \text{ TRUE} \end{array}$$

The ordered pair $(4, 3)$ is the solution.

11. $a = -4b$ (1)

$a + 5b = 5$ (2)

Substitute $-4b$ for a into the second equation and solve for b.

$a + 5b = 5$ (2)

$(-4b) + 5b = 5$

$-4b + 5b = 5$

$b = 5$

Next substitute 5 for b in either equation of the original system and solve for a.

$a = -4b$ (1)

$a = -4(5)$

$a = -20$

Check the ordered pair $(-20, 5)$.

$$\begin{array}{c|c} a = -4b \\ \hline (-20) & -4(5) \\ -20 & -20 \text{ TRUE} \end{array}$$

$$\begin{array}{c|c} a + 5b = 5 \\ \hline (-20) + 5(5) & 5 \\ -20 + 25 & 5 \\ 5 & 5 \text{ TRUE} \end{array}$$

The ordered pair $(-20, 5)$ is the solution.

13. $2x + 3y = 8$ (1)

$x = y - 6$ (2)

Substitute $y - 6$ for x in the first equation and solve for y.

$2x + 3y = 8$ (1)

$2(y - 6) + 3y = 8$

$2y - 12 + 3y = 8$

$5y - 12 = 8$

$5y = 20$

$y = 4$

Next substitute 4 for y in either equation of the original system and solve for x.

$x = y - 6$ (2)

$x = 4 - 6$

$x = -2$

Check the ordered pair $(-2, 4)$.

$$\begin{array}{c|c} 2x + 3y = 8 \\ \hline 2(-2) + 3 \cdot 4 & 8 \\ -4 + 12 \\ 8 & 8 \end{array} \qquad \begin{array}{c|c} x = y - 6 \\ \hline -2 & 4 - 6 \\ -2 & -2 \end{array}$$

TRUE   TRUE

The ordered pair $(-2, 4)$ is the solution.

15. $x = 2y + 1$ (1)

$3x - 6y = 2$ (2)

Substitute $2y + 1$ for x in the second equation and solve for y.

$3x - 6y = 2$ (2)

$3(2y + 1) - 6y = 2$

$6y + 3 - 6y = 2$

$3 = 2$

We get a false equation, or contradiction. The system has no solution.

17. $s + t = -4$ (1)

$s - t = 2$ (2)

Solve the first equation for s.

$s + t = -4$ (1)

$s = -t - 4$ (3)

Substitute $-t - 4$ for s in the second equation and solve for t.

$s - t = 2$ (2)

$(-t - 4) - t = 2$

$-2t - 4 = 2$

$-2t = 6$

$t = -3$

Next substitute -3 for t in Equation (3).

$s = -t - 4$　(3)

$s = -(-3) - 4$

$s = 3 - 4$

$s = -1$

The ordered pair $(-1, -3)$ works in both equations (1) and (2). It is the solution.

19.　$x - y = 5$　(1)

$x + 2y = 7$　(2)

Solve the first equation for x.

$x - y = 5$　　(1)

　$x = y + 5$　(3)

Substitute $y + 5$ for x in the second equation and solve for y.

$x + 2y = 7$　(2)

$(y + 5) + 2y = 7$

$3y + 5 = 7$

$3y = 2$

$y = \dfrac{2}{3}$

Next substitute $\frac{2}{3}$ for y in Equation (3).

$x = y + 5$　(3)

$x = \dfrac{2}{3} + \dfrac{15}{3}$

$x = \dfrac{17}{3}$

The ordered pair $\left(\frac{17}{3}, \frac{2}{3}\right)$ works in both equations (1) and (2). It is the solution.

21.　$x - 2y = 7$　　(1)

$3x - 21 = 6y$　(2)

Solve the first equation for x.

$x - 2y = 7$　　　(1)

　$x = 2y + 7$　(3)

Substitute $2y + 7$ for x in the second equation and solve for y.

$3x - 21 = 6y$　(2)

$3(2y + 7) - 21 = 6y$

$6y + 21 - 21 = 6y$

$6y = 6y$

We have an identity. The equations are dependent, and the solution set is infinite.

$\{(x, y) \mid x = 2y + 7\}$

23.　$y = 2x + 5$　　(1)

$-2y = -4x - 10$　(2)

Substitute $2x + 5$ for y in the second equation and solve for x.

$-2y = -4x - 10$　(2)

$-2(2x + 5) = -4x - 10$

$-4x - 10 = -4x - 10$

We have an identity. The equations are dependent, and the solution set is infinite.

$\{(x, y) \mid y = 2x + 5\}$

25.　$2x + 3y = -2$　(1)

$2x - \ y = 9$　　(2)

Solve the second equation for y.

$2x - y = 9$　　　(2)

　$-y = -2x + 9$

　$y = 2x - 9$　(3)

Substitute $2x - 9$ for y in the first equation and solve for x.

$2x + 3y = -2$　(1)

$2x + 3(2x - 9) = -2$

$2x + 6x - 27 = -2$

$8x - 27 = -2$

$8x = 25$

$x = \dfrac{25}{8}$

Next substitute $\frac{25}{8}$ for x in Equation (3).

$y = 2x - 9$　　　(3)

$y = 2\left(\dfrac{25}{8}\right) - 9$

$y = \dfrac{25}{4} - \dfrac{36}{4}$

$y = \dfrac{-11}{4}$

The ordered pair $\left(\frac{25}{8}, \frac{-11}{4}\right)$ works in both equations (1) and (2). It is the solution.

27.　$a - b = 6$　(1)

$3a - 4b = 18$　(2)

Solve the first equation for a.

$a - b = 6$　　(1)

　$a = b + 6$　(3)

Substitute $b + 6$ for a in the second equation and solve for b.

$3a - 4b = 18$ (2)

$3(b+6) - 4b = 18$

$3b + 18 - 4b = 18$

$-b + 18 = 18$

$-b = 0$

$b = 0$

Next substitute 0 for b in Equation (3) and solve for a.

$a = b + 6$ (3)

$a = 0 + 6$

$a = 6$

The ordered pair (6, 0) works in both equations (1) and (2). It is the solution.

29. $s = \dfrac{1}{2}r$ (1)

$3r - 4s = 10$ (2)

Substitute $\dfrac{1}{2}r$ for s in the second equations and solve for s.

$3r - 4s = 10$ (2)

$3r - 4\left(\dfrac{1}{2}r\right) = 10$

$3r - 2r = 10$

$r = 10$

Next substitute 10 for r in Equation (1).

$s = \dfrac{1}{2}r$ (1)

$s = \dfrac{1}{2}(10)$

$s = 5$

The ordered pair (10, 5) works in both equations (1) and (2). It is the solution.

31. $8x + 2y = 6$ (1)

$y = 3 - 4x$ (2)

Substitute $3 - 4x$ for y in the first equation and solve for x.

$8x + 2y = 6$ (1)

$8x + 2(3 - 4x) = 6$

$8x + 6 - 8x = 6$

$6 = 6$

We have an identity. The equations are dependent, and the solution set is infinite.

$\{(x, y) \mid y = 3 - 4x\}$

33. $x - 2y = 5$ (1)

$2y - 3x = 1$ (2)

Solve the first equation for x.

$x - 2y = 5$ (1)

$x = 2y + 5$ (3)

Substitute $2y + 5$ for x in the second equation and solve for y.

$2y - 3x = 1$ (2)

$2y - 3(2y + 5) = 1$

$2y - 6y - 15 = 1$

$-4y - 15 = 1$

$-4y = 16$

$y = -4$

Next substitute –4 for y in Equation (3).

$x = 2y + 5$ (3)

$x = 2(-4) + 5$

$x = -8 + 5$

$x = -3$

The ordered pair (–3, –4) works in both equations (1) and (2). It is the solution.

35. $2x - y = 0$ (1)

$2x - y = -2$ (2)

$2x - y$ cannot equal 0 [Equation (1)] *and* –2 [Equation (2)]. This system has no solution.

37.   **Familiarize.** Let x = the lesser number and y = the greater number.

Translate.

Greater Number	is	5 more than	Lesser Number
↓	↓	↓	↓
y	=	5 +	x

The sum	is	83
↓	↓	↓
$x + y$	=	83

Carry out. Solve the system of equations.

$y = 5 + x$ (1)

$x + y = 83$ (2)

Substitute $5 + x$ for y in Equation (2).

$x + y = 83$ (2)

$x + (5 + x) = 83$

$2x + 5 = 83$

$2x = 78$

$x = 39$

Substitute 39 for x in Equation (1).

$y = 5 + x$ (1)

$y = 5 + 39$

$y = 44$.

Check. 44 is 5 more than 39, and $44 + 39 = 83$. The numbers check.

State. The two numbers are 39 and 44.

39. **Familiarize.** Let x = one number and y = the second number.

Translate.

The sum is $93 \rightarrow x + y = 93$

The difference is $9 \rightarrow x - y = 9$

Carry out. Solve the system of equations.

$x + y = 93$ (1)

$x - y = 9$ (2)

Solve the second equation for x.

$x - y = 9$ (2)

$x = y + 9$ (3)

Substitute $y + 9$ for x in Equation (1).

$x + y = 93$ (1)

$y + 9 + y = 93$

$2y + 9 = 93$

$2y = 84$

$y = 42$

Substitute 42 for y in Equation (3).

$x = y + 9$

$x = 42 + 9$

$x = 51$

Check. $42 + 51 = 93$ and $51 - 42 = 9$.

The numbers check.

State. The two numbers are 42 and 51.

41. **Familiarize.** Let x = the larger number and y = the smaller number.

Translate.

Difference is $16 \rightarrow x - y = 16$

3 times larger is 7 times smaller

↓   ↓   ↓   ↓   ↓   ↓   ↓

3   · x = 7   · y

Carry out. Solve the system.

$x - y = 16$ (1)

$3x = 7y$ (2)

Solve Equation (1) for x.

$x - y = 16$ (1)

$x = y + 16$ (3)

Substitute $y + 16$ for x in Equation (2).

$3x = 7y$ (2)

$3(y + 16) = 7y$

$3y + 48 = 7y$

$48 = 4y$

$12 = y$

Substitute 12 for y in Equation (3).

$x = y + 16$ (3)

$x = 12 + 16$

$x = 28$

Check. $28 - 12 = 16$ and $3 \cdot 28 = 84 = 7 \cdot 12$.

The numbers check.

State. The numbers are 12 and 28.

43. **Familiarize.** Two angles are supplementary if the sum of their measures is 180°. Let x and y represent the measures of the two angles.

Translate. Since the angles are supplementary, $x + y = 180$.

one angle   is   15° more than twice the other angle

↓   ↓   ↓

y = $2x + 15$

Carry out. Solve the system of equations.

$x + y = 180$ (1)

$y = 2x + 15$ (2)

Substitute $2x + 15$ for y in Equation (1).

$x + y = 180$

$x + (2x + 15) = 180$

$3x + 15 = 180$

$3x = 165$

$x = 55$

Substitute 55 for x in Equation (2).

$y = 2x + 15$ (2)

$y = 2 \cdot 55 + 15$

$y = 110 + 15$

$y = 125$

Check.

$55 + 125 = 180$ and $2 \cdot 55 + 15$

$= 110 + 15 = 125$.

The numbers check.

State. The angles have measures of 55° and 125°.

45. **Familiarize.** Two angles are complementary if the sum of their measures is 90°. Let x and y represent the measures of the two angles.

Translate. Since the angles are complementary $x + y = 90$

$\underbrace{\text{Their difference}}$ is $18°$

$\quad\quad\downarrow\quad\quad\quad\downarrow\quad\downarrow$

$\quad\quad x - y\quad\quad = \quad 18$

Carry out. Solve the system of equations.

$x + y = 90$ (1)

$x - y = 18$ (2)

Solve equation (2) for x.

$x - y = 18 \quad\quad$ (2)

$\quad x = y + 18$ (3)

Substitute $y + 18$ for x in Equation (1).

$\quad\quad x + y = 90$ (1)

$(y + 18) + y = 90$

$\quad 2y + 18 = 90$

$\quad\quad 2y = 72$

$\quad\quad y = 36$

Substitute 36 for y in Equation (3).

$x = y + 18$

$x = 36 + 18$

$x = 54$

Check. $54 + 36 = 90$ and $54 - 36 = 18$.
The numbers check.

State. The angles have measures of $36°$ and $54°$.

47. **Familiarize.** $p = 2l + 2w$, where $l =$ the length of the poster, in inches and $w =$ the width, in inches.

Translate. $\underbrace{\text{Perimeter}}$ is $\underline{100}$

$\quad\quad\quad\downarrow\quad\quad\quad\downarrow\quad\downarrow$

$\quad\quad 2l + 2w\quad = \quad 100$

$\quad\quad\quad\quad\quad\text{6 inches more}$

$\underbrace{\text{Length}}$ is $\underbrace{\text{than the width}}$

$\quad\downarrow\quad\quad\downarrow\quad\quad\quad\downarrow$

$\quad l\quad\quad =\quad\quad w + 6$

Carry out. Solve the system of equations.

$2l + 2w = 100 \quad\quad$ (1)

$\quad l = w + 6$ (2)

Substitute $w + 6$ for l in Equation (1).

$2l + 2w = 100$ (1)

$2(w + 6) + 2w = 100$

$2w + 12 + 2w = 100$

$\quad 4w + 12 = 100$

$\quad\quad 4w = 88$

$\quad\quad w = 22$

Substitute 22 for w in Equation (2) to find l.

$l = w + 6$ (2)

$l = 22 + 6$

$l = 28$

Check. $2 \cdot 28 + 2 \cdot 22 = 56 + 44$

$\quad\quad\quad\quad\quad\quad = 100$

$\quad\quad\text{and } 28 = 22 + 6$

The numbers check.

State. The length is 28 in., and the width is 22 in.

49. **Familiarize.** Perimeter $2l + 2w$, where $l =$ the length and $w =$ the width, in miles.

Translate. $\underbrace{\text{Perimeter}}$ is $\underline{1300}$.

$\quad\quad\quad\downarrow\quad\quad\quad\downarrow\quad\quad\downarrow$

$\quad\quad 2l + 2w\quad = \quad 1300$

$\underbrace{\text{Width}}$ is $\underbrace{\text{110 less than the length}}$

$\quad\downarrow\quad\quad\downarrow\quad\quad\quad\quad\downarrow$

$\quad w\quad\quad =\quad\quad\quad l - 110$

Carry out. Solve the system of equations.

$2l + 2w = 1300 \quad\quad$ (1)

$\quad w = l - 110$ (2)

Substitute $l - 110$ for w in Equation (1).

$\quad\quad 2l + 2w = 1300$ (1)

$2l + 2(l - 110) = 1300$

$\quad 2l + 2l - 220 = 1300$

$\quad\quad\quad 4l = 1520$

$\quad\quad\quad l = 380$

Substitute 380 for l in Equation (2)

$w = l - 110$ (2)

$w = 380 - 110$

$w = 270$

Check.

$2(380) + 2(270) = 760 + 540$

$\quad\quad\quad\quad\quad\quad = 1300 \text{ and } 380 - 110 = 270.$

The numbers check.

State. Colorado is roughly 380 mi long and 270 mi wide.

51. *Familiarize.* $P = 2l + 2w$, l and w in yards.

Translate. Perimeter is $280 \rightarrow 2l + 2w = 280$

$\underline{\text{width}}$ is $\underline{\text{5 more than half the length}}$

$$\downarrow \quad \downarrow \qquad \qquad \downarrow$$
$$w \quad = \qquad \qquad \tfrac{1}{2}l + 5$$

Carry out. Solve the system of equations.

$$2l + 2w = 280 \qquad (1)$$
$$w = \frac{1}{2}l + 5 \quad (2)$$

Substitute $\tfrac{1}{2}l + 5$ for w in Equation (1).

$$2l + 2w = 280 \quad (1)$$
$$2l + 2\left(\frac{1}{2}l + 5\right) = 280$$
$$2l + l + 10 = 280$$
$$3l + 10 = 280$$
$$3l = 270$$
$$l = 90$$

Substitute 90 for l in Equation (2).

$$w = \frac{1}{2}l + 5 \quad (2)$$
$$w = \frac{1}{2} \cdot 90 + 5$$
$$w = 45 + 5$$
$$w = 50$$

Check.
$$2 \cdot 90 + 2 \cdot 50 = 180 + 100$$
$$= 280 \text{ and } 50 = \frac{1}{2} \cdot 90 + 5.$$

The numbers check.

State. The soccer field is 90 yd long and 50 yd wide.

53. *Familiarize.* Let h = height and w = width, both in feet.

Translate. Total is $25 \rightarrow h + w = 25$ and

height is 4 times width $\rightarrow h = 4w$.

Carry out. Solve the system of equations.
$$h + w = 25 \quad (1)$$
$$h = 4w \quad (2)$$

Substitute $4w$ for h in Equation (1).
$$h + w = 25 \quad (1)$$
$$4w + w = 25$$
$$5w = 25$$
$$w = 5$$

Substitute 5 for w in Equation (2).
$$h = 4w \quad (2)$$
$$h = 4 \cdot 5$$
$$h = 20$$

Check. $5 + 20 = 25$ and $4 \cdot 5 = 20$.
The numbers check.

State. The height is 20 ft; and the width is 5 ft.

55. *Thinking and Writing Exercise.*

57. $2(5x + 3y) - 3(5x + 3y) = 10x + 6y - 15x - 9y$
$$= (10 - 15)x + (6 - 9)y$$
$$= -5x - 3y$$

We could also work this problem using the distributive law, since $(5x + 3y)$ is a common factor.
$$2(5x + 3y) - 3(5x + 3y) = (5x + 3y)(2 - 3)$$
$$= (5x + 3y)(-1)$$
$$= -5x - 3y$$

59. $4(5x + 6y) - 5(4x + 7y)$
$$= 20x + 24y - 20x - 35y$$
$$= (20 - 20)x + (24 - 35)y$$
$$= -11y$$

61. $2(5x - 3y) - 5(2x + y)$
$$= 10x - 6y - 10x - 5y$$
$$= (10 - 10)x + (-6 - 5)y$$
$$= -11y$$

63. *Thinking and Writing Exercise.*

65. $\dfrac{1}{6}(a + b) = 1 \quad (1)$

$\dfrac{1}{4}(a - b) = 2 \quad (2)$

Multiply Equation (1) by 6 and Equation (2) by 4 to obtain integer coefficients.

$$6 \cdot \frac{1}{6}(a + b) = 6 \cdot 1$$
$$a + b = 6 \quad (3)$$
$$4 \cdot \frac{1}{4}(a - b) = 4 \cdot 2$$
$$a - b = 8 \quad (4)$$

Solve Equation (4) for a.
$$a - b = 8 \qquad (4)$$
$$a = b + 8 \quad (5)$$

Substitute $b + 4$ for a in Equation (3).

$a+b=6$ (3)

$b+8+b=6$

$2b+8=6$

$2b=-2$

$b=-1$

Substitute -1 for b in Equation (5).

$a=b+8$ (5)

$a=-1+8$

$a=7$

The ordered pair $(7, -1)$ checks in both equations. It is the solution.

67. $y+5.97=2.35x$ (1)

$2.14y-x=4.88$ (2)

Solve Equation (1) for y.

$y+5.97=2.35x$ (1)

$y=2.35x-5.97$ (3)

Substitute $2.35x-5.97$ for y in Equation (2).

$2.14y-x=4.88$ (2)

$2.14(2.35x-5.97)-x=4.88$

$5.029x-12.7758-x=4.88$

$4.029x-12.7758=4.88$

$4.029x=17.6558$

$x\approx 4.38$

Substitute 4.38 for x in Equation (3).

$y=2.35x-5.97$ (3)

$y=2.35(4.38)-5.97$

$y=10.293-5.97$

$y\approx 4.32$

The ordered pair $(4.38, 4.32)$ checks – remember, the answers are approximate values – in both equations. It is the solution.

69. *Familiarize.* Let $x=$ Trudy's age and $y=$ Dennis' age

   Translate.

   $\underline{\text{Trudy's}}$ $\underline{\text{is}}$ $\underbrace{\text{20 years younger than Dennis}}$

   $\quad\downarrow\qquad\downarrow\qquad\qquad\downarrow$

   $\quad x\qquad=\qquad\qquad y-20$

   $\underbrace{\begin{matrix}\text{Trudy's}\\\text{age}\end{matrix}}$ $\underline{\text{is}}$ $\underbrace{\begin{matrix}\text{more than half}\\\text{of Dennis' age}\end{matrix}}$

   $\quad\downarrow\qquad\downarrow\qquad\downarrow$

   $\quad x\qquad=\qquad 7+\frac{1}{2}y$

   Carry out. Solve the system of equations.

$x=y-20$ (1)

$x=7+\dfrac{1}{2}y$ (2)

Substitute $y-20$ for x is Equation (2).

$x=7+\dfrac{1}{2}y$ (2)

$y-20=7+\dfrac{1}{2}y$

$y=27+\dfrac{1}{2}y$

$2\cdot\dfrac{1}{2}y=27\cdot 2$

$y=54$

Substitute 54 for y in Equation (1).

$x=y-20$ (1)

$x=54-20$

$x=34$

Check. $34=\dfrac{1}{2}\cdot 54+7$

$\qquad\qquad = 27+7$ and $54-20=34.$

The numbers check.

State. The youngest age at which Trudy can marry Dennis is 34 yr.

71. $x+y+z=180$ (1)

$x=z-70$ (2)

$2y-z=0$ (3)

Solve Equation (3) for y.

$2y-z=0$ (3)

$2y=z$

$y=\dfrac{1}{2}z$ (4)

Substitute $\frac{1}{2}z$ for y and $z-70$ for x in Equation (1).

$x+y+z=180$ (1)

$(z-70)+\dfrac{1}{2}z+z=180$

$2\dfrac{1}{2}z-70=180$

$\dfrac{2}{5}\cdot\dfrac{5}{2}z=250\cdot\dfrac{2}{5}$

$z=100$

Substitute 100 for z in Equation (2).

$x=z-70$ (2)

$x=100-70$

$x=30$

Substitute 100 for z in Equation (4).

$$y = \frac{1}{2}z \quad (4)$$

$$y = \frac{1}{2} \cdot 100$$

$$y = 50$$

The ordered triple (30, 50, 100) checks in all three equations. It is the solution.

73. *Thinking and Writing Exercise.*

Exercise Set 4.3

---

1. False

3. True

5. $\begin{array}{l} x - y = 6 \\ \underline{x + y = 12} \\ 2x \quad = 18 \quad \text{Adding} \\ \quad x = 9 \end{array}$

   $x + y = 12 \rightarrow \begin{array}{r} 9 + y = 12 \\ y = 3 \end{array}$

   Check.

   $\begin{array}{c|c} \underline{x - y =} & \underline{x + y = 12} \\ 9 - 3\,?\,6 & 9 + 3\ ?\,12 \\ 6\mid 6 & 12 \mid 12 \end{array}$

   TRUE   TRUE

   Since (9, 3) checks, it is the solution.

7. $\begin{array}{l} x + \ y = \ 6 \\ \underline{-x + 3y = -2} \\ \quad 4y = \ 4 \quad \text{Adding} \\ \quad \ y = \ 1 \end{array}$

   $x + y = 6 \rightarrow \begin{array}{r} x + 1 = 6 \\ x = 5 \end{array}$

   Check.

   $\begin{array}{c|c} \underline{x + y = 6} & \underline{-x + 3y = -2} \\ 5 + 1\,?\,6 & -5 + 3(1)\ ?\,-2 \\ 6 \mid 6 & -5 + 3 \,\big|\, \\ & -2 \,\big|\, -2 \end{array}$

   TRUE   TRUE

   Since (5, 1) checks, it is the solution.

9. $\begin{array}{l} 4x - \ y = \ 1 \\ \underline{3x + \ y = 13} \quad \text{Adding} \\ 7x \qquad = 14 \\ \quad x \qquad = \ 2 \end{array}$

   $3x + y = 13 \rightarrow \begin{array}{r} 3 \cdot 2 + y = 13 \\ 6 + y = 13 \\ y = 7 \end{array}$

   The ordered pair (2, 7) checks in both equations; it is the solution.

11. $\begin{array}{l} 5a + 4b = \ 7 \\ \underline{-5a + \ b = \ 8} \\ \quad 5b = 15 \\ \quad \ b = \ 3 \end{array}$

   $5a + 4b = 7 \rightarrow \begin{array}{r} 5a + 4 \cdot 3 = \ 7 \\ 5a + 12 = \ 7 \\ 5a = -5 \\ a = -1 \end{array}$

   The ordered pair (−1, 3) checks in both equations; it is the solution.

13. $\begin{array}{l} 8x - 5y = \ -9 \\ \underline{3x + 5y = \ -2} \\ 11x \qquad = -11 \\ \quad x \qquad = \ -1 \end{array}$

   $3x + 5y = -2 \rightarrow \begin{array}{r} 3(-1) + 5y = -2 \\ 5y = \ 1 \\ y = \dfrac{1}{5} \end{array}$

   The ordered pair $\left(-1, \frac{1}{5}\right)$ checks in both equations; it is the solution.

15. $\begin{array}{l} 3a - 6b = \ 8 \\ \underline{-3a + 6b = -8} \\ \quad 0 = \ 0 \end{array}$

   This is an identity.
   $\{(a, b) \mid 3a - 6b = 8\}$

17. $-x-y=8$ (1)

    $2x-y=-1$ (2)

Multiply Equation (1) by -1 and add the result to Equation (2).

$$\begin{array}{rcl} x + y &=& -8 \\ 2x - y &=& -1 \\ \hline 3x &=& -9 \\ x &=& -3 \end{array}$$

$x+y=-8 \rightarrow -3+y = -8$

$\qquad\qquad\qquad\qquad y = -5$

The ordered pair $(-3, -5)$ checks in both equations; it is the solution.

19. $x+3y=19$ (1)

    $x-y=-1$ (2)

Multiply Equation (2) by -1 and add the result to Equation (1).

$$\begin{array}{rcl} x + 3y &=& 19 \\ -x + y &=& 1 \\ \hline 4y &=& 20 \\ y &=& 5 \end{array}$$

$x-y=-1 \rightarrow x-5 = -1$

$\qquad\qquad\qquad\qquad x = 4$

The ordered pair $(4, 5)$ checks in both equations; it is the solution.

21. $8x-3y=-6$ (1)

    $5x+6y=75$ (2)

Multiply Equation (1) by 2, add the result to Equation (2), and solve for x.

$16x-6y=-12$

$$\begin{array}{rcl} 5x+6y &=& 75 \\ \hline 21x &=& 63 \\ x &=& 3 \end{array}$$

Substitute $x=3$ into equation (1) or (2) to solve for y.

$5x+6y=75$ (2)

$5(3)+6y=75$

$15+6y=75$

$6y=60$

$y=10$

The ordered pair $(3, 10)$ works in both equations (1) and (2). It is the solution.

23. $2w-3z=-1$ (1)

    $-4w+6z=5$ (2)

Multiply Equation (1) by 2 and add the result to Equation (2).

$$\begin{array}{rcl} 4w - 6z &=& -2 \\ -4w + 6z &=& 5 \\ \hline 0 &=& 3 \end{array}$$

This is a contradiction. The system has no solution.

25. $4a+6b=-1$ (1)

    $a-3b=2$ (2)

Multiply Equation (2) by 2, add the result to Equation (1), and solve for a.

$4a+6b=-1$

$$\begin{array}{rcl} 2a-6b &=& 4 \\ \hline 6a &=& 3 \end{array}$$

$$a=\frac{3}{6}=\frac{1}{2}$$

Substitute $a=\frac{1}{2}$ into equation (1) or (2) to solve for b.

$4a+6b=-1$ (1)

$4\left(\frac{1}{2}\right)+6b=-1$

$2+6b=-1$

$6b=-3$

$$b=-\frac{3}{6}=-\frac{1}{2}$$

The ordered pair $\left(\frac{1}{2}, -\frac{1}{2}\right)$ works in both equations (1) and (2). It is the solution.

27. $3y=x \rightarrow x-3y = 0$ (1)

    $5x+14=y \rightarrow 5x-y=-14$ (2)

Multiply Equation (2) by -3 and add the result to Equation (1).

$$\begin{array}{rcl} x-3y &=& 0 \\ -15x+3y &=& 42 \\ \hline -14x &=& 42 \\ x &=& -3 \end{array}$$

$3y=x \rightarrow 3y = -3$

$\qquad\qquad\qquad\quad y = -1$

The ordered pair $(-3, -1)$ checks in both of the original equations; it is the solution.

29. $4x - 10y = 13$ (1)

$-2x + 5y = 8$ (2)

Multiply Equation (2) by 2 and add the result to Equation (1).

$4x - 10y = 13$

$\underline{-4x + 10y = 16}$

$0 = 29$

This is a contradiction. The system has no solution.

31. $8n + 6 - 3m = 0 \rightarrow -3m + 8n = -6$ (1)

$32 = m - n \rightarrow m - n = 32$ (2)

Multiply Equation (2) by 3 and add the result to Equation (1)

$-3m + 8n = -6$

$\underline{3m - 3n = 96}$

$5n = 90$

$n = 18$

$m - n = 32 \rightarrow m - 18 = 32$

$m = 50$

The ordered pair (50, 18) checks in both of the original equations; it is the solution.

33. $3x + 5y = 4$ (1)

$-2x + 3y = 10$ (2)

Multiply Equation (1) by 2, Equation (2) by 3, and add the resulting equations.

$6x + 10y = 8$

$\underline{-6x + 9y = 30}$

$19y = 38$

$y = 2$

$3x + 5y = 4 \rightarrow 3x + 5 \cdot 2 = 4$

$3x + 10 = 4$

$3x = -6$

$x = -2$

The ordered pair (−2, 2) checks in both equations, it is the solution.

35. $0.06x + 0.05y = 0.07$ (1)

$0.4x - 0.3y = 1.1$ (2)

Multiply Equation (1) by 6 and add the result to Equation (2)

$0.36x + 0.30y = 0.42$

$\underline{0.40x - 0.30y = 1.10}$

$0.76x = 1.52$

$x = 2$

$0.4x - 0.3y = 1.1 \rightarrow 0.4 \cdot 2 - 0.3y = 1.1$

$0.8 - 0.3y = 1.1$

$-0.3y = 0.3$

$y = -1$

The ordered pair (2, −1) checks in both equations; it is the solution.

37. $x + \dfrac{9}{2}y = \dfrac{15}{4}$ (1)

$\dfrac{9}{10}x - y = \dfrac{9}{20}$ (2)

Multiply Equation (1) by $\frac{2}{9}$ and add the result to Equation (2)

$\dfrac{2}{9}x + y = \dfrac{5}{6}$

$\dfrac{9}{10}x - y = \dfrac{9}{20}$

$\overline{\left(\dfrac{2}{9} + \dfrac{9}{10}\right)x = \dfrac{5}{6} + \dfrac{9}{20}}$

$\dfrac{20 + 81}{90}x = \dfrac{50 + 27}{60}$

$\dfrac{90}{101} \cdot \dfrac{101}{90}x = \dfrac{77}{60} \cdot \dfrac{90}{101}$

$x = \dfrac{77 \cdot \cancel{90}\,3}{\cancel{60} \cdot 101}$

$x = \dfrac{231}{202}$

$\dfrac{9}{10}x - y = \dfrac{9}{20} \rightarrow \dfrac{9}{10} \cdot \dfrac{231}{202} - y = \dfrac{9}{20}$

$\dfrac{2079}{2020} - y = \dfrac{9}{20}$

$-y = \dfrac{9}{20} - \dfrac{2079}{2020}$

$-y = \dfrac{909 - 2079}{2020}$

$-y = \dfrac{-1170}{2020}$

$y = \dfrac{117}{202}$

The ordered pair $\left(\frac{231}{202}, \frac{117}{202}\right)$ checks in both equations; it is the solution.

39. **Familiarize.** Let x = the number of miles, and c = cost, in dollars.

Translate.

$\underbrace{\text{Cost of}\atop\text{cargo van}}$ is $\$27$ plus $\underbrace{22\text{¢}\cdot\text{miles}}$

$\qquad\downarrow\qquad\downarrow\quad\downarrow\quad\downarrow\qquad\downarrow$

$\quad c_{\text{van}}\quad=\quad27\quad+\quad0.22x$

$\underbrace{\text{Cost of}\atop\text{pickup}}$ is $\$29$ plus $\underbrace{17\text{¢}\cdot\text{miles}}$

$\qquad\downarrow\qquad\downarrow\quad\downarrow\quad\downarrow\qquad\downarrow$

$\quad c_{\text{truck}}\quad=\quad29\quad+\quad0.17x$

Carry out. Solve the system.

$c_{\text{van}}=27+0.22x$ (1)

$c_{\text{truck}}=29+0.17x$ (2)

Set $c_{\text{van}}=c_{\text{truck}}$ and solve for x.

$27+0.22x=29+0.17x$

$(0.22-0.17)x=29-27$

$.05x=2$

$x=40$

Check. The cost of renting the van and driving 40 miles is:

$27+0.22(40)=27+8.8=\$35.80$

The cost of renting the truck and driving 40 miles is:

$29+0.17(40)=29+6.8=\$35.80$

The costs are equal. The answer checks.

State. The cost will be the same for 40 mi.

41. **Familiarize.** Let x and y equal the measures of the two angles, in degrees. Two angles are complementary if the sum of their measures is 90°.

Translate. Two angles are complementary:

$x+y=90°$ (1)

Their difference is 38°:

$x-y=38$ (2)

Carry out. Solve the system.

Add Equation (1) to Equation (2) and solve for x.

$\begin{array}{rcl} x+y & = & 90 \\ x-y & = & 38 \\ \hline 2x & = & 128 \\ x & = & 64 \end{array}$

Then use equation (1) or (2) to find y.

$x+y=90$ (1)

$64+y=90$

$y=26$

Check. $64+26=90$, and

$64-26=38$

The numbers check

State. The two angles measure 26° and 64°.

43. **Familiarize.** Let x = the number of minutes and c = the monthly cost, in dollars.

Translate.

$\underbrace{\text{Monthly}\atop\text{PowerNet cost}}$ is $\$1.99$ plus $\underbrace{43\text{¢}\cdot\text{minutes}}$

$\qquad\downarrow\qquad\qquad\downarrow\quad\downarrow\quad\downarrow\qquad\downarrow$

$\quad c_P\qquad=\quad1.99\quad+\quad0.43x$

$\underbrace{\text{Monthly}\atop\text{AT\&T cost}}$ is $\$3.99$ plus $\underbrace{28\text{¢}\cdot\text{minutes}}$

$\qquad\downarrow\qquad\downarrow\quad\downarrow\quad\downarrow\qquad\downarrow$

$\quad c_A\quad=\quad3.99\quad+\quad0.28x$

Carry out. Set $c_P=c_A$ and solve for x.

$c_P=c_A$

$1.99+0.43x=3.99+0.28x$

$0.43x-0.28x=3.99-1.99$

$0.15x=2$

$x=\dfrac{2}{0.15}=\dfrac{200}{15}=13\dfrac{1}{3}$

Check. PowerNet costs:

$1.99+0.43\left(13\dfrac{1}{3}\right)=1\dfrac{99}{100}+\dfrac{43}{100}\left(\dfrac{40}{3}\right)$

$=\dfrac{199}{100}+\dfrac{43}{100}\left(\dfrac{40}{3}\right)$

$=\dfrac{199}{100}+\dfrac{1720}{300}$

$=\dfrac{597}{300}+\dfrac{1720}{300}=\dfrac{2317}{300}=\$7.72\overline{3}$

AT&T costs:

$$3.99 + 0.28\left(13\frac{1}{3}\right) = 3\frac{99}{100} + \frac{28}{100}\left(\frac{40}{3}\right)$$

$$= \frac{399}{100} + \frac{28}{100}\left(\frac{40}{3}\right)$$

$$= \frac{399}{100} + \frac{1120}{300}$$

$$= \frac{1197}{300} + \frac{1120}{300} = \frac{2317}{300} = \$7.72\overline{3}$$

The costs are equal (when rounded to the nearest penny, \$7.72). The answer checks.
State. The costs will be the same for $13\frac{1}{3}$ min $= 13$ min 20 sec.

45. **Familiarize.** Let x and y equal the measures of the two angles, in degrees. Two angles are supplementary if the sum of their measures is $180°$.
Translate. Supplementary $\rightarrow x + y = 180°$
Their difference is $68° \rightarrow x - y = 68°$
Carry out. Solve the system.

$$x + y = 180 \quad (1)$$
$$x - y = 68 \quad (2)$$

Add the two equations and solve for x,

$$\begin{array}{r} x + y = 180 \\ x - y = 68 \\ \hline 2x = 248 \\ x = 124 \end{array}$$

Use equation (1) or (2) to find y.

$$x + y = 180 \quad (1)$$
$$124 + y = 180$$
$$y = 180 - 124 = 56$$

Check. $124 + 56 = 180$ and $124 - 56 = 68$. The numbers check
State. The measures of the two angles are $124°$ and $56°$.

47. **Familiarize.** Let $x =$ the number of acres of Chardonnay grapes and $y =$ the number of acres of Riesling grapes.
Translate. Total acres equal $820 \rightarrow x + y = 820$

$$\underbrace{140 \text{ more than}}_{} \quad \underbrace{\substack{\text{Riesling} \\ \text{acres}}}_{} \quad \text{is} \quad \underbrace{\substack{\text{Chardonnay} \\ \text{acres}}}_{}$$
$$\downarrow \qquad\qquad \downarrow \quad \downarrow \qquad \downarrow$$
$$140 \quad + \qquad y \quad = \qquad x$$

Carry out. Solve the system.
$$x + y = 820 \quad (1)$$
$$140 + y = x \quad (2)$$
$$x - y = 140 \quad (3)$$

$$\begin{array}{r} x + y = 820 \\ x - y = 140 \\ \hline 2x = 960 \\ x = 480 \end{array}$$

$$140 + y = x \rightarrow 140 + y = 480$$
$$y = 340$$

Check. $480 + 340 = 820$ and $140 + 340 = 480$. The numbers check.
State. South Wind Vineyards plants 340 acres of Riesling grapes and 480 acres of Chardonnay grapes.

49. **Familiarize.** $P = 2l + 2w$, where $l =$ the length and $w =$ the width, in feet
Translate. Perimeter is $18 \rightarrow 18 = 2l + 2w$
Length is twice width $\rightarrow l = 2w$
Carry out. Solve the system.

$$2l + 2w = 18 \quad (1)$$
$$l = 2w \quad (2)$$
$$l - 2w = 0 \quad (3)$$

$$\begin{array}{r} 2l + 2w = 18 \\ l - 2w = 0 \\ \hline 3l = 18 \\ l = 6 \end{array}$$

$$l = 2w \rightarrow 6 = 2w$$
$$3 = w$$

Check. $2 \cdot 6 + 2 \cdot 3 = 12 + 6 = 18$ and $12 = 2 \cdot 6$. The numbers check.
State. The dimensions of the frame should be 6 ft long and 3 ft wide.

51. *Thinking and Writing Exercise.*

53. $12.2\% = \dfrac{12.2}{100} = 0.122$

55. **Translate:**
What percent of 65 is 26?
$$\downarrow \quad\quad \downarrow \quad\quad \downarrow \;\downarrow\;\downarrow\;\downarrow$$
$$x \quad \left(\frac{1}{100}\right) \quad \cdot \quad 65 = 26$$

Solve:

$$x\left(\frac{1}{100}\right)\cdot 65 = 26$$

$$x = \frac{26\cdot 100}{65} = 40$$

State:

26 is 40% of 65.

57. Let x equal the unknown number of liters. Then 12% of the number of liters is:

$$12\cdot\left(\frac{1}{100}\right)\cdot x = 0.12x$$

59. *Thinking and Writing Exercise.*

61. $x+y=7$

$$3(y-x)=9 \to y-x=3$$

$$\begin{array}{r} x+y = 7 \\ -x+y = 3 \\ \hline 2y = 10 \\ y = 5 \end{array}$$

$$x+y=7 \to x+5 = 7$$
$$x \;\;= 2$$

The solution is $(2, 5)$.

63. $2(5a-5b)=10 \to 10a-10b = 10$

$$\underline{-5(2a+6b)=10 \to -10a-30b = 10}$$
$$-40b = 20$$

$$b = -\frac{20}{40} = -\frac{1}{2}$$

$$10a-10b=10 \to a-b = 1$$

$$a-\left(-\frac{1}{2}\right) = 1$$

$$a+\frac{1}{2} = \frac{2}{2}$$

$$a = \frac{1}{2}$$

The solution is $\left(\frac{1}{2}, -\frac{1}{2}\right)$.

65. $y = \dfrac{-2}{7}x+3$

$$y = \frac{4}{5}x+3$$

Since $y = y$, $\dfrac{-2}{7}x+3 = \dfrac{4}{5}x+3$

$$\frac{-2}{7}x = \frac{4}{5}x$$

$$x = 0$$

$$y = \frac{-2}{7}\cdot 0 + 3$$

$$y = 3$$

The solution is $(0, 3)$.

Note: this is the y-intercept of each line.

67. $y = ax+b$ (1)

$$y = x+c \quad\quad (2)$$

Substitute $x+c$ for y in Equation (1).

$$x+c = ax+b$$

$$c = ax-x+b$$

$$c-b = ax-x$$

$$c-b = x(a-1)$$

$$\frac{c-b}{a-1} = x$$

Substitute $\dfrac{c-b}{a-1}$ for x in Equation (2).

$$y = \frac{c-b}{a-1}+c \quad\quad \text{LCD is } (a-1)$$

$$y = \frac{c-b}{a-1}+\frac{c(a-1)}{a-1}$$

$$y = \frac{c-b+ac-c}{a-1}$$

$$y = \frac{ac-b}{a-1}$$

$$x = \frac{c-b}{a-1} \quad \text{and} \quad y = \frac{ac-b}{a-1}.$$

Substitute 1 for y in Equation (1).

$$ax+by+c = 0$$

$$ax+ b+c = 0$$

$$ax = -b-c$$

$$x = \frac{-b-c}{a} \quad \text{and} \quad y = 1.$$

69. **Familiarize.** Let r = the number of rabbits and p = the number of pheasants. Each rabbit has one head and four feet; each pheasant has one head and two feet.

Translate.

Total number of heads	is	No. of rabbits	plus	No. of pheasants
↓	↓	↓	↓	↓
35	=	r	+	p

Total number of feet	is	4 times no. of rabbits	plus	2 times no. of pheasants
↓	↓	↓	↓	↓
94	=	$4r$	+	$2p$

Carry out. Solve the system.

$$r + p = 35 \quad (1)$$
$$4r + 2p = 94 \quad (2)$$

Multiply Equation (1) by -2 and add the result to Equation (2).

$$
\begin{array}{rcr}
-2r - 2p &=& -70 \\
4r + 2p &=& 94 \\
\hline
2r &=& 24 \\
r &=& 12
\end{array}
$$

$$r + p = 35 \rightarrow 12 + p = 35$$
$$p = 23$$

Check. Total number of heads is $12 + 23 = 46$. Total number of feet is $4 \cdot 12 + 2 \cdot 23 = 48 + 46 = 94$. The numbers check.

State. There are 12 rabbits and 23 pheasants.

71. **Familiarize.** Let x = the man's age and y = the daughter's age. 5 years ago, the man's age was $x - 5$, and his daughter's was $y - 5$.

Translate.

Total of man's age plus 5	divided by 5	is	daughter's age
↓	↓	↓	↓
$(x+5)$	$\div 5$	=	y

$$x + 5 = 5y$$
$$x - 5y = -5$$

Five years ago:

Man's age	was	8 times daughter's age
↓	↓	↓
$x - 5$	=	$8(y - 5)$
$x - 5$	=	$8y - 40$
x	=	$8y - 35$

Carry out. Solve the system.

$$x - 5y = -5 \quad (1)$$
$$x = 8y - 35 \quad (2)$$

Substitute $8y - 35$ for x in Equation (1).

$$(8y - 35) - 5y = -5$$
$$3y - 35 = -5$$
$$3y = 30$$
$$y = 10$$

$$
\begin{array}{rcl}
x = 8y - 35 \rightarrow x &=& 8 \cdot 10 - 35 \\
x &=& 80 - 35 \\
x &=& 45
\end{array}
$$

Check. $(45 + 5) \div 5 = 50 \div 5 = 10$ and $(45 - 5) = 8(10 - 5)$. The numbers check.

State. The man is 45 years old, and his daughter is 10 years old.

Mid-Chapter Review

Guided Solutions

1. $2x - 3(x - 1) = 5$ Substituting $x - 1$ for y
   $2x - 3x + 3 = 5$ Using the distributive law
   $-x + 3 = 5$ Combining like terms
   $-x = 2$ Subtracting 3 from both sides
   $x = -2$ Dividing both sides by -1
   $y = x - 1$
   $y = -2 - 1$ Substituting
   $y = -3$
   The solution is $(-2, -3)$.

2.
$$
\begin{array}{rcl}
2x - 5y &=& 1 \\
x + 5y &=& 8 \\
\hline
3x &=& 9 \\
x &=& 3
\end{array}
$$

$x + 5y = 8$

$3 + 5y = 8$ Substituting

$5y = 5$

$y = 1$

The solution is (3, 1).

Mixed Review

1. $x = y$ (1)

    $x + y = 2$ (2)

    This system is easily solved using substitution because equation (1) already has y solved completely in terms of x. Substitute y for x in equation (2) and solve for y. Since $y = x$ finding the value of x is trivial.

    $x + y = 2$ (2)

    $y + y = 2$

    $2y = 2$

    $y = 1$

    $x = y = 1$

    The ordered pair (1, 1) is the solution to the system.

2. $x + y = 10$ (1)

    $x - y = 8$ (2)

    This system is easily solved using elimination because the coefficients of y in equations (1) and (2) are opposites. Add equations (1) and (2) and solve for x. Then substitute the value of x into either equation to find y.

    $x + y = 10$ (1)

    $\underline{x - y = 8 \quad (2)}$

    $2x \quad\;\; = 18$

    $x = 9$

    $x + y = 10$ (1)

    $9 + y = 10$

    $y = 1$

    The ordered pair (9, 1) is the solution to the system.

3. $y = \dfrac{1}{2}x + 1$ (1)

    $y = 2x - 5$ (2)

    This system is easily solved using substitution because both equations already have one variable solved completely in terms of the other. Substitute the expression for y, from

equation (1) into equation (2) and solve for x. Then substitute the value of x into either equation to find y.

$y = 2x - 5$ (2)

$\dfrac{1}{2}x + 1 = 2x - 5$

Multiply both sides of the equation by 2 to clear denominators to find x.

$2\left[\dfrac{1}{2}x + 1\right] = 2\left[2x - 5\right]$

$2 \cdot \dfrac{1}{2}x + 2 \cdot 1 = 2 \cdot 2x - 2 \cdot 5$

$x + 2 = 4x - 10$

$2 + 10 = 4x - x$

$12 = 3x$

$x = 4$

$y = 2x - 5$ (2)

$y = 2(4) - 5 = 8 - 5 = 3$

The ordered pair (4, 3) is the solution to the system.

4. $y = 2x - 3$ (1)

    $x + y = 12$ (2)

    This system is easily solved using substitution because equation (1) already has y solved completely in terms of x. Substitute $2x - 3$ for y into equation (2) and solve for x. Then substitute the value of x into equation (1) to find y.

    $x + y = 12$ (2)

    $x + (2x - 3) = 12$

    $x + 2x - 3 = 12$

    $3x - 3 = 12$

    $3x = 15$

    $x = 5$

    $y = 2x - 3$ (1)

    $= 2(5) - 3 = 10 - 3 = 7$

    The ordered pair (5, 7) is the solution to the system.

5. $x = 5$

    $y = 10$

    The system requires no work to solve. The values of x and y are already given. The ordered pair (5, 10) is the solution.

6. $3x+5y=8$ (1)

$3x-5y=4$ (2)

This system is easily solved using elimination because the coefficients of y in equations (1) and (2) are opposites. Add equations (1) and (2) and solve for x. Then substitute the value of x into either equation to find y.

$3x+5y=8$ (1)

$\underline{3x-5y=4}$ (2)

$6x\qquad =12$

$\qquad x=2$

$3x+5y=8$ (1)

$3(2)+5y=8$

$6+5y=8$

$5y=2$

$y=\dfrac{2}{5}$

The ordered pair $\left(2, \dfrac{2}{5}\right)$ is the solution to the system.

7. $2x-y=1$ (1)

$2y-4x=3$ (2)

This system is easily solved using substitution because equation (1) is easily solved for y in terms of x. Solve equation (1) for y an substitute the expression for y into equation (2) and solve for x. Then substitute the value of x into either equation to find y.

$2x-y=1$ (1)

$-y=-2x+1$

$y=2x-1$

$2y-4x=3$ (2)

$2(2x-1)-4x=3$

$4x-2-4x=3$

$-2=3$

The process led to a contradiction. The system has no solution.

8. $x=2-y$ (1)

$3x+3y=6$ (2)

This system is easily solved using substitution because equation (1) already has x solved completely in terms of y. Substitute the expression for x, from equation (1) into equation (2) and solve for y. Then substitute the value of y into either equation to find x.

$3x+3y=6$ (2)

$3(2-y)+3y=6$

$6-3y+3y=6$

$6=6$

The system has an infinite number of solutions: any solution to the first equation will also be a solution to the second. The solution set is given by the ordered pairs $\{(x,y)\mid x=2-y\}$, or, equivalently, $(2-y,y)$.

9. $x+2y=3$ (1)

$3x=4-y$ (2)

This system is easily solved using substitution because equation (1) is easily solved for x in terms of y. Solve equation (1) for x. Then substitute the result for x into equation (2) and solve for y. Then substitute the value of y into the new equation to find x.

$x+2y=3$ (1)

$x=3-2y$

$3x=4-y$ (2)

$3(3-2y)=4-y$

$9-6y=4-y$

$9-4=-y+6y$

$5=5y$

$y=1$

$x=3-2y$

$=3-2(1)=3-2=1$

The ordered pair (1, 1) is the solution to the system.

10. $9x+8y=0$ (1)

$11x-7y=0$ (2)

This system is more easily solved using elimination because solving either equation for x or y would mean introducing fractions into the system. To eliminate y and find x, multiply equation (1) by 7 and equation (2) by 8 and then add the resulting equations.

$7[9x+8y]=8[0]$

$8[11x-7y]=7[0]$

$63x+56y=0$ (3)

$\underline{88x-56y=0}$ (4)

$151x\qquad =0$

$\qquad x=0$

$9x + 8y = 0$ (1)

$9(0) + 8y = 0$

$0 + 8y = 0$

$8y = 0$

$y = 0$

The ordered pair (0, 0) is the solution to the system.

11. $10x + 20y = 40$ (1)

$x - y = 7$ (2)

This system is easily solved using either method if we begin by noting that every term in equation (1) is divisible by 10. We can then replace the original system with the equivalent system:

$x + 2y = 4$ (3)

$x - y = 7$ (2)

where equation (3) is simply the result of dividing every term in equation (1) by 10. To solve the system using substitution, solve equation (3) for x, substitute the result into equation (2) to find y, then substitute the value of y into equation (3) to find x.

$x + 2y = 4$ (3)

$x = 4 - 2y$

$x - y = 7$ (2)

$(4 - 2y) - y = 7$

$4 - 2y - y = 7$

$4 - 3y = 7$

$-3y = 7 - 4$

$-3y = 3$

$y = -1$

$x = 4 - 2y = 4 - 2(-1)$

$= 4 - (-2) = 4 + 2 = 6$

The ordered pair (6, -1) is the solution to the system.

12. To solve this system, note that when rearranged it reads:

$y - \dfrac{5}{3}x = 7$ (1)

$y - \dfrac{5}{3}x = -8$ (2)

It is impossible for the expression $y - \dfrac{5}{3}x$ to equal two different values. This system has no solution. Graphically, one can see that the

lines have the same slope, but differing y-intercepts. Thus they never cross, confirming the conclusion of no solution.

13. $2x - 5y = 1$ (1)

$3x + 2y = 11$ (2)

This system is more easily solved using elimination because solving either equation for x or y would mean introducing fractions into the system. To eliminate x and find y, multiply equation (1) by 3 and equation (2) by -2 and then add the resulting equations.

$3[2x - 5y] = 3[1]$ (3)

$-2[3x + 2y] = -2[11]$ (4)

$6x - 15y = 3$ (3)

$-6x - 4y = -22$ (4)

$\overline{-19y = -19}$

$y = 1$

$3x + 2y = 11$ (2)

$3x + 2(1) = 11$

$3x + 2 = 11$

$3x = 9$

$x = 3$

The ordered pair (3, 1) is the solution to the system.

14. Begin by clearing denominators in both equations to create an equivalent system of equations: multiply equation (1) by 6, the LCM of 2, 3, and 3, and multiply equation (2) by 20, the LCM of 5, 2, and 4.

$6\left[\dfrac{x}{2} + \dfrac{y}{3}\right] = 6\left[\dfrac{2}{3}\right]$ (3)

$20\left[\dfrac{x}{5} + \dfrac{5y}{2}\right] = 20\left[\dfrac{1}{4}\right]$ (4)

$6 \cdot \dfrac{x}{2} + 6 \cdot \dfrac{y}{3} = 6 \cdot \dfrac{2}{3}$ (3)

$20 \cdot \dfrac{x}{5} + 20 \cdot \dfrac{5y}{2} = 20 \cdot \dfrac{1}{4}$ (4)

$\cancel{2} \cdot 3 \cdot \dfrac{x}{\cancel{2}} + 2 \cdot \cancel{3} \cdot \dfrac{y}{\cancel{3}} = 2 \cdot \cancel{3} \cdot \dfrac{2}{\cancel{3}}$ (3)

$2 \cdot 2 \cdot \cancel{5} \cdot \dfrac{x}{\cancel{5}} + \cancel{2} \cdot 2 \cdot 5 \cdot \dfrac{5y}{\cancel{2}} = \cancel{2} \cdot \cancel{2} \cdot 5 \cdot \dfrac{1}{\cancel{2} \cdot \cancel{2}}$ (4)

$$3 \cdot x + 2 \cdot y = 2 \cdot 2 \quad (3)$$
$$2 \cdot 2 \cdot x + 2 \cdot 5 \cdot 5 y = 5 \cdot 1 \quad (4)$$
$$3x + 2y = 4 \quad (3)$$
$$4x + 50y = 5 \quad (4)$$

The resulting system can then be solved using elimination. Start by finding y by eliminating x. Multiply equation (3) by 4 and equation (4) by -3, then add the results and solve for y.

$$4[3x + 2y] = 4[4]$$
$$-3[4x + 50y] = -3[5]$$
$$12x + 8y = 16$$
$$\underline{-12x - 150y = -15}$$
$$-142y = 1$$

$$y = -\frac{1}{142}$$

Since this value is a little unwieldy to substitute back in to find x, one can solve for x by eliminating y if preferred. Start by multiplying equation (3) by -25 and go from there.

$$-25[3x + 2y] = -25[4]$$
$$4x + 50y = 5$$
$$-75x - 50y = -100$$
$$\underline{4x + 50y = 5}$$
$$-71x = -95$$

$$x = \frac{-95}{-71} = \frac{95}{71}$$

The ordered pair $\left(\dfrac{95}{71}, -\dfrac{1}{142} \right)$ is the solution to the system.

15. One can begin by clearing decimals to create an equivalent system of equations. Since no decimal has a digit beyond the tenths place, multiply both equations by 10.
$$10[1.1x - 0.3y] = 10[0.8]$$
$$10[2.3x + 0.3y] = 10[2.6]$$
$$11x - 3y = 8 \quad (3)$$
$$23x + 3y = 26 \quad (4)$$
Since the coefficients of y are opposites, one can easily find the solution using elimination.
$$11x - 3y = 8 \quad (3)$$
$$\underline{23x + 3y = 26 \quad (4)}$$
$$34x = 34$$
$$x = 1$$

$$11x - 3y = 8 \quad (3)$$
$$11(1) - 3y = 8$$
$$11 - 3y = 8$$
$$-3y = -3$$
$$y = 1$$
The ordered pair (1, 1) is the solution to the system.

16. $y = -3$
$$x = 11$$
The system requires no work to solve. The values of x and y are already given. The ordered pair $(11, -3)$ is the solution.

17. Noting that all the terms in equation (2) are divisible by 3 gives the new system:
$$x - 2y = 5 \quad (1)$$
$$\frac{1}{3}[3x - 15] = \frac{1}{3}[6y] \quad (3)$$
$$x - 2y = 5 \quad (1)$$
$$x - 5 = 2y \quad (3)$$
Now rearrange equation (3) and note that it is exactly the same equation as (1).
$$x - 5 = 2y \quad (3)$$
$$\underline{+5 = +5}$$
$$x = 2y + 5$$
$$\underline{-2y = -2y}$$
$$x - 2y = 5$$
Thus, the two equations are dependent and the system has an infinite number of solutions. (The same conclusion can be reached using substitution or elimination.) Therefore the set of order pairs $\{(x, y) \mid x - 2y = 5\}$ form the solution to the system.

18. $12x - 19y = 13 \quad (1)$
$$8x + 19y = 7 \quad (2)$$
This system is easily solved using elimination because the coefficients of y in equations (1) and (2) are opposites. Add equations (1) and (2) and solve for x. Then substitute the value of x into either equation to find y.
$$12x - 19y = 13 \quad (1)$$
$$\underline{8x + 19y = 7 \quad (2)}$$
$$20x \qquad = 20$$
$$x = 1$$

$12x - 19y = 13$ (1)

$12(1) - 19y = 13$

$12 - 19y = 13$

$-19y = 1$

$y = -\dfrac{1}{19}$

The ordered pair $\left(1, -\dfrac{1}{19}\right)$ is the solution to the system.

19. Clear decimals and use elimination:

$10[0.2x + 0.7y] = 10[1.2]$

$10[0.3x - 0.1y] = 10[2.7]$

$2x + 7y = 12$ (3)

$3x - y = 27$ (4)

$2x + 7y = 12$ (3)

$7[3x - y] = 7[27]$ (5)

$2x + 7y = 12$ (3)

$\underline{21x - 7y = 189}$ (5)

$23x = 201$

$x = \dfrac{201}{23}$

Repeat the process to find y.

$3[2x + 7y] = 3[12]$ (6)

$-2[3x - y] = -2[27]$ (7)

$6x + 21y = 36$ (6)

$\underline{-6x + 2y = -54}$ (7)

$23y = -18$

$y = -\dfrac{18}{23}$

The ordered pair $\left(\dfrac{201}{23}, -\dfrac{18}{23}\right)$ is the solution to the system.

20. Clear denominators and use elimination.

$12\left[\dfrac{1}{4}x\right] - 12\left[\dfrac{1}{3}y\right] = 0$ (3)

$30\left[\dfrac{1}{2}x - \dfrac{1}{15}y\right] = 30[2]$ (4)

$3x - 4y = 0$ (3)

$15x - 2y = 60$ (4)

$3x - 4y = 0$ (3)

$-2[15x - 2y] = -2[60]$ (5)

$3x - 4y = 0$ (3)

$\underline{-30x + 4y = -120}$ (5)

$-27x = -120$

$x = \dfrac{-120}{-27} = \dfrac{\cancel{3} \cdot 40}{\cancel{3} \cdot 9} = \dfrac{40}{9}$

Find y using equation (1).

$\dfrac{1}{4}x = \dfrac{1}{3}y$

$\dfrac{1}{4}\left(\dfrac{40}{9}\right) \cdot 3 = \dfrac{1}{\cancel{3}}y \cdot \cancel{3}$

$\dfrac{1}{\cancel{4}}\left(\dfrac{\cancel{4} \cdot 10}{3 \cdot \cancel{3}}\right) \cdot \cancel{3} = y$

$\dfrac{10}{3} = y$

The ordered pair $\left(\dfrac{40}{9}, \dfrac{10}{3}\right)$ is the solution to the system.

Exercise Set 4.4

1. Let p = the number of endangered plant species, and a = the number of endangered animal species. Then :

$p + a = 1010$ (1)

$p - a = 192$ (2)

The system can be solved using elimination since the coefficients of a are opposites.

$p + a = 1010$

$\underline{p - a = 192}$

$2p \quad\ = 1202$

$p = 601$

$p + a = 1010$ (1)

$601 + a = 1010$

$a = 1010 - 601 = 409$

In 2009 there were 601 endangered plant species and 409 endangered animal species.

3. Let f = the number of Facebook users, in millions, and m = the number of MySpace users, in millions. Then:

$f + m = 160$ (1)

$f = 2m - 8$ (2)

The system can be solved using substitution since f is solved for in equation (2).

$$f + m = 160 \qquad (1)$$
$$f = 2m - 8 \qquad (2)$$
$$(2m - 8) + m = 160$$
$$2m - 8 + m = 160$$
$$3m - 8 = 160$$
$$3m = 160 + 8 = 168$$
$$m = 56$$
$$f = 2m - 8 \quad (2)$$
$$f = 2(56) - 8 = 112 - 8 = 104$$

In 2009 there were 104 million Facebook users and 56 million MySpace users.

5. Let x = the measure of one angle, in degrees, and y = the measure of the second angle, in degrees. Then:
$$x + y = 180 \qquad (1)$$
$$x = 2y - 3 \qquad (2)$$
The system can be solved using substitution since x is solved for in equation (2).
$$x + y = 180 \qquad (1)$$
$$x = 2y - 3 \qquad (2)$$
$$(2y - 3) + y = 180$$
$$2y - 3 + y = 180$$
$$3y - 3 = 180$$
$$3y = 180 + 3 = 183$$
$$y = \frac{183}{3} = 61$$
$$x = 2y - 3 \quad (2)$$
$$x = 2(61) - 3 = 122 - 3 = 119$$
The two angle measures are 119° and 61°.

7. Let x = the number of 3-credit courses and y = the number of 4-credit courses. Then:

Total number of courses   is   48.

$$\downarrow \qquad\qquad \downarrow \quad \downarrow$$
$$x + y \qquad\qquad = \quad 48$$

Total number of credits   is   155.

$$\downarrow \qquad\qquad \downarrow \quad \downarrow$$
$$3x + 4y \qquad\quad = \quad 155$$

Solve the system using elimination:
$$x + y = 48 \quad (1)$$
$$3x + 4y = 155 \quad (2)$$
Multiply Equation (1) by –3 and add the result to Equation (2).

$$-3x - 3y \;=\; -144$$
$$\underline{3x + 4y \;=\; 155}$$
$$y \;=\; 11$$

$$x + y = 48 \quad \rightarrow \quad x + 11 \;=\; 48$$
$$x \;=\; 37$$

Check. $37 + 11 = 48$, and
$$37 \cdot 3 + 11 \cdot 4 = 111 + 44 = 155.$$
The numbers check.
The members of the swim team are taking 37 3-credit courses and 11 4-credit courses.

9. Let x = the number of 5¢ bottles and y = the number of 10¢ bottles. Then:

Total number
of bottles   is   430.

$$\downarrow \qquad\quad \downarrow \quad \downarrow$$
$$x + y \qquad = \quad 430$$

Total value   is   \$26.20

$$\downarrow \qquad\quad \downarrow \quad \downarrow$$
$$0.05x + 0.10y \;=\; \$26.20$$

Solve the system using elimination:
$$x + \quad y = 430 \qquad (1)$$
$$0.05x + 0.10y = 26.50 \qquad (2)$$
Multiply Equation (1) by –0.05 and add the result to Equation (2).

$$-0.05x - 0.05y \;=\; -21.50$$
$$\underline{0.05x + 0.10y \;=\; 26.20}$$
$$0.05y \;=\; 4.7$$
$$y \;=\; 94$$

$$x + y = 430 \quad \rightarrow \quad x + 94 \;=\; 430$$
$$x \;=\; 336$$

Check.
$$336 + 94 = 430 \text{ and } 0.05 \cdot 336 + 0.10 \cdot 94$$
$$= 16.80 + 9.40 = 26.20.$$
The numbers check.
The Daycare collected 336 5¢ bottles and 94 10¢ bottles.

11. Let x = the number of cars and y = the number of motorcycles. Then:

$$x + y = 5950 \qquad (1)$$
$$25x + 20y = 137{,}625 \quad (2)$$

Since only the value of y, the number motorcycles, is asked for, eliminate x by multiplying equation (1) by -25 and add the result to equation (2).

$$-25x - 25y = -25 \cdot 5950$$
$$\underline{25x + 20y = 137{,}625}$$
$$-5y = 137{,}625 - 25 \cdot 5950$$
$$= 137{,}325 - 148{,}750$$
$$= -11{,}125$$
$$y = \frac{-11125}{-5} = 2225$$

2225 motorcycles entered Yellowstone National Park.

13. Let x = the number of sheets of regular papers used, and y = the number of recycled papers used. Then:

$$x + y = 150 \qquad (1)$$
$$1.9x + 2.4y = 341 \qquad (2)$$

Note that the price per page was given in cents, and therefore equation (2) is set to 341¢.

The system can be solved using elimination by multiplying equation (1) by (-1.9) and adding the result to equation (2) to find y, then finding x.

$$-1.9[x + y] = -1.9[150]$$
$$\rightarrow -1.9x - 1.9y = -285$$
$$-1.9x - 1.9y = -285$$
$$\underline{1.9x + 2.4y = 341}$$
$$0.5y = 56$$
$$y = \frac{56}{0.5} = 112$$

Now find x using equation (1).

$$x + y = 150 \qquad (1)$$
$$x + 112 = 150$$
$$x = 150 - 112 = 38$$

38 regular sheets and 112 recycled paper sheets were used.

15. Let x = the number of silver beads purchased, and y = the number of gemstone beads purchased. Then:

$$x + y = 80 \qquad (1)$$
$$40x + 65y = 3900 \qquad (2)$$

Note that the price per bead was given in cents, and therefore equation (2) is set to 3900¢.

The system can be solved using elimination by multiplying equation (1) by (-40) and adding the result to equation (2) to find y, then finding x.

$$-40[x + y] = -40[80] \rightarrow -40x - 40y = -3200$$
$$-40x - 40y = -3200$$
$$\underline{40x + 65y = 3900}$$
$$25y = 700$$
$$y = \frac{700}{25} = 28$$

Now find x using equation (1).

$$x + y = 80 \qquad (1)$$
$$x + 28 = 80$$
$$x = 80 - 28 = 52$$

Alicia purchased 52 silver beads and 28 gemstone beads.

17. Let x = the number of Epson cartridges purchased, and y = the number of HP cartridges purchased. Then:

$$x + y = 50 \qquad (1)$$
$$1699x + 2599y = 98{,}450 \qquad (2)$$

Note all prices in equation (2) are shown in cents.

The system can be solved using elimination by multiplying equation (1) by (-1699) and adding the result to equation (2) to find y, then finding x.

$$-1699[x + y] = -1699[50]$$
$$\rightarrow -1699x - 1699y = -84{,}950$$
$$-1699x - 1699y = -84{,}950$$
$$\underline{1699x + 2599y = 98{,}450}$$
$$9\cancel{0}\cancel{0}y = 13{,}5\cancel{0}\cancel{0}$$
$$y = \frac{135}{9} = 15$$

Now find x using equation (1).

$$x + y = 50 \qquad (1)$$
$$x + 15 = 50$$
$$x = 50 - 15 = 35$$

Office Depot sold 35 Epson cartridges and 15 HP cartridges.

19. Let x = the number of pounds of Mexican coffee used in the blend, and y = the number of pounds of Peruvian used. Then:

$$x + y = 28 \qquad (1)$$
$$13x + 11y = 12 \cdot 28 = 336 \quad (2)$$

The system can be solved using elimination by multiplying equation (1) by (-13) and adding the result to equation (2) to find y, then finding x.

$$-13[x + y] = -13[28] \rightarrow -13x - 13y = -364$$

$$\begin{array}{r} -13x - 13y = -364 \\ 13x + 11y = 336 \\ \hline -2y = -28 \end{array}$$

$$y = \frac{-28}{-2} = 14$$

Now find x using equation (1).

$$x + y = 28 \quad (1)$$
$$x + 14 = 14$$
$$x = 28 - 14 = 14$$

The coffee blend should be made using 14 pounds of Mexican coffee and 14 pounds of Peruvian coffee.

21. Let x = the number of ounces of sumac in the blend, and y = the number of ounces of thyme used. Then:

$$x + y = 20 \qquad (1)$$
$$1.35x + 1.85y = 1.65 \cdot 20 = 33 \qquad (2)$$

The system can be solved using elimination by multiplying equation (1) by (-1.35) and adding the result to equation (2) to find y, then finding x.

$$-1.35[x + y] = -1.35[20]$$
$$\rightarrow -1.35x - 1.35y = -27$$

$$\begin{array}{r} -1.35x - 1.35y = -27 \\ 1.35x + 1.85y = 33 \\ \hline 0.5y = 6 \end{array}$$

$$y = \frac{6}{0.5} = \frac{60}{5} = 12$$

Now find x using equation (1).

$$x + y = 20 \quad (1)$$
$$x + 12 = 20$$
$$x = 20 - 12 = 8$$

The Zahtar seasoning should be made using 8 ounces of sumac and 12 ounces of thyme.

23. Let x = the number of mL of 50%-acid solution and y = the number of mL of the 80% acid solution.
Complete the table with the given information.
Note: Completing the table is part of the exercise.

Type of Solution	50%-Acid	80%-Acid	68%-Acid Mix
Amount of Solution	x	y	200
Percent Acid	50%	80%	68%
Amount of Acid in Solution	$0.5x$	$0.8y$	136

From the table we have an equation for the total mL: $x + y = 200$.

We also have an equation for the total amount of mL of acid:
$$0.5x + 0.8y = 136.$$

Solve the system.

$$x + y = 200 \quad (1)$$
$$0.5x + 0.8y = 136 \quad (2)$$

Multiply Equation (1) by -0.5 and add the result to Equation (2).

$$\begin{array}{r} -0.5x - 0.5y = 100 \\ 0.5x + 0.8y = 136 \\ \hline 0.3y = 36 \end{array}$$

$$y = 120$$

$$x + y = 200 \rightarrow x + 120 = 200$$
$$x = 80$$

Check.
$80 + 120 = 200$, and
$0.5(80) + 0.8(120) = 40 + 96 = 136$.
The numbers check.
State. Jerome should mix 80 mL of the 50%-acid solution and 120 mL of the 80%-acid solution.

25. Let x = the number of pounds of the 50% chocolate mix used in the mix, and y = the number of pounds of the 10% chocolate mix used. Then:

$$x + y = 20 \qquad (1)$$
$$50x + 10y = 25 \cdot 20 = 500 \qquad (2)$$

The system can be solved using elimination by multiplying equation (1) by (-50) and adding the result to equation (2) to find y, then finding x.

$$-50[x+y] = -50[20] \rightarrow -50x - 50y = -1000$$

$$\begin{array}{r} -50x - 50y = -1000 \\ 50x + 10y = 500 \\ \hline -40y = -500 \end{array}$$

$$y = \frac{-500}{-40} = 12\frac{1}{2}$$

Now find x using equation (1).

$$x + y = 20 \quad (1)$$

$$x + 12\frac{1}{2} = 20$$

$$x = 20 - 12\frac{1}{2} = 7\frac{1}{2}$$

The mix should be made using 7.5 pounds of the 50% chocolate mix and 12.5 pounds of the 10% chocolate mix.

27. Let x = the amount borrowed at 6.5%, and y = the amount borrowed at 7.2%. Then:

$$x + y = 12,000 \quad (1)$$

$$0.065x + 0.072y = 811.50 \quad (2)$$

Note that since the interest on the loans (the right side of equation (2)) is not represented as a percent of the total amount borrowed, the percentage rates (shown on the left side of equation (2)) need to be converted to decimals (or the interest could be multiplied by 100: 81,150).
The system can be solved using elimination by multiplying equation (1) by (-0.065) and adding the result to equation (2) to find y, then finding x.

$$-0.065[x+y] = -0.065[12,000]$$

$$\rightarrow -0.065x - 0.065y = -780$$

$$\begin{array}{r} -0.065x - 0.065y = -780 \\ 0.065x + 0.072y = 811.50 \\ \hline 0.007y = 31.50 \end{array}$$

$$y = \frac{31.50}{0.007} = \frac{31,500}{7} = 4500$$

Now find x using equation (1).

$$x + y = 12,000 \quad (1)$$

$$x + 4500 = 12,000$$

$$x = 12,000 - 4500 = 7500$$

Asel borrowed $7500 at 6.5% interest and $4500 at 7.2%.

29. Let x = the number of liters of the 18% alcohol antifreeze used, and y = the number of

liters of the 10% alcohol antifreeze used. Then:

$$x + y = 20 \quad (1)$$

$$18x + 10y = 15 \cdot 20 = 300 \quad (2)$$

The system can be solved using elimination by multiplying equation (1) by (-18) and adding the result to equation (2) to find y, then finding x.

$$-18[x+y] = -18[20] \rightarrow -18x - 18y = -360$$

$$\begin{array}{r} -18x - 18y = -360 \\ 18x + 10y = 300 \\ \hline -8y = -60 \end{array}$$

$$y = \frac{-60}{-8} = 7\frac{1}{2}$$

Now find x using equation (1).

$$x + y = 20 \quad (1)$$

$$x + 7\frac{1}{2} = 20$$

$$x = 20 - 7\frac{1}{2} = 12\frac{1}{2}$$

The mix should be made using 12.5 L of the 18% alcohol solution and 7.5 L of the 10% alcohol solution.

31. Let x = the number of gallons of the 87-octane gasoline used, and y = the number of gallons of the 95-octane gasoline used. Then:

$$x + y = 10 \quad (1)$$

$$87x + 95y = 93 \cdot 10 = 930 \quad (2)$$

The system can be solved using elimination by multiplying equation (1) by (-87) and adding the result to equation (2) to find y, then finding x.

$$-87[x+y] = -87[10] \rightarrow -87x - 87y = -870$$

$$\begin{array}{r} -87x - 87y = -870 \\ 87x + 95y = 930 \\ \hline 8y = 60 \end{array}$$

$$y = \frac{60}{8} = 7\frac{1}{2}$$

Now find x using equation (1).

$$x + y = 10 \quad (1)$$

$$x + 7\frac{1}{2} = 10$$

$$x = 10 - 7\frac{1}{2} = 2\frac{1}{2}$$

The new 93-octane should be made using 2.5 gallons of the 87-octane gasoline and 7.5 gallons of the 95-octane gasoline.

33. Let $x =$ the number of pounds of the whole milk (4% milk fat) used, and $y =$ the number of pounds of the cream (30% milk fat) used. Then:

$$x + y = 200 \qquad (1)$$

$$4x + 30y = 8 \cdot 200 = 1600 \qquad (2)$$

The system can be solved using elimination by multiplying equation (1) by (-4) and adding the result to equation (2) to find y, then finding x.

$$-4[x + y] = -4[200] \rightarrow -4x - 4y = -800$$

$$-4x - 4y = -800$$

$$\underline{4x + 30y = 1600}$$

$$26y = 800$$

$$y = \frac{800}{26} = 30\frac{20}{26} = 30\frac{10}{13}$$

Now find x using equation (1).

$$x + y = 200 \qquad (1)$$

$$x + 30\frac{10}{13} = 200$$

$$x = 200 - 30\frac{10}{13} = 169\frac{3}{13}$$

The cream cheese milk should be made using $169\frac{3}{13}$ pounds of whole milk and $30\frac{10}{13}$ pounds of cream.

35. ***Familiarize.*** We first make a drawing.

Slow train
d kilometers   75 km/h $(t + 2)$ hr

Fast train
d kilometers   125 km/h t hr

From the drawing, we see that the distances are the same. Now complete the chart.

$$d \quad = \quad r \quad \cdot \quad t$$

	Distance	Rate	Time	
Slow train	d	75	$t + 2$	$\rightarrow d = 75(t + 2)$
Fast train	d	125	t	$\rightarrow d = 125t$

Translate. Using $d = rt$ in each row of the table, we get a system of equations.

$$d = 75(t + 2)$$

$$d = 125t$$

Carry out. Solve the system of equations.

$$125t = 75(t + 2) \quad \text{Using substitution}$$

$$125t = 75t + 150$$

$$50t = 150$$

$$t = 3$$

Check. At 125 km/h, in 3 hr the fast train will travel $125 \cdot 3 = 375$ km. At 75 km/h, in $3 + 2$, or 5 hr, the slow train will travel $75 \cdot 5 = 375$ km. The numbers check.

State. The trains will meet 375 km from the station.

37. ***Familiarize.*** We first make a drawing. Let $d =$ the distance and $r =$ the speed of the canoe in still water. Then when the canoe travels downstream its speed is $r + 6$, and its speed upstream is $r - 6$. From the drawing, we see that the distances are the same.

Downstream, 6 km/h current

d km, $r + 6$, 4 hr

Upstream, 6 km/h current

d km, $r - 6$, 10 hr

Organize the information in a table.

$$d \quad = \quad r \quad \cdot \quad t$$

	Distance	Rate	Time
Down-stream	d	$r + 6$	4
Up-stream	d	$r - 6$	10

Translate. Using $d = rt$ in each row of the table, we get a system of equations.

$$d = 4(r + 6)$$

$$d = 10(r - 6)$$

Carry out. Solve the system of equations.

$$4(r + 6) = 10(r - 6) \quad \text{Substitution}$$

$$4r + 24 = 10r - 60$$

$$84 = 6r$$

$$14 = r$$

Check. Going downstream, the speed of the canoe would be $r + 6 = 14 + 6 = 20$ km/h, and in 4 hours, it would travel $4 \cdot 20 = 80$ km. Going upstream, the speed of the canoe would be $r - 6 = 14 - 6 = 8$ km/h, and in 10 hours it would travel $10 \cdot 8 = 80$ km. The numbers check.

State. The speed of the canoe in still water is 14 km/h.

39. ***Familiarize.*** We make a drawing. Note that the plane's speed traveling toward London is 360 + 50, or 410 mph, and the speed traveling toward New York City is 360 − 50, or 310 mph. Also, when the plane is d mi from New York City, it is $3458 - d$ mi from London.

New York City			London
310 mph	t hours	t hours	410 mph

```
|————————— 3458 mi —————————|

|———— d ————|———— 3458 mi −d ————|
```

Organize this information in a table.

	Distance	Rate	Time
Toward NYC	d	310	t
Toward London	$3458 - d$	410	t

Translate. Using $d = rt$ in each row of the table, we get a system of equations.

$$d = 310t \quad (1)$$
$$3458 - d = 410t \quad (2)$$

Carry out. Solve the system of equations.

$$3458 - 310t = 410t \quad \text{Using substitution}$$
$$3458 = 720t$$
$$4.8028 \approx t$$

Substitute 4.8028 for t in (1).

$$d \approx 310(4.8028) \approx 1489$$

Check. If the plane is 1489 mi from New York City, it can return to New York City, flying at 310 mph, in $1489/310 \approx 4.8$ hr. If the plane is $3458 - 1489$, or 1969 mi from London, it can fly to London, traveling at 410 mph in $1969/410 \approx 4.8$ hr. Since the times are the same, the answer checks.

State. The point of no return is about 1489 mi from New York City.

41. Let x = the number of foul shots made, and y = the number of 2-point shots make. Then:

$$x + y = 64 \quad (1)$$
$$1x + 2y = 100 \quad (2)$$

The system can be solved using elimination by multiplying equation (1) by (-1) and adding the result to equation (2) to find y, then finding x.

$$-1[x + y] = -1[64] \rightarrow -x - y = -64$$

$$\begin{array}{r} -x - y = -64 \\ \underline{1x + 2y = 100} \\ y = 36 \end{array}$$

Now find x using equation (1).

$$x + y = 64 \quad (1)$$
$$x + 36 = 64$$
$$x = 64 - 36 = 28$$

Wilt Chamberlain made 28 foul shots and 36 two-point shots.

43. Let x = the number of minutes used for landline calls, and y = the number of minutes used for wireless calls. Then:

$$x + y = 400 \quad (1)$$
$$9x + 15y + 399 = 5889$$
$$9x + 15y = 5889 - 399 = 5490 \quad (2)$$

Note that the price per minute was given in cents, and therefore equation (2) is set to the monthly charge and the bill amount were given in cents in equation (2).

The system can be solved using elimination by multiplying equation (1) by (-9) and adding the result to equation (2) to find y, then finding x.

$$-9[x + y] = -9[400] \rightarrow -9x - 9y = -3600$$

$$\begin{array}{r} -9x - 9y = -3600 \\ \underline{9x + 15y = 5490} \\ 6y = 1890 \end{array}$$

$$y = \frac{1890}{6} = 315$$

Now find x using equation (1).

$$x + y = 400 \quad (1)$$
$$x + 315 = 400$$
$$x = 400 - 315 = 85$$

Kim made 85 minutes worth of landline calls and 315 minutes worth of wireless calls.

45. ***Familiarize.*** Monica tendered $20 to pay for a purchase amounting to $9.25, so she should receive $10.75 in change to be paid back in quarters and fifty-cent pieces. Let x = the number of quarters and y = the number of fifty-cent pieces.

Translate. We organize the information in a table.

	Quarters	Fifty-cent pieces	Total
Number of coins	x	y	30
Value of the coin	$0.25	$0.50	
Total change	$0.25x$	$0.5y$	$10.75

From the "Number of coins" line in the table, we have the equation:

$x + y = 30$.

From the "Total change" line in the table, we have the equation:

$0.25x + 0.50y = 10.75$

After clearing decimals, we have translated to a system of equations:

$$x + y = 30 \quad (1)$$
$$25x + 50y = 1075 \quad (2)$$

Carry out. We can use elimination to solve the system of equations.

$$-25x - 25y = -750 \quad \text{Multiply (1) by } -25$$
$$\underline{25x + 50y = 1075}$$
$$25y = 325$$
$$y = 13$$

Now substitute 13 for y in Equation (1) to solve for x.

$$x + y = 30$$
$$x + 13 = 30$$
$$x = 17$$

Check. There are a total of $17 + 13 = 30$ coins. The total value of the coins is $17(\$0.25) + 13(\$0.50) = \$4.25 + \$6.50 = \$10.75$ The answer checks.

State. Monica will receive 17 quarters and 13 fifty-cent pieces in change amounting to $10.75.

47. *Thinking and Writing Exercise.*

49. $y = 2x - 3$

Slope: 2; y-intercept: $(0, -3)$

We begin by plotting the y-intercept $(0, -3)$ and thinking of the slope as $\frac{2}{1}$, we move up 2 units and right 1 unit to the point $(1, -1)$.

Thinking of the slope as $\frac{2}{1}$ we can begin at $(1, -1)$ and move up 2 units and right 1 unit to

the point $(2, 1)$. We finish by connecting these points to form a line.

51. $y = 2$

We can write this equation as $0 \cdot x + y = 2$. No matter what number we choose for x, we find that y must be 2.

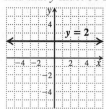

53. $f(x) = -\frac{2}{3}x + 1 \rightarrow y = -\frac{2}{3}x + 1$

Slope: $-\frac{2}{3}$; y-intercept: $(0, 1)$

We begin by plotting the y-intercept $(0, 1)$ and thinking of the slope as $\frac{-2}{3}$, we move down 2 units and right 3 units to the point $(3, -1)$. Thinking of the slope as $\frac{2}{-3}$ we can begin at $(0, 1)$ and move up 2 units and left 3 units to the point $(-3, 3)$. We finish by connecting these points to form a line.

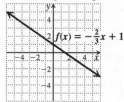

55. *Thinking and Writing Exercise.*

57. ***Familiarize*** In this problem, there is only one unknown amount, the amount of pure silver that must be added to the amount of metal in the coin.

Let x = the number of ounces of pure silver that must be added to the metal in the coin. Then the total amount of metal after the pure silver is added will be: $x + 32$ ounces.

Translate. We organize the information in a table.

	coin silver	pure silver	sterling silver
%	90%	100%	92.5%
amount	32 oz	x	$x+32$
Total silver	$0.90 \cdot 32$	$1.00x$	$0.925(x+32)$

Therefore, we get:

$0.90 \cdot 32 + 1.00x = 0.925(x+32)$

Carry out. Solve the equation for x.

$0.90 \cdot 32 + 1.00x = 0.925(x+32)$

$28.8 + x = 0.925x + 29.6$

$(1 - 0.925)x = 29.6 - 28.8$

$0.075x = 0.8$

$$x = \frac{0.8}{0.075} = \frac{800}{75} = 10\frac{50}{75} = 10\frac{2}{3}$$

Check. There are a total of 32 ounces + $10\frac{2}{3}$ ounces, or $42\frac{2}{3}$ ounces. So

$0.925\left(42\frac{2}{3}\right) \approx 39.4667$ is the total amount

of silver which is $0.90 \cdot 32 + 10\frac{2}{3}$ which is about 39.4667. The answer checks.

State. $10\frac{2}{3}$ ounces of pure silver must be added to the metal in the coin to get a mixture that is sterling silver, or, equivalently, 92.5% pure silver.

59. ***Familiarize.*** Let x = the amount of the original solution that remains after some of the original solution is drained and replaced with pure antifreeze. Let y = the amount of the original solution that is drained and replaced with pure antifreeze. We organize the information in a table. Keep in mind that the table contains information regarding the solution *after* some of the original solution is drained and replaced with pure antifreeze.

Translate. We organize the information in a table.

	Original solution	Pure Anti-freeze	New Mixture
Amount of solution	x	y	6.3 L
Percent of antifreeze	30%	100%	50%
Amount of antifreeze in solution	$0.3x$	$1 \cdot y$, or y	$0.5(6.3)$, or 3.15

We get one equation from the "Amount of solution" row of the table:

$x + y = 6.3$

The last row of the table gives us a second equation:

$0.3x + y = 3.15$

After clearing the decimal we have the problem translated to a system of equations:

$10x + 10y = 63 \quad (1)$

$30x + 100y = 315 \quad (2)$

Carry out. Solve the system of equations using the elimination method.

$$\begin{array}{rl} -30x - 30y = & -189 \quad \text{Multiply (1) by } -3 \\ \underline{30x + 100y = \quad 315} & \\ 70y = & 126 \\ y = & 1.8 \end{array}$$

Now substitute 1.8 for y in Equation (1) to solve for x.

$x + y = 6.3$

$x + 1.8 = 6.3$

$x = 4.5$

Check. The total amount in the mixture is 4.5 L + 1.8 L, or 6.3 L. The amount of antifreeze in the mixture is 30%(4.5) + 100%(1.8) = 1.35 L + 1.8 L = 3.15 L. The answer checks.

State. Michelle should drain 1.8 L of radiator fluid and replace it with 1.8 L of pure antifreeze

61. **Familiarize.** Let x = the number of individual volumes purchased, and y = the number of 3-volume sets purchased. Then:

Translate.
$$x + 3y = 51 \quad (1)$$
$$39x + 88y = 1641 \quad (2)$$

Carry out. y can be found by multiplying equation (1) by (-39) and adding the result to equation (2) to eliminate x.

$$-39[x + 3y] = -39[51]$$
$$\rightarrow -39x - 117y = -1989$$
$$-39x - 117y = -1989$$
$$\underline{39x + 88y = 1641}$$
$$-29y = -348$$
$$y = \frac{-348}{-29} = 12$$

State. 12 three-volume sets were purchased.

63. **Familiarize.** Let x = the number of gallons of pure brown and y = the number of gallons of neutral stain that should be added to the original 0.5 gal. Note that the total of 1 gal of stain needs to be added to bring the amount of stain up to 1.5 gal. The original 0.5 gal of stain contains 20%(0.5 gal), or 0.2(0.5 gal) = 0.1 gal of brown stain. The final solution contains 60%(1.5 gal), or 0.6(1.5 gal) = 0.9 gal and the x gal that are added.

Translate.

The amount of stain added was 1 gal.

$$\downarrow \qquad\qquad \downarrow \quad \downarrow$$
$$x + y \qquad = \quad 1$$

The amount of brown stain in the final solution is 0.9 gal.

$$\downarrow \qquad\qquad \downarrow \quad \downarrow$$
$$0.1 + x \qquad = \quad 0.9$$

We have a system of equations.
$$x + y = 1 \quad (1)$$
$$0.1 + x = 0.9 \quad (2)$$

Carry out. First Solve (2) for x.
$$0.1 + x = 0.9$$
$$x = 0.8$$

Then substitute 0.8 for x in (1) and solve for y.
$$0.8 + y = 1$$
$$y = 0.2$$

Check.

Total amount of stain: $0.5 + 0.8 + 0.2 = 1.5$ gal.

Total amount of brown stain: $0.1 + 0.8 = 0.9$ gal.

Total amount of neutral stain: $0.8(0.5) + 0.2 = 0.4 + 0.2 = 0.6$ gal $= 0.4(1.5$ gal$)$. The answer checks.

State. 0.8 gal of pure brown and 0.2 gal of neutral stain should be added.

65. **Familiarize.** Let x = the number of miles driven in the city and let y = the number of miles driven on the highway.

Translate. We organize this information in a table.

	City driving	Highway driving	
No. of miles	x	y	465
MPG	18	24	
Gallons used	$\dfrac{x}{18}$	$\dfrac{y}{24}$	23

We get one equation from the "No. of miles" row in the table.
$$x + y = 465$$

We get a second equation from the "Gallons used" row in the table.
$$\frac{x}{18} + \frac{y}{24} = 23$$

We have a system of equations:
$$x + y = 465 \quad (1)$$
$$\frac{x}{18} + \frac{y}{24} = 23 \quad (2)$$

Carry out. We use substitution to solve the system of equations. We begin by solving Equation (1) for x.
$$x + y = 465$$
$$x = 465 - y$$

We then substitute $465 - y$ for x in Equation (2) and solve for y.
$$\frac{x}{18} + \frac{y}{24} = 465$$
$$\frac{465 - y}{18} + \frac{y}{24} = 23 \qquad \text{Substituting}$$
$$4(465 - y) + 3y = 72(23) \qquad \text{Mult. by 72}$$
$$1860 - 4y + 3y = 1656$$
$$1860 - y = 1656$$
$$204 = y$$

Substituting 204 for y in (1) and solving for x, we have:

$$x + y = 465$$
$$x + 204 = 465$$
$$x = 261$$

Check. The total number of miles drive is 261 mi + 204 mi, or 465 mi. If the car is driven 261 miles in the city at 18 miles per gallon, the car would use 261/18 = 14.5 gal of fuel. If the car is driven 204 miles on the highway at 24 miles per gallon, it would use 8.5 gal of fuel. The total amount of fuel used would be 14.5 gal + 8.5 gal, or 23 gal. These numbers check.

State. The car was driven 261 miles in the city and 204 miles on the highway.

67. Let x = the number of 2-count pencil packs purchased, and y = the number of 12-count pencil packs purchased. Then:

$$2x + 12y = 138 \qquad (1)$$
$$599x + 749y = 15{,}726 \qquad (2)$$

Note that all prices in equation (2) are given in cents.

y can be found by multiplying equation (1) by $\left(-\dfrac{599}{2}\right)$ and adding the result to equation (2) to eliminate x. Note that since all the coefficients in equation (1) are even, multiplying it by $\left(-\dfrac{599}{2}\right)$ will not introduce any fractions once each term is simplified.

$$\left(-\frac{599}{2}\right)[2x + 12y] = \left(-\frac{599}{2}\right)[138]$$

$$\rightarrow -\frac{599}{\cancel{2}} \cdot \cancel{2}x - \frac{599}{\cancel{2}} \cdot \cancel{2} \cdot 6y = -\frac{599}{\cancel{2}} \cdot \cancel{2} \cdot 69$$

$$\rightarrow -599x - 3594y = -41{,}331$$

$$\begin{array}{r} -599x - 3594y = -41{,}331 \\ 599x + 749y = 15726 \\ \hline -2845y = -25605 \end{array}$$

$$y = \frac{-25605}{-2845} = 9$$

Now find x using equation (1).

$$2x + 12y = 138 \qquad (1)$$
$$2x + 12(9) = 138$$
$$2x + 108 = 138$$
$$2x = 138 - 108 = 30$$
$$x = \frac{30}{2} = 15$$

15 two-count packs and 9 twelve-count packs were purchased.

Exercise Set 4.5

1. e

3. f

5. a

7. $2x - 1 = -5$

The solution is the x-coordinate of the point of intersection of the graph $f(x) = 2x - 1$ and $g(x) = -5$. Inspecting the graph suggests that -2 is the solution.

Check:
$$\begin{array}{c|c} 2x - 1 = -5 & \\ 2(-2) - 1 & -5 \\ -4 - 1 & -5 \\ -5 = -5 & \text{TRUE} \end{array}$$

The solution is -2.

9. $2x + 3 = x - 1$

The solution is the x-coordinate of the point of intersection of the graph $f(x) = 2x + 3$ and $g(x) = x - 1$. Inspecting the graph suggests that -4 is the solution.

Check:
$$\begin{array}{c|c} 2x + 3 = x - 1 & \\ 2(-4) + 3 & -4 - 1 \\ -8 + 3 & -5 \\ -5 = -5 & \text{TRUE} \end{array}$$

The solution is -4.

11. $\frac{1}{2}x + 3 = x - 1$

The solution is the x-coordinate of the point of intersection of the graph $f(x) = \frac{1}{2}x + 3$ and $g(x) = x - 1$. Inspecting the graph suggests that 8 is the solution.

Check: $\dfrac{1}{2}x+3=x-1$

$$\begin{array}{c|c} \frac{1}{2}(8)+3 & 8-1 \\ 4+3 & 7 \\ \hline & 7=7 \qquad \text{TRUE} \end{array}$$

The solution is 8.

13. $f(x)=g(x)$

The solution is the x-coordinate of the point of intersection of the graph $f(x)$ and $g(x)$. Inspecting the graph suggests that 0 is the solution.

15. $y_1=y_2$

The solution is the x-coordinate of the point of intersection of the graph y_1 and y_2. Inspecting the graph suggests that 5 is the solution.

17. The graph of the function crosses the x-axis at $x=-2.-2$ is zero.

19. The graph of the function does not cross the x-axis. There are no zeros.

21. The graph of the function crosses the x-axis at $x=-2$ and $x=2.-2$ and 2 are the zeros.

23. Determine the x-intercept of $f(x)=x-5$. Let $f(x)=0$ and solve for x.
$0=x-5$
$5=x$ 5 is the zero.

25. Determine the x-intercept of $f(x)=\frac{1}{2}x+10$.
Let $f(x)=0$ and solve for x.

$$0=\frac{1}{2}x+10$$

$$-10=\frac{1}{2}x$$

$$2\cdot(-10)=2\cdot\frac{1}{2}x$$

$$-20=x \qquad -20 \text{ is the zero.}$$

27. Determine the x-intercept of $f(x)=2.7-x$.
Let $f(x)=0$ and solve for x.
$0=2.7-x$
$x=2.7$ 2.7 is the zero.

29. Determine the x-intercept of $f(x)=3x+7$.
Let $f(x)=0$ and solve for x.
$0=3x+7$
$-7=3x$

$$-\frac{7}{3}=x \qquad -\frac{7}{3} \text{ is the zero.}$$

31. Solve $x-3=4$.
We graph $f(x)=x-3$ and $g(x)=4$ on the same axes.

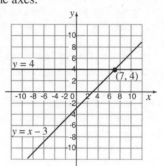

It appears that the lines intersect at $(7,4)$.

$$\begin{array}{c|c} x-3=4 \\ 7-3 & 4 \\ \hline 4 & 4 \quad \text{TRUE} \end{array}$$

The solution is 7.

33. Solve $2x+1=7$.
We graph $f(x)=2x+1$ and $g(x)=7$ on the same axes.

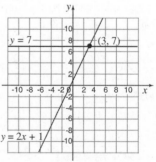

It appears that the lines intersect at $(3,7)$.

$$\begin{array}{c|c} 2x+1=7 \\ 2(3)+1 & 7 \\ \hline 7 & 7 \quad \text{TRUE} \end{array}$$

The solution is 3.

35. Solve $\frac{1}{3}x - 2 = 1$.

We graph $f(x) = \frac{1}{3}x - 2$ and $g(x) = 1$ on the same axes.

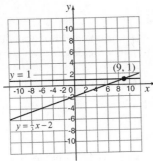

It appears that the lines intersect at $(9,1)$.

$$\frac{1}{3}x - 2 = 1$$

$$\begin{array}{c|c} \frac{1}{3}(9) - 2 & 1 \\ \hline 3 - 2 & 1 \text{ TRUE} \end{array}$$

The solution is 9.

37. Solve $x + 3 = 5 - x$.

We graph $f(x) = x + 3$ and $g(x) = 5 - x$ on the same axes.

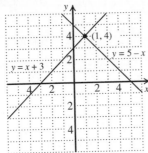

It appears that the lines intersect at $(1,4)$.

$$\begin{array}{c|c} x + 3 = 5 - x \\ \hline 1 + 3 & 5 - 1 \\ 4 & 4 \quad \text{TRUE} \end{array}$$

The solution is 1.

39. Solve $5 - \frac{1}{2}x = x - 4$.

We graph $f(x) = 5 - \frac{1}{2}x$ and $g(x) = x - 4$ on the same axes.

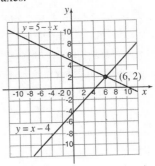

It appears that the lines intersect at $(6,2)$.

$$\begin{array}{c|c} 5 - \frac{1}{2}x = x - 4 \\ \hline 5 - \frac{1}{2}(6) & 6 - 4 \\ 5 - 3 & 2 \\ 2 = 2 & \quad \text{TRUE} \end{array}$$

The solution is 6.

41. Solve $2x - 1 = -x + 3$.

We graph $y_1 = 2x - 1$ and $y_2 = -x + 3$ on the same axes using a graphing calculator. We then use the INTERSECT option from the CALC menu.

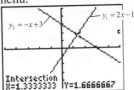

It the lines intersect at $\left(1\frac{1}{3}, 1\frac{2}{3}\right)$. The solution is $1\frac{1}{3}$.

43. **Familiarize.** Let $C =$ the cost to the patient and let $b =$ the hospital bill. The problem asks that we determine how much the patient's hospital bill exceeded $5000 if the patient's cost was $6350.

Translate.

Patient cost is $5000 plus 30% of amount over $5000.

$$C = 5000 + 0.30 \cdot (b - 5000)$$

where $b \geq 5000$ since the bill cannot be negative.

Carry out. To estimate the amount of the hospital bill that would have resulted in the stated cost to the patient, we need to estimate the solution of

$$6350 = 5000 + 0.30(b - 5000),$$

We do this by graphing

$$y_1 = 5000 + 0.30(x - 5000) \text{ and } y_2 = 6350$$

on a graphing calculator, and finding the point of intersection: $(9500, 6350)$.

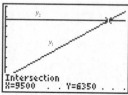

Thus, we estimate that a hospital bill of $9500 would result in a patient's after-insurance bill to be $6350.

Check. We evaluate:

$$C(9500) = 5000 + 0.30(9500 - 5000)$$
$$= 5000 + 0.3(4500)$$
$$= 5000 + 1350$$
$$= 6350$$

Our estimate turns out to be precise.

State. The amount of the hospital bill that resulted in an after-insurance charge of $6350 was a total of $9500. So the excess of that bill over the first $5000 was $4500.

45. **Familiarize.** Let t = the number of months and let C = the total cost of the telephone bill. We are to determine the number of months that would result in cumulative telephone bill of $275.

Translate.

Phone bill is $100 plus $35 · months.

$$\begin{array}{cccccc} \downarrow & \downarrow & \downarrow & \downarrow & \downarrow\downarrow & \downarrow \\ C & = & 100 & + & 35\,· & t \end{array}$$

where $t \geq 0$ since months cannot be negative.

Carry out. To estimate the number of months resulting in a cumulative phone bill of $275, we need to estimate the solution of

$$275 = 100 + 35t,$$

We do this by graphing $y_1 = 100 + 35x$ and

$y_2 = 275$ on a graphing calculator, and we

find the point of intersection at $(5, 275)$.

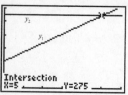

Thus, we estimate that after 5 months, the phone bill under the described plan will be $275.

Check. We evaluate:

$$C(5) = 100 + 35(5)$$
$$= 100 + 175$$
$$= 275$$

Our estimate turns out to be precise.

State. The time required for the cumulative phone bill to equal $275 is 5 months.

47. **Familiarize.** Let x = the number of 15-minute units of time a person is parked. Let F = the parking fee. We are to determine the time a person was parked resulting in a fee of $7.50.

Translate.

Parking fee is $3.00 plus $0.50 per 15-min unit of time.

$$\begin{array}{cccccc} \downarrow & \downarrow & \downarrow & \downarrow & \downarrow & \downarrow \\ F & = & 3.00 & + & 0.50 & ·x \end{array}$$

where $x \geq 0$ since time units cannot be negative.

Rewriting the equation to eliminate decimals we have:

$$F = 300 + 50x$$

where x is in 15-minute units of time and F is parking fee in cents.

Carry out. To estimate the number of 15-minute units of time resulting in a parking fee of $7.50, or 750 cents, we need to estimate the solution of

$$F(x) = 300 + 50x,$$

replacing $F(x)$ with 750. We do this by

graphing $F(x) = 300 + 50x$, and we find the

point of intersection at $(9, 750)$.

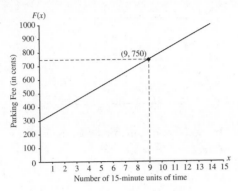

Thus, we estimate that after 9 15-minute units of time, the parking fee will be 750 cents, or $7.50.

Check. We evaluate:

$F(9) = 300 + 50(9)$

$= 300 + 450$

$= 750$

Our estimate turns out to be precise.

State. A parking fee of 750 cents, or $7.50, would result from parking a total of 9 15-minute units of time, or 2 hr and 15 minutes.

49. **Familiarize.** Let p = the weight of the package, in pounds. Let C = the cost to ship the package. We are to determine the weight of a package that costs $325 to ship.

Translate.

$$\underbrace{\text{Shipping charge}}_{} \text{ is } \$130 \text{ plus } \underbrace{\$1.30 \text{ for each pound over 100.}}_{}$$

$$\begin{array}{ccccc} \downarrow & \downarrow & \downarrow & \downarrow & \downarrow \\ C & = & 130 & + & 1.30(p-100) \end{array}$$

where $p \geq 100$ since weight cannot be negative.

Carry out. To estimate the weight of a package that cost $325 to ship, we need to estimate the solution of

$325 = 130 + 1.30(p - 100)$,

We do this by graphing

$y_1 = 130 + 1.30(x - 100)$, $y_2 = 325$ and graphing their intersection on a graphing calculator. We find the point of intersection at $(250, 325)$

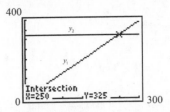

Thus, we estimate that a package weighing 250 lb would cost $325 to ship.

Check.

$C(250) = 130 + 1.30(250 - 100)$

$= 130 + 1.30(150)$

$= 130 + 195$

$= 325$

Our estimate turns out to be precise.

State. The weight of a package that cost $325 to ship is 250 lb.

51. *Thinking and Writing Exercise.*

53. $(-5)^3 = -5^3 = -125$

55. $-2^6 = -64$

57. $2 - (3 - 2^2) + 10 \div 2 \cdot 5$

$= 2 - (3 - 4) + 5 \cdot 5$

$= 2 - (-1) + 25 = 2 + 1 + 25 = 28$

59. *Thinking and Writing Exercise.*

61. $f(x) = g(x)$ is true at those points where the graphs of f and g intersect. The solutions are the first coordinate of each of the points of intersection, or -2, and 2. The solution set is $\{-2, 2\}$.

63. Solve $2x = |x + 1|$. Let $y_1 = 2x$ and $y_2 = |x + 1|$. Graph both in the same window on a graphing calculator and locate the point of intersection.

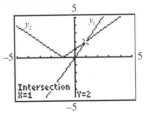

From the graph the solution is 1.

Check.

$$\frac{2x = |x+2|}{2 \cdot 1 \quad | \quad |1+2|}$$

$$\frac{2 \quad | \quad 2}{2 = 2 \text{ is TRUE}}$$

65. Solve $\frac{1}{2}x = 3 - |x|$. Let $y_1 = \frac{1}{2}x$ and

$y_2 = 3 - |x|$. Graph both in the same window on a graphing calculator and locate the points of intersection.

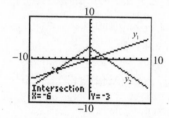

From the graph, the first solution is –6.

Check.

$$\frac{\frac{1}{2}x = 3 - |x|}{\frac{1}{2}(-6) \quad | \quad 3 - |-6|}$$

$$\frac{-3 \quad | \quad 3 - 6}{}$$

$$\frac{-3 \quad | \quad -3}{-3 = -3 \quad \text{is TRUE}}$$

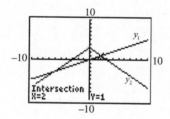

From the graph, the second solution is 2.

Check.

$$\frac{\frac{1}{2}x = 3 - |x|}{\frac{1}{2}(2) \quad | \quad 3 - |2|}$$

$$\frac{1 \quad | \quad 3 - 2}{}$$

$$\frac{1 \quad | \quad 1}{1 = 1 \quad \text{is TRUE}}$$

The two solutions are –6 and 2.

67. Solve $x^2 = x + 2$. Let $y_1 = x^2$ and

$y_2 = x + 2$. Graph both in the same window on a graphing calculator and locate the points of intersection.

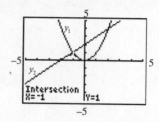

From the graph, the first solution is –1.

Check.

$$\frac{x^2 = x + 2}{(-1)^2 \quad | \quad -1 + 2}$$

$$\frac{1 \quad | \quad 1}{1 = 1 \text{ is TRUE}}$$

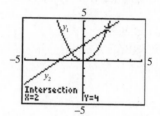

From the graph, the second solution is 2.

Check.

$$\frac{x^2 = x + 2}{2^2 \quad | \quad 2 + 2}$$

$$\frac{4 \quad | \quad 4}{4 = 4 \text{ is TRUE}}$$

The two solutions are –1 and 2.

69.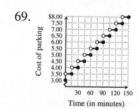

Time (in minutes)

Chapter 4 Study Summary

1. $x - y = 3$ （1)

　　$y = 2x - 5$ （2)

The graph of a line can be made by determining two points that lie on the line using an x-y table. For the first line, defined by $x - y = 3$, substituting $x = 0$ into the equation and solving for y gives the point $(0, -3)$. And substituting $y = 0$ into the equation gives the point $(3, \ 0)$.

For the second line, defined by $y = 2x - 5$, substituting $x = 0$ into the equation and solving for y gives the point $(0, -5)$. Substituting $x = 3$ into the equation gives the point $(3, 1)$.

Using these pairs of points to graph the lines, the intersection appears to be the point $(2, -1)$.

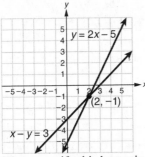

We can verify this by testing the point in each equation.   For the first equation, we have:

$$
\begin{array}{c|c}
x - y & 3 \\
\hline
(2) - (-1) & 3 \\
2 + 1 & 3 \\
3 & 3
\end{array}
$$

For the second equation, we have:

$$
\begin{array}{c|c}
y & 2x - 5 \\
\hline
-1 & 2(2) - 5 \\
-1 & 4 - 5 \\
-1 & -1
\end{array}
$$

The ordered pair $(2, -1)$ is the solution to the system.

2. $\quad x = 3y - 2 \quad (1)$

$\quad y - x = 1 \qquad (2)$

Equation (1) is already solved for x.
Substituting the result into equation (2) gives:

$$y - x = 1 \quad (2)$$
$$y - (3y - 2) = 1$$
$$y - 3y + 2 = 1$$
$$-2y + 2 = 1$$
$$-2y = 1 - 2 = -1$$
$$y = \frac{-1}{-2} = \frac{1}{2}$$

Substituting the value of y back into equation (1) gives the value of x,

$$x = 3y - 2 \quad (1)$$
$$x = 3\left(\frac{1}{2}\right) - 2$$
$$= \frac{3}{2} - 2$$
$$= \frac{3}{2} - \frac{4}{2} = -\frac{1}{2}$$

The ordered pair $\left(-\frac{1}{2}, \frac{1}{2}\right)$ is the solution to the system.

3. $2x - y = 5 \quad (1)$

$\quad x + 3y = 1 \quad (2)$

The system can be solved using elimination by multiplying equation (2) by (-2) and adding the result to equation (1) to find y, then finding x.

$$-2[x + 3y] = -2[1] \rightarrow -2x - 6y = -2$$

$$
\begin{array}{r}
2x - y = 5 \\
-2x - 6y = -2 \\
\hline
-7y = 3
\end{array}
$$

$$y = \frac{3}{-7} = -\frac{3}{7}$$

Now find x using equation (2).

$$x + 3y = 1 \qquad (2)$$
$$x + 3\left(-\frac{3}{7}\right) = 1$$
$$x - \frac{9}{7} = 1$$
$$x = 1 + \frac{9}{7} = 1\frac{9}{7} = \frac{16}{7}$$

The ordered pair $\left(\frac{16}{7}, -\frac{3}{7}\right)$ is the solution to the system.

4. Let $x =$ the number of boxes of Roller Grip™ pens purchased, and $y =$ the number of boxes of GEL™ pens purchased. Then:

$$x + y = 120 \qquad (1)$$
$$1749x + 1649y = 201,080 \qquad (2)$$

Note that all prices shown in equation (2), and the total spent, are in cents.
The system can be solved using elimination by multiplying equation (1) by (-1749) and adding the result to equation (2) to find y, then using equation (1) to find x.

$$-1749[x+y] = -1749[120]$$
$$\rightarrow -1749x - 1749y = -209,880$$
$$-1749x - 1749y = -209,880$$
$$\underline{1749x + 1649y = 201,080}$$
$$-100y = -8800$$
$$y = \frac{-8800}{-100} = 88$$

Now find x using equation (1).
$$x + y = 120 \quad (1)$$
$$x + 88 = 120$$
$$x = 120 - 88 = 32$$

Barlow's Office Supply purchased 32 boxes of Roller Grip™ pens and 88 boxes of GEL™ pens.

5. Let x = the amount, in liters, of 40% nitric acid solution used, and y = the amount, in liters, of 15% nitric acid solution used. Then:
$$x + y = 2 \quad (1)$$
$$40x + 15y = 25 \cdot 2 = 50 \quad (2)$$
The system can be solved using elimination by multiplying equation (1) by (-40) and adding the result to equation (2) to find y, then finding x.
$$-40[x+y] = -40[2] \rightarrow -40x - 40y = -80$$
$$-40x - 40y = -80$$
$$\underline{40x + 15y = 50}$$
$$-25y = -30$$
$$y = \frac{-30}{-25} = 1\frac{5}{25} = 1\frac{1}{5} = 1.2$$

Now find x using equation (1).
$$x + y = 2 \quad (1)$$
$$x + 1\frac{1}{5} = 2$$
$$x = 2 - 1\frac{1}{5} = \frac{4}{5} = 0.8$$

To produce 2 L of a 25% mixture, 0.8 L of 40% solution should be added to 1.2 L of 15% solution.

6. Let x = the speed of Ruth paddles in still water. Then, using the equation $d = rt$ for the time traveling with the current gives:
$$d_{\text{with current}} = (x+2) \cdot 1.5$$
and for the time traveling against the current gives:
$$d_{\text{against current}} = (x-2) \cdot 2.5$$

Since the distance traveled with the current is the same as the distance traveled against it, we have:
$$d_{\text{with current}} = d_{\text{against current}}$$
$$(x+2) \cdot 1.5 = (x-2) \cdot 2.5$$
$$1.5x + 3 = 2.5x - 5$$
$$5 + 3 = 2.5x - 1.5x$$
$$8 = x$$
Ruth paddles at a speed of 8 mph in still water.

7. The "zero" of a function is just the value of the x for which $f(x) = 0$. Therefore:
$$f(x) = 0 \rightarrow 8x - 1 = 0 \rightarrow 8x = 1 \rightarrow x = \frac{1}{8}.$$

8. Let $y_1 = x - 3$ and $y_2 = 5x + 1$
The graph of a line can be made by determining two points that lie on the line using an x-y table. For the first line, defined by $y = x - 3$, substituting $x = 0$ into the equation and solving for y gives the point $(0, -3)$. And substituting $y = 0$ into the equation gives the point $(3, 0)$.

For the second line, defined by $y = 5x + 1$, substituting $x = 0$ into the equation and solving for y gives the point $(0, 1)$. Substituting $x = 2$ into the equation gives the point $(2, 11)$.

Using these pairs of points to graph the lines, the intersection appears to be the point $(-1, -4)$.

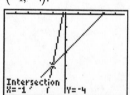

We can verify this by testing the point in each equation. For the first equation, we have:

$x - 3$	-4
$(-1) - 3$	-4
$-1 - 3$	-4
-4	-4

For the second equation, we have:

$$\frac{5x+1 \mid -4}{\begin{array}{c|c} 5(-1)+1 & -4 \\ -5+1 & -4 \\ -4 & -4 \end{array}}$$

The value $x = -1$ is the solution to the equation.

Chapter 4 Review Exercises

1. substitution

2. elimination

3. graphical

4. dependent

5. inconsistent

6. contradiction

7. parallel

8. zero

9. alphabetical

10. x-coordinate

11. Graph each line and determine the point of intersection.

$y = x - 3$ $y = \frac{1}{4}x$

x	y
0	-3
3	0

$m = \frac{1}{4}$

y-intercept is $(0, 0)$

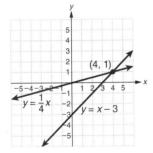

The point of intersection is $(4, 1)$.

12. We first solve each equation for y.

$$16x - 7y = 25 \qquad\qquad 8x + 3y = 19$$
$$-7y = -16x + 25 \qquad\quad 3y = -8x + 19$$
$$y = \frac{16}{7}x - \frac{25}{7} \qquad\quad y = -\frac{8}{3}x + \frac{19}{3}$$

Then we enter and graph $y_1 = \dfrac{16}{7}x - \dfrac{25}{7}$

and $y_2 = -\dfrac{8}{3}x + \dfrac{19}{3}$

Using the INTERSECT option, we determine the point of intersection.

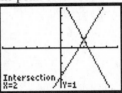

$[-5, 5, \text{Xscl} = 1, -5, 5, \text{Yscl} = 1]$

The point of intersection is $(2, 1)$.

13. $x - y = 8$ (1)

     $y = 3x + 2$ (2)

Substitute $3x + 2$ for y in Equation (1).

$$x - (3x + 2) = 8$$
$$x - 3x - 2 = 8$$
$$-2x = 8 + 2 = 10$$
$$x = \frac{10}{-2} = -5$$

Substitute -5 for x in Equation (1).

$x - y = 8$ (1)

$$(-5) - y = 8$$
$$-5 - y = 8$$
$$-y = 8 + 5 = 13$$
$$y = -13$$

The ordered pair $(-5, -13)$ is the solution to the system.

14. $y = x + 2$ (1)

    $y - x = 8$ (2)

Substitute $x + 2$ for y in equation (2).

$$(x + 2) - x = 8$$
$$2 = 8$$

This is a contradiction. The system has no solution.

15. $x - 3y = -2$ (1)

     $7y - 4x = 6$ (2)

Solve equation (1) for x.

$x - 3y = -2$

$\qquad x = 3y - 2 \quad (3)$

Substitute $3y - 2$ for x in Equation (2).

$7y - 4(3y - 2) = 6$

$\qquad 7y - 12y + 8 = 6$

$\qquad\qquad -5y = -2$

$\qquad\qquad\qquad y = \dfrac{2}{5}$

Substitute $\dfrac{2}{5}$ for y in Equation (3).

$x = 3\left(\dfrac{2}{5}\right) - 2$

$x = \dfrac{6}{5} - \dfrac{10}{5}$

$x = -\dfrac{4}{5}$

The solution is $\left(-\dfrac{4}{5}, \dfrac{2}{5}\right)$.

16. First, rearrange the second equation to align the x's and y's.

$2x - 5y = 11 \quad (1)$

$\qquad y - 2x = 5$

$-2x + y = 5 \quad (2)$

The coefficients of x are already opposites, so add the equations to solve for y. Then solve for x.

$\begin{array}{r} 2x - 5y = 11 \\ -2x + y = 5 \\ \hline -4y = 16 \end{array}$

$\qquad y = \dfrac{16}{-4} = -4$

Now find x using equation (1).

$2x - 5y = 11$

$2x - 5(-4) = 11$

$\quad 2x + 20 = 11$

$\qquad 2x = 11 - 20 = -9$

$\qquad\quad x = \dfrac{-9}{2} = -\dfrac{9}{2}$

The ordered pair $\left(-\dfrac{9}{2}, -4\right)$ is the solution to the system.

17. $4x - 7y = 18 \quad (1)$

$\quad 9x + 14y = 40 \quad (2)$

Multiply equation (1) by 2 and add the result to Equation (2).

$\begin{array}{rcr} 8x - 14y &=& 36 \\ 9x + 14y &=& 40 \\ \hline 17x &=& 76 \\ x &=& \dfrac{76}{17} \end{array}$

Substitute $\dfrac{76}{17}$ for x in Equation (1).

$4\left(\dfrac{76}{17}\right) - 7y = 18$

$\dfrac{304}{17} - 7y = \dfrac{306}{17}$

$\dfrac{-1}{7} \cdot (-7y) = \dfrac{2}{17} \cdot \dfrac{-1}{7}$

$\qquad\qquad y = \dfrac{-2}{119}$

The solution is $\left(\dfrac{76}{17}, -\dfrac{2}{119}\right)$.

18. $3x - 5y = 9 \quad (1)$

$\quad 5x - 3y = -1 \quad (2)$

Multiply Equation (1) by -5, Equation (2) by 3, and add the resulting equations.

$\begin{array}{rcr} -15x + 25y &=& -45 \\ 15x - 9y &=& -3 \\ \hline 16y &=& -48 \\ y &=& -3 \end{array}$

Substitute -3 for y in Equation (2).

$5x - 3(-3) = -1$

$\quad 5x + 9 = -1$

$\qquad 5x = -1 - 9 = -10$

$\qquad\quad x = \dfrac{-10}{5} = -2$

The ordered pair $(-2, -3)$ is the solution to the system.

19. $1.5x - 3 = -2y \rightarrow 1.5x + 2y = 3 \quad (1)$

$\qquad\qquad\qquad\qquad 3x + 4y = 6 \quad (2)$

Multiply Equation (1) by -2 and add the result to Equation (2).

$\begin{array}{rcr} -3x - 4y &=& -6 \\ 3x + 4y &=& 6 \\ \hline 0 &=& 0 \end{array}$

This is an identity.

$\{(x, y) \mid 3x + 4y = 6\}$

20. **Familiarize.** $P = 2l + 2w; l =$ the length and $w =$ the width, both in feet.
Translate. Perimeter is $860 \rightarrow 2l + 2w = 860$
Length is 100 more than width $\rightarrow l = w + 100$
Carryout. Solve the system.
$$2l + 2w = 860 \quad (1)$$
$$l = w + 100 \quad (2)$$
Substitute $w + 100$ for l in Equation (1).
$$2(w + 100) + 2w = 860$$
$$2w + 200 + 2w = 860$$
$$4w + 200 = 860$$
$$4w = 660$$
$$w = 165$$
Substitute 165 for w in Equation (2).
$$l = 165 + 100 = 265$$
Check.: $2 \cdot 265 + 2 \cdot 165 = 530 + 330 = 860$ and $165 = 65 + 100$. The numbers check.
State. The John Hancock building is 265 ft long and 165 ft wide.

21. Let $x =$ the amount, in ounces, of lemon juice used, and $y =$ the amount, in ounces, of linseed oil used. Then:
$$x + y = 32 \quad (1)$$
$$y = 2x \quad (2)$$
Find x by substituting $2x$ for y in equation (1). Then use equation (2) to find y.
$$x + y = 32 \quad (1)$$
$$x + (2x) = 32$$
$$3x = 32$$
$$x = \frac{32}{3} = 10\frac{2}{3}$$
Now find y using equation (2).
$$y = 2x \quad (2)$$
$$y = 2\left(\frac{32}{3}\right) = \frac{64}{3} = 21\frac{1}{3}$$
$10\frac{2}{3}$ ounces of lemon juice and $21\frac{1}{3}$ ounces of linseed oil should be used.

22. **Familiarize.** Let $x =$ the number of students taking private lessons and $y =$ the number of students taking group lessons.
Translate.

Total # of students is $12 \rightarrow x + y = 12$
Total \$ is \$265 $\rightarrow$ \$25x + \$18y = \$265
Carry out. Solve the system.
$$x + y = 12 \quad (1)$$
$$25x + 18y = 265 \quad (2)$$
Multiply equation (1) by -18 and add the result to equation (2)
$$-18x - 18y = -216$$
$$\underline{25x + 18y = 265}$$
$$7x = 49$$
$$x = 7$$
$$x + y = 12 \rightarrow 7 + y = 12$$
$$y = 5$$
Check.
$7 + 5 = 12$, and
$25 \cdot 7 + 18 \cdot 5 = 175 + 90 = 265$.
The numbers check.
State. Jillian had 7 students who took private lessons and 5 students who took group lessons.

23. **Familiarize.** Two angles are supplementary if the sum of their measures is $180°$. Let x and y equal the measures of the two angles in degrees.
Translate. Supplementary $\rightarrow x + y = 180°$

One angle   is $7°$ less than 10 times the other
    $\downarrow$ $\downarrow$ $\downarrow$
    x $=$ $10y - 7$
Carry out. Solve the system.
$$x + y = 180 \quad (1)$$
$$x = 10y - 7 \quad (2)$$
Substitute $10y - 7$ for x in Equation (1).
$$(10y - 7) + y = 180$$
$$11y - 7 = 180$$
$$11y = 187$$
$$y = 17$$
$$x = 10y - 7 \rightarrow x = 10(17) - 7$$
$$x = 170 - 7 = 163$$
Check. $163 + 17 = 180$ and $163 = 10(17) - 7$
The numbers check.
State. The angles measure $163°$ and $17°$.

24. **Familiarize.** We will summarize the information given in a chart and determine the equation using $D = R \times T$.
Translate.

	Distance	Rate	Time
Freight Train	d	44 mph	t
Passenger Train	d	55 mph	$t-1$

We have two equations:
$d = 44t$

$d = 55(t-1)$

Carry out. Solve the system.
$d = 44t$ (1)

$d = 55(t-1) \rightarrow d = 55t - 55$ (2)

Substitute $44t$ for d in Equation (2).
$44t = 55t - 55$

$-11t = -55$

$t = 5$

$t - 1 = 5 - 1 = 4$

Check. The freight train's distance $= 5 \cdot 44 = 220$ m., and the passenger train's

distance $= 4 \cdot 55 = 220$ m. The distances are equal; the numbers check.
State. The passenger train will travel 4 hr before it overtakes the freight train.

25. **Familiarize.** Let $x =$ the number of liters of the 15% juice punch and $y =$ the number of liters of the 8% juice punch.
Translate.
Total liters is $14 \rightarrow x + y = 14$

$\underbrace{\text{Total amount of juice}}$ is $\underbrace{\text{10% of total liters}}$

$\downarrow \qquad\qquad \downarrow$
$0.15x + 0.08y = 0.10 \cdot 14$
$0.15x + 0.08y = 1.4$

Carry out. Solve the system.
$x + y = 14$ (1)
$0.15x + 0.08y = 1.4$ (2)

Multiply Equation (1) by -0.08 and add the result to Equation (2).
$-0.08x - 0.08y = -1.12$
$\underline{0.15x + 0.08y = 1.40}$
$0.07x \qquad\quad = 0.28$
$x \qquad = 4$

$x + y = 14 \rightarrow 4 + y = 14$
$y = 10$

Check.

$4 + 10 = 14$, and

$0.15 \cdot 4 + 0.08 \cdot 10 = 0.6 + 0.8$

$= 1.4 = 0.10 \cdot 14$

The numbers check.
State. D'Andre should purchase 4 L of the 15% juice punch and 10 L of the 8% juice punch.

26. **Familiarize.** Let $x =$ the number of 1300 words/page pages and $y =$ the number of 1850 words/page pages.
Translate.

Total number of pages is $12 \rightarrow x + y = 12$

Total number of words is $18,350 \rightarrow 1300x + 1850y = 18,350$

Carry out. Solve the system.
$x + y = 12$ (1)
$1300x + 1850y = 18,350$ (2)

Multiply Equation (1) by -1300 and add the result to Equation (2).
$-1300x - 1300y = -15,600$
$\underline{1300x + 1850y = 18,350}$
$550y = 2750$
$y = 5$

$x + y = 12 \rightarrow x + 5 = 12$
$x = 7$

Check. $7 + 5 = 12$ and $1300 \cdot 7 + 1850 \cdot 5$
$= 9100 + 9250 = 18,350.$

The numbers check.
State. The typesetter had 7 of the 1300-word pages and 5 of the 1850-word pages.

27. Using the graph, we determine the x-coordinate of the point of intersection.
$x = -2$

28. Using the graph, we determine the x-coordinate of the point where $f(x)$ crosses the x-axis. $x = 1$ 1 is the zero.

29. $f(x) = 4 - 7x$

Determine the x-intercept of $f(x)$. Let $f(x) = 0$ and solve for x.

$$0 = 4 - 7x$$

$$7x = 4$$

$$x = \frac{4}{7} \qquad \frac{4}{7} \text{ is the zero.}$$

30. $3x - 2 = x + 4$

Graph: $y = 3x - 2$ and $y = x + 4$ and determine the x-coordinate of the point of intersection.

$y = 3x - 2$	$y = x + 4$
$m = 3$	$m = 1$
y-intercept is $(0, -2)$	y-intercept is $(0, 4)$

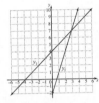

The x-coordinate of the point of intersection is 3; this is the solution.

31. *Thinking and Writing Exercise.* The solution of a system of two equations is an ordered pair that makes both equations true. The graph of an equation represents all ordered pairs that make that equation true. In order for an ordered pair to make both equations true, it must be on both graphs.

32. *Thinking and Writing Exercise.* Both methods involve finding the coordinates of the point of intersection of two graphs. The solution of a system of equations is the ordered pair at the point of intersection; the solution of an equation is the x-coordinate of the point of intersection.

33. Enter and graph $y_1 = x + 2$ and $y_2 = x^2 + 2$. Using the INTERSECT option, determine the points of intersection.

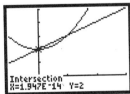

$[-1, 3, \text{Xscl}=1, -1, 5, \text{Yscl}=1]$
Note: 1.947E-14 is extremely close to 0.

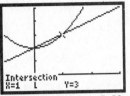

$[-1, 3, \text{Xscl}=1, -1, 5, \text{Yscl}=1]$
The points of intersection are $(0, 2)$ and $(1, 3)$.

34. Let $x = 6$ and $y = 2$.

$$2(6) - D(2) = 6$$

$$12 - 2D = 6$$

$$-2D = -6$$

$$D = 3$$

$$C(6) + 4(2) = 14$$

$$6C + 8 = 14$$

$$6C = 6$$

$$C = 1$$

35. Let $t =$ the tens digit and $u =$ the ones digit. The sum of the digits is 6, so $t + u = 6$. The value of the original number is $10t + u$ and the value of the new number is $10u + t$, using place-value. The new number is 18 more than the original number, so

$$10u + t = 18 + 10t + u$$

$$9u - 9t = 18$$

$$9(u - t) = 9 \cdot 2$$

$$u - t = 2$$

Solve the system.

$$
\begin{array}{r}
t + u = 6 \rightarrow u + t = 6 \\
u - t = 2 \\
\hline
2u \qquad = 8 \\
u \qquad = 4
\end{array}
$$

$$t + u = 6 \rightarrow t + 4 = 6$$

$$t = 2$$

The original number was 24.
Note: $24 + 18 = 42$

36. Let $x =$ the value of the computer, in dollars. 7 mos. is $\frac{7}{12}$ of a year. Since the full value of the computer was included in the prorated package, we have:

$$\frac{7}{12}(42,000+x)=23,750+x$$

$$12\left[\frac{7}{12}(42,000+x)\right]=12\left[23,750+x\right]$$

$$7(42,000+x)=285,000+12x$$

$$294,000+7x=285,000+12x$$

$$294,000-285,000=12x-7x$$

$$9000=5x$$

$$x=\frac{9000}{5}=1800$$

The value of the computer was $1800.

Chapter 4 Test

1. Graph each line and determine the points of intersection.

$$2x+y=8 \qquad y-x=2$$

x	y
0	8
4	0

x	y
−2	0
0	2

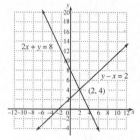

The point of intersection is (2, 4).

2. Graph each line and determine the point of intersection.

$$2y-x=7 \qquad 2x-4y=4$$

x	y
0	$\frac{7}{2}$
−7	0

x	y
0	−1
2	0

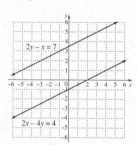

The lines are parallel; they do not intersection. The system has no solution.

3. $x+3y=-8$ (1)

$$4x-3y=23 \quad (2)$$

Solve Equation (2) for $3y$.

$$4x-3y=23$$

$$-3y=-4x+23$$

$$3y=4x-23$$

Substitute $4x-23$ for $3y$ in Equation (1).

$$x+(4x-23)=-8$$

$$5x-23=-8$$

$$5x=15$$

$$x=3$$

$$x+3y=-8\rightarrow 3+3y=-8$$

$$3y=-11$$

$$y=-\frac{11}{3}$$

The solution is $\left(3,-\dfrac{11}{3}\right)$.

4. $2x+4y=-6$ (1)

$$y=3x-9 \quad (2)$$

Substitute $3x-9$ for y in Equation (1).

$$2x+4(3x-9)=-6$$

$$2x+12x-36=-6$$

$$14x=30$$

$$x=\frac{30}{14}=\frac{15}{7}$$

$$y=3x-9\rightarrow y=3\left(\frac{15}{7}\right)-9$$

$$y=\frac{45}{7}-\frac{63}{7}$$

$$y=-\frac{18}{7}$$

The solution is $\left(\dfrac{15}{7},-\dfrac{18}{7}\right)$.

5. $x=5y-10$ (1)

$$15y=3x+30\rightarrow 5y=x+10 \quad (2)$$

Substitute $5y-10$ for x in Equation (2).

$$5y=(5y-10)+10$$

$$5y=5y$$

This is an identity.

$$\{(x,y)\,|\,x=5y-10\}$$

6. $3x - y = 7$ (1)

    $x + y = 1$ (2)

The coefficients of y are opposites. Add the two equations to find x. Then use equation (2) to find y.

$$3x - y = 7$$
$$\underline{x + y = 1}$$
$$4x \quad\quad = 8$$
$$x = \frac{8}{4} = 2$$

Substitute 2 for x in Equation (2) to find y..

$$x + y = 1$$
$$(2) + y = 1$$
$$y = 1 - 2 = -1$$

The ordered pair $(2, -1)$ is the solution to the system.

7. $4y + 2x = 18 \rightarrow 2x + 4y = 18$ (1)

    $3x + 6y = 26$ (2)

Multiply Equation (1) by –3, Equation (2) by 2, and add the resulting equations.

$$-6x - 12y = -54$$
$$\underline{6x + 12y = 52}$$
$$0 = -2$$

This is a contradiction. The system has no solution.

8. $4x - 6y = 3$ (1)

    $6x - 4y = -3$ (2)

Multiply Equation (1) by 3, Equation (2) by –2, and add the resulting equations.

$$12x - 18y = 9$$
$$\underline{-12x + 8y = 6}$$
$$-10y = 15$$
$$y = \frac{-15}{10} = \frac{-3}{2}$$
$$4x - 6y = 3 \rightarrow 4x - 6\left(\frac{-3}{2}\right) = 3$$
$$4x + 9 = 3$$
$$4x = -6$$
$$x = \frac{-6}{4} = -\frac{3}{2}$$

The solution is $\left(-\frac{3}{2}, -\frac{3}{2}\right)$.

9. $4x + 5y = 5$ (1)

$6x + 7y = 7$ (2)

Multiply Equation (1) by 3, Equation (2) by –2, and add the resulting equations.

$$12x + 15y = 15$$
$$\underline{-12x - 14y = -14}$$
$$y = 1$$
$$4x + 5y = 5 \rightarrow 4x + 5 \cdot 1 = 5$$
$$4x = 0$$
$$x = 0$$

The solution is $(0, 1)$.

10. ***Familiarize.*** $P = 2l + 2w$, where $l =$ the length and $w =$ the width, both in feet

Translate.

Perimeter is 66 feet.

$2l + 2w = 66$

Length is 1 foot longer than three times the width $l = 3w + 1$

Carry out. Solve the system.

    $2l + 2w = 66$ (1)

      $l = 3w + 1$ (2)

Substitute $3w + 1$ for l in Equation (1).

$$2(3w + 1) + 2w = 66$$
$$6w + 2 + 2w = 66$$
$$8w + 2 = 66$$
$$8w = 64$$
$$w = \frac{64}{8} = 8$$

Use equation (2) to find l.

$$l = 3w + 1 = 3(8) + 1 = 24 + 1 = 25$$

Check. $2 \cdot 25 + 2 \cdot 8 = 50 + 16 = 66$

      and $25 = 3 \cdot 8 + 1$.

The numbers check.

State. The dimensions of a garden are 25 ft long by 8 ft wide.

11. ***Familiarize.*** Two angles are complementary if the sum of their measures is 90°. Let x and y equal the measures of the angles, in degrees.

Translate. Complementary $\rightarrow x + y = 90$

Sum of first angle and half the second is 64°

$$x + \frac{1}{2}y \qquad\qquad = 64°$$

Carry out. Solve the system.

$$x + \phantom{\frac{1}{2}}y = 90 \quad (1)$$
$$x + \frac{1}{2}y = 64 \quad (2)$$

Multiply Equation (2) by -1 and add the result to Equation (1).

$$\begin{array}{r} x + \phantom{\frac{1}{2}}y = 90 \\ -x - \frac{1}{2}y = -64 \\ \hline 2 \cdot \frac{1}{2}y = 26 \cdot 2 \\ y = 52 \end{array}$$

$$x + y = 90 \rightarrow x + 52 = 90$$
$$x = 38$$

Check. $38 + 52 = 90$ and $38 + \frac{1}{2} \cdot 52$
$$= 38 + 26 = 64.$$

The numbers check.

State. The angles measure $38°$ and $52°$.

12. Let x = the number of Nintendo Wii game machines sold, in millions, and y = the number of PlayStation 3 consoles sold, in millions. Then:

$$x + y = 4.84 \quad (1)$$
$$x = 3y \quad\quad (2)$$

Substitute $3y$ for x in equation (1) to find y, then use equation (2) to find x.

$$x + y = 4.84 \quad (1)$$
$$3y + y = 4.84$$
$$4y = 4.84$$
$$y = \frac{4.84}{4} = 1.21$$

Now find x using equation (2).
$$x = 3y = 3(1.21) = 3.63$$

There were 3.63 million Nintendos Wii game machines and 1.21 million PlayStation 3 consoles sold.

13. **Familiarize.** Let x = the number of hard-backs and y = the number of paperbacks.
Translate.
Total # of books is $23 \rightarrow x + y = 23$

Total cost is $28.25.
$$\rightarrow \$1.75x + \$0.75y = \$28.25$$

Carry out. Solve the system.
$$x + y = 23 \quad\quad (1)$$
$$1.75x + 0.75y = 28.25 \quad (2)$$

Multiply Equation (1) by -0.75 and add the result to Equation (2).

$$\begin{array}{r} -0.75x - 0.75y = -17.25 \\ 1.75x + 0.75y = 28.25 \\ \hline 1.00x = 11 \\ x = 11 \end{array}$$

$$x + y = 23 \rightarrow 11 + y = 23$$
$$y = 12$$

Check. $11 + 12 = 23$, and
$1.75(11) + 0.75(12) = 19.25 + 9 = 28.25.$
These numbers check.
State. Keith purchased 11 hardbacks and 12 paperbacks.

14. **Familiarize.** Let x = the number of grams of Pepperidge Farm® Goldfish and let y = the number of grams of Rold Gold® Pretzels.
Translate. Total grams is 620
$$\rightarrow x + y = 620 .$$

Total fat calories can be found in two ways, thus giving us an equation:

   40% of Goldfish grams + 9% of Pretzel grams, or 15% of total grams
$$0.4x + 0.09y = 0.15(620)$$

Carry out. Solve the system.
$$x + y = 620 \,(1)$$
$$0.4x + 0.09y = 93 \quad (2)$$

Multiply Equation (1) by -0.09, and add the result to Equation (2).

$$\begin{array}{r} -0.09x - 0.09y = -55.8 \\ 0.4x + 0.09y = 93.0 \\ \hline 0.31x = 37.2 \\ x = 120 \end{array}$$

$$x + y = 620 \rightarrow 120 + y = 620 \rightarrow y = 500$$

Check. $120 + 500 = 620$, and $0.4(120) + 0.09(500) = 48 + 45 = 93$. These numbers check.
State. 120 g of Pepperidge Farm® Goldfish and 500 g of Rold Gold® Pretzels should be mixed.

15. **Familiarize.** Let s = speed of boat, in mph.
Translate. We make a table

	Distance	Rate	Time
With current	d	$s+5$	3
Against current	d	$s-5$	5

We have two equations:
$$d = 3(s+5) \rightarrow d = 3s + 15 \quad (1)$$
$$d = 5(s-5) \rightarrow d = 5s - 25 \quad (2)$$

Carry out. Solve the system.

Substitute $3s + 15$ for d in Equation (2).

$3s + 15 = 5s - 25$

$3s + 40 = 5s$

$40 = 2s$

$20 = s$

Check. Downstream distance is

$3(20 + 5) = 3 \cdot 25 = 75$; upstream distance is

$5(20 - 5) = 5 \cdot 15 = 75$. The distances are the same; our solution checks.

State. The speed of the boat is 20 mph.

16. ***Familiarize.*** Let $x =$ the number of 25¢ and $y =$ the number of 5¢.

    Translate.

    Total # of coins

    $13 \rightarrow x + y = 13$

    Total value of coins is

    $\$1.25 \rightarrow 0.25x + 0.05y = \$1.25$

    Carry out. Solve the system.

    $\begin{array}{ll} x + \quad y = 13 & (1) \\ 0.25x + 0.05y = 1.25 & (2) \end{array}$

    Multiply Equation (1) by -0.05 and add the result to Equation (2).

    $\begin{array}{rcl} -0.05x - 0.05y &=& -0.65 \\ 0.25x + 0.05y &=& 1.25 \\ \hline 0.2x \quad\quad &=& 0.6 \\ x &=& 3 \end{array}$

    $x + y = 13 \rightarrow 3 + y = 13$

    $\qquad\qquad\qquad y = 10$

    Check.

    $3 + 10 = 13$, and

    $0.25 \cdot 3 + 0.05 \cdot 10 = 0.75 + 0.50 = 1.25$.

    The numbers check.

    State. There are 3 quarters and 10 nickels in the collection.

17. $2x - 5 = 3x - 3$

    Graph: $y = 2x - 5$ and $y = 3x - 3$ and determine the x-coordinate of the point of intersection.

    $\begin{array}{ll} y = 2x - 5 & y = 3x - 3 \\ m = 2 & m = 3 \end{array}$

    y-intercept is $(0, -5)$; y-intercept is $(0, -3)$.

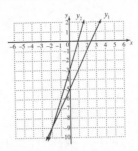

The x-coordinate of the point of intersection is -2; this is the solution.

18. $f(x) = \frac{1}{2}x - 5$

    Let $f(x) = 0$ and solve for x.

    $0 = \frac{1}{2}x - 5$

    $2 \cdot 5 = \frac{1}{2}x \cdot 2$

    $10 = x \qquad$ 10 is the zero.

19. $3(x - y) = 4 + x \qquad x = 5y + 2$

    $3x - 3y = 4 + x$

    $2x - 3y = 4$

    Substitute $5y + 2$ for x.

    $2(5y + 2) - 3y = 4$

    $10y + 4 - 3y = 4$

    $7y = 0$

    $y = 0$

    $x = 5y + 2 \rightarrow x = 5 \cdot 0 + 2 = 2$

    The solution is $(2, 0)$.

20. $\frac{3}{2}x - y = 24$ (1)

    $2x + \frac{3}{2}y = 15$ (2)

    Multiply equation (1) by $\frac{3}{2}$ and add the result to equation (2).

    $\begin{array}{rcl} \frac{9}{4}x - \frac{3}{2}y &=& 36 \\ 2x + \frac{3}{2}y &=& 15 \\ \hline (2 + \frac{9}{4})x &=& 51 \end{array}$

    $\frac{8 + 9}{4}x = 51$

    $\frac{4}{17} \cdot \frac{17}{4}x = 51 \cdot \frac{4}{17}$

    $x = 12$

$$\frac{3}{2}x - y = 24 \rightarrow \frac{3}{2} \cdot 12 - y = 24$$
$$18 - y = 24$$
$$-y = 6$$
$$y = -6$$

The solution is $(12, -6)$.

21. Let $x =$ the number of people behind you, then $x + 2 =$ the number of people ahead of you. There are 3 times the number of people behind, or $3x$ total people, in line. Another way to express the total is $x + (x + 2) + 1$ (the 1 represents you, as you are in line also). So, we have

$$3x = x + x + 2 + 1$$
$$3x = 2x + 3$$
$$x = 3$$

The total number of people in line is $3 \cdot 3$, or 9 people.

22. $f(x) = mx + b$

$$(-1, 3) \rightarrow f(-1) = m(-1) + b$$
$$3 = -m + b$$
$$(-2, -4) \rightarrow f(-2) = m(-2) + b$$
$$-4 = -2m + b$$

Solve the system:

$$3 = -m + b \quad (1)$$
$$-4 = -2m + b \quad (2)$$

Multiply Equation (1) by -2 and add the result to Equation (2).

$$\begin{array}{rcl} -6 &=& 2m - 2b \\ -4 &=& -2m + b \\ \hline -10 &=& -b \end{array}$$
$$10 = b$$

$$3 = -m + b$$
$$3 = -m + 10$$
$$-7 = -m$$
$$7 = m$$

Chapter 5
Polynomials

Exercise Set 5.1

1. e

3. b

5. g

7. c

9. The base of $(5x)^7$ is $5x$. The exponent is 7.

11. The exponential in $8n^0$ is n^0. The base of n^0 is n. The exponent is 0.

13. The exponential in $\dfrac{4y^3}{7}$ is y^3. The base of y^3 is y. The exponent is 3.

15. $d^3 \cdot d^{10} = d^{3+10} = d^{13}$

17. $a^6 \cdot a = a^{6+1} = a^7$

19. $8^4 \cdot 8^7 = 8^{4+7} = 8^{11}$

21. $(3y)^4 (3y)^8 = (3y)^{4+8} = (3y)^{12}$

23. $(7p)^0 (7p)^1 = (7p)^{0+1} = (7p)^1 = 7p$

25. $(x+1)^5 (x+1)^7 = (x+1)^{5+7} = (x+1)^{12}$

27. $(a^2 b^7)(a^3 b^2) = a^2 \cdot a^3 \cdot b^7 \cdot b^2$
$$= a^{2+3} b^{7+2}$$
$$= a^5 b^9$$

29. $r^3 \cdot r^7 \cdot r^0 = r^{3+7+0} = r^{10}$

31. $(mn^5)(m^3 n^4) = mn^5 m^3 n^4$
$$= m^{1+3} n^{5+4}$$
$$= m^4 n^9$$

33. $\dfrac{7^5}{7^2} = 7^{5-2} = 7^3$

35. $\dfrac{t^5}{t} = t^{5-1} = t^4$

37. $\dfrac{(5a)^7}{(5a)^6} = (5a)^{7-6} = (5a)^1 = 5a$

39. $\dfrac{(x+y)^8}{(x+y)^8} = (x+y)^{8-8} = (x+y)^0 = 1$

41. $\dfrac{(r+s)^{12}}{(r+s)^4} = (r+s)^{12-4} = (r+s)^8$

43. $\dfrac{8a^9 b^7}{2a^2 b} = \dfrac{8}{2} a^{9-2} b^{7-1} = 4a^7 b^6$

45. $\dfrac{12d^9}{15d^2} = \dfrac{\cancel{3} \cdot 4}{\cancel{3} \cdot 5} d^{9-2} = \dfrac{4}{5} d^7$

47. $\dfrac{m^9 n^8}{m^0 n^4} = m^{9-0} n^{8-4} = m^9 n^4$

49. When $x = 13, x^0 = 13^0 = 1$

51. $5x^0 = 5(-4)^0 = 5 \cdot 1 = 5$

53. $7^0 + 4^0 = 1 + 1 = 2$

55. $(-3)^1 - (-3)^0 = -3 - (1) = -4$

57. $(x^4)^7 = x^{4 \cdot 7} = x^{28}$

59. $(5^8)^2 = 5^{8 \cdot 2} = 5^{16}$

61. $(t^{20})^4 = t^{20 \cdot 4} = t^{80}$

63. $(7x)^2 = 7^2 \cdot x^2 = 49x^2$

65. $(-2a)^3 = (-2)^3 \cdot a^3 = -8a^3$

67. $\left(-5n^7\right)^2 = (-5)^2(n^7)^2 = 25n^{7 \cdot 2} = 25n^{14}$

69. $(a^2b)^7 = a^{2 \cdot 7} \cdot b^7 = a^{14}b^7$

71. $(x^3y)^2(x^2y^5) = (x^3)^2 y^2 \cdot x^2 \cdot y^5$
$$= x^{3 \cdot 2} \cdot x^2 \cdot y^2 \cdot y^5$$
$$= x^6 x^2 y^2 y^5$$
$$= x^{6+2} y^{2+5}$$
$$= x^8 y^7$$

73. $(2x^5)^3(3x^4) = 2^3 \cdot (x^5)^3 \cdot 3 \cdot x^4$
$$= 8x^{5 \cdot 3} \cdot 3x^4$$
$$= 8 \cdot 3 \cdot x^{15} \cdot x^4$$
$$= 24x^{15+4}$$
$$= 24x^{19}$$

75. $\left(\dfrac{a}{4}\right)^3 = \dfrac{a^3}{4^3} = \dfrac{a^3}{64}$

77. $\left(\dfrac{7}{5a}\right)^2 = \dfrac{7^2}{(5a)^2} = \dfrac{7^2}{5^2 a^2} = \dfrac{49}{25a^2}$

79. $\left(\dfrac{a^4}{b^3}\right)^5 = \dfrac{a^{4 \cdot 5}}{b^{3 \cdot 5}} = \dfrac{a^{20}}{b^{15}}$

81. $\left(\dfrac{x^2 y}{z^3}\right)^4 = \dfrac{(x^2)^4 y^4}{(z^3)^4}$
$$= \dfrac{x^{2 \cdot 4} \cdot y^4}{z^{3 \cdot 4}}$$
$$= \dfrac{x^8 y^4}{z^{12}}$$

83. $\left(\dfrac{a^3}{-2b^5}\right)^4 = \dfrac{a^{3 \cdot 4}}{(-2)^4 b^{5 \cdot 4}} = \dfrac{a^{12}}{16b^{20}}$

85. $\left(\dfrac{5x^7 y}{-2z^4}\right)^3 = \dfrac{5^3 \cdot x^{7 \cdot 3} y^3}{(-2)^3 z^{4 \cdot 3}} = \dfrac{125x^{21} y^3}{-8z^{12}}$
$$= -\dfrac{125x^{21} y^3}{8z^{12}}$$

87. $\left(\dfrac{4x^3 y^5}{3z^7}\right)^0 = 1$

Note: for $x \neq 0$, $y \neq 0$, and $z \neq 0$, we have a nonzero number raised to the 0 power, which is 1.

89. *Thinking and Writing Exercise.*

91. $-10 - 14 = -(10 + 14) = -24$

93. $-16 + 5 = -(16 - 5) = -11$

95. $-8(-10) = 8 \cdot 10 = 80$

97. *Thinking and Writing Exercise.*

99. *Thinking and Writing Exercise.*

101. Choose any number except 0. For example, let $a = 1$. Then
$$(a + 5)^2 = (1 + 5)^2 = 6^2 = 36, \text{ but}$$
$$a^2 + 5^2 = 1^2 + 5^2$$
$$= 1 + 25 = 26.$$

103. Choose any number except $\frac{7}{6}$. For example let $a = 0$. Then $\frac{0+7}{7} = \frac{7}{7} = 1$, but $a = 0$.

105. $a^{10k} \div a^{2k} = a^{10k - 2k} = a^{8k}$

107. $\dfrac{\left(\frac{1}{2}\right)^3 \left(\frac{2}{3}\right)^4}{\left(\frac{5}{6}\right)^3} = \dfrac{\frac{1}{8} \cdot \frac{16}{81}}{\frac{125}{216}} = \dfrac{1}{8} \cdot \dfrac{16}{81} \cdot \dfrac{216}{125}$
$$= \dfrac{1 \cdot 2 \cdot \cancel{8} \cdot \cancel{27} \cdot 8}{\cancel{8} \cdot 3 \cdot \cancel{27} \cdot 125} = \dfrac{16}{375}$$

109. $\dfrac{t^{26}}{t^x} = t^x$
$$t^{26-x} = t^x$$
$$26 - x = x \quad \text{Equating exponents}$$
$$26 = 2x$$
$$13 = x$$
The solution is 13.

111. Since the bases are the same, the expression with the greater exponent is greater. Thus, $4^2 < 4^3$.

113. $4^3 = 64$, $3^4 = 81$, so $4^3 < 3^4$.

115. $25^8 = (5^2)^8 = 5^{16}$
$125^5 = (5^3)^5 = 5^{15}$
$5^{16} > 5^{15}$, or $25^8 > 125^5$.

117. $2^{22} = 2^{10} \cdot 2^{10} \cdot 2^2 \approx 10^3 \cdot 10^3 \cdot 4$
$\approx 1000 \cdot 1000 \cdot 4 \approx 4,000,000.$
Using a calculator, we find that $2^{22} = 4,194,304.$ The difference between the exact value and the approximation is $4,194,304 - 4,000,000,$ or $194,304.$

119. $2^{31} \approx 2^{10} \cdot 2^{10} \cdot 2^{10} \cdot 2 \approx 10^3 \cdot 10^3 \cdot 10^3 \cdot 2$
$\approx 1000 \cdot 1000 \cdot 1000 \cdot 2 = 2,000,000,000.$
Using a calculator, we find that $2^{31} = 2,147,483,648.$ The difference between the exact value and the approximation is $2,147,483,648 - 2,000,000,000$
$= 147,483,648.$

121. $1.5\,\text{MB} = 1500\,\text{K} \approx 1500 \cdot 2^{10} = 1,536,000$ bytes or approximately $1,500,000$ bytes.

Exercise Set 5.2

1. Positive power of 10

3. Negative power of 10

5. Positive power of 10

7. $\left(\dfrac{x^3}{y^2}\right)^{-2} = \left(\dfrac{y^2}{x^3}\right)^2 = \dfrac{y^{2 \cdot 2}}{x^{3 \cdot 2}} = \dfrac{y^4}{x^6}$: c

9. $\left(\dfrac{y^{-2}}{x^{-3}}\right)^{-3} = \left(\dfrac{x^3}{y^2}\right)^{-3} = \left(\dfrac{y^2}{x^3}\right)^3 = \dfrac{y^{2 \cdot 3}}{x^{3 \cdot 3}} = \dfrac{y^6}{x^9}$: a

11. $7^{-2} = \dfrac{1}{7^2} = \dfrac{1}{49}$

13. $(-2)^{-6} = \dfrac{1}{(-2)^6} = \dfrac{1}{64}$

15. $a^{-3} = \dfrac{1}{a^3}$.

17. $\dfrac{1}{5^{-3}} = 5^{-(-3)} = 5^3 = 125,$ or $\dfrac{1}{5^{-3}} = \dfrac{1}{\frac{1}{5^3}} = 5^3$
$= 125$

19. $7^{-1} = \dfrac{1}{7^1} = \dfrac{1}{7}$

21. $8x^{-3} = 8 \cdot \dfrac{1}{x^3} = \dfrac{8}{x^3}$

23. $3a^8 b^{-6} = 3a^8 \cdot \dfrac{1}{b^6} = \dfrac{3a^8}{b^6}$

25. $\dfrac{z^{-4}}{3x^5} = \dfrac{1}{z^4} \cdot \dfrac{1}{3x^5} = \dfrac{1}{3x^5 z^4}$

27. $\dfrac{5x^{-2} y^7}{z^{-4}} = \dfrac{5y^7 z^4}{x^2}$

29. $\left(\dfrac{a}{2}\right)^{-3} = \left(\dfrac{2}{a}\right)^3 = \dfrac{2^3}{a^3} = \dfrac{8}{a^3}$

31. $\dfrac{1}{8^4} = 8^{-4}$

33. $\dfrac{1}{x} = x^{-1}$

35. $x^5 = \dfrac{1}{x^{-5}}$

37. $8^{-2} \cdot 8^{-4} = 8^{-2+(-4)} = 8^{-6} = \dfrac{1}{8^6}$

39. $b^2 \cdot b^{-5} = b^{2+(-5)} = b^{-3} = \dfrac{1}{b^3}$

41. $a^{-3} \cdot a^4 \cdot a = a^{-3+4+1} = a^2$

43. $(5a^{-2}b^{-3})(2a^{-4}b) = 5 \cdot 2 \cdot a^{-2+(-4)}b^{-3+1}$

$\qquad = 10a^{-6}b^{-2}$

$\qquad = 10 \cdot \dfrac{1}{a^6} \cdot \dfrac{1}{b^2}$

$\qquad = \dfrac{10}{a^6 b^2}$

45. $\dfrac{y^4}{y^{-5}} = y^{4-(-5)} = y^9$

47. $\dfrac{2^{-8}}{2^{-5}} = \dfrac{2^5}{2^8} = \dfrac{1}{2^{8-5}} = \dfrac{1}{2^3} = \dfrac{1}{8}$

49. $\dfrac{24a^2}{-8a^3} = -\dfrac{\cancel{8} \cdot 3}{\cancel{8}a^{3-2}} = -\dfrac{3}{a}$

51. $\dfrac{-6a^3b^{-5}}{-3a^7b^{-8}} = \dfrac{\cancel{3} \cdot 2b^8}{\cancel{3}a^{7-3}b^5} = \dfrac{2b^{8-5}}{a^4} = \dfrac{2b^3}{a^4}$

53. $\dfrac{6x^{-2}y^4z^8}{24x^{-5}y^6z^{-3}} = \dfrac{6}{24}x^{-2-(-5)}y^{4-6}z^{8-(-3)}$

$\qquad = \dfrac{1}{4}x^3 y^{-2}z^{11}$

$\qquad = \dfrac{1}{4}x^3 \cdot \dfrac{1}{y^2} \cdot z^{11}$

$\qquad = \dfrac{x^3 z^{11}}{4y^2}$

55. $(n^{-5})^3 = \left(\dfrac{1}{n^5}\right)^3 = \dfrac{1}{n^{5\cdot3}} = \dfrac{1}{n^{15}}$

57. $(t^{-8})^{-5} = t^{-8(-5)} = t^{40}$

59. $(mn)^{-7} = \dfrac{1}{(mn)^7} = \dfrac{1}{m^7 n^7}$

61. $(5r^{-4}t^3)^2 = \left(\dfrac{5t^3}{r^4}\right)^2 = \dfrac{5^2 t^{3\cdot2}}{r^{4\cdot2}} = \dfrac{25t^6}{r^8}$

63. $(3m^5 n^{-3})^{-2} = \left(\dfrac{3m^5}{n^3}\right)^{-2} = \left(\dfrac{n^3}{3m^5}\right)^2$

$\qquad = \dfrac{n^{3\cdot2}}{3^2 m^{5\cdot2}} = \dfrac{n^6}{9m^{10}}$

65. $(a^{-5}b^7c^{-2})(a^{-3}b^{-2}c^6) = a^{-5+(-3)}b^{7+(-2)}c^{-2+6}$

$\qquad = a^{-8}b^5 c^4 = \dfrac{b^5 c^4}{a^8}$

67. $\left(\dfrac{a^4}{3}\right)^{-2} = \dfrac{(a^4)^{-2}}{(3)^{-2}} = \dfrac{a^{-8}}{3^{-2}}$

$\qquad = \dfrac{1}{a^8} \cdot 3^2 = \dfrac{3^2}{a^8} = \dfrac{9}{a^8}$

69. $\left(\dfrac{m^{-1}}{n^{-4}}\right)^3 = \dfrac{m^{-1\cdot3}}{n^{-4\cdot3}} = \dfrac{m^{-3}}{n^{-12}} = \dfrac{n^{12}}{m^3}$

71. $\left(\dfrac{-4x^4 y^{-2}}{5x^{-1}y^4}\right)^{-4} = \left[-\dfrac{4}{5}x^{4-(-1)}y^{-2-4}\right]^{-4}$

$\qquad = \left(-\dfrac{4}{5}\right)^{-4}(x^5)^{-4}(y^{-6})^{-4}$

$\qquad = \left(-\dfrac{5}{4}\right)^4 x^{-20}y^{24}$

$\qquad = \dfrac{5^4 y^{24}}{4^4 x^{20}}$

73. $\left(\dfrac{4a^3 b^{-9}}{6a^{-2}b^5}\right)^0 = 1$

Note: This is a non-zero number raised the zero power.

75. $\dfrac{(2a^3)^3 4a^{-3}}{(a^2)^5} = \dfrac{2^3(a^3)^3 \cdot 4a^{-3}}{(a^2)^5}$

$\qquad = \dfrac{8a^{3\cdot3} \cdot 4a^{-3}}{a^{2\cdot5}}$

$\qquad = \dfrac{8a^9 4a^{-3}}{a^{10}}$

$\qquad = \dfrac{8 \cdot 4 \cdot a^{9+(-3)}}{a^{10}}$

$\qquad = \dfrac{32a^6}{a^{10}}$

$\qquad = 32a^{6-10}$

$\qquad = 32 \cdot a^{-4} = \dfrac{32}{a^4}$

77. $-8^4 = -(8^4) = -4096$

79. $(-2)^{-4} = 0.0625$

81. $3^4 5^{-3} = 81 \cdot 0.008 = 0.648$

83. $4.92 \times 10^5 = 492{,}000$

85. $8.02 \times 10^{-3} = 0.00802$

87. $3.497 \times 10^{-6} = 0.000003497$

89. $9.03 \times 10^{10} = 90{,}300{,}000{,}000$

91. $47{,}000{,}000{,}000 = 4.7 \times 10^{10}$

93. $0.00583 = 5.83 \times 10^{-3}$

95. $407{,}000{,}000{,}000 = 4.07 \times 10^{11}$

97. $0.000000603 = 6.03 \times 10^{-7}$

99. 5.02×10^{18}

101. -3.05×10^{-10}

103. $(2.3 \times 10^6)(4.2 \times 10^{-11})$
$= (2.3 \times 4.2)(10^6 \times 10^{-11})$
$= 9.66 \times 10^{-5}$
$\approx 9.7 \times 10^{-5}$

105. $(2.34 \times 10^{-8})(5.7 \times 10^{-4})$
$= (2.34 \times 5.7)(10^{-8} \times 10^{-4})$
$= 13.338 \times 10^{-12}$
$= 1.3338 \times 10^1 \times 10^{-12}$
$= 1.3338 \times 10^{-11}$
$\approx 1.3 \times 10^{-11}$

107. $(2.0 \times 3.02)(10^6 \times 10^{-6})$
$= 6.04 \times 10^0$
$= 6.04$
≈ 6.0

109. $\dfrac{5.1 \times 10^6}{3.4 \times 10^3} = \dfrac{5.1}{3.4} \times 10^{6-3}$
$= 1.5 \times 10^3$

111. $\dfrac{7.5 \times 10^{-9}}{2.5 \times 10^{-4}} = \dfrac{7.5}{2.5} \times 10^{-9-(-4)}$
$= 3.0 \times 10^{-5}$

113. $\dfrac{1.23 \times 10^8}{6.87 \times 10^{-13}} \approx 1.79 \times 10^{20}$

115. $5.9 \times 10^{23} + 6.3 \times 10^{23}$
$= (5.9 + 6.3) \times 10^{23}$ Distributive Law
$= 12.2 \times 10^{23}$
$= 1.22 \times 10^1 \times 10^{23}$
$= 1.22 \times 10^{24}$
$\approx 1.2 \times 10^{24}$

117. *Familiarize.* We need to divide to determine the average number of megabytes of information generated per person.
Translate. We divide 5 exabytes, or 5×10^{12} megabytes by the worldwide population of 6.3 billion.
$$\frac{5 \times 10^{12}}{6.3 \times 10^9}$$
Carry out. We calculate and write scientific notation for the answer.
$$\frac{5 \times 10^{12}}{6.3 \times 10^9} = \frac{50 \times 10^{11}}{6.3 \times 10^9}$$
$$= \frac{50}{6.3} \times 10^{11-9}$$
$$= 7.936507937 \times 10^2$$
$$\approx 8 \times 10^2$$
Check. We reverse the process (multiply) to check our answer.
$(8 \times 10^2)(6.3 \times 10^9) = (8 \times 6.3)[10^{2+9}]$
$= 50.4 \times 10^{11}$
$\approx 5 \times 10^{12}$
Reminder: We used an approximate value to write the answer with the correct number of significant digits. Our answer checks.
State. The average number of megabytes of information per person is 8×10^2 megabytes.

119. 121 million $\cdot 5.8$ hours

$$= 121 \cdot 10^6 \cdot 5.8 \; \cancel{\text{hr}} \cdot \left(\frac{60 \text{ min}}{1 \; \cancel{\text{hr}}} \right)$$

$$= 121 \cdot 10^6 \cdot 5.8 \; \cdot 60 \text{ min} = 42{,}108 \cdot 10^6 \text{ min}$$

$$= 4.2108 \cdot 10^4 \cdot 10^6 \text{ min} = 4.2108 \cdot 10^{10} \text{ min}$$

121. **Familiarize.** We will divide the diameter by the light year to determine n, the number of light years.

Translate. $n = \dfrac{5.88 \times 10^{17}}{5.88 \times 10^{12}}$

Carry out. We solve the equation.

$$n = \frac{5.88 \times 10^{17}}{5.88 \times 10^{12}}$$

$$= \frac{5.88}{5.88} \times 10^{17-12}$$

$$= 1 \times 10^5$$

Check: $(1 \times 10^5)(5.88 \times 10^{12}) = 5.88 \times 10^{17}$

Our answer checks.

State. It is 1×10^5 light years from one end of the Milky Way to the other end.

123. The cable is like a long cylinder. Let $l =$ the length of a cylinder and $d =$ its diameter. Then the volume of the cylinder is given by the formula: $\pi r^2 l = \pi \left(\dfrac{d}{2} \right)^2 l = \dfrac{\pi d^2 l}{4}$.

In order to calculate a volume, make sure that the units for d and l match. The problem gives l in km and d in cm. Converting both to meters will give an answer in units of m^3.

$$125 \text{ km} = 125 \cdot 10^3 \text{ m} = 1.25 \cdot 10^2 \cdot 10^3 \text{ m}$$

$$= 1.25 \cdot 10^5 \text{ m}$$

and $0.6 \text{ cm} = 0.6 \cdot 10^{-2} \text{ m} = 6 \cdot 10^{-1} \cdot 10^{-2} \text{ m}$

$$= 6 \cdot 10^{-3} \text{ m}$$

Therefore the volume of cable is:

$$\frac{\pi d^2 l}{4} = \frac{\pi \left(6 \cdot 10^{-3} \text{ m} \right)^2 \left(1.25 \cdot 10^5 \text{ m} \right)}{4}$$

$$= \frac{\pi \left(36 \cdot 10^{-6} \text{ m}^2 \right) \left(1.25 \cdot 10^5 \text{ m} \right)}{4}$$

$$= \frac{\pi \left(36 \cdot 1.25 \cdot 10^{-6} \cdot 10^5 \right)}{4} \text{ m}^3$$

$$= \frac{\pi \left(45 \cdot 10^{-1} \right)}{4} \text{ m}^3 = \frac{4.5\pi}{4} \text{ m}^3$$

$$= 1.125\pi \text{ m}^3 \approx 3.5 \text{ m}^3$$

125. **Familiarize.** There are 10 million bacteria per square centimeter of coral in a coral reef. The reefs near the Hawaiian Islands cover 14,000 square kilometers.

Translate. We will multiply after converting the given information to the same units of area. Since

$1 \text{ m}^2 = 100 \text{ cm} \times 100 \text{ cm} = 10^4 \text{ cm}^2$ and

$1 \text{ km}^2 = 1000 \text{ m} \times 1000 \text{ m} = 10^6 \text{ m}^2$, we

simply multiply $\left(10 \times 10^6 \right) \cdot 10^4 \cdot 10^6$ to

convert cm^2 to km^2, obtaining 10×10^{16}

bacteria per km^2.

Carry out. $10 \times 10^{16} \cdot 14{,}000$

$$= 1 \times 10^{17} \cdot 1.4 \times 10^4$$

$$= 1.4 \times 10^{17+4}$$

$$= 1.4 \times 10^{21}$$

Check. We reverse the process to check our answer.

$1.4 \times 10^{21} \div 1.4 \times 10^4 = 1 \times 10^{21-4} = 1 \times 10^{17}$

which is 10×10^{16} bacteria per km^2.

Our answer checks.

State. 1.4×10^{21} bacteria are in Hawaii's coral reef.

127. **Familiarize.** Each side of the house is covered with a. 4-ft high-sheet of 8-mil plastic. These dimensions remain constant for all of the sheets/sides which are rectangular prisms. The volume of a rectangular prism is the area of the base times the height of the prism, or $v = l \cdot w \cdot h$. Two sides have length of 24-ft, and the other two sides have length of 32-ft. 1 ft = 12 inches

Translate. We will convert the measurement in feet to inches.

$24 \text{ ft.} = 24 \cdot 12 = 288 \text{ in.}$

$32 \text{ ft.} = 32 \cdot 12 = 384 \text{ in.}$

$4 \text{ ft.} = 4 \cdot 12 \; = 48 \text{ in.}$

Substitute these values into the volume formula.

Note: 8-mil = .008

$V = 2(288 \times 48 \times .008) + 2(384 \times 48 \times .008)$

Carry out. We solve the equation

$V = 2(288 \times 48 \times .008) + 2(384 \times 48 \times .008)$

$ = 2(110.592) + 2(147.456)$

$ = 221.184 + 294.912$

$ = 516.096$

$ = 5.16096 \times 10^2 \, \text{in.}^3.$

$ \approx 5 \times 10^2 \, \text{in.}^3, \quad \text{or} \quad 3 \times 10^{-1} \, \text{ft}^3$

Note: $1\text{in.}^3 = \left(\dfrac{1}{12}\right)^3 \text{ft}^3, \quad \text{or} \quad \dfrac{1}{1728}\,\text{ft}^3$

Check. We recalculate to check our solution. The answer checks.

State. The volume of plastic used is

$5 \times 10^2 \, \text{in.}^3, \quad \text{or} \quad 3 \times 10^{-1} \, \text{ft}^3.$

129. *Thinking and Writing Exercise.*

131. $9x + 2y - x - 2y = (9-1)x + (2-2)y$

$ = 8x + 0y = 8x$

133. $-3x + (-2) - 5 - (-x) = -3x - 2 - 5 + x$

$ = (-3+1)x - 7$

$ = -2x - 7$

135. $4 + x^3 = 4 + (10)^3 = 4 + 1000 = 1004$

137. *Thinking and Writing Exercise.*

139. *Thinking and Writing Exercise.*

141. $5^0 - 5^{-1} = 1 - \dfrac{1}{5} = \dfrac{5}{5} - \dfrac{1}{5} = \dfrac{4}{5}$

143. $(7^{-12}) \cdot 7^{25} = 7^{-12 \cdot 2} \cdot 7^{25}$

$\phantom{143.\ (7^{-12}) \cdot 7^{25}} = 7^{-24} \cdot 7^{25}$

$\phantom{143.\ (7^{-12}) \cdot 7^{25}} = 7^{-24+25}$

$\phantom{143.\ (7^{-12}) \cdot 7^{25}} = 7^1$

$\phantom{143.\ (7^{-12}) \cdot 7^{25}} = 7$

Note: These steps could be combined, since the bases are both 7.

$(7^{-12})^2 \cdot 7^{25} = 7^{-12 \cdot 2 + 25}$

$\phantom{(7^{-12})^2 \cdot 7^{25}} = 7^{-24+25}$

$\phantom{(7^{-12})^2 \cdot 7^{25}} = 7^1$

$\phantom{(7^{-12})^2 \cdot 7^{25}} = 7$

145. $\dfrac{4.2 \times 10^8 [(2.5 \times 10^{-5}) \div [5.0 \times 10^{-9})]}{3.0 \times 10^{-12}}$

$= \dfrac{4.2 \times 10^8 \left[\frac{2.5}{5.0} \times 10^{-5-(-9)}\right]}{3.0 \times 10^{-12}}$

$= \dfrac{4.2 \times 10^8 [0.5 \times 10^4]}{3.0 \times 10^{-12}}$

$= \dfrac{4.2 \times 10^8 [5 \times 10^{-1} \times 10^4]}{3.0 \times 10^{-12}}$

$= \dfrac{4.2 \times 10^8 [5 \times 10^3]}{3.0 \times 10^{-12}}$

$= \dfrac{(4.2 \times 5) \times (10^8 \times 10^3)}{3.0 \times 10^{-12}}$

$= \dfrac{21.0 \times 10^{11}}{3.0 \times 10^{-12}}$

$= \dfrac{21.0}{3.0} \times 10^{11-(-12)}$

$= 7.0 \times 10^{23}$

147. $81^3 \cdot 27 \div 9^2 = (3^4)^3 \cdot 3^3 \div (3^2)^2$

$ = 3^{12} \cdot 3^3 \div 3^4$

$ = 3^{12+3-4}$

$ = 3^{11}$

149. $\dfrac{1}{8.00 \times 10^{-23}} = \dfrac{1}{8.00} \times 10^{23} = 0.125 \times 10^{23}$

$\phantom{149.\ \dfrac{1}{8.00 \times 10^{-23}}} = 1.25 \times 10^{-1} \times 10^{23}$

$\phantom{149.\ \dfrac{1}{8.00 \times 10^{-23}}} = 1.25 \times 10^{-1+23}$

$\phantom{149.\ \dfrac{1}{8.00 \times 10^{-23}}} = 1.25 \times 10^{22}$

151. Observe that there are 2^{n-1} grains of sand on the nth square of the chessboard. Let g represent this quantity. Recall that a chessboard has 64 squares. Note also that $2^{10} \approx 10^3$.

We write the equation $g = 2^{n-1}$.

To find the number of grains of sand on the last (or 64th) square, substitute 64 for n:

$$g = 2^{64-1}$$

Do the calculations, expressing the result in scientific notation.

$$g = 2^{64-1} = 2^{63} = 2^3 \left(2^{10}\right)^6$$

$$\approx 2^3 \left(10^3\right)^6 \approx 8 \times 10^{18}$$

Approximately 8×10^{18} grains of sand are required for the last square.

Exercise Set 5.3

---

1. b

3. h

5. g

7. a

9. $3x - 7$ can be written as a sum of monomials, so it is a polynomial.

11. $\dfrac{x^2 + x + 1}{x^3 - 7}$ cannot be written as a sum of monomials, so it is not a polynomial.

13. $\dfrac{1}{4}x^{10} - 8.6$ can be written as a sum of monomials, so it is a polynomial.

15. $7x^4 + x^3 - 5x + 8 = 7x^4 + x^3 + (-5x) + 8$. The terms are $7x^4, x^3, -5x$, and 8.

17. $-t^6 + 7t^3 - 3t^2 + 6 = -t^4 + 7t^3 + (-3t^2) + 6$. The terms are $-t^6, 7t^3, -3t^2$, and 6.

19. $4x^5 + 7x$

Term	Coefficient	Degree
$4x^5$	4	5
$7x$	7	1

21. $9t^2 - 3t + 4$

Term	Coefficient	Degree
$9t^2$	9	2
$-3t$	-3	1
4	4	0

23. $x^4 - x^3 + 4x - 3$

Term	Coefficient	Degree
x^4	1	4
$-x^3$	-1	3
$4x$	4	1
-3	-3	0

25. $2a^3 + 7a^5 + a^2$

a.

Term	$2a^3$	$7a^5$	a^2
Degree	3	5	2

b. The term of highest degree is $7a^5$. This is the leading term. Then the leading coefficient is 7.

c. Since the term of highest degree is $7a^5$, the degree of the polynomial is 5.

27. $9x^4 + x^2 + x^7 + 4$

a.

Term	$9x^4$	x^2	x^7	4
Degree	4	2	7	0

b. The term of highest degree is x^7. This is the leading term. Then the leading coefficient is 1.

c. Since the term of highest degree is x^7, the degree of the polynomial is 7.

29. $9a - a^4 + 3 + 2a^3$

a.

Term	$9a$	$-a^4$	3	$2a^3$
Degree	1	4	0	3

b. The term of highest degree is $-a^4$. This is the leading term. Then the leading coefficient is -1.

c. Since the term of highest degree is $-a^4$, the degree of the polynomial is 4.

31. $7x^2 + 8x^5 - 4x^3 + 6 - \dfrac{1}{2}x^4$

Term	Coefficient	Degree of Term	Degree of Polynomial
$8x^5$	8	5	
$-\dfrac{1}{2}x^4$	$-\dfrac{1}{2}$	4	
$-4x^3$	-4	3	5
$7x^2$	7	2	
6	6	0	

33. Three monomials are added, so $x^2 - 23x + 17$ is a trinomial.

35. The polynomial $x^3 - 7x^2 + 2x - 4$ is a polynomial with no special name.

37. Two monomials are added, so $y + 5$ is a binomial.

39. The polynomial 17 is a monomial because it is the product of a constant and a variable raised to a whole number power. (In this case the variable is raised to the power 0.)

41. $7x^2 + 3x + 4x^2 = (7+4)x^2 + 3x$
$\qquad\qquad\qquad\; = 11x^2 + 3x$

43. $3a^4 - 2a + 2a + a^4 = (3+1)a^4 + (-2+2)a$
$\qquad\qquad\qquad\qquad = 4a^4 + 0a = 4a^4$

45. $9t^3 - 11t + 5t + t^2 = 9t^3 + t^2 + (-11+5)t$
$\qquad\qquad\qquad\qquad = 9t^3 + t^2 - 6t$

47. $\quad 4b^3 + 5b + 7b^3 + b^2 - 6b$
$\qquad = (4+7)b^3 + b^2 + (5-6)b$
$\qquad = 11b^3 + b^2 - b$

49. $10x^2 + 2x^3 - 3x^3 - 4x^2 - 6x^2 - x^4$
$\qquad = -x^4 + (2-3)x^3 + (10-4-6)x^2$
$\qquad = -x^4 - x^3$

51. $\dfrac{1}{5}x^4 + 7 - 2x^2 + 3 - \dfrac{2}{15}x^4 + 2x^2$

$\quad = \left(\dfrac{1}{5} - \dfrac{2}{15}\right)x^4 + (-2+2)x^2 + (7+3)$

$\quad = \left(\dfrac{3}{15} - \dfrac{2}{15}\right)x^4 + 0x^2 + 10 = \dfrac{1}{15}x^4 + 10$

53. $5.9x^2 - 2.1x + 6 + 3.4x - 2.5x^2 - 0.5$
$\quad = (5.9 - 2.5)x^2 + (-2.1 + 3.4)x + (6 - 0.5)$
$\quad = 3.4x^2 + 1.3x + 5.5$

55. $-7x + 4 = -7 \cdot 3 + 4 = -21 + 4 = -17$
$\quad -7x + 4 = -7 \cdot (-3) + 4 = 21 + 4 = 25$

57. $2x^2 - 3x + 7 = 2 \cdot 3^2 - 3 \cdot 3 + 7$
$\qquad\qquad\qquad = 18 - 9 + 7$
$\qquad\qquad\qquad = 16$
$\quad 2x^2 - 3x + 7 = 2 \cdot (-3)^2 - 3 \cdot (-3) + 7$
$\qquad\qquad\qquad = 2 \cdot 9 + 9 + 7$
$\qquad\qquad\qquad = 18 + 9 + 7$
$\qquad\qquad\qquad = 34$

59. $-2x^3 - 3x^2 + 4x + 2 = -2 \cdot 3^3 - 3 \cdot 3^2 + 4 \cdot 3 + 2$
$\qquad\qquad\qquad\qquad = -2 \cdot 27 - 3 \cdot 9 + 4 \cdot 3 + 2$
$\qquad\qquad\qquad\qquad = -54 - 27 + 12 + 2$
$\qquad\qquad\qquad\qquad = -67$
$\quad -2x^3 - 3x^2 + 4x + 2 = -2 \cdot (-3)^3 - 3 \cdot (-3)^2$
$\qquad\qquad\qquad\qquad\quad + 4 \cdot (-3) + 2$
$\qquad\qquad\qquad\qquad = -2 \cdot (-27) - 3 \cdot 9$
$\qquad\qquad\qquad\qquad\quad + 4 \cdot (-3) + 2$
$\qquad\qquad\qquad\qquad = 54 - 27 - 12 + 2$
$\qquad\qquad\qquad\qquad = 17$

61. $\dfrac{1}{3}x^4 - 2x^3 = \dfrac{1}{3} \cdot 3^4 - 2 \cdot 3^3$
$\qquad\qquad\qquad = \dfrac{1}{3} \cdot 81 - 2 \cdot 27$
$\qquad\qquad\qquad = 27 - 54$
$\qquad\qquad\qquad = -27$
$\quad \dfrac{1}{3}x^4 - 2x^3 = \dfrac{1}{3} \cdot (-3)^4 - 2 \cdot (-3)^3$
$\qquad\qquad\qquad = \dfrac{1}{3} \cdot 81 - 2 \cdot (-27)$
$\qquad\qquad\qquad = 27 + 54$
$\qquad\qquad\qquad = 81$

63. $-x - x^2 - x^3$

$= -3 - 3^2 - 3^3$

$= -3 - 9 - 27 = -39$

$-x - x^2 - x^3$

$= -(-3) - (-3)^2 - (-3)^3$

$= 3 - (9) - (-27) = 3 - 9 + 27 = 21$

65. $P(x) = 3x^2 - 2x + 7$

$P(4) = 3 \cdot 4^2 - 2 \cdot 4 + 7$

$= 48 - 8 + 7$

$= 47$

$P(0) = 3 \cdot 0^2 - 2 \cdot 0 + 7$

$= 0 - 0 + 7$

$= 7$

67. $P(y) = 8y^3 - 12y - 5$

$P(-2) = 8(-2)^3 - 12(-2) - 5$

$= -64 + 24 - 5$

$= -45$

$P\left(\dfrac{1}{3}\right) = 8\left(\dfrac{1}{3}\right)^3 - 12 \cdot \dfrac{1}{3} - 5$

$= 8 \cdot \dfrac{1}{27} - 4 - 5$

$= \dfrac{8}{27} - 9$

$= \dfrac{8}{27} - \dfrac{243}{27}$

$= -\dfrac{235}{27}$, or $-8\dfrac{19}{27}$

69. $f(x) = -5x^3 + 3x^2 - 4x - 3$

$f(-1) = -5(-1)^3 + 3(-1)^2 - 4(-1) - 3$

$= 5 + 3 + 4 - 3$

$= 9$

71. Use $t = 2$ because 2006 is 2 years after 2004.

$f(t) = 0.4x + 1.13$

$f(2) = 0.4(2) + 1.13$

$= 0.8 + 1.13$

$= 1.93$

The amount spent on shoes for college in 2006 is $1.93 billion.

73. $f(t) = 11.12t^2$

$f(10) = 11.12 \cdot 10^2$

$= 11.12 \cdot 100$

$= 1112$

A skydiver has fallen approximately 1112 ft. 10 sec. after jumping from a plane.

75. $C = 2\pi r$ $\pi \approx 3.14$

$C \approx 2 \cdot 3.14 \cdot 10 \text{ cm}$

$\approx 62.8 \text{ cm}$

77. $A = \pi r^2$ $\pi \approx 3.14$

$\approx 3.14 \cdot (7\,\text{m})^2$

$\approx 3.14 \cdot 49\,\text{m}^2$

$\approx 153.86\,\text{m}^2$

79. $s(t) = 16t^2$

$s(2.9) = 16 \cdot (2.9)^2 = 16 \cdot 8.41 = 134.56$

Approximately 135 feet

81. From the graph, there appears to be about 55 million Web sites in 2004. Using the polynomial, the number is about 54 million:

$w(t) = 4.03t^2 + 6.78t + 42.86$

$w(2004 - 2003) = w(1)$

$= 4.03(1)^2 + 6.78(1) + 42.86 = 53.67$

83. From the graph, 5 years after 2003 appears to have about 175 million Web sites. Using the polynomial, the number is about 178 million:

$w(t) = 4.03t^2 + 6.78t + 42.86$

$w(5) = 4.03(5)^2 + 6.78(5) + 42.86$

$= 4.03(25) + 6.78(5) + 42.86 = 177.51$

85. $N(x) = \dfrac{1}{3}x^3 + \dfrac{1}{2}x^2 + \dfrac{1}{6}x$

$N(3) = \dfrac{1}{3} \cdot 3^3 + \dfrac{1}{2} \cdot 3^2 + \dfrac{1}{6} \cdot 3$

$= \dfrac{1}{3} \cdot 27 + \dfrac{1}{2} \cdot 9 + \dfrac{1}{6} \cdot 3$

$= 9 + \dfrac{9}{2} + \dfrac{1}{2}$

$= 14$

From the diagram we see that the bottom layer contains 9 balls, the second layer contains 4, and the top layer contains 1 for a total of $9+4+1$, or 14

$$N(5) = \frac{1}{3} \cdot 5^3 + \frac{1}{2} \cdot 5^2 + \frac{1}{6} \cdot 5$$

$$= \frac{125}{3} + \frac{25}{2} + \frac{5}{6} \qquad \text{LCD is 6.}$$

$$= \frac{125}{3} \cdot \frac{2}{2} + \frac{25}{2} \cdot \frac{3}{3} + \frac{5}{6}$$

$$= \frac{250 + 75 + 5}{6}$$

$$= \frac{330}{6} = 55 \text{ oranges}$$

87. Locate 2 on the horizontal axis. From there move vertically to the graph and then horizontally to the $C(t)$-axis. This locates a value of about 2.3 mcg/mL.

89. $C(t)$ has a minimum value of 0 and a maximum value of 10, so the range is $[0, 10]$.

91. The function has a maximum value of 3 and no minimum is indicated, so the range is $(-\infty, 3]$.

93. There is no maximum or minimum value indicated by the graph, so the range is $(-\infty, \infty)$.

95. The function has a minimum value of –4 and no maximum is indicated, so the range is $[-4, \infty)$.

97. The function has a minimum value of –65 and no maximum is indicated, so the range is $[-65, \infty)$.

99. We graph $f(x) = x^2 + 2x + 1$ in the standard viewing window.

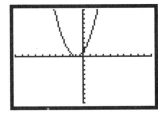

The range appears to be $[0, \infty)$.

101. We graph $q(x) = -2x^2 + 5$ in the standard viewing window.

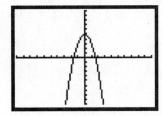

The range appears to be $(-\infty, 5]$.

103. We graph $p(x) = -2x^3 + x + 5$ in the standard viewing window.

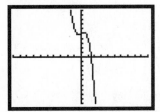

The range appears to be $(-\infty, \infty)$.

105. We graph $g(x) = x^4 + 2x^3 - 5$ in the window $[-5, 5, -10, 10]$.

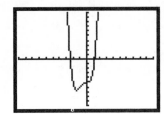

We estimate that the range is $[-6.7, \infty)$.

107. *Thinking and Writing Exercise.*

109. $3x + 7 - (x + 3) = 3x + 7 - x - 3 = 2x + 4$

111. $4a + 11 - (-2a - 9) = 4a + 11 + 2a + 9$
$$= 6a + 20$$

113. $4t^4 + 3t^2 + 8t - (3t^4 + 9t^2 + 8t)$
$$= 4t^4 + 3t^2 + 8t - 3t^4 - 9t^2 - 8t$$
$$= t^4 - 6t^2$$

115. *Thinking and Writing Exercise.*

117. Answer may vary. Use an ax^5-term, where a is an integer, and 3 other terms with different degrees, each less than degree 5, and consecutive even integer coefficients. Three answers are

$$6x^5 + 8x^4 + 10x^2 + 12,$$

$$-8x^5 - 6x^4 - 4x^2 - 2, \text{ and}$$

$$2x^5 + 4x^4 + 6x^2 + 8.$$

119. $(5m^5)^2 = 5^2 m^{5 \cdot 2} = 25 m^{10}$

The degree is 10.

121. $\dfrac{9}{2}x^8 + \dfrac{1}{9}x^2 + \dfrac{1}{2}x^9 + \dfrac{9}{2}x + \dfrac{9}{2}x^9$

$\qquad + \dfrac{8}{9}x^2 + \dfrac{1}{2}x - \dfrac{1}{2}x^8$

$= \left(\dfrac{1}{2} + \dfrac{9}{2}\right)x^9 + \left(\dfrac{9}{2} - \dfrac{1}{2}\right)x^8$

$\quad + \left(\dfrac{1}{9} + \dfrac{8}{9}\right)x^2 + \left(\dfrac{9}{2} + \dfrac{1}{2}\right)x$

$= \dfrac{10}{2}x^9 + \dfrac{8}{2}x^8 + \dfrac{9}{9}x^2 + \dfrac{10}{2}x$

$= 5x^9 + 4x^8 + x^2 + 5x$

123. Let c = the coefficient of x^3. Solve:

$$c + (c - 3) + 3(c - 3) + (c + 2) = -4$$

$$c + c - 3 + 3c - 9 + c + 2 = -4$$

$$6c - 10 = -4$$

$$6c = 6$$

$$c = 1$$

Coefficient of x^3, c : 1

Coefficient of x^2, $c - 3$: $1 - 3$, or -2

Coefficient of x, $3(c - 3)$: $3(1 - 3)$, or -6

Coefficient remaining (constant term), $c + 2$: $1 + 2$, or 3

The polynomial is $x^3 - 2x^2 - 6x + 3$.

125. *Familiarize.* Since q and t represent averages, we will determine the average of each using

$$(\text{Mean}) \text{ Average} = \frac{\text{Sum of scores}}{\text{Number of scores}}.$$

We will substitute those values into the given formula for A.

Translate.

$$\text{Quiz Average } q = \frac{\text{Sum of quizzes}}{\text{Number of quizzes}}$$

$$= \frac{60 + 85 + 72 + 91}{4}$$

$$\text{Test average } t = \frac{\text{Sum of tests}}{\text{Number of tests}}$$

$$= \frac{89 + 93 + 90}{3}$$

Carry out. We calculate q and t. We substitute the values of q, t, f, and h into A and solve the equation.

$$q = \frac{60 + 85 + 72 + 91}{4} = \frac{308}{4} = 77$$

$$t = \frac{89 + 93 + 90}{3} = \frac{272}{3} = 90.\overline{6}$$

$$a = 0.3q + 0.4t + 0.2f + 0.1h$$

$$= 0.3 \cdot 77 + 0.4 \cdot 90.\overline{6} + 0.2 \cdot 84 + 0.1 \cdot 88$$

$$= 23.1 + 36.2\overline{6} + 16.8 + 8.8$$

$$= 84.9\overline{6}$$

$$\approx 85.0$$

Check. We can recalculate to see that we obtain the same answer. The answer checks.

State. Mary Lou's average is 85.0 for the course.

127.

t	$-t^2 + 10t - 18$
3	$-3^2 + 10 \cdot 3 - 18 = -9 + 30 - 18 = 3$
4	$-4^2 + 10 \cdot 4 - 18 = -16 + 40 - 18 = 6$
5	$-5^2 + 10 \cdot 5 - 18 = -25 + 50 - 18 = 7$
6	$-6^2 + 10 \cdot 6 - 18 = -36 + 60 - 18 = 6$
7	$-7^2 + 10 \cdot 7 - 18 = -49 + 70 - 18 = 3$

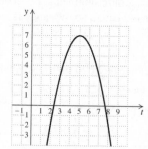

129. Using a calculator, we complete the table.

d	$-0.0064d^2 + 0.8d + 2$
0	2
30	20.24
60	26.96
90	22.16
120	5.84

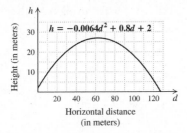

Exercise Set 5.4

1. $(3x^2 + 2) + (6x^2 + 7) = (3 + 6)\boxed{x^2} + (2 + 7)$

3. $(9x^3 - x^2) - (3x^3 + x^2) = 9x^3 - x^2 - 3x^3 \boxed{-} x^2$

5. $(3x + 2) + (x + 7) = 3x + 2 + x + 7$
$$= 4x + 9$$

7. $(2y - 3) + (-9y + 1) = 2y - 3 - 9y + 1$
$$= -7y - 2$$

9. $(-6x + 2) + (x^2 + x - 3) = x^2 + (-6 + 1)x$
$$+ (2 - 3)$$
$$= x^2 - 5x - 1$$

11. $\left(7t^2 - 3t + 6\right) + \left(2t^2 + 8t - 9\right)$
$$= (7 + 2)t^2 + (-3 + 8)t + (6 - 9)$$
$$= 9t^2 + 5t - 3$$

13. $\left(2m^3 - 7m^2 + m - 6\right) + \left(4m^3 + 7m^2 - 4m - 2\right)$
$$= 2m^3 - 7m^2 + m - 6 + 4m^3 + 7m^2 - 4m - 2$$
$$= 6m^3 - 3m - 8$$

15. $\left(3 + 6a + 7a^2 + a^3\right) + \left(4 + 7a - 8a^2 + 6a^3\right)$
$$= 3 + 6a + 7a^2 + a^3 + 4 + 7a - 8a^2 + 6a^3$$
$$= 7a^3 - a^2 + 13a + 7$$

17. $\left(9x^8 - 7x^4 + 2x^2 + 5\right) + \left(8x^7 + 4x^4 - 2x\right)$

$$= 9x^8 + 8x^7 + (-7 + 4)x^4 + 2x^2 - 2x + 5$$
$$= 9x^8 + 8x^7 - 3x^4 + 2x^2 - 2x + 5$$

19. $\left(\dfrac{1}{4}x^4 + \dfrac{2}{3}x^3 + \dfrac{5}{8}x^2 + 7\right) + \left(-\dfrac{3}{4}x^4 + \dfrac{3}{8}x^2 - 7\right)$

$$= \left(\dfrac{1}{4} - \dfrac{3}{4}\right)x^4 + \dfrac{2}{3}x^3 + \left(\dfrac{5}{8} + \dfrac{3}{8}\right)x^2 + (7 - 7)$$
$$= -\dfrac{2}{4}x^4 + \dfrac{2}{3}x^3 + \dfrac{8}{8}x^2 + 0$$
$$= -\dfrac{1}{2}x^4 + \dfrac{2}{3}x^3 + x^2$$

21. $\left(5.3t^2 - 6.4t - 9.1\right) + \left(4.2t^3 - 1.8t^2 + 7.3\right)$
$$= 4.2t^3 + (5.3 - 1.8)t^2 - 6.4t + (-9.1 + 7.3)$$
$$= 4.2t^3 + 3.5t^2 - 6.4t - 1.8$$

23. $-3x^4 + 6x^2 + 2x - 1$
$\underline{\quad - 3x^2 + 2x + 1}$
$-3x^4 + 3x^2 + 4x + 0$
$-3x^4 + 3x^2 + 4x$

25. Rewrite the problem so the coefficient of like terms have the same number of decimal places.

$0.15x^4 + 0.10x^2 - 0.90x^2$
$\quad\quad - 0.01x^3 + 0.01x^2 + x$
$1.25x^4 \quad\quad\quad + 0.11x^2 \quad\quad + 0.01$
$\quad\quad 0.27x^3 \quad\quad\quad\quad\quad + 0.99$
$\underline{-0.35x^4 \quad\quad\quad + 15.00x^2 \quad\quad - 0.03}$
$1.05x^4 + 0.36x^3 + 14.22x^2 + x + 0.97$

27. Two forms of the opposite of $-t^3 + 4t^2 - 9$ are

a. $-\left(-t^3 + 4t^2 - 9\right)$ and

b. $t^3 - 4t^2 + 9$. (Changing the sign of each term).

29. Two forms for the opposite of $12x^4 - 3x^3 + 3$ are

a. $-\left(12x^4 - 3x^3 + 3\right)$ and

b. $-12x^4 + 3x^3 - 3$. (Changing the sign of each term).

31. $-(8x - 9) = -8x + 9$

33. $-\left(3a^4 - 5a^2 + 9\right) = -3a^4 + 5a^2 - 9$

35. $-\left(-4x^4 + 6x^2 + \frac{3}{4}x - 8\right) = 4x^4 - 6x^2 - \frac{3}{4}x + 8$

37. $(7x+4) - (2x+1) = 7x + 4 - 2x - 1$
$= 5x + 3$

39. $(-5t+6) - (t^2 + 3t - 1) = -5t + 6 - t^2 - 3t + 1$
$= -t^2 - 8t + 7$

41. $(8y^2 + y - 11) - (3 - 6y^3 - 8y^2)$
$= 8y^2 + y - 11 - 3 + 6y^3 + 8y^2$
$= 6y^3 + 16y^2 + y - 14$

43. $\left(1.2x^3 + 4.5x^2 - 3.8x\right) - \left(-3.4x^3 - 4.7x^2 + 23\right)$
$= 1.2x^3 + 4.5x^2 - 3.8x + 3.4x^3 + 4.7x^2 - 23$
$= 4.6x^3 + 9.2x^2 - 3.8x - 23$

45. $\left(7x^3 - 2x^2 + 6\right) - \left(7x^3 - 2x^2 + 6\right)$

Observe that we are subtracting the polynomial $7x^3 - 2x^2 + 6$ from itself. The result is 0.

47. $\left(3 + 5a + 3a^2 - a^3\right) - \left(2 + 3a - 4a^2 + 2a^3\right)$
$= 3 + 5a + 3a^2 - a^3 - 2 - 3a + 4a^2 - 2a^3$
$= 1 + 2a + 7a^2 - 3a^3$

49. $\left(\frac{5}{8}x^3 - \frac{1}{4}x - \frac{1}{3}\right) - \left(-\frac{1}{8}x^3 + \frac{1}{4}x - \frac{1}{3}\right)$
$= \frac{5}{8}x^3 - \frac{1}{4}x - \frac{1}{3} + \frac{1}{8}x^3 - \frac{1}{4}x + \frac{1}{3}$
$= \frac{6}{8}x^3 - \frac{2}{4}x$
$= \frac{3}{4}x^3 - \frac{1}{2}x$

51. $(0.07t^3 - 0.03t^2 - 0.25t) - (0.02t^3 + 0.04t^2 - 0.3t)$
$= 0.07t^3 - 0.03t^2 - 0.25t - 0.02t^3 - 0.04t^2 + 0.3t$
$= 0.05t^3 - 0.07t^2 + 0.05t$

53. $\begin{array}{l} x^2 + 5x + 6 \\ \underline{-(x^2 + 2x + 1)} \end{array}$

$\begin{array}{l} x^2 + 5x + 6 \\ \underline{-x^2 - 2x - 1} \\ 3x + 5 \end{array}$ Changing signs and removing parentheses
Adding

55. $\begin{array}{l} 5x^4 + 6x^3 \\ \underline{-(-6x^4 - 6x^3 + x^2)} \end{array}$

$\begin{array}{l} 5x^4 + \ 6x^3 \\ \underline{6x^4 + \ 6x^3 - \ x^2} \\ 11x^4 + 12x^3 - x^2 \end{array}$ Changing signs and removing parentheses
Adding

57. a.

Familiarize. The area of a rectangle is the product of the length and the width.
Translate. The sum of the areas is found as follows:

$\begin{array}{ccccccc} \text{Area} & & \text{Area} & & \text{Area} & & \text{Area} \\ \text{of } A & + & \text{of } B & + & \text{of } C & + & \text{of } D \\ = \ 3x \cdot x & + & x \cdot x & + & 4 \cdot x & + & x \cdot x \end{array}$

Carry out. We collect like terms.
$3x^2 + x^2 + 4x + x^2 = 5x^2 + 4x$
Check. We can recalculate or assign some value to x, say 2, and carry out the computation of the area in two ways.
Sum of areas: $3 \cdot 2 \cdot 2 + 2 \cdot 2 + 4 \cdot 2 + 2 \cdot 2$
$= 12 + 4 + 8 + 4 = 28$
Substituting in the polynomial:
$5(2)^2 + 4 \cdot 2 = 20 + 8 = 28$
Since the results are the same, our solution is probably correct.
State. A polynomial for the sum of the areas is $5x^2 + 4x$.

b. For $x = 5: 5x^2 + 4x = 5 \cdot 5^2 + 4 \cdot 5$
$= 5 \cdot 25 + 4 \cdot 5 = 125 + 20$
$= 145$
When $x = 5$, the sum of the areas is 145 square units.
For $x = 7: 5x^2 + 4x = 5 \cdot 7^2 + 4 \cdot 7$
$= 5 \cdot 49 + 4 \cdot 7 = 245 + 28$
$= 273$
When $x = 7$, the sum of the areas is 273 square units.

59.

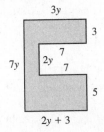

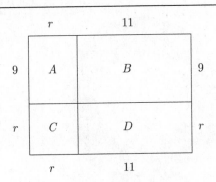

Familiarize. The perimeter is the sum of the lengths of the sides.

Translate. The sum of the lengths is found as follows:

$$3y + 7y + (2y+3) + 5 + 7 + 2y + 7 + 3$$

Carry out. We collect like terms.

$$(3+7+2+2)y + (3+5+7+7+3) = 14y + 25$$

Check. We can recalculate or assign some value to y, say 3, and carry out the computation of the perimeter in two ways.

Sum of lengths:

$$3\cdot3 + 7\cdot3 + (2\cdot3+3) + 5 + 7 + 2\cdot3 + 7 + 3$$

$$= 9 + 21 + 9 + 5 + 7 + 6 + 7 + 3$$

$$= 67$$

Substituting in the polynomial:

$$14\cdot3 + 25 = 42 + 25 = 67$$

Since the results are the same, our solution is probably correct.

State. A polynomial for the perimeter of the figure is $14y + 25$.

The area of the figure can be found by adding the areas of the four rectangles A, B, C, and D. The area of a rectangle is the product of the length and the width.

Area of A	+	Area of B	+	Area of C	+	Area of D
= 9·r	+	11·9	+	r·r	+	11·r
= 9r	+	99	+	r^2	+	11r

An algebraic expression for the area of the figure is $9r + 99 + r^2 + 11r$.

61.

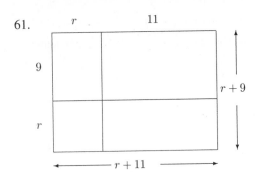

The length and width of the figure can be expressed as $r+11$ and $r+9$, respectively. The area of this figure (a rectangle) is the product of the length and width. An algebraic expression for the area is $(r+11)\cdot(r+9)$.

The algebraic expressions $9r + 99 + r^2 + 11r$ and $(r+11)\cdot(r+9)$ represent the same area.

63.

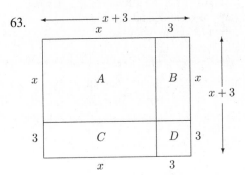

The length and width of the figure can each be expressed as $x+3$. The area can be expressed as

$$(x+3)\cdot(x+3) = x^2 + 3x + 3x + 9, \text{ or } (x+3)^2.$$

Another way to express the area is to find an expression for the sum of the areas of the four rectangles A, B, C, and D. The area of each rectangle is the product of its length and width.

Area of A	+	Area of B	+	Area of C	+	Area of D
= $x\cdot x$	+	3·x	+	3·x	+	3·3
= x^2	+	3x	+	3x	+	9

The algebraic expressions $(x+3)^2$ and $x^2 + 3x + 3x + 9$ represents the same area.

$$(x+3)^2 = x^2 + 3x + 3x + 9$$

65.

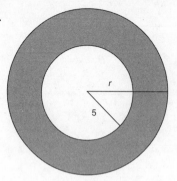

Familiarize. Recall that the area of a circle is the product of π and the square of the radius, r^2.

Translate.

$$\underset{\text{with radius } r}{\text{Area of circle}} - \underset{\text{with radius } 5}{\text{Area of circle}} = \underset{\text{area}}{\text{Shaded}}$$

$$\pi \cdot r^2 \quad - \quad 25\pi \quad = \text{Shaded area}$$

Carry out. We simplify the expression.

Check. We can recalculate or assign some value to r, say 7, and carry out the computation in two ways.

Difference of areas:

$$\pi \cdot 7^2 - \pi \cdot 5^2 = 49\pi - 25\pi = 24\pi$$

Substituting in the polynomial:

$$\pi \cdot 7^2 - 25\pi = 49\pi - 25\pi = 24\pi$$

Since the results are the same, our solution is probably correct.

State. A polynomial for the shaded area is $\pi r^2 - 25\pi$.

67. **Familiarize.** We label the figure with additional information.

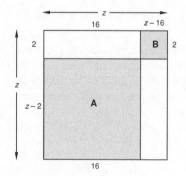

Translate.

Area of shaded sections $=$ Area of A + Area of B

Area of shaded sections

$$= 16(z - 2) + 2(z - 16) = 16z - 32 + 2z - 32$$

$$= 18z - 64.$$

Carry out. We can recalculate or assign some value to z, say 30, and carry out the computation in two ways.

Sum of areas:

$$16 \cdot 28 + 2 \cdot 14 = 448 + 28 = 476$$

Substituting in the polynomial:

$$18 \cdot 30 - 64 = 540 - 64 = 476$$

Since the results are the same, our solution is probably correct.

State. A polynomial for the shaded area is $18z - 64$.

69. The square bathroom has area of x^2 ft^2 and the enclosure has area of 2 ft $\cdot$ 6 ft, or 12 ft^2. The remaining floor area is their difference, or $\left(x^2 - 12\right)$ ft^2.

71. The square garden has an area of z^2 ft^2. The 12 ft-wide round patio's diameter is 12 ft, thus its radius is $\frac{1}{2} \cdot$ 12 ft = 6 ft.

Area of a circle $= \pi r^2$. The area of the patio is $\pi \cdot 6^2$ ft$^2 = 36\pi$ ft^2.

The remaining area of the patio is their difference, or $\left(z^2 - 36\pi\right)$ ft^2.

73. The 12 m by 12 m square mat has area of (12 m)^2, or 144 m^2. The diameter of the circle is d meters, thus its radius $= \frac{1}{2}d$ meters.

Area of a circle $= \pi r^2$. The area of the circle is $\pi \cdot \left(\frac{1}{2}d\right)^2$ m^2, or $\frac{1}{4}\pi d^2$ m^2. The mat area, outside of the wrestling circle is their difference, or $\left(144 - \frac{1}{4}\pi d^2\right)$ m^2 which can be written as $\left(144 - \frac{\pi d^2}{4}\right)$ m^2.

75. Using one of the methods on page 389 in the text, we see that the addition is not correct.

77. Using one of the methods on page 389 in the text, we see that the subtraction is correct.

79. Using one of the methods on page 389 in the text, we see that the subtraction is not correct.

81. *Thinking and Writing Exercise.*

83. $2(x^2 - x + 3) = 2 \cdot x^2 - 2 \cdot x + 2 \cdot 3$

$$= 2x^2 - 2x + 6$$

85. $t^2 t^{11} = t^{2+11} = t^{13}$

87. $2n \cdot n^6 = 2n^1 \cdot n^6 = 2n^{1+6} = 2n^7$

89. *Thinking and Writing Exercise.*

91. $\left(6t^2 - 7t\right) + \left(3t^2 - 4t + 5\right) - (9t - 6)$

$\quad = 6t^2 - 7t + 3t^2 - 4t + 5 - 9t + 6$

$\quad = 9t^2 - 20t + 11$

93. $4(x^2 - x + 3) - 2(2x^2 + x - 1)$

$\quad = 4x^2 - 4x + 12 - 4x^2 - 2x + 2$

$\quad = -6x + 14$

95. $\left(345.099x^3 - 6.178x\right) - \left(94.508x^3 - 8.99x\right)$

$\quad = 345.099x^3 - 6.178x - 94.508x^3 + 8.99x$

$\quad = 250.591x^3 + 2.812x$

For Exercises 96–99, the surface area of a right rectangular solid is $2lw + 2lh + 2wh$, where l = length, w = width, and h = height of the rectangular solid.

97. Surface Area $= 2lw + 2lh + 2wh$

$\quad = 2 \cdot 3 \cdot w + 2 \cdot 3 \cdot 7 + 2 \cdot w \cdot 7$

$\quad = 6w + 42 + 14w$

$\quad = (20w + 42) \text{ units}^2$

99. Surface Area $= 2lw + 2lh + 2wh$

$\quad = 2 \cdot x \cdot x + 2 \cdot x \cdot 5 + 2 \cdot x \cdot 5$

$\quad = 2x^2 + 10x + 10x$

$\quad = \left(2x^2 + 20x\right) \text{ units}^2$

101. a) $P(x) = R(x) - C(x)$

$\quad = 175x - 0.4x^2 - \left(5000 + 0.6x^2\right)$

$\quad = 175x - 0.4x^2 - 5000 - 0.6x^2$

$\quad = -x^2 + 175x - 5000$

b) $P(x) = -x^2 + 175x - 5000$

$\quad P(75) = -(75)^2 + 175(75) - 5000$

$\quad = -5625 + 13,125 - 5000 = 2500$

The total profit on the production and sale of 75 cameras is $2500.

c) $P(x) = -x^2 + 175x - 5000$

$\quad P(120) = -(120)^2 + 175(120) - 5000$

$\quad = -14,400 + 21,000 - 5000$

$\quad = 1600$

The total profit on the production and sale of 120 cameras is $1600.

Exercise Set 5.5

1. $3x^2 \cdot 2x^4 = 3 \cdot 2 \cdot x^{2+4} = 6x^6$:   c

3. $4x^3 \cdot 2x^5 = 4 \cdot 2 \cdot x^{3+5} = 8x^8$:   d

5. $4x^6 + 2x^6 = (4+2)x^6 = 6x^6$:   c

7. $(4x^3)9 = (4 \cdot 9)x^3 = 36x^3$

9. $(-x^2)(-x) = (-1 \cdot x^2)(-1 \cdot x)$

$\quad = (-1 \cdot -1)(x^2 \cdot x)$

$\quad = x^3$

11. $(-x^6)(x^2) = (-1 \cdot x^6)(x^2)$

$\quad = -1 \cdot (x^6 \cdot x^2)$

$\quad = -x^8$

13. $(7t^5)(4t^3) = (7 \cdot 4)(t^5 \cdot t^3)$

$\quad = 28t^8$

15. $(-0.1x^6)(0.2x^4) = (-0.1 \cdot 0.2)(x^6 \cdot x^4)$

$\quad = -0.02x^{10}$

17. $\left(-\dfrac{1}{5}x^3\right)\left(-\dfrac{1}{3}x\right) = \left(-\dfrac{1}{5} \cdot -\dfrac{1}{3}\right)(x^3 \cdot x)$

$\quad = \dfrac{1}{15}x^4$

19. $(-1)(-19t^2) = (-1)(-19)t^2 = 19t^2$

21. $(-4y^5)(6y^2)(-3y^3) = (-4) \cdot 6 \cdot (-3)(y^5 y^2 y^3)$

$\quad = 72y^{10}$

23. $4x(x+1) = 4x \cdot x + 4x \cdot 1$ Distributive Law

$\quad = 4x^2 + 4x$

25. $(a-7)4a = 4a \cdot a + 4a(-7)$
$$= 4a^2 - 28a$$

27. $x^2(x^3+1) = x^2 \cdot x^3 + x^2 \cdot 1$
$$= x^5 + x^2$$

29. $-3n(2n^2 - 8n + 1)$
$$= (-3n)(2n^2) - (-3n)(8n) + (-3n)(1)$$
$$= -6n^3 - (-24n^2) + (-3n)$$
$$= -6n^3 + 24n^2 - 3n$$

31. $-5t^2(3t^3 + 6t)$
$$= (-5t^2)(3t^3) + (-5t^2)(6t)$$
$$= -15t^{2+3} + (-30t^{2+1})$$
$$= -15t^5 - 30t^3$$

33. $\dfrac{2}{3}a^4\left(6a^5 - 12a^3 - \dfrac{5}{8}\right)$
$$= \dfrac{2}{3}a^4(6a^5) - \dfrac{2}{3}a^4(12a^3) - \dfrac{2}{3}a^4\left(\dfrac{5}{8}\right)$$
$$= \dfrac{12}{3}a^9 - \dfrac{24}{3}a^7 - \dfrac{10}{24}a^4$$
$$= 4a^9 - 8a^7 - \dfrac{5}{12}a^4$$

35. $(x+6)(x+1) = (x+6)\cdot x + (x+6)\cdot 1$
$$= x\cdot x + 6\cdot x + x\cdot 1 + 6\cdot 1$$
$$= x^2 + 6x + x + 6$$
$$= x^2 + 7x + 6$$

37. $(x+5)(x-2) = (x+5)\cdot x + (x+5)(-2)$
$$= x\cdot x + 5\cdot x + x\cdot(-2) + 5\cdot(-2)$$
$$= x^2 + 5x - 2x - 10$$
$$= x^2 + 3x - 10$$

39. $(a-0.6)(a-0.7)$
$$= a(a-0.7) - 0.6(a-0.7)$$
$$= a^2 - 0.7a - 0.6a + 0.42$$
$$= a^2 - 1.3a + 0.42$$

41. $(x+3)(x-3) = (x+3)x + (x+3)(-3)$
$$= x\cdot x + 3\cdot x + x(-3) + 3(-3)$$
$$= x^2 + 3x - 3x - 9$$
$$= x^2 - 9$$

43. $(5-x)(5-2x) = (5-x)5 + (5-x)(-2x)$
$$= 5\cdot 5 - x\cdot 5 + 5(-2x) - x(-2x)$$
$$= 25 - 5x - 10x + 2x^2$$
$$= 25 - 15x + 2x^2$$

45. $\left(t+\dfrac{3}{2}\right)\left(t+\dfrac{4}{3}\right) = \left(t+\dfrac{3}{2}\right)t + \left(t+\dfrac{3}{2}\right)\left(\dfrac{4}{3}\right)$
$$= t\cdot t + \dfrac{3}{2}\cdot t + t\cdot \dfrac{4}{3} + \dfrac{3}{2}\cdot\dfrac{4}{3}$$
$$= t^2 + \dfrac{3}{2}t + \dfrac{4}{3}t + 2$$
$$= t^2 + \dfrac{9}{6}t + \dfrac{8}{6}t + 2$$
$$= t^2 + \dfrac{17}{6}t + 2$$

47. $\left(\dfrac{1}{4}a+2\right)\left(\dfrac{3}{4}a-1\right) = \left(\dfrac{1}{4}a+2\right)\left(\dfrac{3}{4}a\right)$
$$+\left(\dfrac{1}{4}a+2\right)(-1)$$
$$= \dfrac{1}{4}a\left(\dfrac{3}{4}a\right) + 2\cdot\dfrac{3}{4}a$$
$$+\dfrac{1}{4}a(-1) + 2(-1)$$
$$= \dfrac{3}{16}a^2 + \dfrac{3}{2}a - \dfrac{1}{4}a - 2$$
$$= \dfrac{3}{16}a^2 + \dfrac{6}{4}a - \dfrac{1}{4}a - 2$$
$$= \dfrac{3}{16}a^2 + \dfrac{5}{4}a - 2$$

49. Illustrate $x(x+5)$ as the area of a rectangle with width x and length $x+5$.

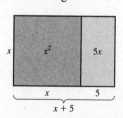

51. Illustrate $(x+1)(x+2)$ as the area of a rectangle with width $x+1$ and length $x+2$.

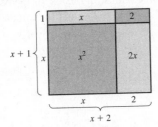

53. Illustrate $(x+5)(x+3)$ as the area of a rectangle with width $x+5$ and length $x+3$.

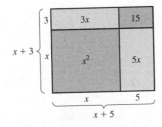

55. $(x^2-x+5)(x+1)$

$= (x^2-x+5)x+(x^2-x+5)1$

$= x^3-x^2+5x+x^2-x+5$

$= x^3+4x+5$

A partial check can be made by selecting a convenient replacement for x, say 1, and comparing the values of the original expression and the result.

$(1^2-1+5)(1+1)$ $1^3+4\cdot1+5$

$= (1-1+5)(1+1)$ $= 1+4+5$

$= 5\cdot2$ $= 10$

$= 10$

Since the value of both expression is 10, the multiplication is very likely correct.

57. $(2a+5)(a^2-3a+2)$

$= (2a+5)a^2-(2a+5)(3a)+(2a+5)2$

$= 2a\cdot a^2+5\cdot a^2-2a\cdot3a-5\cdot3a$

$\quad +2a\cdot2+5\cdot2$

$= 2a^3+5a^2-6a^2-15a+4a+10$

$= 2a^3-a^2-11a+10$

59. $(y^2-7)(2y^3+y+1)$

$= (y^2-7)(2y^3)+(y^2-7)y+(y^2-7)(1)$

$= y^2\cdot2y^3-7\cdot2y^3+y^2\cdot y$

$\quad -7\cdot y+y^2\cdot1-7\cdot1$

$= 2y^5-14y^3+y^3-7y+y^2-7$

$= 2y^5-13y^3+y^2-7y-7$

61. $(3x+2)(5x+4x+7)$

Note: We can simplify the second polynomial before multiplying.

$= (3x+2)(9x+7) = (3x+2)\cdot9x+(3x+2)\cdot7$

$\quad\quad\quad\quad\quad = 3x\cdot9x+2\cdot9x+3x\cdot7+2\cdot7$

$\quad\quad\quad\quad\quad = 27x^2+18x+21x+14$

$\quad\quad\quad\quad\quad = 27x^2+39x+14$

63.

$$
\begin{array}{lll}
& x^2-3x+2 & \text{Line up like terms}\\
& x^2+\ x+1 & \text{in columns}\\
\hline
& x^2-3x+2 & \text{Multiplying by 1}\\
& x^3-3x^2\ +2x & \text{Multiplying by } x\\
x^4-3x^3+2x^2 & & \text{Multiplying by } x^2\\
\hline
x^4-2x^3\quad\quad -x\ +2 &
\end{array}
$$

65.

$$
\begin{array}{lll}
& 2t^2-5t-4 & \\
& 3t^2-\ t+1 & \\
\hline
& 2t^2-5t-4 & \text{Multiplying by 1}\\
& -\ 2t^3+\ 5t^2+4t & \text{Multiplying by } -t\\
6t^4-15t^3-12t^2 & & \text{Multiplying by } 3t^2\\
\hline
6t^4-17t^3-\ 5t^2-t-4 &
\end{array}
$$

67. We will multiply horizontally while still aligning like terms.

$(x+1)(x^3+7x^2+5x+4)$

$$
\begin{array}{ll}
= x^4+7x^3+\ 5x^2+4x & \text{Multiplying by } x\\
\ \ +\ x^3+\ 7x^2+5x+4 & \text{Multiplying by 1}\\
\hline
= x^4+8x^3+12x^2+9x+4 &
\end{array}
$$

69. *Thinking and Writing Exercise.*

71. $(9-3)(9+3)+3^2-9^2$

$= 6\cdot12+9-81$

$= 72+9-81 = 81-81 = 0$

73. $5+\dfrac{7+4+2\cdot5}{7}$

$= 5+\dfrac{11+10}{7} = 5+\dfrac{21}{7} = 5+3 = 8$

75. $(4 + 3 \cdot 5 + 5) \div 3 \cdot 4 = (4 + 15 + 5) \div 3 \cdot 4$

$\qquad = 24 \div 3 \cdot 4 = 8 \cdot 4 = 32$

77. *Thinking and Writing Exercise.*

79. The shaded area is the area of the larger rectangle, $6y(14y - 5)$ less the area of the unshaded rectangle, $3y(3y + 5)$. We have:

$6y(14y - 5) - 3y(3y + 5)$

$\quad = 84y^2 - 30y - 9y^2 - 15y$

$\quad = 75y^2 - 45y$

81. Let $n =$ the missing number. Label the figure with the known areas.

2	2x	2n
x	x^2	nx
	x	n

The area of the figure is $x^2 + 2x + nx + 2n$. This is equivalent to $x^2 + 7x + 10$, so we have $2x + nx = 7x$ and $2n = 10$. Solving either equation for n, we find that the missing number is 5.

83.

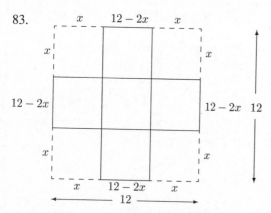

The dimensions, in inches, of the box are $12 - 2x$ by $12 - 2x$ by x. The volume is the product of the dimensions (volume = length × width × height):

Volume $= (12 - 2x)(12 - 2x)x$

$\qquad = (144 - 48x + 4x^2)x$

$\qquad = (144x - 48x^2 + 4x^3)$ in^3, or

$\qquad (4x^3 - 48x^2 + 144x)$ in^3

The outside surface area is the sum of the area of the bottom and the areas of the four sides. The dimensions, in inches, of the bottom are $12 - 2x$ by $12 - 2x$, and the dimensions, in inches, of each sides are x by $12 - 2x$.

Surface area = Area of bottom

$\qquad + 4 \cdot$ Area of each side

$\quad = (12 - 2x)(12 - 2x)$

$\qquad + 4 \cdot x(12 - 2x)$

$\quad = 144 - 24x - 24x$

$\qquad + 4x^2 + 48x - 8x^2$

$\quad = 144 - 48x + 4x^2 + 48x - 8x^2$

$\quad = (144 - 4x^2)$ in^2,

$\qquad$ or $(-4x^2 + 144)$ in^2

85.

The interior dimensions of the open box are $x - 2$ cm by $x - 2$ cm by $x - 1$ cm.

The volume is the product of the dimensions. $(V = l \cdot w \cdot h)$

Interior volume $= (x - 2)(x - 2)(x - 1)$

$[(x - 2)(x - 2)](x - 1)$

$\quad = (x^2 - 4x + 4)(x - 1)$

$\quad = (x^2 - 4x + 4)x - (x^2 - 4x + 4)$

$\quad = x^3 - 4x^2 + 4x - x^2 + 4x - 4$

$\quad = x^3 - 5x^2 + 8x - 4$

The interior volume can be represented by $(x^3 - 5x^2 + 8x - 4)$ cm^3

Exercise Set 5.6

87. We have a rectangular solid with dimensions x m by x m by $x+2$ m with a rectangular solid piece with dimension 6 m by 5 m by 7 m cut out of it.

$$\text{Volume} = \frac{\text{Volume of}}{\text{large solid}} - \frac{\text{Volume of}}{\text{small solid}}$$

$$= (x\,\text{m})(x\,\text{m})(x+2\,\text{m}) - (6\,\text{m})(5\,\text{m})(7\,\text{m})$$

$$= x^2(x+2)\,\text{m}^3 - 210\,\text{m}^3$$

$$= (x^3 + 2x^2 - 210)\,\text{m}^3$$

89. $(x-2)(x-7) - (x-7)(x-2)$

First observe that, by the commutative law of multiplication, $(x-2)(x-7)$ and $(x-7)(x-2)$ are equivalent expressions. Then when we subtract $(x-7)(x-2)$ from $(x-2)(x-7)$, the result is 0.

91. $(x+2)(x+4)(x-5) = [(x+2)(x+4)](x-5)$

$$= [x(x+4) + 2(x+4)](x-5)$$

$$= [x^2 + 4x + 2x + 8](x-5)$$

$$= [x^2 + 6x + 8](x-5)$$

$$= x^2(x-5) + 6x(x-5) + 8(x-5)$$

$$= x^3 - 5x^2 + 6x^2 - 30x + 8x - 40$$

$$= x^3 + x^2 - 22x - 40$$

93. $(x-a)(x-b)\cdots(x-x)(x-y)(x-z)$

$$= (x-a)(x-b)\cdots 0 \cdot (x-y)(x-z)$$

$$= 0$$

Exercise Set 5.6

1. True

3. False

5. $(x+3)(x^2+5)$

$$\quad \text{F} \qquad \text{O} \qquad \text{I} \qquad \text{L}$$

$$= x \cdot x^2 + x \cdot 5 + 3 \cdot x^2 + 3 \cdot 5$$

$$= x^3 \quad + 5x \quad + 3x^2 \quad + 15$$

$$= x^3 \quad + 3x^2 + 5x \quad + 15$$

7. $(t^4 - 2)(t + 7)$

$$\qquad \text{F} \qquad \text{O} \qquad \text{I} \quad \text{L}$$

$$= t^4 \cdot t + \; 7t^4 - 2t - 7 \cdot 2$$

$$= t^5 \quad + 7t^4 \quad - 2t - 14$$

9. $(y+2)(y-3)$

$$\quad \text{F} \qquad \text{O} \qquad \text{I} \qquad \text{L}$$

$$= y \cdot y + \; y \cdot (-3) + 2 \cdot y + 2 \cdot (-3)$$

$$= y^2 \quad - 3y \qquad + 2y \quad - 6$$

$$= y^2 \quad - y \qquad\qquad - 6$$

11. $(3x+2)(3x+5)$

$$\quad \text{F} \qquad \text{O} \qquad \text{I} \qquad \text{L}$$

$$= 3x \cdot 3x + \; 3x \cdot 5 + 2 \cdot 3x + 2 \cdot 5$$

$$= 9x^2 \quad + 15x \quad + 6x \quad + 10$$

$$= 9x^2 \quad + 21x \qquad\qquad + 10$$

13. $(5x-6)(x+2)$

$$\quad \text{F} \qquad \text{O} \qquad \text{I} \qquad \text{L}$$

$$= 5x \cdot x + \; 5x \cdot 2 + (-6) \cdot x + (-6) \cdot 2$$

$$= 5x^2 \quad + 10x \quad - 6x \qquad - 12$$

$$= 5x^2 \quad + 4x \qquad\qquad - 12$$

15. $(1+3t)(2-3t)$

$$\quad \text{F} \qquad \text{O} \qquad \text{I} \qquad \text{L}$$

$$= 1 \cdot 2 + \; 1(-3t) + 3t \cdot 2 + 3t(-3t)$$

$$= 2 \quad - 3t \qquad + 6t \quad - 9t^2$$

$$= 2 \quad + 3t \qquad\qquad - 9t^2$$

17. $(x^2+3)(x^2-7)$

$$\quad \text{F} \qquad \text{O} \qquad \text{I} \qquad \text{L}$$

$$= x^2 \cdot x^2 - \; x^2 \cdot 7 + 3 \cdot x^2 - 3 \cdot 7$$

$$= x^4 \quad - 7x^2 \quad + 3x^2 \quad - 21$$

$$= x^4 \quad - 4x^2 \qquad\qquad - 21$$

19. $\left(p - \dfrac{1}{4}\right)\left(p + \dfrac{1}{4}\right)$

$$\quad \text{F} \qquad \text{O} \qquad \text{I} \qquad\qquad \text{L}$$

$$= p \cdot p + \; p \cdot \frac{1}{4} + \left(-\frac{1}{4}\right) \cdot p + \left(-\frac{1}{4}\right) \cdot \frac{1}{4}$$

$$= p^2 \quad + \frac{1}{4}p \quad - \frac{1}{4}p \qquad - \frac{1}{16}$$

$$= p^2 \qquad\qquad\qquad\qquad - \frac{1}{16}$$

21. $(x-0.1)(x-0.1)$

$$\quad \text{F} \qquad \text{O} \qquad \text{I} \qquad \text{L}$$

$$= x \cdot x - \; x \cdot 0.1 - 0.1 \cdot x + 0.1 \cdot 0.1$$

$$= x^2 \quad - 0.1x \quad - 0.1x \quad + 0.01$$

$$= x^2 \quad - 0.2x \qquad\qquad + 0.01$$

23. $(-3n+2)(n+7)$

$$\qquad\quad \text{F}\qquad\quad\text{O}\qquad\quad\text{I}\qquad\quad\text{L}$$
$$= -3n\cdot n\ -\ 3n\cdot 7\ +\ 2\cdot n\ +\ 2\cdot 7$$
$$= -3n^2\ \ -\ 21n\ \ +\ 2n\ \ +\ 14$$
$$= -3n^2\ \ -\ 19n\ \qquad\ +\ 14$$

25. $(a+9)(a+9)$

$$\qquad\quad\text{F}\quad\ \ \text{O}\quad\ \ \text{I}\quad\ \ \text{L}$$
$$= a^2\ +\ \ 9a\ +\ 9a\ +\ 81$$
$$= a^2\ +\ 18a\qquad\ \ +\ 81$$

27. $(1-3t)(1+5t^2)$

$$\qquad\quad\text{F}\qquad\ \text{O}\qquad\ \text{I}\qquad\ \ \text{L}$$
$$= 1\cdot 1\ +\ 1\cdot 5t^2\ -\ 3t\cdot 1\ -\ 3t\cdot 5t^2$$
$$= 1\ \ +\ 5t^2\ \ -\ 3t\ \ -\ 15t^3$$

29. $\left(x^2+3\right)\left(x^3-1\right)$

$$\qquad\quad\text{F}\qquad\text{O}\qquad\ \text{I}\quad\ \ \text{L}$$
$$= x^5\ -\ \ x^2\ +\ 3x^3\ -\ 3$$
$$= x^5\ +\ 3x^3\ -\ \ x^2\ -\ 3$$

31. $\left(3x^2-2\right)\left(x^4-2\right)$

$$\qquad\quad\text{F}\qquad\ \text{O}\qquad\ \text{I}\quad\ \text{L}$$
$$= 3x^6\ -\ 6x^2\ -\ 2x^4\ +\ 4$$
$$= 3x^6\ -\ 2x^4\ -\ 6x^2\ +\ 4$$

33. $\left(2t^3+5\right)\left(2t^3+3\right)$

$$\qquad\quad\text{F}\qquad\ \text{O}\qquad\ \ \text{I}\quad\ \ \text{L}$$
$$= 4t^6\ +\ 6t^3\ \ +\ 10t^3\ +\ 15$$
$$= 4t^6\ +\ 16t^3\qquad\ \ +\ 15$$

35. $\left(8x^3+5\right)\left(x^2+2\right)$

$$\qquad\quad\text{F}\qquad\ \text{O}\qquad\ \ \text{I}\quad\ \ \text{L}$$
$$= 8x^5\ +\ 16x^3\ +\ 5x^2\ +\ 10$$

37. $(10x^2+3)(10x^2-3)$

$$\qquad\quad\text{F}\qquad\qquad\text{O}\qquad\ \ \text{I}\qquad\ \ \text{L}$$
$$= 10x^2\cdot 10x^2\ -\ 3\cdot 10x^2\ +\ 3\cdot 10x^2\ -\ 3\cdot 3$$
$$= 100x^4\qquad\ -\ 30x^2\ \ +\ 30x^2\ \ -\ 9$$
$$= 100x^4\qquad\qquad\qquad\qquad\ \ -\ 9$$

39. $(x+7)(x-7)=x^2-7^2$　　Product of sum and difference of the same two terms

$$= x^2-49$$

41. $(2x+1)(2x-1)=(2x)^2-1^2$　　Product of sum and difference of the same two terms

$$= 4x^2-1$$

43. $(5m^2-9)(5m^2+9)$　　Product of the sum and difference of the same two terms

$$= (5m^2)^2-9^2$$
$$= 25m^4-81$$

45. $(6a^3+1)(6a^3-1)$　　Product of the sum and difference of the same two terms

$$= (6a^3)^2-1^2$$
$$= 36a^6-1$$

47. $(x^4+0.1)(x^4-0.1)$　　Product of the sum and difference of the same two terms

$$= (x^4)^2-(0.1)^2$$
$$= x^8-0.01$$

49. $\left(t-\dfrac{3}{4}\right)\left(t+\dfrac{3}{4}\right)=t^2-\left(\dfrac{3}{4}\right)^2$　　Product of the sum and difference of the same two terms

$$= t^2-\dfrac{9}{16}$$

51. $(x+2)^2=x^2+2\cdot x\cdot 2+2^2$　　Square of a binomial

$$= x^2+4x+4$$

53. $(3x^5-1)^2=\left(3x^5\right)^2-2\cdot 3x^5\cdot 1+1^2$　　Square of a binomial

$$= 9x^{10}-6x^5+1$$

55. $\left(a-\dfrac{2}{5}\right)^2=a^2-2\cdot a\cdot\dfrac{2}{5}+\left(\dfrac{2}{5}\right)^2$　　Square of a binomial

$$= a^2-\dfrac{4}{5}a+\dfrac{4}{25}$$

57. $\left(x^2+1\right)\left(x^2-x+2\right)$　　Multiplying a binomial and a trinomial

$$= x^4\ -\ x^3\ +\ 2x^2$$
$$\underline{\qquad\qquad +\ \ x^2\ -\ x\ +\ 2}$$
$$= x^4\ -\ x^3\ +\ 3x^2\ -\ x\ +\ 2$$

59. $(2-3x^4)^2 = 2^2 - 2 \cdot 2 \cdot 3x^4 + (3x^4)^2$ — Square of a binomial

$= 4 - 12x^4 + 9x^8$

61. $(5+6t^2)^2 = 5^2 + 2 \cdot 5 \cdot 6t^2 + (6t^2)^2$ — Square of a binomial

$= 25 + 60t^2 + 36t^4$

63. $(7x-0.3)^2 = (7x)^2 - 2 \cdot 7x \cdot 0.3 + 0.3^2$ — Square of a binomial

$= 49x^2 - 4.2x + 0.09$

65. $5a^3(2a^2-1) = 5a^3 \cdot 2a^2 - 5a^3 \cdot 1$ — Multiplying each term of the binomial by the monomial

$= 10a^5 - 5a^3$

67. $(a-3)(a^2+2a-4)$

$\begin{array}{l} = a^3 + 2a^2 - 4a \\ - 3a^2 - 6a + 12 \\ \hline a^3 - a^2 - 10a + 12 \end{array}$ — Multiplying horizontally and aligning like terms

69. $(7-3x^4)(7-3x^4) = (7-3x^4)^2$ — Square of a bionomial

$= 7^2 - 2 \cdot 7 \cdot 3x^4 + (3x^4)^2$

$= 49 - 42x^4 + 9x^8$

71. $-4x(x^2+6x-3)$

$= -4x \cdot x^2 - 4x \cdot 6x + 4x \cdot 3$ — Multiplying each term of a trinomial by a monomial

$= -4x^3 - 24x^2 + 12x$

73. $(-t^3+1)^2 = (-t^3)^2$ — Square of a binomial

$ + 2(-t)^3(1) + 1^2$

$= t^6 - 2t^3 + 1$

75. $3t^2(5t^3-t^2+t)$

$= 3t^2 \cdot 5t^3 + 3t^2(-t^2) + 3t^2 \cdot t$ — Multiplying each term of the trinomial by the monomial

$= 15t^5 - 3t^4 + 3t^3$

77. $(6x^4-3x)^2$ — Square of a binomial

$= (6x^4)^2 - 2 \cdot 6x^4 \cdot 3x + (3x)^2$

$= 36x^8 - 36x^5 + 9x^2$

79. $(9a+0.4)(2a^3+0.5)$ — Product of two binomials; use FOIL

$= 9a \cdot 2a^3 + 9a \cdot 0.5 + 0.4 \cdot 2a^3 + 0.4 \cdot 0.5$

$= 18a^4 + 4.5a + 0.8a^3 + 0.2,$ or

$18a^4 + 0.8a^3 + 4.5a + 0.2$

81. $\left(\dfrac{1}{5}-6x^4\right)\left(\dfrac{1}{5}+6x^4\right)$ — Product of the sum and difference of the same two terms

$= \left(\dfrac{1}{5}\right)^2 - (6x^4)^2$

$= \dfrac{1}{25} - 36x^8,$ or $-36x^8 + \dfrac{1}{25}$

83. $(a+1)(a^2-a+1)$

$\begin{array}{l} = a^3 - a^2 + a \\ \; a^2 - a + 1 \\ \hline a^3 + 1 \end{array}$ — Multiplying horizontally and aligning like terms

85.

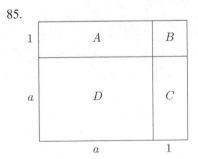

We can find the shaded area in two ways.
Method 1: The figure is a square with side $a+1$, so the area is $(a+1)^2 = a^2 + 2a + 1$.
Method 2: We add the areas of A, B, C, and D.

$1 \cdot a + 1 \cdot 1 + 1 \cdot a + a \cdot a = a + 1 + a + a^2$

$= a^2 + 2a + 1.$

Either way, we find that the total shaded area is $a^2 + 2a + 1.$

87.

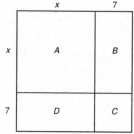

We can find the shaded area in two ways.
Method 1: The figure is a rectangle with
dimensions $x+5$ by $x+2$, so the area is

$$(x+5)(x+2) = x^2 +2x+5x+10$$

$$= x^2 +7x+10.$$

Method 2: We add the areas of A, B, C,
and D.

$$5 \cdot x+2 \cdot 5+2 \cdot x+x \cdot x = 5x+10+2x+x^2$$

$$= x^2 +7x+10.$$

Either way, we find that the area is
$x^2 +7x+10.$

89.

We can find the shaded area in two ways.
Method 1: The figure is a square with side
$x+7$, so the area is $(x+7)^2 = x^2 +14x+49.$
Method 2: We add the areas of A, B, C,
and D.

$$x \cdot x+x \cdot 7+7 \cdot 7+7 \cdot x = x^2 +7x+49+7x$$

$$= x^2 +14x+49.$$

Either way, we find that the total shaded area
is $x^2 +14x+49.$

91.

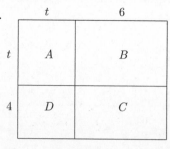

We can find the shaded area in two ways.
Method 1: The figure is a rectangle with
dimensions $t+6$ by $t+4$, so the area is

$$(t+6)(t+4) = t^2 +4t+6t+24$$

$$= t^2 +10t+24.$$

Method 2: We add the areas of A, B, C,
and D.

$$t \cdot t+t \cdot 6+6 \cdot 4+4 \cdot t = t^2 +6t+24+4t$$

$$= t^2 +10t+24.$$

Either way, we find that the total shaded area
is $t^2 +10t+24.$

93.

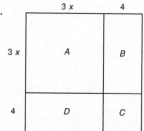

We can find the shaded area in two ways.
Method 1: The figure is a rectangle with
dimensions $t+9$ by $t+4$, so the area is

$$(t+9)(t+4) = t^2 +4t+9t+36$$

$$= t^2 +13t+36$$

Method 2: We add the areas of A, B, C,
and D.

$$9 \cdot t+t \cdot t+4 \cdot t+4 \cdot 9 = 9t+t^2 +4t+36$$

$$= t^2 +13t+36.$$

Either way, we find that the total shaded area
is $t^2 +13t+36.$

95.

We can find the shaded area in two ways.
Method 1: The figure is a square with side
$3x+4$, so the area is

$$(3x+4)^2 = 9x^2 +24x+16.$$

Method 2: We add the areas of A, B, C, and D.

$$3x \cdot 3x+3x \cdot 4+4 \cdot 4+3x \cdot 4 = 9x^2 +12x$$

$$+16+12x$$

$$= 9x^2 +24x+16.$$

Either way, we find that the total shaded area
is $9x^2 +24x+16.$

97. We draw a square with side $x + 5$.

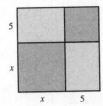

99. We draw a square with side $t + 9$.

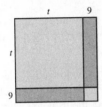

101. We draw a square with side $3 + x$.

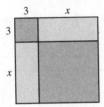

103. *Thinking and Writing Exercise.*

105. ***Familiarize.*** Let t = the number of watts used by the washing machine. Then $21t$ = the number of watts used by the refrigerator, and $11t$ = the number of watts used by the freezer.
Translate.

refrigerator watts	+	freezer watts	+	washer watts	=	Total watts
↓	↓	↓	↓	↓	↓	↓
$21t$	+	$11t$	+	t	=	297

Solve. We solve the equation.
$$21t + 11t + t = 297$$
$$33t = 297$$
$$t = 9$$
The possible solution is:
washer, t: 9 kWh/mo

freezer, $11t$: $11 \cdot 9$, or 99 kWh / mo

refrigerator, $21t$: $21 \cdot 9$, or 189 kWh / mo

Check. The number of kWh/mo for the freezer, 99, is 11 times 9, the number of the washer. The number of kWh/mo for the refrigerator, 189, is 21 times 9, the number

for the washer. Also, $9 + 99 + 189 = 297$, the total kWh/mo.
State. The washer uses 9 kWh/mo, the freezer uses 99 kWh/mo, and the refrigerator uses 189 kWh/mo.

107. $5xy = 8$

$y = \dfrac{8}{5x}$ Dividing both sides by $5x$

109. $ax - by = c$

$ax = by + c$ Adding by to both sides

$x = \dfrac{by + c}{a}$ Dividing both sides by a

111. *Thinking and Writing Exercise.*

113. $18 \times 22 = (20 - 2)(20 + 2)$
$$= 20^2 - 2^2$$
$$= 400 - 4$$
$$= 396$$

115. $(4x^2 + 9)(2x + 3)(2x - 3) = (4x^2 + 9)(4x^2 - 9)$
$$= 16x^4 - 81$$

117. $(3t - 2)^2 (3t + 2)^2 = [(3t - 2)(3t + 2)]^2$
$$= \left(9t^2 - 4\right)^2$$
$$= 81t^4 - 72t^2 + 16$$

119. $\left(t^3 - 1\right)^4 \left(t^3 + 1\right)^4 = [(t^3 - 1)(t^3 + 1)]^4$
$$= \left(t^6 - 1\right)^4$$
$$= [(t^6 - 1)^2]^2$$
$$= \left(t^{12} - 2t^6 + 1\right)^2$$
$$= \left(t^{12} - 2t^6 + 1\right)\left(t^{12} - 2t^6 + 1\right)$$
$$= t^{24} - 2t^{18} + t^{12} - 2t^{18} + 4t^{12}$$
$$\quad - 2t^6 + t^{12} - 2t^6 + 1$$
$$= t^{24} - 4t^{18} + 6t^{12} - 4t^6 + 1$$

121. $(x+2)(x-5) = (x+1)(x-3)$

$\quad x^2 - 5x + 2x - 10 = x^2 - 3x + x - 3$

$\quad\quad x^2 - 3x - 10 = x^2 - 2x - 3$

$\quad\quad\quad -3x - 10 = -2x - 3$ Adding $-x^2$

$\quad\quad\quad -3x + 2x = 10 - 3$ Adding $2x$ and 10

$\quad\quad\quad\quad\quad -x = 7$

$\quad\quad\quad\quad\quad\quad x = -7$

The solution is -7.

123. The areas of the white regions are
$\quad 2(y-2) = 2y - 4,$

$\quad\quad 2 \cdot 2 = 4,$ and

$\quad 2(y-2) = 2y - 4.$

The total area of the white regions is
$\quad 2y - 4 + 4 + 2y - 4 = 4y - 4$

We subtract this total from y^2

$\quad y^2 - (4y - 4) = y^2 - 4y + 4$

125. If $l = $ the length, then $l + 1 = $ the height, and
$l - 1 = $ the width. Recall that the volume of a
rectangular solid is given by length $\times$ width $\times$
height.

Volume $= l(l-1)(l+1) = l(l^2 - 1) = l^3 - l$

127.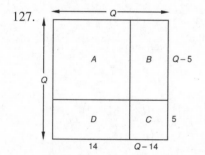

The dimensions of the shaded area, B, are
$Q - 14$ by $Q - 5,$ so one expression is
$(Q-14)(Q-5).$

To find another expression, we find the area
of regions B and C together and subtract the
area of region C. The region consisting of B
and C together has dimensions Q by $Q - 14,$
so its area is $Q(Q-14)$. Region C has
dimensions 5 by $Q - 14,$ so its area is
$5(Q-14).$ Then another expression for the
shaded area, B, is $Q(Q-14) - 5(Q-14).$
It is possible to find other equivalent
expressions also.

129.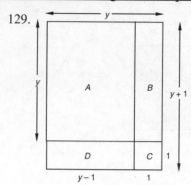

The dimensions of the shaded area, regions A
and D together, are $y+1$ by $y-1$ so the area
is $(y+1)(y-1).$

To find another expression we add the areas
of regions A and D. The dimensions of region
A are y by $y-1,$ and the dimensions of region
D are $y-1$ by 1, so the sum of the areas is
$y(y-1) + (y-1)(1),$ or $y(y-1) + y - 1.$

It is possible to find other equivalent
expressions also.

Mid-Chapter Review

Guided Solutions

1. $\left(2x^2 y^{-5}\right)^{-10} = 2^{-10}\left(x^2\right)^{-10}\left(y^{-5}\right)^{-10}$

$\quad\quad\quad\quad = 2^{-10} x^{-20} y^{50}$

$\quad\quad\quad\quad = \dfrac{y^{50}}{2^{10} x^{20}}$

2. $(x^2 + 7x)(x^2 - 7x) = \left(x^2\right)^2 - \left(7x\right)^2$

$\quad\quad\quad\quad\quad = x^4 - 49x^2$

Mixed Review

1. $(x^2 y^5)^8 = x^{2 \cdot 8} y^{5 \cdot 8} = x^{16} y^{40}$

2. $(4x)^0 = 1.$ Any number or expression raised
to the 0 power is 1.

3. $d^{-10} = \left(\dfrac{1}{d}\right)^{10},$ or $\dfrac{1}{d^{10}}$

4. $\dfrac{3a^{11}}{12a} = \dfrac{\cancel{3} \cdot a^{10} \cdot \cancel{a}}{\cancel{3} \cdot 4 \cdot \cancel{a}} = \dfrac{a^{10}}{4},$ or $\dfrac{1}{4}a^{10}$

5. $(m^{-3}y)^{-1} = \left(\dfrac{y}{m^3}\right)^{-1} = \left(\dfrac{m^3}{y}\right)^{1} = \dfrac{m^3}{y}$

6. $\dfrac{-48ab^{-7}}{18a^{11}b^{-6}} = -\dfrac{48}{18}\cdot\dfrac{a}{a^{11}}\cdot\dfrac{b^6}{b^7}$

$\qquad = -\dfrac{\cancel{6}\cdot 8}{\cancel{6}\cdot 3}\cdot\dfrac{\cancel{a}}{\cancel{a}\cdot a^{10}}\cdot\dfrac{\cancel{b^6}}{b\cdot \cancel{b^6}}$

$\qquad = -\dfrac{8}{3}\cdot\dfrac{1}{a^{10}}\cdot\dfrac{1}{b} = -\dfrac{8}{3a^{10}b}$

7. $(3x^2 - 2x + 6) + (5x - 3)$

$\quad = 3x^2 - 2x + 6 + 5x - 3$

$\quad = 3x^2 + (-2+5)x + 6 - 3$

$\quad = 3x^2 + 3x + 3$

8. $(9x + 6) - (2x - 1)$

$\quad = 9x + 6 - 2x + 1$

$\quad = (9 - 2)x + 6 + 1 = 7x + 7$

9. $6x^3(8x^2 - 7)$

$\quad = 6x^3\cdot 8x^2 - 6x^3\cdot 7$

$\quad = 48x^{3+2} - 42x^3$

$\quad = 48x^5 - 42x^3$

10. $(3x + 2)(2x - 1)$

$\quad = 3x(2x - 1) + 2(2x - 1)$

$\quad = 3x\cdot 2x - 3x\cdot 1 + 2\cdot 2x - 2\cdot 1$

$\quad = 6x^2 - 3x + 4x - 2$

$\quad = 6x^2 + x - 2$

11. $(4x^2 - x - 7) - (10x^2 - 3x + 5)$

$\quad = 4x^2 - x - 7 - 10x^2 + 3x - 5$

$\quad = (4 - 10)x^2 + (-1 + 3)x - 7 - 5$

$\quad = -6x^2 + 2x - 12$

12. $(3x + 8)(3x + 7)$

$\quad = 3x(3x + 7) + 8(3x + 7)$

$\quad = 3x\cdot 3x + 3x\cdot 7 + 8\cdot 3x + 8\cdot 7$

$\quad = 9x^2 + 21x + 24x + 56$

$\quad = 9x^2 + 45x + 56$

13. $(t^9 + 3t^6 - 8t^2) + (5t^7 - 3t^6 + 8t^2)$

$\quad = t^9 + \cancel{3t^6} - \cancel{8t^2} + 5t^7 - \cancel{3t^6} + \cancel{8t^2}$

$\quad = t^9 + 5t^7$

14. $(2m - 1)^2 = (2m)^2 - 2\cdot(2m)\cdot(1) + (1)^2$

$\qquad = 4m^2 - 4m + 1$

15. $(x - 1)(x^2 + x + 1)$

$\quad = x(x^2 + x + 1) - 1(x^2 + x + 1)$

$\quad = x\cdot x^2 + x\cdot x + x\cdot 1 - x^2 - x - 1$

$\quad = x^3 + \cancel{x^2} + \cancel{x} - \cancel{x^2} - \cancel{x} - 1$

$\quad = x^3 - 1$

16. $(c + 3)(c - 3) = c^2 - 3^2 = c^2 - 9$

17. $(4y^3 + 7)^2 = (4y^3)^2 + 2(4y^3)(7) + (7)^2$

$\qquad\qquad = 16y^6 + 56y^3 + 49$

18. $(3a^4 - 9a^3 - 7) - (4a^3 + 13a^2 - 3)$

$\quad = 3a^4 - 9a^3 - 7 - 4a^3 - 13a^2 + 3$

$\quad = 3a^4 - 9a^3 - 4a^3 - 13a^2 - 7 + 3$

$\quad = 3a^4 - 13a^3 - 13a^2 - 4$

19. $(4t^2 - 5)(4t^2 + 5) = (4t^2)^2 - 5^2 = 16t^4 - 25$

20. $(a^4 + 3)(a^4 - 8)$

$\quad = a^4(a^4 - 8) + 3(a^4 - 8)$

$\quad = a^4\cdot a^4 - a^4\cdot 8 + 3\cdot a^4 - 3\cdot 8$

$\quad = a^{4+4} - 8a^4 + 3a^4 - 24$

$\quad = a^8 - 5a^4 - 24$

Exercise Set 5.7

1. coefficient

3. degree

5. binomial

7. like terms

9. We replace x with 5 and y with -2.

$\quad x^2 - 3y^2 + 2xy = 5^2 - 3(-2)^2 + 2\cdot 5(-2)$

$\qquad\qquad\qquad = 25 - 12 - 20 = -7.$

11. We replace x with 2, y with -3, and z with -4.

$$xyz^2 - z = 2(-3)(-4)^2 - (-4)$$
$$= -96 + 4 = -92$$

13. Evaluate the polynomial for $h = 160$ and $A = 50$.

$$0.041h - 0.018A - 2.69$$
$$= 0.041(160) - 0.018(50) - 2.69$$
$$= 6.56 - 0.9 - 2.69$$
$$= 2.97$$

The woman's lung capacity is 2.97 liters.

15. Evaluate the polynomial for $w = 87$, $h = 185$, and $a = 59$.

$$19.18w + 7h - 9.52a + 92.4$$
$$= 19.18(87) + 7(185) - 9.52(59) + 92.4$$
$$= 1668.66 + 1295 - 561.68 + 92.4$$
$$= 2494.38$$

The male author needs approximately 2494 calories each day.

17. Evaluate the polynomial for

$\pi \approx 3.14$, $h = 7$, and $r = 1\dfrac{1}{2} = \dfrac{3}{2}$.

$2\pi rh + \pi r^2$

$$= 2 \cdot (3.14) \cdot \left(\frac{3}{2}\right) \cdot (7) + (3.14)\left(\frac{3}{2}\right)^2$$
$$= \frac{2 \cdot 3.14 \cdot 3 \cdot 7}{2} + (3.14)\left(\frac{9}{4}\right)$$
$$= \frac{131.88}{2} + \frac{3.14 \cdot 9}{4}$$
$$= \frac{263.76}{4} + \frac{28.26}{4}$$
$$= \frac{263.76 + 28.26}{4} = \frac{292.02}{4} = 73.005$$

The surface area is about 73.005 in.^{2}

19. Evaluate the polynomial for $v = 18$, $h = 50$, and $t = 2$.

$$h + vt - 4.9t^2 = 50 + 18 \cdot 2 - 4.9 \cdot 2^2$$
$$= 50 + 18 \cdot 2 - 4.9 \cdot 4$$
$$= 50 + 36 - 19.6 = 66.4$$

The ball will be 66.4 m high 2 sec. After it is thrown.

21. $x^3 y - 2xy + 3x^2 - 5$

Term	Coefficient	Degree	
$x^3 y$	1	4	(Think: $x^3 y = x^3 y^1$)
$-2xy$	-2	2	(Think: $-2xy = -2x^1 y^1$)
$3x^2$	3	2	
-5	-5	0	(Think: $-5 = -5x^0$)

The degree of the polynomial is the degree of the term of highest degree. The term of highest degree is $x^3 y$. Its degree is 4, so the degree of the polynomial is 4.

23. $11 - abc + a^2 b + 0.5ab^2$

Term	Coeff	Degree	
11	11	0	(Think: $11 = 11a^0$)
$-abc$	-1	3	(Think: $-abc$ $= (-1)a^1 b^1 c^1$)
$a^2 b$	1	3	(Think: $a^2 b = (1)a^2 b^1$)
$0.5ab^2$	0.5	3	(Think: $0.5ab^2$ $= 0.5a^1 b^2$)

The polynomial has three terms that are of degree 3 and one of degree 0. The degree of the polynomial equals the degree of the highest order term, 3.

25. $7a + b - 4a - 3b = (7 - 4)a + (1 - 3)b$
$$= 3a - 2b$$

27. $3x^2 y - 2xy^2 + x^2 + 5x$

There are *no* like terms, so none of the terms can be combined.

29. $2u^2 v - 3uv^2 + 6u^2 v - 2uv^2 + 7u^2$
$$= (2 + 6)u^2 v + (-3 - 2)uv^2 + 7u^2$$
$$= 8u^2 v - 5uv^2 + 7u^2$$

31. $5a^2 c - 2ab^2 + a^2 b - 3ab^2 + a^2 c - 2ab^2$
$$= (5 + 1)a^2 c + (-2 - 3 - 2)ab^2 + a^2 b$$
$$= 6a^2 c - 7ab^2 + a^2 b$$

33. $\left(4x^2 - xy + y^2\right) + \left(-x^2 - 3xy + 2y^2\right)$
$$= (4 - 1)x^2 + (-1 - 3)xy + (1 + 2)y^2$$
$$= 3x^2 - 4xy + 3y^2$$

35. $\left(3a^4 - 5ab + 6ab^2\right) - \left(9a^4 + 3ab - ab^2\right)$

$= 3a^4 - 5ab + 6ab^2 - 9a^4 - 3ab + ab^2$

$= (3-9)a^4 + (-5-3)ab + (6+1)ab^2$

$= -6a^4 - 8ab + 7ab^2$

37. $\left(5r^2 - 4rt + t^2\right) + \left(-6r^2 - 5rt - t^2\right)$

$\qquad + \left(-5r^2 + 4rt - t^2\right)$

Observe that the polynomials $5r^2 - 4rt + t^2$ and $-5r^2 + 4rt - t^2$ are opposites. Thus, their sum is 0 and the sum in the exercise is the remaining polynomial, $-6r^2 - 5rt - t^2$.

39. $\left(x^3 - y^3\right) - \left(-2x^3 + x^2y - xy^2 + 2y^3\right)$

$= x^3 - y^3 + 2x^3 - x^2y + xy^2 - 2y^3$

$= (1+2)x^3 + (-1-2)y^3 - x^2y + xy^2$

$= 3x^3 - 3y^3 - x^2y + xy^2,$ or

$= 3x^3 - x^2y + xy^2 - 3y^3$

41. $\left(2y^4x^2 - 5y^3x\right) + \left(5y^4x^2 - y^3x\right)$

$\qquad + \left(3y^4x^2 - 2y^3x\right)$

$= (2+5+3)y^4x^2 + (-5-1-2)y^3x$

$= 10y^4x^2 - 8y^3x$

43. $(4x+5y) + (-5x+6y) - (7x+3y)$

$= 4x + 5y - 5x + 6y - 7x - 3y$

$= (4-5-7)x + (5+6-3)y$

$= -8x + 8y$

45. $\qquad\qquad$ F $\qquad$ O $\qquad$ I $\qquad$ L

$(3z-u)(2z+3u) = 6z^2 + 9uz - 2uz - 3u^2$

$\qquad\qquad\qquad = 6z^2 + 7uz \qquad - 3u^2$

47. $\qquad\qquad$ F $\qquad$ O $\qquad$ I $\qquad$ L

$(xy+7)(xy-4) = x^2y^2 - 4xy + 7xy - 28$

$\qquad\qquad\qquad = x^2y^2 + 3xy \qquad - 28$

49. $(2a-b)(2a+b) \quad (A+B)(A-B) = A^2 - B^2$

$= (2a)^2 - (b)^2$

$= 4a^2 - b^2$

51. $\qquad\qquad$ F $\qquad$ O $\qquad$ I $\qquad$ L

$(5rt-2)(3rt+1) = 15r^2t^2 + 5rt - 6rt - 2$

$\qquad\qquad\qquad = 15r^2t^2 - rt \qquad - 2$

53. $\qquad\qquad$ F $\qquad$ O $\qquad$ I $\qquad$ L

$\left(m^3n+8\right)\left(m^3n-6\right) = m^6n^2 - 6m^3n + 8m^3n - 48$

$\qquad\qquad\qquad = m^6n^2 \qquad + 2m^3n - 48$

55. $\qquad\qquad$ F $\qquad$ O $\qquad$ I $\qquad$ L

$(6x-2y)(5x-3y) = 30x^2 - 18xy - 10xy + 6y^2$

$\qquad\qquad\qquad = 30x^2 - 28xy \qquad + 6y^2$

57. $(aw+0.1)(-aw+0.1) = (0.1+aw)(0.1-aw)$

$\qquad\qquad\qquad = (0.1)^2 - (aw)^2$

$\qquad\qquad\qquad = 0.01 - a^2w^2$

$\qquad (A+B)(A-B) = A^2 - B^2$

59. $(x+h)^2 = x^2 + 2xh + h^2$

$\qquad (A+B)^2 = A^2 + 2AB + B^2$

61. $(4a-5b)^2 = (4a)^2 - 2 \cdot (4a)(5b) + (5b)^2$

$\qquad\qquad = 16a^2 - 40ab + 25b^2$

$\qquad (A-B)^2 = A^2 - 2AB + B^2$

63. $\left(ab+cd^2\right)\left(ab-cd^2\right) = (ab)^2 - \left(cd^2\right)^2$

$\qquad\qquad\qquad = a^2b^2 - c^2d^4$

$\qquad (A+B)(A-B) = A^2 - B^2$

65. $(2xy + x^2y + 3)(xy + y^2)$

$= 2xy(xy + y^2) + x^2y(xy + y^2) + 3(xy + y^2)$

$= 2xy \cdot xy + 2xy \cdot y^2 + x^2y \cdot xy$

$\qquad + x^2y \cdot y^2 + 3 \cdot xy + 3 \cdot y^2$

$= 2x^2y^2 + 2xy^3 + x^3y^2 + x^2y^3 + 3xy + 3y^2$

67. $(a+b-c)(a+b+c)$

$= [(a+b)-c][(a+b)+c]$

$= (a+b)^2 - c^2$

$= a^2 + 2ab + b^2 - c^2$

69. $[a+b+c][a-(b+c)]$
$= [a+(b+c)][a-(b+c)]$
$= a^2 - (b+c)^2$
$= a^2 - (b^2 + 2bc + c^2)$
$= a^2 - b^2 - 2bc - c^2$

71. The figure is a rectangle with dimensions $a+b$ by $a+c$. Its area is
$(a+b)(a+c) = a^2 + ac + ab + bc.$

73. The figure is a parallelogram with base $x+z$ and height $x-z$. Thus the area is
$(x+z)(x-z) = x^2 - z^2.$

75. The figure is a square with side $x+y+z$. Thus the area is
$(x+y+z)^2 = [(x+y)+z]^2$
$\qquad = (x+y)^2 + 2(x+y)(z) + z^2$
$\qquad = x^2 + 2xy + y^2 + 2xz + 2yz + z^2.$

77. The figure is a triangle with base $x+2y$ and height $x-y$. Thus the area is
$\frac{1}{2}(x+2y)(x-y) = \frac{1}{2}\left(x^2 + xy - 2y^2\right)$
$\qquad\qquad\qquad = \frac{1}{2}x^2 + \frac{1}{2}xy - y^2.$

79. We draw a rectangle with dimensions $r+s$ by $u+v$.

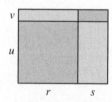

81. We draw a rectangle with dimensions $a+b+c$ by $a+d+f$.

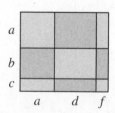

83. a) $f(t-1) = (t-1)^2 + 5$
$\qquad = t^2 - 2t + 1 + 5$
$\qquad = t^2 - 2t + 6$

b) $f(a+h) - f(a) = [(a+h)^2 + 5] - (a^2 + 5)$
$\qquad = a^2 + 2ah + h^2$
$\qquad\qquad + 5 - a^2 - 5$
$\qquad = 2ah + h^2$

c) $f(a) - f(a-h)$
$\qquad = a^2 + 5 - \left[(a-h)^2 + 5\right]$
$\qquad = a^2 + 5 - \left[a^2 - 2ah + h^2 + 5\right]$
$\qquad = a^2 + 5 - a^2 + 2ah - h^2 - 5$
$\qquad = 2ah - h^2$

85. *Thinking and Writing Exercise.*

87. $\begin{array}{r} x^2 - 3x - 7 \\ -(5x - 3) \\ \hline \end{array} = \begin{array}{r} x^2 - 3x - 7 \\ - 5x + 3 \\ \hline x^2 - 8x - 4 \end{array}$

89. $\begin{array}{r} 3x^2 + x + 5 \\ -(3x^2 + 3x) \\ \hline \end{array} = \begin{array}{r} 3x^2 + x + 5 \\ - 3x^2 - 3x \\ \hline - 2x + 5 \end{array}$

91. $\begin{array}{r} 5x^3 - 2x^2 + 1 \\ -(5x^3 - 15x^2) \\ \hline \end{array} = \begin{array}{r} 5x^3 - 2x^2 + 1 \\ - 5x^3 + 15x^2 \\ \hline 13x^2 + 1 \end{array}$

93. *Thinking and Writing Exercise.*

95. The unshaded region is a circle with radius $a-b$. Then the shaded area is the area of a circle with radius a minus the area of a circle with radius $a-b$:
Shaded area $= \pi a^2 - \pi(a-b)^2$
$\qquad = \pi a^2 - \pi(a^2 - 2ab + b^2)$
$\qquad = \pi a^2 - \pi a^2 + 2\pi ab - \pi b^2$
$\qquad = 2\pi ab - \pi b^2$

97. The shaded area is the area of a square with side a less the areas of 4 squares with side b. Thus, the shaded area is
$a^2 - 4 \cdot b^2$, or $a^2 - 4b^2.$

99. We can think of the figure as four rectangular prisms with dimensions
$y \times y \times 2y$, $x \times x \times y$, $x \times y \times y$, and
$x \times x \times x$. The volume of a rectangular prism is the product of its dimensions we determine the volume of each and then add to determine the total volume.
$$y \cdot y \cdot 2y = 2y^3$$
$$x \cdot x \cdot y = x^2 y$$
$$x \cdot y \cdot y = xy^2$$
$$x \cdot x \cdot x = x^3$$
$$V = 2y^3 + x^2 y + xy^2 + x^3, \text{ or}$$
$$x^3 + x^2 y + xy^2 + 2y^3$$

101. The lateral surface area of the outer portion of the solid is the lateral surface area of a right circular cylinder with radius n and height h. The lateral surface area of the inner portion is the lateral surface area of a right circular cylinder with radius m and height h. Recall that the formula for the lateral surface area of a right circular cylinder with radius r and height h is $2\pi rh$.
The surface area of the top is the area of a circle with radius n less the area of a circle with radius m. The surface area of the bottom is the same as the surface area of the top. Thus, the surface area of the solid is
$$2\pi nh + 2\pi mh + 2\pi n^2 - 2\pi m^2.$$

103. *Thinking and Writing Exercise.*

105. $D4 = 2 \cdot 5 + 3 \cdot 10 = 10 + 30 = 40$

107. Replace t with 2 and multiply.
$$P(1+r)^2 = P\left(1 + 2r + r^2\right)$$
$$= P + 2Pr + Pr^2$$

109. Substitute \$ 10,400 for P, 8.5% or 0.085 for r, and 5 for t.
$$P(1+r)^t = \$10,400(1+0.085)^5$$
$$= \$10,400(1.085)^5$$
$$\approx \$10,400(1.50365669)$$
$$\approx \$15,638.03$$

Exercise Set 5.8

1. quotient

3. dividend

5. $\dfrac{32x^5 - 24x}{8} = \dfrac{32x^5}{8} - \dfrac{24x}{8}$
$$= \dfrac{32}{8}x^5 - \dfrac{24}{8}x$$
$$= 4x^5 - 3x$$
To check, we multiply the quotient by 8:
$$\left(4x^5 - 3x\right)8 = 32x^5 - 24x.$$
The answer checks.

7. $\dfrac{u - 2u^2 + u^7}{u} = \dfrac{u}{u} - \dfrac{2u^2}{u} + \dfrac{u^7}{u}$
$$= 1 - 2u + u^6$$
Check: We multiply.
$$1 - 2u + u^6$$
$$\underline{\hspace{2cm} u}$$
$$u - 2u^2 + u^7$$

9. $\left(15t^3 - 24t^2 + 6t\right) \div (3t)$
$$= \dfrac{15t^3 - 24t^2 + 6t}{3t}$$
$$= \dfrac{15t^3}{3t} - \dfrac{24t^2}{3t} + \dfrac{6t}{3t}$$
$$= 5t^2 - 8t + 2$$
Check: We multiply.
$$5t^2 - 8t + 2$$
$$\underline{\hspace{3cm} 3t}$$
$$15t^3 - 24t^2 + 6t$$

11. $\left(24t^5 - 40t^4 + 6t^3\right) \div \left(4t^3\right)$
$$= \dfrac{24t^5}{4t^3} - \dfrac{40t^4}{4t^3} + \dfrac{6t^3}{4t^3}$$
$$= 6t^2 - 10t + \dfrac{3}{2}$$
Check: We multiply.
$$6t^2 - 10t + \dfrac{3}{2}$$
$$\underline{\hspace{4cm} 4t^3}$$
$$24t^5 - 40t^4 + 6t^3$$

13. $\dfrac{15x^7 - 21x^4 - 3x^2}{-3x^2}$

$= \dfrac{15x^7}{-3x^2} + \dfrac{-21x^4}{-3x^2} + \dfrac{-3x^2}{-3x^2}$

$= -5x^5 + 7x^2 + 1$

Check. We multiply

$$
\begin{array}{r}
-5x^5 + 7x^2 + 1 \\
-3x^2 \\
\hline
15x^7 - 21x^4 - 3x^2
\end{array}
$$

15. $\dfrac{8x^2 - 10x + 1}{2x} = \dfrac{8x^2}{2x} - \dfrac{10x}{2x} + \dfrac{1}{2x}$

$= 4x - 5 + \dfrac{1}{2x}$

Check. We multiply.

$$
\begin{array}{r}
4x - 5 + \dfrac{1}{2x} \\
2x \\
\hline
8x^2 - 10x + 1
\end{array}
$$

17. $\dfrac{9r^2s^2 + 3r^2s - 6rs^2}{-3rs} = \dfrac{9r^2s^2}{-3rs} + \dfrac{3r^2s}{-3rs} - \dfrac{6rs^2}{-3rs}$

$= -3rs - r + 2s$

Check: We multiply.

$$
\begin{array}{r}
-3rs - r + 2s \\
- 3rs \\
\hline
9r^2s^2 + 3r^2s - 6rs^2
\end{array}
$$

$= \dfrac{10x^5y^2}{5x^2y} + \dfrac{15x^2y^2}{5x^2y} - \dfrac{5x^2y}{5x^2y}$

$= 2x^3y + 3y - 1$

Check. Multiply

$$
\begin{array}{r}
2x^3y + 3y - 1 \\
5x^2y \\
\hline
10x^5y^2 + 15x^2y^2 - 5x^2y
\end{array}
$$

19. $\left(10x^5y^2 + 15x^2y^2 - 5x^2y\right) \div \left(5x^2y\right)$

$= \dfrac{10x^5y^2}{5x^2y} + \dfrac{15x^2y^2}{5x^2y} - \dfrac{5x^2y}{5x^2y}$

$= 2x^3y + 3y - 1$

Check. Multiply

$$
\begin{array}{r}
2x^3y + 3y - 1 \\
5x^2y \\
\hline
10x^5y^2 + 15x^2y^2 - 5x^2y
\end{array}
$$

21. $\left(x^2 + 10x + 21\right) \div (x + 7) = \dfrac{(x+7)(x+3)}{x+7}$

$= \dfrac{(\cancel{x+7})(x+3)}{\cancel{x+7}}$

$= x + 3$

Check. We multiply.

$(x+3)(x+7) = x^2 + 7x + 3x + 21$

$= x^2 + 10x + 21$

23.
$$
\begin{array}{r}
a - 12 \\
a+4 \overline{\smash{\big)}\, a^2 - 8a - 16} \\
\underline{a^2 + 4a} \\
-12a - 16 \\
\underline{-12a - 48} \\
32
\end{array}
$$
$(a^2 - 8a) - (a^2 + 4a) = -12a$
$(-12a - 16) - (-12a - 48) = 32$

The answer is $a - 12$, R 32, or $a - 12 + \dfrac{32}{a+4}$.

Check. We multiply

$(a+4)\left[(a-12) + \dfrac{32}{a+4}\right]$

$= (a+4)(a-12) + (a+4)\left(\dfrac{32}{a+4}\right)$

$= a^2 - 12a + 4a - 48 + 32$

$= a^2 - 8a - 16$

25.
$$
\begin{array}{r}
2x - 1 \\
x+6 \overline{\smash{\big)}\, 2x^2 + 11x - 5} \\
\underline{-(2x^2 + 12x)} \\
-x - 5 \\
\underline{-(-x - 6)} \\
1
\end{array}
$$

The answer is $2x - 1$, R 1, or $2x - 1 + \dfrac{1}{x+6}$.

Check. Multiply.

$(x+6)\left[2x - 1 + \dfrac{1}{x+6}\right]$

$= (x+6)(2x-1) + (x+6)\left(\dfrac{1}{x+6}\right)$

$= x^2 - x + 12x - 6 + 1$

$= x^2 + 11x - 5$

27. $\left(y^2 - 25\right) \div (y + 5) = \dfrac{y^2 - 25}{y + 5}$

$\qquad\qquad\qquad = \dfrac{(y+5)(y-5)}{y+5}$

$\qquad\qquad\qquad = \dfrac{(\cancel{y+5})(y-5)}{\cancel{y+5}}$

$\qquad\qquad\qquad = y - 5$

We could also find this quotient as follows.

$$y + 5 \overline{)\, y^2 + 0y - 25}$$ with quotient $y - 5$ — Writing in the mising term

$\qquad \dfrac{y^2 + 5y}{-5y - 25}$

$\qquad \dfrac{-5y - 25}{0}$

Check. We multiply.

$(y - 5)(y + 5) = y^2 - 5^2 = y^2 - 25$

29. $a + 2 \overline{)\, a^3 + 0a^2 + 0a + 8}$ with quotient $a^2 - 2a + 4$ — Writing in the missing terms

$\qquad \dfrac{a^3 + 2a^2}{-2a^2 + 0a}$

$\qquad \dfrac{-2a^2 - 4a}{4a + 8}$

$\qquad \dfrac{4a + 8}{0}$

Check. We multiply.

$\qquad a^2 \ - \ 2a \ + \ 4$

$\qquad\qquad\qquad a \ + \ 2$

$\qquad \overline{\ 2a^2 \ - \ 4a \ + \ 8\ }$

$\qquad \dfrac{a^3 \ -2a^2 \ + \ 4a}{a^3 \qquad\qquad + \ 8}$

31. $t - 4 \overline{)\, t^2 + 0t - 13}$ with quotient $t + 4$

$\qquad \dfrac{t^2 - 4t}{4t - 13}$

$\qquad \dfrac{4t - 16}{3}$

The answer is $t + 4$, R 3, or $t + 4 + \dfrac{3}{t - 4}$

Check. We multiply.

$(t - 4)\left[(t + 4) + \dfrac{3}{t - 4}\right]$

$= (t - 4)(t + 4) + (t - 4)\left(\dfrac{3}{t - 4}\right)$

$= t^2 - 4^2 + 3$

$= t^2 - 16 + 3$

$= t^2 - 13$

33. $2t - 3 \overline{)\, 2t^3 - 9t^2 + 11t - 3}$ with quotient $t^2 - 3t + 1$

$\qquad \dfrac{2t^3 - 3t^2}{-6t^2 + 11t}$

$\qquad \dfrac{-6t^2 + 9t}{2t - 3}$

$\qquad \dfrac{2t - 3}{0}$

Check. We multiply.

$\qquad t^2 \ - \ 3t \ + \ 1$

$\qquad\qquad\qquad 2t \ - \ 3$

$\qquad \overline{\ - \ 3t^2 \ + \ 9t \ - \ 3\ }$

$\qquad \dfrac{2t^3 \ - \ 6t^2 \ + \ 2t}{2t^3 \ - \ 9t^2 \ + \ 11t \ - \ 3}$

35. $5x + 1 \overline{)\, 5x^2 - 14x}$ with quotient $x - 3$

$\qquad \dfrac{-(5x^2 + x)}{-15x}$

$\qquad \dfrac{-(-15x - 3)}{3}$

The answer is $x - 3$, R 3, or $x - 3 + \dfrac{3}{5x + 1}$

Check. Multiply.

$(5x + 1)\left[x - 3 + \dfrac{3}{5x + 1}\right]$

$= (5x + 1)(x - 3) + (5x + 1)\left(\dfrac{3}{5x + 1}\right)$

$= 5x^2 - 15x + x - 3 + 3$

$= 5x^2 - 14x$

37. $\left(t^3 + t - t^2 - 1\right) \div (t+1)$,

    or $\left(t^3 - t^2 + t - 1\right) \div (t+1)$

$$
\begin{array}{r}
t^2 - 2t + 3 \\
t+1 \overline{)t^3 - t^2 + t - 1} \\
\underline{t^3 + t^2} \\
-2t^2 + t \\
\underline{-2t^2 - 2t} \\
3t - 1 \\
\underline{3t + 3} \\
-4
\end{array}
$$

The answer is $t^2 - 2t + 3 + \dfrac{-4}{t+1}$.

Check. We multiply.

$(t+1)\left[\left(t^2 - 2t + 3\right) + \dfrac{-4}{t+1}\right]$

$= t^3 - 2t^2 + 3t + t^2 - 2t + 3 - 4$

$= t^3 - t^2 + t - 1$

39. $t^2 + 5 \overline{)t^4 + 0t^3 + 4t^2 + 3t - 6}$

$$
\begin{array}{r}
t^2 - 1 \\
t^2 + 5 \overline{)t^4 + 0t^3 + 4t^2 + 3t - 6} \\
\underline{t^4 + 5t^2} \\
- t^2 + 3t - 6 \\
\underline{- t^2 - 5} \\
3t - 1
\end{array}
$$

The answer is $t^2 - 1 + \dfrac{3t - 1}{t^2 + 5}$.

Check. We multiply.

$\left(t^2 + 5\right)\left[\left(t^2 - 1\right) + \dfrac{3t - 1}{t^2 + 5}\right]$

$= \left(t^2 + 5\right)\left(t^2 - 1\right) + \left(t^2 + 5\right)\left(\dfrac{3t - 1}{t^2 + 5}\right)$

$= t^4 - t^2 + 5t^2 - 5 + 3t - 1$

$= t^4 + 4t^2 + 3t - 6$

41. $\left(4x^4 - 3 - x - 4x^2\right) \div \left(2x^2 - 3\right)$,

    or $\left(4x^4 - 4x^2 - x - 3\right) \div \left(2x^2 - 3\right)$

$$
\begin{array}{r}
2x^2 + 1 \\
2x^2 - 3 \overline{)4x^4 + 0x^3 - 4x^2 - x - 3} \\
\underline{4x^4 - 6x^2} \\
2x^2 - x - 3 \\
\underline{2x^2 - 3} \\
- x
\end{array}
$$

The answer is $2x^2 + 1 + \dfrac{-x}{2x^2 - 3}$.

Check. We multiply.

$\left(2x^2 - 3\right)\left[\left(2x^2 + 1\right) + \dfrac{-x}{2x^2 - 3}\right]$

$= \left(2x^2 - 3\right)\left(2x^2 + 1\right) + \left(2x^2 - 3\right)\left(\dfrac{-x}{2x^2 - 3}\right)$

$= 4x^4 + 2x^2 - 6x^2 - 3 - x$

$= 4x^4 - 4x^2 - x - 3$

43. $\left(x^3 - 2x^2 + 2x - 7\right) \div (x+1)$

$= \left(x^3 - 2x^2 + 2x - 7\right) \div [x - (-1)]$

$$
\begin{array}{r|rrrr}
-1 & 1 & -2 & 2 & -7 \\
 & & -1 & 3 & -5 \\
\hline
 & 1 & -3 & 5 & -12
\end{array}
$$

The answer is $x^2 - 3x + 5$, R -12,

     or $x^2 - 3x + 5 + \dfrac{-12}{x+1}$.

45. $\left(a^2 + 8a + 11\right) \div (a + 3)$

$= \left(a^2 + 8a + 11\right) \div [a - (-3)]$

$$
\begin{array}{r|rrr}
-3 & 1 & 8 & 11 \\
 & & -3 & -15 \\
\hline
 & 1 & 5 & -4
\end{array}
$$

The answer is $a + 5$, R -4, or $a + 5 + \dfrac{-4}{a + 3}$.

47. Begin by writing the dividend in descending order:

$\left(x^3 - 13x - 7x^2 + 3\right) \div (x - 2)$

$= \left(x^3 - 7x^2 - 13x + 3\right) \div (x - 2)$

$$
\begin{array}{r|rrrr}
2 & 1 & -7 & -13 & 3 \\
 & & 2 & -10 & -46 \\
\hline
 & 1 & -5 & -23 & -43
\end{array}
$$

The answer is $x^2 - 5x - 23$, R -43, or

     $x^2 - 5x - 23 + \dfrac{-43}{x - 2}$.

49. $\left(3x^3 + 7x^2 - 4x + 3\right) \div (x+3)$

$= \left(3x^3 + 7x^2 - 4x + 3\right) \div [x - (-3)]$

$$
\begin{array}{r}
-3 \, | \; \begin{array}{rrrr} 3 & 7 & -4 & 3 \\ & -9 & 6 & -6 \\ \hline 3 & -2 & 2 & | \; -3 \end{array}
\end{array}
$$

The answer is $3x^2 - 2x + 2, R - 3$, or

$$3x^2 - 2x + 2 + \frac{-3}{x+3}.$$

51. $\left(x^5 - 32\right) \div (x - 2)$

$= \left(x^5 + 0x^4 + 0x^3 + 0x^2 + 0x - 32\right) \div (x - 2)$

$$
\begin{array}{r}
2 \, | \; \begin{array}{rrrrrr} 1 & 0 & 0 & 0 & 0 & -32 \\ & 2 & 4 & 8 & 16 & 32 \\ \hline 1 & 2 & 4 & 8 & 16 & | \; 0 \end{array}
\end{array}
$$

The answer is $x^4 + 2x^3 + 4x^2 + 8x + 16$.

53. $\left(3x^3 + 1 - x + 7x^2\right) \div \left(x + \frac{1}{3}\right)$

$= \left(3x^3 + 7x^2 - x + 1\right) \div \left[x - \left(-\frac{1}{3}\right)\right]$

$$
\begin{array}{r}
-\frac{1}{3} \, | \; \begin{array}{rrrr} 3 & 7 & -1 & 1 \\ & -1 & -2 & 1 \\ \hline 3 & 6 & -3 & | \; 2 \end{array}
\end{array}
$$

$3x^2 + 6x - 3$, R2, or $3x^2 + 6x - 3 + \dfrac{2}{x + \frac{1}{3}}$.

55. *Thinking and Writing Exercise.*

57. $3x - 4y = 12$. In this equation both the coefficient of x and the coefficient of y go into 12 evenly. It is easy to find the x and y intercepts as a result. Setting $x = 0$ gives the y-intercept:

$3(0) - 4y = 12 : 0 - 4y = 12 : y = \dfrac{12}{-4} = -3$,

$(0, -3)$ is the y- intercept. Setting $y = 0$ gives the x-intercept:

$3x - 4(0) = 12 : 3x - 0 = 12 : x = \dfrac{12}{3} = 4$

$(4, 0)$ is the x-intercept. Plotting the points $(0, -3)$ and $(4, 0)$ and graphing the line passing through them gives the graph of the line defined by $3x - 4y = 12$.

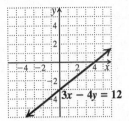

59. $3y - 2 = 7$. There are no x's in the equation for this line. It is therefore a horizontal line. Solving the equation for y gives $y = 3$. The graph of this line is then the horizontal line that passes through the y-axis at $(0, 3)$.

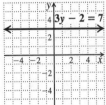

61. $m = \dfrac{y_2 - y_1}{x_2 - x_1} = \dfrac{5 - 2}{-7 - 3} = \dfrac{3}{-10} = -\dfrac{3}{10}$

63. m is -5 and the y-intercept b is $(0, -10)$.

$y = mx + b \rightarrow y = -5x - 10$

65. *Thinking and Writing Exercise.*

67. $\left(10x^{9k} - 32x^{6k} + 28x^{3k}\right) \div \left(2x^{3k}\right)$

$= \dfrac{10x^{9k} - 32x^{6k} + 28x^{3k}}{2x^{3k}}$

$= \dfrac{10x^{9k}}{2x^{3k}} - \dfrac{32x^{6k}}{2x^{3k}} + \dfrac{28x^{3k}}{2x^{3k}}$

$= 5x^{9k-3k} - 16x^{6k-3k} + 14x^{3k-3k}$

$= 5x^{6k} - 16x^{3k} + 14$

69.
$$
\begin{array}{r}
3t^{2h} + 2t^h - 5 \\
2t^h + 3 \overline{) 6t^{3h} + 13t^{2h} - 4t^h - 15} \\
\underline{6t^{3h} + 9t^{2h}} \\
4t^{2h} - 4t^h \\
\underline{4t^{2h} + 6t^h} \\
-10t^h - 15 \\
\underline{-10t^h - 15} \\
0
\end{array}
$$

The answer is $3t^{2h} + 2t^h - 5$.

71.
$$
\begin{array}{r}
a+3 \\
5a^2-7a-2{\overline{\smash{\big)}\,5a^3+8a^2-23a-1}} \\
\underline{5a^3-7a^2-2a} \\
15a^2-21a-1 \\
\underline{15a^2-21a-6} \\
5
\end{array}
$$

The answer is $a+3+\dfrac{5}{5a^2-7a-2}$.

73. $\left(4x^5-14x^3-x^2+3\right)$
$\qquad +\left(2x^5+3x^4+x^3-3x^2+5x\right)$
$= 6x^5+3x^4-13x^3-4x^2+5x+3$

$$
\begin{array}{r}
2x^2+\ x\ -\ 3 \\
3x^3-2x-1{\overline{\smash{\big)}\,6x^5+3x^4-13x^3-4x^2+\ 5x+3}} \\
\underline{6x^5\qquad\ -4x^3\ -2x^2} \\
3x^4-9x^3\ -\ 2x^2+5x \\
\underline{3x^4\qquad\ -2x^2-\ x} \\
-9x^3\qquad\ +6x+3 \\
\underline{-9x^3\qquad\ +6x+3} \\
0
\end{array}
$$

The answer is $2x^2+x-3$.

75.
$$
\begin{array}{r}
x-3 \\
x-1{\overline{\smash{\big)}\,x^2-4x+c}} \\
\underline{x^2-x} \\
-3x+c \\
\underline{-3x+3} \\
c-3
\end{array}
$$

We set the remainder equal to 0.
$c-3=0$
$c=3$
Thus, c must be 3.

77.
$$
\begin{array}{r}
c^2x+(2c+c^2) \\
x-1{\overline{\smash{\big)}\,c^2x^2+2cx+1}} \\
\underline{c^2x^2-c^2x} \\
(2c+c^2)x+1 \\
\underline{(2c+c^2)x-(2c+c^2)} \\
1+(2c+c^2)
\end{array}
$$

We set the remainder equal to 0.
$c^2+2c+1=0$
$(c+1)^2=0$
$c+1=0 \qquad$ or $\quad c+1=0$
$c=-1 \qquad$ or $\qquad c=-1$
Thus, c must be -1.

1. difference

3. evaluate

5. excluding

7. $f(2)+g(2)$
$\qquad f(2)=-3\cdot 2+1=-5$
$\qquad g(2)=2^2+2=6$
$\quad f(2)+g(2)=-5+6=1$
We could also use this method.
$\quad f(x)+g(x)=(-3x+1)+\left(x^2+2\right)$
$\qquad\qquad\qquad = x^2-3x+3$
$\quad (f+g)(2)=2^2-3\cdot 2+3$
$\qquad\qquad\quad = 4-6+3$
$\qquad\qquad\quad = 1$

9. $f(5)-g(5)$
$\qquad f(5)=-3\cdot 5+1=-15+1=-14$
$\qquad g(5)=5^2+2=25+2=27$
$\quad f(5)-g(5)=-14-27=-41$
We could also use this method:
$\quad f(x)-g(x)=(-3x+1)-\left(x^2+2\right)$
$\qquad\qquad\qquad = -3x+1-x^2-2$
$\qquad\qquad\qquad = -x^2-3x-1$
$\quad (f-g)(5)=-5^2-3\cdot 5-1$
$\qquad\qquad\quad = -25-15-1$
$\qquad\qquad\quad = -41$

11. $f(-1)\cdot g(-1)$
$\qquad f(-1)=-3(-1)+1=3+1=4$
$\qquad g(-1)=(-1)^2+2=1+2=3$
$\quad (f\cdot g)(-1)=4\cdot 3=12$
We could also use this method.
$\quad f(x)\cdot g(x)=(-3x+1)\left(x^2+2\right)$
$\qquad\qquad\qquad = -3x^3-6x+x^2+2$
$\qquad\qquad\qquad = -3x^3+x^2-6x+2$
$\quad (f\cdot g)(-1)=-3(-1)^3+(-1)^2-6(-1)+2$
$\qquad\qquad\qquad = -3(-1)+1-6(-1)+2$
$\qquad\qquad\qquad = 3+1+6+2$
$\qquad\qquad\qquad = 12$

13. $f(-4)/g(-4)$

$$f(-4) = -3(-4) + 1 = 12 + 1 = 13$$

$$g(-4) = (-4)^2 + 2 = 16 + 2 = 18$$

$$f(-4)/g(-4) = \frac{13}{18}$$

15. $g(1) - f(1)$

$$g(1) = 1^2 + 2 = 1 + 2 = 3$$

$$f(1) = -3 \cdot 1 + 1 = -3 + 1 = -2$$

$$g(1) - f(1) = 3 - (-2) = 5$$

17. $(f+g)(x) = (-3x+1) + (x^2 + 2)$

$$= x^2 - 3x + 3$$

19. $(f \cdot g)(x) = (-3x+1)(x^2 + 2)$

$$= -3x^3 - 6x + x^2 + 2$$

$$= -3x^3 + x^2 - 6x + 2$$

21. $(F+G)(x) = (x^2 - 2) + (5 - x)$

$$= x^2 - x + 3$$

23. $(F+G)(x) = x^2 - x + 3$

$$(F+G)(-4) = (-4)^2 - (-4) + 3$$

$$= 16 + 4 + 3$$

$$= 23$$

25. $(F-G)(3)$

$$F(3) = 3^2 - 2 = 9 - 2 = 7$$

$$G(3) = 5 - 3 = 2$$

$$(F-G)(3) = 7 - 2 = 5$$

27. $(F \cdot G)(-3)$

$$F(-3) = (-3)^2 - 2 = 9 - 2 = 7$$

$$G(-3) = 5 - (-3) = 5 + 3 = 8$$

$$(F \cdot G)(-3) = 7 \cdot 8 = 56$$

29. $(F/G)(x) = (x^2 - 2) \div (5 - x)$

$$= \frac{x^2 - 2}{5 - x};$$

$$x \neq 5$$

31. $(F-G)(x) = (x^2 - 2) - (5 - x)$

$$= x^2 - 2 - 5 + x$$

$$= x^2 + x - 7$$

33. $(F/G)(-2)$

$$f(-2) = (-2)^2 - 2 = 4 - 2 = 2$$

$$g(-2) = 5 - (-2) = 5 + 2 = 7$$

$$(F/G)(-2) = \frac{2}{7}$$

35. Locate 2 on the horizontal axis then move vertically to the graph of P, the blue graph. $P(2) \approx 26.5$. Locate 2 on the horizontal axis then move vertically to the graph of L, the red graph.
$L(2) \approx 22.5$; $(P-L)(2) \approx 26.5 - 22.5 \approx 4\%$

37. Answers may vary slightly.
Using the graph,
$C(2004) \approx 1.2$ and $B(2004) \approx 2.9$.
$$N(2004) = C(2004) + B(2004)$$
$$\approx 1.2 + 2.9$$
$$= 4.1 \text{ million}$$

39. Answers may vary slightly.
Since the p and r bands are stacked on top of one another, the value of $(p+r)('05)$ can be found by subtracting the value at the bottom of the r band for '05 from the value at the top of the p band for '05. Thus:
$(p+r)('05) \approx 256 - 162 = 94$ million tons.
This value represents the amount of trash in 2005 that was either composted or recycled in the U.S..

41. Answers may vary slightly.
$F('96) \approx 215$ million tons. This value represents the total amount of trash generated in the U.S. in 1996.

43. Answers may vary slightly.
$(F-p)('04)$ is found graphically by determining the value at the bottom of the p band for 2004. Thus:
$(F-p)('04) \approx 231$ million tons. This value represents the amount of trash generated in the U.S. in 2004 that was not composted.

45. $f(x) = x^2$; $g(x) = 7x - 4$

The domain of f is $\{x \mid x$ is a real number $\}$, or $\mathbb{R}$. The domain of g is $\mathbb{R}$. The domains of $f + g$, $f - g$, and $f \cdot g$ are the set of all elements common to both the domain of f and the domain of g. Thus, domain of $f + g$, domain of $f - g$, and domain of $f \cdot g = \{x \mid x$ is a real number$\}$.

47. $f(x) = \dfrac{1}{x-3}$; $g(x) = 4x^3$

The domain of $f(x)$ is $\{x \mid x$ is a real number, and $x \neq 3\}$

The domain $g(x)$ is $\mathbb{R}$. The domains of $f + g$, $f - g$, and $f \cdot g = \{x \mid x$ is a real number, and $x \neq 3\}$.

49. $f(x) = \dfrac{2}{x}$; $g(x) = x^2 - 4$

Domain of $f = \{x \mid x$ is a real number, and $x \neq 0\}$

Domain of g is $\mathbb{R}$.

The domains of $f + g$, $f - g$, and $f \cdot g = \{x \mid x$ is a real number, and $x \neq 0\}$.

51. $f(x) = x + \dfrac{2}{x-1}$; $g(x) = 3x^3$

Domain of $f = \{x \mid x$ is a real number, and $x \neq 1\}$

Domain of g is $\mathbb{R}$. The domain of $f + g$, $f - g$, and $f \cdot g = \{x \mid x$ is a real number and $x \neq 1\}$.

53. $f(x) = \dfrac{x}{2x-9}$; $g(x) = \dfrac{5}{1-x}$

Domain of $f = \{x \mid x$ in $\mathbb{R}, \ x \neq \dfrac{9}{2}\}$

Domain of $g = \{x \mid x$ in $\mathbb{R}, \ x \neq 1\}$.

The domains of $f + g$, $f - g$, and $f \cdot g$ are the set of elements common to both the domain of f and the domain of g, which is

$\{x \mid x$ in $\mathbb{R}, \ x \neq \dfrac{9}{2}$, or $1\}$.

55. $f(x) = x^4$; $g(x) = x - 3$

Domain of f is $\mathbb{R}$; domain of g is $\mathbb{R}$.

$f / g = \dfrac{x^4}{x-3}$

The domain of $f / g = \{x \mid x$ is a real number and $x \neq 3\}$.

57. $f(x) = 3x - 2$; $g(x) = 2x - 8$

Domain of f is $\mathbb{R}$; domain of g is $\mathbb{R}$.

$f / g = \dfrac{3x-2}{2x-8}$

$2x - 8 = 0$
$\quad 2x = 8$
$\quad\ \ x = 4$ So, $x \neq 4$.

Domain of $f / g = \{x \mid x$ is a real number and $x \neq 4\}$

59. $f(x) = \dfrac{3}{x-4}$; $g(x) = 5 - x$

Domain of $f = \{x \mid x$ is a real number and $x \neq 4\}$

Domain of g is $\mathbb{R}$.

$f / g = \dfrac{3}{\frac{x-4}{5-x}}$ So, $x \neq 5$

Domain of $f / g = \{x \mid x$ is a real number and $x \neq 4$ and $x \neq 5\}$.

61. $f(x) = \dfrac{2x}{x+1}$; $g(x) = 2x + 5$

Domain of $f = \{x \mid x$ is a real number and $x \neq -1\}$

Domain of g is $\mathbb{R}$.

$f / g = \dfrac{\frac{2x}{x+1}}{2x+5}$

$2x + 5 = 0$
$\quad 2x = -5$
$\quad\ \ x = \dfrac{-5}{2}$

So, $x \neq \dfrac{-5}{2}$.

Domain of $f / g = \{ x \, | \, x$ is a real number and

$x \neq -1$ and $x \neq \frac{-5}{2} \}$

63. $f(x) = 8x^3 + 27; \quad g(x) = 2x + 3$

$F(x) = \dfrac{8x^3 + 27}{2x + 3}; \quad x \neq \dfrac{-3}{2}$

$$2x + 3 \overline{\smash{\big)}\, 8x^3 + 0x^2 + 0x + 27} \quad \underset{}{4x^2 - 6x + 9}$$

$$\underline{8x^3 + 12x^2}$$
$$-12x^2 + 0x$$
$$\underline{-12x^2 - 18x}$$
$$18x + 27$$
$$\underline{18x + 27}$$
$$0$$

$F(x) = 4x^2 - 6x + 9; \quad x \neq \dfrac{-3}{2}$

65. $f(x) = 6x^2 - 11x - 10; \quad g(x) = 3x + 2$

$F(x) = \dfrac{6x^2 - 11x - 10}{3x + 2}; \quad x \neq \dfrac{-2}{3}$

$$3x + 2 \overline{\smash{\big)}\, 6x^2 - 11x - 10} \quad \underset{}{2x - 5}$$

$$\underline{6x^2 + 4x}$$
$$-15x - 10$$
$$\underline{-15x - 10}$$
$$0$$

$F(x) = 2x - 5; \quad x \neq \dfrac{-2}{3}$

67. From the graph we determine:

$F(5) = 1$ and $G(5) = 3$.

$F(5) + G(5) = 1 + 3 = 4$

$F(7) = -1$ and $G(7) = 4$.

$F(7) + G(7) = -1 + 4 = 3$

69. From the graph we determine:

$G(7) = 4$ and $F(7) = -1$

$G(7) - F(7) = 4 - (-1) = 4 + 1 = 5$

$G(3) = 1$ and $F(3) = 2$

$G(3) - F(3) = 1 - 2 = -1$

71. The x-values of f are from 0 to 9, so the domain of $F = \{ x \, | \, 0 \leq x \leq 9 \}$.

The x-values of g are from 3 to 10, so the domain of $G = \{ x \, | \, 3 \leq x \leq 10 \}$.

The domain of $F + G$ is the set of numbers common to the domains of f and g; so, the domain of $F + G = \{ x \, | \, 3 \leq x \leq 9 \}$.

The domain of F / G is the set of numbers common to the domains of F and G, excluding any number for which $G = 0$ (there are no exclusions, as $G \neq 0$). The domain of $F / G = \{ x \, | \, 3 \leq x \leq 9 \}$.

73. We determine some values of $F + G$ and graph these points. We join these points with a smooth curve.

	$F + G$
3	$2 + 1 = 3$
5	$1 + 3 = 4$
7	$-1 + 4 = 3$
9	$1 + 2 = 3$

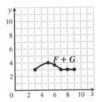

75. *Thinking and Writing Exercise.*

77. $15x + 20y + 5 = 5 \cdot 3x + 5 \cdot 4y + 5 \cdot 1$
$$= 5(3x + 4y + 1)$$

79. $\quad x + 5 = 0$
$\quad x + 5 - 5 = 0 - 5$
$\qquad x = -5$

81. $\quad 3x - 1 = 0$
$\quad 3x - 1 + 1 = 0 + 1$
$\qquad 3x = 1$
$\qquad \dfrac{3x}{3} = \dfrac{1}{3}$
$\qquad x = \dfrac{1}{3}$

83. *Thinking and Writing Exercise.*

85. $f(x) = \dfrac{3x}{2x + 5}; \quad g(x) = \dfrac{x^4 - 1}{3x + 9}$

Domain $f : 2x + 5 = 0$

$$2x = -5$$

$$x = \frac{-5}{2}, \text{ So } x \neq \frac{-5}{2}$$

Domain of $f = \{x \mid x \text{ is a real number and}$

$x \neq \frac{-5}{2}\}$

Domain of $g : 3x + 9 = 0$

$$3x = -9$$

$$x = -3; \text{ So } x \neq -3$$

Domain of $g = \{x \mid x \text{ is a real number and}$

$x \neq -3\}$.

$$f / g = \frac{3x}{2x+5} \Big/ \frac{x^4 - 1}{3x+9}$$

$x^4 - 1 \neq 0,$ since the divisor must be non-zero.

$$x^4 - 1 = 0$$

$$(x^2 + 1)(x^2 - 1) = 0$$

$x^2 + 1 = 0$ or $x^2 - 1 = 0$

No real $(x+1)(x-1) = 0$

values $x = -1$ or $x = 1$

So, $x \neq \pm 1$.

Therefore, the domain of $f / g = \{x \mid x \text{ is a}$

real number and $x \neq \frac{-5}{2}, x \neq -3, \text{ and } x \neq \pm 1\}$

87. Answers may vary.

89. The domain of m is $\{x \mid -1 < x < 5\}$.

$n(x) = 0$ when $2x - 3 = 0$

$$2x = 3$$

$$x = \frac{3}{2}$$

We exclude this value from the domain of
m / n.

The domain of

$m / n = \left\{ x \mid -1 < x < 5 \text{ and } x \neq \frac{3}{2} \right\}$.

91. Answers may vary.

$$f(x) = \frac{1}{x+2}; \ g(x) = \frac{1}{x-5}$$

93. $y_1 = 2.5x + 1.5$ and $y_2 = x - 3$

$$\frac{y_1}{y_2} = \frac{2.5x + 1.5}{x - 3}$$

Domain: $\{x \mid \text{ is any real number } and \ x \neq 3\}$

The CONNECTED mode graph contains the
line $x = 3$, whereas, the DOT mode graph
contains no points having 3 as the first
coordinate. Thus, the DOT mode graph
represents y_3 more accurately.

Chapter 5 Study Summary

1. $6^1 = 6$

2. $(-5)^0 = 1$

3. $x^5 x^{11} = x^{5+11} = x^{16}$

4. $\dfrac{8^9}{8^2} = 8^{9-2} = 8^7$

5. $\left(y^5\right)^3 = y^{5 \cdot 3} = y^{15}$

6. $\left(x^3 y\right)^{10} = x^{3 \cdot 10} y^{1 \cdot 10} = x^{30} y^{10}$

7. $\left(\dfrac{x^2}{7}\right)^5 = \dfrac{x^{2 \cdot 5}}{7^{1 \cdot 5}} = \dfrac{x^{10}}{7^5}$

8. $10^{-1} = \dfrac{1}{10^1} = \dfrac{1}{10}$

9. $\dfrac{x^{-1}}{y^{-3}} = \dfrac{y^3}{x^1} = \dfrac{y^3}{x}$

10. $0.000904 = 9.04 \cdot 10^{-4}$

11. $6.9 \cdot 10^5 = 690,000$

12. $x^2 - 10 + 5x - 8x^6$ has four terms:
    $x^2, -10, \ 5x, \text{ and } -8x^6$

13. $5x = 5x^1$, so the term is of degree 1.

14. $x^2 = 1x^2$, so the coefficient is 1.

15. The polynomial is not written in descending order, but the "leading term" is still the term containing the highest power of x: $-8x^6$

16. The leading coefficient is the coefficient of the leading term: -8.

17. The degree of the polynomial is the degree of the leading term: 6.

18. Three terms: trinomial.

19. $3x^2 + 5x - 10x + x = 3x^2 + (5 - 10 + 1)x$
$$= 3x^2 + (-4)x$$
$$= 3x^2 - 4x$$

20. $2 - 3x - x^2 = 2 - 3(-1) - (-1)^2$ @ $x = -1$
$$= 2 - (-3) - (1) = 2 + 3 - 1 = 4$$

21. $(9x^2 - 3x) + (4x - x^2) = 9x^2 - 3x + 4x - x^2$
$$= (9 - 1)x^2 + (-3 + 4)x$$
$$= 8x^2 + x$$

22. $(9x^2 - 3x) - (4x - x^2) = 9x^2 - 3x - 4x + x^2$
$$= (9 + 1)x^2 + (-3 - 4)x$$
$$= 10x^2 + (-7)x$$
$$= 10x^2 - 7x$$

23. $(x - 1)(x^2 - x - 2)$
$$= x(x^2 - x - 2) - 1(x^2 - x - 2)$$
$$= x \cdot x^2 - x \cdot x - x \cdot 2 - 1 \cdot x^2 + 1 \cdot x + 1 \cdot 2$$
$$= x^3 - x^2 - 2x - x^2 + x + 2$$
$$= x^3 - 2x^2 - x + 2$$

24. $(x + 4)(2x + 3)$
$$= x(2x + 3) + 4(2x + 3)$$
$$= x \cdot 2x + x \cdot 3 + 4 \cdot 2x + 4 \cdot 3$$
$$= 2x^2 + 3x + 8x + 12$$
$$= 2x^2 + 11x + 12$$

25. $(5 + 3x)(5 - 3x) = 5^2 - (3x)^2 = 25 - 9x^2$

26. $(x + 9)^2 = (x)^2 + 2 \cdot (x) \cdot (9) + (9)^2$
$$= x^2 + 18x + 81$$

27. $(8x - 1)^2 = (8x)^2 - 2 \cdot (8x) \cdot (1) + (1)^2$
$$= 64x^2 - 16x + 1$$

28. $xy - y^2 - 4x$
$$= (-2)(3) - (3)^2 - 4(-2) \text{ for } (x, y) = (-2, 3)$$
$$= -6 - 9 - (-8) = -15 + 8 = -7$$

29. $4mn^5 = 4m^1n^5$ so the degree is $1 + 5 = 6$.

30. $(3cd^2 + 2c) + (4cd - 9c) = 3cd^2 + 2c + 4cd - 9c$
$$= 3cd^2 + 4cd - 7c$$

31. $(8pw - p^2w) - (p^2w + 8pw)$
$$= 8pw - p^2w - p^2w - 8pw = -2p^2w$$

32. $(7xy - x^2)^2 = (7xy)^2 - 2 \cdot (7xy) \cdot (x^2) + (x^2)^2$
$$= 49x^2y^2 - 14x^3y + x^4$$

33. $\dfrac{4y^5 - 8y^3 + 16y^2}{4y^2} = \dfrac{4y^5}{4y^2} - \dfrac{8y^3}{4y^2} + \dfrac{16y^2}{4y^2}$
$$= \dfrac{4y^2 \cdot y^3}{4y^2} - \dfrac{4y^2 \cdot 2y}{4y^2} + \dfrac{4y^2 \cdot 4}{4y^2}$$
$$= y^3 - 2y + 4$$

34. $x + 1 \overline{)x^2 - x + 4}$ with quotient $x - 2$
$$-(x^2 + x)$$
$$-2x + 4$$
$$-(-2x - 2)$$
$$6$$

$x - 2$, R 6, or $x - 2 + \dfrac{6}{x + 1}$.

35. $(f + g)(x) = f(x) + g(x)$
$$= (x - 2) + (x - 7)$$
$$= x - 2 + x - 7$$
$$= 2x - 9$$

36. $(f - g)(x) = f(x) - g(x)$
$$= (x - 2) - (x - 7)$$
$$= x - 2 - x + 7$$
$$= 5$$

37. $(f \cdot g)(x) = f(x)g(x)$
$$= (x - 2)(x - 7)$$
$$= x^2 - 7x - 2x + 14$$
$$= x^2 - 9x + 14$$

38. $\left(\dfrac{f}{g}\right)(x) = \dfrac{f(x)}{g(x)} = \dfrac{x-2}{x-7},\ x \neq 7$

Chapter 5 Review Exercises

1. False

2. True

3. True

4. False

5. True

6. False

7. True

8. True

9. $y^7 \cdot y^3 \cdot y = y^{7+3+1} = y^{11}$

10. $(3x)^5 \cdot (3x)^9 = (3x)^{5+9} = (3x)^{14}$

11. $t^8 \cdot t^0 = t^{8+0} = t^8$

12. $\dfrac{4^5}{4^2} = 4^{5-2} = 4^3,\ \text{or } 64$

13. $\dfrac{(a+b)^4}{(a+b)^4} = (a+b)^{4-4} = (a+b)^0 = 1$

    Also note: An expression divided by itself = 1.

14. $\dfrac{-20m^4 n^5}{10mn^3} = -\dfrac{\cancel{10} \cdot 2 \cdot \cancel{m} \cdot m^3 \cdot \cancel{n^3} \cdot n^2}{\cancel{10} \cdot \cancel{m} \cdot \cancel{n^3}}$

    $= -2m^3 n^2$

15. $(-2xy^2)^3 = (-2)^3 x^3 (y^2)^3$

    $= -8x^3 y^6$

16. $(2x^3)(-3x)^2 = 2x^3 \cdot (-3)^2 x^2$

    $= 2 \cdot 9 \cdot x^3 \cdot x^2$

    $= (2 \cdot 9) x^{3+2}$

    $= 18x^5$

17. $(a^2 b)(ab)^5 = a^2 \cdot b \cdot a^5 \cdot b^5$

    $= a^{2+5} b^{1+5}$

    $= a^7 b^6$

18. $\left(\dfrac{3x^2}{2y^3}\right)^2 = \dfrac{3^2 x^{2\cdot 2}}{2^2 y^{3 \cdot 2}} = \dfrac{9x^4}{4y^6}$

19. $m^{-7} = \dfrac{1}{m^7},\ \text{or } \left(\dfrac{1}{m}\right)^7$

20. $7^2 \cdot 7^{-4} = 7^{2+(-4)} = 7^{-2} = \dfrac{1}{7^2} = \dfrac{1}{49}$

21. $\dfrac{\cancel{6} a^{-5} b}{\cancel{6} a^8 b^8} = \dfrac{\cancel{b}}{a^8 a^5 \cancel{b} \cdot b^7} = \dfrac{1}{a^{13} b^7}$

22. $(x^3)^{-4} = x^{3(-4)} = x^{-12} = \dfrac{1}{x^{12}}$

23. $(2x^{-3} y)^{-2} = 2^{-2} (x^{-3})^{-2} y^{-2}$

    $= \dfrac{1}{2^2} x^6 \cdot \dfrac{1}{y^2}$

    $= \dfrac{x^6}{4y^2}$

24. $\left(\dfrac{2x}{y}\right)^{-3} = \left(\dfrac{y}{2x}\right)^3 = \dfrac{y^3}{2^3 x^3} = \dfrac{y^3}{8x^3}$

25. $8.3 \times 10^6 = 8{,}300{,}000$

26. $0.0000328 = 3.28 \times 10^{-5}$

27. $(3.8 \times 10^4)(5.5 \times 10^{-1})$

    $= (3.8 \times 5.5)(10^4 \times 10^{-1})$

    $= 20.9 \times 10^3$

    $= 2.09 \times 10^1 \times 10^3$

    $= 2.09 \times 10^4$

    $= 2.1 \times 10^4$

28. $\dfrac{1.28 \times 10^{-8}}{2.5 \times 10^{-4}} = \dfrac{1.28}{2.5} \times 10^{-8-4}$

$= 0.512 \times 10^{-4}$

$= 5.12 \times 10^{-1} \times 10^{-4}$

$= 5.12 \times 10^{-5}$

$= 5.1 \times 10^{-5}$

29. $0.5 \text{ L} \cdot \dfrac{10^{6} \, \mu\text{L}}{1 \text{ L}} = 5.0 \cdot 10^{5} \, \mu\text{L}$

$5.0 \cdot 10^{5} \, \mu\text{L} \cdot \dfrac{4.5 \cdot 10^{6} \text{ red blood cells}}{1 \, \mu\text{L}}$

$= (5 \cdot 4.5) \cdot (10^{5} \cdot 10^{6}) \text{ red blood cells}$

$= 22.5 \cdot 10^{11} \text{ red blood cells}$

$= 2.25 \cdot 10^{12} \text{ red blood cells}$

$= 2 \cdot 10^{12} \text{ red blood cells}$

30. $3x^{2} + 6x + \dfrac{1}{2}$

The terms are $3x^{2}$, $6x$, and $\frac{1}{2}$.

31. $-4y^{5} + 7y^{2} - 3y - 2$

The terms are $-4y^{5}$, $7y^{2}$, $-3y$, and -2.

32. $7x^{2} - x + 7$

The coefficients are $7, -1,$ and 7.

33. $4x^{3} + x^{2} - 5x + \dfrac{5}{3}$

The coefficients are $4, 1, -5,$ and $\frac{5}{3}$.

34. $4t^{2} + 6 + 15t^{3}$

a)

Term	$4t^{2}$	6	$15t^{3}$
Degree	2	0	3

b) The highest degree term is $15t^{3}$. This is the leading term, and the leading coefficient is 15.

c) Since the highest degree term is $15t^{3}$, the degree of the polynomial is 3.

35. $-2x^{5} + x^{4} - 3x^{2} + x$

a)

Term	$-2x^{5}$	x^{4}	$-3x^{2}$	x
Degree	5	4	2	1

b) The term of highest degree is $-2x^{5}$. This is the leading term. The leading coefficient is -2.

c) Since the term of highest degree is $-2x^{5}$, the degree of the polynomial is 5.

36. Two monomial terms are added, so $4x^{3} - 1$ is a binomial.

37. The polynomial $4 - 9t^{3} - 7t^{4} + 10t^{2}$ has no special name.

38. This is a monomial, since it is a product of a constant and a variable raised to a whole power.

39. $5x - x^{2} + 4x = -x^{2} + (5 + 4)x$

$= -x^{2} + 9x$

40. $\dfrac{3}{4}x^{3} + 4x^{2} - x^{3} + 7 = \left(\dfrac{3}{4} - 1\right)x^{3} + 4x^{2} + 7$

$= -\dfrac{1}{4}x^{3} + 4x^{2} + 7$

41. $-2x^{4} + 16 + 2x^{4} + 9 - 3x^{5}$

$= -3x^{5} + (-2 + 2)x^{4} + (16 + 9)$

$= -3x^{5} + 25$

42. $-x + \dfrac{1}{2} + 14x^{4} - 7x^{2} - 1 - 4x^{4}$

$= (14 - 4)x^{4} - 7x^{2} - x + \left(\dfrac{1}{2} - 1\right)$

$= 10x^{4} - 7x^{2} - x - \dfrac{1}{2}$

43. $P(x) = x^{2} - 3x + 6$

$P(-1) = (-1)^{2} - 3(-1) + 6$

$= 1 + 3 + 6$

$= 10$

44. $3 - 5x$, for $x = -5$

$3 - 5(-5) = 3 + 25 = 28$

45. Graph $p(x) = 2 - x^2$ in a standard viewing window.

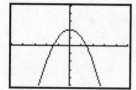

The range appears to be $(-\infty, 2]$

46. Locate 2 on the horizontal axis and move vertically to the graph to determine $m(2) \approx 340$ mg.

47. Locate 4 on the horizontal axis and move vertically to the graph to determine $m(4) \approx 185$ mg.

48. The $m(t)$ values have a minimum of 0 and a maximum of about 345. The range is $0 \le m(t) \le 345$, or $[0, 345]$.

49. The t values begin at $t = 0$ and finish at $t = 6$. The domain is $0 \le t \le 6$, or $[0, 6]$.

50. $(3x^4 - x^3 + x - 4) + (x^5 + 7x^3 - 3x - 5)$

$= x^5 + 3x^4 + (-1 + 7)x^3 + (1 - 3)x + (-4 - 5)$

$= x^5 + 3x^4 + 6x^3 - 2x - 9$

51. $(5x^2 - 4x + 1) - (3x^2 + 7)$

$= 5x^2 - 4x + 1 - 3x^2 - 7$

$= (5 - 3)x^2 - 4x + (1 - 7)$

$= 2x^2 - 4x - 6$

52. $(3x^5 - 4x^4 + 2x^2 + 3)$

     $- (2x^5 - 4x^4 + 3x^3 + 4x^2 - 5)$

$= 3x^5 - 4x^4 + 2x^2 + 3 - 2x^5$

     $+ 4x^4 - 3x^3 - 4x^2 + 5$

$= (3 - 2)x^5 + (-4 + 4)x^4 - 3x^3$

     $+ (2 - 4)x^2 + (3 + 5)$

$= x^5 - 3x^3 - 2x^2 + 8$

53.

$$
\begin{array}{llll}
\dfrac{-3}{4}x^4 & +\dfrac{1}{2}x^3 & & +\dfrac{7}{8} \\[2mm]
& -\dfrac{1}{4}x^3 & -x^2 & -\dfrac{7}{4}x \\[2mm]
+\dfrac{3}{2}x^4 & & +\dfrac{2}{3}x^2 & -\dfrac{1}{2}
\end{array}
$$

$$\left(\dfrac{-3}{4} + \dfrac{3}{2}\right)x^4 + \left(\dfrac{1}{2} - \dfrac{1}{4}\right)x^3 + \left(-1 + \dfrac{2}{3}\right)x^2 - \dfrac{7}{4}x + \left(\dfrac{7}{8} - \dfrac{1}{2}\right)$$

$$\left(\dfrac{-3}{4} + \dfrac{6}{4}\right)x^4 + \left(\dfrac{2}{4} - \dfrac{1}{4}\right)x^3 + \left(\dfrac{-3}{3} + \dfrac{2}{3}\right)x^2 - \dfrac{7}{4}x + \left(\dfrac{7}{8} - \dfrac{4}{8}\right)$$

$$\dfrac{3}{4}x^4 \quad +\dfrac{1}{4}x^3 \quad -\dfrac{1}{3}x^2 \quad -\dfrac{7}{4}x + \dfrac{3}{8}$$

54.

$$
\begin{array}{llll}
2x^5 & -x^3 & +x+3 \\
-(3x^5 - x^4 + 4x^3 + 2x^2 - x + 3) \\
\hline
2x^5 & -x^3 & +x+3 \\
-3x^5 + x^4 - 4x^3 - 2x^2 + x - 3 \\
\hline
-x^5 + x^4 - 5x^3 - 2x^2 + 2x
\end{array}
$$

55.

         $w + 3$

                     w

a. The perimeter is the sum of the lengths of the sides.

$P = (w + 3) + w + (w + 3) + w$

$= 4w + 6$

b. The area of a rectangle is length $\times$ width

$P = (w + 3)w$

$= w^2 + 3w$

56. $3x(-4x^2) = 3(-4)x \cdot x^2$

$= -12x^3$

57. $(7x + 1)^2 = (7x)^2 + 2(7x) \cdot 1 + 1^2$

$= 49x^2 + 14x + 1$

58. $(a - 7)(a + 4) = a^2 + 4a - 7a - 28$

$= a^2 - 3a - 28$

59. $(m + 5)(m - 5) = m^2 - 5^2$

$= m^2 - 25$

60.
$$4x^2 - 5x + 1$$
$$\underline{3x - 2}$$
$$-8x^2 + 10x - 2$$
$$\underline{12x^3 - 15x^2 + 3x}$$
$$12x^3 - 23x^2 + 13x - 2$$

61. $(x-9)^2 = x^2 - 2 \cdot x \cdot 9 + 9^2$
$$= x^2 - 18x + 81$$

62. $3t^2(5t^3 - 2t^2 + 4t) = 15t^5 - 6t^4 + 12t^3$

63. $(3a+8)(3a-8) = (3a)^2 - 8^2 = 9a^2 - 64$

64. $(x-0.3)(x-0.5)$
$$= x^2 - 0.5x - 0.3x + 0.15$$
$$= x^2 - 0.8x + 0.15$$

65.
$$x^4 - 2x + 3$$
$$\underline{x^3 + x - 1}$$
$$-x^4 \qquad + 2x - 3$$
$$x^5 \qquad -2x^2 + 3x$$
$$\underline{x^7 \qquad -2x^4 + 3x^3}$$
$$x^7 + x^5 \quad -3x^4 + 3x^3 - 2x^2 + 5x - 3$$

66. $(3x^4 - 5)^2 = (3x^4)^2 - 2(3x^4)(5) + (5)^2$
$$= 9x^8 - 30x^4 + 25$$

67. $(2t^2 + 3)(t^2 - 7) = 2t^4 - 14t^2 + 3t^2 - 21$
$$= 2t^4 - 11t^2 - 21$$

68. $\left(a - \dfrac{1}{2}\right)\left(a + \dfrac{2}{3}\right) = a^2 + \dfrac{2}{3a} - \dfrac{1}{2a} - \dfrac{1}{3}$
$$= a^2 + \left(\dfrac{4}{6} - \dfrac{3}{6}\right)a - \dfrac{1}{3}$$
$$= a^2 + \dfrac{1}{6a} - \dfrac{1}{3}$$

69. $(-7 + 2n)(7 + 2n) = (2n - 7)(2n + 7)$
$$= (2n)^2 - 7^2$$
$$= 4n^2 - 49$$

70. $2 - 5xy + y^2 - 4xy^3 + x^6$

Replace the x terms with -1 and the y terms with 2 and evaluate.
$$= 2 - 5(-1)(2) + (2)^2 - 4(-1)(2)^3 + (-1)^6$$
$$= 2 - 5(-1)(2) + 4 - 4(-1) \cdot 8 + 1$$
$$= 2 + 10 + 4 + 32 + 1$$
$$= 49$$

71. $x^5y - 7xy + 9x^2 - 8$

Term	Coefficient	Degree
x^5y	1	6
$-7xy$	-7	2
$9x^2$	9	2
-8	-8	0

The term of highest degree is x^5y. Its degree is 6, so the degree of the polynomial is 6.

72. $x^2y^5z^9 - y^{40} + x^{13}z^{10}$

Term	Coefficient	Degree
$x^2y^5z^9$	1	16
$-y^{40}$	-1	40
$x^{13}z^{10}$	1	23

The term of highest degree is $-y^{40}$. Its degree is 40, so the degree of the polynomial is 40.

73. $y + w - 2y + 8w - 5 = (1 - 2)y + (1 + 8)w - 5$
$$= -y + 9w - 5$$

74. $6m^3 + 3m^2n + 4mn^2 + m^2n - 5mn^2$
$$= 6m^3 + (3 + 1)m^2n + (4 - 5)mn^2$$
$$= 6m^3 + 4m^2n - mn^2$$

75. $(5x^2 - 7xy + y^2) + (-6x^2 - 3xy - y^2)$
$$= (5 - 6)x^2 + (-7 - 3)xy + (1 - 1)y^2$$
$$= -x^2 - 10xy$$

76. $(6x^3y^2 - 4x^2y - 6x)$
$$\quad - (-5x^3y^2 + 4x^2y + 6x^2 - 6)$$
$$= 6x^3y^2 - 4x^2y - 6x + 5x^3y^2$$
$$\quad - 4x^2y - 6x^2 + 6$$
$$= (6 + 5)x^3y^2 + (-4 - 4)x^2y$$
$$\quad - 6x^2 - 6x + 6$$
$$= 11x^3y^2 - 8x^2y - 6x^2 - 6x + 6$$

77. $p^2 + pq + q^2$

$\dfrac{p - q}{}$

$-p^2q - pq^2 - q^3$

$\dfrac{p^3 + p^2q + pq^2}{}$

$p^3 \qquad\qquad - q^3$

78. $\left(5ab - cd^2\right)^2 = (5ab)^2 - 2(5ab)(cd^2) + (cd^2)^2$

$= 25a^2b^2 - 10abcd^2 + c^2d^4$

79. The area of a triangle is $\frac{1}{2} \cdot$ base $\cdot$ height.

$A = \dfrac{1}{2} \cdot b \cdot h$

$= \dfrac{1}{2}(x + y)(x - y)$

$= \dfrac{1}{2}(x^2 - y^2)$

$= \dfrac{1}{2}x^2 - \dfrac{1}{2}y^2$

80. $(10x^3 - x^2 + 6x) \div (2x) = \dfrac{10x^3}{2x} - \dfrac{x^2}{2x} + \dfrac{6x}{2x}$

$= 5x^2 - \dfrac{1}{2}x + 3$

81. $(6x^3 - 5x^2 - 13x + 13) \div (2x + 3)$

$$
\begin{array}{r}
3x^2 - 7x + 4 \\
2x+3{\overline{\smash{\big)}\,6x^3 - 5x^2 - 13x + 13}} \\
\underline{6x^3 + 9x^2} \\
-14x^2 - 13x \\
\underline{-14x^2 - 21x} \\
8x + 13 \\
\underline{8x + 12} \\
1
\end{array}
$$

The answer is $3x^2 - 7x + 4, R1$ or

$3x^2 - 7x + 4 + \dfrac{1}{2x + 3}$

82. $(t^4 + t^3 + 2t^2 - t - 3) \div (t + 1)$

$= (t^4 + t^3 + 2t^2 - t - 3) \div [t - (-1)]$

$$
\begin{array}{r|rrrrr}
-1 & 1 & 1 & 2 & -1 & -3 \\
   &   & -1 & 0 & -2 & 3 \\
\hline
   & 1 & 0 & 2 & -3 & 0
\end{array}
$$

The answer is $t^3 + 2t - 3$

83. $(g \cdot h)(4)$

$g(4) = 3 \cdot 4 - 6 = 12 - 6 = 6$

$h(4) = 4^2 + 1 = 16 + 1 = 17$

$(g \cdot h)(4) = 6 \cdot 17 = 102$

84. $(g - h)(-2)$

$g(-2) = 3(-2) - 6 = -6 - 6 = -12$

$h(-2) = (-2)^2 + 1 = 4 + 1 = 5$

$(g - h)(-2) = -12 - 5 = -17$

85. $(g / h)(-1)$

$g(-1) = 3(-1) - 6 = -3 - 6 = -9$

$h(-1) = (-1)^2 + 1 = 1 + 1 = 2$

$(g / h)(-1) = \dfrac{-9}{2}$

86. $(g + h)(x) = g(x) + h(x)$

$= (3x - 6) + (x^2 + 1)$

$= x^2 + 3x - 5$

87. $(g \cdot h)(x) = g(x) \cdot h(x)$

$= (3x - 6) + (x^2 + 1)$

$= 3x^3 + 3x - 6x^2 - 6$

$= 3x^3 - 6x^2 + 3x - 6$

88. $(h / g)(x) = \dfrac{x^2 + 1}{3x - 6}$

Note: $3x - 6 \neq 0$

$3x \neq 6$

$x \neq 2$

89. The domains of $g + h$ and $g \cdot h$ are those numbers common to the domains of g and h. The domain of g is $\mathbb{R}$, and the domain of h is $\mathbb{R}$. The domains of $g + h$ and $g \cdot h$ equal $\mathbb{R}$.

90. The domain of h / g is $\mathbb{R}$, excluding any value(s) which makes/make $g = 0$.

$0 = 3x - 6$

$6 = 3x$

$2 = x$

The domain of $h \mid g = \{x \mid x$ is a real number and $x \neq 2\}$

91. *Thinking and Writing Exercise.*

$5x^3 = 5 \cdot x \cdot x \cdot x$ whereas

$(5x)^3 = 5x \cdot 5x \cdot 5x$

$ = (5 \cdot 5 \cdot 5)(x \cdot x \cdot x)$

$ = 125x^3$

The powers of x are equal, but the coefficients are not.

92. *Thinking and Writing Exercise.* No. There are exactly two binomials that, when squared, give $x^2 - 6x + 9$. $(x-3)^2 = x^2 - 6x + 9$ and $(3-x)^2 = 9 - 6x + x^2 = x^2 - 6x + 9$.

The problem is that $(x-3) = -(3-x)$, but when opposites are squared, they give the same result.

93. a) The degree of a product will be the sum of the degrees of each factor. Therefore the degree of $(x^5 - 6x^2 + 3)(x + x^7 + 11)$ is $5 + 7 = 12$.

   b) The degree of a power will be the degree of the base times the power. Therefore the degree of $(x^3 - 1)^5$ is $3 \cdot 5 = 15$.

94. $-3x^5 \cdot 3x^3 - x^6(2x)^2 + (3x^4)^2$
$\quad + (2x^2)^4 - 40x^2(x^3)^2$

$= -3x^5 \cdot 3x^3 - x^6 \cdot 2^2 \cdot x^2 + 3^2 x^{4 \cdot 2}$
$\quad + 2^4 x^{2 \cdot 4} - 40x^2 x^{3 \cdot 2}$

$= -3x^5 \cdot 3x^3 - x^6 \cdot 4 \cdot x^2 + 9x^8$
$\quad + 16x^8 - 40x^2 x^6$

$= -9x^8 - 4x^8 + 9x^8 + 16x^8 - 40x^8$

$= (-9 - 4 + 9 + 16 - 40)x^8$

$= -28x^8$

95. With the polynomial form –

$Ax^4 + Bx^3 + Cx + D.$ Using the given information we have the following equations

A is 2 times $B \rightarrow A = 2B$

C is A less $3 \rightarrow C = A - 3$

D is C less $7 \rightarrow D = C - 7$

$A + B + C + D = 15$ (Sum is 15).

We will solve using substitution.

$A + B + C + D = 15$

$2B + B + (A - 3) + (C - 7) = 15$

$3B + A + C - 10 = 15$

$3B + 2B + (A - 3) - 10 = 15$

$5B + A - 13 = 15$

$5B + 2B - 13 = 15$

$7B = 28$

$B = 4$

$A = 2B \rightarrow A = 2 \cdot 4 = 8$

$C = A - 3 \rightarrow C = 8 - 3 = 5$

$D = C - 7 \rightarrow D = 5 - 7 = -2$

Note: $8 + 4 + 5 - 2 = 15$

The polynomial is $8x^4 + 4x^3 + 5x - 2$

96. $[(x-5) - 4x^3][(x-5) + 4x^3]$

Note: This is of the form

$(A+B)(A-B) = A^2 - B^2$

$ = (x-5)^2 - (4x^3)^2$

$ = x^2 - 10x + 25 - 16x^6$

$ = -16x^6 + x^2 - 10x + 25$

97. $\quad (x-7)(x+10) = (x-4)(x-6)$

$x^2 + 10x - 7x - 70 = x^2 - 6x - 4x + 24$

$\quad x^2 + 3x - 70 = x^2 - 10x + 24$

$\quad\quad 3x - 70 = -10x + 24$

$\quad\quad 13x - 70 = 24$

$\quad\quad 13x = 94$

$\quad\quad x = \dfrac{94}{13}$

Chapter 5 Test

1. $t^2 \cdot t^5 \cdot t = t^{2+5+1} = t^8$

2. $(x^4)^9 = x^{4 \cdot 9} = x^{36}$

3. $\dfrac{3^5}{3^2} = 3^{5-2} = 3^3 = 27$

4. $\dfrac{(2x)^5}{(2x)^5} = 1$

5. $(5x^4 y)(-2x^5 y)^3 = 5x^4 y(-8x^{15} y^3)$

$ = -40x^{4+15} y^{1+3} = -40x^{19} y^4$

6. $\dfrac{-24a^7b^4}{8a^2b} = -3a^{7-2}b^{4-1} = -3a^5b^3$

7. $5^{-3} = \dfrac{1}{5^3}$, or $\left(\dfrac{1}{5}\right)^3$

8. $t^{-4} \cdot t^{-2} = t^{-4+(-2)} = t^{-6} = \dfrac{1}{t^6}$

9. $\dfrac{12x^3y^2}{15x^8y^{-3}} = \dfrac{3 \cdot 4y^2y^3}{3 \cdot 5x^{8-3}} = \dfrac{4y^5}{5x^5}$

10. $(2a^3b^{-1})^{-4} = 2^{-4}(a^3)^{-4}(b^{-1})^{-4}$
$$= 2^{-4} \cdot a^{-12} \cdot b^4$$
$$= \dfrac{1}{2^4} \cdot \dfrac{1}{a^{12}} \cdot b^4$$
$$= \dfrac{b^4}{16a^{12}}$$

11. $\left(\dfrac{ab}{c}\right)^{-3} = \left(\dfrac{c}{ab}\right)^3 = \dfrac{c^3}{a^3b^3}$

12. $3{,}060{,}000{,}000 = 3.06 \times 10^9$

13. $5 \times 10^{-4} = 0.0005$

14. $\dfrac{5.6 \times 10^6}{3.2 \times 10^{-11}} = \dfrac{5.6}{3.2} \times \dfrac{10^6}{10^{-11}}$
$$= 1.75 \times 10^{6-(-11)}$$
$$= 1.75 \times 10^{17}$$
$$\approx 1.8 \times 10^{17}$$

15. $(2.4 \times 10^5)(5.4 \times 10^{16}) = (2.4 \times 5.4)(10^5 \times 10^{16})$
$$= 12.96 \times 10^{5+16}$$
$$= 1.296 \times 10^1 \times 10^{21}$$
$$= 1.296 \times 10^{1+21}$$
$$\approx 1.3 \times 10^{22}$$

16. 12.4 billion is $12.4 \cdot 10^9 = 1.24 \cdot 10^{10}$. So:
$$1.24 \cdot 10^{10} \cdot 4\sec = 4.96 \cdot 10^{10} \sec$$
$$= 4.96 \cdot 10^{10} \, \cancel{\sec} \cdot \dfrac{1\,\cancel{\min}}{60\,\cancel{\sec}} \cdot \dfrac{1\,\text{hr}}{60\,\cancel{\min}}$$

$$= \dfrac{4.96 \cdot 10^{10}}{3600} \, \text{hr} = \dfrac{4.96 \cdot 10^{10}}{3.6 \cdot 10^3} \, \text{hr}$$
$$= \dfrac{4.96}{3.6} \cdot \dfrac{10^{10}}{10^3} \, \text{hr} = \dfrac{49.6}{36} \cdot 10^{10-3} \, \text{hr}$$
$$\approx 1.38 \cdot 10^7 \, \text{hr}$$

Approximately $1.38 \cdot 10^7$, or 13.8 million, hours are wasted, collectively each day.

17. $6t^2 - 9t$ has two terms. It is a binomial.

18. $\dfrac{1}{3}x^5 - x + 7$

The coefficients are $\frac{1}{3}$, -1, and 7.

19. $2t^3 - t + 7t^5 + 4$

Term	$2t^3$	$-t$	$7t^5$	4
Degree	3	1	5	0

The term of highest degree is $7t^5$. This is the leading term. Then the leading coefficient is 7, and the degree of the polynomial is 5.

20. $p(x) = x^2 + 5x - 1$
$$p(-2) = (-2)^2 + 5(-2) - 1$$
$$= 4 - 10 - 1$$
$$= -7$$

21. $4a^2 - 6 + a^2 = (4+1)a^2 - 6 = 5a^2 - 6$

22. $y^2 - 3y - y + \dfrac{3}{4}y^2 = \left(1 + \dfrac{3}{4}\right)y^2 + (-3-1)y$
$$= \left(\dfrac{4}{4} + \dfrac{3}{4}\right)y^2 + (-3-1)y$$
$$= \dfrac{7}{4}y^2 - 4y$$

23. $3 - x^2 + 2x^3 + 5x^2 - 6x - 2x + x^5$
$$= x^5 + 2x^3 + (-1+5)x^2 + (-6-2)x + 3$$
$$= x^5 + 2x^3 + 4x^2 - 8x + 3$$

24. Graph $f(x) = x^3 - x + 1$ in the standard viewing window.

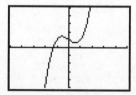

The range appears to be $(-\infty, \infty)$

25. $(3x^5 + 5x^3 - 5x^2 - 3) + (x^5 + x^4 - 3x^2 + 2x - 4)$

$= 3x^5 + 5x^3 - 5x^2 - 3 + x^5 + x^4 - 3x^2 + 2x - 4$

$= (3+1)x^5 + x^4 + 5x^3 - (5+3)x^2 + 2x - 7$

$= 4x^5 + x^4 + 5x^3 - 8x^2 + 2x - 7$

26. $(2x^4 + x^3 - 8x^2 - 6x - 3) - (6x^4 - 8x^2 + 2x)$

$= 2x^4 + x^3 - 8x^2 - 6x - 3 - 6x^4 + 8x^2 - 2x$

$= (2-6)x^4 + x^3 + (-8+8)x^2 + (-6-2)x - 3$

$= -4x^4 + x^3 - 8x - 3$

27. $(t^3 - 0.3t^2 - 20) - (t^4 - 1.5t^3 + 0.3t^2 - 11)$

$(t^3 - 0.3t^2 - 20) - (t^4 - 1.5t^3 + 0.3t^2 - 11)$

$= t^3 - 0.3t^2 - 20 - t^4 + 1.5t^3 - 0.3t^2 + 11$

$= -t^4 + (1+1.5)t^3 - (0.3+0.3)t^2 - 9$

$= -t^4 + 2.5t^3 - 0.6t^2 - 9$

28. $-3x^2(4x^2 - 3x - 5)$

$= -3x^2 \cdot 4x^2 - 3x^2(-3x) - 3x^2(-5)$

$= -12x^4 + 9x^3 + 15x^2$

29. $\left(x - \dfrac{1}{3}\right)^2 = x^2 - 2 \cdot x \cdot \dfrac{1}{3} + \dfrac{1}{3}^2$

$= x^2 - \dfrac{2}{3}x + \dfrac{1}{9}$

30. $(5t - 7)(5t + 7) = (5t)^2 - 7^2$

$= 25t^2 - 49$

31. $(3b + 5)(b - 3) = 3b^2 - 9b + 5b - 15$

$= 3b^2 - 4b - 15$

32. $(x^6 - 4)(x^8 + 4) = x^{14} + 4x^6 - 4x^8 - 16$

$= x^{14} - 4x^8 + 4x^6 - 16$

33. $(8 - y)(6 + 5y) = 48 + 40y - 6y - 5y^2$

$= 48 + 34y - 5y^2$

34. $(2x + 1)(3x^2 - 5x - 3)$

$3x^2 - 5x - 3$

$2x + 1$

$\overline{3x^2 - 5x - 3}$

$\underline{6x^3 - 10x^2 - 6x}$

$6x^3 - 7x^2 - 11x - 3$

35. $(8a + 3)^2 = (8a)^2 + 2(8a) \cdot 3 + 3^2$

$= 64a^2 + 48a + 9$

36. $x^3y - y^3 + xy^3 + 8 - 6x^3y - x^2y^2 + 11$

$= (1 - 6)x^3y - x^2y^2 + xy^3 - y^3 + (8 + 11)$

$= -5x^3y - x^2y^2 + xy^3 - y^3 + 19$

37. $(8a^2b^2 - ab + b^3) - (-6ab^2 - 7ab - ab^3 + 5b^3)$

$= 8a^2b^2 - ab + b^3 + 6ab^2 + 7ab + ab^3 - 5b^3$

$= 8a^2b^2 + (-1 + 7)ab + 6ab^2$

$ + ab^3 + (1 - 5)b^3$

$= 8a^2b^2 + 6ab + 6ab^2 + ab^3 - 4b^3$

38. $(3x^5 - 4y)(3x^5 + 4y) = (3x^5)^2 - (4y)^2$

$= 9x^{10} - 16y^2$

39. $(12x^4 + 9x^3 - 15x^2) \div (3x^2)$

$= \dfrac{12x^4}{3x^2} + \dfrac{9x^3}{3x^2} - \dfrac{15x^2}{3x^2}$

$= 4x^2 + 3x - 5$

40. $(6x^3 - 8x^2 - 14x + 13) \div (x + 2)$

$= (6x^3 - 8x^2 - 14x + 13) \div [x - (-2)]$

$\underline{-2}\,|\;\;6\;\;-8\;\;-14\;\;\;\;13$

$-12\;\;\;\;40\;\;-52$

$\overline{6\;\;-20\;\;\;\;26\;|\;-39}$

The answer is $6x^2 - 20x + 26,\ R - 39$ or

$6x^2 - 20x + 26 + \dfrac{-39}{x + 2}$

41. $(g \cdot h)(3)$

$g(3) = \dfrac{1}{3}$

$h(3) = 2(3) + 1 = 6 + 1 = 7$

$(g \cdot h)(3) = g(3) \cdot h(3) = \dfrac{1}{3} \cdot 7 = \dfrac{7}{3},\ \text{or } 2\dfrac{1}{3}$

42. $(g + h)(x) = g(x) + h(x)$

$= \dfrac{1}{x} + 2x + 1,\ \text{or } \dfrac{2x^2 + x + 1}{x}$

43. The domain of g/h is the set of real numbers that are common to the domains of h and g, minus any values where $h = 0$. The domain of g is: $\{x \mid x \text{ is a real number}, x \ne 0\}$. The domain of h is all real numbers. To find where h is zero, set the expression for h equal to zero and solve for x:

$$2x + 1 = 0$$
$$2x = -1$$
$$x = -\frac{1}{2}$$

The domain of $\left(\dfrac{g}{h}\right)(x)$ is therefore:

$$\left\{ x \mid x \text{ is a real number}, x \ne 0, -\frac{1}{2} \right\}$$

44. **Familiarize.** Let $l = $ the length of the box. Express the height and width in terms of l. $V = $ product of the dimensions.

    Translate.

height	is	length	less 1
↓	↓	↓	↓
h	$=$	l	-1

length	is	width	and 2 more
↓	↓	↓	↓
l	$=$	w	$+2$

    or $l - 2 = w$

    $V = l \cdot w \cdot h$

    $V = l \cdot (l-2) \cdot (l-1) = l \cdot [(l-2)(l-1)]$

    $\quad = l \cdot [l^2 - l - 2l + 2] = l \cdot [l^2 - 3l + 2]$

    $\quad = l^3 - 3l^2 + 2l$

    A polynomial for the volume is $l^3 - 3l^2 + 2l$

45. $2^{-1} - 4^{-1} = \dfrac{1}{2} - \dfrac{1}{4} = \dfrac{2}{4} - \dfrac{1}{4} = \dfrac{1}{4}$

Chapter 6
Polynomial Factorizations and Equations

Exercise Set 6.1

1. False

3. True

5. True

7. True

9. $x^2 + 6x + 9$ has no equals sign, so it is an expression.

11. $3x^2 = 3x$ has an equals sign, so it is an equation.

13. $2x^3 + x^2 = 0$ has an equals sign, so it is an equation.

15. From the graph we see that $f(x) = 0$ when $x = -3$ or $x = 5$. These are the solutions.

17. From the graph we see that $f(x) = 0$ when $x = -2$ or $x = 0$. These are the zeros of the function.

19. From the graph we see that $f(x) = 3$ when $x = -3$ or $x = 1$. These are the solutions.

21. From the graph we see that $f(x) = 0$ when $x = -4$ or $x = 2$. These are the solutions.

23. We can graph $y_1 = x^2$ and $y_2 = 5x$ and use the Intersect feature to find the first coordinates of the points of intersection, or we can begin by rewriting the equation so that one side is 0:
$$x^2 = 5x$$
$x^2 - 5x = 0$ Subtracting $5x$ on both sides

Then graph $y = x^2 - 5x$ and use the Zero feature to find the roots of the equation. In either case, we find that the solutions are 0 and 5.

25. We can graph $y_1 = 4x$ and $y_2 = x^2 + 3$ and use the Intersect feature to find the first coordinates of the points of intersection, or we can begin by rewriting the equation so that one side is 0:
$$4x = x^2 + 3$$
$0 = x^2 - 4x + 3$
Then graph $y = x^2 - 4x + 3$ and use the Zero feature to find the roots of the equation. In either case, we find that the solutions are 1 and 3.

27. We can graph $y_1 = x^2 + 150$ and $y_2 = 25x$ and use the Intersect feature to find the first coordinates of the points of intersection, or we can begin by rewriting the equation so that one side is 0:
$$x^2 + 150 = 25x$$
$x^2 - 25x + 150 = 0$
Then graph $y = x^2 - 25x + 150$ and use the Zero feature to find the roots of the equation. In either case, we find that the solutions are 10 and 15.

29. Graph $y = x^3 - 3x^2 + 2x$ and use the Zero feature to find the roots of the equation. The solutions are 0, 1, and 2.

31. Graph $y = x^3 - 3x^2 - 198x + 1080$ and use the Zero feature to find the roots of the equation. The solutions are -15, 6, and 12.

33. Graph $y = 21x^2 + 2x - 3$ and use the Zero feature to find the roots of the equation. The solutions are approximately -0.42857 and 0.33333.

35. Graph $y = x^2 - 4x - 45$ and use the Zero feature to find the zeros of the function. They are -5 and 9.

37. Graph $y = 2x^2 - 13x - 7$ and use the Zero feature to find the zeros of the function. They are -0.5 and 7.

39. Graph $y = x^3 - 2x^2 - 3x$ and use the Zero feature to find the zeros of the function. They are -1, 0, and 3.

41. We see that $2x - 1 = 0$ when $x = 0.5$ and $3x + 1 = 0$ when $x = -0.\overline{3}$, so Graph III corresponds to the given function.

43. We see that $4 - x = 0$ when $x = 4$ and $2x - 11 = 0$ when $x = 5.5$, so Graph I corresponds to the given function.

45. $2t^2 + 8t = 2t \cdot t + 2t \cdot 4$
$= 2t(t + 4)$

47. $9y^3 - y^2 = y^2 \cdot 9y - y^2 \cdot 1$
$= y^2(9y - 1)$

49. $15x^2 - 5x^4 + 5x = 5x \cdot 3x - 5x \cdot x^3 + 5x \cdot 1$
$= 5x(3x - x^3 + 1)$

51. $4x^2 y - 12xy^2 = 4xy \cdot x - 4xy \cdot 3y$
$= 4xy(x - 3y)$

53. $3y^2 - 3y - 9 = 3 \cdot y^2 - 3 \cdot y - 3 \cdot 3$
$= 3(y^2 - y - 3)$

55. $6ab - 4ad + 12ac = 2a \cdot 3b - 2a \cdot 2d + 2a \cdot 6c$
$= 2a(3b - 2d + 6c)$

57. $72x^3 - 36x^2 + 24x = 12x \cdot 6x^2$
$-12x \cdot 3x + 12x \cdot 2$
$= 12x(6x^2 - 3x + 2)$

59. $x^5 y^5 + x^4 y^3 + x^3 y^3 - xy^2$
$= xy^2 \cdot x^4 y^3 + xy^2 \cdot x^3 y$
$+ xy^2 \cdot x^2 y - xy^2 \cdot 1$
$= xy^2 \cdot (x^4 y^3 + x^3 y + x^2 y - 1)$

61. $9x^3 y^6 z^2 - 12x^4 y^4 z^4 + 15x^2 y^5 z^3$
$= 3x^2 y^4 z^2 \cdot 3xy^2 - 3x^2 y^4 z^2 \cdot 4x^2 z^2$
$+ 3x^2 y^4 z^2 \cdot 5yz$
$= 3x^2 y^4 z^2 (3xy^2 - 4x^2 z^2 + 5yz)$

63. $-5x + 35 = -5(x - 7)$

65. $-2x^2 + 4x - 12 = -2(x^2 - 2x + 6)$

67. $3y - 24x = -3(-y + 8x)$,   or $-3(8x - y)$

69. $-x^2 + 5x - 9 = -(x^2 - 5x + 9)$

71. $-a^4 + 2a^3 - 13a = -a(a^3 - 2a^2 + 13)$

73. $a(b - 5) + c(b - 5) = (b - 5)(a + c)$

75. $(x + 7)(x - 1) + (x + 7)(x - 2)$
$= (x + 7)(x - 1 + x - 2)$
$= (x + 7)(2x - 3)$

77. $a^2(x - y) + 5(y - x)$
$= a^2(x - y) + 5(-1)(x - y)$ Factoring out -1
$= a^2(x - y) - 5(x - y)$ to reverse the
$= (x - y)(a^2 - 5)$ second
                                   subtraction

79. $ac + ad + bc + bd = a(c + d) + b(c + d)$
$= (c + d)(a + b)$

81. $b^3 - b^2 + 2b - 2 = b^2(b - 1) + 2(b - 1)$
$= (b - 1)(b^2 + 2)$

83. $x^3 - x^2 - 2x + 5 = x^2(x - 1) - 1(2x - 5)$
There is no common factor to the polynomial, so the polynomial is not factorable by grouping.

85. $a^3 - 3a^2 + 6 - 2a$
$= a^2(a - 3) + 2(3 - a)$ Factoring out -1
$= a^2(a - 3) + 2(-1)(a - 3)$ to reverse the
$= a^2(a - 3) - 2(a - 3)$ second
$= (a - 3)(a^2 - 2)$ subtraction

87. $x^6 - x^5 - x^3 + x^4 = x^3(x^3 - x^2 - 1 + x)$
$= x^3[x^2(x - 1) - 1 + x]$
$= x^3[x^2(x - 1) + 1(x - 1)]$
$= x^3(x - 1)(x^2 + 1)$

89. $2y^4 + 6y^2 + 5y^2 + 15 = 2y^2(y^2 + 3)$
$$+ 5(y^2 + 3)$$
$$= (y^2 + 3)(2y^2 + 5)$$

91. a. $h(t) = -16t^2 + 72t$
$$h(t) = -8t(2t - 9)$$

b. Using $h(t) = -16t^2 + 72t$:
$$h(1) = -16 \cdot 1^2 + 72 \cdot 1 = -16 \cdot 1 + 72$$
$$= -16 + 72 = 56 \text{ ft}$$
Using $h(t) = -8t(2t - 9)$:
$$h(1) = -8(1)(2 \cdot 1 - 9) = -8(1)(-7) = 56 \text{ ft}$$
The expressions have the same value for $t = 1$, so the factorization is probably correct.

93. $R(n) = n^2 - n$
$$R(n) = n(n - 1)$$

95. $P(x) = x^2 - 3x$
$$P(x) = x(x - 3)$$

97. $R(x) = 280x - 0.4x^2$
$$R(x) = 0.4x(700 - x)$$

99. $N(x) = \frac{1}{6}x^3 + \frac{1}{2}x^2 + \frac{1}{3}x$
$$= \frac{1}{6} \cdot x^3 + \frac{1}{6} \cdot 3x^2 + \frac{1}{6} \cdot 2x$$
$$= \frac{1}{6}\left(x^3 + 3x^2 + 2x\right)$$

101. $H(n) = \frac{1}{2}n^2 - \frac{1}{2}n$
$$= \frac{1}{2}n(n - 1)$$

103. $(x + 3)(x - 4) = 0$
$$x + 3 = 0 \quad \text{or} \quad x - 4 = 0$$
$$x = -3 \quad \text{or} \quad x = 4$$
The solutions are –3 and 4.

105. $x(x + 1) = 0$
$$x = 0 \quad \text{or} \quad x + 1 = 0$$
$$x = 0 \quad \text{or} \quad x = -1$$
The solutions are 0 and –1.

107. $x^2 - 3x = 0$
$$x(x - 3) = 0 \qquad \text{Factoring}$$
$$x = 0 \quad \text{or} \quad x - 3 = 0$$
$$x = 0 \quad \text{or} \qquad x = 3$$
The solutions are 0 and 3.

109. To use the principle of *zero* products, we set the equation equal to *zero*.
$$-5x^2 = 15x$$
$$0 = 5x^2 + 15x$$
$$0 = 5x(x + 3)$$
$$5x = 0 \quad \text{or} \quad x + 3 = 0$$
$$x = 0 \quad \text{or} \qquad x = -3$$
The solutions are –3 and 0.

111. $12x^4 + 4x^3 = 0$
$$4x^3(3x + 1) = 0$$
$$4x^3 = 0 \quad \text{or} \quad 3x + 1 = 0$$
$$x^3 = 0 \quad \text{or} \qquad 3x = -1$$
$$x = 0 \quad \text{or} \qquad x = -\frac{1}{3}$$
The solutions are $-\frac{1}{3}$ and 0.

113. $f(x) = (x - 3)(x + 7)$
If $f(a) = 0$, then
$$0 = (a - 3)(a + 7)$$
$$a - 3 = 0 \quad \text{or} \quad a + 7 = 0$$
$$a = 3 \quad \text{or} \qquad a = -7$$
The solutions are 3 and –7.

115. $f(x) = 2x(5x + 9)$
If $f(a) = 0$, then
$$0 = 2a(5a + 9)$$
$$5a + 9 = 0 \quad \text{or} \quad 2a = 0$$
$$5a = -9$$
$$a = -\frac{9}{5} \quad \text{or} \qquad a = 0$$
The solutions are 0 and $-\frac{9}{5}$.

117. $f(x) = x^3 - 3x^2$

If $f(a) = 0$, then

$$0 = a^3 - 3a^2 = a^2(a - 3)$$

$$a^2 = 0 \quad \text{or} \quad a - 3 = 0$$

$$a = 0 \quad \text{or} \quad a = 3$$

The solutions are 0 and 3.

119. *Thinking and Writing Exercise.*

121. $(x + 2)(x + 7) = x \cdot x + x \cdot 7 + 2 \cdot x + 2 \cdot 7$

$$= x^2 + 7x + 2x + 14$$

$$= x^2 + 9x + 14$$

123. $(x + 2)(x - 7) = x \cdot x + x \cdot (-7) + 2 \cdot x + 2 \cdot (-7)$

$$= x^2 - 7x + 2x - 14$$

$$= x^2 - 5x - 14$$

125. $(a - 1)(a - 3) = a \cdot a + a \cdot (-3)$

$$+ (-1) \cdot a + (-1) \cdot (-3)$$

$$= a^2 - 3a - a + 3$$

$$= a^2 - 4a + 3$$

127. $(t - 5)(t + 10) = t \cdot t + t \cdot 10 + (-5) \cdot t + (-5) \cdot 10$

$$= t^2 + 10t - 5t - 50$$

$$= t^2 + 5t - 50$$

129. *Thinking and Writing Exercise.*

131. We use the principle of zero products in reverse. Since the zeros of $f(x) = x^2 + 2x - 8$ are -4 and 2, we have

$$x = -4 \quad \text{or} \quad x = 2$$

$$x + 4 = 0 \quad \text{or} \quad x - 2 = 0,$$

so $x^2 + 2x - 8 = (x + 4)(x - 2)$.

133. $x^5 y^4 + \underline{\quad} = x^4 y^4 \left(\underline{\quad} + y^2 \right)$

The term that goes in the first blank is the product of $x^4 y^4$ and y^2, or $x^4 y^6$.

The term that goes in the second blank is the expression that is multiplied with $x^4 y^4$ to obtain $x^5 y^4$, or x. Thus, we have

$$x^5 y^4 + x^4 y^6 = x^4 y^4 (x + y^2).$$

135. $x^{-6} + x^{-9} + x^{-3} = x^{-9}(x^3 + 1 + x^6)$

137. $x^{1/3} - 5x^{1/2} + 3x^{3/4}$

$$= x^{4/12} - 5x^{6/12} + 3x^{9/12}$$

$$= x^{4/12}(1 - 5x^{2/12} + 3x^{5/12})$$

$$= x^{1/3}(1 - 5x^{1/6} + 3x^{5/12})$$

139. $5x^5 - 5x^4 + x^3 - x^2 + 3x - 3$

$$= 5x^4(x - 1) + x^2(x - 1) + 3(x - 1)$$

$$= (5x^4 + x^2 + 3)(x - 1)$$

141. $2x^{3a} + 8x^a + 4x^{2a} = 2x^{a+2a} + 8x^a + 4x^{a+a}$

$$= 2x^a \cdot x^{2a} + 2x^a \cdot 4$$

$$+ 2x^a \cdot 2x^a$$

$$= 2x^a(x^{2a} + 4 + 2x^a)$$

143. The shaded area is equal to the difference between the area of the rectangle and the area of the circles and partial circles. Because the diameter $(2x)$ of each circle is equal to the width of the rectangle, and two diameters $(4x)$ are equal to the length of the rectangle, the area of the rectangle (A_r) is

$$A_r = l \cdot w = 4x \cdot 2x = 8x^2.$$

The area of the full circle and the two half-circles equals 2 times the area of one circle, or $A_c = 2(\pi r^2) = 2\pi x^2$.

The area of the shaded region is therefore

$$A_s = A_r - A_c = 8x^2 - 2\pi x^2$$

$$= 2x^2(4 - \pi)$$

Exercise Set 6.2

1. True

3. False

5. True

7. True

9. $x^2 + 8x + 12$

We look for two numbers whose product is 12 and whose sum is 8. Since 12 and 8 are both positive, we need only consider positive factors.

Pairs of Factors	Sum of Factors
1, 12	13
2, 6	8

The numbers we need are 2 and 6. The factorization is $(x+2)(x+6)$.

11. $t^2 + 8t + 15$

Since the constant term is positive and the coefficient of the middle term is also positive, we look for a factorization of 15 in which both factors are positive. Their sum must be 8.

Pair of Factors	Sum of Factors
1, 15	16
3, 5	8

The numbers we need are 3 and 5. The factorization is $(t+3)(t+5)$.

13. $a^2 - 7a + 12$

Since the constant term is positive and the coefficient of the middle term is negative, we look for two negative terms whose product is 12 and whose sum must be −7.

Pair of Factors	Sum of Factors
−12, −1	−13
−6, −2	−8
−4, −3	−7

The numbers we want are −4 and −3. The factorization is $(a-4)(a-3)$.

15. $x^2 - 2x - 15$

Since the constant term is negative, we look for factorization of −15 in which one factor is positive and one factor is negative. Their sum must be −2, so the negative factor must have the larger absolute value. Thus we consider only pairs of factors in which the negative factor has the larger absolute value.

Pair of Factors	Sum of Factors
1, −15	−14
3, −5	−2

The numbers we need are 3 and −5. The factorization is $(x+3)(x-5)$.

17. $x^2 + 2x - 15$

Since the constant term is negative, we look for factorization of −15 in which one factor is positive and one factor is negative. Their sum must be 2, so the positive factor must have the larger absolute value. Thus we consider only pairs of factors in which the positive factor has the larger absolute value.

Pair of Factors	Sum of Factors
−1, 15	14
−3, 5	2

The numbers we need are 5 and −3. The factorization is $(x+5)(x-3)$.

19. $2n^2 - 20n + 50$

$= 2(n^2 - 10n + 25)$ Removing the common factor

We now factor $n^2 - 10n + 25$. We look for two numbers whose product is 25 and whose sum is −10. Since the constant term is positive and the coefficient of the middle term is negative, we look for factorization of 25 in which both factors are negative.

Pair of Factors	Sum of Factors
−1, −25	−26
−5, −5	−10

The numbers we need are −5 and −5. $n^2 - 10n + 25 = (n-5)(n-5)$. We must not forget to include the common factor 2. $2n^2 - 20n + 50 = 2(n-5)(n-5)$, or $2(n-5)^2$

21. $a^3 - a^2 - 72a$

$= a\left(a^2 - a - 72\right)$ Removing the

common factor

We now factor $a^2 - a - 72.$ Since the constant term is negative, we look for a factorization of –72 in which one factor is positive and one factor is negative. We consider only pairs of factors in which the negative factor has the larger absolute value, since the sum of the factors, –1, is negative.

Pair of Factors	Sum of Factors
–72, 1	–71
–36, 2	–34
–18, 4	–14
–9, 8	–1

The numbers we need are –9 and 8.

$a^2 - a - 72 = \left(a - 9\right)\left(a + 8\right)$

We must not forget to include the common factor a.

$a^3 - a^2 - 72a = a\left(a - 9\right)\left(a + 8\right)$

23. $14x + x^2 + 45 = x^2 + 14x + 45$

Since the constant term and the middle term are both positive, we look for a factorization of 45 in which both factors are positive. Their sum must be 14.

Pair of Factors	Sum of Factors
45, 1	46
15, 3	18
9, 5	14

The numbers we need are 9 and 5. The factorization is $\left(x + 9\right)\left(x + 5\right).$

25. $3x + x^2 - 10 = x^2 + 3x - 10$

Since the constant term is negative, we look for a factorization of –10 in which one factor is positive and one factor is negative. We consider only pairs of factors in which the positive factor has the larger absolute value, since the sum of the factors, 3, is positive.

Pair of Factors	Sum of Factors
10, –1	9
5, –2	3

The numbers we need are 5 and –2. The factorization is $\left(x + 5\right)\left(x - 2\right).$

27. $3x^2 - 15x + 18$

$= 3\left(x^2 - 5x + 6\right)$ Removing the

common factor

We now factor $x^2 - 5x + 6.$ Since the constant term is positive and the coefficient of the middle term is negative, we look for two negative terms whose product is 6 and whose sum is –5.

Pair of Factors	Sum of Factors
–1, –6	–7
–2, –3	–5

The numbers we need are –2 and –3.

$x^2 - 5x + 6 = \left(x - 2\right)\left(x - 3\right)$

We must not forget to include the common factor 3.

$3x^2 - 15x + 18 = 3\left(x - 2\right)\left(x - 3\right)$

29. $56 + x - x^2 = -x^2 + x + 56 = -\left(x^2 - x - 56\right)$

We now factor $x^2 - x - 56.$ Since the constant term is negative, we look for a factorization of –56 in which one factor is positive and one factor is negative. We consider only pairs of factors in which the negative factor has the larger absolute value, since the sum of the factors, –1, is negative.

Pair of Factors	Sum of Factors
–56, 1	–55
–28, 2	–26
–14, 4	–10
–8, 7	–1

The numbers we need are –8 and 7. Thus, $x^2 - x - 56 = \left(x - 8\right)\left(x + 7\right).$ We must not forget to include the factor that was factored out earlier:

$56 + x - x^2 = -(x-8)(x+7)$, or

$(-x+8)(x+7)$, or $(8-x)(7+x)$.

31. $32y + 4y^2 - y^3$

There is a common factor, y. We also factor out -1 in order to make the leading coefficient positive.

$32y + 4y^2 - y^3 = -y(-32 - 4y + y^2)$

$= -y(y^2 - 4y - 32)$

Now we factor $y^2 - 4y - 32$. Since the constant term is negative, we look for factorization of -32 in which one factor is positive and one factor is negative. We consider only pairs of factors in which the negative factor has the larger absolute value, since the sum of the factors, -4, is negative.

Pair of Factors	Sum of Factors
$-32, 1$	-31
$-16, 2$	-14
$-8, 4$	-4

The numbers we need are -8 and 4. Thus, $y^2 - 4y - 32 = (y-8)(y+4)$. We must not forget to include the common factor:

$32y + 4y^2 - y^3 = -y(y-8)(y+4)$, or

$y(-y+8)(y+4)$, or $y(8-y)(4+y)$

33. $x^4 + 11x^3 - 80x^2$

$= x^2(x^2 + 11x - 80)$ Removing the
                                  common factor

We now factor $x^2 + 11x - 80$. We look for pairs of factors of -80, one positive and one negative, such that the positive factor has the larger absolute value and the sum of the factors is 11.

Pair of Factors	Sum of Factors
$80, -1$	79
$40, -2$	38
$20, -4$	16
$16, -5$	11
$10, -8$	2

The numbers we need are 16 and -5. Then $x^2 + 11x - 80 = (x+16)(x-5)$. We must not forget to include the common factor: $x^4 + 11x^3 - 80x^2 = x^2(x+16)(x-5)$.

35. $x^2 + 12x + 13$

There are no factors of 13 whose sum is 12. This trinomial is not factorable into binomials with integer coefficients. The polynomial is prime.

37. $p^2 - 5pq - 24q^2$

We look for numbers r and s such that $p^2 - 5pq - 24q^2 = (p + rq)(p + sq)$.

Our thinking is much the same as if we were factoring $p^2 - 5p - 24$. We look for factors of -24 whose sum is -5, one positive and one negative, such that the negative factor has the larger absolute value.

Pair of Factors	Sum of Factors
$-24, 1$	-23
$-12, 2$	-10
$-8, 3$	-5
$-6, 4$	-2

The numbers we need are -8 and 3. The factorization is $(p - 8q)(p + 3q)$.

39. $y^2 + 8yz + 16z^2$

We look for numbers p and q such that $y^2 + 8yz + 16z^2 = (y + pz)(y + qz)$. Our thinking is much the same as if we factor $y^2 + 8y + 16$. Since the constant term is positive and the coefficient of the middle term is positive, we look for a factorization of 16 in which both factors are positive. Their sum must be 8.

Pair of Factors	Sum of Factors
$1, 16$	17
$2, 8$	10
$4, 4$	8

The numbers we need are 4 and 4. The factorization is

$(y + 4z)(y + 4z)$ or $(y + 4z)^2$.

41. $p^4 - 80p^3 + 79p^2$

$= p^2\left(p^2 - 80p + 79\right)$ Factor out p^2

We now factor $p^2 - 80p + 79$. We look for a pair of factors of 79 whose sum is -80. The only negative pair of factors is -1 and -79. These are the numbers we need, so

$p^2 - 80p + 79 = (p - 1)(p - 79)$.

We must not forget to include the common factor:

$p^4 - 80p^3 + 79p^2 = p^2(p - 1)(p - 79)$.

43. $x^2 + 8x + 12 = 0$

$(x + 2)(x + 6) = 0$ From Exercise 9

$x + 2 = 0$ or $x + 6 = 0$ Using the principle of zero products:

$x = -2$ or $x = -6$

The solutions are -2 and -6.

45. $2n^2 + 50 = 20n$

$2n^2 - 20n + 50 = 0$ Subtracting $20n$
                          from both sides

$2(n - 5)(n - 5) = 0$ From Exercise 15

$n - 5 = 0$ or $n - 5 = 0$

Using the principle of zero products:

$n = 5$ or $n = 5$

The solution is 5.

47. The x-intercepts are $(-5, 0)$ and $(1, 0)$, so the solutions are -5 and 1.

Check: For -5:

$$\frac{x^2 + 4x - 5 = 0}{(-5)^2 + 4(-5) - 5 \;\Big|\; 0}$$
$$25 - 20 - 5$$
$$0 \;\Big|\; 0 \quad \text{TRUE}$$

For 1:

$$\frac{x^2 + 4x - 5 = 0}{1^2 + 4 \cdot 1 - 5 \;\Big|\; 0}$$
$$1 + 4 - 5$$
$$0 \;\Big|\; 0 \quad \text{TRUE}$$

Both numbers check, so they are the solutions.

49. The intercepts are $(-3, 0)$ and $(2, 0)$, so the solutions are -3 and 2.

Check: For -3:

$$\frac{x^2 + x - 6 = 0}{(-3)^2 + (-3) - 6 \;\Big|\; 0}$$
$$9 - 3 - 6$$
$$0 \;\Big|\; 0 \quad \text{TRUE}$$

For 2:

$$\frac{x^2 + x - 6 = 0}{2^2 + 2 - 6 \;\Big|\; 0}$$
$$4 + 2 - 6$$
$$0 \;\Big|\; 0 \quad \text{TRUE}$$

Both numbers check, so they are the solutions.

51. The zeros of $f(x) = x^2 - 4x - 45$ are the solutions of the equation $x^2 - 4x - 45 = 0$. We factor and use the principle of zero products.

$x^2 - 4x - 45 = 0$

$(x - 9)(x + 5) = 0$

$x - 9 = 0$ or $x + 5 = 0$

$x = 9$ or $x = -5$

The zeros are 9 and -5.

53. The zeros of $r(x) = x^3 + 4x^2 + 3x$ are the solutions of the equation $x^3 + 4x^2 + 3x = 0$. We factor and use the principle of zero products.

$x^3 + 4x^2 + 3x = 0$

$x\left(x^2 + 4x + 3\right) = 0$

$x(x + 1)(x + 3) = 0$

$x = 0$ or $x + 1 = 0$ or $x + 3 = 0$

$x = 0$ or $x = -1$ or $x = -3$

The zeros are 0, -1, and -3.

55. $x^2 + 4x = 45$

$x^2 + 4x - 45 = 0$

$(x + 9)(x - 5) = 0$

$x + 9 = 0$ or $x - 5 = 0$

$x = -9$ or $x = 5$

The solutions are -9 and 5.

57. $x^2 - 9x = 0$

$x(x-9) = 0$

$x = 0 \ or \ x-9 = 0$

$x = 0 \ or \quad x = 9$

The solutions are 0 and 9.

59. $\qquad a^3 + 40a = 13a^2$

$a^3 - 13a^2 + 40a = 0$

$a(a^2 - 13a + 40) = 0$

$a(a-5)(a-8) = 0$

$a = 0 \ or \ a-5 = 0 \ or \ a-8 = 0$

$a = 0 \ or \quad a = 5 \ or \quad a = 8$

The solutions are 0, 5, and 8.

61. $(x-3)(x+2) = 14$

$x^2 - x - 6 = 14$

$x^2 - x - 20 = 0$

$(x-5)(x+4) = 0$

$x-5 = 0 \ or \ x+4 = 0$

$x = 5 \ or \quad x = -4$

The solutions are 5 and –4.

63. $\qquad 35 - x^2 = 2x$

$35 - 2x - x^2 = 0$

$(7+x)(5-x) = 0$

$7+x = 0 \quad or \ 5-x = 0$

$x = -7 \ or \quad 5 = x$

The solutions are –7 and 5.

65. From the graph we see that the zeros of $f(x) = x^2 + 10x - 264$ are –22 and 12. We also know that –22 is a zero of $g(x) = x + 22$ and 12 is a zero of $h(x) = x - 12$. Using the principle of zero products in reverse, we have $x^2 + 10x - 264 = (x+22)(x-12)$.

67. Graph $y = x^2 + 40x + 384$ and find the zeros. They are –24 and –16. We know that –24 is a zero of $g(x) = x + 24$ and –16 is a zero of $h(x) = x + 16$. Using the principle of zero products in reverse, we have $x^2 + 40x + 384 = (x+24)(x+16)$.

69. Graph $y = x^2 + 26x - 2432$ and find the zeros. They are –64 and 38. We know that –64 is a zero of $g(x) = x + 64$ and 38 is a zero of $h(x) = x - 38$. Using the principle of zero products in reverse, we have $x^2 + 26x - 2432 = (x+64)(x-38)$.

71. We write a linear function for each zero: –1 is a zero of $g(x) = x + 1$; 2 is a zero of $h(x) = x - 2$.

Then $f(x) = (x+1)(x-2)$,

or $f(x) = x^2 - x - 2$.

73. We write a linear function for each zero: –7 is a zero of $g(x) = x + 7$; –10 is a zero of $h(x) = x + 10$.

Then $f(x) = (x+7)(x+10)$, or

$f(x) = x^2 + 17x + 70$.

75. We write a linear function for each zero:

0 is a zero of $g(x) = x$;

1 is a zero of $h(x) = x - 1$;

2 is a zero of $k(x) = x - 2$.

Then $f(x) = x(x-1)(x-2)$, or

$f(x) = x^3 - 3x^2 + 2x$.

77. *Thinking and Writing Exercise.*

79. $(2x+3)(3x+4) = 2x \cdot 3x + 2x \cdot 4 + 3 \cdot 3x + 3 \cdot 4$

$= 6x^2 + 8x + 9x + 12$

$= 6x^2 + 17x + 12$

81. $(2x-3)(3x+4) = 2x \cdot 3x + 2x \cdot 4$

$+ (-3) \cdot 3x + (-3) \cdot 4$

$= 6x^2 + 8x - 9x - 12$

$= 6x^2 - x - 12$

83. $(5x-1)(x-7) = 5x \cdot x + 5x \cdot (-7) + (-1) \cdot x$

$+ (-1) \cdot (-7)$

$= 5x^2 - 35x - x + 7$

$= 5x^2 - 36x + 7$

85. *Thinking and Writing Exercise.*

87. The x-coordinates of the x-intercepts are -1 and 3. These are the solutions of $x^2 - 2x - 3 = 0$. From the graph we see that the x-values for which $f(x) < 5$ are in the interval $(-2, 4)$. We could also express the solution set as $\{x \mid -2 < x < 4\}$.

89. Answers may vary. A polynomial function of lowest degree that meets the given criteria is of the form $f(x) = ax^3 + bx^2 + cx + d$.
Substituting we have

$$a \cdot 2^3 + b \cdot 2^2 + c \cdot 2 + d = 0,$$
$$a(-1)^3 + b(-1)^2 + c(-1) + d = 0,$$
$$a \cdot 3^3 + b \cdot 3^2 + c \cdot 3 + d = 0,$$
$$a \cdot 0^3 + b \cdot 0^2 + c \cdot 0 + d = 30, \text{ or}$$
$$8a + 4b + 2c + d = 0,$$
$$-a + b - c + d = 0,$$
$$27a + 9b + 3c + d = 0,$$
$$d = 30$$

Solving the system of equations, we get $(5, -20, 5, 30)$, so the corresponding function is $f(x) = 5x^3 - 20x^2 + 5x + 30$.

91. Graph $y_1 = -x^2 + 13.80x$ and $y_2 = 47.61$ and use the Intersect feature to find the first coordinate of the point of intersection. The solution is 6.90.

93. Graph $y_1 = x^3 - 3.48x^2 + x$ and $y_2 = 3.48$ and use the Intersect feature to find the first coordinate of the point of intersection. The solution is 3.48.

95. $x^2 + \frac{1}{2}x - \frac{3}{16}$
We look for factors of $-\frac{3}{16}$ whose sum is $\frac{1}{2}$.
The factors are $\frac{3}{4}$ and $-\frac{1}{4}$. The factorization is $x^2 + \frac{1}{2}x - \frac{3}{16} = \left(x + \frac{3}{4}\right)\left(x - \frac{1}{4}\right)$.

97. $x^{2a} + 5x^a - 24$
Substitute u for x^a $\left(\text{and } u^2 \text{ for } x^{2a}\right)$. We factor $u^2 + 5u - 24$. We look for factors of -24 whose sum is 5. The factors are 8 and $-$

3. We have $u^2 + 5u - 24 = (u + 8)(u - 3)$.
Replace u by x^a: $x^{2a} + 5x^a - 24$
$= (x^a + 8)(x^a - 3)$.

99. $(a+1)x^2 + (a+1)3x + (a+1)2$
$= (a+1)(x^2 + 3x + 2)$
We factor $x^2 + 3x + 2$ by looking for factors of 2 whose sum is 3. These factors are 2 and 1, so we have $x^2 + 3x + 2 = (x+2)(x+1)$.
$(a+1)x^2 + (a+1)3x + (a+1)2$
$= (a+1)(x+2)(x+1)$.

101. $(x+3)^2 - 2(x+3) - 35$
We want two factors of -35 whose sum is -2. They are -7 and 5.
$\left[(x+3) - 7\right]\left[(x+3) + 5\right] = (x-4)(x+8)$

103. $x^2 + qx - 32$
All such q are the sums of the factors of -32.

Pair of Factors	Sum of Factors
$32, -1$	31
$-32, 1$	-31
$16, -2$	14
$-16, 2$	-14
$8, -4$	4
$-8, 4$	-4

q can be 31, -31, 14, -14, 4, or -4.

105. The area can be described as the sum of the individual areas:
$x^2 + 5x + 4x + 20 = x^2 + 9x + 20$.
We factor $x^2 + 9x + 20$ by looking for factors of 20 whose sum is 9. These factors are 4 and 5, so we have
$x^2 + 9x + 20 = (x+4)(x+5)$.

Exercise Set 6.3

1. f

3. e

5. g

7. h

9. $2x^2 + 7x - 4$
We will use the FOIL method.
1. There is no common factor (other than 1 or −1).
2. Factor the first term, $2x^2$. The factors are $2x, x$. The only possibility is
$$(2x+\ )(x+\ ).$$
3. Factor the last term, −4. The possibilities are $4(-1)$, $-1 \cdot 4$, $-4 \cdot 1$, $1(-4)$, $-2 \cdot 2$, and $2(-2)$.
4. We need factors for which the sum of the products (the "outer" and "inner" parts of FOIL) is the middle term, $7x$. Try some possibilities and check by multiplying.
$$(2x+4)(x-1) = 2x^2 + 2x - 4$$
We try again.
$$(2x-1)(x+4) = 2x^2 + 7x - 4$$
The factorization is $(2x-1)(x+4)$.

11. $3x^2 - 17x - 6$
We will use the FOIL method.
1. There is no common factor (other than 1 or −1).
2. Factor the first term, $3x^2$. The only factors are $3x, x$. The only possibility is:
$$(3x+\ )(x+\ ).$$
3. Factor the last term, −6. The possibilities are $6(-1)$, $-1 \cdot 6$, $-6 \cdot 1$, $1(-6)$, $-2 \cdot 3$, $3(-2)$, $-3 \cdot 2$, and $2(-3)$,
4. We need factors for which the sum of the products (the "outer" and "inner" parts of FOIL) is the middle term, $-17x$. Try various possibilities, and check by multiplying.
The factorization is $(3x+1)(x-6)$.

13. $15a^2 - 14a + 3$
We will use the FOIL method.
1. There is no common factor (other than 1 or −1).
2. Factor the first term, $15a^2$. The factors are $15a, a$ and $5a, 3a$. We have these

possibilities:
$$(15a+\ )(a+\ ), (5a+\ )(3a+\ ).$$
3. Factor the last term, 3. The possibilities are $1 \cdot 3$, $3 \cdot 1$, $(-1)(-3)$, and $(-3)(-1)$.
4. Look for factors such that the sum of the products is the middle term, $-14a$. Trial and error leads us to the correct factorization:
$$15a^2 - 14a + 3 = (5a-3)(3a-1)$$

15. $6t^2 + 17t + 7$
We will use the FOIL method.
1. There is no common factor (other than 1 or −1).
2. Factor the first term, $6t^2$. The factors are $6t, t$ and $3t, 2t$. We have these possibilities:
$$(6t+\ )(t+\ ), (3t+\ )(2t+\ ).$$
3. Factor the last term, 7. The possibilities are $1 \cdot 7$, $7 \cdot 1$, $(-1)(-7)$, and $(-7)(-1)$.
4. Look for factors such that the sum of the products is the middle term, $17t$. Trial and error leads us to the correct factorization:
$$6t^2 + 17t + 7 = (3t+7)(2t+1)$$

17. $6x^2 - 10x - 4$
We will use the FOIL method.
1. Factor out the common factor, 2:
$$2(3x^2 - 5x - 2)$$
2. Now we factor the trinomial $3x^2 - 5x - 2$. Factor the first term, $3x^2$. The factors are $3x$ and x. We have this possibility: $(3x+\ )(x+\ )$.
3. Factor the last term, −2. The possibilities are $2(-1)$, $-1 \cdot 2$, $-2 \cdot 1$, and $1(-2)$.
4. Look for factors such that the sum of the products is the middle term, $-10x$. Trial and error leads us to the correct factorization:
$$3x^2 - 5x - 2 = (3x+1)(x-2)$$
We must include the common factor to get a factorization of the original trinomial:
$$6x^2 - 10x - 4 = 2(3x+1)(x-2)$$

19. $8x^2 - 16 - 28x = 8x^2 - 28x - 16$

We will use the grouping method.

1. Factor out the common factor, 4:
$$4\left(2x^2 - 7x - 4\right)$$

2. Now we factor the trinomial $2x^2 - 7x - 4$. Multiply the leading coefficient, 2, and the constant, –4:
$$2(-4) = -8$$

3. Factor –8 so the sum of the factors is –7. We need only consider pairs of factors in which the negative factor has the larger absolute value, since their sum is negative.

Pair of Factors	Sum of Factors
–4, 2	–2
–8, 1	–7

4. Split $-7x$ using the results of step (3):
$$-7x = -8x + x$$

5. Factor by grouping:
$$2x^2 - 7x - 4 = 2x^2 - 8x + x - 4$$
$$= 2x(x - 4) + 1(x - 4)$$
$$= (x - 4)(2x + 1)$$

We must include the common factor to get a factorization of the original trinomial:
$$8x^2 - 16 - 28x = 4(x - 4)(2x + 1)$$

21. $14x^4 - 19x^3 - 3x^2$

We will use the grouping method.

1. Factor out the common factor, x^2.
$$x^2\left(14x^2 - 19x - 3\right)$$

2. Now we factor the trinomial $14x^2 - 19x - 3$. Multiply the leading coefficient, 14, and the constant, –3:
$$14(-3) = -42$$

3. Factor –42 so the sum of the factors is –19. We need only consider pairs of factors in which the negative factor has the larger absolute value, since the sum is negative.

Pair of Factors	Sum of Factors
–42, 1	–41
–21, 2	–19
–14, 3	–11
–7, 6	–1

4. Split $-19x$ using the results of step (3):
$$-19x = -21x + 2x$$

5. Factor by grouping:
$$14x^2 - 19x - 3 = 14x^2 - 21x + 2x - 3$$
$$= 7x(2x - 3) + 2x - 3$$
$$= (2x - 3)(7x + 1)$$

We must include the common factor to get a factorization of the original trinomial:
$$14x^4 - 19x^3 - 3x^2 = x^2(2x - 3)(7x + 1)$$

23. $10 - 23x + 12x^2 = 12x^2 - 23x + 10$

We will use the grouping method.

1. There is no common factor (other than 1 or –1).

2. We now factor the trinomial $12x^2 - 23x + 10$. Multiply the leading coefficient and the constant $12 \cdot 10 = 120$.

3. We want factors of 120 whose sum equals –23. Since $-15(-8) = 120$ and $-15 - 8 = -23$, we will split $-23x$ into $-15x$ and $-8x$.

4. Factor by grouping:
$$12x^2 - 23x + 10 = 12x^2 - 8x - 15x + 10$$
$$= 4x(3x - 2) - 5(3x - 2)$$
$$= (4x - 5)(3x - 2)$$

25. $9x^2 + 15x + 4$

We will use the grouping method.

1. There is no common factor (other than 1 or –1).

2. Multiply the leading coefficient and constant: $9(4) = 36$

3. Factor 36 so the sum of the factors is 15. We need only consider pairs of positive factors since 36 and 15 are both positive.

Pair of Factors	Sum of Factors
36, 1	37
18, 2	20
12, 3	15
9, 4	13
6, 6	12

4. Split $15x$ using the results of step (3):
$$15x = 12x + 3x$$

5. Factor by grouping:
$$9x^2 + 15x + 4 = 9x^2 + 12x + 3x + 4$$
$$= 3x(3x + 4) + 3x + 4$$
$$= (3x + 4)(3x + 1)$$

27. $4x^2 + 15x + 9$
We will use the FOIL method.
1. There is no common factor other than 1 or -1)
2. Factor the first term, $4x^2$. The possibilities are $(4x + \quad)(x + \quad)$ and $(2x + \quad)(2x + \quad)$.
3. Factor the last term, 9. We consider only positive factors since the middle term and the last term are positive. The possibilities are $9 \cdot 1$ and $3 \cdot 3$.
4. We need factors for which the sum of products is the middle term, $15x$. Trial and error leads us to the correct factorization: $(4x + 3)(x + 3)$

29. $4 + 6t^2 - 13t = 6t^2 - 13t + 4$
We will use the FOIL method.
1. There is no common factor (other than 1 or -1).
2. Factor the first term, $6t^2$, to get the possibilities $(6t + \quad)(t + \quad)$ and $(3t + \quad)(2t + \quad)$.
3. Factor 4. The possibilities are $1 \cdot 4$, $4 \cdot 1$, $(-4)(-1)$, $(-1)(-4)$, $2 \cdot 2$ and $(-2)(-2)$.
4. Look for factors such that the sum of the products is the middle term, $-13t$. Trial and error indicates that no possible combination of factors will produce the

polynomial $6t^2 - 13t + 4$. Therefore, the polynomial $4 + 6t^2 - 13t$ is prime.

31. $-8t^2 - 8t + 30$
We will use the grouping method.
1. Factor out -2: $-2(4t^2 + 4t - 15)$
2. Now we factor the trinomial $4t^2 + 4t - 15$. Multiply the leading coefficient and the constant:
$$4(-15) = -60$$
3. Factor -60 so the sum of the factors is 4. The desired factorization is $10(-6)$.
4. Split $4t$ using the results of step (3):
$$4t = 10t - 6t$$
5. Factor by grouping:
$$4t^2 + 4t - 15 = 4t^2 + 10t - 6t - 15$$
$$= 2t(2t + 5) - 3(2t + 5)$$
$$= (2t + 5)(2t - 3)$$

We must include the common factor to get a factorization of the original trinomial:
$$-8t^2 - 8t + 30 = -2(2t + 5)(2t - 3)$$

33. $8 - 6z - 9z^2$
We will use the FOIL method.
1. There is no common factor (other than 1 or -1).
2. Factor the first term, 8. The possibilities are $(8 + \quad)(1 + \quad)$ and $(4 + \quad)(2 + \quad)$.
3. Factor the last term, $-9z^2$. The possibilities are $-9z \cdot z, -3z \cdot 3z$, and $9z(-z)$.
4. We need factors for which the sum of products is the middle term, $-6z$. Trial and error leads us to the correct factorization: $(4 + 3z)(2 - 3z)$

35. $18xy^3 + 3xy^2 - 10xy$
We will use the FOIL method.
1. Factor out the common factor, xy.
$$xy(18y^2 + 3y - 10)$$
2. We now factor the trinomial $18y^2 + 3y - 10$. Factor the first term, $18y^2$. The possibilities are

$(18y+\ )(y+\ ), (9y+\ )(2y+\ )$, and

$(6y+\ )(3y+\ )$.

3. Factor the last term, -10. The possibilities are $-10 \cdot 1, -5 \cdot 2, 10(-1)$ and $5(-2)$.

4. We need factors for which the sum of the products is the middle term, $3y$. Trial and error leads us to the correct factorization.

$18y^2 + 3y - 10 = (6y+5)(3y-2)$

We must include the common factor to get a factorization of the original trinomial:

$18xy^3 + 3xy^2 - 10xy$

$\quad = xy(6y+5)(3y-2)$

37. $24x^2 - 2 - 47x = 24x^2 - 47x - 2$

We will use the grouping method.

1. There is no common factor (other than 1 or -1).

2. Multiply the leading coefficient and the constant: $24(-2) = -48$

3. Factor -48 so the sum of the factors is -47. The desired factorization is $-48 \cdot 1$.

4. Split $-47x$ using the results of step (3): $-47x = -48x + x$

5. Factor by grouping:

$24x^2 - 47x - 2 = 24x^2 - 48x + x - 2$

$\quad = 24x(x-2)+(x-2)$

$\quad = (x-2)(24x+1)$

39. $63x^3 + 111x^2 + 36x$

We will use the FOIL method.

1. Factor out the common factor, $3x$.

$3x(21x^2 + 37x + 12)$

2. Now we will factor the trinomial $21x^2 + 37x + 12$. Factor the first term, $21x^2$. The factors are $21x, x$ and $7x, 3x$. We have these possibilities:

$(21x+\ )(x+\ )$ and $(7x+\ )(3x+\ )$.

3. Factor the last term, 12. The possibilities are $12 \cdot 1, (-12)(-1), 6 \cdot 2, (-6)(-2), 4 \cdot 3,$

and $(-4)(-3)$ as well as $1 \cdot 12,$

$(-1)(-12),\ 2 \cdot 6,\ (-2)(-6),\ 3 \cdot 4,$ and

$(-3)(-4)$.

4. Look for factors such that the sum of the products is the middle term, $37x$. Trial and error leads us to the correct factorization: $(7x+3)(3x+4)$

We must include the common factor to get a factorization of the original trinomial:

$63x^3 + 111x^2 + 36x = 3x(7x+3)(3x+4)$

41. $48x^4 + 4x^3 - 30x^2$

We will use the grouping method.

1. We factor out the common factor, $2x^2$.

$2x^2(24x^2 + 2x - 15)$

2. We now factor $24x^2 + 2x - 15$. Multiply the leading coefficient and the constant: $24(-15) = -360$

3. Factor -360 so the sum of the factors is 2. The desired factorization is $-18 \cdot 20$.

4. Split $2x$ using the results of step (3): $2x = -18x + 20x$

5. Factor by grouping:

$24x^2 + 2x - 15 = 24x^2 - 18x + 20x - 15$

$\quad = 6x(4x-3)+5(4x-3)$

$\quad = (4x-3)(6x+5)$

We must not forget to include the common factor:

$48x^4 + 4x^3 - 30x^2 = 2x^2(4x-3)(6x+5)$

43. $12a^2 - 17ab + 6b^2$

We will use the FOIL method. (Our thinking is much the same as if we were factoring $12a^2 - 17a + 6$.)

1. There is no common factor (other than 1 or -1).

2. Factor the first term, $12a^2$. The factors are $12a, a$ and $6a, 2a$ and $4a, 3a$. We have these possibilities: $(12a+\ )(a+\ )$ and $(6a+\ )(2a+\ )$ and $(4a+\ )(3a+\ )$.

3. Factor the last term, $6b^2$. The possibilities are $6b \cdot b, (-6b)(-b), 3b \cdot 2b,$ and $(-3b)(-2b)$ as well as $b \cdot 6b, (-b)(-6b), 2b \cdot 3b,$ and $(-2b)(-3b)$.

4. Look for factors such that the sum of the products is the middle term, $-17ab$. Trial and error leads us to the correct factorization: $(4a-3b)(3a-2b)$

45. $2x^2 + xy - 6y^2$

We will use the grouping method.
1. There is no common factor (other than 1 or -1).
2. Multiply the coefficients of the first and last terms: $2(-6) = -12$
3. Factor -12 so the sum of the factors is 1. The desired factorization is $4(-3)$.
4. Split xy using the results of step (3): $xy = 4xy - 3xy$
5. Factor by grouping:
$$2x^2 + xy - 6y^2$$
$$= 2x^2 + 4xy - 3xy - 6y^2$$
$$= 2x(x+2y) - 3y(x+2y)$$
$$= (x+2y)(2x-3y)$$

47. $8s^2 + 22st + 14t^2$

We will use the FOIL method.
1. We factor out the common factor, 2.
$$2(4s^2 + 11st + 7t^2)$$
2. Now we will factor the trinomial $4s^2 + 11st + 7t^2$. Factor the first term, $4s^2$. The factors are $4s, s$ and $2s, 2s$. We have these possibilities:
$(4s+\quad)(s+\quad)$ and $(2s+\quad)(2s+\quad)$.
3. Factor the last term, $7t^2$. the possibilities are $7t \cdot t$, $(-7t)(-t)$, $t \cdot 7t$, and $(-t)(-7t)$.
4. Look for factors such that the sum of the products is the middle term, $11st$. Trial and error leads us to the correct factorization: $(s+t)(4s+7t)$
We must include the common factor to get a factorization of the original trinomial:
$$8s^2 + 22st + 14t^2 = 2(s+t)(4s+7t)$$

49. $9x^2 - 30xy + 25y^2$

We will use the grouping method.
1. There is no common factor (other than 1 or -1).
2. Multiply the coefficient of the first and last terms: $9(25) = 225$
3. Factor 225 so the sum of the factors is -30. The desired factorization is $-15(-15)$.
4. Split $-30xy$ using the results of step (3): $-30xy = -15xy - 15xy$
5. Factor by grouping:
$$9x^2 - 30xy + 25y^2$$
$$= 9x^2 - 15xy - 15xy + 25y^2$$
$$= 3x(3x - 5y) - 5y(3x - 5y)$$
$$= (3x - 5y)(3x - 5y) \text{ or } (3x - 5y)^2$$

51. $9x^2y^2 + 5xy - 4$

Let $u = xy$ and $u^2 = x^2y^2$. Factor $9u^2 + 5u - 4$. We will use the FOIL method.
1. There is no common factor (other than 1 or -1).
2. Factor the first term, $9u^2$. The factors are $9u, u$ and $3u, 3u$. We have these possibilities: $(9u+\quad)(u+\quad)$ and $(3u+\quad)(3u+\quad)$.
3. Factor the last term, -4. The possibilities are: $-4 \cdot 1, -2 \cdot 2, 2 \cdot -2$ and $-1 \cdot 4$.
4. We need factors for which the sum of the products is the middle term, $5u$. Trial and error leads us to the factorization: $(9u - 4)(u + 1)$. Replace u by xy. We have $9x^2y^2 + 5xy - 4 = (9xy - 4)(xy + 1)$.

53. $9z^2 + 6z = 8$
$$0 = 8 - 6z - 9z^2$$
$$0 = (4 + 3z)(2 - 3z) \quad \text{From Ex. 33}$$
$$4 + 3z = 0 \text{ or } 2 - 3z = 0 \quad \text{Using the principle of zero products}$$
$$3z = -4 \quad or \quad 2 = 3z$$
$$z = -\frac{4}{3} \quad or \quad \frac{2}{3} = z$$

The solutions are $-\frac{4}{3}$ and $\frac{2}{3}$.

55. $63x^3 + 111x^2 + 36x = 0$

$3x(7x+3)(3x+4) = 0$ From Ex. 39

$3x = 0$ or $7x + 3 = 0$ or $3x + 4 = 0$

$x = 0$ or $7x = -3$ or $3x = -4$

$x = 0$ or $x = -\dfrac{3}{7}$ or $x = -\dfrac{4}{3}$

The solutions are $0, -\dfrac{3}{7}, -\dfrac{4}{3}$.

57. $3x^2 - 8x + 4 = 0$

$(3x - 2)(x - 2) = 0$ Factoring

$3x - 2 = 0$ or $x - 2 = 0$

$3x = 2$ or $x = 2$

$x = \dfrac{2}{3}$ or $x = 2$

The solutions are $\dfrac{2}{3}$ and 2.

59. $4t^3 + 11t^2 + 6t = 0$

$t(4t^2 + 11t + 6) = 0$

$t(4t + 3)(t + 2) = 0$

$t = 0$ or $4t + 3 = 0$ or $t + 2 = 0$

$t = 0$ or $4t = -3$ or $t = -2$

$t = 0$ or $t = -\dfrac{3}{4}$ or $t = -2$

The solutions are 0, $-\dfrac{3}{4}$, and -2.

61. $6x^2 = 13x + 5$

$6x^2 - 13x - 5 = 0$

$(2x - 5)(3x + 1) = 0$

$2x - 5 = 0$ or $3x + 1 = 0$

$2x = 5$ or $3x = -1$

$x = \dfrac{5}{2}$ or $x = -\dfrac{1}{3}$

The solutions are $\dfrac{5}{2}$ and $-\dfrac{1}{3}$.

63. $x(5 + 12x) = 28$

$5x + 12x^2 = 28$

$5x + 12x^2 - 28 = 0$

$12x^2 + 5x - 28 = 0$ Rearranging

$(4x + 7)(3x - 4) = 0$

$4x + 7 = 0$ or $3x - 4 = 0$

$4x = -7$ or $3x = 4$

$x = -\dfrac{7}{4}$ or $x = \dfrac{4}{3}$

The solutions are $-\dfrac{7}{4}$ and $\dfrac{4}{3}$.

65.   The zeros of $f(x) = 2x^2 - 13x - 7$ are the roots, or solutions, of the equation $2x^2 - 13x - 7 = 0$.

$2x^2 - 13x - 7 = 0$

$(2x + 1)(x - 7) = 0$

$2x + 1 = 0$ or $x - 7 = 0$

$2x = -1$ or $x = 7$

$x = -\dfrac{1}{2}$ or $x = 7$

The zeros are $-\dfrac{1}{2}$ and 7.

67. $f(x) = x^2 + 12x + 40$

We set $f(a) = 8$.

$a^2 + 12a + 40 = 8$

$a^2 + 12a + 32 = 0$

$(a + 8)(a + 4) = 0$

$a + 8 = 0$ or $a + 4 = 0$

$a = -8$ or $a = -4$

The values of a for which $f(a) = 8$ are -8 and -4.

69. $g(x) = 2x^2 + 5x$

We set $g(a) = 12$.

$2a^2 + 5a = 12$

$2a^2 + 5a - 12 = 0$

$(2a - 3)(a + 4) = 0$

$2a - 3 = 0$ or $a + 4 = 0$

$2a = 3$ or $a = -4$

$a = \dfrac{3}{2}$ or $a = -4$

The values for a for which $g(a) = 12$ are $\dfrac{3}{2}$ and -4.

71. $f(x) = \dfrac{3}{x^2 - 4x - 5}$

$f(x)$ cannot be calculated for any x-value for which the denominator, $x^2 - 4x - 5$, is 0. To find the excluded values, we solve:
$$x^2 - 4x - 5 = 0$$
$$(x - 5)(x + 1) = 0$$
$$x - 5 = 0 \ or \ x + 1 = 0$$
$$x = 5 \ or \quad x = -1$$
The domain of f is { $x | x$ is a real number *and* $x \ne 5$ *and* $x \ne -1$ }.

73. $f(x) = \dfrac{x - 5}{9x - 18x^2}$

$f(x)$ cannot be calculated for any x-value for which the denominator, $9x - 18x^2$, is 0. To find the excluded values, we solve:
$$9x - 18x^2 = 0$$
$$9x(1 - 2x) = 0$$
$$9x = 0 \ or \ 1 - 2x = 0$$
$$x = 0 \ or \ -2x = -1$$
$$x = 0 \ or \qquad x = \dfrac{1}{2}$$
The domain of f is { $x | x$ is a real number *and* $x \ne 0$ *and* $x \ne \dfrac{1}{2}$ }.

75. $f(x) = \dfrac{3x}{2x^2 - 9x + 4}$

$f(x)$ cannot be calculated for any x-value for which the denominator, $2x^2 - 9x + 4$ is 0. To find the excluded values, we solve:
$$2x^2 - 9x + 4 = 0$$
$$(2x - 1)(x - 4) = 0$$
$$2x - 1 = 0 \ or \ x - 4 = 0$$
$$2x = 1$$
$$x = \dfrac{1}{2} \ or \qquad x = 4$$
The domain of f is { $x | x$ is a real number *and* $x \ne \dfrac{1}{2}$ *and* $x \ne 4$ }.

77. $f(x) = \dfrac{7}{5x^3 - 35x^2 + 50x}$

$f(x)$ cannot be calculated for any x-value for which the denominator, $5x^3 - 35x^2 + 50x$, is 0. To find the excluded values, we solve:
$$5x^3 - 35x^2 + 50x = 0$$
$$5x(x^2 - 7x + 10) = 0$$
$$5x(x - 2)(x - 5) = 0$$
$$5x = 0 \ or \ x - 2 = 0 \ or \ x - 5 = 0$$
$$x = 0 \ or \quad x = 2 \ or \quad x = 5$$
The domain of f is { $x | x$ is a real number and $x \ne 0$ and $x \ne 2$ and $x \ne 5$ }.

79. *Thinking and Writing Exercise.*

81. $(x - 2)^2 = x^2 - 2 \cdot x \cdot 2 + 2^2$
$$= x^2 - 4x + 4$$

83. $(x + 2)(x - 2) = x^2 - 2^2$
$$= x^2 - 4$$

85. $(4a + 1)^2 = (4a)^2 + 2 \cdot (4a) \cdot 1 + 1^2$
$$= 16a^2 + 8a + 1$$

87. $(3c - 10)^2 = (3c)^2 - 2 \cdot (3c) \cdot 10 + 10^2$
$$= 9c^2 - 60c + 100$$

89. $(8n + 3)(8n - 3) = (8n)^2 - 3^2$
$$= 64n^2 - 9$$

91. *Thinking and Writing Exercise.*

93. Graph $y = 4x^2 + 120x + 675$ and find the zeros. They are -7.5 and -22.5, or $-\dfrac{15}{2}$ and $-\dfrac{45}{2}$. We know that $-\dfrac{15}{2}$ is a zero of $g(x) = 2x + 15$ and $-\dfrac{45}{2}$ is a zero of $h(x) = 2x + 45$. We have
$$y = 4x^2 + 120x + 675 = (2x + 15)(2x + 45)$$

95. First factor out the largest common factor.

$$3x^3 + 150x^2 - 3672x = 3x(x^2 + 50x - 1224)$$

Now graph $y = x^2 + 50x - 1224$ and find the zeros. They are -68 and 18. We know that -68 is a zero of $g(x) = x + 68$, and 18 is a zero of $h(x) = x - 18$.

We have $x^2 + 50x - 1224 = (x + 68)(x - 18)$,

so $3x^3 + 150x^2 - 3672x = 3x(x + 68)(x - 18)$

97. $(8x + 11)(12x^2 - 5x - 2) = 0$

$(8x + 11)(3x - 2)(4x + 1) = 0$

$8x + 11 = 0 \quad or \quad 3x - 2 = 0 \quad or \quad 4x + 1 = 0$

$8x = -11 \, or \quad 3x = 2 \quad or \quad 4x = -1$

$x = -\dfrac{11}{8} \quad or \quad x = \dfrac{2}{3} \quad or \quad x = -\dfrac{1}{4}$

The solutions are $-\dfrac{11}{8}, \dfrac{2}{3}$, and $-\dfrac{1}{4}$.

99. $\qquad (x - 2)^3 = x^3 - 2$

$x^3 - 6x^2 + 12x - 8 = x^3 - 2$

$0 = 6x^2 - 12x + 6$

$0 = 6(x^2 - 2x + 1)$

$0 = 6(x - 1)(x - 1)$

$x - 1 = 0 \, or \, x - 1 = 0$

$x = 1 \, or \quad x = 1$

The solution is 1.

101. $18a^2b^2 - 3ab - 10$

Let $u = ab$ (and $u^2 = a^2b^2$). Factor $18u^2 - 3u - 10$.

We will use the FOIL method.

1. There is no common factor (other than 1 or -1).
2. Factor the first term, $18u^2$. The factors are $18u, u$ and $9u, 2u$ and $6u, 3u$. We have these possibilities: $(18u + \;)(u + \;)$ and $(9u + \;)(2u + \;)$ and $(6u + \;)(3u + \;)$.
3. Factor the last term, -10. The possibilities are $-10 \cdot 1, -5 \cdot 2,$ $-2 \cdot 5, -1 \cdot 10$.
4. We need factors for which the sum of the products is the middle term, $-3u$. Trial

and error leads us to the factorization: $(6u - 5)(3u + 2)$. Replace u by ab. The factorization is $(6ab - 5)(3ab + 2)$.

103. $16a^2b^3 + 25ab^2 + 9$

There is no common factor (other than 1 or -1).

The trinomial is not of the form $ax^2 + bx + c$, and also is not of the form $au^2 + bu + c$, using substitution. Therefore, $16a^2b^3 + 25ab^2 + 9$ is prime.

105. $25t^{10} - 10t^5 + 1$

Let $u = t^5$ (and $u^2 = t^{10}$). Factor

$25u^2 - 10u + 1$. We will use the grouping method. Multiply the leading coefficient and the constant: $25(1) = 25$. Factor 25 so the sum of the factors is -10. The desired factorization is $-5, -5$. Split the middle term and factor by grouping.

$25u^2 - 10u + 1 = 25u^2 - 5u - 5u + 1$

$\qquad\qquad = 5u(5u - 1) - 1(5u - 1)$

$\qquad\qquad = (5u - 1)(5u - 1)$

$\qquad\qquad or \; (5u - 1)^2$

Replace u by t^5. The factorization is $(5t^5 - 1)^2$.

107. $20x^{2n} + 16x^n + 3$

Let $u = x^n$ (and $u^2 = x^{2n}$). Factor $20u^2 + 16u + 3$. We will use the FOIL method.

1. There is no common factor (other than 1 or -1).
2. Factor the first term, $20u^2$. The factors are $20u, u$ and $10u, 2u$ and $5u, 4u$. We have these possibilities: $(20u + \;)(u + \;)$ and $(10u + \;)(2u + \;)$ and $(5u + \;)(4u + \;)$.
3. Factor the last term, 3. The possibilities are $3 \cdot 1$ and $-3 \cdot -1$.
4. We need factors for which the sum of the products is the middle term, $16u$. Trial and error leads us to the factorization

$(10u+3)(2u+1)$. Replace u by x^n.

The factorization is $(10x^n+3)(2x^n+1)$.

109. $7(t-3)^{2n}+5(t-3)^n-2$

Let $u=(t-3)^n$ $\left(\text{and } u^2=(t-3)^{2n}\right)$. Factor

$7u^2+5u-2$. We will use the FOIL method.

1. There is no common factor (other than 1 or –1).

2. Factor the first term, $7u^2$. The factors are $7u, u$. The possibility is
    $(7u+ \quad)(u+ \quad)$.

3. Factor the last term, –2. The possibilities are $-2\cdot1$ and $-1,2$.

4. We need factors for which the sum of the product is the middle term, $5u$. Trial and error leads us to the factorization
    $(7u-2)(u+1)$. Replace u by $(t-3)^n$.

    The factorization is
    $\left[7(t-3)^n-2\right]\left[(t-3)^n+1\right]$

111. $2a^4b^6-3a^2b^3-20$

Let $u=a^2b^3$ $\left(\text{and } u^2=a^4b^6\right)$. Factor

$2u^2-3u-20$. We will use the FOIL method.

1. There is no common factor (other than 1 or –1).

2. Factor the first term, $2u^2$. The factors are $2u, u$. The possibility is
    $(2u+ \quad)(u+ \quad)$.

3. Factor the last term, –20. The possibilities are $-20\cdot1, -10\cdot2$,
    $-5\cdot4, -4\cdot5, -2\cdot10$, and $-1\cdot20$.

4. We need factors for which the sum of the products is the middle term, $-3u$. Trial and error leads us to the factorization:
    $(2u+5)(u-4)$. Replace u by a^2b^3.

    We have $(2a^2b^3+5)(a^2b^3-4)$.

113. $ax^2+bx+c=(mx+r)(nx+s)$

Let $P=ms$ and $Q=rn$. Then

$ax^2+bx+c=(mx+r)(nx+s)$

$\qquad = mnx^2+msx+rnx+rs$

$\qquad = mnx^2+Px+Qx+rs$

$\qquad = mnx^2+(P+Q)x+rs$

Comparing the terms on the right and left sides of the equation,

$ax^2=mnx^2$, or $a=mn$

$bx=(P+Q)x$, or $b=P+Q$

$\quad c=rs$

and

$ac=(mn)(rs)=(ms)(rn)=PQ$.

Mid-Chapter Review

Guided Solutions

1. $12x^3y-8xy^2+24x^2y$
    $=4xy\left(3x^2-2y+6x\right)$

2. $3a^3-3a^2-90a=3a\left(a^2-a-30\right)$
    $\qquad\qquad\qquad = 3a(a-6)(a+5)$

Mixed Review

1. $6x^5-18x^2=6x^2\cdot x^3-6x^2\cdot3$
    $\qquad\qquad = 6x^2\left(x^3-3\right)$

2. $x^2+10x+16$
    We look for two numbers whose product is 16 and whose sum is 10. Since 16 and 10 are both positive, we need only consider positive factors.

Pairs of Factors	Sum of Factors
1, 16	17
2, 8	10
4, 4	8

    The numbers we need are 2 and 8. The factorization is $(x+2)(x+8)$.

3. $2x^2 + 13x - 7$
     We will use the FOIL method.
     1. There is no common factor (other than 1 or –1).
     2. Factor the first term, $2x^2$. The factors are $2x$ and x. We have the possibility $(2x+\ )(x+\ )$.
     3. Factor the last term, –7. The possibilities are $-1 \cdot 7$, $1 \cdot (-7)$, $-7 \cdot 1$, and $7(-1)$.
     4. Look for factors for which the sum of the products is $13x$. Trial and error leads us to: $(2x-1)(x+7)$.

4. $x^3 + 3x^2 + 2x + 6 = x^2(x+3) + 2(x+3)$
     $= (x^2+2)(x+3)$

5. $5x^2 + 40x - 100$
     $= 5(x^2 + 8x - 20)$ Common factor
     Since the constant term is negative, we look for a factorization of –20 in which one factor is positive and one factor is negative. Their sum must be 8, so the positive factor must have the larger absolute value. Thus we consider only pairs of factors in which the positive factor has the larger absolute value.

Pair of Factors	Sum of Factors
–1, 20	19
–2, 10	8
–4, 5	1

     The numbers we need are 10 and –2. The factorization is $(x+10)(x-2)$.

     We must not forget to include the common factor 5.
     $5x^2 - 40x - 100 = 5(x+10)(x-2)$.

6. $x^2 - 2x - 5$
     We want a positive and negative factor of –5, whose sum is –2. The only possible integer factors whose sum is negative are 1 and –5, and their sum is –4. Therefore, the polynomial cannot be factored with integer coefficients, and thus $x^2 - 2x - 5$ is prime.

7. $7x^2 y - 21xy - 28y$
     $= 7y(x^2 - 3x - 4)$ Common factor
     Since the constant term is negative, we look for a factorization of –4 in which one factor is positive and one factor is negative. Their sum must be –3, so the negative factor must have the larger absolute value. Thus we consider only pairs of factors in which the negative factor has the larger absolute value.

Pair of Factors	Sum of Factors
1, –4	–3
2, –2	0

     The numbers we need are 1 and –4. The factorization is $(x+1)(x-4)$.

     We must not forget to include the common factor $7y$.
     $7x^2 y - 21xy - 28y = 7y(x+1)(x-4)$.

8. $15a^4 - 27a^2 b^2 + 21a^2 b$
     $= 3a^2 \cdot 5a^2 - 3a^2 \cdot 9b^2 + 3a^2 \cdot 7b$
     $= 3a^2(5a^2 - 9b^2 + 7b)$

9. $b^2 - 14b + 49$
     Since the constant term is positive and the coefficient of the middle term is negative, we look for a factorization of 49 in which both factors are negative.

Pair of Factors	Sum of Factors
–1, –49	–50
–7, –7	–14

     The numbers we need are –7 and –7.
     $b^2 - 14b + 49 = (b-7)(b-7)$, or $(b-7)^2$

10. $12x^2 - x - 1$
     We will use the grouping method.
     1. There is no common factor (other than 1 or –1).
     2. Multiply the leading coefficient and the constant $12(-1) = -12$.
     3. We want factors of –12 whose sum is –1. Since $-4 \cdot 3 = -12$ and $-4 + 3 = -1$, we will split $-x$ into $-4x$ and $3x$.
     4. Factor by grouping:

$$12x^2 - x - 1 = 12x^2 - 4x + 3x - 1$$
$$= 4x(3x-1) + 1(3x-1)$$
$$= (4x+1)(3x-1)$$

11. $x^2y^2 - xy - 2$

Substitute u for xy $\left(\text{and } u^2 \text{ for } x^2y^2\right)$. We

factor $u^2 - u - 2$. Look for factors of -2 whose sum is -1. The factors are -2 and 1, so we have $u^2 - u - 2 = (u-2)(u+1)$.

Replace u by xy to obtain the final form of the factored polynomial.

$$x^2y^2 - xy - 2 = (xy-2)(xy+1)$$

12. $2x^2 + 30x - 200$

$$= 2(x^2 + 15x - 100) \qquad \text{Common factor}$$

Since the constant term is negative, we look for a factorization of -100 in which one factor is positive and one factor is negative. Their sum must be 15, so the positive factor must have the larger absolute value. Thus we consider only pairs of factors in which the positive factor has the larger absolute value.

Pair of Factors	Sum of Factors
−1, 100	99
−2, 50	48
−4, 25	21
−5, 20	15
−10, 10	0

The numbers we need are -5 and 20. The factorization is $(x+20)(x-5)$.

We must not forget to include the common factor 2.

$$2x^2 + 30x - 200 = 2(x+20)(x-5).$$

13. $t^2 + t - 10$

Since the constant term is negative, we look for a factorization of -10 in which one factor is positive and one factor is negative. Their sum must be 1, so the positive factor must have the larger absolute value. Thus we consider only pairs of factors in which the positive factor has the larger absolute value.

Pair of Factors	Sum of Factors
10, −1	9
5, −2	3

Neither of these possibilities satisfies the necessary conditions for factoring. Therefore, the polynomial cannot be factored with integer coefficients, and thus $t^2 + t - 10$ is prime.

14. $15d^2 - 30d + 75 = 15 \cdot d^2 - 15 \cdot 2d + 15 \cdot 5$
$$= 15(d^2 - 2d + 5)$$

15. $15p^2 + 16px + 4x^2$

We will use the FOIL method.

1. There is no common factor (other than 1 or −1).

2. Factor the first term, $15p^2$. The factors are $15p, p$ and $5p, 3p$. We have these possibilities: $(15p +)(p +)$ and $(5p +)(3p +)$.

3. Factor the last term, $4x^2$. The possibilities are $4x \cdot x$, $(-4x)(-x)$, $2x \cdot 2x$, and $(-2x)(-2x)$.

4. Look for factors such that the sum of the products is the middle term, $16px$.

   Trial and error leads us to the correct factorization: $(5p + 2x)(3p + 2x)$

16. $-2t^3 - 10t^2 - 12t$

$$= -2t(t^2 + 5t + 6) \qquad \text{Common factor}$$

We look for two numbers whose product is 6 and whose sum is 5. Since 6 and 5 are both positive, we need only consider positive factors.

Pair of Factors	Sum of Factors
1, 6	7
2, 3	5

The numbers we need are 2 and 3. The factorization is $(t+2)(t+3)$.

We must not forget to include the common factor $-2t$.

$$-2t^3 - 10t^2 - 12t = -2t(t+2)(t+3)$$

17. $x^2 + 4x - 77$

Since the constant term is negative, we look for a factorization of –77 in which one factor is positive and one factor is negative. Their sum must be 4, so the positive factor must have the larger absolute value. Thus we consider only pairs of factors in which the positive factor has the larger absolute value.

Pair of Factors	Sum of Factors
11, −7	4

The numbers we need are 11 and −7. The factorization is $(x+11)(x-7)$.

18. $10c^2 + 20c + 10$

$= 10(c^2 + 2c + 1)$ Common factor

We look for two numbers whose product is 1 and whose sum is 2. Since 1 and 2 are both positive, we need only consider positive factors.

Pair of Factors	Sum of Factors
1, 1	2

The numbers we need are 1 and 1. The factorization is $(c+1)(c+1)$.

We must not forget to include the common factor 10.

$10c^2 + 20c + 10 = 10(c+1)(c+1)$, or

$10(c+1)^2$.

19. $5 + 3x - 2x^2 = -2x^2 + 3x + 5$

We will use the grouping method.

1. There is no common factor (other than 1 or −1).
2. Multiply the leading coefficient and the constant $(-2) \cdot 5 = -10$.
3. Factor −10 so the sum of the factors is 3. The desired terms are 5 and −2.
4. Split $3x$ using the results of step (3):
    $3x = 5x - 2x = -2x + 5x$.
5. Factor by grouping:

$-2x^2 + 3x + 5 = -2x^2 - 2x + 5x + 5$
$= -2x(x+1) + 5(x+1)$
$= (-2x+5)(x+1)$
$= -1(2x-5)(x+1)$

20. $2m^3n - 10m^2n - 6mn + 30n$
$= 2n \cdot m^3 - 2n \cdot 5m^2 - 2n \cdot 3m + 2n \cdot 15$
$= 2n(m^3 - 5m^2 - 3m + 15)$

Factoring the polynomial $m^3 - 5m^2 - 3m + 15$
$m^3 - 5m^2 - 3m + 15$
$= m^2 \cdot m - m^2 \cdot 5 - 3 \cdot m + 3 \cdot 5$
$= m^2(m-5) - 3(m-5)$
$= (m^2 - 3)(m-5)$

So the complete factorization is
$2m^3n - 10m^2n - 6mn + 30n$
$= 2n(m^2 - 3)(m-5)$.

Exercise Set 6.4

1. Differences of two squares

3. Perfect-square trinomial

5. None of these

7. Prime polynomial

9. $x^2 + 18x + 81 = x^2 + 2 \cdot 9 \cdot x + 9^2$
This is of the form $A^2 + 2BA + B^2$, which is a perfect-square trinomial.

11. $x^2 - 10x - 25$
The sign of the third term must be "+" for the polynomial to be a perfect-square trinomial. Therefore, $x^2 - 10x - 25$ is not a perfect-square trinomial.

13. $x^2 - 3x + 9$
The second term must equal $-6x$ for $x^2 - 3x + 9$ to be a perfect-square trinomial. Therefore, $x^2 - 3x + 9$ is not a perfect-square trinomial.

15. $9x^2 + 25 - 30x = 9x^2 - 30x + 25$
$$= (3x)^2 - 2 \cdot 5 \cdot 3x + 5^2$$
This is of the form $A^2 - 2BA + B^2$, which is a perfect-square trinomial.

17. $t^2 + 6t + 9 = (t+3)^2$
Find the square terms and write the quantities that were squared with a plus sign between them.

19. $a^2 - 14a + 49 = (a-7)^2$
Find the square terms and write the quantities that were squared with a minus sign between them.

21. $4a^2 - 16a + 16$
$$= 4(a^2 - 4a + 4)$$
Factoring out the common factor
$$= 4(a-2)^2$$
Factoring the perfect-square trinomial

23. $1 - 2t + t^2$
$$= t^2 - 2t + 1$$
Changing order
$$= (t-1)^2, \text{ or } (1-t)^2$$
Factoring the perfect-square trinomial

25. $24a^2 + a^3 + 144a$
$$= a^3 + 24a^2 + 144a$$
Changing order
$$= a(a^2 + 24a + 144)$$
Factoring out the common factor
$$= a(a+12)^2$$
Factoring the perfect-square trinomial

27. $20x^2 + 100x + 125$
$$= 5(4x^2 + 20x + 25)$$
Factoring out the common factor
$$= 5(2x+5)^2$$
Factoring the perfect-square trinomial

29. $1 + 8d^3 + 16d^6 = (1+4d^3)^2 \text{ or } (4d^3+1)^2$
Find the square terms and write the quantities that were squared with an addition sign between them.

31. $-y^3 + 8y^2 - 16y$
$$= -y(y^2 - 8y + 16)$$
Factoring out the common factor
$$= -y(y-4)^2$$
Factoring the perfect-square trinomial

33. $0.25x^2 + 0.30x + 0.09 = (0.5x + 0.3)^2$
Find the square terms and write the quantities that were squared with a plus sign between them. Square this binomial.

35. $x^2 - 2xy + y^2 = (x-y)^2$
Find the square terms and write the quantities that were squared with a minus sign between them. Square this binomial.

37. $25a^6 + 30a^3b^3 + 9b^6 = (5a^3 + 3b^3)^2$
Find the square terms and write the quantities that were squared with a plus sign between them. Square this binomial.

39. $5a^2 - 10ab + 5b^2$
$$= 5(a^2 - 2ab + b^2)$$
Factoring out the common factor
$$= 5(a-b)^2$$
Factoring the perfect-square trinomial

41. $x^2 - 100$ is a difference of squares, because $x^2 = (x)^2$ and $100 = 10^2$, and the terms have different signs.

43. $n^4 + 1$ is not a difference of squares, because the terms have the same sign,.

45. $-1 + 64t^2$ is a difference of squares, because $1 = (1)^2$ and $64t^2 = (8t)^2$, and the terms have different signs.

47. $y^2 - 100 = y^2 - 10^2 = (y+10)(y-10)$

49. $m^2 - 64 = m^2 - 8^2 = (m+8)(m-8)$

51. $-49 + t^2 = t^2 - 7^2 = (t+7)(t-7)$

53. $8x^2 - 8y^2$

$= 8(x^2 - y^2)$

Factoring out the common factor

$= 8(x+y)(x-y)$

Factoring the difference of squares

55. $-80a^6 + 45 = -5(16a^6 - 9)$

$= -5\left[(4a^3)^2 - 3^2\right]$

$= -5(4a^3 + 3)(4a^3 - 3)$

57. $49a^4 + 100$

This expression is the sum of two squares, and therefore cannot be factored any further. It is therefore prime.

59. $t^4 - 1 = (t^2)^2 - 1^2$

$= (t^2 + 1)(t^2 - 1)$

$= (t^2 + 1)(t^2 - 1^2)$

$= (t^2 + 1)(t+1)(t-1)$

61. $9a^4 - 25a^2b^4 = a^2(9a^2 - 25b^4)$

$= a^2\left[(3a)^2 - (5b^2)^2\right]$

$= a^2(3a + 5b^2)(3a - 5b^2)$

63. $16x^4 - y^4 = (4x^2)^2 - (y^2)^2$

$= (4x^2 + y^2)(4x^2 - y^2)$

$= (4x^2 + y^2)((2x)^2 - y^2)$

$= (4x^2 + y^2)(2x + y)(2x - y)$

65. $\dfrac{1}{49} - x^2 = \left(\dfrac{1}{7}\right)^2 - x^2 = \left(\dfrac{1}{7} + x\right)\left(\dfrac{1}{7} - x\right)$

67. $(a+b)^2 - 9 = (a+b)^2 - 3^2$

$= \left[(a+b)+3\right]\left[(a+b)-3\right]$

$= (a+b+3)(a+b-3)$

69. $x^2 - 6x + 9 - y^2$

$= (x^2 - 6x + 9) - y^2$

Grouping as a difference of squares

$= (x-3)^2 - y^2$

$= (x-3+y)(x-3-y)$

71. $t^3 + 8t^2 - t - 8$

$= t^2(t+8) - 1(t+8)$

Factoring by grouping

$= (t+8)(t^2 - 1)$

$= (t+8)(t+1)(t-1)$

Factoring the difference of squares

73. $r^3 - 3r^2 - 9r + 27$

$= r^2(r-3) - 9(r-3)$

Factoring by grouping

$= (r-3)(r^2 - 9)$

$= (r-3)(r+3)(r-3)$ or $(r-3)^2(r+3)$

Factoring the difference of squares

75. $m^2 - 2mn + n^2 - 25$

$= (m^2 - 2mn + n^2) - 25$

Grouping as a difference of squares

$= (m-n)^2 - 5^2$

$= (m-n+5)(m-n-5)$

77. $36 - (x+y)^2 = 6^2 - (x+y)^2$

$= \left[6+(x+y)\right]\left[6-(x+y)\right]$

$= (6+x+y)(6-x-y)$

79. $16 - a^2 - 2ab - b^2$

$= 16 - \left(a^2 + 2ab + b^2\right)$

     Grouping as a difference of squares

$= 4^2 - (a+b)^2$

$= \left[4 + (a+b)\right]\left[4 - (a+b)\right]$

$= (4 + a + b)(4 - a - b)$

81. $a^3 - ab^2 - 2a^2 + 2b^2$

$= a\left(a^2 - b^2\right) - 2\left(a^2 - b^2\right)$

$= \left(a^2 - b^2\right)(a - 2)$

$= (a + b)(a - b)(a - 2)$

83. $a^2 + 1 = 2a$

$a^2 - 2a + 1 = 0$

$(a - 1)(a - 1) = 0$

$a - 1 = 0 \; or \; a - 1 = 0$

$a = 1 \; or \;\;\;\; a = 1$

The solution is 1.

85. $2x^2 - 24x + 72 = 0$

$2\left(x^2 - 12x + 36\right) = 0$

$2(x - 6)(x - 6) = 0$

$x - 6 = 0 \; or \; x - 6 = 0$

$x = 6 \; or \;\;\;\; x = 6$

The solution is 6.

87. $x^2 - 9 = 0$

$(x + 3)(x - 3) = 0$

$x + 3 = 0 \;\; or \; x - 3 = 0$

$x = -3 \; or \;\;\;\; x = 3$

The solutions are -3 and 3.

89. $a^2 = \dfrac{1}{25}$

$a^2 - \dfrac{1}{25} = 0$

$\left(a + \dfrac{1}{5}\right)\left(a - \dfrac{1}{5}\right) = 0$

$a + \dfrac{1}{5} = 0 \;\;\; or \; a - \dfrac{1}{5} = 0$

$a = -\dfrac{1}{5} \; or \;\;\;\; a = \dfrac{1}{5}$

The solutions are $-\dfrac{1}{5}$ and $\dfrac{1}{5}$.

91. $8x^3 + 1 = 4x^2 + 2x$

$8x^3 - 4x^2 - 2x + 1 = 0$

$4x^2(2x - 1) - (2x - 1) = 0$

$(2x - 1)\left(4x^2 - 1\right) = 0$

$(2x - 1)(2x + 1)(2x - 1) = 0$

$2x - 1 = 0 \; or \; 2x + 1 = 0 \;\; or \; 2x - 1 = 0$

$2x = 1 \; or \;\;\;\; 2x = -1 \; or \;\;\;\; 2x = 1$

$x = \dfrac{1}{2} \; or \;\;\;\; x = -\dfrac{1}{2} \; or \;\;\;\; x = \dfrac{1}{2}$

The solutions are $\dfrac{1}{2}$ and $-\dfrac{1}{2}$.

93. $x^3 + 3 = 3x^2 + x$

$x^3 - 3x^2 - x + 3 = 0$

$x^2(x - 3) - (x - 3) = 0$

$(x - 3)\left(x^2 - 1\right) = 0$

$(x - 3)(x + 1)(x - 1) = 0$

$x - 3 = 0 \; or \; x + 1 = 0 \;\; or \; x - 1 = 0$

$x = 3 \; or \;\;\;\; x = -1 \; or \;\;\;\; x = 1$

The solutions are 3, -1, and 1.

95.   The polynomial $x^2 - 3x - 7$ is prime. We solve the equation by graphing $y = x^2 - 3x - 7$ and finding the zeros. They are approximately -1.541 and 4.541. These are the solutions.

97.   The polynomial $2x^2 + 8x + 1$ is prime. We solve the equation by graphing $y = 2x^2 + 8x + 1$ and finding the zeros. They are approximately -3.871 and -0.129. These are the solutions.

99.   The polynomial $x^3 + 3x^2 + x - 1$ is prime. We solve the equation by graphing $y = x^3 + 3x^2 + x - 1$ and finding the zeros. They are approximately -2.414, -1, and approximately 0.414. These are the solutions.

101. $f(x) = x^2 - 12x$

We set $f(a)$ equal to –36.

$$a^2 - 12a = -36$$
$$a^2 - 12a + 36 = 0$$
$$(a-6)(a-6) = 0$$
$$a - 6 = 0 \text{ or } a - 6 = 0$$
$$a = 6 \text{ or } \quad a = 6$$

The value of a for which $f(a) = -36$ is 6.

103. To find the zeros of $f(x) = x^2 - 16$, we find

the roots of the equation $x^2 - 16 = 0$.

$$x^2 - 16 = 0$$
$$(x+4)(x-4) = 0$$
$$x + 4 = 0 \quad \text{or } x - 4 = 0$$
$$x = -4 \text{ or } \quad x = 4$$

The zeros are –4 and 4.

105. To find the zeros of $f(x) = 2x^2 + 4x + 2$, we

find the roots of the equation

$$2x^2 + 4x + 2 = 0$$
$$2(x^2 + 2x + 1) = 0$$
$$2(x+1)(x+1) = 0$$
$$x + 1 = 0 \quad \text{or } x + 1 = 0$$
$$x = -1 \text{ or } \quad x = -1$$

The zero is –1.

107. To find the zeros of $f(x) = x^3 - 2x^2 - x + 2$,

we find the roots of the equation

$x^3 - 2x^2 - x + 2 = 0$.

$$x^3 - 2x^2 - x + 2 = 0$$
$$x^2(x-2) - (x-2) = 0$$
$$(x-2)(x^2 - 1) = 0$$
$$(x-2)(x+1)(x-1) = 0$$
$$x - 2 = 0 \quad \text{or } x + 1 = 0 \quad \text{or } x - 1 = 0$$
$$x = 2 \text{ or } \quad x = -1 \text{ or } \quad x = 1$$

The zeros are 2, –1, and 1.

109. *Thinking and Writing Exercise.*

111. $\left(2x^2 y^4\right)^3 = 2^3 \left(x^2\right)^3 \left(y^4\right)^3$

$$= 8 \cdot x^6 \cdot y^{12}$$
$$= 8x^6 y^{12}$$

113. $(x+1)(x+1)(x+1)$

$$= (x+1)\left(x^2 + 2 \cdot x \cdot 1 + 1\right)$$
$$= (x+1)\left(x^2 + 2x + 1\right)$$
$$= x\left(x^2 + 2x + 1\right) + \left(x^2 + 2x + 1\right)$$
$$= x^3 + 2x^2 + x + x^2 + 2x + 1$$
$$= x^3 + 2x^2 + x^2 + 2x + x + 1$$
$$= x^3 + 3x^2 + 3x + 1$$

115. $(m+n)^3$

$$= (m+n)(m+n)(m+n)$$
$$= (m+n)\left(m^2 + 2 \cdot m \cdot n + n^2\right)$$
$$= (m+n)\left(m^2 + 2mn + n^2\right)$$
$$= m\left(m^2 + 2mn + n^2\right) + n\left(m^2 + 2mn + n^2\right)$$
$$= m^3 + 2m^2 n + mn^2 + nm^2 + 2mn^2 + n^3$$
$$= m^3 + 2m^2 n + nm^2 + 2mn^2 + mn^2 + n^3$$
$$= m^3 + 3m^2 n + 3mn^2 + n^3$$

117. *Thinking and Writing Exercise .*

119. $x^8 - 2^8 = \left(x^4\right)^2 - \left(2^4\right)^2$

$$= \left(x^4 + 2^4\right)\left(x^4 - 2^4\right)$$
$$= \left(x^4 + 2^4\right)\left(\left(x^2\right)^2 - \left(2^2\right)^2\right)$$
$$= \left(x^4 + 2^4\right)\left(x^2 + 2^2\right)\left(x^2 - 2^2\right)$$
$$= \left(x^4 + 2^4\right)\left(x^2 + 2^2\right)(x+2)(x-2)$$
$$\text{or } \left(x^4 + 16\right)\left(x^2 + 4\right)(x+2)(x-2)$$

121. $3x^2 - \dfrac{1}{3} = 3\left(x^2 - \dfrac{1}{9}\right)$

$$= 3\left(x^2 - \left(\frac{1}{3}\right)^2\right)$$
$$= 3\left(x + \frac{1}{3}\right)\left(x - \frac{1}{3}\right)$$

123. $0.09x^8 + 0.48x^4 + 0.64 = \left(0.3x^4 + 0.8\right)^2$,

$\quad$ or $\dfrac{1}{100}\left(3x^4 + 8\right)^2$

125. $r^2 - 8r - 25 - s^2 - 10s + 16$

$\quad = \left(r^2 - 8r + 16\right) - \left(s^2 + 10s + 25\right)$

$\quad = (r-4)^2 - (s+5)^2$

$\quad = \left[(r-4) + (s+5)\right]\left[(r-4) - (s+5)\right]$

$\quad = (r - 4 + s + 5)(r - 4 - s - 5)$

$\quad = (r + s + 1)(r - s - 9)$

127. $x^{4a} - 49y^{2a} = \left(x^{2a}\right)^2 - \left(7y^a\right)^2$

$\quad = \left(x^{2a} + 7y^a\right)\left(x^{2a} - 7y^a\right)$

129. $3(x+1)^2 + 12(x+1) + 12$

$\quad = 3\left[(x+1)^2 + 4(x+1) + 4\right]$

$\quad = 3\left[(x+1) + 2\right]^2$

$\quad = 3(x+3)^2$

131. $9x^{2n} - 6x^n + 1 = \left(3x^n\right)^2 - 6x^n + 1 = \left(3x^n - 1\right)^2$

133. $s^2 - 4st + 4t^2 + 4s - 8t + 4$

$\quad = \left(s^2 - 4st + 4t^2\right) + (4s - 8t) + 4$

$\quad = (s - 2t)^2 + 4(s - 2t) + 2^2$

$\quad = (s - 2t + 2)^2$

135. If $P(x) = x^4$, then

$\quad P(a+h) - P(a)$

$\quad = (a+h)^4 - a^4$

$\quad = \left[(a+h)^2 + a^2\right]\left[(a+h)^2 - a^2\right]$

$\quad = \left[(a+h)^2 + a^2\right]\left[(a+h) + a\right]\left[(a+h) - a\right]$

$\quad = \left(a^2 + 2ah + h^2 + a^2\right)(2a + h)(h)$

$\quad = h(2a+h)\left(2a^2 + 2ah + h^2\right)$

Exercise Set 6.5

1. Difference of cubes

3. Difference of squares

5. Sum of cubes

7. None of these

9. Difference of cubes

11. $x^3 + 64 = x^3 + 4^3$

$\quad = (x+4)\left(x^2 - 4x + 16\right)$

$\quad A^3 + B^3 = (A+B)\left(A^2 - AB + B^2\right)$

13. $z^3 - 1 = z^3 - 1^3$

$\quad = (z-1)\left(z^2 + z + 1\right)$

$\quad A^3 - B^3 = (A - B)\left(A^2 + AB + B^2\right)$

15. $t^3 - 1000 = t^3 - 10^3$

$\quad = (t - 10)\left(t^2 + 10t + 100\right)$

$\quad A^3 - B^3 = (A - B)\left(A^2 + AB + B^2\right)$

17. $27x^3 + 1 = (3x)^3 + 1^3$

$\quad = (3x + 1)\left(9x^2 - 3x + 1\right)$

19. $64 - 125x^3 = (4)^3 - (5x)^3$

$\quad = (4 - 5x)\left(16 + 20x + 25x^2\right)$

21. $8y^3 + 64 = 8\left(y^3 + 8\right)$

$\quad = 8\left(y^3 + 2^3\right)$

$\quad = 8(y + 2)\left(y^2 - 2y + 4\right)$

23. $x^3 - y^3 = (x - y)\left(x^2 + xy + y^2\right)$

25. $a^3 + \dfrac{1}{8} = a^3 + \left(\dfrac{1}{2}\right)^3$

$\quad = \left(a + \dfrac{1}{2}\right)\left(a^2 - \dfrac{1}{2}a + \dfrac{1}{4}\right)$

27. $8t^3 - 8 = 8\left(t^3 - 1\right) = 8(t - 1)\left(t^2 + t + 1\right)$

29. $y^3 - \dfrac{1}{1000} = y^3 - \left(\dfrac{1}{10}\right)^3$

$$= \left(y - \dfrac{1}{10}\right)\left(y^2 + \dfrac{1}{10}y + \dfrac{1}{100}\right)$$

31. $ab^3 + 125a = a\left(b^3 + 125\right)$

$$= a\left(b^3 + 5^3\right)$$

$$= a(b+5)\left(b^2 - 5b + 25\right)$$

33. $5x^3 - 40z^3 = 5\left(x^3 - 8z^3\right)$

$$= 5\left[x^3 - (2z)^3\right]$$

$$= 5(x - 2z)\left(x^2 + 2xz + 4z^2\right)$$

35. $x^3 + 0.001 = x^3 + 0.1^3$

$$= (x + 0.1)\left(x^2 - 0.1x + 0.01\right)$$

37. $64x^6 - 8t^6 = 8\left(8x^6 - t^6\right)$

$$= 8\left[\left(2x^2\right)^3 - \left(t^2\right)^3\right]$$

$$= 8\left(2x^2 - t^2\right)\left(4x^4 + 2x^2t^2 + t^4\right)$$

39. $2y^4 - 128y = 2y\left(y^3 - 64\right)$

$$= 2y\left(y^3 - 4^3\right)$$

$$= 2y(y - 4)\left(y^2 + 4y + 16\right)$$

41. $z^6 - 1$

$= \left(z^3\right)^2 - 1^2$ Writing as a difference

                of squares

$= \left(z^3 + 1\right)\left(z^3 - 1\right)$ Factoring as a

              difference of squares

$= (z + 1)\left(z^2 - z + 1\right)(z - 1)\left(z^2 + z + 1\right)$

              Factoring a sum and

              a difference of cubes

43. $t^6 + 64y^6$

$$= \left(t^2\right)^3 + \left(4y^2\right)^3$$

        Writing as a sum of cubes

$$= \left(t^2 + 4y^2\right)\left(t^4 - 4t^2y^2 + 16y^4\right)$$

        Factoring a sum of cubes

45. $x^{12} - y^3z^{12} = \left(x^4\right)^3 - \left(yz^4\right)^3$

        Writing as a difference of cubes

$$= \left(x^4 - yz^4\right)\left(x^8 + x^4yz^4 + y^2z^8\right)$$

        Factoring a difference of cubes

47. $\qquad x^3 + 1 = 0$

$\qquad (x + 1)\left(x^2 - x + 1\right) = 0$

$\qquad x + 1 = 0 \quad or \quad x^2 - x + 1 = 0$

$\qquad\qquad x = -1$

We cannot factor $x^2 - x + 1$. The only real-number solution is -1.

49. $\qquad\qquad 8x^3 = 27$

$\qquad\qquad 8x^3 - 27 = 0$

$\qquad (2x - 3)\left(4x^2 + 6x + 9\right) = 0$

$\qquad 2x - 3 = 0 \quad or \quad 4x^2 + 6x + 9 = 0$

$\qquad\qquad 2x = 3$

$\qquad\qquad x = \dfrac{3}{2}$

We cannot factor $4x^2 + 6x + 9$. The only real-number solution is $\dfrac{3}{2}$.

51. $\qquad\qquad 2t^3 - 2000 = 0$

$\qquad\qquad 2\left(t^3 - 1000\right) = 0$

$\qquad 2(t - 10)\left(t^2 + 10t + 100\right) = 0$

$\qquad t - 10 = 0 \quad or \quad t^2 + 10t + 100 = 0$

$\qquad\qquad t = 10$

We cannot factor $t^2 + 10t + 100$. The only real-number solution is 10.

53. *Thinking and Writing Exercise.*

55. We can use any two points on the line such as $(-2,-5)$ and $(3,-6)$.

$$m = \frac{\text{change in } y}{\text{change in } x}$$

$$= \frac{-6-(-5)}{3-(-2)} = \frac{-6+5}{3+2} = \frac{-1}{5} = -\frac{1}{5}$$

57. $2x - 5y = 10$

To find the y-intercept, we let $x = 0$ and solve for y:

$2 \cdot 0 - 5y = 10$

$\qquad -5y = 10$

$\qquad\quad y = -2$

Thus, the y-intercept is $(0,-2)$.

To find the x-intercept, we let $y = 0$ and solve for x:

$2x - 5 \cdot 0 = 10$

$\qquad 2x = 10$

$\qquad\quad x = 5$

Thus, the x-intercept is $(5,0)$.

Before drawing the line, we plot a third point as a check. We substitute any convenient value for x and solve for y. If we let $x = -5$, then

$2 \cdot (-5) - 5y = 10$

$\qquad -10 - 5y = 10$

$\qquad\quad -5y = 20$

$\qquad\qquad y = -4$

The point $(-5,-4)$ appears to line up with the intercepts. To finish, we draw and label the line.

59. $y = \frac{2}{3}x - 1$

Slope: $\frac{2}{3}$; y-intercept: $(0,-1)$

First we plot the y-intercept $(0,-1)$. We can start at the y-intercept and using the slope, $\frac{2}{3}$, find another point. We move up 2

units and right 3 units to get a new point $(3,1)$. Thinking of the slope as $\frac{-2}{-3}$ we can start at $(0,-1)$ and move down 2 units and left 3 units to get another point $(-3,-3)$.

61. *Thinking and Writing Exercise.*

63. $x^{6a} - y^{3b} = \left(x^{2a}\right)^3 - \left(y^b\right)^3$

$\qquad = \left(x^{2a} - y^b\right)\left(x^{4a} + x^{2a}y^b + y^{2b}\right)$

65. $(x+5)^3 + (x-5)^3$ Sum of cubes

$= \left[(x+5) + (x-5)\right] \cdot$

$\quad \left[(x+5)^2 - (x+5)(x-5) + (x-5)^2\right]$

$= 2x\left[\left(x^2 + 10x + 25\right) - \left(x^2 - 25\right)\right.$

$\quad \left. + \left(x^2 - 10x + 25\right)\right]$

$= 2x\left(x^2 + 10x + 25 - x^2 + 25 + x^2 - 10x + 25\right)$

$= 2x\left(x^2 + 75\right)$

67. $5x^3y^6 - \frac{5}{8}$

$= 5\left(x^3y^6 - \frac{1}{8}\right)$

$= 5\left(xy^2 - \frac{1}{2}\right)\left(x^2y^4 + \frac{1}{2}xy^2 + \frac{1}{4}\right)$

69. $x^{6a} - \left(x^{2a} + 1\right)^3$

$= \left[x^{2a} - \left(x^{2a} + 1\right)\right]\left[x^{4a} + x^{2a}\left(x^{2a} + 1\right) + \left(x^{2a} + 1\right)^2\right]$

$= \left(x^{2a} - x^{2a} - 1\right)\left(x^{4a} + x^{4a} + x^{2a} + x^{4a} + 2x^{2a} + 1\right)$

$= -\left(3x^{4a} + 3x^{2a} + 1\right)$

71. $t^4 - 8t^3 - t + 8$

$= t^3(t-8) - (t-8)$

$= (t-8)(t^3 - 1)$

$= (t-8)(t-1)(t^2 + t + 1)$

73. If $Q(x) = x^6$, then

$Q(a+h) - Q(a)$

$= (a+h)^6 - a^6$

$= \left[(a+h)^3 + a^3\right]\left[(a+h)^3 - a^3\right]$

$= \left[(a+h) + a\right] \cdot \left[(a+h)^2 - (a+h)a + a^2\right] \cdot$

$\quad \left[(a+h) - a\right] \cdot \left[(a+h)^2 + (a+h)a + a^2\right]$

$= (2a+h) \cdot \left(a^2 + 2ah + h^2 - a^2 - ah + a^2\right) \cdot$

$\quad (h) \cdot \left(a^2 + 2ah + h^2 + a^2 + ah + a^2\right)$

$= h(2a+h)\left(a^2 + ah + h^2\right)\left(3a^2 + 3ah + h^2\right)$

Exercise Set 6.6

---

1. common factor

3. grouping

5. $10a^2 - 640 = 10(a^2 - 64)$　　　10 is a common

$\qquad = 10(a+8)(a-8)$　　factor $A^2 - B^2 = (A+B)(A-B)$

7. $y^2 + 49 - 14y = y^2 - 14y + 49$

$\qquad = y^2 - 2 \cdot y \cdot 7 + 7^2$　Perfect-square

$\qquad = (y-7)^2$　　　　trinomial

9. $2t^2 + 11t + 12$

There is no common factor (other than 1), and this is not a perfect square trinomial. $2t^2$ factors into $2t \cdot t$ and since all signs are positive, we consider only positive factors.

$(2t + \underline{\quad})(t + \underline{\quad})$

We factor the last term, 12.

Possibilities are 1, 12; 2, 6; 3, 4; 4, 3; 6, 2; and 12, 1. The pair 3, 4 gives us a sum of 11 for the middle term coefficient.

$(2t + 3)(t + 4)$

11. $x^3 - 18x^2 + 81x$

$= x(x^2 - 18x + 81)$　　Common factor

$= x(x^2 - 2 \cdot x \cdot 9 + 9^2)$　Perfect-square trinomial

$= x(x-9)^2$

13. $x^3 - 5x^2 - 25x + 125$

$= (x^3 - 5x^2) - (25x - 125)$　Grouping

$= x^2(x-5) - 25(x-5)$　Common factors (of each pair)

$= (x-5)(x^2 - 25)$　Common factor

$= (x-5)(x+5)(x-5)$　Difference of squares

$= (x-5)^2(x+5)$

15. $27t^3 - 3t = 3t(9t^2 - 1)$　Common factor

$\qquad = 3t(3t+1)(3t-1)$　Difference of squares

17. $9x^3 + 12x^2 - 45x$

$= 3x(3x^2 + 4x - 15)$　Common factor

We now factor $3x^2 + 4x - 15$, which is not a perfect-square trinomial.

Factor by grouping: multiply the leading coefficient, 3, and the constant, -15.

$3 \cdot (-15) = -45$ factor -45, so the sum of the factors is 4. The factors are 9 and -5, since $9 \cdot (-5) = -45$ and $9 + (-5) = 4$. Rewrite the middle term, $4x$, as $9x - 5x$. Factor by grouping.

$3x^2 + 4x - 15 = 3x^2 + 9x - 5x - 15$

$\qquad = (3x^2 + 9x) - (5x + 15)$

$\qquad = 3x(x+3) - 5(x+3)$

$\qquad = (x+3)(3x-5)$

Remember to include the common factor, $3x$.

$9x^3 + 12x^2 - 45x = 3x(3x-5)(x+3)$

19. $t^2 + 25$ is the *sum* of squares, which is prime (in $\mathbb{R}$).

21. $6x^2 + 3x - 45 = 3(2x^2 + x - 15)$

$\qquad = 3(2x - 5)(x + 3)$

23. $-2a^6 + 8a^5 - 8a^4 = -2a^4(a^2 - 4a + 4)$
$$= -2a^4(a-2)^2$$

25. $5x^5 - 80x = 5x(x^4 - 16)$
$$= 5x(x^2 + 4)(x^2 - 4)$$
Note: we can factor $(x^2 - 4)$

$$5x(x^2 + 4)(x^2 - 4) = 5x(x^2 + 4)(x+2)(x-2)$$

27. $t^4 - 9 = (t^2 + 3)(t^2 - 3)$

29. $-x^6 + 2x^5 - 7x^4 = -x^4(x^2 - 2x + 7)$

31. $x^3 - y^3 = (x - y)$
$$\quad\quad\quad \cdot (x^2 + xy + y^2) \quad \text{Difference of cubes}$$

33. $ax^2 + ay^2 = a(x^2 + y^2)$

35. $80cd^2 - 36c^2d + 4c^3 = 4c(20d^2 - 9cd + c^2)$
$$= 4c(4d - c)(5d - c)$$

37. $2\pi rh + 2\pi r^2 = 2\pi r(h + r)$

39. $(a+b)5a + (a+b)3b = (a+b)(5a + 3b)$

41. $x^2 + x + xy + y = (x^2 + x) + (xy + y)$
$$= x(x+1) + y(x+1)$$
$$= (x+1)(x+y)$$

43. $n^2 - 10n + 25 - 9m^2$
Note: There are $\underline{3}$ squared terms: n^2, 5^2 and
$\quad\quad (3m)^2$. Group into a difference of
squares.
$$(n^2 - 10n + 25) - 9m^2$$
$$= (n-5)^2 - (3m)^2$$
$$= [(n-5) + 3m][(n-5) - 3m]$$
$$= (n - 5 + 3m)(n - 5 - 3m)$$

45. $3x^2 + 13xy - 10y^2$
$3 \cdot (-10) = -30 \quad \text{and} \quad 15 \cdot (-2)$
$$= -30, 15 + (-2) = 13$$
$3x^2 + 15xy - 2xy - 10y^2$
$$= (3x^2 + 15xy) - (2xy + 10y^2)$$
$$= 3x(x + 5y) - 2y(x + 5y)$$
$$= (x + 5y)(3x - 2y)$$

47. $4b^2 + a^2 - 4ab = a^2 - 4ab + 4b^2$
$$= a^2 - 2 \cdot a \cdot 2b + (2b)^2$$
$$= (a - 2b)^2$$

49. $16x^2 + 24xy + 9y^2 = (4x)^2 + 2 \cdot 4x \cdot 3y + (3y)^2$
$$= (4x + 3y)^2$$

51. $t^2 - 8t + 10 \quad$ Prime
Note: $1 \cdot 10 = 10$. There is not a pair of factors
of 10, whose sum is -8.

53. $64t^6 - 1 = (8t^3)^2 - 1^2 \quad\quad\quad\quad$ Difference
$$= (8t^3 + 1)(8t^3 - 1) \quad\quad \text{of squares}$$
$$= [(2t)^3 + 1^3][(2t)^3 - 1^3] \quad \text{Sum \&}$$
$$= (2t + 1)[(2t)^2 - 2t \cdot 1 + 1^2] \quad \text{difference}$$
$$\quad \cdot (2t - 1)[(2t)^2 + 2t \cdot 1 + 1^2] \quad \text{of cubes}$$
$$= (2t + 1)(4t^2 - 2t + 1)(2t - 1)$$
$$\quad \cdot (4t^2 + 2t + 1)$$

55. $8m^3n - 32m^2n^2 + 24mn = 8mn(m^2 - 4mn + 3)$

57. $3b^2 + 17ab - 6a^2 = (3b - a)(b + 6a)$

59. $-12 - x^2y^2 - 8xy = -(x^2y^2 + 8xy + 12)$
Note: Look for a pair of factors of 12, whose
sum is 8. They are $\underline{6}$ and $\underline{2}$.
$$-(x^2y^2 + 8xy + 12) = -(xy + \underline{6})(xy + \underline{2})$$

61. $t^8 - s^{10} - 12s^5 - 36$
Note: There are 3 squared terms and one
additional term with s. We can group
into a   difference of squares.
$$t^8 - s^{10} - 12s^5 - 36$$

$$= t^8 - (s^{10} + 12s^5 + 36)$$
$$= t^8 - [(s^5)^2 + 2 \cdot s^5 \cdot 6 + 6^2]$$
$$= (t^4)^2 - (s^5 + 6)^2$$
$$= [t^4 + (s^5 + 6)][t^4 - (s^5 + 6)]$$
$$= (t^4 + s^5 + 6)(t^4 - s^5 - 6)$$

63. $54a^4 + 16ab^3$
$$= 2a(27a^3 + 8b^3)$$
$$= 2a[(3a)^3 + (2b)^3] \quad \text{Sum of cubes}$$
$$= 2a(3a + 2b)[(3a)^2 - 3a \cdot 2b + (2b)^2]$$
$$= 2a(3a + 2b)(9a^2 - 6ab + 4b^2)$$

65. $x^6 + x^5 y - 2x^4 y^2 = x^4(x^2 + xy - 2y^2)$
$$= x^4(x + 2y)(x - y)$$

67. $36a^2 - 15a + \dfrac{25}{16} = (6a)^2 - 2 \cdot 6a \cdot \dfrac{5}{4} + \left(\dfrac{5}{4}\right)^2$
$$= \left(6a - \dfrac{5}{4}\right)^2$$

69. $\dfrac{1}{81}x^2 - \dfrac{8}{27}x + \dfrac{16}{9} = \left(\dfrac{1}{9}x\right)^2 - 2 \cdot \dfrac{1}{9}x \cdot \dfrac{4}{3} + \left(\dfrac{4}{3}\right)^2$
$$= \left(\dfrac{1}{9}x - \dfrac{4}{3}\right)^2$$

71. $1 - 16x^{12} y^{12}$
$$= 1^2 - (4x^6 y^6)^2 \qquad \text{Difference of squares}$$
$$= (1 + 4x^6 y^6)(1 - 4x^6 y^6)$$
$$= (1 + 4x^6 y^6)[1^2 - (2x^3 y^3)^2] \qquad \text{Differences of squares}$$
$$= (1 + 4x^6 y^6)(1 + 2x^3 y^3)$$
$$\cdot (1 - 2x^3 y^3)$$

73. $4a^2 b^2 + 12ab + 9 = (2ab)^2 + 2 \cdot 2ab \cdot 3 + 3^2$
$$= (2ab + 3)^2$$

75. $a^4 + 8a^2 + 8a^3 + 64a$

$$= a^4 + 8a^3 + 8a^2 + 64a$$
$$= a(a^3 + 8a^2 + 8a + 64)$$
$$= a[(a^3 + 8a^2) + (8a + 64)]$$
$$= a[a^2(a + 8) + 8(a + 8)]$$
$$= a[(a + 8)(a^2 + 8)]$$
$$= a(a^2 + 8)(a + 8)$$

77.   *Thinking and Writing Exercise.*

79.   Let m and n represent the numbers. The square of the sum of two numbers is $(m + n)^2$.

81.   Let x represent the first integer. Then $x + 1$ represents the second integer. The product of the two consecutive integers is therefore $x(x + 1)$.

83.   *Familiarize.* Let $x =$ the measure of the second angle, in degrees. Then the measure of the first angle is $4x$ degrees, and the third angle measures $x - 30$ degrees. Recall that the sum of the measures of the angles of any triangle is 180°.
Translate.
$$\underbrace{\text{The sum of the measures of the three angles}} \quad \text{is 180 degrees.}$$
$$\downarrow \qquad\qquad \downarrow \ \downarrow$$
$$x + 4x + (x - 30) \qquad = 180$$
Carry out. We solve the equation.
$$x + 4x + (x - 30) = 180$$
$$6x - 30 = 180 \qquad \text{Combining like terms}$$
$$6x = 210 \qquad \text{Adding 30}$$
$$x = 35 \qquad \text{Dividing by 6}$$
If the second angle measures 35°, then the first angle measures $4 \cdot 35°$, or 140°, and the third angle measures $35° - 30°$, or 5°.
Check. The sum of the measures of the three angles is 140° + 35° + 5°, or 180°. These results check.
State. The angles measure 140°, 35°, and 5°, respectively.

85.   *Thinking and Writing Exercise.*

87. $-(x^5 + 7x^3 - 18x) = -x(x^4 + 7x^2 - 18)$
$$= -x(x^2 + 9)(x^2 - 2)$$

89. $-3a^4 + 15a^2 - 12 = -3(a^4 - 5a^2 + 4)$

$= -3(a^2 - 1)(a^2 - 4)$

$= -3(a+1)(a-1)(a+2)(a-2)$

91. $y^2(y+1) - 4y(y+1) - 21(y+1)$

$= (y+1)(y^2 - 4y - 21)$

$= (y+1)(y-7)(y+3)$

93. $6(x-1)^2 + 7y(x-1) - 3y^2$

$= [2(x-1) + 3y][3(x-1) - y]$

$= (2x + 3y - 2)(3x - y - 3), \quad \text{or}$

$\quad (2x - 2 + 3y)(3x - 3 - y)$

95. $2(a+3)^4 - (a+3)^3(b-2) - (a+3)^2(b-2)^2$

$= (a+3)^2[2(a+3)^2 - (a+3)(b-2) - (b-2)^2]$

$= (a+3)^2[2(a+3) + (b-2)][(a+3) - (b-2)]$

$= (a+3)^2(2a + 6 + b - 2)(a + 3 - b + 2)$

$= (a+3)^2(2a + b + 4)(a - b + 5)$

97. $49x^4 + 14x^2 + 1 - 25x^6$

$= \left(49x^4 + 14x^2 + 1\right) - 25x^6$

$= \left(7x^2 + 1\right)^2 - \left(5x^3\right)^2$

$= \left[\left(7x^2 + 1\right) + 5x^3\right]\left[\left(7x^2 + 1\right) - 5x^3\right]$

$= \left(7x^2 + 1 + 5x^3\right)\left(7x^2 + 1 - 5x^3\right)$

Exercise Set 6.7

1. ***Familiarize.*** Let x represent the number.

Translate.

Square of

number plus number is 132

$\underbrace{\qquad}$ $\quad$ $\underbrace{\qquad}$

$\downarrow$ $\quad$ $\downarrow$ $\quad$ $\downarrow$ $\quad$ $\downarrow$ $\downarrow$

$\quad x^2 \quad + \quad x \quad = 132$

Carry out. We solve the equation:

$x^2 + x = 132$

$x^2 + x - 132 = 0$

$(x + 12)(x - 11) = 0$

$x + 12 = 0 \quad or \; x - 11 = 0$

$\quad x = -12 \; or \quad\quad x = 11$

Check. The square of -12, which is 144, plus -12 is 132. The square of 11, which is 121, plus 11 is 132. Both numbers check.

State. The number is -12 or 11.

3. ***Familiarize.*** Let x represent the number of the first parking space, and $x + 1$ the number of the next parking space.

Translate.

First space $\quad\quad$ second space

number times number is 110

$\underbrace{\qquad}$ $\quad$ $\underbrace{\qquad}$ $\quad\downarrow\quad\downarrow$

$\downarrow$ $\qquad$ $\downarrow$ $\qquad$ $\downarrow$

$\quad x \quad\quad \times \quad\quad (x+1) \quad = 110$

Carry out. We solve the equation:

$x(x+1) = 110$

$x^2 + x - 110 = 0$

$(x + 11)(x - 10) = 0$

$x + 11 = 0 \quad or \; x - 10 = 0$

$\quad x = -11 \; or \quad\quad x = 10$

Check. The number -11 is not a solution, because consecutive numbering of objects uses natural numbers. The product of 10 and $10 + 1 = 11$ is 110. The answer checks.

State. The parking spaces are numbered 10 and 11.

5. ***Familiarize.*** We let w represent the width and $5w$ represent the length. Recall that the formula for the area of a rectangle is $A =$ length $\times$ width.

Translate.

Area $\quad$ is $\quad$ 180 ft^2

$\downarrow$ $\qquad$ $\downarrow$ $\qquad$ $\downarrow$

$w(5w) \quad = \quad\quad 180$

Carry out. We solve the equation:

$w(5w) = 180$

$5w^2 = 180$

$w^2 = 36$

$w = -6 \quad or \quad w = 6$

Check. The number -6 is not a solution, because width cannot be negative. If the width is 6 ft, and the length is $5 \cdot 6$, or 30 ft, then the area is $6 \cdot 30 = 180$ ft^2. The answer checks.

State. The length is 30 ft, and the width is 6 ft.

7. ***Familiarize.*** We let w represent the width and $w + 5$ represent the length. We make a drawing and label it.

Recall that the formula for the area of a rectangle is $A = $ length $\times$ width.

Translate.

$$\underset{\downarrow}{\underline{\text{Area}}} \quad \underset{\downarrow}{\underline{\text{is}}} \quad \underset{\downarrow}{84 \text{ cm}^2}$$

$$w(w+5) = \quad 84$$

Carry out. We solve the equation:

$$w(w+5) = 84$$
$$w^2 + 5w = 84$$
$$w^2 + 5w - 84 = 0$$
$$(w+12)(w-7) = 0$$
$$w+12 = 0 \quad or \quad w-7 = 0$$
$$w = -12 \; or \qquad w = 7$$

Check. The number -12 is not a solution , because width cannot be negative. If the width is 7 cm and the length is 5 cm more, or 12 cm, then the area is $12 \cdot 7$, or 84 cm^2. This is a solution.

State. The length is 12 cm, and the width is 7 cm.

9. **Familiarize.** Let x represent the length of the foot of the sail, and $x+5$ represents the height. Recall that the formula for the area of a triangle is $A = \dfrac{1}{2} \times$ base $\times$ height.

Translate. The area is 42 ft^2.

$$\underset{\downarrow}{\underline{\text{the area}}} \quad \underset{\downarrow}{\underline{\text{is}}} \quad \underset{\downarrow}{\underline{42 \text{ ft}^2}}.$$

$$\frac{1}{2}x(x+5) = \quad 42$$

Carry out. We solve the equation:

$$\frac{1}{2}x(x+5) = 42$$
$$x(x+5) = 84$$
$$x^2 + 5x = 84$$
$$x^2 + 5x - 84 = 0$$
$$(x+12)(x-7) = 0$$
$$x+12 = 0 \quad or \quad x-7 = 0$$
$$x = -12 \; or \qquad x = 7$$

Check. We check only 7, since the length of the foot cannot be negative. If the base is 7 ft, the height is $7 + 5$, or 12 ft, and the area is $\dfrac{1}{2} \cdot 7 \cdot 12$, or 42 ft^2. The answer checks.

State. The foot is 7 ft, and the height is 12 ft.

11. **Familiarize.** Let h represent the height of the triangle, and $\dfrac{1}{2}h$ represents the width of the base. Recall that the formula for the area of a triangle is $A = \dfrac{1}{2} \times$ base $\times$ height.

Translate. The area is 64 ft^2.

$$\underset{\downarrow}{\underline{\text{the area}}} \quad \underset{\downarrow}{\underline{\text{is}}} \quad \underset{\downarrow}{\underline{64 \text{ ft}^2}}.$$

$$\frac{1}{2}\left(\frac{h}{2}\right)h = \quad 64$$

Carry out. We solve the equation:

$$\frac{1}{2}\left(\frac{h}{2}\right)h = 64$$
$$\frac{h^2}{4} = 64$$
$$h^2 = 256$$
$$h = -16 \quad or \quad h = 16$$

Check. We check only 16, since the height cannot be negative. If the height is 16 ft, the width of the base is $\dfrac{16}{2}$, or 8 ft, and the area is $\dfrac{1}{2} \cdot 8 \cdot 16$, or 64 ft^2. The answer checks.

State. The base is 8 ft, and the height is 16 ft.

13. **Familiarize.** We use the function $x^2 - x = N$ and set N equal to 240 games.

Translate.

$$\underset{\downarrow}{\underline{\text{Number of games}}} \quad \underset{\downarrow}{\underline{\text{equals}}} \quad \underset{\downarrow}{\underline{240}}$$

$$x^2 - x \qquad = \quad 240$$

Carry out. We solve the equation:

$$x^2 - x = 240$$
$$x^2 - x - 240 = 0$$
$$(x+15)(x-16) = 0$$
$$x+15 = 0 \quad or \quad x-16 = 0$$
$$x = -15 \; or \qquad x = 16$$

Check. We check only 16 since the number of teams cannot be negative. If 16 teams are in the league, the number of games played is $16^2 - 16 = 256 - 16 = 240$. The answer checks.

State. The number of teams in the league is 16.

15. ***Familiarize.*** We use the function $A = -50t^2 + 200t$ and set A equal to 150 micrograms.

Translate.

$$\underbrace{\text{Size of dose}} \quad \underbrace{\text{equals}} \quad \underbrace{150 \ \mu g}$$
$$\downarrow \qquad\qquad \downarrow \qquad\quad \downarrow$$
$$-50t^2 + 200t \quad = \qquad 150$$

Carry out. We solve the equation:
$$-50t^2 + 200t = 150$$
$$-50t^2 + 200t - 150 = 0$$
$$-50\left(t^2 - 4t + 3\right) = 0$$
$$t^2 - 4t + 3 = 0$$
$$(t-3)(t-1) = 0$$
$$t - 3 = 0 \quad or \quad t - 1 = 0$$
$$t = 3 \quad or \qquad t = 1$$

Check. If the time is 3 minutes, the amount of Albuterol in the bloodstream is
$$-50(3)^2 + 200 \cdot 3 = -50 \cdot 9 + 600$$
$$= -450 + 600$$
$$= 150$$

If the time is 1 minute, the amount of Albuterol in the bloodstream is
$$-50(1)^2 + 200 \cdot 1 = -50 + 200$$
$$= 150$$

Both answers check.

State. The time at which 150 micrograms of Albuterol is present in the bloodstream is either 1 minute or 3 minutes.

17. ***Familiarize.*** We use the function $H = 0.006x^2 + 0.6x$ and set H equal to 6.6 ft.

Translate.

$$\underbrace{\text{Height of wave}} \quad \underbrace{\text{equals}} \quad \underbrace{6.6 \ \text{ft}}$$
$$\downarrow \qquad\qquad \downarrow \qquad\quad \downarrow$$
$$0.006x^2 + 0.6x \quad = \qquad 6.6$$

Carry out. We solve the equation:

$$0.006x^2 + 0.6x = 6.6$$
$$0.006x^2 + 0.6x - 6.6 = 0$$
$$0.006\left(x^2 + 100x - 1100\right) = 0$$
$$x^2 + 100x - 1100 = 0$$
$$(x+110)(x-10) = 0$$
$$x + 110 = 0 \qquad or \quad x - 10 = 0$$
$$x = -110 \quad or \qquad\quad x = 10$$

Check. We check only 10 since the wind speed cannot be negative. If the wind speed is 10 knots, the height of the wave is
$$0.006(10)^2 + 0.6 \cdot 10 = 0.006 \cdot 100 + 6$$
$$= 0.6 + 6$$
$$= 6.6$$

The answer checks.

State. The speed of the wind required to produce a wave 6.6 ft high is 10 knots.

19. ***Familiarize.*** We use the function $h(t) = -15t^2 + 75t + 10$. Note that t cannot be negative since it represents time after launch.

Translate. We need to determine the value of t for which $h(t) = 70$ ft:

$$-15t^2 + 75t + 10 = 70$$

Carry out. We solve the equation.
$$-15t^2 + 75t + 10 = 70$$
$$-15t^2 + 75t = 60$$
$$-15\left(t^2 - 5t\right) = 60$$
$$t^2 - 5t = -4$$
$$t^2 - 5t + 4 = 0$$
$$(t-4)(t-1) = 0$$
$$t - 4 = 0 \quad or \quad t - 1 = 0$$
$$t = 4 \quad or \qquad t = 1$$

Check. We reject 1, since this represents the moment of launch when the tee shirt starts traveling upwards.
$$h(4) = -15 \cdot 4^2 + 75 \cdot 4 + 10$$
$$= -240 + 300 + 10 = 70$$

State. The tee shirt was airborne for 4 sec before it was caught.

21. ***Familiarize.*** We will use the given formula, $h(t) = -16t^2 + 64t + 80$. Note that t cannot be negative since it represents time after launch.

Translate. We need to find the value of t for which $h(t) = 0$. We have:

$$-16t^2 + 64t + 80 = 0$$

Carry out. We solve the equation.

$$-16t^2 + 64t + 80 = 0$$
$$-16\left(t^2 - 4t - 5\right) = 0$$
$$-16(t - 5)(t + 1) = 0$$
$$t - 5 = 0 \ or \ t + 1 = 0$$
$$t = 5 \ or \ \ \ \ t = -1$$

Check. Since t cannot be negative, we check only 5. $h(5) = -16 \cdot 5^2 + 64 \cdot 5 + 80$

$= -400 + 320 + 80 = 0$. The number 5 checks.

State. The cardboard shell will reach the ground in 5 sec after it is launched.

23. **Familiarize.** We make a drawing and label it. We let x represent the length of a side of the original square, in meters.

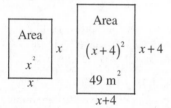

Translate.

Area of
$\underbrace{\text{new square}}$ is $49 \ m^2$
       $\downarrow$ $\downarrow$ $\downarrow$

$\ \ \ \ (x + 4)^2 \ \ \ \ = \ \ \ 49$

Carry out. We solve the equation.

$$(x + 4)^2 = 49$$
$$x^2 + 8x + 16 = 49$$
$$x^2 + 8x - 33 = 0$$
$$(x - 3)(x + 11) = 0$$
$$x - 3 = 0 \ or \ x + 11 = 0$$
$$x = 3 \ or \ \ \ \ x = -11$$

Check. We check only 3 since the length of a side cannot be negative. If we increase the length by 4, the new length is $3 + 4$, or 7 m. Then the new area is $7 \cdot 7$, or $49 \ m^2$. We have a solution.

State. The length of a side of the original square is 3 m.

25. **Familiarize.** We make a drawing and label it with both known and unknown information. We let x represent the width of the frame. (We assume the frame is of uniform width.) The dimensions of the picture that show are represented by $20 - 2x$ and $12 - 2x$. The area of the picture that shows is $84 \ cm^2$.

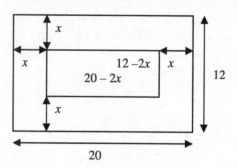

Translate. Using the formula for the area of a rectangle $A = l \cdot w$, we have

$$84 = (20 - 2x)(12 - 2x).$$

Carry out. We solve the equation:

$$84 = 240 - 64x + 4x^2$$
$$84 = 4\left(60 - 16x + x^2\right)$$
$$21 = 60 - 16x + x^2$$
$$0 = x^2 - 16x + 39$$
$$0 = (x - 3)(x - 13)$$
$$x - 3 = 0 \ or \ x - 13 = 0$$
$$x = 3 \ or \ \ \ \ x = 13$$

Check. We see that 13 is not a solution because when $x = 13$, $20 - 2x = -6$ and $12 - 26 = -14$, and the length and width of the picture cannot be negative. We check 3. When $x = 3$, $20 - 2x = 14$ and $12 - 2x = 6$ and $14 \cdot 6 = 84$. The area is 84. The value checks.

State. The width of the frame is 3 cm.

27. ***Familiarize.*** We let x represent the width of the sidewalk. We make a drawing and label it with both the known and unknown information.

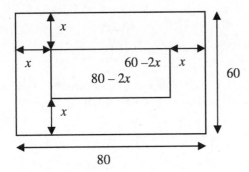

The area of the new lawn is
$(80 - 2x)(60 - 2x)$.

Translate.

$$\underbrace{\text{Area of new lawn}}_{(80-2x)(60-2x)} \quad \underset{=}{\text{is}} \quad \underset{2400}{2400 \, ft^2}$$

Carry out. We solve the equation:

$(80 - 2x)(60 - 2x) = 2400$

$4800 - 280x + 4x^2 = 2400$

$4x^2 - 280x + 2400 = 0$

$x^2 - 70x + 600 = 0$ Dividing by 4

$(x - 10)(x - 60) = 0$

$x - 10 = 0 \quad or \quad x - 60 = 0$

$x = 10 \; or \qquad x = 60$

Check. If the sidewalk is 10 ft wide, the length of the new lawn will be $80 - 2 \cdot 10$, or 60 ft, and its width will be $60 - 2 \cdot 10$, or 40ft. Then the area of the new lawn will be $60 \cdot 40$, or 2400 ft². This answer checks. If the sidewalk is 60 ft wide, the length of the new lawn will be $80 - 2 \cdot 60$, or –40 ft. Since the length cannot be negative, 60 is not a solution.

State. The sidewalk is 10 ft wide.

29. ***Familiarize.*** Let h = the height of the tower braces. We use the Pythagorean theorem $a^2 + b^2 = c^2$.

Translate. $\quad a^2 \; + \; b^2 \; = \; c^2$
$$12^2 \; + \; h^2 \; = \; 15^2$$

Carry out. We solve the equation:

$12^2 + h^2 = 15^2$

$h^2 = 15^2 - 12^2$

$h^2 = 225 - 144$

$h^2 = 81$

$h = -9 \quad or \quad h = 9$

Check. We check only 9 since the height of a side cannot be negative. Using the Pythagorean theorem to find the length of the diagonal, $12^2 + 9^2 = 144 + 81 = 225 = 15^2$. The answer checks.

State. The brace is 9 ft high.

31. ***Familiarize.*** Let w = the width of Main Street. We use the Pythagorean theorem $a^2 + b^2 = c^2$.

Translate. $\quad a^2 \; + \; b^2 \; = \; c^2$
$$24^2 \; + \; w^2 \; = \; 40^2$$

Carry out. We solve the equation:

$24^2 + w^2 = 40^2$

$w^2 = 40^2 - 24^2$

$w^2 = 1600 - 576$

$w^2 = 1024$

$w = -32 \quad or \quad w = 32$

Check. We check only 32 since the width cannot be negative. Using the Pythagorean theorem to find the diagonal across Main and Elliott Streets, $24^2 + 32^2 = 576 + 1024 = 1600 = 40^2$. The answer checks.

State. Main Street is 32 ft across.

33. ***Familiarize.*** Let x represent the unknown side of the right triangle and $x + 200$ represent the hypotenuse. We use the Pythagorean Theorem $a^2 + b^2 = c^2$.

Translate. $\quad a^2 \; + \; b^2 \; = \; c^2$
$$400^2 + \; x^2 \; = (x + 200)^2$$

Carry out. We solve the equation:

$400^2 + x^2 = (x + 200)^2$

$400^2 + x^2 = x^2 + 400x + 200^2$

$400^2 - 200^2 = 400x$

$160,000 - 40,000 = 400x$

$120,000 = 400x$

$300 = x$

Check. Using the Pythagorean theorem to find the length of the garden's diagonal:

$$400^2 + 300^2 = 160,000 + 90,000$$
$$= 250,000$$
$$= 500^2$$
$$= (300 + 200)^2$$

The answer checks.

State. The sides are 300 ft and 400 ft in length, and the diagonal is 500 ft long.

35. Familiarize. Let d represent the base of the right triangle and $d + 4$ represent the height. These are the legs of the right triangle and 20 is the length of the hypotenuse. Recall: Pythagorean Theorem $a^2 + b^2 = c^2$.

Translate.
$$\underset{\downarrow}{a^2} + \underset{\downarrow}{b^2} = \underset{\downarrow}{c^2}$$

$$d^2 + (d+4)^2 = 20^2$$

Carry out. We solve the equation:

$$d^2 + (d+4)^2 = 20^2$$
$$d^2 + d^2 + 8d + 16 = 400$$
$$2d^2 + 8d - 384 = 0$$
$$2(d^2 + 4d - 192) = 0$$
$$d^2 + 4d - 192 = 0$$
$$(d+16)(d-12) = 0$$
$$d + 16 = 0 \quad or \quad d - 12 = 0$$
$$d = -16 \quad or \quad d = 12$$

Check. Since measure cannot be negative, we know -16 is not a solution. If $d = 12$, $d + 4 = 16$ and

$$12^2 + 16^2 = 20^2$$
$$144 + 256 = 400$$
$$400 = 400$$

The answer checks.

State. The distance is 12 ft, and the height of the tower is 16 ft.

37. Familiarize. Let x represent the base of the right triangle, $x + 1$ represent the other leg, and $x + 2$ represent the hypotenuse. Recall: Pythagorean Theorem $a^2 + b^2 = c^2$.

Translate.
$$\underset{\downarrow}{a^2} + \underset{\downarrow}{b^2} = \underset{\downarrow}{c^2}$$

$$x^2 + (x+1)^2 = (x+2)^2$$

Carry out. We solve the equation:

$$x^2 + (x+1)^2 = (x+2)^2$$
$$x^2 + x^2 + 2x + 1 = x^2 + 4x + 4$$
$$x^2 + 2x - 4x + 1 - 4 = 0$$
$$x^2 - 2x - 3 = 0$$
$$(x-3)(x+1) = 0$$
$$x - 3 = 0 \quad or \quad x + 1 = 0$$
$$x = 3 \quad or \quad x = -1$$

Check. Since measure cannot be negative, we know -1 is not a solution. If $x = 3$, $x + 1 = 4$ and

$$3^2 + 4^2 = (3+2)^2$$
$$9 + 16 = 5^2$$
$$25 = 25$$

State. The lengths of the sides are 3, 4 and 5 units.

39. Familiarize. Let w represent the width and $w + 10$ represent the length.

$$Area = length \times width$$

Translate.
$$\underset{\downarrow}{\underline{Area}} \quad \underset{\downarrow}{is} \quad \underset{\downarrow}{264 \text{ ft}^2}$$

$$w(w+10) = 264$$

Carry out. We solve the equation:

$$w(w+10) = 264$$
$$w^2 + 10w = 264$$
$$w^2 + 10w - 264 = 0$$
$$(w+22)(w-12) = 0$$
$$w + 22 = 0 \quad or \quad w - 12 = 0$$
$$w = -22 \quad or \quad w = 12$$

Check. -22 cannot be a solution because width cannot be negative. If $w = 12$ then the length is $w + 10$, which is 22. The area is $12 \cdot 22$ which is 264 ft^2.

State. The dimensions for the total space will be 12 ft by 22 ft. This will divide as 12 ft by 12 ft for the dining room, and 12 ft by 10 ft for the kitchen.

41.
a) Enter the data in the graphing calculator, letting x represent the number of years after 1990. Then use the quartic regression feature to find the desired function $P(x) = 0.01823x^4 - 0.77199x^3 + 11.62153x^2 - 73.65807x + 179.76190$.

b) In 2005, $x = 15$.

$P(15) \approx 7\%$

c) We solve $P(x) = 12$ graphically. Graph $y_1 = P(x)$ and $y_2 = 12$. Both $x \approx 13$ years and $x \approx 17$ years after 1990 satisfy the equation. These values correspond to the years 2003 and 2007.

43. a) Enter the data in the graphing calculator, letting x represent the number of years after 1900. Then use the cubic regression feature to find the desired function $F(x) = -0.03587x^3 + 10.35169x^2 - 871.97543x + 23,423.45189$.

b) In 2012, $x = 112$.
$F(112) \approx 5220$ athletes

c) We solve $F(x) = 6000$ graphically.
Graph $y_1 = F(x)$ and $y_2 = 6000$. Both $x \approx 122$ years and $x \approx 137$ after 1900 satisfy the equation. These correspond to the years 2022 and 2037; however, since the Summer Olympics occur every four years, we estimate these years as 2024 and 2036.

45. *Thinking and Writing Exercise.*

47. $-\dfrac{3}{5} \cdot \dfrac{4}{7} = -\dfrac{3 \cdot 4}{5 \cdot 7} = -\dfrac{12}{35}$

49. $-\dfrac{5}{6} - \dfrac{1}{6} = \dfrac{-5-1}{6} = \dfrac{-(5+1)}{6} = \dfrac{-6}{6} = -1$

51. $-\dfrac{3}{8} \cdot \left(-\dfrac{10}{15}\right) = -\left(-\dfrac{3 \cdot 10}{8 \cdot 15}\right)$

$= \dfrac{3 \cdot 10}{8 \cdot 15}$

$= \dfrac{1 \cdot 3 \cdot 2 \cdot 5}{2 \cdot 4 \cdot 3 \cdot 5}$

$= \dfrac{1}{4}$

53. $\dfrac{5}{24} + \dfrac{3}{28} = \dfrac{5}{4 \cdot 6} + \dfrac{3}{4 \cdot 7}$

$= \dfrac{5 \cdot 7}{4 \cdot 6 \cdot 7} + \dfrac{3 \cdot 6}{4 \cdot 7 \cdot 6}$

$= \dfrac{35}{168} + \dfrac{18}{168}$

$= \dfrac{35 + 18}{168}$

$= \dfrac{53}{168}$

55. *Thinking and Writing Exercise.*

57. *Familiarize.* From the drawing in the text, we note that the roof has two sides of equal area. The length of each side is 32 ft, while the width must be found using the Pythagorean Theorem. The width equals the diagonal of a right triangle whose height equals the difference between the heights of the peak of the house (25 ft) and the base of the roof (16 ft), while the base of the right triangle equals half of the house width (24 ft). The number of squares of shingles equals the total area, in square feet, divided by 100.

Translate.

$\underbrace{\text{Area}}\ \underbrace{\text{is}}\ \underbrace{\text{twice}}\ \underbrace{\text{length}}\ \underbrace{\text{times}}\quad\ \underbrace{\text{width}}.$

$A = 2 \left[32 \cdot \sqrt{\left(\dfrac{24}{2}\right)^2 + (25-16)^2} \right]$

$\underbrace{\text{Number of squares}}\ \underbrace{\text{is}}\ \underbrace{\text{area}}\ \underbrace{\text{divided by}}\ \underbrace{100}.$

$N = A \div 100$

Carry out. We solve the equations.

$A = 2 \left[32 \cdot \sqrt{\left(\dfrac{24}{2}\right)^2 + (25-16)^2} \right]$

$= 2 \left[32 \cdot \sqrt{12^2 + 9^2} \right]$

$= 2 \left[32 \cdot \sqrt{144 + 81} \right]$

$= 2 \left[32 \cdot \sqrt{225} \right]$

$= 2 \left[32 \cdot 15 \right]$

$= 2 \cdot 480$

$= 960$

$N = \dfrac{A}{100} = \dfrac{960}{100} = 9.6 \approx 10$

Check. Find the height of the roof by taking the approximate area for either side,

$\dfrac{10\cdot100\text{ ft}^2}{2}$, or 500 ft^2, and applying the area

formula in reverse.

$w = \dfrac{500}{32} = 15.625$

$h = \sqrt{15.625^2 - \left(\dfrac{24}{2}\right)^2} \approx \sqrt{244^2 - 144^2} \approx 10$

The approximate roof height of 10 ft is close to the actual height of $25-16$, or 9 ft. The answer checks.

State. A total of 10 squares of shingles will be needed to cover the area of the roof.

59. We solve $N(t) = 18$ graphically. Graph $y_1 = N(t)$ and $y_2 = 18$. Both $t = 2$ hours and $t \approx 4.2$ hours satisfy the equation.

61. We solve for the time at which $N(t)$ is at a maximum by graphing $y_1 = N(t)$ and using the "maximum" feature to find the peak value $y_1 = 19.683$. This values is reached at $t = 3$ hours.

63. **Familiarize.** Using the labels on the drawing in the text, we let x represent the width of the piece of tin and $2x$ represent the length. Then the width and length of the base of the box are represented by $x-4$ and $2x-4$, respectively. Recall that the formula for the volume of a rectangular solid with length l, width w, and height h is $l\cdot w\cdot h$.

Translate.   The volume is $480cm^3$.

$(2x-4)(x-4)(2) = 480$

Carry out. We solve the equation.

$(2x-4)(x-4)(2) = 480$

$(2x-4)(x-4) = 240$ Dividing by 2

$2x^2 - 12x + 16 = 240$

$2x^2 - 12x - 224 = 0$

$x^2 - 6x - 112 = 0$ Dividing by 2

$(x+8)(x-14) = 0$

$x+8 = 0 \ \ or \ x - 14 = 0$

$x = -8 \ or \ \ \ \ x = 14$

Check. We check only 14 since the width cannot be negative. If the width of the piece of tin is 14 cm, then its length is $2\cdot14$, or 28 cm, and the dimensions of the base of the box are $14-4$, or 10 cm by $28-4$, or 24 cm. The volume of the box is $24\cdot10\cdot2$, or 480 cm^3. The answer checks.

State. The dimensions of the piece of tin are 14 cm by 28 cm.

65. Graph $y_1 = 11.12(x+1)^2$ and $y_2 = 15.4x^2$ in
a   window that shows the point of intersection of the graphs. The window $[0,10,0,1000]$, Xscl = 1, Yscl = 100 is one good choice. Then find the first coordinate of the point of intersection. It is approximately 5.7, so it
will   take the camera about 5.7 sec to catch up
to   the skydiver.

Chapter 6 Study Summary

1. $12x^4 - 18x^3 + 30x$

$= 6x(2x^3 - 3x^2 + 5)$

2. $2x^3 - 6x^2 - x + 3 = 2x^2(x-3) - 1(x-3)$

$= (2x^2 - 1)(x-3)$

3. $8x = 6x^2$

$6x^2 - 8x = 0$

$2x(3x-4) = 0$

$2x = 0 \ \ \ or \ \ \ 3x - 4 = 0$

$x = 0 \ \ \ or \ \ \ \ \ \ \ x = \dfrac{4}{3}$

4. $x^2 - 7x - 18$

Since the constant term is negative, we look for factorization of −18 in which one factor is positive and one factor is negative. Their sum must be −7, so the negative factor must have the larger absolute value. Thus we consider only pairs of factors in which the positive factor has the larger absolute value.

Pair of Factors	Sum of Factors
1, −18	−17
2, −9	−7
3, −6	−3

The numbers we need are 2 and −9. The factorization is $(x-9)(x+2)$.

5. $6x^2 + x - 2$
We will use the FOIL method.
1. There is no common factor (other than 1 or −1).
2. Factor the first term, $6x^2$. The factors are $6x$ and x, and $3x$ and $2x$. We have the possibilities $(6x+\ )(x+\ )$ and $(3x+\ )(2x+\ )$.
3. Factor the last term, −2. The possibilities are $-1\cdot2$, $1\cdot(-2)$, $-2\cdot1$, and $2(-1)$.
4. Look for factors for which the sum of the products is x. Trial and error leads us to: $(3x+2)(2x-1)$.

6. $8x^2 - 22x + 15$
We will use the grouping method.
1. There is no common factor (other than 1 or −1).
2. Multiply the leading coefficient and the constant $8(15) = 120$.
3. We want factors of 120 whose sum is −22.

Pair of Factors	Sum of Factors
−1, −120	−121
−2, −60	−62
−3, −40	−43
−4, −30	−34
−5, −24	−29
−6, −20	−26
−8, −15	−23
−10, −12	−22

Since $-10\cdot(-12) = 120$ and $-10-12 = -22$, we will split $-22x$ into $-10x$ and $-12x$.

4.   Factor by grouping:
$$8x^2 - 22x + 15 = 8x^2 - 10x - 12x + 15$$
$$= 2x(4x-5) - 3(4x-5)$$
$$= (2x-3)(4x-5)$$

7. $100n^2 + 81 + 180n = 100n^2 + 180n + 81$
$$= 100n^2 + 2\cdot9\cdot10n + 9^2$$
This is of the form $A^2 + 2BA + B^2$, which is a perfect-square trinomial $(A+B)^2$, so
$$100n^2 + 81 + 180n = (10n+9)^2$$

8. $144t^2 - 25$
This is of the form $A^2 - B^2$ (difference of squares) because $144t^2 = (12t)^2$ and $25 = 5^2$, and the terms have different signs.
$$144t^2 - 25 = (12t)^2 - 5^2 = (12t+5)(12t-5)$$

9. $a^3 - 1 = a^3 - 1^3$
$$= (a-1)(a^2 + a + 1)$$
$$A^3 - B^3 = (A-B)(A^2 + AB + B^2)$$

10. $3x^4 - 24x + 5x^3 - 40$
$$= 3x^4 + 5x^3 - 24x - 40$$
$$= x^3(3x+5) - 8(3x+5)$$
$$= (x^3 - 8)(3x+5)$$
$$= (x^3 - 2^3)(3x+5)$$
$$= (x-2)(x^2 + 2x + 4)(3x+5)$$

11. $a^2\ +\ b^2\ =\ c^2$
$\downarrow\qquad\downarrow\qquad\downarrow$
$$5^2\ +\ x^2 = (x+1)^2$$
$$5^2 + x^2 = (x+1)^2$$
$$25 + x^2 = x^2 + 2x + 1$$
$$25 - 1 = 2x$$
$$24 = 2x$$
$$x = 12$$
The lengths of the sides are 5, 12 and 13 units.

Chapter 6 Review Exercises

1. False

2. True

3. True

4. False

5. False

6. True

7. True

8. False

9. $7x^2 + 6x$

$= x \cdot 7x + x \cdot 6$

$= x(7x + 6)$

10. $18y^4 - 6y^2$

$= 6y^2 \cdot 3y^2 + 6y^2 \cdot -1$

$= 6y^2(3y^2 - 1)$

11. $100t^2 - 1 = (10t)^2 - 1^2$ Difference of squares

$= (10t + 1)(10t - 1)$ $A^2 - B^2 = (A+B)(A-B)$

12. $a^2 - 12a + 27$

We want two negative factors of 27, whose sum is –12. They are –9 and –3.

$= (a - 9)(a - 3)$

13. $3m^2 + 14m + 8$

Using the grouping method, multiply the leading coefficient, 3 and the constant, 8. $3 \cdot 8 = 24$. Factor 24 so the sum of factors is 14; both factors will be positive since $14m$ and 8 are both positive. They are 12 and 2. Split the middle term using $14m = 12m + 2m$ and factor by grouping.

$3m^2 + 14m + 8 = 3m^2 + 12m + 2m + 8$

$= 3m(m + 4) + 2(m + 4)$

$= (m + 4)(3m + 2)$

14. $25x^2 + 20x + 4 = (5x)^2 + 2(5x)2 + 2^2$

$= (5x + 2)^2$

Find the square terms and write the quantities that were squared with a plus sign between them, i.e., factoring the perfect-square trinomial.

15. $4y^2 - 16$

$= 4(y^2 - 4)$ Common Factor

$= 4(y^2 - 2^2)$ Difference of two squares

$= 4(y + 2)(y - 2)$ $A^2 - B^2 = (A+B)(A-B)$

16. $5x^2 + x^3 - 14x = x^3 + 5x^2 - 14x$

$= x(x^2 + 5x - 14)$

Common Factor

To factor $x^2 + 5x - 14$, we want to find two factors of –14 whose sum is 5. They are 7 and –2:

$= x(x + 7)(x - 2)$

17. $ax + 2bx - ay - 2by$

$= x(a + 2b) - y(a + 2b)$ Factor by grouping

$= (a + 2b)(x - y)$

18. $3y^3 + 6y^2 - 5y - 10$

$= 3y^2(y + 2) - 5(y + 2)$ Factor by grouping

$= (y + 2)(3y^2 - 5)$

19. $81a^4 - 1 = (9a^2)^2 - 1^2$ Difference of squares

$= (9a^2 + 1)(9a^2 - 1)$ $A^2 - B^2 = (A+B)(A-B)$

$= (9a^2 + 1)((3a)^2 - 1^2)$ Difference of squares

$= (9a^2 + 1)(3a + 1)(3a - 1)$ $A^2 - B^2 =$
$(A+B)(A-B)$

20. $48t^2 - 28t + 6$

$= 2(24t^2 - 14t + 3)$ Common Factor

21. $27x^3 - 8 = (3x)^3 - 2^3$ Difference of 2 cubes

$= (3x - 2)\left[(3x)^2 + 3x \cdot 2 + 2^2\right]$

$= (3x - 2)(9x^2 + 6x + 4)$

$A^3 - B^3 = (A - B)(A^2 + AB + B^2)$

22. $-t^3 + t^2 + 42t$

$= -t(t^2 - t - 42)$ Common factor

$= -t(t - 7)(t + 6)$

23. $a^2 b^4 - 64$

$= \left(ab^2\right)^2 - 8^2$ Difference of squares

$= (ab^2 + 8)(ab^2 - 8)$

$A^2 - B^2 = (A + B)(A - B)$

24. $3x + x^2 + 5 = x^2 + 3x + 5$

We want two positive factors of 5, whose sum is 3. This is not possible with integer coefficients, so $3x + x^2 + 5$ is prime.

25. $2z^8 - 16z^6 = 2z^6\left(z^2 - 8\right)$ Common Factor

26. $54x^6 y - 2y$

$= 2y \cdot 27x^6 - 2y \cdot 1$ Common Factor

$= 2y\left(27x^6 - 1\right)$

Factor $27x^6 - 1$.

$27x^6 - 1 = \left(3x^2\right)^3 - 1^3$ Difference of 2 cubes

$= (3x^2 - 1)(9x^4 + 3x^2 + 1)$

$A^3 - B^3 = (A - B)(A^2 + AB + B^2)$

Therefore,

$54x^6 y - 2y = 2y(3x^2 - 1)(9x^4 + 3x^2 + 1)$

27. $75 + 12x^2 - 60x$

$= 3\left(4x^2 - 20x + 25\right)$ Common Factor

$= 3(2x - 5)^2$ Perfect-square trinomial

28. $6t^2 + 17pt + 5p^2$

Factors of $6t^2$ are $6t$, t and $3t$, $2t$.

Positive factors of $5p^2$ are $5p$, p.

To get $17pt$ we determine the factorization is $(3t + p)(2t + 5p)$.

$6t^2 + 17pt + 5p^2 = (3t + p)(2t + 5p)$

29. $x^3 + 2x^2 - 9x - 18$

$= x^2(x + 2) - 9(x + 2)$ Factor by grouping

$= (x + 2)(x^2 - 9)$

$= (x + 2)(x + 3)(x - 3)$ Factor difference of two squares

30. $a^2 - 2ab + b^2 - 4t^2$

$= (a^2 - 2ab + b^2) - 4t^2$ Grouping as a difference of squares

$= (a - b)^2 - (2t)^2$

$= (a - b + 2t)(a - b - 2t)$

31. From the graph we see that $p(x) = 0$ when $x = -2$, $x = 1$, or $x = 5$. These are the solutions.

32. The zeros of $f(x)$ are the values of x which make $f(x) = 0$. We solve $0 = x^2 - 11x + 28$.

$0 = x^2 - 11x + 28$

$0 = (x - 7)(x - 4)$

$x - 7 = 0 \ or \ x - 4 = 0$ Principle of

$x = 7 \ or \ \ \ \ \ x = 4$ Zero Products

The solutions are 7 and 4.

33. $(x - 9)(x + 11) = 0$

$x - 9 = 0 \ \ or \ \ x + 11 = 0$

$x = 9 \ \ or \ \ \ \ \ \ x = -11$

The solutions are 9 and −11.

34. $6b^2 - 13b + 6 = 0$

Factor and use Principle of Zero Products

$(2b - 3)(3b - 2) = 0$

$2b - 3 = 0 \ or \ 3b - 2 = 0$

$b = \dfrac{3}{2} \ or \ \ \ \ b = \dfrac{2}{3}$

The solutions are $\dfrac{3}{2}$ and $\dfrac{2}{3}$.

35. $8t^2 = 14t$

Factor and use the principle of zero products.

$$8t^2 = 14t$$

$$8t^2 - 14t = 0$$

$$2t(4t - 7) = 0$$

$$2t = 0 \text{ or } 4t - 7 = 0$$

$$t = 0 \text{ or } \quad t = \frac{7}{4}$$

The solutions are 0 and $\frac{7}{4}$.

36. $x^2 - 20x = -100$

$$x^2 - 20x + 100 = 0$$

$$(x - 10)^2 = 0 \quad \text{Factor the}$$
$$\text{perfect-square}$$
$$\text{trinomial}$$

$$x - 10 = 0 \quad \text{or } x - 10 = 0 \quad \text{Principle of}$$

$$x = 10 \text{ or } \quad x = 10 \quad \text{Zero Products}$$

The solution is 10.

37. $r^2 = 16$

Factor and use the principle of zero products.

$$r^2 = 16$$

$$r^2 - 16 = 0$$

$$(r + 4)(r - 4) = 0$$

$$r + 4 = 0 \quad \text{or } r - 4 = 0$$

$$r = -4 \text{ or } \quad r = 4$$

The solutions are –4 and 4, or ±4.

38. $a^3 = 4a^2 + 21a$

$$a^3 - 4a^2 - 21a = 0$$

Factor and use the principle of zero products.

$$a^3 - 4a^2 - 21a = a(a^2 - 4a - 21)$$

$$= a(a - 7)(a + 3)$$

$$a(a - 7)(a + 3) = 0$$

$$a = 0 \text{ or } a - 7 = 0 \text{ or } a + 3 = 0$$

$$a = 0 \text{ or } a = 7 \quad \text{or} \quad a = -3$$

The solutions are –3, 0, and 7.

39. $x(x - 1) = 20$

$$x^2 - x = 20$$

$$x^2 - x - 20 = 0$$

Factor and use the principle of zero products.

$$(x - 5)(x + 4) = 0$$

$$x - 5 = 0 \quad \text{or } x + 4 = 0$$

$$x = 5 \quad \text{or} \quad x = -4$$

The solutions are –4 and 5.

40. $x^3 - 5x^2 - 16x + 80 = 0$

$$x^2(x - 5) - 16(x - 5) = 0 \quad \text{Factor by}$$
$$\text{grouping}$$

$$(x - 5)(x^2 - 16) = 0$$

$$(x - 5)(x + 4)(x - 4) = 0 \quad \text{Factor the}$$
$$\text{perfect-square}$$
$$\text{trinomial}$$

$$x - 5 = 0 \text{ or } x + 4 = 0 \quad \text{or } x - 4 = 0$$

$$x = 5 \text{ or } \quad x = -4 \quad \text{or} \quad x = 4$$

Principle of Zero Products

The solutions are –4, 4, and 5.

41. $x^2 + 180 = 27x$

$$x^2 - 27x + 180 = 0$$

Factor and use the principle of zero products.

$$(x - 15)(x - 12) = 0$$

$$x - 15 = 0 \quad \text{or } x - 12 = 0$$

$$x = 15 \text{ or } \quad x = 12$$

The solutions are 12 and 15.

42.   Graph $y = x^2 - 2x - 6$ and find the zeros.
Using the zero feature we find $x \approx -1.646$
and $x \approx 3.646$ which are the solutions.

43. $f(x) = x^2 - 7x - 40$. Let $f(a) = 4$ and
solve the resulting equation by factoring and
using the principle of zero products.

$$4 = x^2 - 7x - 40$$

$$0 = x^2 - 7x - 44$$

$$0 = (x - 11)(x + 4)$$

$$x - 11 = 0 \quad \text{or } x + 4 = 0$$

$$x = 11 \text{ or } \quad x = -4$$

The solutions are –4 and 11.

44. $f(x) = \dfrac{x - 3}{3x^2 + 19x - 14}$

To find the excluded values solve
$3x^2 + 19x - 14 = 0$, by factoring and using the
principle of zero products.

$$3x^2 + 19x - 14 = 0$$

$$(3x - 2)(x + 7) = 0$$

$3x - 2 = 0$ or $x + 7 = 0$

$$x = \frac{2}{3} \quad or \quad x = -7$$

The domain is

$\left\{ x \mid x \text{ is a real number and } x \neq -7 \text{ and } x \neq \frac{2}{3} \right\}$

45. **Familiarize.** We use the function $x^2 - x = N$ and set N equal to 90 games.
   Translate.

   $\underbrace{\text{Number of games}}_{\downarrow} \quad \underbrace{\text{equals}}_{\downarrow} \quad \underbrace{90}_{\downarrow}$

   $x^2 - x \qquad = \qquad 90$

   Carry out. We solve the equation:

   $x^2 - x = 90$

   $x^2 - x - 90 = 0$

   $(x + 9)(x - 10) = 0$

   $x + 9 = 0 \quad or \quad x - 10 = 0$

   $x = -9 \quad or \qquad x = 10$

   Check. We check only 10 since the number of teams cannot be negative. If 10 teams are in the league, the number of games played is $10^2 - 10 = 100 - 10 = 90$. The answer checks.
   State. The number of teams in the league is 10.

46. **Familiarize.** Let w represent the width of the gable, and $\frac{3}{4} w$ represents the height. Recall that the formula for the area of a triangle is $A = \frac{1}{2} \times \text{base} \times \text{height}$.
   Translate. The area is 216 m^2.

   $\underbrace{\text{the area}}_{\downarrow} \quad \underbrace{\text{is}}_{\downarrow} \quad \underbrace{216 \text{ m}^2,}_{\downarrow}$

   $\frac{1}{2} w \left(\frac{3w}{4} \right) = 216$

   Carry out. We solve the equation:

   $\frac{1}{2} w \left(\frac{3w}{4} \right) = 216$

   $\frac{3w^2}{8} = 216$

   $w^2 = \frac{216 \cdot 8}{3}$

   $w^2 = 576$

   $w = -24 \quad or \quad w = 24$

Check. We check only 24, since the width cannot be negative. If the width is 24 m, the height is $\frac{3}{4}(24)$, or 18 m, and the area is $\frac{1}{2} \cdot 24 \cdot 18$, or 216 m^2. The answer checks.
State. The gable is 24 m wide and 18 m high.

47. **Familiarize.** Let $x =$ the width of the photograph and $x + 3$ the length.

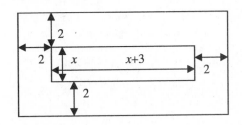

Translate.
Area of photograph

$\underbrace{\text{and border}}_{\downarrow} \quad \underbrace{\text{is}}_{\downarrow} \quad \underbrace{108 \text{ in}^2}_{\downarrow}$

$(x + 7)(x + 4) \quad = \quad 108$

Carry out. We solve the equation.

$(x + 7)(x + 4) = 108$

$x^2 + 11x - 80 = 0$

$(x + 16)(x - 5) = 0$

$x + 16 = 0 \quad or \quad x - 5 = 0$

$x = -16 \quad or \quad x = 5$

Check. Since length and width cannot be negative, we only check $x = 5$. The length of the frame is $(5 + 3) + 4$, or 12, and the width is $5 + 4$, or 9. So, the area of the frame is $12 \cdot 9 = 108$.
State. The width is 5 inches and the length is 8 inches.

48. **Familiarize.** Draw a triangle to represent the known and unknown information.

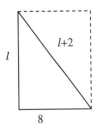

Translate. Use the Pythagorean Theorem to obtain the equation:

$$l^2 + 8^2 = (l+2)^2$$

Carry out. Solve the equation.

$$l^2 + 64 = l^2 + 4l + 4$$
$$60 = 4l$$
$$l = 15$$

Check. If $l = 15, l + 2 = 17$ and

$$8^2 + 15^2 = 64 + 225 = 289 = 17^2$$

State. The path is 17 ft.

49. a) Let x = the number of years after 1970. Enter the data in a graphing calculator. Use the quadratic regression feature to obtain $P(x) = 0.021174x^2 -$

   $0.944163x + 26.650798$

   b) In 2012, $x = 42$. $P(42) \approx 24.3\%$

   c) Solve $P(x) = 25$ graphically. Graph

   Both $x \approx 1.8$ years (approximately 2 years after 1970, or 1972) and $x \approx 42.8$ years (approximately 43 years after 1970), or 2013.

50. *Thinking and Writing Exercise.* The zeros of a polynomial function are the x-coordinates of the points at which the graph of the function crosses the x-axis.

51. *Thinking and Writing Exercise.* If the factors of the quadratic polynomial are different, there will be two different solutions. If the polynomial is a perfect square, the factors will be the same, and there will not be two different solutions.

52. $128x^6 - 2y^6$
$= 2(64x^6 - y^6)$ Common factor
$= 2[(8x^3)^2 - (y^3)^2]$ Factor as difference

of squares
$= 2(8x^3 + y^3)(8x^3 - y^3)$
$= 2[(2x)^3 + y^3][(2x)^3 - y^3]$
$= 2(2x + y)(4x^2 - 2xy + y^2)$
$(2x - y)(4x^2 + 2xy + y^2)$

Factoring a sum and difference of cubes

53. $(x-1)^3 - (x+1)^3$

$$(x-1)^3 = (x-1)(x-1)(x-1)$$
$$= (x-1)(x^2 - 2x + 1)$$
$$= x^3 - 2x^2 + x - x^2 + 2x - 1$$
$$= x^3 - 3x^2 + 3x - 1$$
$$(x+1)^3 = (x+1)(x+1)(x+1)$$
$$= (x+1)(x^2 + 2x + 1)$$
$$= x^3 + 2x^2 + x + x^2 + 2x + 1$$
$$= x^3 + 3x^2 + 3x + 1$$
$$(x-1)^3 - (x+1)^3 = (x^3 - 3x^2 + 3x - 1) -$$
$$(x^3 + 3x^2 + 3x + 1)$$
$$= x^3 - 3x^2 + 3x - 1 - x^3 -$$
$$3x^2 - 3x - 1$$
$$= -6x^2 - 2$$
$$= -2(3x^2 + 1)$$

54. $(x+1)^3 = x^2(x+1)$

Factor and use the principle of zero products.

$$(x+1)^3 - x^2(x+1) = 0$$
$$(x+1)[(x+1)^2 - x^2] = 0$$
Common factor
$$(x+1)[x^2 + 2x + 1 - x^2] = 0$$
$$(A+B)^2 = A^2 + 2AB + B^2$$
$$(x+1)(2x+1) = 0$$

$$x + 1 = 0 \quad or \quad 2x + 1 = 0$$
$$x = -1 \, or \qquad x = -\frac{1}{2}$$

The solutions are -1 and $-\frac{1}{2}$.

55. $x^2 + 100 = 0$
$$x^2 = -100$$
There is no real solution to this equation.

Chapter 6 Test

1. $x^2 - 10x + 25 = x^2 - 5x - 5x + 25$

    $\qquad = x(x-5) - 5(x-5)$

    $\qquad = (x-5)^2$

    Factor by grouping

2. $y^3 + 5y^2 - 4y - 20$

    $= y^2(y+5) - 4(y+5)$

    Factor by grouping

    $= (y+5)(y^2-4)$

    $= (y+5)(y+2)(y-2)$

    Factor the difference of squares

3. $p^2 - 12p - 28$

    We want two factors of -28 whose sum is -12. They are -14 and 2.

    $p^2 - 12p - 28 = (p-14)(p+2)$

4. $t^7 - 3t^5 = t^5(t^2-3)$

    Factor out the common factor

5. $12m^2 + 20m + 3$

    The factors of $12m^2$ are $12m, m$ and $6m, 2m$ and $4m, 3m$. Since $20m$ is positive, we want two positive factors of 3. They are 3 and 1. We want the sum of the products to equal $20m$. The factors are $(6m+1)(2m+3)$.

6. $9y^2 - 25 = (3y)^2 - 5^2$

    Write as difference of squares

    $= (3y+5)(3y-5)$

    $A^2 - B^2 = (A+B)(A-B)$

7. $3r^3 - 3 = 3(r^3-1)$

    Factor out the common factor

    $= 3(r-1)(r^2+r+1)$

    Factor the difference of two

    cubes using

    $A^3 - B^3 = (A-B)(A^2+AB+B^2)$

8. $45x^2 + 20 + 60x = 45x^2 + 60x + 20$

    $= 5(9x^2 + 12x + 4)$

    Factor out the common factor

    $= 5(9x^2 + 6x + 6x + 4)$

    $= 5[3x(3x+2) + 2(3x+2)]$

    $= 5(3x+2)(3x+2)$

    Factor by grouping

    $= 5(3x+2)^2$

9. $3x^4 - 48y^4 = 3(x^4 - 16y^4)$

    Factor out the common factor

    $= 3\left[(x^2)^2 - (4y^2)^2\right]$

    Write as difference of two squares

    $= 3(x^2 + 4y^2)(x^2 - 4y^2)$

    Factor the difference of two squares

    $= 3(x^2 + 4y^2)\left[x^2 - (2y)^2\right]$

    Repeat procedure

    $= 3(x^2 + 4y^2)(x+2y)(x-2y)$

10. $y^2 + 8y + 16 - 100t^2$

    $= (y^2 + 8y + 16) - 100t^2$

    Group as difference of two squares

    $= (y+4)^2 - (10t)^2$

    Factor perfect-square trinomial

    $= [(y+4) + 10t][(y+4) - 10t]$

    Factor difference of two squares

    $= (y+4+10t)(y+4-10t)$

11. $x^2 + 3x + 6$

    We want two positive factors of 6, whose sum is 3. This is not possible with integer coefficients, so $x^2 + 3x + 6$ is prime.

12. $20a^2 - 5b^2$

    $= 5(4a^2 - b^2)$ Common Factor

    $= 5(2a+b)(2a-b)$ Difference of 2 squares

13. $24x^2 - 46x + 10$

$= 2(12x^2 - 23x + 5)$ Common Factor

Factor $12x^2 - 23x + 5$ by the grouping method. The product of the leading coefficient, 12 and constant, 5, is $12 \cdot 5 = 60$. We want two factors of 60 whose sum is –23. They are –20 and –3. Split $-23x$ into $-20x + (-3x)$, and factor by grouping.

$12x^2 - 23x + 5 = 12x^2 - 20x + (-3x) + 5$

$\qquad\qquad\qquad = 4x(3x - 5) - 1(3x - 5)$

$\qquad\qquad\qquad = (3x - 5)(4x - 1)$

The original trinomial has a factorization that includes the common factor.

$24x^2 - 46x + 10 = 2(3x - 5)(4x - 1)$

14. $3m^2 - 9mn - 30n^2$

$= 3(m^2 - 3mn - 10n^2)$ Common Factor

Factor $m^2 - 3mn - 10n^2$ by the grouping method. The product of the leading coefficient, 1 and constant, –10, is $1 \cdot (-10) = -10$. We want two factors of –10 whose sum is –3. They are –5 and 2. Split $-3mn$ into $-5mn + 2mn$, and factor by grouping.

$3m^2 - 9mn - 30n^2$

$= 3(m^2 - 3mn - 10n^2)$

$= 3(m^2 - 5mn + 2mn - 10n^2)$

$= 3[m(m - 5n) + 2n(m - 5n)]$

$= 3(m + 2n)(m - 5n)$

15. From the graph we see $p(x) = 0$ when $x = -3$, $x = -1$, $x = 2$, and $x = 4$. These are the zeros of p.

16. Let $f(x) = 0$ and solve the resulting equation using factoring and the principle of zero products.

$0 = 2x^2 - 11x - 40$

$0 = (2x + 5)(x - 8)$

$2x + 5 = 0$ or $x - 8 = 0$

$x = -\dfrac{5}{2}$ or $x = 8$

The zeros are $-\dfrac{5}{2}$ and 8.

17. $x^2 - 3x - 18 = 0$

$(x + 3)(x - 6) = 0$

$x + 3 = 0$ or $x - 6 = 0$

$x = -3$ or $x = 6$

The solutions are –3 and 6.

18. $5t^2 = 125$

$5t^2 - 125 = 0$

$5(t^2 - 25) = 0$

$t^2 - 25 = 0$ Multiplying by $\dfrac{1}{5}$

$(t + 5)(t - 5) = 0$

$t + 5 = 0$ or $t - 5 = 0$

$t = -5$ or $t = 5$

The solutions are ±5.

19. $2x^2 + 21 = -17x$

$2x^2 + 17x + 21 = 0$

$(2x + 3)(x + 7) = 0$

$2x + 3 = 0$ or $x + 7 = 0$

$x = -\dfrac{3}{2}$ or $x = -7$

The solutions are –7 and $-\dfrac{3}{2}$.

20. $9x^2 + 3x = 0$

$3x(3x + 1) = 0$

$3x = 0$ or $3x + 1 = 0$

$x = 0$ or $x = -\dfrac{1}{3}$

The solutions are $-\dfrac{1}{3}$ and 0.

21. $x^2 + 81 = 18x$

$x^2 - 18x + 81 = 0$

$(x - 9)^2 = 0$

$x - 9 = 0$ or $x - 9 = 0$

$x = 9$ or $x = 9$

The solution is 9.

22. $x^2(x+1) = 8x$

$x^3 + x^2 - 8x = 0$

Graph $y = x^3 + x^2 - 8x$ and find the zeros.

They are $x \approx -3.372$, $x = 0$, and

$x \approx 2.372$, which are the solutions.

23. $f(x) = 3x^2 - 15x + 11$

Let $f(a) = 11$ and solve the equation.

$11 = 3a^2 - 15a + 11$

$0 = 3a(a-5)$

$3a = 0 \text{ or } a - 5 = 0$

$a = 0 \text{ or } \quad a = 5$

$f(0) = 11 \text{ and } f(5) = 11$

24. $f(x) = \dfrac{3-x}{x^2 + 2x + 1}$

To find the excluded values solve

$x^2 + 2x + 1 = 0$.

$x^2 + 2x + 1 = 0$

$(x+1)^2 = 0$

$(x+1)(x+1) = 0 \rightarrow x = -1$

The domain of $f(x)$ is $\{x | x \text{ is a real}$

number and $x \neq -1\}$

25. **Familiarize.** Let x = width of the photograph

and $x + 3$ the length.

Translate. $\underbrace{\text{Area}}$ $\underset{\downarrow}{\text{is}}$ $\underset{\downarrow}{40cm^2}$

$x(x+3) = \quad 40$

Carry out. Solve the equation.

$x(x+3) = 40$

$x^2 + 3x - 40 = 0$

$(x+8)(x-5) = 0$

$x + 8 = 0 \text{ or } x - 5 = 0$

$x = -8 \text{ or } \quad x = 5$

Check. Since measure cannot be negative,

we check only $x = 5$. When $x = 5$,

$x + 3 = 5 + 3 = 8$. The area is $5 \cdot 8 = 40$.

State. The photo is 5 cm wide and 8 cm long.

26. Let $h(t) = 0$ and solve using factoring and

the principle of zero products.

$h(t) = -16t^2 + 64t + 36$

$0 = -4(4t^2 - 16t - 9)$

$0 = 4t^2 - 16t - 9$ Multiplying by $-\dfrac{1}{4}$

$0 = (2t-9)(2t+1)$

$2t - 9 = 0 \quad \text{or } 2t + 1 = 0$

$t = \dfrac{9}{2} \text{ or } \qquad t = -\dfrac{1}{2}$

Since time cannot be negative, we check

$t = \dfrac{9}{2}$.

$h\left(\dfrac{9}{2}\right) = -16 \cdot \left(\dfrac{9}{2}\right)^2 + 64 \cdot \dfrac{9}{2} + 36$

$= -324 + 288 + 36 = 0$

The shell will reach the water in $4\dfrac{1}{2}$ sec.

27. **Familiarize.** Let h = the height the ladder

reaches on the wall. The ladder is 2 ft longer,

so $h + 2$ represents its length.

Translate. $a^2 + b^2 = \quad c^2$

$h^2 + 10^2 = (h+2)^2$

Carry out. We solve the equation:

$h^2 + 100 = h^2 + 4h + 4$

$96 = 4h$

$h = 24$

Check. If $h = 24$, then $h + 2 = 26$

$24^2 + 10^2 = 26^2$

State. The ladder reaches 24 ft up the wall.

28. a) Let x = the number of years after 2000.

Enter the data in a graphing calculator

and use the quadratic regression feature

to obtain: $E(x) = 0.50625x^2 + 0.245x$

$+ 32.86375$.

b) In 2009, $x = 9$. $E(9) \approx \$76.1$ billion.

c) Solve $E(x) = 100$ graphically. Graph

$y_1 = E$ and $y_2 = 100$. Determine the

point of intersection for which x is

positive, which is $x \approx 11.3$ years. Eleven

years after 2000 is 2011.

29. $(a+3)^2 - 2(a+3) - 35$

Let $u = a+3$. Using substitution, we have
$u^2 - 2u - 35$, which factors as
$u^2 - 2u - 35 = (u-7)(u+5)$

Substituting $u = a+3$ produces
$$(a+3)^2 - 2(a+3) - 35 = (a+3-7)(a+3+5)$$
$$= (a-4)(a+8)$$

30. $20x(x+2)(x-1) = 5x^3 - 24x - 14x^2$

$20x(x^2 + x - 2) = 5x^3 - 24x - 14x^2$

$20x^3 + 20x^2 - 40x = 5x^3 - 24x - 14x^2$

$20x^3 - 5x^3 + 20x^2 + 14x^2 - 40x + 24x = 0$

$15x^3 + 34x^2 - 16x = 0$

$x(15x^2 + 34x - 16) = 0$

Factor $15x^2 + 34x - 16$ by the grouping method. The product of the leading coefficient, 15 and constant, -16, is $15 \cdot (-16) = -240$. We want two factors of -240 whose sum is 34. They are -6 and 40. Split $34x$ into $-6x + 40x$, and factor by grouping into factors of the form $(5x+\ )(3x+\ )$.

$x(15x^2 + 34x - 16)$

$= x(15x^2 - 6x + 40x - 16)$

$= x[3x(5x-2) + 8(5x-2)]$

$= x(3x+8)(5x-2)$

$3x + 8 = 0 \quad or \quad 5x - 2 = 0 \quad or \quad x = 0$

$3x = -8 \quad or \qquad 5x = 2$

$x = -\dfrac{8}{3} \quad or \qquad x = \dfrac{2}{5} \quad or \quad x = 0$

The solutions are $-\dfrac{8}{3}, \dfrac{2}{5}$, and 0.

Chapters 1 – 6
Cumulative Review

1. $x = -8$
   $-1 \cdot x = -1 \cdot (-8)$
   $-x = 8$

2. $x = -8$
   $-1 \cdot x = -1 \cdot (-8)$
   $-(-x) = -1 \cdot -1 \cdot x$
   $\quad\quad = -1 \cdot (-1) \cdot (-8)$
   $\quad\quad = -8$

3. $f(x) = -0.1x^3 - 3x^2 + 10x + 0.5$
   $f(-2) = -0.1(-2)^3 - 3(-2)^2 + 10(-2) + 0.5$
   $\quad\quad = -0.1(-8) - 3 \cdot 4 + 10(-2) + 0.5$
   $\quad\quad = 0.8 - 12 - 20 + 0.5$
   $\quad\quad = -30.7$

4. $-2 + (20 \div 4)^2 - 6 \cdot (-1)^3$
   $= -2 + (5)^2 - 6 \cdot (-1 \cdot -1 \cdot -1)$
   $= -2 + 25 - 6 \cdot (-1)$
   $= -2 + 25 + 6$
   $= 29$

5. $\left(3x^2 y^3\right)^{-2} = \dfrac{1}{\left(3x^2 y^3\right)^2} = \dfrac{1}{9x^4 y^6}$

6. $\left(3x^4 - 2x^2 + x - 7\right) + \left(5x^3 + 2x^2 - 3\right)$
   $= 3x^4 - 2x^2 + x - 7 + 5x^3 + 2x^2 - 3$
   $= 3x^4 + 5x^3 - 2x^2 + 2x^2 + x - 7 - 3$
   $= 3x^4 + 5x^3 + x - 10$

7. $\left(a^2 b - 2ab^2 + 3b^3\right) - \left(4a^2 b - ab^2 + b^3\right)$
   $= a^2 b - 2ab^2 + 3b^3 - 4a^2 b + ab^2 - b^3$
   $= a^2 b - 4a^2 b - 2ab^2 + ab^2 + 3b^3 - b^3$
   $= -3a^2 b - ab^2 + 2b^3$

8. $\dfrac{3t^3 s^{-1}}{12t^{-5}s} = \dfrac{3t^3 \cdot t^5}{12s \cdot s} = \dfrac{t^8}{4s^2}$

9. $\left(\dfrac{-2x^2 y}{3z^4}\right)^3 = \dfrac{(-2)^3 x^{2 \cdot 3} y^{1 \cdot 3}}{3^3 z^{4 \cdot 3}} = -\dfrac{8x^6 y^3}{27z^{12}}$

10. $-4t^5 \left(t^3 - 2t - 5\right)$
    $= -4t^5 \cdot t^3 - 4t^5 \cdot (-2t) - 4t^5 \cdot (-5)$
    $= -4t^8 + 8t^6 + 20t^5$

11. $(6x - 5y)^2 = (6x)^2 - 2 \cdot 6x \cdot 5y + (5y)^2$
    $\quad\quad\quad\quad = 36x^2 - 60xy + 25y^2$

12. $\left(10x^5 + 1\right)\left(10x^5 - 1\right)$
    $= 10x^5 \cdot 10x^5 - 10x^5 + 10x^5 + 1 \cdot (-1)$
    $= 100x^{10} - 1$

13. $(x - 1)\left(x^2 - x - 1\right)$
    $= x\left(x^2 - x - 1\right) - 1 \cdot \left(x^2 - x - 1\right)$
    $= x^3 - x^2 - x - x^2 + x + 1$
    $= x^3 - 2x^2 + 1$

14. $\dfrac{15x^4 - 12x^3 + 6x^2 + 2x + 18}{3x^2}$
    $= \dfrac{15x^4}{3x^2} - \dfrac{12x^3}{3x^2} + \dfrac{6x^2}{3x^2} + \dfrac{2x}{3x^2} + \dfrac{18}{3x^2}$
    $= 5x^2 - 4x + 2 + \dfrac{2}{3x} + \dfrac{6}{x^2}$

15. $(x^4 + 2x^3 + 6x^2 + 2x + 18) \div (x + 3)$
    $= (x^4 + 2x^3 + 6x^2 + 2x + 18) \div [x - (-3)]$

    $\underline{-3}\,|\ 1\quad 2\quad 6\quad\quad 2\quad 18$
    $\quad\quad\quad\quad -3\quad 3\ \ -27\ \ 75$
    $\quad\quad\overline{1\ \ -1\quad 9\ \ -25\,|\ 93}$

    The answer is $x^3 - x^2 + 9x - 25$, R93 or
    $x^3 - x^2 + 9x - 25 + \dfrac{93}{x + 3}$

16. $c^2 - 1 = c^2 - 1^2 = (c + 1)(c - 1)$

17. $5x + 5y + 10x^2 + 10xy$
    $= 5\left(x + y + 2x^2 + 2xy\right)$
    $= 5\left[1 \cdot (x + y) + 2x(x + y)\right]$
    $= 5(1 + 2x)(x + y)$

18. $6x - 2x^2 - 24x^4 = 2x(3 - x - 12x^3)$

19. $16x^2 - 81 = (4x)^2 - 9^2 = (4x + 9)(4x - 9)$

20. $t^2 - 10t + 24 = (t - 6)(t - 4)$

21. $8x^2 + 10x + 3 = (4x + 3)(2x + 1)$

22. $6x^2 - 28x + 16 = 2(3x^2 - 14x + 8)$
$\qquad\qquad\qquad\quad = 2(3x - 2)(x - 4)$

23. $2x^3 + 250 = 2(x^3 + 125)$
$\qquad\qquad\quad = 2(x^3 + 5^3)$
$\qquad\qquad\quad = 2(x + 5)(x^2 - x \cdot 5 + 5^2)$
$\qquad\qquad\quad = 2(x + 5)(x^2 - 5x + 25)$

24. $16x^2 + 40x + 25 = (4x)^2 + 2 \cdot 4x \cdot 5 + 5^2$
$\qquad\qquad\qquad\qquad = (4x + 5)^2$

25. $5(x - 2) = 40$
$\quad 5x - 10 = 40$
$\qquad\quad 5x = 50$
$\qquad\quad\; x = 10$
The solution is 10.

26. $-4x < -18$
$\qquad x > \dfrac{-18}{-4}$

Note: We must reverse the inequality when we multiply or divide by a negative number.
$x > \dfrac{18}{4}$
$x > \dfrac{9}{2}$
The solution $\left\{ x \mid x > \frac{9}{2} \right\}$, or $\left(\frac{9}{2}, \infty \right)$.

27. $(x - 1)(x + 3) = 0$
$\quad x - 1 = 0 \quad or \quad x + 3 = 0$
$\qquad\; x = 1 \quad or \qquad x = -3$
The solutions are 1 and –3.

28. $\qquad\quad x^2 + 10 = 11x$
$\quad x^2 - 11x + 10 = 0$
$\quad (x - 10)(x - 1) = 0$
$\qquad x - 10 = 0 \qquad or \quad x - 1 = 0$
$\qquad\qquad x = 10 \qquad or \qquad x = 1$
The solutions are 10 and 1.

29. $\quad \frac{1}{3}x - \frac{2}{9} = \frac{2}{3} + \frac{4}{9}x \qquad$ LCD is 9
$\quad 9\left(\frac{1}{3}x - \frac{2}{9} \right) = 9\left(\frac{2}{3} + \frac{4}{9}x \right)$
$\qquad 3x - 2 = 6 + 4x$
$\qquad\quad -2 = 6 + x$
$\qquad\quad -8 = x$
The solution is –8.

30. $3x + 2y = 5$
$\quad x - 3y = 9$
Multiply $x - 3y = 9$ by –3 and add the result to $3x + 2y = 5$ to eliminate x.

$-3(x - 3y = 9) \rightarrow -3x \;+\; 9y \;=\; -27$
$\qquad\qquad\qquad\qquad \underline{3x \;+\; 2y \;=\;\quad 5}$
$\qquad\qquad\text{Add} \qquad\qquad 11y \;=\; -22$
$\qquad\qquad\qquad\qquad\qquad\quad y = \dfrac{-22}{11}$
$\qquad\qquad\qquad\qquad\qquad\quad y = -2$

Substitute –2 for y into any equation with both x and y to determine x.
$\qquad x - 3y = 9$
$\quad x - 3(-2) = 9$
$\qquad\; x + 6 = 9$
$\qquad\qquad x = 9 - 6$
$\qquad\qquad x = 3$
The solution is $(3, -2)$.

31. $\qquad 2x^2 + 7x = 4$
$\quad 2x^2 + 7x - 4 = 0$
$\quad (2x - 1)(x + 4) = 0$
$\qquad 2x - 1 = 0 \quad or \quad x + 4 = 0$
$\qquad\qquad 2x = 1 \quad or \quad x = -4$
$\qquad\qquad\; x = \dfrac{1}{2} \; or \; x = -4$
The solutions are –4 and $\frac{1}{2}$.

32. $4(x+7) < 5(x-3)$

 $4x + 28 < 5x - 15$

 $28 < x - 15$

 $43 < x, \quad \text{or} \quad x > 43$

 The solution is $\{x \mid x > 43\}$, or $(43, \infty)$

33. $2y = 4x - 3$

 $4x = 1 - y$

 Substitute $1 - y$ for $4x$ into the equation

 $2y = 4x - 3$

 $2y = 1 - y - 3$

 $2y = -y - 2$

 $3y = -2$

 $y = -\dfrac{2}{3}$

 Substitute $-\frac{2}{3}$ for y into any equation

 with x and y to determine x.

 $4x = 1 - y$

 $4x = 1 - \left(-\dfrac{2}{3}\right)$

 $4x = \dfrac{3}{3} + \dfrac{2}{3}$

 $4x = \dfrac{5}{3}$

 $\dfrac{1}{4} \cdot 4x = \dfrac{1}{4} \cdot \dfrac{5}{3}$

 $x = \dfrac{5}{12}$

 The solution is $\left(\frac{5}{12}, -\frac{2}{3}\right)$.

34. $a = bc + dc$

          $a = c(b + d)$

 $\dfrac{a}{b+d} = c$

35. $x + 2y - 2z + \frac{1}{2}x - z = \left(1 + \frac{1}{2}\right)x$

 $+ 2y + (-2 - 1)z$

 $= \frac{3}{2}x + 2y - 3z$

36. $2x^3 - 7 + \frac{3}{7}x^2 - 6x^3 - \frac{4}{7}x^2 + 5$

 $= (2 - 6)x^3 + \left(\frac{3}{7} - \frac{4}{7}\right)x^2 + (-7 + 5)$

 $= -4x^3 - \frac{1}{7}x^2 - 2$

37. $y = 1 - \frac{1}{2}x$

 Determine at least 2 points. Plot the points and draw the line which contains them.

x	y
0	1
4	−1

 Let $x = 0$ $y = 1 - \dfrac{1}{2} \cdot 0 = 1$

 Let $x = 4$ $y = 1 - \dfrac{1}{2}(4)$

 $y = 1 - 2 = -1$

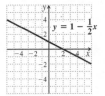

38. $x = -3$

 This is of the form $x = c$, where c is a constant. This is a vertical line with x-intercept $(-3, 0)$.

39. $x - 6y = 6$

 Determine at least 2 points. Plot these points and draw the line which contains them. We will find the intercepts.

x	y
0	−1
6	0

 Let $x = 0$ $0 - 6y = 6$

 $-6y = 6$

 $y = -1$

 Let $y = 0$ $x - 6 \cdot 0 = 6$

 $x = 6$

40. $y = 6$

This is of the form $y = c$, where c is a constant. This is a horizontal line with y-intercept $(0, 6)$.

41. $y = x^2 - 4$

Graph $y = x^2 - 4$ using a standard viewing window.

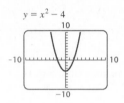

42. $(1, 5)$ and $(2, 3)$.

$$m = \frac{y_2 - y_1}{x_2 - x_1}$$

$$m = \frac{3 - 5}{2 - 1} = \frac{-2}{1} = -2$$

43. $3x + 4y = 8$

We will solve for y, which is slope-intercept form.

$$3x + 4y = 8$$

$$4y = -3x + 8$$

$$y = \frac{-3}{4}x + \frac{8}{4}$$

$$y = -\frac{3}{4}x + 2$$

$$y = mx + b$$

The slope is $-\frac{3}{4}$, and the y-intercept is $(0, 2)$.

44. Use slope intercept form.

$$y = mx + b$$

$$y = \frac{1}{2}x - 7$$

45. $y = x + 3$ has slope of 1 and y-intercept $(0, 3)$. The equation of the line perpendicular to $y = x + 3$ has of -1, since the negative reciprocal of 1 is $-\frac{1}{1} = -1$.

We use slope-intercept form:

$$y = mx + b$$

$$y = -x + 3$$

46. ***Familiarize.*** Let $x = $ number of 3G subscribers, in millions. $80\% = 0.80$
Translate.

		80%		
Number of 3G Subscribers	plus	Increase in 3G Subscribers	is	64.2 million
↓	↓	↓	↓	↓
x	$+$	$0.80\,x$	$=$	64.2

Carry out. We solve for x.

$$x + 0.80x = 64.2$$

$$1.80x = 64.2$$

$$x = \frac{64.2}{1.80}$$

$$x = 35.6\overline{6}$$

Check. 80% of $35.67 = 0.80\,(35.67)$
$$\approx 28.53$$
$35.67 + 28.53 = 64.2$
The answer checks.

State. The number of U.S. subscribers with 3G devices in June 2007 was about 35.7 million.

47. ***Familiarize.*** Let $x = $ royalty income for Andrew; then $2x = $ royalty income for Emily, both in dollars.
Translate.

Andrew's royalties	plus	Emily's royalties	equal	$1260.
↓	↓	↓	↓	↓
x	$+$	$2x$	$=$	1260

Carry out. Solve for x.

$$x + 2x = 1260$$

$$3x = 1260$$

$$x = 420$$

$$2x = 2 \cdot 420 = 840$$

Check. $420 + 840 = 1260$. Our answer checks.

State. Emily should receive $840, and Andrew should receive $420.

48. The number of people, in thousands, who are on an organ-transplant waiting list is

$N(t) = 2.38t + 77.38$,

where t = number years since 2000. For the year 2010, $t = 10$, and

$N(10) = 2.38(10) + 77.38$

$\qquad = 23.8 + 77.38$

$\qquad = 101.18$

The number of people in the United States on organ-transplant waiting lists in 2010 was about 101,180.

49. **Familiarize.** Chamberlain scored 100 total points and made a total of 64 shots. Each two-point shot was worth 2-points, and each foul shot was worth 1-point. Let x = the number of 2-point shots, then $64 - x$ = the number of foul shots.

Translate.

Points from 2-point shots	plus	Points from foul shots	Equal	Total Points
↓	↓	↓	↓	↓
$2x$	$+$	$1(64 - x)$	$=$	100

Carry out. We solve for x.

$2x + 1(64 - x) = 100$

$\quad 2x + 64 - x = 100$

$\qquad\quad x + 64 = 100$

$\qquad\qquad\quad x = 36$

$64 - x = 64 - 36 = 28$

Check. $2 \cdot 36 + 1 \cdot 28 = 72 + 28 = 100$ Our answer checks.

State. With Chamberlain made 36 two-point shots and 28 foul shots in his 100-point game.

50. **Familiarize.** Use the drawing provided. Let x represent the distance along the ground from the bottom of the ladder to the building, and $x + 7$ represent the height of the ladder. These are the legs of the right triangle and 13 is the length of the hypotenuse. Recall: Pythagorean Theorem $a^2 + b^2 = c^2$.

Translate.

$$\begin{array}{ccccc} a^2 & + & b^2 & = & c^2 \\ \downarrow & & \downarrow & & \downarrow \\ x^2 & + & (x+7)^2 & = & 13^2 \end{array}$$

Carry out. We solve the equation:

$x^2 + (x + 7)^2 = 13^2$

$x^2 + x^2 + 14x + 49 = 169$

$\quad 2x^2 + 14x - 120 = 0$

$\quad 2(x^2 + 7x - 60) = 0$

$\qquad x^2 + 7x - 60 = 0$

$\qquad (x + 12)(x - 5) = 0$

$x + 12 = 0 \quad or \quad x - 5 = 0$

$\qquad x = -12 \quad or \qquad x = 5$

Check. Since measure cannot be negative, we know -12 is not a solution. If $x = 5$,

$x + 7 = 12$ and

$5^2 + 12^2 = 13^2$

$25 + 144 = 169$

$\qquad 169 = 169$

The answer checks.

State. The distance from the building is 5 ft, and the height of the ladder is 12 ft.

51. **Familiarize.** We let w represent the width and $6w$ represent the length. Recall that the formula for the area of a rectangle is $A =$ length × width.

Translate.

Area	is	24 ft^2
↓	↓	↓
$w(6w)$	$=$	24

Carry out. We solve the equation:

$w(6w) = 24$

$\quad 6w^2 = 24$

$\qquad w^2 = 4$

$w = -2 \quad or \quad w = 2$

Check. The number -2 is not a solution, because width cannot be negative. If the width is 2 ft, and the length is $6 \cdot 2$, or 12 ft, then the area is $2 \cdot 12 = 24$ ft^2. The answer checks.

State. The length is 12 ft, and the width is 2 ft.

52. a. From the given information, we have (55, 7.6) and (70, 6.1). We plot these points and draw the line which contains them.

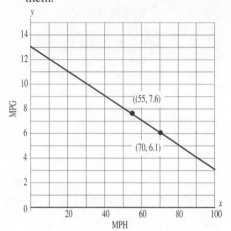

To determine an equation, we will use the points (55, 7.6) and (70, 6.1)

$$\text{Slope} = m = \frac{y_2 - y_1}{x_2 - x_1} = \frac{6.1 - 7.6}{70 - 55}$$

$$= \frac{-1.5}{15} = -0.1$$

Choose either point and substitute into point-slope form.

$$y - y_1 = m(x - x_1)$$

$$y - 7.6 = -0.1(x - 55)$$

$$y - 7.6 = -0.1x + 5.5$$

$$y = -0.1x + 13.1$$

b. $x = 60$ $y = -0.1x + 13.1$

$$y = -0.1(60) + 13.1$$

$$y = -6 + 13.1$$

$$y = 7.1 \text{ mpg.}$$

53. The data increases, decreases, and then increases. A linear function would not be an appropriate model.

54. The data appears to increase at a steady (0.3) rate. A linear function would be an appropriate model.

55. Using the bar graph from Exercise 54, we obtain the following data:

Years after 2002	College enrollment (in millions)
0	16.6
1	16.9
2	17.3
3	17.5
4	17.8
5	18.0
6	18.2

We enter the data into a graphing calculator and choose the Linear Regression feature. We have:

$f(x) = 0.26786x + 16.66786,$ where x is the number of years after 2002 and f is in millions of students.

56. *Domain:* $\mathbb{R}$; Range: the function has a constant value of 3; $\{3\}$.

57. *Domain:* $\mathbb{R}$; Range: $\mathbb{R}$.

58. *Domain:* $\mathbb{R}$; Range: the minimum function value is –3; there is no maximum. $\{y \mid y \geq -3\}$, or $[-3, \infty)$.

59. *Domain:* $\mathbb{R}$; Range: the minimum function value is –4; there is no maximum. $\{y \mid y \geq -4\}$, or $[-4, \infty)$.

60. $(x + 7)(x - 4) - (x + 8)(x - 5)$

$$= x^2 - 4x + 7x - 28 - (x^2 - 5x + 8x - 40)$$

$$= x^2 + 3x - 28 - (x^2 + 3x - 40)$$

$$= x^2 + 3x - 28 - x^2 - 3x + 40$$

$$= (1 - 1)x^2 + (3 - 3)x + (-28 + 40)$$

$$= 12$$

61. $2a^{32} - 13,122b^{40} = 2(a^{32} - 6561b^{40})$

$$= 2[(a^{16})^2 - (81b^{20})^2]$$

$$= 2(a^{16} + 81b^{20})(a^{16} - 81b^{20})$$

$$= 2(a^{16} + 81b^{20})$$

$$\cdot [(a^8)^2 - (9b^{10})^2]$$

$$= 2(a^{16} + 81b^{20})(a^8 + 9b^{10})$$

$$\cdot (a^8 - 9b^{10})$$

$$= 2(a^{16} + 81b^{20})(a^8 + 9b^{10})$$

$$\cdot [(a^4)^2 - (3b^5)^2]$$

$$= 2(a^{16} + 81b^{20})(a^8 + 9b^{10})$$

$$\cdot (a^4 + 3b^5)(a^4 - 3b^5)$$

62. $-\left| 0.875 - \left(-\dfrac{1}{8} \right) - 8 \right|$

$-\left| 0.875 + 0.125 - 8 \right|$

$-\left| -7 \right|$

-7

63. $f(x) = x^4 - 34x^2 + 225$

$0 = (x^2 - 25)(x^2 - 9)$

$0 = (x + 5)(x - 5)(x + 3)(x - 3)$

$x + 5 = 0$ or $x - 5 = 0$ or

$x + 3 = 0$ or $x - 3 = 0$

$x = -5$ or $x = 5$ or

$x = -3$ or $x = 3$

The solutions are –5, –3, 3, and 5.

Chapter 7

Rational Expressions, Equations, and Functions

1. e

3. i

5. d

7. $H(t) = \dfrac{t^2 + 3t}{2t + 3}$

$H(5) = \dfrac{5^2 + 3 \cdot 5}{2 \cdot 5 + 3} = \dfrac{25 + 15}{10 + 3}$

$= \dfrac{40}{13}$ hr, or $3\dfrac{1}{13}$ hr

9. $v(t) = \dfrac{4t^2 - 5t + 2}{t + 3}$

$v(0) = \dfrac{4 \cdot 0^2 - 5 \cdot 0 + 2}{0 + 3} = \dfrac{0 - 0 + 2}{0 + 3} = \dfrac{2}{3}$

$v(-2) = \dfrac{4(-2)^2 - 5(-2) + 2}{-2 + 3} = \dfrac{16 + 10 + 2}{-2 + 3} = 28$

$v(7) = \dfrac{4 \cdot 7^2 - 5 \cdot 7 + 2}{7 + 3} = \dfrac{196 - 35 + 2}{7 + 3} = \dfrac{163}{10}$

11. $g(x) = \dfrac{2x^3 - 9}{x^2 - 4x + 4}$

$g(0) = \dfrac{2 \cdot 0^3 - 9}{0^2 - 4 \cdot 0 + 4} = \dfrac{0 - 9}{0 - 0 + 4} = -\dfrac{9}{4}$

$g(2) = \dfrac{2 \cdot 2^3 - 9}{2^2 - 4 \cdot 2 + 4} = \dfrac{16 - 9}{4 - 8 + 4} = \dfrac{7}{0}$

Since division by zero is not defined, $g(2)$ does not exist.

$g(-1) = \dfrac{2(-1)^3 - 9}{(-1)^2 - 4(-1) + 4} = \dfrac{-2 - 9}{1 + 4 + 4} = -\dfrac{11}{9}$

13. $\dfrac{25}{-7x}$

We find the real number(s) that make the denominator 0. To do so, we set the denominator equal to 0 and solve for x:

$-7x = 0$

$x = 0$

The expression is undefined for $x = 0$.

15. $\dfrac{t - 3}{t + 8}$

Set the denominator equal to 0 and solve for t:

$t + 8 = 0$

$t = -8$

The expression is undefined for $t = -8$.

17. $\dfrac{a}{3a - 12}$

Set the denominator equal to 0 and solve for a:

$3a - 12 = 0$

$3a = 12$

$a = 4$

The expression is undefined for $a = 4$.

19. $\dfrac{x^2 - 16}{x^2 - 3x - 28}$

Set the denominator equal to 0 and solve for x:

$x^2 - 3x - 28 = 0$

$(x - 7)(x + 4) = 0$

$x - 7 = 0$ or $x + 4 = 0$

$x = 7$ or $x = -4$

The expression is undefined for $x = 7$ and $x = -4$.

21. $\dfrac{m^3 - 2m}{m^2 - 25}$

Set the denominator equal to 0 and solve for m:

$m^2 - 25 = 0$

$(m + 5)(m - 5) = 0$

$m + 5 = 0$ or $m - 5 = 0$

$m = -5$ or $m = 5$

The expression is undefined for $m = -5$ and $m = 5$.

23. $\dfrac{15x}{5x^2} = \dfrac{5x \cdot 3}{5x \cdot x}$ Factoring; the greatest common factor is $5x$.

$= \dfrac{5x}{5x} \cdot \dfrac{3}{x}$ Factoring the rational expression

$= 1 \cdot \dfrac{3}{x} \qquad \dfrac{5x}{5x} = 1$

$= \dfrac{3}{x} \qquad$ Removing a factor equal to 1

25. $\dfrac{18t^3 w^2}{27t^7 w} = \dfrac{9t^3 w \cdot 2w}{9t^3 w \cdot 3t^4}$ Factor the numerator and denominator

$= \dfrac{9t^3 w}{9t^3 w} \cdot \dfrac{2w}{3t^4}$ Factor the rational expression

$= \dfrac{2w}{3t^4} \qquad$ Remove a factor of 1

27. $\dfrac{2a - 10}{2} = \dfrac{2}{2} \cdot \dfrac{a - 5}{1}$ Factoring the rational expression

$= \dfrac{a - 5}{1} = a - 5$ Removing a factor equal to 1

29. $\dfrac{3x^2 - 12x}{3x^2 + 15x} = \dfrac{3x(x - 4)}{3x(x + 5)} = \dfrac{3x}{3x} \cdot \dfrac{x - 4}{x + 5} = \dfrac{x - 4}{x + 5}$

31. $\dfrac{6a^2 - 3a}{7a^2 - 7a} = \dfrac{3a(2a - 1)}{7a(a - 1)}$

$= \dfrac{a}{a} \cdot \dfrac{3(2a - 1)}{7(a - 1)}$

$= 1 \cdot \dfrac{3(2a - 1)}{7(a - 1)}$

$= \dfrac{3(2a - 1)}{7(a - 1)}$

33. $\dfrac{3a - 1}{2 - 6a} = \dfrac{3a - 1}{2(1 - 3a)}$

$= \dfrac{-1(1 - 3a)}{2(1 - 3a)}$ Factoring out -1 in the numerator reverses the subtraction

$= \dfrac{-1}{2} \cdot \dfrac{1 - 3a}{1 - 3a}$

$= -\dfrac{1}{2}$

35. $\dfrac{3a^2 + 9a - 12}{6a^2 - 30a + 24} = \dfrac{3(a^2 + 3a - 4)}{6(a^2 - 5a + 4)}$

$= \dfrac{3(a + 4)(a - 1)}{3 \cdot 2(a - 4)(a - 1)}$

$= \dfrac{3(a - 1)}{3(a - 1)} \cdot \dfrac{a + 4}{2(a - 4)}$

$= 1 \cdot \dfrac{a + 4}{2(a - 4)}$

$= \dfrac{a + 4}{2(a - 4)}$

37. $\dfrac{x^2 + 8x + 16}{x^2 - 16} = \dfrac{(x + 4)(x + 4)}{(x + 4)(x - 4)}$

$= \dfrac{x + 4}{x + 4} \cdot \dfrac{x + 4}{x - 4}$

$= 1 \cdot \dfrac{x + 4}{x - 4}$

$= \dfrac{x + 4}{x - 4}$

39. $\dfrac{t^2 - 1}{t + 1} = \dfrac{(t + 1)(t - 1)}{t + 1}$

$= \dfrac{t + 1}{t + 1} \cdot \dfrac{t - 1}{1}$

$= 1 \cdot \dfrac{t - 1}{1}$

$= t - 1$

41. $\dfrac{y^2 + 4}{y + 2}$ can not be simplified.

Neither the numerator nor the denominator can be factored.

43. $\dfrac{5x^2 - 20}{10x^2 - 40} = \dfrac{5(x^2 - 4)}{10(x^2 - 4)}$

$= \dfrac{1 \cdot \cancel{5} \cdot \cancel{(x^2 - 4)}}{2 \cdot \cancel{5} \cdot \cancel{(x^2 - 4)}}$

$= \dfrac{1}{2}$

45. $\dfrac{x - 8}{8 - x} = \dfrac{x - 8}{-(x - 8)}$

$= \dfrac{1}{-1} \cdot \dfrac{x - 8}{x - 8}$

$= \dfrac{1}{-1} \cdot 1$

$= -1$

47. $\dfrac{2t-1}{1-4t^2} = \dfrac{2t-1}{(1+2t)(1-2t)}$

$\qquad = \dfrac{-1(1-2t)}{(1+2t)(1-2t)}$ Factoring out -1 in the numerator reverses the subtraction

$\qquad = \dfrac{-1}{1+2t} \cdot \dfrac{1-2t}{1-2t}$

$\qquad = -\dfrac{1}{1+2t}$

49. $\dfrac{a^2-25}{a^2+10a+25} = \dfrac{(a+5)(a-5)}{(a+5)(a+5)}$

$\qquad = \dfrac{a+5}{a+5} \cdot \dfrac{a-5}{a+5}$

$\qquad = \dfrac{a-5}{a+5}$

51. $\dfrac{7s^2-28t^2}{28t^2-7s^2}$

Note that the numerator and denominator are opposites. Thus, we have an expression divided by its opposite, so the result is -1.

53. $\dfrac{x^3-1}{x^2-1} = \dfrac{(x-1)(x^2+x+1)}{(x+1)(x-1)}$

$\qquad = \dfrac{x-1}{x-1} \cdot \dfrac{x^2+x+1}{x+1}$

$\qquad = \dfrac{x^2+x+1}{x+1}$

55. $\dfrac{3y^3+24}{y^2-2y+4} = \dfrac{3(y^3+8)}{y^2-2y+4}$

$\qquad = \dfrac{3(y+2)(y^2-2y+4)}{y^2-2y+4}$

$\qquad = \dfrac{y^2-2y+4}{y^2-2y+4} \cdot \dfrac{3(y+2)}{1}$

$\qquad = 3(y+2)$

57. $f(x) = \dfrac{3x+21}{x^2+7x}$

Set the denominator equal to 0 and solve for x:

$\qquad x^2+7x = 0$

$\qquad x(x+7) = 0$

$x = 0 \quad \text{or} \quad x+7 = 0$

$x = 0 \quad \text{or} \qquad x = -7$

The domain of f is $\{x \mid x$ is a real number and $x \neq -7$ and $x \neq 0\}$

$\dfrac{3x+21}{x^2+7x} = \dfrac{3(x+7)}{x(x+7)}$

$\qquad = \dfrac{x+7}{x+7} \cdot \dfrac{3}{x}$

$\qquad = \dfrac{3}{x}$

$f(x) = \dfrac{3}{x}, \ x \neq -7, 0$

59. $g(x) = \dfrac{x^2-9}{5x+15}$

$5x+15 = 0$

$\qquad 5x = -15$

$\qquad x = -3$

The domain of g is $\{x \mid x$ is a real number and $x \neq -3\}$.

$\dfrac{x^2-9}{5x+15} = \dfrac{(x+3)(x-3)}{5(x+3)}$

$\qquad = \dfrac{x+3}{x+3} \cdot \dfrac{x-3}{5}$

$\qquad = \dfrac{x-3}{5}$

$g(x) = \dfrac{x-3}{5}, \ x \neq -3$

61. $h(x) = \dfrac{4-x}{5x-20}$

$5x - 20 = 0$

$5x = 20$

$x = 4$

The domain of h is $\{x \mid x \text{ is a real number and}$ $x \neq 4\}$.

$\dfrac{4-x}{5x-20} = \dfrac{-1(x-4)}{5(x-4)}$

$= \dfrac{-1}{5} \cdot \dfrac{x-4}{x-4}$

$= -\dfrac{1}{5}$

$h(x) = -\dfrac{1}{5}, \; x \neq 4$

63. $f(t) = \dfrac{t^2 - 9}{t^2 + 4t + 3}$

$t^2 + 4t + 3 = 0$

$(t+3)(t+1) = 0$

$t + 3 = 0 \quad \text{or} \quad t + 1 = 0$

$t = -3 \quad \text{or} \quad t = -1$

The domain of f is $\{t \mid t \text{ is a real number and}$ $t \neq -1 \text{ or } -3\}$

$f(t) = \dfrac{t^2 - 9}{t^2 + 4t + 3} = \dfrac{\cancel{(t+3)}(t-3)}{\cancel{(t+3)}(t+1)}$

$= \dfrac{(t-3)}{(t+1)}, \; t \neq -3, \; t \neq -1$

65. $g(t) = \dfrac{21 - 7t}{3t - 9}$

$3t - 9 = 0$

$3t = 9$

$t = 3$

The domain of g is $\{t \mid t \text{ is a real number and}$ $t \neq 3\}$.

$\dfrac{21 - 7t}{3t - 9} = \dfrac{-7(t-3)}{3(t-3)}$

$= \dfrac{-7}{3} \cdot \dfrac{t-3}{t-3}$

$= -\dfrac{7}{3}$

$g(t) = -\dfrac{7}{3}, \; t \neq 3$

67. $h(t) = \dfrac{t^2 + 5t + 4}{t^2 - 8t - 9}$

$t^2 - 8t - 9 = 0$

$(t-9)(t+1) = 0$

$t - 9 = 0 \quad \text{or} \quad t + 1 = 0$

$t = 9 \quad \text{or} \qquad t = -1$

The domain of h is $\{t \mid t \text{ is a real number and}$ $t \neq -1 \text{ and } t \neq 9\}$.

$\dfrac{t^2 + 5t + 4}{t^2 - 8t - 9} = \dfrac{(t+4)(t+1)}{(t-9)(t+1)}$

$= \dfrac{t+4}{t-9} \cdot \dfrac{t+1}{t+1}$

$= \dfrac{t+4}{t-9}$

$h(t) = \dfrac{t+4}{t-9}, \; t \neq -1, 9$

69. $f(x) = \dfrac{9x^2 - 4}{3x - 2}$

$3x - 2 = 0$

$3x = 2$

$x = \dfrac{2}{3}$

The domain of f is $\{x \mid x \text{ is a real number and}$ $x \neq \tfrac{2}{3}\}$.

$\dfrac{9x^2 - 4}{3x - 2} = \dfrac{(3x+2)(3x-2)}{1 \cdot (3x-2)}$

$= \dfrac{3x+2}{1} \cdot \dfrac{3x-2}{3x-2}$

$= \dfrac{3x+2}{1}$

$= 3x + 2$

$f(x) = 3x + 2, \; x \neq \dfrac{2}{3}$

71. First we simplify the rational expression describing the function.

$\dfrac{3x - 12}{3x + 15} = \dfrac{3(x-4)}{3(x+5)} = \dfrac{3}{3} \cdot \dfrac{x-4}{x+5} = \dfrac{x-4}{x+5}$

$x + 5 = 0$ when $x = -5$. Thus, the vertical asymptote is $x = -5$.

73. First we simplify the rational expression describing the function.

$$\frac{12-6x}{5x-10} = \frac{-6(-2+x)}{5(x-2)}$$

$$= \frac{-6(x-2)}{5(x-2)}$$

$$= \frac{-6}{5} \cdot \frac{x-2}{x-2}$$

$$= -\frac{6}{5}$$

The denominator of the simplified expression is not equal to 0 for any value of x, so there are no vertical asymptotes.

75. First we simplify the rational expression describing the function.

$$\frac{x^3+3x^2}{x^2+6x+9} = \frac{x^2(x+3)}{(x+3)(x+3)}$$

$$= \frac{x^2}{x+3} \cdot \frac{x+3}{x+3} = \frac{x^2}{x+3}$$

$x+3=0$ when $x=-3$. Thus, the vertical asymptote is $x=-3$.

77. First we simplify the rational expression describing the function.

$$\frac{x^2-x-6}{x^2-6x+8} = \frac{(x-3)(x+2)}{(x-4)(x-2)}$$

We cannot remove a factor equal to 1. Observe that $x-4=0$ when $x=4$ and $x-2=0$ when $x=2$. Thus, the vertical asymptotes are $x=4$ and $x=2$.

79. The vertical asymptote of $h(x)=\frac{1}{x}$ is $x=0$. Observe that $h(x)>0$ for $x>0$ and $h(x)<0$ for $x<0$. Thus, graph (b) corresponds to this function.

81. The vertical asymptote of $f(x)=\frac{x}{x-3}$ is $x=3$. Thus, graph (f) corresponds to this function.

83. $\dfrac{4x-2}{x^2-2x+1} = \dfrac{2(2x-1)}{(x-1)(x-1)}$

The vertical asymptote of $r(x)$ is $x=1$. Thus, graph (a) corresponds to this function.

85. *Thinking and Writing Exercise.*

87. $-\dfrac{2}{15} \cdot \dfrac{10}{7} = -\dfrac{2}{3 \cdot 5} \cdot \dfrac{2 \cdot 5}{7} = -\dfrac{4}{21}$

89. $\dfrac{5}{8} \div \left(-\dfrac{1}{6}\right) = -\dfrac{5}{8} \cdot \dfrac{6}{1} = -\dfrac{5}{2 \cdot 4} \cdot \dfrac{2 \cdot 3}{1} = -\dfrac{15}{4}$

91. $\dfrac{7}{9} - \dfrac{2}{3} \cdot \dfrac{6}{7} = \dfrac{7}{9} - \dfrac{2}{3} \cdot \dfrac{3 \cdot 2}{7} = \dfrac{7}{9} - \dfrac{4}{7}$

$$= \dfrac{7}{9} \cdot \dfrac{7}{7} - \dfrac{4}{7} \cdot \dfrac{9}{9} = \dfrac{49-36}{63} = \dfrac{13}{63}$$

93. *Thinking and Writing Exercise.*

95. $(a, f(a))$ and $(a+h, f(a+h))$

$$m = \frac{y_2-y_1}{x_2-x_1} = \frac{f(a+h)-f(a)}{a+h-a}$$

$$= \frac{f(a+h)-f(a)}{h}$$

Substituting: $f(a)=a^2+5$ and

$$f(a+h)=(a+h)^2+5$$

$$= a^2+2ah+h^2+5$$

we have:

$$\frac{(a^2+2ah+h^2+5)-(a^2+5)}{h}$$

$$= \frac{a^2+2ah+h^2+5-a^2-5}{h}$$

$$= \frac{2ah+h^2}{h} = \frac{h\cdot(2a+h)}{h\cdot 1}$$

$$= 2a+h$$

97. $\dfrac{x^4-y^4}{(y-x)^4} = \dfrac{(x^2+y^2)(x^2-y^2)}{[-(x-y)]^4}$

$$= \frac{(x^2+y^2)(x+y)(x-y)}{(-1)^4(x-y)(x-y)^3}$$

$$= \frac{(x^2+y^2)(x+y)}{(x-y)^3}$$

99. $\dfrac{(x-1)(x^4-1)(x^2-1)}{(x^2+1)(x-1)^2(x^4-2x^2+1)}$

$$= \dfrac{(x-1)(x^4-1)(x^2-1)}{(x^2+1)(x-1)^2(x^2-1)^2}$$

$$= \dfrac{(x-1)(x^2+1)(x^2-1)(x+1)(x-1)}{(x^2+1)(x-1)(x-1)(x^2-1)(x^2-1)}$$

$$= \dfrac{(x-1)(x^2+1)(x^2-1)(x+1)(x-1)\cdot 1}{(x^2+1)(x-1)(x-1)(x^2-1)(x+1)(x-1)}$$

$$= \dfrac{1}{x-1}$$

101. $\dfrac{a^3-2a^2+2a-4}{a^3-2a^2-3a+6} = \dfrac{(a^3-2a^2)+(2a-4)}{(a^3-2a^2)-(3a-6)}$

$$= \dfrac{a^2(a-2)+2(a-2)}{a^2(a-2)-3(a-2)}$$

$$= \dfrac{(a^2+2)(a-2)}{(a^2-3)(a-2)}$$

$$= \dfrac{(a^2+2)(a-2)}{(a^2-3)(a-2)}$$

$$= \dfrac{a^2+2}{a^2-3}$$

103. $\dfrac{(t+2)^3(t^2+2t+1)(t+1)}{(t+1)^3(t^2+4t+4)(t+2)}$

$$= \dfrac{(t+2)^3(t+1)^2(t+1)}{(t+1)^3(t+2)^2(t+2)}$$

$$= \dfrac{(t+2)^3(t+1)^3}{(t+1)^3(t+2)^3} = 1$$

105. From the graph we see that the domain consists of all real numbers except –2 and 1, so the domain is $(-\infty,-2)\cup(-2,1)\cup(1,\infty)$.
We also see that the range consists of all real numbers except 2 and 3, so the range is $(-\infty,2)\cup(2,3)\cup(3,\infty)$.

107. From the graph we see that the domain consists of all real numbers except –1 and 1, so the domain is $(-\infty,-1)\cup(-1,1)\cup(1,\infty)$.
We also see that the range consists of all real numbers less than or equal to –1 or greater than 0. Thus, the range is $(-\infty,-1]\cup(0,\infty)$.

109. *Thinking and Writing Exercise.*

Exercise Set 7.2

1. d

3. a

5. b

7. $\dfrac{7x}{5}\cdot\dfrac{x-5}{2x+1} = \dfrac{7x(x-5)}{5(2x+1)}$

9. $\dfrac{a-4}{a+6}\cdot\dfrac{a+2}{a+6} = \dfrac{(a-4)(a+2)}{(a+6)^2}$

11. $\dfrac{2x+3}{4}\cdot\dfrac{x+1}{x-5} = \dfrac{(2x+3)(x+1)}{4(x-5)}$

13. $\dfrac{5a^4}{6a}\cdot\dfrac{2}{a} = \dfrac{5a^4\cdot 2}{6a\cdot a}$ Multiplying the numerators and the denominators

$$= \dfrac{5\cdot a\cdot a\cdot a\cdot a\cdot 2}{2\cdot 3\cdot a\cdot a}$$ Factoring the numerator and the denominator

$$= \dfrac{5\cdot a\cdot a\cdot a\cdot a\cdot 2}{2\cdot 3\cdot a\cdot a}$$ Removing a factor equal to 1

$$= \dfrac{5a^2}{3}$$ Simplifying

15. $\dfrac{3c}{d^2}\cdot\dfrac{8d}{6c^3} = \dfrac{3c\cdot 8d}{d^2\cdot 6c^3}$ Multiplying the numerators and the denominators

$$= \dfrac{3\cdot c\cdot 2\cdot 4\cdot d}{d\cdot d\cdot 3\cdot 2\cdot c\cdot c\cdot c}$$ Factoring the numerator and the denominator

$$= \dfrac{3\cdot c\cdot 2\cdot 4\cdot d}{d\cdot d\cdot 3\cdot 2\cdot c\cdot c\cdot c}$$

$$= \dfrac{4}{dc^2}$$

17. $\dfrac{y^2-16}{4y+12}\cdot\dfrac{y+3}{y-4}=\dfrac{(y^2-16)(y+3)}{(4y+12)(y-4)}$

$\qquad\qquad=\dfrac{(y+4)(y-4)(y+3)}{4(y+3)(y-4)}$

$\qquad\qquad=\dfrac{(y+4)(\cancel{y-4})(\cancel{y+3})}{4(\cancel{y+3})(\cancel{y-4})}$

$\qquad\qquad=\dfrac{y+4}{4}$

19. $\dfrac{x^2-3x-10}{(x-2)^2}\cdot\dfrac{x-2}{x-5}=\dfrac{(x^2-3x-10)(x-2)}{(x-2)^2(x-5)}$

$\qquad\qquad=\dfrac{(x-5)(x+2)(x-2)}{(x-2)(x-2)(x-5)}$

$\qquad\qquad=\dfrac{(\cancel{x-5})(x+2)(\cancel{x-2})}{(\cancel{x-2})(x-2)(\cancel{x-5})}$

$\qquad\qquad=\dfrac{x+2}{x-2}$

21. $\dfrac{y^2-y}{y^2+5y+4}\cdot(y+4)=\dfrac{y(y-1)}{(y+1)\cancel{(y+4)}}\cdot\cancel{(y+4)}$

$\qquad\qquad=\dfrac{y(y-1)}{(y+1)}$

23. $\dfrac{a^2-9}{a^2}\cdot\dfrac{5a}{a^2+a-12}=\dfrac{(a+3)(a-3)\cdot5\cdot a}{a\cdot a(a+4)(a-3)}$

$\qquad\qquad=\dfrac{(a+3)(\cancel{a-3})\cdot5\cdot\cancel{a}}{\cancel{a}\cdot a(a+4)(\cancel{a-3})}$

$\qquad\qquad=\dfrac{5(a+3)}{a(a+4)}$

25. $\dfrac{4a^2}{3a^2-12a+12}\cdot\dfrac{3a-6}{2a}$

$\qquad=\dfrac{4a^2(3a-6)}{(3a^2-12a+12)2a}$

$\qquad=\dfrac{2\cdot2\cdot a\cdot a\cdot3\cdot(a-2)}{3\cdot(a-2)\cdot(a-2)\cdot2\cdot a}$

$\qquad=\dfrac{\cancel{2}\cdot2\cdot\cancel{a}\cdot a\cdot\cancel{3}\cdot(\cancel{a-2})}{\cancel{3}\cdot(\cancel{a-2})\cdot(a-2)\cdot\cancel{2}\cdot\cancel{a}}$

$\qquad=\dfrac{2a}{a-2}$

27. $\dfrac{t^2+2t-3}{t^2+4t-5}\cdot\dfrac{t^2-3t-10}{t^2+5t+6}$

$\qquad=\dfrac{(t^2+2t-3)(t^2-3t-10)}{(t^2+4t-5)(t^2+5t+6)}$

$\qquad=\dfrac{(t+3)(t-1)(t-5)(t+2)}{(t+5)(t-1)(t+3)(t+2)}$

$\qquad=\dfrac{(\cancel{t+3})(\cancel{t-1})(t-5)(\cancel{t+2})}{(t+5)(\cancel{t-1})(\cancel{t+3})(\cancel{t+2})}$

$\qquad=\dfrac{t-5}{t+5}$

29. $\dfrac{5a^2-180}{10a^2-10}\cdot\dfrac{20a+20}{2a-12}$

$\qquad=\dfrac{(5a^2-180)(20a+20)}{(10a^2-10)(2a-12)}$

$\qquad=\dfrac{5(a+6)(a-6)(2)(10)(a+1)}{10(a+1)(a-1)(2)(a-6)}$

$\qquad=\dfrac{5(a+6)(\cancel{a-6})(\cancel{2})(\cancel{10})(\cancel{a+1})}{\cancel{10}(\cancel{a+1})(a-1)(\cancel{2})(\cancel{a-6})}$

$\qquad=\dfrac{5(a+6)}{a-1}$

31. $\dfrac{x^2+4x+4}{(x-1)^2}\cdot\dfrac{x^2-2x+1}{(x+2)^2}=\dfrac{(x+2)^2(x-1)^2}{(x-1)^2(x+2)^2}=1$

33. $\dfrac{t^2+8t+16}{(t+4)^3}\cdot\dfrac{(t+2)^3}{t^2+4t+4}$

$\qquad=\dfrac{(t+4)^2(t+2)^3}{(t+4)^3(t+2)^2}$

$\qquad=\dfrac{(t+4)^2(t+2)^2(t+2)}{(t+4)^2(t+4)(t+2)^2}$

$\qquad=\dfrac{(t+4)^2(t+2)^2}{(t+4)^2(t+2)^2}\cdot\dfrac{t+2}{t+4}$

$\qquad=1\cdot\dfrac{t+2}{t+4}$

$\qquad=\dfrac{t+2}{t+4}$

35. $\dfrac{7a-14}{4-a^2}\cdot\dfrac{5a^2+6a+1}{35a+7}$

$=\dfrac{(7a-14)(5a^2+6a+1)}{(4-a^2)(35a+7)}$

$=\dfrac{7(a-2)(5a+1)(a+1)}{(2+a)(2-a)(7)(5a+1)}$

$=\dfrac{7(-1)(2-a)(5a+1)(a+1)}{(2+a)(2-a)(7)(5a+1)}$

$=\dfrac{\cancel{7}(-1)(\cancel{2-a})(\cancel{5a+1})(a+1)}{(2+a)(\cancel{2-a})(\cancel{7})(\cancel{5a+1})}$

$=\dfrac{-1(a+1)}{2+a}$

$=\dfrac{-(a+1)}{2+a},\quad\text{or}\quad-\dfrac{a+1}{2+a}$

37. $(10x^2-x-2)\cdot\dfrac{4x^2-8x+3}{10x^2-11x-6}$

$=\cancel{(5x+2)}(2x-1)\cdot\dfrac{\cancel{(2x-3)}(2x-1)}{\cancel{(5x+2)}\,\cancel{(2x-3)}}=(2x-1)^2$

39. $\dfrac{c^3+8}{c^5-4c^3}\cdot\dfrac{c^6-4c^5+4c^4}{c^2-2c+4}$

$=\dfrac{(c^3+8)(c^6-4c^5+4c^4)}{(c^5-4c^3)(c^2-2c+4)}$

$=\dfrac{(c+2)(c^2-2c+4)(c^4)(c-2)(c-2)}{c^3(c+2)(c-2)(c^2-2c+4)}$

$=\dfrac{c^3(c+2)(c^2-2c+4)(c-2)}{c^3(c+2)(c^2-2c+4)(c-2)}\cdot\dfrac{c(c-2)}{1}$

$=c(c-2)$

41. The reciprocal of $\frac{3x}{7}$ is $\frac{7}{3x}$ because

$\dfrac{3x}{7}\cdot\dfrac{7}{3x}=1$

43. The reciprocal of a^3-8a is $\frac{1}{a^3-8a}$

because $a^3-8a\cdot\frac{1}{a^3-8a}=1$

45. $\dfrac{5}{8}\div\dfrac{3}{7}=\dfrac{5}{8}\cdot\dfrac{7}{3}$ Multiplying by the
reciprocal of the divisor

$=\dfrac{5}{8}\cdot\dfrac{7}{3}$

$=\dfrac{35}{24}$

47. $\dfrac{x}{4}\div\dfrac{5}{x}=\dfrac{x}{4}\cdot\dfrac{x}{5}$ Multiplying by the
reciprocal of the divisor

$=\dfrac{x\cdot x}{4\cdot 5}$

$=\dfrac{x^2}{20}$

49. $\dfrac{a^5}{b^4}\div\dfrac{a^2}{b^7}=\dfrac{a^5}{b^4}\cdot\dfrac{b^7}{a^2}$ Multiplying by
the reciprocal
of the divisor

$=a^3b^3$

51. $\dfrac{y+5}{4}\div\dfrac{y}{2}=\dfrac{(y+5)}{4}\cdot\dfrac{2}{y}$

$=\dfrac{(y+5)(2)}{4\cdot y}$

$=\dfrac{(y+5)(\cancel{2})}{\cancel{2}\cdot 2y}$

$=\dfrac{y+5}{2y}$

53. $\dfrac{4y-8}{y+2}\div\dfrac{y-2}{y^2-4}=\dfrac{4y-8}{y+2}\cdot\dfrac{y^2-4}{y-2}$

$=\dfrac{(4y-8)(y^2-4)}{(y+2)(y-2)}$

$=\dfrac{4(\cancel{y-2})(\cancel{y+2})(y-2)}{(\cancel{y+2})(\cancel{y-2})(1)}$

$=4(y-2)$

55. $\dfrac{a}{a-b}\div\dfrac{b}{b-a}=\dfrac{a}{a-b}\cdot\dfrac{b-a}{b}$

$=\dfrac{a(b-a)}{(a-b)(b)}$

$=\dfrac{a(-1)(\cancel{a-b})}{(\cancel{a-b})(b)}$

$=\dfrac{-a}{b}=-\dfrac{a}{b}$

57. $(y^2 - 9) \div \dfrac{y^2 - 2y - 3}{y^2 + 1} = \dfrac{(y^2 - 9)}{1} \cdot \dfrac{y^2 + 1}{y^2 - 2y - 3}$

$= \dfrac{(y^2 - 9)(y^2 + 1)}{y^2 - 2y - 3}$

$= \dfrac{(y + 3)(y - 3)(y^2 + 1)}{(y - 3)(y + 1)}$

$= \dfrac{(y + 3)\cancel{(y - 3)}(y^2 + 1)}{\cancel{(y - 3)}(y + 1)}$

$= \dfrac{(y + 3)(y^2 + 1)}{y + 1}$

59. $\dfrac{-3 + 3x}{16} \div \dfrac{x - 1}{5} = \dfrac{-3 + 3x}{16} \cdot \dfrac{5}{x - 1}$

$= \dfrac{(-3 + 3x) \cdot 5}{16(x - 1)}$

$= \dfrac{3(x - 1) \cdot 5}{16(x - 1)}$

$= \dfrac{3\cancel{(x - 1)} \cdot 5}{16\cancel{(x - 1)}}$

$= \dfrac{15}{16}$

61. $\dfrac{a + 2}{a - 1} \div \dfrac{3a + 6}{a - 5} = \dfrac{a + 2}{a - 1} \cdot \dfrac{a - 5}{3a + 6}$

$= \dfrac{(a + 2)(a - 5)}{(a - 1)(3a + 6)}$

$= \dfrac{(a + 2)(a - 5)}{(a - 1) \cdot 3 \cdot (a + 2)}$

$= \dfrac{\cancel{(a + 2)}(a - 5)}{(a - 1) \cdot 3 \cdot \cancel{(a + 2)}}$

$= \dfrac{a - 5}{3(a - 1)}$

63. $\dfrac{25x^2 - 4}{x^2 - 9} \div \dfrac{2 - 5x}{x + 3} = \dfrac{25x^2 - 4}{x^2 - 9} \cdot \dfrac{x + 3}{2 - 5x}$

$= \dfrac{25x^2 - 4}{x^2 - 9} \cdot \dfrac{x + 3}{2 - 5x}$

$= \dfrac{(25x^2 - 4)(x + 3)}{(x^2 - 9)(2 - 5x)}$

$= \dfrac{(5x + 2)(5x - 2)(x + 3)}{(x + 3)(x - 3)(-1)(5x - 2)}$

$= \dfrac{(5x + 2)\cancel{(5x - 2)}\cancel{(x + 3)}}{\cancel{(x + 3)}(x - 3)(-1)\cancel{(5x - 2)}}$

$= \dfrac{5x + 2}{-x + 3}, \text{ or } -\dfrac{5x + 2}{x - 3}$

65. $(2x - 1) \div \dfrac{2x^2 - 11x + 5}{4x^2 - 1}$

$= \dfrac{2x - 1}{1} \cdot \dfrac{4x^2 - 1}{2x^2 - 11x + 5}$

$= \dfrac{(2x - 1)(4x^2 - 1)}{1 \cdot (2x^2 - 11x + 5)}$

$= \dfrac{(2x - 1)(2x + 1)(2x - 1)}{1 \cdot (2x - 1)(x - 5)}$

$= \dfrac{\cancel{(2x - 1)}(2x + 1)(2x - 1)}{1 \cdot \cancel{(2x - 1)}(x - 5)}$

$= \dfrac{(2x - 1)(2x + 1)}{x - 5}$

67. $\dfrac{w^2 - 14w + 49}{2w^2 - 3w - 14} \div \dfrac{3w^2 - 20w - 7}{w^2 - 6w - 16}$

$= \dfrac{w^2 - 14w + 49}{2w^2 - 3w - 14} \cdot \dfrac{w^2 - 6w - 16}{3w^2 - 20w - 7}$

$= \dfrac{(w - 7)\cancel{(w - 7)}}{(2w - 7)\cancel{(w + 2)}} \cdot \dfrac{(w - 8)\cancel{(w + 2)}}{(3w + 1)\cancel{(w - 7)}}$

$= \dfrac{(w - 7)(w - 8)}{(2w - 7)(3w + 1)}$

69. $\dfrac{c^2+10c+21}{c^2-2c-15} \div (5c^2+32c-21)$

$= \dfrac{c^2+10c+21}{c^2-2c-15} \cdot \dfrac{1}{5c^2+32c-21}$

$= \dfrac{(c^2+10c+21)\cdot 1}{(c^2-2c-15)(5c^2+32c-21)}$

$= \dfrac{(c+7)(c+3)}{(c-5)(c+3)(5c-3)(c+7)}$

$= \dfrac{(c+7)(c+3)}{(c+7)(c+3)} \cdot \dfrac{1}{(c-5)(5c-3)}$

$= \dfrac{1}{(c-5)\left(5c-3\right)}$

71. $\dfrac{x^3-64}{x^3+64} \div \dfrac{x^2-16}{x^2-4x+16}$

$= \dfrac{x^3-64}{x^3+64} \cdot \dfrac{x^2-4x+16}{x^2-16}$

$= \dfrac{(x^3-64)(x^2-4x+16)}{(x^3+64)(x^2-16)}$

$= \dfrac{(x-4)(x^2+4x+16)(x^2-4x+16)}{(x+4)(x^2-4x+16)(x+4)(x-4)}$

$= \dfrac{(x-4)(x^2-4x+16)}{(x-4)(x^2-4x+16)} \cdot \dfrac{x^2+4x+16}{(x+4)(x+4)}$

$= \dfrac{x^2+4x+16}{(x+4)(x+4)}, \quad \text{or} \quad \dfrac{x^2+4x+16}{(x+4)^2}$

73. $\dfrac{8a^3+b^3}{2a^2+3ab+b^2} \div \dfrac{8a^2-4ab+2b^2}{4a^2+4ab+b^2}$

$= \dfrac{8a^3+b^3}{2a^2+3ab+b^2} \cdot \dfrac{4a^2+4ab+b^2}{8a^2-4ab+2b^2}$

$= \dfrac{(8a^3+b^3)(4a^2+4ab+b^2)}{(2a^2+3ab+b^2)(8a^2-4ab+2b^2)}$

$= \dfrac{(2a+b)(4a^2-2ab+b^2)(2a+b)(2a+b)}{(2a+b)(a+b)(2)(4a^2-2ab+b^2)}$

$= \dfrac{(2a+b)(4a^2-2ab+b^2)}{(2a+b)(4a^2-2ab+b^2)} \cdot \dfrac{(2a+b)(2a+b)}{(a+b)(2)}$

$= \dfrac{(2a+b)(2a+b)}{2(a+b)}, \quad \text{or} \quad \dfrac{(2a+b)^2}{2(a+b)}$

75. *Thinking and Writing Exercise.*

77. $\dfrac{3}{4}+\dfrac{5}{6} = \dfrac{3}{4}\cdot\dfrac{3}{3}+\dfrac{5}{6}\cdot\dfrac{2}{2}$

$= \dfrac{9}{12}+\dfrac{10}{12}$

$= \dfrac{19}{12}$

79. $\dfrac{2}{9}-\dfrac{1}{6} = \dfrac{2}{9}\cdot\dfrac{2}{2}-\dfrac{1}{6}\cdot\dfrac{3}{3}$

$= \dfrac{4}{18}-\dfrac{3}{18}$

$= \dfrac{1}{18}$

81. $2x^2-x+1-(x^2-x-2)$

$= 2x^2-x+1-x^2+x+2$

$= (2-1)x^2+(-1+1)x+1+2$

$= x^2+3$

83. *Thinking and Writing Exercise.*

85. $2\dfrac{1}{3}x = \dfrac{7}{3}x = \dfrac{7x}{3} \to \dfrac{3}{7x}$

87. $\dfrac{3x-y}{2x+y} \div \dfrac{3x-y}{2x+y}$

We have the rational expression $\dfrac{3x-y}{2x+y}$

divided by itself: Thus, the result is 1 for
$2x+y \neq 0$, and $3x-y \neq 0.$.

89. $(x-2a) \div \dfrac{a^2x^2-4a^4}{a^2x+2a^3} = \dfrac{x-2a}{1} \cdot \dfrac{a^2x+2a^3}{a^2x^3-4a^4}$

$= \dfrac{(x-2a)(a^2x+2a^3)}{a^2x^2-4a^4}$

$= \dfrac{(x-2a)(a^2)(x+2a)}{a^2(x+2a)(x-2a)}$

$= \dfrac{(x-2a)(a^2)(x+2a)}{a^2(x+2a)(x+2a)}$

$= 1$

91. $\dfrac{a^2-3b}{a^2+2b} \cdot \dfrac{a^2-2b}{a^2+3b} \cdot \dfrac{a^2+2b}{a^2-3b}$

Note that $\dfrac{a^2-3b}{a^2+2b} \cdot \dfrac{a^2+2b}{a^2-3b}$ is the product of

reciprocals and thus is equal to 1. Then the
product in the original exercise is the

remaining factor, $\dfrac{a^2-2b}{a^2+3b}.$

93. $\left[\dfrac{r^2-4s^2}{r+2s}\div(r+2s)\right]\cdot\dfrac{2s}{r-2s}$

$=\left[\dfrac{r^2-4s^2}{r+2s}\cdot\dfrac{1}{r+2s}\right]\cdot\dfrac{2s}{r-2s}$

$=\dfrac{(r^2-4s^2)(2s)}{(r+2s)(r+2s)(r-2s)}$

$=\dfrac{(r+2s)(r-2s)(2s)}{(r+2s)(r+2s)(r-2s)}$

$=\dfrac{(\cancel{r+2s})(\cancel{r-2s})(2s)}{(\cancel{r+2s})(r+2s)(\cancel{r-2s})}$

$=\dfrac{2s}{r+2s}$

95. $g(x)=\dfrac{2x+3}{4x-1}$

(a) $g(x+h)=\dfrac{2(x+h)+3}{4(x+h)-1}=\dfrac{2x+2h+3}{4x+4h-1}$

(b)

$g(2x-2)\cdot g(x)=\dfrac{2(2x-2)+3}{4(2x-2)-1}\cdot\dfrac{2x+3}{4x-1}$

$=\dfrac{4x-1}{8x-9}\cdot\dfrac{2x+3}{4x-1}$

$=\dfrac{2x+3}{8x-9}$

(c) $g\left(\dfrac{1}{2}x+1\right)\cdot g(x)=\dfrac{2(\frac{1}{2}x+1)+3}{4(\frac{1}{2}x+1)-1}\cdot\dfrac{2x+3}{4x-1}$

$=\dfrac{x+5}{2x+3}\cdot\dfrac{2x+3}{4x-1}$

$=\dfrac{x+5}{4x-1}$

Exercise Set 7.3

1. numerators; denominator

3. least common denominator; LCD

5. $\dfrac{6}{x}+\dfrac{4}{x}=\dfrac{10}{x}$ Adding numerators

7. $\dfrac{x}{12}+\dfrac{2x+5}{12}=\dfrac{3x+5}{12}$ Adding numerators

9. $\dfrac{4}{a+3}+\dfrac{5}{a+3}=\dfrac{9}{a+3}$ Adding numerators

11. $\dfrac{11}{4x-7}-\dfrac{3}{4x-7}=\dfrac{8}{4x-7}$ Subtracting numerators

13. $\dfrac{3y+8}{2y}-\dfrac{y+1}{2y}$

$=\dfrac{3y+8-(y+1)}{2y}$ Subtracting numerators

$=\dfrac{3y+8-y-1}{2y}$ Removing parentheses

$=\dfrac{2y+7}{2y}$

15. $\dfrac{7x+8}{x+1}+\dfrac{4x+3}{x+1}=\dfrac{11x+11}{x+1}$ Adding numerators

$=\dfrac{11(x+1)}{x+1}$ Factoring

$=\dfrac{11(\cancel{x+1})}{\cancel{x+1}}$ Removing a factor equal to 1

$=11$

17. $\dfrac{7x+8}{x+1}-\dfrac{4x+3}{x+1}=\dfrac{7x+8-(4x+3)}{x+1}$

$=\dfrac{7x+8-4x-3}{x+1}$

$=\dfrac{3x+5}{x+1}$

19. $\dfrac{a^2}{a-4}+\dfrac{a-20}{a-4}=\dfrac{a^2+a-20}{a-4}$

$=\dfrac{(a+5)(a-4)}{a-4}$

$=\dfrac{(a+5)(\cancel{a-4})}{\cancel{a-4}}$

$=a+5$

21. $\dfrac{x^2}{x-2} - \dfrac{6x-8}{x-2} = \dfrac{x^2 - (6x-8)}{x-2}$

$\qquad = \dfrac{x^2 - 6x + 8}{x-2}$

$\qquad = \dfrac{(x-4)(x-2)}{x-2}$

$\qquad = \dfrac{(x-4)(\cancel{x-2})}{\cancel{x-2}}$

$\qquad = x-4$

23. $\dfrac{t^2 - 5t}{t-1} + \dfrac{5t - t^2}{t-1}$

Note that the numerators are opposites, so their sum is 0. Then we have $\frac{0}{t-1}$, or 0.

25. $\dfrac{x-6}{x^2+5x+6} + \dfrac{9}{x^2+5x+6} = \dfrac{x+3}{x^2+5x+6}$

$\qquad\qquad = \dfrac{x+3}{(x+3)(x+2)}$

$\qquad\qquad = \dfrac{\cancel{x+3}}{(\cancel{x+3})(x+2)}$

$\qquad\qquad = \dfrac{1}{x+2}$

27. $\dfrac{3a^2+14}{a^2+5a-6} - \dfrac{13a}{a^2+5a-6} = \dfrac{3a^2 - 13a + 14}{a^2+5a-6}$

$\qquad\qquad = \dfrac{(3a-7)(a-2)}{(a+6)(a-1)}$

(No simplification is possible.)

29. $\dfrac{t^2-3t}{t^2+6t+9} + \dfrac{2t-12}{t^2+6t+9} = \dfrac{t^2-t-12}{t^2+6t+9}$

$\qquad\qquad = \dfrac{(t-4)(t+3)}{(t+3)^2}$

$\qquad\qquad = \dfrac{(t-4)(\cancel{t+3})}{(t+3)(\cancel{t+3})}$

$\qquad\qquad = \dfrac{t-4}{t+3}$

31. $\dfrac{2x^2+x}{x^2-8x+12} - \dfrac{x^2-2x+10}{x^2-8x+12}$

$\qquad = \dfrac{2x^2+x - (x^2-2x+10)}{x^2-8x+12}$

$\qquad = \dfrac{2x^2+x - x^2 + 2x - 10}{x^2-8x+12}$

$\qquad = \dfrac{x^2+3x-10}{x^2-8x+12}$

$\qquad = \dfrac{(x+5)(x-2)}{(x-6)(x-2)}$

$\qquad = \dfrac{(x+5)(\cancel{x-2})}{(x-6)(\cancel{x-2})}$

$\qquad = \dfrac{x+5}{x-6}$

33. $\dfrac{3-2x}{x^2-6x+8} + \dfrac{7-3x}{x^2-6x+8}$

$\qquad = \dfrac{10-5x}{x^2-6x+8}$

$\qquad = \dfrac{5(2-x)}{(x-4)(x-2)}$

$\qquad = \dfrac{5(-1)(x-2)}{(x-4)(x-2)}$

$\qquad = \dfrac{5(-1)(\cancel{x-2})}{(x-4)(\cancel{x-2})}$

$\qquad = \dfrac{-5}{x-4}, \quad\text{or}\quad -\dfrac{5}{x-4}, \quad\text{or}\quad \dfrac{5}{4-x}$

35. $\dfrac{x-9}{x^2+3x-4} - \dfrac{2x-5}{x^2+3x-4} = \dfrac{x-9-(2x-5)}{x^2+3x-4}$

$\qquad\qquad = \dfrac{x-9-2x+5}{x^2+3x-4}$

$\qquad\qquad = \dfrac{-x-4}{(x+4)(x-1)}$

$\qquad\qquad = \dfrac{-(x+4)}{(x+4)(x-1)}$

$\qquad\qquad = \dfrac{-(\cancel{x+4})}{(\cancel{x+4})(x-1)}$

$\qquad\qquad = \dfrac{-1}{x-1}, \quad\text{or}\quad -\dfrac{1}{x-1},$

$\qquad\qquad\qquad \text{or}\quad \dfrac{1}{1-x}$

37. $\quad 15 = 3 \cdot 5$
$\qquad 27 = 3 \cdot 3 \cdot 3$
$\qquad$ LCM $= 3 \cdot 3 \cdot 3 \cdot 5$, or 135

39. $\qquad 8 = 2 \cdot 2 \cdot 2$
$\qquad 9 = 3 \cdot 3$
$\qquad$ LCM $= 2 \cdot 2 \cdot 2 \cdot 3 \cdot 3$, or 72

41. $\qquad 6 = 2 \cdot 3$
$\qquad 12 = 2 \cdot 2 \cdot 3$
$\qquad 15 = 3 \cdot 5$
$\qquad$ LCM $= 2 \cdot 2 \cdot 3 \cdot 5 = 60$

43. $12x^2 = 2 \cdot 2 \cdot 3 \cdot x \cdot x$
$\qquad 6x^3 = 2 \cdot 3 \cdot x \cdot x \cdot x$
$\qquad$ LCM $= 2 \cdot 2 \cdot 3 \cdot x \cdot x \cdot x$, or $12x^3$

45. $15a^4b^7 = 3 \cdot 5 \cdot a \cdot a \cdot a \cdot a \cdot b \cdot b \cdot b \cdot b \cdot b \cdot b \cdot b$
$\qquad 10a^2b^8 = 2 \cdot 5 \cdot a \cdot a \cdot b \cdot b \cdot b \cdot b \cdot b \cdot b \cdot b \cdot b$
$\qquad$ LCM $= 2 \cdot 3 \cdot 5 \cdot a \cdot a \cdot a \cdot a \cdot b \cdot b \cdot b \cdot b$
$\qquad\qquad \cdot b \cdot b \cdot b$, or $30a^4b^8$

47. $2(y-3) = 2 \cdot (y-3)$
$\qquad 6(y-3) = 2 \cdot 3 \cdot (y-3)$
$\qquad$ LCM $= 2 \cdot 3 \cdot (y-3)$, or $6(y-3)$

49. $\qquad x^2 - 4 = (x+2)(x-2)$
$\qquad x^2 + 5x + 6 = (x+3)(x+2)$
$\qquad$ LCM $= (x+2)(x-2)(x+3)$

51. $t^3 + 4t^2 + 4t = t(t^2 + 4t + 4)$
$\qquad\qquad = t(t+2)(t+2)$
$\qquad t^2 - 4t = t(t-4)$
$\qquad$ LCM $= t(t+2)(t+2)(t-4)$
$\qquad\qquad = t(t+2)^2(t-4)$

53. $10x^2y = 2 \cdot 5 \cdot x \cdot x \cdot y$
$\qquad 6y^2z = 2 \cdot 3 \cdot y \cdot y \cdot z$
$\qquad 5xz^3 = 5 \cdot x \cdot z \cdot z \cdot z$
$\qquad$ LCM $= 2 \cdot 3 \cdot 5 \cdot x \cdot x \cdot y \cdot y \cdot z \cdot z \cdot z$
$\qquad\qquad = 30x^2y^2z^3$

55. $\qquad a + 1 = a + 1$
$\qquad (a-1)^2 = (a-1)(a-1)$
$\qquad a^2 - 1 = (a+1)(a-1)$
$\qquad$ LCM $= (a+1)(a-1)(a-1)$
$\qquad\qquad = (a+1)(a-1)^2$

57. $\quad 2n^2 + n - 1 = (2n-1)(n+1)$
$\qquad 2n^2 + 3n - 2 = (2n-1)(n+2)$
$\qquad\qquad$ LCM $= (2n-1)(n+2)(n+1)$

59. $t - 3, t + 3, t^2 - 9$
$\qquad t^2 - 9 = (t+3)(t-3)$
$\qquad$ LCM $= t^2 - 9$

61. $6x^3 - 24x^2 + 18x = 6x(x^2 - 4x + 3)$
$\qquad\qquad = 2 \cdot 3 \cdot x(x-1)(x-3)$
$\qquad 4x^5 - 24x^4 + 20x^3 = 4x^3(x^2 - 6x + 5)$
$\qquad\qquad = 2 \cdot 2 \cdot x \cdot x \cdot x(x-1)(x-5)$
$\qquad$ LCM $= 2 \cdot 2 \cdot 3 \cdot x \cdot x \cdot x(x-1)$
$\qquad\qquad \cdot (x-3)(x-5)$
$\qquad\qquad = 12x^3(x-1)(x-3)(x-5)$

63. $2t^3 - 2 = 2(t^3 - 1) = 2(t-1)(t^2 + t + 1)$
$\qquad t^2 - 1 = (t+1)(t-1)$
$\qquad$ LCM $= 2(t-1)(t^2 + t + 1)(t+1)$

65. $\quad 6x^5 = 2 \cdot 3 \cdot x \cdot x \cdot x \cdot x \cdot x$
$\qquad 12x^3 = 2 \cdot 2 \cdot 3 \cdot x \cdot x \cdot x$
$\qquad$ The LCD is $2 \cdot 2 \cdot 3 \cdot x \cdot x \cdot x \cdot x \cdot x$, or $12x^5$.

The factor of the LCD that is missing from the first denominator is 2. We multiply by 1 using $2/2$:

$$\frac{5}{6x^5} \cdot \frac{2}{2} = \frac{10}{12x^5}$$

The second denominator is missing two factors of x, or x^2. We multiply by 1 using x^2/x^2:

$$\frac{y}{12x^3} \cdot \frac{x^2}{x^3} = \frac{x^2y}{12x^5}$$

67. $2a^2b = 2 \cdot a \cdot a \cdot b$

$8ab^2 = 2 \cdot 2 \cdot 2 \cdot a \cdot b \cdot b$

The LCD is $2 \cdot 2 \cdot 2 \cdot a \cdot a \cdot b \cdot b$, or $8a^2b^2$.

We multiply the first expression by $\frac{4b}{4b}$ to obtain the LCD:

$$\frac{3}{2a^2b} \cdot \frac{4b}{4b} = \frac{12b}{8a^2b^2}$$

We multiply the second expression by a/a to obtain the LCD:

$$\frac{7}{8ab^2} \cdot \frac{a}{a} = \frac{7a}{8a^2b^2}$$

69. The LCD is $(x+2)(x-2)(x+3)$.

$$\frac{2x}{x^2-4} = \frac{2x}{(x+2)(x-2)} \cdot \frac{x+3}{x+3}$$

$$= \frac{2x(x+3)}{(x+2)(x-2)(x+3)}$$

$$\frac{4x}{x^2+5x+6} = \frac{4x}{(x+3)(x+2)} \cdot \frac{x-2}{x-2}$$

$$= \frac{4x(x-2)}{(x+3)(x+2)(x-2)}$$

71. *Thinking and Writing Exercise.*

73. $-\dfrac{5}{8} = \dfrac{-5}{8} = \dfrac{5}{-8}$

75. $-(x-y) = -x+y = y-x$

77. $-1(2x-7) = -2x+7 = 7-2x$

79. *Thinking and Writing Exercise.*

81. $\dfrac{6x-1}{x-1} + \dfrac{3(2x+5)}{x-1} + \dfrac{3(2x-3)}{x-1}$

$= \dfrac{6x-1+6x+15+6x-9}{x-1}$

$= \dfrac{18x+5}{x-1}$

83. $\dfrac{x^2}{3x^2-5x-2} - \dfrac{2x}{3x+1} \cdot \dfrac{1}{x-2}$

$= \dfrac{x^2}{(3x+1)(x-2)} - \dfrac{2x}{(3x+1)(x-2)}$

$= \dfrac{x^2-2x}{(3x+1)(x-2)}$

$= \dfrac{x\cancel{(x-2)}}{(3x+1)\cancel{(x-2)}}$

$= \dfrac{x}{3x+1}$

85. The smallest number of strands that can be used is the LCM of 10 and 3.

$10 = 2 \cdot 5$

$3 = 3$

$\text{LCM} = 2 \cdot 5 \cdot 3 = 30$

87. If the number of strands must also be a multiple of 4, we find the smallest multiple of 30 that is also a multiple of 4.

$1 \cdot 30 = 30$, not a multiple of 4

$2 \cdot 30 = 60 = 15 \cdot 4$, a multiple of 4

The smallest number of strands that can be used is 60.

89. $4x^2 - 25 = (2x-5)(2x+5)$

$6x^2 - 7x - 20 = (2x-5)(3x+4)$

$(9x^2+24x+16)^2 = [(3x+4)^2]^2 = (3x+4)^4$

$\text{LCM} = (2x-5)(2x+5)(3x+4)^4$

91. If the Sharp DX-C400 prints 40 pages per minute, then it prints one page in $\frac{1}{40}$th of a minute. Similarly, the Canon Image Runner 1025N can print one page in $\frac{1}{25}$th of a minute. The LCM of $\frac{1}{40}$ and $\frac{1}{25}$ is the LCM of $\frac{1}{40} \cdot \frac{5}{5} = \frac{5}{200}$ and $\frac{1}{25} \cdot \frac{8}{8} = \frac{8}{200}$, or $\frac{40}{200} = \frac{1}{5}$. $\frac{1}{5}$th of a minute is 12 seconds. Thus, every 12 seconds, both copiers begin printing new pages.

93. The number of minutes after 5:00 A.M. when the shuttles will first: leave at the same time again is the LCM of their departure intervals, 25 minutes and 35 minutes.

$$25 = 5 \cdot 5$$
$$35 = 5 \cdot 7$$

LCM $= 5 \cdot 5 \cdot 7$, or 175

Thus, the shuttles will leave at the same time 175 minutes after 5:00 A.M., or at 7:55 A.M.

95. *Thinking and Writing Exercise.*

Exercise Set 7.4

1. LCD

3. numerators; LCD

5. LCD $= x \cdot x$, or x^2

$$\frac{4}{x} + \frac{9}{x^2} = \frac{4}{x} \cdot \frac{x}{x} + \frac{9}{x^2}$$
$$= \frac{4x+9}{x^2}$$

7. $\left.\begin{array}{l} 6r = 2 \cdot 3 \cdot r \\ 8r = 2 \cdot 2 \cdot 2 \cdot r \end{array}\right\}$ LCD $= 2 \cdot 2 \cdot 2 \cdot 3 \cdot r = 24r$

$$\frac{1}{6r} - \frac{3}{8r} = \frac{1}{6r} \cdot \frac{4}{4} - \frac{3}{8r} \cdot \frac{3}{3}$$
$$= \frac{4-9}{24r}$$
$$= -\frac{5}{24r}$$

9. $\left.\begin{array}{l} c^2 d = c \cdot c \cdot d \\ cd^3 = c \cdot d \cdot d \cdot d \end{array}\right\}$ LCD $= c \cdot c \cdot d \cdot d \cdot d = c^2 d^3$

$$\frac{2}{c^2 d} + \frac{7}{cd^3} = \frac{2}{c^2 d} \cdot \frac{d^2}{d^2} + \frac{7}{cd^3} \cdot \frac{c}{c}$$
$$= \frac{2d^2 + 7c}{c^2 d^3}$$

11. $\left.\begin{array}{l} 3xy^2 = 3 \cdot x \cdot y \cdot y \\ x^2 y^3 = x \cdot x \cdot y \cdot y \cdot y \end{array}\right\}$ LCD $= 3 \cdot x \cdot x \cdot y \cdot y \cdot y = 3x^2 y^3$

$$\frac{-2}{3xy^2} - \frac{6}{x^2 y^3} = \frac{-2}{3xy^2} \cdot \frac{xy}{xy} - \frac{6}{x^2 y^3} \cdot \frac{3}{3}$$
$$= \frac{-2xy - 18}{3x^2 y^3}$$

13. $\left.\begin{array}{l} 9 = 3 \cdot 3 \\ 6 = 2 \cdot 3 \end{array}\right\}$ LCD $= 2 \cdot 3 \cdot 3 = 18$

$$\frac{x-4}{9} + \frac{x+5}{6} = \frac{x-4}{9} \cdot \frac{2}{2} + \frac{x+5}{6} \cdot \frac{3}{3}$$
$$= \frac{2(x-4) + 3(x+5)}{18}$$
$$= \frac{2x - 8 + 3x + 15}{18}$$
$$= \frac{5x + 7}{18}$$

15. $\left.\begin{array}{l} 2 = 2 \\ 4 = 2 \cdot 2 \end{array}\right\}$ LCD $= 4$

$$\frac{a+2}{2} - \frac{a-4}{4} = \frac{a+2}{2} \cdot \frac{2}{2} - \frac{a-4}{4}$$
$$= \frac{2a+4}{4} - \frac{a-4}{4}$$
$$= \frac{2a+4 - (a-4)}{4}$$
$$= \frac{2a+4 - a + 4}{4}$$
$$= \frac{a+8}{4}$$

17. $\left.\begin{array}{l} 6a^2 = 3 \cdot a \cdot a \\ 9a = 3 \cdot 3 \cdot a \end{array}\right\}$ LCD $= 3 \cdot 3 \cdot a \cdot a$, or $9a^2$

$$\frac{2a-1}{3a^2} + \frac{5a+1}{9a} = \frac{2a-1}{3a^2} \cdot \frac{3}{3} + \frac{5a+1}{9a} \cdot \frac{a}{a}$$
$$= \frac{6a-3}{9a^2} + \frac{5a^2 + a}{9a^2}$$
$$= \frac{5a^2 + 7a - 3}{9a^2}$$

19. $\left. \begin{array}{l} 4x = 4 \cdot x \\[4pt] x = x \end{array} \right\}$ LCD $= 4x$

$$\frac{x-1}{4x} - \frac{2x+3}{x} = \frac{x-1}{4x} - \frac{2x+3}{x} \cdot \frac{4}{4}$$
$$= \frac{x-1}{4x} - \frac{8x+12}{4x}$$
$$= \frac{x-1-(8x+12)}{4x}$$
$$= \frac{x-1-8x-12}{4x}$$
$$= \frac{-7x-13}{4x}$$

21. $\left. \begin{array}{l} c^2 d = c \cdot c \cdot d \\[4pt] cd^2 = c \cdot d \cdot d \end{array} \right\}$ LCD $= c \cdot c \cdot d \cdot d$, or $c^2 d^2$

$$\frac{2c-d}{c^2 d} + \frac{c+d}{cd^2} = \frac{2c-d}{c^2 d} \cdot \frac{d}{d} + \frac{c+d}{cd^2} \cdot \frac{c}{c}$$
$$= \frac{d(2c-d) + c(c+d)}{c^2 d^2}$$
$$= \frac{2cd - d^2 + c^2 + cd}{c^2 d^2}$$
$$= \frac{c^2 + 3cd - d^2}{c^2 d^2}$$

23. $\left. \begin{array}{l} 2x^2 y = 2 \cdot x \cdot x \cdot y \\[4pt] xy^2 = x \cdot y \cdot y \end{array} \right\}$ LCD $= 2 \cdot x \cdot x \cdot y \cdot y$, or $2x^2 y^2$

$$\frac{5x+3y}{2x^2 y} - \frac{3x+4y}{xy^2} = \frac{5x+3y}{2x^2 y} \cdot \frac{y}{y}$$
$$\qquad - \frac{3x+4y}{xy^2} \cdot \frac{2x}{2x}$$
$$= \frac{5xy + 3y^2}{2x^2 y^2} - \frac{6x^2 + 8xy}{2x^2 y^2}$$
$$= \frac{5xy + 3y^2 - (6x^2 + 8xy)}{2x^2 y^2}$$
$$= \frac{5xy + 3y^2 - 6x^2 - 8xy}{2x^2 y^2}$$
$$= \frac{3y^2 - 3xy - 6x^2}{2x^2 y^2}$$

(Although $3y^2 - 3xy - 6x^2$ can be factored, doing so will not enable us to simplify the result further.)

25. The denominators cannot be factored, so the LCD is their product, $(x-1)(x+1)$.

$$\frac{5}{x-1} + \frac{5}{x+1} = \frac{5}{x-1} \cdot \frac{x+1}{x+1} + \frac{5}{x+1} \cdot \frac{x-1}{x-1}$$
$$= \frac{5(x+1) + 5(x-1)}{(x-1)(x+1)}$$
$$= \frac{5x+5+5x-5}{(x-1)(x+1)}$$
$$= \frac{10x}{(x-1)(x+1)}$$

27. The denominators cannot be factored, so the LCD is their product, $(z-1)(z+1)$.

$$\frac{4}{z-1} - \frac{2}{z+1} = \frac{4}{z-1} \cdot \frac{z+1}{z+1} - \frac{2}{z+1} \cdot \frac{z-1}{z-1}$$
$$= \frac{4z+4}{(z-1)(z+1)} - \frac{2z-2}{(z-1)(z+1)}$$
$$= \frac{4z+4-(2z-2)}{(z-1)(z+1)}$$
$$= \frac{4z+4-2z+2}{(z-1)(z+1)}$$
$$= \frac{2z+6}{(z-1)(z+1)}$$

(Although $2z+6$ can be factored, doing so will not enable us to simplify the result further.)

29. $\left. \begin{array}{l} x+5 = x+5 \\[4pt] 4x = 4 \cdot x \end{array} \right\}$ LCD $= 4x(x+5)$

$$\frac{2}{x+5} + \frac{3}{4x} = \frac{2}{x+5} \cdot \frac{4x}{4x} + \frac{3}{4x} \cdot \frac{x+5}{x+5}$$
$$= \frac{2 \cdot 4x + 3(x+5)}{4x(x+5)}$$
$$= \frac{8x + 3x + 15}{4x(x+5)}$$
$$= \frac{11x + 15}{4x(x+5)}$$

31. $\begin{array}{r} 3t^2 - 15t = 3t(t-5) \\ 2t - 10 = 2(t-5) \end{array} \bigg\}$ LCD $= 6t(t-5)$

$\dfrac{8}{3t(t-5)} - \dfrac{3}{2(t-5)} = \dfrac{8}{3t(t-5)} \cdot \dfrac{2}{2}$

$\qquad\qquad\qquad - \dfrac{3}{2(t-5)} \cdot \dfrac{3t}{3t}$

$\qquad\qquad = \dfrac{16}{6t(t-5)} - \dfrac{9t}{6t(t-5)}$

$\qquad\qquad = \dfrac{16 - 9t}{6t(t-5)}$

33. $\dfrac{4x}{x^2 - 25} + \dfrac{x}{x+5} = \dfrac{4x}{(x+5)(x-5)} + \dfrac{x}{x+5}$

$\qquad\qquad$ LCD $= (x+5)(x-5)$

$\qquad\qquad = \dfrac{4x + x(x-5)}{(x+5)(x-5)}$

$\qquad\qquad = \dfrac{4x + x^2 - 5x}{(x+5)(x-5)}$

$\qquad\qquad = \dfrac{x^2 - x}{(x+5)(x-5)}$

(Although $x^2 - x$ can be factored, doing so will not enable us to simplify the result further.)

35. $\dfrac{t}{t-3} - \dfrac{5}{4t - 12} = \dfrac{t}{t-3} - \dfrac{5}{4(t-3)}$

$\qquad\qquad$ LCD $= 4(t-3)$

$\qquad\qquad = \dfrac{t}{t-3} \cdot \dfrac{4}{4} - \dfrac{5}{4(t-3)}$

$\qquad\qquad = \dfrac{4t - 5}{4(t-3)}$

37. $\dfrac{2}{x+3} + \dfrac{4}{(x+3)^2}$ LCD $= (x+3)^2$

$\qquad = \dfrac{2}{x+3} \cdot \dfrac{x+3}{x+3} + \dfrac{4}{(x+3)^2}$

$\qquad = \dfrac{2(x+3) + 4}{(x+3)^2}$

$\qquad = \dfrac{2x + 6 + 4}{(x+3)^2}$

$\qquad = \dfrac{2x + 10}{(x+3)^2}$

(Although $2x + 10$ can be factored, doing so will not enable us to simplify the result further.)

39. $\dfrac{t-3}{t^3 - 1} - \dfrac{2}{1 - t^3} = \dfrac{t-3}{t^3 - 1} - \dfrac{2}{-(t^3 - 1)} = \dfrac{t-3}{t^3 - 1} + \dfrac{2}{t^3 - 1}$

$\qquad\qquad = \dfrac{t - 3 + 2}{t^3 - 1} = \dfrac{\cancel{t-1}}{(\cancel{t-1})(t^2 + t + 1)}$

$\qquad\qquad = \dfrac{1}{t^2 + t + 1}$

41. $\dfrac{3a}{4a - 20} + \dfrac{9a}{6a - 30} = \dfrac{3a}{2 \cdot 2(a-5)} + \dfrac{9a}{2 \cdot 3(a-5)}$

$\qquad\qquad$ LCD $= 2 \cdot 2 \cdot 3(a-5)$

$\qquad\qquad = \dfrac{3a}{2 \cdot 2(a-5)} \cdot \dfrac{3}{3}$

$\qquad\qquad\quad + \dfrac{9a}{2 \cdot 3(a-5)} \cdot \dfrac{2}{2}$

$\qquad\qquad = \dfrac{9a + 18a}{2 \cdot 2 \cdot 3(a-5)}$

$\qquad\qquad = \dfrac{27a}{2 \cdot 2 \cdot 3(a-5)}$

$\qquad\qquad = \dfrac{\cancel{3} \cdot 9 \cdot a}{2 \cdot 2 \cdot \cancel{3}(a-5)}$

$\qquad\qquad = \dfrac{9a}{4(a-5)}$

43. $\dfrac{x}{x-5} + \dfrac{x}{5-x} = \dfrac{x}{x-5} + \dfrac{x}{5-x} \cdot \dfrac{-1}{-1}$

$\qquad\qquad = \dfrac{x}{x-5} + \dfrac{-x}{x-5}$

$\qquad\qquad = 0$

45. $\dfrac{6}{a^2+a-2}+\dfrac{4}{a^2-4a+3}$

$=\dfrac{6}{(a+2)(a-1)}+\dfrac{4}{(a-3)(a-1)}$

$\text{LCD}=(a+2)(a-1)(a-3)$

$=\dfrac{6}{(a+2)(a-1)}\cdot\dfrac{a-3}{a-3}$

$\quad+\dfrac{4}{(a-3)(a-1)}\cdot\dfrac{a+2}{a+2}$

$=\dfrac{6(a-3)+4(a+2)}{(a+2)(a-1)(a-3)}$

$=\dfrac{6a-18+4a+8}{(a+2)(a-1)(a-3)}$

$=\dfrac{10a-10}{(a+2)(a-1)(a-3)}$

$=\dfrac{10(a-1)}{(a+2)(a-1)(a-3)}=\dfrac{10}{(a+2)(a-3)}$

47. $\dfrac{x}{x^2+9x+20}-\dfrac{4}{x^2+7x+12}$

$=\dfrac{x}{(x+4)(x+5)}-\dfrac{4}{(x+3)(x+4)}$

$\text{LCD}=(x+3)(x+4)(x+5)$

$=\dfrac{x}{(x+4)(x+5)}\cdot\dfrac{x+3}{x+3}$

$\quad-\dfrac{4}{(x+3)(x+4)}\cdot\dfrac{x+5}{x+5}$

$=\dfrac{x(x+3)-4(x+5)}{(x+3)(x+4)(x+5)}$

$=\dfrac{x^2+3x-4x-20}{(x+3)(x+4)(x+5)}$

$=\dfrac{x^2-x-20}{(x+3)(x+4)(x+5)}$

$=\dfrac{(x+4)(x-5)}{(x+3)(x+4)(x+5)}$

$=\dfrac{x-5}{(x+3)(x+5)}$

49. $\dfrac{3z}{z^2-4z+4}+\dfrac{10}{z^2+z-6}$

$=\dfrac{3z}{(z-2)^2}+\dfrac{10}{(z+3)(z-2)}$

$\text{LCD}=(z-2)^2(z+3)$

$=\dfrac{3z}{(z-2)^2}\cdot\dfrac{z+3}{z+3}+\dfrac{10}{(z+3)(z-2)}\cdot\dfrac{z-2}{z-2}$

$=\dfrac{3z^2+9z}{(z-2)^2(z+3)}+\dfrac{10z-20}{(z-2)^2(z+3)}$

$=\dfrac{3z^2+9z+10z-20}{(z-2)^2(z+3)}$

$=\dfrac{3z^2+19z-20}{(z-2)^2(z+3)}$

51. $\dfrac{-5}{x^2+17x+16}-\dfrac{0}{x^2+9x+8}$

Note that $\frac{0}{x^2+9x+8}=0$, so the difference is

$\dfrac{-5}{x^2+17x+16}$.

53. $\dfrac{4x}{5}+\dfrac{x+3}{-5}=\dfrac{4x}{5}-\dfrac{x+3}{5}=\dfrac{4x-(x+3)}{5}$

$\qquad=\dfrac{4x-x-3}{5}=\dfrac{3x-3}{5}$

55. $\dfrac{y^2}{y-3}-\dfrac{9}{3-y}=\dfrac{y^2}{y-3}+\dfrac{9}{y-3}$

$\qquad=\dfrac{y^2+9}{y-3}$

57. $\dfrac{c-5}{c^2-64}-\dfrac{5-c}{64-c^2}=\dfrac{c-5}{c^2-64}+\dfrac{5-c}{c^2-64}$

$\qquad=\dfrac{c-5+5-c}{c^2-64}=\dfrac{0}{c^2-64}=0$

59. $\dfrac{y+2}{y-7}+\dfrac{3-y}{49-y^2}=\dfrac{y+2}{y-7}+\dfrac{3-y}{(7+y)(7-y)}$

$$=\dfrac{y+2}{y-7}+\dfrac{3-y}{(7+y)(7-y)}\cdot\dfrac{-1}{-1}$$

$$=\dfrac{y+2}{y-7}+\dfrac{y-3}{(y+7)(y-7)}$$

$$\text{LCD}=(y+7)(y-7)$$

$$=\dfrac{y+2}{y-7}\cdot\dfrac{y+7}{y+7}$$

$$+\dfrac{y-3}{(y+7)(y-7)}$$

$$=\dfrac{y^2+9y+14+y-3}{(y+7)(y-7)}$$

$$=\dfrac{y^2+10y+11}{(y+7)(y-7)}$$

61. $\dfrac{x}{x-4}-\dfrac{3}{16-x^2}=\dfrac{x}{x-4}+\dfrac{3}{x^2-16}$

$$=\dfrac{x}{x-4}\cdot\dfrac{x+4}{x+4}+\dfrac{3}{(x-4)(x+4)}$$

$$=\dfrac{x(x+4)+3}{(x-4)(x+4)}=\dfrac{x^2+4x+3}{(x-4)(x+4)}$$

$$=\dfrac{(x+1)(x+3)}{(x-4)(x+4)}$$

63. $\dfrac{3x+2}{3x+6}+\dfrac{x}{4-x^2}=\dfrac{3x+2}{3(x+2)}+\dfrac{x}{(2+x)(2-x)}$

$$\text{LCD}=3(x+2)(2-x)$$

$$=\dfrac{3x+2}{3(x+2)}\cdot\dfrac{2-x}{2-x}$$

$$+\dfrac{x}{(2+x)(2-x)}\cdot\dfrac{3}{3}$$

$$=\dfrac{(3x+2)(2-x)+x\cdot3}{3(x+2)(2-x)}$$

$$=\dfrac{-3x^2+4x+4+3x}{3(x+2)(2-x)}$$

$$=\dfrac{-3x^2+7x+4}{3(x+2)(2-x)},\text{ or}$$

$$\dfrac{3x^2-7x-4}{3(x+2)(x-2)}$$

65. $\dfrac{4-a^2}{a^2-9}-\dfrac{a-2}{3-a}=\dfrac{4-a^2}{(a+3)(a-3)}-\dfrac{a-2}{3-a}$

$$=\dfrac{4-a^2}{(a+3)(a-3)}-\dfrac{a-2}{3-a}\cdot\dfrac{-1}{-1}$$

$$=\dfrac{4-a^2}{(a+3)(a-3)}-\dfrac{2-a}{a-3}$$

$$\text{LCD}=(a+3)(a-3)$$

$$=\dfrac{4-a^2}{(a+3)(a-3)}-\dfrac{2-a}{a-3}\cdot\dfrac{a+3}{a+3}$$

$$=\dfrac{4-a^2-(2a+6-a^2-3a)}{(a+3)(a-3)}$$

$$=\dfrac{4-a^2-2a-6+a^2+3a}{(a+3)(a-3)}$$

$$=\dfrac{a-2}{(a+3)(a-3)}$$

67. $\dfrac{x-3}{2-x}-\dfrac{x+3}{x+2}+\dfrac{x+6}{4-x^2}$

$$=\dfrac{x-3}{2-x}-\dfrac{x+3}{x+2}+\dfrac{x+6}{(2+x)(2-x)}$$

$$\text{LCD}=(2+x)(2-x)$$

$$=\dfrac{x-3}{2-x}\cdot\dfrac{2+x}{2+x}-\dfrac{x+3}{x+2}\cdot\dfrac{2-x}{2-x}$$

$$+\dfrac{x+6}{(2+x)(2-x)}$$

$$=\dfrac{(x-3)(2+x)-(x+3)(2-x)+(x+6)}{(2+x)(2-x)}$$

$$=\dfrac{x^2-x-6-(-x^2-x+6)+x+6}{(2+x)(2-x)}$$

$$=\dfrac{x^2-x-6+x^2+x-6+x+6}{(2+x)(2-x)}$$

$$=\dfrac{2x^2+x-6}{(2+x)(2-x)}$$

$$=\dfrac{(2x-3)(x+2)}{(2+x)(2-x)}$$

$$=\dfrac{2x-3}{2-x}$$

69. $\dfrac{x+5}{x+3}+\dfrac{x+7}{x+2}-\dfrac{7x+19}{(x+3)(x+2)}$

$\qquad\qquad$ LCD is $(x+3)(x+2)$

$\quad = \dfrac{x+5}{x+3}\cdot\dfrac{x+2}{x+2}+\dfrac{x+7}{x+2}\cdot\dfrac{x+3}{x+3}-\dfrac{7x+19}{(x+3)(x+2)}$

$\quad = \dfrac{(x+5)(x+2)+(x+7)(x+3)-(7x+19)}{(x+3)(x+2)}$

$\quad = \dfrac{x^2+7x+10+x^2+10x+21-7x-19}{(x+3)(x+2)}$

$\quad = \dfrac{2x^2+10x+12}{(x+3)(x+2)}$

$\quad = \dfrac{2\cancel{(x+3)}\cancel{(x+2)}}{\cancel{(x+3)}\cancel{(x+2)}}$

$\quad = 2$

71. $\dfrac{1}{x+y}+\dfrac{1}{x-y}-\dfrac{2x}{x^2-y^2}$

$\quad$ LCD $=(x+y)(x-y)$

$\qquad = \dfrac{1}{x+y}\cdot\dfrac{x-y}{x-y}+\dfrac{1}{x-y}\cdot\dfrac{x+y}{x+y}$

$\qquad\quad -\dfrac{2x}{(x+y)(x-y)}$

$\qquad = \dfrac{(x-y)+(x+y)-2x}{(x+y)(x-y)}$

$\qquad = 0$

73. $f(x)=2+\dfrac{x}{x-3}-\dfrac{18}{x^2-9}$

$\quad = \dfrac{2}{1}+\dfrac{x}{x-3}-\dfrac{18}{(x+3)(x-3)}$

$\quad$ Note: $x \ne 3$ and $x \ne -3$

$\quad$ LCD $=(x-3)(x+3)$

$\qquad = \dfrac{2}{1}\cdot\dfrac{(x-3)(x+3)}{(x-3)(x+3)}+\dfrac{x}{x-3}\cdot\dfrac{x+3}{x+3}$

$\qquad\quad -\dfrac{18}{(x+3)(x-3)}$

$\qquad = \dfrac{2(x-3)(x+3)+x(x+3)-18}{(x-3)(x+3)}$

$\qquad = \dfrac{2x^2-18+x^2+3x-18}{(x-3)(x+3)}$

$\qquad = \dfrac{3x^2+3x-36}{(x-3)(x+3)}$

$\qquad = \dfrac{3(x+4)\cancel{(x-3)}}{\cancel{(x-3)}(x+3)}$

$\qquad = \dfrac{3(x+4)}{x+3}$

$\quad f(x)=\dfrac{3(x+4)}{x+3},\ \ x \ne \pm 3$

75. $f(x) = \dfrac{3x-1}{x^2+2x-3} - \dfrac{x+4}{x^2-16}$

$= \dfrac{3x-1}{(x+3)(x-1)} - \dfrac{x+4}{(x+4)(x-4)}$

Note: $x \neq -3, 1, -4, 4$

$= \dfrac{3x-1}{(x+3)(x-1)} - \dfrac{\cancel{(x+4)}\cdot 1}{\cancel{(x+4)}(x-4)}$

$\text{LCD} = (x+3)(x-1)(x-4)$

$= \dfrac{3x-1}{(x+3)(x-1)} \cdot \dfrac{x-4}{x-4}$

$- \dfrac{1}{x-4} \cdot \dfrac{(x+3)(x-1)}{(x+3)(x-1)}$

$= \dfrac{(3x-1)(x-4)}{(x+3)(x-1)(x-4)}$

$- \dfrac{(x+3)(x-1)}{(x+3)(x-1)(x-4)}$

$= \dfrac{3x^2-13x+4}{(x+3)(x-1)(x-4)}$

$- \dfrac{x^2+2x-3}{(x+3)(x-1)(x-4)}$

$= \dfrac{3x^2-13x+4-(x^2+2x-3)}{(x+3)(x-1)(x-4)}$

$= \dfrac{3x^2-13x+4-x^2-2x+3}{(x+3)(x-1)(x-4)}$

$= \dfrac{2x^2-15x+7}{(x+3)(x-1)(x-4)}$

$f(x) = \dfrac{2x^2-15x+7}{(x+3)(x-1)(x-4)}, \; x \neq -4,$

$-3, 1, 4$

77. $f(x) = \dfrac{1}{x^2+5x+6} - \dfrac{2}{x^2+3x+2}$

$- \dfrac{1}{x^2+5x+6}$

Observe that $\dfrac{1}{x^2+5x+6} - \dfrac{1}{x^2+5x+6} = 0.$

Thus the remaining expression

$- \dfrac{2}{x^2+3x+2}, \;$ or $\; \dfrac{-2}{x^2+3x+2} \;$ is the answer.

We must remember to exclude values that

make the expression x^2+5x+6 equal 0

(and also x^2+3x+2).

$x^2+5x+6=0 \qquad x^2+3x+2=0$

$(x+3)(x+2)=0 \qquad (x+2)(x+1)=0$

$x = -3, \text{ or } x = -2 \qquad x = -2 \text{ or } x = -1$

$f(x) = \dfrac{-2}{x^2+3x+2}, \; x \neq -3, -2, -1$

79. *Thinking and Writing Exercise.*

81. $-\dfrac{3}{8} \div \dfrac{11}{4} = -\dfrac{3}{8} \cdot \dfrac{4}{11} = -\dfrac{3}{\cancel{4}\cdot 2} \cdot \dfrac{\cancel{4}}{11} = -\dfrac{3}{22}$

83. $\dfrac{\frac{3}{4}}{\frac{5}{6}} = \dfrac{3}{4} \cdot \dfrac{6}{5} = \dfrac{3}{\cancel{2}\cdot 2} \cdot \dfrac{\cancel{2}\cdot 3}{5} = \dfrac{9}{10}$

85. $\dfrac{2x+6}{x-1} \div \dfrac{3x+9}{x-1} = \dfrac{2x+6}{\cancel{x-1}} \cdot \dfrac{\cancel{x-1}}{3x+9}$

$= \dfrac{2\cancel{(x+3)}}{3\cancel{(x+3)}} = \dfrac{2}{3} \text{ for } x \neq 1, -3$

87. *Thinking and Writing Exercise.*

89. $P = 2\left(\dfrac{3}{x+4}\right) + 2\left(\dfrac{2}{x-5}\right)$

$= \dfrac{6}{x+4} + \dfrac{4}{x-5}$

$\text{LCD} = (x+4)(x-5)$

$= \dfrac{6}{x+4} \cdot \dfrac{x-5}{x-5} + \dfrac{4}{x-5} \cdot \dfrac{x+4}{x+4}$

$= \dfrac{6x-30+4x+16}{(x+4)(x-5)}$

$= \dfrac{10x-14}{(x+4)(x-5)}, \;$ or $\; \dfrac{10x-14}{x^2-x-20}$

$A = \left(\dfrac{3}{x+4}\right)\left(\dfrac{2}{x-5}\right)$

$= \dfrac{6}{(x+4)(x-5)}, \;$ or $\; \dfrac{6}{x^2-x-20}$

91. $\dfrac{2x+11}{x-3}\cdot\dfrac{3}{x+4}+\dfrac{2x+1}{4+x}\cdot\dfrac{3}{3-x}$

$=\dfrac{6x+33}{(x-3)(x+4)}+\dfrac{6x+3}{(4+x)(3-x)}$

$=\dfrac{6x+33}{(x-3)(x+4)}+\dfrac{6x+3}{(4+x)(3-x)}\cdot\dfrac{-1}{-1}$

$=\dfrac{6x+33}{(x-3)(x+4)}+\dfrac{-6x-3}{(x+4)(x-3)}$

$=\dfrac{6x+33-6x-3}{(x-3)(x+4)}$

$=\dfrac{30}{(x-3)(x+4)}$

93. $\left(\dfrac{x}{x+7}-\dfrac{3}{x+2}\right)\left(\dfrac{x}{x+7}+\dfrac{3}{x+2}\right)$

$=\dfrac{x^2}{(x+7)^2}-\dfrac{9}{(x+2)^2}$

$LCD=(x+7)^2(x+2)^2$

$=\dfrac{x^2}{(x+7)^2}\cdot\dfrac{(x+2)^2}{(x+2)^2}$

$\quad-\dfrac{9}{(x+2)^2}\cdot\dfrac{(x+7)^2}{(x+7)^2}$

$=\dfrac{x^2(x+2)^2-9(x+7)^2}{(x+7)^2(x+2)^2}$

$=\dfrac{x^2(x^2+4x+4)-9(x^2+14x+49)}{(x+7)^2(x+2)^2}$

$=\dfrac{x^4+4x^3+4x^2-9x^2-126x-441}{(x+7)^2(x+2)^2}$

$=\dfrac{x^4+4x^3-5x^2-126x-441}{(x+7)^2(x+2)^2}$

95. $\dfrac{2x^2+5x-3}{2x^2-9x+9}+\dfrac{x+1}{3-2x}+\dfrac{4x^2+8x+3}{x-3}\cdot\dfrac{x+3}{9-4x^2}$

$=\dfrac{2x^2+5x-3}{(2x-3)(x-3)}+\dfrac{x+1}{3-2x}$

$\quad+\dfrac{(4x^2+8x+3)(x+3)}{(x-3)(3+2x)(3-2x)}$

$=\dfrac{2x^2+5x-3}{(2x-3)(x-3)}\cdot\dfrac{-1}{-1}+\dfrac{x+1}{3-2x}$

$\quad+\dfrac{4x^3+20x^2+27x+9}{(x-3)(3+2x)(3-2x)}$

$=\dfrac{-2x^2-5x+3}{(3-2x)(x-3)}+\dfrac{x+1}{3-2x}$

$\quad+\dfrac{4x^3+20x^2+27x+9}{(x-3)(3+2x)(3-2x)}$

$LCD=(x-3)(3+2x)(3-2x)$

$=\dfrac{-2x^2-5x+3}{(3-2x)(x-3)}\cdot\dfrac{3+2x}{3+2x}$

$\quad+\dfrac{x+1}{3-2x}\cdot\dfrac{(x-3)(3+2x)}{(x-3)(3+2x)}$

$\quad+\dfrac{4x^3+20x^2+27x+9}{(x-3)(3+2x)(3-2x)}$

$=[(-4x^3-16x^2-9x+9+2x^3-x^2$

$\quad-12x-9+4x^3+20x^2+27x+9)]/$

$\quad[(x-3)(3+2x)(3-2x)]$

$=\dfrac{2x^3+3x^2+6x+9}{(x-3)(3+2x)(3-2x)}$

$=\dfrac{x^2(2x+3)+3(2x+3)}{(x-3)(3+2x)(3-2x)}$

$=\dfrac{(2x+3)(x^2+3)}{(x-3)(3+2x)(3-2x)}$

$=\dfrac{x^2+3}{(x-3)(3-2x)},\ \text{or}\ \dfrac{-x^2-3}{(x-3)(2x-3)}$

97. $5(x-3)^{-1}+4(x+3)^{-1}-2(x+3)^{-2}$

$=\dfrac{5}{x-3}+\dfrac{4}{x+3}-\dfrac{2}{(x+3)^2}$

$\left[\text{LCD is } (x-3)(x+3)^2.\right]$

$=\dfrac{5(x+3)^2+4(x-3)(x+3)-2(x-3)}{(x-3)(x+3)^2}$

$=\dfrac{5x^2+30x+45+4x^2-36-2x+6}{(x-3)(x+3)^2}$

$=\dfrac{9x^2+28x+15}{(x-3)(x+3)^2}$

99. Answers may vary. $\dfrac{a}{a-b}+\dfrac{3b}{b-a}$

101. $(f+g)(x)=\dfrac{x^3}{x^2-4}+\dfrac{x^2}{x^2+3x-10}$

$=\dfrac{x^3}{(x+2)(x-2)}+\dfrac{x^2}{(x+5)(x-2)}$

$=\dfrac{x^3(x+5)+x^2(x+2)}{(x+2)(x-2)(x+5)}$

$=\dfrac{x^4+5x^3+x^3+2x^2}{(x+2)(x-2)(x+5)}$

$=\dfrac{x^4+6x^3+2x^2}{(x+2)(x-2)(x+5)}$

103. $(f\cdot g)(x)=\dfrac{x^3}{x^2-4}\cdot\dfrac{x^2}{x^2+3x-10}$

$=\dfrac{x^5}{(x^2-4)(x^2+3x-10)}$

105. The denominator of $f+g$ is 0 when $x=-2$, $x=2$, or $x=-5$. Thus the domain of $f+g$ is $\{x\,|\,x$ is a real number and $x\ne -2$ and $x\ne 2$ and $x\ne -5\,\}$, or $(-\infty,-5)\cup(-5,-2)\cup(-2,2)\cup(2,\infty)$.

107.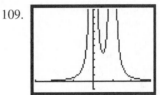

From the graph (shown in the standard window) we see that the domain of the function consists of all real numbers except -1, so the domain of f is $\{x\,|\,x$ is a real number $and\ x\ne -1\}$, or $(-\infty,-1)\cup(-1,\infty)$.

We also see that the range consists of all real numbers except 3, so the range of f is $\{y\,|\,y$ is a real number and $y\ne 3\}$, or $(-\infty,3)\cup(3,\infty)$.

109.

From the graph (shown in the window $[-3,3,-2,20]$, $\text{Yscl}=2$), we see that the domain consists of all real numbers except 0 and 1, so the domain of r is $\{x\,|\,x$ is a real number $and\ x\ne 0\ and\ x\ne 1\}$, or $(-\infty,0)\cup (0,1)\cup(1,\infty)$. We also see that the range consists of all real numbers greater than 0, so the range of r is $\{y\,|\,y>0\}$, or $(0,\infty)$.

Mid-Chapter Review

Guided Solutions:

1. $\dfrac{a^2}{a-10}\div\dfrac{a^2+5a}{a^2-100}=\dfrac{a^2}{a-10}\cdot\dfrac{\boxed{a^2-100}}{\boxed{a^2+5a}}$

$=\dfrac{a\cdot a\cdot(a+10)\cdot\boxed{a-10}}{(a-10)\cdot a\cdot\boxed{a+5}}$

$=\dfrac{\boxed{a(a-10)}}{\boxed{a(a-10)}}\cdot\dfrac{a(a+10)}{a+5}$

$=\dfrac{a(a+10)}{a+5}$

2. $\dfrac{2}{x} + \dfrac{1}{x^2 + x} = \dfrac{2}{x} + \dfrac{1}{x\boxed{(x+1)}}$

$= \dfrac{2}{x} \cdot \dfrac{\boxed{(x+1)}}{\boxed{(x+1)}} + \dfrac{1}{x(x+1)}$

$= \dfrac{\boxed{2x+2}}{x(x+1)} + \dfrac{1}{x(x+1)}$

$= \dfrac{\boxed{2x+3}}{x(x+1)}$

Mixed Review:

1. $\dfrac{3}{5x} + \dfrac{2}{x^2} = \dfrac{3}{5x} \cdot \dfrac{x}{x} + \dfrac{2}{x^2} \cdot \dfrac{5}{5}$

$\quad = \dfrac{3x}{5x^2} + \dfrac{10}{5x^2} = \dfrac{3x+10}{5x^2}$

2. $\dfrac{3}{5x} \cdot \dfrac{2}{x^2} = \dfrac{3 \cdot 2}{5x \cdot x^2} = \dfrac{6}{5x^3}$

3. $\dfrac{3}{5x} \div \dfrac{2}{x^2} = \dfrac{3}{5x} \cdot \dfrac{x^2}{2} = \dfrac{3}{5 \cdot \cancel{x}} \cdot \dfrac{x \cdot \cancel{x}}{2} = \dfrac{3x}{10}$ for $x \neq 0$

4. $\dfrac{3}{5x} - \dfrac{2}{x^2} = \dfrac{3}{5x} \cdot \dfrac{x}{x} - \dfrac{2}{x^2} \cdot \dfrac{5}{5}$

$\quad = \dfrac{3x}{5x^2} - \dfrac{10}{5x^2} = \dfrac{3x-10}{5x^2}$

5. $\dfrac{2x-6}{5x+10} \cdot \dfrac{x+2}{6x-12} = \dfrac{\cancel{2} \cdot (x-3)}{5 \cdot \cancel{(x+2)}} \cdot \dfrac{\cancel{x+2}}{\cancel{2} \cdot 3 \cdot (x-2)}$

$\quad = \dfrac{x-3}{15(x-2)}$ for $x \neq -2$

6. $\dfrac{2}{x+3} \cdot \dfrac{3}{x+4} = \dfrac{3 \cdot 2}{(x+3)(x+4)} = \dfrac{6}{(x+3)(x+4)}$

7. $\dfrac{2}{x-5} \div \dfrac{6}{x-5} = \dfrac{\cancel{2}}{\cancel{x-5}} \cdot \dfrac{\cancel{x-5}}{\cancel{2} \cdot 3} = \dfrac{1}{3}$ for $x \neq 5$

8. $\dfrac{x}{x+2} - \dfrac{1}{x-1} = \dfrac{x}{x+2} \cdot \dfrac{x-1}{x-1} - \dfrac{1}{x-1} \cdot \dfrac{x+2}{x+2}$

$\quad = \dfrac{x(x-1)-(x+2)}{(x+2)(x-1)} = \dfrac{x^2-x-x-2}{(x+2)(x-1)}$

$\quad = \dfrac{x^2-2x-2}{(x+2)(x-1)}$

9. $\dfrac{2}{x+3} + \dfrac{3}{x+4} = \dfrac{2}{x+3} \cdot \dfrac{x+4}{x+4} + \dfrac{3}{x+4} \cdot \dfrac{x+3}{x+3}$

$= \dfrac{2(x+4)+3(x+3)}{(x+3)(x+4)} = \dfrac{2x+8+3x+9}{(x+3)(x+4)}$

$= \dfrac{5x+17}{(x+3)(x+4)}$

10. $\dfrac{5}{2x-1} + \dfrac{10x}{1-2x} = \dfrac{5}{2x-1} - \dfrac{10x}{2x-1} = \dfrac{5-10x}{2x-1}$

$= \dfrac{5(1-2x)}{2x-1} = -\dfrac{5 \cdot \cancel{(2x-1)}}{\cancel{(2x-1)}} = -5$ for $x \neq \dfrac{1}{2}$

11. $\dfrac{3}{x-4} - \dfrac{2}{4-x} = \dfrac{3}{x-4} + \dfrac{2}{x-4} = \dfrac{5}{x-4}$

12. $= \dfrac{(x-2)(2x+3)}{(x+1)(x-5)} \div \dfrac{(x-2)(x+1)}{(x-5)(x+3)}$

$= \dfrac{\cancel{(x-2)}(2x+3)}{(x+1)\cancel{(x-5)}} \cdot \dfrac{\cancel{(x-5)}(x+3)}{\cancel{(x-2)}(x+1)}$

$= \dfrac{(2x+3)(x+3)}{(x+1)^2}$ for $x \neq 2, 5, -3$

13. $\dfrac{a}{6a-9b} - \dfrac{b}{4a-6b}$

$= \dfrac{a}{3(2a-3b)} \cdot \dfrac{2}{2} - \dfrac{b}{2(2a-3b)} \cdot \dfrac{3}{3}$

$= \dfrac{2a}{6(2a-3b)} - \dfrac{3b}{6(2a-3b)} = \dfrac{\cancel{2a-3b}}{6\cancel{(2a-3b)}}$

$= \dfrac{1}{6}$ for $2a \neq 3b$

14. $= \dfrac{x^2-16}{x^2-x} \cdot \dfrac{x^2}{x^2-5x+4}$

$= \dfrac{(x+4)\cancel{(x-4)}}{\cancel{x}(x-1)} \cdot \dfrac{\cancel{x} \cdot x}{\cancel{(x-4)}(x-1)}$

$= \dfrac{x(x+4)}{(x-1)^2}$ for $x \neq 0, 4$

15. $= \dfrac{x+1}{x^2-7x+10} + \dfrac{3}{x^2-x-2}$

$= \dfrac{x+1}{(x-5)(x-2)} \cdot \dfrac{x+1}{x+1} + \dfrac{3}{(x-2)(x+1)} \cdot \dfrac{x-5}{x-5}$

$= \dfrac{(x+1)^2 + 3(x-5)}{(x-5)(x-2)(x+1)} = \dfrac{x^2+2x+1+3x-15}{(x-5)(x-2)(x+1)}$

$= \dfrac{x^2+5x-14}{(x-5)(x-2)(x+1)} = \dfrac{(x+7)\cancel{(x-2)}}{(x-5)\cancel{(x-2)}(x+1)}$

$= \dfrac{(x+7)}{(x-5)(x+1)}$ for $x \neq 2$

16.
$$= \frac{3u^2-3}{4} \div \frac{4u+4}{3} = \frac{3u^2-3}{4}\cdot\frac{3}{4u+4}$$
$$= \frac{3(u^2-1)}{4}\cdot\frac{3}{4(u+1)} = \frac{3(u+1)(u-1)}{4}\cdot\frac{3}{4(u+1)}$$
$$= \frac{9(u-1)}{16} \text{ for } u \neq -1$$

17.
$$\frac{t+2}{10} + \frac{2t+1}{15} = \frac{t+2}{2\cdot 5}\cdot\frac{3}{3} + \frac{2t+1}{3\cdot 5}\cdot\frac{2}{2}$$
$$= \frac{3(t+2)+2(2t+1)}{30} = \frac{3t+6+4t+2}{30}$$
$$= \frac{7t+8}{30}$$

18.
$$(t^2+t-20)\cdot\frac{t+5}{t-4} = \frac{(t+5)(t-4)}{1}\cdot\frac{t+5}{(t-4)}$$
$$= (t+5)^2 \text{ for } t \neq 4$$

19.
$$\frac{a^2-2a+1}{a^2-4}\div(a^2-3a+2)$$
$$= \frac{a^2-2a+1}{a^2-4}\cdot\frac{1}{a^2-3a+2}$$
$$= \frac{(a-1)(a-1)}{(a+2)(a-2)}\cdot\frac{1}{(a-2)(a-1)}$$
$$= \frac{(a-1)}{(a+2)(a-2)^2} \text{ for } a \neq 1$$

20.
$$\frac{2x-7}{x} - \frac{3x-5}{2} = \frac{2x-7}{x}\cdot\frac{2}{2} - \frac{3x-5}{2}\cdot\frac{x}{x}$$
$$= \frac{2(2x-7)-x(3x-5)}{2x} = \frac{4x-14-3x^2+5x}{2x}$$
$$= \frac{-3x^2+9x-14}{2x}$$

Exercise Set 7.5

1. a

3. b

5.
$$\frac{\frac{1}{2}+\frac{1}{3}}{\frac{1}{4}-\frac{1}{6}} = \frac{\frac{1}{2}+\frac{1}{3}}{\frac{1}{4}-\frac{1}{6}}\cdot\frac{12}{12} = \frac{12\cdot\frac{1}{2}+12\cdot\frac{1}{3}}{12\cdot\frac{1}{4}-12\cdot\frac{1}{6}}$$
$$= \frac{6\cdot 2\cdot\frac{1}{2}+4\cdot 3\cdot\frac{1}{3}}{3\cdot 4\cdot\frac{1}{4}-2\cdot 6\cdot\frac{1}{6}}$$
$$= \frac{6\cdot 1+4\cdot 1}{3\cdot 1-2\cdot 1} = \frac{6+4}{3-2} = \frac{10}{1} = 10$$

7.
$$\frac{1+\frac{1}{4}}{2+\frac{3}{4}} = \frac{1+\frac{1}{4}}{2+\frac{3}{4}}\cdot\frac{4}{4} = \frac{4\cdot 1 + 4\cdot\frac{1}{4}}{4\cdot 2 + 4\cdot\frac{3}{4}}$$
$$= \frac{4\cdot 1+1\cdot 1}{4\cdot 2+1\cdot 3} = \frac{4+1}{8+3} = \frac{5}{11}$$

9.
$$\frac{\frac{x}{4}+x}{\frac{4}{x}+x} = \frac{\frac{x}{4}+x}{\frac{4}{x}+x}\cdot\frac{4x}{4x} = \frac{4x\cdot\frac{x}{4}+4x\cdot x}{4x\cdot\frac{4}{x}+4x\cdot x}$$
$$= \frac{x\cdot x+4x\cdot x}{4\cdot 4+4x\cdot x} = \frac{x^2+4x^2}{16+4x^2}$$
$$= \frac{5x^2}{16+4x^2}$$

11.
$$\frac{\frac{x+2}{x-1}}{\frac{x+4}{x-3}} = \frac{x+2}{x-1}\cdot\frac{x-3}{x+4} = \frac{(x+2)(x-3)}{(x-1)(x+4)}$$

13.
$$\frac{\frac{5}{a}-\frac{4}{b}}{\frac{2}{a}+\frac{3}{b}} = \frac{\frac{5}{a}-\frac{4}{b}}{\frac{2}{a}+\frac{3}{b}}\cdot\frac{ab}{ab} = \frac{ab\cdot\frac{5}{a}-ab\cdot\frac{4}{b}}{ab\cdot\frac{2}{a}+ab\cdot\frac{3}{b}}$$
$$= \frac{b\cdot 5-a\cdot 4}{b\cdot 2+a\cdot 3} = \frac{5b-4a}{2b+3a}$$

15.
$$\frac{\frac{1}{t}-\frac{1}{5}}{t-5} = \frac{\frac{1}{t}-\frac{1}{5}}{t-5}\cdot\frac{5t}{5t} = \frac{5t\cdot\frac{1}{t}-5t\cdot\frac{1}{5}}{5t\cdot\frac{t-5}{t}} = \frac{5\cdot 1-t\cdot 1}{5(t-5)}$$
$$= \frac{5-t}{5(t-5)} = -\frac{t-5}{5(t-5)} = -\frac{1}{5}$$

17.
$$\frac{\frac{a^2-b^2}{ab}}{\frac{a-b}{b}} = \frac{a^2-b^2}{ab}\cdot\frac{b}{a-b} = \frac{(a-b)(a+b)}{ab}\cdot\frac{b}{a-b}$$
$$= \frac{a+b}{a}$$

19. $\dfrac{1-\dfrac{2}{3x}}{x-\dfrac{4}{9x}} = \dfrac{1-\dfrac{2}{3x}}{x-\dfrac{4}{9x}}\cdot\dfrac{9x}{9x} = \dfrac{9x\cdot1-9x\cdot\dfrac{2}{3x}}{9x\cdot x-9x\cdot\dfrac{4}{9x}}$

$= \dfrac{9x\cdot1-3\cdot\cancel{3x}\cdot\dfrac{2}{\cancel{3x}}}{9x\cdot x-\cancel{9x}\cdot\dfrac{4}{\cancel{9x}}} = \dfrac{9x\cdot1-3\cdot2}{9x\cdot x-1\cdot4}$

$= \dfrac{9x-6}{9x^2-4} = \dfrac{3\,\cancel{(3x-2)}}{(3x+2)\,\cancel{(3x-2)}}$

$= \dfrac{3}{(3x+2)}$

21. $\dfrac{x^{-1}+y^{-1}}{x^2-y^2} = \dfrac{\dfrac{1}{x}+\dfrac{1}{y}}{\dfrac{x^2-y^2}{xy}} = \dfrac{\dfrac{1}{x}+\dfrac{1}{y}}{\dfrac{x^2-y^2}{xy}}\cdot\dfrac{xy}{xy}$

$= \dfrac{\cancel{x}y\cdot\dfrac{1}{\cancel{x}}+x\cancel{y}\cdot\dfrac{1}{\cancel{y}}}{\cancel{xy}\cdot\dfrac{x^2-y^2}{\cancel{xy}}} = \dfrac{y\cdot1+x\cdot1}{1\cdot(x^2-y^2)}$

$= \dfrac{\cancel{y+x}}{\cancel{(x+y)}(x-y)} = \dfrac{1}{x-y}$

23. $\dfrac{\dfrac{1}{a-h}-\dfrac{1}{a}}{h}$

$= \dfrac{\dfrac{1}{a-h}\cdot\dfrac{a}{a}-\dfrac{1}{a}\cdot\dfrac{a-h}{a-h}}{h}$ $\quad$ Combining terms in the numerator

$= \dfrac{\dfrac{a-a+h}{a(a-h)}}{h} = \dfrac{\dfrac{h}{a(a-h)}}{h}$

$= \dfrac{h}{a(a-h)}\cdot\dfrac{1}{h}$ $\quad$ Multiplying by the reciprocal of the divisor

$= \dfrac{1\cdot\cancel{h}\cdot1}{a(a-h)\cdot\cancel{h}}$

$= \dfrac{1}{a(a-h)}$

25. $\dfrac{\dfrac{a^2-4}{a^2+3a+2}}{\dfrac{a^2-5a-6}{a^2-6a-7}}$

$= \dfrac{a^2-4}{a^2+3a+2}\cdot\dfrac{a^2-6a-7}{a^2-5a-6}$ $\quad$ Multiplying by the reciprocal of the divisor

$= \dfrac{(a+2)(a-2)}{(a+2)(a+1)}\cdot\dfrac{(a+1)(a-7)}{(a+1)(a-6)}$

$= \dfrac{\cancel{(a+2)}(a-2)\,\cancel{(a+1)}(a-7)}{\cancel{(a+2)}\,\cancel{(a+1)}(a+1)(a-6)}$

$= \dfrac{(a-2)(a-7)}{(a+1)(a-6)}$

27. $\dfrac{\dfrac{x}{x^2+3x-4}-\dfrac{1}{x^2+3x-4}}{\dfrac{x}{x^2+6x+8}+\dfrac{3}{x^2+6x+8}}$

$= \dfrac{\dfrac{x-1}{x^2+3x-4}}{\dfrac{x+3}{x^2+6x+8}}$ $\quad$ Combining terms in the numerator and denominator

$= \dfrac{x-1}{x^2+3x-4}\cdot\dfrac{x^2+6x+8}{x+3}$

$= \dfrac{(x-1)(x+4)(x+2)}{(x+4)(x-1)(x+3)}$

$= \dfrac{\cancel{(x-1)}\,\cancel{(x+4)}(x+2)}{\cancel{(x+4)}\,\cancel{(x-1)}(x+3)} = \dfrac{x+2}{x+3}$

29. $\dfrac{y+y^{-1}}{y-y^{-1}} = \dfrac{y+\dfrac{1}{y}}{y-\dfrac{1}{y}}$ $\quad$ Rewriting with positive exponents

$= \dfrac{y+\dfrac{1}{y}}{y-\dfrac{1}{y}}\cdot\dfrac{y}{y}$ $\quad$ Multiply by 1, using the LCD

$= \dfrac{y\cdot y+\dfrac{1}{y}\cdot y}{y\cdot y-\dfrac{1}{y}\cdot y} = \dfrac{y^2+1}{y^2-1}$

(Although the denominator can be factored, doing so does not lead to further simplification.)

31. $\dfrac{\dfrac{x^2}{x^2-y^2}}{\dfrac{x}{x+y}} = \dfrac{x^2}{x^2-y^2} \cdot \dfrac{y+x}{x} = \dfrac{x \cdot \cancel{x}}{\cancel{(x+y)}(x-y)} \cdot \dfrac{\cancel{y+x}}{\cancel{x}}$

$= \dfrac{x}{x-y}$

33. $\dfrac{\dfrac{x}{5y^3}+\dfrac{3}{10y}}{\dfrac{3}{10y}+\dfrac{x}{5y^3}} = \dfrac{\dfrac{x}{5y^3}+\dfrac{3}{10y}}{\dfrac{x}{5y^3}+\dfrac{3}{10y}} = 1$

35. $\dfrac{\dfrac{3}{ab^4}+\dfrac{4}{a^3b}}{ab} = \dfrac{\dfrac{3}{ab^4}+\dfrac{4}{a^3b}}{ab} \cdot \dfrac{a^3b^4}{a^3b^4}$

$= \dfrac{a^3b^4 \cdot \dfrac{3}{ab^4}+a^3b^4 \cdot \dfrac{4}{a^3b}}{a^3b^4 \cdot ab}$

$= \dfrac{a^2 \cdot \cancel{ab^4} \cdot \dfrac{3}{\cancel{ab^4}}+b^3 \cdot \cancel{a^3b} \cdot \dfrac{4}{\cancel{a^3b}}}{a^3b^4 \cdot ab}$

$= \dfrac{a^2 \cdot 3+b^3 \cdot 4}{a^4b^5} = \dfrac{3a^2+4b^3}{a^4b^5}$

37. $\dfrac{\dfrac{x-y}{1}}{\dfrac{1}{x^3}-\dfrac{1}{y^3}} = \dfrac{x-y}{\dfrac{1}{x^3}-\dfrac{1}{y^3}} \cdot \dfrac{x^3y^3}{x^3y^3}$

$= \dfrac{x^3y^3(x-y)}{\cancel{x^3}y^3 \cdot \dfrac{1}{\cancel{x^3}}-x^3\cancel{y^3} \cdot \dfrac{1}{\cancel{y^3}}}$

$= \dfrac{x^3y^3(x-y)}{y^3-x^3} = \dfrac{x^3y^3(x-y)}{(y-x)(y^2+xy+x^2)}$

$= -\dfrac{x^3y^3\cancel{(x-y)}}{\cancel{(x-y)}(y^2+xy+x^2)}$

$= -\dfrac{x^3y^3}{y^2+xy+x^2}$

39. $\dfrac{a(a+3)^{-1}-2(a-1)^{-1}}{a(a+3)^{-1}-(a-1)^{-1}}$

$= \dfrac{\dfrac{a}{a+3}-\dfrac{2}{a-1}}{\dfrac{a}{a+3}-\dfrac{1}{a-1}}$

$= \dfrac{\dfrac{a}{a+3}-\dfrac{2}{a-1}}{\dfrac{a}{a+3}-\dfrac{1}{a-1}} \cdot \dfrac{(a+3)(a-1)}{(a+3)(a-1)}$

Multiplying by 1, using the LCD

$= \dfrac{\dfrac{a}{a+3} \cdot (a+3)(a-1)-\dfrac{2}{a-1} \cdot (a+3)(a-1)}{\dfrac{a}{a+3} \cdot (a+3)(a-1)-\dfrac{1}{a-1} \cdot (a+3)(a-1)}$

$= \dfrac{a(a-1)-2(a+3)}{a(a-1)-(a+3)}$

$= \dfrac{a^2-a-2a-6}{a^2-a-a-3} = \dfrac{a^2-3a-6}{a^2-2a-3}$

(Although the denominator can be factored, doing so does not lead to further simplification.)

41. $\dfrac{\dfrac{2}{a^2-1}+\dfrac{1}{a+1}}{\dfrac{3}{a^2-1}+\dfrac{2}{a-1}}$

$= \dfrac{\dfrac{2}{(a+1)(a-1)}+\dfrac{1}{a+1}}{\dfrac{3}{(a+1)(a-1)}+\dfrac{2}{a-1}}$

$= \dfrac{\dfrac{2}{(a+1)(a-1)}+\dfrac{1}{a+1}}{\dfrac{3}{(a+1)(a-1)}+\dfrac{2}{a-1}} \cdot \dfrac{(a+1)(a-1)}{(a+1)(a-1)}$

Multiplying by 1, using the LCD

$= \dfrac{\dfrac{2}{(a+1)(a-1)}(a+1)(a-1)+\dfrac{1}{a+1}(a+1)(a-1)}{\dfrac{3}{(a+1)(a-1)}(a+1)(a-1)+\dfrac{2}{a-1}(a+1)(a-1)}$

$= \dfrac{2+a-1}{3+2(a+1)} = \dfrac{a+1}{3+2a+2} = \dfrac{a+1}{2a+5}$

43. $\dfrac{\dfrac{y^2}{y^2-25}-\dfrac{y}{y-5}}{\dfrac{y}{y^2-25}-\dfrac{1}{y+5}}=\dfrac{\dfrac{y^2}{(y+5)(y-5)}-\dfrac{y}{y-5}}{\dfrac{y}{(y+5)(y-5)}-\dfrac{1}{y+5}}$

$=\dfrac{\dfrac{y^2}{(y+5)(y-5)}-\dfrac{y}{y-5}}{\dfrac{y}{(y+5)(y-5)}-\dfrac{1}{y+5}}\cdot\dfrac{(y+5)(y-5)}{(y+5)(y-5)}$

Multiplying by 1, using the LCD

$=\dfrac{\dfrac{y^2}{(y+5)(y-5)}\cdot(y+5)(y-5)-\dfrac{y}{y-5}\cdot(y+5)(y-5)}{\dfrac{y}{(y+5)(y-5)}\cdot(y+5)(y-5)-\dfrac{1}{y+5}\cdot(y+5)(y-5)}$

$=\dfrac{y^2-y(y+5)}{y-(y-5)}=\dfrac{y^2-y^2-5y}{y-y+5}=\dfrac{-5y}{5}=\dfrac{-1\cdot\cancel{5}y}{\cancel{5}}=-y$

45. $\dfrac{\dfrac{3}{a^2-4a+3}+\dfrac{3}{a^2-5a+6}}{\dfrac{3}{a^2-3a+2}+\dfrac{3}{a^2+3a-10}}$

$=\dfrac{\dfrac{3}{(a-1)(a-3)}+\dfrac{3}{(a-2)(a-3)}}{\dfrac{3}{(a-1)(a-2)}+\dfrac{3}{(a+5)(a-2)}}$

$=\dfrac{\dfrac{3}{(a-1)(a-3)}+\dfrac{3}{(a-2)(a-3)}}{\dfrac{3}{(a-1)(a-2)}+\dfrac{3}{(a+5)(a-2)}}\cdot$

$\dfrac{(a-1)(a-3)(a-2)(a+5)}{(a-1)(a-3)(a-2)(a+5)}$

Multiplying by 1, using the LCD

$=\dfrac{3(a-2)(a+5)+3(a-1)(a+5)}{3(a-3)(a+5)+3(a-1)(a-3)}$

$=\dfrac{3[(a-2)(a+5)+(a-1)(a+5)]}{3[(a-3)(a+5)+(a-1)(a-3)]}$

$=\dfrac{\cancel{3}[(a-2)(a+5)+(a-1)(a+5)]}{\cancel{3}[(a-3)(a+5)+(a-1)(a-3)]}$

$=\dfrac{a^2+3a-10+a^2+4a-5}{a^2+2a-15+a^2-4a+3}$

$=\dfrac{2a^2+7a-15}{2a^2-2a-12}$ or $\dfrac{(2a-3)(a+5)}{2(a-3)(a+2)}$

47. $\dfrac{t+5+\dfrac{3}{t}}{t+2+\dfrac{1}{t}}=\dfrac{t+5+\dfrac{3}{t}}{t+2+\dfrac{1}{t}}\cdot\dfrac{t}{t}=\dfrac{t\cdot t+t\cdot5+\cancel{t}\cdot\dfrac{3}{\cancel{t}}}{t\cdot t+t\cdot2+\cancel{t}\cdot\dfrac{1}{\cancel{t}}}$

$=\dfrac{t^2+5t+3}{t^2+2t+1}$ for $t\ne0$

49. $\dfrac{x-2-\dfrac{1}{x}}{x-5-\dfrac{4}{x}}=\dfrac{x-2-\dfrac{1}{x}}{x-5-\dfrac{4}{x}}\cdot\dfrac{x}{x}=\dfrac{x\cdot x-x\cdot2-\cancel{x}\cdot\dfrac{1}{\cancel{x}}}{x\cdot x-x\cdot5-\cancel{x}\cdot\dfrac{4}{\cancel{x}}}$

$=\dfrac{x^2-2x-1}{x^2-5x-4}$ for $x\ne0$

51. *Thinking and Writing Exercise.*

53. $3x-5+2(4x-1)=12x-3$

$3x-5+8x-2=12x-3$

$11x-7=12x-3$

$-7+3=12x-11x$

$-4=x$

55. $\dfrac{3}{4}x-\dfrac{5}{8}=\dfrac{3}{8}x+\dfrac{7}{4}$

$8\cdot\left[\dfrac{3}{4}x-\dfrac{5}{8}\right]=8\cdot\left[\dfrac{3}{8}x+\dfrac{7}{4}\right]$

$8\cdot\dfrac{3}{4}x-8\cdot\dfrac{5}{8}=8\cdot\dfrac{3}{8}x+8\cdot\dfrac{7}{4}$

$\cancel{8}\cdot2\cdot\dfrac{3}{\cancel{4}}x-\cancel{8}\cdot\dfrac{5}{\cancel{8}}=\cancel{8}\cdot\dfrac{3}{\cancel{8}}x+\cancel{8}\cdot2\cdot\dfrac{7}{\cancel{4}}$

$2\cdot3\cdot x-1\cdot5=1\cdot3\cdot x+2\cdot7$

$6x-5=3x+14$

$6x-3x=14+5$

$3x=19$

$x=\dfrac{19}{3}$

57. $x^2-7x+12=0$

$(x-3)(x-4)=0$

$\Rightarrow x-3=0$ or $x-4=0$

$\Rightarrow x=3$ or $x=4$

59. *Thinking and Writing Exercise.*

61. The function is undefined when the numerator is undefined, at $x=6$, when the denominator is undefined, at $x=8$, and when the denominator is 0, at $x=7$.

$x\ne6,7,8$

63. The function is undefined when the numerator is undefined, which occurs when

$5x + 4 = 0 \Rightarrow 5x = -4 \Rightarrow x = -\dfrac{4}{5}$, and when the

denominator is 0, which occurs when

$\dfrac{3}{7} - \dfrac{x^2}{21} = 0 \Rightarrow 21 \cdot \left[\dfrac{3}{7} - \dfrac{x^2}{21}\right] = 21 \cdot 0$

$\Rightarrow 21 \cdot \dfrac{3}{7} - 21 \cdot \dfrac{x^2}{21} = 21 \cdot 0 \Rightarrow \cancel{7} \cdot 3 \cdot \dfrac{3}{\cancel{7}} - \cancel{21} \cdot \dfrac{x^2}{\cancel{21}} = 21 \cdot 0$

$\Rightarrow 3 \cdot 3 - 1 \cdot x^2 = 21 \cdot 0 \Rightarrow 9 - x^2 = 0 \Rightarrow (3 - x)(3 + x) = 0$

$\Rightarrow x = 3 \text{ or } x = -3$

$x \neq -3, -\dfrac{4}{5}, 3$

65. $\dfrac{\dfrac{A}{B}}{\dfrac{C}{D}} = \dfrac{\dfrac{A}{B}}{\dfrac{C}{D}} \cdot \dfrac{BD}{BD} = \dfrac{\cancel{B}D \cdot \dfrac{A}{\cancel{B}}}{B\cancel{D} \cdot \dfrac{C}{\cancel{D}}}$

$= \dfrac{D \cdot A}{B \cdot C} = \dfrac{A \cdot D}{B \cdot C} = \dfrac{A}{B} \cdot \dfrac{D}{C}$

67. Substitute $\dfrac{c}{4}$ for both v_1 and v_2.

$\dfrac{\dfrac{c}{4} + \dfrac{c}{4}}{1 + \dfrac{c}{4} \cdot \dfrac{c}{4}} = \dfrac{\dfrac{2c}{4}}{1 + \dfrac{c^2}{16}} = \dfrac{\dfrac{c}{2}}{1 + \dfrac{1}{16}} = \dfrac{\dfrac{c}{2}}{\dfrac{17}{16}} = \dfrac{c}{2} \cdot \dfrac{16}{17} = \dfrac{8c}{17}$

The observed speed is $\dfrac{8c}{17}$, or $\dfrac{8}{17}$ the speed of light.

69. $\dfrac{\dfrac{x-1}{x-1} - 1}{\dfrac{x+1}{x-1} + 1} = \dfrac{\dfrac{x-1}{x-1} - 1}{\dfrac{x+1}{x-1} + 1} \cdot \dfrac{x-1}{x-1}$

$= \dfrac{(x-1)\dfrac{x-1}{x-1} - (x-1)1}{(x-1)\dfrac{x+1}{x-1} + (x-1)1} = \dfrac{x-1-(x-1)}{x+1+x-1}$

$= \dfrac{x-1-x+1}{x+1+x-1} = \dfrac{0}{2x} = 0 \text{ for } x \neq 0, 1$

Therefore, for $x \neq 0$ or 1, :

$\left[\dfrac{\dfrac{x-1}{x-1} - 1}{\dfrac{x+1}{x-1} + 1}\right]^5 = 0^5 = 0$

71. $\dfrac{\dfrac{z}{1 - \dfrac{z}{2+2z}} - 2z}{\dfrac{2z}{5z-2} - 3} = \dfrac{\dfrac{z}{1 - \dfrac{z}{2+2z}} \cdot \dfrac{2+2z}{2+2z} - 2z}{\dfrac{2z}{5z-2} - 3}$

$= \dfrac{\dfrac{z(2+2z)}{2+2z-z} - 2z}{\dfrac{2z}{5z-2} - 3} = \dfrac{\dfrac{z(2+2z)}{2+z} - 2z}{\dfrac{2z}{5z-2} - 3}$

$= \dfrac{\dfrac{z(2+2z)}{2+z} - 2z}{\dfrac{2z}{5z-2} - 3} \cdot \dfrac{(2+z)(5z-2)}{(2+z)(5z-2)}$

$= \dfrac{z(2+2z)(5z-2) - 2z(2+z)(5z-2)}{2z(2+z) - 3(2+z)(5z-2)}$

$= \dfrac{z(5z-2)[(2+2z) - 2(2+z)]}{(2+z)[2z - 3(5z-2)]}$

$= \dfrac{z(5z-2)[2+2z-4-2z]}{(2+z)[2z-15z+6]} = \dfrac{-2z(5z-2)}{(2+z)(6-13z)}$

$= \dfrac{2z(5z-2)}{(z+2)(13z-6)}$

73. $f(x) = \dfrac{3}{x}, f(x+h) = \dfrac{3}{x+h}$

$= \dfrac{f(x+h) - f(x)}{h} = \dfrac{\dfrac{3}{x+h} - \dfrac{3}{x}}{h}$

$= \dfrac{\dfrac{3x - 3(x+h)}{x(x+h)}}{h}$

$= \dfrac{3x - 3(x+h)}{x(x+h)} \cdot \dfrac{1}{h}$

$= \dfrac{3x - 3x - 3h}{xh(x+h)}$

$= \dfrac{-3h}{xh(x+h)}$

$= \dfrac{-3\cancel{h}}{x\cancel{h}(x+h)}$

$= \dfrac{-3}{x(x+h)}$

75. To avoid division by zero in $\dfrac{1}{x}$ and $\dfrac{8}{x^2}$ we must exclude 0 from the domain of F. To avoid division by zero in the complex fraction we solve:

$$2 - \frac{8}{x^2} = 0$$

$$2x^2 - 8 = 0$$

$$2\left(x^2 - 4\right) = 0$$

$$2(x+2)(x-2) = 0$$

$$x + 2 = 0 \ \ or \ x - 2 = 0$$

$$x = -2 \ or \quad x = 2$$

The domain of $F = \left\{x \middle| x \text{ is a real number and }\right.$ $x \neq 0$ and $x \neq -2$ and $\left. x \neq 2\right\}$.

77. $\dfrac{30{,}000 \cdot \dfrac{0.075}{12}}{\left(1 + \dfrac{0.075}{12}\right)^{120} - 1} = \dfrac{30{,}000\left(0.00625\right)}{\left(1 + 0.00625\right)^{120} - 1}$

$$= \dfrac{187.5}{\left(1 + 0.00625\right)^{120} - 1}$$

$$\approx \dfrac{187.5}{2.112064637 - 1}$$

$$\approx \dfrac{187.5}{1.112064637}$$

$$\approx 168.61$$

Alexis' monthly investment is $168.61.

Exercise Set 7.6

---

1. Equation

3. Expression

5. Equation

7. Equation

9. Expression

11. $\qquad \dfrac{3}{5} - \dfrac{2}{3} = \dfrac{x}{6}$, LCD is 30

$$30 \cdot \frac{3}{5} - 30 \cdot \frac{2}{3} = 30 \cdot \frac{x}{6}$$

$$\not{5} \cdot 6 \cdot \frac{3}{\not{5}} - \not{3} \cdot 10 \cdot \frac{2}{\not{3}} = \not{6} \cdot 5 \cdot \frac{x}{\not{6}}$$

$$18 - 20 = 5x$$

$$5x = -2$$

$$x = -\frac{2}{5}$$

Check.

$\dfrac{3}{5} - \dfrac{2}{3} = \dfrac{x}{6}$	
$\dfrac{3}{5} \cdot \dfrac{3}{3} - \dfrac{2}{3} \cdot \dfrac{5}{5}$	$-\dfrac{2}{5}$
$\dfrac{9-10}{15}$	$\dfrac{-\dfrac{2}{5}}{6}$
$-\dfrac{1}{15}$	$-\dfrac{2}{5} \cdot \dfrac{1}{6}$
	$-\dfrac{\not{2}}{5} \cdot \dfrac{1}{\not{2} \cdot 3}$
	$-\dfrac{1}{15}$
	TRUE

13. $\dfrac{1}{8} + \dfrac{1}{12} = \dfrac{1}{t}$, LCD is $24t$

Because $\dfrac{1}{t}$ is undefined at $t = 0$, note at the onset that $t \neq 0$.

$$\frac{1}{8} + \frac{1}{12} = \frac{1}{t}$$

$$24t \cdot \frac{1}{8} + 24t \cdot \frac{1}{12} = 24t \cdot \frac{1}{t}$$

$$\not{8} \cdot 3t \cdot \frac{1}{\not{8}} + \not{12} \cdot 2t \cdot \frac{1}{\not{12}} = 24 \cdot \not{t} \cdot \frac{1}{\not{t}}$$

$$3t + 2t = 24$$

$$5t = 24$$

$$t = \frac{24}{5} \neq 0$$

Check.

$$\frac{1}{8}+\frac{1}{12}=\frac{1}{t}$$

$\dfrac{1}{8}+\dfrac{1}{12}$	$\dfrac{1}{24}$
$\dfrac{1}{8}\cdot\dfrac{3}{3}+\dfrac{1}{12}\cdot\dfrac{2}{2}$	$\dfrac{5}{24}$
$\dfrac{3+2}{24}$	$1\cdot\dfrac{5}{24}$
$\dfrac{5}{24}$	$\dfrac{5}{24}$
	TRUE

15. $\dfrac{x}{6}-\dfrac{6}{x}=0$, LCD is $6x$

Because $\dfrac{6}{x}$ is undefined at $x=0$, note at the

onset that $x\neq 0$.

$$\frac{x}{6}-\frac{6}{x}=0$$

$$6x\cdot\frac{x}{6}-6x\cdot\frac{6}{x}=0$$

$$x^2-36=0$$

$$(x+6)(x-6)=0$$

$$x=-6,6\neq 0$$

Check

$$\frac{x}{6}-\frac{6}{x}=\frac{6}{6}-\frac{6}{6}=0\ \surd$$

$$\frac{x}{6}-\frac{6}{x}=\frac{-6}{6}-\frac{6}{-6}=-\frac{6}{6}+\frac{6}{6}=0\ \surd$$

17. $\dfrac{2}{3}-\dfrac{1}{t}=\dfrac{7}{3t}$

Because $\dfrac{1}{t}$ is undefined when t is 0, we note

at the outset that $t\neq 0$. Then we multiply

both sides by the LCD, $3\cdot t=3t$.

$$3t\left(\frac{2}{3}-\frac{1}{t}\right)=3t\cdot\frac{7}{3t}$$

$$3t\cdot\frac{2}{3}-3t\cdot\frac{1}{t}=3t\cdot\frac{7}{3t}$$

$$2t-3=7$$

$$2t=10$$

$$t=5$$

Check.

$$\frac{2}{3}-\frac{1}{t}=\frac{7}{3t}$$

$\dfrac{2}{3}-\dfrac{1}{5}$	$\dfrac{7}{3\cdot5}$
$\dfrac{10}{15}-\dfrac{3}{15}$	$\dfrac{7}{15}$
$\dfrac{7}{15}$	$\dfrac{7}{15}$ TRUE

The solution is 5.

19. $\dfrac{n+2}{n-6}=\dfrac{1}{2}$, LCD is $2(n-6)$

Because $\dfrac{n+2}{n-6}$ is undefined at $n=6$, note at

the onset that $n\neq 6$.

$$\frac{n+2}{n-6}=\frac{1}{2}$$

$$2(n-6)\frac{n+2}{n-6}=2(n-6)\frac{1}{2}$$

$$2(n+2)=n-6$$

$$2n+4=n-6$$

$$2n-n=-6-4$$

$$n=-10\neq 6$$

Check

$$\frac{(-10)+2}{(-10)-6}=\frac{-8}{-16}=\frac{1}{2}\ \surd$$

21. $\dfrac{12}{x}=\dfrac{x}{3}\Rightarrow 3x\cdot\dfrac{12}{x}=3x\cdot\dfrac{x}{3}\Rightarrow 36=x^2$

$$\Rightarrow x^2-36=0\Rightarrow(x+6)(x-6)=0$$

$$\Rightarrow x=-6,6$$

23. $\dfrac{2}{6}+\dfrac{1}{2x}=\dfrac{1}{3}$

Because $\dfrac{1}{2x}$ is undefined when $x=0$, we

note at the outset that $x\neq 0$. We multiply

both sides by the LCD, $6x$.

$$6x\left(\frac{2}{6}+\frac{1}{2x}\right)=6x\cdot\frac{1}{3}$$

$$6x\cdot\frac{2}{6}+6x\cdot\frac{1}{2x}=6x\cdot\frac{1}{3}$$

$$2x+3=2x$$

$$3=0$$

We get a false equation. The given equation

has no solution.

25. $y + \dfrac{4}{y} = -5$

Because $\dfrac{4}{y}$ is undefined when y is 0, we note at the outset that $y \neq 0$. Then we multiply both sides by the LCD, y.

$$y\left(y + \dfrac{4}{y}\right) = y(-5)$$

$$y \cdot y + y \cdot \dfrac{4}{y} = -5y$$

$$y^2 + 4 = -5y$$

$$y^2 + 5y + 4 = 0$$

$$(y+1)(y+4) = 0$$

$$y + 1 = 0 \ \ or \ \ y + 4 = 0$$

$$y = -1 \ or \ \ \ \ \ y = -4$$

Both values check. The solutions are -1 and -4.

27. $x - \dfrac{12}{x} = 4$

Because $\dfrac{12}{x}$ is undefined when $x = 0$, we note that $x \neq 0$. LCD $= x$.

$$x \cdot \left(x - \dfrac{12}{x}\right) = x \cdot 4$$

$$x \cdot x - x \cdot \dfrac{12}{x} = x \cdot 4$$

$$x^2 - 12 = 4x$$

$$x^2 - 4x - 12 = 0$$

$$(x-6)(x+2) = 0$$

$$x - 6 = 0 \ or \ x + 2 = 0$$

$$x = 6 \ or \ \ \ \ x = -2$$

Both values check. The solutions are -2 and 6.

29. $\dfrac{y+3}{y-3} = \dfrac{6}{y-3}$

$$(y-3)\dfrac{y+3}{y-3} = (y-3)\dfrac{6}{y-3}$$

$$y + 3 = 6$$

$$y = 3$$

But $y = 3$ is not a valid solution, because the expressions in the original equation are undefined at $y = 3$. Therefore, since the equation had no other solutions, the equation has no solution.

31. $\dfrac{x}{x-5} = \dfrac{25}{x^2 - 5x}$

To assure that neither denominator is 0, we note that $x \neq 5$ and $x \neq 0$. We then multiply both sides by the LCD, $x(x-5)$.

$$x(x-5) \cdot \dfrac{x}{x-5} = x(x-5) \cdot \dfrac{25}{x(x-5)}$$

$$x^2 = 25$$

$$x^2 - 25 = 0$$

$$(x+5)(x-5) = 0$$

$$x + 5 = 0 \ \ or \ x - 5 = 0$$

$$x = -5 \ or \ \ \ \ x = 5$$

Recall that, because of the restriction above, 5 cannot be a solution . A check confirms this. **Check** $x = 5$.

$\dfrac{x}{x-5} = \dfrac{25}{x^2 - 5x}$	
5	25
$\dfrac{5}{5-5}$	$\dfrac{25}{5^2 - 5 \cdot 5}$
$\dfrac{5}{0}$	$\dfrac{5}{0}$ UNDEFINED

Also, Check $x = -5$.

$\dfrac{x}{x-5} = \dfrac{25}{x^2 - 5x}$	
-5	25
$\dfrac{-5}{-5-5}$	$\dfrac{25}{(-5)^2 - 5(-5)}$
$\dfrac{-5}{-10}$	$\dfrac{25}{25 + 25}$
$\dfrac{1}{2}$	$\dfrac{25}{50}$
	$\dfrac{1}{2}$ TRUE

The solution is -5.

33.
$$\frac{n+1}{n+2} = \frac{n-3}{n+1}$$

$$(n+1)(n+2)\frac{n+1}{n+2} = (n+1)(n+2)\frac{n-3}{n+1}$$

$$(n+1)^2 = (n+2)(n-3)$$

$$n^2 + 2n + 1 = n^2 - n - 6$$

$$2n + n = -6 - 1$$

$$3n = -7$$

$$n = -\frac{7}{3}$$

35.
$$\frac{x^2+4}{x-1} = \frac{5}{x-1}$$

To assure that neither denominator is 0, we note at the outset that $x \neq 1$. Then we multiply both sides by the LCD, $x-1$.

$$(x-1)\cdot\frac{x^2+4}{x-1} = (x-1)\cdot\frac{5}{x-1}$$

$$x^2 + 4 = 5$$

$$x^2 - 1 = 0$$

$$(x+1)(x-1) = 0$$

$$x+1 = 0 \ or \ x-1 = 0$$

$$x = -1 \ or \quad x = 1$$

Recall that, because of the restriction above, 1 cannot be a solution. The number -1 checks and is the solution.

We might also observe that since the denominators are the same, the numerators must be equal. Solving $x^2 + 4 = 5$, we get $x = -1$ or $x = 1$ as shown above. Again because of the restriction $x \neq 1$, only -1 is a solution of the equation.

37.
$$\frac{6}{a+1} = \frac{a}{a-1}$$

To assure that neither denominator is 0, we note at the outset that $x \neq -1$ and $x \neq 1$. Then we multiply both sides by the LCD, $(a+1)(a-1)$.

$$(a+1)(a-1)\cdot\frac{6}{a+1} = (a+1)(a-1)\cdot\frac{a}{a-1}$$

$$6(a-1) = a(a+1)$$

$$6a - 6 = a^2 + a$$

$$0 = a^2 - 5a + 6$$

$$0 = (a-2)(a-3)$$

$$a-2 = 0 \ or \ a-3 = 0$$

$$a = 2 \ or \quad a = 3$$

Both values check. The solutions are 2 and 3.

39.
$$\frac{60}{t-5} - \frac{18}{t} = \frac{40}{t}$$

To assure that none of the denominators are 0, we note at the outset that $t \neq 5$ and $t \neq 0$. Then we multiply on both sides by the LCD, $t(t-5)$.

$$t(t-5)\left(\frac{60}{t-5} - \frac{18}{t}\right) = t(t-5)\cdot\frac{40}{t}$$

$$60t - 18(t-5) = 40(t-5)$$

$$60t - 18t + 90 = 40t - 200$$

$$2t = -290$$

$$t = -145$$

This value checks. The solution is -145.

41.
$$\frac{3}{x-3} + \frac{5}{x+2} = \frac{5x}{x^2-x-6}$$

To assure that none of the denominators are 0, we note that $x \neq 3$ and $x \neq -2$. LCD is $(x-3)(x+2)$.

$$(x-3)(x+2)\left(\frac{3}{x-3} + \frac{5}{x+2}\right) =$$

$$(x-3)(x+2)\cdot\frac{5x}{x^2-x-6}$$

$$3(x+2) + 5(x-3) = 5x$$

$$3x + 6 + 5x - 15 = 5x$$

$$3x = 9$$

$$x = 3$$

Recall that, because of the restriction above, 3 cannot be a solution. The equation has no solution.

43.
$$\frac{3}{x} + \frac{x}{x+2} = \frac{4}{x^2+2x}$$

$$\frac{3}{x} + \frac{x}{x+2} = \frac{4}{x(x+2)}$$

To assure that none of the denominators are 0, we note at the outset that $x \neq 0$ and $x \neq -2$. Then we multiply both sides by the LCD, $x(x+2)$.

$$x(x+2)\left(\frac{3}{x}+\frac{x}{x+2}\right)=x(x+2)\cdot\frac{4}{x(x+2)}$$

$$3(x+2)+x\cdot x=4$$

$$3x+6+x^2=4$$

$$x^2+3x+2=0$$

$$(x+1)(x+2)=0$$

$$x+1=0 \ \ or \ x+2=0$$

$$x=-1 \ or \ \ \ x=-2$$

Recall that, because of the restrictions above, -2 cannot be a solution. The number -1 checks. The solution is -1.

45. $\dfrac{5}{x+2}-\dfrac{3}{x-2}=\dfrac{2x}{4-x^2}$

$$\frac{5}{x+2}-\frac{3}{x-2}=\frac{2x}{(2+x)(2-x)}$$

$$\frac{5}{x+2}+\frac{3}{2-x}=\frac{2x}{(2+x)(2-x)}$$

$$\boxed{-\frac{3}{x-2}=\frac{3}{2-x}}$$

First note that $x\neq-2$ and $x\neq 2$. Then multiply on both sides by the LCD, $(2+x)(2-x)$.

$$(2+x)(2-x)\left(\frac{5}{x+2}+\frac{3}{2-x}\right)=$$

$$(2+x)(2-x)\cdot\frac{2x}{(2+x)(2-x)}$$

$$5(2-x)+3(2+x)=2x$$

$$10-5x+6+3x=2x$$

$$16-2x=2x$$

$$16=4x$$

$$4=x$$

This value checks. The solution is 4.

47. $\dfrac{3}{x^2-6x+9}+\dfrac{x-2}{3x-9}=\dfrac{x}{2x-6}$

$$\frac{3}{(x-3)(x-3)}+\frac{x-2}{3(x-3)}=\frac{x}{2(x-3)}$$

Note that $x\neq 3$.

$$6(x-3)(x-3)\left(\frac{3}{(x-3)(x-3)}+\frac{x-2}{3(x-3)}\right)=$$

$$6(x-3)(x-3)\cdot\frac{x}{2(x-3)}$$

$$6\cdot 3+2(x-3)(x-2)=3x(x-3)$$

$$18+2x^2-10x+12=3x^2-9x$$

$$0=x^2+x-30$$

$$0=(x+6)(x-5)$$

$$x=-6 \ or \ x=5$$

Both values check. The solutions are -6 and 5.

49. We find all values of a for which

$$2a-\frac{15}{a}=7. \ \ \text{First note that} \ \ a\neq 0. \ \ \text{Then}$$

multiply on both sides by the LCD, a.

$$a\left(2a-\frac{15}{a}\right)=a\cdot 7$$

$$a\cdot 2a-a\cdot\frac{15}{a}=7a$$

$$2a^2-15=7a$$

$$2a^2-7a-15=0$$

$$(2a+3)(a-5)=0$$

$$a=-\frac{3}{2} \ or \ a=5$$

Both values check. The solutions are $-\dfrac{3}{2}$ and 5.

51. We find all values of a for which $\dfrac{a-5}{a+1}=\dfrac{3}{5}$.

First note that $a\neq-1$. Then multiply on both sides by the LCD, $5(a+1)$.

$$5(a+1)\cdot\frac{a-5}{a+1}=5(a+1)\cdot\frac{3}{5}$$

$$5(a-5)=3(a+1)$$

$$5a-25=3a+3$$

$$2a=28$$

$$a=14$$

This value checks. The solution is 14.

53. We find all values of a for which
$\dfrac{12}{a} - \dfrac{12}{2a} = 8$. First note that $a \neq 0$. then
multiply on both sides by the LCD, $2a$.

$$2a\left(\dfrac{12}{a} - \dfrac{12}{2a}\right) = 2a \cdot 8$$

$$2a \cdot \dfrac{12}{a} - 2a \cdot \dfrac{12}{2a} = 16a$$

$$24 - 12 = 16a$$

$$12 = 16a$$

$$\dfrac{3}{4} = a$$

This value checks. The solution is $\dfrac{3}{4}$.

55.
$$f(a) = g(a)$$

$$\dfrac{a+1}{3} - 1 = \dfrac{a-1}{2}$$

$$6 \cdot \dfrac{a+1}{3} - 6 \cdot 1 = 6 \cdot \dfrac{a-1}{2}$$

$$\cancel{6} \cdot 2 \cdot \dfrac{a+1}{\cancel{3}} - 6 \cdot 1 = \cancel{2} \cdot 3 \cdot \dfrac{a-1}{\cancel{2}}$$

$$2(a+1) - 6 = 3(a-1)$$

$$2a + 2 - 6 = 3a - 3$$

$$2a - 4 = 3a - 3$$

$$-4 + 3 = 3a - 2a$$

$$-1 = a$$

57. First note that $g(x)$ can easily be written as a single term:

$$g(x) = \dfrac{4}{x-3} + \dfrac{2x}{x-3} = \dfrac{4+2x}{x-3} = \dfrac{2x+4}{x-3}.$$

Therefore:

$$f(a) = g(a)$$

$$\dfrac{12}{a^2 - 6a + 9} = \dfrac{2a+4}{a-3}$$

$$\dfrac{12}{(a-3)^2} = \dfrac{2a+4}{a-3}$$

$$(a-3)^2 \dfrac{12}{(a-3)^2} = (a-3)^2 \dfrac{2a+4}{a-3}$$

$$\cancel{(a-3)^2} \dfrac{12}{\cancel{(a-3)^2}} = (a-3)\cancel{(a-3)} \dfrac{2a+4}{\cancel{a-3}}$$

$$12 = (a-3)(2a+4)$$

$$(a-3)(2a+4) - 12 = 0$$

$$2a^2 + 4a - 6a - 12 - 12 = 0$$

$$2a^2 - 2a - 24 = 0$$

$$(2a+6)(a-4) = 0$$

$$2a + 6 = 0 \quad \text{or} \quad a - 4 = 0$$

$$2a = -6 \qquad\qquad a = 4$$

$$a = -3$$

Both -3 and 4 are in the domains of f and g, so both are valid solutions.

59. *Thinking and Writing Exercise.* Note that the domain of f is all real numbers while the domain of g is all real numbers except 2. Then we see that the graph on the right represents f, and the graph on the left represents g.

61. Let $x =$ the smallest number. Then $x + 2 =$ the larger number. Then:

$$x + (x+2) = 276$$

$$x + x + 2 = 276$$

$$2x + 2 = 276$$

$$2x = 274$$

$$x = \dfrac{274}{2} = 137.$$

The two numbers are $x = 137$, and $x + 2 = 139$.

63. The height of the triangle is defined in terms of its base. Therefore let b = the base, in cm, of the triangle. Then $b + 3$ = the height, in cm, of the triangle. Then:

$$A = \frac{1}{2}bh$$

$$\frac{1}{2}b(b+3) = 54$$

$$2 \cdot \frac{1}{2}b(b+3) = 2 \cdot 54$$

$$b(b+3) = 108$$

$$b^2 + 3b - 108 = 0$$

$$(b+12)(b-9) = 0$$

$$b = -12 \text{ or } b = 9$$

Since the base of the triangle cannot be a negative number, the base of the triangle is 9 cm, and the height is $9 + 3 = 12$ cm.

65. $\dfrac{0.9 \text{ cm}}{24 - 9 \text{ day}} = \dfrac{0.9 \text{ cm}}{15 \text{ day}} = 0.06$ cm per day

67. *Thinking and Writing Exercise.*

69. $$f(a) = g(a)$$

$$\frac{a - \frac{2}{3}}{a + \frac{1}{2}} = \frac{a + \frac{2}{3}}{a - \frac{3}{2}}$$

$$\frac{a - \frac{2}{3}}{a + \frac{1}{2}} \cdot \frac{6}{6} = \frac{a + \frac{2}{3}}{a - \frac{3}{2}} \cdot \frac{6}{6}$$

$$\frac{6a - \frac{2}{3} \cdot 6}{6a + \frac{1}{2} \cdot 6} = \frac{6a + \frac{2}{3} \cdot 6}{6a - \frac{3}{2} \cdot 6}$$

$$\frac{6a - 4}{6a + 3} = \frac{6a + 4}{6a - 9}$$

$$\frac{6a - 4}{3(2a+1)} = \frac{6a + 4}{3(2a-3)}$$

To assure that neither denominator is 0, we note at the outset that $a \neq -\frac{1}{2}$ and $a \neq \frac{3}{2}$.

Then we multiply both sides by the LCD, $3(2a+1)(2a-3)$.

$$3(2a+1)(2a-3) \cdot \frac{6a-4}{3(2a+1)} =$$

$$3(2a+1)(2a-3) \cdot \frac{6a+4}{3(2a-3)}$$

$$(2a-3)(6a-4) = (2a+1)(6a+4)$$

$$12a^2 - 26a + 12 = 12a^2 + 14a + 4$$

$$-26a + 12 = 14a + 4$$

$$-40a + 12 = 4$$

$$-40a = -8$$

$$a = \frac{1}{5}$$

This number checks. For $a = \frac{1}{5}, f(a) = g(a).$

71. $$f(a) = g(a)$$

$$\frac{a+3}{a+2} - \frac{a+4}{a+3} = \frac{a+5}{a+4} - \frac{a+6}{a+5}$$

Note that $a \neq -2$ and $a \neq -3$ and $a \neq -4$ and $a \neq -5$.

$$(a+2)(a+3)(a+4)(a+5)\left(\frac{a+3}{a+2} - \frac{a+4}{a+3}\right) =$$

$$(a+2)(a+3)(a+4)(a+5)\left(\frac{a+5}{a+4} - \frac{a+6}{a+5}\right)$$

$$(a+3)(a+4)(a+5)(a+3) -$$
$$(a+2)(a+4)(a+5)(a+4) =$$
$$(a+2)(a+3)(a+5)(a+5) -$$
$$(a+2)(a+3)(a+4)(a+6)$$

$$a^4 + 15a^3 + 83a^2 + 201a + 180 -$$
$$\left(a^4 + 15a^3 + 82a^2 + 192a + 160\right) =$$
$$a^4 + 15a^3 + 81a^2 + 185a + 150 -$$
$$\left(a^4 + 15a^3 + 80a^2 + 180a + 144\right)$$

$$a^2 + 9a + 20 = a^2 + 5a + 6$$

$$4a = -14$$

$$a = -\frac{7}{2}$$

This value checks. When $a = -\frac{7}{2}, f(a) = g(a).$

73.
$$1 + \frac{x-1}{x-3} = \frac{2}{x-3} - x$$

$$(x-3)\cdot 1 + (x-3)\cdot \frac{x-1}{x-3}$$

$$= (x-3)\cdot \frac{2}{x-3} - (x-3)\cdot x$$

$$x-3+x-1 = 2 - x(x-3)$$

$$2x-4 = 2 - x^2 + 3x$$

$$x^2 + 2x - 3x - 4 - 2 = 0$$

$$x^2 - x - 6 = 0$$

$$(x-3)(x+2) = 0$$

$$x = 3 \text{ or } x = -2$$

By clearing denominators we find two solutions, 3 and −2. However, 3 is not in the domain of the original equation and we must discard this solution as extraneous. Therefore, the equation has one solution: $x = -2$.

75.
$$\frac{5-3a}{a^2+4a+3} - \frac{2a+2}{a+3} = \frac{3-a}{a+1}$$

$$\frac{5-3a}{(a+3)(a+1)} - \frac{2(a+1)}{a+3} = \frac{3-a}{a+1}$$

$$(a+3)(a+1)\frac{5-3a}{(a+3)(a+1)}$$

$$- (a+3)(a+1)\frac{2(a+1)}{a+3}$$

$$= (a+3)(a+1)\frac{3-a}{a+1}$$

$$\Rightarrow 5 - 3a - 2(a+1)^2 = (a+3)(3-a)$$

$$\Rightarrow 5 - 3a - 2(a^2+2a+1) = 3a - a^2 + 9 - 3a$$

$$\Rightarrow 5 - 3a - 2a^2 - 4a - 2 = 9 - a^2$$

$$\Rightarrow 3 - 7a - 2a^2 = 9 - a^2 \Rightarrow 9 - a^2 = 3 - 7a - 2a^2$$

$$\Rightarrow 9 - a^2 - 3 + 7a + 2a^2 = 0 \Rightarrow a^2 + 7a + 6 = 0$$

$$\Rightarrow (a+6)(a+1) = 0 \Rightarrow a = -6 \text{ or } a = -1$$

By clearing denominators we find two solutions, −6 and −1. However, −1 is not in the domain of the original equation and we must discard this solution as extraneous. Therefore, the equation has one solution: $a = -6$.

77.
$$\frac{\frac{1}{3}}{x} = \frac{1 - \frac{1}{x}}{x}$$

$$\frac{1}{3}\cdot\frac{1}{x} = \frac{1-\frac{1}{x}}{x}\cdot\frac{x}{x}$$

$$\frac{1}{3x} = \frac{x\cdot 1 - x\cdot\frac{1}{x}}{x\cdot x}$$

$$\frac{1}{3x} = \frac{x-1}{x^2}$$

$$3x^2\cdot\frac{1}{3x} = 3x^2\cdot\frac{x-1}{x^2}$$

$$3x\cdot x\cdot\frac{1}{3x} = 3x^2\cdot\frac{x-1}{x^2}$$

$$x = 3(x-1)$$

$$x = 3x - 3$$

$$3x - 3 = x$$

$$3x - x = 3$$

$$2x = 3$$

$$x = \frac{3}{2}$$

Exercise Set 7.7

1. $\dfrac{1 \text{ cake}}{2 \text{ hr}} = \dfrac{1}{2}$ cake per hour

3. $\dfrac{1}{2}$ cake per hour $+ \dfrac{1}{3}$ cake per hour

$= \left(\dfrac{1}{2} + \dfrac{1}{3}\right)$ cake per hour $= \dfrac{5}{6}$ cake per hour

5. $\dfrac{1 \text{ yard}}{3 \text{ hr}} = \dfrac{1}{3}$ lawn per hour

7. *Familiarize.* The job takes Trey 8 hours working alone and Matt 6 hours working alone. Then in 1 hour, Trey does $\dfrac{1}{8}$ of the job, and Matt does $\dfrac{1}{6}$ of the job. Working together they can do $\dfrac{1}{8} + \dfrac{1}{6}$ of the job in 1 hour. Let t represent the number of hours for Trey and Matt, working together, to do the job.

Translate. We want to find t such that

$t\left(\dfrac{1}{8}\right) + t\left(\dfrac{1}{6}\right) = 1$, or $\dfrac{t}{8} + \dfrac{t}{6} = 1$, where 1 represents one entire job.

Carry out. We solve the equation.

$$\dfrac{t}{8} + \dfrac{t}{6} = 1, \text{ LCD is } 24$$

$$24\left(\dfrac{t}{8} + \dfrac{t}{6}\right) = 24 \cdot 1$$

$$3t + 4t = 24$$

$$7t = 24$$

$$t = \dfrac{24}{7}$$

Check. In $\dfrac{24}{7}$ hours, Trey will do

$\dfrac{1}{8} \cdot \dfrac{24}{7}$, or $\dfrac{3}{7}$ of the job, and Matt will do

$\dfrac{1}{6} \cdot \dfrac{24}{7}$, or $\dfrac{4}{7}$ of the job. Together they do

$\dfrac{3}{7} + \dfrac{4}{7}$, or 1 entire job. The answer checks.

State. It will take $\dfrac{24}{7}$ hr., or $3\dfrac{3}{7}$ hr., for Trey and Matt working together, to do the job.

9. **Familiarize.** The pool can be filled in 12 hours by only the pipe and in 30 hours with only the hose. Then in 1 hour the pipe fills $\dfrac{1}{12}$ of the pool, and the hose fills $\dfrac{1}{30}$.

Working together, they fill $\dfrac{1}{12} + \dfrac{1}{30}$ of the pool in an hour. Let t equal the number of hours it takes them to fill the pool together.

Translate. We want to find t such that

$t\left(\dfrac{1}{12}\right) + t\left(\dfrac{1}{30}\right) = 1$, or $\dfrac{t}{12} + \dfrac{t}{30} = 1$

Carry out. We solve the equation. LCD = 60.

$$60\left(\dfrac{t}{12} + \dfrac{t}{30}\right) = 60 \cdot 1$$

$$5t + 2t = 60$$

$$7t = 60$$

$$t = \dfrac{60}{7}$$

Check. The pipe fills $\dfrac{1}{12} \cdot \dfrac{60}{7}$, or $\dfrac{5}{7}$, and the hose fills $\dfrac{1}{30} \cdot \dfrac{60}{7}$, or $\dfrac{2}{7}$. Working together, they fill $\dfrac{5}{7} + \dfrac{2}{7} = 1$, or the entire pool in $\dfrac{60}{7}$ hrs.

State. Working together, the pipe and hose can fill the pool in $\dfrac{60}{7}$ hrs., or $8\dfrac{4}{7}$ hr.

11. **Familiarize.** The $\dfrac{1}{4}$ HP does $\dfrac{1}{70}$ of the job in 1 minute, and the $\dfrac{1}{3}$ HP does $\dfrac{1}{30}$ of the job. Working together, they can do $\dfrac{1}{70} + \dfrac{1}{30}$ of the job in 1 minute. Let t equal the number of minutes it takes them working together.

Translate. We want to find t such that

$t\left(\dfrac{1}{70}\right) + t\left(\dfrac{1}{30}\right) = 1$, or $\dfrac{t}{70} + \dfrac{t}{30} = 1$

Carry out. We solve the equation. LCD is 210.

$$210\left(\dfrac{t}{70} + \dfrac{t}{30}\right) = 210 \cdot 1$$

$$3t + 7t = 210$$

$$10t = 210$$

$$t = \dfrac{210}{10} = 21$$

Check. In 21 minutes the $\dfrac{1}{4}$ HP empties

$\dfrac{1}{70} \cdot 21 = \dfrac{1}{\cancel{7} \cdot 10} \cdot \cancel{7} \cdot 3 = \dfrac{3}{10}$ of the basement,

and the $\dfrac{1}{3}$ HP empties

$\dfrac{1}{30} \cdot 21 = \dfrac{1}{\cancel{3} \cdot 10} \cdot \cancel{3} \cdot 7 = \dfrac{7}{10}$ of the basement.

Together they remove water from

$\dfrac{3}{10} + \dfrac{7}{10} = \dfrac{10}{10} = 1$ basement. The answer checks.

State. It will take 21 minutes for the machines working together to remove the water.

13. *Familiarize.* Let t = the time, in minutes, it takes the MP C7500 to copy the proposal. Then $3t$ = the time, in minutes, it takes the MP C2500 to copy the proposal. In 1 minute, the C7500 does $\frac{1}{t}$ of the job, and the C2500 does $\frac{1}{3t}$ of the job.

Translate. Together, they can do the entire job in 1.5 min, so we want to find t such that
$$1.5 \cdot \frac{1}{t} + 1.5 \cdot \frac{1}{3t} = 1$$

Carry out. Solve the equation.
$$1.5 \cdot \frac{1}{t} + 1.5 \cdot \frac{1}{3t} = 1$$
$$\Rightarrow 10 \cdot 1.5 \cdot \frac{1}{t} + 10 \cdot 1.5 \cdot \frac{1}{3t} = 10 \cdot 1$$
$$\Rightarrow 15 \cdot \frac{1}{t} + \cancel{3} \cdot 5 \cdot \frac{1}{\cancel{3}t} = 10 \Rightarrow \frac{15}{t} + \frac{5}{t} = 10$$
$$\Rightarrow \cancel{t} \cdot \frac{15}{\cancel{t}} + \cancel{t} \cdot \frac{5}{\cancel{t}} = t \cdot 10 \Rightarrow 15 + 5 = 10t$$
$$\Rightarrow t = \frac{20}{10} = 2$$

Check. The C7500 does the job in 2 min; in 1.5 min, it does
$$1.5 \cdot \frac{1}{2} = \frac{1.5}{2} = \frac{1.5 \cdot 10}{2 \cdot 10} = \frac{15}{20} = \frac{3}{4}$$ of the job.
The C2500 does the job in $3 \cdot 2 = 6$ min; in 1.5 min. it does
$$1.5 \cdot \frac{1}{6} = \frac{1.5}{6} = \frac{1.5 \cdot 10}{6 \cdot 10} = \frac{15}{60} = \frac{5}{20} = \frac{1}{4}$$ of the job. Together, they do $\frac{3}{4} + \frac{1}{4} = 1$ job in 1.5 min. The answer checks.
State. The C7500 can do the entire job in 2 min., and the C2500 in 6 min.

15. *Familiarize.* Let t = the time, in minutes, it takes the Airgle 750 to purify the air. Then $t + 20$ the time, in minutes, it takes the Healthmate 400 to copy the proposal. In 1 minute, the Airgle 750 does $\frac{1}{t}$ of the job, and the Healthmate 400 does $\frac{1}{t+20}$ of the job.

Translate. Together, they can do the entire job in 10.5 min, so we want to find t such that
$$10.5 \cdot \frac{1}{t} + 10.5 \cdot \frac{1}{t+20} = 1$$

Carry out. Solve the equation.
$$10.5 \cdot \frac{1}{t} + 10.5 \cdot \frac{1}{t+20} = 1$$
$$\Rightarrow 10 \cdot 10.5 \cdot \frac{1}{t} + 10 \cdot 10.5 \cdot \frac{1}{t+20} = 10 \cdot 1$$
$$\Rightarrow \frac{105}{t} + \frac{105}{t+20} = 10$$
$$\Rightarrow \cancel{t}(t+20) \cdot \frac{105}{\cancel{t}} + t\cancel{(t+20)} \cdot \frac{105}{\cancel{t+20}} = t(t+20) \cdot 10$$
$$\Rightarrow 105(t+20) + 105t = 10t(t+20)$$
$$\Rightarrow 105t + 2100 + 105t = 10t^2 + 200t$$
$$\Rightarrow 10t^2 + 200t - 105t - 105t - 2100 = 0$$
$$\Rightarrow 10t^2 - 10t - 2100 = 0 \Rightarrow t^2 - t - 210 = 0$$
$$\Rightarrow (t+14)(t-15) \Rightarrow t = -14 \text{ or } t = 15$$

Check. Since $t > 0$ the Airgle 750 does the job in 15 min; in 10.5 min, it does
$$10.5 \cdot \frac{1}{15} = \frac{10.5}{15} = \frac{10.5 \cdot 10}{15 \cdot 10} = \frac{105}{150} = \frac{21}{30} = \frac{7}{10}$$ of the job. The Healthmate 400 does the job in $15 + 20 = 35$ min; in 10.5 min. it does
$$10.5 \cdot \frac{1}{35} = \frac{10.5}{35} = \frac{10.5 \cdot 10}{35 \cdot 10} = \frac{105}{350} = \frac{21}{70} = \frac{3}{10}$$ of the job. Together, they do $\frac{7}{10} + \frac{3}{10} = 1$ job in 1.5 min. The answer checks.
State. The Airgle 750 can do the entire job in 15 min., and the Healthmate 400 in 35 min.

17. *Familiarize.* Let t represent the number of hours that the Erickson takes, working alone to douse the fire. Then $4t$ represents the number of hours the S-58T requires. In 1 hour, the Erickson does $\frac{1}{t}$ of the job, and the S-58T does $\frac{1}{4t}$ of the job.

Translate. Together they can do the job in 8 hr, so we want to find t such that

$$8 \cdot \frac{1}{t} + 8 \cdot \frac{t}{4t} = 1, \text{ or } \frac{8}{t} + \frac{2}{t} = 1$$

Carry out. We solve the equation.

$$\frac{8}{t} + \frac{2}{t} = 1, \text{ LCD is } t$$

$$t\left(\frac{8}{t} + \frac{2}{t}\right) = t \cdot 1$$

$$8 + 2 = t$$

$$t = 10$$

Check. The Erickson takes 10 hrs, working alone; it can do $8 \cdot \frac{1}{10} = \frac{4}{5}$ of the job in 8 hrs.

The S-58T does $8 \cdot \frac{1}{40} = \frac{1}{5}$ of the job in 8 hrs. Together, they do $\frac{4}{5} + \frac{1}{5}$, the entire job.

State. It takes the Erickson 10 hrs, and the S-58T 40 hrs, working alone.

19. ***Familiarize.*** We will convert hours to minutes:

$$2 \text{ hr } = 2 \cdot 60 \text{ min} = 120 \text{ min}$$

2 hr 55 min = 120 min + 55 min = 175 min
Let t = the number of minutes it takes Deb to do the job alone. Then $t + 120$ = the number of minutes it takes Dawn alone. In 1 hour (60 minutes) Deb does $\frac{1}{t}$ and Dawn does $\frac{1}{t+120}$ of the job.

Translate. In 175 min Dawn and Deb will complete one entire job, so we have

$$175\left(\frac{1}{t}\right) + 175\left(\frac{1}{t+120}\right) = 1,$$

or $\dfrac{175}{t} + \dfrac{175}{t+120} = 1$.

Carry out. We solve the equation. Multiply on both sides by the LCD, $t(t+120)$.

$$t(t+120) \cdot \left(\frac{175}{t} + \frac{175}{t+120}\right) = t(t+120)(1)$$

$$175(t+120) + 175t = t^2 + 120t$$

$$175t + 21{,}000 + 175t = t^2 + 120t$$

$$0 = t^2 - 230t - 21{,}000$$

$$0 = (t-300)(t+70)$$

$$t = 300 \text{ or } t = -70$$

Check. Since negative time has no meaning in this problem, -70 is not a solution of the original problem. If the job takes Deb 300 min and it take Dawn $300 + 120 = 420$ min, then in 175 min they would complete

$$175\left(\frac{1}{300}\right) + 175\left(\frac{1}{420}\right) = \frac{7}{12} + \frac{5}{12} = 1 \text{ job.}$$

The results check.
State. It would take Deb 300 min, or 5 hr, to do the job alone.

21. ***Familiarize***. Let r = the speed of the AMTRAK train in km/h. Then $r - 14 =$ the speed of the B&M train in km/h. We complete the table.

	Distance =	Rate ·	Time
	Distance (in km)	Speed (in km/h)	Time (in hours)
B&M	330	$r-14$	$\dfrac{330}{r-14}$
AMTRAK	400	r	$\dfrac{400}{r}$

Note: "Filling In" this specific chart is part of the exercise.
Translate. Since the times are the same (equal), we have the equation:

$$\frac{330}{r-14} = \frac{400}{r}$$

Carry out. We solve the equation.

$$\frac{330}{r-14} = \frac{400}{r}, \text{ LCD is } r(r-14)$$

$$r(r-14) \cdot \frac{330}{r-14} = r(r-14) \cdot \frac{400}{r}$$

$$330r = 400(r-14)$$

$$330r = 400r - 5600$$

$$5600 = 70r$$

$$80 = r$$

Check. If the AMTRAK train's speed is 80 km/h, then the B&M train's speed is $80 - 14$, or 66 km/h. Traveling 400 km at 80 km/h takes $\frac{400}{80} = 5$ hr., and traveling 330 km. at 66 km/h. takes $\frac{330}{66} = 5$ hr. Since the times are equal, the answer checks.
State. The speed of the AMTRAK train is 80 km/h and the speed of the B&M train is 66 km/h.

23. *Familiarize.* We first make a drawing. Let r = the kayak's speed in still water in mph. Then $r - 3$ = the speed upstream and $r + 3$ = the speed downstream.

$$\xrightarrow{\text{Upstream 4 miles } r-3 \text{ mph}}$$

$$\xleftarrow{\text{10 miles} \qquad r+3 \text{ mph} \qquad \text{Downstream}}$$

We organize the information in a table. The time is the same both upstream and downstream so we use t for each time.

	Distance	Speed	Time
Upstream	4	$r-3$	t
Downstream	10	$r+3$	t

Translate. Using the formula Time = Distance/Rate in each row of the table and the fact that the times are the same, we can write an equation.

$$\frac{4}{r-3} = \frac{10}{r+3}$$

Carry out. We solve the equation.

$$\frac{4}{r-3} = \frac{10}{r+3}, \text{ LCD is } (r-3)(r+3)$$

$$(r-3)(r+3) \cdot \frac{4}{r-3} = (r-3)(r+3) \cdot \frac{10}{r+3}$$

$$4(r+3) = 10(r-3)$$

$$4r + 12 = 10r - 30$$

$$42 = 6r$$

$$7 = r$$

Check. If $r = 7$ mph, then $r - 3$ is 4 mph and $r + 3$ is 10 mph. the time upstream is $\frac{4}{4}$, or 1 hour. The time downstream is $\frac{10}{10}$, or 1 hour. Since the times are the same, the answer checks.

State. The speed of the kayak in still water is 7 mph.

25. Note that 38 mi is 7 mi less than 45 mi and that the local bus travels 7 mph slower than the express. Then the express travels 45 mi in one hr, or 45 mph, and the local bus travels 38 mi in one hr, or 38 mph.

27. *Familiarize.* We first make a drawing. Let r = Kaitlyn's speed on a non-moving sidewalk in ft/sec. Then his speed moving forward on the moving sidewalk is $r + 1.7$ and his speed in the opposite direction is $r - 1.7$.

$$\xrightarrow{\text{Forward} \qquad r+1.7 \qquad 120 \text{ ft}}$$

$$\xleftarrow{\text{52 ft. } r-1.7 \text{ Opposite Direction}}$$

We organize the information in a table. The time is the same both forward and in the opposite direction, so we use t for each time.

	Distance	Speed	Time
Forward	120	$r+1.7$	t
Opposite Direction	52	$r-1.7$	t

Translate. Using the formula $T = D/R$ in each row of the table and the fact the times are the same (equal), we have:

$$\frac{120}{r+1.7} = \frac{52}{r-1.7}$$

Carry out. We solve the equation.

$$\frac{120}{r+1.7} = \frac{52}{r-1.7}; \text{ LCD is } (r+1.7)(r-1.7)$$

$$(r+1.7)(r-1.7) \cdot \frac{120}{r+1.7} =$$

$$(r+1.7)(r-1.7) \cdot \frac{52}{r-1.7}$$

$$120(r-1.7) = 52(r+1.7)$$

$$120r - 204 = 52r + 88.4$$

$$68r = 292.4$$

$$r = 4.3$$

Check. If Kaitlyn's speed on a non-moving sidewalk is 4.3 ft/sec, then his speed moving forward is 4.3 + 1.7, or 6 ft/sec, and his speed moving in the opposite direction on the sidewalk is 4.3 – 1.7, or 2.6 ft/sec. Moving 120 ft at 6 ft/sec takes $\frac{120}{6} = 20$ sec. Moving 52 ft at 2.6 ft/sec takes $\frac{52}{2.6} = 20$ sec. Since the times are the same, the answer checks.

State. Kaitlyn would be walking 4.3 ft/sec on a non-moving sidewalk.

29. Let t = the time, in hours, it takes Caledonia to drive to town. Then $t + 1$ = the time, in hours, it takes Manley to drive to town. Then:

$$r_M = r_C$$

$$\frac{d_M}{t_M} = \frac{d_C}{t_C}$$

$$\frac{20}{t+1} = \frac{15}{t}$$

$$20t = 15(t+1)$$

$$20t = 15t + 15$$

$$20t - 15t = 15$$

$$5t = 15$$

$$t = 3$$

It takes Caledonia 3 hours to drive to town.

31. **Familiarize.** Let r = speed of the river in km/h. Then $15 + r$ = the speed downstream, and $15 - r$ = the speed upstream. Using a table to organize the information, we have:

	Distance	Speed	Time
Downstream	140	$15 + r$	t
Upstream	35	$15 - r$	t

Translate. Using T = D/R in each row of the table and the fact that the times are the same, we can write the equation.

$$\frac{140}{15+r} = \frac{35}{15-r}$$

Carry out. We solve the equation.

$$\frac{140}{15+r} = \frac{35}{15-r}, \text{ LCD is } (15+r)(15-r)$$

$$(15+r)(15-r) \cdot \frac{140}{15+r} =$$

$$(15+r)(15-r) \cdot \frac{35}{15-r}$$

$$140(15-r) = 35(15+r)$$

$$2100 - 140r = 525 + 35r$$

$$1575 = 175r$$

$$9 = r$$

Check. The speed downstream is 15 + 9, or 24 km/h and the speed upstream is 15 − 9, or 6 km/h. Traveling 140 km at 24 km/h takes $\frac{140}{24} = 5\frac{5}{6}$ hr and traveling 35 km at 6 km/h takes $\frac{35}{6} = 5\frac{5}{6}$ hr. Since the times are the same, the number checks.

State. The speed of the river is 9 km/h.

33. **Familiarize.** Let c = the speed of the current, in km/h. Then $7 + c$ = the speed downriver and $7 - c$ = the speed upriver. We organize the information in a table.

	Distance	Speed	Time
Downriver	45	$7 + c$	t_1
Upriver	45	$7 - c$	t_2

Translate. Using the formula Time = Distance/Rate we see that $t_1 = \dfrac{45}{7+c}$ and $t_2 = \dfrac{45}{7-c}$. The total time upriver and back is 14 hr, so $t_1 + t_2 = 14$, or $\dfrac{45}{7-c} + \dfrac{45}{7+c} = 14$

Carry out. We solve the equation. Multiply both sides by the LCD, $(7+c)(7-c)$.

$$(7+c)(7-c)\left(\frac{45}{7+c} + \frac{45}{7-c}\right) =$$
$$(7+c)(7-c)14$$

$$45(7-c) + 45(7+c) = 14(49 - c^2)$$

$$315 - 45c + 315 + 45c = 686 - 14c^2$$

$$14c^2 - 56 = 0$$

$$14(c+2)(c-2) = 0$$

$$c + 2 = 0 \ \text{ or } c - 2 = 0$$

$$c = -2 \ \text{or} \quad c = 2$$

Check. Since speed cannot be negative in this problem, −2 cannot be a solution of the original problem. If the speed of the current is 2 km/h, the barge travels upriver at 7 − 2, or 5 km/h. At this rate it takes $\dfrac{45}{5}$, or 9 hr, to travel 45 km. The barge travels downriver at 7 + 2, or 9 km/h. At this rate it takes $\dfrac{45}{9}$, or 5 hr, to travel 45 km.

The total travel time is 9 + 5, or 14 hr. The answer checks.

State. The speed of the current is 2 km/h.

35. ***Familiarize.*** Let r = the speed at which the train actually traveled in mph, and let t = the actual travel time in hours. We organize the information in a table.

	Distance	Speed	Time
Actual speed	120	r	t
Faster speed	120	$r+10$	$t-2$

Translate. From the first row of the table we have $120 = rt$, and from the second row we have $120 = (r+10)(t-2)$. Solving the first equation for t, we have $t = \dfrac{120}{r}$. Substituting for t in the second equation, we have

$$120 = (r+10)\left(\dfrac{120}{r} - 2\right).$$

Carry out. We solve the equation.

$$120 = (r+10)\left(\dfrac{120}{r} - 2\right)$$

$$120 = 120 - 2r + \dfrac{1200}{r} - 20$$

$$20 = -2r + \dfrac{1200}{r}$$

$$r \cdot 20 = r\left(-2r + \dfrac{1200}{r}\right)$$

$$20r = -2r^2 + 1200$$

$$2r^2 + 20r - 1200 = 0$$

$$2\left(r^2 + 10r - 600\right) = 0$$

$$2(r+30)(r-20) = 0$$

$$r = -30 \ or \ r = 20$$

Check. Since speed cannot be negative in this problem, -30 cannot be a solution of the original problem. If the speed is 20 mph, it takes $\dfrac{120}{20}$, or 6 hr, to travel 120 mi. If the speed is 10 mph faster, or 30 mph, it takes $\dfrac{120}{30}$, or 4 hr, to travel 120 mi. Since 4 hr is 2 hr less time than 6 hr, the answer checks.

State. The speed was 20 mph.

37. Write a proportion and then solve it.

$$\dfrac{b}{6} = \dfrac{7}{4}$$

$$b = \dfrac{7}{4} \cdot 6$$

$$b = \dfrac{42}{4}, \ \ or \ \ 10.5$$

$\left(\text{Note that the proportions } \frac{6}{b} = \frac{4}{7}, \frac{b}{7} = \frac{6}{4},\right.$

$\left. \text{or } \frac{7}{b} = \frac{4}{6} \text{ could also be used.}\right)$

39. We write a proportion and then solve it.

$$\dfrac{4}{f} = \dfrac{6}{4}$$

$$4f \cdot \dfrac{4}{f} = 4f \cdot \dfrac{6}{4}$$

$$16 = 6f$$

$$\dfrac{8}{3} = f \qquad \text{Simplifying}$$

$\left(\text{One of the following proportions could}\right.$
$\left. \text{also be used:}\right.$

$$\dfrac{f}{4} = \dfrac{4}{6}, \dfrac{4}{f} = \dfrac{9}{6}, \dfrac{f}{4}$$

$$\left. = \dfrac{6}{9}, \dfrac{4}{9} = \dfrac{f}{6}, \dfrac{9}{4} = \dfrac{6}{f}\right)$$

41. We write a proportion and then solve it.

$$\dfrac{\text{Inches}}{\text{Feet}} \qquad \dfrac{P}{15} = \dfrac{1}{4}$$

$$P = \dfrac{1}{4} \cdot 15$$

$$P = \dfrac{15}{4}, \ \ or \ \ 3\dfrac{3}{4} \ \text{in.}$$

43. We write a proportion and then solve it.

$$\dfrac{\text{Inches}}{\text{Feet}} \qquad \dfrac{5}{r} = \dfrac{1}{4}$$

$$4r \cdot \dfrac{5}{r} = 4r \cdot \dfrac{1}{4}$$

$$20 = r$$

$$r = 20 \, \text{ft}$$

45. We write a proportion and then solve it.

$$\frac{l}{10} = \frac{6}{4}$$

$$l = \frac{6}{4} \cdot 10$$

$$l = \frac{60}{4} = 15\,\text{ft}$$

47. We write a proportion and then solve it.

$$\frac{5}{7} = \frac{9}{r}$$

$$7r \cdot \frac{5}{7} = 7r \cdot \frac{9}{r}$$

$$5r = 63$$

$$r = \frac{63}{5}, \quad \text{or} \quad 12.6$$

49. Write a proportion and then solve it.

$$\frac{n}{30} = \frac{384}{8}$$

$$n = \frac{384}{8} \cdot 30 = 96 \cdot 15$$

$$n = 1440 \text{ messages}$$

51. Let x = number of photos taken. Then:

$$\frac{234 \text{ photos}}{14 \text{ days}} = \frac{x \text{ photos}}{42 \text{ days}}$$

$$x = \frac{234 \cdot 42}{14} = 702 \text{ photos}$$

53. Let x = width of the wing in cm. Then:

$$\frac{24 \text{ cm}}{180 \text{ cm}} = \frac{x \text{ cm}}{200 \text{ cm}}$$

$$x = \frac{24 \cdot 200}{180} = \frac{24 \cdot 20\cancel{0}}{18\cancel{0}} = \frac{480}{18} = \frac{80}{3} = 26\frac{2}{3} \text{ cm}$$

55. Let x = number of defective drives.

Then: $\dfrac{7}{150} = \dfrac{x}{2700} \Rightarrow x = \dfrac{7 \cdot 2700}{150} = 126$

At the same rate 126 defective drives would be expected.

57. Let x = the ounces of water needed. Then:

$\dfrac{12}{8} = \dfrac{x}{5} \Rightarrow x = \dfrac{12 \cdot 5}{8} = 7\dfrac{1}{2}$

At the same rate the Bolognese would need $7\dfrac{1}{2}$ ounces of water.

59. Let x = the size of the pod. Then:

$\dfrac{x}{27} = \dfrac{40}{12} \Rightarrow x = \dfrac{40 \cdot 27}{12} = 90$

The pod has about 90 whales.

61. *Familiarize.* The ratio of the weight of an object on the moon to the weight of an object on Earth is 0.16 to 1.

a) We wish to find how much a 12-ton rocket would weigh on the moon.

b) We wish to find how much a 180-lb astronaut would weigh on the moon.

We can determine ratios.

$$\frac{0.16}{1} \qquad \frac{R}{12} \qquad \frac{A}{180}$$

Translate. Assuming the ratios are the same, we can translate to proportions.

a) $\dfrac{\text{Weight on moon}}{\text{Weight on Earth}} \quad \dfrac{0.16}{1} = \dfrac{R}{12}$

b) $\dfrac{\text{Weight on moon}}{\text{Weight on Earth}} \quad \dfrac{0.16}{1} = \dfrac{A}{180}$

Carry out. We solve each proportion.

a) $\qquad \dfrac{0.16}{1} = \dfrac{R}{12}$

$$12(0.16) = R$$

$$1.92 = R$$

b) $\qquad \dfrac{0.16}{1} = \dfrac{A}{180}$

$$120(0.16) = A$$

$$28.8 = A$$

Check. $\dfrac{0.16}{1} = 0.16, \quad \dfrac{1.92}{12} = 0.16,$

$$\text{and} \quad \dfrac{28.8}{180} = 0.16.$$

The ratios are the same.

State.

a) A 12-ton rocket would weigh 1.92 tons on the moon.

b) A 180-lb astronaut would weigh 28.8-lb. on the moon.

63. *Thinking and Writing Exercise.*

65. $a = \dfrac{b}{c} \Rightarrow a \cdot c = \dfrac{b}{\cancel{c}} \cdot \cancel{c} \Rightarrow b = ac$

67. $2x - 5y = 10 \Rightarrow 2x - 5y + 5y - 10 = 10 + 5y - 10$

$\Rightarrow 2x - 10 = 5y \Rightarrow \dfrac{1}{5} \cdot 2x - \dfrac{1}{5} \cdot 10 = \dfrac{1}{5} \cdot 5y$

$\Rightarrow \dfrac{2}{5}x - 2 = y$

69. $an + b = a \Rightarrow an + b - an = a - an$

$\Rightarrow b = a(1 - n) \Rightarrow b \cdot \dfrac{1}{1-n} = a(1-n) \cdot \dfrac{1}{1-n}$

$\Rightarrow a = \dfrac{b}{1-n}$

71. *Thinking and Writing Exercise.* Yes, in the time it takes the slower steamroller to do half the job alone, the faster steamroller can do more than half of the job.

73. *Familiarize.* If the drainage gate is closed, $\dfrac{1}{9}$ of the bog is filled in 1 hr. If the bog is not being filled, $\dfrac{1}{11}$ of the bog is drained in 1 hr. If the bog is being filled with the drainage gate left open, $\dfrac{1}{9} - \dfrac{1}{11}$ of the bog is filled in 1 hr. Let t = the time it takes to fill the bog with the drainage gate left open.
Translate. We want to find t such that

$$t\left(\dfrac{1}{9} - \dfrac{1}{11}\right) = 1, \text{ or } \dfrac{t}{9} - \dfrac{t}{11} = 1.$$

Carry out. We solve the equation. First we multiply by the LCD, 99.

$$99\left(\dfrac{t}{9} - \dfrac{t}{11}\right) = 99 \cdot 1$$

$$11t - 9t = 99$$

$$2t = 99$$

$$t = \dfrac{99}{2}$$

Check. In $\dfrac{99}{2}$ hr, we have

$$\dfrac{99}{2}\left(\dfrac{1}{9} - \dfrac{1}{11}\right) = \dfrac{11}{2} - \dfrac{9}{2} = \dfrac{2}{2} = 1 \text{ full bog.}$$

State. It will take $\dfrac{99}{2}$, or $49\dfrac{1}{2}$ hr, to fill the bog.

75. First let t = the time in hours it takes Julia and Tristan working together to complete one batch:

$$\dfrac{t}{3} + \dfrac{t}{4} = 1 \Rightarrow 4t + 3t = 12 \Rightarrow 7t = 12 \Rightarrow t = \dfrac{12}{7}$$

In $\dfrac{12}{7}$ ths of an hour, Julia completes:

$$\dfrac{1}{3} \cdot \dfrac{12}{7} = \dfrac{4}{7} \text{ths of the batch.} \quad \dfrac{4}{7} \approx 0.571 = 57.1\%$$

77. *Familiarize.* Let p = the number of people per hour moved by the 60-cm-wide escalator. Then $2p$ = the number of people per hour moved by the 100-cm-wide escalator. We convert 1575 people per 14 minutes to people per hour:

$$\dfrac{1575 \text{ people}}{14 \text{ min}} \cdot \dfrac{60 \text{ min}}{1 \text{ hr}} = 6750 \text{ people/hr}$$

Translate. We use the information that together the escalators move 6750 people per hour to write an equation.

$$p + 2p = 6750$$

Carry out. We solve the equation.

$$p + 2p = 6750$$

$$3p = 6750$$

$$p = 2250$$

Check. If the 60 cm-wide escalator moves 2250 people per hour, then the 100 cm-wide escalator moves $2 \cdot 2250$, or 4500 people per hour. Together, they move $2250 + 4500$, or 6750 people per hour. The answer checks.
State. The 60 cm-wide escalator moves 2250 people per hour.

79. *Familiarize.* Let d = the distance, in miles, the paddleboat can cruise upriver before it is time to turn around. The boat's speed upriver is $12 - 5$, or 7 mph, and its speed downriver is $12 + 5$, or 17 mph. We organize the information in a table.

	Distance	Speed	Time
Upriver	d	7	t_1
Downriver	d	17	t_2

Translate. Using the formula Time = Distance/Rate we see that

$t_1 = \dfrac{d}{7}$ and $t_2 = \dfrac{d}{17}$. The time upriver and back is 3 hr, so $t_1 + t_2 = 3$, or $\dfrac{d}{7} + \dfrac{d}{17} = 3$

Carry out. We solve the equation.

$$7 \cdot 17 \left(\frac{d}{7} + \frac{d}{17} \right) = 7 \cdot 17 \cdot 3$$

$$17d + 7d = 357$$

$$24d = 357$$

$$d = \frac{119}{8}$$

Check. Traveling $\frac{119}{8}$ mi upriver at a speed

of 7 mph takes $\frac{119/8}{7} = \frac{17}{8}$ hr. Traveling

$\frac{119}{8}$ mi downriver at a speed of 17 mph takes

$\frac{119/8}{17} = \frac{7}{8}$ hr. The total time is

$\frac{17}{8} + \frac{7}{8} = \frac{24}{8} = 3$ hr. The answer checks.

State. The pilot can go $\frac{119}{8}$, or $14\frac{7}{8}$ mi

upriver before it is time to turn around.

81. **Familiarize.** The Admissions Office printer

can do $\frac{1}{50}$ of the job in 1 min, and the

Business Office printer can do $\frac{1}{40}$ of the job

in 1 min. Let $t =$ the time they work together.
Translate. The work of the Admission Office

printer is $t \cdot \frac{1}{50}$, and the work of the Business

Office printer is $t \cdot \frac{1}{40}$. Working together, to

do one entire job gives us the equation:

$$t \cdot \frac{1}{50} + t \cdot \frac{1}{40} = 1, \text{ or } \frac{t}{50} + \frac{t}{40} = 1$$

Carry out. We solve the equation.

$$\frac{t}{50} + \frac{t}{40} = 1, \text{ LCD is } 200$$

$$200 \left(\frac{t}{50} + \frac{t}{40} \right) = 200 \cdot 1$$

$$4t + 5t = 200$$

$$9t = 200$$

$$t = \frac{200}{9} = 22\frac{2}{9}$$

It will require $22\frac{2}{9}$ min to complete the job,

if they work together. To determine on what page they will meet, we will find the amount of work done by the faster Business Office machine. (Note: we could also find the amount of work done by the slower machine and do a similar computation.) The Business

Office machine does $\frac{200}{9} \cdot \frac{1}{40}$ or $\frac{5}{9}$ of the

job. To determine the page number, we take

$\frac{5}{9}$ of $500 \approx 277.8$, or page 278, since this

machine begins on page 1.
Check. We already have determined that the

Business Office machine does $\frac{5}{9}$ of the job.

Similarly, the Admissions Office machine

does $\frac{200}{9} \cdot \frac{1}{50}$, or $\frac{4}{9}$ of the job. Since

$\frac{5}{9} + \frac{4}{9} = 1$, the number checks.

State. Working together, the machines will meet on page 278.

83. **Familiarize.** Express the position of the hands in terms of minute units on the face of the clock. At 10:30 the hour hand is at

$\frac{10.5}{12}$ hr $\times \frac{60 \text{ min}}{1 \text{ hr}}$, or 52.5 minutes, and the

minute hand is at 30 minutes. The rate of the minute hand is 12 times the rate of the hour hand. (When the minute hand moves 60 minutes, the hour hand moves 5 minutes.) Let $t =$ the number of minutes after 10:30 that the hands will first be perpendicular. After t minutes the minute hand has moved t units,

and the hour hand has moved $\frac{t}{12}$ units. The

position of the hour hand will be 15 units "ahead" of the position of the minute hand when they are first perpendicular.

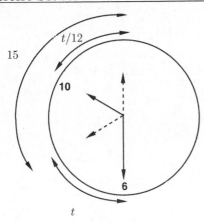

Translate.

Position of   position of
hour hand   is minute hand plus 15 min
after t min   after t min

$$52.5 + \frac{t}{12} = 30 + t + 15$$

Solve. We solve the equation.

$$52.5 + \frac{t}{12} = 30 + t + 15$$

$$52.5 + \frac{t}{12} = 45 + t, \text{ LCM is } 12$$

$$12\left(52.5 + \frac{t}{12}\right) = 12(45 + t)$$

$$630 + t = 540 + 12t$$

$$90 = 11t$$

$$\frac{90}{11} = t \ or \ t = 8\frac{2}{11}$$

Check. At $\frac{90}{11}$ min after 10:30, the position

of the hour hand is at $52.5 + \frac{90/11}{12}$, or $53\frac{2}{11}$

min. The minute hand is at $30 + \frac{90}{11}$, or $38\frac{2}{11}$

min. The hour hand is 15 minutes ahead of
the minute hand so the hands are
perpendicular. The answer checks.

State. After 10:30 the hands of a clock will

first be perpendicular in $8\frac{2}{11}$ min. The time

is $10:38\frac{2}{11}$, or $21\frac{9}{11}$ min before 11:00.

85. **Familiarize.** Let r = the speed in mph Liam
would have to travel for the last half of the
trip in order to average a speed of 45 mph for
the entire trip. We organize the information
in a table.

	Distance	Speed	Time
First half	50	40	t_1
Last half	50	r	t_2

The total distance is 50 + 50, or 100 mi. The

total time is $t_1 + t_2$, or $\frac{50}{40} + \frac{50}{r}$, or $\frac{5}{4} + \frac{50}{r}$.

The average speed is 45 mph.

Translate. Average speed = $\dfrac{\text{Total distance}}{\text{Total time}}$

$$45 = \frac{100}{\dfrac{5}{4} + \dfrac{50}{r}}$$

Carry out. We solve the equation.

$$45 = \frac{100}{\dfrac{5}{4} + \dfrac{50}{r}}$$

$$45 = \frac{100}{\dfrac{5r + 200}{4r}}$$

$$45 = 100 \cdot \frac{4r}{5r + 200}$$

$$45 = \frac{400r}{5r + 200}$$

$$(5r + 200)(45) = (5r + 200)\frac{400r}{5r + 200}$$

$$225r + 9000 = 400r$$

$$9000 = 175r$$

$$\frac{360}{7} = r$$

Check. Traveling 50 mi at 40 mph takes

$\frac{50}{40}$, or $\frac{5}{4}$ hr. Traveling 50 mi at $\frac{360}{7}$ mph

takes $\frac{50}{360/7}$, or $\frac{35}{36}$ hr. Then the total time

is $\frac{5}{4} + \frac{35}{36} = \frac{80}{36} = \frac{20}{9}$ hr and the average speed

is $\frac{100}{20/9} = 45$ mph. The answer checks.

State. Liam would have to travel at a speed

of $\frac{360}{7}$, or $51\frac{3}{7}$ mph for the last half of the trip so that the average speed for the entire trip would be 45 mph.

Exercise Set 7.8

1. Clear denominators in an equation using the LCD. : d

3. $y = \dfrac{k}{x} \Rightarrow xy = k$ The product xy is constant. : e

5. If $y = kx$, then by definition, y varies directly as x. : a

7. Inverse

9. Direct

11. Inverse

13. $f = \dfrac{L}{d}$ Solve for d.

     $df = L$ Multiplying by d

     $d = \dfrac{L}{f}$ Dividing by f

15. $s = \dfrac{(v_1 + v_2)t}{2}$ Solve for v_1.

     $2s = (v_1 + v_2)t$ Multiplying by 2

     $\dfrac{2s}{t} = v_1 + v_2$ Dividing by t

$\dfrac{2s}{t} - v_2 = v_1$

This result can also be expressed as

$v_1 = \dfrac{2s - tv_2}{t}$.

17. $\dfrac{t}{a} + \dfrac{t}{b} = 1$ Solve for b.

$ab\left(\dfrac{t}{a} + \dfrac{t}{b}\right) = ab \cdot 1$ Multiplying by the LCD

     $bt + at = ab$

       $at = ab - bt$

       $at = b(a - t)$ Factoring out b

    $\dfrac{at}{a-t} = b$ Multiplying by $\dfrac{1}{a-t}$

19. $I = \dfrac{2V}{R + 2r}$ Solve for R.

$I(R + 2r) = \dfrac{2V}{R + 2r} \cdot (R + 2r)$ Multiplying by the LCD

$I(R + 2r) = 2V$

     $R + 2r = \dfrac{2V}{I}$

       $R = \dfrac{2V}{I} - 2r$, or $\dfrac{2V - 2Ir}{I}$

21. $R = \dfrac{gs}{g + s}$ Solve for g.

$(g + s) \cdot R = (g + s) \cdot \dfrac{gs}{g + s}$ Multiplying by the LCD

     $Rg + Rs = gs$

       $Rs = gs - Rg$

       $Rs = g(s - R)$ Factoring out g

    $\dfrac{Rs}{s - R} = g$ Multiplying by $\dfrac{1}{s - R}$

23. $\dfrac{1}{p} + \dfrac{1}{q} = \dfrac{1}{f}$ Solve for q.

$pqf\left(\dfrac{1}{p} + \dfrac{1}{q}\right) = pqf \cdot \dfrac{1}{f}$ Multiplying by the LCD

     $qf + pf = pq$

       $pf = pq - qf$

       $pf = q(p - f)$

    $\dfrac{pf}{p - f} = q$

25. $S = \dfrac{H}{m(t_1 - t_2)}$ Solve for t_1.

$(t_1 - t_2)S = \dfrac{H}{m}$ Multiplying by $t_1 - t_2$

     $t_1 - t_2 = \dfrac{H}{Sm}$ Dividing by S

       $t_1 = \dfrac{H}{Sm} + t_2$, or $\dfrac{H + Smt_2}{Sm}$

27. $\dfrac{E}{e} = \dfrac{R+r}{r}$ Solve for r

$er \cdot \dfrac{E}{e} = er \cdot \dfrac{R+r}{r}$ Mult. by LCD

$Er = e(R+r)$

$Er = eR + er$

$Er - er = eR$

$r(E-e) = eR$

$r = \dfrac{eR}{E-e}$

29. $S = \dfrac{a}{1-r}$ Solve for r.

$(1-r)S = a$ Multiplying by the LCD, $1-r$

$1-r = \dfrac{a}{S}$ Dividing by S

$1 - \dfrac{a}{S} = r$ Adding r and $-\dfrac{a}{S}$

This result can also be expressed as

$r = \dfrac{S-a}{S}$.

31. $c = \dfrac{f}{(a+b)c}$ Solve for $(a+b)$.

$\dfrac{a+b}{c} \cdot c = \dfrac{a+b}{c} \cdot \dfrac{f}{(a+b)c}$

$a+b = \dfrac{f}{c^2}$

33. $P = \dfrac{A}{1+r}$ Solve for r.

$P(1+r) = \dfrac{A}{1+r} \cdot (1+r)$

$P(1+r) = A$

$1+r = \dfrac{A}{P}$

$r = \dfrac{A}{P} - 1$, or $\dfrac{A-P}{P}$

35. $v = \dfrac{d_2 - d_1}{t_2 - t_1}$ Solve for t_2.

$(t_2 - t_1)v = (t_2 - t_1) \cdot \dfrac{d_2 - d_1}{t_2 - t_1}$

$(t_2 - t_1)v = d_2 - d_1$

$t_2 - t_1 = \dfrac{d_2 - d_1}{v}$

$t_2 = \dfrac{d_2 - d_1}{v} + t_1$, or $\dfrac{d_2 - d_1 + t_1 v}{v}$

37. $\dfrac{1}{t} = \dfrac{1}{a} + \dfrac{1}{b}$ Solve for t.

$tab \cdot \dfrac{1}{t} = tab\left(\dfrac{1}{a} + \dfrac{1}{b}\right)$

$ab = tb + ta$

$ab = t(b+a)$

$\dfrac{ab}{b+a} = t$

39. $A = \dfrac{2Tt + Qq}{2T + Q}$ Solve for Q.

$(2T+Q) \cdot A = (2T+Q) \cdot \dfrac{2Tt + Qq}{2T + Q}$

$2AT + AQ = 2Tt + Qq$

Adding $-2AT$ and $-Qq$

$AQ - Qq = 2Tt - 2At$

$Q(A-q) = 2Tt - 2AT$

$Q = \dfrac{2Tt - 2AT}{A-q}$

41. $p = \dfrac{-98.42 + 4.15c - 0.082w}{w}$

$\Rightarrow w \cdot p = \not{w} \cdot \dfrac{-98.42 + 4.15c - 0.082w}{\not{w}}$

$\Rightarrow pw = -98.42 + 4.15c - 0.082w$

$\Rightarrow pw + 0.082w = -98.42 + 4.15c$

$\Rightarrow w(p + 0.082) = 4.15c - 98.42$

$\Rightarrow w = \dfrac{4.15c - 98.42}{p + 0.082}$

43. $y = kx$

$28 = k \cdot 4$ Substituting

$7 = k$

The variation constant is 7.

The equation of variation is $y = 7x$.

45. $y = kx$

$3.4 = k \cdot 2$ Substituting

$1.7 = k$

The variation constant is 1.7

The equation of variation is $y = 1.7x$.

47. $y = kx$

$2 = k \cdot \dfrac{1}{3}$ Substituting

$6 = k$

The variation constant is 6.

The equation of variation is $y = 6x$.

49. $y = \dfrac{k}{x}$

$3 = \dfrac{k}{20}$ Substituting

$60 = k$

The variation constant is 60.

The equation of variation is $y = \dfrac{60}{x}$.

51. $y = \dfrac{k}{x}$

$11 = \dfrac{k}{6}$ Substituting

$66 = k$

The variation constant is 66.

The equation of variation is $y = \dfrac{66}{x}$.

53. $y = \dfrac{k}{x}$

$27 = \dfrac{k}{\dfrac{1}{3}}$ Substituting

$9 = k$

The variation constant is 9.

The equation of variation is $y = \dfrac{9}{x}$.

55.   **Familiarize.** Because of the phrase "d . . . varies directly as . . . m," we express the distance as a function of the mass. Thus we have $d(m) = km$. We know that $d(3) = 20$.

Translate. We find the variation constant and then find the equation of variation.

$d(m) = km$

$d(3) = k \cdot 3$ Replacing m with 3

$20 = k \cdot 3$ Replacing $d(3)$ with 20

$\dfrac{20}{3} = k$ Variation constant

The equation of variation is $d(m) = \dfrac{20}{3}m$.

Carry out. We compute $d(5)$.

$d(m) = \dfrac{20}{3}m$

$d(5) = \dfrac{20}{3} \cdot 5$ Replacing m with 5

$d(5) = \dfrac{100}{3}$, or $33\dfrac{1}{3}$

Check. Reexamine the calculations. Note that the answer seems reasonable since

$\dfrac{3}{20}$ and $\dfrac{5}{100/3}$ are equal.

State. The spring is stretched $33\dfrac{1}{3}$ cm by a hanging object with mass 5 kg.

57.   **Familiarize.** Because T varies inversely as P, we write $T(P) = k / P$. We know that $T(7) = 5$.

Translate. We find the variation constant and the equation of variation.

$T(P) = \dfrac{k}{P}$

$T(7) = \dfrac{k}{7}$ Replacing P with 7

$5 = \dfrac{k}{7}$ Replacing $T(P)$ with 5

$35 = k$ Variation constant

$T(P) = \dfrac{35}{P}$ Equation of variation

Carry out. We find $T(10)$.

$T(10) = \dfrac{35}{10}$

$= 3.5$

Check. Reexamine the calculations.

State. It would take 3.5 hr for 10 volunteers to complete the job.

59.
$$W = kS$$
$$\frac{W}{S} = k$$
$$\frac{W_1}{S_1} = \frac{W_2}{S_2}$$
$$\frac{16.8 \text{ cm}}{150 \text{ cm}} = \frac{W_2}{500 \text{ in}}$$
$$W_2 = \frac{16.8}{150} \cdot 500 \text{ in} = 56 \text{ in}$$

61. Since the number of kilograms of water varies directly as the mass, and a 96-kg person contains 64 kg of water, a 48-kg person (weighing 1/2 of 96-kg), would contain half of 64 kg of water, or $\frac{1}{2} \cdot 64 = 32kg$.

63.
$$F = \frac{k}{L}$$
$$FL = k$$
$$F_1L_1 = F_2L_2$$
$$F_2 = \frac{F_1L_1}{L_2}$$
$$\frac{F_1L_1}{L_2} = \frac{33 \text{ cm} \cdot 260 \text{ Hz}}{30 \text{ cm}} = 286 \text{ Hz}$$

65. *Familiarize.* Because of the phrase "*t* varies inversely as . . . *u*," we write $t = k/u$. We know that $70 = \frac{k}{4}$.

Translate. We find the variation constant and then we find the equation of variation.
$$70 = \frac{k}{4} \Rightarrow 70 \cdot 4 = k \Rightarrow 280 = k \Rightarrow t = \frac{280}{u}$$

Carry out. We find *t* when *u* = 14.
$$t = \frac{280}{14} = 20$$

Check. Reexamine the calculations. Note that, as expected, as the UV rating increases, the time it takes to burn goes down.

State. It will take 20 min to burn when the UV rating is 14.

67. *Familiarize.* Let *A* = the amount of carbon monoxide released, in tons, and *P* = the population. Then:
$$A = kP \Rightarrow \frac{A}{P} = k \Rightarrow \frac{A_1}{P_1} = \frac{A_2}{P_2} \Rightarrow A_2 = \frac{A_1}{P_1} \cdot P_2$$
$$= \frac{0.65 \text{ tons}}{2.6 \text{ people}} \cdot 308,000,000 \text{ people}$$
$$= 77,000,000 \text{ tons}$$

Check. Reexamine the calculations.
State. The U.S. released approx. 77 million tons of carbon monoxide in 1 year.

69. $y = kx^2$
$$6 = k \cdot 3^2 \qquad \text{Substituting}$$
$$6 = 9k$$
$$\frac{6}{9} = k$$
$$\frac{2}{3} = k \qquad \text{Variation constant}$$

The equation of variation is $y = \frac{2}{3}x^2$.

71. $y = \frac{k}{x^2}$
$$6 = \frac{k}{3^2} \qquad \text{Substituting}$$
$$6 = \frac{k}{9}$$
$$6 \cdot 9 = k$$
$$54 = k \qquad \text{Variation constant}$$

The equation of variation is $y = \frac{54}{x^2}$.

73. $y = kxz^2$
$$105 = k \cdot 14 \cdot 5^2 \qquad \begin{array}{l}\text{Substituting 105 for } y, \\ 14 \text{ for } x, \text{ and } 5 \text{ for } z.\end{array}$$
$$105 = 350k$$
$$\frac{105}{350} = k$$
$$0.3 = k \qquad \text{Variation constant}$$

The equation of variation is $y = 0.3xz^2$.

75. $y = k \cdot \dfrac{wx^2}{z}$

$49 = k \cdot \dfrac{3 \cdot 7^2}{12}$ Substituting

$4 = k$ Variation constant

The equation of variation is $y = \dfrac{4wx^2}{z}$.

77. **Familiarize.** Because d varies directly as r^2, we write $d = kr^2$. We know that $d = 138$ ft when $r = 60$ mph.
Translate. Determine k and the equation of variation.

$d = kr^2$

$138 = k \cdot 60^2$

$\dfrac{138}{3600} = k$

$\dfrac{23}{600} = k$

$d = \dfrac{23}{600} \cdot r^2$ Equation of variation

Carry out. Substitute 40 for r and solve for d.

$d = \dfrac{23}{600} \cdot 40^2$

$d = 61\dfrac{1}{3}$

Check. Reexamine the calculations.
State. A car traveling at 40 mph will require $61\dfrac{1}{3}$ ft to stop.

79. **Familiarize.** Because V varies directly as T and inversely as P, we write $V = \dfrac{kT}{P}$. We know that $V = 231$ when $T = 300$ and $P = 20$.
Translate. Determine k and the equation of variation.

$V = \dfrac{kT}{P}$

$231 = \dfrac{k \cdot 300}{20}$

$\dfrac{20}{300} \cdot 231 = k$

$15.4 = k$

$V = \dfrac{15.4T}{P}$ Equation of variation

Carry out. Substitute 320 for T and 16 for P to determine V.

$V = \dfrac{15.4 \cdot 320}{16}$

$V = 308 \text{ cm}^3$

Check. Reexamine the calculations.
State. The volume is 308 cm^3 when $T = 320°k$ and $P = 16\text{lb/cm}^2$.

81. **Familiarize.** The drag W varies jointly as the surface area A and velocity v, so we write $W = kAv$. We know that $W = 222$ when $A = 37.8$ and $v = 40$.
Translate. Find k.

$W = kAv$

$222 = k(37.8)(40)$

$\dfrac{222}{37.8(40)} = k$

$\dfrac{37}{252} = k$

$W = \dfrac{37}{252}Av$ Equation of variation

Carry out. Substitute 51 for A and 430 for W and solve for v.

$430 = \dfrac{37}{252} \cdot 51 \cdot v$

$57.42 \text{ mph} \approx v$

(If we had used the rounded value 0.1468 for k, the resulting speed would have been approximately 57.43 mph.)
Check. Reexamine the calculations.
State. The car must travel about 57.42 mph.

83. a) From the data table we see that as the population density increases the annual VMT per household decreases. Therefore the population P is inversely to proportional the VMT, V.

b) $P = \dfrac{k}{V} \Rightarrow 25 = \dfrac{k}{12,000}$

$\Rightarrow k = 25 \cdot 12,000 = 300,000$

$\Rightarrow P = \dfrac{300,000}{V}$

As a check, we note that $PV = k$ for all pairs of points (P, V).

$50 \cdot 6000 = 300,000 \quad \checkmark$

$100 \cdot 3000 = 300,000 \quad \checkmark$

$200 \cdot 1500 = 300,000 \quad \checkmark$

c) $P = \dfrac{300{,}000}{V}$

$V = \dfrac{300{,}000}{P} = \dfrac{300{,}000}{10} = 30{,}000 \text{ VMT}$

85. a)   From the data table we see that, as the selling price increases, the seller's commission also increases. Therefore the commission C is directly proportional to the selling price P.

b) $C = kP \Rightarrow 111.85 = k \cdot 200$

$\Rightarrow k = \dfrac{111.85}{200} \approx 0.56$

$\Rightarrow C = 0.56P$

As a check, we note that $\dfrac{C}{P} = k$ for all

pairs of points (P, C).

$\dfrac{41.42}{75} \approx 0.552$

$\dfrac{55.50}{100} = 0.5550$

$\dfrac{111.85}{200} = 0.55925$

$\dfrac{240.55}{400} \approx 0.601$

We see that the "constant" of variation is not exactly constant. Over the range of selling prices, there is a rise of about 0.008 or 0.8% in the commission. However, we will use the equation determined above in accordance with the directions to use the point (200, 111.85).

c) $C = 0.56P = 0.56(150) = 84$ The seller receives approximately $84.00.

87. *Thinking and Writing Exercise.*

89. $3 - x < 3x + 5$

$3 - 5 < 3x + x$

$-2 < 4x$

$-\dfrac{2}{4} = -\dfrac{1}{2} < x$

$x > -\dfrac{1}{2}$

$\left\{ x \mid x > -\dfrac{1}{2} \right\}$, or $\left(-\dfrac{1}{2}, \infty \right)$

91. $2(x - 1) \ge 4(x + 1)$

$2x - 2 \ge 4x + 4$

$-2 - 4 \ge 4x - 2x$

$-6 \ge 2x$

$\dfrac{-6}{2} \ge \dfrac{2x}{2}$

$-3 \ge x \Rightarrow x \le -3$

$\{ x \mid x \le -3 \}$, or $(-\infty, -3]$

93. $\dfrac{2x + 3}{4} \le 5$

$\cancel{4} \cdot \dfrac{2x + 3}{\cancel{4}} \le 4 \cdot 5$

$2x + 3 \le 20$

$2x \le 20 - 3 = 17$

$x \le \dfrac{17}{2}$

$\left\{ x \mid x \le \dfrac{17}{2} \right\}$, or $\left(-\infty, \dfrac{17}{2} \right]$

95. *Thinking and Writing Exercise.*

97. Use the result of Example 2.

$h = \dfrac{2R^2 g}{V^2} - R$

We have $V = 6.5$ mi/sec, $R = 3960$ mi, and $g = 32.2$ ft/sec^2. We must convert 32.2 ft/sec^2 to mi/sec^2 so all units of length are the same.

$32.2 \dfrac{\cancel{ft}}{\sec^2} \cdot \dfrac{1 \text{ mi}}{5280 \cancel{ft}} \approx 0.0060985 \dfrac{\text{mi}}{\sec^2}$

Now we substitute and compute.

$h = \dfrac{2(3960)^2 (0.0060985)}{(6.5)^2} - 3960$

$h \approx 567$

The satellite is about 567 mi from the surface of the earth.

99. $c = \dfrac{a}{a+12} \cdot d$

$c = \dfrac{2a}{2a+12} \cdot d$ Doubling a

$= \dfrac{\cancel{2}a}{\cancel{2}(a+6)} \cdot d$

$= \dfrac{a}{a+6} \cdot d$ Simplifying

The ratio of the larger dose to the smaller dose is

$\dfrac{\dfrac{a}{a+6} \cdot d}{\dfrac{a}{a+12} \cdot d} = \dfrac{\dfrac{ad}{a+6}}{\dfrac{ad}{a+12}}$

$= \dfrac{ad}{a+6} \cdot \dfrac{a+12}{ad}$

$= \dfrac{\cancel{ad}(a+12)}{(a+6)\cancel{ad}}$

$= \dfrac{a+12}{a+6}$

The amount by which the dosage increases is

$\dfrac{a}{a+6} \cdot d - \dfrac{a}{a+12} \cdot d$

$= \dfrac{ad}{a+6} - \dfrac{ad}{a+12}$

$= \dfrac{ad}{a+6} \cdot \dfrac{a+12}{a+12} - \dfrac{ad}{a+12} \cdot \dfrac{a+6}{a+6}$

$= \dfrac{ad(a+12) - ad(a+6)}{(a+6)(a+12)}$

$= \dfrac{a^2 d + 12ad - a^2 d - 6ad}{(a+6)(a+12)}$

$= \dfrac{6ad}{(a+6)(a+12)}$

Then the percent by which the dosage increases is

$\dfrac{\dfrac{6ad}{(a+6)(a+12)}}{\dfrac{a}{a+12} \cdot d} = \dfrac{\dfrac{6ad}{(a+6)(a+12)}}{\dfrac{ad}{a+12}}$

$= \dfrac{6ad}{(a+6)(a+12)} \cdot \dfrac{a+12}{ad}$

$= \dfrac{6 \cdot \cancel{ad} \cdot \cancel{(a+12)}}{(a+6)\cancel{(a+12)} \cdot \cancel{ad}}$

$= \dfrac{6}{a+6}$

This is a decimal representation for the percent of increase. To give the result in percent notation we multiply by 100 and use a percent symbol. We have

$\dfrac{6}{a+6} \cdot 100\%$, or $\dfrac{600}{a+6}\%$

101. $a = \dfrac{\dfrac{d_4 - d_3}{t_4 - t_3} - \dfrac{d_2 - d_1}{t_2 - t_1}}{t_4 - t_2}$

$a(t_4 - t_2) = \dfrac{d_4 - d_3}{t_4 - t_3} - \dfrac{d_2 - d_1}{t_2 - t_1}$

Multiplying by $t_4 - t_2$

$a(t_4 - t_2)(t_4 - t_3)(t_2 - t_1) =$
$(d_4 - d_3)(t_2 - t_1) - (d_2 - d_1)(t_4 - t_3)$

Multiplying by $(t_4 - t_3)(t_2 - t_1)$

$a(t_4 - t_2)(t_4 - t_3)(t_2 - t_1) - (d_4 - d_3)(t_2 - t_1) =$
$\qquad -(d_2 - d_1)(t_4 - t_3)$

$(t_2 - t_1)[a(t_4 - t_2)(t_4 - t_3) - (d_4 - d_3)] =$
$\qquad -(d_2 - d_1)(t_4 - t_3)$

$t_2 - t_1 = \dfrac{-(d_2 - d_1)(t_4 - t_3)}{a(t_4 - t_2)(t_4 - t_3) - (d_4 - d_3)}$

$t_2 + \dfrac{(d_2 - d_1)(t_4 - t_3)}{a(t_4 - t_2)(t_4 - t_3) + d_3 - d_4} = t_1$

103. Write I as a function of w, the wattage and d, the distance, and then double both w and d.

$$I = \frac{w}{d^2}$$

$$I = \frac{2w}{(2d)^2}$$

$$I = \frac{2w}{4d^2}$$

$$I = \frac{1}{2} \cdot \frac{w}{d^2}$$

Comparing $I = \frac{w}{d^2}$ and $I = \frac{1}{2} \cdot \frac{w}{d^2}$, we see that the intensity is halved.

105. **Familiarize.** We write $T = kml^2 f^2$. We know that $T = 100$ when $m = 5$, $l = 2$, and $f = 80$.
Translate. Find k.

$$T = kml^2 f^2$$

$$100 = k(5)(2)^2 (80)^2$$

$$0.00078125 = k$$

$$T = 0.00078125ml^2 f^2$$

Carry out. Substitute 72 for T, 5 for m, and 80 for f and solve for l.

$$72 = 0.00078125(5)l^2 (80)^2$$

$$2.88 = l^2$$

$$1.697 \approx l$$

Check. Recheck the calculations.
State. The string should be about 1.697 m long.

107. **Familiarize.** Because d varies inversely as s, we write $d(s) = k / s$. We know that

$$d(0.56) = 50.$$

Translate.

$$d(s) = \frac{k}{s}$$

$$d(0.56) = \frac{k}{0.56} \quad \text{Replacing } s \text{ with } 0.56$$

$$50 = \frac{k}{0.56} \quad \text{Replacing } d(56) \text{ with } 50$$

$$28 = k$$

$$d(s) = \frac{28}{s} \quad \text{Equation of variation}$$

Carry out. Find $d(0.40)$.

$$d(0.40) = \frac{28}{0.40}$$

$$= 70$$

Check. Reexamine the calculations. Also serve that, as expected, when d decreases, then s increases.
State. The equation of variation is

$$d(s) = \frac{28}{s}. \quad \text{The distance is 70 yd.}$$

Chapter 7 Study Summary

1. The expression is undefined where
   $(t+7)^2 = 0 \Rightarrow t+7 = \pm\sqrt{0} \Rightarrow t+7 = 0 \Rightarrow t = -7$

2. $\dfrac{y^2 - 5y}{y^2 - 25} = \dfrac{y(y-5)}{(y+5)(y-5)} = \dfrac{y}{y+5}$ for $y \neq 5$

3. $\dfrac{6x-12}{2x^2 + 3x - 2} \cdot \dfrac{x^2 - 4}{8x - 8}$

   $= \dfrac{6(x-2)}{(2x-1)(x+2)} \cdot \dfrac{(x+2)(x-2)}{8(x-1)}$

   $= \dfrac{6(x-2)^2}{8(x-1)(2x-1)}$

   $= \dfrac{2 \cdot 3(x-2)^2}{2 \cdot 4(x-1)(2x-1)} = \dfrac{3(x-2)^2}{4(x-1)(2x-1)}$

4. $\dfrac{t-3}{6} \div \dfrac{t+1}{15} = \dfrac{t-3}{6} \cdot \dfrac{15}{t+1} = \dfrac{t-3}{3 \cdot 2} \cdot \dfrac{3 \cdot 5}{t+1}$

   $= \dfrac{5(t-3)}{2(t+1)}$

5. $\dfrac{5x+4}{x+3} + \dfrac{4x+1}{x+3} = \dfrac{(5x+4)+(4x+1)}{x+3} = \dfrac{9x+5}{x+3}$

6. $\dfrac{5x+4}{x+3} - \dfrac{4x+1}{x+3} = \dfrac{(5x+4)-(4x+1)}{x+3}$

   $= \dfrac{5x+4-4x-1}{x+3} = \dfrac{x+3}{x+3} = 1$ for $x \neq -3$

7. $x^2 - 2x - 15 = (x-5)(x+3)$, and
   $x^2 - 9 = (x+3)(x-3)$.
   $\text{LCM} = (x-5)(x+3)(x-3)$

8. $\dfrac{t}{t-1} - \dfrac{t-2}{t+1} = \dfrac{t}{t-1} \cdot \dfrac{t+1}{t+1} - \dfrac{t-2}{t+1} \cdot \dfrac{t-1}{t-1}$

$= \dfrac{t(t+1) - (t-2)(t-1)}{(t+1)(t-1)} = \dfrac{t^2 + t - (t^2 - 2t - t + 2)}{(t+1)(t-1)}$

$= \dfrac{\cancel{t^2} + t \cancel{-t^2} + 2t + t - 2}{(t+1)(t-1)} = \dfrac{4t-2}{(t+1)(t-1)}$

9. $\dfrac{\dfrac{4}{x} - 4}{\dfrac{7}{x} - 7} = \dfrac{\dfrac{4}{x} - 4}{\dfrac{7}{x} - 7} \cdot \dfrac{x}{x} = \dfrac{\cancel{x} \cdot \dfrac{4}{\cancel{x}} - x \cdot 4}{\cancel{x} \cdot \dfrac{7}{\cancel{x}} - x \cdot 7} = \dfrac{4 - 4x}{7 - 7x}$

$= \dfrac{4(1-x)}{7(1-x)} = \dfrac{4}{7}$ for $x \neq 0, 1$

10. $\dfrac{3}{x+4} = \dfrac{1}{x-1} \Rightarrow 3(x-1) = 1(x+4)$

$\Rightarrow 3x - 3 = x + 4 \Rightarrow 3x - x = 4 + 3$

$\Rightarrow 2x = 7 \Rightarrow x = \dfrac{7}{2}$

11. Jackson can sand $\dfrac{1}{12}$th of the floors per hour, and Charis can sand $\dfrac{1}{9}$th. Let $t =$ the time it takes for them to sand the floors (1 job) working together. Then:

$\left(\dfrac{1}{12} + \dfrac{1}{9}\right) t = 1 \Rightarrow \dfrac{1}{12} + \dfrac{1}{9} = \dfrac{1}{t}$

$\Rightarrow 36t \cdot \dfrac{1}{12} + 36t \cdot \dfrac{1}{9} = 36t \cdot \dfrac{1}{t}$

$\Rightarrow \cancel{12} \cdot 3t \cdot \dfrac{1}{\cancel{12}} + \cancel{9} \cdot 4t \cdot \dfrac{1}{\cancel{9}} = 36 \cdot \cancel{t} \cdot \dfrac{1}{\cancel{t}}$

$\Rightarrow 3t + 4t = 36 \Rightarrow 7t = 36 \Rightarrow t = \dfrac{36}{7} = 5\dfrac{1}{7}$

They can do the job in $5\dfrac{1}{7}$ hours working together.

12. Let $b =$ the speed of the boat in still water. We are told that the time going downstream is the same as the time going upstream. Thus:

$t_{DOWNSTREAM} = t_{UPSTREAM} \Rightarrow$

$\left(\dfrac{d}{r}\right)_{DOWNSTREAM} = \left(\dfrac{d}{r}\right)_{UPSTREAM}$

$\Rightarrow \dfrac{35}{b+4} = \dfrac{15}{b-4} \Rightarrow 35(b-4) = 15(b+4)$

$\Rightarrow 35b - 140 = 15b + 60 \Rightarrow 35b - 15b = 60 + 140$

$\Rightarrow 20b = 200 \Rightarrow b = \dfrac{200}{20} = 10$

Drew's speed in still water is 10 mph.

13. $\dfrac{m}{16} = \dfrac{15}{10} = \dfrac{3}{2} \Rightarrow m = \dfrac{3}{2} \cdot 16 = \dfrac{3}{\cancel{2}} \cdot \cancel{2} \cdot 8 = 24$

14. $y = kx \Rightarrow k = \dfrac{y}{x} = \dfrac{10}{0.2} = \dfrac{10 \cdot 10}{0.2 \cdot 10} = \dfrac{100}{2} = 50$

$\Rightarrow y = 50x$

15. $y = \dfrac{k}{x} \Rightarrow k = xy = 5 \cdot 8 = 40$

$\Rightarrow y = \dfrac{40}{x}$

16. $y = kxz \Rightarrow k = \dfrac{y}{xz} = \dfrac{2}{5 \cdot 4} = \dfrac{2}{20} = \dfrac{\cancel{2}}{\cancel{2} \cdot 10} = \dfrac{1}{10}$

$\Rightarrow y = \dfrac{1}{10} xz$, or $\dfrac{xz}{10}$

Chapter 7 Review Exercises

1. False

2. True

3. False

4. False

5. False

6. True

7. True

8. False

9. False

10. True

11. $f(t) = \dfrac{t^2 - 3t + 2}{t^2 - 9}$

a) $f(0) = \dfrac{0^2 - 3 \cdot 0 + 2}{0^2 - 9}$

$= \dfrac{-2}{9}$

b) $f(-1) = \dfrac{(-1)^2 - 3 \cdot (-1) + 2}{(-1)^2 - 9}$

$= \dfrac{6}{-8} = -\dfrac{3}{4}$

c) $f(2) = \dfrac{2^2 - 3 \cdot 2 + 2}{2^2 - 9}$

$\quad = \dfrac{0}{-5} = 0$

12. $\dfrac{17}{-x^2}$ We find the number which makes the denominator 0. To do so, we set the denominator equal to 0 and solve for x.

$-x^2 = 0$

$\quad x = 0$

The expression is undefined for $x = 0$.

13. $\dfrac{9}{a-4}$ We find the number which makes the denominator 0. To do so, we set the denominator equal to 0 and solve for a.

$a - 4 = 0$

$\quad a = 4$

The expression is undefined for $a = 4$.

14. $\dfrac{x-5}{x^2-36}$ We find the numbers which make the denominator 0. To do so, we set the denominator equal to 0 and solve for x.

$$x^2 - 36 = 0$$

$$(x+6)(x-6) = 0$$

$x + 6 = 0 \quad \text{or} \quad x - 6 = 0$

$\quad x = -6 \quad \text{or} \quad \quad x = 6$

The expression is undefined for $x = \pm 6$.

15. $\dfrac{x^2 + 3x + 2}{x^2 + x - 30}$

We find the numbers which make the denominator 0. To do so, we set the denominator equal to 0 and solve for x.

$$x^2 + x - 30 = 0$$

$$(x+6)(x-5) = 0$$

$x + 6 = 0 \quad \text{or} \quad x - 5 = 0$

$\quad x = -6 \quad \text{or} \quad \quad x = 5$

The expression is undefined for $x = -6$ and $x = 5$.

16. $\dfrac{4x^2 - 8x}{4x^2 + 4x} = \dfrac{4x(x-2)}{4x(x+1)}$

$\quad = \dfrac{4x}{4x} \cdot \dfrac{x-2}{x+1} = \dfrac{x-2}{x+1}$

17. $\dfrac{14x^2 - x - 3}{2x^2 - 7x + 3} = \dfrac{(7x+3)\,\cancel{(2x-1)}}{\cancel{(2x-1)}\,(x-3)} = \dfrac{7x+3}{x-3}$

18. $\dfrac{5x^2 - 20y^2}{2y - x} = \dfrac{5(x^2 - 4y^2)}{-(x-2y)}$

$\quad = \dfrac{5(x+2y)(x-2y)}{-(x-2y)}$

$\quad = \dfrac{5(x+2y)}{-1} \cdot \dfrac{x-2y}{x-2y}$

$\quad = -5(x+2y)$

19. $\dfrac{a^2 - 36}{10a} \cdot \dfrac{2a}{a+6} = \dfrac{(a+6)(a-6)}{5 \cdot 2 \cdot a} \cdot \dfrac{2 \cdot a}{a+6}$

$\quad = \dfrac{(a+6)(a-6) \cdot 2 \cdot a}{5 \cdot 2 \cdot a(a+6)}$

$\quad = \dfrac{(\cancel{a+6})(a-6)\cancel{2} \cdot \cancel{a}}{5 \cdot \cancel{2} \cdot \cancel{a} \cdot (\cancel{a+6})}$

$\quad = \dfrac{a-6}{5}$

20. $\dfrac{8t+8}{2t^2 + t - 1} \cdot \dfrac{t^2 - 1}{t^2 - 2t + 1}$

$\quad = \dfrac{8(t+1)}{(2t-1)(t+1)} \cdot \dfrac{(t+1)(t-1)}{(t-1)(t-1)}$

$\quad = \dfrac{8(t+1)(t+1)(t-1)}{(2t-1)(t+1)(t-1)(t-1)}$

$\quad = \dfrac{8(\cancel{t+1})(t+1)(\cancel{t-1})}{(2t-1)(\cancel{t+1})(\cancel{t-1})(t-1)}$

$\quad = \dfrac{8(t+1)}{(2t-1)(t-1)}$

21. $\dfrac{16 - 8t}{3} \div \dfrac{t-2}{12t} = \dfrac{16 - 8t}{3} \cdot \dfrac{12t}{t-2}$

$\quad = \dfrac{-8(-2+t)}{3} \cdot \dfrac{3 \cdot 4 \cdot t}{t-2}$

$\quad = \dfrac{-8(t-2) \cdot 3 \cdot 4 \cdot t}{3(t-2)}$

$\quad = \dfrac{-8(\cancel{t-2})\cancel{3} \cdot 4 \cdot t}{\cancel{3}(\cancel{t-2})}$

$\quad = -32t$

22. $\dfrac{4x^4}{x^2-1} \div \dfrac{2x^3}{x^2-2x+1}$

$= \dfrac{4x^4}{x^2-1} \cdot \dfrac{x^2-2x+1}{2x^3}$

$= \dfrac{4x^4}{(x+1)(x-1)} \cdot \dfrac{(x-1)(x-1)}{2x^3}$

$= \dfrac{2 \cdot 2 \cdot x^3 \cdot x(x-1)(x-1)}{(x+1)(x-1) \cdot 2x^3}$

$= \dfrac{\cancel{2} \cdot 2 \cdot \cancel{x^3} \cdot x \cdot (\cancel{x-1})(x-1)}{(x+1)(\cancel{x-1}) \cdot \cancel{2} \, \cancel{x^3}}$

$= \dfrac{2x(x-1)}{x+1}$

23. $\dfrac{x^2+1}{x-2} \cdot \dfrac{2x+1}{x+1} = \dfrac{(x^2+1)(2x+1)}{(x-2)(x+1)}$

24. $(t^2+3t-4) \div \dfrac{t^2-1}{t+4}$

$= \dfrac{t^2+3t-4}{1} \cdot \dfrac{t+4}{t^2-1}$

$= \dfrac{(t+4)(t-1)}{1} \cdot \dfrac{t+4}{(t+1)(t-1)}$

$= \dfrac{(t+4)(t-1)(t+4)}{(t+1)(t-1)}$

$= \dfrac{(t+4)(\cancel{t-1})(t+4)}{(t+1)(\cancel{t-1})}$

$= \dfrac{(t+4)^2}{t+1}$

25. $10a^3b^8 = 2 \cdot 5 \cdot a^3 \cdot b^8,\ 12a^5b = 2^2 \cdot 3 \cdot a^5 \cdot b$

$\text{LCM} = 2^2 \cdot 3 \cdot 5 \cdot a^5 \cdot b^8 = 60a^5b^8$

26. $x^2 - x = x(x-1)$

$x^5 - x^3 = x^3(x^2-1) = x^3(x-1)(x+1)$

$x^4 = x^4$

$\text{LCM} = x^4(x-1)(x+1)$

27. $y^2 - y - 2 = (y-2)(y+1),\ y^2-4 = (y+2)(y-2)$

$\text{LCM} = (y-2)(y+1)(y+2)$

28. $\dfrac{x+6}{x+3} + \dfrac{9-4x}{x+3} = \dfrac{x+6+9-4x}{x+3} = \dfrac{-3x+15}{x+3}$

29. $\dfrac{6x-3}{x^2-x-12} - \dfrac{2x-15}{x^2-x-12}$

$= \dfrac{(6x-3)-(2x-15)}{x^2-x-12}$

$= \dfrac{6x-3+(-2x)+15}{x^2-x-12}$

$= \dfrac{4x+12}{x^2-x-12}$

$= \dfrac{4(x+3)}{(x-4)(x+3)}$

$= \dfrac{4}{x-4} \cdot \dfrac{x+3}{x+3}$

$= \dfrac{4}{x-4}$

30. $\dfrac{3x-1}{2x} - \dfrac{x-3}{x} \quad \text{LCD is } 2x$

$= \dfrac{3x-1}{2x} - \dfrac{x-3}{x} \cdot \dfrac{2}{2}$

$= \dfrac{3x-1}{2x} - \dfrac{2(x-3)}{2x}$

$= \dfrac{3x-1-2(x-3)}{2x}$

$= \dfrac{3x-1-2x+6}{2x} = \dfrac{x+5}{2x}$

31. $\dfrac{x+5}{x-2} - \dfrac{x}{2-x} = \dfrac{x+5}{x-2} - \dfrac{x}{-1(x-2)}$

$= \dfrac{x+5}{x-2} + \dfrac{x}{x-2}$

$= \dfrac{x+5+x}{x-2} = \dfrac{2x+5}{x-2}$

32. $\dfrac{2a}{a+1} - \dfrac{4a}{1-a^2} = \dfrac{2a}{a+1} - \dfrac{4a}{-1(a^2-1)}$

$= \dfrac{2a}{a+1} + \dfrac{4a}{(a^2-1)}$

$= \dfrac{2a}{a+1} + \dfrac{4a}{(a+1)(a-1)}$

$\text{LCD is } (a+1)(a-1)$

$= \dfrac{2a}{a+1} \cdot \dfrac{a-1}{a-1} + \dfrac{4a}{(a+1)(a-1)}$

$= \dfrac{2a(a-1)+4a}{(a+1)(a-1)}$

$= \dfrac{2a^2-2a+4a}{(a+1)(a-1)}$

$$= \frac{2a^2 + 2a}{(a+1)(a-1)}$$

$$= \frac{2a(a+1)}{(a+1)(a-1)}$$

$$= \frac{2a(\cancel{a+1})}{(\cancel{a+1})(a-1)}$$

$$= \frac{2a}{a-1}$$

33. $\dfrac{d^2}{d-2} + \dfrac{4}{2-d} = \dfrac{d^2}{d-2} - \dfrac{4}{d-2} = \dfrac{d^2-4}{d-2}$

$= \dfrac{(d+2)(\cancel{d-2})}{\cancel{d-2}} = d+2$ for $d \neq 2$

34. $\dfrac{1}{x^2-25} - \dfrac{x-5}{x^2-4x-5}$

$= \dfrac{1}{(x+5)(x-5)}$

$\quad - \dfrac{x-5}{(x-5)(x+1)}$

LCD is $(x+5)$

$\times (x-5)(x+1)$

$= \dfrac{1}{(x+5)(x-5)} \cdot \dfrac{x+1}{x+1}$

$\quad - \dfrac{x-5}{(x-5)(x+1)} \cdot \dfrac{x+5}{x+5}$

$= \dfrac{x+1}{(x+5)(x-5)(x+1)}$

$\quad - \dfrac{(x-5)(x+5)}{(x+5)(x-5)(x+1)}$

$= \dfrac{x+1 - (x^2-25)}{(x+5)(x-5)(x+1)}$

$= \dfrac{x+1 - x^2 + 25}{(x+5)(x-5)(x+1)}$

$= \dfrac{-x^2 + x + 26}{(x+5)(x-5)(x+1)}$

35. $\dfrac{2}{5x} + \dfrac{3}{2x+4} = \dfrac{2}{5x} + \dfrac{3}{2(x+2)}$

LCD is $5 \cdot 2 \cdot x \cdot (x+2)$

$= 10x(x+2)$

$= \dfrac{2}{5x} \cdot \dfrac{2(x+2)}{2(x+2)} + \dfrac{3}{2(x+2)} \cdot \dfrac{5x}{5x}$

$= \dfrac{4(x+2)}{10x(x+2)} + \dfrac{15x}{10x(x+2)}$

$= \dfrac{4x + 8 + 15x}{10x(x+2)}$

$= \dfrac{19x + 8}{10x(x+2)}$

36. $f(x) = \dfrac{14x^2 - x - 3}{2x^2 - 7x + 3}$

$2x^2 - 7x + 3 = 0$

$(2x-1)(x-3) = 0$

$2x - 1 = 0 \quad$ or $\quad x - 3 = 0$

$x = \dfrac{1}{2} \quad$ or $\qquad x = 3$

Thus $x \neq \frac{1}{2}$ and $x \neq 3$ since these values
make the denominator 0.

$f(x) = \dfrac{14x^2 - x - 3}{2x^2 - 7x + 3}$

$= \dfrac{(7x+3)(2x-1)}{(2x-1)(x-3)}$

$= \dfrac{(7x+3)(\cancel{2x-1})}{(\cancel{2x-1})(x-3)}$

$f(x) = \dfrac{7x+3}{x-3}; \quad x \neq \dfrac{1}{2}, \quad x \neq 3$

37. $f(x) = \dfrac{3x}{x+2} - \dfrac{x}{x-2} + \dfrac{8}{x^2-4}$

$= \dfrac{3x}{x+2} - \dfrac{x}{x-2} + \dfrac{8}{(x+2)(x-2)}$

Note: $x \neq -2;\ x \neq 2$

LCD is $(x+2)(x-2)$

$f(x) = \dfrac{3x}{x+2} \cdot \dfrac{x-2}{x-2} - \dfrac{x}{x-2} \cdot \dfrac{x+2}{x+2}$

$\qquad + \dfrac{8}{(x+2)(x-2)}$

$= \dfrac{3x(x-2)}{(x+2)(x-2)} - \dfrac{x(x+2)}{(x+2)(x-2)}$

$\qquad + \dfrac{8}{(x+2)(x-2)}$

$= \dfrac{3x^2 - 6x}{(x+2)(x-2)} - \dfrac{x^2 + 2x}{(x+2)(x-2)}$

$\qquad + \dfrac{8}{(x+2)(x-2)}$

$= \dfrac{3x^2 - 6x - x^2 - 2x + 8}{(x+2)(x-2)}$

$= \dfrac{2x^2 - 8x + 8}{(x+2)(x-2)}$

$= \dfrac{2(x-2)(x-2)}{(x+2)(x-2)}$

$= \dfrac{2(x-2)(\cancel{x-2})}{(x+2)(\cancel{x-2})}$

$f(x) = \dfrac{2(x-2)}{x+2},\ \ x \neq \pm 2$

38. $\dfrac{\dfrac{1}{z}+1}{\dfrac{1}{z^2}-1} = \dfrac{\dfrac{1}{z}+1}{\dfrac{1}{z^2}-1} \cdot \dfrac{z^2}{z^2} = \dfrac{z^2 \cdot \dfrac{1}{z} + z^2 \cdot 1}{z^2 \cdot \dfrac{1}{z^2} - z^2 \cdot 1}$

$= \dfrac{\cancel{z} \cdot z \cdot \dfrac{1}{\cancel{z}} + z^2 \cdot 1}{\cancel{z^2} \cdot \dfrac{1}{\cancel{z^2}} - z^2 \cdot 1} = \dfrac{z \cdot 1 + z^2 \cdot 1}{1 \cdot 1 - z^2 \cdot 1} = \dfrac{z + z^2}{1 - z^2}$

$= \dfrac{z(1+z)}{(1+z)(1-z)} = \dfrac{z}{1-z}$ for $z \neq -1$

39. $\dfrac{\dfrac{5}{2x^2}}{\dfrac{3}{4x} + \dfrac{4}{x^3}} = \dfrac{\dfrac{5}{2x^2}}{\dfrac{3}{4x} + \dfrac{4}{x^3}} \cdot \dfrac{4x^3}{4x^3} = \dfrac{4x^3 \cdot \dfrac{5}{2x^2}}{4x^3 \cdot \dfrac{3}{4x} + 4x^3 \cdot \dfrac{4}{x^3}}$

$= \dfrac{\cancel{2x^2} \cdot 2x \cdot \dfrac{5}{\cancel{2x^2}}}{\cancel{4x} \cdot x^2 \cdot \dfrac{3}{\cancel{4x}} + \cancel{x^3} \cdot 4 \cdot \dfrac{4}{\cancel{x^3}}} = \dfrac{2x \cdot 5}{x^2 \cdot 3 + 4 \cdot 4}$

$= \dfrac{10x}{3x^2 + 16}$

40. $\dfrac{\dfrac{y^2 + 4y - 77}{y^2 - 10y + 25}}{\dfrac{y^2 - 5y - 14}{y^2 - 25}}$

$= \dfrac{y^2 + 4y - 77}{y^2 - 10y + 25} \cdot \dfrac{y^2 - 25}{y^2 - 5y - 14}$

Multiplying by the reciprocal of the divisor

$= \dfrac{(y+11)(y-7)}{(y-5)^2} \cdot \dfrac{(y+5)(y-5)}{(y-7)(y+2)}$

$= \dfrac{(y+11)(y-7)(y+5)(y-5)}{(y-5)(y-5)(y-7)(y+2)}$

$= \dfrac{(y+11)\cancel{(y-7)}(y+5)\cancel{(y-5)}}{\cancel{(y-5)}(y-5)\cancel{(y-7)}(y+2)}$

$= \dfrac{(y+11)(y+5)}{(y-5)(y+2)}$

41. $\dfrac{\dfrac{5}{x^2-9}-\dfrac{3}{x+3}}{\dfrac{4}{x^2+6x+9}+\dfrac{2}{x-3}}$

$=\dfrac{\dfrac{5}{(x+3)(x-3)}-\dfrac{3}{x+3}}{\dfrac{4}{(x+3)^2}+\dfrac{2}{x-3}}$

$\left[\text{LCD is } (x+3)^2(x-3)\right]$

$=\dfrac{(x+3)^2(x-3)}{(x+3)^2(x-3)}\cdot\dfrac{\dfrac{5}{(x+3)(x-3)}-\dfrac{3}{x+3}}{\dfrac{4}{(x+3)^2}+\dfrac{2}{x-3}}$

$=\dfrac{(x+3)^2(x-3)\cdot\left[\dfrac{5}{(x+3)(x-3)}-\dfrac{3}{x+3}\right]}{(x+3)^2(x-3)\cdot\left[\dfrac{4}{(x+3)^2}+\dfrac{2}{x-3}\right]}$

$=\dfrac{5(x+3)-3(x+3)(x-3)}{4(x-3)+2(x+3)^2}$

$=\dfrac{5x+15-3(x^2-9)}{4x-12+2(x^2+6x+9)}$

$=\dfrac{5x+15-3x^2+27}{4x-12+2x^2+12x+18}$

$=\dfrac{-3x^2+5x+42}{2x^2+16x+6}$

We can factor the numerator and denominator, but doing so leads to no further simplification.

42. $\dfrac{3}{x}+\dfrac{7}{x}=5$ LCD is x; $x\neq 0$

$x\left(\dfrac{3}{x}+\dfrac{7}{x}\right)=x\cdot 5$ Multiplying both

$3+7=5x$ sides by x

$10=5x$

$2=x$

The solution is 2.

43. $\dfrac{5}{3x+2}=\dfrac{3}{2x}$

LCD is $2x(3x+2)$

$x\neq-\dfrac{3}{2}$ and $x\neq 0$

$2x(3x+2)\cdot\dfrac{5}{3x+2}=2x(3x+2)\cdot\dfrac{3}{2x}$

$2x\cdot 5=(3x+2)\cdot 3$

$10x=9x+6$

$x=6$

The solution is 6.

44. $x+\dfrac{6}{x}=-7\Rightarrow x\cdot x+\cancel{x}\cdot\dfrac{6}{\cancel{x}}=x\cdot(-7)$

$\Rightarrow x^2+6=-7x\Rightarrow x^2+7x+6=0$

$\Rightarrow (x+6)(x+1)=0\Rightarrow x=-6,-1$

Both solutions are valid.

45. $\dfrac{x+6}{x^2+x-6}+\dfrac{x}{x^2+4x+3}=\dfrac{x+2}{x^2-x-2}$

$\dfrac{x+6}{(x+3)(x-2)}+\dfrac{x}{(x+3)(x+1)}=$

$\dfrac{x+2}{(x-2)(x+1)}$

Note: $x\neq-3, x\neq 2, x\neq-1$

$\left[\text{LCD is } (x+3)(x-2)(x+1)\right]$

$(x+3)(x-2)(x+1)\left(\dfrac{x+6}{(x+3)(x-2)}+\right.$

$\left.\dfrac{x}{(x+3)(x+1)}\right)=$

$(x+3)(x-2)(x+1)\cdot\dfrac{x+2}{(x-2)(x+1)}$

$(x+1)(x+6)+(x-2)x=(x+3)(x+2)$

$x^2+7x+6+x^2-2x=x^2+5x+6$

$x^2=0$

$x=0$

46. $f(x) = \dfrac{2}{x-1} + \dfrac{2}{x+2}$

$f(a) = \dfrac{2}{a-1} + \dfrac{2}{a+2}$ Substitute a for x

$1 = \dfrac{2}{a-1} + \dfrac{2}{a+2}$ Substitute 1 for $f(a)$

Note: $a \neq 1$ and $a \neq -2$

LCD is $(a-1)(a+2)$

$(a-1)(a+2) \cdot 1 = (a-1)(a+2)\left(\dfrac{2}{a-1} + \dfrac{2}{a+2} \right)$

$a^2 + a - 2 = 2(a+2) + 2(a-1)$

$a^2 + a - 2 = 2a + 4 + 2a - 2$

$a^2 - 3a - 4 = 0$

$(a-4)(a+1) = 0$

$a - 4 = 0 \ or \ a + 1 = 0$

$a = 4 \ or \quad a = -1$

The solutions are -1 and 4.

47. **Familiarize.** Megan can do $\dfrac{1}{9}$ of the job in 1

hour, and Kelly can do $\dfrac{1}{12}$ of the same job in

1 hour. Let $t =$ time worked together.
Translate. Rate $\times$ Time = Work. We have
the equation.

$\dfrac{1}{9} \cdot t + \dfrac{1}{12} \cdot t = 1$, or $\dfrac{t}{9} + \dfrac{t}{12} = 1$

Carry out. Solve the equation.

$\dfrac{t}{9} + \dfrac{t}{12} = 1$, LCD = 36

$36\left(\dfrac{t}{9} + \dfrac{t}{12} \right) = 36 \cdot 1$

$36 \cdot \dfrac{t}{9} + 36 \cdot \dfrac{t}{12} = 36 \cdot 1$

$4t + 3t = 36$

$7t = 36$

$t = \dfrac{36}{7}$, or $5\dfrac{1}{7}$

Check. Megan's work is $\dfrac{1}{9} \cdot \dfrac{36}{7} = \dfrac{4}{7}$, and

Kelly's work is $\dfrac{1}{12} \cdot \dfrac{36}{7} = \dfrac{3}{7}$. Together they

do $\dfrac{4}{7} + \dfrac{3}{7}$, or 1 entire job. The number

checks.

State. Megan and Kelly, working together

can arrange the books in $5\dfrac{1}{7}$ hrs.

48. **Familiarize.** Let $t =$ the time it takes the Core
2 Quad to process a data file. Then $t + 15 =$
the time it takes the Core 2 Duo to process a
data file.

Translate. Using $R = \dfrac{W}{T}$, the rate of the

Core 2 Quad is $\dfrac{1}{t}$ and the rate of the Core 2

Duo is $\dfrac{1}{t+15}$. Knowing it takes them 18 sec

when working together, we write the
equation.

$\dfrac{1}{t} \cdot 18 + \dfrac{1}{t+15} \cdot 18 = 1$, or $\dfrac{18}{t} + \dfrac{18}{t+15} = 1$

Carry out. Solve the equation.

$\dfrac{18}{t} + \dfrac{18}{t+15} = 1$, LCD is $t(t+15)$

$t(t+15)\left(\dfrac{18}{t} + \dfrac{18}{t+15} \right) = t(t+15) \cdot 1$

$18(t+15) + 18t = t(t+15)$

$18t + 270 + 18t = t^2 + 15t$

$0 = t^2 - 21t - 270$

$0 = (t-30)(t+9)$

$t - 30 = 0 \ or \ t + 9 = 0$

$t = 30 \ or \ t = -9$

Check. Since time cannot be negative, use
only $t = 30$. The Core 2 Quad does

$\dfrac{18}{30}$, or $\dfrac{3}{5}$ of the job, and the Core 2 Duo

does $\dfrac{18}{30+15} = \dfrac{18}{45}$, or $\dfrac{2}{5}$ of the job.

Together they do $\dfrac{3}{5} + \dfrac{2}{5}$, or one complete job.

State. The Core 2 Quad can process the data
file in 30 seconds and the Core 2 Duo takes
45 seconds.

49. **Familiarize.** The rate of the current is 6 mph.
Let $R =$ the speed of the boat in still water.
The rate of the boat going downstream, with
the current, is $R + 6$, and the rate of the boat
going upstream, against the current, is $R - 6$.

Translate. Using the fact that the times are the same and $T = \dfrac{D}{R}$, we have the equation

$$\frac{50}{R+6} = \frac{30}{R-6}$$

Carry out. We solve the equation.

$$\frac{50}{R+6} = \frac{30}{R-6}, \text{ LCD is } (R+6)(R-6)$$

$$(R+6)(R-6) \cdot \frac{50}{R+6} = (R+6)(R-6) \cdot \frac{30}{R-6}$$

$$50(R-6) = 30(R+6)$$

$$50R - 300 = 30R + 180$$

$$20R = 480$$

$$R = 24 \text{ mph}$$

Check. The time downstream is

$$\frac{50}{24+6} = \frac{50}{30}, \text{ or } \frac{5}{3} \text{ and the time upstream is}$$

$$\frac{30}{24-6} = \frac{30}{18}, \text{ or } \frac{5}{3}. \text{ The times are equal; the}$$

number checks.

State. The speed of the boat in still water is 24 mph.

50. *Familiarize.* Let R = speed of the motorcycle and $R + 8$ = speed of the car. The times are equal. $T = \dfrac{D}{R}$

Translate. The time of the car is $\dfrac{105}{R+8}$, and

the time of the motorcycle is $\dfrac{93}{R}$. Since the

times are the same, we have the equation.

$$\frac{105}{R+8} = \frac{93}{R}$$

Carry out. Solve the equation.

$$\text{LCD} = (R+8)R$$

$$(R+8)R \cdot \frac{105}{R+8} = (R+8)R \cdot \frac{93}{R}$$

$$105R = 93R + 744$$

$$12R = 744$$

$$R = 62$$

Check. The time of the car is

$$\frac{105}{62+8} = \frac{105}{70}, \text{ or } \frac{3}{2}; \text{ the time of the}$$

motorcycle is $\dfrac{93}{62}$, or $\dfrac{3}{2}$. The times are

equal; the number checks.

State. The speed of the motorcycle is 62 mph, and the speed of the car is 70 mph.

51. The ratio of seals tagged to the total number of seals in the pond S is $\dfrac{33}{S}$. Of the 40 seals sampled later, 24 were tagged. The ratio of tagged seals to seals caught is $\dfrac{24}{40} = \dfrac{3}{5}$.

Therefore:

$$\frac{33}{S} = \frac{3}{5} \Rightarrow \frac{S}{33} = \frac{5}{3}$$

$$S = \frac{5}{3} \cdot 33 = \frac{5}{\cancel{3}} \cdot \cancel{3} \cdot 11 = 55 \text{ seals}$$

52. We write a proportion and then solve it.

$$\frac{x}{2.4} = \frac{8.5}{3.4}$$

$$x = 2.4 \cdot \frac{8.5}{3.4}$$

$$x = 6$$

53.
$$R = \frac{gs}{g+s}, \text{ LCD is } g+s$$

$$(g+s)R = (g+s) \cdot \frac{gs}{g+s}$$

$$gR + sR = gs$$

$$Rg = gs - sR$$

$$Rg = s(g-R)$$

$$\frac{Rg}{g-R} = s$$

54. $S = \dfrac{H}{m(t_1 - t_2)}$

$$m \cdot S = m \left[\frac{H}{m(t_1 - t_2)} \right] \quad \begin{array}{l}\text{Multiplying both}\\ \text{sides of the}\\ \text{equation by } m\end{array}$$

$$mS = \frac{H}{t_1 - t_2}$$

$$\frac{1}{S} \cdot mS = \frac{1}{S} \cdot \frac{H}{t_1 - t_2} \quad \begin{array}{l}\text{Multiplying both}\\ \text{sides of the}\\ \text{equation by } \dfrac{1}{S}\end{array}$$

$$m = \frac{H}{S(t_1 - t_2)}$$

55. $\dfrac{1}{ac} = \dfrac{2}{ab} - \dfrac{3}{bc}$, LCD is abc

$$abc \cdot \dfrac{1}{ac} = abc\left(\dfrac{2}{ab} - \dfrac{3}{bc}\right)$$

$$\dfrac{abc}{ac} = \dfrac{abc \cdot 2}{ab} - \dfrac{abc \cdot 3}{bc}$$

$$b = 2c - 3a$$

$$b + 3a = 2c$$

$$\dfrac{b + 3a}{2} = c$$

56. $T = \dfrac{A}{v(t_2 - t_1)}$

$$(t_2 - t_1)T = (t_2 - t_1)\dfrac{A}{v(t_2 - t_1)}$$

$$Tt_2 - Tt_1 = \dfrac{A}{v}$$

$$-Tt_1 = \dfrac{A}{v} - Tt_2$$

$$-\dfrac{1}{T} \cdot -Tt_1 = -\dfrac{1}{T}\left(\dfrac{A}{v} - Tt_2\right)$$

$$t_1 = -\dfrac{A}{vT} + t_2, \text{ or } \dfrac{-A + vTt_2}{vT}$$

57. W varies directly as P, where W is the pounds of waste generated daily, and P is the number of people in the household.

$$W = kP$$

$$\dfrac{W}{P} = k$$

$$\dfrac{W_1}{P_1} = \dfrac{W_2}{P_2}$$

$$W_2 = \dfrac{W_1}{P_1} \cdot P_2 = \dfrac{11.96}{2.6} \cdot 5 = 23$$

At the same rate a family of 5 would generate 23 pounds of waste daily.

58.   The volume V of the dye used varies directly as the square of the diameter, or d^2. We write $V = kd^2$. Using $V = 4$ and $d = 10$, we substitute to determine k and the equation of variation.

$$4 = k \cdot 10^2$$

$$.04 = k \qquad \text{Variation constant}$$

$$V = .04d^2$$

Let $d = 40$ and determine V.

$$V = .04 \cdot 40^2$$

$$V = 64L$$

64 L of dye is needed for 40-m wide circle.

59. $y = \dfrac{k}{x}$

Substitute $y = 3$ and $x = \dfrac{1}{4}$ to determine k.

$$3 = \dfrac{k}{\dfrac{1}{4}}$$

$$\dfrac{3}{4} = k \qquad \text{Variation constant}$$

$$y = \dfrac{\dfrac{3}{4}}{x}$$

$$y = \dfrac{3}{4x} \qquad \text{Equation of variation}$$

60.   a.   Inverse

     b.   Using $(12, 2)$, $x = 12$ and $y = 2$.

$$y = \dfrac{k}{x}$$

$$2 = \dfrac{k}{12}$$

$$24 = k \qquad \text{Variation constant}$$

$$y = \dfrac{24}{x} \qquad \text{Equation of variation}$$

     c. $x = 8$

$$y = \dfrac{24}{8}$$

$$y = 3 \text{ oz}$$

61.   The LCD was used to add and subtract rational expressions, to simplify complex rational expressions, and to solve rational equations.

62.   A rational expression is a quotient of two polynomials and can be simplified, multiplied, or added, but cannot be solved for the variable. A rational equation is an equality containing rational expressions, and we can solve for a variable.

63. $\dfrac{5}{x-13} - \dfrac{5}{x} = \dfrac{65}{x^2 - 13x}$

$$\dfrac{5}{x-13} - \dfrac{5}{x} = \dfrac{65}{x(x-13)}$$

Note: $x \neq 13$ and $x \neq 0$

$$\left[\text{LCD is } x(x-13)\right]$$

$$x(x-13)\left(\dfrac{5}{x-13} - \dfrac{5}{x}\right) = x(x-13) \cdot \dfrac{65}{x(x-13)}$$

$$5x - 5(x-13) = 65$$

$$5x - 5x + 65 = 65$$

$$65 = 65$$

Since $65 = 65$ (Reflexive Property of Equality), we have an identity. We must remember to exclude the values which make the denominators equal zero; i.e., the solution set is the domain $\{x \mid x \text{ is a real number, and } x \neq 0 \text{ and } x \neq 13\}$.

64. $\dfrac{\dfrac{x}{x^2-25} + \dfrac{2}{x-5}}{\dfrac{3}{x-5} - \dfrac{4}{x^2-10x+25}} = 1$

Note: $x \neq \pm 5$

$$\dfrac{x}{x^2-25} + \dfrac{2}{x-5} = \dfrac{3}{x-5} - \dfrac{4}{x^2-10x+25}$$

Multiplying both sides of the equation by the denominator

$$\left[\text{LCD is } (x-5)^2(x+5)\right]$$

$$(x-5)^2(x+5)\left[\dfrac{x}{(x-5)(x+5)} + \dfrac{2}{x-5}\right] =$$

$$(x-5)^2(x+5)\left[\dfrac{3}{x-5} - \dfrac{4}{(x-5)^2}\right]$$

$$x(x-5) + 2(x-5)(x+5) =$$
$$\qquad 3(x-5)(x+5) - 4(x+5)$$

$$x^2 - 5x + 2x^2 - 50 = 3x^2 - 75 - 4x - 20$$

$$45 = x$$

The solution is 45.

65. $\dfrac{2a^2+5a-3}{a^2} \cdot \dfrac{5a^3+30a^2}{2a^2+7a-4} \div \dfrac{a^2+6a}{a^2+7a+12}$

$$= \dfrac{2a^2+5a-3}{a^2} \cdot \dfrac{5a^3+30a^2}{2a^2+7a-4} \cdot \dfrac{a^2+7a+12}{a^2+6a}$$

$$= \dfrac{(2a-1)(a+3)}{a^2} \cdot \dfrac{5a^2(a+6)}{(2a-1)(a+4)} \cdot \dfrac{(a+3)(a+4)}{a(a+6)}$$

$$= \dfrac{5(a+3)^2}{a} \quad \left(\text{for } a \neq -6, -4, -3, \tfrac{1}{2}\right)$$

66. $\dfrac{5(x-y)}{(x-y)(x+2y)} - \dfrac{5(x-3y)}{(x+2y)(x-3y)}$

$$= \dfrac{5(x-y)}{(x-y)(x+2y)} - \dfrac{5(x-3y)}{(x+2y)(x-3y)}$$

$$= \dfrac{5}{x+2y} - \dfrac{5}{x+2y} = 0 \quad \left(\text{for } x \neq y, x \neq 3y\right)$$

Chapter 7 Test

1. $\dfrac{2-x}{5x}$

We find the number which makes the denominator 0.

$$5x = 0$$

$$x = 0$$

The expression is undefined for $x = 0$.

2. $\dfrac{x^2+x-30}{x^2-3x+2}$

We find the numbers which make the denominator 0.

$$x^2 - 3x + 2 = 0$$

$$(x-2)(x-1) = 0$$

$$x - 2 = 0 \quad \text{or} \quad x - 1 = 0$$

$$x = 2 \quad \text{or} \qquad x = 1$$

The expression is undefined for $x = 1$ and $x = 2$.

3. $\dfrac{6x^2-7x-5}{3x^2-2x-5} = \dfrac{(2x+1)(3x-5)}{(x+1)(3x-5)}$

$$= \dfrac{2x+1}{x+1} \quad \left(\text{for } x \neq \tfrac{5}{3}\right)$$

4. $\dfrac{a^2-25}{9a}\cdot\dfrac{6a}{5-a}=\dfrac{(a+5)(a-5)}{9a}\cdot\dfrac{6a}{-1(a-5)}$

$=\dfrac{(a+5)(a-5)\cdot2\cdot3\cdot a}{3\cdot3\cdot a\cdot(-1)(a-5)}$

$=\dfrac{(a+5)(a-5)\cdot2\cdot3\cdot a}{3\cdot3\cdot a(-1)(a-5)}$

$=\dfrac{2(a+5)}{-3}=\dfrac{-2(a+5)}{3}$

5. $\dfrac{25y^2-1}{9y^2-6y}\div\dfrac{5y^2+9y-2}{3y^2+y-2}$

$=\dfrac{25y^2-1}{9y^2-6y}\cdot\dfrac{3y^2+y-2}{5y^2+9y-2}$

$=\dfrac{(5y+1)(5y-1)}{3y(3y-2)}\cdot\dfrac{(3y-2)(y+1)}{(5y-1)(y+2)}$

$=\dfrac{(5y+1)(5y-1)(3y-2)(y+1)}{3y(3y-2)(5y-1)(y+2)}$

$=\dfrac{(5y+1)(y+1)}{3y(y+2)}$

6. $\dfrac{4x^2-1}{x^2-2x+1}\div\dfrac{x-2}{x^2+1}=\dfrac{4x^2-1}{x^2-2x+1}\cdot\dfrac{x^2+1}{x-2}$

$=\dfrac{(2x+1)(2x-1)}{(x-1)^2}\cdot\dfrac{x^2+1}{x-2}$

$=\dfrac{(2x+1)(2x-1)(x^2+1)}{(x-1)^2(x-2)}$

7. $(x^2+6x+9)\cdot\dfrac{(x-3)^2}{x^2-9}$

$=\dfrac{(x+3)(x+3)}{1}\cdot\dfrac{(x-3)(x-3)}{(x+3)(x-3)}$

$=\dfrac{(x+3)(x+3)(x-3)(x-3)}{(x+3)(x-3)}$

$=(x+3)(x-3)$

8. $y^2-9,\ y^2+10y+21,\ y^2+4y-21$

$y^2-9=(y+3)(y-3)$

$y^2+10y+21=(y+7)(y+3)$

$y^2+4y-21=(y+7)(y-3)$

$\text{LCM}=(y+3)(y-3)(y+7)$

9. $\dfrac{2+x}{x^3}+\dfrac{7-4x}{x^3}=\dfrac{2+x+7-4x}{x^3}=\dfrac{-3x+9}{x^3}$

10. $\dfrac{5-t}{t^2+1}-\dfrac{t-3}{t^2+1}=\dfrac{5-t-(t-3)}{t^2+1}$

$=\dfrac{5-t-t+3}{t^2+1}$

$=\dfrac{-2t+8}{t^2+1}$

11. $\dfrac{x-4}{x-3}+\dfrac{x-1}{3-x}=\dfrac{x-4}{-1(3-x)}+\dfrac{x-1}{3-x}$

$=\dfrac{-(x-4)}{3-x}+\dfrac{x-1}{3-x}$

$=\dfrac{-x+4+x-1}{3-x}$

$=\dfrac{3}{3-x}\ \text{or}\ -\dfrac{3}{x-3}$

12. $\dfrac{x-4}{x-3}-\dfrac{x-1}{3-x}=\dfrac{x-4}{x-3}-\dfrac{x-1}{-1(x-3)}$

$=\dfrac{x-4}{x-3}+\dfrac{x-1}{x-3}$

$=\dfrac{x-4+x-1}{x-3}$

$=\dfrac{2x-5}{x-3}$

13. $\dfrac{7}{t-2}+\dfrac{4}{t}\quad\text{LCD is }t(t-2)$

$=\dfrac{7}{t-2}\cdot\dfrac{t}{t}+\dfrac{4}{t}\cdot\dfrac{t-2}{t-2}$

$=\dfrac{7t}{t(t-2)}+\dfrac{4(t-2)}{t(t-2)}$

$=\dfrac{7t+4t-8}{t(t-2)}$

$=\dfrac{11t-8}{t(t-2)}$

14. $\dfrac{4}{x^2-16} - \dfrac{x-1}{x^2+5x+4}$

$= \dfrac{4}{(x+4)(x-4)} - \dfrac{x-1}{(x+4)(x+1)}$

$= \dfrac{4}{(x+4)(x-4)} \cdot \dfrac{x+1}{x+1} - \dfrac{x-1}{(x+4)(x+1)} \cdot \dfrac{x-4}{x-4}$

$= \dfrac{4(x+1)-(x-1)(x-4)}{(x+1)(x+4)(x-4)} = \dfrac{4x+4-(x^2-5x+4)}{(x+1)(x+4)(x-4)}$

$= \dfrac{4x\cancel{+4}-x^2+5x\cancel{-4}}{(x+1)(x+4)(x-4)} = \dfrac{-x^2+9x}{(x+1)(x+4)(x-4)}$

The numerator can be factored, but the expression cannot be reduced.

15. $f(x) = \dfrac{6x^2+17x+7}{2x^2+7x+3}$

Since $2x^2+7x+3 = (2x+1)(x+3)$, the

denominator is zero if $x=-3$ or $x=-\tfrac{1}{2}$.

$f(x) = \dfrac{(3x+7)(2x+1)}{(2x+1)(x+3)}$

$= \dfrac{(3x+7)\cancel{(2x+1)}}{\cancel{(2x+1)}(x+3)}$

$f(x) = \dfrac{3x+7}{x+3}$ $x \ne -3$ and $x \ne -\tfrac{1}{2}$

16. $f(x) = \dfrac{4}{x+3} - \dfrac{x}{x-2} + \dfrac{x^2+4}{x^2+x-6}$

$= \dfrac{4}{x+3} - \dfrac{x}{x-2} + \dfrac{x^2+4}{(x+3)(x-2)}$

Note: $x \ne -3$ and $x \ne 2$.

LCD is $(x+3)(x-2)$.

$f(x) = \dfrac{4}{x+3} \cdot \dfrac{x-2}{x-2} - \dfrac{x}{x-2} \cdot \dfrac{x+3}{x+3}$

$+ \dfrac{x^2+4}{(x+3)(x-2)}$

$= \dfrac{4(x-2)}{(x+3)(x-2)} - \dfrac{x(x+3)}{(x-2)(x+3)}$

$+ \dfrac{x^2+4}{(x+3)(x-2)}$

$= \dfrac{4x-8-(x^2+3x)+x^2+4}{(x+3)(x-2)}$

$= \dfrac{4x-8-x^2-3x+x^2+4}{(x+3)(x-2)}$

$= \dfrac{x-4}{(x+3)(x-2)}$, $x \ne -3$ and $x \ne 2$.

17. $\dfrac{9-\dfrac{1}{y^2}}{3-\dfrac{1}{y}} = \dfrac{9-\dfrac{1}{y^2}}{3-\dfrac{1}{y}} \cdot \dfrac{y^2}{y^2} = \dfrac{y^2 \cdot 9 - y^2 \cdot \dfrac{1}{y^2}}{y^2 \cdot 3 - y^2 \cdot \dfrac{1}{y}}$

$= \dfrac{y^2 \cdot 9 - \cancel{y^2} \cdot \dfrac{1}{\cancel{y^2}}}{y^2 \cdot 3 - \cancel{y} \cdot y \cdot \dfrac{1}{\cancel{y}}} = \dfrac{y^2 \cdot 9 - 1 \cdot 1}{y^2 \cdot 3 - y \cdot 1} = \dfrac{9y^2-1}{3y^2-y}$

$= \dfrac{(3y+1)(3y-1)}{y(3y-1)} = \dfrac{3y+1}{y}$ for $y \ne \dfrac{1}{3}$

18. $\dfrac{\dfrac{x}{8}-\dfrac{8}{x}}{\dfrac{1}{8}+\dfrac{1}{x}} = \dfrac{\dfrac{x}{8}-\dfrac{8}{x}}{\dfrac{1}{8}+\dfrac{1}{x}} \cdot \dfrac{8x}{8x} = \dfrac{8x \cdot \dfrac{x}{8} - 8x \cdot \dfrac{8}{x}}{8x \cdot \dfrac{1}{8} + 8x \cdot \dfrac{1}{x}}$

$= \dfrac{x \cdot x - 8 \cdot 8}{x \cdot 1 + 8 \cdot 1} = \dfrac{x^2-64}{x+8}$

$= \dfrac{(x+8)(x-8)}{x+8} = x-8$ for $x \ne -8, 0$

19. $\dfrac{\dfrac{x^2-5x-36}{x^2-36}}{\dfrac{x^2+x-12}{x^2-12x+36}}$

$= \dfrac{x^2-5x-36}{x^2-36} \div \dfrac{x^2+x-12}{x^2-12x+36}$

$= \dfrac{x^2-5x-36}{x^2-36} \cdot \dfrac{x^2-12x+36}{x^2+x-12}$

$= \dfrac{(x-9)(x+4)}{(x+6)(x-6)} \cdot \dfrac{(x-6)^2}{(x+4)(x-3)}$

$= \dfrac{(x-9)\,\cancel{(x+4)}\,\cancel{(x-6)}\,(x-6)}{(x+6)\,\cancel{(x-6)}\,\cancel{(x+4)}\,(x-3)}$

$= \dfrac{(x-9)(x-6)}{(x+6)(x-3)}$

20. $\dfrac{1}{t}+\dfrac{1}{3t}=\dfrac{1}{2} \Rightarrow 6t\cdot\dfrac{1}{t}+6t\cdot\dfrac{1}{3t}=6t\cdot\dfrac{1}{2}$

$\Rightarrow \cancel{t}\cdot 6\cdot\dfrac{1}{\cancel{t}}+\cancel{3}t\cdot 2\cdot\dfrac{1}{\cancel{3t}}=\cancel{2}\cdot 3t\cdot\dfrac{1}{\cancel{2}}$

$\Rightarrow 6+2=3t \Rightarrow 3t=8 \Rightarrow t=\dfrac{8}{3}$

21. $\dfrac{4}{2x-5}=\dfrac{6}{5x+3}$ Note: $x\neq\dfrac{5}{2},-\dfrac{3}{5}$

$\big[\text{LCD is }(2x-5)(5x+3)\big]$

$(2x-5)(5x+3)\cdot\dfrac{4}{2x-5}=$

$\qquad\qquad (2x-5)(5x+3)\cdot\dfrac{6}{5x+3}$

$4(5x+3)=6(2x-5)$

$20x+12=12x-30$

$8x=-42$

$x=-\dfrac{42}{8}=-\dfrac{21}{4}$

The solution is $-\dfrac{21}{4}$.

22. $\dfrac{15}{x}-\dfrac{15}{x-2}=-2$

$\Rightarrow \cancel{x}(x-2)\cdot\dfrac{15}{\cancel{x}}-x\cancel{(x-2)}\cdot\dfrac{15}{\cancel{x-2}}$

$= x(x-2)\cdot(-2)$

$\Rightarrow 15(x-2)-15x=-2x(x-2)$

$\Rightarrow \cancel{15x}-30\cancel{-15x}=-2x^2+4x$

$\Rightarrow 2x^2-4x-30=0 \Rightarrow x^2-2x-15=0$

$\Rightarrow (x-5)(x+3)=0 \Rightarrow x=-3,5$

Both solutions are valid.

23. $f(x)=\dfrac{x+3}{x-1}$

$f(2)=\dfrac{2+3}{2-1}=\dfrac{5}{1}=5$

$f(-3)=\dfrac{-3+3}{-3-1}=\dfrac{0}{-4}=0$

24. $f(a)=\dfrac{a+3}{a-1},$ Note: $a\neq 1$

$7=\dfrac{a+3}{a-1},$ LCD is $a-1$

$(a-1)\cdot 7=(a-1)\cdot\dfrac{a+3}{a-1}$

$7a-7=a+3$

$6a=10$

$a=\dfrac{10}{6}=\dfrac{5}{3}$

$f\left(\dfrac{5}{3}\right)=7$

25. $A=\dfrac{h(b_1+b_2)}{2}$

$2A=h(b_1+b_2)$

$\dfrac{2A}{h}=b_1+b_2$

$\dfrac{2A}{h}-b_2=b_1$

$b_1=\dfrac{2A}{h}-b_2,$ or $\dfrac{2A-b_2h}{h}$

26. Let r = speed of the wind. Emma's speed with the wind is $12 + r$, and her speed against the wind is $12 - r$. Since the times are the same, we solve for time $\left(T = \dfrac{D}{R}\right)$ and write the equation.

$$\frac{14}{12+r} = \frac{8}{12-r}, \text{ LCD is } (12+r)(12-r)$$

$$(12+r)(12-r) \cdot \frac{14}{12+r} =$$

$$\qquad (12+r)(12-r) \cdot \frac{8}{12-r}$$

$$14(12-r) = 8(12+r)$$

$$168 - 14r = 96 + 8r$$

$$72 = 22r$$

$$\frac{72}{22} = r$$

$$\frac{36}{11} = r$$

With the wind, $t = \dfrac{14}{12 + \dfrac{36}{11}} = \dfrac{14}{\dfrac{168}{11}} = \dfrac{154}{168} = \dfrac{11}{12}$

Against the wind,

$t = \dfrac{8}{12 - \dfrac{36}{11}} = \dfrac{8}{\dfrac{96}{11}} = \dfrac{88}{96} = \dfrac{11}{12}$

Since the times are equal, the number checks. The speed of the wind is

$\dfrac{36}{11}$ mph, or $3\dfrac{3}{11}$ mph.

27. Kyla can do $\dfrac{1}{3.5}$, or $\dfrac{2}{7}$ of the job in one hour, and Brock can do $\dfrac{1}{4.5}$, or $\dfrac{2}{9}$ of the job in one hour.

Let t = time together and write the equation.

$t \cdot \dfrac{2}{7} + t \cdot \dfrac{2}{9} = 1$, or $\dfrac{2t}{7} + \dfrac{2t}{9} = 1$

Solve the equation. LCD is 63.

$$63\left(\frac{2t}{7} + \frac{2t}{9}\right) = 63 \cdot 1$$

$$18t + 14t = 63$$

$$32t = 63$$

$$t = \frac{63}{32} \text{ hr.}$$

Kyla: $\dfrac{2}{7} \cdot \dfrac{63}{32} = \dfrac{9}{16}$

Brock: $\dfrac{2}{9} \cdot \dfrac{63}{32} = \dfrac{7}{16}$

Together: $\dfrac{9}{16} + \dfrac{7}{16}$, or one entire job.

Working together, they can install the vinyl in $\dfrac{63}{32}$ hr., or $1\dfrac{31}{32}$ hr.

28. Let t_P = the time in hours it would take Pe'rez to mulch the flower beds working alone, and t_E = the time in hours it would take Ellia to mulch the flower beds working alone. Then:

$$t_P = t_E + 6 \quad (1)$$

$$\left(\frac{1}{t_P} + \frac{1}{t_E}\right)\left(\frac{20}{7}\right) = 1 \quad (2)$$

Substitute the expression for t_P in (1) into (2) and solve for t_E.

$$\left(\frac{1}{t_P} + \frac{1}{t_E}\right)\left(\frac{20}{7}\right) = 1 \quad (2)$$

$$\Rightarrow \frac{1}{t_P} + \frac{1}{t_E} = \frac{7}{20} \Rightarrow \frac{1}{t_E + 6} + \frac{1}{t_E} = \frac{7}{20}$$

$$\Rightarrow 20t_E(t_E + 6)\left[\frac{1}{t_E + 6} + \frac{1}{t_E}\right]$$

$$= 20t_E(t_E + 6)\left[\frac{7}{20}\right]$$

$$\Rightarrow 20t_E + 20(t_E + 6) = 7t_E(t_E + 6)$$

$$\Rightarrow 7t_E{}^2 + 2t_E - 120 = 0$$

$$\Rightarrow (7t_E + 30)(t_E - 4)$$

$$\Rightarrow t_E = 4 \text{ since } t_E > 0.$$

Use (1) and the value for t_E to find t_P.

$$t_P = t_E + 6 \quad (1)$$

$$\Rightarrow t_P = 4 + 6 = 10$$

It takes Ellia 4 hr working alone and Pe'rez 10 hr.

29. *Familiarize.* $3\frac{1}{2}c$ of whole wheat flour are used with $1\frac{1}{4}c$ of warm water. This gives us the ratio $\dfrac{3\frac{1}{2}}{1\frac{1}{4}}$. We want to determine the amount of water, w, to be used with $6c$ of

whole wheat flour. We have a second ratio, $\frac{6}{w}$.

Translate. Assuming the two ratios are equal, we have a proportion.

$$\frac{\text{Flour}}{\text{Water}} \quad \frac{3\frac{1}{2}}{1\frac{1}{4}} = \frac{6}{w}$$

Carry out. We solve the proportion.

$$\frac{3\frac{1}{2}}{1\frac{1}{4}} = \frac{6}{w}$$

$$1\frac{1}{4} w \cdot \frac{3\frac{1}{2}}{1\frac{1}{4}} = 1\frac{1}{4} w \cdot \frac{6}{w}$$

$$3\frac{1}{2} w = 1\frac{1}{4} \cdot 6$$

$$\frac{7}{2} w = \frac{15}{2}$$

$$\frac{2}{7} \cdot \frac{7}{2} w = \frac{2}{7} \cdot \frac{15}{2}$$

$$w = \frac{15}{7}, \quad \text{or} \quad 2\frac{1}{7}$$

Check. $\frac{3\frac{1}{2}}{1\frac{1}{4}} = 2.8 \quad \frac{6}{2\frac{1}{7}} = 2.8$

The ratios are the same.

State. $2\frac{1}{7}c$ of warm water should be used with $6c$ of whole wheat flour.

30. N varies inversely as t, so $N = \frac{k}{t}$.

Let $N = 25$ and $t = 6$, and determine k, the variation constant and the equation of variation.

$$25 = \frac{k}{6}$$

$$150 = k \qquad \text{Variation constant}$$

$$N = \frac{150}{t} \qquad \text{Equation of variation}$$

Let $t = 5$ and solve for N.

$$N = \frac{150}{5}$$

$$N = 30$$

It will take 30 workers to clean the stadium in 5 hours.

31. Surface area, SA varies directly as the square of the radius, R^2.

The equation is $SA = kR^2$

Let $SA = 325$ and $R = 5$, and determine the variation constant and the equation of variation.

$$325 = k \cdot 5^2$$

$$13 = k \qquad \text{Variation constant}$$

$$SA = 13r^2 \qquad \text{Equation of variation}$$

Let $r = 7$ to determine SA for 7 in.

$$SA = 13 \cdot 7^2$$

$$SA = 13 \cdot 49$$

$$SA = 637$$

When the radius is 7 in., the surface area of the balloon is 637 in^2.

32.
$$\frac{6}{x-15} - \frac{6}{x} = \frac{90}{x^2 - 15}$$

Note: $x \neq 15$, $x \neq 0$.

$$\left[\text{LCD is } x(x-15) \right]$$

$$x(x-15)\left(\frac{6}{x-15} - \frac{6}{x} \right) = x(x-15) \cdot \frac{90}{x^2 - 15}$$

$$6x - 6(x-15) = 90$$

$$6x - 6x + 90 = 90$$

$$90 = 90$$

Since $90 = 90$ (Reflexive Property of Equality), we have an identity. We must remember to exclude the values which make the denominators equal zero. i.e., the solution is the domain.

$$\{ x | x \text{ is a real number, and } x \neq 0 \text{ and } x \neq 15 \}$$

33. $1 - \dfrac{1}{1 - \dfrac{1}{1 - \dfrac{1}{a}}} = 1 - \dfrac{1}{1 - \dfrac{1}{1 - \dfrac{1}{a}} \cdot \dfrac{a}{a}} = 1 - \dfrac{1}{1 - \dfrac{a}{a-1}}$

$= 1 - \dfrac{1}{1 - \dfrac{a}{a-1}} \cdot \dfrac{a-1}{a-1} = 1 - \dfrac{a-1}{\cancel{a-1} \cancel{-a}} = 1 - \dfrac{a-1}{-1}$

$= 1 + \dfrac{a-1}{1} = 1 + a - 1 = a \text{ for } a \neq 0,1$

34. The ratio of the number of lawns Andy mowed to the number of lawns Chad mowed is $\dfrac{4}{3}$. Let $x =$ the number of lawns Andy mowed, then $98 - x =$ number of lawns Chad mowed. Set the ratios equal to one another and solve for x.

$$\frac{4}{3} = \frac{x}{98 - x}, \text{ LCD is } 3(98 - x)$$

$$3(98 - x) \cdot \frac{4}{3} = 3(98 - x) \cdot \frac{x}{98 - x}$$

$$4(98 - x) = 3x$$

$$392 - 4x = 3x$$

$$392 = 7x$$

$$56 = x$$

Substituting, $x = 56,$ so $98 - 56 = 42.$

$\dfrac{56}{42} = \dfrac{4}{3}$ and $56 + 42 = 98,$ the number checks. Andy mowed 56 lawns and Chad mowed 42 lawns.

Chapter 8

Inequalities

Exercise Set 8.1

1. h

3. f

5. e

7. b

9. c

11. $f(x) \geq g(x)$ when the graph of f lies above, or is at the same height as the graph of g. From the graph, for all values greater than or equal to 2, $f(x) \geq g(x)$. The solution is $\{x \mid x \geq 2\}$, or $[2, \infty)$.

13. $y_1 < y_2$ when the graph of y_1 lies below the graph of y_2. From the graph, for all values less than 3, $y_1 < y_2$. The solution is $\{x \mid x < 3\}$, or $(-\infty, 3)$.

15. a) $2x + 1 \leq x - 1$
    We can rewrite this inequality as $f(x) \leq h(x)$. This inequality is true when the graph of f lies at the same height or below the graph of $h(x)$. This occurs for all x values less than or equal to -2. The solution is $\{x \mid x \leq -2\}$, or $(-\infty, -2]$.

    b) $x - 1 > -\frac{1}{2}x + 3$
    We can rewrite this inequality as $h(x) > g(x)$. This inequality is true when the graph of h lies above the graph of g. This occurs for all x values greater than $\frac{8}{3}$. The solution is $\{x \mid x > \frac{8}{3}\}$, or $(\frac{8}{3}, \infty)$.

    c) $-\frac{1}{2}x + 3 < 2x + 1$
    We can rewrite this inequality as $g(x) < f(x)$. This inequality is true when the graph of g lies below the graph of $f(x)$. This occurs for all x values greater than $\frac{4}{5}$. The solution is $\{x \mid x > \frac{4}{5}\}$, or $(\frac{4}{5}, \infty)$.

17. $x - 3 < 4$
    We graph $f(x) = x - 3$ and $g(x) = 4$ on the same axes.

    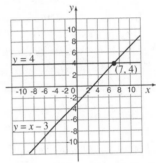

    $f(x) < g(x)$ when the graph of f lies below the graph of g. From the graph, for all values less than 7, $f(x) < g(x)$. The solution is $\{x \mid x < 7\}$, or $(-\infty, 7)$.

19. $2x - 3 \geq 1$
    We graph $f(x) = 2x - 3$ and $g(x) = 1$ on the same axes.

    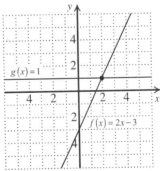

    $f(x) \geq g(x)$ when the graph of f lies above, or is at the same height as the graph of g. From the graph, for all values greater than or equal to 2, $f(x) \geq g(x)$. The solution is $\{x \mid x \geq 2\}$, or $[2, \infty)$.

21. $x + 3 > 2x - 5$

We graph $f(x) = x + 3$ and $g(x) = 2x - 5$ on the same axes.

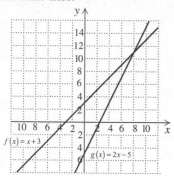

$f(x) > g(x)$ when the graph of f lies above, the graph of g. From the graph, for all values less than 8, $f(x) > g(x)$. The solution is $\{x \mid x < 8\}$, or $(-\infty, 8)$.

23. $\frac{1}{2}x - 2 \leq 1 - x$

We graph $f(x) = \frac{1}{2}x - 2$ and $g(x) = 1 - x$ on the same axes.

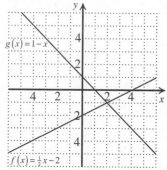

$f(x) \leq g(x)$ when the graph of f lies below, or is at the same height as the graph of g. From the graph, for all values less than or equal to 2, $f(x) \leq g(x)$. The solution is $\{x \mid x \leq 2\}$, or $(-\infty, 2]$.

25. $4x + 7 \leq 3 - 5x$

We graph $y_1 = 4x + 7$ and $y_2 = 3 - 5x$ on a graphing calculator in the same window. We can also graph $y_3 = y_1 \leq y_2$.

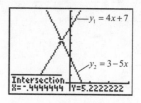

$y_1 \leq y_2$ when the graph of y_1 lies below, or is at the same height as the graph of y_2. From the graph, for all values less than or equal to $-\frac{4}{9}$, $y_1 \leq y_2$. The solution is $\{x \mid x \leq -\frac{4}{9}\}$, or $(-\infty, -\frac{4}{9}]$.

27. ***Familiarize.*** Let n = the number of people who must attend and let R = the amount the band will receive. We are to determine the number of people who must attend in order for the band to be paid at least $1200.
Translate.

Band pay	is at least	$750 plus 15% of	receipts in excess of $750.
↓	↓	↓ ↓ ↓ ↓	↓
R	$\leq$	$750 + 0.15 \cdot$	$(6n - 750)$

where $n \geq 125$ since number of people attending cannot be negative.
Carry out. To estimate the number of people attending the band's concert in order for the band to receive at least $1200, we need to estimate the solution of

$R(n) \leq 750 + 0.15(6n - 750)$

replacing $R(n)$ with 1200. We do this by letting $y_1 = 1200$,

$y_2 = 750 + 0.15(6x - 750)$ and graphing their intersection on a graphing calculator. We find the point of intersection at $(625, 1200)$ and note that y_2 is greater than or equal to y_1 for all x values greater than or equal to 625.

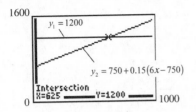

Thus, we estimate that at least 625 people will have to attend the show for the band to receive at least $1200.

Check. If $n = 624$, then the band would receive $750 + 0.15[$6(624) − $750], or $1199.10. However, when $n = 626$, the band would receive $750 + 0.15[$6(626) − $750], or $1200.90. We cannot check all possible numbers so we stop here.

State. In order for the band to receive at least $1200, at least 625 people must attend the show.

29. Let x = the number of years after 2000 and let r = the estimated annual ridership at amusement parks in billions. Enter the data in the table on a graphing calculator and use the linear regression feature to determine the linear regression equation. The equation is $r(x) = -0.04x + 2.0233$.

Create a scatter plot along with the graph of the regression equation by letting y_1 equal the regression equation. Let $y_2 = 1.5$ and plot these equations in the same window. Determine the intersection of y_1 and y_2. Note that $y_1 < y_2$ when $x > 13$, or 13 years after 2000. After 2013 there will be fewer than 1.5 billion riders.

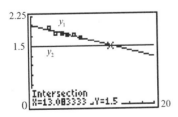

31. Let x = the number of years after 2006 and let n = the number of hours per person per year spent reading newspapers. Let v = the number of hours per person per year playing video games. Enter the data in two tables on a graphing calculator and use the linear regression feature to determine the linear

regression equations for both. The equations are:
$$n(x) = -6.4x + 94.9$$
$$t(x) = 3.7x - 10.2$$
Create a scatter plot along with the graph of the regression equations by letting y_1 equal the regression equation for $n(x)$. Let y_2 equal the regression equation for $t(x)$ and plot both equations in the same window. Determine the intersection of y_1 and y_2. To determine those years for which advertising revenue for the Internet will *exceed* that for newspapers use $y_2 > y_1$. The solution is any year after 2010 (2006 + 4.4).

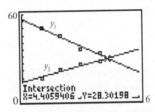

33. Let x = the number of years after 1900 and let f = the world record for the women's 100 meter freestyle, in seconds. Enter the data in a table on a graphing calculator and use the linear regression feature to determine the linear regression equation. The equation is:
$$f(x) = -0.08259x + 54.541.$$
Create a scatter plot along with the graph of the regression equation by letting y_1 equal the regression equation for $f(x)$. Let $y_2 = 50$ and plot both equations in the same window. Determine the intersection of y_1 and y_2. To determine the years in which the world record will be less than 50 seconds, we note those x-values for which $y_1 < y_2$. The solution is any year after 2045 (1990 + 55).

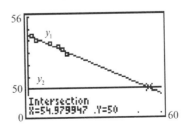

35. $\{5,9,11\} \cap \{9,11,18\}$

The numbers 9 and 11 are common to both sets, so they intersection is $\{9,11\}$.

37. $\{0,5,10,15\} \cup \{5,15,20\}$

The numbers 0, 5, 10, 15, and 20 are in either or both sets, so the union is $\{0,5,10,15,20\}$.

39. $\{a,b,c,d,e,f\} \cap \{b,d,f\}$

The letters b, d, and f are common to both sets, so the intersection is $\{b,d,f\}$.

41. $\{r,s,t\} \cup \{r,u,t,s,v\}$

The letters r, s, t, u, and v, are in either or both sets, so the union is $\{r,s,t,u,v\}$.

43. $\{3,6,9,12\} \cap \{5,10,15\}$

There are no numbers common to both sets, so the solution has no members. It is $\varnothing$.

45. $\{3,5,7\} \cup \varnothing$

The members in either or both sets are 3, 5, and 7, so the union is $\{3,5,7\}$.

47. $3 < x < 7$

$(3,7)$

49. $-6 \le y \le 0$

$[-6,0]$

51. $x < -1$ or $x > 4$

$(-\infty,-1) \cup (4,\infty)$

53. $x \le -2$ or $x > 1$

$(-\infty,-2] \cup (1,\infty)$

55. $x > -2$ and $x < 4$

$(-2,4)$

57. $-4 \le -x < 2$

$(-2,4]$

59. $5 > a$ or $a > 7$

$(-\infty,5) \cup (7,\infty)$

61. $x \ge 5$ or $-x \ge 4$

$(-\infty,-4] \cup [5,\infty)$

63. $7 > y$ and $y \ge -3$

$[-3,7)$

65. $x < 7$ and $x \ge 3$

$[3,7)$

67. $t < 2$ or $t < 5$

$(-\infty,5)$

69. $-2 < t+1 < 8$

$-3 < t < 7$

The solution set is $\{t \mid -3 < t < 7\}$, or $(-3,7)$.

71. $4 < x+4$ and $x-1 < 3$

$0 < x$ and $\qquad x < 4$

We can abbreviate the answer as $0 < x < 4$.

The solution set is $\{x \mid 0 < x < 4\}$, or $(0,4)$.

73. $-7 \le 2a-3$ or $3a+1 > 7$

$-4 \le 2a \qquad$ or $\qquad 3a > 6$

$-2 \le a \qquad$ or $\qquad a > 2$

The solution set is $\{a \mid a \ge -2 \text{ or } a > 2\}$, or $[-2,\infty)$.

75. $x+7 \le -2$ or $x+7 \ge -3$

Observe that any real number is either less than or equal to –9 or greater than or equal to –10. Then the solution set is $\{x \mid x \text{ is any real number}\}$, or $(-\infty,\infty)$.

77. $-7 \le 4x + 5 \le 13$

$-12 \le 4x \le 8$

$-3 \le x \le 2$

The solution set is $\{x \mid -3 \le x \le 2\}$, or

$[-3, 2]$.

79. $5 > \dfrac{x-3}{4} > 1$

$20 > x - 3 > 4$

$23 > x > 7$

The solution set is $\{x \mid 7 < x < 23\}$, or

$(7, 23)$.

81. $-2 \le \dfrac{x+2}{-5} \le 6$

$10 \ge x + 2 \ge -30$

$8 \ge x \ge -32$

The solution set is $\{x \mid -32 \le x \le 8\}$, or

$[-32, 8]$.

83. $2 \le 3x - 1 \le 8$

$3 \le 3x \le 9$

$1 \le x \le 3$

The solution set is $\{x \mid 1 \le x \le 3\}$, or $[1, 3]$.

85. $-21 \le -2x - 7 < 0$

$-14 \le -2x < 7$

$7 \ge x > -\dfrac{7}{2}$, or

$-\dfrac{7}{2} < x \le 7$

The solution set is $\left\{ x \,\middle|\, -\dfrac{7}{2} < x \le 7 \right\}$, or

$\left(-\dfrac{7}{2}, 7 \right]$.

87. $5t + 3 < 3 \ \ or \ \ 5t + 3 > 8$

$5t < 0 \ \ or \ \ \ \ \ 5t > 5$

$t < 0 \ \ or \ \ \ \ \ \ t > 1$

The solution set is $\{t \mid t < 0 \ or \ t > 1\}$, or

$(-\infty, 0) \cup (1, \infty)$.

89. $6 > 2a - 1 \ \ or \ \ -4 \le -3a + 2$

$7 > 2a \ \ \ \ or \ \ -6 \le -3a$

$\dfrac{7}{2} > a \ \ \ \ or \ \ \ \ \ 2 \ge a$

The solution set is $\left\{ a \,\middle|\, \dfrac{7}{2} > a \right\} \cup \{a \mid 2 \ge a\} =$

$\left\{ a \,\middle|\, \dfrac{7}{2} > a \right\}$, or $\left\{ a \,\middle|\, a < \dfrac{7}{2} \right\}$, or $\left(-\infty, \dfrac{7}{2} \right)$.

91. $a + 3 < -2 \ \ and \ \ 3a - 4 < 8$

$a < -5 \ \ and \ \ \ \ \ \ 3a < 12$

$a < -5 \ \ and \ \ \ \ \ \ \ \ a < 4$

The solution set is

$\{a \mid a < -5\} \cap \{a \mid a < 4\} = \{a < -5\}$, or

$(-\infty, -5)$.

93. $3x + 2 < 2 \ \ and \ \ 3 - x < 1$

$3x < 0 \ \ and \ \ \ -x < -2$

$x < 0 \ \ and \ \ \ \ \ \ x > 2$

The solution set is $\{x \mid x < 0\} \cap \{x \mid x > 2\} =$

the empty set, or $\varnothing$.

95. $2t - 7 \le 5 \ \ or \ \ 5 - 2t > 3$

$2t \le 12 \ or \ \ -2t > -2$

$t \le 6 \ \ or \ \ \ \ \ \ t < 1$

The solution set is

$\{t \mid t \le 6\} \cup \{t \mid t < 1\} = \{t \mid t \le 6\}$, or $(-\infty, 6]$.

97. From the graph, we observe that the values of x for which $2x - 5 > -7$ and $2x - 5 < 7$ are $\{x \mid -1 < x < 6\}$, or $(-1, 6)$.

99. $f(x) = \dfrac{9}{x+8}$

$f(x)$ is undefined when $x + 8 = 0 \Rightarrow x = -8$.

Therefore, the domain of f is the set

$(-\infty, -8) \cup (-8, \infty)$.

101. $f(x) = \dfrac{-8}{x}$

$f(x)$ is undefined when $x = 0$. Therefore,

the domain of f is the set $(-\infty, 0) \cup (0, \infty)$.

103. $f(x) = \sqrt{x-6}$

The expression $\sqrt{x-6}$ is not a real number when $x-6$ is negative. Thus, the domain of f is the set of all x-values for which $x-6 \geq 0$. Since $x-6 \geq 0$ is equivalent to $x \geq 6$, we have the domain of $f = [6, \infty)$.

105. $f(x) = \sqrt{2x+7}$

The expression $\sqrt{2x+7}$ is not a real number when $2x+7$ is negative. Thus, the domain of f is the set of all x-values for which $2x+7 \geq 0$. Since $2x+7 \geq 0$ is equivalent to $x \geq -\dfrac{7}{2}$, we have the domain of $f = \left[-\dfrac{7}{2}, \infty \right)$.

107. $f(x) = \sqrt{8-2x}$

The expression $\sqrt{8-2x}$ is not a real number when $8-2x$ is negative. Thus, the domain of f is the set of all x-values for which $8-2x \geq 0$. Since $8-2x \geq 0$ is equivalent to $x \leq 4$, we have the domain of $f = (-\infty, 4]$.

109. $f(x) = \sqrt{x-5}$, $g(x) = \sqrt{\frac{1}{2}x+1}$

The domain of f is the set of all x-values for which $x-5 \geq 0$, or $[5, \infty)$. The domain of g is the set of all x-values for which $\frac{1}{2}x+1 \geq 0$, or $[-2, \infty)$. The intersection of the domains is $[5, \infty)$.

111. $f(x) = \sqrt{3-x}$, $g(x) = \sqrt{3x-2}$

The domain of f is the set of all x-values for which $3-x \geq 0$, or $(-\infty, 3]$. The domain of g is the set of all x-values for which $3x-2 \geq 0$, or $\left[\frac{2}{3}, \infty\right)$. The intersection of the domains is $\left[\frac{2}{3}, 3\right]$.

113. *Thinking and Writing Exercise.* .

115. Graph: $g(x) = 2x$
We make a table of values, plot points, and draw the graph.

x	$g(x)$
-2	-4
-1	-2
0	0
1	2
2	4

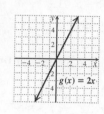

117. Graph: $g(x) = -3$

The graph of any constant function $y = c$ is a horizontal line that crosses the vertical axis at $(0, c)$. Thus, the graph of $g(x) = -3$ is a horizontal line that crosses the vertical axis at $(0, -3)$.

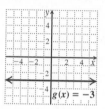

119. $f(x) = x+4$
$g(x) = 3$
Graph both functions.

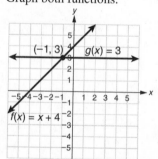

The point of intersection appears to be $(-1, 3)$, so the solution is apparently -1.

$$\begin{array}{c|c} x+4 = 3 & \\ \hline -1+4 & 3 \\ 3 & 3 \text{ TRUE} \end{array}$$

The solution is -1.

121. *Thinking and Writing Exercise.*

123. Solve $18,000 \le s(t) \le 21,000$, or
$18,000 \le 500t + 16,500 \le 21,000$.
$$18,000 \le 500t + 16,500 \le 21,000$$
$$1500 \le 500t \le 4500$$
$$3 \le t \le 9$$
Thus, from 3 through 9 years after 2000, from 2003 to 2009 the number of student visits to the counseling center is between 18,000 and 21,000.

125. Solve $32 < f(x) < 46$, or
$32 < 2(x+10) < 46$.
$$32 < 2(x+10) < 46$$
$$32 < 2x + 20 < 46$$
$$12 < 2x < 26$$
$$6 < x < 13$$
For U.S. dress sizes between 6 and 13, dress size in Italy will be between 32 and 46.

127. Solve $25 \le F(d) \le 31..$
$$25 \le (4.95/d - 4.50) \times 100 \le 31$$
$$\Rightarrow 0.25 \le 4.95/d - 4.50 \le 0.31$$
$$\Rightarrow 4.75 \le 4.95/d \le 4.81$$
$$\Rightarrow 4.75 \le 4.95/d \le 4.81$$
$$\Rightarrow 0.9596 \le 1/d \le 0.9717$$
$$\Rightarrow 1.029 \le d \le 1.042 \text{ since } \frac{1}{x} \le \frac{1}{y} \Rightarrow y \le x$$
Thus, densities considered acceptable for a woman are from 1.03 kg/L to 1.04 kg/L.

129. $4m - 8 > 6m + 5 \quad or \quad 5m - 8 < -2$
$$-13 > 2m \quad or \quad 5m < 6$$
$$-\frac{13}{2} > m \quad or \quad m < \frac{6}{5}$$
The solution set is $\left\{ m \mid m < \frac{6}{5} \right\}$, or $\left(-\infty, \frac{6}{5} \right)$.

131. $3x < 4 - 5x < 5 + 3x$
$$0 < 4 - 8x < 5$$
$$-4 < -8x < 1$$
$$\frac{1}{2} > x > -\frac{1}{8}$$

The solution set is $\left\{ x \mid -\frac{1}{8} < x < \frac{1}{2} \right\}$, or $\left(-\frac{1}{8}, \frac{1}{2} \right)$.

133. If $-b < -a$, then $-1(-b) > -1(-a)$, or $b > a$, or $a < b$. The statement is true.

135. Let $a = b$, $c = 12$, and $b = 2$. Then $a < c$ and $b < c$, but $a \not< b$. The given statement is false.

137. $f(x) = \dfrac{\sqrt{3-4x}}{x+7}$
The expression $\sqrt{3-4x}$ is not a real number when $3 - 4x$ is negative. Then for $3 - 4x \ge 0$, or for $x \le \dfrac{3}{4}$, the numerator of $f(x)$ is a real number. In addition, $f(x)$ cannot be computed when the denominator is 0. Since $x + 7 = 0$ is equivalent to $x = -7$, we have the domain of
$$f = \left\{ x \mid x \le \frac{3}{4} \text{ and } x \ne -7 \right\}, \text{ or}$$
$$(-\infty, -7) \cup \left(-7, \frac{3}{4} \right].$$

139. For the student to complete.

Exercise Set 8.2

1. True

3. True

5. True

7. False

9. g

11. d

13. a

15. The solutions of $|x+2|=3$ are the first coordinates of the points of intersection of $y_1 = abs(x+2)$ and $y_2 = 3$. They are –5 and 1, so the solution set is $\{-5,1\}$.

17. The graph of $y_1 = abs(x+2)$ lies below the graph of $y_2 = 3$ for $\{x \mid -5 < x < -1\}$, or on $(-5,1)$.

19. The graph of $y_1 = abs(x+2)$ lies on or above the graph of $y_2 = 3$ for $\{x \mid x \leq -5 \ or \ x \geq 1\}$, or on $(\infty, -5] \cup [1, \infty)$.

21. $|x| = 7$

    $x = -7 \ or \ x = 7$

    The solution set is $\{-7,7\}$.

23. $|x| = -6$

    The absolute value of a number is always nonnegative. Therefore, the solution is $\varnothing$.

25. $|p| = 0$

    The only number whose absolute value is 0 is 0. The solution set is $\{0\}$.

27. $|2x-3| = 4$

    $2x-3 = -4 \ or \ 2x-3 = 4$

    $2x = -1 \ or \ \ \ \ \ \ \ 2x = 7$

    $x = -\dfrac{1}{2} \ or \ \ \ \ \ \ x = \dfrac{7}{2}$

    The solution set is $\left\{-\dfrac{1}{2}, \dfrac{7}{2}\right\}$.

29. $|3x-5| = -8$

    The absolute value of a number is always nonnegative. Therefore, the solution is $\varnothing$.

31. $|x-2| = 6$

    $x-2 = -6 \ or \ x-2 = 6$

    $x = -4 \ or \ \ \ \ \ \ x = 8$

    The solution set is $\{-4,8\}$.

33. $|x-5| = 3$

    $x-5 = -3 \ or \ x-5 = 3$

    $x = 2 \ or \ \ \ \ \ \ x = 8$

    The solution set is $\{2,8\}$.

35. $|t| + 1.1 = 6.6 \Rightarrow |t| = 5.5$

    $t = -5.5 \ or \ t = 5.5$

    The solution set is $\{-5.5, 5.5\}$.

37. $|5x| - 3 = 37$

    $|5x| = 40$

    $5x = -40 \ or \ 5x = 40$

    $x = -8 \ or \ \ \ x = 8$

    The solution set is $\{-8,8\}$.

39. $7|q| - 2 = 9$

    $7|q| = 11$ Adding 2

    $|q| = \dfrac{11}{7}$ Multiplying by $\dfrac{1}{7}$

    $q = -\dfrac{11}{7} \ or \ \dfrac{11}{7}$

    The solution set is $\left\{-\dfrac{11}{7}, \dfrac{11}{7}\right\}$.

41. $\left|\dfrac{2x-1}{3}\right| = 4$

    $\dfrac{2x-1}{3} = -4 \ or \ \dfrac{2x-1}{3} = 4$

    $2x-1 = -12 \ or \ 2x-1 = 12$

    $2x = -11 \ or \ 2x = 13$

    $x = -\dfrac{11}{2} \ or \ x = \dfrac{13}{2}$

    The solution set is $\left\{-\dfrac{11}{2}, \dfrac{13}{2}\right\}$

43. $|5-m| + 9 = 16 \Rightarrow |5-m| = 7$

    $5-m = -7 \ or \ 5-m = 7$

    $-m = -12 \ or \ -m = 2$

    $m = 12 \ or \ \ \ \ \ m = -2$

    The solution set is $\{-2,12\}$.

45. $5 - 2|3x - 4| = -5$

$-2|3x - 4| = -10$

$|3x - 4| = 5$

$3x - 4 = -5 \ or \ 3x - 4 = 5$

$3x = -1 \ or \ 3x = 9$

$x = -\dfrac{1}{3} \ or \ x = 3$

The solution set is $\left\{ -\dfrac{1}{3}, 3 \right\}$.

47. $f(x) = |2x + 6|; \ f(x) = 8$

$|2x + 6| = 8$

$2x + 6 = -8 \ or \ 2x + 6 = 8$

$2x = -14 \ or \ 2x = 2$

$x = -7 \ or \ x = 1$

The solution set is $\{-7, 1\}$.

49. $f(x) = |x| - 3; \ f(x) = 5.7$

$|x| - 3 = 5.7$

$|x| = 8.7$

$x = -8.7 \ or \ x = 8.7$

The solution set is $\{-8.7, 8.7\}$.

51. $\left| \dfrac{3x - 2}{5} \right| = 2$

$\dfrac{3x - 2}{5} = -2 \quad or \quad \dfrac{3x - 2}{5} = 2$

$3x - 2 = -10 \ or \ 3x - 2 = 10$

$3x = -8 \quad or \qquad 3x = 12$

$x = -\dfrac{8}{3} \quad or \qquad x = 4$

The solution set is $\left\{ -\dfrac{8}{3}, 4 \right\}$.

53. $|x + 4| = |2x - 7|$

$x + 4 = 2x - 7 \ or \ x + 4 = -(2x - 7)$

$4 = x - 7 \ or \ x + 4 = -2x + 7$

$11 = x \qquad or \ 3x + 4 = 7$

$3x = 3$

$x = 1$

The solution set is $\{1, 11\}$.

55. $|x + 4| = |x - 3|$

$x + 4 = x - 3 \ or \ x + 4 = -(x - 3)$

$4 = -3 \quad or \ x + 4 = -x + 3$

$\text{False} \qquad 2x = -1$

$x = -\dfrac{1}{2}$

The solution set is $\left\{ -\dfrac{1}{2} \right\}$.

57. $|3a - 1| = |2a + 4|$

$3a - 1 = 2a + 4 \ or \ 3a - 1 = -(2a + 4)$

$a - 1 = 4 \qquad or \ 3a - 1 = -2a - 4$

$a = 5 \qquad or \ 5a - 1 = -4$

$5a = -3$

$a = -\dfrac{3}{5}$

The solution set is $\left\{ -\dfrac{3}{5}, 5 \right\}$.

59. $|n - 3| = |3 - n|$

$n - 3 = 3 - n \ or \ n - 3 = -(3 - n)$

$2n - 3 = 3 \ or \ n - 3 = -3 + n$

$2n = 6 \ or \quad -3 = -3$

$n = 3 \qquad \text{True for all real}$

$\text{values of } n$

The solution set is the set of all real numbers.

61. $|7 - 4a| = |4a + 5|$

$7 - 4a = 4a + 5 \quad or \ 7 - 4a = -(4a + 5)$

$7 = 8a + 5 \ or \quad 7 - 4a = -4a - 5$

$2 = 8a \qquad or \qquad 7 = -5$

$\dfrac{1}{4} = a \qquad\qquad \text{False}$

The solution set is $\left\{ \dfrac{1}{4} \right\}$.

63. $|a| \le 9$

$-9 \le a \le 9$

The solution set is $\{a \, | \, -9 \le a \le 9\}$, or

$[-9, 9]$.

65. $|t| > 0$

$t < 0 \ or \ 0 < t$

The solution set is $\{t \mid t < 0 \ or \ t > 0\}$, or $\{t \mid t \neq 0\}$, or $(-\infty, 0) \cup (0, \infty)$.

67. $|x - 1| < 4$

$-4 < x - 1 < 4$

$-3 < x < 5$ Adding 1

The solution set is $\{x \mid -3 < x < 5\}$, or $(-3, 5)$.

69. $|x + 2| \leq 6$

$-6 \leq x + 2 \leq 6$

$-8 \leq x \leq 4$ Adding -2

The solution set is $x \mid -8 \leq x \leq 4$, or $[-8, 4]$.

71. $|x - 3| + 2 > 7$

$|x - 3| > 5$ Adding -2

$x - 3 < -5 \ or \ 5 < x - 3$

$x < -2 \ or \ 8 < x$

The solution set is $\{x \mid x < -2 \ or \ x > 8\}$, or $(-\infty, -2) \cup (8, \infty)$.

73. $|2y - 9| > -5$

Since the absolute value is never negative, any value of $2y - 9$, and hence any value of y, will satisfy the inequality. The solution set is the set of all real numbers, or $(-\infty, \infty)$.

75. $|3a - 4| + 2 \geq 8$

$|3a - 4| \geq 6$ Adding -2

$3a - 4 \leq -6 \ or \ 6 \leq 3a - 4$

$3a \leq -2 \ or \ 10 \leq 3a$

$a \leq -\dfrac{2}{3} \ or \ \dfrac{10}{3} \leq a$

The solution set is $\left\{a \mid a \leq -\dfrac{2}{3} \ or \ a \geq \dfrac{10}{3}\right\}$,

or $\left(-\infty, -\dfrac{2}{3}\right] \cup \left[\dfrac{10}{3}, \infty\right)$.

77. $|y - 3| < 12$

$-12 < y - 3 < 12$

$-9 < y < 15$ Adding 3

The solution set is $\{y \mid -9 < y < 15\}$, or $(-9, 15)$.

79. $9 - |x + 4| \leq 5$

$-|x + 4| \leq -4$

$|x + 4| \geq 4$ Multiplying by -1

$x + 4 \leq -4 \ or \ 4 \leq x + 4$

$x \leq -8 \ or \ 0 \leq x$

The solution set is $\{x \mid x \leq -8 \ or \ x \geq 0\}$, or $(-\infty, -8] \cup [0, \infty)$.

81. $6 + |3 - 2x| > 10$

$|3 - 2x| > 4$

$3 - 2x < -4 \ or \ 4 < 3 - 2x$

$-2x < -7 \ or \ 1 < -2x$

$x > \dfrac{7}{2} \ or \ -\dfrac{1}{2} > x$

The solution set is $\left\{x \mid x < -\dfrac{1}{2} \ or \ x > \dfrac{7}{2}\right\}$, or

$\left(-\infty, -\dfrac{1}{2}\right) \cup \left(\dfrac{7}{2}, \infty\right)$.

83. $|5 - 4x| < -6$

Absolute value is always nonnegative, so the inequality has no solution. The solution set is $\varnothing$.

85. $\left|\dfrac{2 - 5x}{4}\right| \geq \dfrac{2}{3}$

$\dfrac{2 - 5x}{4} \leq -\dfrac{2}{3} \ or \ \dfrac{2}{3} \leq \dfrac{2 - 5x}{4}$

$2 - 5x \leq -\dfrac{8}{3} \ or \ \dfrac{8}{3} \leq 2 - 5x$

$-5x \leq -\dfrac{14}{3} \ or \ \dfrac{2}{3} \leq -5x$

$x \geq \dfrac{14}{15} \ or \ -\dfrac{2}{15} \geq x$

The solution set is $\left\{x \mid x \le -\dfrac{2}{15} \ or \ x \ge \dfrac{14}{15}\right\}$,

or $\left(-\infty, -\dfrac{2}{15}\right] \cup \left[\dfrac{14}{15}, \infty\right)$.

87. $|m+3| + 8 \le 14$

$|m+3| \le 6$

$-6 \le m + 3 \le 6$

$-9 \le m \le 3$

The solution set is $\{m \mid -9 \le m \le 3\}$, or

$[-9, 3]$.

89. $25 - 2|a+3| > 19$

$-2|a+3| > -6$

$|a+3| < 3$

$-3 < a + 3 < 3$

$-6 < a < 0$

The solution set is $\{a \mid -6 < a < 0\}$, or

$(-6, 0)$.

91. $|2x-3| \le 4$

$-4 \le 2x - 3 \le 4$

$-1 \le 2x \le 7$

$-\dfrac{1}{2} \le x \le \dfrac{7}{2}$

The solution set is $\left\{x \mid -\dfrac{1}{2} \le x \le \dfrac{7}{2}\right\}$, or

$\left[-\dfrac{1}{2}, \dfrac{7}{2}\right]$.

93. $5 + |3x-4| \ge 16$

$|3x-4| \ge 11$

$3x - 4 \le -11 \ or \ 11 \le 3x - 4$

$3x \le -7 \ or \ 3x \ge 15$

$x \le -\dfrac{7}{3} \ or \ x \ge 5$

The solution set is $\left\{x \mid x \le -\dfrac{7}{3} \ or \ x \ge 5\right\}$, or

$\left(-\infty, -\dfrac{7}{3}\right] \cup [5, \infty)$.

95. $7 + |2x-1| < 16$

$|2x-1| < 9$

$-9 < 2x - 1 < 9$

$-8 < 2x < 10$

$-4 < x < 5$

The solution set is $\{x \mid -4 < x < 5\}$, or

$(-4, 5)$.

97. *Thinking and Writing Exercise.*

99. $3x - y = 6$

To find the y-intercept, let $x = 0$ and solve for y:

$3 \cdot 0 - y = 6$

$y = -6$

The y-intercept is $(0, -6)$.

To find the x-intercept, let $y = 0$ and solve for x:

$3x - 0 = 6$

$3x = 6$

$x = 2$

The x-intercept is $(2, 0)$.

Before drawing the line, plot a third point as a check. Substitute any convenient value for x and solve for y. For $x = 1$,

$3 \cdot 1 - y = 6$

$3 - y = 6$

$y = -3$

The point $(1, -3)$ appears to line up with the intercepts. To finish, draw and label the line.

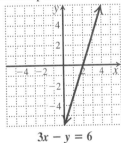

$3x - y = 6$

101. $x = -2$

We can write this equation as $x + 0 \cdot y = -2$.
No matter what number is chosen for y, x must be -2. Consider the following table.

x	y	(x, y)
-2	-4	$(-2, -4)$
-2	0	$(-2, 0)$
-2	4	$(-2, 4)$

Plot the ordered pairs and connect the points.

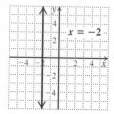

103. $x - 3y = 8$ (1)

$2x + 3y = 4$ (2)

Solve the first equation for x.

$x - 3y = 8$

$x = 3y + 8$ (3)

Substitute $3y + 8$ for x in the second equation and solve for y.

$2x + 3y = 4$ (2)

$2(3y + 8) + 3y = 4$

$6y + 16 + 3y = 4$

$9y + 16 = 4$

$9y = -12$

$y = -\dfrac{4}{3}$

Next substitute $-\dfrac{4}{3}$ for y in Equation (3).

$x = 3y + 8$ (3)

$x = 3 \cdot \left(-\dfrac{4}{3}\right) + 8$

$x = -4 + 8$

$x = 4$

The ordered pair $\left(4, -\dfrac{4}{3}\right)$ checks in both equations; it is the solution.

105. $y = 1 - 5x$ (1)

$2x - y = 4$ (2)

Substitute $1 - 5x$ for y in the second equation and solve for y.

$2x - y = 4$ (2)

$2x - (1 - 5x) = 4$

$2x - 1 + 5x = 4$

$7x - 1 = 4$

$7x = 5$

$x = \dfrac{5}{7}$

Next substitute $\dfrac{5}{7}$ for x in Equation (1).

$y = 1 - 5x$ (1)

$y = 1 - 5\left(\dfrac{5}{7}\right)$

$y = \dfrac{7}{7} - \dfrac{25}{7}$

$y = -\dfrac{18}{7}$

The ordered pair $\left(\dfrac{5}{7}, -\dfrac{18}{7}\right)$ checks in both equations; it is the solution.

107. *Thinking and Writing Exercise.*

109. From the definition of absolute value,
$|3t - 5| = 3t - 5$ only when $3t - 5 \geq 0$. Solve

$3t - 5 \geq 0$

$3t \geq 5$

$t \geq \dfrac{5}{3}$

The solution set is $\left\{ t \middle| t \geq \dfrac{5}{3} \right\}$, or $\left[\dfrac{5}{3}, \infty \right)$.

111. $|x + 2| > x$

The inequality is true for all $x < 0$ (because absolute value must be nonnegative). The solution set in this case is $\{x \mid x < 0\}$. If $x = 0$, we have $|0 + 2| > 0$, which is true.

The solution set in this case is $\{0\}$. If $x > 0$, we have the following:

$x + 2 < -x \;\; or \;\; x < x + 2$

$2x < -2 \;\; or \;\; 0 < 2$

$x < -1$

Although $x > 0$ *and* $x < -1$ yields no solutions, $x > 0$ and $2 > 0$ (true for all x) yields the solution set $\{x \mid x > 0\}$ in this case. The solution set for the inequality is $\{x \mid x < 0\} \cup \{0\} \cup \{x \mid x > 0\}$, or $\{x \mid x \text{ is a real number}\}$, or $(-\infty, \infty)$.

113. $|5t - 3| = 2t + 4$

From the definition of absolute value, we know that $2t + 4 \geq 0$, or $t \geq -2$. So we have $t \geq -2$ *and*

$$5t - 3 = -(2t + 4) \quad or \quad 5t - 3 = 2t + 4$$
$$5t - 3 = -2t - 4 \quad or \quad 3t = 7$$
$$7t = -1 \quad or \quad t = \frac{7}{3}$$
$$t = -\frac{1}{7} \quad or \quad t = \frac{7}{3}$$

Since $-\frac{1}{7} \geq -2$ and $\frac{7}{3} \geq -2$, the solution set is $\left\{ -\frac{1}{7}, \frac{7}{3} \right\}$.

115. Using part (b) we find that $-3 < x < 3$ is equivalent to $|x| < 3$.

117. $x < -8 \ or \ 2 < x$
$x + 3 < -5 \ or \ 5 < x + 3$ Adding 3
$|x + 3| > 5$ Using part (c)

$-8 \ -3 \ 2$

119. The distance from x to 7 is $|x - 7|$ or $|7 - x|$, so we have $|x - 7| < 2$, or $|7 - x| < 2$.

121. The length of the segment from -1 to 7 is $|-1 - 7| = |-8| = 8$ units. The midpoint of the segment is $\frac{-1 + 7}{2} = \frac{6}{2} = 3$. Thus, the interval extends 8/2, or 4 units, on each side of 3. An inequality for which the closed interval is the solution set is then $|x - 3| \leq 4$.

123. The length of the segment from -7 to -1 is $|-7 - (-1)| = |-6| = 6$ units. The midpoint of

the segment is $\frac{-7 + (-1)}{2} = \frac{-8}{2} = -4$. Thus, the interval extends 6/2, or 3 units on each side of -4. An inequality for which the open interval is the solution set is $|x - (-4)| < 3$, or $|x + 4| < 3$.

125. Let d = the distance above the river. This distance must satisfy the inequality $|d - 60| \leq 10$. First solve for d.
$$|d - 60| \leq 10$$
$$-10 \leq d - 60 \leq 10$$
$$50 \leq d \leq 70$$
Since the bridge is 150 ft from the river, the bungee jumper will, at any point in time, be $150 - d$ feet from the bridge, so the jumper will be a maximum of 150 ft – 50 ft, or 100 ft from the bridge *and* a minimum of 150 ft – 70 ft, or 80 ft from the bridge. The solution is between 80 ft and 100 ft.

Mid-Chapter Review

Guided Solutions

1. $-3 < x - 5 < 6$
   $2 < x < 11$
   The solution is $(2, 11)$.

2. $|x - 1| > 9$
   $x - 1 < -9 \ or \ 9 < x - 1$
   $x < -8 \ or \ 10 < x$
   The solution is $(-\infty, -8) \cup (10, \infty)$.

Mixed Review

1. $|x| = 15$
   $x = -15 \ or \ x = 15$
   The solution set is $\{-15, 15\}$.

2. $|t| < 10$
   $-10 < t < 10$
   The solution set is $\{t \mid -10 < t < 10\}$, or $(-10, 10)$.

3. $|p| > 15$

   $p < -15 \ or \ p > 15$

   The solution set is $\{p \mid p < -15 \ or \ p > 15\}$,

   or $(-\infty, -15) \cup (15, \infty)$.

4. $|2x + 1| = 7$

   $2x + 1 = -7 \ or \ 2x + 1 = 7$

   $2x = -8 \ or \quad 2x = 6$

   $x = -4 \ or \quad x = 3$

   The solution set is $\{-4, 3\}$.

5. $-3 < x - 5 < 6$

   $2 < x < 11$

   The solution set is $\{x \mid 2 < x < 11\}$, or $(2, 11)$.

6. $5|t| < 20$

   $|t| < 4$

   $-4 < t < 4$

   The solution set is $\{t \mid -4 < t < 4\}$, or $(-4, 4)$.

7. $x + 8 < 2 \ or \ x - 4 > 9$

   $x < -6 \ or \quad x > 13$

   The solution set is $\{x \mid x < -6 \ or \ x > 13\}$, or

   $(-\infty, -6) \cup (13, \infty)$.

8. $|x + 2| \le 5$

   $-5 \le x + 2 \le 5$

   $-7 \le x \le 3$

   The solution set is $\{x \mid -7 \le x \le 3\}$, or

   $[-7, 3]$.

9. $2 + |3x| = 10$

   $|3x| = 8$ Subtracting 2

   $3x = -8 \ or \ 3x = 8$

   $x = -\dfrac{8}{3} \ or \quad x = \dfrac{8}{3}$

   The solution set is $\left\{-\dfrac{8}{3}, \dfrac{8}{3}\right\}$.

10. $|x - 3| \le 10$

    $-10 \le x - 3 \le 10$

    $-7 \le x \le 13$

The solution set is $\{x \mid -7 \le x \le 13\}$, or

$[-7, 13]$.

11. $-12 < 2n + 6 \ and \ 3n - 1 \le 7$

    $-18 < 2n \quad and \quad 3n \le 8$

    $-9 < n \quad\quad and \quad n \le \dfrac{8}{3}$

    The solution set is $\left\{n \mid -9 < n \le \dfrac{8}{3}\right\}$, or

    $\left(-9, \dfrac{8}{3}\right]$.

12. $|t| < 0$

    $0 < t < 0$ FALSE

    Absolute value is always nonnegative, so there is no value for t that can be negative, thus there is no solution. The solution set is $\varnothing$.

13. $|2x + 5| + 1 \ge 13$

    $|2x + 5| \ge 12$

    $2x + 5 \le -12 \ or \ 12 \le 2x + 5$

    $2x \le -17 \ or \quad 7 \le 2x$

    $x \le -\dfrac{17}{2} \ or \quad \dfrac{7}{2} \le x$

    The solution set is $\left\{x \mid x \le -\dfrac{17}{2} \ or \ x \ge \dfrac{7}{2}\right\}$, or

    $\left(-\infty, -\dfrac{17}{2}\right] \cup \left[\dfrac{7}{2}, \infty\right)$.

14. $5x + 1 < 1 \ and \ 7 - x < 2$

    $5x < 0 \quad and \quad -x < -5$

    $x < 0 \quad and \quad\quad x > 5$

    No real number is less than 0 and greater than 5. The solution set is the empty set, or $\varnothing$.

15. $|m + 6| - 8 < 10 \Rightarrow$

    $|m + 6| < 18$

    $-18 < m + 6 < 18$

    $-24 < m < 12$

    The solution set is $\{x \mid -24 < x < 12\}$, or

    $(-24, 12)$.

16. $\left|\dfrac{x+2}{5}\right| = 8$

$\dfrac{x+2}{5} = -8 \quad or \quad \dfrac{x+2}{5} = 8$

$x+2 = -40 \quad or \quad x+2 = 40$

$x = -42 \quad or \qquad x = 38$

The solution set is $\{-42, 38\}$.

17. $4 - |7 - t| \le 1$

$-|7 - t| \le -3$

$|7 - t| \ge 3$

$7 - t \le -3 \quad or \quad 3 \le 7 - t$

$-t \le -10 \quad or \quad -4 \le -t$

$t \ge 10 \quad or \quad 4 \ge t$

The solution set is $\{t \mid t \le 4 \ or \ t \ge 10\}$, or

$(-\infty, 4] \cup [10, \infty)$.

18. $|8x - 11| + 6 < 2$

$|8x - 11| < -4$

Absolute values are always nonnegative, so there is no value for x that can make the expression in the absolute value symbol negative. There is no solution, $\varnothing$.

19. $8 - 5|a + 6| > 3$

$-5|a + 6| > -5$

$|a + 6| < 1$

$-1 < a + 6 < 1$

$-7 < a < -5$

The solution set is $\{a \mid -7 < a < -5\}$, or

$(-7, -5)$.

20. $|5x + 7| + 9 \ge 4$

$|5x + 7| \ge -5$

Since absolute values are never negative, any value of x will make $|5x + 7| \ge 0$. The solution set is the set of all real numbers, or $(-\infty, \infty)$.

Exercise Set 8.3

1. e

3. d

5. b

7. We replace x with -4 and y with 2.

$2x + 3y < -1$

$$\begin{array}{c|c} 2(-4) + 3 \cdot 2 & -1 \\ \hline -8 + 6 & -1 \\ -2 & -1 \ \text{TRUE} \end{array}$$

Since $-2 < -1$ is true, $(-4, 2)$ is a solution.

9. We replace x with 8 and y with 14.

$2y - 3x \ge 9$

$$\begin{array}{c|c} 2 \cdot 14 - 3 \cdot 8 & 9 \\ \hline 28 - 24 & 9 \\ 4 & 9 \ \text{FALSE} \end{array}$$

Since $4 \ge 9$ is false, $(8, 14)$ is not a solution.

11. Graph: $y \ge \dfrac{1}{2}x$

We first graph the boundary line $y = \dfrac{1}{2}x$.

We draw the line solid since the inequality symbol is $\ge$. To determine which half-plane to shade, test a point not on the line. We try $(0, 1)$:

$y \ge \dfrac{1}{2}x$

$$\begin{array}{c|c} 1 & \dfrac{1}{2} \cdot 0 \\ \hline 1 & 0 \ \text{TRUE} \end{array}$$

Since $1 \ge 0$ is true, $(0, 1)$ is a solution as are all of the points in the half-plane containing $(0, 1)$. We shade that half-plane and obtain the graph.

13. Graph: $y > x - 3$

We first graph the boundary line $y = x - 3$.

We draw the line dashed since the inequality symbol is $>$. To determine which half-plane to shade, test a point not on the line. We try $(0, 0)$:

$$y > x - 3$$

$$\begin{array}{c|c} 0 & 0 - 3 \\ \hline 0 & -3 \quad \text{TRUE} \end{array}$$

Since $0 > -3$ is true, $(0,0)$ is a solution as are all of the points in the half-plane containing $(0,0)$. We shade that half-plane and obtain the graph.

15. Graph: $y \leq x + 5$

We first graph the boundary line $y = x + 5$. We draw the line solid since the inequality symbol is $\leq$. To determine which half-plane to shade, test a point not on the line. We try $(0,0)$:

$$y \leq x + 5$$

$$\begin{array}{c|c} 0 & 0 + 5 \\ \hline 0 & 5 \quad \text{TRUE} \end{array}$$

Since $0 \leq 5$ is true, $(0,0)$ is a solution as are all of the points in the half-plane containing $(0,0)$. We shade that half-plane and obtain the graph.

17. Graph: $x - y \leq 4$

We first graph the boundary line $x - y = 4$. We draw the line solid since the inequality symbol is $\leq$. To determine which half-plane to shade, test a point not on the line. We try $(0,0)$:

$$x - y \leq 4$$

$$\begin{array}{c|c} 0 - 0 & 4 \\ \hline 0 & 4 \quad \text{TRUE} \end{array}$$

Since $0 \leq 4$ is true, $(0,0)$ is a solution as are all of the points in the half-plane containing $(0,0)$. We shade that half-plane and obtain the graph.

19. Graph: $2x + 3y > 6$

We first graph the boundary line $2x + 3y = 6$. We draw the line dashed since the inequality symbol is $>$. To determine which half-plane to shade, test a point not on the line. We try $(0,0)$:

$$2x + 3y > 6$$

$$\begin{array}{c|c} 2 \cdot 0 + 3 \cdot 0 & 6 \\ \hline 0 & 6 \quad \text{FALSE} \end{array}$$

Since $0 > 6$ is false, $(0,0)$ is not a solution nor are any of the points in the half-plane containing $(0,0)$. The points in the other half-plane are the solutions so we shade that half-plane and obtain the graph.

21. Graph: $2y - x \leq 4$

We first graph the boundary line $2y - x \leq 4$. We draw the line solid since the inequality symbol is $\leq$. To determine which half-plane to shade, test a point not on the line. We try $(0,0)$:

$$2y - x \leq 4$$

$$\begin{array}{c|c} 2 \cdot 0 - 0 & 4 \\ \hline 0 & 4 \quad \text{TRUE} \end{array}$$

Since $0 \leq 4$ is true, $(0,0)$ is a solution as are all of the points in the half-plane containing $(0,0)$. We shade that half-plane and obtain the graph.

23. Graph: $2x - 2y \geq 8 + 2y$

$$2x - 4y \geq 8$$

We first graph the boundary line $2x - 4y = 8$. We draw the line solid since the inequality symbol is $\geq$. To determine which half-plane to shade, test a point not on the line. We try $(0,0)$:

$$2x - 4y \geq 8$$

$2 \cdot 0 - 4 \cdot 0$	8
0	8 FALSE

Since $0 \geq 8$ is false, $(0,0)$ is not a solution nor are any of the points in the half-plane containing $(0,0)$. The points in the other half-plane are the solutions so we shade that half-plane and obtain the graph.

$2x - 2y \geq 8 + 2y$

25. Graph: $x > -2$

We first graph the boundary line $x = -2$. We draw the line dashed since the inequality symbol is >. To determine which half-plane to shade, test a point not on the line. We try $(0,0)$:

$$x > -2$$

0	-2
0	-2 TRUE

Since $0 > -2$ is true, $(0,0)$ is a solution as are all of the points in the half-plane containing $(0,0)$. We shade that half-plane and obtain the graph.

27. Graph: $y \leq 6$

We first graph the boundary line $y = 6$. We draw the line solid since the inequality symbol is $\leq$. To determine which half-plane to shade, test a point not on the line. We try $(0,0)$:

$$y \leq 6$$

0	6
0	6 TRUE

Since $0 < 6$ is true, $(0,0)$ is a solution as are all of the points in the half-plane containing $(0,0)$. We shade that half-plane and obtain the graph.

29. Graph: $-2 < y < 7$

This is a system of inequalities:

$$-2 < y$$
$$y < 7$$

We graph the boundary line $-2 = y$ and see that the graph of $-2 < y$ is the half-plane above the line $-2 = y$. We also graph the boundary line $y = 7$ and see that the graph of $y < 7$ is the half-plane below the line $y = 7$.

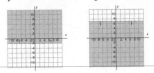

Finally, we shade the intersection of these graphs.

31. Graph $-4 \leq x \leq 2$

This is a system of inequalities.

$$-4 \leq x$$
$$x \leq 2$$

We graph the equations $-4 = x$ and see that the graph of $-4 \leq x$ is the half-plane right of the line $-4 = x$. We also graph $x = 2$ and see that the graph of $x \leq 2$ is the half-plane left of the line $x = 2$.

Finally, we shade the intersection of these graphs.

$-4 \le x \le 2$

33. Graph: $0 \le y \le 3$

This is a system of inequalities:

$0 \le y$

$y \le 3$

We graph the equations $0 = y$ and see that the graph of $0 \le y$ is the half-plane above the line $0 = y$. We also graph $y = 3$ and see that the graph of $y \le 3$ is the half-plane below the line $y = 3$. Finally, we shade the intersection of these graphs.

35. $y > x + 3.5$

37. First get y alone on one side of the inequality.

$8x - 2y < 11$

$-2y < -8x + 11$

$y > \dfrac{-8x + 11}{-2}$

39. Graph: $y > x$

$y < -x + 3$

We graph the boundary lines $y = x$ and $y = -x + 3$, using dashed lines. Note where the regions overlap and shade the region of solutions.

41. Graph: $y \le x$

$y \le 2x - 5$

We graph the boundary lines $y = x$ and $y = 2x - 5$, using solid lines. Note where the regions overlap and shade the region of solutions.

43. Graph: $y \le -3$

$x \ge -1$

We graph the boundary lines $y = -3$ and $x = -1$, using solid lines. We indicate the region for each inequality by the arrows at the ends of the lines. Note where the regions overlap and shade the region of solutions.

45. Graph: $x > -4$

$y < -2x + 3$

We graph the lines $x = -4$ and $y = -2x + 3$ using dashed lines. We indicate the region for each inequality by the arrows at the ends of the lines. Note where the regions overlap and shade the region of solutions.

47. Graph: $y \le 5$

$y \ge -x + 4$

We graph the lines $y = 5$ and $y = -x + 4$, using solid lines. We indicate the region for each inequality by the arrows at the ends of the lines. Note where the regions overlap and shade the region of solutions.

49. Graph: $x + y \le 6$

$\qquad x - y \le 4$

We graph the lines $x + y = 6$ and $x - y = 4$, using solid lines. We indicate the region for each inequality by the arrows at the ends of the lines. Note where the regions overlap and shade the region of solutions.

51. Graph: $y + 3x > 0$

$\qquad y + 3x < 2$

We graph the lines $y + 3x = 0$ and $y + 3x = 2$, using dashed lines. We indicate the region for each inequality by the arrows at the ends of the lines. Note where the regions overlap and shade the region of solutions.

53. Graph: $y \le 2x - 3$ $\qquad$ (1)

$\qquad y \ge -2x + 1$ $\qquad$ (2)

$\qquad x \le 5$ $\qquad$ (3)

Graph the lines $y = 2x - 3$, $y = -2x + 1$, and $x = 5$ using solid lines. Indicate the region for each inequality by arrows, and shade the region where they overlap.

To find the vertices we solve three different systems of related equations.

From (1) and (2) we have $y = 2x - 3$

$\qquad\qquad\qquad\qquad y = -2x + 1.$

Solving, we obtain the vertex $(1, -1)$.

From (1) and (3) we have $y = 2x - 3$

$\qquad\qquad\qquad\qquad x = 5.$

Solving, we obtain the vertex $(5, 7)$.

From (2) and (3) we have $y = -2x + 1$

$\qquad\qquad\qquad\qquad x = 5.$

Solving, we obtain the vertex $(5, -9)$.

55. Graph: $x + 2y \le 12$ $\qquad$ (1)

$\qquad 2x + y \le 12$ $\qquad$ (2)

$\qquad x \ge 0$ $\qquad$ (3)

$\qquad y \ge 0$ $\qquad$ (4)

Graph the lines $x + 2y = 12$, $2x + y = 12$, $x = 0$, and $y = 0$ using solid lines. Indicate the region for each inequality by arrows, and shade the region where they overlap.

To find the vertices we solve four different systems of related equations.

From (1) and (2) we have $x + 2y = 12$

$\qquad\qquad\qquad\qquad 2x + y = 12.$

Solving, we obtain the vertex $(4, 4)$.

From (1) and (3) we have $x + 2y = 12$

$\qquad\qquad\qquad\qquad x = 0.$

Solving, we obtain the vertex $(0, 6)$.

From (2) and (4) we have $2x + y = 12$

$\qquad\qquad\qquad\qquad y = 0.$

Solving, we obtain the vertex $(6, 0)$.

From (3) and (4) we have $x = 0$

$\qquad\qquad\qquad\qquad y = 0.$

Solving, we obtain the vertex $(0, 0)$.

57. Graph: $8x + 5y \le 40$ $\qquad$ (1)

$\qquad x + 2y \le 8$ $\qquad$ (2)

$\qquad x \ge 0$ $\qquad$ (3)

$\qquad y \ge 0$ $\qquad$ (4)

Graph the lines $8x + 5y = 40$, $x + 2y = 8$, $x = 0$, and $y = 0$ using solid lines. Indicate the region for each inequality by arrows, and shade the region where they overlap.

To find the vertices we solve four different systems of related equations.

From (1) and (2) we have $8x + 5y = 40$

$$x + 2y = 8.$$

Solving, we obtain the vertex $\left(\dfrac{40}{11}, \dfrac{24}{11}\right)$.

From (1) and (4) we have $8x + 5y = 40$

$$y = 0.$$

Solving, we obtain the vertex $(5, 0)$.

From (2) and (3) we have $x + 2y = 8$

$$x = 0.$$

Solving, we obtain the vertex $(0, 4)$.

From (3) and (4) we have $x = 0$

$$y = 0.$$

Solving, we obtain the vertex $(0, 0)$.

59. Graph: $y - x \geq 2$ (1)

$$y - x \leq 4 \quad (2)$$

$$2 \leq x \leq 5 \quad (3)$$

Think of (3) as two inequalities:

$$2 \leq x \quad (4)$$

$$x \leq 5 \quad (5)$$

Graph the lines $y - x = 2$, $y - x = 4$, $2 = x$, and $x = 5$ using solid lines. Indicate the region for each inequality by arrows, and shade the region where they overlap.

To find the vertices we solve three different systems of related equations.
From (1) and (4) we have $y - x = 2$

$$2 = x.$$

Solving, we obtain the vertex $(2, 4)$.

From (1) and (5) we have $x - y = 2$

$$x = 5.$$

Solving, we obtain the vertex $(5, 7)$.

From (2) and (5) we have $y - x = 4$

$$x = 5.$$

Solving, we obtain the vertex $(5, 9)$.

From (2) and (4) we have $y - x = 4$

$$x = 2.$$

Solving, we obtain the vertex $(2, 6)$.

61. *Thinking and Writing Exercise.*

63. $x^2 - 1 = 0$

$$(x + 1)(x - 1) = 0$$

$$x + 1 = 0 \ \ or \ \ x - 1 = 0$$

$$x = -1 \ \ or \ \ x = 1$$

The solutions are -1 and 1.

65. $10x^3 - 30x^2 + 20x = 0$

$$10x\left(x^2 - 3x + 2\right) = 0$$

$$10x(x - 1)(x - 2) = 0$$

$$10x = 0 \ \ or \ \ x - 1 = 0 \ \ or \ \ x - 2 = 0$$

$$x = 0 \ \ or \ \ x = 1 \ \ or \ \ x = 2$$

The solutions are 0, 1, and 2.

67. $\dfrac{x - 3}{x + 4} = 5$

To assure that none of the denominators are 0, we note at the outset that $x \neq -4$. Then we multiply on both sides by $(x + 4)$.

$$(x + 4)\dfrac{x - 3}{x + 4} = 5(x + 4)$$

$$x - 3 = 5x + 20$$

$$-23 = 4x$$

$$-\dfrac{23}{4} = x$$

This value checks. The solution is $-\dfrac{23}{4}$.

69. $\dfrac{x}{(x - 3)(x + 7)} = 0$

To assure that none of the denominators are 0, we note at the outset that $x \neq 3$ and $x \neq -7$. Then we multiply on both sides by $(x - 3)(x + 7)$.

$$(x - 3)(x + 7)\dfrac{x}{(x - 3)(x + 7)} = 0 \cdot (x - 3)(x + 7)$$

$$x = 0$$

This value checks. The solution is 0.

71. *Thinking and Writing Exercise.*

73. Graph: $x + y > 8$

$x + y \le -2$

Graph the line $x + y = 8$ using a dashed line and graph $x + y = -2$ using a solid line. Indicate the region for each inequality by arrows. The regions do not overlap (the solution set is $\varnothing$), so we do not shade any portion of the graph.

75. Graph: $x - 2y \le 0$

$-2x + y \le 2$

$x \le 2$

$y \le 2$

$x + y \le 4$

Graph the lines $x - 2y = 0$, $-2x + y = 2$, $x = 2$, $y = 2$, and $x + y = 4$ using solid lines. Indicate the regions for each inequality by arrows. Note where the regions overlap and shade the region of solutions.

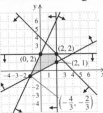

77. $w > 0$

$h > 0$

$w + h + 30 \le 62$, or

$w + h \le 32$

$2w + 2h + 30 \le 130$, or

$w + h \le 50$

79. Graph: $h \le 2w$

$w \le 1.5h$

$h \le 3200$

$h \ge 0$

$w \ge 0$

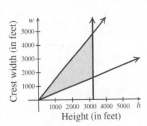

81. Graph: $q + v \ge 1150$

$q \ge 700$

$q \le 800$

$v \ge 400$

$v \le 800$

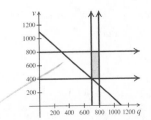

83. The shaded region lies below the graphs of $y = x$ and $y = 2$, and both lines are solid. Thus, we have

$y \le x$

$y \le 2$

85. The shaded region lies below the graphs of $y = x + 2$ and $y = -x + 4$, and above $y = 0$. All of the lines are solid. Thus, we have

$y \le x + 2$

$y \le -x + 4$

$y \ge 0$

Exercise Set 8.4

1. True

3. True

5. False

7. We see that $p(x) = 0$ at $x = -4$ and $x = \dfrac{3}{2}$,

and $p(x) < 0$ between -4 and $\dfrac{3}{2}$. The

solution set of the inequality is $\left[-4, \dfrac{3}{2}\right]$, or

$\left\{x \mid -4 \le x \le \dfrac{3}{2}\right\}$.

9. $x^4 + 12x > 3x^3 + 4x^2$

$x^4 - 3x^3 - 4x^2 + 12x > 0$

From the graph we see that $p(x) > 0$ on

$(-\infty, -2) \cup (0, 2) \cup (3, \infty)$. This union, or

$\{x \mid x < -2 \ or \ 0 < x < 2 \ or \ x > 3\}$, is the

solution set of the inequality.

11. $\dfrac{x-1}{x+2} < 3$

$\dfrac{x-1}{x+2} - 3 < 0$

$\dfrac{x-1}{x+2} - \dfrac{3(x+2)}{x+2} < 0$

$\dfrac{x-1-3(x+2)}{x+2} < 0$

$\dfrac{-2x-7}{x+2} < 0$

$\dfrac{-2x-7}{x+2} < 0$ on $\left(-\infty, -\dfrac{7}{2}\right) \cup (-2, \infty)$. This

union, or $\left\{x \mid x < -\dfrac{7}{2} \ or \ x > -2\right\}$, is the

solution set of the inequality.

13. $(x+4)(x-3) < 0$

We solve the related equation.

$(x+4)(x-3) = 0$

$x + 4 = 0 \quad or \quad x - 3 = 0$

$x = -4 \ or \qquad x = 3$

The numbers -4 and 3 divide the number line
into 3 intervals.

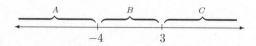

We graph $p(x) = (x+4)(x-3)$ in the

window $[-10, 10, -15, 5]$ and determine the

sign of the function in each interval.

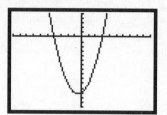

We see that $p(x) < 0$ in interval B, or in

$(-4, 3)$. Thus, the solution set of the inequality

is $(-4, 3)$, or $\{x \mid -4 < x < 3\}$.

15. $(x+7)(x-2) \ge 0$

The solutions of $(x+7)(x-2) = 0$ are -7 and

2. They divide the number line into three

intervals as shown:

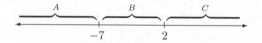

We graph $p(x) = (x+7)(x-2)$ in the

window $[-10, 10, -25, 5]$, Yscl = 5.

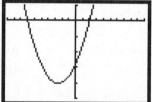

We see that $p(x) \ge 0$ in intervals A and C, or

in $(-\infty, -7) \cup (2, \infty)$. We also know that

$p(-7) = 0$ and $p(2) = 0$. Thus, the solution

set of the inequality is $(-\infty, -7] \cup [2, \infty)$, or

$\{x \mid x \le -7 \ or \ x \ge 2\}$.

17.　$x^2 - x - 2 > 0$

$(x-2)(x+1) > 0$

The solutions of $(x-2)(x+1) = 0$ are 2 and –1. We graph $p(x) = (x-2)(x+1)$ in the standard window.

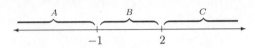

We see that $p(x) > 0$ on $(-\infty, -1)$ and $(2, \infty)$. The solution set of the inequality is $(-\infty, -1) \cup (2, \infty)$, or $\{x | x < -1 \text{ or } x > 2\}$.

19.　$x^2 + 4x + 4 < 0$

$(x+2)^2 < 0$

Observe that $(x+2)^2 \geq 0$ for all values of x. Thus, the solution set is $\emptyset$. The graph of $p(x) = x^2 + 4x + 4$ confirms this.

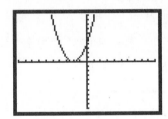

21.　$x^2 - 4x < 12$

$x^2 - 4x - 12 < 0$

$(x-6)(x+2) < 0$

The solutions of $(x-6)(x+2) = 0$ are 6 and –2. Graph $p(x) = x^2 - 4x - 12$ in the window $[-10, 10, -20, 5]$, Yscl $= 5$.

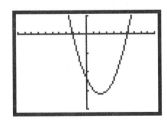

We see that $p(x) < 0$ on the interval $(-2, 6)$ or $\{x | -2 < x < 6\}$. This is the solution set of the inequality.

23.　$3x(x+2)(x-2) < 0$

The solutions of $3x(x+2)(x-2) = 0$ are 0, –2, and 2. Graph $p(x) = 3x(x+2)(x-2)$ in the window $[-5, 5, -10, 10]$.

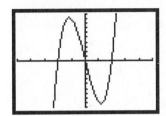

We see that $p(x) < 0$ on $(-\infty, -2) \cup (0, 2)$, or $\{x | x < -2 \text{ or } 0 < x < 2\}$. This is the solution set for the inequality.

25.　$(x-1)(x+2)(x-4) \geq 0$

The solutions of $(x-1)(x+2)(x-4) = 0$ are 1, –2. and 4. Graph $p(x) = (x-1)(x+2)(x-4)$ in the window $[-5, 5, -12, 12]$.

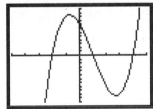

We see that $p(x) \geq 0$ on $[-2, 1]$ and $[4, \infty)$. the solution set of the inequality includes the zeros; it is $[-2, 1] \cup [4, \infty)$ or $\{x | -2 \leq x \leq 1 \text{ or } x \geq 4\}$.

27.　$4.32x^2 - 3.54x - 5.34 \leq 0$

Graph $p(x) = 4.32x^2 - 3.54x - 5.34$ in the window $[-5, 5, -10, 10]$.

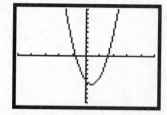

Using the Zero feature we find that $p(x) = 0$ when $x \approx -0.78$ and when $x \approx 1.59$. Also observe that $p(x) < 0$ in the interval $(-0.78, 1.59)$. Thus, the solution set of the inequality is $[-0.78, 1.59]$, or $\{x \mid -0.78 \le x \le 1.59\}$.

29. $x^3 - 2x^2 - 5x + 6 < 0$
Graph $p(x) = x^3 - 2x^2 - 5x + 6$ in the window $[-5, 5, -10, 10]$.

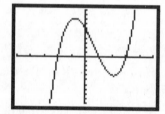

Using the Zero feature we find that $p(x) = 0$ when $x = -2$, when $x = 1$, and when $x = 3$. Then we see that $p(x) < 0$ on $(-\infty, -2) \cup (1, 3)$, or $\{x \mid x < -2 \text{ or } 1 < x < 3\}$. This is the solution set of the inequality.

31. $f(x) \ge 3$
$7 - x^2 \ge 3$
$4 - x^2 \ge 0$
The solutions of $(x + 2)(x - 2) = 0$ are -2 and 2. Graph $p(x) = 4 - x^2$ in a standard window.

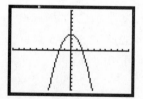

We see that $p(x) \ge 0$, or $f(x) \ge 3$, on $(-2, 2)$. The solution set includes the zeros. $[-2, 2]$ or $\{x \mid -2 \le x \le 2\}$.

33. $(x - 2)(x - 3)(x + 1) > 0$
The solutions of $(x - 2)(x - 3)(x + 1) = 0$ are 2, 3, and -1. Graph $g(x) = (x - 2)(x - 3)(x + 1)$ in the window $[-5, 5, -10, 10]$.

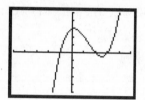

We see that $g(x) > 0$ on $(-1, 2) \cup (3, \infty)$, or $\{x \mid -1 < x < 2 \text{ or } x > 3\}$.   This is the solution set of the inequality.

35. $x^3 - 7x^2 + 10x \le 0$
$x(x - 5)(x - 2) \le 0$
The solutions of $x(x - 5)(x - 2) = 0$ are 0, 5, and 2. Graph $F(x) = x^3 - 7x^2 + 10x$.

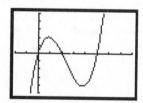

We see that $F(x) < 0$ on $(-\infty, 0) \cup (2, 5)$. The solution set, which includes the zeros is
$(-\infty, 0] \cup [2, 5]$, or
$\{x \mid x \le 0 \text{ or } 2 \le x \le 5\}$.

37. $\dfrac{1}{x+5} < 0$

The related equation $\dfrac{1}{x+5} = 0$ has no

solution. Also $x+5 = 0$ when $x = -5$.

Choose a test number from each interval using substitution or a graphing calculator.

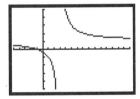

We see that $y < 0$ when $x < -5$, so the solution

set is $(-\infty, -5)$, or $\{x | x < -5\}$.

39. $\dfrac{x+1}{x-3} \geq 0$

Find the values that make the denominator 0.
$x - 3 = 0$ when $x = 3$

Graph $r(x) = \dfrac{x+1}{x-3}$ in the window [–5, 15,

–5, 5]. Using the Zero feature we find that
$r(x) = 0$ when $x = -1$. We include the –1 in
the solution.

The solution is $(-\infty, -1] \cup (3, \infty)$ or

$\{x | x \leq -1 \text{ or } x > 3\}$.

41. $\dfrac{x+1}{x+6} \geq 1$

$\dfrac{x+1}{x+6} - 1 \geq 0$

If $r(x) = \dfrac{x+1}{x+6} - 1$, the solution set of the

inequality is all values of x for which
$r(x) \geq 0$.

First we solve $r(x) = 0$.

$$\dfrac{x+1}{x+6} - 1 = 0$$

$$(x+6)\left(\dfrac{x+1}{x+6} - 1\right) = (x+6) \cdot 0$$

$$(x+6)\left(\dfrac{x+1}{x+6}\right) - (x+6) \cdot 1 = 0$$

$$x + 1 - x - 6 = 0$$

$$-5 = 0$$

This equation has no solution. Find the
values that make the denominator 0.

$$x + 6 = 0$$

$$x = -6$$

Use –6 to divide the number line into
intervals.

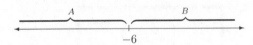

Enter $y = r(x)$ on a graphing calculator and

evaluate a test number in each interval. We
test –7 and 0.

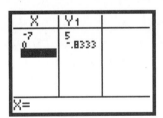

We see that $r(x) > 0$ in interval A. The

solution set is $(-\infty, -6)$, or $\{x | x < -6\}$.

43. $\dfrac{(x-2)(x+1)}{x-5} \le 0$

Solve the related equation.

$\dfrac{(x-2)(x+1)}{x-5} = 0$

$(x-2)(x+1) = 0$

$x = 2 \ or \ x = -1$

Find the values that make the denominator 0.

$x - 5 = 0$

$x = 5$

Use the numbers 2, −1, and 5 to divide the number line into intervals as shown:

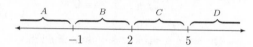

Enter $r(x) = \dfrac{(x-2)(x+1)}{x-5}$ and evaluate a test number in each interval. We test −2, 0, 3, and 6.

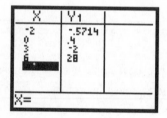

We see that $r(x) < 0$ in intervals A and C.

From above we also know that $r(x) = 0$ when $x = 2$ or $x = -1$. Thus, the solution set is $(-\infty, -1] \cup [2, 5)$, or $\{x | x \le -1 \text{ or } 2 \le x < 5\}$.

45. $\dfrac{x}{x+3} \ge 0$

Graph $r(x) = \dfrac{x}{x+3}$ using DOT mode in the window $[-10, 10, -5, 5]$.

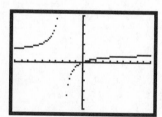

Using the Zero feature we find that $r(x) = 0$ when $x = 0$. Also observe that $r(x) > 0$ in the interval $(-\infty, -3)$ and in $(0, \infty)$. Then the solution set is $(-\infty, -3) \cup [0, \infty)$, or $\{x | x < -3 \text{ or } x \ge 0\}$.

47. $\dfrac{x-5}{x} < 1$

$\dfrac{x-5}{x} - 1 < 0$

Let $r(x) = \dfrac{x-5}{x} - 1$ and solve $r(x) = 0$.

$\dfrac{x-5}{x} - 1 = 0$

$x\left(\dfrac{x-5}{x} - 1\right) = x \cdot 0$

$x\left(\dfrac{x-5}{x}\right) - x \cdot 1 = 0$

$x - 5 - x = 0$

$-5 = 0$

This equation has no solution. Find the values that make the denominator 0.

$x = 0$

Use the number 0 to divide the number line into two intervals as shown.

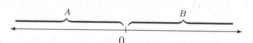

Enter $y = r(x)$ in a graphing calculator and evaluate a test number in each interval. We test −1 and 1.

We see that $r(x) < 0$ in interval B. Thus, the solution set is $(0, \infty)$, or $\{x | x > 0\}$.

49. $\dfrac{x-1}{(x-3)(x+4)} \le 0$

Solve the related equation.

$\dfrac{x-1}{(x-3)(x+4)} = 0$

$x-1 = 0$

$x = 1$

Find the values that make the denominator 0.

$(x-3)(x+4) = 0$

$x = 3 \ or \ x = -4$

Use the numbers 1, 3, and –4 to divide the number line into intervals as shown.

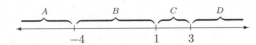

Enter $r(x) = \dfrac{x-1}{(x-3)(x+4)}$ in a graphing

calculator and evaluate a test point in each interval. We test –5, 0, 2, and 4.

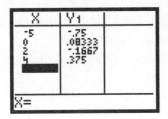

We see that $r(x) < 0$ in intervals A and C.

From above we also know that $r(x) = 0$

when $x = 1$. Thus, the solution set is

$(-\infty, -4) \cup [1,3)$, or $\left\{x \mid x < -4 \ or \ 1 \le x < 3\right\}$.

51. $\dfrac{5-2x}{4x+3} \ge 0$

Solve the related equation.

$\dfrac{5-2x}{4x+3} = 0$

$5-2x = 0$

$\dfrac{5}{2} = x$

Find the values that make the denominator 0.

$4x+3 = 0$

$x = -\dfrac{3}{4}$

Now graph $r(x) = \dfrac{5-2x}{4x+3}$ in the window

$[-10, 10, -5, 5]$.

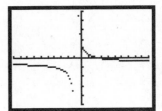

We see $r(x) > 0$ on $\left(-\dfrac{3}{4}, \dfrac{5}{2}\right)$. We include

the zero in the solution set:

$\left(-\dfrac{3}{4}, \dfrac{5}{2}\right]$ or $\left\{x \mid -\dfrac{3}{4} < x \le \dfrac{5}{2}\right\}$.

53. $\dfrac{1}{x-2} \le 1$

$\dfrac{1}{x-2} - 1 \le 0$

$\dfrac{1}{x-2} - \dfrac{x-2}{x-2} \le 0$

$\dfrac{3-x}{x-2} \le 0$

Solve the related equation.

$\dfrac{3-x}{x-2} = 0$

$3-x = 0$

$3 = x$

Find the values that make the denominator 0.

$x - 2 = 0$ when $x = 2$.

Use these numbers to divide the number line into intervals.

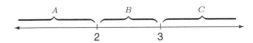

Choose a test point from each interval to determine where the original inequality is True i.e. the value is negative. We determine this to be intervals A and C.

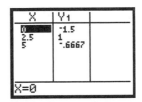

We include the zero in the solution.

$(-\infty, 2) \cup [3, \infty)$ or $\{x | x < 2 \text{ or } x \geq 3\}$

The solution is $\left(-\dfrac{3}{8}, \dfrac{1}{8}\right)$.

55. *Thinking and Writing Exercise.*

63. *Thinking and Writing Exercise.*

57. $3x - 2y = 7$

$\dfrac{x + 2y = 1}{4x \quad\quad = 8}$ Adding

$x = 2$

$x + 2y = 1 \rightarrow 2 + 2y = 1$

$2y = -1$

$y = -\dfrac{1}{2}$

The solution is $\left(2, -\dfrac{1}{2}\right)$.

65. $x^4 + x^2 < 0$

Note that when we raise any number to the fourth power or second power, the result is non-negative. If we add two non-negative numbers, the result is also non-negative. Thus $x^4 + x^2 \geq 0$, so $x^4 + x^2 < 0$ has no solution. The solution set is $\varnothing$.

67. $x^4 + 3x^2 \leq 0$

$x^2\left(x^2 + 3\right) \leq 0$

$x^2 = 0$ for $x = 0$, $x^2 > 0$ for $x \neq 0$, $x^2 + 3 > 0$ for all x. The solution set is $\{0\}$.

59. $3x - 5y = 1$ (1)

$2x + 3y = 7$ (2)

Multiply equation (1) by 2, equation (2) by -3, and add the resulting equations.

$6x - 10y = 2$

$\dfrac{-6x - 9y = -21}{-19y = -19}$

$y = 1$

$3x - 5y = 1 \rightarrow 3x - 5 \cdot (1) = 1$

$3x - 5 = 1$

$3x = 6$

$x = 2$

The solution is (2, 1).

69. a) $-3x^2 + 630x - 6000 > 0$

$x^2 - 210x + 2000 < 0$ Multiply by $-\dfrac{1}{3}$

$(x - 200)(x - 10) < 0$

The solutions of

$f(x) = (x - 200)(x - 10) = 0$

are 200 and 10. They divide the number line as shown:

Enter $p(x) = x^2 - 210x + 2000$ in a graphing calculator and evaluate a test point in each interval. We test 9, 11, and 201. Note that only nonnegative values of x have meaning in this problem.

61. $y - 5x = 2 \rightarrow -5x + y = 2$ (1)

$3x + y = -1 \rightarrow 3x + y = -1$ (2)

Multiply equation (2) by -1, and add the resulting equation to equation (1).

$-5x + y = 2$

$\dfrac{-3x - y = 1}{-8x = 3}$

$x = -\dfrac{3}{8}$

$3x + y = -1 \rightarrow 3 \cdot \left(-\dfrac{3}{8}\right) + y = -1$

$-\dfrac{9}{8} + y = -1$

$y = \dfrac{1}{8}$

We see that $p(x) < 0$ in interval B, so the company makes a profit for values of x such that $10 < x < 200$, or for values of x in the interval (10, 200), or in the set $\{x | 10 < x < 200\}$.

b) See part (a). Keep in mind that x must be nonnegative since negative numbers have no meaning in this application.
The company loses money for values of x such that $0 \le x < 10$ or $x > 200$, or for values of x in the interval $[0,10) \cup (200,\infty)$, or in the set $\{x | 0 \le x < 10 \text{ or } x > 200\}$.

71. We find values of n such that $N \ge 66$ and $N \le 300$.
For $N \ge 66$

$$\frac{n(n-1)}{2} \ge 66$$

$$n(n-1) \ge 132$$

$$n^2 - n - 132 \ge 0$$

$$(n-12)(n+11) \ge 0$$

The solutions of $f(n) = (n-12)(n+11) = 0$ are 12 and –11. They divide the number line as shown:

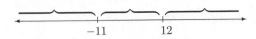

However, only positive values of n have meaning in this exercise so we need only consider the intervals shown below:

Enter $p(x) = x^2 - x - 132$ in a graphing calculator and evaluate a test point in each interval. We test 1 and 13.

X	Y1	
1	-132	
13	24	

X=

We see that $p(x) > 0$ in interval B. From above we also know that $p(12) = 0$, so the solution set for this inequality is $[12,\infty)$.
For $N \le 300$

$$\frac{n(n-1)}{2} \le 300$$

$$n(n-1) \le 600$$

$$n^2 - n - 600 \le 0$$

$$(n-25)(n+24) \le 0$$

The solutions of $f(n) = (n-25)(n+24) = 0$ are 25 and –24. They divide the number line as shown:

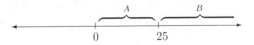

However, only positive values of n have meaning in this exercise so we need only consider the intervals shown below:

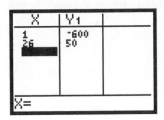

Enter $p(x) = x^2 - x - 600$ in a graphing calculator and evaluate a test point in each interval. We test 1 and 26.

X	Y1	
1	-600	
26	50	

X=

We see that $p(x) < 0$ in interval A. From above we also know that $p(25) = 0$, so the solution set for this inequality is $(0,25]$.
Then $66 \le N \le 300$ for $[12,\infty) \cap (0,25]$, or $[12, 25]$. We can express the solution set as $\{n | n \text{ is an integer and } 12 \le n \le 25\}$.

73. From the graph we determine the following:

$f(x)$ has no zeros.

The solutions of

$f(x) < 0$ are $(-\infty, 0)$ or $\{x | x < 0\}$.

The solutions of

$f(x) > 0$ are $(0, \infty)$ or $\{x | x > 0\}$.

75. From the graph we determine the following:

The solutions of $f(x) = 0$ are $-1, 0$

The solution of

$f(x) < 0$ is $(-\infty, -3) \cup (-1, 0)$ or

$\{x | x < -3 \text{ or } -1 < x < 0\}$

The solution of

$f(x) > 0$ is $(-3, -1) \cup (0, 2) \cup (2, \infty)$ or

$\{x | -3 < x < -1 \text{ or } 0 < x < 2 \text{ or } x > 2\}$

77. The domain of $f(x)$ are all x-values that make the expression under the square root symbol non-negative. Factoring the polynomial gives:

$x^2 - 4x - 45 \geq 0$

$(x + 5)(x - 9) \geq 0$

$x \leq -5$ or $x \geq 9$. So the domain of f is

$(-\infty, -5] \cup [9, \infty)$, or $\{x | x \leq -5 \text{ or } x \geq 9\}$

79. The domain of $f(x)$ are all x-values that make the expression under the square root symbol non-negative. Factoring the polynomial gives:

$x^2 + 8x \geq 0$.

$x(x + 8) \geq 0$

$x \leq -8$ or $x \geq 0$. So the domain of f is

$(-\infty, -8] \cup [0, \infty)$, or $\{x | x \leq -8 \text{ or } x \geq 0\}$.

Chapter 8 Study Summary

1. $2x - 1 < x$

    On a graphing calculator, let $y_1 = 2x - 1$ and $y_2 = x$. We graph the equations in the window $[-5, 5, -5, 5]$.

    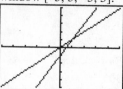

    We can see from the graph that the point of intersection is $(1, 1)$, and that $y_1 < y_2$ when $x < 1$. The solution set is $\{x | x < 1\}$, or $(-\infty, 1)$.

2. $-5 < 4x + 3 \leq 0$

    $-8 < 4x \leq -3$

    $-2 < x \leq -\dfrac{3}{4}$

    The solution set is $\left\{x | -2 < x \leq -\dfrac{3}{4}\right\}$, or $\left(-2, -\frac{3}{4}\right]$.

3. $x - 3 \leq 10$ *or* $25 - x < 3$

    $x \leq 13$ *or* $-x < -22$

    $x \leq 13$ *or* $x > 22$

    The solution set is $\{x | x \leq 13 \text{ or } x > 22\}$, or $(-\infty, 13] \cup (22, \infty)$.

4. $|4x - 7| = 11$

    $4x - 7 = -11$ *or* $4x - 7 = 11$

    $4x = -4$ *or* $4x = 18$

    $x = -1$ *or* $x = \dfrac{9}{2}$

    The solution set is $\left\{-1, \dfrac{9}{2}\right\}$.

5. $|x - 12| \leq 1$

    $-1 \leq x - 12 \leq 1$

    $11 \leq x \leq 13$

    The solution set is $\{x | 11 \leq x \leq 13\}$, or $[11, 13]$.

6. $|2x+3|>7$

$2x+3<-7$ or $7<2x+3$

$2x<-10$ or $4<2x$

$x<-5$ or $2<x$

The solution set is $\{x|x<-5$ or $x>2\}$, or $(-\infty,-5)\cup(2,\infty)$.

7. $2x-y<5$

First graph the boundary line $2x-y=5$ using a dashed line. Since the test point $(0,0)$ is a solution and $(0,0)$ is above the line, we shade the half-plane above the line.
The solution is the region which is shaded.

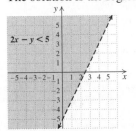

8. $x^2-11x-12<0$

We find the zeros of the relation equation.

$x^2-11x-12=0$

$(x+1)(x-12)=0$

$x+1=0$ or $x-12=0$

$x=-1$ or $x=12$

These zeros divide the number line into three intervals, A, B, and C.

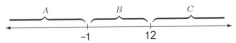

We choose a convenient test point from each interval to determine the sign of the interval. Using the table feature we see the signs on either side of the zeros.

X	Y₁			X	Y₁	
-3	30			9	-30	
-2	14			10	-22	
-1	0			11	-12	
0	-12			12	0	
1	-22			13	14	
2	-30			14	30	
3	-36			15	48	
X=-1				X=12		

Since $x^2-11x-12<0$, we want the interval where the sign is negative, or interval B.

Our solution is $\{x|-1<x<12\}$ or $(-1,12)$.

1. True

2. False

3. True

4. True

5. False

6. True

7. True

8. False

9. False

10. True

11. $4-3x>1$

Let $y_1=4-3x$, $y_2=1$ and graph y_1 and y_2 in the standard viewing window. We see the point of intersection, where $y_1=y_2$, is $(1,1)$.
$y_1>y_2$ to the left of the intersection, so the solution set is $\{x|x<1\}$, or $(-\infty,1)$

Note: We could also manually graph each linear equation, using various graphing techniques.

12. $x-3<3x+5$

Let $y_1=x-3$ and $y_2=3x+5$ and graph y_1 and y_2 in the standard viewing window. We see that the point of intersection, where $y_1=y_2$, is $(-4,-7)$. $y_1<y_2$ to the right of the intersection, thus the solution set is $\{x|x>-4\}$, or $(-4,\infty)$.

Note: We could also manually graph each linear equation, using various graphing techniques.

13. $x+1 \geq \frac{1}{2}x - 2$

Let $y_1 = x+1$, $y_2 = \frac{1}{2}x - 2$ and graph y_1 and y_2 in the standard viewing window. We see the point of intersection, where $y_1 = y_2$, is $(-6, -5)$. $y_1 \geq y_2$ to the right of the intersection, so the solution set is $\{x \mid x \geq -6\}$, or $[-6, \infty)$.

Note: We could also manually graph each linear equation, using various graphing techniques.

14.   Let $y_1 = 3x + 2$ and $y_2 = 10 - x$ and graph y_1 and y_2 in the standard viewing window. We see that the point of intersection, where $y_1 = y_2$, is $(2, 8)$. $y_1 \leq y_2$ to the left of the intersection, thus the solution set is $\{x \mid x \leq 2\}$, or $(-\infty, 2]$

15. $\{1, 2, 5, 6, 9\} \cap \{1, 3, 5, 9\}$

The numbers 1, 5, and 9 are common to both sets, so the intersection is $\{1, 5, 9\}$.

16   $\{1, 2, 5, 6, 9\} \cup \{1, 3, 5, 9\}$

The numbers 1,2,3,5,6,and 9 are in either or both sets, so the union is $\{1, 2, 3, 5, 6, 9\}$.

17. $x \leq 2$ and $x > -3$

This conjunction can be abbreviated as $-3 < x \leq 2$.

Interval notation: $(-3, 2]$

18. $x \leq 3$ or $x > -5$

Interval notation: $(-\infty, \infty)$

19. $-4 < x + 8 \leq 5$

$-12 < x \leq -3$

Set builder notation: $\{x \mid -12 < x \leq -3\}$

Interval notation: $(-12, -3]$

20. $-15 < -4x - 5 < 0$

$-10 < -4x < 5$

$\frac{5}{2} > x > -\frac{5}{4}$

Set builder notation: $\left\{ x \mid -\frac{5}{4} < x < \frac{5}{2} \right\}$

Interval notation: $\left(-\frac{5}{4}, \frac{5}{2} \right)$

21. $3x < -9$ or $-5x < -5$

$x < -3$ or $x > 1$

Set builder notation: $\{x \mid x < -3 \ or \ x > 1\}$

Interval notation: $(-\infty, -3) \cup (1, \infty)$

22. $2x + 5 < -17$ or $-4x + 10 \leq 34$

$2x < -22$ or $\quad -4x \leq 24$

$x < -11$ or $\quad x \geq -6$

Set builder notation: $\{x \mid x < -11 \ or \ x \geq -6\}$

Interval notation: $(-\infty, -11) \cup [-6, \infty)$

23. $2x + 7 \leq -5$ or $x + 7 \geq 15$

$2x \leq -12$ or $\quad x \geq 8$

$x \leq -6$ or $\quad x \geq 8$

Set builder notation: $\{x \mid x \leq -6 \ or \ x \geq 8\}$

Interval notation: $(-\infty, -6] \cup [8, \infty)$

24. $f(x) < -5$ or $f(x) > 5$

$3 - 5x < -5$ or $3 - 5x > 5$

$8 < 5x$ or $-2 > 5x$

$\dfrac{8}{5} < x$ or $-\dfrac{2}{5} > x$

Set builder notation: $\left\{ x \middle| x < -\dfrac{2}{5} \ or \ x > \dfrac{8}{5} \right\}$

Interval notation: $\left(-\infty, -\dfrac{2}{5} \right) \cup \left(\dfrac{8}{5}, \infty \right)$

25. $f(x) = \dfrac{2x}{x-8}$

$f(x)$ cannot be computed when the denominator is 0. Since $x - 8 = 0$ is equivalent for $x = 8$, we have Domain of $f = \{ x \mid x$ is any real number $and \ x \ne 8 \}$

$= (-\infty, 8) \cup (8, \infty)$.

26. $f(x) = \sqrt{x+5}$

The expression $\sqrt{x+5}$ is not a real number when $x + 5$ is negative. Thus, the domain of f is the set of all x-values for which $x + 5 \ge 0$. Since $x + 5 \ge 0$ is equivalent to $x \ge -5$, we have the domain of $f = [-5, \infty)$.

27. $f(x) = \sqrt{8 - 3x}$

The expression $\sqrt{8 - 3x}$ is not a real number when $8 - 3x$ is negative. Thus, the domain of f is the set of all x-values for which $8 - 3x \ge 0$. Since $8 - 3x \ge 0$ is equivalent to $\frac{8}{3} \ge x$, we have the domain of $f = \left(-\infty, \frac{8}{3} \right]$.

28. $|x| = 5$

$x = -5$ or $x = 5$

The solution set is $\{ -5, 5 \}$.

29. $|t| \ge 21$

$t \le -21$ or $t \ge 21$

The solution set is $\{ t \mid t \le -21 \ or \ t \ge 21 \}$, or $(-\infty, -21] \cup [21, \infty)$.

30. $|x - 3| = 7$

$x - 3 = -7$ or $x - 3 = 7$

$x = -4$ or $x = 10$

The solution set is $\{ -4, 10 \}$.

31. $|4a + 3| < 11$

$-11 < 4a + 3 < 11$

$-14 < 4a < 8$

$-\dfrac{7}{2} < a < 2$

The solution set is $\left\{ a \middle| -\dfrac{7}{2} < a < 2 \right\}$, or

$\left(-\dfrac{7}{2}, 2 \right)$.

32. $|3x - 4| - 6 \ge 9$

$|3x - 4| \ge 15$

$-15 \ge 3x - 4$ or $15 \le 3x - 4$

$-11 \ge 3x$ or $19 \le 3x$

$-\dfrac{11}{3} \ge x$ or $\dfrac{19}{3} \le x$

The solution set is $\left\{ x \middle| x \le -\dfrac{11}{3} \ or \ x \ge \dfrac{19}{3} \right\}$,

or $\left(-\infty, -\dfrac{11}{3} \right] \cup \left[\dfrac{19}{3}, \infty \right)$.

33. $|2x + 5| = |x - 9|$

$2x + 5 = x - 9$ or $2x + 5 = -(x - 9)$

$x = -14$ or $2x + 5 = -x + 9$

$x = -14$ or $3x = 4$

$x = -14$ or $x = \dfrac{4}{3}$

The solution set is $\left\{ -14, \dfrac{4}{3} \right\}$.

34. $|5n + 6| = -8$

The absolute value is always nonnegative, so the equation has not solution. The solution set is $\varnothing$.

35. $\left|\dfrac{x+4}{6}\right| \le 2$

$-2 \le \dfrac{x+4}{6} \le 2$

$-12 \le x+4 \le 12$

$-16 \le x \le 8$

The solution set is $\{x \mid -16 \le x \le 8\}$, or $[-16, 8]$.

36. $2|x-5|-7 > 3$

$2|x-5| > 10$

$|x-5| > 5$

$x-5 < -5 \;\; or \;\; x-5 > 5$

$x < 0 \;\; or \;\;\;\;\; x > 10$

The solution set is $\{x \mid x < 0 \; or \; x > 10\}$, or $(-\infty, 0) \cup (10, \infty)$.

37. $f(x) = |3x-5|$

$f(x) < 0$

$|3x-5| < 0$

The absolute value is always nonnegative, so the equation has no solution. The solution set is $\varnothing$.

38. Graph: $x - 2y \ge 6$

First graph the line $x - 2y = 6$. Draw it solid since the inequality symbol is $\ge$. Test the point $(0,0)$ to determine if it is a solution.

$$\begin{array}{c|c} x-2y \ge 6 & \\ \hline 0 - 2 \cdot 0 & 6 \\ 0 & 6 \;\; \text{False} \end{array}$$

Since $0 \ge 6$ is false, $(0,0)$ is not a solution, nor are any points in the half-plane containing $(0,0)$. The points in the other half-plane are solutions, so we shade that half-plane and obtain the graph.

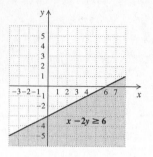

39. Graph: $x + 3y > -1$

$x + 3y < 4$

We graph the lines $x + 3y = -1$ and $x + 3y = 4$, using dashed lines. We indicate the region for each inequality by arrows at the ends of the lines. Note where the regions overlap and shade the region of solutions.

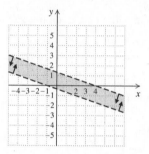

40. Graph: $x - 3y \le 3$

$x + 3y \ge 9$

$y \le 6$

We graph the lines $x - 3y = 3$, $x + 3y = 9$, and $y = 6$ using solid lines. We indicate the region for each inequality by arrows at the ends of the lines. Note where the regions overlap and shade the region of solutions.

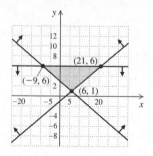

41. $x^3 - 3x > 2x^2$

$x^3 - 2x^2 - 3x > 0$
Solve the related equation.

$x^3 - 2x^2 - 3x = 0$

$x(x^2 - 2x - 3) = 0$

$x(x - 3)(x + 1) = 0$

$x = 0$ or $x - 3 = 0$ or $x + 1 = 0$

$x = 0$ or $x = 3$ or $x = -1$

The zeros are -1, 0, and 3. These zeros divide the number line into four intervals: A, B, C, and D.

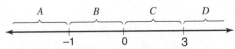

We will select a convenient test point from each interval to determine the sign of $x^3 - 2x^2 - 3x$ for that interval.

Interval A: Let $x = -5$:
$$(-5)^3 - 2(-5)^2 - 3(-5)$$
$$= -125 - 2(25) - 3(-5)$$
$$= -125 - 50 + 15$$
$$= -160$$

Interval B: Let $x = -0.5$:
$$(-0.5)^3 - 2(-0.5)^2$$
$$- 3(-0.5)$$
$$= -0.125 - 2(0.25) - 3(-0.5)$$
$$= -0.125 - 0.5 + 1.5 = 0.875$$

Interval C: Let $x = 1$:
$$(1)^3 - 2(1)^2 - 3(1)$$
$$= 1 - 2(1) - 3(1)$$
$$= 1 - 2 - 3 = -4$$

Interval D: Let $x = 4$:
$$4^3 - 2 \cdot 4^2 - 3 \cdot 4$$
$$= 64 - 2 \cdot 16 - 3 \cdot 4$$
$$= 64 - 32 - 12 = 20$$

We indicate our results on the number line.

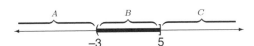

We are looking for the intervals where $x^3 - 2x^2 - 3x > 0$, or the $+$ intervals.

The solution set is $(-1, 0) \cup (3, \infty)$ or $\{x \mid -1 < x < 0 \text{ or } x > 3\}$

42. $\dfrac{x - 5}{x + 3} \le 0; \quad x \ne -3$

Solve the related equation.

$$\frac{x - 5}{x + 3} = 0 \qquad \text{LCD is } x + 3$$

$$(x + 3)\frac{x - 5}{x + 3} = (x + 3) \cdot 0$$

$$x - 5 = 0$$

$$x = 5$$

We must also find any values which make the denominator 0.

$$x + 3 = 0$$

$$x = -3$$

Use 5 and -3 to divide the number line into intervals.

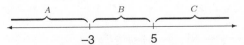

We will select a convenient test point from each interval to determine the sign of

$$\frac{x - 5}{x + 3}$$

Interval A: Let $x = -4$

$$\frac{-4 - 5}{-4 + 3} = \frac{-9}{-1} = 9$$

Interval B: Let $x = 0$

$$\frac{0 - 5}{0 + 3} = \frac{-5}{3}$$

Interval C: Let $x = 10$

$$\frac{10 - 5}{10 + 3} = \frac{5}{13}$$

We indicate our results on the number line.

We are looking for intervals where $\frac{x-5}{x+3}$ is less than 0. We also want the value which makes the expression equal 0, or $x = 5$. The solution is $(-3, 5]$, or $\{x \mid -3 < x \le 5\}$.

43. *Thinking and Writing Exercise.* The equation $|X| = p$ has two solutions when p is positive because X can be either p or $-p$. The same equation has no solution when p is negative because no number has a negative absolute value.

44. *Thinking and Writing Exercise.* The solution set of a system of inequalities is all ordered pairs that make *all* the individual inequalities true. This consists of ordered pairs that are common to all the individual solution sets, or the intersection of the graphs.

45. $|2x+5| \le |x+3|$

This inequality says that the distance from $x+3$ to zero is greater than or equal than the distance from $2x+5$ to zero.

Case 1: $2x+5 \ge 0$ and $x+3 \ge 0$, which
imply that $x \ge -\dfrac{5}{2}$ and $x \ge -3$.

$$2x+5 \le x+3$$
$$x \le -2$$

The intersection of these three
intervals is $\left[-\dfrac{5}{2}, -2\right]$.

Case 2: $2x+5 < 0$ and $x+3 < 0$, which
imply that $x < -\dfrac{5}{2}$ and $x < -3$.

$$-(2x+5) \le -(x+3)$$
$$-2x-5 \le -x-3$$
$$-2 \le x$$

This is impossible so we rule it out.

Case 3: $2x+5 < 0$ and $x+3 \ge 0$, which
imply that $x < -\dfrac{5}{2}$ and $x \ge -3$.

$$-(2x+5) \le x+3$$
$$-2x-5 \le x+3$$
$$-8 \le 3x$$
$$-\dfrac{8}{3} \le x$$

The intersection of these three
intervals is $\left[-\dfrac{8}{3}, -\dfrac{5}{2}\right]$.

Case 4: $2x+5 \ge 0$ and $x+3 < 0$, which
imply that $x \ge -\dfrac{5}{2}$ and $x < -3$.

$$2x+5 \le -(x+3)$$
$$2x+5 \le -x-3$$
$$3x \le -8$$
$$x \le -\dfrac{8}{3}$$

This case is impossible so we rule it out.

The union of the intervals for Cases 1 and 3 is
$$\left[-\dfrac{5}{2}, -2\right] \cup \left[-\dfrac{8}{3}, -\dfrac{5}{2}\right] = \left[-\dfrac{8}{3}, -2\right], \text{ or}$$
$$\left\{x \middle| -\dfrac{8}{3} \le x \le -2\right\}.$$

46. The thickness of the paper ranges from 18 thousandths to 25 thousandths of an inch, or $|18-25| = |-7| = 7$ thousandths. The midpoint of the thickness is $\dfrac{18+25}{2} = \dfrac{43}{2} = 21.5$. Thus, the thickness ranges from 7/2, or 3.5, thousandths of an inch on each side of 21.5 thousandths of an inch. An inequality for which the closed interval is the solution set is $|t - 21.5| \le 3.5$, where t is in thousandths of an inch.

Chapter 8 Test

1. $3-x < 2$

On a graphing calculator, let $y_1 = 3-x$ and $y_2 = 2$. We graph the equations in the window $[-5, 5, -5, 5]$.

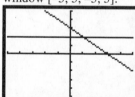

We can see from the graph that the point of intersection is $(1, 2)$, and that $y_1 < y_2$ when $x > 1$. The solution set is $\{x \mid x > 1\}$, or $(1, \infty)$.

2. $2x-3 \ge x+1$

On a graphing calculator, let $y_1 = 2x-3$ and $y_2 = x+1$. We graph the equations in the window $[-10, 10, -10, 10]$.

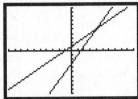

We can see from the graph that the point of intersection is (4, 5), and that $y_1 \geq y_2$ when $x \geq 4$. The solution set is $\{x \mid x \geq 4\}$, or $[4, \infty)$.

3. $\{1,3,5,7,9\} \cap \{3,5,11,13\}$

The numbers 3 and 5 are common to both sets, so the intersection is $\{3,5\}$.

4. $\{1,3,5,7,9\} \cup \{3,5,11,13\}$

The numbers 1,3,5,7,9,11, and 13 are in either or both sets, so the union is $\{1,3,5,7,9,11,13\}$.

5. $f(x) = \sqrt{6-3x}$

The expression $\sqrt{6-3x}$ is not a real number when $6-3x$ is negative. Thus, the domain of f is the set of all x-values for which $6-3x \geq 0$. Since $6-3x \geq 0$ is equivalent $x \leq 2$, we have the domain of $f = (-\infty, 2]$.

6. $f(x) = \dfrac{x}{x-7}$

Note: We exclude numbers which make the denominator equal 0.

$x - 7 = 0$

$\quad x = 7$

The domain is $x \in \mathbb{R}, x \neq 7$, or $(-\infty, 7) \cup (7, \infty)$.

7. $-2 < x - 3 < 5$

$\quad 1 < x < 8$

The solution set is $\{x \mid 1 < x < 8\}$, or $(1,8)$.

8. $-11 \leq -5t - 2 < 0$

$\quad -9 \leq -5t < 2$

$\quad \dfrac{9}{5} \geq t > -\dfrac{2}{5}$

The solution set is $\left\{t \mid -\dfrac{2}{5} < t \leq \dfrac{9}{5}\right\}$, or

$\left(-\dfrac{2}{5}, \dfrac{9}{5}\right]$.

9. $3x - 2 < 7 \ or \ x - 2 > 4$

$\quad 3x < 9 \ or \ x > 6$

$\quad x < 3 \ or \ x > 6$

The solution set is $\{x \mid x < 3 \ or \ x > 6\}$, or $(-\infty, 3) \cup (6, \infty)$.

10. $-3x > 12 \ or \ 4x > -10$

$\quad x < -4 \ or \ x > -\frac{5}{2}$

The solution set is $\left\{x \mid x < -4 \ or \ x > -\dfrac{5}{2}\right\}$,

or $(-\infty, -4) \cup \left(-\dfrac{5}{2}, \infty\right)$.

11. $|x| = 13$

$\quad x = -13 \ or \ x = 13$

The solution set is $\{-13, 13\}$.

12. $|a| > 7$

$\quad a < -7 \ or \ a > 7$

The solution set is $\{a \mid a < -7 \ or \ a > 7\}$, or $(-\infty, -7) \cup (7, \infty)$.

13. $|3x - 1| < 7$

$\quad -7 < 3x - 1 < 7$

$\quad -6 < 3x < 8$

$\quad -2 < x < \dfrac{8}{3}$

The solution set is $\left\{x \mid -2 < x < \dfrac{8}{3}\right\}$, or

$\left(-2, \dfrac{8}{3}\right)$.

14. $|-5t - 3| \geq 10$

$\quad -5t - 3 \leq -10 \ or \ 10 \leq -5t - 3$

$\quad -5t \leq -7 \ or \ 13 \leq -5t$

$\quad t \geq \dfrac{7}{5} \ or \ -\dfrac{13}{5} \geq t$

The solution set is $\left\{ t \middle| t \le -\dfrac{13}{5} \ or \ t \ge \dfrac{7}{5} \right\}$, or

$\left(-\infty, -\dfrac{13}{5} \right] \cup \left[\dfrac{7}{5}, \infty \right)$.

15. $|2 - 5x| = -12$

Absolute value is always nonnegative, so the equation has no solution. The solution set is $\varnothing$.

16. $g(x) < -3 \ or \ g(x) > 3; \ g(x) = 4 - 2x$

$4 - 2x < -3 \ or \ 4 - 2x > 3$

$\qquad -2x < -7 \ or \ -2x > -1$

$\qquad x > \dfrac{7}{2} \quad or \quad x < \dfrac{1}{2}$

The solution set is $\left\{ x \middle| x < \dfrac{1}{2} \ or \ x > \dfrac{7}{2} \right\}$, or

$\left(-\infty, \dfrac{1}{2} \right) \cup \left(\dfrac{7}{2}, \infty \right)$.

17. $f(x) = |x + 10| \ and \ g(x) = |x - 12|$

$f(x) = g(x)$

$|x + 10| = |x - 12|$

$x + 10 = x - 12 \ or \ x + 10 = -(x - 12)$

$10 = -12 \quad or \quad x + 10 = -x + 12$

False - $\qquad\qquad 2x = 2$

yields no $\qquad\qquad x = 1$

solution

The solution set is $\{1\}$.

18. $y \le 2x + 1$

First graph the boundary line $y = 2x + 1$ as a solid line. Since the test point $(0,0)$ is a solution and $(0,0)$ is below the line, we shade the half-plane below the line. The solution is the region which is shaded.

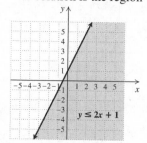

19. $x + y \ge 3$ (1)

$\quad x - y \ge 5$ (2)

First sketch the graph of $x + y = 3$ using a solid line. Since the test point $(0,0)$ is not a solution and $(0,0)$ is below the line, we shade the half-plane above the line (note the arrows pointing up-and-right).

Next, sketch the graph of $x - y = 5$ using a solid line. Since the test point $(0,0)$ is not a solution and $(0,0)$ is above the line, we shade the half-plane below the line (note the arrows pointing down-and-right). The solution is the region which is shaded.

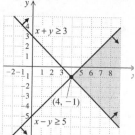

20. $2y - x \ge -7$

$\quad 2y + 3x \le 15$

$\quad\quad y \le 0$

$\quad\quad x \le 0$

First sketch the graph of $2y - x = -7$ using a solid line. Since the test point $(0,0)$ is a solution and $(0,0)$ is above the line, we shade the half-plane above the line (note the arrows pointing up-and-left).

Next, sketch the graph of $2y + 3x = 15$ using a solid line. Since the test point $(0,0)$ is a solution and $(0,0)$ is below the line, we shade the half-plane below the line (note the arrows pointing down-and-left).

We graph the equation $y = 0$ and note that the graph of $y \le 0$ is the half-plane below the x-axis.

We graph the equation $x = 0$ and note that the graph of $x \le 0$ is the half-plane left of the y-axis.

The solution is the region which is shaded which is where all the regions overlap.

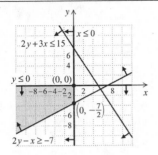

21. $x^2 + 5x \le 6$

$x^2 + 5x - 6 \le 0$

We find the zeros of the relation equation.

$x^2 + 5x - 6 = 0$

$(x+6)(x-1) = 0$

$x + 6 = 0$ or $x - 1 = 0$

$x = -6$ or $x = 1$

These zeros divide the number line into three intervals, A, B, and C.

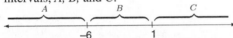

We choose a convenient test point from each interval to determine the sign of the interval.

X	Y₁	
-10	44	
0	-6	
3	18	
X=-10		

Since $x^2 + 5x - 6 \le 0$, we want the interval where the sign is negative, or interval B. We also want the zeros since $x^2 + 5x - 6 = 0$. Our solution is $\{x \mid -6 \le x \le 1\}$ or $[-6, 1]$.

22. $x - \dfrac{1}{x} > 0,\ x \ne 0$

We find the zeros of the related equation.

$x - \dfrac{1}{x} = 0,$ LCD is x

$x\left(x - \dfrac{1}{x} \right) = 0$

$x^2 - 1 = 0$

$(x+1)(x-1) = 0$

$x = -1$ or $x = 1$

As already noted, $x \ne 0$.

These three numbers divide the number line into four intervals: A, B, C, and D.

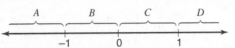

We choose a convenient test point from each interval to determine the sign of the interval.

X	Y₁	
-4	-3.75	
-.5	1.5	
.5	-1.5	
3	2.6667	
X=-4		

Since $x - \dfrac{1}{x} > 0$, we want the interval(s) where the sign is positive, so we choose intervals B and D.

Our solution is $\{x \mid -1 < x < 0,\ \text{or}\ x > 1\}$, or $(-1, 0) \cup (1, \infty)$.

23. $|2x - 5| \le 7\ \ and\ \ |x - 2| \ge 2$

For the inequality $|2x - 5| \le 7$:

$-7 \le 2x - 5 \le 7$

$-2 \le 2x \le 12$

$-1 \le x \le 6$

For the inequality $|x - 2| \ge 2$:

$x - 2 \le -2\ \ or\ \ 2 \le x - 2$

$x \le 0\ \ \ or\ \ 4 \le x$

The solution set is $[-1, 0] \cup [4, 6]$.

24. $7x < 8 - 3x < 6 + 7x$

For the inequality $7x < 8 - 3x$:

$7x < 8 - 3x$

$10x < 8$

$x < \dfrac{4}{5}$

For the inequality $8 - 3x < 6 + 7x$:

$-10x < -2$

$x > \dfrac{1}{5}$

The solution is $\left(\dfrac{1}{5}, \dfrac{4}{5} \right)$.

25. The length of the segment from -8 to 2 is
$\left|-8-2\right| = \left|-10\right| = 10$ units. The midpoint of

the segment is $\dfrac{-8+2}{2} = \dfrac{-6}{2} = -3$. Thus, the

interval extends 10/2, or 5, units on each side
of -3. An inequality for which the closed
interval is the solution set is $\left|x-(-3)\right| \le 5$, or
$\left|x+3\right| \le 5$.

Chapter 9

More on Systems

1. True

3. False

5. True

7. Substitute $(2,-1,-2)$ into the three equations, using alphabetical order.

$$x + y - 2z = 5$$
$$\frac{2+(-1)-2(-2)}{\;} \mid 5$$
$$2-1+4$$
$$\qquad\qquad 5 \mid 5 \quad \text{True}$$

$$2x - y - z = 7$$
$$\frac{2\cdot 2-(-1)-(-2)}{\;} \mid 7$$
$$4+1+2$$
$$\qquad\qquad 7 \mid 7 \quad \text{True}$$

$$-x - 2y - 3z = 6$$
$$\frac{-2-2(-1)-3(-2)}{\;} \mid 6$$
$$-2+2+6$$
$$\qquad\qquad 6 \mid 6 \quad \text{True}$$

The triple $(2,-1,-2)$ is a solution to the system.

9. $\qquad x - y - z = 0 \quad (1)$
$\qquad 2x - 3y + 2z = 7 \quad (2)$
$\qquad -x + 2y + z = 1 \quad (3)$

1., 2. The equations are already in standard form with no fractions or decimals.

3. Use Equations (1) and (3) to eliminate x:

$$x - y - z = 0 \quad (1)$$
$$\frac{-x + 2y + z = 1}{\;} \quad (3)$$
$$y = 1 \qquad\qquad (4)$$

4. Use a different pair of equations and eliminate x:

$$2x - 3y + 2z = 7 \quad (2)$$
$$\frac{-2x + 4y + 2z = 2}{\;} \quad \text{Mult. (3) by 2}$$
$$y + 4z = 9 \qquad\qquad (5)$$

5. Now solve the system of Equations (4) and (5).

$$y = 1 \qquad (4)$$
$$y + 4z = 9 \quad (5)$$

Substitute $y = 1$ into (5):

$$1 + 4z = 9$$
$$4z = 8$$
$$z = 2$$

6. Substitute in one of the original equations to find x.

$$x - 1 - 2 = 0 \quad \text{Substituting in } (1)$$
$$x - 3 = 0$$
$$x = 3$$

We obtain $(3,1,2)$. This checks, so it is the solution.

11. $\qquad x - y - z = 1 \quad (1)$
$\qquad 2x + y + 2z = 4 \quad (2)$
$\qquad x + y + 3z = 5 \quad (3)$

1., 2. The equations are already in standard form with no fractions or decimals.

3. We eliminate y from two different pairs of equations.

$$x - y - z = 1 \quad (1)$$
$$\frac{2x + y + 2z = 4}{\;} \quad (2)$$
$$3x \qquad + z = 5 \quad (4) \quad \text{Adding}$$

4. Use a different pair of equations and eliminate y:

$$x - y - z = 1 \quad (1)$$
$$\frac{x + y + 3z = 5}{\;} \quad (3)$$
$$2x \qquad + 2z = 6 \quad (5)$$

5. Now solve the system of Equations (4) and (5).

$$3x + z = 5 \quad (4)$$
$$2x + 2z = 6 \quad (5)$$

$$-6\dot{x}-2z=-10 \quad \text{Mult. } (4) \text{ by} -2$$
$$\underline{2x+2z=6 \quad (5)}$$
$$-4x=-4$$
$$x=1$$

$$3(1)+z=5 \quad \text{Substituting in } (4)$$
$$3+z=5$$
$$z=2$$

6. Substitute in one of the original equations to find y.
$$1-y-2=1 \quad \text{Substituting in } (1)$$
$$-y-1=1$$
$$-y=2$$
$$y=-2$$

We obtain $(1,-2,2)$. This checks, so it is the solution.

13. $3x +4y-3z = 4 \quad (1)$
$5x - y +2z = 3 \quad (2)$
$x + 2y - z = -2 \quad (3)$

1., 2. The equations are already in standard form with no fractions or decimals.

3., 4. We eliminate y from two different pairs of equations.
$$3x +4y-3z = 4 \quad (1)$$
$$\underline{-2x-4y+2z = 4 \quad \text{Mult. } (3) \text{ by} -2}$$
$$x- \quad z = 8 \quad (4) \text{ Adding}$$

Use a different pair of equations and eliminate y:
$$10x -2y+4z = 6 \quad \text{Mult. } (2) \text{ by } 2$$
$$\underline{x +2y-z = -2 \quad (3)}$$
$$11x \quad +3z = 4 \quad (5) \text{ Adding}$$

5. Now solve the system of Equations (4) and (5).
$$x-z=8 \quad (4)$$
$$11x+3z = 4 \quad (5)$$

$$3x-3z = 24 \quad \text{M. } (4) \text{ by } 3$$
$$\underline{11x+3z = 4 \quad (5)}$$
$$14x = 28$$
$$x=2$$

$$2-z=8 \quad \text{Substituting in } (4)$$
$$-z=6$$
$$z=-6$$

6. Substitute in one of the original equations to find y.
$$2+2y-(-6)=-2 \quad \text{Substituting in } (3)$$
$$2y+8=-2$$
$$2y=-10$$
$$y=-5$$

We obtain $(2,-5,-6)$. This checks, so it is the solution.

15. $x + y +z = 0 \quad (1)$
$2x +3y+ 2z = -3 \quad (2)$
$-x -2y -z = 1 \quad (3)$

1., 2. The equations are already in standard form with no fractions or decimals.

3., 4. We eliminate z from two different pairs of equations.
$$-2x-2y-2z = 0 \quad \text{Mult. } (1) \text{ by} -2$$
$$\underline{2x+3y+2z = -3 \quad (2)}$$
$$y = -3 \quad (4) \text{ Adding}$$

Use a different pair of equations and eliminate x:
$$x+ y+z = 0 \quad (1)$$
$$\underline{-x-2y-z = 1 \quad (3)}$$
$$-y = 1 \quad \text{Adding}$$
$$y = -1 \quad (5)$$

5. Now solve the system of Equations (4) and (5).
$$y = -3 \quad (4)$$
$$y = -1 \quad (5)$$

$$y = -3 \quad (4)$$
$$\underline{-y = 1 \quad \text{Mult. } (5) \text{ by} -1}$$
$$0 = -2$$

We get a false equation, or a contradiction. There is no solution.

17. $2x-3y-z = -9 \quad (1)$
$2x+5y+z = 1 \quad (2)$
$x-y+z = 3 \quad (3)$

1., 2. The equations are already in standard form with no fractions or decimals.

3., 4. We eliminate z from two different pairs of equations.

$$2x - 3y - z = -9 \quad (1)$$
$$\underline{2x + 5y + z = 1 \quad (2)}$$
$$4x + 2y = -8 \quad (4) \text{ Adding}$$

Use a different pair of equations and eliminate z:

$$2x - 3y - z = -9 \quad (1)$$
$$\underline{x - y + z = 3 \quad (3)}$$
$$3x - 4y = -6 \quad (5) \text{ Adding}$$

5. Now solve the system of Equations (4) and (5).

$$4x + 2y = -8 \quad (4)$$
$$3x - 4y = -6 \quad (5)$$

$$8x + 4y = -16 \quad \text{M. } (4) \text{ by } 2$$
$$\underline{3x - 4y = -6 \quad (5)}$$
$$11x = -22$$
$$x = -2$$

$$3(-2) - 4y = -6 \quad \text{Subst. in } (5)$$
$$-6 - 4y = -6$$
$$-4y = 0$$
$$y = 0$$

6. Substitute in one of the original equations to find z.

$$x - y + z = 3 \quad \text{Sub. } (3)$$
$$-2 - 0 + z = 3$$
$$z = 5$$

We obtain $(-2, 0, 5)$. This checks, so it is the solution.

19. $a + b + c = 5 \quad (1)$
$$2a + 3b - c = 2 \quad (2)$$
$$2a + 3b - 2c = 4 \quad (3)$$

1., 2. The equations are already in standard form with no fractions or decimals.
3., 4. We eliminate a from two different pairs of equations.

$$-2a - 2b - 2c = -10 \quad \text{Mult. } (1) \text{ by } -2$$
$$\underline{2a + 3b - 2c = 4 \quad (2)}$$
$$b - 4c = -6 \quad (4) \text{ Adding}$$

Use a different pair of equations and eliminate a:

$$2a + 3b - c = 2 \quad (2)$$
$$\underline{-2a - 3b + 2c = -4 \quad \text{M. } (3) \text{ by } -1}$$
$$c = -2 \quad (5) \text{ Adding}$$

5. Now solve the system of Equations (4) and (5).

$$b - 4c = -6 \quad (4)$$
$$c = -2 \quad (5)$$

Substitute $c = -2$ into (4):

$$b - 4(-2) = -6$$
$$b = -14$$

6. Substitute in one of the original equations to find a.

$$a + (-14) + (-2) = 5 \quad \text{Sub. in } (1)$$
$$a - 16 = 5$$
$$a = 21$$

We obtain $(21, -14, -2)$. This checks, so it is the solution.

21. $-2x + 8y + 2z = 4 \quad (1)$
$$x + 6y + 3z = 4 \quad (2)$$
$$3x - 2y + z = 0 \quad (3)$$

1., 2. The equations are already in standard form with no fractions or decimals.
3., 4. We eliminate z from two different pairs of equations.

$$-2x + 8y + 2z = 4 \quad (1)$$
$$\underline{-6x + 4y - 2z = 0 \quad \text{Mult. } (3) \text{ by } -2}$$
$$-8x + 12y = 4 \quad (4) \text{ Adding}$$

Use a different pair of equations and eliminate z:

$$x + 6y + 3z = 4 \quad (2)$$
$$\underline{-9x + 6y - 3z = 0 \quad \text{M. } (3) \text{ by } -3}$$
$$-8x + 12y = 4 \quad (5) \text{ Adding}$$

5. Now solve the system of Equations (4) and (5).

$$-8x + 12y = 4 \quad (4)$$
$$-8x + 12y = 4 \quad (5)$$

$$-8x + 12y = 4 \quad (4)$$
$$\underline{8x - 12y = -4 \quad \text{M. } (5) \text{ by } -1}$$
$$0 = 0 \quad (6)$$

Equation (6) indicates that Equation (1), (2), and (3) are dependent. (Note that if Equation

(1) is subtracted from Equation (2), the result is Equation (3).) We could also have concluded that the equations are dependent by observing that Equations (4) and (5) are identical.

23. $2u - 4v - w = 8$ (1)

$3u + 2v + w = 6$ (2)

$5u - 2v + 3w = 2$ (3)

1., 2. The equations are already in standard form with no fractions or decimals.

3., 4. We eliminate w from two different pairs of equations.

$$2u - 4v - w = 8 \quad (1)$$
$$\underline{3u + 2v + w = 6 \quad (2)}$$
$$5u - 2v = 14 \quad (4) \text{ Adding}$$

Use a different pair of equations and eliminate w:

$$6u - 12v - 3w = 24 \quad \text{M. } (1) \text{ by } 3$$
$$\underline{5u - 2v + 3w = 2 \quad (3)}$$
$$11u - 14v = 26 \quad (5) \text{ Adding}$$

5. Now solve the system of Equations (4) and (5).

$$5u - 2v = 14 \quad (4)$$
$$11u - 14v = 26 \quad (5)$$

$$-35u + 14v = -98 \quad \text{M. } (4) \text{ by } -7$$
$$\underline{11u - 14v = 26 \quad \quad (5)}$$
$$-24u \quad \quad = -72$$
$$u = 3$$

$$5(3) - 2v = 14$$
$$15 - 2v = 14 \quad \text{Subst. in } (4)$$
$$-2v = -1$$
$$v = \frac{1}{2}$$

6. Substitute in one of the original equations to find w.

$$2(3) - 4\left(\frac{1}{2}\right) - w = 8 \quad \text{Sub. in } (1)$$
$$-w = 4$$
$$w = -4$$

We obtain $\left(3, \dfrac{1}{2}, -4\right)$. This checks, so it is the solution.

25. $r + \dfrac{3}{2}s + 6t = 2$ (1)

$2r - 3s + 3t = 0.5$ (2)

$r + s + t = 1$ (3)

1. The equations are already in standard form.

2. Multiply the first equation by 2 to clear the fraction. Also, multiply the second equation by 10 to clear the decimal.

$2r + 3s + 12t = 4$ (1)

$20r - 30s + 30t = 5$ (2)

$r + s + t = 1$ (3)

3., 4. We eliminate s from two different pairs of equations.

$$20r + 30s + 120t = 40 \quad \text{Mult. } (1) \text{ by } 10$$
$$\underline{20r - 30s + 30t = 5 \quad (2)}$$
$$40r \quad \quad + 150t = 45 \quad (4) \text{ Adding}$$

Use a different pair of equations and eliminate s:

$$20r - 30s + 30t = 5 \quad (2)$$
$$\underline{30r + 30s + 30t = 30 \quad \text{M. } (3) \text{ by } 30}$$
$$50r \quad \quad + 60t = 35 \quad (5) \text{ Adding}$$

5. Now solve the system of Equations (4) and (5).

$$40r + 150t = 45 \quad (4)$$
$$50r + 60t = 35 \quad (5)$$

$$200r + 750t = 225 \quad \text{M. } (4) \text{ by } 5$$
$$\underline{-200r - 240t = -140 \quad \text{M. } (5) \text{ by } -4}$$
$$510t = 85$$
$$t = \frac{85}{510}$$
$$t = \frac{1}{6}$$

$$40r + 150\left(\frac{1}{6}\right) = 45 \quad \text{Subst. in } (4)$$
$$40r + 25 = 45$$
$$40r = 20$$
$$r = \frac{1}{2}$$

6. Substitute in one of the original equations to find s.

$$\frac{1}{2} + s + \frac{1}{6} = 1 \quad \text{Sub. in } (3)$$

$$s + \frac{2}{3} = 1$$

$$s = \frac{1}{3}$$

We obtain $\left(\frac{1}{2}, \frac{1}{3}, \frac{1}{6}\right)$. This checks, so it is the solution.

27. $\quad 4a + 9b \qquad = 8 \quad (1)$

$\quad\; 8a \qquad\; + 6c = -1 \quad (2)$

$\qquad\quad 6b + 6c = -1 \quad (3)$

1., 2. The equations are already in standard form with no fractions or decimals.

3., 4. Note that there is no c in Equation (1). We will use Equations (2) and (3) to obtain another equation with no c–term.

$$8a \qquad + 6c = -1 \quad (2)$$

$$\underline{\qquad -6b - 6c = \;\; 1 \quad \text{Mult. } (3) \text{ by } -1}$$

$$8a - 6b \qquad = \;\; 0 \quad (4) \text{ Adding}$$

5. Now solve the system of Equations (1) and (4).

$$-8a - 18b = -16 \quad \text{Mult. } (1) \text{ by } -2$$

$$\underline{\;\;8a - \;\;6b = \quad 0}$$

$$-24b = -16$$

$$b = \frac{2}{3}$$

$$8a - 6\left(\frac{2}{3}\right) = 0 \quad \text{Subst. in } (4)$$

$$8a - 4 = 0$$

$$8a = 4$$

$$a = \frac{1}{2}$$

6. Substitute in Equations (2) or (3) to find c.

$$8\left(\frac{1}{2}\right) + 6c = -1 \quad \text{Sub. in } (2)$$

$$4 + 6c = -1$$

$$6c = -5$$

$$c = -\frac{5}{6}$$

We obtain $\left(\frac{1}{2}, \frac{2}{3}, -\frac{5}{6}\right)$. This checks, so it is the solution.

29. $\quad x + y + z = 57 \quad (1)$

$\;\; -2x + y \qquad = \;\; 3 \quad (2)$

$\qquad x \quad\;\; - z = \;\; 6 \quad (3)$

1., 2. The equations are already in standard form with no fractions or decimals.

3., 4. Note that there is no z in Equation (2). We will use Equations (1) and (3) to obtain another equation with no z–term.

$$x + y + z = 57 \quad (1)$$

$$\underline{\;\;x \qquad - z = \;\; 6 \quad (3)}$$

$$2x \qquad + y = 63 \quad (4) \text{ Adding}$$

5. Now solve the system of Equations (2) and (4).

$$-2x + y = \;\; 3 \quad (2)$$

$$2x + y = 63 \quad (4)$$

$$-2x + y = \;\; 3 \quad (2)$$

$$\underline{\;\;2x + y = 63 \quad (4)}$$

$$2y = 66$$

$$y = 33$$

$$2x + 33 = 63 \quad \text{Subst. in } (4)$$

$$2x = 30$$

$$x = 15$$

6. Substitute in Equations (1) or (3) to find z.

$$15 - z = 6 \quad \text{Sub. 15 for } x \text{ in } (3)$$

$$9 = z$$

We obtain $(15, 33, 9)$. This checks, so it is the solution.

31.

$$a \qquad -3c = 6 \quad (1)$$
$$b + 2c = 2 \quad (2)$$
$$7a - 3b - 5c = 14 \quad (3)$$

1., 2. The equations are already in standard form with no fractions or decimals.

3., 4. Note that there is no b in Equation (1). We will use Equations (2) and (3) to obtain another equation with no b –term.

$$3b + 6c = 6 \quad \text{Mult. (2) by 3}$$
$$\underline{7a - 3b - 5c = 14 \quad (3)}$$
$$7a \qquad + c = 20 \quad (4) \text{ Adding}$$

5. Now solve the system of Equations (1) and (4).

$$a - 3c = 6 \quad (1)$$
$$7a + c = 20 \quad (4)$$

$$a - 3c = 6 \quad (1)$$
$$\underline{21a + 3c = 60 \quad \text{Mult. (4) by 3}}$$
$$22a = 66$$
$$a = 3$$

$$3 - 3c = 6 \quad \text{Subst. in (1)}$$
$$-3c = 3$$
$$c = -1$$

6. Substitute in Equations (2) or (3) to find b.

$$b + 2(-1) = 2 \quad \text{Sub. in (2)}$$
$$b - 2 = 2$$
$$b = 4$$

We obtain $(3, 4, -1)$. This checks, so it is the solution.

33.

$$x + y + z = 83 \quad (1)$$
$$y = 2x + 3 \quad (2)$$
$$z = 40 + x \quad (3)$$

1., 2. Equations (2) and (3) are not in standard form. However, both equations (2) and (3) have variables y and z written in terms of the variable x. Substitute $2x + 3$ for y in Equation (1), and substitute $40 + x$ for z in Equation (1).

$$x + y + z = 83$$
$$x + 2x + 3 + 40 + x = 83$$
$$4x + 43 = 83$$
$$4x = 40$$
$$x = 10$$

3. Substitute 10 for x in Equation (2) to solve for y.

$$y = 2x + 3$$
$$y = 2 \cdot 10 + 3$$
$$y = 23$$

Substitute 10 for x in Equation (3) to solve for z.

$$z = 40 + x$$
$$z = 40 + 10$$
$$z = 50$$

We obtain $(10, 23, 50)$. This checks, so it is the solution.

35.

$$x \qquad + z = 0 \quad (1)$$
$$x + y + 2z = 3 \quad (2)$$
$$y + z = 2 \quad (3)$$

1., 2. The equations are already in standard form with no fractions or decimals.

3., 4. Note that there is no y in Equation (1). We will use Equations (2) and (3) to obtain another equation with no y –term.

$$x + y + 2z = 3 \quad (2)$$
$$\underline{-y - z = -2 \quad \text{Mult. (3) by } -1}$$
$$x \qquad + z = 1 \quad (4) \text{ Adding}$$

5. Now solve the system of Equations (1) and (4).

$$x + z = 0 \quad (1)$$
$$x + z = 1 \quad (4)$$

$$x + z = 0 \quad (1)$$
$$\underline{-x - z = -1 \quad \text{Mult. (4) by } -1}$$
$$0 = -1 \quad \text{Adding}$$

We get a false equation, or contradiction. There is no solution.

37. $\quad x + y + z = 1 \quad (1)$

$\quad -x + 2y + z = 2 \quad (2)$

$\quad 2x - y = -1 \quad (3)$

1., 2. The equations are already in standard form with no fractions or decimals.

3. Note that there is no z in Equation (3). We will use Equations (1) and (2) to eliminate z:

$\quad x + y + z = 1 \quad (1)$

$\quad \underline{x - 2y - z = -2} \quad$ Mult. (2) by -1

$\quad 2x - y = -1 \quad (4)$ Adding

Equations (3) and (4) are identical, so Equations (1), (2), and (3) are dependent. (We have seen that if Equation (2) is multiplied by –1 and added to Equation (1), the result is Equation (3).)

39. *Thinking and Writing Exercise.*

41. One number is half another.
Let x and y represent the two numbers, then translate.

$$x = \frac{1}{2}y$$

43. The sum of three consecutive numbers is 100.
Let x represent the first number, so $(x + 1)$ and $(x + 2)$ represent the other numbers, then translate.
$$x + (x + 1) + (x + 2) = 100$$

45. The product of two numbers is five times a third number.
Let x, y, and z represent the three numbers, then translate.
$$xy = 5z$$

47. *Thinking and Writing Exercise.*

49. $\dfrac{x+2}{3} - \dfrac{y+4}{2} + \dfrac{z+1}{6} = 0$

$\dfrac{x-4}{3} + \dfrac{y+1}{4} - \dfrac{z-2}{2} = -1$

$\dfrac{x+1}{2} + \dfrac{y}{2} + \dfrac{z-1}{4} = \dfrac{3}{4}$

1., 2. We clear fractions and write each equation in standard form.

To clear fractions we multiply both sides of each equation by the LCM of its

denominators. The LCMs are 6, 12, and 4, respectively.

$$6\left(\frac{x+2}{3} - \frac{y+4}{2} + \frac{z+1}{6}\right) = 6 \cdot 0$$

$$2(x+2) - 3(y+4) + (z+1) = 0$$

$$2x + 4 - 3y - 12 + z + 1 = 0$$

$$2x - 3y + z = 7$$

$$12\left(\frac{x-4}{3} + \frac{y+1}{4} - \frac{z-2}{2}\right) = 12 \cdot (-1)$$

$$4(x-4) + 3(y+1) - 6(z-2) = -12$$

$$4x - 16 + 3y + 3 - 6z + 12 = -12$$

$$4x + 3y - 6z = -11$$

$$4\left(\frac{x+1}{2} + \frac{y}{2} + \frac{z-1}{4}\right) = 4 \cdot \frac{3}{4}$$

$$2(x+1) + 2(y) + (z-1) = 3$$

$$2x + 2 + 2y + z - 1 = 3$$

$$2x + 2y + z = 2$$

The resulting system is

$2x - 3y + z = 7 \quad (1)$

$4x + 3y - 6z = -11 \quad (2)$

$2x + 2y + z = 2 \quad (3)$

3., 4. We eliminate z from two different pairs of equations.

$12x - 18y + 6z = 42 \quad$ Mult. (1) by 6

$\underline{4x + 3y - 6z = -11} \quad (2)$

$16x - 15y = 31 \quad (4)$ Adding

$2x - 3y + z = 7 \quad (1)$

$\underline{-2x - 2y - z = -2} \quad$ Mult. (3) by -1

$-5y = 5 \quad (5)$ Adding

5. Solve (5) for y: $\quad -5y = 5$

$$y = -1$$

Substitute –1 for y in (4):

$16x - 15(-1) = 31$

$16x + 15 = 31$

$16x = 16$

$x = 1$

6. Substitute 1 for x and –1 for y in (1):

$2 \cdot 1 - 3(-1) + z = 7$

$5 + z = 7$

$z = 2$

We obtain $(1,-1,2)$. This checks, so it is the solution.

51.
$$
\begin{array}{rcl}
w+\ x-\ y+\ z &=&\ 0\quad(1)\\
w-2x-2y-\ z &=&-5\quad(2)\\
w-3x-\ y+\ z &=&\ 4\quad(3)\\
2w-\ x-\ y+3z &=&\ 7\quad(4)
\end{array}
$$

The equations are already in standard form with no fractions or decimals.
Start by eliminating z from three different pairs of equations.

$$
\begin{array}{rcl}
w+\ x-\ y+z &=&\ 0\quad(1)\\
w-2x-2y-z &=&-5\quad(2)\\
\hline
2w\ -x-3y\ \ \ \ \ &=&-5\quad(5)\ \text{Adding}
\end{array}
$$

$$
\begin{array}{rcl}
w-2x-2y-z &=&-5\quad(2)\\
w-3x-\ y+z &=&\ 4\quad(3)\\
\hline
2w-5x-3y\ \ \ \ \ &=&-1\quad(6)\ \text{Adding}
\end{array}
$$

$$
\begin{array}{rcl}
3w-6x-6y-3z &=&-15\quad\text{Mult. (2) by 3}\\
2w-\ x-\ y+3z &=&\ 7\quad(4)\\
\hline
5w-7x-7y\ \ \ \ \ \ &=&-8\quad(7)\ \text{Adding}
\end{array}
$$

Now solve the system of equations (5), (6), and (7).

$$
\begin{array}{rcl}
2w-x-3y &=&-5\quad(5)\\
2w-5x-3y &=&-1\quad(6)\\
5w-7x-7y &=&-8\quad(7)
\end{array}
$$

$$
\begin{array}{rcl}
2w-\ x-3y &=&-5\quad(5)\\
-2w+5x+3y &=&\ 1\quad\text{Mult. (6) by }-1\\
\hline
4x\ \ \ \ \ \ &=&-4\\
x\ \ \ \ \ &=&-1
\end{array}
$$

Substituting -1 for x in (5) and (7) and simplifying, we have
$$
\begin{array}{rcl}
2w-3y &=&-6\quad(8)\\
5w-7y &=&-15\quad(9)
\end{array}
$$

Now solve the system of Equations (8) and (9).

$$
\begin{array}{rcl}
10w-15y &=&-30\quad\text{Mult. (8) by 5}\\
-10w+14y &=&\ 30\quad\text{Mult. (9) by }-2\\
\hline
-y &=&\ 0\\
y &=&\ 0
\end{array}
$$

Substitute 0 for y in Equation (8) or (9) and solve for w.

$$
\begin{array}{rcl}
2w-3\cdot0 &=&-6\quad\text{Subst. in (8)}\\
2w &=&-6\\
w &=&-3
\end{array}
$$

Substitute in one of the original equations to find z.

$$
\begin{array}{rcl}
-3-1-0+z &=&0\quad\text{Subst. in (1)}\\
-4+z &=&0\\
z &=&4
\end{array}
$$

We obtain $(-3,-1,0,4)$. This checks, so it is the solution.

53.
$$
\begin{array}{rcl}
\dfrac{2}{x}+\dfrac{2}{y}-\dfrac{3}{z} &=&\ 3\quad(1)\\[2mm]
\dfrac{1}{x}-\dfrac{2}{y}-\dfrac{3}{z} &=&\ 9\quad(2)\\[2mm]
\dfrac{7}{x}-\dfrac{2}{y}+\dfrac{9}{z} &=&-39\quad(3)
\end{array}
$$

Let u represent $\dfrac{1}{x}$, v represent $\dfrac{1}{y}$, and w represent $\dfrac{1}{z}$. Substituting, we have
$$
\begin{array}{rcl}
2u+2v-3w &=&\ 3\quad(1)\\
u-2v-3w &=&\ 9\quad(2)\\
7u-2v+9w &=&-39\quad(3)
\end{array}
$$

1., 2. The equations in u, v, and w are in standard form with no fractions or decimals.

3., 4. We eliminate v from two different pairs of equations.

$$
\begin{array}{rcl}
2u+2v-3w &=&3\quad(1)\\
u-2v-3w &=&9\quad(2)\\
\hline
3u\ \ \ \ \ \ -6w &=&12\quad(4)\ \text{Adding}
\end{array}
$$

$$
\begin{array}{rcl}
2u+2v-3w &=&\ 3\quad(1)\\
7u-2v+9w &=&-39\quad(3)\\
\hline
9u\ \ \ \ \ \ +6w &=&-36\quad(5)\ \text{Adding}
\end{array}
$$

5. Now solve the system of Equations (4) and (5).

$$3u - 6w = 12 \quad (4)$$
$$9u + 6w = -36 \quad (5)$$
$$\overline{12u = -24}$$
$$u = -2$$

$$3(-2) - 6w = 12 \quad \text{Subst. in } (4)$$
$$-6 - 6w = 12$$
$$-6w = 18$$
$$w = -3$$

6. Substitute in Equation (1), (2), or (3) to find v.

$$2(-2) + 2v - 3(-3) = 3 \quad \text{Subst. in } (1)$$
$$2v + 5 = 3$$
$$2v = -2$$
$$v = -1$$

Solve for x, y, and z. We substitute -2 for u, -1 for v and -3 for w.

$$u = \frac{1}{x} \qquad v = \frac{1}{y} \qquad w = \frac{1}{z}$$

$$-2 = \frac{1}{x} \qquad -1 = \frac{1}{y} \qquad -3 = \frac{1}{z}$$

$$x = -\frac{1}{2} \qquad y = -1 \qquad z = -\frac{1}{3}$$

We obtain $\left(-\dfrac{1}{2}, -1, -\dfrac{1}{3}\right)$. This checks, so it is the solution.

55.
$$5x - 6y + kz = -5 \quad (1)$$
$$x + 3y - 2z = 2 \quad (2)$$
$$2x - y + 4z = -1 \quad (3)$$

1., 2. The equations in x, y, and z are in standard form with no fractions or decimals.

3., 4. We eliminate y from two different pairs of equations.

$$5x - 6y + kz = -5 \quad (1)$$
$$2x + 6y - 4z = 4 \quad \text{Mult. } (2) \text{ by } 2$$
$$\overline{7x + (k-4)z = -1} \quad (4) \text{ Adding}$$

$$x + 3y - 2z = 2 \quad (2)$$
$$6x - 3y + 12z = -3 \quad \text{Mult. } (3) \text{ by } 3$$
$$\overline{7x + 10z = -1} \quad (5)$$

5. Now solve the system of equations (4) and (5).

$$7x + (k-4)z = -1 \quad (4)$$
$$7x + 10z = -1 \quad (5)$$

$$-7x + (k-4)z = -1 \quad \text{M. } (4) \text{ by } -1$$
$$7x + 10z = -1 \quad (5)$$
$$\overline{(-k+14)z = 0} \quad (6)$$

The system is dependent for the value of k that makes Equation (6) true. This occurs when $-k+14$ is 0. We solve for k:

$$-k + 14 = 0$$
$$14 = k$$

57. $z = b - mx - ny$

Three solutions are $(1,1,2)$, $(3,2,-6)$, and $\left(\dfrac{3}{2},1,1\right)$. We substitute for x, y, and z and then solve for b, m, and n.

$$2 = b - m - n$$
$$-6 = b - 3m - 2n$$
$$1 = b - \frac{3}{2}m - n$$

1., 2. Write the equations in standard form. Also clear the fractions in the last equation.

$$b - m - n = 2 \quad (1)$$
$$b - 3m - 2n = -6 \quad (2)$$
$$2b - 3m - 2n = 2 \quad (3)$$

3., 4. Eliminate b from two different pairs of equations.

$$b - m - n = 2 \quad (1)$$
$$-b + 3m + 2n = 6 \quad \text{Mult. } (2) \text{ by } -1$$
$$\overline{2m + n = 8} \quad (4)$$

$$-2b + 2m + 2n = -4 \quad \text{Mult. } (1) \text{ by } -2$$
$$2b - 3m - 2n = 2 \quad (3)$$
$$\overline{-m = -2} \quad (5) \text{ Adding}$$

5. We solve Equation (5) for m:

$$-m = -2$$
$$m = 2$$

Substitute in Equation (4) and solve for n.

$$2 \cdot 2 + n = 8$$
$$4 + n = 8$$
$$n = 4$$

6. Substitute in one of the original equations to find b.

$$b - 2 - 4 = 2 \quad \text{Subst. 2 for } m \text{ and}$$
$$\qquad\qquad\qquad 4 \text{ for } n \text{ in (1)}$$
$$b - 6 = 2$$
$$b = 8$$

The solution is $(8, 2, 4)$, so the equation is $z = 8 - 2x - 4y$.

Exercise Set 9.2

---

1. ***Familiarize.*** Let x = the first number, y = the second number, and z = the third number.

    Translate.

    The sum of three numbers is 57.

    $$x + y + z \qquad = \quad 57$$

    The second is 3 more than the first.

    $$y \qquad = 3 \quad + \qquad x$$

    The third is 6 more than the first.

    $$z \qquad = 6 \quad + \qquad x$$

    We now have a system of equations.
    $$\begin{array}{ll} x + y + z = 57 & \text{or} \quad x + y + z = 57 \\ y = 3 + x & \quad\quad -x + y = 3 \\ z = 6 + x & \quad\quad -x + z = 6 \end{array}$$

    Carry out. Solving the system we get $(16, 19, 22)$.

    Check. The sum of the three numbers is $16 + 19 + 22$, or 57. The second number, 19, is three more than the first number, 16. The third number, 22, is 6 more than the first number, 16. The numbers check.

    State. The numbers are 16, 19, and 22.

3. ***Familiarize.*** Let x = the first number, y = the second number, and z = the third number.

    Translate.

    The sum of three numbers is 26.

    $$x + y + z \qquad = \quad 26$$

    Twice the first minus the second is the third less 2.

    $$2x \quad - \quad y \quad = \quad z \quad - \quad 2$$

    The third is the second minus 3 times the first.

    $$z \quad = \quad y \quad - \quad 3x$$

    We now have a system of equations.
    $$\begin{array}{ll} x + y + z = 26 & \text{or} \quad x + y + z = 26 \\ 2x - y = z - 2 & \quad\quad 2x - y - z = -2 \\ z = y - 3x & \quad\quad 3x - y + z = 0 \end{array}$$

    Carry out. Solving the system we get $(8, 21, -3)$.

    Check. The sum of the three numbers is $8 + 21 + (-3)$, or 26. Twice the first minus the second is $2 \cdot 8 - 21$, or -5, which is 2 less than the third. The second minus three times the first is $21 - 3 \cdot 8$, or -3, which is the third. The numbers check.

    State. The numbers are 8, 21, and -3.

5. ***Familiarize.*** We first make a drawing.

    We let x, y, and z represent the measures of angles A, B, and C, respectively The measures of the angles of a triangle add up to $180°$.

    Translate.

    The sum of the measures is $180°$.

    $$x + y + z \qquad = 180$$

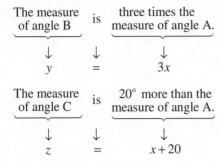

$\downarrow$ $\downarrow$ $\downarrow$
y $=$ $3x$

The measure   is $20°$ more than the
of angle C   measure of angle A.

$\downarrow$ $\downarrow$ $\downarrow$
z $=$ $x+20$

We now have a system of equations.
$x+y+z=180$

$y=3x$

$z=x+20$

Carry out. Solving the system we get
$(32,96,52)$.

Check. The sum of the measures is $32° + 96°$
$+ 52°$, or $180°$. Three times the measure of
angle A is $3\cdot32°$, or $96°$, the measure of
angle B. $20°$ more than the measure of angle
A is $32° + 20°$, or $52°$, the measure of angle
C. The numbers check.

State. The measures of the angles A, B, and
C, are $32°$, $96°$, and $52°$, respectively.

7. Familiarize.
Let x = the average score in math, y = the
average score in reading, and
$z =$ the average score in writing. Then
Translate.
$x+y+z=1509$ (1)

$x=y+14$ (2)

$z=y-8$ (3)

Carry out. Equations (2) and (3) respectively
define x and z in terms of y. Through
substitution, we rewrite equation (1) entirely in
terms of y and solve for y.
$x+y+z=1509$ (1)

$(y+14)+y+(y-8)=1509$

$y+14+y+y-8=1509$

$3y+6=1509$

$3y=1503$

$y=501$

Substitute the result into equations (2) and (3)

to find x and z:
$x=y+14$ (2)

$x=501+14=515$

$z=y-8$ (3)

$z=501-8=493$

Check.
$515+501+493=1509$ (1)$\checkmark$

$515=501+14$ (2)$\checkmark$

$493=501-8$ (3)$\checkmark$

State. On the 2009 SAT, the average math
score was 515, the average reading score was
501, and the average writing score was 493.

9. **Familiarize.** Let x = the number of grams of
fiber in a bran muffin. Let y = the number of
grams of fiber in a banana. Let z = the
number of grams of fiber in 1-cup of serving
of Wheaties®.

Translate.
We can summarize this information in a
table.

Bran Muffin	Banana	Wheaties 1-cup	Fiber
$2x$	y	z	9
x	$2y$	z	10.5
$2x$		z	6

We now have a system of equations.
$2x+y+z=9$

$x+2y+z=10.5$

$2x+z=6$

Carry out. Solving the system we get
$(1.5,3,3)$.

Check. A breakfast of 2 bran muffins (1.5
grams of fiber each), one banana (3 grams of
fiber), and a 1-cup serving of Wheaties (3
grams of fiber) provides $2(1.5)+3+3$, or 9
grams of fiber. A breakfast of 1 bran muffin,
2 bananas, and a 1-cup serving of Wheaties
provides $1.5+2\cdot3+3$, or 10.5 grams of
fiber. A breakfast of 2 bran muffins and a 1-
cup serving of Wheaties provides $2(1.5)+3$,
or 6 grams of fiber. These numbers check.

State. Each bran muffin contains 1.5 g of
fiber, each banana contains 3 g of fiber, and
each 1-cup serving of Wheaties contains 3 g
of fiber.

11. **Familiarize.**

Let x = the price of the basic model in 2010, y = the price of the car cover, and z = the price of satellite radio. Then:

Translate.

$$x + y = 24{,}030 \quad (1)$$
$$x + y + z = 24{,}340 \quad (2)$$
$$x + z = 24{,}110 \quad (3)$$

Carry out.

Subtracting equation (1) from equation (2) will allow us to find the value of z.

$$x + y + z = 24{,}340 \quad (2)$$
$$x + y \phantom{{}+z} = 24{,}030 \quad (1)$$
$$z = 24{,}340 - 24{,}030 = 310$$

Use the value of z and equation (3) to find x:

$$x + z = 24{,}110 \quad (3)$$
$$x = 24{,}110 - z = 24{,}110 - 310 = 23{,}800$$

Use the value of x and equation (1) to find y:

$$x + y = 24{,}030 \quad (1)$$
$$y = 24{,}030 - x = 24{,}030 - 23{,}800 = 230$$

Check.

$$23{,}800 + 230 = 24{,}030 \quad (1)$$
$$23{,}800 + 230 + 310 = 24{,}340 \quad (2)$$
$$23{,}800 + 310 = 24{,}110 \quad (3)$$

State.

In 2010, the basic price of a Honda Civic Hybrid was \$23,800. The car cover cost \$230, and satellite radio for the car cost \$310.

13. **Familiarize.** Let x = the number of 12 oz cups sold. Let y = the number of 16 oz cups sold, and let z = the number of 20 oz cups sold. 12, 16, and 20 oz cups sell for \$1.65, \$1.85, and \$1.95, respectively. If Reba empties six 144-oz "brewers," then a total of $6 \cdot 144 = 864$ oz of coffee was sold.

Translate.

	12 oz	16 oz	20 oz	Total
Cups sold	x	y	z	55
Oz sold	$12x$	$16y$	$20z$	864
Sales	$165x$	$185y$	$195z$	9965

Now we have a system of equations.

$$x + y + z = 55 \quad (1)$$
$$12x + 16y + 20z = 864 \quad (2)$$
$$165x + 185y + 195z = 9965 \quad (3)$$

Note that equation (3) is in cents. We can reduce the coefficients in equations (2) and (3) if we note that all the coefficients in equation (2) are divisible by 4, and all the coefficients in equation (3) are divisible by 5.

$$12x + 16y + 20z = 864$$
$$\frac{1}{4} \cdot [12x + 16y + 20z] = \frac{1}{4} \cdot [864]$$
$$3x + 4y + 5z = 216 \quad (5)$$
$$165x + 185y + 195z = 9965$$
$$\frac{1}{5} \cdot [165x + 185y + 195z] = \frac{1}{5} \cdot [9965]$$
$$33x + 37y + 39z = 1993 \quad (6)$$

We can now solve the new system:

$$x + y + z = 55 \quad (1)$$
$$3x + 4y + 5z = 216 \quad (5)$$
$$33x + 37y + 39z = 1993 \quad (6)$$

Carry out.

Adding $-3 \cdot (1)$ to (5), and $-33(1)$ to (6) will give two equations in y and z only.

$$-3x - 3y - 3z = -165 \quad -3 \cdot (1)$$
$$3x + 4y + 5z = 216 \quad (5)$$
$$y + 2z = 51 \quad (7)$$

$$-33x - 33y - 33z = -1815 \quad -33 \cdot (1)$$
$$33x + 37y + 39z = 1993 \quad (6)$$
$$4y + 6z = 178$$

$$\frac{1}{2} \cdot [4y + 6z] = \frac{1}{2} \cdot [178]$$
$$2y + 3z = 89 \quad (8)$$

Adding $-2 \cdot (7)$ to (8) will give an equation in z only:

$$-2y - 4z = -102 \quad -2 \cdot (7)$$
$$2y + 3z = 89 \quad (8)$$
$$-z = -13 \Rightarrow z = 13$$

Knowing z, use equation (7) to find y and then equation (1) to find x.

$$y + 2z = 51 \quad (7)$$
$$y = 51 - 2z = 51 - 2(13) = 51 - 26 = 25$$
$$x + y + z = 55 \quad (1)$$
$$x = 55 - y - z = 55 - 25 - 13 = 17$$

Check.

$$17 + 13 + 25 = 55 \quad (1) \checkmark$$
$$12(17) + 16(25) + 20(13) = 864 \quad (2) \checkmark$$
$$165(17) + 185(25) + 195(13) = 9965 \quad (3) \checkmark$$

State. Reba sold 17 twelve-oz coffees, 25 sixteen-oz coffees, and 13 twenty-oz coffees.

15. ***Familiarize.*** Let x = amount borrowed at 8%, y = amount borrowed at 5%, and z = amount borrowed at 4%.

Translate.

$$x + y + z = 120,000 \qquad (1)$$

$$0.08x + 0.05y + 0.04z = 5750$$

$$\text{or } 8x + 5y + 4z = 575,000 \qquad (2)$$

$$0.04z = 0.08x + 1600$$

$$\text{or } 8x - 4z = -160,000$$

$$\text{or } 2x - z = -40,000 \qquad (3)$$

Carry out.

Using $(1) + (3)$ and $(2) + 4 \cdot (3)$ will generate a two-variable system in x and y.

$$\begin{array}{ll} x + y + z = 120,000 & (1) \\ 2x \quad - z = -40,000 & (2) \\ \hline 3x + y \quad = 80,000 & (4) \end{array}$$

and

$$\begin{array}{ll} 8x + 5y + 4z = 575,000 & (2) \\ 8x \quad - 4z = -160,000 & -4 \cdot (3) \\ \hline 16x + 5y \quad = 415,000 & (5) \end{array}$$

Using $-5 \cdot (4) + (5)$ we can find x and then back-substitute into equations (4) and (1) to find y and z.

$$\begin{array}{ll} -15x - 5y \quad = -400,000 & -5 \cdot (4) \\ 16x + 5y \quad = 415,000 & (5) \\ \hline x = 15,000 \end{array}$$

So: $3x + y \quad = 80,000 \qquad (4)$

$$y = 80,000 - 3x = 80,000 - 3(15,000)$$

$$y = 80,000 - 45,000 = 35,000$$

and: $x + y + z = 120,000 \qquad (1)$

$$z = 120,000 - x - y$$

$$z = 120,000 - 15,000 - 35,000 = 70,000$$

Check.

$$15,000 + 35,000 + 70,000 = 120,000 \quad (1) \checkmark$$

$$8(15,000) + 5(35,000) + 4(70,000) = 575,000 \quad (2) \checkmark$$

$$2(15,000) - 70,000 = -40,000 \quad (3) \checkmark$$

State. Chelsea borrowed $15,000 at 8%, $35,000 at 5%, and $70,000 at 4% interest.

17. ***Familiarize.*** Let x = cost per gram of gold, y = cost per gram of silver, and z = cost per gram of copper. Note that x% of 100 grams equals x.

Translate.

$$75x + 5y + 20z = 2265.40$$

$$\text{or } 15x + y + 4z = 453.08 \qquad (1)$$

$$75x + 12.5y + 12.5z = 2287.75$$

$$\text{or } 6x + y + z = 183.02 \qquad (2)$$

$$37.5x + 62.5y = 1312.50$$

$$\text{or } 3x + 5y = 105 \qquad (3)$$

Carry out.

Using $4 \cdot (2) - (1)$ and (3) will generate a two-variable system in x and y.

$$\begin{array}{ll} 24x + 4y + 4z = 732.08 & 4 \cdot (2) \\ -15x \;\; - y - 4z = -453.08 & -1 \cdot (1) \\ \hline 9x + 3y \quad = 279 \end{array}$$

$$\text{or } 3x + y = 93 \qquad (5)$$

Using $(3) - 1 \cdot (5)$ we can find y and then back-substitute into equations (5) and (2) to find x and z.

$$\begin{array}{ll} 3x + 5y = 105 & (3) \\ -3x - y = -93 & -1 \cdot (5) \\ \hline 4y = 12 \end{array}$$

$$y = \frac{12}{4} = 3$$

So: $3x + y \quad = 93 \qquad (5)$

$$3x = 93 - y = 93 - 3 = 90$$

$$x = \frac{90}{3} = 30$$

and: $6x + y + z = 183.02 \qquad (2)$

$$z = 183.02 - 6x - y = 183.02 - 6(30) - 3$$

$$z = 183.02 - 180 - 3 = 0.02$$

Check.

$$75(30) + 5(3) + 20(0.02) = 2265.40 \quad (1) \checkmark$$

$$75(30) + 12.5(3) + 12.5(0.02) = 2287.75 \quad (2) \checkmark$$

$$37.5(30) + 62.5(3) = 1312.50 \quad (3) \checkmark$$

State. Gold is $30 a gram, silver is $3 a gram and copper is $0.02 a gram.

19. ***Familiarize.*** Let r = the number of servings of roast beef, p = the number of baked potatoes, and b = the number of servings of broccoli. Then r servings of roast beef contain $300r$ Calories, $20r$ g of protein, and no vitamin C. In p baked potatoes there are $100p$ Calories, $5p$ g of protein, and $20p$ mg of vitamin C. And b servings of broccoli contain $50b$ Calories, $5b$ g of protein, and $100b$ mg of vitamin C. The patient requires

800 Calories, 55 g of protein, and 220 mg of vitamin C.

Translate. Write equations for the total number of calories, the total amount of protein, and the total amount of vitamin C.

$$300r + 100p + 50b = 800 \quad \text{(Calories)}$$
$$20r + 5p + 5b = 55 \quad \text{(Protein)}$$
$$20p + 100b = 220 \quad \text{(Vitamin C)}$$

We now have a system of equations.

Carry out. Solving the system we get $(2, 1, 2)$.

Check. Two servings of roast beef provide 600 Calories, 40 g of protein, and no vitamin C. One baked potato provides 100 Calories, 5 g of protein, and 20 gm of vitamin C. And 2 servings of broccoli provide 100 Calories, 10 g of protein, and 200 mg of vitamin C. Together, then, they provide 800 Calories, 55 g of protein, and 220 mg of vitamin C. The value check.

State. The dietician should prepare 2 servings of roast beef, 1 baked potato, and 2 servings of broccoli.

21. **Familiarize.** Let x = number of main floor tickets sold, y = number of first mezzanine tickets sold, and z = number of second mezzanine tickets sold.
   Translate.
   $$x + y + z = 40 \quad (1)$$
   $$x + y = z$$
   $$\text{or } x + y - z = 0 \quad (2)$$
   $$38x + 52y + 28z = 1432$$
   $$\text{or } 19x + 26y + 14z = 716 \quad (3)$$
   Carry out.
   Using $(1) + (2)$ and $(3) - 14 \cdot (1)$ will generate a two-variable system in x and y.
   $$x + y + z = 40 \quad (1)$$
   $$\underline{x + y - z = 0 \quad (2)}$$
   $$2x + 2y \quad = 40$$
   $$\text{or } x + y = 20 \quad (5)$$

   This result actually allows us to find z right away using equation (2)
   $$x + y - z = 0 \quad (2)$$
   $$z = x + y = 20$$
   Back to $(3) - 14 \cdot (1)$ to find x and y.

$$19x + 26y + 14z = 716 \quad (3)$$
$$\underline{-14x - 14y - 14z = -560 \quad -14 \cdot (1)}$$
$$5x + 12y = 156 \quad (6)$$

Using $(6) - 5 \cdot (5)$ we can find y and then back-substitute into equation (5) to find x.
$$5x + 12y = 156 \quad (6)$$
$$\underline{-5x - 5y = -100 \quad -5 \cdot (5)}$$
$$7y = 56$$

$$y = \frac{56}{7} = 8$$

So: $x + y = 20 \quad (5)$
$$x = 20 - y = 20 - 8 = 12$$

Check.
$$12 + 8 + 20 = 40 \quad (1) \checkmark$$
$$12 + 8 = 20 \quad (2) \checkmark$$
$$38(12) + 52(8) + 28(20) = 1432 \quad (3) \checkmark$$

State. 12 floor seats were sold, 8 first mezzanine seats were sold, and 20 second mezzanine seats were sold.

23. **Familiarize.** Let x = the number of two-point field goals, y = the number of 3-point field goals, and z = the number of 1-point foul shots.

Translate.

Total number of baskets is 50.
$$x + y + z = 50$$

Total number of points is 92.
$$2x + 3y + z = 92$$

Number of 2-pointers is 19 more than the number of foul shots.
$$x = 19 + z$$

Now we have a system of equations.
$$x + y + z = 50$$
$$2x + 3y + z = 92$$
$$x = 19 + z$$

Carry out. Solving the system, we get $(32, 5, 13)$.

Check. The total number of baskets made was $32 + 5 + 13$, or 50. The total number of points made was $32 \cdot 2 + 5 \cdot 3 + 13 \cdot 1$, or $64 + 15 + 13 = 92$. The number of 2-pointers, 32, was 19 more than the number of foul shots, 13. These numbers check.

State. The number of two-point field goals, three-point field goals, and the number of foul shots was 32, 5, and 13, respectively.

25. *Thinking and Writing Exercise.*

27. $-2(2x - 3y) = -4x + 6y$

29. $-6(x - 2y) + (6x - 5y)$
$= -6x + 12y + 6x - 5y = 7y$

31. $-(2a - b - 6c) = -2a + b + 6c$

33. $-2(3x - y + z) + 3(-2x + y - 2z)$
$= -6x + 2y - 2z - 6x + 3y - 6z$
$= -12x + 5y - 8z$

35. *Thinking and Writing Exercise.*

37. **Familiarize.** Let x = the cost for the applicant, y = the cost for the spouse, z = the cost for the first child, and w = the cost for the second child.

Translate.

Monthly cost for an applicant and spouse is 135.

$$x + y = 135$$

Monthly cost for an applicant and spouse and one child is 154.

$$x + y + z = 154$$

Monthly cost for an applicant and spouse and two children is 173.

$$x + y + z + w = 173$$

Monthly cost for an applicant, one child is 102.

$$x + z = 102$$

We now have a system of equations.

$$x + y = 135$$
$$x + y + z = 154$$
$$x + y + z + w = 173$$
$$x + z = 102$$

Carry out. Solving the system we get $(83, 52, 19, 19)$.

Check. The cost for an applicant and his or her spouse is $\$83 + \$52 = \$135$. The cost when just one child is added is: $\$83 + \$52 + \$19 = \$154$. When an additional child is added, the cost is $\$83 + \$52 + \$19 + \$19 = \$173$. Finally, the cost for an applicant and just one child is $\$83 + \$19 = \$102$. These numbers check.

State. The monthly costs for an applicant, spouse, first child, and second child are $\$83$, $\$52$, $\$19$, and $\$19$, respectively.

39. **Familiarize.** Let t = Tammy's age, let c = Carmen's age, let d = Dennis's age, and let m = Mark's age.

Translate.

Tammy's age is the sum of Carmen and Dennis's ages.

$$t = c + d$$

Carmen's age is 2 more than the sum of Dennis and Mark's ages.

$$c = 2 + d + m$$

Dennis's age is four times Mark's age.

$$d = 4m$$

Sum of all four ages is 42.

$$t + c + d + m = 42$$

Now we have a system of equations.

$$t = c + d$$
$$c = 2 + d + m$$
$$d = 4m$$
$$t + c + d + m = 42$$

Carry out. We are only asked to determine Tammy's age, but we will solve the whole system in order to check our work.   Solving the system we get $(20, 12, 8, 2)$.

Check. The sum of Carmen's and Dennis's ages are 12 + 8, or 20, which is Tammy's age. Carmen's age, 12, is 2 more than the sum of Dennis and Mark's age, 8 + 2 = 10. Dennis's age is 8, which is 4 times Mark's age. The sum of all four ages is 20 + 12 + 8 + 2, or 42. These numbers check.

State. Tammy is 20 years old. In addition, Carmen is 12, Dennis is 8, and Mark is 2.

41. **Familiarize.** Let *T, G,* and *H* represent the number of tickets Tom, Gary, and Hal begin with respectively.

Translate. After Hal gives tickets to Tom and Gary, each has the following number of tickets:

Tom: $T + T$, or $2T$

Gary: $G + G$, or $2G$

Hal: $H - T - G$

After Tom gives tickets to Gary and Hal, each has the following number of tickets:

Gary:

$2G + 2G$, or $4G$

Hal:

$(H - T - G) + (H - T - G)$, or

$2(H - T - G)$

Tom:

$2T - 2G - (H - T - G)$, or

$3T - H - G$

After Gary gives tickets to Hal and Tom, each has the following number of tickets:

Hal:

$2(H - T - G) + 2(H - T - G)$, or

$4(H - T - G)$

Tom:

$(3T - H - G) + (3T - H - G)$, or

$2(3T - H - G)$

Gary:

$4G - 2(H - T - G) - (3T - H - G)$, or

$7G - H - T$

Since Hal, Tom, and Gary each finish with 40 tickets, we write the following system of equations:

$4(H - T - G) = 40$

$2(3T - H - G) = 40$

$7G - H - T = 40$

Carry out. Solving the system we get $(35, 20, 65)$.

Check. Hal has 65 tickets to start with and gives Tom 35 tickets and gives Gary 20 tickets. As a result, Tom has 70 tickets and Gary has 40 tickets. Hal's supply of tickets has been reduced by 35 + 20, or 55, leaving him with 10 tickets. So, Tom, Gary, and Hal now have 70, 40, and 10 tickets, respectively. Tom then gives Hal 10 tickets and Gary 40 tickets. As a result, Hal has 20 tickets and Gary has 80 tickets. Tom's supply of tickets has been reduced by 10 + 40, or 50, leaving him with 20 tickets. So, Tom, Gary, and Hal now have 20, 80, 20 tickets, respectively. Finally, Gary gives Hal 20 tickets and gives Tom 20 tickets, so they now each have 40 tickets. Gary had 80 and gave away 20 + 20, or 40, so he has 40 left. So, all three ended up with 40 tickets apiece.

State. Tom had 35 tickets to start with. Incidentally, Gary had 20, and Hal had 65.

Exercise Set 9.3

1. matrix

3. entry

5. rows

7. $x + 2y = 11$

$3x - y = 5$

Write a matrix using only the constants.

$\begin{bmatrix} 1 & 2 & | & 11 \\ 3 & -1 & | & 5 \end{bmatrix}$

Multiply row 1 by –3 and add it to row 2.

$\begin{bmatrix} 1 & 2 & | & 11 \\ 0 & -7 & | & -28 \end{bmatrix}$ New R2 = $-3(R1) + R2$

Reinserting the variables, we have

$x + 2y = 11$ (1)

$-7y = -28$ (2)

Solve Equation (2) for *y*.

$-7y = -28$

$y = 4$

Substitute 4 for y in Equation (1) and solve for x.

$x + 2y = 11 \Rightarrow x = 11 - 2y = 11 - 2(4) = 3$

The solution is $(3, 4)$.

9. $x + 4y = 8$

$3x + 5y = 3$

Write a matrix using only the constants.

$$\begin{bmatrix} 1 & 4 & | & 8 \\ 3 & 5 & | & 3 \end{bmatrix}$$

Multiply the first row by -3 and add it to the second row.

$$\begin{bmatrix} 1 & 4 & | & 8 \\ 0 & -7 & | & -21 \end{bmatrix}$$ New Row 2 $= -3(\text{Row } 1)$ $+ \text{Row } 2$

Reinserting the variables, we have

$x + 4y = 8 \quad (1)$

$-7y = -21 \quad (2)$

Solve Equation (2) for y.

$-7y = -21$

$y = 3$

Substitute 3 for y in Equation (1) and solve for x.

$x + 4 \cdot 3 = 8$

$x + 12 = 8$

$x = -4$

The solution is $(-4, 3)$.

11. $6x - 2y = 4$

$7x + y = 13$

Write a matrix using only the constants.

$$\begin{bmatrix} 6 & -2 & | & 4 \\ 7 & 1 & | & 13 \end{bmatrix}$$

Multiply the second row by 6 to make the first number in row 2 a multiple of 6.

$$\begin{bmatrix} 6 & -2 & | & 4 \\ 42 & 6 & | & 78 \end{bmatrix}$$ New Row 2 $= 6(\text{Row } 2)$

Multiply the first row by -7 and add it to the second row.

$$\begin{bmatrix} 6 & -2 & | & 4 \\ 0 & 20 & | & 50 \end{bmatrix}$$ New Row 2 $= 7(\text{Row } 1) +$ Row 2

Reinserting the variables, we have

$6x - 2y = 4 \quad (1)$

$20y = 50 \quad (2)$

Solve Equation (2) for y.

$20y = 50$

$y = \dfrac{5}{2}$

Substitute $\dfrac{5}{2}$ for y in Equation (1) and solve for x.

$6x - 2y = 4$

$6x - 2\left(\dfrac{5}{2}\right) = 4$

$6x - 5 = 4$

$6x = 9$

$x = \dfrac{3}{2}$

The solution is $\left(\dfrac{3}{2}, \dfrac{5}{2}\right)$.

13. $3x + 2y + 2z = 3$

$x + 2y - z = 5$

$2x - 4y + z = 0$

Write a matrix using only the constants.

$$\begin{bmatrix} 3 & 2 & 2 & | & 3 \\ 1 & 2 & -1 & | & 5 \\ 2 & -4 & 1 & | & 0 \end{bmatrix}$$

First interchange rows 1 and 2 so that each number below the first number in the first row is a multiple of that number.

$$\begin{bmatrix} 1 & 2 & -1 & | & 5 \\ 3 & 2 & 2 & | & 3 \\ 2 & -4 & 1 & | & 0 \end{bmatrix}$$

Multiply row 1 by -3 and add it to row 2. Multiply row 1 by -2 and add it to row 3.

$$\begin{bmatrix} 1 & 2 & -1 & | & 5 \\ 0 & -4 & 5 & | & -12 \\ 0 & -8 & 3 & | & -10 \end{bmatrix}$$

Multiply row 2 by -2 and add it to row 3.

$$\begin{bmatrix} 1 & 2 & -1 & | & 5 \\ 0 & -4 & 5 & | & -12 \\ 0 & 0 & -7 & | & 14 \end{bmatrix}$$

Reinserting the variables, we have

$x + 2y - z = 5 \quad (1)$

$-4y + 5z = -12 \quad (2)$

$-7z = 14 \quad (3)$

Solve Equation (3) for z.

$$-7z = 14$$
$$z = -2$$

Substitute -2 for z in Equation (2) and solve for y.

$$-4y + 5(-2) = -12$$
$$-4y - 10 = -12$$
$$-4y = -2$$
$$y = \frac{1}{2}$$

Substitute $\frac{1}{2}$ for y and -2 for z in (1) and solve for x.

$$x + 2 \cdot \frac{1}{2} - (-2) = 5$$
$$x + 1 + 2 = 5$$
$$x + 3 = 5$$
$$x = 2$$

The solution is $\left(2, \frac{1}{2}, -2\right)$.

15. $a - 2b - 3c = 3$
$$2a - b - 2c = 4$$
$$4a + 5b + 6c = 4$$

Write a matrix using only the constants.

$$\begin{bmatrix} 1 & -2 & -3 & | & 3 \\ 2 & -1 & -2 & | & 4 \\ 4 & 5 & 6 & | & 4 \end{bmatrix}$$

Multiply row 1 by -2 and add it to row 2.
Multiply row 1 by -4 and add it to row 3.

$$\begin{bmatrix} 1 & -2 & -3 & | & 3 \\ 0 & 3 & 4 & | & -2 \\ 0 & 13 & 18 & | & -8 \end{bmatrix} \quad \begin{array}{l} \text{New R2} = -2\text{R1} + \text{R2} \\ \\ \text{New R3} = -4\text{R1} + \text{R3} \end{array}$$

Multiply row 3 by 3.

$$\begin{bmatrix} 1 & -2 & -3 & | & 3 \\ 0 & 3 & 4 & | & -2 \\ 0 & 39 & 54 & | & -24 \end{bmatrix} \quad \text{New R3} = 3\text{R3}$$

Multiply row 2 by -13 and add it to row 3.

$$\begin{bmatrix} 1 & -2 & -3 & | & 3 \\ 0 & 3 & 4 & | & -2 \\ 0 & 0 & 2 & | & 2 \end{bmatrix} \quad \text{New R3} = -13\text{R2} + \text{R3}$$

Reinserting the variables, we have

$$a - 2b - 3c = 3 \quad (1)$$
$$3b + 4c = -2 \quad (2)$$
$$2c = 2 \quad (3)$$

Solve Equation (3) for c.

$$2c = 2 \Rightarrow c = 1$$

Substitute $c = 1$ into equation (2) and solve for b.

$$3b + 4 \cdot 1 = -2$$
$$3b + 4 = -2$$
$$3b = -6$$
$$b = -2$$

Substitute $b = -2$ and $c = 1$ into (1) and solve for a.

$$a - 2(-2) - 3 \cdot 1 = 3$$
$$a + 4 - 3 = 3$$
$$a + 1 = 3$$
$$a = 2$$

The solution is $(2, -2, 1)$.

17. $3u + 2w = 11$
$$v - 7w = 4$$
$$u - 6v = 1$$

Write a matrix using only the constants.

$$\begin{bmatrix} 3 & 0 & 2 & | & 11 \\ 0 & 1 & -7 & | & 4 \\ 1 & -6 & 0 & | & 1 \end{bmatrix}$$

Interchange row 1 and row 3.

$$\begin{bmatrix} 1 & -6 & 0 & | & 1 \\ 0 & 1 & -7 & | & 4 \\ 3 & 0 & 2 & | & 11 \end{bmatrix} \quad \text{Interchange R1 and R3}$$

Multiply row 1 by -3 and add it to row 3.

$$\begin{bmatrix} 1 & -6 & 0 & | & 1 \\ 0 & 1 & -7 & | & 4 \\ 0 & 18 & 2 & | & 8 \end{bmatrix} \quad \text{New R3} = -3\text{R1} + \text{R3}$$

Multiply row 2 by -18 and add it to row 3.

$$\begin{bmatrix} 1 & -6 & 0 & | & 1 \\ 0 & 1 & -7 & | & 4 \\ 0 & 0 & 128 & | & -64 \end{bmatrix} \quad \text{New R3} = -18\text{R2} + \text{R3}$$

Reinserting the variables, we have

$$u - 6v = 1 \quad (1)$$
$$v - 7w = 4 \quad (2)$$
$$128w = -64 \quad (3)$$

Solve Equation (3) for w.
$$128w = -64$$
$$w = -\frac{1}{2}$$

Substitute $w = -\frac{1}{2}$ into equation (2) and solve for v.

$$v - 7w = 4$$
$$v - 7\left(-\frac{1}{2}\right) = 4$$
$$v + \frac{7}{2} = 4$$
$$v = \frac{1}{2}$$

Substitute $v = \frac{1}{2}$ in (1) and solve for u.

$$u - 6 \cdot \frac{1}{2} = 1$$
$$u - 3 = 1$$
$$u = 4$$

The solution is $\left(4, \frac{1}{2}, -\frac{1}{2}\right)$.

19. We will rewrite the equations with the variables in alphabetical order:
$$-2w + 2x + 2y - 2z = -10$$
$$w + x + y + z = -5$$
$$3w + x - y + 4z = -2$$
$$w + 3x - 2y + 2z = -6$$

Write a matrix using only the constants.
$$\begin{bmatrix} -2 & 2 & 2 & -2 & | & -10 \\ 1 & 1 & 1 & 1 & | & -5 \\ 3 & 1 & -1 & 4 & | & -2 \\ 1 & 3 & -2 & 2 & | & -6 \end{bmatrix}$$

Multiply row 1 by $\frac{1}{2}$.
$$\begin{bmatrix} -1 & 1 & 1 & -1 & | & -5 \\ 1 & 1 & 1 & 1 & | & -5 \\ 3 & 1 & -1 & 4 & | & -2 \\ 1 & 3 & -2 & 2 & | & -6 \end{bmatrix}$$
New Row 1 = $\frac{1}{2}$(Row 1)

Add row 1 to row 2.

Add 3 times row 1 to row 3.
Add row 1 to row 4.
$$\begin{bmatrix} -1 & 1 & 1 & -1 & | & -5 \\ 0 & 2 & 2 & 0 & | & -10 \\ 0 & 4 & 2 & 1 & | & -17 \\ 0 & 4 & -1 & 1 & | & -11 \end{bmatrix}$$
New Row 2 = Row 1 + Row 2
New Row 3 = 3(Row 1) + Row 3
New Row 4 = Row 1 + Row 4

Multiply row 2 by -2 and add it to row 3.
Multiply row 2 by -2 and add it to row 4.
$$\begin{bmatrix} -1 & 1 & 1 & -1 & | & -5 \\ 0 & 2 & 2 & 0 & | & -10 \\ 0 & 0 & -2 & 1 & | & 3 \\ 0 & 0 & -5 & 1 & | & 9 \end{bmatrix}$$
New Row 3 = -2(Row 2) + Row 3
New Row 4 = -2(Row 2) + Row 4

Multiply row 4 by 2.
$$\begin{bmatrix} -1 & 1 & 1 & -1 & | & -5 \\ 0 & 2 & 2 & 0 & | & -10 \\ 0 & 0 & -2 & 1 & | & 3 \\ 0 & 0 & -10 & 2 & | & 18 \end{bmatrix}$$
New Row 4 = 2(Row 4)

Multiply row 3 by -5 and add it to row 4.
$$\begin{bmatrix} -1 & 1 & 1 & -1 & | & -5 \\ 0 & 2 & 2 & 0 & | & -10 \\ 0 & 0 & -2 & 1 & | & 3 \\ 0 & 0 & 0 & -3 & | & 3 \end{bmatrix}$$
New Row 4 = -5(Row 3) + Row 4

Reinserting the variables, we have
$$-w + x + y - z = -5 \quad (1)$$
$$2x + 2y = -10 \quad (2)$$
$$-2y + z = 3 \quad (3)$$
$$-3z = 3 \quad (4)$$

Solve (4) for z.
$$-3z = 3$$
$$z = -1$$

Substitute -1 for z in (3) and solve for y.
$$-2y + (-1) = 3$$
$$-2y = 4$$
$$y = -2$$

Substitute -2 for y in (2) and solve for x.

$$2x + 2(-2) = -10$$
$$2x - 4 = -10$$
$$2x = -6$$
$$x = -3$$

Substitute -3 for x, -2 for y, and -1 for z and solve for w.

$$-w + (-3) + (-2) - (-1) = -5$$
$$-w - 3 - 2 + 1 = -5$$
$$-w - 4 = -5$$
$$-w = -1$$
$$w = 1$$

The solution is $(1, -3, -2, -1)$.

21. **Familiarize.** Let d = the number of dimes and n = the number of nickels. The value of the dimes is $\$0.10d$. The value of the nickels is $\$0.05n$.
Translate.

Total number of coins is 42.

$$\downarrow \qquad\qquad \downarrow\ \downarrow$$
$$d + n \qquad\qquad = 42$$

Total value of the coins is $3.00.

$$\downarrow \qquad\qquad\quad \downarrow\ \downarrow$$
$$0.10d + 0.05n \qquad = 3.00$$

After clearing decimals, we have this system.
$$d + n = 42$$
$$10d + 5n = 300$$

Carry out. Solve using matrices.

$$\begin{bmatrix} 1 & 1 & | & 42 \\ 10 & 5 & | & 300 \end{bmatrix}$$

$$\begin{bmatrix} 1 & 1 & | & 42 \\ 0 & -5 & | & -120 \end{bmatrix}$$ New Row 2 $= -10$(Row 1) $+$ Row 2

Reinserting the variables, we have
$$d + n = 42 \qquad (1)$$
$$-5n = -120 \qquad (2)$$

Solve (2) for n.
$$-5n = -120$$
$$n = 24$$
$$d + n = 42 \text{ Substituting in } (2)$$
$$d + 24 = 42$$
$$d = 18$$

Check. The sum of the two numbers is 42. The total value is $\$0.10(18) + \$0.05(24) = \$1.80 + \$1.20 = \$3.00$. The answer checks.
State. There are 18 dimes and 24 nickels.

23. **Familiarize.** Let $x =$ pounds of dried-fruit used, and $y =$ pounds of macadamia nuts used. Then:
Translate.

$$x + y = 15 \qquad\qquad (1)$$
$$580x + 1475y = 938 \cdot 15 = 14,070$$
or $116x + 295y = 2814 \qquad (2)$
or

$$\begin{bmatrix} 1 & 1 & | & 15 \\ 116 & 295 & | & 2814 \end{bmatrix}$$

Carry out.
Multiply row 1 by -116 and add to row 2.

$$\begin{bmatrix} 1 & 1 & | & 15 \\ 0 & 179 & | & 1074 \end{bmatrix}$$

Reinsert variables and solve for x and y.
$$x + y = 15 \qquad (1)$$
$$179y = 1074 \qquad (2)$$
$$y = \frac{1074}{179} = 6$$
$$x + 6 = 15 \Rightarrow x = 9$$
Check.
$$9 + 6 = 15 \qquad (1)\checkmark$$
$$580(9) + 1475(6) = 14,070 \qquad (2)\checkmark$$
State. The snack mix should be made with 9 pounds of dried fruit and 6 pounds of macadamia nuts.

25. **Familiarize.** We let x, y, and z represent the amounts invested at 7%, 8%, and 9%, respectively. Recall the formula for simple interest.

Interest = Principal $\times$ Rate $\times$ Time

Translate. We organize the information in a table.

	First Investment	Second Investment	Third Investment	Total
P	x	y	z	$2500
R	7%	8%	9%	
T	1 yr	1 yr	1 yr	
I	$0.07x$	$0.08y$	$0.09z$	$212

The first row give us one equation:

$x + y + z = 2500$

The last row gives a second equation:

$0.07x + 0.08y + 0.09z = 212$

Amount invested is at 9%	$1100	more than	amount invested at 8%

$$z = \$1100 + y$$

After clearing decimals, we have this system:

$x + y + z = 2500$

$7x + 8y + 9z = 21,200$

$-y + z = 1100$

Carry out. Solve using matrices.

$$\begin{bmatrix} 1 & 1 & 1 & | & 2500 \\ 7 & 8 & 9 & | & 21,200 \\ 0 & -1 & 1 & | & 1100 \end{bmatrix}$$

$$\begin{bmatrix} 1 & 1 & 1 & | & 2500 \\ 0 & 1 & 2 & | & 3700 \\ 0 & -1 & 1 & | & 1100 \end{bmatrix}$$ New Row 2 = $-7(\text{Row } 1) + \text{Row } 2$

$$\begin{bmatrix} 1 & 1 & 1 & | & 2500 \\ 0 & 1 & 2 & | & 3700 \\ 0 & 0 & 3 & | & 4800 \end{bmatrix}$$ New Row 3 = Row 2 + Row 3

Reinserting the variables, we have

$x + y + z = 2500$ (1)

$y + 2z = 3700$ (2)

$3z = 4800$ (3)

Solve (3) for z.

$3z = 4800$

$z = 1600$

Substitute 1600 for z in (2) and solve for y.

$y + 2 \cdot 1600 = 3700$

$y + 3200 = 3700$

$y = 500$

Substitute 500 for y and 1600 for z in (1) and solve for x.

$x + 500 + 1600 = 2500$

$x + 2100 = 2500$

$x = 400$

Check. The total investment is $400 + $500 + $1600, or $2500. The total interest is $0.07(\$400) + 0.08(\$500) + 0.09(\$1600) = \$28 + \$40 + \$144 = \$212$. The amount invested

at 9%, $1600, is $1100 more than the amount invested at 8%, $500. The numbers check.

State. The amounts invested at 7%, 8%, and 9%, are $400, $500, and $1600, respectively.

27. *Thinking and Writing Exercise.*

29. $5(-3) - (-7)4 = -15 - (-28) = -15 + 28 = 13$

31. $-2(5 \cdot 3 - 4 \cdot 6) - 3(2 \cdot 7 - 15) + 4(3 \cdot 8 - 5 \cdot 4)$

$= -2(15 - 24) - 3(14 - 15) + 4(24 - 20)$

$= -2(-9) - 3(-1) + 4(4)$

$= 18 + 3 + 16$

$= 21 + 16$

$= 37$

33. *Thinking and Writing Exercise.*

35. **Familiarize.** Let w, x, y, and z represent the thousand's, hundred's, ten's, and one's digits, respectively.

Translate.

Total number of the digits is 10.

$$w + x + y + z = 10$$

Twice the sum of the thousand's and ten's digits	is	the sum of the hundred's and one's digits	less one.

$$2(w + y) = x + z \quad -1$$

The ten's digit	is twice	the thousand's digit.

$$y = 2 \cdot w$$

The one's digit	equals	the sum of the thousand's and hundred's digits.

$$z = w + x$$

We have a system of equations which can be written as

$w + x + y + z = 10$

$2w - x + 2y - z = -1$

$-2w + y = 0$

$w + x - z = 0$

Carry out. We can use matrices to solve the system. We get $(1, 3, 2, 4)$.

Check. The sum of the digits is 10. Twice the sum of 1 and 2 is 6. This is one less than the sum of 3 and 4. The ten's digit, 2, is twice the thousand's digit, 1. The one's digit, 4, equals $1 + 3$. The numbers check.

State. The number is 1324.

Mid-Chapter Review

Guided Solutions

1. $x - y + z = 4$

$\underline{x + y - 2z = 3}$

$2x - z = 7$

$x + y - 2z = 3$

$\underline{2x - y - z = 9}$

$3x - 3z = 12$

$2x - z = 7$

$3x - 3z = 12$

$x = 3, z = -1$

$x + y - 2z = 3$

$3 + y - 2(-1) = 3$

$y = -2$

The solution is $(3, -2, -1)$

2. $\begin{bmatrix} 2 & 3 & | & 6 \\ 4 & -5 & | & 1 \end{bmatrix}$

$\begin{bmatrix} 2 & 3 & | & 6 \\ 0 & -11 & | & -11 \end{bmatrix}$

$2x + 3y = 6$

$-11y = -11$

$y = 1$

$2x + 3(1) = 6$

$x = \dfrac{3}{2}$

The solution is $\left(\dfrac{3}{2}, 1 \right)$.

Mixed Review

1. $x + 2y + z = 4$ (1)

$2x - y - z = 1$ (2)

$x - 2y + z = 8$ (3)

Add (1) and (2) to generate an equation in x and y. Then add (2) and (3) to generate a second equation in x and y.

$x + 2y + z = 4$ (1)

$\underline{2x - y - z = 1}$ (2)

$3x + y\ \ \ \ \ = 5$ (4)

$x - 2y + z = 8$ (3)

$\underline{2x - y - z = 1}$ (2)

$3x - 3y\ \ \ = 9$ (5)

Add $-1 \cdot (5)$ to (4) to eliminate x and find y, then back substitute to find x and z.

$3x + y = 5$ (4)

$\underline{-3x + 3y = -9}$ $-1 \cdot (5)$

$4y = -4$

$y = \dfrac{-4}{4} = -1$

Use (4) and the value of y to find x.

$3x + y = 5$ (4)

$3x = 5 - y = 5 - (-1) = 6$

$x = \dfrac{6}{3} = 2$

Use (1) and the values of x and y to find z.

$x + 2y + z = 4$ (1)

$z = 4 - 2y - x = 4 - 2(-1) - 2 = 4$

$(x, y, z) = (2, -1, 4)$

2. $x + 2y - z = 3$ (1)

$x - y + z = -7$ (2)

$2x + 3y - z = 2$ (3)

Add (1) and (2) to generate an equation in x and y. Then add (2) and (3) to generate a second equation in x and y.

$x + 2y - z = 3$ (1)

$\underline{x - y + z = -7}$ (2)

$2x + y\ \ \ \ \ = -4$ (4)

$2x + 3y - z = 2$ (3)

$\underline{x - y + z = -7}$ (2)

$3x + 2y\ \ \ \ = -5$ (5)

Add $-2 \cdot (4)$ to (5) to eliminate y and find x, then back substitute to find y and z.

$$-4x - 2y = 8 \quad -2 \cdot (4)$$
$$\underline{3x + 2y = -5 \quad (5)}$$
$$-x \qquad = 3$$

$x = -3$

Use (4) and the value of x to find y.

$2x + y = -4 \quad (4)$

$y = -4 - 2x = -4 - 2(-3) = 2$

Use (1) and the values of x and y to find z.

$x + 2y - z = 3 \quad (1)$

$z = x + 2y - 3 = (-3) + 2(2) - 3 = -2$

$(x, y, z) = (-3, 2, -2)$

3. $\quad 5a - 2b - c = 0 \quad (1)$

$\quad 3a + 4b + c = 8 \quad (2)$

$\quad 9a - 6b - 2c = -1 \quad (3)$

Add (1) and (2), to generate an equation in a and b. Then add $2 \cdot (2)$ and (3), to generate a second equation in a and b.

$$5a - 2b - c = 0 \quad (1)$$
$$\underline{3a + 4b + c = 8 \quad (2)}$$
$$8a + 2b \qquad = 8$$

or $4a + b = 4 \quad (4)$

$$6a + 8b + 2c = 16 \quad 2 \cdot (2)$$
$$\underline{9a - 6b - 2c = -1 \quad (2)}$$
$$15a + 2b \qquad = 15 \quad (5)$$

Add $-2 \cdot (4)$ to (5) to eliminate b and find a, then back substitute to find b and c.

$$-8a - 2b = -8 \quad -2 \cdot (4)$$
$$\underline{15a + 2b = 15 \quad (5)}$$
$$7a \qquad = 7$$

$a = 1$

Use (4) and the value of a to find b.

$4a + b = 4 \quad (4)$

$b = 4 - 4a = 4 - 4(1) = 0$

Use (2) and the values of a and b to find c.

$3a + 4b + c = 8 \quad (2)$

$c = 8 - 3a - 4b = 8 - 3(1) - 4(0) = 5$

$(a, b, c) = (1, 0, 5)$

4. $\quad 2u + v - w = 3 \quad (1)$

$\quad u + 5v + 2w = 2 \quad (2)$

$\quad 2u - v - w = 4 \quad (3)$

Add $-1 \cdot (1)$ and $2 \cdot (2)$ to generate an equation in v and w. Then add $2 \cdot (2)$ and $-1 \cdot (3)$, to generate a second equation in v and w.

$$-2u - v + w = -3 \quad -1 \cdot (1)$$
$$\underline{2u + 10v + 4w = 4 \quad 2 \cdot (2)}$$
$$9v + 5w = 1 \quad (4)$$

$$2u + 10v + 4w = 4 \quad 2 \cdot (2)$$
$$\underline{-2u + v + w = -4 \quad -1 \cdot (3)}$$
$$11v + 5w = 0 \quad (5)$$

Add $-1 \cdot (4)$ to (5) to eliminate w and find v, then back substitute to find w and u.

$$-9v - 5w = -1 \quad -1 \cdot (4)$$
$$\underline{11v + 5w = 0 \quad (5)}$$
$$2v \qquad = -1$$

$v = -\dfrac{1}{2}$

Use (4) and the value of v to find w.

$9v + 5w = 1 \quad (4)$

$5w = 1 - 9v = 1 - 9\left(-\dfrac{1}{2}\right) = \dfrac{11}{2}$

$w = \dfrac{1}{5} \cdot \dfrac{11}{2} = \dfrac{11}{10}$

Use (2) and the values of v and w to find u.

$u + 5v + 2w = 2 \quad (2)$

$u = 2 - 5v - 2w = 2 - 5\left(-\dfrac{1}{2}\right) - 2\left(\dfrac{11}{10}\right)$

$u = \dfrac{20}{10} + \dfrac{25}{10} - \dfrac{22}{10} = \dfrac{23}{10}$

$(u, v, w) = \left(\dfrac{23}{10}, -\dfrac{1}{2}, \dfrac{11}{10}\right)$

5. $\quad y + 3z = 2 \quad (1)$

$\quad x - 2z = 4 \quad (2)$

$\quad 2x - y = 0 \quad (3)$

Add $2 \cdot (1)$ and $3 \cdot (2)$ to generate an equation in x and y. (3) is already an equation in x and y.

$$2y + 6z = 4 \quad 2 \cdot (1)$$
$$\underline{3x - 6z = 12 \quad 3 \cdot (2)}$$
$$3x + 2y \qquad = 16 \quad (4)$$

Add $2 \cdot (3)$ to (4) to eliminate y and find x, then back substitute to find y and z.

$$4x - 2y = 0 \quad 2 \cdot (3)$$
$$\underline{3x + 2y = 16 \quad (4)}$$
$$7x \quad\quad = 16$$

$$x = \frac{16}{7}$$

Use (3) and the value of x to find y.

$$2x - y = 0 \quad (3)$$

$$y = 2x = 2\left(\frac{16}{7}\right) = \frac{32}{7}$$

Use (2) and the value of x to find z.

$$x - 2z = 4 \quad (2)$$

$$2z = x - 4 = \frac{16}{7} - 4 = \frac{16}{7} - 4 \cdot \frac{7}{7}$$

$$2z = \frac{16}{7} - \frac{28}{7} = -\frac{12}{7}$$

$$z = -\frac{12}{7} \cdot \frac{1}{2} = -\frac{6}{7}$$

$$(x, y, z) = \left(\frac{16}{7}, \frac{32}{7}, -\frac{6}{7}\right)$$

6. $\quad 2x \quad\quad - z = 1 \quad (1)$
$\quad\quad x - 2y + 3z = 2 \quad (2)$
$\quad\quad x + 2y - 4z = -2 \quad (3)$

Add (2) and (3) to generate an equation in x and z. (1) is already an equation in x and z.

$$x - 2y + 3z = 2 \quad (2)$$
$$\underline{x + 2y - 4z = -2 \quad (3)}$$
$$2x \quad\quad - z = 0 \quad (4)$$

Add $-1 \cdot (1)$ to (4) to eliminate x and find z.

$$-2x + z = -1 \quad 2 \cdot (3)$$
$$\underline{2x - z = 0 \quad (4)}$$
$$0 = -1$$

The result is a contradiction. The system has no solution.

7. $\quad 2x + 4y = -1 \quad (1)$
$\quad\quad x + 3y = 2 \quad (2)$

Create the augmented coefficient matrix.

$$\begin{bmatrix} 2 & 4 & | & -1 \\ 1 & 3 & | & 2 \end{bmatrix}$$

Replace R2 with R1 + (-2)R2 to get a leading 0 in R2.

$$\begin{array}{rrrl} 2 & 4 & -1 & \text{R1} \\ -2 & -6 & -4 & (-2)\text{R2} \\ \hline 0 & -2 & -5 & \text{New R2} \end{array}$$

$$\begin{bmatrix} 2 & 4 & | & -1 \\ 0 & -2 & | & -5 \end{bmatrix}$$

Reinsert the variables in R2 to find y.

$$-2y = -5 \Rightarrow y = \frac{-5}{-2} = \frac{5}{2}$$

Reinsert the variables into R1 and use the value of y to find x.

$$2x + 4y = -1 \Rightarrow 2x = -1 - 4y$$

$$= -1 - 4\left(\frac{5}{2}\right) = -1 - 10 = -11$$

$$x = -\frac{11}{2}$$

$$(x, y) = \left(-\frac{11}{2}, \frac{5}{2}\right)$$

8. $\quad 4x - y = 5 \quad (1)$
$\quad\quad 2x + 3y = -1 \quad (2)$

Create the augmented coefficient matrix.

$$\begin{bmatrix} 4 & -1 & | & 5 \\ 2 & 3 & | & -1 \end{bmatrix}$$

Replace R2 with R1 + (-2)R2 to get a leading 0 in R2.

$$\begin{array}{rrrl} 4 & -1 & 5 & \text{R1} \\ -4 & -6 & 2 & (-2)\text{R2} \\ \hline 0 & -7 & 7 & \text{New R2} \end{array}$$

$$\begin{bmatrix} 4 & -1 & | & 5 \\ 0 & -7 & | & 7 \end{bmatrix}$$

Reinsert the variables in R2 to find y.

$$-7y = 7 \Rightarrow y = \frac{7}{-7} = -1$$

Reinsert the variables into R1 and use the value of y to find x.

$$4x - y = 5$$

$$4x = 5 + y = 5 + (-1) = 4$$

$$x = \frac{4}{4} = 1$$

$$(x, y) = (1, -1)$$

9. $2x + y + 2z = 5$ (1)

$x + 2y + 4z = 6$ (2)

$2x + 3y + 5z = 8$ (3)

Create the augmented coefficient matrix.

$$\begin{bmatrix} 2 & 1 & 2 & | & 5 \\ 1 & 2 & 4 & | & 6 \\ 2 & 3 & 5 & | & 8 \end{bmatrix}$$

Replace R2 with (-1) R1 + 2R2 to get a leading 0 in R2. And replace R3 with (-1) R1 + R3 to get a leading 0 in R3.

-2	-1	-2	-5	(-1)R1
2	4	8	12	2R2
0	3	6	7	New R2

-2	-1	-2	-5	(-1)R1
2	3	5	8	R3
0	2	3	3	New R3

$$\begin{bmatrix} 2 & 1 & 2 & | & 5 \\ 0 & 3 & 6 & | & 7 \\ 0 & 2 & 3 & | & 3 \end{bmatrix}$$

Replace R3 with 2R2 + (-3) R3 to get a second leading 0 in R3.

0	6	12	14	2R2
0	-6	-9	-9	(-3)R3
0	0	3	5	New R3

Reinsert the variables in R3 to find z.

$$3z = 5 \Rightarrow z = \frac{5}{3}$$

Reinsert the variables into R2 and use the value of z to find y.

$$3y + 6z = 7 \Rightarrow 3y = 7 - 6z = 7 - 6\left(\frac{5}{3}\right) = -3$$

$$y = \frac{-3}{3} = -1$$

Reinsert the variables into R1 and use the value of z and y to find x.

$2x + y + 2z = 5 \Rightarrow 2x = 5 - y - 2z$

$$= 5 - (-1) - 2\left(\frac{5}{3}\right) = 6 - \frac{10}{3}$$

$$= \frac{18}{3} - \frac{10}{3} = \frac{8}{3}$$

$$x = \frac{8}{3} \cdot \frac{1}{2} = \frac{4}{3}$$

$$(x, y, z) = \left(\frac{4}{3}, -1, \frac{5}{3}\right)$$

10. $-2a + 8b + c = -2$ (1)

$4a - 2b - 6c = 1$ (2)

$2a - b - c = 2$ (3)

Create the augmented coefficient matrix.

$$\begin{bmatrix} -2 & 8 & 1 & | & -2 \\ 4 & -2 & -6 & | & 1 \\ 2 & -1 & -1 & | & 2 \end{bmatrix}$$

Replace R2 with 2R1 + R2 to get a leading 0 in R2. And replace R3 with R1 + R3 to get a leading 0 in R3.

-4	16	2	-4	2R1
4	-2	-6	1	R2
0	14	-4	-3	New R2

-2	8	1	-2	R1
2	-1	-1	2	R3
0	7	0	0	New R3

$$\begin{bmatrix} -2 & 8 & 1 & | & -2 \\ 0 & 14 & -4 & | & -3 \\ 0 & 7 & 0 & | & 0 \end{bmatrix}$$

Since we can see from R3 that $b = 0$, we will leave that row alone by interchanging R2 and R3.

$$\begin{bmatrix} -2 & 8 & 1 & | & -2 \\ 0 & 7 & 0 & | & 0 \\ 0 & 14 & -4 & | & -3 \end{bmatrix}$$

Now replace R3 with $2R2 + (-1)R3$ to get a second leading 0 in R3.

0	14	0	0	2R2
0	-14	4	3	(-1)R3
0	0	4	3	New R3

$$\begin{bmatrix} -2 & 8 & 1 & | & -2 \\ 0 & 7 & 0 & | & 0 \\ 0 & 0 & 4 & | & 3 \end{bmatrix}$$

Reinsert the variables into R3 to find c.

$$4c = 3 \Rightarrow c = \frac{3}{4}$$

Reinsert the variables into R1 and use the value of b and c to find a.

$$-2a + 8b + c = -2$$
$$-2a = -2 - 8b - c$$
$$-2a = -2 - 8(0) - \frac{3}{4} = -\frac{11}{4}$$
$$a = -\frac{1}{2} \cdot \left(-\frac{11}{4}\right) = \frac{11}{8}$$
$$(a, b, c) = \left(\frac{11}{8}, 0, \frac{3}{4}\right)$$

11. Let x = the first number, y = the second number, and z = the third number. Then:

$$x + y + z = 15 \quad (1)$$
$$x - 2y = 2z$$
or $x - 2y - 2z = 0 \quad (2)$
$$x - y + z = 19 \quad (3)$$

Add $-1 \cdot (1)$ to (3), giving $-2y = 4$. So, $y = -2$.

Substituting the value of y into (1) and (2) gives:

$$x + z = 17$$
$$x - 2z = -4$$

Eliminating x gives:

$3z = 21$, so $z = 7$.

Substituting the value of y and z into (1) gives $x = 10$. The numbers are:

$$(x, y, z) = (10, -2, 7).$$

12. Let x = the measure of angle A, y = the measure of angle B, and z = the measure of angle C. Then:

$$y = 2x \quad (1)$$
$$z = y + 40 \quad (2)$$
$$x + y + z = 180 \quad (3)$$

Substituting $y = 2x$ into equations (2) and (3) will give a two variable system in x and z.

$$z = y + 40 \quad (2)$$
$$z = 2x + 40 \quad (4)$$
$$x + y + z = 180 \quad (3)$$
$$x + 2x + z = 180$$
or $3x + z = 180 \quad (5)$

Substituting $z = 2x + 40$ into (5) allows us to find x.

$$3x + z = 180 \quad (5) \text{ Substituting the}$$
$$3x + (2x + 40) = 180$$
$$3x + 2x = 180 - 40$$
$$5x = 140$$
$$x = \frac{140}{5} = 28$$

value of x into (4) allows us to find z.

$$z = 2x + 40 \quad (4)$$
$$= 2(28) + 40 = 96$$

Substituting the value of x into (1) allows us to find y.

$$y = 2x \quad (1)$$
$$= 2(28) = 56$$
$$(x, y, z) = (28, 56, 96)$$

13. Let x = the number of brownies sold, y = the number of bags of chips sold, and z = the number of hot dogs sold. Then:

$$x + y + z = 125 \quad (1)$$
$$75x + 100y + 200z = 16375$$
or $3x + 4y + 8z = 655 \quad (2)$
$$z = x + y - 25 \quad (3)$$

Substituting $z = x + y - 25$ into equations (1) and (2) will give a two variable system in x and y.

$$x + y + z = 125 \quad (1)$$
$$x + y + (x + y - 25) = 125$$
or $2x + 2y = 150$
or $y = 75 - x \quad (4)$
$$3x + 4y + 8z = 655 \quad (2)$$
$$3x + 4y + 8(x + y - 25) = 655$$
or $11x + 12y = 855 \quad (5)$

Substituting $y = 75 - x$ into (5) allows us to find x.

$$11x + 12y = 855 \quad (5)$$
$$11x + 12(75 - x) = 855$$
$$-x = 855 - 12 \cdot 75 = -45$$
$$x = 45$$

Substituting the value of x into (4) allows us to find y.

$$y = 75 - x \quad (4)$$
$$= 75 - 45 = 30$$

Substituting the values of x and y into (3) allows us to find z.

$$z = x + y - 25 \quad (1)$$
$$= 30 + 45 - 25 = 50$$
$$(x, y, z) = (45, 30, 50)$$

14. Let x = the balance on the 2% card, y = the balance on the 1% card, and z = the balance on the 1.5% card. Then:
$$x + y + z = 10,300 \quad (1)$$
$$0.02x + 0.01y + 0.015z = 151$$
$$\text{or } 20x + 10y + 15z = 151,000$$
$$\text{or } 4x + 2y + 3z = 30,200 \quad (2)$$
$$x + z = y + 3300$$
$$\text{or } y = x + z - 3300 \quad (3)$$

Substituting $y = x + z - 3300$ into equations (1) and (2) will give a two variable system in x and z.

$$x + y + z = 10,300 \quad (1)$$
$$x + (x + z - 3300) + z = 10,300$$
$$\text{or } 2x + 2z = 13,600$$
$$\text{or } x + z = 6800$$
$$\text{or } z = 6800 - x \quad (4)$$
$$4x + 2(x + z - 3300) + 3z = 30,200 \quad (2)$$
$$\text{or } 6x + 5z = 36,800 \quad (5)$$

Substituting $z = 6800 - x$ into (5) allows us to find x.

$$6x + 5z = 36,800 \quad (5)$$
$$6x + 5(6800 - x) = 36,800$$
$$x = 36,800 - 5 \cdot 6800 = 2800$$

Substituting the value of x into (4) allows us to find z.

$$z = 6800 - x \quad (4)$$
$$= 6800 - 2800 = 4000$$

Substituting the values of x and z into (3) allows us to find z.

$$y = x + z - 3300 \quad (3)$$
$$= 4000 + 2800 - 3300 = 3500$$
$$(x, y, z) = (2800, 3500, 4000)$$

1. True

3. True

5. False

7. $\begin{vmatrix} 5 & 1 \\ 2 & 4 \end{vmatrix} = 5 \cdot 4 - 2 \cdot 1 = 20 - 2 = 18$

9. $\begin{vmatrix} 10 & 8 \\ -5 & -9 \end{vmatrix} = 10 \cdot (-9) - (-5) \cdot 8 = -90 + 40$
$$= -50$$

11. $\begin{vmatrix} 1 & 4 & 0 \\ 0 & -1 & 2 \\ 3 & -2 & 1 \end{vmatrix}$

$$= 1 \begin{vmatrix} -1 & 2 \\ -2 & 1 \end{vmatrix} - 0 \begin{vmatrix} 4 & 0 \\ -2 & 1 \end{vmatrix} + 3 \begin{vmatrix} 4 & 0 \\ -1 & 2 \end{vmatrix}$$
$$= 1(-1 + 4) - 0(4 + 0) + 3(8 + 0)$$
$$= 1(3) - 0(4) + 3(8)$$
$$= 3 - 0 + 24$$
$$= 27$$

13. $\begin{vmatrix} -4 & -2 & 3 \\ -3 & 1 & 2 \\ 3 & 4 & -2 \end{vmatrix}$

$$= -4 \begin{vmatrix} 1 & 2 \\ 4 & -2 \end{vmatrix} - (-3) \begin{vmatrix} -2 & 3 \\ 4 & -2 \end{vmatrix} + 3 \begin{vmatrix} -2 & 3 \\ 1 & 2 \end{vmatrix}$$
$$= -4(-2 - 8) + 3(4 - 12) + 3(-4 - 3)$$
$$= -4(-10) + 3(-8) + 3(-7)$$
$$= 40 - 24 - 21$$
$$= -5$$

15. $5x + 8y = 1$
$$3x + 7y = 5$$

We have

$$x = \frac{\begin{vmatrix} 1 & 8 \\ 5 & 7 \end{vmatrix}}{\begin{vmatrix} 5 & 8 \\ 3 & 7 \end{vmatrix}} = \frac{1 \cdot 7 - 5 \cdot 8}{5 \cdot 7 - 3 \cdot 8} = \frac{7 - 40}{35 - 24} = \frac{-33}{11}$$
$$= -3$$

and

$$y = \frac{\begin{vmatrix} 5 & 1 \\ 3 & 5 \end{vmatrix}}{\begin{vmatrix} 5 & 8 \\ 3 & 7 \end{vmatrix}} = \frac{5 \cdot 5 - 3 \cdot 1}{11} = \frac{25 - 3}{11} = \frac{22}{11} = 2$$

The solution is $(-3, 2)$ which checks.

17. $5x - 4y = -3$

   $7x + 2y = 6$

   We have

$$x = \frac{\begin{vmatrix} -3 & -4 \\ 6 & 2 \end{vmatrix}}{\begin{vmatrix} 5 & -4 \\ 7 & 2 \end{vmatrix}} = \frac{(-3) \cdot 2 - 6(-4)}{5 \cdot 2 - 7 \cdot (-4)}$$

$$= \frac{-6 + 24}{10 + 28} = \frac{18}{38} = \frac{9}{19}$$

and

$$y = \frac{\begin{vmatrix} 5 & -3 \\ 7 & 6 \end{vmatrix}}{\begin{vmatrix} 5 & -4 \\ 7 & 2 \end{vmatrix}} = \frac{5 \cdot 6 - 7 \cdot (-3)}{38} = \frac{30 + 21}{38} = \frac{51}{38}$$

The solution is $\left(\frac{9}{19}, \frac{51}{38}\right)$ which checks.

19. $3x - y + 2z = 1$

    $x - y + 2z = 3$

   $-2x + 3y + z = 1$

   We compute D, D_x, D_y, and D_z.

$$D = \begin{vmatrix} 3 & -1 & 2 \\ 1 & -1 & 2 \\ -2 & 3 & 1 \end{vmatrix}$$

$$= 3\begin{vmatrix} -1 & 2 \\ 3 & 1 \end{vmatrix} - 1\begin{vmatrix} -1 & 2 \\ 3 & 1 \end{vmatrix} - 2\begin{vmatrix} -1 & 2 \\ -1 & 2 \end{vmatrix}$$

$$= 3(-1 - 6) - 1(-1 - 6) - 2(-2 + 2)$$

$$= 3(-7) - 1(-7) - 2 \cdot 0 = -21 + 7 - 0 = -14$$

$$D_x = \begin{vmatrix} 1 & -1 & 2 \\ 3 & -1 & 2 \\ 1 & 3 & 1 \end{vmatrix}$$

$$= 1\begin{vmatrix} -1 & 2 \\ 3 & 1 \end{vmatrix} - 3\begin{vmatrix} -1 & 2 \\ 3 & 1 \end{vmatrix} + 1\begin{vmatrix} -1 & 2 \\ -1 & 2 \end{vmatrix}$$

$$= 1(-1 - 6) - 3(-1 - 6) + 1(-2 + 2)$$

$$= 1(-7) - 3(-7) + 1(0)$$

$$= -7 + 21 + 0 = 14$$

$$D_y = \begin{vmatrix} 3 & 1 & 2 \\ 1 & 3 & 2 \\ -2 & 1 & 1 \end{vmatrix}$$

$$= 3\begin{vmatrix} 3 & 2 \\ 1 & 1 \end{vmatrix} - 1\begin{vmatrix} 1 & 2 \\ 1 & 1 \end{vmatrix} - 2\begin{vmatrix} 1 & 2 \\ 3 & 2 \end{vmatrix}$$

$$= 3(3 - 2) - 1(1 - 2) - 2(2 - 6)$$

$$= 3 \cdot 1 - 1(-1) - 2(-4)$$

$$= 3 + 1 + 8 = 12$$

$$D_z = \begin{vmatrix} 3 & -1 & 1 \\ 1 & -1 & 3 \\ -2 & 3 & 1 \end{vmatrix}$$

$$= 3\begin{vmatrix} -1 & 3 \\ 3 & 1 \end{vmatrix} - 1\begin{vmatrix} -1 & 1 \\ 3 & 1 \end{vmatrix} - 2\begin{vmatrix} -1 & 1 \\ -1 & 3 \end{vmatrix}$$

$$= 3(-1 - 9) - 1(-1 - 3) - 2(-3 + 1)$$

$$= 3(-10) - 1(-4) - 2(-2)$$

$$= -30 + 4 + 4 = -22$$

$$x = \frac{D_x}{D} = \frac{14}{-14} = -1$$

$$y = \frac{D_y}{D} = \frac{12}{-14} = -\frac{6}{7}$$

$$z = \frac{D_z}{D} = \frac{-22}{-14} = \frac{11}{7}$$

The solution is $\left(-1, -\frac{6}{7}, \frac{11}{7}\right)$ which checks.

21. $2x - 3y + 5z = 27$

    $x + 2y - z = -4$

    $5x - y + 4z = 27$

   We compute D, D_x, D_y, and D_z.

$$D = \begin{vmatrix} 2 & -3 & 5 \\ 1 & 2 & -1 \\ 5 & -1 & 4 \end{vmatrix}$$

$$= 2\begin{vmatrix} 2 & -1 \\ -1 & 4 \end{vmatrix} - 1\begin{vmatrix} -3 & 5 \\ -1 & 4 \end{vmatrix} + 5\begin{vmatrix} -3 & 5 \\ 2 & -1 \end{vmatrix}$$

$$= 2(8 - 1) - 1(-12 + 5) + 5(3 - 10)$$

$$= 2 \cdot 7 - 1(-7) + 5(-7)$$

$$= 14 + 7 - 35$$

$$= -14$$

$$D_x = \begin{vmatrix} 27 & -3 & 5 \\ -4 & 2 & -1 \\ 27 & -1 & 4 \end{vmatrix}$$

$$= 27 \begin{vmatrix} 2 & -1 \\ -1 & 4 \end{vmatrix} - (-4) \begin{vmatrix} -3 & 5 \\ -1 & 4 \end{vmatrix} + 27 \begin{vmatrix} -3 & 5 \\ 2 & -1 \end{vmatrix}$$

$$= 27(8-1) + 4(-12+5) + 27(3-10)$$

$$= 27(7) + 4(-7) + 27(-7)$$

$$= 189 - 28 - 189$$

$$= -28$$

$$D_y = \begin{vmatrix} 2 & 27 & 5 \\ 1 & -4 & -1 \\ 5 & 27 & 4 \end{vmatrix}$$

$$= 2 \begin{vmatrix} -4 & -1 \\ 27 & 4 \end{vmatrix} - 1 \begin{vmatrix} 27 & 5 \\ 27 & 4 \end{vmatrix} + 5 \begin{vmatrix} 27 & 5 \\ -4 & -1 \end{vmatrix}$$

$$= 2(-16+27) - 1(108-135)$$
$$\qquad\qquad + 5(-27+20)$$

$$= 2(11) - 1(-27) + 5(-7)$$

$$= 22 + 27 - 35$$

$$= 14$$

$$D_z = \begin{vmatrix} 2 & -3 & 27 \\ 1 & 2 & -4 \\ 5 & -1 & 27 \end{vmatrix}$$

$$= 2 \begin{vmatrix} 2 & -4 \\ -1 & 27 \end{vmatrix} - 1 \begin{vmatrix} -3 & 27 \\ -1 & 27 \end{vmatrix} + 5 \begin{vmatrix} -3 & 27 \\ 2 & -4 \end{vmatrix}$$

$$= 2(54-4) - 1(-81+27) + 5(12-54)$$

$$= 2(50) - 1(-54) + 5(-42)$$

$$= 100 + 54 - 210$$

$$= -56$$

$$x = \frac{D_x}{D} = \frac{-28}{-14} = 2$$

$$y = \frac{D_y}{D} = \frac{39}{-14} = \frac{14}{-14} = -1$$

$$z = \frac{D_z}{D} = \frac{-56}{-14} = 4$$

The solution is $(2, -1, 4)$ which checks.

23. $r - 2s + 3t = 6$
  $2r - s - t = -3$
  $r + s + t = 6$

We compute D, D_r, D_s, and D_t.

$$D = \begin{vmatrix} 1 & -2 & 3 \\ 2 & -1 & -1 \\ 1 & 1 & 1 \end{vmatrix}$$

$$= 1 \begin{vmatrix} -1 & -1 \\ 1 & 1 \end{vmatrix} - 2 \begin{vmatrix} -2 & 3 \\ 1 & 1 \end{vmatrix} + 1 \begin{vmatrix} -2 & 3 \\ -1 & -1 \end{vmatrix}$$

$$= 1(-1+1) - 2(-2-3) + 1(2+3)$$

$$= 1(0) - 2(-5) + 1(5)$$

$$= 0 + 10 + 5$$

$$= 15$$

$$D_r = \begin{vmatrix} 6 & -2 & 3 \\ -3 & -1 & -1 \\ 6 & 1 & 1 \end{vmatrix}$$

$$= 6 \begin{vmatrix} -1 & -1 \\ 1 & 1 \end{vmatrix} - (-3) \begin{vmatrix} -2 & 3 \\ 1 & 1 \end{vmatrix} + 6 \begin{vmatrix} -2 & 3 \\ -1 & -1 \end{vmatrix}$$

$$= 6(-1+1) + 3(-2-3) + 6(2+3)$$

$$= 6(0) + 3(-5) + 6(5)$$

$$= 0 - 15 + 30$$

$$= 15$$

$$D_s = \begin{vmatrix} 1 & 6 & 3 \\ 2 & -3 & -1 \\ 1 & 6 & 1 \end{vmatrix}$$

$$= 1 \begin{vmatrix} -3 & -1 \\ 6 & 1 \end{vmatrix} - 2 \begin{vmatrix} 6 & 3 \\ 6 & 1 \end{vmatrix} + 1 \begin{vmatrix} 6 & 3 \\ -3 & -1 \end{vmatrix}$$

$$= 1(-3+6) - 2(6-18) + 1(-6+9)$$

$$= 1(3) - 2(-12) + 1(3)$$

$$= 3 + 24 + 3$$

$$= 30$$

$$D_t = \begin{vmatrix} 1 & -2 & 6 \\ 2 & -1 & -3 \\ 1 & 1 & 6 \end{vmatrix}$$

$$= 1\begin{vmatrix} -1 & -3 \\ 1 & 6 \end{vmatrix} - 2\begin{vmatrix} -2 & 6 \\ 1 & 6 \end{vmatrix} + 1\begin{vmatrix} -2 & 6 \\ -1 & -3 \end{vmatrix}$$

$$= 1(-6+3) - 2(-12-6) + 1(6+6)$$

$$= 1(-3) - 2(-18) + 1(12)$$

$$= -3 + 36 + 12$$

$$= 45$$

$$r = \frac{D_r}{D} = \frac{15}{15} = 1$$

$$s = \frac{D_s}{D} = \frac{30}{15} = 2$$

$$t = \frac{D_t}{D} = \frac{45}{15} = 3$$

The solution is $(1, 2, 3)$ which checks.

25. *Thinking and Writing Exercise.*

27. $f(90) = 80(90) + 2500 = 7200 + 2500 = 9700$

29. $(g - f)(10) = 70(10) - 2500 = 700 - 2500$
$$= -1800$$

31. $g(x) = f(x)$
$$\Rightarrow 150x = 80x + 2500 \Rightarrow 70x = 2500$$
$$\Rightarrow x = \frac{250\cancel{0}}{7\cancel{0}} = \frac{250}{\cdot 7} = 35\frac{5}{7}$$

33. *Thinking and Writing Exercise.*

35. $\begin{vmatrix} y & -2 \\ 4 & 3 \end{vmatrix} = 44$

Evaluating the determinant, we have
$$3y + 8 = 44$$
$$3y = 36$$
$$y = 12$$

37. $\begin{vmatrix} m+1 & -2 \\ m-2 & 1 \end{vmatrix} = 27$

First evaluate the determinant.

$$\begin{vmatrix} m+1 & -2 \\ m-2 & 1 \end{vmatrix} = (m+1)(1) - (m-2)(-2)$$
$$= m + 1 + 2m - 4$$
$$= 3m - 3$$

Next set $3m - 3$ equal to 27 and solve the resulting equation.
$$3m - 3 = 27$$
$$3m = 30$$
$$m = 10$$
The solution is 10.

Exercise Set 9.5

1. b

3. e

5. h

7. g

9. $C(x) = 45x + 300,000 \quad R(x) = 65x$

a) $P(x) = R(x) - C(x)$
$$= 65x - (45x + 300,000)$$
$$= 65x - 45x - 300,000$$
$$= 20x - 300,000$$

b) To find the break-even point we solve the system:
$$R(x) = 65x$$
$$C(x) = 45x + 300,000$$

Since $R(x) = C(x)$ at the break-even point, we can rewrite the system:
$$R(x) = 65x \qquad (1)$$
$$R(x) = 45x + 300,000 \quad (2)$$

We solve using substitution.
$$65x = 45x + 300,000 \quad \text{Substituting } 65x$$
$$\text{for } R(x) \text{ in } (2)$$
$$20x = 300,000$$
$$x = 15,000$$

Thus, 15,000 units must be produced and sold in order to break even. Also, $R(15,000) = C(15,000) = 975,000$, so the breakeven point is (15,000 units, $975,000).

11. $C(x) = 15x + 3100$ $R(x) = 40x$

a) $P(x) = R(x) - C(x)$

$\qquad = 40x - (15x + 3100)$

$\qquad = 25x - 3100$

b) To find the break-even point we solve the system, set $P(x) = 0$.

$P(x) = 0 \Rightarrow 25x - 3100 = 0$

$\Rightarrow x = \dfrac{3100}{25} = 124$

Thus, 124 units must be produced and sold in order to break even.

$C(124) = R(124) = 40(124) = \$4960$

So the break-even point is (124 units, \$4960).

13. $C(x) = 40x + 22{,}500$ $R(x) = 85x$

a) $P(x) = R(x) - C(x)$

$\qquad = 85x - (40x + 22{,}500)$

$\qquad = 85x - 40x - 22{,}500$

$\qquad = 45x - 22{,}500$

b) To find the break-even point we solve the system:

$R(x) = 85x$

$C(x) = 40x + 22{,}500$

Since $R(x) = C(x)$ at the break-even point, we can rewrite the system:

$R(x) = 85x \qquad (1)$

$R(x) = 40x + 22{,}500 \quad (2)$

We solve using substitution.

$85x = 40x + 22{,}500$ Substituting $85x$
for $R(x)$ in (2)

$45x = 22{,}500$

$x = 500$

Thus, 500 units must be produced and sold in order to break even. Also,

$R(500) = C(500) = 42{,}500$, so

the breakeven point is (500 units, \$42,500).

15. $C(x) = 24x + 50{,}000$ $R(x) = 40x$

a) $P(x) = R(x) - C(x)$

$\qquad = 40x - (24x + 50{,}000)$

$\qquad = 16x - 50{,}000$

b) To find the break-even point we solve the system, set $P(x) = 0$.

$P(x) = 0 \Rightarrow 16x - 50{,}000 = 0$

$\Rightarrow x = \dfrac{50{,}000}{16} = 3125$

3125 units must be produced and sold in order to break even.

$C(3125) = R(3125) = 40(3125) = \$125{,}000$

17. $C(x) = 75x + 100{,}000$ $R(x) = 125x$

a) $P(x) = R(x) - C(x)$

$\qquad = 125x - (75x + 100{,}000)$

$\qquad = 125x - 75x - 100{,}000$

$\qquad = 50x - 100{,}000$

b) To find the break-even point we solve the system:

$R(x) = 125x$

$C(x) = 75x + 100{,}000$

Since $R(x) = C(x)$ at the break-even point, we can rewrite the system:

$R(x) = 125x \qquad (1)$

$R(x) = 75x + 100{,}000 \quad (2)$

We solve using substitution.

$125x = 75x + 100{,}000$ Substituting
$125x$ for $R(x)$
in (2)

$50x = 100{,}000$

$x = 2000$

Thus, 2000 units must be produced and sold in order to break even. Also,

$R(2000) = C(2000) = 250{,}000$, so the

breakeven point is (2000 units, \$250,000).

19. $D(p) = 1000 - 10p$

$S(p) = 230 + p$

Since both demand and supply are quantities, the system can be written:

$q = 1000 - 10p$ (1)

$q = 230 + p$ (2)

Substitute $1000 - 10p$ for q in (2) and solve.

$1000 - 10p = 230 + p$

$770 = 11p$

$70 = p$

The equilibrium price is \$70 per unit. To find the equilibrium quantity we substitute \$70 into either $D(p)$ or $S(p)$.

$D(70) = 1000 - 10 \cdot 70 = 1000 - 700 = 300$

The equilibrium quantity is 300 units.

The equilibrium point is $(\$70, 300 \text{ units})$.

21. $D(p) = 760 - 13p$

$S(p) = 430 + 2p$

Since both demand and supply are quantities, the system can be written:

$q = 760 - 13p$ (1)

$q = 430 + 2p$ (2)

Substitute $760 - 13p$ for q in (2) and solve.

$760 - 13p = 430 + 2p$

$330 = 15p$

$22 = p$

The equilibrium price is \$22 per unit. To find the equilibrium quantity we substitute \$22 into either $D(p)$ or $S(p)$.

$S(22) = 430 + 2(22) = 430 + 44 = 474$

The equilibrium quantity is 474 units.

The equilibrium point is $(\$22, 474 \text{ units})$.

23. $D(p) = 7500 - 25p$

$S(p) = 6000 + 5p$

Since both demand and supply are quantities, the system can be written:

$q = 7500 - 25p$ (1)

$q = 6000 + 5p$ (2)

Substitute $7500 - 25p$ for q in (2) and solve.

$7500 - 25p = 6000 + 5p$

$1500 = 30p$

$50 = p$

The equilibrium price is \$50 per unit. To find the equilibrium quantity we substitute \$50 into either $D(p)$ or $S(p)$.

$D(50) = 7500 - 25(50) = 7500 - 1250 = 6250$

The equilibrium quantity is 6250 units.

The equilibrium point is $(\$50, 6250 \text{ units})$.

25. $D(p) = 1600 - 53p$

$S(p) = 320 + 75p$

Since both demand and supply are quantities, the system can be written:

$q = 1600 - 53p$ (1)

$q = 320 + 75p$ (2)

Substitute $1600 - 53p$ for q in (2) and solve.

$1600 - 53p = 320 + 75p$

$1280 = 128p$

$10 = p$

The equilibrium price is \$10 per unit. To find the equilibrium quantity we substitute \$10 into either $D(p)$ or $S(p)$.

$S(10) = 320 + 75(10) = 320 + 750 = 1070$

The equilibrium quantity is 1070 units.

The equilibrium point is $(\$10, 1070 \text{ units})$.

27. **a)** $C(x) = $ Fixed costs + Variable Costs

$C(x) = 45,000 + 40x$

where x is the number of MP3 phones produced

b) Each MP3 phone sells for \$130. The total revenue is 130 times the number of MP3 phones sold. We assume that all MP3 phones produced are sold.

$R(x) = 130x$

c) $P(x) = R(x) - C(x)$

$P(x) = 130x - (45,000 + 40x)$

$= 90x - 45,000$

d) $P(3000) = 90(3000) - 45,000 = 225,000$

The company will realize a \$225,000 profit when 3000 MP3 phones are produced and sold.

$P(400) = 90(400) - 45,000 = -9000$

The company will lose \$9000 if only 400 MP3 phones are produced and sold.

e) Set $P(x) = 0$

$$\Rightarrow 90x - 45,000 = 0$$

$$\Rightarrow x = \frac{45,00\cancel{0}}{9\cancel{0}} = 500$$

The break-even point is 500 MP3 phones.

$C(500) = R(500) = 130(500) = \$65,000$

29. a) $C(x) = $ Fixed costs + Variable Costs

$C(x) = 10,000 + 30x$

where x is the number of car seats produced

b) Each car seat sells for \$80. The total revenue is 80 times the number of car seats sold. We assume that all car seats are sold.

$R(x) = 80x$

c) $P(x) = R(x) - C(x)$

$P(x) = 80x - (10,000 + 30x)$

$= 50x - 10,000$

d) $P(2000) = 50(2000) - 10,000 = 90,000$

The company will make a \$90,000 profit if 2000 car seats are produced and sold.

$P(50) = 50(50) - 10,000 = -7500$

The company will lose \$7500 if only 50 car seats are produced and sold.

e) Set $P(x) = 0$

$$\Rightarrow 50x - 10,000 = 0$$

$$\Rightarrow x = \frac{10,00\cancel{0}}{5\cancel{0}} = 200$$

The break-even point is 200 car seats.

$C(200) = R(200) = 80(200) = \$16,000$

31. a) $D(p) = -14.97p + 987.35$

$S(p) = 98.55p - 5.13$

Rewrite the system:

$q = -14.97p + 987.35$ (1)

$q = 98.55p - 5.13$ (2)

Substitute $-14.97p + 987.35$ for q in (2) and solve.

$-14.97p + 987.35 = 98.55p - 5.13$

$992.48 = 113.52p$

$8.74 \approx p$

The equilibrium price is \$8.74 per unit. A price of \$8.74 per unit should be charged in order to have equilibrium between supply and demand.

b) $R(x) = 8.74x$

$C(x) = 5.15x + 87,985$

Rewrite the system:

$d = 8.74x$ (1)

$d = 5.15x + 87,985$ (2)

We solve using substitution.

$8.74x = 5.15x + 87,985$ Substituting $8.74x$ for d in (2)

$3.59x = 87,985$

$x = \dfrac{87,985}{3.59} = 24,509$ Rounding up

Thus 24,509 units must be sold in order to break even.

33. *Thinking and Writing Exercise.*

35. $f(a) + h = 4a - 7 + h$

37. $2x + 1 = 0 \Rightarrow x = -\dfrac{1}{2}$. Therefore the domain of f

is $\left\{ x \mid x \neq -\dfrac{1}{2} \right\}$

39. $2x + 8 \geq 0 \Rightarrow 2x \geq -8 \Rightarrow x \geq -4$ Therefore the

domain of f is $\{ x \mid x \geq -4 \}$

41. *Thinking and Writing Exercise.*

43. The supply function contains the points $(\$2, 100)$ and $(\$8, 500)$. We find its equation:

$$m = \frac{500 - 100}{8 - 2} = \frac{400}{6} = \frac{200}{3}$$

$y - y_1 = m(x - x_1)$ Point-slope form

$y - 100 = \dfrac{200}{3}(x - 2)$

$$y - 100 = \frac{200}{3}x - \frac{400}{3}$$

$$y = \frac{200}{3}x - \frac{100}{3}$$

We can equivalently express supply S as a function of price p:

$$S(p) = \frac{200}{3}p - \frac{100}{3}$$

The demand function contains the points $(\$1, 500)$ and $(\$9, 100)$. We find its equation:

$$m = \frac{100 - 500}{9 - 1} = \frac{-400}{8} = -50$$

$$y - y_1 = m(x - x_1)$$

$$y - 500 = -50(x - 1)$$

$$y - 500 = -50x + 50$$

$$y = -50x + 550$$

We can equivalently express demand D as a function of price p:

$$D(p) = -50p + 550$$

We have a system of equations

$$S(p) = \frac{200}{3}p - \frac{100}{3}$$

$$D(p) = -50p + 550$$

Rewrite the system:

$$q = \frac{200}{3}p - \frac{100}{3} \quad (1)$$

$$q = -50p + 550 \quad (2)$$

Substitute $\frac{200}{3}p - \frac{100}{3}$ for q in (2) and solve.

$$\frac{200}{3}p - \frac{100}{3} = -50p + 550$$

$$200p - 100 = -150p + 1650 \quad \text{Multiplying by 3 to clear fractions}$$

$$350p - 100 = 1650$$

$$350p = 1750$$

$$p = 5$$

The equilibrium price is $5 per unit.
To find the equilibrium quantity, we substitute $5 into either $S(p)$ or $D(p)$.

$$D(5) = -50(5) + 550 = -250 + 550 = 300$$

The equilibrium quantity is 300 units.
The equilibrium point is $(\$5, 300 \text{ yo-yo's})$.

45.   a)   Enter the data and use the linear regression feature to get
$$S(p) = 15.97p - 1.05 .$$

    b)   Enter the data and use the linear regression feature to get
$$D(p) = -11.26p + 41.16 .$$

    c)   Find the point of intersection of the graphs of the functions found in parts (a) and (b).

We see that the equilibrium point is $(\$1.55, 23.7 \text{ million jars})$.

Chapter 9 Study Summary

1. $x - 2y - z = 8 \quad (1)$
      $2x + 2y - z = 8 \quad (2)$
      $x - 8y + z = 1 \quad (3)$

Add (1) and (3) to generate an equation in x and y. Then add (2) and (3) to generate a second equation in x and y.

$$\begin{array}{ll} x - 2y - z = 8 & (1) \\ x - 8y + z = 1 & (3) \\ \hline 2x - 10y \phantom{{}= 0} = 9 & (4) \end{array}$$

$$\begin{array}{ll} 2x + 2y - z = 8 & (2) \\ x - 8y + z = 1 & (3) \\ \hline 3x - 6y \phantom{{}=0} = 9 & \\ \text{or } x - 2y = 3 & (5) \end{array}$$

Add $-1 \cdot (4)$ to $2(5)$ to eliminate x and find y. Then back-substitute to find x and z.

$$\begin{array}{l} -2x + 10y = -9 \quad -1 \cdot (4) \\ \underline{2x - 4y = 6 \quad\quad (5)} \\ 6y = -3 \end{array}$$

$$\Rightarrow y = \frac{-3}{6} = -\frac{1}{2}$$

Use (5) and the value of y to find x.
$$x - 2y = 3 \quad (5)$$

$$\Rightarrow x = 3 + 2y = 3 + 2\left(-\frac{1}{2}\right) = 2$$

Use (3) and the values of x and y to find z.

$x - 8y + z = 1$ (3)

$\Rightarrow z = 1 - x + 8y = 1 - 2 + 8\left(-\dfrac{1}{2}\right) = -5$

$(x, y, z) = \left(2, -\dfrac{1}{2}, -5\right)$

2.   Let x = the first number, y = the second number, and z = the third number. Then

$x + y + z = 9$ (1)

$z = \dfrac{1}{2}(x + y)$ (2)

$y = x + z - 2$ (3)

Note that equation (2) implies that $x + y = 2z$

Substituting this result into (1) allows us to find z.

$x + y + z = 9 \Rightarrow (x + y) + z = 9$

$\Rightarrow (2z) + z = 9 \Rightarrow 3z = 9 \Rightarrow z = 3$

Use this result to simplify equation (1) and (3).

$x + y + z = 9$ (1)

$\Rightarrow x + y + 3 = 9$

$\Rightarrow x + y = 6$ (4)

$y = x + z - 2$ (3)

$\Rightarrow y = x + 3 - 2 = x + 1$

$\Rightarrow x - y = -1$ (5)

Add (4) to (5) to eliminate y and find x. Then back-substitute to find y.

$x + y = 6$ (4)

$\underline{x - y = -1 \quad (5)}$

$2x = 5$

$\Rightarrow x = \dfrac{5}{2}$

Use (4) and the value of x to find y.

$x + y = 6$ (4)

$\Rightarrow y = 6 - x = 6 - \dfrac{5}{2} = \dfrac{12}{2} - \dfrac{5}{2} = \dfrac{7}{2}$

$(x, y, z) = \left(\dfrac{5}{2}, \dfrac{7}{2}, 3\right)$

3. $3x - 2y = 10$ (1)

     $x + y = 5$ (2)

Write the augmented coefficient matrix,

$\begin{bmatrix} 3 & -2 & | & 10 \\ 1 & 1 & | & 5 \end{bmatrix}$

Since R2 is easy to work with, switch R1 and

R2.

$\begin{bmatrix} 1 & 1 & | & 5 \\ 3 & -2 & | & 10 \end{bmatrix}$

Now add 3R1 to (-1)R2 to get a new R2

$\begin{array}{rrrl} 3 & 3 & 15 & 3R1 \\ -3 & 2 & -10 & -R2 \\ \hline 0 & 5 & 5 & \text{New R2} \end{array}$

$\begin{bmatrix} 1 & 1 & | & 5 \\ 0 & 5 & | & 5 \end{bmatrix}$

Reinsert the variables to find x and y.

$R2 \Rightarrow 5y = 5 \Rightarrow y = 1$

$R1 \Rightarrow x + y = 5 \Rightarrow x = 5 - y = 5 - 1 = 4$

$(x, y) = (4, 1)$

4. $\begin{vmatrix} 3 & -5 \\ 2 & 6 \end{vmatrix} = 3 \cdot 6 - (-5)(2) = 18 + 10 = 28$

5. $\begin{vmatrix} 1 & 2 & -1 \\ 2 & 0 & 3 \\ 0 & 1 & 5 \end{vmatrix} = 1\begin{vmatrix} 0 & 3 \\ 1 & 5 \end{vmatrix} - 2\begin{vmatrix} 2 & -1 \\ 1 & 5 \end{vmatrix} + 0\begin{vmatrix} 2 & -1 \\ 0 & 3 \end{vmatrix}$

$= (0 - 3) - 2(10 - (-1)) + 0$

$= -3 - 22 = -25$

6. $3x - 5y = 12$ (1)

    $2x + 6y = 1$ (2)

$D = \begin{vmatrix} 3 & -5 \\ 2 & 6 \end{vmatrix} = 18 + 10 = 28$

$D_X = \begin{vmatrix} 12 & -5 \\ 1 & 6 \end{vmatrix} = 72 + 5 = 77$

$D_Y = \begin{vmatrix} 3 & 12 \\ 2 & 1 \end{vmatrix} = 3 - 24 = -21$

$x = \dfrac{D_X}{D} = \dfrac{77}{28} = \dfrac{11}{4}$

$y = \dfrac{D_Y}{D} = \dfrac{-21}{28} = -\dfrac{3}{4}$

7. $C(x) = 15x + 9000$

    $R(x) = 90x$

a) $P(x) = R(x) - C(x)$

        $= 90x - (15x + 9000)$

        $= 75x - 9000$

b) $P(x) = 0 \Rightarrow 75x - 9000 = 0$

$\Rightarrow x = \dfrac{9000}{75} = 120$

$C(120) = R(120) = 90(120) = \$10,800$

8. $S(p) = 60 + 9p$

$D(p) = 195 - 6p$

$S(p) = D(p) \Rightarrow 60 + 9p = 195 - 6p$

$\Rightarrow 15p = 135 \Rightarrow p = \dfrac{135}{15} = 9$

$S(9) = D(9) = 60 + 9(9) = 60 + 81 = 141$

Chapter 9 Review Exercises

1. Elimination

2. Consistent

3. $180°$

4. 1

5. Square

6. Determinant

7. Total profit

8. Fixed

9. Equilibrium point

10. Zero

11. $x + 4y + 3z = 2 \quad (1)$

$2x + y + z = 10 \quad (2)$

$-x + y + 2z = 8 \quad (3)$

1., 2. The equations are already in standard form with no fractions or decimals.

3., 4. We eliminate x from two different pairs of equations.

$\begin{array}{l} x + 4y + 3z = 2 \quad (1) \\ \underline{-x + y + 2z = 8} \quad (3) \\ 5y + 5z = 10 \quad (4) \end{array}$

$\begin{array}{l} 2x + y + z = 10 \quad (2) \\ \underline{-2x + 2y + 4z = 16} \quad \text{Mult. (3) by 2} \\ 3y + 5z = 26 \quad (5) \end{array}$

5. Now solve the system of equations (4) and (5)

$5y + 5z = 10 \quad (4)$

$3y + 5z = 26 \quad (5)$

$\begin{array}{l} 5y + 5z = 10 \quad (4) \\ \underline{-3y - 5z = -26} \quad \text{Mult. (5) by } -1 \\ 2y = -16 \\ y = -8 \end{array}$

$5y + 5z = 10 \quad \text{Substituting in (4)}$

$5(-8) + 5z = 10$

$-40 + 5z = 10$

$5z = 50$

$z = 10$

6. Substitute in one of the original equations to find x.

$x + 4y + 3z = 2$

$x + 4(-8) + 3(10) = 2$

$x - 32 + 30 = 2$

$x - 2 = 2$

$x = 4$

We obtain $(4, -8, 10)$. This checks, so it is the solution.

12. $4x + 2y - 6z = 34 \quad (1)$

$2x + y + 3z = 3 \quad (2)$

$6x + 3y - 3z = 37 \quad (3)$

1., 2. The equations are already in standard form with no fractions or decimals.

3., 4. We eliminate z from two different pairs of equations.

$\begin{array}{l} 4x + 2y - 6z = 34 \quad (1) \\ \underline{4x + 2y + 6z = 6} \quad \text{Mult. (2) by 2} \\ 8x + 4y = 40 \quad (4) \end{array}$

$\begin{array}{l} 2x + y + 3z = 3 \quad (2) \\ \underline{6x + 3y - 3z = 37} \quad (3) \\ 8x + 4y = 40 \quad (5) \end{array}$

5. Now solve the system of equations (4) and (5)

$\begin{array}{l} 8x + 4y = 40 \quad (4) \\ \underline{-8x - 4y = -40} \quad \text{Mult (5) by } -1 \\ 0 = 0 \quad (6) \end{array}$

Equation (6) indicates that equations (1), (2) and (3) are dependent.

13. $2x - 5y - 2z = -4$ (1)

$7x + 2y - 5z = -6$ (2)

$-2x + 3y + 2z = 4$ (3)

1., 2. The equations are already in standard form with no fractions or decimals.

3., 4. We eliminate x from two different pairs of equations.

$2x - 5y - 2z = -4$ (1)

$-2x + 3y + 2z = 4$ (3)

$\overline{ -2y = 0}$ (4)

$-14x + 35y + 14z = 28$ M. (1) by -7

$14x + 4y - 10z = -12$ M. (2) by 2

$\overline{ 39y + 4z = 16}$ (5)

5. From equation (4) we have

$-2y = 0$

$y = 0$

Substituting 0 for y in equation (5) we have

$39y + 4z = 16$

$39(0) + 4z = 16$

$0 + 4z = 16$

$4z = 16$

$z = 4$

6. Substitute in one of the original equations to find x.

$2x - 5y - 2z = -4$

$2x - 5(0) - 2(4) = -4$

$2x - 0 - 8 = -4$

$2x - 8 = -4$

$2x = 4$

$x = 2$

We obtain $(2, 0, 4)$. This checks, so it is the solution.

14. $2x - 3y + z = 1$ (1)

$x - y + 2z = 5$ (2)

$3x - 4y + 3z = -2$ (3)

Multiply Equation (1) by -2 and add the result to Equation (2) to Eliminate z.

$-2(2x - 3y + z = 1) \rightarrow -4x + 6y - 2z = -2$

$x - y + 2z = 5$

$\overline{-3x + 5y = 3}$ (4)

Multiply Equation (1) by -3 and add the result to Equation (3) to Eliminate z.

$-3(2x - 3y + z = 1) \rightarrow -6x + 9y - 3z = -3$

$3x - 4y + 3z = -2$

$\overline{-3x + 5y = -5}$ (5)

We solve the resulting system:

$-3x + 5y = 3$ (4)

$-(-3x + 5y = -5)$ (5)

$\overline{ 0 = 8}$

We have a contradiction. The system has no solution.

15. $3x + y = 2$ (1)

$x + 3y + z = 0$ (2)

$x + z = 2$ (3)

1., 2. The equations are already in standard form with no fractions or decimals.

3., 4. We eliminate x from equations (2) and (3).

$x + 3y + z = 0$ (2)

$-x - z = -2$ Mult. (3) by -1

$\overline{ 3y = -2}$ (4)

Solve equation (4), we have

$3y = -2$.

$y = -\dfrac{2}{3}$

Now substitute $-\dfrac{2}{3}$ for y in equation (1) and solve for x.

$3x + y = 2$

$3x + \left(-\dfrac{2}{3}\right) = 2$

$3x = 2 + \dfrac{2}{3}$

$3x = \dfrac{8}{3}$

$x = \dfrac{8}{9}$

Now substitute in equation (3) to solve for z.

$x + z = 2$

$\dfrac{8}{9} + z = 2$

$z = 2 - \dfrac{8}{9} = \dfrac{10}{9}$

We obtain $\left(\dfrac{8}{9}, -\dfrac{2}{3}, \dfrac{10}{9}\right)$. This checks, so it is the solution.

16. *Familiarize.* Let x = the measure of angle A, y = the measure of angle B, and z = the measure of angle C.

Translate.

$\underbrace{\text{The sum of the measures of the angles}}$ $\underbrace{\text{is}}$ $180°$.

$\downarrow$ $\qquad$ $\downarrow$ $\downarrow$

$x + y + z$ $\qquad$ $= 180$

$\underbrace{\text{Measure of angle } A}$ $\underbrace{\text{is}}$ $\underbrace{4 \text{ times}}$ $\underbrace{\text{the measure of angle } C.}$

$\downarrow$ $\quad$ $\downarrow$ $\quad$ $\downarrow$ $\qquad$ $\downarrow$

x $\quad$ $=$ $\quad$ $4 \cdot$ $\qquad$ z

$\underbrace{\text{Measure of angle } B}$ $\underbrace{\text{is}}$ $\underbrace{45° \text{ more than}}$ $\underbrace{\text{the measure of angle } C}$

$\downarrow$ $\quad$ $\downarrow$ $\quad$ $\downarrow$ $\qquad$ $\downarrow$

y $\quad$ $=$ $\quad$ $45 +$ $\qquad$ z

We now have a system of equations.

$x + y + z = 180 \quad (1)$

$x = 4z \qquad\quad (2)$

$y = 45 + z \qquad (3)$

Carry out. Substituting $4z$ for x and $45 + z$ for y in equation (1), we solve for z.

$x + y + z = 180$

$4z + (45 + z) + z = 180$

$6z + 45 = 180$

$6z + 45 = 180$

$6z = 135$

$z = 22.5$

Substituting 22.5 for z in equation (2), we can solve for x.

$x = 4z$

$x = 4(22.5)$

$x = 90$

Substituting in equation (1), we can solve for y.

$x + y + z = 180$

$90 + y + 22.5 = 180$

$y + 112.5 = 180$

$y = 67.5$

Check. The sum of the measures of the three angles is $90° + 67.5° + 22.5°$, or $180°$. The measure of angle A, which is $90°$, is 4 times the measure of angle C, which is 22.5. The measure of angle B is equal to the measure of angle C, plus $45°$, or $22.5° + 45° = 67.5°$. These numbers check.

State. Angles A, B, and C measure $90°$, $67.5°$, and $22.5°$, respectively.

17. *Familiarize.* Let x = the average times a man cries each month, y = the average times a woman cries each month, and z = the average times a one-year-old cries each month. Then:

Translate.

$x + y + z = 56.7 \qquad (1)$

$y = 3.9x \qquad\qquad (2)$

$z = x + y + 43.3 \quad (3)$

Carry out. Use (2) to rewrite (1) and (3) in terms of x and z.

$x + y + z = 56.7 \qquad (1)$

$x + 3.9x + z = 56.7$

$4.9x + z = 56.7 \qquad (4)$

$z = x + y + 43.3 \qquad (3)$

$z = x + 3.9x + 43.3$

$z = 4.9x + 43.3 \qquad (5)$

Use (5) to rewrite (4) in terms of x and solve for x.

$4.9x + z = 56.7 \qquad (4)$

$4.9x + (4.9x + 43.3) = 56.7$

$9.8x = 56.7 - 43.3 = 13.4$

$x = \dfrac{13.4}{9.8} \approx 1.367$

Use (2) and the value of x to find y.

$y = 3.9x \quad (5)$

$y \approx 3.9(1.367) \approx 5.33$

Use (5) and the value of x to find z.

$z = 4.9x + 43.3 \quad (5)$

$z \approx 4.9(1.367) + 43.3 \approx 50$

State. Rounding to the nearest tenth gives: $(x, y, z) = (1.4, 5.3, 50)$.

18. $3x + 4y = -13$

$5x + 6y = 8$

Write a matrix using only the constants.

$$\begin{bmatrix} 3 & 4 & | & -13 \\ 5 & 6 & | & 8 \end{bmatrix}$$

Multiply the second row by 3 to make the first number in row 2 a multiply of 3.

$$\begin{bmatrix} 3 & 4 & | & -13 \\ 15 & 18 & | & 24 \end{bmatrix} \text{New Row 2} = 3(\text{Row 2})$$

Now multiply the first row by –5 and add it to the second row.

$$\begin{bmatrix} 3 & 4 & | & -13 \\ 0 & -2 & | & 89 \end{bmatrix} \text{Row 2} = -5(\text{Row 1}) + \text{Row 2}$$

Reinserting the variables, we have

$3x + 4y = -13$ (1)

$-2y = 89$ (2)

Solve Equation (2) for y.

$-2y = 89$

$$y = -\frac{89}{2}$$

Substituting $-\dfrac{89}{2}$ for y in Equation (1) and solve for x.

$3x + 4y = -13$

$3x + 4\left(-\dfrac{89}{2}\right) = -13$

$3x - 178 = -13$

$3x = 165$

$x = 55$

The solution is $\left(55, -\dfrac{89}{2}\right)$.

19. $3x - y + z = -1$

$2x + 3y + z = 4$

$5x + 4y + 2z = 5$

We first write a matrix using only the constants.

$$\begin{bmatrix} 3 & -1 & 1 & | & -1 \\ 2 & 3 & 1 & | & 4 \\ 5 & 4 & 2 & | & 5 \end{bmatrix}$$

Multiply the first row by 10.
Multiply the second row by –15.
Multiply the third row by –6.

$$\begin{bmatrix} 30 & -10 & 10 & | & -10 \\ -30 & -45 & -15 & | & -60 \\ -30 & -24 & -12 & | & -30 \end{bmatrix} \begin{array}{l} \text{Multiply by 10.} \\ \text{Multiply by } -15. \\ \text{Multiply by } -6 \end{array}$$

Now add row 1 to row 2 and add row 1 to row 3.

$$\begin{bmatrix} 30 & -10 & 10 & | & -10 \\ 0 & -55 & -5 & | & -70 \\ 0 & -34 & -2 & | & -40 \end{bmatrix} \begin{array}{l} \text{R2 = R1 + R2.} \\ \text{R3 = R1 + R3} \end{array}$$

Multiply row 2 by –34 and multiply row 3 by 55.

$$\begin{bmatrix} 30 & -10 & 10 & | & -10 \\ 0 & 1870 & 170 & | & 2380 \\ 0 & -1870 & -110 & | & -2200 \end{bmatrix} \begin{array}{l} \text{M. by } -34. \\ \text{M. by 55.} \end{array}$$

Add row 2 to row 3.

$$\begin{bmatrix} 30 & -10 & 10 & | & -10 \\ 0 & 1870 & 170 & | & 2380 \\ 0 & 0 & 60 & | & 180 \end{bmatrix} \text{R3 = R2 + R3}$$

Reinserting the variables, we have

$30x - 10y + 10z = -10$ (1)

$1870y + 170z = 2380$ (2)

$60z = 180$ (3)

Solve (3) for z.

$60z = 180$

$z = 3$

Substitute 3 for z in (2) and solve for y.

$1870y + 170z = 2380$

$1870y + 170(3) = 2380$

$1870y + 510 = 2380$

$1870y = 1870$

$y = 1$

Substitute 1 for y and 3 for z in (1) and solve for x.

$30x - 10y + 10z = -10$

$30x - 10(1) + 10(3) = -10$

$30x - 10 + 30 = -10$

$30x + 20 = -10$

$30x = -30$

$x = -1$

The solution is $(-1, 1, 3)$.

20. $\begin{vmatrix} -2 & -5 \\ 3 & 10 \end{vmatrix} = -2(10) - 3(-5) = -20 + 15 = -5$

21. $\begin{vmatrix} 2 & 3 & 0 \\ 1 & 4 & -2 \\ 2 & -1 & 5 \end{vmatrix}$

$= 2\begin{vmatrix} 4 & -2 \\ -1 & 5 \end{vmatrix} - 1\begin{vmatrix} 3 & 0 \\ -1 & 5 \end{vmatrix} + 2\begin{vmatrix} 3 & 0 \\ 4 & -2 \end{vmatrix}$

$= 2(20-2)-1(15+0)+2(-6-0)$

$= 2(18)-1(15)+2(-6)$

$= 36-15-12$

$= 9$

22. $2x+3y=6$

$x-4y=14$

$x = \dfrac{\begin{vmatrix} 6 & 3 \\ 14 & -4 \end{vmatrix}}{\begin{vmatrix} 2 & 3 \\ 1 & -4 \end{vmatrix}} = \dfrac{-24-42}{-8-3} = \dfrac{-66}{-11} = 6$

$y = \dfrac{\begin{vmatrix} 2 & 6 \\ 1 & 14 \end{vmatrix}}{\begin{vmatrix} 2 & 3 \\ 1 & -4 \end{vmatrix}} = \dfrac{28-6}{-8-3} = \dfrac{22}{-11} = -2$

The solution is $(6,-2)$.

23. $2x+y+z=-2$

$2x-y+3z=6$

$3x-5y+4z=7$

First find D, D_x, D_y and D_z.

$D = \begin{vmatrix} 2 & 1 & 1 \\ 2 & -1 & 3 \\ 3 & -5 & 4 \end{vmatrix}$

$= 2\begin{vmatrix} -1 & 3 \\ -5 & 4 \end{vmatrix} - 2\begin{vmatrix} 1 & 1 \\ -5 & 4 \end{vmatrix} + 3\begin{vmatrix} 1 & 1 \\ -1 & 3 \end{vmatrix}$

$= 2(-4+15)-2(4+5)+3(3+1)$

$= 2(11)-2(9)+3(4)$

$= 22-18+12$

$= 16$

$D_x = \begin{vmatrix} -2 & 1 & 1 \\ 6 & -1 & 3 \\ 7 & -5 & 4 \end{vmatrix}$

$= -2\begin{vmatrix} -1 & 3 \\ -5 & 4 \end{vmatrix} - 6\begin{vmatrix} 1 & 1 \\ -5 & 4 \end{vmatrix} + 7\begin{vmatrix} 1 & 1 \\ -1 & 3 \end{vmatrix}$

$= -2(-4+15)-6(4+5)+7(3+1)$

$= -2(11)-6(9)+7(4)$

$= -22-54+28$

$= -48$

$D_y = \begin{vmatrix} 2 & -2 & 1 \\ 2 & 6 & 3 \\ 3 & 7 & 4 \end{vmatrix}$

$= 2\begin{vmatrix} 6 & 3 \\ 7 & 4 \end{vmatrix} - 2\begin{vmatrix} -2 & 1 \\ 7 & 4 \end{vmatrix} + 3\begin{vmatrix} -2 & 1 \\ 6 & 3 \end{vmatrix}$

$= 2(24-21)-2(-8-7)+3(-6-6)$

$= 2(3)-2(-15)+3(-12)$

$= 6+30-36$

$= 0$

$D_z = \begin{vmatrix} 2 & 1 & -2 \\ 2 & -1 & 6 \\ 3 & -5 & 7 \end{vmatrix}$

$= 2\begin{vmatrix} -1 & 6 \\ -5 & 7 \end{vmatrix} - 2\begin{vmatrix} 1 & -2 \\ -5 & 7 \end{vmatrix} + 3\begin{vmatrix} 1 & -2 \\ -1 & 6 \end{vmatrix}$

$= 2(-7+30)-2(7-10)+3(6-2)$

$= 2(23)-2(-3)+3(4)$

$= 46+6+12$

$= 64$

$x = \dfrac{D_x}{D} = \dfrac{-48}{16} = -3$

$y = \dfrac{D_y}{D} = \dfrac{0}{16} = 0$

$z = \dfrac{D_z}{D} = \dfrac{64}{16} = 4$

The solution is $(-3, 0, 4)$.

24. a. $P(x) = R(x) - C(x)$

$\quad = (50x) - (30x + 15,800)$

$\quad = 50x - 30x - 15,800$

$\quad P(x) = 20x - 15,800$

b. $R(x) = C(x)$

$\quad 50x = 30x + 15,800$

$\quad 20x = 15,800$

$\quad x = 790$

$\quad R(x) = 50x$

$\quad R(790) = 50 \cdot 790 = \$39,500$

The break-even point is $(790, \$39,500)$ which represents 790 units and revenue of $39,500.

25. $D(p) = 120 - 13p$

$S(p) = 60 + 7p$

Rewrite the system:

$q = 120 - 13p \quad (1)$

$q = 60 + 7p \quad (2)$

Substitute $120 - 13p$ for q in (2) and solve.

$120 - 13p = 60 + 7p$

$\quad 60 = 20p$

$\quad 3 = p$

The equilibrium price is $3 per unit.
To find the equilibrium quantity we substitute $3 into either $D(p)$ or $S(p)$.

$S(3) = 60 + 7(3)$

$\quad = 60 + 21$

$\quad = 81$

The equilibrium quantity is 81 units.
The equilibrium point is $(\$3, 81)$.

26. Let x= the number of pints of honey produced and sold.

a) $C(x) = 4.75x + 54,000$

b) $R(x) = 9.25x$

c) $P(x) = R(x) - C(x)$

$\quad = 9.25x - (4.75x + 54,000)$

$\quad = 4.50x - 54,000$

d)   If 5000 pints are produced and sold:

$\quad P(5000) = 4.5(5000) - 54,000$

$\quad = -31,500$

there is a loss of $31,500.

If 15,000 pints are produced and sold:

$\quad P(15,000) = 4.5(15,000) - 54,000$

$\quad = 13,500$

there is a profit of $13,500.

e) $P(x) = 0 \Rightarrow 4.5x - 54,000 = 0$

$\quad \Rightarrow x = \dfrac{54,000}{4.5} = 12,000$

$\quad C(12,000) = R(12,000) = 9.25(12,000)$

$\quad = \$111,000$

27. *Thinking and Writing Exercise.* To solve a problem involving four variables, go through the *Familiarize* and *Translate* steps as usual. The resulting system of equations can be solved using the elimination method just as for three variables but likely with more steps.

28. *Thinking and Writing Exercise.* A system of equations can be both dependent and inconsistent if it is equivalent to a system with fewer equations that has no solution. An example is a system of three equations in three unknowns in which two of the equations represent the same plane and the third represents a parallel plane.

29. Danae must make a profit producing and selling honey equal to the pay of her previous job, or $36,000, so $P(x) = 36,000$. We must solve for x to determine the number of units of honey she needs to produce and sell.

$P(x) = 4.5x - 54,000 = 36,000$

$\quad 4.5x = 90,000$

$\quad x = \dfrac{90,000}{4.5} = 20,000$

Danae must produce and sell 20,000 pints of honey.

30. $f(x) = ax^2 + bx + c$ contains the points $(-2,3)$, $(1,1)$, and $(0,3)$. Therefore, these points must satisfy the given equation.

$(-2,3): \quad f(x) = a(-2)^2 + b(-2) + c$

$\quad = 4a - 2b + c = 3$

$(1,1): \quad f(x) = a(1)^2 + b(1) + c$

$\quad = a + b + c = 1$

$(0,3): \quad f(x) = a(0)^2 + b(0) + c$

$\quad = c = 3$

Substituting 3 for c in the first two equations
$4a - 2b + c = 3$ and $a + b + c = 1$

$4a - 2b + 3 = 3$ $\qquad a + b + 3 = 1$

$4a - 2b = 0$ $\qquad a + b = -2$

gives the system:

$4a - 2b = 0$ $\quad$ (1)

$a + b = -2$ $\quad$ (2)

Using elimination, we solve the system:

$4a - 2b = 0$

$\underline{2a + 2b = -4}$ $\quad$ Multiply eq. (2) by 2

$\quad 6a = -4$

$\qquad a = -\dfrac{2}{3}$

Substituting $-\dfrac{2}{3}$ for a in equation (2), we

solve for b.

$a + b = -2$

$-\dfrac{2}{3} + b = -2$

$\qquad b = -2 + \dfrac{2}{3} = -\dfrac{4}{3}$

The values for a, b, and c, are $-\dfrac{2}{3}$, $-\dfrac{4}{3}$ and

3, respectively. Therefore the function is

$f(x) = -\dfrac{2}{3}x^2 - \dfrac{4}{3}x + 3$.

Chapter 9 Test

1. $-3x + y - 2z = 8$ $\quad$ (1)

$\quad -x + 2y - z = 5$ $\quad$ (2)

$\quad 2x + y + z = -3$ $\quad$ (3)

1., 2. $\quad$ The equations are already in standard form with no fractions or decimals.

3., 4. $\quad$ We eliminate x from equations (1) and (2).

$-3x \ + y - 2z = \ \ \ 8$

$\underline{3x - 6y + 3z = -15}$ $\quad$ M. eq. 2 by -3

$\quad -5y + z = -7$ $\quad$ (4)

We eliminate x from equations (2) and (3).

$-2x + 4y - 2z = 10$ $\quad$ M. eq. 2 by 2

$\underline{2x \ + y \ + z = -3}$

$\quad 5y \ - z = 7$ $\quad$ (5)

Solving equations (4) and (5), we have:

$-5y + z = -7$ $\quad$ (4)

$\underline{5y - z = 7}$ $\quad$ (5)

$\quad 0 = 0$ $\quad$ Adding

Because $0 = 0$ is a true statement, the system is dependent.

2. $\quad 6x + 2y - 4z = 15$ $\quad$ (1)

$\quad -3x - 4y + 2z = -6$ $\quad$ (2)

$\quad 4x - 6y + 3z = 8$ $\quad$ (3)

1., 2. $\quad$ The equations are already in standard form with no fractions or decimals.

3., 4. $\quad$ We eliminate x from equations (1) and (2).

$6x + 2y - 4z = \ \ \ 15$

$\underline{-6x - 8y + 4z = -12}$ $\quad$ M. eq. 2 by 2

$\quad -6y = 3$ $\quad$ (4)

$\qquad y = -\dfrac{1}{2}$

We eliminate x from equations (2) and (3).

$-12x - 16y + 8z = -24$ $\quad$ M. eq. 2 by 4

$\underline{12x - 18y + 9z = 24}$ $\quad$ M. eq. 3 by 3

$\quad -34y + 17z = 0$ $\quad$ (5)

Substituting $-\dfrac{1}{2}$ for y in equation (5)

and solving for z, we have:

$-34y + 17z = 0$

$-34\left(-\dfrac{1}{2}\right) + 17z = 0$

$\quad 17 + 17z = 0$

$\quad 17z = -17$

$\quad z = -1$

Substituting $-\dfrac{1}{2}$ for y and -1 for z in

any of the original equations, we can solve for x. We choose equation (1).

$$6x + 2y - 4z = 15$$

$$6x + 2\left(-\frac{1}{2}\right) - 4(-1) = 15$$

$$6x - 1 + 4 = 15$$

$$6x + 3 = 15$$

$$6x = 12$$

$$x = 2$$

The solution is $\left(2, -\frac{1}{2}, -1\right)$ which

checks.

3. $2x + 2y = 0 \;\; (1)$

$$4x + 4z = 4 \;\; (2)$$

$$2x + y + z = 2 \;\; (3)$$

1., 2.   The equations are already in standard form with no fractions or decimals.

3., 4.   We can simplify equations (1) and (2) by multiplying by $\frac{1}{2}$ and $\frac{1}{4}$, respectively.

$$x + y = 0 \;\; (4)$$

$$x + z = 1 \;\; (5)$$

$$2x + y + z = 2 \;\; (3)$$

We can eliminate the variable x from equations (4) and (5).

$$\begin{array}{l} x + y = 0 \\ \underline{-x - z = -1} \;\; \text{M. eq. 5 by } -1 \\ y - z = -1 \;\; (6) \end{array}$$

We can eliminate the variable x from equations (5) and (3).

$$\begin{array}{l} -2x - 2z = -2 \;\; \text{M. eq. 5 by } -2 \\ \underline{2x + y + z = 2} \\ y - z = 0 \;\; (7) \end{array}$$

Solving equations (6) and (7), we have:

$$\begin{array}{l} -y + z = 1 \;\; \text{M. eq. 6 by } -1 \\ \underline{y - z = 0} \\ 0 = 1 \end{array}$$

Because the statement $0 = 1$ is false, the system has no solution.

4. $3x + 3z = 0 \;\; (1)$

$$2x + 2y = 2 \;\; (2)$$

$$3y + 3z = 3 \;\; (3)$$

1., 2.   The equations are already in standard form with no fractions or decimals.

3., 4.   We can simplify all three equations by multiplying by $\frac{1}{3}$, $\frac{1}{2}$, and $\frac{1}{3}$, respectively.

$$x + z = 0 \;\; (4)$$

$$x + y = 1 \;\; (5)$$

$$y + z = 1 \;\; (6)$$

We can eliminate the variable x from equations (4) and (5).

$$\begin{array}{l} x + z = 0 \\ \underline{-x - y = -1} \;\; \text{M. eq. 5 by } -1 \\ -y + z = -1 \;\; (7) \end{array}$$

Now we can solve equations (6) and (7).

$$\begin{array}{l} y + z = 1 \;\; (6) \\ \underline{-y + z = -1} \;\; (7) \\ 2z = 0 \\ z = 0 \end{array}$$

We can substitute 0 for z in equation (6) and solve for y.

$$y + z = 1$$

$$y + 0 = 1$$

$$y = 1$$

We can substitute 0 for z in equation (4) and solve for x.

$$x + z = 0$$

$$x + 0 = 0$$

$$x = 0$$

The solution is $(0, 1, 0)$, which checks.

5. $4x + y = 12 \;\; (1)$

$$3x + 2y = 2 \;\; (2)$$

Write the augmented coefficient matrix,

$$\begin{bmatrix} 4 & 1 & | & 12 \\ 3 & 2 & | & 2 \end{bmatrix}$$

Add (-3) R1 to 4R2 to get a new R2

$$\begin{array}{llll} -12 & -3 & -36 & -3\text{R1} \\ \underline{12} & \underline{8} & \underline{8} & \underline{4\text{R2}} \\ 0 & 5 & -28 & \text{New R2} \end{array}$$

$$\begin{bmatrix} 4 & 1 & | & 12 \\ 0 & 5 & | & -28 \end{bmatrix}$$

Reinsert the variables to find x and y.

$R2 \Rightarrow 5y = -28 \Rightarrow y = -\dfrac{28}{5}$

$R1 \Rightarrow 4x + y = 12 \Rightarrow 4x = 12 - y$

$\qquad = 12 - \left(-\dfrac{28}{5}\right) = \dfrac{60}{5} + \dfrac{28}{5} = \dfrac{88}{5}$

$x = \dfrac{1}{4} \cdot \dfrac{88}{5} = \dfrac{22}{5}$

$(x,\ y) = \left(\dfrac{22}{5}, -\dfrac{28}{5}\right)$

6. $x + 3y - 3z = 12$

$3x - y + 4z = 0$

$-x + 2y - z = 1$

We first write a matrix with only constants.

$$\begin{bmatrix} 1 & 3 & -3 & | & 12 \\ 3 & -1 & 4 & | & 0 \\ -1 & 2 & -1 & | & 1 \end{bmatrix}$$

Multiply row 1 by −3 and add it to row 2.
Add row 1 to row 3.

$$\begin{bmatrix} 1 & 3 & -3 & | & 12 \\ 0 & -10 & 13 & | & -36 \\ 0 & 5 & -4 & | & 13 \end{bmatrix}$$

Reverse rows 2 and 3.

$$\begin{bmatrix} 1 & 3 & -3 & | & 12 \\ 0 & 5 & -4 & | & 13 \\ 0 & -10 & 13 & | & -36 \end{bmatrix}$$

Multiply row 2 by 2 and add it to row 3.

$$\begin{bmatrix} 1 & 3 & -3 & | & 12 \\ 0 & 5 & -4 & | & 13 \\ 0 & 0 & 5 & | & -10 \end{bmatrix}$$

Reinserting the variables, we have

$x + 3y - 3z = 12 \quad (1)$

$5y - 4z = 13 \quad (2)$

$5z = -10 \quad (3)$

Solve (3) for z.

$5z = -10$

$z = -2$

Substitute −2 for z in (2) and solve for y.

$5y - 4z = 13$

$5y - 4(-2) = 13$

$5y + 8 = 13$

$5y = 5$

$y = 1$

Substitute −2 for z and 1 for y in equation (1) and solve for x.

$x + 3y - 3z = 12$

$x + 3(1) - 3(-2) = 12$

$x + 3 + 6 = 12$

$x + 9 = 12$

$x = 3$

The solution is $(3, 1, -2)$ which checks.

7. $\begin{vmatrix} 4 & -2 \\ 3 & -5 \end{vmatrix} = 4(-5) - 3(-2) = -20 + 6 = -14$

8. $\begin{vmatrix} 3 & 4 & 2 \\ -2 & -5 & 4 \\ 0 & 5 & -3 \end{vmatrix}$

$= 3\begin{vmatrix} -5 & 4 \\ 5 & -3 \end{vmatrix} + 2\begin{vmatrix} 4 & 2 \\ 5 & -3 \end{vmatrix} + 0\begin{vmatrix} 4 & 2 \\ -5 & 4 \end{vmatrix}$

$= 3(15 - 20) + 2(-12 - 10) + 0$

$= -59$

9. $3x + 4y = -1$

$5x - 2y = 4$

$D = \begin{vmatrix} 3 & 4 \\ 5 & -2 \end{vmatrix} = -6 - 20 = -26$

$D_x = \begin{vmatrix} -1 & 4 \\ 4 & -2 \end{vmatrix} = 2 - 16 = -14$

$D_Y = \begin{vmatrix} 3 & -1 \\ 5 & 4 \end{vmatrix} = 12 + 5 = 17$

$x = \dfrac{D_X}{D} = \dfrac{-14}{-26} = \dfrac{7}{13}$

$y = \dfrac{D_Y}{D} = \dfrac{17}{-26} = -\dfrac{17}{26}$

10. **Familiarize.** Let x, y, and z equal the measures of
angles A, B, an C (in degrees)
Respectively. The sum of the measures of the angles of a triangle $= 180°$.
Translate.

$\underbrace{\text{Sum of Measures}}$ is $\underline{180°}$
$\qquad\qquad\downarrow\qquad\quad\downarrow\quad\downarrow$
$\qquad x+y+z\quad = \quad 180$

$\underbrace{\text{Measure of}\atop \sphericalangle B}$ is $\underbrace{\text{3 times}}$ $\underbrace{\text{Measure of}\atop \sphericalangle A}$ less $5°$

$\qquad\downarrow\qquad\qquad\downarrow$
$\qquad y\qquad = \quad 3\cdot\qquad x\qquad - \quad 5$

$\underbrace{\text{Measure of}\atop \sphericalangle C}$ is $\underbrace{\text{3 times}}$ $\underbrace{\text{Measure of}\atop \sphericalangle B}$ $+\,5°$

$\quad\downarrow\qquad\downarrow\quad\downarrow\qquad\downarrow\qquad\downarrow\downarrow$
$\quad z\quad = \quad 3\cdot\qquad y\quad + \quad 5$

Carry out.
We have the system:
$x+y+z=180$ (1)
$y=3x-5$ (2)
$z=3y+5$ (3)
We will substitute $3x-5$ for y in Equation (3)
$z=3y+5$ (1)
$z=3(3x-5)+5$
$z=9x-15+5$
$z=9x-10$ (4)
We now substitute $9x-10$ for z and $3x-5$ for y in Equation (1).
$\qquad\qquad x+y+z=180$ (1)
$x+(3x-5)+(9x-10)=180$
$\qquad\qquad\qquad 13x-15=180$
$\qquad\qquad\qquad 13x=195$
$\qquad\qquad\qquad x=15$
Substitute 15 for x in Equation (2).
$y=3x-5$ (2)
$y=3\cdot15-5$
$y=45-5$
$y=40$
Substitute 40 for y in Equation (3).

$z=3y+5$ (3)
$z=3\cdot40+5$
$z=120+5$
$z=125$
Check. $15+40+125=180$
$\qquad\qquad 40=3(15)-5$
$\qquad\qquad 125=3(40)+5$
The numbers check.
State. In triangle ABC,
$m\sphericalangle A=15°$, $m\sphericalangle B=40°$, and $m\sphericalangle C=125°$.

11. **Familiarize.** Let x = the number of hours the electrician worked, y = the number of hours the carpenter worked, and z = the number of hours the plumber worked.

Translate.
$\underbrace{\text{The total number of hours worked}}$ is 21.5.
$\qquad\qquad\downarrow\qquad\qquad\downarrow\downarrow$
$\qquad\quad x+y+z\qquad = 21.5$

$\underbrace{\text{The total pay for all three}}$ was \$673

$\qquad\qquad\downarrow\qquad\qquad\downarrow\qquad\downarrow$
$30x+28.50y+34z\quad = \quad 673$

$\underbrace{\text{Hours the}\atop\text{plumber}\atop\text{worked}}$ is $\underbrace{\text{2 more than}}$ $\underbrace{\text{the hours}\atop\text{worked by}\atop\text{carpenter.}}$

$\quad\downarrow\qquad\downarrow\qquad\downarrow\qquad\qquad\downarrow$
$\quad z\quad = \quad 2\quad + \qquad y$
We now have a system of equations.
$\qquad\qquad x+y+z=21.5$
$30x+28.50y+34z=673$
$\qquad\qquad\qquad z=y+2$

Carry out. Clearing decimals we have:
$\qquad 10x+10y+10z=215$
$\qquad$ or $2x+2y+2z=43$ (1)
$300x+285y+340z=6730$
$\qquad$ or $60x+57y+68z=1346$ (2)
$\qquad\qquad\qquad\qquad z=y+2$ (3)
Substitute $y+2$ for z in (1) and (2):

$2x + 2y + 2z = 43$ (1)

$\Rightarrow 2x + 2y + 2(2 + y) = 43$

$\Rightarrow 2x + 4y = 39$ (4)

$60x + 57y + 68z = 1346$ (2)

$\Rightarrow 60x + 57y + 68(2 + y) = 1346$

$\Rightarrow 60x + 125y = 1210$

or $12x + 25y = 242$ (5)

Solve the system:

$2x + 4y = 39$ (4)

$12x + 25y = 242$ (5)

Add $-6 \cdot$ (4) to (5) to eliminate x and solve for y.

$\begin{array}{l} -12x - 24y = -234 \quad \text{Multiply (4) by } -6 \\ \underline{12x + 25y = 242} \qquad\qquad (5) \\ y = 8 \end{array}$

$z = y + 2$ (3)

$\Rightarrow z = 8 + 2 = 10$

Use (1) and the values of y and z to solve for x.

$x + y + z = 21.5$ (1)

$\Rightarrow x = 21.5 - y - z = 21.5 - 8 - 10 = 3.5$

State. The electrician worked 3.5 hours; the carpenter worked 8 hours, and the plumber worked 10 hours.

12. $D(p) = 79 - 8p$

$S(p) = 37 + 6p$

Rewrite the system.

$q = 79 - 8p$ (1)

$q = 37 + 6p$ (2)

Substitute $79 - 8p$ for q in (2) and solve.

$79 - 8p = 37 + 6p$

$79 - 37 = 6p + 8p$

$42 = 14p$

$3 = p$

The equilibrium price is \$3 per unit.
To find the equilibrium quantity we substitute \$3 into either $D(p)$ or $S(p)$.

$D(3) = 79 - 8(3)$

$= 79 - 24$

$= 55$

The equilibrium quantity is 55 units.
The equilibrium point is $(\$3, \, 55 \text{ units})$.

13. a) $C(x) = 25x + 44{,}000$

b) $R(x) = 80x$

c) $P(x) = R(x) - C(x)$

$= 80x - (25x + 44{,}000)$

$= 55x - 44{,}000$

d)   For 300 hammocks,

$P(300) = 55(300) - 44{,}000 = -27{,}500$

the company will have a loss of \$27,500.
For 900 hammocks,

$P(900) = 55(900) - 44{,}000 = 5500$

the company will realize a \$5500 profit.

e) $P(x) = 0 \Rightarrow 55x - 44{,}000 = 0$

$\Rightarrow x = \dfrac{44{,}000}{55} = 800$

$C(800) = R(800) = 80(800) = 64{,}000$

14. $2w - x + 5y - z = 1$ (1)

$w + x - 2y + z = 2$ (2)

$w + 3x - y + z = 0$ (3)

$3w + x + y - z = 0$ (4)

Start by Eliminating z from three different pairs of Equations.

$\begin{array}{l} 2w - x + 5y - z = 1 \qquad (1) \\ \underline{w + x - 2y + z = 2} \qquad (2) \\ 3w + 3y = 3 \qquad (5) \end{array}$

$\begin{array}{l} 2w - x + 5y - z = 1 \qquad (1) \\ \underline{w + 3x - y + z = 0} \qquad (3) \\ 3w + 2x + 4y = 1 \qquad (6) \end{array}$

$\begin{array}{l} w + x - 2y + z = 2 \qquad (2) \\ \underline{3w + x + y - z = 0} \qquad (4) \\ 4w + 2x - y = 2 \qquad (7) \end{array}$

Now solve the system of equations (5), (6), and (7)

$3w + 3y = 3$ (5)

$3w + 2x + 4y = 1$ (6)

$4w + 2x - y = 2$ (7)

$\begin{array}{l} -3w - 2x - 4y = -1 \quad \text{Multiply (6) by} -1 \\ \underline{4w + 2x - y = 2} \quad (7) \\ w - 5y = 1 \quad (8) \end{array}$

$\begin{array}{l} -3w + 15y = -3 \qquad \text{Multiply (8) by} -3 \\ \underline{3w + 3y = 3} \qquad (5) \\ 18y = 0 \\ y = 0 \end{array}$

Substitute 0 for y in Equation (8).

$w - 5y = 1$

$w - 5 \cdot 0 = 1$

$w = 1$

Substitute 1 for w and 0 for y in Equation (6).

$3w + 2x + 4y = 1$

$3 \cdot 1 + 2x + 4 \cdot 0 = 1$

$3 + 2x = 1$

$2x = -2$

$x = -1$

Substitute 1 for w, -1 for x, and 0 for y in Equation (2)

$w + x - 2y + z = 2$

$1 + (-1) - 2 \cdot 0 + z = 2$

$z = 2$

We check $w = 1$, $x = -1$, $y = 0$, and $z = 2$ in the four original Equations. The numbers check. $(1, -1, 0, 2)$ is the solution.

15. **Familiarize.** Let x = the number of adult tickets,
y = the number of senior citizen's tickets, and
z = the number of children's tickets.

Translate.

Number of adult and senior tickets sold	was	30 more than	the number of children's tickets sold.
↓	↓	↓	↓
$x + y$	$=$	$30 +$	z

Number of adult tickets sold	was	6 more than	4 times	the number of senior tickets sold.
↓	↓	↓	↓	↓
x	$=$	$6 +$	$4 \cdot$	y

Total receipts	were	$11,219.50.
↓	↓	↓
$5.5x + 4y + 1.5z$	$=$	$11,219.50$

We now have a system of equations.

$x + y = 30 + z$

$x = 6 + 4y$

$5.5x + 4y + 1.5z = 11,219.50$

After clearing decimals and writing in standard form, we have the system:

$x + y - z = 30 \qquad (1)$

$x - 4y = 6 \qquad (2)$

$55x + 40y + 15z = 112,195 \qquad (3)$

Eliminating the x variable from equations (1) and (2), we have:

$x + y - z = 30$

$\underline{-x + 4y = -6} \quad \text{Multiply (2) by } -1$

$5y - z = 24 \qquad (4)$

Eliminating the x variable from equations (2) and (3), we have:

$-55x + 220y = -330 \quad \text{M. (2) by } -55$

$\underline{55x + 40y + 15z = 112,195}$

$260y + 15z = 111,865 \qquad (5)$

Now we have the system:

$5y - z = 24 \qquad (4)$

$260y + 15z = 111,865 \quad (5)$

Solving this system we have:

$-260y + 52z = -1248 \quad \text{M. (4) by } -52$

$\underline{260y + 15z = 111,865}$

$67z = 110,617$

$z = 1651$

Substitute 1651 for z in equation (4) and solve for y.

$5y - z = 24$

$5y - 1651 = 24$

$5y = 1675$

$y = 335$

Finally, substitute 1651 for z and 335 for y in equation (1) and solve for x.

$x + y - z = 30$

$x + 335 - 1651 = 30$

$x - 1316 = 30$

$x = 1346$

Check. The number of adult's and senior tickets sold, $1346 + 335$, or 1681, was 30 more than the number of children's tickets, 1651. The number of adult tickets sold, 1346, was 6 more than four times the number of senior tickets sold, $6 + 4 \cdot 335 = 6 + 1340 = 1346$. The total receipts amounted to

$1346(\$5.50) + 335(\$4.00) + 1651(\$1.50)$

$= \$7403 + \$1340 + \$2476.50 = \$11,219.50$. These numbers check.

State. The number of adult tickets sold was 1346; the number of senior citizen tickets sold was 335, and the number of children's tickets sold was 1651.

Chapters 1 – 9

Cumulative Review

1. $3 + 24 \div 2^2 \cdot 3 - (6 - 7)$
   $= 3 + 24 \div 4 \cdot 3 - (-1)$
   $= 3 + 6 \cdot 3 + 1 = 3 + 18 + 1 = 22$

2. $3c - [8 - 2(1 - c)]$
   $= 3c - [8 - 2 + 2c]$
   $= 3c - [6 + 2c]$
   $= 3c - 6 - 2c$
   $= c - 6$

3. $-10^{-2} = -\dfrac{1}{10^2} = -\dfrac{1}{100}$

4. $(3xy^{-4})(-2x^3 y) = -6x^{1+3} y^{-4+1} = -6x^4 y^{-3}$
   $= -\dfrac{6x^4}{y^3}$

5. $\left(\dfrac{18a^2 b^{-1}}{12a^{-1}b}\right)^2 = \left(\dfrac{\cancel{6} \cdot 3a^2 a^1}{\cancel{6} \cdot 2bb^1}\right)^2 = \left(\dfrac{3a^3}{2b^2}\right)^2 = \dfrac{9a^{3 \cdot 2}}{b^{2 \cdot 2}}$
   $= \dfrac{9a^6}{b^4}$

6. $\dfrac{2x - 10}{x^3 - 125} = \dfrac{2x - 10}{x^3 - 5^3} = \dfrac{2\cancel{(x - 5)}}{\cancel{(x - 5)}(x^2 + 5x + 25)}$
   $= \dfrac{2}{x^2 + 5x + 25}$ for $x \neq 5$

7. $(x - 5)(x + 5) = x^2 - 5^2 = x^2 - 25$

8. $(3n - 2)(5n + 7) = 3n(5n + 7) - 2(5n + 7)$
   $= 15n^2 + 21n - 10n - 14 = 15n^2 + 11n - 14$

9. $\dfrac{1}{x + 5} - \dfrac{1}{x - 5} = \dfrac{1}{x + 5} \cdot \dfrac{x - 5}{x - 5} - \dfrac{1}{x - 5} \cdot \dfrac{x + 5}{x + 5}$
   $= \dfrac{x - 5 - (x + 5)}{(x - 5)(x + 5)} = \dfrac{x - 5 - x - 5}{(x - 5)(x + 5)}$
   $= -\dfrac{10}{(x - 5)(x + 5)}$

10. $\dfrac{x^2 - 3x}{2x^2 - x - 3} \div \dfrac{x^3}{x^2 - 2x - 3}$
    $= \dfrac{x^2 - 3x}{2x^2 - x - 3} \cdot \dfrac{x^2 - 2x - 3}{x^3}$
    $= \dfrac{\cancel{x}(x - 3)}{(2x - 3)\cancel{(x + 1)}} \cdot \dfrac{(x - 3)\cancel{(x + 1)}}{\cancel{x} \cdot x^2}$
    $= \dfrac{(x - 3)^2}{x^2(2x - 3)}$ for $x \neq -1, -\dfrac{3}{2}, 3, 0$

11. $y + \dfrac{2}{3y} = y \cdot \dfrac{3y}{3y} + \dfrac{2}{3y} = \dfrac{3y^2 + 2}{3y}$

12. $4x^3 + 18x^2 = 2x^2(2x + 9)$

13. $x^2 + 8x - 84 = (x + 14)(x - 6)$

14. $16y^2 - 81 = (4y)^2 - 9^2$
    $= (4y + 9)(4y - 9)$

15. $64x^3 + 8 = 8(8x^3 + 1)$
    $= 8(2x + 1)(4x^2 - 2x + 1)$

16. $t^2 - 16t + 64 = (t - 8)^2$

17. $x^6 - x^2 = x^2(x^4 - 1)$
    $= x^2(x^2 + 1)(x^2 - 1)$
    $= x^2(x^2 + 1)(x + 1)(x - 1)$

18. $0.027b^3 - 0.008c^3$
    $= (0.3b)^3 - (0.2c)^3$
    $= (0.3b - 0.2c)(0.09b^2 + 0.06bc + 0.04c^2)$

19. $20x^2 + 7x - 3$
    $= (5x + 3)(4x - 1)$

20. $3(x - 2) = 14 - x \Rightarrow 3x - 6 = 14 - x$
    $\Rightarrow 3x + x = 14 + 6 \Rightarrow 4x = 20$
    $\Rightarrow x = \dfrac{20}{4} = 5$

21. $x - 2 < 6$, or $2x + 1 > 5 \Rightarrow x < 8$, or $x > 2$
Since every real number is either greater than 2 or less than 8, possibly both, the solution set is all real numbers, $(-\infty, \infty)$

22. $x^2 - 2x - 3 = 5 \Rightarrow x^2 - 2x - 8 = 0$
$\Rightarrow (x - 4)(x + 2) = 0$
$\Rightarrow x - 4 = 0$, or $x + 2 = 0$
$\Rightarrow x = -2$, or 4

23. $\dfrac{3}{x + 1} = \dfrac{x}{4} \Rightarrow 4 \cdot 3 = x \cdot (x + 1) \Rightarrow 12 = x^2 + x$
$\Rightarrow x^2 + x - 12 = 0 \Rightarrow (x + 4)(x - 3)$
$\Rightarrow x + 4 = 0$, or $x - 3 = 0$
$\Rightarrow x = -4$, or 3

24. $y = \dfrac{1}{2}x - 7$ (1)
$2x - 4y = 3$ (2)
(1) already has y solved for completely in terms of x, so substitute the expression into (2) and solve for x, then back-substitute to find y.
$2x - 4y = 3$ (1) $\Rightarrow 2x - 4\left(\dfrac{1}{2}x - 7\right) = 3$
$\Rightarrow 2x - 2x + 28 = 3 \Rightarrow 28 = 3$
Since the resulting equation is a contradiction, the system has no solution.

25. $x + 3y = 8$ (1)
$2x - 3y = 7$ (2)
The coefficients of y are opposites, so add (1) and (2) to eliminate y and find x, then back-substitute to find y.
$x + 3y = 8$ (1)
$\underline{2x - 3y = 7 \quad (2)}$
$3x \quad\quad = 15$
$\Rightarrow x = \dfrac{15}{3} = 5$
Use equation (1) and the value of x to find y.
$x + 3y = 8$ (1)
$\Rightarrow 3y = 8 - x = 8 - 5 = 3$
$\Rightarrow y = \dfrac{3}{3} = 1$

26. $-3x + 4y + z = -5$ (1)
$x - 3y - z = 6$ (2)
$2x + 3y + 5z = -8$ (3)
Add equations (1) and (2) to eliminate variable z.
$-3x + 4y + z = -5$ (1)
$\underline{x - 3y - z = 6 \quad (2)}$
$-2x + y = 1$ (4)
Add 5 times equation (2) to equation (3).
$5x - 15y - 5z = 30$ Mult. 5 times (2)
$\underline{2x + 3y + 5z = -8 \quad (3)}$
$7x - 12y = 22$ (5)
Multiply 12 times equation (4) and add it to equation (5).
$-24x + 12y = 12$ Mult. 12 times (4)
$\underline{7x - 12y = 22 \quad (5)}$
$-16x = 32$
$x = -2$
Substitute -2 for x in equation (4) and solve for y.
$-2x + y = 1$
$-2(-2) + y = 1$
$4 + y = 1$
$y = -3$
Substitute -2 for x and -3 for y in equation (2) and solve for z.
$x - 3y - z = 6$
$-2 - 3(-3) - z = 6$
$-2 + 9 - 6 = z$
$1 = z$
The solution is $(-2, -3, 1)$.

27. $|2x - 1| = 8$
$\Rightarrow 2x - 1 = 8$, or $2x - 1 = -8$
$\Rightarrow x = \dfrac{9}{2}$, or $x = -\dfrac{7}{2}$

28. $9(x - 3) - 4x < 2 - (3 - x)$
$\Rightarrow 9x - 27 - 4x < 2 - 3 + x$
$\Rightarrow 5x - 27 < x - 1 \Rightarrow 5x - x < -1 + 27$
$\Rightarrow 4x < 26 \Rightarrow x < \dfrac{26}{4} = \dfrac{13}{2}$
$x < \dfrac{13}{2}$, or $\left(-\infty, \dfrac{13}{2}\right)$

29. $|4t| > 12 \Rightarrow 4t > 12,\ or\ 4t < -12$

$\Rightarrow t > \dfrac{12}{4} = 3,\ or\ t < \dfrac{-12}{4} = -3$

$\{t \mid t > 3,\ or\ t < -3\},\ or\ (-\infty, -3) \cup (3, \infty)$

30. $|3x - 2| \le 8 \Rightarrow -8 \le 3x - 2 \le 8$

$\Rightarrow -8 + 2 \le 3x \le 8 + 2 \Rightarrow -6 \le 3x \le 10$

$\Rightarrow \dfrac{-6}{3} \le x \le \dfrac{10}{3} \Rightarrow -2 \le x \le \dfrac{10}{3}$

$\left\{ x \mid -2 \le x \le \dfrac{10}{3} \right\},\ or\ \left[-2, \dfrac{10}{3} \right]$

31. b

32. a

33. d

34. c

35. $y = \dfrac{2}{3}x - 4$: The equation is linear and is in slope-intercept form. Therefore one point on the graph is $(0, -4)$. We can use this point and the slope to find a second point.
$(0, -4) \rightarrow (0 + 3, -4 + 2) = (3, -2)$ Therefore, plot the points $(0, -4)$ and $(3, -2)$ and draw the line passing through them.

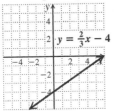

36. $x = -3$ This equation is a graph of a vertical line passing through -3 on the x-axis.

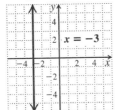

37. $3x - y = 3$: This equation is linear and is easy to put in slope-intercept form:
$3x - y = 3 \Rightarrow y = 3x - 3$ Therefore one point on the graph is $(0, -3)$. We can use this point and the slope to find a second point.
$(0, -3) \rightarrow (0 + 1, -3 + 3) = (1, 0)$ Therefore, plot the points $(0, -3)$ and $(1, 0)$ and draw the line passing through them.

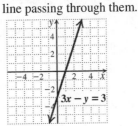

38. $x + y \ge -2 \Rightarrow y \ge -x - 2$: Begin by plotting the solid boundary line $y = -x - 2$. This line goes through the points:
$(0, -2)$ and $(0 + 1, -2 - 1) = (1, -3)$. Since when y is isolated in $y \ge -x - 2$ with the $\ge$ sign, we shade above the above the boundary line.

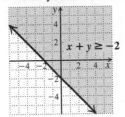

39. Graphically, $f(x) = -x + 1$ is equivalent to $y = -x + 1$: The equation is linear and is in slope-intercept form. Therefore one point on the graph is $(0, 1)$. We can use this point and the slope to find a second point.
$(0, 1) \rightarrow (0 + 1, 1 - 1) = (1, 0)$ Therefore, plot the points $(0, 1)$ and $(1, 0)$ and draw the line passing through them.

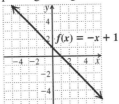

40. $x - 2y > 4$ (1)

$x + 2y \geq -2$ (2)

To plot the solution to this system of linear inequalities plot them one at a time and look for the region of overlap. Beginning with (1), start by plotting the dashed boundary line $x - 2y = 4$. By first substituting in 0 for x, and then substituting in 0 for y, locate the x- and y-intercepts.

$x - 2y = 4 \Rightarrow 0 - 2y = 4 \Rightarrow y = -2 \Rightarrow (0, -2)$

$x - 2y = 4 \Rightarrow x - 2 \cdot 0 = 4 \Rightarrow x - 4 \Rightarrow (4, 0)$

After plotting the dashed boundary line through the points $(0, -2)$ and $(4, 0)$ use the point $(0, 0)$ as a test point to determine which way to shade:

$x - 2y > 4 \Rightarrow 0 - 2 \cdot 0 > 4 \Rightarrow 0 > 4$. Since the result is false, shade in the direction that does not include the point $(0,0)$.

In similar fashion $x + 2y \geq -2$ has a solid boundary line passing through the point $(0, -1)$ and $(-2, 0)$ Since the test point $(0, 0)$ leads to the true statement $0 \geq -2$ shade in a direction that includes the point $(0, 0)$.

The region of overlap is the solution set for the system.

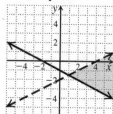

41. To find the slope and y-intercept of the line defined by the equation $4x - 9y = 18$, rearrange the equation to put it in the slope-intercept form ($y = mx + b$) form and read off the requested quantities:

$4x - 9y = 18$

$4x - 18 = 9y$

$\dfrac{4x}{9} - \dfrac{18}{9} = y$

$y = \dfrac{4}{9}x - 2$

Therefore $m = \dfrac{4}{9}$ and $b = -2$

42. As long as the slope is not undefined, any line can be put in the slope-intercept form: $y = mx + b$. To complete the problem determine the values of m and b and insert them into the equation. m is given, so only b needs to be determined. Use the given slope and the given point to determine b.

$y = -7x + b \Rightarrow -4 = -7(-3) + b$

$\Rightarrow -4 = 21 + b \Rightarrow b = -4 - 21 = -25$

Therefore the slope-intercept form of the equation is $y = -7x - 25$.

43. As long as its slope is not undefined, any line can be put in the slope-intercept form: $y = mx + b$. To complete the problem determine the values of m and b, and insert them into the equation. In this problem b is given, so only m needs to be determined.

Use the fact that if two lines are perpendicular then the product of their slopes is -1.($l_1 \perp l_2 \Rightarrow m_1 m_2 = -1$). Find the slope of the line defined by $3x + 2y = 1$.

$3x + 2y = 1 \Rightarrow 2y = -3x + 1 \Rightarrow y = -\dfrac{3}{2}x + 1$

$\Rightarrow m_1 = -\dfrac{3}{2} \Rightarrow -\dfrac{3}{2}m_2 = -1 \Rightarrow m_2 = \dfrac{2}{3}$

Therefore the slope-intercept form of the equation is $y = \dfrac{2}{3}x + 4$.

44. The function is defined for all values of x; its domain is $(-\infty, \infty)$. The function takes on a minimum value of -2 at its vertex and no maximum value; its range is $[-2, \infty)$.

45. The function is defined for all values of x except those that make the denominator 0.

$2x + 5 = 0 \Rightarrow x = -\dfrac{5}{2}$. The domain of $f(x)$ is

all real number except $-\dfrac{5}{2}$: $\left\{ x \mid x \neq -\dfrac{5}{2} \right\}$.

46. $g(-2) = 3(-2)^2 - 5(-2) = 3 \cdot 4 + 10 = 22$

47. $(f - g)(x) = f(x) - g(x)$

$= x^2 + 3x - (9 - 3x) = x^2 + 3x - 9 + 3x$

$= x^2 + 6x - 9$

48. $-3 \le f(x) \le 2 \Rightarrow -3 \le 1 - x \le 2$

$\Rightarrow -3 - 1 \le -x \le 2 - 1 \Rightarrow -4 \le -x \le 1$

$\Rightarrow 4 \ge x \ge -1,$ or $-1 \le x \le 4$

49. The domain of h/g is the set of values for which both h and g are well-defined and $g \ne 0$. The domain of h is $\{x \mid x \ne 0\}$ and the domain of g is all real numbers.

$g(x) = 0 \Rightarrow 3x - 1 = 0 \Rightarrow x = \dfrac{1}{3}.$

Therefore the domain of h/g is $\left\{ x \mid x \ne 0, \dfrac{1}{3} \right\}.$

50. $at - dt = c \Rightarrow t(a - d) = c \Rightarrow t = \dfrac{c}{a - d}$

51. The two lines intersect at the point $(1, 5)$, so $x = 1$.

52. From the graph, we see that the function values of f are less than the function values of g when $x \ge 1$. Because the function values are represented by the vertical position, we want the values where the graph of f is below that of g. The solution is the set $\{x \mid x \ge 1\}$, or $[1, \infty)$.

53. The solution is the point of intersection of the graphs of the two linear functions, or $(2, -3)$.

54 The solution is the x-value, or x-coordinate of the point of intersection of the two lines, or $x = 2$.

55. The solution is the x-value, or x-coordinate of the point where the parabola intersects the x-axis, or $x = 1$.

56. Let $x =$ equal the number of snow cannons needed for 1448 acres and build a proportion.

$$\frac{549 \text{ snow cannons}}{4344 \text{ acres}} = \frac{x}{1448 \text{ acres}}$$

$$\Rightarrow \frac{549 \text{ snow cannons}}{4344 \text{ acres}} \cdot 1448 \text{ acres}$$

$$= \frac{x}{1448 \text{ acres}} \cdot 1448 \text{ acres}$$

$$\Rightarrow x = \frac{549 \cdot 1448}{4344} \text{ snow cannons}$$

$$= 183 \text{ snow cannons}$$

57. Let $x =$ the gallons of water needed to produce a pound of beef, and $y =$ the gallons of water needed to produce a pound of wheat. Then:

$x + y = 11,600$ (1)

$\quad x = y + 7000$ (2)

Substituting $y + 7000$ for x in (1) will allow us to find y.

$x + y = 11,600$ (1)

$(y + 7000) + y = 11,600 \Rightarrow 2y + 7000 = 11,600$

$\Rightarrow 2y = 11,600 - 7000 = 4600$

$\Rightarrow y = \dfrac{4600}{2} = 2300$

Use (2) and the value of y to find x.

$x = y + 7000$ (2)

$\Rightarrow x = 2300 + 7000 = 9300$

It takes 9300 gallons of water to make one pound of beef, and 2300 gallons to make one pound of wheat.

58. Let $x =$ the cost of each dinner.
a) The rental fee will be waived if

$$150x > 6000 \Rightarrow x > \frac{6000}{150} = 40$$

For the catering fee to be waived, each dinner should more than $40.

b) The total cost is the cost per person times the number of people and the $1500 fee. Therefore:

$150x + 1500 > 6000$

$150x > 4500$

$x > \dfrac{4500}{150}$

$x > 30$

The total cost will exceed $6000 when the costs per person are greater than $30.

59. Let l = the length of the rectangle, in cm, and let w = its width, in cm. Then:

$$2l + 2w = 32$$

or $l + w = 16$ (1)

$$5w = 3l$$

or $3l - 5w = 0$ (2)

Add $5 \cdot$ (1) to (2) to eliminate w and find l.

$$5l + 5w = 80 \quad 5 \cdot (1)$$
$$\underline{3l - 5w = 0 \qquad (2)}$$
$$8l = 80$$

$$\Rightarrow l = \frac{80}{8} = 10$$

Use (1) and the value of l to find w.

$l + w = 16$ (1)

$$\Rightarrow w = 16 - l = 16 - 10 = 6$$
$$(l, w) = (10 \text{ cm}, 6 \text{ cm})$$

60. Let x = the cost of electricity, y = the cost of rent, and z = the cost of the cell phone. Then:

$$x + y + z = 920 \qquad (1)$$

$$x = \frac{1}{4} y$$

or $y = 4x$ (2)

$$z = x - 40 \qquad (3)$$

(2) and (3) respectively define both y and z in terms of x. Substitute $4x$ for y, and $x - 40$ for z, into (1) and solve for x.

$$x + y + z = 920 \Rightarrow x + (4x) + (x - 40) = 920$$
$$\Rightarrow 6x - 40 = 920 \Rightarrow 6x = 960$$

$$\Rightarrow x = \frac{960}{6} = 160$$

Use (2) and the value of x to find y.

$y = 4x$ (2)

$$\Rightarrow y = 4(160) = 640$$

The rent is $640.

61. If t = the number of years since 2006, $t = 0$ in 2006, $t = 1$ in 2007, and so on. Therefore the problem states that $f(0) = 19$ and $f(3) = 14$.

a) Since we are asked to find a linear function, we can assume $f(t) = mt + b$.

We need to find m and b. Since we are given the point $(0, 9)$, $b = 19$. To find m use the formula

$$m = \frac{y_2 - y_1}{x_2 - x_1}$$

and the points $(0, 19)$ and $(3, 1)$

$$m = \frac{y_2 - y_1}{x_2 - x_1} = \frac{14 - 19}{3 - 0} = -\frac{5}{3}$$

Therefore $f(t) = -\dfrac{5}{3}t + 19$

b) $t = 2012 - 2006 = 6$

$$f(6) = -\frac{5}{3}(6) + 19 = -10 + 19 = 9$$

c) We are asked to find t such that $f(t) = 4$

$$f(t) = 4 \Rightarrow -\frac{5}{3}t + 19 = 4 \Rightarrow -\frac{5}{3}t = 4 - 19 = -15$$

$$\Rightarrow \frac{5}{3}t = 15 \Rightarrow t = \frac{3}{5} \cdot 15 = 9$$

$t = 9$ corresponds to the year $2006 + 9 = 2015$.

62. If t = the number of years since 2006, $t = 0$ in 2006, $t = 1$ in 2007, and so on

a) Use linear regression to find the equation of line for the points $(0, 672)$, $(1, 767)$, and $(2, 859)$: The resulting equation is $C(t) = 93.5t + 672.5$

b) For 2011, $t = 2011 - 2007 = 4$.
$C(4) = 93.5(4) + 672.5 = 1046.5$.

63. For the function to be well-defined the expression under the radical must be greater than or equal to zero, and the denominator must not be zero:
$x + 4 \geq 0 \Rightarrow x \geq -4$ Since the denominator is just x, $x \neq 0$. The domain of f is $[-4, 0) \cup (0, \infty)$.

64.
$$\frac{2^{a-1} \cdot 2^{4a}}{2^{3(-2a+5)}} = \frac{2^{a-1+4a}}{2^{-6a+15}} = \frac{2^{5a-1}}{2^{-6a+15}}$$
$$= 2^{5a-1-(-6a+15)} = 2^{5a-1+6a-15} = 2^{11a-16}$$

65. To find the roots we set $f(x) = 0$, and factor f to solve.

$$x^4 - 34x^2 + 225 = (x^2)^2 - 34x^2 + 225$$

$$= (x^2 - 25)(x^2 - 9) = (x+5)(x-5)(x+3)(x-3)$$

Therefore, $f(x) = 0$

$\Rightarrow (x+5)(x-5)(x+3)(x-3) = 0$

$\Rightarrow x+5 = 0 \Rightarrow x = -5,$

or $x - 5 = 0 \Rightarrow x = 5,$

or $x + 3 = 0 \Rightarrow x = -3,$

or $x - 3 = 0 \Rightarrow x = 3$

The roots are $\{-5, -3, 3, 5\}$

66. $4 \leq |3 - x| \leq 6$

$\Rightarrow \quad 4 \leq 3 - x \leq 6 \qquad (1),$ or

$\quad -6 \leq 3 - x \leq -4 \quad (2)$

(1) gives us one solution set

$4 \leq 3 - x \leq 6 \quad (1)$

$\Rightarrow 4 - 3 \leq -x \leq 6 - 3 \Rightarrow 1 \leq -x \leq 3$

$\Rightarrow -1 \geq x \geq -3,$ or $-3 \leq x \leq -1$

(2) gives us a second solution set

$-6 \leq 3 - x \leq -4 \quad (2)$

$\Rightarrow -6 - 3 \leq -x \leq -4 - 3 \Rightarrow -9 \leq -x \leq -7$

$\Rightarrow 9 \geq x \geq 7,$ or $7 \leq x \leq 9$

In interval notation, the solution is

$[-3, -1] \cup [7, 9]$

67. $\dfrac{18}{x-9} + \dfrac{10}{x+5} = \dfrac{28x}{x^2 - 4x - 45}$

$\Rightarrow \dfrac{18}{x-9} + \dfrac{10}{x+5} = \dfrac{28x}{(x-9)(x+5)}$

$\Rightarrow (x-9)(x+5)\left[\dfrac{18}{x-9} + \dfrac{10}{x+5}\right]$

$= (x-9)(x+5)\left[\dfrac{28x}{(x-9)(x+5)}\right]$

$\Rightarrow 18(x+5) + 10(x-9) = 28x$ for $x \neq -5, 9$

$\Rightarrow 18x + 90 + 10x - 90 = 28x$

$\Rightarrow 28x = 28x \Rightarrow 0 = 0$

The equation is an identity and is true for all values of x for which the equation is defined. That is for all real numbers other than -5 and 9. In setbuilder notation, the solution set is $\{x \mid x \neq -5, 9\}$.

Chapter 10

Exponents and Radical Functions

1. Two

3. Positive

5. Irrational

7. Nonnegative

9. The square roots of 49 are 7 and –7, because
$7^2 = 49$ and $(-7)^2 = 49$.

11. The square roots of 144 are 12 and –12,
because $12^2 = 144$ and $(-12)^2 = 144$.

13. The square roots of 400 are 20 and –20,
because $20^2 = 400$ and $(-20)^2 = 400$.

15. The square roots of 900 are 30 and –30,
because $30^2 = 900$ and $(-30)^2 = 900$.

17. $\sqrt{49} = \sqrt{7 \cdot 7} = 7$

19. $-\sqrt{16} = -4$ since $\sqrt{16} = 4$.

21. $\sqrt{\dfrac{36}{49}} = \sqrt{\dfrac{6 \cdot 6}{7 \cdot 7}} = \dfrac{6}{7}$

23. $-\sqrt{\dfrac{16}{81}} = -\dfrac{4}{9}$ since $\sqrt{\dfrac{16}{81}} = \dfrac{4}{9}$.

25. $\sqrt{0.04} = 0.2$, since $(0.2)^2 = 0.04$.

27. $\sqrt{0.0081} = 0.09$

29. $5\sqrt{p^2} + 4$
The radicand is the expression written under
the radical sign, p^2.
Since the index is not written, it is understood
to be 2.

31. $xy\sqrt[5]{\dfrac{x}{y+4}}$
The radicand is the expression written under
the radical sign, $\dfrac{x}{y+4}$.
The index is 5.

33. $f(t) = \sqrt{5t - 10}$
$f(3) = \sqrt{5 \cdot 3 - 10} = \sqrt{5}$
$f(2) = \sqrt{5 \cdot 2 - 10} = \sqrt{0} = 0$
$f(1) = \sqrt{5 \cdot 1 - 10} = \sqrt{-5}$
Since negative numbers do not have real-
number square roots, $f(1)$ does not exist.
$f(-1) = \sqrt{5(-1) - 10} = \sqrt{-15}$
Since negative numbers do not have real-
number square roots, $f(-1)$ does not exist.

35. $t(x) = -\sqrt{2x^2 - 1}$
$t(5) = -\sqrt{2(5)^2 - 1} = -\sqrt{49} = -7$
$t(0) = -\sqrt{2(0)^2 - 1} = -\sqrt{-1}$;
$\quad t(0)$ does not exist
$t(-1) = -\sqrt{2(-1)^2 - 1} = -\sqrt{1} = -1$
$t\left(-\dfrac{1}{2}\right) = -\sqrt{2\left(-\dfrac{1}{2}\right)^2 - 1} = -\sqrt{-\dfrac{1}{2}}$;
$\quad t\left(-\dfrac{1}{2}\right)$ does not exist

37. $f(t) = \sqrt{t^2 + 1}$
$f(0) = \sqrt{0^2 + 1} = \sqrt{1} = 1$
$f(-1) = \sqrt{(-1)^2 + 1} = \sqrt{2}$
$f(-10) = \sqrt{(-10)^2 + 1} = \sqrt{101}$

39. $\sqrt{64x^2} = \sqrt{(8x)^2} = |8x| = 8|x|$

Since x might be negative, absolute-value notation is necessary.

41. $\sqrt{(-4b)^2} = |-4b| = |-4| \cdot |b| = 4|b|$

Since b might be negative, absolute-value notation is necessary.

43. $\sqrt{(8-t)^2} = |8-t|$

Since $8-t$ might be negative, absolute-value notation is necessary.

45. $\sqrt{y^2 + 16y + 64} = \sqrt{(y+8)^2} = |y+8|$

47. $\sqrt{4x^2 + 28x + 49} = \sqrt{(2x+7)^2} = |2x+7|$

49. $-\sqrt[4]{256} = -4$ since $4^4 = 256$

51. $\sqrt[5]{-1} = -1$ since $-1^5 = -1$

53. $-\sqrt[5]{-\dfrac{32}{243}} = -\left(-\dfrac{2}{3}\right) = \dfrac{2}{3}$ since $\left(-\dfrac{2}{3}\right)^5 = -\dfrac{32}{243}$

55. $\sqrt[6]{x^6} = |x|$ The index is even. Use absolute-value notation since x could have a negative value.

57. $\sqrt[9]{t^9} = t$ The index is odd.

59. $\sqrt[4]{(6a)^4} = |6a| = 6|a|$ The index is even.

61. $\sqrt[10]{(-6)^{10}} = |-6| = 6$

63. $\sqrt[414]{(a+b)^{414}} = |a+b|$ The index is even.

65. $\sqrt{a^{22}} = |a^{11}|$ Since $\left(a^{11}\right)^2 = a^{22}$; a^{11} could have a negative value.

67. $\sqrt{-25}$ Cannot be simplified.

69. $\sqrt{16x^2} = \sqrt{(4x)^2} = 4x$ Assuming $x \ge 0$.

71. $-\sqrt{(3t)^2} = -3t$ Assuming $t \ge 0$.

73. $\sqrt{(a+1)^2} = a+1$ Assuming $a+1 \ge 0$.

75. $\sqrt{9t^2 - 12t + 4} = \sqrt{(3t-2)^2} = 3t-2$

77. $\sqrt[3]{27a^3} = \sqrt[3]{(3a)^3} = 3a$

79. $\sqrt[4]{16x^4} = \sqrt[4]{(2x)^4} = 2x$

81. $\sqrt[5]{(x-1)^5} = x-1$

83. $-\sqrt[3]{-125y^3} = -\sqrt[3]{(-5y)^3} = -(-5y) = 5y$

85. $\sqrt{t^{18}} = \sqrt{\left(t^9\right)^2} = t^9$

87. $\sqrt{(x-2)^8} = \sqrt{\left[(x-2)^4\right]^2} = (x-2)^4$

89. $f(x) = \sqrt[3]{x+1}$

$f(7) = \sqrt[3]{7+1} = \sqrt[3]{8} = 2$

$f(26) = \sqrt[3]{26+1} = \sqrt[3]{27} = 3$

$f(-9) = \sqrt[3]{-9+1} = \sqrt[3]{-8} = -2$

$f(-65) = \sqrt[3]{-65+1} = \sqrt[3]{-64} = -4$

91. $g(t) = \sqrt[4]{t-3}$

$g(19) = \sqrt[4]{19-3} = \sqrt[4]{16} = 2$

$g(-13) = \sqrt[4]{-13-3} = \sqrt[4]{-16};$

$\quad g(-13)$ does not exist

$g(1) = \sqrt[4]{1-3} = \sqrt[4]{-2};$

$\quad g(1)$ does not exist

$g(84) = \sqrt[4]{84-3} = \sqrt[4]{81} = 3$

93. $f(x) = \sqrt{x-6}$

Since the index is even, the radicand, $x-6$, must be non-negative. We solve the inequality:

$$x - 6 \geq 0$$

$$x \geq 6$$

Domain of $f = \{x | x \geq 6\}$, or $[6, \infty)$

95. $g(t) = \sqrt[4]{t+8}$

Since the index is even, the radicand, $t+8$, must be non-negative. We solve the inequality:

$$t + 8 \geq 0$$

$$t \geq -8$$

Domain of $g = \{t | t \geq -8\}$, or $[-8, \infty)$

97. $g(x) = \sqrt[4]{2x-10}$

$$2x - 10 \geq 0$$

$$2x \geq 10$$

$$x \geq 5$$

Domain of $g = \{x | x \geq 5\}$, or $[5, \infty)$

99. $f(t) = \sqrt[5]{8-3t}$

Since the index is odd, the radicand can be any real number.

Domain of $f = \{t | t \text{ is a real number}\}$,

or $(-\infty, \infty)$

101. $h(z) = -\sqrt[6]{5z+2}$

$$5z + 2 \geq 0$$

$$5z \geq -2$$

$$z \geq -\frac{2}{5}$$

Domain of $h = \left\{z \middle| z \geq -\frac{2}{5}\right\}$, or $\left[-\frac{2}{5}, \infty\right)$

103. $f(t) = 7 + \sqrt[8]{t^8}$

Since we can compute $7 + \sqrt[8]{t^8}$ for any real number t, the domain is the set of real numbers, or $\{t | t \text{ is a real number}\}$,

or $(-\infty, \infty)$.

105. $f(x) = \sqrt{5-x}$

Find all values of x for which the radicand is non-negative.

$$5 - x \geq 0$$

$$5 \geq x$$

The domain is $\{x | x \leq 5\}$, or $(-\infty, 5]$.

We graph the function in the standard window.

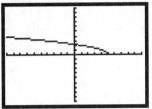

The range appears to be $\{y | y \geq 0\}$, or $[0, \infty)$.

107. $f(x) = 1 - \sqrt{x+1}$

Find all values of x for which the radicand is non-negative.

$$x + 1 \geq 0$$

$$x \geq -1$$

The domain is $\{x | x \geq -1\}$, or $[-1, \infty)$.

We graph the function in the standard window.

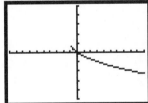

The range appears to be $\{y | y \leq 1\}$, or $(-\infty, 1]$.

109. $g(x) = 3 + \sqrt{x^2 + 4}$

Since $x^2 + 4$ is positive for all values of x, the domain is $\{x | x \text{ is a real number}\}$,

or $(-\infty, \infty)$.

We graph the function in the standard window.

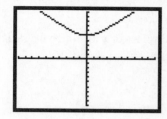

The range appears to be $\{y|y \geq 5\}$, or $[5, \infty)$.

111. For $f(x) = \sqrt{x-4}$, the domain is $[4, \infty)$ and all of the function values are non-negative. Graph (c) corresponds to this function.

113. For $h(x) = \sqrt{x^2 + 4}$, the domain is $(-\infty, \infty)$ Graph (d) corresponds to this function.

For problems 115 – 120, use a scatter plot of the data (for each problem, separately) and determine whether it could be modeled with a radical function.

115. Yes

117. Yes

119. No

121. $f(x) = 118.8\sqrt{x}$

    $f(50) = 118.8\sqrt{50}$

    $f(50) \approx 840$ GPM

    $f(175) = 118.8\sqrt{175}$

    $f(175) \approx 1572$ GPM

123. *Thinking and Writing Exercise.*

125. $(a^2 b)(a^4 b) = a^{2+4} b^{1+1} = a^6 b^2$

127. $(5x^2 y^{-3})^3 = 5^3 \cdot x^{2 \cdot 3} y^{-3 \cdot 3}$

$$= 125 \cdot x^6 \cdot y^{-9} = \frac{125 x^6}{y^9}$$

129. $\left(\dfrac{10 x^{-1} y^5}{5 x^2 y^{-1}}\right)^{-1} = \left(\dfrac{2 y^6}{x^3}\right)^{-1} = \dfrac{x^3}{2 y^6}$

131. *Thinking and Writing Exercise.*

133. *Thinking and Writing Exercise.*

135. $N = 2.5\sqrt{A}$

    a. $N = 2.5\sqrt{25} = 2.5(5) = 12.5 \approx 13$

    b. $N = 2.5\sqrt{36} = 2.5(6) = 15$

    c. $N = 2.5\sqrt{49} = 2.5(7) = 17.5 \approx 18$

    d. $N = 2.5\sqrt{64} = 2.5(8) = 20$

137. $\{x|x \geq -5\}$, or $[-5, \infty)$

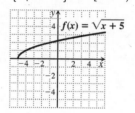

139. $\{x|x \geq 0\}$, or $[0, \infty)$

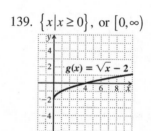

141. $g(x) = \dfrac{\sqrt[4]{5-x}}{\sqrt[6]{x+4}}$

In the numerator we must have $5 - x \geq 0$, or $x \leq 5$, and in the denominator we must have $x + 4 > 0$, or $x > -4$. Thus we have $x \leq 5$ and $x > -4$, so

Domain of $g = \{x|-4 < x \leq 5\}$, or $(-4, 5]$.

143. Cubic

Exercise Set 10.2

1. g

3. e

5. a

7. b

9. $x^{1/6} = \sqrt[6]{x}$

11. $(16)^{1/2} = \sqrt{16} = 4$

13. $32^{1/5} = \sqrt[5]{32} = 2$

15. $9^{1/2} = \sqrt{9} = 3$

17. $(xyz)^{1/2} = \sqrt{xyz}$

19. $(a^2 b^2)^{1/5} = \sqrt[5]{a^2 b^2}$

21. $t^{2/5} = \sqrt[5]{t^2}$

23. $16^{3/4} = \sqrt[4]{16^3} = \left(\sqrt[4]{16}\right)^3 = 2^3 = 8$

25. $27^{4/3} = \sqrt[3]{27^4} = \left(\sqrt[3]{3^3}\right)^4 = 3^4 = 81$

27. $(81x)^{3/4} = \sqrt[4]{(81x)^3} = \sqrt[4]{81^3 x^3}$, or $\sqrt[4]{81^3} \cdot \sqrt[4]{x^3}$
$= \left(\sqrt[4]{81}\right)^3 \cdot \sqrt[4]{x^3} = 3^3 \sqrt[4]{x^3} = 27\sqrt[4]{x^3}$

29. $(25x^4)^{3/2} = \sqrt{(25x^4)^3} = \sqrt{25^3 \cdot x^{12}}$
$= \sqrt{25^3} \cdot \sqrt{x^{12}} = \left(\sqrt{25}\right)^3 x^6$
$= 5^3 x^6 = 125x^6$

31. $\sqrt[3]{20} = 20^{1/3}$

33. $\sqrt{17} = 17^{1/2}$

35. $\sqrt{x^3} = x^{3/2}$

37. $\sqrt[5]{m^2} = m^{2/5}$

39. $\sqrt[4]{cd} = (cd)^{1/4}$ Parentheses are required.

41. $\sqrt[5]{xy^2 z} = \left(xy^2 z\right)^{1/5}$

43. $\left(\sqrt{3mn}\right)^3 = (3mn)^{3/2}$

45. $\left(\sqrt[7]{8x^2 y}\right)^5 = \left(8x^2 y\right)^{5/7}$

47. $\dfrac{2x}{\sqrt[3]{z^2}} = \dfrac{2x}{z^{2/3}}$

49. $8^{-1/3} = \dfrac{1}{8^{1/3}} = \dfrac{1}{\left(2^3\right)^{1/3}} = \dfrac{1}{2}$

51. $(2rs)^{-3/4} = \dfrac{1}{(2rs)^{3/4}}$

53. $\left(\dfrac{1}{16}\right)^{-3/4} = \dfrac{16^{3/4}}{1} = \left(2^4\right)^{3/4} = 2^3 = 8$

55. $\dfrac{2c}{a^{-3/5}} = 2c \cdot a^{3/5} = 2a^{3/5}c$

57. $5x^{-2/3} y^{4/5} z = 5 \cdot \dfrac{1}{x^{2/3}} \cdot y^{4/5} z = \dfrac{5 y^{4/5} z}{x^{2/3}}$

59. $3^{-5/2} a^3 b^{-7/3} = \dfrac{1}{3^{5/2}} \cdot a^3 \cdot \dfrac{1}{b^{7/3}} = \dfrac{a^3}{3^{5/2} b^{7/3}}$

61. $\left(\dfrac{2ab}{3c}\right)^{-5/6} = \left(\dfrac{3c}{2ab}\right)^{5/6}$
Find the reciprocal of the base and change the sign of the exponent.

63. $\dfrac{6a}{\sqrt[4]{b}} = \dfrac{6a}{b^{1/4}}$

65. $f(x) = \sqrt[4]{x+7} = (x+7)^{1/4}$
Enter $y = (x+7) \wedge (1/4)$, or
$y = (x+7) \wedge 0.25$.
Since the index is even, the domain of the function is the set of all x for which the radicand is non-negative, or $[-7, \infty)$. One good choice of a viewing window is $[-10, 25, -1, 5]$, Xscl $= 5$.

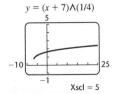

$y = (x+7)\wedge(1/4)$
Xscl = 5

67. $r(x) = \sqrt[7]{3x-2} = (3x-2)^{1/7}$

Enter $y = (3x-2)\wedge(1/7)$. Since the index is odd the domain of the function is $(-\infty, \infty)$. One good choice of a viewing window is $[-10, 10, -5, 5]$.

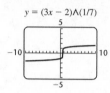

$y = (3x - 2)\wedge(1/7)$

69. $f(x) = \sqrt[6]{x^3} = (x^3)^{1/6} = \dot{x}^{3/6}$

Enter $y = x\wedge(3/6)$. The function is defined only for non-negative values of x, so the domain is $[0, \infty)$. One good choice of a window is $[-5, 25, -1, 5]$, Xscl $= 5$.

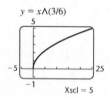

$y = x\wedge(3/6)$

Xscl = 5

71. $\sqrt[5]{9} = 9^{1/5} = 9\wedge(1/5) \approx 1.552$

73. $\sqrt[4]{10} = 10^{1/4} = 10\wedge(1/4) \approx 1.778$

75. $\sqrt[3]{(-3)^5} = (-3)^{5/3} = (-3)\wedge(5/3) \approx -6.240$

77. $7^{\frac{3}{4}} \cdot 7^{\frac{1}{8}} = 7^{\frac{3}{4}+\frac{1}{8}} = 7^{\frac{6}{8}+\frac{1}{8}} = 7^{\frac{7}{8}}$

We added exponents after finding their common denominator.

79. $\dfrac{3^{\frac{5}{8}}}{3^{-\frac{1}{8}}} = 3^{\frac{5}{8}-\left(-\frac{1}{8}\right)} = 3^{\frac{6}{8}} = 3^{\frac{3}{4}}$

81. $\dfrac{5.2^{-\frac{1}{6}}}{5.2^{-\frac{2}{3}}} = 5.2^{-\frac{1}{6}-\left(-\frac{2}{3}\right)} = 5.2^{-\frac{1}{6}+\frac{4}{6}} = 5.2^{\frac{3}{6}} = 5.2^{\frac{1}{2}}$

We subtracted exponents after finding a common denominator.

83. $\left(10^{\frac{3}{5}}\right)^{\frac{2}{5}} = 10^{\frac{3}{5}\cdot\frac{2}{5}} = 10^{\frac{6}{25}}$

85. $a^{\frac{2}{3}} \cdot a^{\frac{5}{4}} = a^{\frac{2}{3}+\frac{5}{4}} = a^{\frac{8}{12}+\frac{15}{12}} = a^{\frac{23}{12}}$

87. $\left(64^{\frac{3}{4}}\right)^{\frac{4}{3}} = 64^{\frac{3}{4}\cdot\frac{4}{3}} = 64^{1} = 64$

89. $\left(m^{\frac{2}{3}}n^{-\frac{1}{4}}\right)^{\frac{1}{2}} = m^{\frac{2}{3}\cdot\frac{1}{2}}n^{-\frac{1}{4}\cdot\frac{1}{2}} = m^{\frac{1}{3}}n^{-\frac{1}{8}} = \dfrac{m^{\frac{1}{3}}}{n^{\frac{1}{8}}}$

91. $\sqrt[8]{x^4} = x^{4/8}$ Convert to exponential notation

   $= x^{1/2}$ Simplifying the exponent

   $= \sqrt{x}$ Returning to radical notation

93. $\sqrt[4]{a^{12}} = a^{12/4}$ Convert to exponential notation

   $= a^3$ Simplifying the exponent

95. $\sqrt[12]{y^8} = y^{8/12} = y^{2/3} = \sqrt[3]{y^2}$

97. $\left(\sqrt[7]{xy}\right)^{14} = (xy)^{14/7}$ Convert to exponential notation

   $= (xy)^2$ Simplifying the exponent

   $= x^2y^2$ Using the law of exponents

99. $\sqrt[4]{(7a)^2} = (7a)^{2/4}$ Convert to exponential notation

   $= (7a)^{1/2}$ Simplifying the exponent

   $= \sqrt{7a}$ Returning to radical notation

101. $\left(\sqrt[8]{2x}\right)^6 = (2x)^{6/8} = (2x)^{3/4}$

   $= \sqrt[4]{(2x)^3} = \sqrt[4]{8x^3}$

103. $\sqrt{\sqrt[5]{m}} = \sqrt{m^{1/5}}$ Convert to exponential notation

   $= \left(m^{1/5}\right)^{1/2}$

   $= m^{1/10}$ Using a law of exponents

   $= \sqrt[10]{m}$ Returning to radical notation

105. $\sqrt[4]{(xy)^{12}} = (xy)^{12/4} = (xy)^3 = x^3y^3$

107. $\left(\sqrt[5]{(a^2b^4)}\right)^{15} = (a^2b^4)^{15/5} = (a^2b^4)^3 = a^6b^{12}$

109. $\sqrt[3]{\sqrt[4]{xy}} = \sqrt[3]{(xy)^{1/4}} = \left[(xy)^{1/4}\right]^{1/3}$
$= (xy)^{1/12} = \sqrt[12]{xy}$

111. *Thinking and Writing Exercise.*

113. $(x+5)(x-5)$
$= x \cdot x + 5x - 5x + 5(-5)$
$= x^2 - 25$

115. $4x^2 + 20x + 25 = (2x)^2 + 2(2x)(5) + 5^2$
$= (2x+5)^2$

117. $5t^2 - 10t + 5 = 5(t^2 - 2t + 1)$
$= 5(t^2 - 2(t)(1) + 1^2)$
$= 5(t-1)^2$

119. *Thinking and Writing Exercise.*

121. $\sqrt{x\sqrt[3]{x^2}} = \sqrt{x \cdot x^{2/3}} = \left(x^{5/3}\right)^{1/2} = x^{5/6} = \sqrt[6]{x^5}$

123. $\sqrt[12]{p^2 + 2pq + q^2} = \sqrt[12]{(p+q)^2} = \left[(p+q)^2\right]^{1/12}$
$= (p+q)^{2/12} = (p+q)^{1/6}$
$= \sqrt[6]{p+q}$

125. $f(x) = k \cdot 2^{x/12}$
$f(24) = 440 \cdot 2^{24/12}$
$= 440 \cdot 2^2$
$= 1760 \text{ cycles per second}$

127. $2^{4/12} \approx 1.2599 \approx 1.25$ so the C sharp that is 4 half steps above concert A has a frequency that is 125% of, or 25% greater than, that of concert A.

129. $L = \dfrac{0.000169d^{2.27}}{h}$

a. $L = \dfrac{(0.000169)60^{2.27}}{1} \approx 1.8 \text{ m}$

b. $L = \dfrac{(0.000169)75^{2.27}}{0.9906} \approx 3.1\,\text{m}$

c. $L = \dfrac{(0.000169)80^{2.27}}{2.4} \approx 1.5 \text{ m}$

d. $L = \dfrac{(0.000169)100^{2.27}}{1.1} \approx 5.3 \text{ m}$

131. $T = 0.936d^{1.97}h^{0.85}$
$= 0.936 \cdot 3^{1.97}80^{0.85}$
$\approx 338 \text{ ft}^3$

133. $y_1 = x^{1/2},\ y_2 = 3x^{2/5},$
$y_3 = x^{4/7},\ y_4 = \frac{1}{5}x^{3/4}$

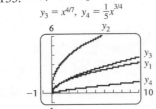

Exercise Set 10.3

1. True

3. False

5. True

7. $\sqrt{5} \cdot \sqrt{7} = \sqrt{5 \cdot 7} = \sqrt{35}$

9. $\sqrt[3]{3} \cdot \sqrt[3]{2} = \sqrt[3]{3 \cdot 2} = \sqrt[3]{6}$

11. $\sqrt[4]{6} \cdot \sqrt[4]{3} = \sqrt[4]{6 \cdot 3} = \sqrt[4]{18}$

13. $\sqrt{2x} \cdot \sqrt{13y} = \sqrt{2x \cdot 13y} = \sqrt{26xy}$

15. $\sqrt[5]{8y^3} \cdot \sqrt[5]{10y} = \sqrt[5]{8y^3 \cdot 10y} = \sqrt[5]{80y^4}$

17. $\sqrt{y-b} \cdot \sqrt{y+b} = \sqrt{(y-b)(y+b)} = \sqrt{y^2 - b^2}$

19. $\sqrt[3]{0.7y} \cdot \sqrt[3]{0.3y} = \sqrt[3]{0.7y \cdot 0.3y} = \sqrt[3]{0.21y^2}$

21. $\sqrt[5]{x-2} \cdot \sqrt[5]{(x-2)^2} = \sqrt[5]{(x-2)(x-2)^2}$

$\qquad\qquad\qquad\qquad = \sqrt[5]{(x-2)^3}$

23. $\sqrt{\dfrac{3}{t}} \cdot \sqrt{\dfrac{7s}{11}} = \sqrt{\dfrac{3}{t} \cdot \dfrac{7s}{11}} = \sqrt{\dfrac{21s}{11t}}$

25. $\sqrt[7]{\dfrac{x-3}{4}} \cdot \sqrt[7]{\dfrac{5}{x+2}} = \sqrt[7]{\dfrac{x-3}{4} \cdot \dfrac{5}{x+2}} = \sqrt[7]{\dfrac{5x-15}{4x+8}}$

27. $\sqrt{18}$

$= \sqrt{9 \cdot 2}$ 9 is the greatest perfect
$\qquad\qquad\qquad$ square of 18

$= \sqrt{9} \cdot \sqrt{2}$

$= 3\sqrt{2}$

29. $\sqrt{27}$

$= \sqrt{9 \cdot 3}$ 9 is the greatest perfect
$\qquad\qquad\qquad$ square of 27

$= \sqrt{9} \cdot \sqrt{3}$

$= 3\sqrt{3}$

31. $\sqrt{8x^9} = \sqrt{4x^8 \cdot 2x} = \sqrt{4x^8} \cdot \sqrt{2x} = 2x^4\sqrt{2x}$

33. $\sqrt{120} = \sqrt{4 \cdot 30} = \sqrt{4} \cdot \sqrt{30} = 2\sqrt{30}$

35. $\sqrt{36a^4 b}$

$= \sqrt{36a^4 \cdot b}$ $36a^4$ is a perfect square

$= \sqrt{36a^4} \cdot \sqrt{b}$ Factoring into 2 radicals

$= 6a^2\sqrt{b}$ Taking the square root of $36a^4$

37. $\sqrt[3]{8x^3 y^2}$

$= \sqrt[3]{8x^3 \cdot y^2}$ $8x^3$ is a perfect cube

$= \sqrt[3]{8x^3} \cdot \sqrt[3]{y^2}$ Factoring into 2 radicals

$= 2x\sqrt[3]{y^2}$ Taking the cube root of $8x^3$

39. $\sqrt[3]{-16x^6}$

$= \sqrt[3]{-8x^6 \cdot 2}$ $-8x^6$ is a perfect cube

$= \sqrt[3]{-8x^6} \cdot \sqrt[3]{2}$

$= -2x^2\sqrt[3]{2}$ Taking the cube root of $-8x^6$

41. $f(x) = \sqrt[3]{125x^5}$

$= \sqrt[3]{125x^3 \cdot x^2}$

$= \sqrt[3]{125x^3} \cdot \sqrt[3]{x^2}$

$= 5x\sqrt[3]{x^2}$

43. $f(x) = \sqrt{49(x-3)^2}$ $49(x-3)^2$ is a
$\qquad\qquad\qquad\qquad$ perfect square

$= \left|7(x-3)\right|$, or $7|x-3|$

45. $f(x) = \sqrt{5x^2 - 10x + 5}$

$= \sqrt{5(x^2 - 2x + 1)}$

$= \sqrt{5(x-1)^2}$

$= \sqrt{(x-1)^2} \cdot \sqrt{5}$

$= |x-1|\sqrt{5}$

47. $\sqrt{a^6 b^7}$

$= \sqrt{a^6 \cdot b^6 \cdot b}$ Identifying the greatest
$\qquad\qquad\qquad$ even powers of a and b

$= \sqrt{a^6} \cdot \sqrt{b^6} \cdot \sqrt{b}$ Factoring into radicals

$= a^3 b^3 \sqrt{b}$

49. $\sqrt[3]{x^5 y^6 z^{10}}$

$= \sqrt[3]{x^3 \cdot x^2 \cdot y^6 \cdot z^9 \cdot z}$ Identifying the
greatest perfect-cube powers of x,
y, and z.

$= \sqrt[3]{x^3} \cdot \sqrt[3]{y^6} \cdot \sqrt[3]{z^9} \cdot \sqrt[3]{x^2 z}$

Factoring into radicals

$= xy^2 z^3 \sqrt[3]{x^2 z}$

51. $\sqrt[4]{16x^5 y^{11}}$

$= \sqrt[4]{2^4 \cdot x^4 \cdot x \cdot y^8 \cdot y^3}$

$= \sqrt[4]{2^4} \cdot \sqrt[4]{x^4} \cdot \sqrt[4]{y^8} \cdot \sqrt[4]{xy^3}$

$= 2xy^2 \sqrt[4]{xy^3}$

53. $\sqrt[5]{x^{13}y^8z^{17}}$

$= \sqrt[5]{x^{10} \cdot x^3 \cdot y^5 \cdot y^3 \cdot z^{15} \cdot z^2}$

$= \sqrt[5]{x^{10}} \cdot \sqrt[5]{y^5} \cdot \sqrt[5]{z^{15}} \cdot \sqrt[5]{x^3 y^3 z^2}$

$= x^2 yz^3 \sqrt[5]{x^3 y^3 z^2}$

55. $\sqrt[3]{-80a^{14}} = \sqrt[3]{-8 \cdot 10 \cdot a^{12} a^2}$

$\qquad = \sqrt[3]{(-2)^3} \cdot \sqrt[3]{a^{12}} \cdot \sqrt[3]{10a^2}$

$\qquad = -2a^4 \sqrt[3]{10a^2}$

57. $\sqrt{6} \cdot \sqrt{3} = \sqrt{18} = \sqrt{9 \cdot 2} = 3\sqrt{2}$

59. $\sqrt{10} \cdot \sqrt{14} = \sqrt{140} = \sqrt{4 \cdot 35} = 2\sqrt{35}$

61. $\sqrt[3]{9} \cdot \sqrt[3]{3} = \sqrt[3]{27} = \sqrt[3]{3^3} = 3$

63. $\sqrt{18a^3} \cdot \sqrt{18a^3} = \left(\sqrt{18a^3}\right)^2 = 18a^3$

65. $\sqrt[3]{5a^2} \cdot \sqrt[3]{2a} = \sqrt[3]{10a^3} = \sqrt[3]{a^3 \cdot 10} = a\sqrt[3]{10}$

67. $3\sqrt{2x^5} \cdot 4\sqrt{10x^2} = 12\sqrt{20x^7} = 12\sqrt{4x^6 \cdot 5x}$

$\qquad\qquad = 12 \cdot 2x^3 \sqrt{5x} = 24x^3 \sqrt{5x}$

69. $\sqrt[3]{s^2 t^4} \cdot \sqrt[3]{s^4 t^6} = \sqrt[3]{s^6 t^{10}} = \sqrt[3]{s^6 t^9 t} = s^2 t^3 \sqrt[3]{t}$

71. $\sqrt[3]{(x+5)^2} \cdot \sqrt[3]{(x+5)^4} = \sqrt[3]{(x+5)^6} = (x+5)^2$

73. $\sqrt[4]{20a^3 b^7} \cdot \sqrt[4]{4a^2 b^5} = \sqrt[4]{80a^5 b^{12}}$

$\qquad\qquad = \sqrt[4]{16 \cdot 5 \cdot a^4 ab^{12}}$

$\qquad\qquad = 2ab^3 \sqrt[4]{5a}$

75. $\sqrt[5]{x^3 (y+z)^6} \cdot \sqrt[5]{x^3 (y+z)^4} = \sqrt[5]{x^6 (y+z)^{10}}$

$\qquad\qquad = \sqrt[5]{x^5 (y+z)^{10} \cdot x}$

$\qquad\qquad = x(y+z)^2 \sqrt[5]{x}$

77. *Thinking and Writing Exercise.*

79. $\dfrac{15a^2 x}{8b} \cdot \dfrac{24b^2 x}{5a} = \dfrac{15a^2 x \cdot 24b^2 x}{8b \cdot 5a}$

$\qquad\qquad = \dfrac{15a^2 x}{5a} \cdot \dfrac{24b^2 x}{8b}$

$\qquad\qquad = \dfrac{3a \cdot 5a \cdot x}{5a} \cdot \dfrac{3b \cdot 8b \cdot x}{8b}$

$\qquad\qquad = 3ax \cdot 3bx$

$\qquad\qquad = 9abx^2$

81. $\dfrac{x-3}{2x-10} - \dfrac{3x-5}{x^2-25} = \dfrac{x-3}{2(x-5)} - \dfrac{3x-5}{(x+5)(x-5)}$

LCD is $2(x+5)(x-5)$

$\dfrac{x-3}{2x-10} - \dfrac{3x-5}{x^2-25}$

$= \dfrac{(x-3)(x+5)}{2(x-5)(x+5)} - \dfrac{2(3x-5)}{2(x+5)(x-5)}$

$= \dfrac{(x-3)(x+5) - 2(3x-5)}{2(x-5)(x+5)}$

$= \dfrac{x^2 - 3x + 5x - 15 - 6x + 10}{2(x-5)(x+5)}$

$= \dfrac{x^2 - 4x - 5}{2(x-5)(x+5)}$

$= \dfrac{(x-5)(x+1)}{2(x-5)(x+5)}$

$= \dfrac{(x+1)}{2(x+5)}$

83. $\dfrac{a^{-1} + b^{-1}}{ab} = \dfrac{\dfrac{1}{a} + \dfrac{1}{b}}{ab}$

$\qquad\qquad = \dfrac{ab\left(\dfrac{1}{a} + \dfrac{1}{b}\right)}{ab(ab)}$

$\qquad\qquad = \dfrac{b+a}{a^2 b^2}$

85. *Thinking and Writing Exercise.*

87. $R(x) = \dfrac{1}{2} \sqrt[4]{\dfrac{x \cdot 3.0 \times 10^6}{\pi^2}}$

$R(5 \times 10^4) = \dfrac{1}{2} \sqrt[4]{\dfrac{(5 \times 10^4) \cdot (3.0 \times 10^6)}{\pi^2}}$

$\qquad\qquad \approx 175.6 \text{ mi}$

89. $T_w = 33 - \dfrac{\left(10.45 + 10\sqrt{v} - v\right)(33 - T)}{22}$

   a. $T_w = 33 - \dfrac{\left(10.45 + 10\sqrt{8} - 8\right)(33 - 7)}{22}$

      $\approx -3.3°C$

   b. $T_w = 33 - \dfrac{\left(10.45 + 10\sqrt{12} - 12\right)(33 - 0)}{22}$

      $\approx -16.6°C$

   c. $T_w = 33 - \dfrac{\left(10.45 + 10\sqrt{14} - 14\right)(33 - -5)}{22}$

      $\approx -25.5 °C$

   d. $T_w = 33 - \dfrac{\left(10.45 + 10\sqrt{15} - 15\right)(33 - -23)}{22}$

      $\approx -54.0 °C$

91. $\left(\sqrt[3]{25x^4}\right)^4 = \sqrt[3]{\left(25x^4\right)^4} = \sqrt[3]{25^4 x^{16}}$

$= \sqrt[3]{25^3 \cdot 25 \cdot x^{15} \cdot x}$

$= \sqrt[3]{25^3} \cdot \sqrt[3]{x^{15}} \cdot \sqrt[3]{25x}$

$= 25x^5 \sqrt[3]{25x}$

93. $\left(\sqrt{a^3 b^5}\right)^7 = \sqrt{\left(a^3 b^5\right)^7} = \sqrt{a^{21} b^{35}}$

$= \sqrt{a^{20} \cdot a \cdot b^{34} \cdot b}$

$= a^{10} b^{17} \sqrt{ab}$

95.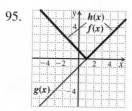

We see that $f(x) = h(x)$ and $f(x) \neq g(x)$.

97. $g(x) = \sqrt{x^2 - 6x + 8}$

We must have $x^2 - 6x + 8 \geq 0$, or $(x - 2)(x - 4) \geq 0$. We graph $y = x^2 - 6x + 8$.

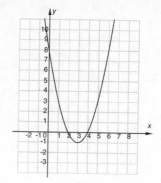

From the graph we see that $y \geq 0$ for $x \leq 2$ or $x \geq 4$, so the domain of g is $\{x | x \leq 2 \ or \ x \geq 4\}$, or $(-\infty, 2] \cup [4, \infty)$.

99. $\sqrt[5]{4a^{3k+2}} \, \sqrt[5]{8a^{6-k}} = 2a^4$

$\sqrt[5]{4a^{3k+2} \cdot 8a^{6-k}} = 2a^4$

$\sqrt[5]{32a^{2k+8}} = 2a^4$

$2\sqrt[5]{a^{2k+8}} = 2a^4$

$\sqrt[5]{a^{2k+8}} = a^4$

$\left(a^{2k+8}\right)^{1/5} = a^4$

$a^{\frac{2k+8}{5}} = a^4$

Since the base is the same, the exponents must be equal. We have:

$\dfrac{2k + 8}{5} = 4$

$2k + 8 = 20$

$2k = 12$

$k = 6$

101. *Thinking and Writing Exercise.*

Exercise Set 10.4

1. e

3. f

5. h

7. a

9. $\sqrt{\dfrac{36}{25}} = \dfrac{\sqrt{36}}{\sqrt{25}} = \dfrac{6}{5}$

11. $\sqrt[3]{\dfrac{64}{27}} = \dfrac{\sqrt[3]{64}}{\sqrt[3]{27}} = \dfrac{4}{3}$

13. $\sqrt{\dfrac{49}{y^2}} = \dfrac{\sqrt{49}}{\sqrt{y^2}} = \dfrac{7}{y}$

15. $\sqrt{\dfrac{36y^3}{x^4}} = \dfrac{\sqrt{36y^3}}{\sqrt{x^4}} = \dfrac{\sqrt{36y^2 \cdot y}}{\sqrt{x^4}} = \dfrac{6y\sqrt{y}}{x^2}$

17. $\sqrt[3]{\dfrac{27a^4}{8b^3}} = \dfrac{\sqrt[3]{27a^3 \cdot a}}{\sqrt[3]{8b^3}} = \dfrac{3a\sqrt[3]{a}}{2b}$

19. $\sqrt[4]{\dfrac{32a^4}{2b^4c^8}} = \sqrt[4]{\dfrac{16a^4}{b^4c^8}} = \dfrac{\sqrt[4]{16a^4}}{\sqrt[4]{b^4c^8}} = \dfrac{2a}{bc^2}$

21. $\sqrt[4]{\dfrac{a^5b^8}{c^{10}}} = \dfrac{\sqrt[4]{a^4 \cdot b^8 \cdot a}}{\sqrt[4]{c^8 \cdot c^2}} = \dfrac{ab^2\sqrt[4]{a}}{c^2\sqrt[4]{c^2}}$, or $\dfrac{ab^2}{c^2}\sqrt[4]{\dfrac{a}{c^2}}$

23. $\sqrt[5]{\dfrac{32x^6}{y^{11}}} = \dfrac{\sqrt[5]{32x^5 \cdot x}}{\sqrt[5]{y^{10} \cdot y}} = \dfrac{2x\sqrt[5]{x}}{y^2\sqrt[5]{y}}$, or $\dfrac{2x}{y^2}\sqrt[5]{\dfrac{x}{y}}$

25. $\sqrt[6]{\dfrac{x^6y^8}{z^{15}}} = \dfrac{\sqrt[6]{x^6y^6 \cdot y^2}}{\sqrt[6]{z^{12} \cdot z^3}} = \dfrac{xy\sqrt[6]{y^2}}{z^2\sqrt[6]{z^3}}$,

   or $\dfrac{xy}{z^2}\sqrt[6]{\dfrac{y^2}{z^3}}$

27. $\dfrac{\sqrt{18y}}{\sqrt{2y}} = \sqrt{\dfrac{18y}{2y}} = \sqrt{9} = 3$

29. $\dfrac{\sqrt[3]{26}}{\sqrt[3]{13}} = \sqrt[3]{\dfrac{26}{13}} = \sqrt[3]{2}$

31. $\dfrac{\sqrt{40xy^3}}{\sqrt{8x}} = \sqrt{\dfrac{40xy^3}{8x}} = \sqrt{5y^3} = \sqrt{y^2 \cdot 5y}$

   $= \sqrt{y^2}\sqrt{5y} = y\sqrt{5y}$

33. $\dfrac{\sqrt[3]{96a^4b^2}}{\sqrt[3]{12a^2b}} = \sqrt[3]{\dfrac{96a^4b^2}{12a^2b}} = \sqrt[3]{8a^2b} = 2\sqrt[3]{a^2b}$

35. $\dfrac{\sqrt{100ab}}{5\sqrt{2}} = \dfrac{1}{5} \cdot \dfrac{\sqrt{100ab}}{\sqrt{2}} = \dfrac{1}{5}\sqrt{\dfrac{100ab}{2}}$

   $= \dfrac{1}{5}\sqrt{50ab} = \dfrac{1}{5}\sqrt{25 \cdot 2ab}$

   $= \dfrac{1}{5} \cdot 5\sqrt{2ab} = \sqrt{2ab}$

37. $\dfrac{\sqrt[4]{48x^9y^{13}}}{\sqrt[4]{3xy^{-2}}} = \sqrt[4]{\dfrac{48x^9y^{13}}{3xy^{-2}}} = \sqrt[4]{16x^8y^{15}}$

   $= \sqrt[4]{16x^8y^{12} \cdot y^3} = 2x^2y^3\sqrt[4]{y^3}$

39. $\dfrac{\sqrt[3]{x^3 - y^3}}{\sqrt[3]{x - y}} = \sqrt[3]{\dfrac{x^3 - y^3}{x - y}}$

   $= \sqrt[3]{\dfrac{(x - y)(x^2 + xy + y^2)}{x - y}}$

   $= \sqrt[3]{\dfrac{\cancel{(x - y)}(x^2 + xy + y^2)}{\cancel{x - y}}}$

   $= \sqrt[3]{x^2 + xy + y^2}$

41. $\sqrt{\dfrac{3}{2}} = \sqrt{\dfrac{3}{2} \cdot \dfrac{2}{2}} = \sqrt{\dfrac{6}{4}} = \dfrac{\sqrt{6}}{\sqrt{4}} = \dfrac{\sqrt{6}}{2}$

43. $\dfrac{2\sqrt{5}}{7\sqrt{3}} = \dfrac{2\sqrt{5}}{7\sqrt{3}} \cdot \dfrac{\sqrt{3}}{\sqrt{3}} = \dfrac{2\sqrt{15}}{7 \cdot 3} = \dfrac{2\sqrt{15}}{21}$

45. $\sqrt[3]{\dfrac{5}{4}} = \sqrt[3]{\dfrac{5}{2^2} \cdot \dfrac{2}{2}} = \sqrt[3]{\dfrac{10}{2^3}} = \dfrac{\sqrt[3]{10}}{\sqrt[3]{2^3}} = \dfrac{\sqrt[3]{10}}{2}$

47. $\dfrac{\sqrt[3]{3a}}{\sqrt[3]{5c}} = \dfrac{\sqrt[3]{3a}}{\sqrt[3]{5c}} \cdot \dfrac{\sqrt[3]{5^2c^2}}{\sqrt[3]{5^2c^2}} = \dfrac{\sqrt[3]{75ac^2}}{\sqrt[3]{5^3c^3}} = \dfrac{\sqrt[3]{75ac^2}}{5c}$

49. $\dfrac{\sqrt[4]{5y^6}}{\sqrt[4]{9x}} = \dfrac{y\sqrt[4]{5y^2}}{\sqrt[4]{3^2x^1}} \cdot \dfrac{\sqrt[4]{3^2x^3}}{\sqrt[4]{3^2x^3}} = \dfrac{y\sqrt[4]{45x^3y^2}}{\sqrt[4]{3^4x^4}}$

   $= \dfrac{y\sqrt[4]{45x^3y^2}}{3x}$

51. $\sqrt[3]{\dfrac{2}{x^2y}} = \sqrt[3]{\dfrac{2}{x^2y} \cdot \dfrac{xy^2}{xy^2}} = \sqrt[3]{\dfrac{2xy^2}{x^3y^3}}$

   $= \dfrac{\sqrt[3]{2xy^2}}{\sqrt[3]{x^3y^3}} = \dfrac{\sqrt[3]{2xy^2}}{xy}$

53. $\sqrt{\dfrac{7a}{18}} = \sqrt{\dfrac{7a}{18} \cdot \dfrac{2}{2}} = \sqrt{\dfrac{14a}{36}} = \dfrac{\sqrt{14a}}{6}$

55. $\sqrt{\dfrac{9}{20x^2 y}} = \sqrt{\dfrac{9}{20x^2 y} \cdot \dfrac{5y}{5y}}$

$= \sqrt{\dfrac{9 \cdot 5y}{100x^2 y^2}} = \dfrac{3\sqrt{5y}}{10xy}$

57. $\sqrt{\dfrac{10ab^2}{72a^3 b}} = \sqrt{\dfrac{5b}{36a^2}} = \dfrac{\sqrt{5b}}{6a}$

59. $\sqrt{\dfrac{5}{11}} = \dfrac{\sqrt{5}}{\sqrt{11}} = \dfrac{\sqrt{5}}{\sqrt{11}} \cdot \dfrac{\sqrt{5}}{\sqrt{5}} = \dfrac{\sqrt{25}}{\sqrt{55}} = \dfrac{5}{\sqrt{55}}$

61. $\dfrac{2\sqrt{6}}{5\sqrt{7}} = \dfrac{2\sqrt{6} \cdot \sqrt{6}}{5\sqrt{7} \cdot \sqrt{6}} = \dfrac{2 \cdot 6}{5\sqrt{42}} = \dfrac{12}{5\sqrt{42}}$

63. $\dfrac{\sqrt{8}}{2\sqrt{3x}} = \dfrac{\cancel{2}\sqrt{2}}{\cancel{2}\sqrt{3x}} \cdot \dfrac{\sqrt{2}}{\sqrt{2}} = \dfrac{2}{\sqrt{6x}}$

65. $\dfrac{\sqrt[3]{7}}{\sqrt[3]{2}} = \dfrac{\sqrt[3]{7}}{\sqrt[3]{2}} \cdot \dfrac{\sqrt[3]{7^2}}{\sqrt[3]{7^2}} = \dfrac{\sqrt[3]{7^3}}{\sqrt[3]{98}} = \dfrac{7}{\sqrt[3]{98}}$

67. $\sqrt{\dfrac{7x}{3y}} = \sqrt{\dfrac{7x \cdot 7x}{3y \cdot 7x}} = \dfrac{\sqrt{(7x)^2}}{\sqrt{21xy}} = \dfrac{7x}{\sqrt{21xy}}$

69. $\sqrt[3]{\dfrac{2a^5}{5b}} = \sqrt[3]{\dfrac{2a^5}{5b} \cdot \dfrac{4a}{4a}} = \sqrt[3]{\dfrac{8a^6}{20ab}} = \dfrac{2a^2}{\sqrt[3]{20ab}}$

71. $\sqrt{\dfrac{x^3 y}{2}} = \sqrt{\dfrac{x^3 y}{2} \cdot \dfrac{xy}{xy}} = \sqrt{\dfrac{x^4 y^2}{2xy}}$

$= \dfrac{\sqrt{x^4 y^2}}{\sqrt{2xy}} = \dfrac{x^2 y}{\sqrt{2xy}}$

73. *Thinking and Writing Exercise.*

75. $3x - 8xy + 2xz = x(3 - 8y + 2z)$

77. $(a + b)(a - b) = a^2 - b^2$

79. $(8 + 3x)(7 - 4x)$

$= 8 \cdot 7 - 8 \cdot 4x + 3x \cdot 7 - 3x \cdot 4x$

$= 56 - 32x + 21x - 12x^2$

$= 56 - 11x - 12x^2$

81. *Thinking and Writing Exercise.*

83. a. $T = 2\pi \sqrt{\dfrac{65}{980}} \approx 1.62 \,\text{sec}$

    b. $T = 2\pi \sqrt{\dfrac{98}{980}} \approx 1.99 \,\text{sec}$

    c. $T = 2\pi \sqrt{\dfrac{120}{980}} \approx 2.20 \,\text{sec}$

85. $\dfrac{\left(\sqrt[3]{81mn^2}\right)^2}{\left(\sqrt[3]{mn}\right)^2} = \dfrac{\sqrt[3]{\left(81mn^2\right)^2}}{\sqrt[3]{\left(mn\right)^2}} = \dfrac{\sqrt[3]{6561m^2 n^4}}{\sqrt[3]{m^2 n^2}}$

$= \sqrt[3]{\dfrac{6561m^2 n^4}{m^2 n^2}} = \sqrt[3]{6561n^2}$

$= \sqrt[3]{729 \cdot 9n^2} = \sqrt[3]{729}\sqrt[3]{9n^2}$

$= 9\sqrt[3]{9n^2}$

87. $\sqrt{a^2 - 3} - \dfrac{a^2}{\sqrt{a^2 - 3}}$

$= \sqrt{a^2 - 3} - \dfrac{a^2}{\sqrt{a^2 - 3}} \cdot \dfrac{\sqrt{a^2 - 3}}{\sqrt{a^2 - 3}}$

$= \sqrt{a^2 - 3} - \dfrac{a^2 \sqrt{a^2 - 3}}{a^2 - 3}$

$= \sqrt{a^2 - 3} \cdot \dfrac{a^2 - 3}{a^2 - 3} - \dfrac{a^2 \sqrt{a^2 - 3}}{a^2 - 3}$

$= \dfrac{a^2 \sqrt{a^2 - 3} - 3\sqrt{a^2 - 3} - a^2 \sqrt{a^2 - 3}}{a^2 - 3}$

$= \dfrac{-3\sqrt{a^2 - 3}}{a^2 - 3}, \text{ or } \dfrac{-3}{\sqrt{a^2 - 3}}$

89. Step 1: $\sqrt[n]{x} = x^{1/n}$, by definition;

Step 2: $\left(\dfrac{x}{y}\right)^{1/n} = \dfrac{x^{1/n}}{y^{1/n}}$, raising a quotient to a power;

Step 3: $x^{1/n} = \sqrt[n]{x}$, by definition.

91. $f(x) = \sqrt{18x^3}, g(x) = \sqrt{2x}$

$$(f/g)(x) = \frac{f(x)}{g(x)} = \frac{\sqrt{18x^3}}{\sqrt{2x}}$$

$$= \sqrt{\frac{18x^3}{2x}} = \sqrt{9x^2} = 3x$$

$\sqrt{2x}$ is defined for $2x \geq 0$, or $x \geq 0$. To avoid division by 0, we must exclude 0 from the domain. Thus, the domain of $f/g = \{x | x \text{ is a real number and } x > 0\}$, or $(0, \infty)$.

93. $f(x) = \sqrt{x^2 - 9}, g(x) = \sqrt{x - 3}$

$$(f/g)(x) = \frac{f(x)}{g(x)} = \frac{\sqrt{x^2 - 9}}{\sqrt{x - 3}}$$

$$= \sqrt{\frac{x^2 - 9}{x - 3}}$$

$$= \sqrt{\frac{(x+3)(x-3)}{x - 3}}$$

$$= \sqrt{x + 3}$$

$\sqrt{x-3}$ is defined for $x - 3 \geq 0$, or $x \geq 3$. To avoid division by 0, we must exclude 3 from the domain. Thus, the domain of $f/g = \{x | x \text{ is a real number and } x > 3\}$, or $(3, \infty)$.

Exercise Set 10.5

1. Radicands; indices

3. Bases

5. Numerator; conjugate

7. $2\sqrt{5} + 7\sqrt{5} = (2 + 7)\sqrt{5} = 9\sqrt{5}$

9. $7\sqrt[3]{4} - 5\sqrt[3]{4} = (7 - 5)\sqrt[3]{4} = 2\sqrt[3]{4}$

11. $\sqrt[3]{y} + 9\sqrt[3]{y} = (1 + 9)\sqrt[3]{y} = 10\sqrt[3]{y}$

13. $8\sqrt{2} - \sqrt{2} + 5\sqrt{2} = (8 - 1 + 5)\sqrt{2} = 12\sqrt{2}$

15. $9\sqrt[3]{7} - \sqrt{3} + 4\sqrt[3]{7} + 2\sqrt{3}$
$= (9 + 4)\sqrt[3]{7} + (-1 + 2)\sqrt{3}$
$= 13\sqrt[3]{7} + \sqrt{3}$

17. $4\sqrt{27} - 3\sqrt{3} = 4\sqrt{9 \cdot 3} - 3\sqrt{3}$
$= 4 \cdot 3\sqrt{3} - 3\sqrt{3}$
$= 12\sqrt{3} - 3\sqrt{3}$
$= (12 - 3)\sqrt{3} = 9\sqrt{3}$

19. $3\sqrt{45} - 8\sqrt{20} = 3\sqrt{9 \cdot 5} - 8\sqrt{4 \cdot 5}$
$= 3 \cdot 3\sqrt{5} - 8 \cdot 2\sqrt{5}$
$= 9\sqrt{5} - 16\sqrt{5} = -7\sqrt{5}$

21. $3\sqrt[3]{16} + \sqrt[3]{54} = 3\sqrt[3]{8 \cdot 2} + \sqrt[3]{27 \cdot 2}$
$= 3 \cdot 2\sqrt[3]{2} + 3\sqrt[3]{2} = 6\sqrt[3]{2} + 3\sqrt[3]{2}$
$= 9\sqrt[3]{2}$

23. $\sqrt{a} + 3\sqrt{16a^3} = \sqrt{a} + 3\sqrt{16a^2 \cdot a}$
$= \sqrt{a} + 3 \cdot 4a\sqrt{a}$
$= \sqrt{a} + 12a\sqrt{a}$
$= (1 + 12a)\sqrt{a}$

25. $\sqrt[3]{6x^4} - \sqrt[3]{48x} = \sqrt[3]{x^3 \cdot 6x} - \sqrt[3]{8 \cdot 6x}$
$= x\sqrt[3]{6x} - 2\sqrt[3]{6x}$
$= (x - 2)\sqrt[3]{6x}$

27. $\sqrt{4a - 4} + \sqrt{a - 1} = \sqrt{4 \cdot (a - 1)} + \sqrt{a - 1}$
$= 2\sqrt{a - 1} + \sqrt{a - 1}$
$= 3\sqrt{a - 1}$

29. $\sqrt{x^3 - x^2} + \sqrt{9x - 9} = \sqrt{x^2(x - 1)} + \sqrt{9(x - 1)}$
$= x\sqrt{x - 1} + 3\sqrt{x - 1}$
$= (x + 3)\sqrt{x - 1}$

31. $\sqrt{3}(4 + \sqrt{3}) = \sqrt{3} \cdot 4 + \sqrt{3} \cdot \sqrt{3} = 4\sqrt{3} + 3$

33. $3\sqrt{5}(\sqrt{5} - \sqrt{2}) = 3\sqrt{5} \cdot \sqrt{5} - 3\sqrt{5} \cdot \sqrt{2}$
$= 3 \cdot 5 - 3\sqrt{10} = 15 - 3\sqrt{10}$

35. $\sqrt{2}\left(3\sqrt{10}-2\sqrt{2}\right)=\sqrt{2}\cdot 3\sqrt{10}-\sqrt{2}\cdot 2\sqrt{2}$

$\qquad =3\sqrt{20}-2\cdot 2=3\sqrt{4\cdot 5}-4$

$\qquad =3\cdot 2\sqrt{5}-4=6\sqrt{5}-4$

37. $\sqrt[3]{3}\left(\sqrt[3]{9}-4\sqrt[3]{21}\right)=\sqrt[3]{3}\cdot\sqrt[3]{9}-\sqrt[3]{3}\cdot 4\sqrt[3]{21}$

$\qquad =\sqrt[3]{27}-4\sqrt[3]{63}=3-4\sqrt[3]{63}$

39. $\sqrt[3]{a}\left(\sqrt[3]{a^2}+\sqrt[3]{24a^2}\right)=\sqrt[3]{a}\cdot\sqrt[3]{a^2}+\sqrt[3]{a}\cdot\sqrt[3]{24a^2}$

$\qquad =\sqrt[3]{a^3}+\sqrt[3]{24a^3}$

$\qquad =a+\sqrt[3]{8a^3\cdot 3}$

$\qquad =a+2a\sqrt[3]{3}$

41. $\left(2+\sqrt{6}\right)\left(5-\sqrt{6}\right)$

$\qquad =2\cdot 5-2\sqrt{6}+5\sqrt{6}-\sqrt{6}\cdot\sqrt{6}$

$\qquad =10+3\sqrt{6}-6=4+3\sqrt{6}$

43. $\left(\sqrt{2}+\sqrt{7}\right)\left(\sqrt{3}-\sqrt{7}\right)$

$\qquad =\sqrt{2}\cdot\sqrt{3}-\sqrt{2}\cdot\sqrt{7}+\sqrt{7}\cdot\sqrt{3}-\sqrt{7}\cdot\sqrt{7}$

$\qquad =\sqrt{6}-\sqrt{14}+\sqrt{21}-7$

45. $\left(3-\sqrt{5}\right)\left(3+\sqrt{5}\right)=3^2-\left(\sqrt{5}\right)^2=9-5=4$

47. $\left(\sqrt{6}+\sqrt{8}\right)\left(\sqrt{6}-\sqrt{8}\right)=\left(\sqrt{6}\right)^2-\left(\sqrt{8}\right)^2$

$\qquad\qquad =6-8=-2$

49. $\left(3\sqrt{7}+2\sqrt{5}\right)\left(2\sqrt{7}-4\sqrt{5}\right)$

$\qquad =3\sqrt{7}\cdot 2\sqrt{7}-3\sqrt{7}\cdot 4\sqrt{5}+2\sqrt{5}\cdot 2\sqrt{7}-$

$\qquad\qquad\qquad 2\sqrt{5}\cdot 4\sqrt{5}$

$\qquad =3\cdot 2\cdot 7-12\sqrt{35}+4\sqrt{35}-2\cdot 4\cdot 5$

$\qquad =42-8\sqrt{35}-40=2-8\sqrt{35}$

51. $\left(2+\sqrt{3}\right)^2=2^2+2\cdot 2\cdot\sqrt{3}+\left(\sqrt{3}\right)^2$

$\qquad =4+4\sqrt{3}+3=7+4\sqrt{3}$

53. $\left(\sqrt{3}-\sqrt{2}\right)^2=\left(\sqrt{3}\right)^2-2\sqrt{3}\sqrt{2}+\left(\sqrt{2}\right)^2$

$\qquad =3-2\sqrt{6}+2=5-2\sqrt{6}$

55. $\left(\sqrt{2t}+\sqrt{5}\right)^2=\left(\sqrt{2t}\right)^2+2\sqrt{2t}\sqrt{5}+\left(\sqrt{5}\right)^2$

$\qquad =2t+2\sqrt{10t}+5$

$\qquad =2t+5+2\sqrt{10t}$

57. $\left(3-\sqrt{x+5}\right)^2=3^2-2\cdot 3\cdot\sqrt{x+5}+\left(\sqrt{x+5}\right)^2$

$\qquad =9-6\sqrt{x+5}+x+5$

$\qquad =14+x-6\sqrt{x+5}$

59. $\left(2\sqrt[4]{7}-\sqrt[4]{6}\right)\left(3\sqrt[4]{9}+2\sqrt[4]{5}\right)$

$\qquad =2\sqrt[4]{7}\cdot 3\sqrt[4]{9}+2\sqrt[4]{7}\cdot 2\sqrt[4]{5}-\sqrt[4]{6}\cdot 3\sqrt[4]{9}-$

$\qquad\qquad\qquad\qquad\qquad \sqrt[4]{6}\cdot 2\sqrt[4]{5}$

$\qquad =6\sqrt[4]{63}+4\sqrt[4]{35}-3\sqrt[4]{54}-2\sqrt[4]{30}$

61. $\dfrac{6}{3-\sqrt{2}}=\dfrac{6}{3-\sqrt{2}}\cdot\dfrac{3+\sqrt{2}}{3+\sqrt{2}}=\dfrac{18+6\sqrt{2}}{9-2}$

$\qquad =\dfrac{18+6\sqrt{2}}{7}$

63. $\dfrac{2+\sqrt{5}}{6+\sqrt{3}}=\dfrac{2+\sqrt{5}}{6+\sqrt{3}}\cdot\dfrac{6-\sqrt{3}}{6-\sqrt{3}}$

$\qquad =\dfrac{12-2\sqrt{3}+6\sqrt{5}-\sqrt{15}}{36-3}$

$\qquad =\dfrac{12-2\sqrt{3}+6\sqrt{5}-\sqrt{15}}{33}$

65. $\dfrac{\sqrt{a}}{\sqrt{a}+\sqrt{b}}=\dfrac{\sqrt{a}}{\sqrt{a}+\sqrt{b}}\cdot\dfrac{\sqrt{a}-\sqrt{b}}{\sqrt{a}-\sqrt{b}}$

$\qquad =\dfrac{\sqrt{a}\left(\sqrt{a}-\sqrt{b}\right)}{\left(\sqrt{a}+\sqrt{b}\right)\left(\sqrt{a}-\sqrt{b}\right)}=\dfrac{a-\sqrt{ab}}{a-b}$

67. $\dfrac{\sqrt{7}-\sqrt{3}}{\sqrt{3}-\sqrt{7}}=\dfrac{-1\left(\sqrt{3}-\sqrt{7}\right)}{\sqrt{3}-\sqrt{7}}$

$\qquad =-1\cdot\dfrac{\sqrt{3}-\sqrt{7}}{\sqrt{3}-\sqrt{7}}=-1\cdot 1=-1$

69. $\dfrac{3\sqrt{2}-\sqrt{7}}{4\sqrt{2}+2\sqrt{5}} = \dfrac{3\sqrt{2}-\sqrt{7}}{4\sqrt{2}+2\sqrt{5}} \cdot \dfrac{4\sqrt{2}-2\sqrt{5}}{4\sqrt{2}-2\sqrt{5}}$

$= \dfrac{24-6\sqrt{10}-4\sqrt{14}+2\sqrt{35}}{32-20}$

$= \dfrac{\cancel{2}\left(12-3\sqrt{10}-2\sqrt{14}+\sqrt{35}\right)}{\cancel{2}\cdot 6}$

$= \dfrac{12-3\sqrt{10}-2\sqrt{14}+\sqrt{35}}{6}$

71. $\dfrac{\sqrt{5}+1}{4} = \dfrac{\sqrt{5}+1}{4}\cdot\dfrac{\sqrt{5}-1}{\sqrt{5}-1} = \dfrac{5-1}{4\left(\sqrt{5}-1\right)}$

$= \dfrac{4}{4\left(\sqrt{5}-1\right)} = \dfrac{1}{\sqrt{5}-1}$

73. $\dfrac{\sqrt{6}-2}{\sqrt{3}+7} = \dfrac{\sqrt{6}-2}{\sqrt{3}+7}\cdot\dfrac{\sqrt{6}+2}{\sqrt{6}+2}$

$= \dfrac{6-4}{\sqrt{18}+2\sqrt{3}+7\sqrt{6}+14}$

$= \dfrac{2}{3\sqrt{2}+2\sqrt{3}+7\sqrt{6}+14}$

75. $\dfrac{\sqrt{x}-\sqrt{y}}{\sqrt{x}+\sqrt{y}} = \dfrac{\sqrt{x}-\sqrt{y}}{\sqrt{x}+\sqrt{y}}\cdot\dfrac{\sqrt{x}+\sqrt{y}}{\sqrt{x}+\sqrt{y}}$

$= \dfrac{x-y}{x+2\sqrt{xy}+y}$

77. $\dfrac{\sqrt{a+h}-\sqrt{a}}{h} = \dfrac{\sqrt{a+h}-\sqrt{a}}{h}\cdot\dfrac{\sqrt{a+h}+\sqrt{a}}{\sqrt{a+h}+\sqrt{a}}$

$= \dfrac{(a+h)-a}{h\left(\sqrt{a+h}+\sqrt{a}\right)}$

$= \dfrac{\cancel{h}\cdot 1}{\cancel{h}\left(\sqrt{a+h}+\sqrt{a}\right)}$

$= \dfrac{1}{\sqrt{a+h}+\sqrt{a}}$

79. $\sqrt[3]{a}\,\sqrt[6]{a}$

$= a^{1/3}\cdot a^{1/6}$

Converting to exponential notation

$= a^{3/6}$ Adding exponents

$= a^{1/2}$ Simplifying exponent

$= \sqrt{a}$ Returning to radical notation

81. $\sqrt[5]{b^2}\,\sqrt{b^3}$

$= b^{2/5}\cdot b^{3/2}$

Converting to exponential notation

$= b^{19/10}$ Adding exponents

$= b^{1+9/10}$ Writing 19/10 as a mixed number

$= b\cdot b^{9/10}$ Factoring

$= b\sqrt[10]{b^9}$ Returning to radical notation

83. $\sqrt{xy^3}\,\sqrt[3]{x^2 y} = \left(xy^3\right)^{1/2}\left(x^2 y\right)^{1/3}$

$= \left(xy^3\right)^{3/6}\left(x^2 y\right)^{2/6}$

$= \left[\left(xy^3\right)^3\left(x^2 y\right)^2\right]^{1/6}$

$= \sqrt[6]{x^3 y^9 \cdot x^4 y^2}$

$= \sqrt[6]{x^7 y^{11}}$

$= \sqrt[6]{x^6 y^6 \cdot xy^5}$

$= xy\sqrt[6]{xy^5}$

85. $\sqrt[4]{9ab^3}\,\sqrt{3a^4 b} = \left(9ab^3\right)^{1/4}\left(3a^4 b\right)^{1/2}$

$= \left(9ab^3\right)^{1/4}\left(3a^4 b\right)^{2/4}$

$= \left[\left(9ab^3\right)\left(3a^4 b\right)^2\right]^{1/4}$

$= \sqrt[4]{9ab^3 \cdot 9a^8 b^2}$

$= \sqrt[4]{81a^9 b^5}$

$= \sqrt[4]{81a^8 b^4 \cdot ab}$

$= 3a^2 b\sqrt[4]{ab}$

87. $\sqrt{a^4b^3c^4}\,\sqrt[3]{ab^2c} = \left(a^4b^3c^4\right)^{1/2}\left(ab^2c\right)^{1/3}$

$\qquad\qquad = \left(a^4b^3c^4\right)^{3/6}\left(ab^2c\right)^{2/6}$

$\qquad\qquad = \left[\left(a^4b^3c^4\right)^3\left(ab^2c\right)^2\right]^{1/6}$

$\qquad\qquad = \sqrt[6]{a^{12}b^9c^{12}\cdot a^2b^4c^2}$

$\qquad\qquad = \sqrt[6]{a^{14}b^{13}c^{14}}$

$\qquad\qquad = \sqrt[6]{a^{12}b^{12}c^{12}\cdot a^2bc^2}$

$\qquad\qquad = a^2b^2c^2\sqrt[6]{a^2bc^2}$

89. $\dfrac{\sqrt[3]{a^2}}{\sqrt[4]{a}}$

$= \dfrac{a^{2/3}}{a^{1/4}}$ Converting to exponential notation

$= a^{2/3-1/4}$ Subtracting exponents

$= a^{5/12}$ Converting back

$= \sqrt[12]{a^5}$ to radical notation

91. $\dfrac{\sqrt[4]{x^2y^3}}{\sqrt[3]{xy}}$

$= \dfrac{\left(x^2y^3\right)^{1/4}}{\left(xy\right)^{1/3}}$ Converting to exponential notation

$= \dfrac{x^{2/4}y^{3/4}}{x^{1/3}y^{1/3}}$ Using the power and product rule

$= x^{2/4-1/3}\,y^{3/4-1/3}$ Subtracting exponents

$= x^{2/12}y^{5/12}$ Converting back

$= \sqrt[12]{x^2y^5}$ to radical notation

93. $\dfrac{\sqrt{ab^3}}{\sqrt[5]{a^2b^3}}$

$= \dfrac{\left(ab^3\right)^{1/2}}{\left(a^2b^3\right)^{1/5}}$ Converting to exponential notation

$= \dfrac{a^{1/2}b^{3/2}}{a^{2/5}b^{3/5}}$ Using the power rule

$= a^{1/10}b^{9/10}$ Subtracting exponents

$= \left(ab^9\right)^{1/10}$ Converting back

$= \sqrt[10]{ab^9}$ to radical notation

95. $\dfrac{\sqrt{(7-y)^3}}{\sqrt[3]{(7-y)^2}}$

$= \dfrac{(7-y)^{3/2}}{(7-y)^{2/3}}$ Converting to exponential notation

$= (7-y)^{3/2-2/3}$ Subtracting exponents

$= (7-y)^{5/6}$ Converting back

$= \sqrt[6]{(7-y)^5}$ to radical notation

97. $\dfrac{\sqrt[4]{(5+3x)^3}}{\sqrt[3]{(5+3x)^2}}$

$= \dfrac{(5+3x)^{3/4}}{(5+3x)^{2/3}}$ Converting to exponential notation

$= (5+3x)^{3/4-2/3}$ Subtracting exponents

$= (5+3x)^{1/12}$ Converting back

$= \sqrt[12]{5+3x}$ to radical notation

99. $\sqrt[3]{x^2y}\left(\sqrt{xy}-\sqrt[5]{xy^3}\right)$

$= \left(x^2y\right)^{1/3}\left[\left(xy\right)^{1/2}-\left(xy^3\right)^{1/5}\right]$

$= x^{2/3}y^{1/3}\left(x^{1/2}y^{1/2}-x^{1/5}y^{3/5}\right)$

$= x^{2/3}y^{1/3}x^{1/2}y^{1/2}-x^{2/3}y^{1/3}x^{1/5}y^{3/5}$

$= x^{2/3+1/2}y^{1/3+1/2}-x^{2/3+1/5}y^{1/3+3/5}$

$= x^{7/6}y^{5/6}-x^{13/15}y^{14/15}$

$= x^{1\frac{1}{6}}y^{\frac{5}{6}}-x^{13/15}y^{14/15}$

Writing a mixed numeral

$= x\cdot x^{\frac{1}{6}}y^{\frac{5}{6}}-x^{13/15}y^{14/15}$

$= x\left(xy^5\right)^{1/6}-\left(x^{13}y^{14}\right)^{1/15}$

$= x\sqrt[6]{xy^5}-\sqrt[15]{x^{13}y^{14}}$

101. $\left(m+\sqrt[3]{n^2}\right)\left(2m+\sqrt[4]{n}\right)$

$=\left(m+n^{2/3}\right)\left(2m+n^{1/4}\right)$

   Converting to exponential notation

$=2m^2+mn^{1/4}+2mn^{2/3}+n^{2/3}n^{1/4}$

   Using FOIL

$=2m^2+mn^{1/4}+2mn^{2/3}+n^{2/3+1/4}$

   Adding exponents

$=2m^2+mn^{1/4}+2mn^{2/3}+n^{11/12}$

$=2m^2+m\sqrt[4]{n}+2m\sqrt[3]{n^2}+\sqrt[12]{n^{11}}$

   Converting back to radical notation

103. $f(x)=\sqrt[4]{x},\ g(x)=2\sqrt{x}-\sqrt[3]{x^2}$

$(f\cdot g)(x)=\sqrt[4]{x}\left(2\sqrt{x}-\sqrt[3]{x^2}\right)$

$=\sqrt[4]{x}\left(2\sqrt{x}\right)-\left(\sqrt[4]{x}\right)\sqrt[3]{x^2}$

$=2x^{\frac{1}{4}+\frac{1}{2}}-x^{\frac{1}{4}+\frac{2}{3}}=2x^{\frac{3}{4}}-x^{\frac{11}{12}}$

$=2\sqrt[4]{x^3}-\sqrt[12]{x^{11}}$

105. $f(x)=x+\sqrt{7},\ g(x)=x-\sqrt{7}$

$(f\cdot g)(x)=\left(x+\sqrt{7}\right)\left(x-\sqrt{7}\right)$

$=x^2-\left(\sqrt{7}\right)^2$

$=x^2-7$

107. $f(x)=x^2$

$f\left(5+\sqrt{2}\right)=\left(5+\sqrt{2}\right)^2$

$=5^2+2\cdot5\sqrt{2}+\left(\sqrt{2}\right)^2$

$=25+10\sqrt{2}+2$

$=27+10\sqrt{2}$

109. $f(x)=x^2$

$f\left(\sqrt{3}-\sqrt{5}\right)=\left(\sqrt{3}-\sqrt{5}\right)^2$

$=\left(\sqrt{3}\right)^2-2\cdot\sqrt{3}\cdot\sqrt{5}+\left(\sqrt{5}\right)^2$

$=3-2\sqrt{15}+5$

$=8-2\sqrt{15}$

111. *Thinking and Writing Exercise.*

113. $3x-1=125$

$3x=125+1=126$

$x=\dfrac{126}{3}=42$

115. $x^2+2x+1=22-2x$

$x^2+2x+2x-22=0$

$x^2+4x-21=0$

$(x+7)(x-3)=0$

$x+7=0\quad or\quad x-3=0$

$x=-7\quad or\quad x=3$

117. $\dfrac{1}{x}+\dfrac{1}{2}=\dfrac{1}{6}$

$6x\cdot\dfrac{1}{x}+6x\cdot\dfrac{1}{2}=6x\cdot\dfrac{1}{6}$

$6+3x=x$

$2x=-6$

$x=-3$

119. *Thinking and Writing Exercise.*

121. $f(x)=$

$=\sqrt{x^3-x^2}+\sqrt{9x^3-9x^2}-\sqrt{4x^3-4x^2}$

$=\sqrt{x^2(x-1)}+\sqrt{9x^2(x-1)}-\sqrt{4x^2(x-1)}$

$=x\sqrt{x-1}+3x\sqrt{x-1}-2x\sqrt{x-1}$

$=2x\sqrt{x-1}$

123. $f(x)=\sqrt[4]{x^5-x^4}+3\sqrt[4]{x^9-x^8}$

$=\sqrt[4]{x^4(x-1)}+3\sqrt[4]{x^8(x-1)}$

$=\sqrt[4]{x^4}\cdot\sqrt[4]{x-1}+3\sqrt[4]{x^8}\cdot\sqrt[4]{x-1}$

$=x\sqrt[4]{x-1}+3x^2\sqrt[4]{x-1}$

$=\left(x+3x^2\right)\sqrt[4]{x-1}$

125. $7x\sqrt{(x+y)^3} - 5xy\sqrt{x+y} - 2y\sqrt{(x+y)^3}$

$= 7x\sqrt{(x+y)^2(x+y)} - 5xy\sqrt{x+y} -$
$\qquad\qquad 2y\sqrt{(x+y)^2(x+y)}$

$= 7x(x+y)\sqrt{x+y} - 5xy\sqrt{x+y} -$
$\qquad\qquad 2y(x+y)\sqrt{x+y}$

$= \left[7x(x+y) - 5xy - 2y(x+y)\right]\sqrt{x+y}$

$= \left(7x^2 + 7xy - 5xy - 2xy - 2y^2\right)\sqrt{x+y}$

$= \left(7x^2 - 2y^2\right)\sqrt{x+y}$

127. $\sqrt{8x(y+z)^5}\,\sqrt[3]{4x^2(y+z)^2}$

$= \left[8x(y+z)^5\right]^{1/2}\left[4x^2(y+z)^2\right]^{1/3}$

$= \left[8x(y+z)^5\right]^{3/6}\left[4x^2(y+z)^2\right]^{2/6}$

$= \left\{\left[2^3 x(y+z)^5\right]^3\left[2^2 x^2(y+z)^2\right]^2\right\}^{1/6}$

$= \sqrt[6]{2^9 x^3(y+z)^{15}\cdot 2^4 x^4(y+z)^4}$

$= \sqrt[6]{2^{13} x^7(y+z)^{19}}$

$= \sqrt[6]{2^{12} x^6(y+z)^{18}\cdot 2x(y+z)}$

$= 2^2 x(y+z)^3\sqrt[6]{2x(y+z)}$, or

$\quad 4x(y+z)^3\sqrt[6]{2x(y+z)}$

129. $\dfrac{\frac{1}{\sqrt{w}}-\sqrt{w}}{\frac{\sqrt{w}+1}{\sqrt{w}}} = \dfrac{\frac{1}{\sqrt{w}}-\sqrt{w}}{\frac{\sqrt{w}+1}{\sqrt{w}}}\cdot\dfrac{\sqrt{w}}{\sqrt{w}} = \dfrac{1-w}{\sqrt{w}+1}$

$= \dfrac{1-w}{\sqrt{w}+1}\cdot\dfrac{\sqrt{w}-1}{\sqrt{w}-1}$

$= \dfrac{\sqrt{w}-1-w\sqrt{w}+w}{w-1}$

$= \dfrac{(w-1)-\sqrt{w}(w-1)}{w-1}$

$= \dfrac{(w-1)(1-\sqrt{w})}{w-1} = 1-\sqrt{w}$

131. $x-5 = \left(\sqrt{x}\right)^2 - \left(\sqrt{5}\right)^2 = \left(\sqrt{x}+\sqrt{5}\right)\left(\sqrt{x}-\sqrt{5}\right)$

133. $x-a = \left(\sqrt{x}\right)^2 - \left(\sqrt{a}\right)^2$

$= \left(\sqrt{x}+\sqrt{a}\right)\left(\sqrt{x}-\sqrt{a}\right)$

135. $\left(\sqrt{x+2}-\sqrt{x-2}\right)^2$

$= \left(\sqrt{x+2}\right)^2 - 2\sqrt{x+2}\sqrt{x-2} + \left(\sqrt{x-2}\right)^2$

$= x+2 - 2\sqrt{(x+2)(x-2)} + x - 2$

$= 2x - 2\sqrt{x^2-4}$

Mid-Chapter Review

Guided Solutions

1. $\sqrt{6x^9}\cdot\sqrt{2xy} = \sqrt{6x^9\cdot 2xy}$

$= \sqrt{12x^{10}y}$

$= \sqrt{4x^{10}\cdot 3y}$

$= \sqrt{4x^{10}}\cdot\sqrt{3y}$

$= 2x^5\sqrt{3y}$

2. $\sqrt{12}-3\sqrt{75}+\sqrt{8} = 2\sqrt{3}-3\cdot 5\sqrt{3}+2\sqrt{2}$

$= 2\sqrt{3}-15\sqrt{3}+2\sqrt{2}$

$= -13\sqrt{3}+2\sqrt{2}$

Mixed Review

1. $\sqrt{81} = \sqrt{9\cdot 9} = 9$

2. $-\sqrt{\dfrac{9}{100}} = -\sqrt{\dfrac{3\cdot 3}{10\cdot 10}} = -\dfrac{3}{10}$

3. $\sqrt{64t^2} = \sqrt{(8t)^2} = |8t| = 8|t|$

4. $\sqrt[5]{x^5} = x$

5. $f(x) = \sqrt[3]{12x-4}$

$f(-5) = \sqrt[3]{12(-5)-4}$

$= \sqrt[3]{-60-4}$

$= \sqrt[3]{-64} = -4$

6. $g(x) = \sqrt[4]{10-x}$

    Since the index is even, the radicand, $10-x$,
    must be non-negative.

    $$10 - x \geq 0$$
    $$10 \geq x$$
    $$x \leq 10$$

    Domain of $f = \{x | x \leq 10\}$, or $(-\infty, 10]$

7. $8^{2/3} = \left(\sqrt[3]{8}\right)^2 = 2^2 = 4$

8. $\sqrt[6]{\sqrt{a}} = \sqrt[6]{a^{1/2}} = \left(a^{1/2}\right)^{1/6} = a^{1/12} = \sqrt[12]{a}$

9. $\sqrt[3]{y^{24}} = y^{24/3} = y^8$

10. $\sqrt{(t+5)^2} = t+5 \ \left(\text{for } t+5 \geq 0\right)$

11. $\sqrt[3]{-27a^{12}} = \sqrt[3]{\left(-3a^4\right)^3} = -3a^4$

12. $\sqrt{6x}\sqrt{15x} = \sqrt{6x \cdot 15x}$
    $$= \sqrt{90x^2}$$
    $$= \sqrt{9x^2 \cdot 10}$$
    $$= 3x\sqrt{10} \ \left(\text{for } x \geq 0\right)$$

13. $\dfrac{\sqrt{20y}}{\sqrt{45y}} = \sqrt{\dfrac{20y}{45y}} = \sqrt{\dfrac{4 \cdot 5y}{9 \cdot 5y}} = \sqrt{\dfrac{4}{9}} = \dfrac{2}{3}$

14. $\sqrt{15t} + 4\sqrt{15t} = (1+4)\sqrt{15t} = 5\sqrt{15t}$

15. $\sqrt[5]{a^5 b^{10} c^{11}} = \sqrt[5]{a^5} \sqrt[5]{b^{10}} \sqrt[5]{c^{11}}$
    $$= a\sqrt[5]{\left(b^2\right)^5} \sqrt[5]{c^{10} \cdot c}$$
    $$= ab^2 \sqrt[5]{\left(c^2\right)^5} \sqrt[5]{c}$$
    $$= ab^2 c^2 \sqrt[5]{c}$$

16. $\sqrt{6}\left(\sqrt{10} - \sqrt{33}\right) = \sqrt{6} \cdot \sqrt{10} - \sqrt{6} \cdot \sqrt{33}$
    $$= \sqrt{60} - \sqrt{198}$$
    $$= \sqrt{4 \cdot 15} - \sqrt{9 \cdot 22}$$
    $$= 2\sqrt{15} - 3\sqrt{22}$$

17. $\dfrac{\sqrt{t}}{\sqrt[8]{t^3}} = \dfrac{t^{1/2}}{t^{3/8}} = t^{4/8 - 3/8} = t^{1/8} = \sqrt[8]{t}$

18. $\sqrt[5]{\dfrac{3a^{12}}{96a^2}} = \sqrt[5]{\dfrac{a^{10}}{32}} = \sqrt[5]{\dfrac{\left(a^2\right)^5}{2^5}} = \dfrac{a^2}{2}$

19. $2\sqrt{3} - 5\sqrt{12} = 2\sqrt{3} - 5\sqrt{4 \cdot 3}$
    $$= 2\sqrt{3} - 5 \cdot 2\sqrt{3}$$
    $$= 2\sqrt{3} - 10\sqrt{3}$$
    $$= -8\sqrt{3}$$

20. $\left(\sqrt{5}+3\right)\left(\sqrt{5}-3\right) = \left(\sqrt{5}\right)^2 - 3^2 = 5 - 9 = -4$

21. $\left(\sqrt{15} + \sqrt{10}\right)^2 = \left(\sqrt{15}\right)^2 + 2\sqrt{15}\sqrt{10} + \left(\sqrt{10}\right)^2$
    $$= 15 + 2\sqrt{150} + 10$$
    $$= 25 + 2\sqrt{25 \cdot 6}$$
    $$= 25 + 2 \cdot 5\sqrt{6}$$
    $$= 25 + 10\sqrt{6}$$

22. $\sqrt{25x - 25} - \sqrt{9x - 9}$
    $$= \sqrt{25(x-1)} - \sqrt{9(x-1)}$$
    $$= 5\sqrt{x-1} - 3\sqrt{x-1}$$
    $$= 2\sqrt{x-1}$$

23. $\sqrt{x^3 y} \sqrt[5]{xy^4} = x\sqrt{xy} \sqrt[5]{xy^4} = x \cdot x^{\frac{1}{2}} y^{\frac{1}{2}} \cdot x^{\frac{1}{5}} y^{\frac{4}{5}}$
    $$= x \cdot x^{\frac{1}{2} + \frac{1}{5}} y^{\frac{1}{2} + \frac{4}{5}} = x \cdot x^{\frac{7}{10}} y^{\frac{13}{10}}$$
    $$= x \cdot x^{\frac{7}{10}} \cdot y^1 \cdot y^{\frac{3}{10}} = xy\sqrt[10]{x^7 y^3}$$

24. $\sqrt[3]{5000} + \sqrt[3]{625} = \sqrt[3]{1000 \cdot 5} + \sqrt[3]{125 \cdot 5}$
    $$= 10\sqrt[3]{5} + 5\sqrt[3]{5}$$
    $$= 15\sqrt[3]{5}$$

25. $\sqrt[3]{12x^2 y^5} \sqrt[3]{18x^7 y} = y\sqrt[3]{12x^2 y^2} \cdot x^2 \sqrt[3]{18xy}$
    $$= x^2 y\sqrt[3]{\left(2^2 \cdot 3x^2 y^2\right)\left(2 \cdot 3^2 xy\right)} = x^2 y\sqrt[3]{2^3 3^3 x^3 y^3}$$
    $$= x^2 y \cdot 2 \cdot 3 \cdot xy = 6x^3 y^2$$

1. False

3. True

5. True

7. $\sqrt{5x+1} = 4$

$\left(\sqrt{5x+1}\right)^2 = 4^2$ Principle of powers (squaring)

$5x+1 = 16$

$5x = 15$

$x = \dfrac{15}{5} = 3$

Check:

$$\begin{array}{c|c} \sqrt{5x+1} = 4 & \\ \hline \sqrt{5\cdot 3+1} & 4 \\ \sqrt{15+1} & \\ \sqrt{16} & \\ 4 & 4 \quad \text{TRUE} \end{array}$$

The solution is 3.

9. $\sqrt{3x}+1 = 6$

$\sqrt{3x} = 5$ Sub. to isolate the radical

$\left(\sqrt{3x}\right)^2 = 5^2$ Principle of powers (squaring)

$3x = 25$

$x = \dfrac{25}{3}$

Check:

$$\begin{array}{c|c} \sqrt{3x}+1 = 6 & \\ \hline \sqrt{3\cdot\dfrac{25}{3}+1} & 6 \\ 5+1 & \\ 6 & 6 \quad \text{TRUE} \end{array}$$

The solution is $\dfrac{25}{3}$.

11. $\sqrt{y+1}-5 = 8$

$\sqrt{y+1} = 13$

Adding to isolate the radical

$\left(\sqrt{y+1}\right)^2 = 13^2$

Principle of powers (squaring)

$y+1 = 169$

$y = 168$

Check:

$$\begin{array}{c|c} \sqrt{y+1}-5 = 8 & \\ \hline \sqrt{168+1}-5 & 8 \\ 13-5 & \\ 8 & 8 \quad \text{TRUE} \end{array}$$

The solution is 168.

13. $\sqrt{8-x}+7 = 10$

$\sqrt{8-x} = 3$ Isolate the radical

$\left(\sqrt{8-x}\right)^2 = 3^2$ Principle of powers

(squaring)

$8-x = 9$

$-x = 1$

$x = -1$

Check:

$$\begin{array}{c|c} \sqrt{8-x}+7 = 10 & \\ \hline \sqrt{8-(-1)}+7 & 10 \\ \sqrt{9}+7 & \\ 3+7 & \\ 10 & 10 \quad \text{TRUE} \end{array}$$

The solution is -1.

15. $\sqrt[3]{x+5} = 2$

$\left(\sqrt[3]{x+5}\right)^3 = 2^3$

$x+5 = 8$

$x = 3$

Check:

$$\begin{array}{c|c} \sqrt[3]{x+5} = 2 & \\ \hline \sqrt[3]{3+5} & 2 \\ \sqrt[3]{8} & \\ 2 & 2 \quad \text{TRUE} \end{array}$$

The solution is 3.

17. $\sqrt[4]{y-1} = 3$

$$\left(\sqrt[4]{y-1}\right)^4 = 3^4$$

$$y - 1 = 81$$

$$y = 82$$

Check:

$$\frac{\sqrt[4]{y-1} = 3}{\begin{array}{c|c} \sqrt[4]{82-1} & 3 \\ \sqrt[4]{81} & \\ 3 & 3 \end{array}} \quad \text{TRUE}$$

The solution is 82.

19. $3\sqrt{x} = x$

$$\left(3\sqrt{x}\right)^2 = x^2$$

$$9x = x^2$$

$$0 = x^2 - 9x$$

$$0 = x(x-9)$$

$$x = 0 \ or \ x = 9$$

Check:

For 0: $\dfrac{3\sqrt{x} = x}{\begin{array}{c|c} 3\sqrt{0} & 0 \\ 3\cdot 0 & \\ 0 & 0 \end{array}}$ TRUE

For 9: $\dfrac{3\sqrt{x} = x}{\begin{array}{c|c} 3\sqrt{9} & 9 \\ 3\cdot 3 & \\ 9 & 9 \end{array}}$ TRUE

The solutions are 0 and 9.

21. $2y^{1/2} - 13 = 7$

$$2\sqrt{y} - 13 = 7$$

$$2\sqrt{y} = 20$$

$$\sqrt{y} = 10$$

$$\left(\sqrt{y}\right)^2 = 10^2$$

$$y = 100$$

Check:

$$\frac{2y^{1/2} - 13 = 7}{\begin{array}{c|c} 2\cdot 100^{1/2} - 13 & 7 \\ 2\cdot 10 - 13 & \\ 7 & 7 \end{array}} \quad \text{TRUE}$$

The solution is 100.

23. $\sqrt[3]{x} = -3$

$$\left(\sqrt[3]{x}\right)^3 = (-3)^3$$

$$x = -27$$

Check:

$$\frac{\sqrt[3]{x} = -3}{\begin{array}{c|c} \sqrt[3]{-27} & -3 \\ -3 & -3 \end{array}} \quad \text{TRUE}$$

The solution is −27.

25. $z^{1/4} + 8 = 10$

$$z^{1/4} = 2$$

$$\left(z^{1/4}\right)^4 = 2^4$$

$$z = 16$$

Check:

$$\frac{z^{1/4} + 8 = 10}{\begin{array}{c|c} 16^{1/4} + 8 & 10 \\ 2 + 8 & \\ 10 & 10 \end{array}} \quad \text{TRUE}$$

The solution is 16.

27. $\sqrt{n} = -2$

This equation has no solution, since the principal square root is never negative.

29. $\sqrt[4]{3x+1} - 4 = -1$

$$\sqrt[4]{3x+1} = 3$$

$$\left(\sqrt[4]{3x+1}\right)^4 = 3^4$$

$$3x + 1 = 81$$

$$3x = 80$$

$$x = \frac{80}{3}$$

Check:

$$\frac{\sqrt[4]{3x+1}-4=-1}{\begin{array}{c|c} \sqrt[4]{3\cdot\dfrac{80}{3}+1}-4 & -1 \\ \sqrt[4]{81}-4 & \\ 3-4 & \\ -1 & -1 \quad \text{TRUE} \end{array}}$$

The solution is $\dfrac{80}{3}$.

31. $\quad (21x+55)^{1/3}=10$

$$\left[(21x+55)^{1/3}\right]^3=10^3$$

$$21x+55=1000$$

$$21x=945$$

$$x=\frac{945}{21}=45$$

Check:

$$\frac{(21x+55)^{1/3}=10}{\begin{array}{c|c} (21\cdot45+55)^{1/3} & 10 \\ (945+55)^{1/3} & \\ 1000^{1/3} & \\ 10 & 10 \quad \text{TRUE} \end{array}}$$

The solution is 45.

33. $\quad \sqrt[3]{3y+6}+7=8$

$$\sqrt[3]{3y+6}=1$$

$$\left(\sqrt[3]{3y+6}\right)^3=1^3$$

$$3y+6=1$$

$$3y=-5$$

$$y=-\frac{5}{3}$$

Check:

$$\frac{\sqrt[3]{3y+6}+7=8}{\begin{array}{c|c} \sqrt[3]{3\left(-\dfrac{5}{3}\right)+6}+7 & 8 \\ \sqrt[3]{1}+7 & \\ 1+7 & \\ 8 & 8 \quad \text{TRUE} \end{array}}$$

The solution is $-\dfrac{5}{3}$.

35. $\quad \sqrt{3t+4}=\sqrt{4t+3}$

$$\left(\sqrt{3t+4}\right)^2=\left(\sqrt{4t+3}\right)^2$$

$$3t+4=4t+3$$

$$4=t+3$$

$$1=t$$

Check:

$$\frac{\sqrt{3t+4}=\sqrt{4t+3}}{\begin{array}{c|c} \sqrt{3\cdot1+4} & \sqrt{4\cdot1+3} \\ \sqrt{7} & \sqrt{7} \quad \text{TRUE} \end{array}}$$

The solution is 1.

37. $\quad 3(4-t)^{1/4}=6^{1/4}$

$$\left[3(4-t)^{1/4}\right]^4=\left(6^{1/4}\right)^4$$

$$81(4-t)=6$$

$$324-81t=6$$

$$-81t=-318$$

$$t=\frac{106}{27}$$

The number $\dfrac{106}{27}$ checks and is the solution.

39. $\quad 3+\sqrt{5-x}=x$

$$\sqrt{5-x}=x-3$$

$$\left(\sqrt{5-x}\right)^2=(x-3)^2$$

$$5-x=x^2-6x+9$$

$$0=x^2-5x+4$$

$$0=(x-1)(x-4)$$

$$x-1=0 \; or \; x-4=0$$

$$x=1 \; or \quad x=4$$

Check:

For 1:
$$\frac{3+\sqrt{5-x}=x}{\begin{array}{c|c} 3+\sqrt{5-1} & 1 \\ 3+\sqrt{4} & \\ 3+2 & \\ 5 & 1 \quad \text{FALSE} \end{array}}$$

For 4:

$$\frac{3+\sqrt{5-x}=x}{3+\sqrt{5-4}\ \bigg|\ 4}$$

$$\begin{array}{c|c} 3+\sqrt{1} & \\ 3+1 & \\ 4 & 4 \quad \text{TRUE} \end{array}$$

Since 4 checks but 1 does not, the solution is 4.

41. $\sqrt{4x-3}=2+\sqrt{2x-5}$ One radical is already isolated.

$\left(\sqrt{4x-3}\right)^2=\left(2+\sqrt{2x-5}\right)^2$ Squaring both sides

$4x-3=4+4\sqrt{2x-5}+2x-5$

$2x-2=4\sqrt{2x-5}$

$x-1=2\sqrt{2x-5}$

$x^2-2x+1=8x-20$

$x^2-10x+21=0$

$(x-7)(x-3)=0$

$x-7=0 \ or \ x-3=0$

$x=7 \ or \ \ \ x=3$

Both numbers check. The solutions are 7 and 3.

43. $\sqrt{20-x}+8=\sqrt{9-x}+11$

$\sqrt{20-x}=\sqrt{9-x}+3$ Isolating one radical

$\left(\sqrt{20-x}\right)^2=\left(\sqrt{9-x}+3\right)^2$ Squaring both sides

$20-x=9-x+6\sqrt{9-x}+9$

$2=6\sqrt{9-x}$ Isolating the remaining Radical

$1=3\sqrt{9-x}$ Multiplying by $\frac{1}{2}$

$1^2=\left(3\sqrt{9-x}\right)^2$ Squaring both sides

$1=9(9-x)$

$1=81-9x$

$-80=-9x$

$\frac{80}{9}=x$

The number $\frac{80}{9}$ checks and is the solution.

45. $\sqrt{x+2}+\sqrt{3x+4}=2$

$\sqrt{x+2}=2-\sqrt{3x+4}$ Isolating one radical

$\left(\sqrt{x+2}\right)^2=\left(2-\sqrt{3x+4}\right)^2$

$x+2=4-4\sqrt{3x+4}+3x+4$

$-2x-6=-4\sqrt{3x+4}$ Isolating the remaining radical

$x+3=2\sqrt{3x+4}$ Multiplying by $-\frac{1}{2}$

$(x+3)^2=\left(2\sqrt{3x+4}\right)^2$

$x^2+6x+9=4(3x+4)$

$x^2+6x+9=12x+16$

$x^2-6x-7=0$

$(x-7)(x+1)=0$

$x-7=0 \ or \ x+1=0$

$x=7 \ or \ \ \ x=-1$

Check:

For 7:

$$\frac{\sqrt{x+2}+\sqrt{3x+4}=2}{\sqrt{7+2}+\sqrt{3\cdot 7+4}\ \bigg|\ 2}$$

$$\begin{array}{c|c} \sqrt{9}+\sqrt{25} & \\ 8 & 2 \quad \text{FALSE} \end{array}$$

For -1:

$$\frac{\sqrt{x+2}+\sqrt{3x+4}=2}{\sqrt{-1+2}+\sqrt{3\cdot(-1)+4}\ \bigg|\ 2}$$

$$\begin{array}{c|c} \sqrt{1}+\sqrt{1} & \\ 2 & 2 \quad \text{TRUE} \end{array}$$

Since -1 checks but 7 does not, the solution is -1.

47. We must have $f(x) = 1$ or $\sqrt{x} + \sqrt{x-9} = 1$.

$$\sqrt{x} + \sqrt{x-9} = 1$$

$$\sqrt{x-9} = 1 - \sqrt{x} \quad \text{Isolating one radical term.}$$

$$\left(\sqrt{x-9}\right)^2 = \left(1 - \sqrt{x}\right)^2$$

$$x - 9 = 1 - 2\sqrt{x} + x$$

$$-10 = -2\sqrt{x} \quad \text{Isolating the remaining radical term}$$

$$5 = \sqrt{x}$$

$$25 = x$$

This value does not check. There is no solution, so there is no value of x for which $f(x) = 1$.

49. We must have $f(t) = -3$ or

$$\sqrt{t-2} - \sqrt{4t+1} = -3.$$

$$\sqrt{t-2} - \sqrt{4t+1} = -3$$

$$\sqrt{t-2} = \sqrt{4t+1} - 3$$

$$\left(\sqrt{t-2}\right)^2 = \left(\sqrt{4t+1} - 3\right)^2$$

$$t - 2 = 4t + 1 - 6\sqrt{4t+1} + 9$$

$$-3t - 12 = -6\sqrt{4t+1}$$

$$t + 4 = 2\sqrt{4t+1}$$

$$\left(t+4\right)^2 = \left(2\sqrt{4t+1}\right)^2$$

$$t^2 + 8t + 16 = 4(4t+1)$$

$$t^2 + 8t + 16 = 16t + 4$$

$$t^2 - 8t + 12 = 0$$

$$(t-2)(t-6) = 0$$

$$t - 2 = 0 \text{ or } t - 6 = 0$$

$$t = 2 \text{ or } \quad t = 6$$

Both numbers check, so we have $f(t) = -3$ when $t = 2$ and when $t = 6$.

51. We must have $\sqrt{2x-3} = \sqrt{x+7} - 2$.

$$\sqrt{2x-3} = \sqrt{x+7} - 2$$

$$\left(\sqrt{2x-3}\right)^2 = \left(\sqrt{x+7} - 2\right)^2$$

$$2x - 3 = x + 7 - 4\sqrt{x+7} + 4$$

$$x - 14 = -4\sqrt{x+7}$$

$$\left(x-14\right)^2 = \left(-4\sqrt{x+7}\right)^2$$

$$x^2 - 28x + 196 = 16(x+7)$$

$$x^2 - 28x + 196 = 16x + 112$$

$$x^2 - 44x + 84 = 0$$

$$(x-2)(x-42) = 0$$

$$x = 2 \text{ or } x = 42$$

Since 2 checks but 42 does not, we have $f(x) = g(x)$ when $x = 2$.

53.

$$4 - \sqrt{t-3} = (t+5)^{1/2}$$

$$\left(4 - \sqrt{t-3}\right)^2 = \left[(t+5)^{1/2}\right]^2$$

$$16 - 8\sqrt{t-3} + t - 3 = t + 5$$

$$-8\sqrt{t-3} = -8$$

$$\sqrt{t-3} = 1$$

$$\left(\sqrt{t-3}\right)^2 = 1^2$$

$$t - 3 = 1$$

$$t = 4$$

The number 4 checks, so we have $f(t) = g(t)$ when $t = 4$.

55. *Thinking and Writing Exercise.*

57. *Familiarize.* We let x represent the width of the sign, and $13x + 5$ represent the length. Recall that the perimeter is given by the formula $P = 2w + 2l$.

Translate.

$$\underbrace{\text{The perimeter}} \quad \text{is} \quad 430 \text{ ft.}$$

$$2x + 2(13x + 5) \;=\; 430$$

Carry out. We solve the equation.

$$2x + 2(13x + 5) = 430$$
$$2x + 26x + 10 = 430$$
$$28x = 420 \quad \text{Combining like terms}$$
$$x = 15 \quad \text{Dividing by 28}$$

If the width of the sign is 15 ft, then the length is $13 \cdot 15 + 5 = 195 + 5 = 200$ ft.

Check The length of the sign 200 ft. is 13 times the width, 15 ft, plus 5 ft. The perimeter is $2 \cdot 200 + 2 \cdot 15 = 400 + 30 = 430$ ft. These results check.

State. The width of the sign is 15 ft., and its length is 200 ft.

59. *Familiarize.* Let x = the width of the photograph and $x + 4$ the length.

Translate.

$$\underbrace{\text{Area of photograph}} \quad \text{is} \quad 140 \text{ in}^2$$

$$x(x+4) \quad = \quad 140$$

Carry out. We solve the equation.

$$x(x+4) = 140$$
$$x^2 + 4x - 140 = 0$$
$$(x+14)(x-10) = 0$$
$$x + 14 = 0 \quad \text{or } x - 10 = 0$$
$$x = -14 \text{ or} \quad x = 10$$

Check. Since length and width cannot be negative, we only check $x = 10$. The length of the photograph is $10 + 4 = 14$ in^2. The area of the photograph is $10 \cdot 14 = 140$ in^2. These results check.

State. The width is 10 inches and the length is 14 inches.

61. *Familiarize.* Let x represent the base of the right triangle, $x + 2$ represent the other leg, and $x + 4$ represent the hypotenuse. Thus, each side represents one of three consecutive integers. Recall: Pythagorean Theorem $a^2 + b^2 = c^2$.

Translate.

$$a^2 + \quad b^2 \quad = \quad c^2$$

$$x^2 + (x+2)^2 = (x+4)^2$$

Carry out. We solve the equation:

$$x^2 + (x+2)^2 = (x+4)^2$$
$$x^2 + x^2 + 4x + 4 = x^2 + 8x + 16$$
$$x^2 + 4x - 8x + 4 - 16 = 0$$
$$x^2 - 4x - 12 = 0$$
$$(x-6)(x+2) = 0$$
$$x - 6 = 0 \quad \text{or} \quad x + 2 = 0$$
$$x = 6 \quad \text{or} \quad x = -2$$

Check. Since measure cannot be negative, we know –2 is not a solution. If $x = 6$, $x + 2 = 8$, and $x + 4 = 10$.

$$6^2 + 8^2 = 10^2$$
$$36 + 64 = 100$$
$$100 = 100$$

The results check.

State. The lengths of the sides are 6, 8, and 10 units.

63. *Thinking and Writing Exercise.*

65. $v(p) = 12.1\sqrt{p}$

Substitute 100 for $v(p)$ and solve for p.

$$100 = 12.1\sqrt{p}$$
$$\sqrt{p} = \frac{100}{12.1}$$
$$p = \left(\frac{100}{12.1}\right)^2$$
$$p \approx 68.3$$

The pressure is about 68 psi.

67. $f(T) = k\sqrt{T}$

To find the value of k, substitute 260 for $f(T)$ and 28 for T.

$260 = k\sqrt{28}$

$k = \dfrac{260}{\sqrt{28}} = \dfrac{130}{\sqrt{7}}$

When $T = 32$ N,

$f(T) = \dfrac{130}{\sqrt{7}}\sqrt{32} = 260\sqrt{\dfrac{8}{7}} \approx 277.95$

The new frequency is about 278 Hz.

69. $S(t) = 1087.7\sqrt{\dfrac{9t+2617}{2457}}$

Substitute 1880 for $S(t)$ and solve for t.

$1880 = 1087.7\sqrt{\dfrac{9t+2617}{2457}}$

$1.7284 \approx \sqrt{\dfrac{9t+2617}{2457}}$

$2.9874 \approx \dfrac{9t+2617}{2457}$

$7340.0418 \approx 9t+2617$

$4723.0418 \approx 9t$

$524.7824 \approx t$

The temperature is about 524.8°C.

71. $S(t) = 1087.7\sqrt{\dfrac{9t+2617}{2457}}$

$\dfrac{S}{1087.7} = \sqrt{\dfrac{9t+2617}{2457}}$

$\left(\dfrac{S}{1087.7}\right)^2 = \left(\sqrt{\dfrac{9t+2617}{2457}}\right)^2$

$\dfrac{S^2}{1087.7^2} = \dfrac{9t+2617}{2457}$

$\dfrac{2457 S^2}{1087.7^2} = 9t+2617$

$\dfrac{2457 S^2}{1087.7^2} - 2617 = 9t$

$\dfrac{1}{9}\left(\dfrac{2457 S^2}{1087.7^2} - 2617\right) = t$

73. $d(n) = 0.75\sqrt{2.8n}$

Substitute 84 for $d(n)$ and solve for n.

$84 = 0.75\sqrt{2.8n}$

$112 = \sqrt{2.8n}$

$(112)^2 = \left(\sqrt{2.8n}\right)^2$

$12544 = 2.8n$

$4480 \approx n$

Approximately 4480 rpm will produce peak performance.

75. $v = \sqrt{2gr}\sqrt{\dfrac{h}{r+h}}$

$v^2 = 2gr \cdot \dfrac{h}{r+h}$ Squaring both sides

$v^2(r+h) = 2grh$ Multiplying by $r+h$

$v^2 r + v^2 h = 2grh$

$v^2 h = 2grh - v^2 r$

$v^2 h = r(2gh - v^2)$

$\dfrac{v^2 h}{2gh - v^2} = r$

77. $\dfrac{x+\sqrt{x+1}}{x-\sqrt{x+1}} = \dfrac{5}{11}$

$11\left(x+\sqrt{x+1}\right) = 5\left(x-\sqrt{x+1}\right)$

$11x + 11\sqrt{x+1} = 5x - 5\sqrt{x+1}$

$16\sqrt{x+1} = -6x$

$8\sqrt{x+1} = -3x$

$\left(8\sqrt{x+1}\right)^2 = (-3x)^2$

$64(x+1) = 9x^2$

$64x + 64 = 9x^2$

$x = 9x^2 - 64x - 64$

$0 = (9x+8)(x-8)$

$9x+8 = 0$ or $x-8 = 0$

$9x = -8$ or $x = 8$

$x = -\dfrac{8}{9}$ or $x = 8$

Since $-\dfrac{8}{9}$ checks but 8 does not, the solution

is $-\dfrac{8}{9}$.

79. $\left(z^2+17\right)^{3/4}=27$

$$\left[\left(z^2+17\right)^{3/4}\right]^{4/3}=\left(3^3\right)^{4/3}$$

$$z^2+17=3^4$$

$$z^2+17=81$$

$$z^2-64=0$$

$$\left(z+8\right)\left(z-8\right)=0$$

$z=-8 \; or \; z=8$

Both –8 and 8 check. They are the solutions.

81. $\sqrt{8-b}=b\sqrt{8-b}$

$$\left(\sqrt{8-b}\right)^2=\left(b\sqrt{8-b}\right)^2$$

$$\left(8-b\right)=b^2\left(8-b\right)$$

$$0=b^2\left(8-b\right)-\left(8-b\right)$$

$$0=\left(8-b\right)\left(b^2-1\right)$$

$$0=\left(8-b\right)\left(b+1\right)\left(b-1\right)$$

$8-b=0 \; or \; b+1=0 \; or \; b-1=0$

$8=b \; or \quad b=-1 \; or \quad b=1$

Since the number 8 and 1 check but –1 does not, 8 and 1 are the solutions.

83. We find the values of x for which $g\left(x\right)=0$.

$$6x^{1/2}+6x^{-1/2}-37=0$$

$$6\sqrt{x}+\frac{6}{\sqrt{x}}=37$$

$$\left(6\sqrt{x}+\frac{6}{\sqrt{x}}\right)^2=37^2$$

$$36x+72+\frac{36}{x}=1369$$

$36x^2+72x+36=1369x$ Multiplying by x

$$36x^2-1297x+36=0$$

$$\left(36x-1\right)\left(x-36\right)=0$$

$36x-1=0 \; or \; x-36=0$

$36x=1 \quad or \qquad x=36$

$x=\dfrac{1}{36} \; or \qquad x=36$

Both numbers check. The x-intercepts are $\left(\dfrac{1}{36},0\right)$ and $\left(36,0\right)$.

1. d

3. e

5. f

7. $a^2+b^2=c^2$ Pythagorean equation

$5^2+3^2=c^2$ Substituting

$25+9=c^2$

$34=c^2$

$c=\sqrt{34}$ Exact answer

$c\approx 5.831$ Approximation

9. Since $a=b$, this is an isosceles right triangle. The hypotenuse = length of a leg $\cdot\sqrt{2}$, or $9\sqrt{2}$, or approximately, 12.728.

11. $a^2+b^2=c^2$

$a^2+12^2=13^2$

$a^2+144=169$

$a^2=25$

$a=\sqrt{25}=5$

13. $a^2+b^2=c^2; c=8$, and $a=4\sqrt{3}$ (or $b\cdot 4\sqrt{3}$, since it can be the length of either leg.)

$$\left(4\sqrt{3}\right)^2+b^2=8^2$$

$$48+b^2=64$$

$$b^2=16$$

$$b=\sqrt{16}$$

$$b=4 \; m$$

15. $a^2+b^2=c^2; c=\sqrt{20}, \; a=1$

$$1^2+b^2=\left(\sqrt{20}\right)^2$$

$$1+b^2=20$$

$$b^2=19$$

$$b=\sqrt{19} \; in$$

$$b\approx 4.359 \; in$$

17.　$a^2 + b^2 = c^2; a = 1,\ c = \sqrt{2}$

$$1^2 + b^2 = \left(\sqrt{2}\right)^2$$

$$1 + b^2 = 2$$

$$b^2 = 1$$

$$b = 1\text{ m}$$

Also, we might have observed that the hypotenuse is $\sqrt{2}\cdot$ length of the leg, 1. So, we know the right triangle is isosceles, thus $a = b = 1$.

19.

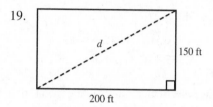

Let d = the distance, in feet, across the parking lot. A right triangle is formed in which the length of one leg is 200 ft., and the other is 150 ft.

We substitute these values into the Pythagorean equation to determine d.

$$d^2 = 200^2 + 150^2$$

$$d^2 = 40,000 + 22,500$$

$$d^2 = 62,500$$

$$d = \sqrt{62,500}$$

$$d = 250$$

The distance is 250 ft.

21.

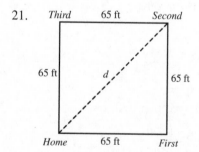

Let d = the distance, in feet, from home plate to second base.

Since the triangle formed is an isosceles right triangle, we know the hypotenuse, d, is $\sqrt{2}\cdot$ the length of a leg, 65. Thus,

$$d = 65\sqrt{2}$$

$$d \approx 91.924$$

The distance is $65\sqrt{2}$ ft, or approximately 91.924 ft.

23.　Let h = height, in inches of the TV set. We substitute the given values into the Pythagorean equation, $w^2 + h^2 = d^2$, to determine h.

$$45^2 + h^2 = 51^2$$

$$2025 + h^2 = 2601$$

$$h^2 = 576$$

$$h = \sqrt{576} = 24$$

The height is 24 inches.

25.　First, find the distance from corner to corner of the room.

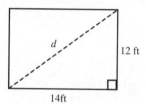

We use the Pythagorean equation to find d.

$$12^2 + 14^2 = d^2$$

$$144 + 196 = d^2$$

$$340 = d^2$$

$$d = \sqrt{340} = 2\sqrt{85}$$

Since 4 ft of slack is required on each end, we add $2\cdot4$, or 8 to d, or $2\sqrt{85} + 8$. The length of wire is $\left(2\sqrt{85} + 8\right)$ ft, or approximately 26.439 ft.

27.

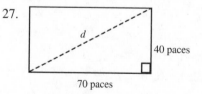

We use the Pythagorean equation to find d, the number of paces on the diagonal path.

$$d^2 = 40^2 + 70^2$$

$$d^2 = 1600 + 4900$$

$$d^2 = 6500$$

$$d = \sqrt{6500} = 10\sqrt{65}$$

If Marissa does not use the diagonal path, she will walk 40 + 70, or 110 paces. Marissa will save $\left(110-10\sqrt{65}\right)$ paces, or about 29.377 paces.

29. Since one acute angle is 45°, this is an isosceles right triangle with $b = 5$. Then $a = 5$ also. We substitute to find c.

$$c = a\sqrt{2}$$
$$c = 5\sqrt{2}$$

Exact answer: $a = 5; c = 5\sqrt{2}$

Approximation: $c \approx 7.071$

31. This is a 30-60-90 right triangle with $c = 14$.
$$c = 2a$$
$$14 = 2a$$
$$7 = a$$
$$b = a\sqrt{3}$$
$$b = 7\sqrt{3}$$

Exact answer: $a = 7, b = 7\sqrt{3}$

Approximation: $b \approx 12.124$

33. This is a 30-60-90 right triangle with $b = 15$.
$$b = a\sqrt{3}$$
$$15 = a\sqrt{3}$$
$$\frac{15}{\sqrt{3}} = a$$
$$\frac{15\sqrt{3}}{3} = a \quad \text{Rationalizing the denominator}$$
$$5\sqrt{3} = a \quad \text{Simplifying}$$
$$c = 2a$$
$$c = 2 \cdot 5\sqrt{3}$$
$$c = 10\sqrt{3}$$

Exact answer: $a = 5\sqrt{3}, c = 10\sqrt{3}$

Approximation: $a \approx 8.660, c \approx 17.321$

35. This is an isosceles right triangle with $c = 13$.
$$a = \frac{c\sqrt{2}}{2}$$
$$a = \frac{13\sqrt{2}}{2}$$

Since $a = b$, we have $b = \frac{13\sqrt{2}}{2}$ also.

Exact answer: $a = \frac{13\sqrt{2}}{2}, b = \frac{13\sqrt{2}}{2}$

Approximation $a \approx 9.192, b \approx 9.192$.

37. This is a 30-60-90 right triangle with $a = 14$.
$$b = a\sqrt{3} \qquad c = 2a$$
$$b = 14\sqrt{3} \qquad c = 2 \cdot 14$$
$$c = 28$$

Exact answer: $b = 14\sqrt{3}, c = 28$

Approximation: $b \approx 24.249$

39.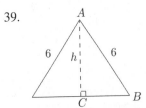

This is an equilateral triangle, so all the angles are 60°. The altitude bisects one angle and one side. Then triangle ABC is a 30-60-90 right triangle with the shorter leg of length 6/2, or 3, and hypotenuse of length 6. We substitute to find the length of the other leg.

$$b = a\sqrt{3}$$
$$h = 3\sqrt{3} \quad \text{Substituting } h \text{ for } b \text{ and 3 for } a$$

Exact answer: $h = 3\sqrt{3}$

Approximation: $h \approx 5.196$

41.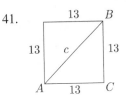

Triangle ABC is an isosceles right triangle with $a = 13$. We substitute to find c.

$$c = a\sqrt{2}$$
$$c = 13\sqrt{2}$$

Exact answer: $c = 13\sqrt{2}$

Approximation: $c \approx 18.385$

43.

Triangle ABC is an isosceles right triangle with $c = 19$. We substitute to find a.

$$a = \frac{c\sqrt{2}}{2}$$

$$a = \frac{19\sqrt{2}}{2}$$

Exact answer: $a = \dfrac{19\sqrt{2}}{2}$

Approximation: $a \approx 13.435$

45. We will express all distances in feet. Recall that 1 mi = 5280 ft.

We use the Pythagorean equation to find h.

$$h^2 + (5280)^2 = (5281)^2$$

$$h^2 + 27,878,400 = 27,888,961$$

$$h^2 = 10,561$$

$$h = \sqrt{10,561}$$

$$h \approx 102.767$$

The height of the bulge is $\sqrt{10,561}$ ft., or about 102.767 ft.

47. The lodge is an equilateral triangle, so all the angles are 60°. The "height" bisects one angle and one side, so the the building can be depicted as two identical 30-60-90 right triangles, each with the shorter legs (a) of length 33/2 ft and hypotenuses (c) of length 33 ft. We substitute to find h, the length of the remaining side.

$$b = a\sqrt{3}$$

$$h = \left(\frac{33}{2}\right)\sqrt{3} \quad \text{Substituting } h \text{ for } b$$

$$\text{and } \frac{33}{2} \text{ for } a$$

The area of the lodge is the area of the equilateral triangle, which is $A = \dfrac{1}{2}ch$.

$$A = \frac{1}{2}(33)\left(\frac{33\sqrt{3}}{2}\right)$$

$$= \frac{1089\sqrt{3}}{4} \text{ ft}^2 \approx 471.551 \text{ ft}^2$$

49.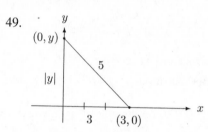

$$|y|^2 + 3^2 = 5^2$$

$$y^2 + 9 = 25$$

$$y^2 = 16$$

$$y = \pm 4$$

The points are $(0, -4)$ and $(0, 4)$.

51. Substitute the coordinates into the distance formula.

$$d = \sqrt{(x_2 - x_1)^2 + (y_2 - y_1)^2}$$

$$= \sqrt{(7 - 4)^2 + (1 - 5)^2}$$

$$= \sqrt{3^2 + (-4)^2}$$

$$= \sqrt{25}$$

$$= 5$$

53. Substitute the coordinates into the distance formula.

$$d = \sqrt{(x_2 - x_1)^2 + (y_2 - y_1)^2}$$

$$= \sqrt{(1 - 0)^2 + (-2 - (-5))^2}$$

$$= \sqrt{1^2 + 3^2}$$

$$= \sqrt{10}$$

$$\approx 3.162$$

55. Substitute the coordinates into the distance formula.

$$d = \sqrt{(x_2 - x_1)^2 + (y_2 - y_1)^2}$$
$$= \sqrt{(6-(-4))^2 + (-6-4)^2}$$
$$= \sqrt{10^2 + (-10)^2}$$
$$= \sqrt{200}$$
$$\approx 14.142$$

57. Substitute the coordinates into the distance formula.

$$d = \sqrt{(x_2 - x_1)^2 + (y_2 - y_1)^2}$$
$$= \sqrt{(-9.2-8.6)^2 + (-3.4-(-3.4))^2}$$
$$= \sqrt{(-17.8)^2 + 0^2}$$
$$= \sqrt{17.8^2}$$
$$= 17.8$$

59. Substitute the coordinates into the distance formula.

$$d = \sqrt{(x_2 - x_1)^2 + (y_2 - y_1)^2}$$
$$= \sqrt{\left(\frac{5}{6} - \frac{1}{2}\right)^2 + \left(-\frac{1}{6} - \frac{1}{3}\right)^2}$$
$$= \sqrt{\left(\frac{5}{6} - \frac{3}{6}\right)^2 + \left(-\frac{1}{6} - \frac{2}{6}\right)^2}$$
$$= \sqrt{\left(\frac{2}{6}\right)^2 + \left(-\frac{3}{6}\right)^2}$$
$$= \sqrt{\frac{4+9}{36}}$$
$$= \frac{\sqrt{13}}{6}$$
$$\approx 0.601$$

61. Substitute the coordinates into the distance formula.

$$d = \sqrt{(x_2 - x_1)^2 + (y_2 - y_1)^2}$$
$$= \sqrt{\left(0-(-\sqrt{6})\right)^2 + \left(0-\sqrt{6}\right)^2}$$
$$= \sqrt{\left(\sqrt{6}\right)^2 + \left(-\sqrt{6}\right)^2}$$
$$= \sqrt{12}$$
$$\approx 3.464$$

63. Substitute the coordinates into the distance formula.

$$d = \sqrt{(x_2 - x_1)^2 + (y_2 - y_1)^2}$$
$$= \sqrt{(-2-(-1))^2 + (-40-(-30))^2}$$
$$= \sqrt{(-1)^2 + (-10)^2}$$
$$= \sqrt{101}$$
$$\approx 10.050$$

65. Substitute the coordinates into the midpoint formula.

$$\left(\frac{x_1 + x_2}{2}, \frac{y_1 + y_2}{2}\right)$$
$$\left(\frac{-2+8}{2}, \frac{5+3}{2}\right), \text{ or } \left(\frac{6}{2}, \frac{8}{2}\right), \text{ or } (3,4)$$

67. Substitute the coordinates into the midpoint formula.

$$\left(\frac{x_1 + x_2}{2}, \frac{y_1 + y_2}{2}\right)$$
$$\left(\frac{2+5}{2}, \frac{-1+8}{2}\right), \text{ or } \left(\frac{7}{2}, \frac{7}{2}\right)$$

69. Substitute the coordinates into the midpoint formula.

$$\left(\frac{x_1 + x_2}{2}, \frac{y_1 + y_2}{2}\right)$$
$$\left(\frac{-8+6}{2}, \frac{-5+(-1)}{2}\right), \text{ or } \left(\frac{-2}{2}, \frac{-6}{2}\right),$$
or $(-1,-3)$

71. Substitute the coordinates into the midpoint formula.

$$\left(\frac{x_1 + x_2}{2}, \frac{y_1 + y_2}{2}\right)$$
$$\left(\frac{-3.4+4.8}{2}, \frac{8.1+(-8.1)}{2}\right), \text{ or } \left(\frac{1.4}{2}, \frac{0}{2}\right),$$
or $(0.7, 0)$

73. Substitute the coordinates into the midpoint formula.

$$\left(\frac{x_1 + x_2}{2}, \frac{y_1 + y_2}{2}\right)$$

$$\left(\frac{\frac{1}{6} + \left(-\frac{1}{3}\right)}{2}, \frac{-\frac{3}{4} + \frac{5}{6}}{2}\right), \text{ or } \left(\frac{-\frac{1}{6}}{2}, \frac{\frac{1}{12}}{2}\right),$$

$$\text{or } \left(-\frac{1}{12}, \frac{1}{24}\right)$$

75. Substitute the coordinates into the midpoint formula.

$$\left(\frac{x_1 + x_2}{2}, \frac{y_1 + y_2}{2}\right)$$

$$\left(\frac{\sqrt{2} + \sqrt{3}}{2}, \frac{-1 + 4}{2}\right), \text{ or } \left(\frac{\sqrt{2} + \sqrt{3}}{2}, \frac{3}{2}\right)$$

77. *Thinking and Writing Exercise.*

79. $y = 2x - 3$

Slope is 2, y-intercept is –3. Graph the point $(0,-3)$ and use the slope to determine a second point, from $(0,-3)$ go right 1 unit and up 2, or $(1,-1)$.

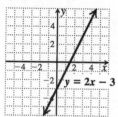

81. $8x - 4y = 8$

Rewrite the equation in slope-intercept form.

$8x - 4y = 8$

$8x - 8 = 4y$

$2x - 2 = y$

Slope is 2, y-intercept is –2. Graph the point $(0,-2)$ and use the slope to determine a second point, from $(0,-2)$ go right 1 unit and up 2, to $(1,0)$.

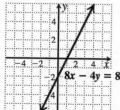

83. $x \geq 1$

Graph the boundary line $x = 1$ as a solid line. Use a point of substitution (for instance, $(0, 1)$) to determine which half-plane to shade.

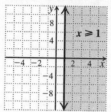

85. *Thinking and Writing Exercise.*

87. A regular hexagon has 6 sides of equal length, thus the length of each side is $72 \div 6$, or 12. The shaded region has a base of 12, and using 1/2 of the shaded region, we determine the height (it is a 30°-60°-90° triangle). The height is $6\sqrt{3}$.

Area of triangle is	$\frac{1}{2}$	×	length of the base	×	length of the height
↓	↓	↓	↓		↓
A	= $\frac{1}{2}$	·	12	·	$6\sqrt{3}$

$A = 36\sqrt{3}$

The area of the shaded region is $36\sqrt{3}$ cm², or approximately 62.354 cm².

89.
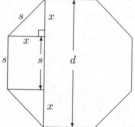

To determine x, use the Pythagorean equation on the right triangle, and solve.

$$x^2 + x^2 = s^2$$

$$2x^2 = s^2$$

$$x^2 = \frac{s^2}{2}$$

$$x = \sqrt{\frac{s^2}{2}}$$

$$x = \frac{s}{\sqrt{2}}, \text{ or } \frac{s\sqrt{2}}{2}$$

Thus, $d = x + s + x$

$$d = 2x + s$$

$$d = 2\left(\frac{s\sqrt{2}}{2}\right) + s$$

$$d = s\sqrt{2} + s$$

91.

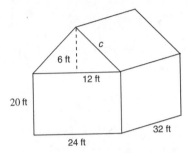

The area to be painted consists of two 20 ft by 24 ft rectangles, two 20 ft by 32 ft rectangles, and two triangles with height 6 ft and base 24 ft. The area of the two 20 ft by 24 ft rectangles is $2 \cdot 20$ ft $\cdot 24$ ft $= 960$ ft^2. The area of the two 20 ft by 32 ft rectangles is $2 \cdot 20$ ft $\cdot 32$ ft $= 1280$ ft^2. The area of the two triangles is $2 \cdot \frac{1}{2} \cdot 24$ ft $\cdot 6$ ft $= 144$ ft^2. Thus, the total area to be painted is 960 ft$^2 +$ 1280 ft$^2 + 144$ ft$^2 = 2384$ ft^2.

One gallon of paint covers 500 ft^2, so we divide to determine how many gallons of paint are required: $\frac{2384}{500} \approx 4.8$. Thus, 5 gallons of paint should be bought to paint the house.

93. First we find the radius of a circle with an area of 6160 ft^2.

$$A = \pi r^2$$

$$6160 = \pi r^2$$

$$\frac{6160}{\pi} = r^2$$

$$\sqrt{\frac{6160}{\pi}} = r$$

$$44.28 \approx r$$

Now we make a drawing. Let $s =$ the length of the side of the room.

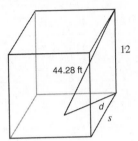

Using the Pythagorean equation, find d.

$$d^2 + 12^2 = 44.28^2$$

$$d^2 = 44.28^2 - 12^2$$

$$d \approx 42.63$$

$2 \cdot d$ is the length of the diagonal across the floor, and $s =$ the length of each side of the square room. Again, using the Pythagorean equations, we find s.

$$s^2 + s^2 = (2 \cdot 42.63)^2$$

$$2s^2 = 85.26^2$$

$$s^2 = \frac{85.26^2}{2}$$

$$s = \sqrt{\frac{85.26^2}{2}}$$

$$s \approx 60.28$$

The dimensions of the room are 60.28 ft by 60.28 ft.

95. i)
$$\sqrt{\left(\frac{x_1 + x_2}{2} - x_1\right)^2 + \left(\frac{y_1 + y_2}{2} - y_1\right)^2}$$

$$\overset{?}{=} \sqrt{\left(x_2 - \frac{x_1 + x_2}{2}\right)^2 + \left(y_2 - \frac{y_1 + y_2}{2}\right)^2}$$

$$\sqrt{\left(\frac{x_1 + x_2 - 2x_1}{2}\right)^2 + \left(\frac{y_1 + y_2 - 2y_1}{2}\right)^2}$$

$$\overset{?}{=} \sqrt{\left(\frac{2x_2 - x_1 - x_2}{2}\right)^2 + \left(\frac{2y_2 - y_1 - y_2}{2}\right)^2}$$

$$\sqrt{\left(\frac{x_2 - x_1}{2}\right)^2 + \left(\frac{y_2 - y_1}{2}\right)^2}$$

$$= \sqrt{\left(\frac{x_2 - x_1}{2}\right)^2 + \left(\frac{y_2 - y_1}{2}\right)^2}$$

ii) The sum of the lengths of the short segments must equal the length of the long segment, $\sqrt{(x_2-x_1)^2+(y_2-y_1)^2}$.

$$\sqrt{\left(\frac{x_1+x_2}{2}-x_1\right)^2+\left(\frac{y_1+y_2}{2}-y_1\right)^2}$$

$$+\sqrt{\left(x_2-\frac{x_1+x_2}{2}\right)^2+\left(y_2-\frac{y_1+y_2}{2}\right)^2}$$

$$=\sqrt{\left(\frac{x_1+x_2-2x_1}{2}\right)^2+\left(\frac{y_1+y_2-2y_1}{2}\right)^2}$$

$$+\sqrt{\left(\frac{2x_2-x_1-x_2}{2}\right)^2+\left(\frac{2y_2-y_1-y_2}{2}\right)^2}$$

$$=\sqrt{\left(\frac{x_2-x_1}{2}\right)^2+\left(\frac{y_2-y_1}{2}\right)^2}$$

$$+\sqrt{\left(\frac{x_2-x_1}{2}\right)^2+\left(\frac{y_2-y_1}{2}\right)^2}$$

$$=2\sqrt{\left(\frac{x_2-x_1}{2}\right)^2+\left(\frac{y_2-y_1}{2}\right)^2}$$

$$=2\sqrt{\frac{1}{4}(x_2-x_1)^2+\frac{1}{4}(y_2-y_1)^2}$$

$$=2\left(\frac{1}{2}\right)\sqrt{(x_2-x_1)^2+(y_2-y_1)^2}$$

$$=\sqrt{(x_2-x_1)^2+(y_2-y_1)^2}$$

For the sum of the short segments to equal the long segment, the three points must be on the same line, and the short segments are equal in length. Thus, the midpoint formula is correct.

Exercise Set 10.8

1. False

3. True

5. False

7. False

9. $\sqrt{-100}=\sqrt{100\cdot-1}=\sqrt{100}\cdot\sqrt{-1}=10i$

11. $\sqrt{-13}=\sqrt{-1\cdot13}=\sqrt{-1}\cdot\sqrt{13}$
$\phantom{\sqrt{-13}}=i\sqrt{13}$ or $\sqrt{13}i$

13. $\sqrt{-8}=\sqrt{4\cdot2\cdot-1}=\sqrt{4}\cdot\sqrt{2}\cdot\sqrt{-1}=2\sqrt{2}i$,
or $2i\sqrt{2}$

15. $-\sqrt{-3}=-\sqrt{3\cdot-1}=-\sqrt{3}i$, or $-i\sqrt{3}$

17. $-\sqrt{-81}=-\sqrt{81\cdot-1}=-9i$

19. $-\sqrt{-300}=-\sqrt{100\cdot3\cdot-1}$
$\phantom{-\sqrt{-300}}=-\sqrt{100}\cdot\sqrt{3}\cdot\sqrt{-1}=-10\sqrt{3}i,$
or $-10i\sqrt{3}$

21. $6-\sqrt{-84}=6-\sqrt{4\cdot21\cdot-1}$
$\phantom{6-\sqrt{-84}}=6-\sqrt{4}\cdot\sqrt{21}\cdot\sqrt{-1}$
$\phantom{6-\sqrt{-84}}=6-2\sqrt{21}i,$
or $6-2i\sqrt{21}$

23. $-\sqrt{-76}+\sqrt{-125}=-\sqrt{4\cdot19\cdot-1}+\sqrt{25\cdot5\cdot-1}$
$\phantom{-\sqrt{-76}+\sqrt{-125}}=-\sqrt{4}\cdot\sqrt{19}\cdot\sqrt{-1}+$
$\phantom{-\sqrt{-76}+\sqrt{-125}=}\sqrt{25}\cdot\sqrt{5}\cdot\sqrt{-1}$
$\phantom{-\sqrt{-76}+\sqrt{-125}}=-2\sqrt{19}i+5\sqrt{5}i$
$\phantom{-\sqrt{-76}+\sqrt{-125}}=\left(-2\sqrt{19}+5\sqrt{5}\right)i$

25. $\sqrt{-18}-\sqrt{-100}=\sqrt{9\cdot2\cdot-1}-\sqrt{100\cdot-1}$
$\phantom{\sqrt{-18}-\sqrt{-100}}=\sqrt{9}\cdot\sqrt{2}\cdot\sqrt{-1}-\sqrt{100}\cdot\sqrt{-1}$
$\phantom{\sqrt{-18}-\sqrt{-100}}=3\sqrt{2}i-10i=\left(3\sqrt{2}-10\right)i$

27. $(6+7i)+(5+3i)=(6+5)+(7+3)i$
$=11+10i$

29. $(9+8i)-(5+3i)=(9-5)+(8-3)i=4+5i$

31. $(7-4i)-(5-3i)=(7-5)+\left[-4-(-3)\right]i$
$=2-i$

33. $(-5-i)-(7+4i)=(-5-7)+(-1-4)i$
$=-12-5i$

35. $7i\cdot6i=42i^2=-42\quad\left[i^2=-1\right]$

37. $(-4i)(-6i)=24i^2=-24$

39. $\sqrt{-36}\cdot\sqrt{-9} = \sqrt{36\cdot-1}\cdot\sqrt{9\cdot-1}$
$= 6i\cdot 3i = 18i^2 = -18$

41. $\sqrt{-5}\cdot\sqrt{-2} = \sqrt{5}i\cdot\sqrt{2}i = \sqrt{10}i^2 = -\sqrt{10}$

43. $\sqrt{-6}\cdot\sqrt{-21} = \sqrt{6}i\cdot\sqrt{21}i = \sqrt{126}i^2$
$= -\sqrt{9\cdot 14} = -3\sqrt{14}$

45. $5i(2+6i)$
$= 5i\cdot 2 + 5i\cdot 6i \quad$ Distributive law
$= 10i + 30i^2$
$= 10i + 30(-1) \quad i^2 = -1$
$= -30 + 10i \quad$ Standard form

47. $-7i(3-4i)$
$= (-7i\cdot 3) + (-7i)(-4i)$
$= -21i + 28i^2$
$= -28 - 21i$

49. $(1+i)(3+2i)$
$= 1\cdot 3 + 1\cdot 2i + i\cdot 3 + i\cdot 2i \quad$ Using FOIL
$= 3 + 5i + 2i^2$
$= 3 + 5i - 2 \qquad\qquad i^2 = -1$
$= 1 + 5i$

51. $(6-5i)(3+4i)$
$= 18 + 9i - 20i^2$
$= 18 + 9i + 20$
$= 38 + 9i$

53. $(7-2i)(2-6i)$
$= 14 - 46i + 12i^2$
$= 14 - 46i - 12$
$= 2 - 46i$

55. $(3+8i)(3-8i) = 3^2 - (8i)^2$
$= 9 - 64i^2 = 9 - 64(-1) = 73$

57. $(-7+i)(-7-i) = (-7)^2 - (i)^2$
$= 49 - (-1) = 50$

59. $(4-2i)^2$
$= 4^2 - 2\cdot 4\cdot 2i + (2i)^2 \quad$ Squaring a binomial
$= 16 - 16i + 4i^2$
$= 12 - 16i$

61. $(2+3i)^2$
$= 2^2 + 2\cdot 2\cdot 3i + (3i)^2$
$= 4 + 12i + 9i^2$
$= 4 + 12i - 9$
$= -5 + 12i$

63. $(-2+3i)^2$
$= (-2^2) + 2\cdot-2\cdot 3i + (3i)^2$
$= 4 - 12i - 9$
$= -5 - 12i$

65. $\dfrac{10}{3+i}$
$= \dfrac{10}{3+i}\cdot\dfrac{3-i}{3-i} \quad$ Multiplying by 1, using the conjugate
$= \dfrac{30-10i}{9-i^2} \quad$ Multiplying
$= \dfrac{30-10i}{9-(-1)} \quad i^2 = -1$
$= \dfrac{30-10i}{10}$
$= \dfrac{30}{10} - \dfrac{10}{10}i \quad$ Standard form
$= 3 - i$

67. $\dfrac{2}{3-2i}$
$= \dfrac{2}{3-2i}\cdot\dfrac{3+2i}{3+2i} \quad$ Multiplying by 1, using the conjugate
$= \dfrac{6+4i}{9-4i^2} \quad$ Multiplying
$= \dfrac{6+4i}{9-4(-1)} \quad i^2 = -1$
$= \dfrac{6+4i}{13}$
$= \dfrac{6}{13} + \dfrac{4}{13}i \quad$ Standard form

69.
$$\frac{2i}{5+3i}$$
$$=\frac{2i}{5+3i}\cdot\frac{5-3i}{5-3i}$$
$$=\frac{10i-6i^2}{25-9i^2}$$
$$=\frac{10i+6}{25+9}\quad i^2=-1$$
$$=\frac{6+10i}{34}=\frac{\cancel{2}(3+5i)}{\cancel{2}\cdot17}$$
$$=\frac{3}{17}+\frac{5}{17}i$$

71. $\dfrac{5}{6i}=\dfrac{5}{6i}\cdot\dfrac{i}{i}=\dfrac{5i}{6i^2}=\dfrac{5i}{-6}=-\dfrac{5}{6}i$

73.
$$\frac{5-3i}{4i}=\frac{5-3i}{4i}\cdot\frac{i}{i}=\frac{5i-3i^2}{4i^2}$$
$$=\frac{5i+3}{-4}=-\frac{3}{4}-\frac{5}{4}i$$

75.
$$\frac{7i+14}{7i}=\frac{7i}{7i}+\frac{14}{7i}=1+\frac{2}{i}=1+\frac{2}{i}\cdot\frac{i}{i}$$
$$=1+\frac{2i}{i^2}=1+\frac{2i}{-1}=1-2i$$

77.
$$\frac{4+5i}{3-7i}=\frac{4+5i}{3-7i}\cdot\frac{3+7i}{3+7i}$$
$$=\frac{12+28i+15i+35i^2}{9-49i^2}$$
$$=\frac{12+43i-35}{9+49}=\frac{-23+43i}{58}$$
$$=-\frac{23}{58}+\frac{43}{58}i$$

79.
$$\frac{2+3i}{2+5i}=\frac{2+3i}{2+5i}\cdot\frac{2-5i}{2-5i}$$
$$=\frac{4-10i+6i-15i^2}{4-25i^2}$$
$$=\frac{4-4i+15}{4+25}=\frac{19-4i}{29}$$
$$=\frac{19}{29}-\frac{4}{29}i$$

81.
$$\frac{3-2i}{4+3i}=\frac{3-2i}{4+3i}\cdot\frac{4-3i}{4-3i}=\frac{12-9i-8i+6i^2}{16-9i^2}$$
$$=\frac{12-17i-6}{16+9}=\frac{6-17i}{25}=\frac{6}{25}-\frac{17}{25}i$$

83. $i^7=i^6\cdot i=\left(i^2\right)^3\cdot i=\left(-1\right)^3\cdot i=-1\cdot i=-i$

85. $i^{32}=\left(i^2\right)^{16}=\left(-1\right)^{16}=1$

87. $i^{42}=\left(i^2\right)^{21}=\left(-1\right)^{21}=-1$

89. $i^9=\left(i^2\right)^4\cdot i=\left(-1\right)^4\cdot i=1\cdot i=i$

91. $\left(-i\right)^6=i^6=\left(i^2\right)^3=\left(-1\right)^3=-1$

93. $\left(5i\right)^3=5^3\cdot i^3=125\cdot i^2\cdot i$
$$=125\left(-1\right)\left(i\right)=-125i$$

95. $i^2+i^4=-1+\left(i^2\right)^2=-1+\left(-1\right)^2=-1+1=0$

97. *Thinking and Writing Exercise.*

99. $x^2-x-6=0$
$$\left(x-3\right)\left(x+2\right)=0$$
$$x-3=0\;\;or\;\;x+2=0$$
$$x=3\quad or\quad x=-2$$
The solutions are -2 and 3.

101. $t^2=100$
$$t=\pm\sqrt{100}=\pm10$$
The solutions are -10 and 10.

103. $15x^2=14x+8$
$$15x^2-14x-8=0$$
$$\left(5x+2\right)\left(3x-4\right)=0\quad\text{Factoring}$$
$$5x+2=0\;\;or\;\;3x-4=0$$
$$5x=-2\;\;or\;\;\;\;3x=4$$
$$x=-\frac{2}{5}\;or\;\;\;\;\;x=\frac{4}{3}$$
The solutions are $-\dfrac{2}{5}$ and $\dfrac{4}{3}$.

105. *Thinking and Writing Exercise.*

107.

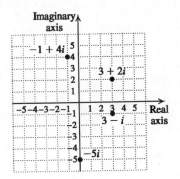

109. $|3+4i| = \sqrt{3^2 + 4^2} = \sqrt{9+16} = \sqrt{25} = 5$

111. $|-1+i| = \sqrt{(-1)^2 + 1^2} = \sqrt{1+1} = \sqrt{2}$

113. First we simplify $g(z)$.

$$g(z) = \frac{z^4 - z^2}{z-1} = \frac{z^2(z+1)(z-1)}{z-1}$$
$$= z^2(z+1) \text{ for } z \neq 1$$

Now we substitute.

$$g(3i) = (3i)^2(3i+1) = (9i^2)(3i+1)$$
$$= (-9)(3i+1) = -27i-9$$
$$= -9-27i$$

115. We use the simplified form of $g(z)$ found in Exercise 113.

$$g(5i-1) = (5i-1)^2(5i-1+1)$$
$$= (25i^2 - 10i + 1)(5i)$$
$$= (-25 - 10i + 1)(5i)$$
$$= (-24 - 10i)(5i)$$
$$= -120i - 50i^2$$
$$= 50 - 120i$$

117. $\dfrac{1}{w-w^2} = \dfrac{1}{\left(\dfrac{1-i}{10}\right) - \left(\dfrac{1-i}{10}\right)^2} = \dfrac{1}{\left(\dfrac{1-i}{10}\right) - \dfrac{(1-i^2)}{100}}$

$$= \frac{1 \cdot 100}{\left(\dfrac{1-i}{10}\right) \cdot 100 - \dfrac{(1-i)^2}{100} \cdot 100}$$

$$= \frac{100}{(1-i)10 - (1-i)^2} = \frac{100}{10 - 10i - (1 - 2i + i^2)}$$

$$= \frac{100}{10 - 10i - (1 - 2i - 1)} = \frac{100}{10 - 8i} = \frac{50}{5 - 4i}$$

$$= \frac{50}{5-4i} \cdot \frac{5+4i}{5+4i} = \frac{50(5+4i)}{25+16}$$

$$= \frac{250 + 200i}{41} = \frac{250}{41} + \frac{200i}{41}$$

119. $(1-i)^3(1+i)^3$
$$= (1-i)(1+i) \cdot (1-i)(1+i) \cdot (1-i)(1+i)$$
$$= (1-i^2)(1-i^2)(1-i^2)$$
$$= (1+1)(1+1)(1+1) = 2 \cdot 2 \cdot 2 = 8$$

121. $\dfrac{6}{1+\dfrac{3}{i}} = \dfrac{6}{\dfrac{i+3}{i}} = \dfrac{6i}{i+3} \cdot \dfrac{-i+3}{-i+3} = \dfrac{-6i^2 + 18i}{-i^2 + 9}$

$$= \frac{6 + 18i}{10} = \frac{6}{10} + \frac{18}{10}i = \frac{3}{5} + \frac{9}{5}i$$

123. $\dfrac{i - i^{38}}{1+i} = \dfrac{i - (i^2)^{19}}{1+i} = \dfrac{i - (-1)^{19}}{1+i}$

$$= \frac{i - (-1)}{1+i} = \frac{i+1}{i+1} = 1$$

Chapter 10 Study Summary

1. $-\sqrt{81} = -9$ since $\sqrt{81} = 9, -\sqrt{81} = -9$.

2. $\sqrt[3]{-1} = -1$ since $(-1)^3 = -1$

3. $\sqrt{36x^2} = \sqrt{(6x)^2} = |6x| = 6|x|$

Since x might be negative, absolute-value notation is necessary.

4. $\sqrt[4]{x^4} = |x| = x$, for $x \geq 0$

5. $100^{-1/2} = \dfrac{1}{100^{1/2}} = \dfrac{1}{\left(10^2\right)^{1/2}} = \dfrac{1}{10}$

6. $\sqrt{7x} \cdot \sqrt{3y} = \sqrt{7x \cdot 3y} = \sqrt{21xy}$

7. $\sqrt{200x^5 y^{18}} = \sqrt{100 \cdot 2 \cdot x^4 \cdot x \cdot y^{18}}$
$= \sqrt{100 x^4 y^{18} \cdot 2x}$
$= 10x^2 \left|y\right|^9 \sqrt{2x}$

The expression is undefined when $x < 0$, but y might be negative.

8. $\sqrt{\dfrac{12x^3}{25}} = \dfrac{\sqrt{4 \cdot 3 \cdot x^2 \cdot x}}{\sqrt{25}} = \dfrac{\sqrt{4x^2 \cdot 3x}}{5} = \dfrac{2x\sqrt{3x}}{5}$

Since the original expression is undefined when $x < 0$, there is no need to use the absolute value.

9. $\sqrt{\dfrac{2x}{3y^2}} = \dfrac{\sqrt{2x}}{\left|y\right|\sqrt{3}} \cdot \dfrac{\sqrt{3}}{\sqrt{3}} = \dfrac{\sqrt{6x}}{3\left|y\right|}$

10. $5\sqrt{8} - 3\sqrt{50} = 5\sqrt{4 \cdot 2} - 3\sqrt{25 \cdot 2}$
$= 5 \cdot 2\sqrt{2} - 3 \cdot 5\sqrt{2}$
$= 10\sqrt{2} - 15\sqrt{2}$
$= -5\sqrt{2}$

11. $\left(2 - \sqrt{3}\right)\left(5 - 7\sqrt{3}\right) =$
$= 2 \cdot 5 - 2 \cdot 7\sqrt{3} - \sqrt{3} \cdot 5 + 7\sqrt{3} \cdot \sqrt{3}$
$= 10 - 14\sqrt{3} - 5\sqrt{3} + 7 \cdot 3$
$= 31 - 19\sqrt{3}$

12. $\dfrac{\sqrt{15}}{3 + \sqrt{5}} = \dfrac{\sqrt{15}}{3 + \sqrt{5}} \cdot \dfrac{3 - \sqrt{5}}{3 - \sqrt{5}} = \dfrac{3\sqrt{15} - \sqrt{15} \cdot \sqrt{5}}{9 - 5}$
$= \dfrac{3\sqrt{15} - \sqrt{75}}{4} = \dfrac{3\sqrt{15} - \sqrt{25 \cdot 3}}{4}$
$= \dfrac{3\sqrt{15} - 5\sqrt{3}}{4}$

13. $\dfrac{\sqrt{x^5}}{\sqrt[3]{x}}$

$= \dfrac{x^{5/2}}{x^{1/3}}$ Converting to exponential notation

$= x^{5/2 - 1/3}$ Subtracting exponents

$= x^{13/6}$ Converting back

$= \sqrt[6]{x^{13}}$ to radical notation

14. $\sqrt{2x + 3} = x$
$\left(\sqrt{2x + 3}\right)^2 = x^2$
$2x + 3 = x^2$
$x^2 - 2x - 3 = 0$
$(x + 1)(x - 3) = 0$
$x + 1 = 0 \;\; or \;\; x - 3 = 0$
$x = -1 \; or \;\;\;\; x = 3$

Check:

For -1:

$\sqrt{2x + 3} = x$	
$\sqrt{-2 + 3}$	-1
$\sqrt{1}$	
1	-1 FALSE

For 3:

$\sqrt{2x + 3} = x$	
$\sqrt{6 + 3}$	3
$\sqrt{9}$	
3	3 TRUE

Since 3 checks but -1 does not, the solution is 3.

15. $a^2 + b^2 = c^2$ Pythagorean equation
$a^2 + 7^2 = 10^2$ Substituting
$a^2 + 49 = 100$
$a^2 = 100 - 49 = 51$
$a = \sqrt{51}$ Exact answer
$a \approx 7.141$ Approximation

16. Because this is an isosceles right triangle, the hypotenuse = length of a leg $\cdot \sqrt{2}$.
$6\sqrt{2} = a\sqrt{2}$
$6 = a$

17. For a 30-60-90 triangle, $c = 2a$, $b = \sqrt{3}a$
$a = 5$, $b = 5\sqrt{3} \approx 8.660$, $c = 2a = 10$

18. Substitute the coordinates into the distance formula.

$$d = \sqrt{(x_2 - x_1)^2 + (y_2 - y_1)^2}$$
$$= \sqrt{(6 - (-2))^2 + (-10 - 1)^2}$$
$$= \sqrt{8^2 + (-11)^2}$$
$$= \sqrt{64 + 121}$$
$$= \sqrt{185}$$
$$\approx 13.601$$

19. Substitute the coordinates into the midpoint formula.

$$\left(\frac{x_1 + x_2}{2}, \frac{y_1 + y_2}{2} \right) = \left(\frac{-2 + 6}{2}, \frac{1 + (-10)}{2} \right)$$
$$= \left(\frac{4}{2}, \frac{-9}{2} \right) = \left(2, -\frac{9}{2} \right)$$

20. $(5 - 3i) + (-8 - 9i) = (5 - 8) + (-3 - 9)i$
$$= -3 - 12i$$

21. $(2 - i) - (-1 + i) = 2 - i + 1 - i = 3 - 2i$

22. $(1 - 7i)(3 - 5i) = 1 \cdot 3 - 1 \cdot 5i - 7i \cdot 3 + 7 \cdot 5i^2$
$$= 3 - 5i - 21i - 35$$
$$= -32 - 26i$$

23. $\dfrac{1 + i}{1 - i} = \dfrac{1 + i}{1 - i} \cdot \dfrac{1 + i}{1 + i} = \dfrac{1 + 2i + i^2}{1 - i^2} = \dfrac{1 + 2i + (-1)}{1 - (-1)}$
$$= \frac{2i}{2} = i$$

Chapter 10 Review Exercises

1. True

2. False

3. False

4. True

5. True

6. True

7. True

8. False

9. $\sqrt{\dfrac{49}{9}} = \dfrac{\sqrt{49}}{\sqrt{9}} = \dfrac{\sqrt{7^2}}{\sqrt{3^2}} = \dfrac{7}{3}$

10. $-\sqrt{0.25} = -\sqrt{(0.5^2)} = -0.5$

11. $f(x) = \sqrt{2x - 7}$
$$f(16) = \sqrt{2 \cdot 16 - 7} = \sqrt{32 - 7}$$
$$= \sqrt{25} = \sqrt{5^2} = 5$$

12. The domain of f are the values of x which make $2x - 7$ non-negative.
$$2x - 7 \geq 0$$
$$2x \geq 7$$
$$x \geq \frac{7}{2}$$
The domain is $\left\{ x \mid x \geq \dfrac{7}{2} \right\}$, or $\left[\dfrac{7}{2}, \infty \right)$

13. $M = -5 + \sqrt{6.7x - 444}$

    a. $x = 300$, so $M = -5 + \sqrt{6.7(300) - 444}$
    $$M \approx 34.6 \text{ lb}$$

    b. $x = 100$, so $M = -5 + \sqrt{6.7(100) - 444}$
    $$M \approx 10.0 \text{ lb}$$

    c. $x = 200$, so $M = -5 + \sqrt{6.7(200) - 444}$
    $$M \approx 24.9 \text{ lb}$$

    d. $x = 400$, so $M = -5 + \sqrt{6.7(400) - 444}$
    $$M \approx 42.3 \text{ lb}$$

14. $\sqrt{25t^2} = \sqrt{25} \cdot \sqrt{t^2} = 5|t|$

15. $\sqrt{(c + 8)^2} = |c + 8|$

16. $\sqrt{4x^2 + 4x + 1} = \sqrt{(2x + 1)^2} = |2x + 1|$

17. $\sqrt[5]{-32} = \sqrt[5]{(-2)^5} = -2$

18. $\left(\sqrt[3]{5ab}\right)^4 = (5ab)^{4/3}$

19. $\left(16a^6\right)^{3/4} = \left(2^4\right)^{3/4} \cdot \left(a^6\right)^{3/4} = 2^3 \cdot a^{9/2}$

$\qquad = 8 \cdot a^{\frac{8}{2}+\frac{1}{2}} = 8a^4\sqrt{a}$

20. $\sqrt{x^6 y^{10}} = \left(x^6 y^{10}\right)^{1/2} = \left(x^6\right)^{1/2}\left(y^{10}\right)^{1/2} = x^3 y^5$

21. $\left(\sqrt[6]{x^2 y}\right)^2 = \left[\left(x^2\right)^{1/6} y^{1/6}\right]^2 = \left[x^{1/3} y^{1/6}\right]^2$

$\qquad = \left(x^{1/3}\right)^2 \left(y^{1/6}\right)^2 = x^{2/3} y^{1/3}$

$\qquad = \left(x^2 y\right)^{1/3} = \sqrt[3]{x^2 y}$

22. $\left(x^{-2/3}\right)^{3/5} = x^{-2/3 \cdot 3/5} = x^{-2/5} = \dfrac{1}{x^{2/5}}$

23. $\dfrac{7^{-1/3}}{7^{-1/2}} = 7^{-1/3-(-1/2)} = 7^{-2/6-(-3/6)} = 7^{1/6}$

24. $f(x) = \sqrt{25(x-6)^2} = \sqrt{5^2(x-6)^2}$

$\qquad f(x) = 5|x-6|$

25. $\sqrt[4]{16x^{20}y^8} = \sqrt[4]{\left(2x^5 y^2\right)^4} = 2x^5 y^2$

26. $\sqrt{250x^3 y^2} = \sqrt{25x^2 y^2 \cdot 10x} = 5xy\sqrt{10x}$

27. $\sqrt{2x} \cdot \sqrt{3y} = \sqrt{2x \cdot 3y} = \sqrt{6xy}$

28. $\sqrt[3]{3x^4 b}\sqrt[3]{9xb^2} = \sqrt[3]{3x^4 b \cdot 9xb^2} = \sqrt[3]{27x^5 b^3}$

$\qquad = \sqrt[3]{27x^3 b^3 \cdot x^2} = 3xb\sqrt[3]{x^2}$

29. $\sqrt[3]{-24x^{10}y^8} \cdot \sqrt[3]{18x^7 y^4} = \sqrt[3]{-432x^{17}y^{12}}$

$\qquad = \sqrt[3]{-216x^{15}y^{12} \cdot 2x^2}$

$\qquad = -6x^5 y^4 \sqrt[3]{2x^2}$

30. $\dfrac{\sqrt[3]{60xy^3}}{\sqrt[3]{10x}} = \sqrt[3]{\dfrac{60xy^3}{10x}} = \sqrt[3]{6y^3} = y\sqrt[3]{6}$

31. $\dfrac{\sqrt{75x}}{2\sqrt{3}} = \dfrac{\sqrt{75x}}{2\sqrt{3}} \cdot \dfrac{\sqrt{3}}{\sqrt{3}} = \dfrac{\sqrt{225x}}{2\cdot 3} = \dfrac{15\sqrt{x}}{6} = \dfrac{5\sqrt{x}}{2}$

32. $\sqrt[4]{\dfrac{48a^{11}}{c^8}} = \dfrac{\sqrt[4]{48a^{11}}}{\sqrt[4]{c^8}} = \dfrac{\sqrt[4]{16a^8 \cdot 3a^3}}{\sqrt[4]{c^8}} = \dfrac{2a^2\sqrt[4]{3a^3}}{c^2}$

33. $5\sqrt[3]{x} + 2\sqrt[3]{x} = (5+2)\sqrt[3]{x} = 7\sqrt[3]{x}$

34. $2\sqrt{75} - 9\sqrt{3} = 2\sqrt{25\cdot 3} - 9\sqrt{3}$

$\qquad = 2\cdot 5\sqrt{3} - 9\sqrt{3} = 10\sqrt{3} - 9\sqrt{3}$

$\qquad = (10-9)\sqrt{3} = \sqrt{3}$

35. $\sqrt[3]{8x^4} + \sqrt[3]{xy^6} = \sqrt[3]{8x^3 \cdot x} + \sqrt[3]{y^6 \cdot x}$

$\qquad = 2x\sqrt[3]{x} + y^2\sqrt[3]{x}$

$\qquad = \left(2x + y^2\right)\sqrt[3]{x}$

36. $\sqrt{50} + 2\sqrt{18} + \sqrt{32}$

$\qquad = \sqrt{25\cdot 2} + 2\sqrt{9\cdot 2} + \sqrt{16\cdot 2}$

$\qquad = 5\sqrt{2} + 2\cdot 3\sqrt{2} + 4\sqrt{2}$

$\qquad = (5+6+4)\sqrt{2} = 15\sqrt{2}$

37. $\left(3+\sqrt{10}\right)\left(3-\sqrt{10}\right) = (3)^2 - \left(\sqrt{10}\right)^2$

$\qquad = 9 - 10 = -1$

38. $\left(\sqrt{3} - 3\sqrt{8}\right)\left(\sqrt{5} + 2\sqrt{8}\right)$

$\qquad = \sqrt{3}\cdot\sqrt{5} + \sqrt{3}\cdot 2\sqrt{8} - 3\sqrt{8}\cdot\sqrt{5} - 3\sqrt{8}\cdot 2\sqrt{8}$

$\qquad = \sqrt{15} + 2\sqrt{24} - 3\sqrt{40} - 6\cdot 8$

$\qquad = \sqrt{15} + 2\sqrt{4\cdot 6} - 3\sqrt{4\cdot 10} - 48$

$\qquad = \sqrt{15} + 2\cdot 2\sqrt{6} - 3\cdot 2\sqrt{10} - 48$

$\qquad = \sqrt{15} + 4\sqrt{6} - 6\sqrt{10} - 48$

39. $\sqrt[4]{x} \cdot \sqrt{x} = x^{1/4} \cdot x^{1/2} = x^{1/4+1/2} = x^{1/4+2/4}$

$\qquad = x^{3/4} = \sqrt[4]{x^3}$

40. $\dfrac{\sqrt[3]{x^2}}{\sqrt[4]{x}} = \dfrac{x^{2/3}}{x^{1/4}} = x^{2/3-1/4} = x^{8/12-3/12} = x^{5/12} = \sqrt[12]{x^5}$

41. $\qquad f(x) = x^2$

$\qquad f\left(a - \sqrt{2}\right) = \left(a - \sqrt{2}\right)^2$

$\qquad\qquad = a^2 - 2\cdot a\cdot\sqrt{2} + \left(\sqrt{2}\right)^2$

$\qquad\qquad = a^2 - 2a\sqrt{2} + 2$

42. $\sqrt{\dfrac{x}{8y}} = \dfrac{\sqrt{x}}{2\sqrt{2y}} = \dfrac{\sqrt{x}}{2\sqrt{2y}} \cdot \dfrac{\sqrt{2y}}{\sqrt{2y}} = \dfrac{\sqrt{2xy}}{4y}$

43. $\dfrac{4\sqrt{5}}{\sqrt{2}+\sqrt{3}} = \dfrac{4\sqrt{5}}{\sqrt{2}+\sqrt{3}} \cdot \dfrac{\sqrt{2}-\sqrt{3}}{\sqrt{2}-\sqrt{3}}$

$= \dfrac{4\sqrt{5}\cdot\sqrt{2} - 4\sqrt{5}\cdot\sqrt{3}}{\left(\sqrt{2}\right)^2 - \left(\sqrt{3}\right)^2}$

$= \dfrac{4\sqrt{10} - 4\sqrt{15}}{2-3} = -4\sqrt{10} + 4\sqrt{15}$

44. $\dfrac{4\sqrt{5}}{\sqrt{2}+\sqrt{3}} = \dfrac{4\sqrt{5}}{\sqrt{2}+\sqrt{3}} \cdot \dfrac{\sqrt{5}}{\sqrt{5}}$

$= \dfrac{4\cdot\left(\sqrt{5}\right)^2}{\sqrt{2}\cdot\sqrt{5} + \sqrt{3}\cdot\sqrt{5}}$

$= \dfrac{4\cdot 5}{\sqrt{10}+\sqrt{15}} = \dfrac{20}{\sqrt{10}+\sqrt{15}}$

45. $\sqrt{y+6} - 2 = 3$

$\sqrt{y+6} = 5$

$\left(\sqrt{y+6}\right)^2 = 5^2$

$y+6 = 25$

$y = 19$

Check: $\sqrt{19+6} - 2 = \sqrt{25} - 2 = 5 - 2 = 3$
The solution is 19.

46. $\sqrt{x} = x - 6$

$\left(\sqrt{x}\right)^2 = (x-6)^2$

$x = x^2 - 12x + 36$

$0 = x^2 - 13x + 36$

$0 = (x-4)(x-9)$

$x - 4 = 0 \quad or \ x - 9 = 0$

$x = 4 \quad or \qquad x = 9$

Check: $\sqrt{4} = 4 - 6$

$2 = -2 \quad$ FALSE

$\sqrt{9} = 9 - 6$

$3 = 3 \qquad$ TRUE

The solution is 9.

47. $(x+1)^{1/3} = -5$

$\left((x+1)^{1/3}\right)^3 = (-5)^3$

$x + 1 = -125$

$x = -126$

Check: $(-126+1)^{1/3} = (-125)^{1/3} = -5$

The solution is −126.

48. $1 + \sqrt{x} = \sqrt{3x-3}$

$\left(1+\sqrt{x}\right)^2 = \left(\sqrt{3x-3}\right)^2$

$1 + 2\sqrt{x} + x = 3x - 3$

$2\sqrt{x} = 2x - 4$

$2\sqrt{x} = 2(x-2)$

$\sqrt{x} = x - 2$

$\left(\sqrt{x}\right)^2 = (x-2)^2$

$x = x^2 - 4x + 4$

$0 = x^2 - 5x + 4$

$0 = (x-4)(x-1)$

$x - 4 = 0 \ or \ x - 1 = 0$

$x = 4 \ or \qquad x = 1$

Check: $x = 4$:

$1+\sqrt{4}$	$\sqrt{3(4)-3}$
$1+2$	$\sqrt{12-3}$
3	$\sqrt{9}$
3	3

$TRUE$

$x = 41$:

$1+\sqrt{1}$	$\sqrt{3(1)-3}$
$1+1$	$\sqrt{3-3}$
2	$\sqrt{0}$
2	0

$FALSE$

The solution is 4.

49. $f(x) = \sqrt[4]{x+2}; f(a) = 2.$ By substitution:

$2 = \sqrt[4]{x+2}$

$2^4 = \left(\sqrt[4]{x+2}\right)^4$

$16 = x + 2$

$14 = x$

Check: $\sqrt[4]{14+2} = \sqrt[4]{16} = \sqrt[4]{2^4} = 2$
The solution is 14.

50. Let $x =$ the length of a side of the square.
Using the Pythagorean equations, solve for x.
$$x^2 + x^2 = 10^2$$
$$2x^2 = 100$$
$$x^2 = 50$$
$$x = \sqrt{50} = \sqrt{25 \cdot 2} = 5\sqrt{2}$$
$$x \approx 7.071$$
The length of a side of the square is $5\sqrt{2}$ cm, or approximately 7.071 cm.

51. Using the Pythagorean equation, determine b, the length of the base.
$$b^2 + 2^2 = 6^2$$
$$b^2 + 4 = 36$$
$$b^2 = 32$$
$$b = \sqrt{32} = \sqrt{16 \cdot 2} = 4\sqrt{2}$$
$$b \approx 5.657$$
The base is $4\sqrt{2}$ ft, or about 5.657 ft.

52. This is a 30°-60°-90° triangle. The length of short leg is $1/2 \times$ the length of the hypotenuse, or $\dfrac{1}{2} \cdot 20 = 10$. The length of the long leg is $\sqrt{3} \times$ the length of the short leg, or $\sqrt{3} \cdot 10 = 10\sqrt{3} \approx 17.321$.

53. Substitute the coordinates into the distance formula.
$$d = \sqrt{(x_2 - x_1)^2 + (y_2 - y_1)^2}$$
$$= \sqrt{(-1-(-6))^2 + (5-4)^2}$$
$$= \sqrt{5^2 + 1^2}$$
$$= \sqrt{25+1}$$
$$= \sqrt{26}$$
$$\approx 5.099$$

54. Substitute the coordinates into the midpoint formula.
$$\left(\frac{x_1 + x_2}{2}, \frac{y_1 + y_2}{2} \right)$$
$$\left(\frac{-7+3}{2}, \frac{-2+(-1)}{2} \right), \text{ or } \left(\frac{-4}{2}, \frac{-3}{2} \right),$$
$$\text{or } \left(-2, -\frac{3}{2} \right)$$

55. $\sqrt{-45} = \sqrt{9 \cdot 5 \cdot -1} = \sqrt{9} \cdot \sqrt{5} \cdot \sqrt{-1} = 3\sqrt{5}i$, or $3i\sqrt{5}$

56. $(-4+3i)+(2-12i) = (-4+2) + \left[3+(-12) \right]i$
$$= -2 - 9i$$

57. $(9-7i)-(3-8i) = (9-3) + \left[-7-(-8) \right]i$
$$= 6 + i$$

58. $(2+5i)(2-5i) = 2^2 - (5i)^2 = 4 - 25i^2$
$$= 4 + 25 = 29$$

59. $i^{18} = \left(i^2 \right)^9 = \left(-1 \right)^9 = -1$

60. $(6-3i)(2-i)$
$$= 6 \cdot 2 + 6 \cdot -i + -3i \cdot 2 + -3i \cdot -i$$
$$= 12 - 6i - 6i + 3i^2$$
$$= 12 - 12i + 3(-1)$$
$$= 9 - 12i$$

61. $\dfrac{7-2i}{3+4i} \cdot \dfrac{3-4i}{3-4i} = \dfrac{21-34i+8i^2}{9-16i^2}$
$$= \frac{21-34i-8}{9+16} = \frac{13-34i}{25}$$
$$= \frac{13}{25} - \frac{34}{25}i$$

62. *Thinking and Writing Exercise.* A complex number $a + bi$ is real when $b = 0$; it is imaginary when $a = 0$.

63. *Thinking and Writing Exercise.* An absolute-value sign must be used to simplify $\sqrt[n]{x^n}$ when n is even, since x may be negative. If x is negative, while n is even, the radical expression cannot be simplified to x, since $\sqrt[n]{x^n}$ represents the principal, or non-negative root. When n is odd, there is only one root, and it will be the same sign (positive or negative) as the radicand; thus, no absolute-value sign is needed when n is odd.

64.
$$\sqrt{11x+\sqrt{6+x}}=6$$
$$\left(\sqrt{11x+\sqrt{6+x}}\right)^2=6^2$$
$$11x+\sqrt{6+x}=36$$
$$\left(\sqrt{6+x}\right)^2=\left(36-11x\right)^2$$
$$6+x=1296-792x+121x^2$$
$$0=121x^2-793x+1290$$
$$0=(121x-430)(x-3)$$
$$121x-430=0 \quad or \quad x-3=0$$
$$x=\frac{430}{121} \quad or \quad x=3$$

Check: $x=\frac{430}{121}$; $\sqrt{11\left(\frac{430}{121}\right)+\sqrt{6+\frac{430}{121}}}\overset{?}{=}6$

$$\sqrt{\frac{430}{11}+\sqrt{\frac{1156}{121}}}$$
$$\sqrt{\frac{430}{11}+\frac{34}{11}}$$
$$\sqrt{\frac{464}{11}}\neq 6$$

$x=3$; $\sqrt{11(3)+\sqrt{6+3}}\overset{?}{=}6$
$$\sqrt{33+\sqrt{9}}$$
$$\sqrt{33+3}$$
$$\sqrt{36}$$
$$6=6 \quad \text{True}$$

The solution is 3

65.
$$\frac{2}{1-3i}-\frac{3}{4+2i}$$
$$=\frac{2}{1-3i}\cdot\frac{1+3i}{1+3i}-\frac{3}{4+2i}\cdot\frac{4-2i}{4-2i}$$
$$=\frac{2+6i}{1+9}-\left(\frac{12-6i}{16+4}\right)$$
$$=\frac{2+6i}{10}-\left[\frac{2(6-3i)}{2\cdot 10}\right]$$
$$=\frac{2+6i-(6-3i)}{10}$$
$$=\frac{(2-6)+\left[6-(-3)\right]i}{10}$$
$$=\frac{-4+9i}{10}=\frac{-4}{10}+\frac{9}{10}i \text{ or } -\frac{2}{5}+\frac{9}{10}i$$

66. Variable answers. Possible form $\frac{ai}{bi}$ where a and b are integers (for instance, $\frac{2i}{3i}$).

67. The isosceles triangle has 2 legs of equal length, which equal the length of the hypotenuse $\div\sqrt{2}$, or $\frac{6}{\sqrt{2}}\cdot\frac{\sqrt{2}}{\sqrt{2}}=\frac{6\sqrt{2}}{2}=3\sqrt{2}$.

The area of the isosceles triangle is
$$\frac{1}{2}\cdot 3\sqrt{2}\cdot 3\sqrt{2}=9.$$

The 30°-60°-90° triangle's shorter leg is $\frac{1}{2}$ the length of the hypotenuse, or $\frac{1}{2}\cdot 6=3$. The length of the longer leg is $\sqrt{3}\cdot$ the length of the shorter leg, or $3\sqrt{3}$. The area of the 30°-60°-90° triangle is
$$\frac{1}{2}\cdot 3\cdot 3\sqrt{3}=\frac{9}{2}\sqrt{3}\approx 7.794$$

The isosceles right triangle is greater by $9-7.794$, or 1.206 ft^2.

Chapter 10 Test

1. $\sqrt{50}=\sqrt{25\cdot 2}=\sqrt{25}\cdot\sqrt{2}=5\sqrt{2}$

2. $\sqrt[3]{-\dfrac{8}{x^6}} = \dfrac{\sqrt[3]{-8}}{\sqrt[3]{x^6}} = -\dfrac{2}{x^2}$

3. $\sqrt{81a^2} = \sqrt{81 \cdot a^2} = \sqrt{81} \cdot \sqrt{a^2} = 9|a|$

4. $\sqrt{x^2 - 8x + 16} = \sqrt{(x-4)^2} = |x-4|$

5. $\sqrt{7xy} = (7xy)^{1/2}$

6. $(4a^3b)^{5/6} = \sqrt[6]{(4a^3b)^5}$

7. $f(x) = \sqrt{2x - 10}$

   The domain of f are those values of x which make $2x - 10$ non-negative.
   $$2x - 10 \geq 0$$
   $$2x \geq 10$$
   $$x \geq 5$$
   The domain of f is $\{x | x \geq 5\}$, or $[5, \infty)$

8. If $f(x) = x^2$, substitute to determine $f(5 + \sqrt{2})$.
   $$f(5 + \sqrt{2}) = (5 + \sqrt{2})^2$$
   $$= 5^2 + 2 \cdot 5 \cdot \sqrt{2} + (\sqrt{2})^2$$
   $$= 25 + 10\sqrt{2} + 2$$
   $$= 27 + 10\sqrt{2}$$

9. $\sqrt[5]{32x^{16}y^{10}} = \sqrt[5]{32x^{15}y^{10} \cdot x} = 2x^3y^2\sqrt[5]{x}$

10. $\sqrt[3]{4w} \cdot \sqrt[3]{4v^2} = \sqrt[3]{4w \cdot 4v^2} = \sqrt[3]{16wv^2}$
    $$= \sqrt[3]{8 \cdot 2wv^2} = 2\sqrt[3]{2wv^2}$$

11. $\sqrt{\dfrac{100a^4}{9b^6}} = \dfrac{\sqrt{100a^4}}{\sqrt{9b^6}} = \dfrac{10a^2}{3b^3}$

12. $\dfrac{\sqrt[5]{48x^6y^{10}}}{\sqrt[5]{16x^2y^9}} = \sqrt[5]{\dfrac{48x^6y^{10}}{16x^2y^9}} = \sqrt[5]{3x^4y}$

13. $\sqrt[4]{x^3}\sqrt{x} = x^{3/4} \cdot x^{1/2} = x^{3/4+1/2} = x^{5/4}$
    $$= \sqrt[4]{x^5} = \sqrt[4]{x^4 \cdot x} = x\sqrt[4]{x}$$

14. $\dfrac{\sqrt{y}}{\sqrt[10]{y}} = \dfrac{y^{1/2}}{y^{1/10}} = y^{1/2 - 1/10} = y^{5/10 - 1/10} = y^{4/10}$
    $$= y^{2/5} = \sqrt[5]{y^2}$$

15. $8\sqrt{2} - 2\sqrt{2} = (8-2)\sqrt{2} = 6\sqrt{2}$

16. $\sqrt{x^4y} + \sqrt{9y^3} = \sqrt{x^4 \cdot y} + \sqrt{9y^2 \cdot y}$
    $$= x^2\sqrt{y} + 3y\sqrt{y}$$
    $$= (x^2 + 3y)\sqrt{y}$$

17. $(7 + \sqrt{x})(2 - 3\sqrt{x})$
    $$= 7 \cdot 2 + 7 \cdot (-3\sqrt{x}) + \sqrt{x} \cdot 2 + \sqrt{x} \cdot (-3\sqrt{x})$$
    $$= 14 - 21\sqrt{x} + 2\sqrt{x} - 3x$$
    $$= 14 - 19\sqrt{x} - 3x$$

18. $\dfrac{\sqrt[3]{x}}{\sqrt[3]{4y}} = \dfrac{\sqrt[3]{x}}{\sqrt[3]{2^2y}} = \dfrac{\sqrt[3]{x}}{\sqrt[3]{2^2y}} \cdot \dfrac{\sqrt[3]{2y^2}}{\sqrt[3]{2y^2}} = \dfrac{\sqrt[3]{2xy^2}}{2y}$

19. $6 = \sqrt{x-3} + 5 \Rightarrow 1 = \sqrt{x-3}$
    $$\Rightarrow 1^2 = (\sqrt{x-3})^2$$
    $$\Rightarrow 1 = x - 3 \Rightarrow x = 4$$
    The value checks, $x = 4$.

20. $$x = \sqrt{3x+3} - 1$$
    $$x + 1 = \sqrt{3x+3}$$
    $$(x+1)^2 = (\sqrt{3x+3})^2$$
    $$x^2 + 2x + 1 = 3x + 3$$
    $$x^2 - x - 2 = 0$$
    $$(x+1)(x-2) = 0$$
    $$x + 1 = 0 \quad or \quad x - 2 = 0$$
    $$x = -1 \quad or \qquad x = 2$$
    Check: $x = -1 \qquad -1 \overset{?}{=} \sqrt{3(-1)+3} - 1$
    $$= \sqrt{-3+3} - 1 = -1$$
    $$x = 2 \qquad 2 \overset{?}{=} \sqrt{3(2)+3} - 1$$
    $$= \sqrt{6+3} - 1 = \sqrt{9} - 1$$
    $$= 3 - 1 = 2$$
    The solutions are -1 and 2.

21.
$$\sqrt{2x} = \sqrt{x+1} + 1$$
$$\sqrt{2x} - 1 = \sqrt{x+1}$$
$$\left(\sqrt{2x} - 1\right)^2 = \left(\sqrt{x+1}\right)^2$$
$$2x - 2\sqrt{2x} + 1 = x + 1$$
$$x - 2\sqrt{2x} = 0$$
$$x = 2\sqrt{2x}$$
$$x^2 = \left(2\sqrt{2x}\right)^2$$
$$x^2 = 4 \cdot 2x$$
$$x^2 - 8x = 0$$
$$x(x-8) = 0$$
$$x = 0 \quad or \quad x - 8 = 0$$
$$x = 0 \quad or \quad x = 8$$

Check: $x = 0$ $\quad \sqrt{2(0)} = 0 \overset{?}{=} \sqrt{0+1} + 1$
$$= \sqrt{1} + 1 = 1 + 1 = 2$$
The equation is false, so $x = 0$ is not a solution.

$x = 8$ $\quad \sqrt{2(8)} \overset{?}{=} \sqrt{8+1} + 1$
$$\sqrt{16} \overset{?}{=} \sqrt{9} + 1$$
$$4 \overset{?}{=} 3 + 1$$
$$4 = 4$$
The solution is 8.

22. Let $d =$ the diagonal distance. Using the given information and the Pythagorean equation, we solve for d.
$$d^2 = 50^2 + 90^2$$
$$d^2 = 2500 + 8100$$
$$d^2 = 10,600$$
$$d = \sqrt{10,600} = \sqrt{100 \cdot 106} = 10\sqrt{106}$$
$$d \approx 102.956$$
She jogged $10\sqrt{106}$ ft., or approximately 102.956 ft.

23. If the length of the hypotenuse is 10 cm, the length of the shorter leg is half as long (5 cm), and the length of the longer leg is $5\sqrt{3}$ cm, or approximately 8.660 cm.

24. Substitute the coordinates into the distance formula.
$$d = \sqrt{(x_2 - x_1)^2 + (y_2 - y_1)^2}$$
$$= \sqrt{(-1-3)^2 + (8-7)^2}$$
$$= \sqrt{(-4)^2 + 1^2}$$
$$= \sqrt{16 + 1}$$
$$= \sqrt{17}$$
$$\approx 4.123$$

25. Substitute the coordinates into the midpoint formula.
$$\left(\frac{x_1 + x_2}{2}, \frac{y_1 + y_2}{2}\right)$$
$$\left(\frac{2+1}{2}, \frac{-5+(-7)}{2}\right), \text{ or } \left(\frac{3}{2}, \frac{-12}{2}\right),$$
or $\left(\frac{3}{2}, -6\right)$

26. $\sqrt{-50} = \sqrt{50 \cdot -1} = \sqrt{25 \cdot 2 \cdot -1}$
$$= \sqrt{25} \cdot \sqrt{2} \cdot \sqrt{-1} = 5\sqrt{2}i$$
or $5i\sqrt{2}$

27. $(9+8i) - (-3+6i) = [9 - (-3)] + [8-6]i$
$$= 12 + 2i$$

28. $\sqrt{-16} \cdot \sqrt{-36} = \sqrt{16 \cdot -1} \cdot \sqrt{36 \cdot -1}$
$$= 4i \cdot 6i = 24i^2 = -24$$

29. $(4-i)^2 = 4^2 - 2 \cdot 4 \cdot i + i^2 = 16 - 8i + i^2$
$$= 16 - 8i - 1 = 15 - 8i$$

30. $\dfrac{-2+i}{3-5i} = \dfrac{-2+i}{3-5i} \cdot \dfrac{3+5i}{3+5i} = \dfrac{-6-7i-5}{9+25}$
$$= \dfrac{-11-7i}{34} = -\dfrac{11}{34} - \dfrac{7}{34}i$$

31. $i^{37} = i^{36} \cdot i = \left(i^2\right)^{18} \cdot i = (-1)^{18} \cdot i = 1 \cdot i = i$

32. $\sqrt{2x-2}+\sqrt{7x+4}=\sqrt{13x+10}$

$\left(\sqrt{2x-2}+\sqrt{7x+4}\right)^2=\left(\sqrt{13x+10}\right)^2$

$(2x-2)+2\sqrt{2x-2}\cdot\sqrt{7x+4}+$

$\qquad\qquad\qquad(7x+4)=13x+10$

$9x+2+2\sqrt{(2x-2)(7x+4)}=13x+10$

$2\sqrt{14x^2-6x-8}=4x+8$

$2\sqrt{14x^2-6x-8}=2(2x+4)$

$\sqrt{14x^2-6x-8}=2x+4$

$\left(\sqrt{14x^2-6x-8}\right)^2=(2x+4)^2$

$14x^2-6x-8=4x^2+16x+16$

$10x^2-22x-24=0$

$2(5x^2-11x-12)=0$

$(5x+4)(x-3)=0$

$5x+4=0\quad or\quad x-3=0$

$\qquad x=-\dfrac{4}{5}\ or\qquad x=3$

Check: $x=-\dfrac{4}{5}$

$$\sqrt{2\left(-\frac{4}{5}\right)-2}=\sqrt{-\frac{8}{5}-\frac{10}{5}}=\sqrt{-\frac{18}{5}}$$

Since we cannot take the principle square of a

negative number, using real number, $-\dfrac{4}{5}$ is

not a solution.

Check: $x=3$

$$\sqrt{2(3)-2}+\sqrt{7(3)+4}\overset{?}{=}\sqrt{13(3)+10}$$

$\sqrt{6-2}+\sqrt{21+4}\qquad\sqrt{39+10}$

$\sqrt{4}+\sqrt{25}\qquad\qquad\sqrt{49}$

$\qquad\qquad 2+5=7$

The solution is 3.

33. $\dfrac{1-4i}{4i(1+4i)^{-1}}=\dfrac{1-4i}{4i\cdot\dfrac{1}{1+4i}}=\dfrac{1-4i}{\dfrac{4i}{1+4i}}$

$=(1-4i)\div\dfrac{4i}{1+4i}$

$=(1-4i)\cdot\dfrac{1+4i}{4i}$

$=\dfrac{(1-4i)(1+4i)}{4i}=\dfrac{1-16i^2}{4i}$

$=\dfrac{17}{4i}\cdot\dfrac{-i}{-i}=\dfrac{-17i}{-4i^2}=\dfrac{-17i}{4}$

$=\dfrac{-17}{4}i$

34. $D(h)=180\Rightarrow 1.2\sqrt{h}=180$

$\Rightarrow 1.2\sqrt{h}=180\Rightarrow\sqrt{h}=\dfrac{180}{1.2}=150$

$\Rightarrow\sqrt{h}^2=150^2\Rightarrow h=22,500$

The pilot must fly 22,500 ft above sea level.

Chapter 11
Quadratic Functions and Equations

1. $\sqrt{k}; -\sqrt{k}$

3. $t+3; t+3$

5. $25; 5$

7. There are 2 x-intercepts, so there are 2 real-number solutions.

9. There is 1 x-intercept, so there is 1 real-number solution.

11. There are no x-intercepts, so there are no real-number solutions.

13. $x^2 = 100 \Rightarrow x = \pm\sqrt{100} = \pm 10$

15. $p^2 - 50 = 0 \Rightarrow p^2 = 50 \Rightarrow p = \pm\sqrt{50}$
$= \pm\sqrt{25 \cdot 2} = \pm\sqrt{25} \cdot \sqrt{2} = \pm 5\sqrt{2}$

17. $4x^2 = 20$

$x^2 = 5$ Multiplying by $\dfrac{1}{4}$

$x = \sqrt{5}$ or $x = -\sqrt{5}$ Using the principle of square roots

The solutions are $\sqrt{5}$ and $-\sqrt{5}$, or $\pm\sqrt{5}$.

19. $x^2 = -4 \Rightarrow x = \pm\sqrt{-4} = \pm\sqrt{4}i = \pm 2i$

21. $9x^2 - 16 = 0 \Rightarrow 9x^2 = 16 \Rightarrow x^2 = \dfrac{16}{9}$

$\Rightarrow x = \pm\sqrt{\dfrac{16}{9}} = \pm\dfrac{\sqrt{16}}{\sqrt{9}} = \pm\dfrac{4}{3}$

23. $5t^2 - 3 = 4 \Rightarrow 5t^2 = 7$

$t^2 = \dfrac{7}{5}$

$t = \sqrt{\dfrac{7}{5}}$ or $t = -\sqrt{\dfrac{7}{5}}$ Principle of square roots

$t = \sqrt{\dfrac{7}{5} \cdot \dfrac{5}{5}}$ or $t = -\sqrt{\dfrac{7}{5} \cdot \dfrac{5}{5}}$ Rationalizing denominators

$t = \dfrac{\sqrt{35}}{5}$ or $t = -\dfrac{\sqrt{35}}{5}$

The solutions are $\pm\sqrt{\dfrac{7}{5}}$ or $\pm\dfrac{\sqrt{35}}{5}$.

25. $4d^2 + 81 = 0 \Rightarrow 4d^2 = -81 \Rightarrow d^2 = -\dfrac{81}{4}$

$\Rightarrow d = \pm\sqrt{-\dfrac{81}{4}} = \pm\sqrt{\dfrac{81}{4}}i = \pm\dfrac{\sqrt{81}}{\sqrt{4}}i = \pm\dfrac{9}{2}i$

27. $(x-1)^2 = 49$

$x - 1 = \pm 7$ Principle of square roots

$x - 1 = 7$ or $x - 1 = -7$

$x = 8$ or $x = -6$

The solutions are 8 and –6.

29. $(a-13)^2 = 18$

$a - 13 = \pm\sqrt{18}$

$a - 13 = -\sqrt{9 \cdot 2}$ or $a - 13 = \sqrt{9 \cdot 2}$

$a = 13 - 3\sqrt{2}$ or $a = 13 + 3\sqrt{2}$

The solutions are $13 \pm 3\sqrt{2}$.

31. $(x+1)^2 = -9$

$x + 1 = \pm\sqrt{-9}$

$x + 1 = -\sqrt{9 \cdot -1}$ or $x + 1 = \sqrt{9 \cdot -1}$

$x = -1 - 3i$ or $x = -1 + 3i$

The solutions are $-1 \pm 3i$.

33. $\left(y+\dfrac{3}{4}\right)^2 = \dfrac{17}{16}$

$$y + \dfrac{3}{4} = \pm\sqrt{\dfrac{17}{16}}$$

$$y + \dfrac{3}{4} = -\dfrac{\sqrt{17}}{\sqrt{16}} \ or \ y + \dfrac{3}{4} = \dfrac{\sqrt{17}}{\sqrt{16}}$$

$$y = -\dfrac{3}{4} - \dfrac{\sqrt{17}}{4} \ or \ y = -\dfrac{3}{4} + \dfrac{\sqrt{17}}{4}$$

$$y = -\dfrac{3-\sqrt{17}}{4} \ or \ y = -\dfrac{3+\sqrt{17}}{4}$$

The solutions are $\dfrac{-3 \pm \sqrt{17}}{4}$.

35. $x^2 - 10x + 25 = 64$

$$\left(x-5\right)^2 = 64$$

$$x - 5 = \pm 8$$

$$x = 5 \pm 8$$

$$x = 13 \ or \ x = -3$$

The solutions are 13 and –3.

37. $f(x) = 19 \Rightarrow x^2 = 19 \Rightarrow x = \pm\sqrt{19}$

39. $f(x) = 16$

$$\left(x-5\right)^2 = 16 \quad \text{Substituting}$$

$$x - 5 = 4 \ or \ x - 5 = -4$$

$$x = 9 \ or \quad x = 1$$

The solutions are 9 and 1.

41. $F(t) = 13$

$$\left(t+4\right)^2 = 13 \quad \text{Substituting}$$

$$t + 4 = \sqrt{13} \qquad or \ t + 4 = -\sqrt{13}$$

$$t = -4 + \sqrt{13} \ or \quad t = -4 - \sqrt{13}$$

The solutions are $-4 + \sqrt{13}$ and $-4 - \sqrt{13}$ or $-4 \pm \sqrt{13}$.

43. $g(x) = x^2 + 14x + 49$

Observe first that $g(0) = 49$. Also observe that when $x = -14$, then $x^2 + 14x = (-14)^2$

$-(14)(14) = (14)^2 - (14)^2 = 0$, so

$g(-14) = 49$ as well. Thus, we have $x = 0$ or

$x = -14$. We can also do this problem as follows.

$$g(x) = 49$$

$$x^2 + 14x + 49 = 49 \quad \text{Substituting}$$

$$\left(x+7\right)^2 = 49$$

$$x + 7 = 7 \ or \ x + 7 = -7$$

$$x = 0 \ or \quad x = -14$$

The solutions are 0 and –14.

45. $x^2 + 16x$

Take half of the coefficient of x and square it: half of 16 is 8 and $8^2 = 64$. Add 64.

$$x^2 + 16x + 64, \ \left(x+8\right)^2$$

47. $t^2 - 10t$

Take half of the coefficient of t and square it: half of –10 is –5, and $\left(-5\right)^2 = 25$. Add 25.

$$t^2 - 10t + 25; \ \left(t-5\right)^2$$

49. $t^2 - 2t$

Take half of the coefficient of t and square it: half of –2 is –1, and $\left(-1\right)^2 = 1$. Add 1.

$$t^2 - 2t + 1; \ \left(t-1\right)^2$$

51. $x^2 + 3x$

Take half the coefficient of x and square it:

half of 3 is $\dfrac{3}{2}$, and $\left(\dfrac{3}{2}\right)^2 = \dfrac{9}{4}$. Add $\dfrac{9}{4}$.

$$x^2 + 3x + \dfrac{9}{4} = \left(x+\dfrac{3}{2}\right)^2$$

53. $x^2 + \dfrac{2}{5}x$

$$\dfrac{1}{2} \cdot \dfrac{2}{5} = \dfrac{1}{5}; \ \left(\dfrac{1}{5}\right)^2 = \dfrac{1}{25}; \ \text{add} \ \dfrac{1}{25}$$

$$x^2 + \dfrac{2}{5}x + \dfrac{1}{25}; \ \left(x+\dfrac{1}{5}\right)^2$$

55. $t^2 - \dfrac{5}{6}t$

$\dfrac{1}{2} \cdot \dfrac{-5}{6} = \dfrac{-5}{12};\ \left(\dfrac{-5}{12}\right)^2 = \dfrac{25}{144};$ add $\dfrac{25}{144}$

$t^2 - \dfrac{5}{6}t + \dfrac{25}{144};\ \left(t - \dfrac{5}{12}\right)^2$

57. $\quad x^2 + 6x = 7$

$x^2 + 6x + 9 = 7 + 9 \quad$ Adding 9 to both sides
to complete the square

$(x+3)^2 = 16 \quad$ Factoring

$x + 3 = \pm 4 \quad$ Principle of square roots

$x = -3 \pm 4$

$x = -3 + 4\ or\ x = -3 - 4$

$x = 1 \qquad or\ x = -7$

The solutions are 1 and –7.

59. $\quad t^2 - 10t = -23$

$t^2 - 10t + 25 = -23 + 25$

$(t - 5)^2 = 2$

$t - 5 = \pm \sqrt{2}$

$t = 5 \pm \sqrt{2}$

The solutions are $5 + \sqrt{2}$ and $5 - \sqrt{2}$.

61. $x^2 + 12x + 32 = 0$

$x^2 + 12x \quad = -32$

$x^2 + 12x + 36 = -32 + 36$

$(x + 6)^2 = 4$

$x + 6 = \pm 2$

$x = -6 \pm 2$

$x = -6 + 2\ or\ x = -6 - 2$

$x = -4 \quad or\ x = -8$

The solutions are –4 and –8.

63. $\quad t^2 + 8t - 3 = 0$

$t^2 + 8t \quad = 3$

$t^2 + 8t + 16 = 3 + 16$

$(t + 4)^2 = 19$

$t + 4 = \pm \sqrt{19}$

$t = -4 \pm \sqrt{19}$

The solutions are $-4 \pm \sqrt{19}$.

65. $f(x) = x^2 + 6x + 7$

To determine the x-intercepts, substitute
$f(x) = 0,$ and solve for x.

$x^2 + 6x + 7 = 0$

$x^2 + 6x \quad = -7$

$x^2 + 6x + 9 = -7 + 9$

$(x + 3)^2 = 2$

$x + 3 = \pm \sqrt{2}$

$x = -3 \pm \sqrt{2}$

The intercepts are $\left(-3 - \sqrt{2}, 0\right)$ and

$\left(-3 + \sqrt{2}, 0\right).$

67. $x^2 + 9x - 25 = 0$

$x^2 + 9x \quad = 25$

$x^2 + 9x + \left(\dfrac{9}{2}\right)^2 = 25 + \left(\dfrac{9}{2}\right)^2 = 25 + \dfrac{81}{4}$

$\left(x + \dfrac{9}{2}\right)^2 = \dfrac{100}{4} + \dfrac{81}{4} = \dfrac{181}{4}$

$x + \dfrac{9}{2} = \pm \sqrt{\dfrac{181}{4}} = \pm \dfrac{\sqrt{181}}{2}$

$x = -\dfrac{9}{2} \pm \dfrac{\sqrt{181}}{2}$

The intercepts are $\left(\dfrac{-9 + \sqrt{181}}{2}, 0\right)$ and

$\left(\dfrac{-9 - \sqrt{181}}{2}, 0\right).$

69. $x^2 - 10x - 22 = 0$

$x^2 - 10x \quad = 22$

$x^2 - 10x + 25 = 22 + 25$

$(x - 5)^2 = 47$

$x = 5 \pm \sqrt{47}$

The intercepts are $\left(5 - \sqrt{47}, 0\right)$ and

$\left(5 + \sqrt{47}, 0\right).$

71. $9x^2 + 18x = -8$

$x^2 + 2x = -\dfrac{8}{9}$

$x^2 + 2x + 1 = -\dfrac{8}{9} + 1$

$(x+1)^2 = \dfrac{1}{9}$

$x + 1 = \pm\dfrac{1}{3}$

$x = -1 \pm \dfrac{1}{3}$

$x = -1 - \dfrac{1}{3}$ or $x = -1 + \dfrac{1}{3}$

$x = -\dfrac{4}{3}$ or $x = -\dfrac{2}{3}$

The solutions are $-\dfrac{4}{3}$ and $-\dfrac{2}{3}$.

73. $3x^2 - 5x - 2 = 0$

$3x^2 - 5x = 2$

$x^2 - \dfrac{5}{3}x = \dfrac{2}{3}$

$x^2 - \dfrac{5}{3}x + \dfrac{25}{36} = \dfrac{2}{3} + \dfrac{25}{36}$

$\left(x - \dfrac{5}{6}\right)^2 = \dfrac{49}{36}$

$x - \dfrac{5}{6} = \pm\dfrac{7}{6}$

$x = \dfrac{5}{6} \pm \dfrac{7}{6}$

$x = \dfrac{5}{6} - \dfrac{7}{6}$ or $x = \dfrac{5}{6} + \dfrac{7}{6}$

$x = -\dfrac{2}{6}$ or $x = \dfrac{12}{6}$

$x = -\dfrac{1}{3}$ or $x = 2$

The solutions are 2 and $-\dfrac{1}{3}$.

75. $5x^2 + 4x - 3 = 0$

$5x^2 + 4x = 3$

$x^2 + \dfrac{4}{5}x = \dfrac{3}{5}$

$x^2 + \dfrac{4}{5}x + \dfrac{4}{25} = \dfrac{3}{5} + \dfrac{4}{25}$

$\left(x + \dfrac{2}{5}\right)^2 = \dfrac{19}{25}$

$x = -\dfrac{2}{5} \pm \dfrac{\sqrt{19}}{5}$

$x = -\dfrac{2 \pm \sqrt{19}}{5}$

The solutions are $\dfrac{-2 \pm \sqrt{19}}{5}$.

77. $4x^2 + 2x - 3 = 0$

$x^2 + \dfrac{1}{2}x = \dfrac{3}{4}$

$x^2 + \dfrac{1}{2}x + \dfrac{1}{16} = \dfrac{3}{4} + \dfrac{1}{16}$

$\left(x + \dfrac{1}{4}\right)^2 = \dfrac{13}{16}$

$x + \dfrac{1}{4} = \pm\dfrac{\sqrt{13}}{4}$

$x = -\dfrac{1}{4} \pm \dfrac{\sqrt{13}}{4}$

The x-intercepts are

$\left(\dfrac{-1 - \sqrt{13}}{4}, 0\right)$ and $\left(\dfrac{-1 + \sqrt{13}}{4}, 0\right)$.

79. $2x^2 - 3x - 1 = 0$

$x^2 - \dfrac{3}{2}x = \dfrac{1}{2}$

$x^2 - \dfrac{3}{2}x + \dfrac{9}{16} = \dfrac{1}{2} + \dfrac{9}{16}$

$\left(x - \dfrac{3}{4}\right)^2 = \dfrac{17}{16}$

$x - \dfrac{3}{4} = \pm\dfrac{\sqrt{17}}{4}$

$x = \dfrac{3}{4} \pm \dfrac{\sqrt{17}}{4};$ or $\dfrac{3 \pm \sqrt{17}}{4}$

The x-intercepts are

$\left(\dfrac{3 - \sqrt{17}}{4}, 0\right)$ and $\left(\dfrac{3 + \sqrt{17}}{4}, 0\right)$.

81. **Familiarize.** We are already familiar with the compound-interest formula.
Translate. We substitute into the formula.

$$A = P(1+r)^t$$

$$2420 = 2000(1+r)^2$$

Carry out. We solve for r.

$$2420 = 2000(1+r)^2$$

$$\frac{2420}{2000} = (1+r)^2$$

$$\frac{121}{100} = (1+r)^2$$

$$\pm\sqrt{\frac{121}{100}} = 1+r$$

$$\pm\frac{11}{10} = 1+r$$

$$-\frac{10}{10} \pm \frac{11}{10} = r$$

$$\frac{1}{10} = r \text{ or } -\frac{21}{10} = r$$

Check. Since the interest rate cannot be negative, we need only check $\frac{1}{10}$, or 10%. If $2000 were invested at 10% interest, compounded annually, then in 2 years it would grow to $2000(1.1)^2$, or $2420. The number 10% checks.
State. The interest rate is 10%.

83. **Familiarize.** We are already familiar with the compound-interest formula.
Translate. We substitute into the formula.

$$A = P(1+r)^t$$

$$6760 = 6250(1+r)^2$$

Carry out. We solve for r.

$$6760 = 6250(1+r)^2$$

$$\frac{6760}{6250} = (1+r)^2$$

$$\frac{676}{625} = (1+r)^2$$

$$\pm\frac{26}{25} = 1+r$$

$$-\frac{25}{25} \pm \frac{26}{25} = r$$

$$\frac{1}{25} = r \text{ or } -\frac{51}{25} = r$$

Check. Since the interest rate cannot be negative, we need only check $\frac{1}{25}$ or 4%. If $6250 were invested at 4% interest, compounded annually, then in 2 years it would grow to $6250(1.04)^2$, or $6760. The number 4% checks.
State. The interest rate is 4%.

85. Use $s = 16t^2$. We substitute 4000 for s, and solve for t.

$$4000 = 16t^2 \Rightarrow t^2 = \frac{4000}{16} = 250$$

$$t = +\sqrt{250} = \sqrt{25 \cdot 10} = 5\sqrt{10}$$

$$t \approx 15.8$$

It will take about 15.8 seconds.

87. Use $s = 16t^2$. We substitute 2063 for s, and solve for t.

$$2063 = 16t^2 \Rightarrow t^2 = \frac{2063}{16}$$

$$t = +\sqrt{\frac{2063}{16}} = \frac{\sqrt{2063}}{4}$$

$$t \approx 11.355$$

It will take about 11.4 seconds.

89. *Thinking and Writing Exercise.*

91. $b^2 - 4ac = 2^2 - 4(3)(-5) = 4 - (-60) = 64$

93. $\sqrt{200} = \sqrt{100 \cdot 2} = \sqrt{100}\sqrt{2} = 10\sqrt{2}$

95. $\sqrt{-4} = \sqrt{4}i = 2i$

97. $\sqrt{-8} = \sqrt{8}i = \sqrt{4}\sqrt{2}i = 2\sqrt{2}i$

99. *Thinking and Writing Exercise.*

101. In order for $x^2 + bx + 81$ to be a square, the following must be true:

$$\left(\frac{b}{2}\right)^2 = 81$$

$$\frac{b^2}{4} = 81$$

$$b^2 = 324$$

$$b = 18 \text{ or } b = -18$$

103. We see that x is a factor of each term, so x is also a factor of $f(x)$. We have

$$f(x) = x\left(2x^4 - 9x^3 - 66x^2 + 45x + 280\right).$$

Since $x^2 - 5$ is a factor of $f(x)$ it is also a factor of $2x^4 - 9x^3 - 66x^2 + 45x + 280$. We divide to find another factor.

$$
\begin{array}{r}
2x^2 - 9x - 56 \\
x^2-5\overline{)2x^4 - 9x^3 - 66x^2 + 45x + 280} \\
\underline{2x^4 \qquad\quad -10x^2} \\
-9x^3 - 56x^2 + 45x \\
\underline{-9x^3 \qquad\quad + 45x} \\
-56x^2 \qquad\quad + 280 \\
\underline{-56x^2 \qquad\quad + 280} \\
0
\end{array}
$$

Then we have $f(x) = x\left(x^2 - 5\right)$ $\left(2x^2 - 9x - 56\right)$, or $f(x) = x\left(x^2 - 5\right)$ $(2x+7)(x-8)$. Now we find the values of a for which $f(a) = 0$.

$$f(a) = 0$$

$$a\left(a^2 - 5\right)(2a+7)(a-8) = 0$$

$a = 0$ or $a^2 - 5 = 0$

or $2a+7 = 0$ or $a - 8 = 0$

$\Rightarrow a = 0$ or $a^2 = 5$

or $2a = -7$ or $a = 8$

$\Rightarrow a = 0$ or $a = \pm\sqrt{5}$

or $a = -\dfrac{7}{2}$ or $a = 8$

The solutions are $0, \sqrt{5}, -\sqrt{5}, -\dfrac{7}{2}$, and 8.

105. **Familiarize.** It is helpful to list information in a chart and make a drawing. Let r represent the speed of the fishing boat. Then $r - 7$ represents the speed of the barge.

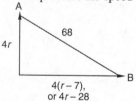

Boat	r	t	d
Fishing	r	4	$4r$
Barge	$r-7$	4	$4(r-7)$

Translate. We use the Pythagorean equation:

$$a^2 + b^2 = c^2$$

$$(4r - 28)^2 + (4r)^2 = 68^2$$

Carry out.

$$(4r - 28)^2 + (4r)^2 = 68^2$$

$$16r^2 - 224r + 784 + 16r^2 = 4624$$

$$32r^2 - 224r - 3840 = 0$$

$$r^2 - 7r - 120 = 0$$

$$(r+8)(r-15) = 0$$

$r + 8 = 0$ or $r - 15 = 0$

$r = -8$ or $r = 15$

Check. We check only 15 since the speeds of the boats cannot be negative. If the speed of the fishing boat is 15 km/h, then the speed of the barge is $15 - 7$, or 8 km/h, and the distances they travel are $4 \cdot 15$ (or 60) and $4 \cdot 8$ (or 32).

$$60^2 + 32^2 = 3600 + 1024 = 4624 = 68^2$$

The values check.

State. The speed of the fishing boat is 15 km/h, and the speed of the barge is 8 km/h.

Exercise Set 11.2

1. True

3. False

5. False

7. $2x^2 + 3x - 5 = 0 \Rightarrow (2x+5)(x-1) = 0$

$\Rightarrow 2x + 5 = 0 \Rightarrow 2x = -5 \Rightarrow x = -\dfrac{5}{2}$

or $x - 1 = 0 \Rightarrow x = 1$

The solutions are $x = -\dfrac{5}{2}$ and $x = 1$.

9. $u^2 + 2u - 4 = 0$

$a = 1,\ b = 2,\ c = -4$

$u = \dfrac{-2 \pm \sqrt{2^2 - 4 \cdot 1 \cdot (-4)}}{2 \cdot 1} = \dfrac{-2 \pm \sqrt{4 + 16}}{2}$

$= \dfrac{-2 \pm \sqrt{20}}{2} = \dfrac{-2 \pm \sqrt{4 \cdot 5}}{2} = \dfrac{-2 \pm 2\sqrt{5}}{2}$

$= \dfrac{\cancel{2}\left(-1 \pm \sqrt{5}\right)}{\cancel{2}} = -1 \pm \sqrt{5}$

The solutions are $u = -1 + \sqrt{5}$ and $u = -1 - \sqrt{5}$.

11. $3p^2 = 18p - 6$

$3p^2 - 18p + 6 = 0$

$p^2 - 6p + 2 = 0$ Dividing by 3

$a = 1,\ b = -6,\ c = 2$

$p = \dfrac{-b \pm \sqrt{b^2 - 4ac}}{2a}$

$p = \dfrac{-(-6) \pm \sqrt{(-6)^2 - 4 \cdot 1 \cdot 2}}{2 \cdot 1} = \dfrac{6 \pm \sqrt{36 - 8}}{2}$

$p = \dfrac{6 \pm \sqrt{28}}{2} = \dfrac{6 \pm 2\sqrt{7}}{2}$

$p = \dfrac{2\left(3 \pm \sqrt{7}\right)}{2} = 3 \pm \sqrt{7}$

The solutions are $3 + \sqrt{7}$ and $3 - \sqrt{7}$.

13. $h^2 + 4 = 6h$

$h^2 - 6h + 4 = 0$

$a = 1,\ b = -6,\ c = 4$

$x = \dfrac{-(-6) \pm \sqrt{(-6)^2 - 4 \cdot 1 \cdot 4}}{2 \cdot 1} = \dfrac{6 \pm \sqrt{36 - 16}}{2}$

$x = \dfrac{6 \pm \sqrt{20}}{2} = \dfrac{6 \pm \sqrt{4 \cdot 5}}{2} = \dfrac{6 \pm 2\sqrt{5}}{2}$

$x = 3 \pm \sqrt{5}$

The solutions are $3 + \sqrt{5}$ and $3 - \sqrt{5}$.

15. $x^2 = 3x + 5 \Rightarrow x^2 - 3x - 5 = 0$

$\Rightarrow a = 1,\ b = -3,\ c = -5$

$x = \dfrac{-(-3) \pm \sqrt{(-3)^2 - 4 \cdot 1 \cdot (-5)}}{2 \cdot 1} = \dfrac{3 \pm \sqrt{9 + 20}}{2}$

$= \dfrac{3 \pm \sqrt{29}}{2} = \dfrac{3}{2} \pm \dfrac{\sqrt{29}}{2}$

The solutions are

$x = \dfrac{3}{2} + \dfrac{\sqrt{29}}{2}$ and $x = \dfrac{3}{2} - \dfrac{\sqrt{29}}{2}$.

17. $3t(t + 2) = 1 \Rightarrow 3t^2 + 6t - 1 = 0$

$\Rightarrow a = 3,\ b = 6,\ c = -1$

$t = \dfrac{-6 \pm \sqrt{6^2 - 4 \cdot 3 \cdot (-1)}}{2 \cdot 3} = \dfrac{-6 \pm \sqrt{36 + 12}}{6}$

$= \dfrac{-6 \pm \sqrt{48}}{6} = \dfrac{-6 \pm \sqrt{16 \cdot 3}}{6} = \dfrac{-6 \pm 4\sqrt{3}}{6}$

$= \dfrac{\cancel{2}\left(-3 \pm 2\sqrt{3}\right)}{\cancel{2} \cdot 3} = \dfrac{-3 \pm 2\sqrt{3}}{3}$

The solutions are

$t = \dfrac{-3 + 2\sqrt{3}}{3}$ and $t = \dfrac{-3 - 2\sqrt{3}}{3}$.

19. $\dfrac{1}{x^2} - 3 = \dfrac{8}{x}$ LCD is x^2

$x^2\left(\dfrac{1}{x^2} - 3\right) = x^2 \cdot \dfrac{8}{x}$

$1 - 3x^2 = 8x$

$0 = 3x^2 + 8x - 1$

$a = 3,\ b = 8,\ c = -1$

$x = \dfrac{-8 \pm \sqrt{8^2 - 4 \cdot 3 \cdot (-1)}}{2 \cdot 3} = \dfrac{-8 \pm \sqrt{64 + 12}}{6}$

$x = \dfrac{-8 \pm \sqrt{76}}{6} = \dfrac{-8 \pm 2\sqrt{19}}{6}$

$x = \dfrac{2\left(-4 \pm \sqrt{19}\right)}{6} = \dfrac{-4 \pm \sqrt{19}}{3}$

The solutions are $-\dfrac{4}{3} \pm \dfrac{\sqrt{19}}{3}$.

21. $t^2 + 10 = 6t \Rightarrow t^2 - 6t + 10 = 0$

$\Rightarrow a = 1,\ b = -6,\ c = 10$

$t = \dfrac{-(-6) \pm \sqrt{(-6)^2 - 4 \cdot 1 \cdot 10}}{2 \cdot 1} = \dfrac{6 \pm \sqrt{36 - 40}}{2}$

$= \dfrac{6 \pm \sqrt{-4}}{2} = \dfrac{6 \pm \sqrt{4}i}{2} = \dfrac{6 \pm 2i}{2}$

$= \dfrac{\cancel{2}\left(3 \pm i\right)}{\cancel{2}} = 3 \pm i$

The solutions are $t = 3 + i$ and $t = 3 - i$.

23. $x^2 + 4x + 6 = 0 \Rightarrow a = 1,\, b = 4,\, c = 6$

$$x = \frac{-4 \pm \sqrt{4^2 - 4 \cdot 1 \cdot 6}}{2 \cdot 1} = \frac{-4 \pm \sqrt{16 - 24}}{2}$$

$$= \frac{-4 \pm \sqrt{-8}}{2} = \frac{-4 \pm \sqrt{8}i}{2} = \frac{-4 \pm 2\sqrt{2}i}{2}$$

$$= \frac{2\left(-2 \pm \sqrt{2}i\right)}{2} = -2 \pm \sqrt{2}i$$

The solutions are

$x = -2 + \sqrt{2}i$ and $x = -2 - \sqrt{2}i$.

25. $12t^2 + 17t = 40 \Rightarrow 12t^2 + 17t - 40 = 0$

$\Rightarrow a = 12,\, b = 17,\, c = -40$

$$t = \frac{-17 \pm \sqrt{17^2 - 4 \cdot 12 \cdot (-40)}}{2 \cdot 12}$$

$$= \frac{-17 \pm \sqrt{289 + 1920}}{24}$$

$$= \frac{-17 \pm \sqrt{2209}}{24} = \frac{-17 \pm 47}{24}$$

The solutions are

$$t = \frac{-17 + 47}{24} = \frac{30}{24} = \frac{5}{4} \text{ and}$$

$$t = \frac{-17 - 47}{24} = \frac{-64}{24} = -\frac{8}{3}.$$

27. $25x^2 - 20x + 4 = 0$

$(5x - 2)(5x - 2) = 0$

$5x - 2 = 0 \text{ or } 5x - 2 = 0$

$$x = \frac{2}{5} \text{ or } \quad x = \frac{2}{5}$$

The solution is $\frac{2}{5}$.

29. $7x(x + 2) + 5 = 3x(x + 1)$

$7x^2 + 14x + 5 = 3x^2 + 3x$

$4x^2 + 11x + 5 = 0$

$$x = \frac{-11 \pm \sqrt{11^2 - 4 \cdot 4 \cdot 5}}{2 \cdot 4}$$

$$x = \frac{-11 \pm \sqrt{41}}{8}$$

The solutions are

$$\frac{-11 + \sqrt{41}}{8} \text{ and } \frac{-11 - \sqrt{41}}{8}.$$

31. $14(x - 4) - (x + 2) = (x + 2)(x - 4)$

$14x - 56 - x - 2 = x^2 - 2x - 8$

$13x - 58 = x^2 - 2x - 8$

$0 = x^2 - 15x + 50$

$0 = (x - 10)(x - 5)$

$x - 10 = 0 \text{ or } x - 5 = 0$

$x = 10 \text{ or } \quad x = 5$

The solutions are 10 and 5.

33. $\qquad 5x^2 = 13x + 17$

$5x^2 - 13x - 17 = 0$

$$x = \frac{-(-13) \pm \sqrt{(-13)^2 - 4(5)(-17)}}{2 \cdot 5}$$

$$x = \frac{13 \pm \sqrt{169 + 340}}{10} = \frac{13 \pm \sqrt{509}}{10}$$

The solutions are $\dfrac{13 + \sqrt{509}}{10}$ and $\dfrac{13 - \sqrt{509}}{10}$.

35. $x(x - 3) = x - 9 \Rightarrow x^2 - 4x + 9 = 0$

$\Rightarrow a = 1,\, b = -4,\, c = 9$

$$x = \frac{-(-4) \pm \sqrt{(-4)^2 - 4 \cdot 1 \cdot 9}}{2 \cdot 1} = \frac{4 \pm \sqrt{16 - 36}}{2}$$

$$= \frac{4 \pm \sqrt{-20}}{2} = \frac{4 \pm \sqrt{20}i}{2} = \frac{4 \pm \sqrt{4 \cdot 5}i}{2}$$

$$= \frac{4 \pm 2\sqrt{5}i}{2} = \frac{2\left(2 \pm \sqrt{5}i\right)}{2} = 2 \pm \sqrt{5}i$$

The solutions are

$x = 2 + \sqrt{5}i$ and $x = 2 - \sqrt{5}i$.

37. $\qquad x^3 - 8 = 0$

$x^3 - 2^3 = 0$

$(x - 2)(x^2 + 2x + 4) = 0$

$x - 2 = 0 \text{ or } x^2 + 2x + 4 = 0$

$$x = 2 \text{ or } x = \frac{-2 \pm \sqrt{2^2 - 4 \cdot 1 \cdot 4}}{2 \cdot 1}$$

$$x = 2 \text{ or } x = \frac{-2 \pm \sqrt{-12}}{2} = \frac{-2 \pm 2i\sqrt{3}}{2}$$

$x = 2 \text{ or } x = -1 \pm i\sqrt{3}$

The solutions are $2,\ -1 + i\sqrt{3}$ and $-1 - i\sqrt{3}$.

39. $g(x) = 0$

$4x^2 - 2x - 3 = 0$ Substituting

$x = \dfrac{-(-2) \pm \sqrt{(-2)^2 - 4 \cdot 4 \cdot (-3)}}{2 \cdot 4} = \dfrac{2 \pm \sqrt{52}}{8}$

$x = \dfrac{2 \pm 2\sqrt{13}}{8} = \dfrac{1 \pm \sqrt{13}}{4}$

The solutions are $\dfrac{1}{4} \pm \dfrac{\sqrt{13}}{4}$.

41. $g(x) = 1$

$\dfrac{2}{x} + \dfrac{2}{x+3} = 1$ Substituting

$2x + 6 + 2x = x^2 + 3x$

$\qquad$ Multiplying by $x(x+3)$

$0 = x^2 - x - 6$

$0 = (x-3)(x+2)$

$x - 3 = 0 \ or \ x + 2 = 0$

$x = 3 \ or \qquad x = -2$

The solutions are 3 and –2.

43. $F(x) = G(x)$

$\dfrac{x+3}{x} = \dfrac{x-4}{3}$ Substituting

$3x\left(\dfrac{x+3}{x}\right) = 3x\left(\dfrac{x-4}{3}\right)$ $\begin{array}{l}\text{Multiplying} \\ \text{by the LCD}\end{array}$

$3x + 9 = x^2 - 4x$

$0 = x^2 - 7x - 9$

$x = \dfrac{-(-7) \pm \sqrt{(-7)^2 - 4 \cdot 1 \cdot (-9)}}{2 \cdot 1}$

$x = \dfrac{7 \pm \sqrt{49 + 36}}{2} = \dfrac{7 \pm \sqrt{85}}{2}$

The solutions are $\dfrac{7 + \sqrt{85}}{2}$ and $\dfrac{7 - \sqrt{85}}{2}$.

45. $x^2 + 4x - 7 = 0$

$x = \dfrac{-4 \pm \sqrt{4^2 - 4 \cdot 1 \cdot (-7)}}{2 \cdot 1}$

$= \dfrac{-4 \pm \sqrt{44}}{2}$

$x = \dfrac{-4 + \sqrt{44}}{2} \approx 1.317$ and

$x = \dfrac{-4 - \sqrt{44}}{2} \approx -5.317$

47. $x^2 - 6x + 4 = 0$

$x = \dfrac{-(-6) \pm \sqrt{(-6)^2 - 4 \cdot 1 \cdot 4}}{2 \cdot 1}$

$= \dfrac{6 \pm \sqrt{20}}{2}$

$x = \dfrac{6 + \sqrt{20}}{2} \approx 5.236$ and

$x = \dfrac{6 - \sqrt{20}}{2} \approx 0.764$

49. $2x^2 - 3x - 7 = 0$

$x = \dfrac{-(-3) \pm \sqrt{(-3)^2 - 4 \cdot 2 \cdot (-7)}}{2 \cdot 2}$

$= \dfrac{3 \pm \sqrt{65}}{4}$

$x = \dfrac{3 + \sqrt{65}}{4} \approx 2.766$ and

$x = \dfrac{3 - \sqrt{65}}{4} \approx -1.266$

51. *Thinking and Writing Exercise.*

53. $(x - 2i)(x + 2i) = x^2 - (2i)^2 = x^2 - 4i^2$

$= x^2 - 4(-1) = x^2 + 4$

55. $(x - (2 - \sqrt{7}))(x - (2 + \sqrt{7}))$

$= (x - 2 + \sqrt{7})(x - 2 - \sqrt{7})$

$= ((x - 2) - \sqrt{7})((x - 2) + \sqrt{7})$

$= (x - 2)^2 - (\sqrt{7})^2 = (x^2 - 4x + 4) - 7$

$= x^2 - 4x - 3$

57. $\dfrac{-6 \pm \sqrt{(-4)^2 - 4(2)(2)}}{2(2)} = \dfrac{-6 \pm \sqrt{16-16}}{4}$

$= \dfrac{-6 \pm 0}{4} = \dfrac{-6}{4} = -\dfrac{3}{2}$

59. *Thinking and Writing Exercise.*

61. $f(x) = \dfrac{x^2}{x-2} + 1$

To find the x-coordinates of the x-intercepts of the graph of f, we solve $f(x) = 0$.

$\dfrac{x^2}{x-2} + 1 = 0$

$x^2 + x - 2 = 0$ Multiplying by $x-2$

$(x+2)(x-1) = 0$

$x = -2 \; or \; x = 1$

The x-intercepts are $(-2, 0)$ and $(1, 0)$.

63. $f(x) = g(x)$

$\dfrac{x^2}{x-2} + 1 = \dfrac{4x-2}{x-2} + \dfrac{x+4}{2}$

Substituting

$2(x-2)\left(\dfrac{x^2}{x-2} + 1\right) = 2(x-2)\left(\dfrac{4x-2}{x-2} + \dfrac{x+4}{2}\right)$

Multiplying by the LCD

$2x^2 + 2(x-2) = 2(4x-2) + (x-2)(x+4)$

$2x^2 + 2x - 4 = 8x - 4 + x^2 + 2x - 8$

$2x^2 + 2x - 4 = x^2 + 10x - 12$

$x^2 - 8x + 8 = 0$

$a = 1, \; b = -8, \; c = 8$

$x = \dfrac{-(-8) \pm \sqrt{(-8)^2 - 4 \cdot 1 \cdot 8}}{2 \cdot 1} = \dfrac{8 \pm \sqrt{64-32}}{2}$

$x = \dfrac{8 \pm \sqrt{32}}{2} = \dfrac{8 \pm \sqrt{16 \cdot 2}}{2} = \dfrac{8 \pm 4\sqrt{2}}{2}$

$x = 4 \pm 2\sqrt{2}$

The solutions are $4 + 2\sqrt{2}$ and $4 - 2\sqrt{2}$.

65. $z^2 + 0.84z - 0.4 = 0$

$z = \dfrac{-0.84 \pm \sqrt{(0.84)^2 - 4 \cdot 1 \cdot (-0.4)}}{2 \cdot 1}$

$z = \dfrac{-0.84 \pm \sqrt{2.3056}}{2}$

$z = \dfrac{-0.84 + \sqrt{2.3056}}{2} \approx 0.339$

$z = \dfrac{-0.84 - \sqrt{2.3056}}{2} \approx -1.179$

67. $\sqrt{2}x^2 + 5x + \sqrt{2} = 0$

$x = \dfrac{-5 \pm \sqrt{5^2 - 4 \cdot \sqrt{2} \cdot \sqrt{2}}}{2\sqrt{2}} = \dfrac{-5 \pm \sqrt{17}}{2\sqrt{2}}$, or

$x = \dfrac{-5 \pm \sqrt{17}}{2\sqrt{2}} \cdot \dfrac{\sqrt{2}}{\sqrt{2}} = \dfrac{-5\sqrt{2} \pm \sqrt{34}}{4}$

The solutions are $\dfrac{-5\sqrt{2} \pm \sqrt{34}}{4}$.

69. $kx^2 + 3x - k = 0$

$k(-2)^2 + 3(-2) - k = 0$

Substituting -2 for

$4k - 6 - k = 0$

$3k = 6$

$k = 2$

$2x^2 + 3x - 2 = 0$

Substituting 2 for k

$(2x-1)(x+2) = 0$

$2x - 1 = 0 \; or \; x + 2 = 0$

$x = \dfrac{1}{2} \; or \quad x = -2$

The other solution is $\dfrac{1}{2}$.

71. *Thinking and Writing Exercise.*

Exercise Set 11.3

1. Discriminant

3. Two

5. Rational

7. $a = 1$, $b = -7$, $c = 5$

$b^2 - 4ac = (-7)^2 - 4 \cdot 1 \cdot 5 = 29$

Since the discriminant is a positive number that is not a perfect square, there are two irrational solutions.

9. $a = 1$, $b = 0$, $c = 3$

$b^2 - 4ac = 0^2 - 4 \cdot 1 \cdot 3 = -12$

Since the discriminant is a negative number, there are two imaginary solutions.

11. $a = 1$, $b = 0$, $c = -5$

$b^2 - 4ac = 0^2 - 4 \cdot 1 \cdot (-5) = 20$

Two irrational solutions.

13. $a = 4$, $b = 8$, $c = -5$

$b^2 - 4ac = 8^2 - 4 \cdot 4 \cdot (-5) = 144$

Since the discriminant is a positive number that is a perfect square, there are two rational solutions.

15. $a = 1$, $b = 4$, $c = 6$

$b^2 - 4ac = 4^2 - 4 \cdot 1 \cdot 6 = -8$

Two imaginary solutions.

17. $a = 9$, $b = -48$, $c = 64$

$b^2 - 4ac = (-48)^2 - 4 \cdot 9 \cdot 64 = 0$

One rational solution.

19. Since $9t^2 - 3t$ is factorable, $3t(3t - 1)$, we know there are two rational solutions.

21. $x^2 + 4x - 8 = 0$ Standard form

$a = 1$, $b = 4$, $c = -8$

$b^2 - 4ac = 4^2 - 4 \cdot 1 \cdot (-8) = 48$

Two irrational solutions.

23. $2a^2 - 3a + 5 = 0$

$a = 2$, $b = -3$, $c = 5$

$b^2 - 4ac = (-3)^2 - 4 \cdot 2 \cdot 5 = -31$

Two imaginary solutions.

25. $7x^2 = 19x \Rightarrow 7x^2 - 19x = 0$

$\Rightarrow a = 7$, $b = -19$, $c = 0$

$\Rightarrow b^2 - 4ac = (-19)^2 - 4 \cdot 7 \cdot 0 = 19^2$

Two rational roots.

27. $y^2 - 4y + \dfrac{9}{4} = 0$

$a = 1$, $b = -4$, $c = \dfrac{9}{4}$

$b^2 - 4ac = (-4)^2 - 4 \cdot 1 \cdot \dfrac{9}{4} = 7$

Two irrational solutions.

29. The solutions are -7, 3

$x = -7$ or $x = 3$

$x + 7 = 0$ or $x - 3 = 0$

$(x + 7)(x - 3) = 0$

$x^2 + 4x - 21 = 0$

31. 3 is the only solution, so $x = 3$ "twice"

$x = 3$

$x - 3 = 0$

$(x - 3)^2 = 0$

$x^2 - 6x + 9 = 0$

Note: This is a perfect-square trinomial.

33. $x = -1$ or $x = -3$

$x + 1 = 0$ or $x + 3 = 0$

$(x + 1)(x + 3) = 0$

$x^2 + 4x + 3 = 0$

35. $x = 5$ or $x = \dfrac{3}{4}$

$x - 5 = 0$ or $x - \dfrac{3}{4} = 0$

$x - 5 = 0$ or $4x - 3 = 0$

$(x - 5)(4x - 3) = 0$

$4x^2 - 23x + 15 = 0$

37. $\left(x + \dfrac{1}{4}\right)\left(x + \dfrac{1}{2}\right) = 0$

$(4x + 1)(2x + 1) = 0$

$8x^2 + 6x + 1 = 0$

39. $(x - 2.4)(x + 0.4) = 0$

$x^2 - 2x - 0.96 = 0$

41. $\left(x + \sqrt{3}\right)\left(x - \sqrt{3}\right) = 0$

$x^2 - 3 = 0$

43. $\left(x-2\sqrt{5}\right)\left(x+2\sqrt{5}\right)=0$

$$x^2-\left(2\sqrt{5}\right)^2=0$$

$$x^2-20=0$$

45. $(x-4i)(x+4i)=0$

$$x^2-(4i)^2=0$$

$$x^2-16i^2=0$$

$$x^2+16=0$$

Reminder: $i^2=-1$

47. $\qquad x=2-7i$ or $\qquad x=2+7i$

$x-2+7i=0\qquad$ or $x-2-7i=0$

$$\left[x+(-2+7i)\right]\left[x+(-2-7i)\right]=0$$

$$x^2-2x-7xi-2x+4+14i$$

$$+7xi-14i-49i^2=0$$

$$x^2-4x+53=0$$

49. $x=3-\sqrt{14}\qquad$ or $\ x=3+\sqrt{14}$

$x-3+\sqrt{14}=0\ $ or $x-3-\sqrt{14}=0$

$$\left[x-\left(3-\sqrt{14}\right)\right]\left[x-\left(3+\sqrt{14}\right)\right]=0$$

$$x^2-3x-x\sqrt{14}-3x+9+$$

$$3\sqrt{14}+x\sqrt{14}-3\sqrt{14}-14=0$$

$$x^2-6x-5=0$$

51.

$$x=1-\frac{\sqrt{21}}{3}\ \text{or}\ x=1+\frac{\sqrt{21}}{3}$$

$$3x=3-\sqrt{21}\ \text{or}\ 3x=3+\sqrt{21}$$

$$3x-3+\sqrt{21}=0$$

$$\text{or}\ 3x-3-\sqrt{21}=0$$

$$\left(3x-3+\sqrt{21}\right)\left(3x-3-\sqrt{21}\right)=0$$

$$9x^2-9x-3x\sqrt{21}-9x+9+3\sqrt{21}+$$

$$3x\sqrt{21}-3\sqrt{21}-21=0$$

$$9x^2-18x-12=0$$

$$3\left(3x^2-6x-4\right)=0$$

$$3x^2-6x-4=0$$

53. $\qquad (x+2)(x-1)(x-5)=0$

$$\left(x^2+x-2\right)(x-5)=0$$

$$x^3+x^2-2x-5x^2-5x+10=0$$

$$x^3-4x^2-7x+10=0$$

55. $(x+1)(x)(x-3)=0$

$$\left(x^2+x\right)(x-3)=0$$

$$x^3-3x^2+x^2-3x=0$$

$$x^3-2x^2-3x=0$$

57. *Thinking and Writing Exercise.*

59. $\dfrac{c}{d}=c+d\Rightarrow \not{d}\cdot\dfrac{c}{\not{d}}=d(c+d)$

$$\Rightarrow c=cd+d^2\Rightarrow c-cd=d^2$$

$$\Rightarrow c(1-d)=d^2\Rightarrow c=\frac{d^2}{1-d}$$

61. $x=\dfrac{3}{1-y}\Rightarrow x(1-y)=3\Rightarrow x-xy=3$

$$\Rightarrow x-3=xy\Rightarrow y=\frac{x-3}{x}$$

63. Let $x=$ Jamal's walking speed in mph, and $y=$ Kade's walking speed in mph. Then:

$x=y+1.5\quad$ (1)

$$\frac{7}{x}=\frac{4}{y}$$

or $7y=4x\quad$ (2)

This system can be solved using substitution. Substitute $y+1.5$ for x in (2) and solve for y. Then use (1) and the value of y to find x.

$7y=4x\quad$ (1)

$\Rightarrow 7y=4(y+1.5)\Rightarrow 7y=4y+6$

$\Rightarrow 3y=6\Rightarrow y=2$ mph

$x=y+1.5\quad$ (1)

$\Rightarrow x=2+1.5=3.5$ mph

Jamal walks at 3.5 mph, and Kade walks at 2 mph.

65. *Thinking and Writing Exercise.*

67. The graph includes the points $(-3,0),(0,-3)$, and $(1,0)$. Substituting in $y = ax^2 + bx + c$, we have three equations.

$0 = 9a - 3b + c,$

$-3 = c,$

$0 = a + b + c$

The solution of this system of equations is $a = 1$, $b = 2$, $c = -3$.

69. a) $kx^2 - 2x + k = 0$; one solution is -3
We first find k by substituting -3 for x.

$$k(-3)^2 - 2(-3) + k = 0$$
$$9k + 6 + k = 0$$
$$10k = -6$$
$$k = -\frac{6}{10}$$
$$k = -\frac{3}{5}$$

b) Now substitute $-\frac{3}{5}$ for k in the original equation.

$$-\frac{3}{5}x^2 - 2x + \left(-\frac{3}{5}\right) = 0$$
$$3x^2 + 10x + 3 = 0$$
$$(3x + 1)(x + 3) = 0$$
$$x = -\frac{1}{3} \text{ or } x = -3$$

The other solution is $-\frac{1}{3}$.

71. a) $x^2 - (6 + 3i)x + k = 0$; one solution is 3.
We first find k by substituting 3 for x.

$$3^2 - (6 + 3i)3 + k = 0$$
$$9 - 18 - 9i + k = 0$$
$$-9 - 9i + k = 0$$
$$k = 9 + 9i$$

b) Now we substitute $9 + 9i$ for k in the original equation.

$$x^2 - (6 + 3i)x + (9 + 9i) = 0$$
$$x^2 - (6 + 3i)x + 3(3 + 3i) = 0$$
$$\left[x - (3 + 3i)\right]\left[x - 3\right] = 0$$
$$x = 3 + 3i \text{ or } x = 3$$

The other solution is $3 + 3i$.

73. The solutions of $ax^2 + bx + c = 0$ are

$x = \dfrac{-b \pm \sqrt{b^2 - 4ac}}{2a}$. When there is just one solution,

$b^2 - 4ac = 0$, so $x = \dfrac{-b \pm 0}{2a} = -\dfrac{b}{2a}$.

75. We substitute $(-3,0), \left(\dfrac{1}{2},0\right)$, and $(0,-12)$ in $f(x) = ax^2 + bx + c$ and get three equations.

$0 = 9a - 3b + c,$

$0 = \dfrac{1}{4}a + \dfrac{1}{2}b + c,$

$-12 = c$

The solution of this system of equations is $a = 8$, $b = 20$, $c = -12$.

77. The only way an equation can have $\sqrt{2}$ as a solution is if $x - \sqrt{2}$ is a factor of the factored form of the equation $a(x - \sqrt{2})(x - k) = 0$. If we select k so that the second factor is the conjugate of $x - \sqrt{2}$, and a to be an integer, then the left side of the equation will have integer coefficients:

$a(x - \sqrt{2})(x + \sqrt{2}) = a(x^2 - \sqrt{2}^2) = ax^2 - 2a.$

Therefore any equation of the form:
$ax^2 - 2a = 0$, where a is an integer will work. Letting $a = 1$ gives the simplest equation:
$x^2 - 2 = 0.$

79. If $1 - \sqrt{5}$ and $3 + 2i$ are two solutions, then $1 + \sqrt{5}$ and $3 - 2i$ are also solutions. The equation of lowest degree that has these solutions is found as follows:

$$\left[x - \left(1 - \sqrt{5}\right)\right]\left[x - \left(1 + \sqrt{5}\right)\right]\left[x - \left(3 + 2i\right)\right]$$
$$\left[x - \left(3 - 2i\right)\right] = 0$$
$$\left(x^2 - 2x - 4\right)\left(x^2 - 6x + 13\right) = 0$$
$$x^4 - 8x^3 + 21x^2 - 2x - 52 = 0$$

Exercise Set 11.4

1. **Familiarize.** Let r represent the speed and t the time for the first part of the trip.

Trip	Distance	Speed	Time
1st part	120	r	t
2nd part	100	$r-10$	$4-t$

Translate. Using $r = \dfrac{d}{t}$, we get two equations from the table, $r = \dfrac{120}{t}$ and

$$r - 10 = \frac{100}{4 - t}.$$

Carry out. We substitute $\dfrac{120}{t}$ for r in the second equation and solve for t.

$$\frac{120}{t} - 10 = \frac{100}{4 - t}$$

LCD is $t(4 - t)$

$$t(4-t)\left(\frac{120}{t} - 10\right) = t(4-t) \cdot \frac{100}{4-t}$$

$$120(4-t) - 10t(4-t) = 100t$$

$$480 - 120t - 40t + 10t^2 = 100t$$

$$10t^2 - 260t + 480 = 0$$

$$t^2 - 26t + 48 = 0$$

$$(t - 24)(t - 2) = 0$$

$t = 24$ or $t = 2$

Check. Since time cannot be negative (If $t = 24$, $4 - 24 = -20$), we check only 2. If $t = 2$, then $4 - 2 = 2$. The speed of the first part is $\dfrac{120}{2}$, or 60 mph. The speed of the second part is $\dfrac{100}{2}$, or 50 mph. The speed of the second part is 10 mph slower than the first part. The value checks.
State. The speed of the first part was 60 mph, and the speed of the second part was 50 mph.

3. **Familiarize.** Let r represent the speed and t the time of the slower trip.

Trip	Distance	Speed	Time
Slower	200	r	t
Faster	200	$r+10$	$t-1$

Translate. Using $t = \dfrac{d}{r}$, we get two equations from the table, $t = \dfrac{200}{r}$ and

$$t - 1 = \frac{200}{r + 10}.$$

Carry out. We substitute $\dfrac{200}{r}$ for t in the second equation and solve for r.

$$\frac{200}{r} - 1 = \frac{200}{r + 10}$$

LCD is $r(r + 10)$

$$r(r+10)\left(\frac{200}{r} - 1\right) = r(r+10) \cdot \frac{200}{r+10}$$

$$200(r+10) - r(r+10) = 200r$$

$$200r + 2000 - r^2 - 10r = 200r$$

$$0 = r^2 + 10r - 2000$$

$$0 = (r + 50)(r - 40)$$

$r = -50$ or $r = 40$

Check. Since speed cannot be negative, we check only 40 mph. If $r = 40$, then $r + 10 = 50$. The time for the slower trip is $\dfrac{200}{40}$, or 5 hrs. The time for the faster trip is $\dfrac{200}{50}$, or 4 hrs. The faster trip is 1 hr. less. The value checks.
State. The speed is 40 mph.

5. **Familiarize.** We let r = the speed and t = the time of the Cessna.

Plane	Distance	Speed	Time
Cessna	600	r	t
Beechcraft	1000	$r+50$	$t+1$

Translate. Using $t = d/r$, we get two equations from the table, $t = \dfrac{600}{r}$ and

$$t + 1 = \frac{1000}{r + 50}.$$

Carry out. We substitute $\dfrac{600}{r}$ for t in the second equation and solve for r.

$$\frac{600}{r}+1=\frac{1000}{r+50}$$

LCD is $r(r+50)$

$$r(r+50)\left(\frac{600}{r}+1\right)=r(r+50)\cdot\frac{1000}{r+50}$$

$$600(r+50)+r(r+50)=1000r$$

$$600r+30{,}000+r^2+50r=1000r$$

$$r^2-350r+30{,}000=0$$

$$(r-150)(r-200)=0$$

$r=150$ *or* $r=200$

Check. If $r=150$, then the Cessna's time is $\frac{600}{150}$, or 4 hr and the Beechcraft's time is $\frac{1000}{150+50}$ or $\frac{1000}{200}$, or 5 hr. If $r=200$, then the Cessna's times is $\frac{600}{200}$, or 3 hr and the Beechcraft's time is $\frac{1000}{200+50}$, or $\frac{1000}{250}$, or 4 hr. Since the Beechcraft's time is 1 hr longer in each case, both values check. There are two solutions.

State. The speed of the Cessna is 150 mph and the speed of the Beechcraft is 200 mph; or the speed of the Cessna is 200 mph and the speed of the Beechcraft is 250 mph.

7. **Familiarize.** We let r represent the speed and t the time of the trip to Hillsboro.

Trip	Distance	Speed	Time
To Hillsboro	40	r	t
Return	40	$r-6$	$14-t$

Translate. Using $t=\frac{d}{r}$, we get two equations from the table,

$$t=\frac{40}{r}\text{ and }14-t=\frac{40}{r-6}.$$

Carry out. We substitute $\frac{40}{r}$ for t in the second equation and solve for r.

$$14-\frac{40}{r}=\frac{40}{r-6}$$

LCD is $r(r-6)$

$$r(r-6)\left(14-\frac{40}{r}\right)=r(r-6)\cdot\frac{40}{r-6}$$

$$14r(r-6)-40(r-6)=40r$$

$$14r^2-84r-40r+240=40r$$

$$14r^2-164r+240=0$$

$$7r^2-82r+120=0$$

$$(7r-12)(r-10)=0$$

$r=\dfrac{12}{7}$ *or* $r=10$

Check. Since negative speed has no meaning in this problem (If $r=\dfrac{12}{7}$, then $r-6=-\dfrac{30}{7}$.) we check only 10 mph. If $r=10$, then the time of the trip to Hillsboro is $\frac{40}{10}$, or 4 hr. The speed of the return trip is $10-6$, or 4 mph and the time is $\frac{40}{4}$, or 10 hr. The total time for the round trip is 4 hr + 10 hr, or 14 hr. The value checks.

State. Naoki's speed on the trip to Hillsboro was 10 mph and it was 4 mph on the return trip.

9. **Familiarize.** Let r represent the speed of the boat in still water and let t represent the time of the trip upriver.

Trip	Distance	Speed	Time
Upriver	60	$r-3$	t
Downriver	60	$r+3$	$9-t$

Translate. Using $t=\frac{d}{r}$, we get

$$t=\frac{60}{r-3}\text{ and }9-t=\frac{60}{r+3}.$$

Carry out. Substitute $\frac{60}{r-3}$ for t in the second equation and solve for r.

$$9 - \frac{60}{r-3} = \frac{60}{r+3}$$

LCD is $(r-3)(r+3)$

$$(r-3)(r+3)\left(9 - \frac{60}{r-3}\right)$$

$$= (r-3)(r+3) \cdot \frac{60}{r+3}$$

$$9(r-3)(r+3) - 60(r+3) = 60(r-3)$$

$$9r^2 - 81 - 60r - 180 = 60r - 180$$

$$9r^2 - 120r - 81 = 0$$

$$3r^2 - 40r - 27 = 0$$

We use the quadratic formula.

$$r = \frac{-(-40) \pm \sqrt{(-40)^2 - 4 \cdot 3 \cdot (-27)}}{2 \cdot 3}$$

$$r = \frac{40 \pm \sqrt{1924}}{6}$$

$r \approx 14$ or $r \approx -0.6$ Since speed cannot be negative, we check $r \approx 14$.

Check. Using the approximation, the speed upriver is $14 - 3$, or 11, and the time is about $\frac{60}{11}$, or 5.5 hrs. The speed downriver is

$14 + 3$, or 17, and the time is about $\frac{60}{17}$, or

3.5. The total time is $5.5 + 3.5$, or 9 hrs. The value checks.

State. The speed of the boat in still water is approximately 14 mph.

11. **Familiarize.** Let x represent the time it takes the spring to fill the pool. Then $x - 6$ represents the time it takes the well to fill the pool. It takes them 4 hr to fill the pool when both wells are working together, so they can

fill $\frac{1}{4}$ of the pool in 1 hr. The spring will fill

$\frac{1}{x}$ of the pool in 1 hr, and the well will fill

$\frac{1}{x-6}$ of the pool in 1 hr.

Translate. We have an equation.

$$\frac{1}{x} + \frac{1}{x-6} = \frac{1}{4}$$

Carry out. We solve the equation.
We multiply by the LCD, $4x(x-6)$.

$$4x(x-6)\left(\frac{1}{x} + \frac{1}{x-6}\right) = 4x(x-6) \cdot \frac{1}{4}$$

$$4(x-6) + 4x = x(x-6)$$

$$4x - 24 + 4x = x^2 - 6x$$

$$0 = x^2 - 14x + 24$$

$$0 = (x-2)(x-12)$$

$x = 2$ or $x = 12$

Check. Since negative time has no meaning in this problem, 2 is not a solution $(2 - 6 = -4)$. We check only 12 hr. This is the time it would take the spring working alone. Then the well would take $12 - 6$, or 6 hr working alone. The well would fill

$4\left(\frac{1}{6}\right)$, or $\frac{2}{3}$, of the pool in 4 hr, and the

spring would fill $4\left(\frac{1}{12}\right)$, or $\frac{1}{3}$, of the pool

in 4 hr. Thus in 4 hr they would fill $\frac{2}{3} + \frac{1}{3}$ of

the pool. This is all of it, so the numbers check.

State. It takes the spring, working alone, 12 hr to fill the pool.

13. **Familiarize.** We let r represent Antonio's speed in still water. Then $r - 2$ is the speed upstream and $r + 2$ is the speed downstream.

Using $t = \dfrac{d}{r}$, we let $\dfrac{1}{r-2}$ represent the time

upstream and $\dfrac{1}{r+2}$ represent the time

downstream.

Trip	Distance	Speed	Time
Upriver	1	$r-2$	$\dfrac{1}{r-2}$
Downriver	1	$r+2$	$\dfrac{1}{r+2}$

Translate. The time for the round trip is 1 hour. We now have an equation.

$$\frac{1}{r-2} + \frac{1}{r+2} = 1$$

Carry out. We solve the equation. We multiply by the LCD, $(r-2)(r+2)$.

$$(r-2)(r+2)\left(\frac{1}{r-2}+\frac{1}{r+2}\right)$$

$$=(r-2)(r+2)\cdot 1$$

$$(r+2)+(r-2)=(r-2)(r+2)$$

$$2r=r^2-4$$

$$0=r^2-2r-4$$

$$r=\frac{-(-2)\pm\sqrt{(-2)^2-4\cdot 1(-4)}}{2\cdot 1}$$

$$r=\frac{2\pm\sqrt{4+16}}{2}=\frac{2\pm\sqrt{20}}{2}$$

$$r=\frac{2\pm 2\sqrt{5}}{2}=1\pm\sqrt{5}$$

$$1+\sqrt{5}\approx 1+2.236\approx 3.24$$

$$1-\sqrt{5}\approx 1-2.236=-1.24$$

Check. Since negative speed has no meaning in this problem, we check only 3.24 mph. If $r\approx 3.24$, then $r-2\approx 1.24$ and $r+2\approx 5.24$. The time it takes to travel upstream is approx. $\frac{1}{1.24}$, or 0.806 hr, and the time it takes to travel downstream is approx. $\frac{1}{5.24}$, or 0.191 hr. The total time is 0.997 which is approximately 1 hour. The value checks.

State. Antonio's speed in still water is approximately 3.24 mph.

15. $$A=4\pi r^2$$

$$\frac{A}{4\pi}=r^2 \quad \text{Dividing by } 4\pi$$

$$\frac{1}{2}\sqrt{\frac{A}{\pi}}=r \quad \text{Taking the positive square root}$$

17. $$A=2\pi r^2+2\pi rh$$

$$0=2\pi r^2+2\pi rh-A \quad \text{Standard form}$$

$$a=2\pi,\ b=2\pi h,\ c=-A$$

$$r=\frac{-2\pi h\pm\sqrt{(2\pi h)^2-4\cdot 2\pi\cdot(-A)}}{2\cdot 2\pi}$$

Using the quadratic formula

$$r=\frac{-2\pi h\pm\sqrt{4\pi^2 h^2+8\pi A}}{4\pi}$$

$$r=\frac{-2\pi h\pm 2\sqrt{\pi^2 h^2+2\pi A}}{4\pi}$$

$$r=\frac{-\pi h\pm\sqrt{\pi^2 h^2+2\pi A}}{2\pi}$$

Since taking the negative square root would result in a negative answer, we take the positive one.

$$r=\frac{-\pi h+\sqrt{\pi^2 h^2+2\pi A}}{2\pi}$$

19. $$F=\frac{Gm_1 m_2}{r^2}$$

$$Fr^2=Gm_1 m_2$$

$$r^2=\frac{Gm_1 m_2}{F}$$

$$r=\sqrt{\frac{Gm_1 m_2}{F}}$$

21. $$c=\sqrt{gH}\ \Rightarrow\ (c)^2=\left(\sqrt{gH}\right)^2\ \Rightarrow\ c^2=gH$$

$$\Rightarrow\ H=\frac{c^2}{g}$$

23. $$w=\frac{l\,g^2}{800}\ \Rightarrow\ \frac{800w}{l}=g^2\ \Rightarrow\ \sqrt{g^2}=\sqrt{\frac{800w}{l}}$$

$$|g|=\sqrt{\frac{800w}{l}}.\ \text{However, if we assume } g>0,$$

we can drop the absolute value bars and get.

$$g=\sqrt{\frac{800w}{l}}.$$

25. $$a^2+b^2=c^2$$

$$b^2=c^2-a^2$$

$$b=\sqrt{c^2-a^2}$$

27. $s = v_0 t + \dfrac{gt^2}{2}$

$0 = \dfrac{gt^2}{2} + v_0 t - s$

$a = \dfrac{g}{2}, \ b = v_0, \ c = -s$

$t = \dfrac{-v_0 \pm \sqrt{v_0^2 - 4\left(\dfrac{g}{2}\right)(-s)}}{2\left(\dfrac{g}{2}\right)}$

$t = \dfrac{-v_0 \pm \sqrt{v_0^2 + 2gs}}{g}$

Since taking the negative square root would result in a negative answer, we take the positive one.

$t = \dfrac{-v_0 + \sqrt{v_0^2 + 2gs}}{g}$

29. $N = \dfrac{1}{2}\left(n^2 - n\right)$

$N = \dfrac{1}{2}n^2 - \dfrac{1}{2}n$

$0 = \dfrac{1}{2}n^2 - \dfrac{1}{2}n - N$

$a = \dfrac{1}{2}, \ b = -\dfrac{1}{2}, \ c = -N$

$n = \dfrac{-\left(-\dfrac{1}{2}\right) \pm \sqrt{\left(-\dfrac{1}{2}\right)^2 - 4 \cdot \dfrac{1}{2} \cdot (-N)}}{2\left(\dfrac{1}{2}\right)}$

$n = \dfrac{1}{2} \pm \sqrt{\dfrac{1}{4} + 2N}$

$n = \dfrac{1}{2} \pm \sqrt{\dfrac{1 + 8N}{4}}$

$n = \dfrac{1}{2} \pm \dfrac{1}{2}\sqrt{1 + 8N}$

Since taking the negative square root would result in a negative answer, we take the positive one.

$n = \dfrac{1}{2} + \dfrac{1}{2}\sqrt{1 + 8N}$, or $\dfrac{1 + \sqrt{1 + 8N}}{2}$

31. $T = 2\pi\sqrt{\dfrac{l}{g}}$

$\dfrac{T}{2\pi} = \sqrt{\dfrac{l}{g}}$

$\dfrac{T^2}{4\pi^2} = \dfrac{l}{g}$

$gT^2 = 4\pi^2 l$

$g = \dfrac{4\pi^2 l}{T^2}$

33. $at^2 + bt + c = 0$

The quadratic formula gives the result.

$t = \dfrac{-b \pm \sqrt{b^2 - 4ac}}{2a}$

35. a) From example 4, we know

$t = \dfrac{-v_0 + \sqrt{v_0^2 + 19.6s}}{9.8}$

Substituting 500 for s and 0 for v_0, we

have $t = \dfrac{0 + \sqrt{0^2 + 19.6(500)}}{9.8}$

$t \approx 10.1$

It takes about 10.1 sec to reach the ground.

b) $t = \dfrac{-v_0 + \sqrt{v_0^2 + 19.6s}}{9.8}$

Substitute 500 for s and 30 for v_0.

$t = \dfrac{-30 + \sqrt{30^2 + 19.6(500)}}{9.8}$

$t \approx 7.49$

It takes about 7.49 sec to reach the ground.

c) We will use the formula in Example 4,
$s = 4.9t^2 + v_0 t$.

Substitute 5 for t and 30 for v_0.

$s = 4.9(5)^2 + 30(5) = 272.5$

The object will fall 272.5 m.

37. From Example 4, we know

$$t = \frac{-v_0 + \sqrt{v_0{}^2 + 19.6s}}{9.8}$$

Substituting 40 for s and 0 for v_0, we have

$$= \frac{0 + \sqrt{0^2 + 19.6(40)}}{9.8}$$

$t \approx 2.9$

He will be falling for about 2.9 sec.

39. From Example 3, we know $T = \dfrac{\sqrt{3V}}{12}$

Substituting 44 for V, we have

$$T = \frac{\sqrt{3 \cdot 44}}{12}$$

$T \approx 0.957$

LeBron's hang time is about 0.957 sec.

41. $s = 4.9t^2 + v_0 t$

Solve the formula for v_0

$s - 4.9t^2 = v_0 t$

$\dfrac{s - 4.9t^2}{t} = v_0$

Substitute 51.6 for s and 3 for t.

$\dfrac{51.6 - 4.9(3)^2}{3} = v_0$

$2.5 = v_0$

The initial velocity is 2.5 m/sec.

43. $A = P_1(1+r)^2 + P_2(1+r)$ where $A =$ total amount in the account after 2 years, P is the amount of the original deposit, P_2 is the amount deposited at the beginning of the second year, and r is the annual interest rate. We are given $P_1 = 3000$, $P_2 = 1700$, and $A = 5253.70$.

Substitute to determine r.

$$r = -1 + \frac{-1700 + \sqrt{(1700)^2 + 4(3000)5253.70}}{2(3000)}$$

Using a calculator, we have $r = 0.07$. The annual interest rate is 0.07, or 7%.

45. *Thinking and Writing Exercise.* .

47. $(m^{-1})^2 = m^{-1 \cdot 2} = m^{-2} = \dfrac{1}{m^2}$

49. $(y^{1/6})^2 = y^{2(1/6)} = y^{2/6} = y^{1/3}$

51. $t^{-1} = \dfrac{1}{2} \Rightarrow \left(t^{-1}\right)^{-1} = \left(\dfrac{1}{2}\right)^{-1} \Rightarrow t^{(-1)(-1)} = \left(\dfrac{2}{1}\right)^1$

$\Rightarrow t = 2$

53. *Thinking and Writing Exercise.*

55. $\qquad A = 6.5 - \dfrac{20.4t}{t^2 + 36}$

$$(t^2 + 36)A = (t^2 + 36)\left(6.5 - \frac{20.4t}{t^2 + 36}\right)$$

$$At^2 + 36A = (t^2 + 36)(6.5) - (t^2 + 36)\left(\frac{20.4t}{t^2 + 36}\right)$$

$$At^2 + 36A = 6.5t^2 + 234 - 20.4t$$

$$At^2 - 6.5t^2 + 20.4t + 36A - 234 = 0$$

$$(A - 6.5)t^2 + 20.4t + (36A - 234) = 0$$

$$a = A - 6.5, \ b = 20.4, \ c = 36A - 234$$

$$t = \frac{-20.4 \pm \sqrt{(20.4)^2 - 4(A - 6.5)(36A - 234)}}{2(A - 6.5)}$$

$$t = \frac{-20.4 \pm \sqrt{416.16 - 144A^2 + 1872A - 6084}}{2(A - 6.5)}$$

$$t = \frac{-20.4 \pm \sqrt{-144A^2 + 1872A - 5667.84}}{2(A - 6.5)}$$

$$t = \frac{-20.4 \pm \sqrt{144(-A^2 + 13A - 39.36)}}{2(A - 6.5)}$$

$$t = \frac{-20.4 \pm 12\sqrt{-A^2 + 13A - 39.36}}{2(A - 6.5)}$$

$$t = \frac{2\left(-10.2 \pm 6\sqrt{-A^2 + 13A - 39.36}\right)}{2(A - 6.5)}$$

$$t = \frac{-10.2 \pm 6\sqrt{-A^2 + 13A - 39.36}}{A - 6.5}$$

57. Let a = the number. Then $a - 1$ is 1 less than

a and the reciprocal of that number is $\dfrac{1}{a-1}$.

Also 1 more than the number is $a + 1$.

$$\frac{1}{(a-1)} = a+1$$

We solve the equation.

$$\frac{1}{a-1} = a+1, \text{ LCD is } a-1$$

$$(a-1)\cdot\frac{1}{a-1} = (a-1)(a+1)$$

$$1 = a^2 - 1$$

$$2 = a^2$$

$$\pm\sqrt{2} = a$$

Both numbers check. The numbers are

$\sqrt{2}$ and $-\sqrt{2}$, or $\pm\sqrt{2}$.

59.
$$\frac{w}{l} = \frac{l}{w+l}$$

$$l(w+l)\cdot\frac{w}{l} = l(w+l)\cdot\frac{l}{w+l}$$

$$w(w+l) = l^2$$

$$w^2 + lw = l^2$$

$$0 = l^2 - lw - w^2$$

Use the quadratic formula with $a = 1$, $b = -w$, and $c = -w^2$.

$$l = \frac{-(-w)\pm\sqrt{(-w)^2 - 4\cdot1(-w^2)}}{2\cdot1}$$

$$l = \frac{w\pm\sqrt{w^2 + 4w^2}}{2} = \frac{w\pm\sqrt{5w^2}}{2}$$

$$l = \frac{w\pm w\sqrt{5}}{2}$$

Since $\dfrac{w-w\sqrt{5}}{2}$ is negative we use the

positive square root: $l = \dfrac{w+w\sqrt{5}}{2}$

61. $mn^4 - r^2pm^3 - r^2n^2 + p = 0$

Let $u = n^2$. Substitute and rearrange.

$mu^2 - r^2u - r^2pm^3 + p = 0$

$a = m$, $b = -r^2$, $c = -r^2pm^3 + p$

$$u = \frac{-(-r^2)\pm\sqrt{(-r^2)^2 - 4\cdot m(-r^2pm^3+p)}}{2\cdot m}$$

$$u = \frac{r^2\pm\sqrt{r^4 + 4m^4r^2p - 4mp}}{2m}$$

$$n^2 = \frac{r^2\pm\sqrt{r^4 + 4m^4r^2p - 4mp}}{2m}$$

$$n = \sqrt{\frac{r^2\pm\sqrt{r^4 + 4m^4r^2p - 4mp}}{2m}}$$

63. Let s represent a length of a side of the cube, let S represent the surface area of the cube, let A represent the surface area of the sphere. Then the diameter of the sphere is s, so the radius r is $s/2$. From Exercise 15, we know, $A = 4\pi r^2$, so when $r = s/2$ we have

$$A = 4\pi\left(\frac{s}{2}\right)^2 = 4\pi\cdot\frac{s^2}{4} = \pi s^2. \text{ From the}$$

formula for the surface area of a cube (See Exercise 16.) we know that $S = 6s^2$, so

$\dfrac{S}{6} = s^2$ and then $A = \pi\cdot\dfrac{S}{6}$, or $A(S) = \dfrac{\pi S}{6}$.

Exercise Set 11.5

1. $x^6 = (x^3)^2$. Let $u = x^3$: f

3. $x^8 = (x^4)^2$. Let $u = x^4$: h

5. $x^{4/3} = (x^{2/3})^2$. Let $u = x^{2/3}$: g

7. $x^{-4/3} = (x^{-2/3})^2$. Let $u = x^{-2/3}$: e

9. $3p - 4\sqrt{p} + 6 = 3\left(\sqrt{p}\right)^2 - 4\sqrt{p} + 6$.

Let $u = \sqrt{p}$.

11. $(x^2 + 3)^2 + (x^2 + 3) - 7$

Let $u = x^2 + 3$.

13. $(1+t)^4 + (1+t)^2 + 4 = [(1+t)^2]^2 + (1+t)^2 + 4$

Let $u = (1+t)^2$.

15. $x^4 - 5x^2 + 4 = 0$

Let $u = x^2$ and $u^2 = x^4$.

$u^2 - 5u + 4 = 0$ Substituting u for x^2

$(u-1)(u-4) = 0$

$u - 1 = 0$ or $u - 4 = 0$

$u = 1$ or $u = 4$

Now replace u with x^2 and solve these equations.

$x^2 = 1$ or $x^2 = 4$

$x = \pm 1$ or $x = \pm 2$

The numbers 1, −1, 2, and −2 check. They are the solutions.

17. $x^4 - 9x^2 + 20 = 0$

Let $u = x^2$ and $u^2 = x^4$.

$u^2 - 9u + 20 = 0$

$(u-4)(u-5) = 0$

$u = 4$ or $u = 5$

$x^2 = 4$ or $x^2 = 5$

$x = \pm 2$ or $x = \pm\sqrt{5}$

All four numbers check.

19. $4t^4 - 19t^2 + 12 = 0$

Let $u = t^2$ and $u^2 = t^4$.

$4u^2 - 19u + 12 = 0$ Substituting

$(4u-3)(u-4) = 0$

$4u - 3 = 0$ or $u - 4 = 0$

$u = \dfrac{3}{4}$ or $u = 4$

Now replace u with t^2 and solve these equations:

$t^2 = \dfrac{3}{4}$ or $t^2 = 4$

$t = \pm\dfrac{\sqrt{3}}{2}$ or $t = \pm 2$

The numbers $\dfrac{\sqrt{3}}{2}$, $-\dfrac{\sqrt{3}}{2}$, 2, and −2 check.

They are the solutions.

21. $w + 4\sqrt{w} - 12 = 0 \Rightarrow \left(\sqrt{w}\right)^2 + 4\sqrt{w} - 12 = 0$

$\Rightarrow u^2 + 4u - 12 = 0$ for $u = \sqrt{w}$

$\Rightarrow (u+6)(u-2) = 0 \Rightarrow u = -6$, or $u = 2$

$\Rightarrow \sqrt{w} = -6 \Rightarrow$ no solution

or $\sqrt{w} = 2 \Rightarrow w = 2^2 = 4$

23. $\left(x^2 - 7\right)^2 - 3\left(x^2 - 7\right) + 2 = 0$

Let $u = x^2 - 7$ and $u^2 = \left(x^2 - 7\right)^2$.

$u^2 - 3u + 2 = 0$ Substituting

$(u-1)(u-2) = 0$

$u = 1$ or $u = 2$

$x^2 - 7 = 1$ or $x^2 - 7 = 2$ Replacing u with $x^2 - 7$

$x^2 = 8$ or $x^2 = 9$

$x = \pm\sqrt{8}$ or $x = \pm\sqrt{9}$

$x = \pm 2\sqrt{2}$ or $x = \pm 3$

The numbers $2\sqrt{2}$, $-2\sqrt{2}$, 3, and −3 check. They are the solutions.

25. $r - 2\sqrt{r} - 6 = 0$

Let $u = \sqrt{r}$ and $u^2 = r$

$u^2 - 2u - 6 = 0$

$u = \dfrac{-(-2) \pm \sqrt{(-2)^2 - 4 \cdot 1 \cdot (-6)}}{2 \cdot 1}$

$u = \dfrac{2 \pm \sqrt{28}}{2} = \dfrac{2 \pm 2\sqrt{7}}{2}$

$u = 1 \pm \sqrt{7}$

$\sqrt{r} = 1 + \sqrt{7}$ or $\sqrt{r} = 1 - \sqrt{7}$

$r = 1 + 2\sqrt{7} + 7$ No solution: $1 - \sqrt{7} < 0$

$r = 8 + 2\sqrt{7}$

The number $8 + 2\sqrt{7}$ checks and is the solution.

27. $\left(1+\sqrt{x}\right)^2 + 5\left(1+\sqrt{x}\right) + 6 = 0$

    Let $u = 1 + \sqrt{x}$ and $u^2 = \left(1+\sqrt{x}\right)^2$.

    $u^2 + 5u + 6 = 0$ Substituting

    $\left(u+3\right)\left(u+2\right) = 0$

    $u = -3 \ or \qquad u = -2$

    $1 + \sqrt{x} = -3 \ or \ 1 + \sqrt{x} = -2$ Replacing u with $1 + \sqrt{x}$

    $\sqrt{x} = -4 \ or \qquad \sqrt{x} = -3$

    Since the principal square root cannot be negative, this equation has no solution.

29. $x^{-2} - x^{-1} - 6 = 0$

    Let $u = x^{-1}$ and $u^2 = x^{-2}$.

    $u^2 - u - 6 = 0$ Substituting

    $\left(u-3\right)\left(u+2\right) = 0$

    $u = 3 \ or \ u = -2$

    Now we replace u with x^{-1} and solve these equations.

    $x^{-1} = 3 \ or \ x^{-1} = -2$

    $\dfrac{1}{x} = 3 \ or \ \dfrac{1}{x} = -2$

    $\dfrac{1}{3} = x \ or \ -\dfrac{1}{2} = x$

    Both $\dfrac{1}{3}$ and $-\dfrac{1}{2}$ check. They are the solutions.

31. $4y^{-2} - 3y^{-1} - 1 = 0$

    Let $u = y^{-1}$, then $u^2 = \left(y^{-1}\right)^2 = y^{-2}$.

    $4u^2 - 3u - 1 = 0$

    $\left(4u+1\right)\left(u-1\right) = 0$

    $u = -\dfrac{1}{4} \ or \ u = 1$

    Replace u with y^{-1} and solve the equations.

    $y^{-1} = -\dfrac{1}{4} \ or \ y^{-1} = 1$

    $\dfrac{1}{y} = -\dfrac{1}{4} \ or \ \dfrac{1}{y} = 1$

    $y = -4 \ or \ y = 1$

33. $t^{2/3} + t^{1/3} - 6 = 0$

    Let $u = t^{1/3}$ and $u^2 = t^{2/3}$.

    $u^2 + u - 6 = 0$ Substituting

    $\left(u+3\right)\left(u-2\right) = 0$

    $u = -3 \ or \ u = 2$

    Now we replace u with $t^{1/3}$ and solve these equations.

    $t^{1/3} = -3 \qquad or \ t^{1/3} = 2$

    $t = \left(-3\right)^3 \ or \qquad t = 2^3$

    $t = -27 \quad or \qquad t = 8$

    Both –27 and 8 check. They are the solutions.

35. $y^{1/3} - y^{1/6} - 6 = 0$

    Let $u = y^{1/6}$ and $u^2 = y^{1/3}$.

    $u^2 - u - 6 = 0$ Substituting

    $\left(u-3\right)\left(u+2\right) = 0$

    $u = 3 \ or \ u = -2$

    Now we replace u with $y^{1/6}$ and solve these equations.

    $y^{1/6} = 3 \ or \ y^{1/6} = -2$

    $\sqrt[6]{y} = 3 \ or \ \sqrt[6]{y} = -2$

    $y = 3^6$ This equation has no solution since principal

    $y = 729$ sixth roots are never negative.

    The number 729 checks and is the solution.

37. $t^{1/3} + 2t^{1/6} = 3$

    $t^{1/3} + 2t^{1/6} - 3 = 0$

    Let $u = t^{1/6}$ and $u^2 = t^{2/6} = t^{1/3}$.

    $u^2 + 2u - 3 = 0$ Substituting

    $\left(u+3\right)\left(u-1\right) = 0$

    $u = -3 \ or \ u = 1$

    $t^{1/6} = -3 \ or \ t^{1/6} = 1$ Substituting $t^{1/6}$ for u

    No solution $t = 1$

    The number 1 checks and is the solution.

39. $\left(10-\sqrt{x}\right)^{2}-2\left(10-\sqrt{x}\right)-35=0$

Let $u=10-\sqrt{x}$, then $u^{2}=\left(10-\sqrt{x}\right)^{2}$.

$u^{2}-2u-35=0\Rightarrow(u-7)(u+5)=0$

$\Rightarrow u=7$ or $u=-5$

$\Rightarrow 10-\sqrt{x}=7\Rightarrow\sqrt{x}=3\Rightarrow x=9$

or $10-\sqrt{x}=-5\Rightarrow\sqrt{x}=15\Rightarrow x=225$

The solutions are $x=9$ and $x=225$.

41. $16\left(\dfrac{x-1}{x-8}\right)^{2}+8\left(\dfrac{x-1}{x-8}\right)+1=0$

Let $u=\dfrac{x-1}{x-8}$ and $u^{2}=\left(\dfrac{x-1}{x-8}\right)^{2}$.

$16u^{2}+8u+1=0$ Substituting

$\left(4u+1\right)\left(4u+1\right)=0$

$$u=-\dfrac{1}{4}$$

Now we replace u with $\dfrac{x-1}{x-8}$ and solve this

equation:

$\dfrac{x-1}{x-8}=-\dfrac{1}{4}$

$4x-4=-x+8$ Multiplying by $4(x-8)$

$5x=12$

$x=\dfrac{12}{5}$

The number $\dfrac{12}{5}$ checks and is the solution.

43. $x^{4}+5x^{2}-36=0$

$\Rightarrow u^{2}+5u-36=0$ for $u=x^{2}$

$\Rightarrow(u+9)(u-4)=0$

$\Rightarrow u=-9\Rightarrow x^{2}=-9\Rightarrow x=\pm\sqrt{-9}=\pm3i$

or $u=4\Rightarrow x^{2}=4\Rightarrow x=\pm\sqrt{4}=\pm2$

The solutions are $x=-2,\ 2,\ 3i,$ and $-3i$

45. $(n^{2}+6)^{2}-7(n^{2}+6)+10=0$

$\Rightarrow u^{2}-7u+10=0$ for $u=n^{2}+6$

$\Rightarrow(u-5)(u-2)=0$

$\Rightarrow u=5\Rightarrow n^{2}+6=5\Rightarrow n^{2}=-1$

$\Rightarrow n=\pm\sqrt{-1}=\pm i$

or $u=2\Rightarrow n^{2}+6=2\Rightarrow n^{2}=-4$

$\Rightarrow n=\pm\sqrt{-4}=\pm2i$

The solutions are $n=-i,\ i,\ -2i,$ and $2i$

47. The x-intercepts occur where $f(x)=0$.

Thus, we must have $5x+13\sqrt{x}-6=0$.

Let $u=\sqrt{x}$ and $u^{2}=x$.

$5u^{2}+13u-6=0$

$\left(5u-2\right)\left(u+3\right)=0$

$u=\dfrac{2}{5}$ or $u=-3$

Now replace u with $\sqrt{x}$ and solve these

equations:

$\sqrt{x}=\dfrac{2}{5}$ or $\sqrt{x}=-3$

$x=\dfrac{4}{25}$ No Solution

The number $\dfrac{4}{25}$ checks. Thus, the

x-intercept is $\left(\dfrac{4}{25},0\right)$.

49. Solve: $\left(x^{2}-3x\right)^{2}-10\left(x^{2}-3x\right)+24=0$

Let $u=x^{2}-3x$ and $u^{2}=\left(x^{2}-3x\right)^{2}$.

$u^{2}-10u+24=0$

$\left(u-6\right)\left(u-4\right)=0$

$u=6$ or $u=4$

Now replace u with $x^{2}-3x$ and solve these

equations:

$x^{2}-3x=6$ or $x^{2}-3x=4$

$x^{2}-3x-6=0$ or $x^{2}-3x-4=0$

$x=\dfrac{-(-3)\pm\sqrt{(-3)^{2}-4(1)(-6)}}{2\cdot1}$ or

$\hspace{3cm}\left(x-4\right)\left(x+1\right)=0$

$x=\dfrac{3\pm\sqrt{33}}{2}$ or $x=4$ or $x=-1$

All four numbers check. Thus, the x-intercepts are $\left(\dfrac{3+\sqrt{33}}{2}, 0\right)$, $\left(\dfrac{3-\sqrt{33}}{2}, 0\right)$, $(4, 0)$, and $(-1, 0)$.

51. Solve: $x^{2/5} + x^{1/5} - 6 = 0$

Let $u = x^{1/5}$ and $u^2 = x^{2/5}$

$u^2 + u - 6 = 0$ Substituting

$(u+3)(u-2) = 0$

$u = -3$ or $u = 2$

$x^{1/5} = -3$ or $x^{1/5} = 2$ Replacing u with $x^{1/5}$

$x = -243$ or $x = 32$ Raising to the fifth power

Both -243 and 32 check. Thus, the x-intercepts are $(-243, 0)$ and $(32, 0)$.

53. $f(x) = \left(\dfrac{x^2+2}{x}\right)^4 + 7\left(\dfrac{x^2+2}{x}\right)^2 + 5$

Observe that, for all real numbers x, each term is positive. Thus, there are no real-number values of x for which $f(x) = 0$ and hence no x-intercepts.

55. *Thinking and Writing Exercise.*

57. $f(x) = x$

The function is linear. The graph is a line that goes through the points:

$(0, f(0)) = (0, 0)$ and $(2, f(2)) = (2, 2)$

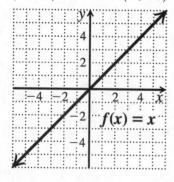

59. $f(x) = x - 2$

The function is linear. The graph is a line that goes through the points:

$(0, f(0)) = (0, -2)$ and $(2, f(2)) = (2, 0)$

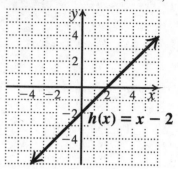

61. $f(x) = x^2 + 2$

The function is quadratic and the leading coefficient is positive. The graph is a parabola that opens upward with vertex at

$$\left(-\frac{b}{2a}, f\left(-\frac{b}{2a}\right)\right) = \left(-\frac{0}{2\cdot 1}, f\left(-\frac{0}{2\cdot 1}\right)\right) = (0, 2)$$

and that goes through the points:

$(-2, f(-2)) = (-2, 6)$, $(2, f(2)) = (2, 6)$,

$(-1, f(-1)) = (-1, 3)$, and $(1, f(1)) = (1, 3)$

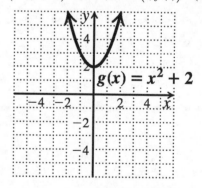

63. *Thinking and Writing Exercise.*

65. $5x^4 - 7x^2 + 1 = 0$

Let $u = x^2$ and $u^2 = x^4$

$5u^2 - 7u + 1 = 0$ Substituting

$$u = \frac{-(-7) \pm \sqrt{(-7)^2 - 4 \cdot 5 \cdot 1}}{2 \cdot 5}$$

$$u = \frac{7 \pm \sqrt{29}}{10}$$

$$x^2 = \frac{7 \pm \sqrt{29}}{10} \quad \text{Replacing } u \text{ with } x^2$$

$$x = \pm\sqrt{\frac{7 \pm \sqrt{29}}{10}}$$

All four numbers check and are the solutions.

67. $\left(x^2 - 4x - 2\right)^2 - 13\left(x^2 - 4x - 2\right) + 30 = 0$

Let $u = x^2 - 4x - 2$ and $u^2 = \left(x^2 - 4x - 2\right)^2$

$u^2 - 13u + 30 = 0$ Substituting

$(u - 3)(u - 10) = 0$

$u = 3 \text{ or} \qquad u = 10$

$x^2 - 4x - 2 = 3 \text{ or} \quad x^2 - 4x - 2 = 10$

Replacing u with $x^2 - 4x - 2$

$x^2 - 4x - 5 = 0 \text{ or} \quad x^2 - 4x - 12 = 0$

$(x - 5)(x + 1) = 0 \text{ or } (x - 6)(x + 2) = 0$

$x = 5 \text{ or } x = -1 \text{ or } x = 6 \text{ or } x = -2$

All four numbers check and are the solutions.

69. $\dfrac{x}{x-1} - 6\sqrt{\dfrac{x}{x-1}} - 40 = 0$

Let $u = \sqrt{\dfrac{x}{x-1}}$ and $u^2 = \dfrac{x}{x-1}$.

$u^2 - 6u - 40 = 0$ Substituting

$(u - 10)(u + 4)$

$u = 10 \quad \text{or} \qquad u = -4$

$\sqrt{\dfrac{x}{x-1}} = 10 \quad \text{or} \quad \sqrt{\dfrac{x}{x-1}} = -4$

$\dfrac{x}{x-1} = 100 \text{ or} \quad \text{No solution}$

$x = 100x - 100$

$100 = 99x$

$\dfrac{100}{99} = x$

The number $\dfrac{100}{99}$ checks. It is the solution.

71. $a^5\left(a^2 - 25\right) + 13a^3\left(25 - a^2\right) + 36a\left(a^2 - 25\right) = 0$

$a^5\left(a^2 - 25\right) - 13a^3\left(a^2 - 25\right) + 36a\left(a^2 - 25\right) = 0$

$a\left(a^2 - 25\right)\left(a^4 - 13a^2 + 36\right) = 0$

$a\left(a^2 - 25\right)\left(a^2 - 4\right)\left(a^2 - 9\right) = 0$

$a = 0 \text{ or } a^2 - 25 = 0 \text{ or } a^2 - 4 = 0 \text{ or } a^2 - 9 = 0$

$a = 0 \text{ or} \qquad a^2 = 25 \text{ or} \qquad a^2 = 4 \text{ or} \qquad a^2 = 9$

$a = 0 \text{ or} \qquad a = \pm 5 \text{ or} \qquad a = \pm 2 \text{ or} \qquad a = \pm 3$

All seven numbers check. The solutions are $0, 5, -5, 2, -2, 3,$ and -3.

73. $x^6 - 28x^3 + 27 = 0$

Let $u = x^3$.

$u^2 - 28u + 27 = 0$

$(u - 27)(u - 1) = 0$

$u = 27 \text{ or} \quad u = 1$

$x^3 = 27 \text{ or } x^3 = 1$

$\left(x^3 - 27\right) = (x - 3)\left(x^2 + 3x + 9\right)$

$\left(x^3 - 1\right) = (x - 1)\left(x^2 + x + 1\right)$

Using the quadratic formula on $x^2 + 3x + 9$ and $x^2 + x + 1$ gives

$-\dfrac{1}{2} \pm \dfrac{\sqrt{3}}{2}i$ and $-\dfrac{3}{2} \pm \dfrac{3\sqrt{3}}{2}i$, 1 and 3.

Mid-Chapter Review

Guided Solutions

1. $(x - 7)^2 = 5$

$x - 7 = \pm\sqrt{5}$

$x = 7 \pm \sqrt{5}$

The solutions re $7 + \sqrt{5}$ and $7 - \sqrt{5}$.

2. $x^2 - 2x - 1 = 0$

$a = 1,\ b = -2,\ c = -1$

$x = \dfrac{-(-2) \pm \sqrt{(-2)^2 - 4 \cdot 1 \cdot (-1)}}{2 \cdot 1}$

$= \dfrac{2 \pm \sqrt{8}}{2}$

$= \dfrac{2}{2} + \dfrac{2\sqrt{2}}{2}$

$= 1 \pm \sqrt{2}$

The solutions are $1 + \sqrt{2}$ and $1 - \sqrt{2}$.

Mixed Review

1. $x^2 - 3x - 10 = 0 \Rightarrow (x-5)(x+2) = 0$

$\Rightarrow x - 5 = 0 \Rightarrow x = 5$

or $x + 2 = 0 \Rightarrow x = -2$

2. $x^2 = 121 \Rightarrow x = \pm\sqrt{121} = \pm 11$

3. $x^2 + 6x = 10 \Rightarrow x^2 + 6x - 10 = 0$

$\Rightarrow a = 1,\ b = 6,\ c = -10$

$x = \dfrac{-6 \pm \sqrt{6^2 - 4 \cdot 1 \cdot (-10)}}{2 \cdot 1} = \dfrac{-6 \pm \sqrt{36 + 40}}{2 \cdot 1}$

$= \dfrac{-6 \pm \sqrt{76}}{2} = \dfrac{-6 \pm \sqrt{4 \cdot 19}}{2} = \dfrac{-6 \pm 2\sqrt{19}}{2}$

$= \dfrac{2(-3 \pm \sqrt{19})}{2} = -3 \pm \sqrt{19}$

4. $x^2 + x - 3 = 0 \Rightarrow a = 1,\ b = 1,\ c = -3$

$x = \dfrac{-1 \pm \sqrt{1^2 - 4 \cdot 1 \cdot (-3)}}{2 \cdot 1} = \dfrac{-1 \pm \sqrt{1 + 12}}{2}$

$= \dfrac{-1 \pm \sqrt{13}}{2} = -\dfrac{1}{2} \pm \dfrac{\sqrt{13}}{2}$

5. $(x+1)^2 = 2 \Rightarrow x + 1 = \pm\sqrt{2}$

$\Rightarrow x = -1 \pm \sqrt{2}$

6. $x^2 - 10x + 25 = 0 \Rightarrow (x-5)^2 = 0$

$\Rightarrow x - 5 = 0 \Rightarrow x = 5$

7. $4t^2 = 11 \Rightarrow t^2 = \dfrac{11}{4} \Rightarrow t = \pm\sqrt{\dfrac{11}{4}} = \pm\dfrac{\sqrt{11}}{2}$

8. $2t^2 + 1 = 3t \Rightarrow 2t^2 - 3t + 1 = 0$

$\Rightarrow (2t-1)(t-1) = 0$

$\Rightarrow 2t - 1 = 0 \Rightarrow 2t = 1 \Rightarrow t = \dfrac{1}{2}$

or $t - 1 = 0 \Rightarrow t = 1$

9. $16c^2 = 7c \Rightarrow 16c^2 - 7c = 0 \Rightarrow c(16c - 7) = 0$

$\Rightarrow c = 0$

or $16c - 7 = 0 \Rightarrow 16c = 7 \Rightarrow c = \dfrac{7}{16}$

10. $y^2 - 2y + 8 = 0 \Rightarrow a = 1,\ b = -2,\ c = 8$

$y = \dfrac{-(-2) \pm \sqrt{(-2)^2 - 4 \cdot 1 \cdot 8}}{2 \cdot 1} = \dfrac{2 \pm \sqrt{4 - 32}}{2}$

$= \dfrac{2 \pm \sqrt{-28}}{2} = \dfrac{2 \pm \sqrt{28}i}{2} = \dfrac{2 \pm 2\sqrt{7}i}{2}$

$= \dfrac{2(1 + \sqrt{7}i)}{2} = 1 \pm \sqrt{7}i$

11. $x^4 - 10x^2 + 9 = 0$

$\Rightarrow u^2 - 10u + 9 = 0$ for $u = x^2$

$\Rightarrow (u-9)(u-1) = 0$

$\Rightarrow u = 9 \Rightarrow x^2 = 9 \Rightarrow x = \pm\sqrt{9} = \pm 3$

or $u = 1 \Rightarrow x^2 = 1 \Rightarrow x = \pm\sqrt{1} = \pm 1$

12. $x^4 - 8x^2 - 9 = 0 \Rightarrow u^2 - 8u - 9 = 0$ for $u = x^2$

$\Rightarrow (u-9)(u+1) = 0$

$\Rightarrow u = 9 \Rightarrow x^2 = 9 \Rightarrow x = \pm\sqrt{9} = \pm 3$

or $u = -1 \Rightarrow x^2 = -1 \Rightarrow x = \pm\sqrt{-1} = \pm i$

13. $(t+4)(t-3) = 18 \Rightarrow t^2 + t - 30 = 0$

$\Rightarrow (t+6)(t-5) = 0$

$\Rightarrow t + 6 = 0 \Rightarrow t = -6$

or $t - 5 = 0 \Rightarrow t = 5$

14. $m^{-4} - 5m^{-2} + 6 = 0$

$\Rightarrow u^2 - 5u + 6 = 0$ for $u = m^{-2}$

$\Rightarrow (u-3)(u-2) = 0$

$\Rightarrow u = 3 \Rightarrow m^{-2} = 3 \Rightarrow \dfrac{1}{m^2} = 3 \Rightarrow m^2 = \dfrac{1}{3}$

$\Rightarrow m = \pm\sqrt{\dfrac{1}{3}} = \pm\dfrac{\sqrt{3}}{3}$

or $u = 2 \Rightarrow m^{-2} = 2 \Rightarrow \dfrac{1}{m^2} = 2 \Rightarrow m^2 = \dfrac{1}{2}$

$\Rightarrow m = \pm\sqrt{\dfrac{1}{2}} = \pm\dfrac{\sqrt{2}}{2}$

15. $x^2 - 8x + 1 = 0$

$\Rightarrow b^2 - 4ac = (-8)^2 - 4 \cdot 1 \cdot 1 = 60$

The coefficients of the quadratic equation are integers, the discriminant is positive, but not a perfect square. There are two irrational, real solutions.

16. $3x^2 = 4x + 7 \Rightarrow 3x^2 - 4x - 7 = 0$

$\Rightarrow b^2 - 4ac = (-4)^2 - 4 \cdot 3 \cdot (-7) = 100$

The coefficients of the quadratic equation are integers, the discriminant is a positive perfect square. There are two rational, real solutions.

17. $5x^2 - x + 6 = 0$

$\Rightarrow b^2 - 4ac = (-1)^2 - 4 \cdot 5 \cdot 6 = -119$

The coefficients of the quadratic equation are integers, the discriminant is negative. There are two imaginary solutions.

18. $F = \dfrac{Av^2}{400} \Rightarrow 400 \cdot F = \cancel{400} \cdot \dfrac{Av^2}{\cancel{400}}$

$\Rightarrow 400F = Av^2 \Rightarrow \dfrac{400F}{A} = \dfrac{\cancel{A}v^2}{\cancel{A}}$

$\Rightarrow v^2 = \dfrac{400F}{A} \Rightarrow \sqrt{v^2} = \sqrt{\dfrac{400F}{A}}$

$v = \sqrt{\dfrac{400F}{A}}$ or $20\sqrt{\dfrac{F}{A}}$ Since we are assuming all

variables are ≥ 0, the negative square root is not shown.

19. $D^2 - 2Dd - 2hd = 0$

$\Rightarrow a = 1,\ b = -2d,\ c = -2hd$

$D = \dfrac{-(-2d) \pm \sqrt{(-2d)^2 - 4 \cdot 1 \cdot (-2hd)}}{2 \cdot 1}$

$= \dfrac{2d \pm \sqrt{4d^2 + 8hd}}{2} = \dfrac{2d \pm \sqrt{4(d^2 + 2hd)}}{2}$

$= \dfrac{2d \pm 2\sqrt{d^2 + 2hd}}{2} = \dfrac{\cancel{2}(d \pm \sqrt{d^2 + 2hd})}{\cancel{2}}$

$= d \pm \sqrt{d^2 + 2hd}$

Since $d^2 + 2hd \geq d$ when $h, d \geq 0$, we discard the difference and give the solution as

$D = d + \sqrt{d^2 + 2hd}$

Exercise Set 11.6

---

1. The quadratic expression is in vertex form. The vertex is located at $(1, 3)$ and the leading coefficient is positive, so the parabola opens upward : h

3. The quadratic expression is in vertex form. The vertex is located at $(-1, 3)$ and the leading coefficient is positive, so the parabola opens upward : f

5. The quadratic expression is in vertex form. The vertex is located at $(-1, 3)$ and the leading coefficient is negative, so the parabola opens downward: b

7. The quadratic expression is in vertex form. The vertex is located at $(-1, -3)$ and the leading coefficient is positive, so the parabola opens upward : e

9. a) The parabola opens upward, so a is positive.

   b) The vertex is at $(3, 1)$.

   c) The axis of symmetry is $x = 3$.

   d) The range is $[1, \infty)$.

11. a) The parabola opens downward, so a is negative.

   b) The vertex is $(-2, -3)$

   c) The axis of symmetry is $x = -2$

   d) The range is $(-\infty, -3]$

13. a) The parabola opens upward, so a is positive.

 b) The vertex is at $(-3, 0)$

 c) The axis of symmetry is $x = -3$

 d) The range is $[0, \infty)$

15.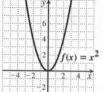

17.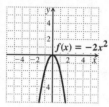

19.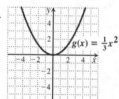

21.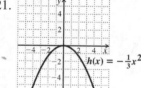

23.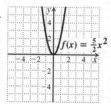

25. $g(x) = (x+1)^2 = [x-(-1)]^2$

 Vertex: $(-1, 0)$, Axis of symmetry: $x = -1$

27. $f(x) = (x-2)^2$

 Vertex: $(2, 0)$, Axis of symmetry: $x = 2$

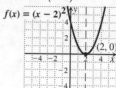

29. $f(x) = -(x+1)^2 = -[x-(-1)]^2$

 Vertex: $(-1, 0)$, Axis of symmetry: $x = -1$

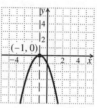

31. $g(x) = -(x-2)^2$

 Vertex: $(2, 0)$, Axis of symmetry: $x = 2$

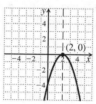

33. $f(x) = 2(x+1)^2 = 2[x-(-1)]^2$

 Vertex: $(-1, 0)$, Axis of symmetry: $x = -1$

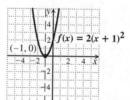

35. $g(x) = 3(x-4)^2$

Vertex: $(4,0)$, Axis of symmetry: $x = 4$

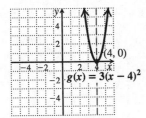

37. $h(x) = -\frac{1}{2}(x-4)^2$

Vertex: $(4,0)$, Axis of symmetry: $x = 4$

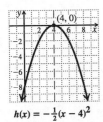

39. $f(x) = \frac{1}{2}(x-1)^2$

Vertex: $(1,0)$, Axis of symmetry: $x = 1$

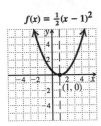

41. $f(x) = -2(x+5)^2 = -2[x-(-5)]^2$

Vertex: $(-5,0)$, Axis of symmetry: $x = -5$

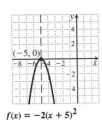

43. $h(x) = -3\left(x - \frac{1}{2}\right)^2$

Vertex: $\left(\frac{1}{2},0\right)$, Axis of symmetry: $x = \frac{1}{2}$

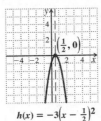

45. $f(x) = (x-5)^2 + 2$

$a > 0$, so the function takes on a minimum at $x = 5$, of $f(5) = 2$, the y value of the vertex.

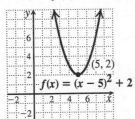

The range of $f(x)$ is $[2, \infty)$.

47. $f(x) = -(x+2)^2 - 1$

$a < 0$, so the function takes on a maximum at $x = -2$, of $f(-2) = -1$, the y value of the vertex.

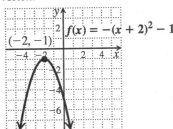

The range of $f(x)$ is $(-\infty, -1]$.

49. $g(x) = \frac{1}{2}(x+4)^2 + 3$

$a > 0$, so the function takes on a minimum at $x = -4$, of $g(-4) = 3$, the y value of the vertex.

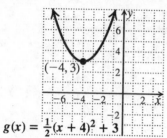

$g(x) = \frac{1}{2}(x+4)^2 + 3$

The range of $g(x)$ is $[3, \infty)$.

51. $h(x) = -2(x-1)^2 - 3$

$a < 0$, so the function takes on a maximum at $x = 1$, of $h(1) = -3$, the y value of the vertex.

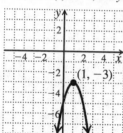

$h(x) = -2(x-1)^2 - 3$

The range of $h(x)$ is $(-\infty, -3]$.

53. $f(x) = (x+1)^2 - 3 = [x - (-1)]^2 + (-3)$

Vertex: $(-1, -3)$

Axis of symmetry: $x = -1$

Minimum: -3

Range: $[-3, \infty)$

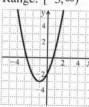

$f(x) = (x+1)^2 - 3$

55. $g(x) = -(x+3)^2 + 5$

Vertex: $(-3, 5)$

Axis of symmetry: $x = -3$

Maximum: 5

Range: $(-\infty, 5]$

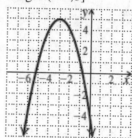

$g(x) = -(x+3)^2 + 5$

57. $f(x) = \frac{1}{2}(x-2)^2 + 1$

Vertex: $(2, 1)$

Axis of symmetry: $x = 2$

Minimum: 1

Range: $[1, \infty)$

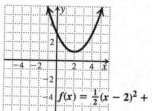

$f(x) = \frac{1}{2}(x-2)^2 + 1$

59. $h(x) = -2(x-1)^2 - 3$

Vertex: $(1, -3)$

Axis of symmetry: $x = 1$

Maximum: -3

Range: $(-\infty, -3]$

$h(x) = -2(x-1)^2 - 3$

61. $f(x) = 2(x+4)^2 + 1$

Vertex: $(-4, 1)$

Axis of symmetry: $x = -4$

Minimum: 1

Range: $[1, \infty)$

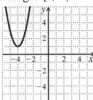

$f(x) = 2(x + 4)^2 + 1$

63. $g(x) = -\frac{3}{2}(x-1)^2 + 4$

Vertex: $(1, 4)$

Axis of symmetry: $x = 1$

Maximum: 4

Range: $(-\infty, 4]$

$g(x) = -\frac{3}{2}(x - 1)^2 + 4$

65. $f(x) = 6(x-8)^2 + 7$

Vertex: $(8, 7)$

Axis of symmetry: $x = 8$

Minimum: 7

67. $h(x) = -\frac{2}{7}(x+6)^2 + 11 = -\frac{2}{7}\left[x - (-6)\right]^2 + 11$

Vertex: $(-6, 11)$

Axis of symmetry: $x = -6$

Maximum: 11

69. $f(x) = \left(x - \frac{7}{2}\right)^2 - \frac{29}{4}$

Vertex: $\left(\frac{7}{2}, -\frac{29}{4}\right)$

Axis of symmetry: $x = \frac{7}{2}$

Minimum: $-\frac{29}{4}$

71. $f(x) = \sqrt{2}(x+4.58)^2 + 65\pi$

$\quad = \sqrt{2}\left[x - (-4.58)\right]^2 + 65\pi$

Vertex: $(-4.58, 65\pi)$

Axis of symmetry: $x = -4.58$

Minimum: 65π

73. *Thinking and Writing Exercise.*

75. $8x - 6y = 24$

To find the x-intercept, let $y = 0$:

$8x - 6(0) = 24$

$\quad\quad 8x = 24$

$\quad\quad\quad x = 3$

To find the y-intercept, let $x = 0$:

$8(0) - 6y = 24$

$\quad\quad -6y = 24$

$\quad\quad\quad y = -4$

The x-intercept is $(3, 0)$, and the y-intercept is $(0, -4)$.

77. $y = x^2 + 8x + 15$

To find the x-intercepts, let $y = 0$:

$0 = x^2 + 8x + 15$

$0 = (x+3)(x+5)$

$x = -5, -3$

The x-intercepts are $(-5, 0)$ and $(-3, 0)$.

79. $x^2 - 14x + \underline{\quad} = (x - \underline{\quad})^2$

$\left(\frac{14}{2}\right)^2 = 7^2 = 49$

$x^2 - 14x + 49 = (x - 7)^2$

81. *Thinking and Writing Exercise.*

83. The equation will be of the form

$f(x) = \frac{3}{5}(x - h)^2 + k$ with $h = 4$ and $k = 1$:

$f(x) = \frac{3}{5}(x - 4)^2 + 1$

85. The equation will be of the form

$f(x) = \frac{3}{5}(x-h)^2 + k$ with $h = 3$ and $k = -1$:

$f(x) = \frac{3}{5}(x-3)^2 + (-1)$ or

$f(x) = \frac{3}{5}(x-3)^2 - 1$

87. The equation will be of the form

$f(x) = \frac{3}{5}(x-h)^2 + k$ with $h = -2$ and $k = -5$:

$f(x) = \frac{3}{5}[x-(-2)]^2 + (-5)$ or

$f(x) = \frac{3}{5}(x+2)^2 - 5$

89. Since there is a minimum at $(2,0)$, the parabola will have the same shape as $f(x) = 2x^2$. It will be of the form

$f(x) = 2(x-h)^2 + k$ with $h = 2$ and $k = 0$:

$f(x) = 2(x-2)^2$

91. Since there is a maximum at $(0,3)$, the parabola will have the same shape as $g(x) = -2x^2$. It will be of the form

$g(x) = -2(x-h)^2 + k$ with $h = 0$ and $k = 3$:

$g(x) = -2x^2 + 3$

93. If h is increased, the x coordinate of the vertex will increase. The graph will be shifted right.

95. Since the parabola is opening downward, the original value of a is negative. Replacing a with $-a$ will then make the leading coefficient positive and the parabola will open upward, but the vertex will remain at (h,k).

97. The maximum value of $g(x)$ is 1 and occurs at the point $(5,1)$, so for $F(x)$ we have $h = 5$ and $k = 1$. $F(x)$ has the same shape as $f(x)$ and has a minimum, so $a = 3$. Thus:

$F(x) = 3(x-5)^2 + 1$.

99. The graph of $y = f(x-1)$ looks like the graph of $y = f(x)$ moved 1 unit to the right.

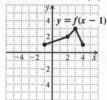

101. The graph of $y = f(x) + 2$ looks like the graph of $y = f(x)$ moved up 2 units.

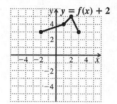

103. The graph of $y = f(x+3) - 2$ looks like the graph of $y = f(x)$ moved 3 units to the left and also moved down 2 units.

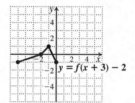

Exercise Set 11.7

---

1. True : The leading coefficient, 3, is positive.

3. True : The quadratic is in vertex form, $a(x-h)^2 + k$ with $(h,k) = (3,7)$.

5. False : The axis of symmetry is the line $x = h$. $g(x)$ is in vertex form with $h = \frac{3}{2}$.

7. False : The x-coordinate of the y-intercept is always 0. The graph goes through (0, 7), not (7, 0).

9. $x^2 - 8x + 2 = x^2 - 8x + \left(\frac{8}{2}\right)^2 + 2 - \left(\frac{8}{2}\right)^2$

$= (x-4)^2 + 2 - 4^2 = (x-4)^2 + (-14)$

11. $x^2 + 3x - 5 = x^2 + 3x + \left(\frac{3}{2}\right)^2 - 5 - \left(\frac{3}{2}\right)^2$

$= \left(x + \frac{3}{2}\right)^2 - 5 - \frac{9}{4} = \left(x - \left(-\frac{3}{2}\right)\right)^2 + \left(-\frac{29}{4}\right)$

13. $3x^2 + 6x - 2 = 3\left(x^2 + 2x + \left(\frac{2}{2}\right)^2\right) - 2 - 3 \cdot \left(\frac{2}{2}\right)^2$

$= 3(x+1)^2 - 2 - 3 = 3(x - (-1))^2 + (-5)$

15. $-x^2 - 4x - 7 = -\left(x^2 + 4x + \left(\frac{4}{2}\right)^2\right) - 7 - (-1) \cdot \left(\frac{4}{2}\right)^2$

$= -(x+2)^2 - 7 + 4 = -(x - (-2))^2 + (-3)$

17. $2x^2 - 5x + 10 = 2\left(x^2 - \frac{5}{2}x + \left(\frac{5}{4}\right)^2\right) + 10 - 2 \cdot \left(\frac{5}{4}\right)^2$

$= 2\left(x - \frac{5}{4}\right)^2 + 10 - \frac{25}{8} = 2\left(x - \frac{5}{4}\right)^2 + \frac{55}{8}$

19. $f(x) = x^2 + 4x + 5$

$= (x^2 + 4x + 4) + (-4 + 5)$

$= (x+2)^2 + 1$

Vertex: $(-2, 1)$, Axis of symmetry: $x = -2$

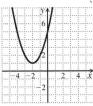

$f(x) = x^2 + 4x + 5$

21. $f(x) = x^2 + 8x + 20$

$= (x^2 + 8x + 16) + (-16 + 20)$

$= (x+4)^2 + 4$

Vertex: $(-4, 4)$, Axis of symmetry: $x = -4$

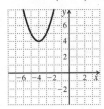

$f(x) = x^2 + 8x + 20$

23. $h(x) = 2x^2 - 16x + 25$

$= 2(x^2 - 8x + 16) - 2 \cdot 16 + 25$

$= 2(x-4)^2 - 7$

Vertex: $(4, -7)$, Axis of symmetry: $x = 4$

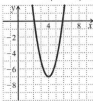

$h(x) = 2x^2 - 16x + 25$

25. $f(x) = -x^2 + 2x + 5$

$= -(x^2 - 2x - 5)$

$= -(x^2 - 2x + 1 - 1 - 5)$

$= -(x-1)^2 + 6$

Vertex: $(1, 6)$, Axis of symmetry: $x = 1$

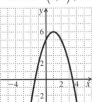

$f(x) = -x^2 + 2x + 5$

27. $g(x) = x^2 + 3x - 10$

$$= \left(x^2 + 3x + \frac{9}{4}\right) - \frac{9}{4} - 10$$

$$= \left(x + \frac{3}{2}\right)^2 - \frac{49}{4}$$

Vertex: $\left(-\frac{3}{2}, -\frac{49}{4}\right)$,

Axis of symmetry: $x = -\frac{3}{2}$

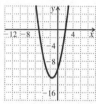

$g(x) = x^2 + 3x - 10$

29. $h(x) = x^2 + 7x$

$$= \left(x^2 + 7x + \frac{49}{4}\right) - \frac{49}{4}$$

$$= \left(x + \frac{7}{2}\right)^2 - \frac{49}{4}$$

Vertex: $\left(-\frac{7}{2}, -\frac{49}{4}\right)$

Axis of symmetry: $x = -\frac{7}{2}$

$h(x) = x^2 + 7x$

31. $f(x) = -2x^2 - 4x - 6$

$$= -2(x^2 + 2x) - 6$$

$$= -2(x^2 + 2x + 1) - 2(-1) - 6$$

$$= -2(x + 1)^2 - 4$$

Vertex: $(-1, -4)$

Axis of symmetry: $x = -1$

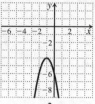

$f(x) = -2x^2 - 4x - 6$

33. (a)

$g(x) = x^2 - 6x + 13 \Rightarrow a = 1,\ b = -6$

$$\Rightarrow -\frac{b}{2a} = -\frac{-6}{2 \cdot 1} = 3$$

$g(3) = (3)^2 - 6(3) + 13$

$= 9 - 18 + 13 = 4$

Vertex: $(3, g(3)) = (3, 4)$

Axis of symmetry: $x = 3$

Minimum of g: $y = 4$

(b)

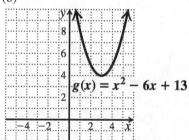

$g(x) = x^2 - 6x + 13$

35. (a)

$g(x) = 2x^2 - 8x + 3 \Rightarrow a = 2,\ b = -8$

$$\Rightarrow -\frac{b}{2a} = -\frac{-8}{2 \cdot 2} = 2$$

$g(2) = 2(2)^2 - 8(2) + 3$

$= 8 - 16 + 3 = -5$

Vertex: $(2, g(2)) = (2, -5)$

Axis of symmetry: $x = 2$

Minimum of g: $y = -5$

(b)

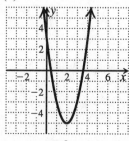

$g(x) = 2x^2 - 8x + 3$

37. (a)

$$f(x) = 3x^2 - 24x + 50 \Rightarrow a = 3, \ b = -24$$

$$\Rightarrow -\frac{b}{2a} = -\frac{-24}{2 \cdot 3} = 4$$

$$f(4) = 3(4)^2 - 24(4) + 50$$

$$= 48 - 96 + 50 = 2$$

Vertex : $(4, f(4)) = (4, 2)$

Axis of symmetry: $x = 4$

Minimum of f: $y = 2$

(b)

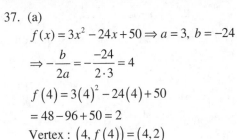

39. (a)

$$f(x) = -3x^2 + 5x - 2 \Rightarrow a = -3, \ b = 5$$

$$\Rightarrow -\frac{b}{2a} = -\frac{5}{2 \cdot (-3)} = \frac{5}{6}$$

$$f\left(\frac{5}{6}\right) = -3\left(\frac{5}{6}\right)^2 + 5\left(\frac{5}{6}\right) - 2$$

$$= -\frac{25}{12} + \frac{25}{6} - 2 = \frac{-25 + 2 \cdot 25 - 2 \cdot 12}{12} = \frac{1}{12}$$

Vertex : $\left(\frac{5}{6}, f\left(\frac{5}{6}\right)\right) = \left(\frac{5}{6}, \frac{1}{12}\right)$

Axis of symmetry: $x = \frac{5}{6}$

Maximum of f: $y = \frac{1}{12}$

(b)

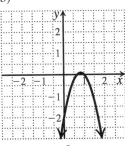

$$f(x) = -3x^2 + 5x - 2$$

41. (a)

$$h(x) = \frac{1}{2}x^2 + 4x + \frac{19}{3} \Rightarrow a = \frac{1}{2}, \ b = 4$$

$$\Rightarrow -\frac{b}{2a} = -\frac{4}{2 \cdot \left(\frac{1}{2}\right)} = -4$$

$$h(-4) = \frac{1}{2}(-4)^2 + 4(-4) + \frac{19}{3}$$

$$= 8 - 16 + \frac{19}{3} = \frac{-8 \cdot 3 + 19}{3} = -\frac{5}{3}$$

Vertex : $(-4, h(-4)) = \left(-4, -\frac{5}{3}\right)$

Axis of symmetry: $x = -4$

Minimum of h: $y = -\frac{5}{3}$

(b)

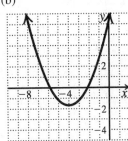

$$h(x) = \frac{1}{2}x^2 + 4x + \frac{19}{3}$$

43. $f(x) = x^2 + x - 6$

The coefficient of x^2 is positive so the graph opens upward and the function has a minimum value. Graph the function in a window that shows the vertex. The standard window is one good choice. Then use the Minimum feature from the CALC menu to find that the vertex is $(-0.5, -6.25)$.

45. $f(x) = 5x^2 - x + 1$

The coefficient of x^2 is positive so the graph opens upward and the function has a minimum value. Graph the function in a window that shows the vertex. The standard window is one good choice. Then use the Minimum feature from the CALC menu to find that the vertex is $(0.1, 0.95)$.

47. $f(x) = -0.2x^2 + 1.4x - 6.7$

The coefficient of x^2 is negative so the graph opens downward and the function has a maximum value. Graph the function in a window that shows the vertex. The standard window is one good choice. Then use the Maximum feature from the CALC menu to find that the vertex is $(3.5, -4.25)$.

49. $f(x) = x^2 - 6x + 3$

To find the x-intercepts, solve the equation $0 = x^2 - 6x + 3$.
Use the quadratic formula.
$$x = \frac{-(-6) \pm \sqrt{(-6)^2 - 4 \cdot 1 \cdot 3}}{2 \cdot 1}$$
$$x = \frac{6 \pm \sqrt{24}}{2} = \frac{6 \pm 2\sqrt{6}}{2} = 3 \pm \sqrt{6}$$
The x-intercepts are $\left(3 - \sqrt{6}, 0\right)$ and $\left(3 + \sqrt{6}, 0\right)$.
The y-intercept is $\left(0, f(0)\right)$, or $(0, 3)$.

51. $g(x) = -x^2 + 2x + 3$

To find the x-intercepts, solve the equation $0 = -x^2 + 2x + 3$. We factor.
$0 = -x^2 + 2x + 3$
$0 = x^2 - 2x - 3$
$0 = (x - 3)(x + 1)$
$x = 3 \; or \; x = -1$
The x-intercepts are $(-1, 0)$ and $(3, 0)$. The y-intercepts are $\left(0, g(0)\right)$, or $(0, 3)$.

53. $f(x) = x^2 - 9x$

To find the x-intercepts, solve the equation.
$0 = x^2 - 9x$
$0 = x(x - 9)$
$x = 0 \; or \; x = 9$
The x-intercepts are $(0, 0)$ and $(9, 0)$. Since $(0, 0)$ is an x-intercept, we observe that $(0, 0)$ is also the y-intercept.

55. $h(x) = -x^2 + 4x - 4$

To find the x-intercepts, solve the equation.
$0 = -x^2 + 4x - 4$
$0 = x^2 - 4x + 4$
$0 = (x - 2)(x - 2)$
$x = 2 \; or \; x = 2$
The x-intercept is $(2, 0)$.
The y-intercept is $\left(0, h(0)\right)$, or $(0, -4)$.

57. $g(x) = x^2 + x - 5$

To find the x-intercepts, solve the equation
$g(x) = 0 \Rightarrow x^2 + x - 5 = 0$
$\Rightarrow a = 1, \; b = 1, \; c = -5$
Use the quadratic formula.
$$x = \frac{-1 \pm \sqrt{1^2 - 4 \cdot 1 \cdot (-5)}}{2 \cdot 1} = \frac{-1 \pm \sqrt{21}}{2}$$
The x-intercepts are $\left(\dfrac{-1 - \sqrt{21}}{2}, 0\right)$ and $\left(\dfrac{-1 + \sqrt{21}}{2}, 0\right)$.
The y-intercept is $\left(0, g(0)\right) = (0, -5)$.

59. $f(x) = 2x^2 - 4x + 6$

To find the x-intercepts, solve the equation $0 = 2x^2 - 4x + 6$. We use the quadratic formula.
$$x = \frac{-(-4) \pm \sqrt{(-4)^2 - 4 \cdot 2 \cdot 6}}{2 \cdot 2}$$
$$x = \frac{4 \pm \sqrt{-32}}{4} = \frac{4 \pm 4i\sqrt{2}}{4} = 1 \pm i\sqrt{2}$$
There are no real-number solutions, so there is no x-intercept. The y-intercept is $\left(0, f(0)\right)$, or $(0, 6)$.

61. *Thinking and Writing Exercise.*

63. $x + y + z = 3$ (1)

$x - y + z = 1$ (2)

$-x - y + z = -1$ (3)

Adding (1) to (3) will generate an equation in y and z. Adding (2) to (3) will generate a second equation in y and z.

$x + y + z = 3$ (1)

$-x - y + z = -1$ (3)

$\overline{2z = 2}$ (4)

$x - y + z = 1$ (2)

$-x - y + z = -1$ (3)

$\overline{-2y + 2z = 0}$

or $y - z = 0$

or $y = z$ (5)

Solving (4) gives $z = 1$. (5) then gives $y = 1$ also. Use equation (1), and the values of y and z, to find x.

$x + y + z = 3$ (1)

$\Rightarrow x = 3 - y - z = 3 - 1 - 1 = 1$

$(x, y, z) = (1, 1, 1)$

65. $z = 8$ (1)

$x + y + z = 23$ (2)

$2x + y - z = 17$ (3)

Substitute 8 for z into (2) and (3) to generate a two-variable system in x and y.

$\begin{array}{l} x + y + z = 23 \quad (2) \\ 2x + y - z = 17 \quad (3) \end{array} \Rightarrow \begin{array}{l} x + y + 8 = 23 \\ 2x + y - 8 = 17 \end{array}$

$\Rightarrow \quad x + y = 15$ (4)

$2x + y = 25$ (5)

Add $-1 \cdot (4)$ and (5) to eliminate y and solve for x.

$-x - y = -15 \quad -1 \cdot (4)$

$2x + y = 25 \quad (5)$

$\overline{x = 10}$

Use (4) and the value of x to find y.

$x + y = 15$ (4)

$\Rightarrow y = 15 - x = 15 - 10 = 5$

$(x, y, z) = (10, 5, 8)$

67. $c = 1.5$ (1)

$25a + 5b + c = 52.5$ (2)

$4a + 2b + c = 7.5$ (3)

Substitute 1.5 for c into (2) and (3) to

generate a two-variable system in a and b.

$\begin{array}{l} 25a + 5b + c = 52.5 \quad (2) \\ 4a + 2b + c = 7.5 \quad\;\; (3) \end{array} \Rightarrow \begin{array}{l} 25a + 5b + 1.5 = 52.5 \\ 4a + 2b + 1.5 = 7.5 \end{array}$

$\Rightarrow 25a + 5b = 51$ (4)

$4a + 2b = 6$

or $2a + b = 3$ (5)

Add (4) to $-5 \cdot (5)$ to eliminate b and solve for a.

$25a + 5b = 51$ (4)

$-10a - 5b = -15 \quad -5 \cdot (5)$

$\overline{15a = 36}$

$\Rightarrow x = \dfrac{36}{15} = \dfrac{12}{5} = 2.4$

Use (5) and the value of a to find b.

$2a + b = 3$ (5)

$\Rightarrow b = 3 - 2a = 3 - 2(2.4) = 3 - 4.8 = -1.8$

$(a, b, c) = (2.4, -1.8, 1.5)$

69. *Thinking and Writing Exercise.* No; the graphs could open in different directions and have different vertices. Consider the graphs of $f(x) = x^2 - 4$ and $g(x) = 4 - x^2$, for example. Both have x-intercepts of $(-2, 0)$ and $(2, 0)$, but the vertex of $f(x)$ is $(0, -4)$ while the vertex of $g(x)$ is $(0, 4)$.

71. $f(x) = 2.31x^2 - 3.135x - 5.89$

a) The coefficient of x^2 is positive so the graph opens upward and the function has a minimum value. Graph the function in a window that shows the vertex. The standard window is one good choice. Then use the Minimum feature from the CALC menu to find that the minimum value is ≈ -6.95.

b) To find the first coordinates of the x-intercepts we use the Zero feature from the CALC menu to find the zeros of the function. They are about -1.06 and 2.41, so the x-intercepts are $(-1.06, 0)$ and $(2.41, 0)$. The y-intercept is $(0, f(0))$, or $(0, -5.89)$.

73. $g(x) = -1.25x^2 + 3.42x - 2.79$

   a) The coefficient of x^2 is negative so the graph opens downward and the function has a maximum value. Graph the function in a window that shows the vertex. The standard window is one good choice. Then use the Maximum feature from the CALC menu to find that the minimum value is about –0.45.

   b) The graph has no x-intercepts. The y-intercept is $(0, f(0))$, or $(0, -2.79)$.

75. $f(x) = x^2 - x - 6$

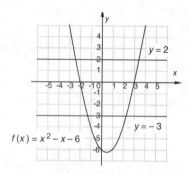

$f(x) = x^2 - x - 6$

   a) The solutions of $x^2 - x - 6 = 2$ are the first coordinates of the points of the intersection of the graphs of $f(x) = x^2 - x - 6$ and $y = 2$. From the graph we see that the solutions are approximately –2.4 and 3.4.

   b) The solutions $x^2 - x - 6 = -3$ are the first coordinates of the points of intersection of the graphs of $f(x) = x^2 - x - 6$ and $y = -3$. From the graph we see that the solutions are approximately –1.3 and 2.3.

77. $f(x) = mx^2 - nx + p$

$$= m\left(x^2 - \frac{n}{m}x\right) + p$$

$$= m\left(x^2 - \frac{n}{m}x + \frac{n^2}{4m^2} - \frac{n^2}{4m^2}\right) + p$$

$$= m\left(x - \frac{n}{2m}\right)^2 - \frac{n^2}{4m} + p$$

$$= m\left(x - \frac{n}{2m}\right)^2 + \frac{-n^2 + 4mp}{4m}, \text{ or}$$

$$m\left(x - \frac{n}{2m}\right)^2 + \frac{4mp - n^2}{4m}$$

79. Since the vertex is given as (3, –5), the function must have the form:

$$f(x) = a(x-3)^2 + (-5) = a(x-3)^2 - 5$$

Use the point $(-1, f(-1)) = (-1, 0)$ to find a.

$$f(-1) = a(-1-3)^2 - 5 = 0$$

$$\Rightarrow a(-4)^2 = 5 \Rightarrow a = \frac{5}{16}$$

$$\Rightarrow f(x) = \frac{5}{16}(x-3)^2 - 5$$

81. $f(x) = |x^2 - 1|$

We plot some points and draw the curve. Note that it will lie entirely on or above the x-axis since absolute value is never negative.

x	$f(x)$
–3	8
–2	3
–1	0
0	1
1	0
2	3
3	8

$f(x) = |x^2 - 1|$

83. $f(x) = |2(x-3)^2 - 5|$

We plot some points and draw the curve. Note that it will lie entirely on or above the *x*-axis since absolute value is never negative.

x	$f(x)$
−1	27
0	13
1	3
2	3
3	5
4	3
5	3
6	13

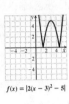

$f(x) = |2(x-3)^2 - 5|$

Exercise Set 11.8

---

1. e

3. c

5. d

7. $P(x) = 0.2x^2 - 2.8x + 9.8$

$\Rightarrow a = 0.2, \ b = -2.8, \ c = 9.8$

$P(x)$ is quadratic with a positive leading coefficient. Its minimum occurs at its vertex point.

$$-\frac{b}{2a} = -\frac{-2.8}{2 \cdot 0.2} = 7$$

7 corresponds to the month of July. The inches of precipitation is given by $P(7)$:

$$P(7) = 0.2(7)^2 - 2.8(7) + 9.8$$

$$= 9.8 - 19.6 + 9.8 = 0$$

It did not rain at all in the month of July.

9. Using $P(x) = R(x) - C(x)$, we have:

$$P(x) = 1000x - x^2 - (3000 + 20x)$$

$$= -x^2 + 980x - 3000$$

We determine the maximum of $P(x)$ by first

finding $-\frac{b}{2a}: -\frac{b}{2a} = -\frac{-980}{2 \cdot -1} = 490$

We now find the maximum value of the function $P(490)$.

$$P(490) = -490^2 + 980(490) - 3000$$

$$= 237,100$$

The maximum profit is $237,100 when $x = 490$.

11. $P = 2l + 2w = 128$

$A = l \cdot w$

Solve the first equation for l, $l = 64 - w$ and substitute into the second equation.

$$A = (64 - w)w$$

$$A = 64w - w^2$$

$$A = -(w^2 - 64w)$$

Complete the square to get:

$$A = -(w^2 - 64w + 1024) + 1024$$

$$A = -(w - 32)^2 + 1024$$

The maximum function value is 1024, when $w = 32$, $l = 64$–32, or 32. The maximum area occurs when the dimensions are 32 in. by 32 in.

13. Since one side is the house, we have:

$$P = l + 2w = 60$$

$A = l \cdot w$

Solving P for l and substituting into A, we have:

$$A = (60 - 2w)w$$

$$A = -2w^2 + 60w$$

$$A = -2(w^2 - 30w + 225) - (-2) \cdot 225$$

$$A = -2(w - 15)^2 + 450$$

The maximum function value of 450 occurs when $w = 15$; $l = 60 - 2(15)$, or 30. Maximum area of 450 ft²; dimensions 15 ft by 30 ft.

15. Let x = height of the file and y = width. We have two equations.

$$2x + y = 14$$
$$V = 8xy$$

Solve $2x + y = 14$ for y, $y = 14 - 2x$, and substitute into the second equation.

$$V = 8x(14 - 2x)$$

$$V = -16x^2 + 112x$$

$$V = -16(x^2 - 7x)$$

$$V = -16\left(x^2 - 7x + \frac{49}{4}\right) - (-16) \cdot \frac{49}{4}$$

$$V = -16\left(x - \frac{7}{2}\right)^2 + 196$$

The maximum of 196 occurs when $x = \frac{7}{2}$.

When $x = \frac{7}{2}$, $y = 14 - 2 \cdot \frac{7}{2} = 7$, the file should be $\frac{7}{2}$ in., or $3\frac{1}{2}$ in. tall.

17. Let x and y represent the numbers.

$$x + y = 18$$
$$P = xy$$

Solve the first equation for y and substitute into the second equation.

$$P = x(18 - x)$$

$$P = -x^2 + 18x$$

$$P = -(x^2 - 18x + 81) - (-81)$$

$$P = -(x - 9)^2 + 81$$

The maximum function value is 81 when $x = 9$. If $x = 9$, $y = 18 - 9$, or 9. The maximum product of 81 occurs for the numbers 9 and 9.

19. Let x and y represent the numbers.

$$x - y = 8$$
$$P = xy$$

Solve the first equation for x and substitute into the second equation.

$$P = (y + 8)y$$

$$P = y^2 + 8y$$

$$P = (y^2 + 8y + 16) - 16$$

$$P = (y + 4)^2 - 16$$

The minimum function value is −16 when

$y = -4$. If $y = -4$, then $x = -4 + 8$, or 4. The minimum product of −16 occurs for the numbers 4 and −4.

21. From the results of Exercise 17 and 18, we might observe that the numbers are −5 and −5 and that the maximum product is 25. We could also solve this problem using the same method(s) as in Exercise 17 and 18.

23. The data points appear nearly linear.

$$f(x) = mx + b$$

25. The data points rise then fall. This appears to represent a parabola that opens downward.

$$f(x) = ax^2 + bx + c, a < 0$$

27. The data points fall then rise. This appears to represent a parabola that opens upward.

$$f(x) = ax^2 + bx + c, a > 0$$

29. The data points rise nonlinearly. This appears to represent a parabola that opens upward.

$$f(x) = ax^2 + bx + c, a > 0$$

31. The data appears linear over the years 1989-1195, but nonlinear after that. The points are neither linear or quadratic over the range 1989-2007.

33. Look for a function of the form $f(x) = ax^2 + bx + c$.

Substituting the data points, we get

$$4 = a(1)^2 + b(1) + c,$$

$$-2 = a(-1)^2 + b(-1) + c,$$

$$13 = a(2)^2 + b(2) + c,$$

or

$$4 = a + b + c,$$

$$-2 = a - b + c,$$

$$13 = 4a + 2b + c$$

Solving this system, we get

$a = 2$, $b = 3$, and $c = -1$.

Therefore the function we are looking for is

$$f(x) = 2x^2 + 3x - 1.$$

35. We look for a function of the form
    $f(x) = ax^2 + bx + c$. Substituting the data
    points, we get
    $$0 = a(2)^2 + b(2) + c,$$
    $$3 = a(4)^2 + b(4) + c,$$
    $$-5 = a(12)^2 + b(12) + c,$$
    or
    $$0 = 4a + 2b + c,$$
    $$3 = 16a + 4b + c,$$
    $$-5 = 144a + 12b + c$$
    Solving this system, we get
    $$a = -\frac{1}{4}, \ b = 3, \text{ and } c = -5.$$
    Therefore the function we are looking for is
    $$f(x) = -\frac{1}{4}x^2 + 3x - 5.$$

37. a) $A(s) = as^2 + bs + c$, where $A(s)$
    represents the number of nighttime
    accidents (for every 200 million km) and s
    represents the travel speed (in
    km/h).
    $$400 = a(60)^2 + b(60) + c,$$
    $$250 = a(80)^2 + b(80) + c,$$
    $$250 = a(100)^2 + b(100) + c,$$
    or
    $$400 = 3600a + 60b + c,$$
    $$250 = 6400a + 80b + c,$$
    $$250 = 10,000a + 100b + c.$$
    Solving the system of equations, we get
    $$a = \frac{3}{16}, b = -\frac{135}{4}, c = 1750.$$
    $$A(s) = \frac{3}{16}s^2 - \frac{135}{4}s + 1750 \text{ fits the data.}$$
    b) Find $A(50)$
    $$A(50) = \frac{3}{16}(50)^2 - \frac{135}{4}(50) + 1750 - 531.25$$

    About 531 accidents occur at 50 km/h.

39. Think of a coordinate system placed on the
    drawing in the text with the origin at the point
    where the arrow is released. Then three
    points on the arrow's parabolic path are $(0,0)$,
    $(63,27)$, and $(126,0)$. We look for a function

of the form $h(d) = ad^2 + bd + c$, where
$h(d)$ represents the arrow's height and d
represents the distance the arrow has traveled
horizontally.
$$0 = a \cdot 0^2 + b \cdot 0 + c,$$
$$27 = a \cdot 63^2 + b \cdot 63 + c,$$
$$0 = a \cdot 126^2 + b \cdot 126 + c,$$
or
$$0 = c,$$
$$27 = 3969a + 63b + c,$$
$$0 = 15,876a + 126b + c.$$
Solving the system of equations, we get
$$a \approx -0.0068, \ b \approx 0.8571, \text{ and } c = 0.$$

41. a) Enter the data and use the quadratic
    regression feature. We have $D(x) =$
    $-0.0083x^2 + 0.8243x + 0.2122$
    b) $D(70) \approx 17.243$, so we estimate that the
    river is about 17.243 ft deep 70 ft from
    the left bank.

43. a) Enter the data and then use the quadratic
    regression feature. We have
    $$t(x) = 18.125x^2 + 78.15x + 24,613$$
    b) In 2017, $x = 2017 - 2005 = 12$
    $$t(12) = 18.125(12)^2 + 78.15(12) + 24,613$$
    $$\approx 28,161 \text{ teachers}$$

45. *Thinking and Writing Exercise.*

47. The graph looks exactly like the graph of
    x^3, but shifted down 2 units.

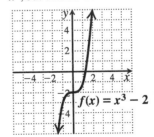

$f(x) = x^3 - 2$

49. $f(x) = x + 7 \Rightarrow f\left(\frac{1}{a^2}\right) = \frac{1}{a^2} + 7$, or $\frac{1 + 7a^2}{a^2}$

51. $g(x) = x^2 + 2 \Rightarrow g(2a + 5) = (2a + 5)^2 + 2$
    $$= 4a^2 + 20a + 25 + 2 = 4a^2 + 20a + 27$$

53. *Thinking and Writing Exercise.*

55. Position the bridge on a coordinate system as shown with the vertex of the parabola at (0, 30).

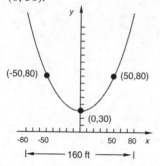

We find a function of the form $y = ax^2 + bx + c$ which represents the parabola. Since (0, 30), (–50, 80), and (50, 80) are on the parabola we know

$30 = a \cdot 0^2 + b \cdot 0 + c,$

$80 = a(-50)^2 + b(-50) + c,$

$80 = a(50)^2 + b(50) + c,$

or

$30 = c,$

$80 = 2500a - 50b + c,$

$80 = 2500a + 50b + c.$

Solving, we get $a = 0.02$, $b = 0$, $c = 30$. The function $y = 0.02x^2 + 30$ represents the parabola. The longest vertical cables occur at $x = -80$ and $x = 80$. For $x \pm 80$,

$y = 0.02(\pm 80)^2 + 30$

$\quad = 128 + 30$

$\quad = 158\,ft$

57.

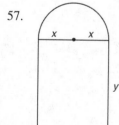

The perimeter of the semicircular portion of the window is $\dfrac{1}{2} \cdot 2\pi x$, or πx. The perimeter of the rectangular portion is $y + 2x + y$, or

$2x + 2y$. The area of the semicircular portion of the window is $\dfrac{1}{2} \cdot \pi x^2$, or $\dfrac{\pi}{2}x^2$.

The area of the rectangular portion is $2xy$. We have two equations, one giving the perimeter of the window and the other giving the area.

$\pi x + 2x + 2y = 24,$

$A = \dfrac{\pi}{2}x^2 + 2xy$

Solve the first equation for y, and substitute into the second equation.

$A = \dfrac{\pi}{2}x^2 + 2x\left(12 - \dfrac{\pi x}{2} - x\right)$

$A = \dfrac{\pi}{2}x^2 + 24x - \pi x^2 - 2x^2$

$A = -2x^2 - \dfrac{\pi}{2}x^2 + 24x$

$A = -\left(2 + \dfrac{\pi}{2}\right)x^2 + 24x$

Completing the square, we get

$A = -\left(2 + \dfrac{\pi}{2}\right)\left(x^2 + \dfrac{24}{-\left(2 + \dfrac{\pi}{2}\right)}x\right)$

$A = -\left(2 + \dfrac{\pi}{2}\right)\left(x^2 - \dfrac{48}{4 + \pi}x\right)$

$A = -\left(2 + \dfrac{\pi}{2}\right)\left(x - \dfrac{24}{4 + \pi}\right)^2 + \left(\dfrac{24}{4 + \pi}\right)^2$

The maximum function value occurs when

$x = \dfrac{24}{4 + \pi}$. When $x = \dfrac{24}{4 + \pi}$,

$y = 12 - \dfrac{\pi}{2}\left(\dfrac{24}{4 + \pi}\right) - \dfrac{24}{4 + \pi} =$

$\dfrac{48 + 12\pi}{4 + \pi} - \dfrac{12\pi}{4 + \pi} - \dfrac{24}{4 + \pi} = \dfrac{24}{4 + \pi}$

The radius of the circular portion of the window and the height of the rectangular portion should each be $\dfrac{24}{4 + \pi}$ ft.

59. Let x represent the number of 25¢ increases in the admission price. Then $10 + 0.25x$ represents the admission price, and $80 - x$ represents the corresponding average attendance. Let R represent the total revenue.

$$R(x) = (10 + 0.25x)(80 - x)$$
$$= -0.25x^2 + 10x + 800$$
$$R(x) = -0.25(x - 20)^2 + 900$$

The maximum function value of 900 occurs when
$x = 20$. The owner should charge
$10 + \$0.25(20)$, or \$15.

Chapter 11 Study Summary

1. $x^2 - 12x + 11 = 0 \Rightarrow (x - 11)(x - 1) = 0$
 $\Rightarrow x = 1, \text{ or } x = 11$

2. $x^2 - 18x + 81 = 5 \Rightarrow (x - 9)^2 = 5$
 $\Rightarrow x - 9 = \pm\sqrt{5} \Rightarrow x = 9 \pm \sqrt{5}$

3. $x^2 + 20x = 21$
 $\Rightarrow x^2 + 20x + \left(\dfrac{20}{2}\right)^2 = 21 + \left(\dfrac{20}{2}\right)^2$
 $\Rightarrow x^2 + 20x + (10)^2 = 21 + (10)^2$
 $\Rightarrow (x + 10)^2 = 121 \Rightarrow x + 10 = \pm\sqrt{121} = \pm 11$
 $\Rightarrow x = -10 \pm 11 \Rightarrow x = -21 \text{ or } x = 1$

4. $2x^2 - 3x - 9 = 0 \Rightarrow (2x + 3)(x - 3) = 0$
 $\Rightarrow 2x + 3 = 0 \Rightarrow 2x = -3 \Rightarrow x = -\dfrac{3}{2}$
 or $x - 3 = 0 \Rightarrow x = 3$

5. $2x^2 + 5x + 9 = 0 \Rightarrow a = 2, \ b = 5, \ c = 9$
 $\Rightarrow b^2 - 4ac = 5^2 - 4 \cdot 2 \cdot 9 = 25 - 72 < 0$
 The discriminant is negative. The equation has two imaginary solutions.

6. $a = n^2 + 1 \Rightarrow n^2 = a - 1 \Rightarrow n = \pm\sqrt{a - 1}$

7. $x - \sqrt{x} - 30 = 0 \Rightarrow u^2 - u - 30 = 0$ for $u = \sqrt{x}$
 $\Rightarrow (u - 6)(u + 5) = 0 \Rightarrow u = 6 \text{ or } u = -5$
 $\Rightarrow \sqrt{x} = 6 \Rightarrow x = 6^2 = 36$
 or $\sqrt{x} = -5$ which has no solution.

8. The function is quadratic with a positive leading coefficient. Its graph is a parabola opening upwards. It takes on a minimum value at its vertex point.

$$f(x) = 2x^2 - 12x + 3 \Rightarrow a = 2, \ b = -12$$
$$\Rightarrow -\frac{b}{2a} = -\frac{-12}{2 \cdot 2} = 3$$
$$f(3) = 2(3)^2 - 12(3) + 3 = -15$$

Vertex : $(3, f(3)) = (3, -15)$

Axis of symmetry: $x = 3$
Minimum of h: $y = -15$

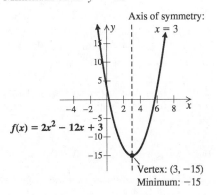

9. Let l = the length of the garden in feet, and w = its width, in feet. Then based on the amount of fencing she can afford, we have:
 $2l + 2w = 120 \Rightarrow l + w = 60$
 $\Rightarrow w = 60 - l$
 Use this result to write the formula for the area in terms of one variable, l, then back substitute to find w.
 $A = lw = l(60 - l) = -l^2 + 60l$
 The formula for the area is quadratic with a negative leading coefficient. It takes on its maximum value at its vertex. We can use
 $-\dfrac{b}{2a}$ to find the l-coordinate of the vertes.
 However, note that in its factored form,
 $A = l(60 - l)$ it is easy to see that the graph has x-intercepts at $l = 0$ and $l = 60$. Since the graph is symmetric, the l-value of the vertex must be halfway between 0 and 60 at $l = 30$. Therefore the garden will have a maximum area when $l = 30$ ft, and $w = 60 - 30 = 30$ ft also. (This method of finding the vertex will work whenever $c = 0$ in $f(x) = ax^2 + bx + c$).

Chapter 11 Review Exercises

1. False : Some quadratics have repeated roots, or, equivalently, one solution.

2. False : Many quadratic equations have two imaginary-number solutions and no real-number solution.

3. True : The quadratic formula is derived by completing the square, so either method applies to all quadratic equations.

4. True : The discriminant is found under the square root term in the quadratic formula, and the square root of a negative number is imaginary.

5. True : The correct substitution can place many rational and radical equations into quadratic form.

6. False : The vertex is at $(-3, -4)$.

7. True : $b = 0 \Rightarrow -\dfrac{b}{2a} = 0$, so the axis of symmetry is $x = 0$.

8. True : When the leading coefficient is negative, the parabola opens downward, and, thus, takes on no minimum, but extends to $-\infty$.

9. True : Find the zeros by setting the function to zero. $x^2 - 9 = 0 \Rightarrow x = \pm 3$.

10. False : If a parabola's vertex is above the x-axis and it opens upward, or if its vertex is below the x-axis and it opens downward, then it will not cross the x-axis.

11. a) 2 : The graph crosses the x-axis in two places.

   b) Positive : The parabola opens upward.

   c) -3 : The minimum occurs at the vertex and is given by the y-coordinate.

12. $9x^2 - 2 = 0 \Rightarrow 9x^2 = 2 \Rightarrow x^2 = \dfrac{2}{9}$

$$\Rightarrow x = \pm\sqrt{\dfrac{2}{9}} = \pm\dfrac{\sqrt{2}}{3}$$

The solutions are $\pm\dfrac{\sqrt{2}}{3}$.

13. $8x^2 + 6x = 0$

$2x(4x + 3) = 0$

$2x = 0 \; or \; 4x + 3 = 0$

$x = 0 \; or \qquad x = -\dfrac{3}{4}$

The solutions are $-\dfrac{3}{4}$ and 0.

14. $x^2 - 12x + 36 = 9$

$(x - 6)^2 = 9$

$x - 6 = \pm\sqrt{9}$

$x - 6 = -3 \; or \; x - 6 = 3$

$x = 3 \quad or \qquad x = 9$

The solutions are 3 and 9.

15. $x^2 - 4x + 8 = 0$

$a = 1, \, b = -4, \, c = 8$

$x = \dfrac{-(-4) \pm \sqrt{(-4)^2 - 4 \cdot 1 \cdot 8}}{2 \cdot 1}$

$x = \dfrac{4 \pm \sqrt{16 - 32}}{2}$

$x = \dfrac{4 \pm \sqrt{16 \cdot -1}}{2} = \dfrac{4 \pm 4i}{2}$

$x = 2 \pm 2i$

The solutions are $2 \pm 2i$.

16. $x(3x + 4) = 4x(x - 1) + 15$

$3x^2 + 4x = 4x^2 - 4x + 15$

$0 = x^2 - 8x + 15$

$0 = (x - 5)(x - 3)$

$x - 5 = 0 \; or \; x - 3 = 0$

$x = 5 \; or \qquad x = 3$

The solutions are 3 and 5.

17. $x^2 + 9x = 1$
$x^2 + 9x - 1 = 0$
$a = 1,\ b = 9,\ c = -1$
$x = \dfrac{-9 \pm \sqrt{9^2 - 4 \cdot 1 \cdot (-1)}}{2 \cdot 1}$
$x = \dfrac{-9 \pm \sqrt{81 + 4}}{2}$
$x = \dfrac{-9 \pm \sqrt{85}}{2}$

The solutions are $\dfrac{-9 \pm \sqrt{85}}{2}$, or $-\dfrac{9}{2} \pm \dfrac{\sqrt{85}}{2}$

18. $x^2 - 5x - 2 = 0$
$a = 1,\ b = -5,\ c = -2$
$x = \dfrac{-(-5) \pm \sqrt{(-5)^2 - 4 \cdot 1 \cdot (-2)}}{2 \cdot 1}$
$x = \dfrac{5 \pm \sqrt{25 + 8}}{2} = \dfrac{5 \pm \sqrt{33}}{2}$
$x = \dfrac{5 + \sqrt{33}}{2} \approx 5.3722813233$
$x = \dfrac{5 - \sqrt{33}}{2} \approx -0.3722813233$

The solutions rounded to 3 decimal places are 5.372 and –0.372.

19. Let $f(x) = 0$ and solve.
$0 = 4x^2 - 3x - 1$
$0 = (4x + 1)(x - 1)$
$4x + 1 = 0 \quad \text{or}\ x - 1 = 0$
$x = -\dfrac{1}{4}$ or $\quad x = 1$

The solutions are $-\dfrac{1}{4}$ and 1.

20. $\dfrac{1}{2} \cdot 12 = 6; 6^2 = 36.\ \ 36; 6$

21. $\dfrac{1}{2} \cdot \dfrac{3}{5} = \dfrac{3}{10}; \left(\dfrac{3}{10}\right)^2 = \dfrac{9}{100}.\ \ \dfrac{9}{100}; \dfrac{3}{10}$

22. $x^2 - 6x + 1 = 0$
$x^2 - 6x + 3^2 = -1 + 3^2$
$(x - 3)^2 = 8$
$x - 3 = \pm\sqrt{8}$
$x = 3 \pm 2\sqrt{2}$
The solutions are $3 \pm 2\sqrt{2}$.

23. $A = P(1 + r)^t$
$2500(1 + r)^2 = 2704 \Rightarrow (1 + r)^2 = \dfrac{2704}{2500} = \dfrac{676}{625}$
$\Rightarrow 1 + r = \pm\sqrt{\dfrac{676}{625}} = \pm\dfrac{26}{25}$
Since r is positive, ignore the negative root.
$\Rightarrow r = -1 + \dfrac{26}{25} = \dfrac{1}{25} = 0.04 = 4\%$

24. $s = 16t^2 \Rightarrow 541 = 16t^2 \Rightarrow t^2 = \dfrac{541}{16}$
$\Rightarrow t = \pm\sqrt{\dfrac{541}{16}}$
Since the time to fall is positive, ignore the negative root.
$\Rightarrow t = \sqrt{\dfrac{541}{16}} = \dfrac{\sqrt{541}}{4} \approx 5.8$ seconds

25. $b^2 - 4ac = 3^2 - 4 \cdot 1 \cdot (-6) = 33$
Two irrational real solutions.

26. $b^2 - 4ac = 2^2 - 4 \cdot 1 + 5 = -16$
Two imaginary solutions.

27. $(x - 3i)(x + 3i) = 0 \Rightarrow x^2 - (3i)^2 = 0$
$\Rightarrow x^2 - 9i^2 = 0 \Rightarrow x^2 + 9 = 0$

28. $x = -4\ or\ x = -4$
$(x + 4)^2 = 0$
$x^2 + 8x + 16 = 0$

29. Let x = the speed of the plane in still air, in mph. Then:

$$t_{\text{TO PLANT}} + t_{\text{BACK}} = 4 \Rightarrow \frac{300}{x+20} + \frac{300}{x-20} = 4$$

$$300(x+20) + 300(x-20)$$
$$= 4(x+20)(x-20)$$

$$300x \cancel{+6000} + 300x \cancel{-6000}$$
$$= 4x^2 - 1600$$

$$4x^2 - 600x - 1600 = 0$$

$$x^2 - 150x - 400 = 0$$

$$a = 1,\ b = -150,\ c = -400$$

$$x = \frac{-(-150) \pm \sqrt{(-150)^2 - 4 \cdot 1 \cdot (-400)}}{2 \cdot 1}$$

$$= \frac{150 \pm \sqrt{22{,}500 + 1600}}{2} = \frac{150 \pm \sqrt{24{,}100}}{2}$$

$$= \frac{150 \pm 10\sqrt{241}}{2} = 75 \pm 5\sqrt{241}$$

Since $75 - 5\sqrt{241} < 0$, the answer is:
$$75 + 5\sqrt{241} \approx 152.6 \text{ mph}$$

30. Let x = the time it takes Shawna to answer all the emails working alone, in hours, and y = the time it takes Erica to answer all the emails working alone, in hours. Then:

$$y = x + 6 \qquad (1)$$

$$\frac{1}{x} \cdot 4 + \frac{1}{y} \cdot 4 = 1 \quad (2)$$

Substitute $x + 6$ for y in (2) to find x.

$$\frac{1}{x} \cdot 4 + \frac{1}{y} \cdot 4 = 1 \quad (2)$$

$$\frac{4}{x} + \frac{4}{x+6} = 1$$

$$4(x+6) + 4x = 1 \cdot x(x+6)$$

$$4x + 24 + 4x = x^2 + 6x$$

$$x^2 - 2x - 24 = 0 \Rightarrow$$

$$(x-6)(x+4) = 0$$

$$x = -4 \text{ or } 6$$

Since x is greater than zero, Shawna can answer all the emails working alone in 6 hours.

31. Let $f(x) = 0$

$$0 = x^4 - 13x^2 + 36$$

$$0 = (x^2 - 9)(x^2 - 4)$$

$$0 = (x+3)(x-3)(x+2)(x-2)$$

$$x + 3 = 0 \ \text{ or } x - 3 = 0 \text{ or } x + 2 = 0 \ \text{ or } x - 2 = 0$$

$$x = -3 \text{ or } \quad x = 3 \text{ or } \quad x = -2 \text{ or } \quad x = 2$$

The x-intercepts are $(-3,0)$, $(-2,0)$, $(2,0)$, and $(3,0)$.

32. $15x^{-2} - 2x^{-1} - 1 = 0$

Let $u = x^{-1}$ $\left(\text{and } u^2 = x^{-2}\right)$

$$15u^2 - 2u - 1 = 0$$

$$(5u + 1)(3u - 1) = 0$$

$$5u + 1 = 0 \ \text{ or } 3u - 1 = 0$$

$$u = -\frac{1}{5} \text{ or } \qquad u = \frac{1}{3}$$

$u = x^{-1}$; substitute and solve for x.

$$x^{-1} = -\frac{1}{5} \text{ or } x^{-1} = \frac{1}{3}$$

$$x = -5 \ \text{ or } \quad x = 3$$

The solutions are -5 and 3.

33. $\left(x^2 - 4\right)^2 - \left(x^2 - 4\right) - 6 = 0$

Let $u = x^2 - 4$ $\left[\text{and } u^2 = \left(x^2 - 4\right)^2\right]$

$$u^2 - u - 6 = 0$$

$$(u - 3)(u + 2) = 0$$

$$u = 3 \text{ or } u = -2$$

$u = x^2 - 4$; substitute and solve for x.

$$x^2 - 4 = 3 \quad \text{ or } x^2 - 4 = -2$$

$$x^2 = 7 \quad \text{ or } \quad x^2 = 2$$

$$x = \pm\sqrt{7} \text{ or } \quad x = \pm\sqrt{2}$$

All four values check. The solutions are $\pm\sqrt{7}$ and $\pm\sqrt{2}$.

34. a) $f(x) = -3(x+2)^2 + 4$

$f(x) = -3(x + 2)^2 + 4$
Maximum: 4

b) Label the vertex $(-2,4)$

c) Draw the axis of symmetry $x = -2$

d) Maximum value is 4.

35. a) $f(x) = 2x^2 - 12x + 23$

$= 2(x^2 - 6x) + 23$

$= 2(x^2 - 6x + 9) - 2 \cdot 9 + 23$

$= 2(x - 3)^2 + 5$

Vertex: $(3,5)$
Axis of symmetry: $x = 3$

b)

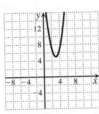

$f(x) = 2x^2 - 12x + 23$

36. $f(x) = x^2 - 9x + 14$

To determine the x-intercepts, let $f(x) = 0$
and solve the equations.

$0 = x^2 - 9x + 14$

$0 = (x - 7)(x - 2)$

$x = 7 \ or \ x = 2$

The x-intercepts are $(7,0)$ and $(2,0)$; the
y-intercept is $(0, f(0))$, or $(0,14)$.

37. $N = 3\pi\sqrt{\dfrac{1}{p}}$

$\dfrac{N}{3\pi} = \sqrt{\dfrac{1}{p}}$

$\left(\dfrac{N}{3\pi}\right)^2 = \dfrac{1}{p}$

$\dfrac{9\pi^2}{N^2} = p$

38. $2A + T = 3T^2$

$0 = 3T^2 - T - 2A$

This is a quadratic equation; use the quadratic
formula to solve. $a = 3$, $b = -1$, $c = -2A$

$T = \dfrac{-(-1) \pm \sqrt{(-1)^2 - 4 \cdot 3 \cdot (-2A)}}{2 \cdot 3}$

$T = \dfrac{1 \pm \sqrt{1 + 24A}}{6}$

39. The data looks linear over the period 2000 to
2004, but then deviates sharply from the line.
The data is neither quadratic nor linear.

40. The data points look like the right side of a
graph of a parabola which opens upward.
$f(x) = ax^2 + bx + c$, $a > 0$

41. The data points are almost linear.
$f(x) = mx + b$

42. Since we have two sides of fencing,
$30 = l + w$, where l is the length and w is the
width. The area of the rectangular area can
be expressed as the equation $A = l \cdot w$. Solve
the system of equations using substitution.
$30 = l + w \rightarrow l = 30 - w$

$A = (30 - w)w$

$A = -w^2 + 30w$

$A = -(w^2 - 30w + 225) - (-225)$

$A = -(w - 15)^2 + 225$

The maximum function value/area is $225 \, ft^2$,
when $w = 15 ft$ and $l = 30 - 15$, or $15 ft$.

43. a)
Find a function of the form
$f(x) = ax^2 + bx + c$ that satisfies the three
points listed. Substitute the given values for x
and $f(x)$ to get

$a(0)^2 + b(0) + c = 1$

$or \ c = 1$ \qquad (1)

$a(20)^2 + b(20) + c = 1000$

$or \ 400a + 20b + c = 1000$ \qquad (2)

$a(60)^2 + b(60) + c = 32,000$

$or \ 3600a + 60b + c = 32,000$ \quad (3)

Substitute $c = 1$ into (2) and (3) to generate
and two-variable system in a and b,

$400a + 20b + 1 = 1000$

or $400a + 20b = 999$ (4)

$3600a + 60b + 1 = 32,000$

or $3600a + 60b = 31,999$ (5)

Add $-3 \cdot (4)$ to (5) to eliminate b and find a.

$-1200a - 60b = -2997$ $-3 \cdot (4)$

$\underline{3600a + 60b = 31,999}$ (5)

$2400a = 29,002$

$\Rightarrow a = \dfrac{29,002}{2400} = \dfrac{14,501}{1200}$

Use (4) and the values of a to find b.

$400a + 20b = 999$ (4)

$\Rightarrow 20b = 999 - 400a = 999 - 400\left(\dfrac{14,501}{1200}\right)$

$= 999 - \dfrac{14,501}{3} = \dfrac{3 \cdot 999 - 14,501}{3}$

$\Rightarrow b = -\dfrac{11,504}{3 \cdot 20} = -\dfrac{\cancel{4} \cdot 2876}{3 \cdot \cancel{4} \cdot 5} = -\dfrac{2876}{15}$

$M(x) = \dfrac{14,501}{1200}x^2 - \dfrac{2876}{15}x + 1$

b) In 2020, $x = 2020 - 1948 = 72$

$M(72) = \dfrac{14,501}{1200}(72)^2 - \dfrac{2876}{15}(72) + 1$

$= \dfrac{14,501}{\cancel{24} \cdot \cancel{2} \cdot 25} \cdot \cancel{24} \cdot 3 \cdot \cancel{2} \cdot 36$

$\quad - \dfrac{2876}{\cancel{3} \cdot 5} \cdot \cancel{3} \cdot 24 + 1$

$= \dfrac{14,501 \cdot 3 \cdot 36 - 5 \cdot 2876 \cdot 24 + 25 \cdot 1}{25}$

$= \dfrac{1,221,013}{25} \approx 48,841$

44. a) $M(x) = 12.6207x^2 - 242.2557x + 706.6461$

    b) $M(72)$

$= 12.6207(72)^2 - 242.2557(72) + 706.6461$

$\approx 48,690$

45. *Thinking and Writing Exercise.* Completing the square was used to solve quadratic equations and to graph functions by rewriting the function in the form

$f(x) = a(x - h)^2 + k$.

46. *Thinking and Writing Exercise.* The model found in Exercise 44 predicts 4363 more restaurants in 2010 than the model from Exercise 43. The greater prediction seems to fit the pattern better.

47. *Thinking and Writing Exercise.* The most solutions a polynomial can have is equal to the degree of the polynomial. Thus the most solutions an equation of the form

$ax^4 + bx^2 + c = 0$ can have is four. One way to see this is to remember that if k is a root, then $x - k$ must be a factor of the polynomial. Multiplying more than four linear factors of this type together would yield a polynomial of degree equal to the number of factors.

48. The x-intercepts are $(-3,0)$ and $(5,0)$, and the y-intercept is $(0,-7)$. Substituting these ordered pairs into the equation

$f(x) = ax^2 + bx + c$ gives a system of equations.

$0 = a \cdot (-3)^2 + b(-3) + c$

$0 = a \cdot 5^2 + b \cdot 5 + c$

$-7 = a \cdot 0 + b \cdot 0 + c$

      or

$0 = 9a - 3b + c$

$0 = 25a + 5b + c$

$-7 = c$

Solving this equation gives us:

$a = \dfrac{7}{15}, b = -\dfrac{14}{15}, c = -7.$

The equation is $f(x) = \dfrac{7}{15}x^2 - \dfrac{14}{15}x - 7$

49. From section 8.4, we know the sum of the solutions of $ax^2 + bx + c = 0$ is $-\dfrac{b}{a}$, and the product is $\dfrac{c}{a}$.

$3x^2 - hx + 4k = 0$

$a = 3, \ b = -h, \ \text{and} \ c = 4k$

Substituting:

$-\dfrac{b}{a} : \dfrac{-(-h)}{3} = 20, h = 60$

$\dfrac{c}{a} : \dfrac{4k}{3} = 80, k = 60$

50. Let x and y represent two positive integers. Since one of the numbers is the square root of the other, we let $y = \sqrt{x}$. To find their average, we find their sum and divide by 2.

$$\frac{x + \sqrt{x}}{2} = 171$$

$$x + \sqrt{x} = 342$$

$$x + \sqrt{x} - 342 = 0$$

Let $u = \sqrt{x}$, $\left(\text{and } u^2 = x\right)$.

$$u^2 + u - 342 = 0$$

$$(u + 19)(u - 18) = 0$$

$$u = -19 \text{ or } u = 18$$

Substituting: $\sqrt{x} = -19$ or $\sqrt{x} = 18$

We use only $\sqrt{x} = 18$

$$x = 324$$

The numbers are 18 and 324.

Chapter 11 Test

1. a) 0 : The graph does not cross the x- axis.

   b) Negative : The parabola opens downward.

   c) -1 : The maximum value of the function is given by the y-coordinate of the vertex.

2. $25x^2 - 7 = 0 \Rightarrow 25x^2 = 7 \Rightarrow x^2 = \dfrac{7}{25}$

   $\Rightarrow x = \pm\sqrt{\dfrac{7}{25}} = \pm\dfrac{\sqrt{7}}{5}$

3. $4x(x - 2) - 3x(x + 1) = -18$

   $$4x^2 - 8x - 3x^2 - 3x = -18$$

   $$x^2 - 11x + 18 = 0$$

   $$(x - 9)(x - 2) = 0$$

   $$x - 9 = 0 \text{ or } x - 2 = 0$$

   $$x = 9 \text{ or } \quad x = 2$$

   The solutions are 9 and 2.

4. $x^2 + 2x + 3 = 0 \Rightarrow a = 1,\ b = 2,\ c = 3$

   $$x = \frac{-2 \pm \sqrt{2^2 - 4 \cdot 1 \cdot 3}}{2 \cdot 1} = \frac{-2 \pm \sqrt{-8}}{2}$$

   $$= \frac{-2 \pm 2\sqrt{2}i}{2} = \frac{\cancel{2}(-1 \pm \sqrt{2}i)}{\cancel{2}} = -1 \pm \sqrt{2}i$$

   The solutions are $-1 \pm \sqrt{2}i$.

5. $2x + 5 = x^2$

   $$0 = x^2 - 2x - 5$$

   $$a = 1,\ b = -2,\ c = 5$$

   $$x = \frac{-(-2) \pm \sqrt{(-2)^2 - 4 \cdot 1 \cdot (-5)}}{2 \cdot 1}$$

   $$x = \frac{2 \pm \sqrt{24}}{2} = \frac{2 \pm 2\sqrt{6}}{2} = \frac{2(1 \pm \sqrt{6})}{2}$$

   $$x = 1 \pm \sqrt{6}$$

   The solutions are $1 \pm \sqrt{6}$.

6. $x^{-2} - x^{-1} = \dfrac{3}{4}$

   Let $u = x^{-1}$ $\left(\text{and } u^2 = x^{-2}\right)$

   $$u^2 - u - \frac{3}{4} = 0$$

   $$4u^2 - 4u - 3 = 0$$

   $$(2u - 3)(2u + 1) = 0$$

   $$u = \frac{3}{2} \text{ or } u = -\frac{1}{2}$$

   $u = x^{-1}$, so $x^{-1} = \dfrac{3}{2}$ or $x^{-1} = -\dfrac{1}{2}$

   $$x = \frac{2}{3} \text{ or } \quad x = -2$$

   The solutions are -2 and $\dfrac{2}{3}$.

7. $x^2 + 3x = 5$

   $$x^2 + 3x - 5 = 0$$

   $$a = 1,\ b = 3,\ c = -5$$

   $$x = \frac{-3 \pm \sqrt{3^2 - 4 \cdot 1 \cdot (-5)}}{2 \cdot 1} = \frac{-3 \pm \sqrt{29}}{2}$$

   $$x = \frac{-3 - \sqrt{29}}{2} \approx -4.193$$

   $$x = \frac{-3 + \sqrt{29}}{2} \approx 1.193$$

8. Let $f(x) = 0$ and solve for x.

$$0 = 12x^2 - 19x - 21$$
$$0 = (4x + 3)(3x - 7)$$
$$x = -\frac{3}{4} \text{ or } x = \frac{7}{3}$$

The solutions are $-\frac{3}{4}$ and $\frac{7}{3}$.

9. $x^2 - 20x + 100 = (x - 10)^2$

10. $\frac{1}{2} \cdot \frac{2}{7} = \frac{1}{7}; \ \frac{1}{7}^2 = \frac{1}{49}. \ \frac{1}{49}; \frac{1}{7}$

11. $x^2 + 10x + 15 = 0$

$$x^2 + 10x = -15$$
$$x^2 + 10x + 25 = -15 + 25$$
$$(x + 5)^2 = 10$$
$$x + 5 = \pm\sqrt{10}$$
$$x = -5 \pm \sqrt{10}$$

The solutions are $-5 \pm \sqrt{10}$.

12. $x^2 + 2x + 5 = 0$

$\Rightarrow b^2 - 4ac = 2^2 - 4 \cdot 1 \cdot 5 < 0$

Because the discriminant is negative, the equation will have two imaginary solutions.

13. $(x - \sqrt{11})(x - (-\sqrt{11})) = 0$

$\Rightarrow (x - \sqrt{11})(x + \sqrt{11}) = 0$

$\Rightarrow x^2 - 11 = 0$

14. Let r = speed of the boat in still water. Since the rate of the river is 4km/h, the rate upstream is $r - 4$ and the rate downstream is $r + 4$. Using $T = \frac{d}{r}$, we have time upstream is $\frac{60}{r - 4}$ and the time downstream is $\frac{60}{r + 4}$.

The total time is 8 hr. We have the equation:

$$\frac{60}{r - 4} + \frac{60}{r + 4} = 8 \ \text{ LCD is } (r - 4)(r + 4)$$

$$(r - 4)(r + 4) \cdot \left(\frac{60}{r - 4} + \frac{60}{r + 4}\right) = (r - 4)(r + 4) \cdot 8$$

$$60(r + 4) + 60(r - 4) = 8(r^2 - 16)$$

$$60r + 240 + 60r - 240 = 8r^2 - 128$$

$$0 = 8r^2 - 120r - 128$$

$$0 = r^2 - 15r - 16$$

$$0 = (r - 16)(r + 1)$$

$$r = 16 \text{ or } r = -1$$

Since the speed cannot be negative, the speed of the boat in still water is 16 km/hr.

15. Let x = the number of hours it takes Dal alone and

$x + 4$ = the number of hours it takes Kim.

Dal can do $\frac{1}{x}$ of the job in 1 hr, and Kim can do $\frac{1}{x + 4}$. They work together for

$1\frac{1}{2}$ hrs., or $\frac{3}{2}$ hrs. We solve the equation

$$\frac{1}{x} \cdot \frac{3}{2} + \frac{1}{x + 4} \cdot \frac{3}{2} = 1, \text{ or } \frac{3}{2x} + \frac{3}{2(x + 4)} = 1$$

for x

LCD is $2x(x + 4)$

$$2x(x + 4) \cdot \left(\frac{3}{2x} + \frac{3}{2(x + 4)}\right) = 2x(x + 4) \cdot 1$$

$$3(x + 4) + 3x = 2x^2 + 8x$$

$$3x + 12 + 3x = 2x^2 + 8x$$

$$0 = 2x^2 + 2x - 12$$

$$0 = x^2 + x - 6$$

$$0 = (x + 3)(x - 2)$$

$x = -3 \text{ or } x = 2$

Since negative time has no meaning for this problem, Dal can assemble the swing set in 2 hrs.

16. $f(x) = 0 \Rightarrow x^4 - 15x^2 - 16 = 0$

$\Rightarrow u^2 - 15u - 16 = 0$ for $u = x^2$

$\Rightarrow (u - 16)(u + 1) = 0$

$\Rightarrow u = 16 \Rightarrow x^2 = 16 \Rightarrow x = \pm\sqrt{16} = \pm 4$

or $u = -1 \Rightarrow x^2 = -1 \Rightarrow x = \pm\sqrt{-1} = \pm i$

The function only has two real roots. Those correspond to the x-intercepts: $(-4, 0)$, $(4, 0)$.

17. a)

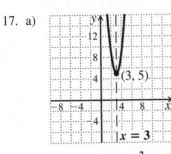

$$f(x) = 4(x - 3)^2 + 5$$
Minimum: 5

b) Label the vertex $(3,5)$

c) Draw the axis of symmetry $x = 3$

d) Minimum function value of 5.

18. a) $f(x) = 2x^2 + 4x - 6$

$= 2(x^2 + 2x) - 6$

$= 2(x^2 + 2x + 1) - 2 \cdot 1 - 6$

$= 2(x + 1)^2 - 8$

Vertex: $(-1, -8)$
Axis of symmetry: $x = -1$

b)

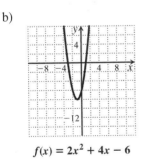

$$f(x) = 2x^2 + 4x - 6$$

19. To find the x-intercepts, set $f(x) = 0$.

$0 = x^2 - x - 6$

$0 = (x - 3)(x + 2)$

$x = 3$ or $x = -2$

The x-intercepts are $(-2, 0)$ and $(3, 0)$ and the y-intercept is $(0, f(0))$, or $(0, -6)$.

20. $V = \frac{1}{3}\pi(R^2 + r^2) \Rightarrow \frac{3V}{\pi} = R^2 + r^2$

$\Rightarrow r^2 = \frac{3V}{\pi} - R^2 \Rightarrow r = \sqrt{\frac{3V}{\pi} - R^2}$

Ignore the negative root since we are assuming $r > 0$.

21. The data points rise then fall; this appears to represent a parabola which opens downward. A quadratic function.

22. $C(x) = 0.2x^2 - 1.3x + 3.4025$

$C(x) = 0.2(x^2 - 6.5x) + 3.4025$

$C(x) = 0.2(x^2 - 6.5x + 10.5625) - 0.2(10.5625) + 3.4025$

$C(x) = 0.2(x - 3.25)^2 + 1.29$

A minimum of $1.29 hundred, or $129 when 3.25 hundred, or 325 cabinets are built.

23. We look for a function of the form $f(x) = ax^2 + bx + c$. Substituting the data points, we have:

$35 = a(0)^2 + b(0) + c$

$310 = a(4)^2 + b(4) + c$

$200 = a(6)^2 + b(6) + c$

$\quad$ or

$35 = c$

$310 = 16a + 4b + c$

$200 = 36a + 6b + c$

Solving this system we get:

$a = -\frac{165}{8}, \ b = \frac{605}{4}, \ c = 35$

The function is $p(x) = -\frac{165}{8}x^2 + \frac{605}{4}x + 35$

24. Enter the data and then use the quadratic regression feature. We have

$p(x) = -23.5417x^2 + 162.2583x + 57.617$

25. $kx^2 + 3x - k = 0$; one solution is -2.
We first find k by substituting -2 for x.

$$k(-2)^2 + 3(-2) - k = 0$$
$$4k - 6 - k = 0$$
$$3k = 6$$
$$k = 2$$

We now substitute 2 for k in the original equation.

$$2x^2 + 3x - 2 = 0$$
$$(2x - 1)(x + 2) = 0$$
$$2x - 1 = 0 \text{ or } x + 2 = 0$$
$$x = \frac{1}{2} \text{ or } x = -2$$

The other solution is $\frac{1}{2}$.

26. If $2i$ is a solution, then $-2i$ must also be a solution since the imaginary solutions of $P(x) = 0$, where $P(x)$ is a polynomial, always come in conjugate pairs.

Thus $x - 2i$, $x + 2i$, and $x + \sqrt{3}$ must be factors of the polynomial. In order for the coefficients of the equation to be integers, $x - \sqrt{3}$ must also be a factor. Since the product of these factors will be a fourth degree polynomial, these are the only factors we need.

$$(x - 2i)(x + 2i)(x + \sqrt{3})(x - \sqrt{3}) = 0$$
$$\Rightarrow (x^2 + 4)(x^2 - 3) = 0$$
$$\Rightarrow x^4 + x^2 - 12 = 0$$

The left side of the equation could also be multiplied by any constant and retain the same roots.

27. $x^4 - 4x^2 - 1 = 0$

$$\Rightarrow u^2 - 4u - 1 = 0 \text{ for } u = x^2$$

$$\Rightarrow u = \frac{4 \pm \sqrt{16 + 4}}{2} = \frac{4 \pm 2\sqrt{5}}{2} = 2 \pm \sqrt{5}$$

$$x^2 = 2 + \sqrt{5} \Rightarrow x = \pm\sqrt{2 + \sqrt{5}}$$

$$\text{or } x^2 = 2 - \sqrt{5} \Rightarrow x = \pm\sqrt{2 - \sqrt{5}}$$

However, since $\sqrt{5} > 2$, these last two solutions are imaginary. Using $2 - \sqrt{5} = -(\sqrt{5} - 2)$, we can write these solutions in terms of i.

$$x = \pm\sqrt{2 - \sqrt{5}} = \pm\sqrt{-(\sqrt{5} - 2)} = \pm\sqrt{\sqrt{5} - 2}\, i$$

Thus the four solutions are:

$$x = \sqrt{2 + \sqrt{5}}, \ x = -\sqrt{2 + \sqrt{5}}. \ x = \sqrt{\sqrt{5} - 2}\, i$$
$$\text{and } x = -\sqrt{\sqrt{5} - 2}\, i$$

Chapter 12

Exponentials and Logarithmic Functions

Exercise Set 12.1

1. True

3. False

5. False

7. True

9. a) $(f \circ g)(1) = f(g(1)) = f(1-3)$
$= f(-2) = (-2)^2 + 1 = 5$

   b) $(g \circ f)(1) = g(f(1)) = g(1^2+1)$
$= g(2) = 2-3 = -1$

   c) $(f \circ g)(x) = f(g(x)) = f(x-3)$
$= (x-3)^2 + 1$
$= x^2 - 6x + 9 + 1$
$= x^2 - 6x + 10$

   d) $(g \circ f)(x) = g(f(x)) = g(x^2+1)$
$= (x^2+1) - 3$
$= x^2 + 1 - 3$
$= x^2 - 2$

11. a) $(f \circ g)(1) = f(g(1)) = f(2 \cdot 1^2 - 7)$
$= f(2-7) = f(-5)$
$= 5(-5) + 1 = -25 + 1 = -24$

   b) $(g \circ f)(1) = g(f(1)) = g(5 \cdot 1 + 1)$
$= g(6) = 2(6)^2 - 7$
$= 2 \cdot 36 - 7$
$= 72 - 7 = 65$

   c) $(f \circ g)(x) = f(g(x)) = f(2x^2 - 7)$
$= 5(2x^2 - 7) + 1$
$= 10x^2 - 35 + 1$
$= 10x^2 - 34$

   d) $(g \circ f)(x) = g(f(x)) = g(5x+1)$
$= 2(5x+1)^2 - 7$
$= 2(25x^2 + 10x + 1) - 7$
$= 50x^2 + 20x + 2 - 7$
$= 50x^2 + 20x - 5$

13. a) $(f \circ g)(1) = f(g(1)) = f\left(\dfrac{1}{1^2}\right)$
$= f(1) = 1 + 7 = 8$

   b) $(g \circ f)(1) = g(f(1)) = g(1+7)$
$= g(8) = \dfrac{1}{8^2} = \dfrac{1}{64}$

   c) $(f \circ g)(x) = f(g(x))$
$= f\left(\dfrac{1}{x^2}\right) = \dfrac{1}{x^2} + 7$

   d) $(g \circ f)(x) = g(f(x))$
$= g(x+7) = \dfrac{1}{(x+7)^2}$

15. a) $(f \circ g)(1) = f(g(1)) = f(1+3)$
$= f(4) = \sqrt{4} = 2$

   b) $(g \circ f)(1) = g(f(1)) = g(\sqrt{1})$
$= g(1) = 1 + 3 = 4$

   c) $(f \circ g)(x) = f(g(x)) = f(x+3)$
$= \sqrt{x+3}$

   d) $(g \circ f)(x) = g(f(x)) = g(\sqrt{x})$
$= \sqrt{x} + 3$

17. a) $(f \circ g)(1) = f(g(1)) = f\left(\frac{1}{1}\right)$

$= f(1) = \sqrt{4 \cdot 1} = 2$

b) $(g \circ f)(1) = g(f(1)) = g(\sqrt{4 \cdot 1})$

$= g(2) = \frac{1}{2}$

c) $(f \circ g)(x) = f(g(x)) = f\left(\frac{1}{x}\right) = \sqrt{\frac{4}{x}}$

d) $(g \circ f)(x) = g(f(x)) = g(\sqrt{4x}) = \frac{1}{\sqrt{4x}}$

19. a) $(f \circ g)(1) = f(g(1)) = f(\sqrt{1-1})$

$= f(0) = 0^2 + 4 = 4$

b) $(g \circ f)(1) = g(f(1)) = g(1^2 + 4)$

$= g(5) = \sqrt{5-1} = \sqrt{4} = 2$

c) $(f \circ g)(x) = f(g(x)) = f(\sqrt{x-1})$

$= (\sqrt{x-1})^2 + 4$

$= x - 1 + 4 = x + 3$

d) $(g \circ f)(x) = g(f(x)) = g(x^2 + 4)$

$= \sqrt{x^2 + 4 - 1}$

$= \sqrt{x^2 + 3}$

21. Since $(y_1 \circ y_2)(-3) = y_1(y_2(-3))$, we first find $y_2(-3)$. Locate -3 in the x-column and then move across to the y_2-column to find that $y_2(-3) = 1$. Now we have $y_1(y_2(-3))$ $= y_1(1)$. Locate 1 in the x-column and then move across to the y_1-column to find that $y_1(1) = 8$. Thus, $(y_1 \circ y_2)(-3) = 8$.

23. Since $(y_1 \circ y_2)(-1) = y_1(y_2(-1))$, we first find $y_2(-1)$. Locate -1 in the x-column and then move across to the y_2-column to find that $y_2(-1) = -3$. Now we have $y_1(y_2(-1))$ $= y_1(-3)$. Locate -3 in the x-column and then move across to the y_1-column to find that $y_1(-3) = -4$.

Thus $(y_1 \circ y_2)(-1) = -4$

25. Since $(y_2 \circ y_1)(1) = y_2(y_1(1))$, we first find $y_1(1)$. Locate 1 in the x-column and then move across to the y_1-column to find that $y_1(1) = 8$. Now we have $y_2(y_1(1)) = y_2(8)$. However, y_2 is not defined for $x = 8$, so $(y_2 \circ y_1)(1)$ is not defined.

27. Since $(f \circ g)(2) = f(g(2))$, we first find $g(2)$. Locate 2 in the x-column and then move across to the $g(x)$-column to find that $g(2) = 5$. Now we have $f(g(2)) = f(5)$. Locate 5 in the x-column and then move across to the $f(x)$- column to find that $f(5) = 4$. Thus, $(f \circ g)(2) = 4$.

29. To find $f(g(3))$ we first find $g(3)$. Locate 3 in the x-column and then move across to the $g(x)$-column to find that $g(3) = 8$. Now we have $f(g(3)) = f(8)$. However, $f(x)$ is not defined for $x = 8$, so $f(g(3))$ is not defined.

31. $h(x) = (3x - 5)^4$

This is $3x - 5$ raised to the fourth power, so the two most obvious functions are $f(x) = x^4$ and $g(x) = 3x - 5$.

33. $h(x) = \sqrt{2x + 7}$

We have $2x + 7$ and take the square root of this expression.

$f(x) = \sqrt{x}$ and $g(x) = 2x + 7$

35. $h(x) = \frac{2}{x - 3}$

This is 2 divided by $x - 3$, so we can use the functions $f(x) = \frac{2}{x}$ and $g(x) = x - 3$.

37. $h(x) = \frac{1}{\sqrt{7x + 2}}$

This is the reciprocal of the square root of $7x + 2$. The two functions can be $f(x) = \frac{1}{\sqrt{x}}$ and $g(x) = 7x + 2$.

39. $h(x) = \dfrac{1}{\sqrt{3x}} + \sqrt{3x}$

This is the reciprocal of the square root of $3x$ plus the square root of $3x$. Two functions that can be used are

$f(x) = \dfrac{1}{x} + x$ and $g(x) = \sqrt{3x}$.

41. The graph of $f(x) = x - 5$ is shown below.

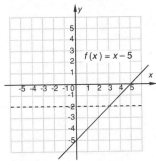

Since there is no horizontal line that crosses the graph more than once, the function is one-to-one.

43. The graph of $f(x) = x^2 + 1$ is shown below.

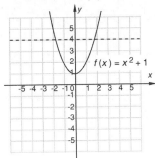

Observe that the graph of this function is a parabola that opens up. There are many horizontal lines that cross the graph more than once. In particular, the line $y = 4$ crosses the graph more than once. The function is not one-to-one.

45. Since no horizontal line crosses the graph more than once, the function is one-to-one.

47. Since we can draw at least one horizontal line that crosses the graph more than once, the function is not one-to-one.

49. a. The function $f(x) = x + 4$ is a linear function that is not constant, so it passes the horizontal-line test. Thus, f is one-to-one.

    b. Replace $f(x)$ by y: $y = x + 4$

    Interchange x and y: $x = y + 4$

    Solve for y: $x - 4 = y$

    Replace y by $f^{-1}(x)$: $f^{-1}(x) = x - 4$

51. a. The function $f(x) = 2x$ is a linear function that is not constant, so it passes the horizontal-line test. Thus, f is one-to-one.

    b. Replace $f(x)$ by y: $y = 2x$

    Interchange x and y: $x = 2y$

    Solve for y: $\dfrac{x}{2} = y$

    Replace y by $f^{-1}(x)$: $f^{-1}(x) = \dfrac{x}{2}$

53. a. The function $g(x) = 3x - 1$ is a linear function that is not constant, so it passes the horizontal-line test. Thus, g is one-to-one.

    b. Replace $g(x)$ by y: $y = 3x - 1$

    Interchange x and y: $x = 3y - 1$

    Solve for y: $\dfrac{x+1}{3} = y$

    Replace y by $g^{-1}(x)$: $g^{-1}(x) = \dfrac{x+1}{3}$

55. a. The function $f(x) = \frac{1}{2}x + 1$ is a linear function that is not constant, so it passes the horizontal-line test. Thus, f is one-to-one.

    b. Replace $f(x)$ by y: $y = \dfrac{1}{2}x + 1$

    Interchange x and y: $x = \dfrac{1}{2}y + 1$

    Solve for y: $2x - 2 = y$

    Replace y by $f^{-1}(x)$: $f^{-1}(x) = 2x - 2$

57. $g(x) = x^2 + 5$

    a. The graph of this function is a parabola which opens upward. This does not pass the horizontal-line test. The function is not one-to-one.

59. a. The function $h(x) = -10 - x$ is a linear function that is not constant, so it passes the horizontal-line test. Thus, h his one-to-one.

   b. Replace $h(x)$ by y: $y = -10 - x$

      Interchange x and y: $x = -10 - y$

      Solve for y: $y = -x - 10$

      Replace y by $h^{-1}(x)$: $h^{-1}(x) = -x - 10$, or $h^{-1}(x) = -10 - x$

61. a. The graph of $f(x) = \dfrac{1}{x}$ is shown below.

      It passes the horizontal-line test, so the function is one-to-one.

      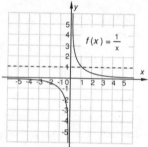

   b. Replace $f(x)$ by y: $y = \dfrac{1}{x}$

      Interchange x and y: $x = \dfrac{1}{y}$

      Solve for y: $xy = 1$

      $$y = \frac{1}{x}$$

      Replace y by $f^{-1}(x)$: $f^{-1}(x) = \dfrac{1}{x}$

63. $G(x) = 4$

   a. The graph of this function is a horizontal line, so the function is not one-to-one.

65. a. The function $f(x) = \dfrac{2x+1}{3} = \dfrac{2}{3}x + \dfrac{1}{3}$ is a linear function that is not constant, so it passes the horizontal-line test. Thus, f is one-to-one.

   b. Replace $f(x)$ by y: $y = \dfrac{2x+1}{3}$

      Interchange x and y: $x = \dfrac{2y+1}{3}$

      Solve for y: $3x = 2y + 1$

$$3x - 1 = 2y$$

$$\frac{3x - 1}{2} = y$$

Replace y by $f^{-1}(x)$: $f^{-1}(x) = \dfrac{3x-1}{2}$

67. a. The graph of $f(x) = x^3 - 5$ is shown below. It passes the horizontal-line test, so the function is one-to-one.

   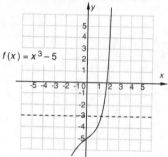

   b. Replace $f(x)$ by y: $y = x^3 - 5$

      Interchange x and y: $x = y^3 - 5$

      Solve for y: $x + 5 = y^3$

      $$\sqrt[3]{x+5} = y$$

      Replace y by $f^{-1}(x)$: $f^{-1}(x) = \sqrt[3]{x+5}$

69. a. The graph of $g(x) = (x-2)^3$ is shown below. It passes the horizontal-line test, so the function is one-to-one.

   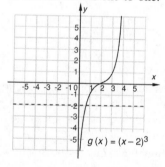

   b. Replace $g(x)$ by y: $y = (x-2)^3$

      Interchange x and y: $x = (y-2)^3$

      Solve for y: $\sqrt[3]{x} = y - 2$

      $$\sqrt[3]{x} + 2 = y$$

      Replace y by $g^{-1}(x)$: $g^{-1}(x) = \sqrt[3]{x} + 2$

71. a. The graph of $f(x) = \sqrt{x}$ is shown below. It passes the horizontal-line test, so the function is one-to-one.

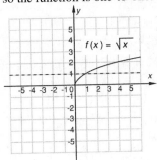

b. Replace $f(x)$ by y: $y = \sqrt{x}$

$\left(\text{Note that } f(x) \geq 0.\right)$

Interchange x and y: $x = \sqrt{y}$

Solve for y: $x^2 = y$

Replace y by $f^{-1}(x)$: $f^{-1}(x) = x^2$, $x \geq 0$

73. First graph $f(x) = \frac{2}{3}x + 4$. Then graph the inverse function by reflecting the graph of $f(x) = \frac{2}{3}x + 4$ across the line $y = x$. The graph of the inverse function can also be found by first finding a formula for the inverse, substituting to find function values, and then plotting points.

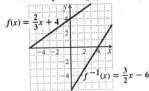

75. Follow the procedure in Exercise 73 to graph the function $f(x) = x^3 + 1$ and its inverse.

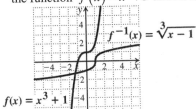

77. Follow the procedure in Exercise 73 to graph the function $g(x) = \frac{1}{2}x^3$ and its inverse.

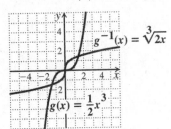

79. Follow the procedure in Exercise 73 to graph the function $F(x) = -\sqrt{x}$ and its inverse.

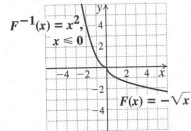

81. Follow the procedure in Exercise 73 to graph the function $f(x) = -x^2$, $x \geq 0$ and its inverse.

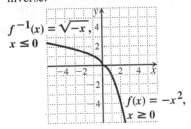

83. We check to see that $f^{-1} \circ f(x) = x$ and $f \circ f^{-1}(x) = x$.

$f^{-1} \circ f(x) = f^{-1}(f(x)) = f^{-1}\left(\sqrt[3]{x-4}\right)$

$= \left(\sqrt[3]{x-4}\right)^3 + 4$

$= x - 4 + 4 = x$

$f \circ f^{-1}(x) = f(f^{-1}(x)) = f(x^3 + 4)$

$= \sqrt[3]{x^3 + 4 - 4} = \sqrt[3]{x^3} = x$

85. We check to see that $f^{-1} \circ f(x) = x$ and $f \circ f^{-1}(x) = x$.

$$f^{-1} \circ f(x) = f^{-1}(f(x)) = f^{-1}\left(\frac{1-x}{x}\right)$$

$$= \frac{1}{\frac{1-x}{x}+1} = \frac{1}{\frac{1-x}{x}+1} \cdot \frac{x}{x}$$

$$= \frac{x}{1-x+x} = \frac{x}{1} = x$$

$$f \circ f^{-1}(x) = f(f^{-1}(x)) = f\left(\frac{1}{x+1}\right)$$

$$= \frac{1-\frac{1}{x+1}}{\frac{1}{x+1}} = \frac{1-\frac{1}{x+1}}{\frac{1}{x+1}} \cdot \frac{x+1}{x+1}$$

$$= \frac{x+1-1}{1} = \frac{x}{1} = x$$

87. Let $y_1 = f(x)$, $y_2 = g(x)$, $y_3 = y_1(y_2)$, and $y_4 = y_2(y_1)$. A table of values shows that $y_3 \neq x$ nor is $y_4 = x$, so $f(x)$ and $g(x)$ are not inverses of each other.

89. Let $y_1 = f(x)$, $y_2 = g(x)$, $y_3 = y_1(y_2)$, and $y_4 = y_2(y_1)$. A table of values shows that $y_3 = x$ and $y_4 = x$ for any value of x, so $f(x)$ and $g(x)$ are inverse of each other.

91. (1) C; (2) D; (3) B; (4) A

93. a. $f(8) = 8 + 32 = 40$

     Size 40 in France corresponds to size 8 in the U.S.

     $f(10) = 10 + 32 = 42$

     Size 42 in France corresponds to size 10 in the U.S.

     $f(14) = 14 + 32 = 46$

     Size 46 in France corresponds to size 14 in the U.S.

     $f(18) = 18 + 32 = 50$

     Size 50 in France corresponds to size 18 in the U.S.

   b.   The function $f(x) = x + 32$ is a linear function that is not constant, so it passes the horizontal-line test. Thus, f is one-to-one and has an inverse that is a function. We now find a formula for the inverse.

Replace $f(x)$ by y: $y = x + 32$

Interchange x and y: $x = y + 32$

Solve for y: $x - 32 = y$

Replace y by $f^{-1}(x)$: $f^{-1}(x) = x - 32$

   c. $f^{-1}(40) = 40 - 32 = 8$

     Size 8 in the U.S. corresponds to size 40 in France.

     $f^{-1}(42) = 42 - 32 = 10$

     Size 10 in the U.S. corresponds to size 42 in France.

     $f^{-1}(46) = 46 - 32 = 14$

     Size 14 in the U.S. corresponds to size 46 in France.

     $f^{-1}(50) = 50 - 32 = 18$

     Size 18 in the U.S. corresponds to size 50 in France.

95. *Thinking and Writing Exercise.*

97. $2^{-3} = \left(\frac{1}{2}\right)^3 = \frac{1}{2} \cdot \frac{1}{2} \cdot \frac{1}{2} = \frac{1}{8}$

99. $4^{5/2} = \left(4^{1/2}\right)^5 = 2^5 = 32$

101. $y = x^3$

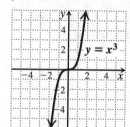

103. *Thinking and Writing Exercise.*

105.

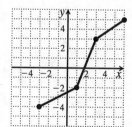

107. From Exercise 94(b), we know that a function that converts dress sizes in Italy to those in the United States is $g(x) = \dfrac{x}{2} - 12$. From Exercise 93(a), we know that a function that converts dress sizes in the United States to those in France is $f(x) = x + 32$. Then a function that converts dress sizes in Italy to those in France is

$$h(x) = (f \circ g)(x)$$

$$h(x) = f\left(\frac{x}{2} - 12\right)$$

$$h(x) = \frac{x}{2} - 12 + 32$$

$$h(x) = \frac{x}{2} + 20.$$

109. *Thinking and Writing Exercise.*

111. $\left(h \circ \left(g^{-1} \circ f^{-1}\right)\right)(x)$

$$= \left((f \circ g) \circ \left(g^{-1} \circ f^{-1}\right)\right)(x)$$

$$= \left(\left(f \circ \left(g \circ g^{-1}\right)\right) \circ f^{-1}\right)(x)$$

$$= \left((f \circ I) \circ f^{-1}\right)(x)$$

$$= \left(f \circ f^{-1}\right)(x) = x$$

Therefore, $\left(g^{-1} \circ f^{-1}\right)(x) = h^{-1}(x)$.

113. *Thinking and Writing Exercise.*

115. $(c \circ f)(n)$ represents the cost of mailing n copies of the book.

117. $R(10) \approx 18$ and $P(18) \approx 22$, so $P(R(10)) \approx 22$ mm of mercury.

119. Locate 20 on the vertical axis of the second graph, move across to the curve, and then move down to the horizontal axis to find that $P^{-1}(20) \approx 15$ liters per minute.

Exercise Set 12.2

1. True

3. True

5. False

7. The function values increase as x increases, so $a > 1$.

9. The function values decrease as x increases, so $0 < a < 1$.

11. Graph: $y = f(x) = 3^x$

We compute some function values and keep the results in a table.

$$f(0) = 3^0 = 1$$

$$f(1) = 3^1 = 3$$

$$f(2) = 3^2 = 9$$

$$f(-1) = 3^{-1} = \frac{1}{3^1} = \frac{1}{3}$$

$$f(-2) = 3^{-2} = \frac{1}{3^2} = \frac{1}{9}$$

x	y, or $f(x)$
0	1
1	3
2	9
-1	$\dfrac{1}{3}$
-2	$\dfrac{1}{9}$

Next we plot these points and connect them with a smooth curve.

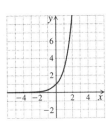

$y = f(x) = 3^x$

13. Graph: $y = 5^x$

We compute some function values, thinking of y as $f(x)$, and keep the results in a table.

$$f(0) = 5^0 = 1$$

$$f(1) = 5^1 = 5$$

$$f(2) = 5^2 = 25$$

$$f(-1) = 5^{-1} = \frac{1}{5^1} = \frac{1}{5}$$

$$f(-2) = 5^{-2} = \frac{1}{5^2} = \frac{1}{25}$$

x	y, or $f(x)$
0	1
1	5
2	25
−1	$\frac{1}{5}$
−2	$\frac{1}{25}$

Next we plot these points and connect them with a smooth curve.

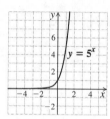

15. Graph: $y = 2^x + 3$

We compute some function values, thinking of y as $f(x)$, and keep the results in a table.

$$f(-4) = 2^{-4} + 3 = \frac{1}{2^4} + 3 = \frac{1}{16} + 3 = 3\frac{1}{16}$$

$$f(-2) = 2^{-2} + 3 = \frac{1}{2^2} + 3 = \frac{1}{4} + 3 = 3\frac{1}{4}$$

$$f(0) = 2^0 + 3 = 1 + 3 = 4$$

$$f(1) = 2^1 + 3 = 2 + 3 = 5$$

$$f(2) = 2^2 + 3 = 4 + 3 = 7$$

x	y, or $f(x)$
−4	$3\frac{1}{16}$
−2	$3\frac{1}{4}$
0	4
1	5
2	7

Next we plot these points and connect them with a smooth curve.

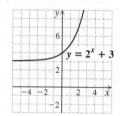

17. Graph: $y = 3^x - 1$

We compute some function values, thinking of y as $f(x)$, and keep the results in a table.

$$f(-3) = 3^{-3} - 1 = \frac{1}{3^3} - 1 = \frac{1}{27} - 1 = -\frac{26}{27}$$

$$f(-1) = 3^{-1} - 1 = \frac{1}{3} - 1 = -\frac{2}{3}$$

$$f(0) = 3^0 - 1 = 1 - 1 = 0$$

$$f(1) = 3^1 - 1 = 3 - 1 = 2$$

$$f(2) = 3^2 - 1 = 9 - 1 = 8$$

x	y, or $f(x)$
−3	$-\frac{26}{27}$
−1	$-\frac{2}{3}$
0	0
1	2
2	8

Next we plot these points and connect them with a smooth curve.

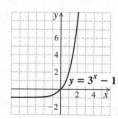

19. Graph: $y = 2^{x-3}$

We compute some function values, thinking of y as $f(x)$, and keep the results in a table.

$$f(-1) = 2^{-1-3} = 2^{-4} = \frac{1}{2^4} = \frac{1}{16}$$

$$f(0) = 2^{0-3} = 2^{-3} = \frac{1}{2^3} = \frac{1}{8}$$

$$f(1) = 2^{1-3} = 2^{-2} = \frac{1}{2^2} = \frac{1}{4}$$

$$f(2) = 2^{2-3} = 2^{-1} = \frac{1}{2^1} = \frac{1}{2}$$

$$f(3) = 2^{3-3} = 2^0 = 1$$

$$f(4) = 2^{4-3} = 2^1 = 2$$

$$f(5) = 2^{5-3} = 2^2 = 4$$

x	y, or $f(x)$
-1	$\frac{1}{16}$
0	$\frac{1}{8}$
1	$\frac{1}{4}$
2	$\frac{1}{2}$
3	1
4	2
5	4

Next we plot these points and connect them with a smooth curve.

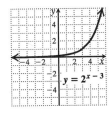

21. Graph: $y = 2^{x+3}$

We compute some function values, thinking of y as $f(x)$, and keep the results in a table.

$$f(-4) = 2^{-4+3} = 2^{-1} = \frac{1}{2}$$

$$f(-2) = 2^{-2+3} = 2$$

$$f(-1) = 2^{-1+3} = 2^2 = 4$$

$$f(0) = 2^{0+3} = 2^3 = 8$$

x	y, or $f(x)$
-4	$\frac{1}{2}$
-2	2
-1	4
0	8

Next we plot these points and connect them with a smooth curve.

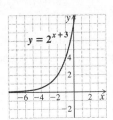

23. Graph: $y = \left(\frac{1}{5}\right)^x$

We compute some function values, thinking of y as $f(x)$, and keep the results in a table.

$$f(0) = \left(\frac{1}{5}\right)^0 = 1$$

$$f(1) = \left(\frac{1}{5}\right)^1 = \frac{1}{5}$$

$$f(2) = \left(\frac{1}{5}\right)^2 = \frac{1}{25}$$

$$f(-1) = \left(\frac{1}{5}\right)^{-1} = \frac{1}{\frac{1}{5}} = 5$$

$$f(-2) = \left(\frac{1}{5}\right)^{-2} = \frac{1}{\frac{1}{25}} = 25$$

x	y, or $f(x)$
0	1
1	$\dfrac{1}{5}$
2	$\dfrac{1}{25}$
-1	5
-2	25

Next we plot these points and connect them with a smooth curve.

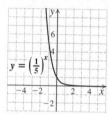

25. Graph: $y = \left(\dfrac{1}{10}\right)^x$

We compute some function values, thinking of y as $f(x)$, and keep the results in a table.

$$f(0) = \left(\frac{1}{10}\right)^0 = 1$$

$$f(1) = \left(\frac{1}{10}\right)^1 = \frac{1}{10}$$

$$f(2) = \left(\frac{1}{10}\right)^2 = \frac{1}{100}$$

$$f(-1) = \left(\frac{1}{10}\right)^{-1} = \frac{1}{\left(\frac{1}{10}\right)^1} = \frac{1}{\frac{1}{10}} = 10$$

$$f(-2) = \left(\frac{1}{10}\right)^{-2} = \frac{1}{\left(\frac{1}{10}\right)^2} = \frac{1}{\frac{1}{100}} = 100$$

x	y, or $f(x)$
0	1
1	$\dfrac{1}{10}$
2	$\dfrac{1}{100}$
-1	10
-2	100

Next we plot these points and connect them with a smooth curve.

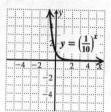

27. Graph: $y = 2^{x-3} - 1$

We compute some function values, thinking of y as $f(x)$, and keep the results in a table.

$$f(0) = 2^{0-3} - 1 = 2^{-3} - 1 = \frac{1}{8} - 1 = -\frac{7}{8}$$

$$f(1) = 2^{1-3} - 1 = 2^{-2} - 1 = \frac{1}{4} - 1 = -\frac{3}{4}$$

$$f(2) = 2^{2-3} - 1 = 2^{-1} - 1 = \frac{1}{2} - 1 = -\frac{1}{2}$$

$$f(3) = 2^{3-3} - 1 = 2^0 - 1 = 1 - 1 = 0$$

$$f(4) = 2^{4-3} - 1 = 2^1 - 1 = 2 - 1 = 1$$

$$f(5) = 2^{5-3} - 1 = 2^2 - 1 = 4 - 1 = 3$$

$$f(6) = 2^{6-3} - 1 = 2^3 - 1 = 8 - 1 = 7$$

x	y, or $f(x)$
0	$-\dfrac{7}{8}$
1	$-\dfrac{3}{4}$
2	$-\dfrac{1}{2}$
3	0
4	1
5	3
6	7

Next we plot these points and connect them with a smooth curve.

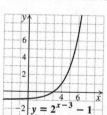

29. Graph: $y = 1.7^x$

We use a graphing calculator.

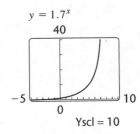

31. Graph: $y = 0.15^x$

We use a graphing calculator.

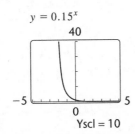

33. Graph: $x = 3^y$

We can find ordered pairs by choosing values for y and then computing values for x.

For $y = 0$, $x = 3^0 = 1$.

For $y = 1$, $x = 3^1 = 3$.

For $y = 2$, $x = 3^2 = 9$.

For $y = 3$, $x = 3^3 = 27$.

For $y = -1$, $x = 3^{-1} = \dfrac{1}{3^1} = \dfrac{1}{3}$.

For $y = -2$, $x = 3^{-2} = \dfrac{1}{3^2} = \dfrac{1}{9}$.

For $y = -3$, $x = 3^{-3} = \dfrac{1}{3^3} = \dfrac{1}{27}$.

x	y
1	0
3	1
9	2
27	3
$\dfrac{1}{3}$	−1
$\dfrac{1}{9}$	−2
$\dfrac{1}{27}$	−3

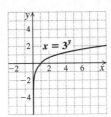

     (1)   Choose values for y.

     (2)   Compute values for x.

We plot the points and connect them with a smooth curve.

35. Graph: $x = 2^{-y} = \left(\dfrac{1}{2}\right)^y$

We can find ordered pairs by choosing values for y and then computing values for x. Then we plot these points and connect them with a smooth curve.

For $y = 0$, $x = \left(\dfrac{1}{2}\right)^0 = 1$.

For $y = 1$, $x = \left(\dfrac{1}{2}\right)^1 = \dfrac{1}{2}$.

For $y = 2$, $x = \left(\dfrac{1}{2}\right)^2 = \dfrac{1}{4}$.

For $y = 3$, $x = \left(\dfrac{1}{2}\right)^3 = \dfrac{1}{8}$.

For $y = -1$, $x = \left(\dfrac{1}{2}\right)^{-1} = \dfrac{1}{\frac{1}{2}} = 2$.

For $y = -2$, $x = \left(\dfrac{1}{2}\right)^{-2} = \dfrac{1}{\frac{1}{4}} = 4$.

For $y = -3$, $x = \left(\dfrac{1}{2}\right)^{-3} = \dfrac{1}{\frac{1}{8}} = 8$.

x	y
1	0
$\dfrac{1}{2}$	1
$\dfrac{1}{4}$	2
$\dfrac{1}{8}$	3
2	−1
4	−2
8	−3

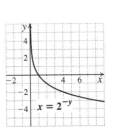

37. Graph: $x = 5^y$

We can ordered pairs by choosing values for y and then computing values for x. Then we plot these points and connect them with a smooth curve.

For $y = 0$, $x = 5^0 = 1$.

For $y = 1$, $x = 5^1 = 5$.

For $y = 2$, $x = 5^2 = 25$.

For $y = -1$, $x = 5^{-1} = \dfrac{1}{5}$.

For $y = -2$, $x = 5^{-2} = \dfrac{1}{25}$.

x	y
1	0
5	1
25	2
$\dfrac{1}{5}$	-1
$\dfrac{1}{25}$	-2

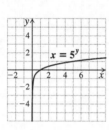

39. Graph: $x = \left(\dfrac{3}{2}\right)^y$

We can find ordered pairs by choosing values for y and then computing values for x. Then we plot these points and connect them with a smooth curve.

For $y = 0$, $x = \left(\dfrac{3}{2}\right)^0 = 1$.

For $y = 1$, $x = \left(\dfrac{3}{2}\right)^1 = \dfrac{3}{2}$.

For $y = 2$, $x = \left(\dfrac{3}{2}\right)^2 = \dfrac{9}{4}$.

For $y = 3$, $x = \left(\dfrac{3}{2}\right)^3 = \dfrac{27}{8}$.

For $y = -1$, $x = \left(\dfrac{3}{2}\right)^{-1} = \dfrac{1}{\frac{3}{2}} = \dfrac{2}{3}$.

For $y = -2$, $x = \left(\dfrac{3}{2}\right)^{-2} = \dfrac{1}{\frac{9}{4}} = \dfrac{4}{9}$.

For $y = -3$, $x = \left(\dfrac{3}{2}\right)^{-3} = \dfrac{1}{\frac{27}{8}} = \dfrac{8}{27}$.

x	y
1	0
$\dfrac{3}{2}$	1
$\dfrac{9}{4}$	2
$\dfrac{27}{8}$	3
$\dfrac{2}{3}$	-1
$\dfrac{4}{9}$	-2
$\dfrac{8}{27}$	-3

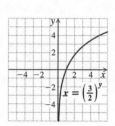

41. Graph $y = 3^x$ (see Exercise 11) and $x = 3^y$ (see Exercise 33) using the same set of axes.

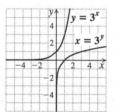

43. Graph $y = \left(\tfrac{1}{2}\right)^x$ and $x = \left(\tfrac{1}{2}\right)^y$ using the same set of axes.

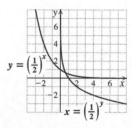

45. $y = \left(\tfrac{5}{2}\right)^x$ is an exponential function of the form $y = a^x$ with $a > 1$, so y-values will increase as x-values increase. Also, observe that when $x = 0$, $y = 1$. Thus, graph (d) corresponds to this equation.

47. For $x = \left(\tfrac{5}{2}\right)^y$, when $y = 0$, $x = 1$. The only graph that contains the point (1, 0) is (f). This graph corresponds to the given equation.

49. $y = \left(\frac{2}{5}\right)^{x-2}$ is an exponential function of the form $y = a^x$ with $0 < a < 1$, so y-values will decrease as x-values increase. Also, observe that when $x = 2$, $y = 1$. Thus, graph (c) corresponds to the given equation.

51. a. In 2006, $t = 2006 - 2003 = 3$
$$M(3) = 0.353(1.244)^3 \approx 0.680$$
The number of tracks downloaded in 2006 was about 0.68 billion.
In 2008, $t = 2008 - 2003 = 5$
$$M(5) = 0.353(1.244)^5 \approx 1.052$$
The number of tracks downloaded in 2008 was about 1.052 billion.
In 2012, $t = 2012 - 2003 = 9$
$$M(9) = 0.353(1.244)^9 \approx 2.519$$
The number of tracks downloaded in 2012 was about 2.519 billion.

b.

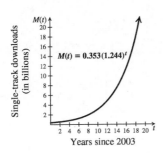

53. a. Since initial time is 0, we substitute the time(s) into the function
$$P(t) = 21.4(0.914)^t$$
$$P(1) = 21.4(0.914)^1 \approx 19.6$$
$$P(3) = 21.4(0.914)^3 \approx 16.3$$
$$P(12) = 21.4(0.914)^{12} \approx 7.3$$
Of the smokers who receive phone counseling, about 19.6% are able to quit for 1 month, about 16.3% are able to quit for 3 months, and about 7.3% are able to quit for 1 year.

b.

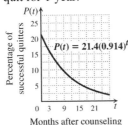

55. a. In 1930, $t = 1930 - 1900 = 30$.
$$P(t) = 150(0.960)^t$$
$$P(30) = 150(0.960)^{30}$$
$$\approx 44.079$$
In 1930, about 44.079 thousand, or 44,079, humpback whales were alive.
In 1960, $t = 1960 - 1900 = 60$.
$$P(t) = 150(0.960)^t$$
$$P(60) = 150(0.960)^{60}$$
$$\approx 12.953$$
In 1960, about 12.953 thousand, or 12,953, humpback whales were alive.

b. Plot the points found in part (a), (30, 44, 079) and (60, 12,953) and additional points as needed and graph the function.

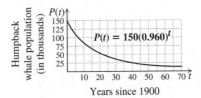

57. a. In 1992, $t = 1992 - 1982 = 10$
$$P(10) = 5.5(1.08)^{10} \approx 11.874$$
In 1992, there were about 11,874 humpback whales.
In 2006, $t = 2006 - 1982 = 24$
$$P(24) = 5.5(1.08)^{24} \approx 34.876$$
In 2006, there were about 34,876 humpback whales.

b.

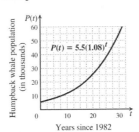

59. a. Since initial time is 0, we substitute the time(s) into the function
$$R(t) = 2(1.75)^t$$
For $t = 10$,
$$R(10) = 2(1.75)^{10} \approx 539$$
After 10 years, there will be about 539 ruffe in the lake.

For $t = 15$,

$$R(15) = 2(1.75)^{15} \approx 8843$$

After 15 years, there will be about 8843 ruffe in the lake.

b. Plot the points found in part (a), (10, 539) and (15, 8843) and additional points as needed and graph the function.

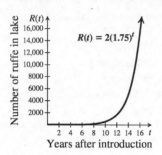

61. *Thinking and Writing Exercise.*

63. $3x^2 - 48 = 3(x^2 - 16) = 3(x^2 - 4^2)$

$$= 3(x + 4)(x - 4)$$

65. $6x^2 + x - 12$

Use the FOIL method to factor the trinomial $(3x + \ )(2x + \ )$.

Factor the last term, −12. The correct pair of factors is −4 and 3.

$(3x - 4)(2x + 3)$

$6x^2 + x - 12 = (3x - 4)(2x + 3)$

67. $6y^2 + 36y - 240 = 6(y^2 + 6y - 40)$

Factoring $y^2 + 6y - 40$, we want two factors of −40, whose sum is 6. They are 10 and −4.

$6y^2 + 36y - 240 = 6(y + 10)(y - 4)$

69. *Thinking and Writing Exercise.*

71. Since the bases are the same, the one with the larger exponent is the larger number. Thus $\pi^{2.4}$ is larger.

73. Graph: $y = 2^x + 2^{-x}$

Construct a table of values, thinking of y as $f(x)$. Then plot these points and connect them with a curve.

$$f(0) = 2^0 + 2^{-0} = 1 + 1 = 2$$

$$f(1) = 2^1 + 2^{-1} = 2 + \frac{1}{2} = 2\frac{1}{2}$$

$$f(2) = 2^2 + 2^{-2} = 4 + \frac{1}{4} = 4\frac{1}{4}$$

$$f(3) = 2^3 + 2^{-3} = 8 + \frac{1}{8} = 8\frac{1}{8}$$

$$f(-1) = 2^{-1} + 2^{-(-1)} = \frac{1}{2} + 2 = 2\frac{1}{2}$$

$$f(-2) = 2^{-2} + 2^{-(-2)} = \frac{1}{4} + 4 = 4\frac{1}{4}$$

$$f(-3) = 2^{-3} + 2^{-(-3)} = \frac{1}{8} + 8 = 8\frac{1}{8}$$

x	y, or $f(x)$
0	2
1	$2\frac{1}{2}$
2	$4\frac{1}{4}$
3	$8\frac{1}{8}$
−1	$2\frac{1}{2}$
−2	$4\frac{1}{4}$
−3	$8\frac{1}{8}$

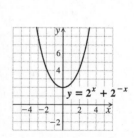

75. Graph: $y = \left|2^x - 2\right|$

We construct a table of values, thinking of y as $f(x)$. Then plot these points and connect them with a curve.

$$f(0) = \left| 2^0 - 2 \right| = \left| 1 - 2 \right| = \left| -1 \right| = 1$$

$$f(1) = \left| 2^1 - 2 \right| = \left| 2 - 2 \right| = \left| 0 \right| = 0$$

$$f(2) = \left| 2^2 - 2 \right| = \left| 4 - 2 \right| = \left| 2 \right| = 2$$

$$f(3) = \left| 2^3 - 2 \right| = \left| 8 - 2 \right| = \left| 6 \right| = 6$$

$$f(-1) = \left| 2^{-1} - 2 \right| = \left| \frac{1}{2} - 2 \right| = \left| -\frac{3}{2} \right| = \frac{3}{2}$$

$$f(-3) = \left| 2^{-3} - 2 \right| = \left| \frac{1}{8} - 2 \right| = \left| -\frac{15}{8} \right| = \frac{15}{8}$$

$$f(-5) = \left| 2^{-5} - 2 \right| = \left| \frac{1}{32} - 2 \right| = \left| -\frac{63}{32} \right| = \frac{63}{32}$$

x	y, or $f(x)$
0	1
1	0
2	2
3	6
−1	$\frac{3}{2}$
−3	$\frac{15}{8}$
−5	$\frac{63}{32}$

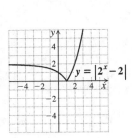

77. Graph: $y = \left| 2^{x^2} - 1 \right|$

We construct a table of values, thinking of y as $f(x)$. Then we plot these points and connect them with a curve.

$$f(0) = \left| 2^{0^2} - 1 \right| = \left| 1 - 1 \right| = 0$$

$$f(1) = \left| 2^{1^2} - 1 \right| = \left| 2 - 1 \right| = 1$$

$$f(2) = \left| 2^{2^2} - 1 \right| = \left| 16 - 1 \right| = 15$$

$$f(-1) = \left| 2^{(-1)^2} - 1 \right| = \left| 2 - 1 \right| = 1$$

$$f(-2) = \left| 2^{(-2)^2} - 1 \right| = \left| 16 - 1 \right| = 15$$

x	y, or $f(x)$
0	0
1	1
2	15
−1	1
−2	15

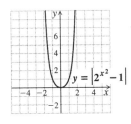

79. Determine a table of values for each function and graph.

$y = 3^{-(x-1)}$

x	y
0	3
1	1
2	$\frac{1}{3}$
3	$\frac{1}{9}$
−1	9

$x = 3^{-(y-1)}$

x	y
3	0
1	1
$\frac{1}{3}$	2
$\frac{1}{9}$	3
9	−1

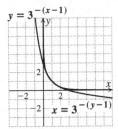

81. Enter the data points (0, 0.5), (4, 4) and (8, 50) and use the exponential regression feature to find an exponential function that models the data.

$$N(t) = 0.4642(1.7783)^t$$

In 2012, $x = 2012 - 2000 = 12$

$$N(12) = 0.4642(1.7783)^{12}$$

$$\approx 464.26$$

In 2012, about 464 million GPS systems will be in use.

83. *Thinking and Writing Exercise.*

85.

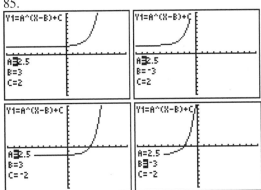

Exercise Set 12.3

1. g

3. a

5. b

7. e

9. $\log_{10} 1000$ is the power/exponent to which we raise 10 to get 1000.
Since $10^3 = 1000$, $\log_{10} 1000 = 3$

11. $\log_2 16$ is the power/exponent to which we raise 2 to get 16.
Since $2^4 = 16$, $\log_2 16 = 4$

13. Since $3^4 = 81$, $\log_3 81 = 4$

15. Since $4^{-2} = \dfrac{1}{16}$, $\log_4 \dfrac{1}{16} = -2$

17. Since $7^{-1} = \dfrac{1}{7}$, $\log_7 \dfrac{1}{7} = -1$

19. Since $5^4 = 625$, $\log_5 625 = 4$

21. Since $8^1 = 8$, $\log_8 8 = 1$

23. Since $8^0 = 1$, $\log_8 1 = 0$

25. Since $9^5 = 9^5$, $\log_9 9^5 = 5$

27. Since $10^{-2} = 0.01$, $\log_{10} 0.01 = -2$

29. Since $9^{\frac{1}{2}} = 3\left[\left(3^2\right)^{\frac{1}{2}} = 3\right]$, $\log_9 3 = \dfrac{1}{2}$

31. Since $9 = 3^2$ and $\left(3^2\right)^{3/2} = 27$, $\log_9 27 = \dfrac{3}{2}$

33. Since $1000 = 10^3$ and $\left(10^3\right)^{2/3} = 10^2 = 100$,

$\log_{1000} 100 = \dfrac{2}{3}$

35. Since $\log_5 7$ is the power to which we raise 5 to get 7, then 5 raised to this power is 7. That is, $5^{\log_5 7} = 7$.

37. Graph: $y = \log_{10} x$

The equation $y = \log_{10} x$ is equivalent to $10^y = x$. We can find ordered pairs by choosing values for y and computing the corresponding x-values.

For $y = 0$, $x = 10^0 = 1$.
For $y = 1$, $x = 10^1 = 10$.
For $y = 2$, $x = 10^2 = 100$.
For $y = -1$, $x = 10^{-1} = \dfrac{1}{10}$.
For $y = -2$, $x = 10^{-2} = \dfrac{1}{100}$.

x, or 10^y	y
1	0
10	1
100	2
$\dfrac{1}{10}$	-1
$\dfrac{1}{100}$	-2

↑   ⌐   (1) Select y.

⌐   (2) Compute x.

We plot the set of ordered pairs and connect the points with a smooth curve.

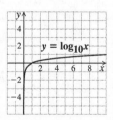

39. Graph: $y = \log_3 x$

The equation $y = \log_3 x$ is equivalent to $3^y = x$. We can find ordered pairs by choosing values for y and computing the corresponding x-values.

For $y = 0$, $x = 3^0 = 1$.

For $y = 1$, $x = 3^1 = 3$.

For $y = 2$, $x = 3^2 = 9$.

For $y = -1$, $x = 3^{-1} = \dfrac{1}{3}$.

For $y = -2$, $x = 3^{-2} = \dfrac{1}{9}$.

x, or 3^y	y
1	0
3	1
9	2
$\dfrac{1}{3}$	−1
$\dfrac{1}{9}$	−2

We plot the set of ordered pairs and connect the points with a smooth curve.

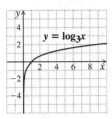

41. Graph: $f(x) = \log_6 x$

Think of $f(x)$ as y. Then $y = \log_6 x$ is equivalent to $6^y = x$. We find ordered pairs by choosing values for y and computing the corresponding x-values. Then we plot the points and connect them with a smooth curve.

For $y = 0$, $x = 6^0 = 1$.

For $y = 1$, $x = 6^1 = 6$.

For $y = 2$, $x = 6^2 = 36$.

For $y = -1$, $x = 6^{-1} = \dfrac{1}{6}$.

For $y = -2$, $x = 6^{-2} = \dfrac{1}{36}$.

x, or 6^y	y
1	0
6	1
36	2
$\dfrac{1}{6}$	−1
$\dfrac{1}{36}$	−2

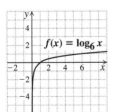

43. Graph: $f(x) = \log_{2.5} x$

Think of $f(x)$ as y. Then $y = \log_{2.5} x$ is equivalent to $2.5^y = x$. We construct a table of values, plot these points and connect them with a smooth curve.

For $y = 0$, $x = 2.5^0 = 1$.

For $y = 1$, $x = 2.5^1 = 2.5$.

For $y = 2$, $x = 2.5^2 = 6.25$.

For $y = 3$, $x = 2.5^3 = 15.625$.

For $y = -1$, $x = 2.5^{-1} = 0.4$.

For $y = -2$, $x = 2.5^{-2} = 0.16$.

x, or 2.5^y	y
1	0
2.5	1
6.25	2
15.625	3
0.4	−1
0.16	−2

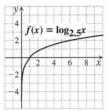

45. Graph $f(x) = 3^x$ (see Exercise Set 12.2, Exercise 11) and $f^{-1}(x) = \log_3 x$ (see Exercise 39 above) on the same set of axes.

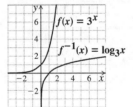

47. $\log 4 \approx 0.6021$

49. $\log 13{,}400 \approx 4.1271$

51. $\log 0.527 \approx -0.2782$

53. $10^{2.3} \approx 199.5262$

55. $10^{-2.9523} \approx 0.0011$

57. $10^{0.0012} \approx 1.0028$

59. $y = \log(x + 2)$

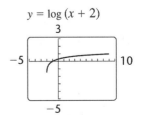

$y = \log(x + 2)$

61. $y = \log(2x) - 3$

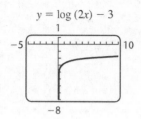

63. $y = \log(x^2)$

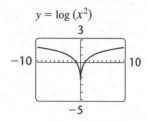

65. $x = \log_{10} 8 \Rightarrow 10^x = 8$

The base remains the same.

The logarithm is the exponent.

67. $\log_9 9 = 1 \Rightarrow 9^1 = 9$

The logarithm is the exponent.

The base remains the same.

69. $\log_{10} 0.1 = -1 \Rightarrow 10^{-1} = 0.1$

71. $\log_{10} 7 = 0.845 \Rightarrow 10^{0.845} = 7$

73. $\log_c m = 8 \Rightarrow c^8 = m$

75. $\log_t Q = r \Rightarrow t^r = Q$

77. $\log_e 0.25 = -1.3863 \Rightarrow e^{-1.3863} = 0.25$

79. $\log_r T = -x \Rightarrow r^{-x} = T$

81. $10^2 = 100 \Rightarrow 2 = \log_{10} 100$

The exponent is the logarithm.

The base remains the same.

83. $4^{-5} = \frac{1}{1024} \Rightarrow -5 = \log_4 \frac{1}{1024}$

The exponent is the logarithm.

The base remains the same.

85. $16^{3/4} = 8 \Rightarrow \frac{3}{4} = \log_{16} 8$

87. $10^{0.4771} = 3 \Rightarrow 0.4771 = \log_{10} 3$

89. $z^m = 6 \Rightarrow m = \log_z 6$

91. $p^m = V \Rightarrow m = \log_p V$

93. $e^3 = 20.0855 \Rightarrow 3 = \log_e 20.0855$

95. $e^{-4} = 0.0183 \Rightarrow -4 = \log_c 0.0183$

97. $\log_3 x = 2$

$3^2 = x$ Converting to an exponential equation

$9 = x$ Computing 3^2

99. $\log_5 125 = x$

$5^x = 125$ Converting to an exponential equation

$5^x = 5^3$

$x = 3$ The exponents must be equal/the same.

101. $\log_x 16 = 4$

$x^4 = 16$ Converting to an exponential equation

$x = \sqrt[4]{16}$

$x = 2$ Computing $\sqrt[4]{16}$

103. $\log_x 7 = 1$

$x^1 = 7$ Converting to an exponential equation

$x = 7$ Simplifying x^1

105. $\log_x 9 = \frac{1}{2}$

$\quad\quad x^{1/2} = 9$ Converting to an
$\quad\quad\quad\quad\quad\quad$ exponential equation

$\quad\quad x = 9^2$
$\quad\quad x = 81$ Computing 9^2

107. $\log_3 x = -2$

$\quad\quad 3^{-2} = x$ Converting to an
$\quad\quad\quad\quad\quad\quad$ exponential equation

$\quad\quad \left(\frac{1}{3}\right)^2 = x$

$\quad\quad \frac{1}{9} = x$ Computing $\left(\frac{1}{3}\right)^2$

109. $\log_{32} x = \frac{2}{5}$

$\quad\quad 32^{2/5} = x$ Converting to an
$\quad\quad\quad\quad\quad\quad$ exponential equation

$\quad\quad \left(2^5\right)^{2/5} = x$

$\quad\quad\quad 4 = x$ Simplifying

111. *Thinking and Writing Exercise.*

113. $a^{12} \cdot a^6 = a^{12+6} = a^{18}$

115. $\dfrac{x^{12}}{x^4} = x^{12-4} = x^8$

117. $\left(y^3\right)^5 = y^{3\cdot5} = y^{15}$

119. $x^2 \cdot x^3 = x^{2+3} = x^5$

121. *Thinking and Writing Exercise.*

123. Graph: $y = \left(\dfrac{3}{2}\right)^x$ Graph: $y = \log_{3/2} x$, or

$\quad\quad\quad\quad\quad\quad\quad\quad\quad\quad\quad x = \left(\dfrac{3}{2}\right)^y$

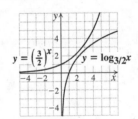

x	y or $\left(\dfrac{3}{2}\right)^x$	x, or $\left(\dfrac{3}{2}\right)^y$ 2^y	y
0	1	1	0
1	$\dfrac{3}{2}$	$\dfrac{3}{2}$	1
2	$\dfrac{9}{4}$	$\dfrac{9}{4}$	2
3	$\dfrac{27}{8}$	$\dfrac{27}{8}$	3
−1	$\dfrac{2}{3}$	$\dfrac{2}{3}$	−1
−2	$\dfrac{4}{9}$	$\dfrac{4}{9}$	−2

125. Graph: $y = \log_3 |x+1|$

Choose values of x and determine corresponding y-values. Plot these points and connect them with a smooth curve.

For $x = 0$, $y = \log_3 |0+1| = 0$

For $x = 2$, $y = \log_3 |2+1| = 1$

For $x = 8$, $y = \log_3 |8+1| = 2$

For $x = -2$, $y = \log_3 |-2+1| = 0$

For $x = -4$, $y = \log_3 |-4+1| = 1$

For $x = -10$, $y = \log_3 |-10+1| = 2$

x	y
0	0
2	1
8	2
−2	0
−4	1
−10	2

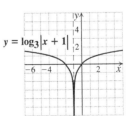

127. $\log_4(3x-2) = 2$

$$4^2 = 3x - 2$$
$$16 = 3x - 2$$
$$18 = 3x$$
$$6 = x$$

129. $\log_{10}(x^2 + 21x) = 2$

$$10^2 = x^2 + 21x$$
$$0 = x^2 + 21x - 100$$
$$0 = (x + 25)(x - 4)$$
$$x = -25 \quad \text{or} \quad x = 4$$

131. Let $\log_{1/5} 25 = x$. Then

$$\left(\frac{1}{5}\right)^x = 25$$
$$\left(5^{-1}\right)^x = 25$$
$$5^{-x} = 5^2$$
$$-x = 2$$
$$x = -2.$$

Thus, $\log_{1/5} 25 = -2$.

133. $\log_{10}\left(\log_4\left(\log_3 81\right)\right)$

$$= \log_{10}\left(\log_4 4\right) \quad \left(\log_3 81 = 4\right)$$
$$= \log_{10} 1 \qquad \left(\log_4 4 = 1\right)$$
$$= 0$$

135. Let $b = 0$, $x = 1$, and $y = 2$. Then $0^1 = 0^2$, but $1 \neq 2$. Let $b = 1, x = 1$, and $y = 2$. Then $1^1 = 1^2$, but $1 \neq 2$.

Exercise Set 12.4

1. e

3. a

5. c

7. $\log_3(81 \cdot 27) = \log_3 81 + \log_3 27$
Using the product rule.

9. $\log_4(64 \cdot 16) = \log_4 64 + \log_4 16$
Using the product rule.

11. $\log_c(rst) = \log_c r + \log_c s + \log_c t$
Using the product rule.

13. $\log_a 5 + \log_a 14 = \log_a(5 \cdot 14)$, or $\log_a 70$
Using the product rule.

15. $\log_c t + \log_c y = \log_c(t \cdot y)$
Using the product rule.

17. $\log_a r^8 = 8 \log_a r$
Using the power rule.

19. $\log_2 y^{1/3} = \frac{1}{3} \log_2 y$
Using the power rule.

21. $\log_b C^{-3} = -3 \log_b C$
Using the power rule.

23. $\log_2 \frac{25}{13} = \log_2 25 - \log_2 13$
Using the quotient rule.

25. $\log_b \frac{m}{n} = \log_b m - \log_b n$
Using the quotient rule.

27. $\log_a 17 - \log_a 6 = \log_a \frac{17}{6}$
Using the quotient rule.

29. $\log_b 36 - \log_b 4 = \log_b \frac{36}{4} = \log_b 9$
Using the quotient rule.

31. $\log_a x - \log_a y = \log_a \frac{x}{y}$
Using the quotient rule.

33. $\log_a(xyz) = \log_a x + \log_a y + \log_a z$
Using the product rule.

35. $\log_a(x^3 z^4) = \log_a x^3 + \log_a z^4$
Using the product rule.
$= 3 \log_a x + 4 \log_a z$
Using the power rule.

37. $\log_a(x^2 y^{-2} z) = \log_a x^2 + \log_a y^{-2} + \log_a z$
Using the product rule
$= 2\log_a x - 2\log_a y + \log_a z$
Using the power rule.

39. $\log_a \dfrac{x^4}{y^3 z} = \log_a x^4 - \log_a y^3 z$
Using the quotient rule.
$= \log_a x^4 - (\log_a y^3 + \log_a z)$
Using the product rule.
$= 4\log_a x - 3\log_a y - \log_a z$
Using the power rule.

41. $\log_b \dfrac{xy^2}{wz^3} = \log_b xy^2 - \log_b wz^3$
Using the quotient rule.
$= \log_b x + \log_b y^2 - (\log_b w + \log_b z^3)$
Using the product rule.
$= \log_b x + 2\log_b y - \log_b w - 3\log_b z$
Using the power rule.

43. $\log_a \sqrt{\dfrac{x^7}{y^5 z^8}}$

$= \log_a \left(\dfrac{x^7}{y^5 z^5}\right)^{1/2}$

$= \dfrac{1}{2}\log_a \dfrac{x^7}{y^5 z^8}$ Using the power rule

$= \dfrac{1}{2}(\log_a x^7 - \log_a y^5 z^8)$ Using the quotient rule

$= \dfrac{1}{2}\left[\log_a x^7 - \left(\log_a y^5 + \log_a z^8\right)\right]$ Using the product rule

$= \dfrac{1}{2}(7\log_a x - 5\log_a y - 8\log_a z)$ Using the power rule

45. $\log_a \sqrt[3]{\dfrac{x^6 y^3}{a^2 z^7}}$

$= \log_a \left(\dfrac{x^6 y^3}{a^2 z^7}\right)^{1/3}$

$= \dfrac{1}{3}\log_a \left(\dfrac{x^6 y^3}{a^2 z^7}\right)$ Using the power rule

$= \dfrac{1}{3}\left(\log_a x^6 y^3 - \log_a a^2 z^7\right)$ Using the quotient rule

$= \dfrac{1}{3}\left[\log_a x^6 + \log_a y^3 - \left(\log_a a^2 + \log_a z^7\right)\right]$ Using the product rule

$= \dfrac{1}{3}\left(\log_a x^6 + \log_a y^3 - 2 - \log_a z^7\right)$ 2 is the number to which we raise a to get a^2.

$= \dfrac{1}{3}(6\log_a x + 3\log_a y - 2 - 7\log_a z)$

47. $8\log_a x + 3\log_a z$
$= \log_a x^8 + \log_a z^3$ Using the power rule
$= \log_a x^8 z^3$ Using the product rule

49. $\log_a x^2 - 2\log_a \sqrt{x}$
$= \log_a x^2 - \log_a (\sqrt{x})^2$ Using the power rule
$= \log_a x^2 - \log_a x$ $(\sqrt{x})^2 = x$
$= \log_a \dfrac{x^2}{x}$ Using the quotient rule
$= \log_a x$ Simplifying

51. $\dfrac{1}{2}\log_a x + 5\log_a y - 2\log_a x$
$= \log_a x^{1/2} + \log_a y^5 - \log_a x^2$ Using the power rule
$= \log_a x^{1/2} y^5 - \log_a x^2$ Using the product rule
$= \log_a \dfrac{x^{1/2} y^5}{x^2}$ Using the quotient rule
The result can also be expressed as
$\log_a \dfrac{\sqrt{x}\,y^5}{x^2}$ or as $\log_a \dfrac{y^5}{x^{3/2}}$.

53. $\log_a(x^2 - 4) - \log_a(x + 2)$

$= \log_a \dfrac{x^2 - 4}{x + 2}$ Using the quotient rule

$= \log_a \dfrac{(x + 2)(x - 2)}{(x + 2)}$ Simplifying

$= \log_a(x - 2)$

55. $\log_b 15 = \log_b(3 \cdot 5)$

$= \log_b 3 + \log_b 5$ Using the product rule

$= 0.792 + 1.161$

$= 1.953$

57. $\log_b \dfrac{3}{5} = \log_b 3 - \log_b 5$ Using the quotient rule

$= 0.792 - 1.161$

$= -0.369$

59. $\log_b \dfrac{1}{5} = \log_b 1 - \log_b 5$ Using the quotient rule

$= 0 - 1.161$ $(\log 1 = 0)$

$= -1.161$

61. $\log_b \sqrt{b^3} = \log_b b^{3/2} = \dfrac{3}{2}$ (log is the exponent)

63. $\log_b 8$ since 8 cannot be expressed using the numbers 1, 3, and 5 (as factors/quotients), we cannot calculate $\log_b 8$ using the given information.

65. $\log_t t^7 = 7$ 7 is the power/exponent/logarithm

67. $\log_e e^m = m$ m is the power/exponent/logarithm

69. $\log_5 125 = 3$ and $\log_5 625 = 4$, so

$\log_5(125 \cdot 625) = \log_5 125 + \log_5 625$

$= 3 + 4 = 7$

71. $\log_2 16 = 4$, so

$\log_2 16^5 = 5 \cdot \log_2 16 = 5 \cdot 4 = 20$

73.   *Thinking and Writing Exercise.*

75.   Graph $f(x) = \sqrt{x} - 3$.

We construct a table of values, plot points, and connect them with a smooth curve. Note that we must choose nonnegative values of x in order for $\sqrt{x}$ to be a real number.

x	$f(x)$
0	-3
1	-2
4	-1
9	0

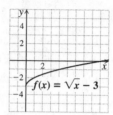

77.   Graph: $g(x) = x^3 + 2$

We construct a table of values, plot points, and connect them with a smooth curve.

x	$f(x)$
-2	-6
-1	1
0	2
1	3
2	10

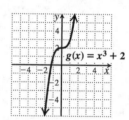

79. $f(x) = \dfrac{x - 3}{x + 7}$

$f(x)$ cannot be calculated for any x-value for which the denominator, $x + 7$, is 0. To find the excluded values, we solve:

$x + 7 = 0$

$x = -7$

The domain of f is $\{\, x | x$ is a real number *and* $x \neq -7 \,\}$, or $(-\infty, -7) \cup (-7, \infty)$.

81. $g(x) = \sqrt{10 - x}$

Since the index is even, the radicand, $10 - x$, must be non-negative. We solve the inequality:

$10 - x \geq 0$

$-x \geq -10$

$x \leq 10$

Domain of $g = \{x | x \leq 10\}$, or $(-\infty, 10]$.

83.   *Thinking and Writing Exercise.*

85. $\log_a(x^8 - y^8) - \log_a(x^2 + y^2)$

$$= \log_a \frac{x^8 - y^8}{x^2 + y^2}$$

$$= \log_a \frac{(x^4 + y^4)(x^2 + y^2)(x+y)(x-y)}{x^2 + y^2}$$

$$= \log_a[(x^4 + y^4)(x^2 - y^2)] \quad \text{Simplifying}$$

$$= \log_a(x^6 - x^4 y^2 + x^2 y^4 - y^6)$$

87. $\log_a \sqrt{1 - s^2}$

$$= \log_a(1 - s^2)^{1/2}$$

$$= \frac{1}{2}\log_a(1 - s^2)$$

$$= \frac{1}{2}\log_a[(1-s)(1+s)]$$

$$= \frac{1}{2}\log_a(1-s) + \frac{1}{2}\log_a(1+s)$$

89. $\log_a \dfrac{\sqrt[3]{x^2 z}}{\sqrt[3]{y^2 z^{-2}}}$

$$= \log_a\left(\frac{x^2 z^3}{y^2}\right)^{1/3}$$

$$= \frac{1}{3}(\log_a x^2 z^3 - \log_a y^2)$$

$$= \frac{1}{3}(2\log_a x + 3\log_a z - 2\log_a y)$$

$$= \frac{1}{3}[2\cdot 2 + 3\cdot 4 - 2\cdot 3]$$

$$= \frac{1}{3}(10)$$

$$= \frac{10}{3}$$

91. $\log_a x = 2$, so $a^2 = x$.

Let $\log_{1/a} x = n$ and solve for n.

$\log_{1/a} a^2 = n$ Substituting a^2 for x

$$\left(\frac{1}{a}\right)^n = a^2$$

$$(a^{-1})^n = a^2$$

$$a^{-n} = a^2$$

$$-n = 2$$

$$n = -2$$

Thus, $\log_{1/a} x = -2$ when $\log_a x = 2$.

93. $\log_2 80 + \log_2 x = 5$

Evaluating $\log_2 80$,

$$\log_2 80 = \log_2(32 \cdot 2.5)$$

$$= \log_2 32 + \log_2 2.5$$

$$= 5 + \log_2 2.5$$

Substituting this expression into the original equation,

$$\log_2 80 + \log_2 x = 5$$

$$5 + \log_2 2.5 + \log_2 x = 5$$

$$\log_2 2.5 + \log_2 x = 5 - 5$$

$$\log_2 x = -\log_2 2.5$$

$$\log_2 x = \log_2 2.5^{-1}$$

$$x = 2.5^{-1}$$

$$x = \frac{1}{2.5} = \frac{2}{5}$$

95. True; $\log_a(Q + Q^2) = \log_a[Q(1+Q)]$

$$= \log_a Q + \log_a(1+Q) = \log_a Q + \log_a(Q+1).$$

Mid-Chapter Review

Guided Solutions

1. $\quad y = 2x - 5$

$$x = 2y - 5$$

$$x + 5 = 2y$$

$$\frac{x+5}{2} = y$$

$$f^{-1}(x) = \frac{x+5}{2}$$

2. $\log_4 x = 1 \Rightarrow x = 4^1$

$$x = 4$$

Mixed Review

1. $(f \circ g)(x) = f(g(x)) = f(x-5)$

$$= (x-5)^2 + 1$$

$$= x^2 - 10x + 25 + 1$$

$$= x^2 - 10x + 26$$

2. $h(x) = \sqrt{5x-3}$

We have $5x - 3$ and take the square root of this expression.

$f(x) = \sqrt{x}$ and $g(x) = 5x - 3$

3. Replace $g(x)$ by y: $y = 6 - x$

Interchange x and y: $x = 6 - y$

Solve for y: $x - 6 = -y$

$\qquad\qquad -x + 6 = y$

Replace y by $g^{-1}(x)$: $g^{-1}(x) = -x + 6$

or $g^{-1}(x) = 6 - x$

4. Graph: $y = 2^x + 3$

We compute some function values, thinking of y as $f(x)$, and keep the results in a table.

$f(-4) = 2^{-4} + 3 = \dfrac{1}{2^4} + 3 = \dfrac{1}{16} + 3 = 3\dfrac{1}{16}$

$f(-2) = 2^{-2} + 3 = \dfrac{1}{2^2} + 3 = \dfrac{1}{4} + 3 = 3\dfrac{1}{4}$

$f(0) = 2^0 + 3 = 1 + 3 = 4$

$f(1) = 2^1 + 3 = 2 + 3 = 5$

$f(2) = 2^2 + 3 = 4 + 3 = 7$

x	y, or $f(x)$
-4	$3\dfrac{1}{16}$
-2	$3\dfrac{1}{4}$
0	4
1	5
2	7

Next we plot these points and connect them with a smooth curve.

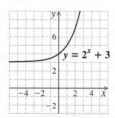

5. $\log_4 16$ is the power/exponent to which we raise 4 to get 16.

Since $4^2 = 16$, $\log_4 16 = 2$

6. Since $5^{-1} = \dfrac{1}{5}$, $\log_5 \dfrac{1}{5} = -1$

7. $\log_{100} 10$ is the power/exponent to which we raise 100 to get 10.

Since $100^{1/2} = 10$, $\log_{100} 10 = \dfrac{1}{2}$

8. $\log_b b^1 = 1 \cdot \log_b b = 1 \cdot 1 = 1$

9. $\log_8 8^{19} = 19 \cdot \log_8 8 = 19 \cdot 1 = 19$

10. $\log_t 1 = 0$

0 is the power/exponent to which t is raised to yield 1 ($t^0 = 1$).

11. $\log_x 3 = m$

$x^m = 3$, or $x = \sqrt[m]{3}$

12. $\log_2 1024 = 10$

$2^{10} = 1024$

13. $e^t = x \Rightarrow t = \log_e x$

14. $64^{2/3} = 16 \Rightarrow \dfrac{2}{3} = \log_{64} 16$

15. $\log \sqrt{\dfrac{x^2}{yz^3}}$

$= \log \left(\dfrac{x^2}{yz^3} \right)^{1/2}$

$= \dfrac{1}{2} \log \dfrac{x^2}{yz^3}$ Using the power rule

$= \dfrac{1}{2} (\log x^2 - \log yz^3)$ Using the quotient rule

$= \dfrac{1}{2} \Big[\log x^2 \\ \qquad - (\log y + \log z^3) \Big]$ Using the product rule

$= \dfrac{1}{2} (2\log x - \log y \\ \qquad - 3\log z)$ Using the power rule

$= \log x - \tfrac{1}{2} \log y - \tfrac{3}{2} \log z$ Simplify

16. $\log a - 2\log b - \log c$

$= \log a - \log b^2 - \log c$ Using the power rule

$= \log \dfrac{a}{b^2} - \log c$ Using the quotient rule

$= \log \dfrac{a}{b^2 c}$ Using the quotient rule

17. $\log_x 64 = 3$

$x^3 = 64$ Converting to an exponential equation

$x = \sqrt[3]{64}$

$x = 4$ Computing $\sqrt[3]{64}$

18. $\log_3 x = -1$

$3^{-1} = x$ Converting to an exponential equation

$\dfrac{1}{3} = x$

19. $\log x = 5$

$10^5 = x$ Converting to an exponential equation

$100,000 = x$ Computing 10^5

20. $\log_x 2 = \tfrac{1}{2}$

$x^{1/2} = 2$ Converting to an exponential equation

$x = 2^2$

$x = 4$ Computing 2^2

Exercise Set 12.5

1. True

3. True

5. True

7. True

9. True

11. 1.6094

13. −5.0832

15. 96.7583

17. 15.0293

19. 0.0305

21. 0.8451

23. 13.0014

25. −0.4260

27. 4.9459

29. We will use common logarithms for the conversion. Let $a = 10, b = 6,$ and $M = 92$ substitute in the change-of-base formula.

$\log_b M = \dfrac{\log_a M}{\log_a b}$

$\log_6 92 = \dfrac{\log_{10} 92}{\log_{10} 6} \approx \dfrac{1.963787827}{0.7781512504} \approx 2.5237$

31. We will use common logarithms for the conversion. Let $a = 10, b = 2,$ and $M = 100$ and substitute in the change-of-base formula.

$\log_2 100 = \dfrac{\log_{10} 100}{\log_{10} 2}$

$\approx \dfrac{2}{0.3010299957}$

≈ 6.6439

33. We will use natural logarithms for the conversion. Let $a = e, b = 0.5,$ and $M = 5$ and substitute in the change-of-base formula.

$\log_{0.5} 5 = \dfrac{\ln 5}{\ln 0.5} \approx \dfrac{1.609437912}{-0.6931471806} \approx -2.3219$

35. We will use common logarithms for the conversion. Let $a = 10, b = 2,$ and $M = 0.2$ and substitute in the change-of-base formula.

$\log_2 0.2 = \dfrac{\log_{10} 0.2}{\log_{10} 2}$

$\approx \dfrac{-0.6989700043}{0.3010299957}$

≈ -2.3219

37. We will use natural logarithms for the conversion. Let $a = e, b = \pi,$ and $M = 58$ and substitute in the change-of-base formula.

$$\log_{\pi} 58 = \frac{\ln 58}{\ln \pi} \approx \frac{4.060443011}{1.144729886} \approx 3.5471$$

39. Graph: $f(x) = e^x$

We find some function values with a calculator. We use these values to plot points and draw the graph.

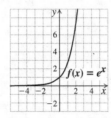

x	e^x
0	1
1	2.7
2	7.4
3	20.1
-1	0.4
-2	0.1

The domain is the set of real numbers and the range is $(0, \infty)$.

41. Graph: $f(x) = e^x + 3$

We find some function values with a calculator. We use these values to plot points and draw the graph.

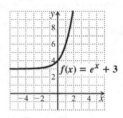

x	$e^x + 3$
0	4
1	5.72
2	10.39
-1	3.37
-2	3.14

The domain is the set of real numbers and range is $(3, \infty)$.

43. Graph: $f(x) = e^x - 2$

We find some function values, plot points, and draw the graph.

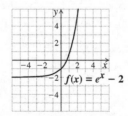

x	$e^x - 2$
0	-1
1	0.72
2	5.4
-1	-1.6
-2	-1.9

The domain is the set of real numbers and the range is $(-2, \infty)$.

45. Graph: $f(x) = 0.5e^x$

We find some function values, plot points, and draw the graph.

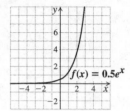

x	$0.5e^x$
0	0.5
1	1.36
2	3.69
-1	0.18
-2	0.07

The domain is the set of real numbers and the range is $(0, \infty)$.

47. Graph: $f(x) = 0.5e^{2x}$

We find some function values, plot points, and draw the graph.

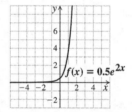

x	$0.5e^{2x}$
0	0.5
1	3.7
2	27.3
-1	0.07
-2	0.009

The domain is the set of real numbers and the range is $(0, \infty)$.

49. Graph: $f(x) = e^{x-3}$

We find some function values, plot points, and draw the graph.

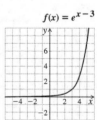

x	e^{x-3}
5	7.4
3	1
0	0.05
-1	0.02

The domain is the set of real numbers and the rage is $(0, \infty)$.

51. Graph: $f(x) = e^{x+2}$

We find some function values, plot points, and draw the graph.

x	e^{x+2}
1	20.1
0	7.4
−1	2.7
−2	1
−3	0.4

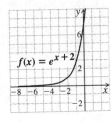

The domain is the set of real numbers and the range is $(0, \infty)$.

53. $f(x) = -e^x$

We find some function values, plot points, and draw the graph.

x	$-e^x$
0	−1
1	−2.7
2	−7.4
3	−20.1
−1	−0.4
−2	−0.1

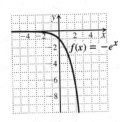

The domain is the set of real number and the range is $(-\infty, 0)$.

55. Graph: $g(x) = \ln x + 1$

Remember : $x > 0$

x	$\ln x + 1$
0.5	0.3
1	1
2	1.7
3	2.1
4	2.4

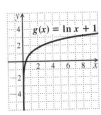

The domain is $(0, \infty)$ and the range is the set of real numbers.

57. Graph: $g(x) = \ln x - 2$

x	$\ln x - 2$
1	−2
2	−1.31
3	−0.90
4	−0.61
5	−0.39

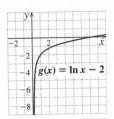

The domain is $(0, \infty)$ and the range is the set of real numbers.

59. Graph: $g(x) = 2 \ln x$

x	$2 \ln x$
0.5	−1.4
1	0
2	1.4
4	2.8
6	3.6

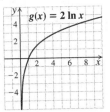

The domain is $(0, \infty)$ and the range is the set of real numbers.

61. Graph: $g(x) = -2 \ln x$

x	$-2 \ln x$
0.5	1.4
1	0
2	−1.4
3	−2.2
4	−2.8

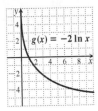

The domain is $(0, \infty)$ and the range is the set of real numbers.

63. Graph: $g(x) = \ln(x + 2)$

x	$\ln(x + 2)$
−1.9	−2.3
−1	0
0	0.7
1	1.1
2	1.4

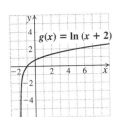

The domain is $(-2, \infty)$ and the range is the set of real numbers.

65. Graph: $g(x) = \ln(x-1)$

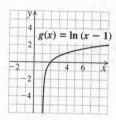

x	$\ln(x-1)$
1.1	-2.3
2	0
3	0.7
4	1.1
5	1.4

The domain is $(1, \infty)$ and the range is the set of real numbers.

67. We use the change-of-base formula:

$$f(x) = \frac{\log x}{\log 5} \text{ or } f(x) = \frac{\ln x}{\ln 5}$$

$y = \log (x)/\log (5)$, or
$y = \ln (x)/\ln (5)$

69. We use the change of base formula.

$$f(x) = \frac{\log(x-5)}{\log 2} \text{ or } f(x) = \frac{\ln(x-5)}{\ln 2}$$

$y = \log (x - 5)/\log (2)$, or
$y = \ln (x - 5)/\ln (2)$

71. We use the change of base formula.

$$f(x) = \frac{\log x}{\log 3} + x \text{ or } f(x) = \frac{\ln x}{\ln 3} + x$$

$y = \log (x)/\log (3) + x$, or
$y = \ln (x)/\ln (3) + x$

73. *Thinking and Writing Exercise.*

75. $x^2 - 3x - 28 = 0$

The trinomial can be factored by finding two numbers whose product is -28 and whose sum is -3. These factors are 4 and -7, and the factorization is $(x+4)(x-7)$.

$$(x+4)(x-7) = 0$$
$$x + 4 = 0 \quad or \quad x - 7 = 0$$
$$x = -4 \quad or \quad x = 7$$

The solutions are -4 and 7.

77. $17x - 15 = 0$
$$17x = 15$$
$$x = \frac{15}{17}$$

The solution is $\frac{15}{17}$.

79. $(x-5) \cdot 9 = 11$
$$9x - 45 = 11$$
$$9x = 45 + 11$$
$$9x = 56$$
$$x = \frac{56}{9}$$

The solution is $\frac{56}{9}$.

81. $x^{1/2} - 6x^{1/4} + 8 = 0$

The trinomial can be factored by finding two numbers whose product is 8 and whose sum is -6. These factors are -2 and -4, and the factorization is $(x^{1/4} - 2)(x^{1/4} - 4)$.

$$(x^{1/4} - 2)(x^{1/4} - 4) = 0$$
$$x^{1/4} - 2 = 0 \quad or \quad x^{1/4} - 4 = 0$$
$$x^{1/4} = 2 \quad or \quad x^{1/4} = 4$$
$$x = 2^4 \quad or \quad x = 4^4$$
$$x = 16 \quad or \quad x = 256$$

The solutions are 16 and 256.

83. *Thinking and Writing Exercise.* $\ln x = 1.5$ is equivalent to $x = e^{1.5} = \text{constant}$. Since $x = \text{constant}$ is a vertical line, the graph on Emma's calculator must be of a different function.

85. We use the change-of-base formula.

$$\log_6 81 = \frac{\log 81}{\log 6}$$

$$= \frac{\log 3^4}{\log(2 \cdot 3)}$$

$$= \frac{4 \log 3}{\log 2 + \log 3}$$

$$\approx \frac{4(0.477)}{0.301 + 0.477}$$

$$\approx 2.452$$

87. We use the change-of-base formula.

$$\log_{12} 36 = \frac{\log 36}{\log 12}$$

$$= \frac{\log(2 \cdot 3)^2}{\log(2^2 \cdot 3)}$$

$$= \frac{2 \log(2 \cdot 3)}{\log 2^2 + \log 3}$$

$$= \frac{2(\log 2 + \log 3)}{2 \log 2 + \log 3}$$

$$\approx \frac{2(0.301 + 0.477)}{2(0.301) + 0.477}$$

$$\approx 1.442$$

89. Use the change-of-base formula with $a = e$ and $b = 10$. We obtain

$$\log M = \frac{\ln M}{\ln 10}.$$

91. $\log(492x) = 5.728$

$$10^{5.728} = 492x$$

$$\frac{10^{5.728}}{492} = x$$

$$1086.5129 \approx x$$

93. $\log 692 + \log x = \log 3450$

$\log x = \log 3450 - \log 692$

$$\log x = \log \frac{3450}{692}$$

$$x = \frac{3450}{692}$$

$$x \approx 4.9855$$

95. a. Domain: $\{x \mid x > 0\}$, or $(0, \infty)$; range: $\{y \mid y < 0.5135\}$, or $(-\infty, 0.5135)$;

b. $[-1, 5, -10, 5]$;

c.
$$y = 3.4 \ln x - 0.25 e^x$$

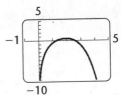

97. a. Domain $\{x \mid x > 0\}$, or $(0, \infty)$; range: $\{y \mid y > -0.2453\}$, or $(-0.2453, \infty)$

b. $[-1, 5, -1, 10]$;

c.
$$y = 2x^3 \ln x$$

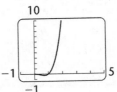

Exercise Set 12.6

1. e

3. f

5. b

7. g

9. $3^{2x} = 81 \Rightarrow 3^{2x} = 3^4 \Rightarrow 2x = 4 \Rightarrow x = 2$

11. $4^x = 32 \Rightarrow \left(2^2\right)^x = 2^5 \Rightarrow 2^{2x} = 2^5$

$$\Rightarrow 2x = 5 \Rightarrow x = \frac{5}{2}$$

13. $2^x = 10 \Rightarrow \log(2^x) = \log(10)$

$$\Rightarrow x \log(2) = 1 \Rightarrow x = \frac{1}{\log(2)} \approx 3.322$$

15. $2^{x+5} = 16 \Rightarrow 2^{x+5} = 2^4 \Rightarrow x + 5 = 4 \Rightarrow x = -1$

17. $8^{x-3} = 19 \Rightarrow \log(8^{x-3}) = \log(19)$

$\Rightarrow (x-3)\log(8) = \log(19)$

$\Rightarrow x-3 = \dfrac{\log(19)}{\log(8)} \Rightarrow x = \dfrac{\log(19)}{\log(8)} + 3 \approx 4.416$

19. $e^{t} = 50 \Rightarrow \ln(e^{t}) = \ln(50) \Rightarrow t\ln(e) = \ln(50)$

$\Rightarrow t = \ln(50) \approx 3.912$

21. $e^{-0.02t} = 8 \Rightarrow \ln(e^{-0.02t}) = \ln(8)$

$\Rightarrow -0.02t\ln(e) = \ln(8)$

$\Rightarrow -0.02t = \ln(8) \Rightarrow t = -\dfrac{\ln(8)}{0.02} \approx -103.972$

23. $5 = 3^{x+1}$

$\log 5 = \log 3^{x+1} \Rightarrow \log 5 = (x+1)\log 3$

$\Rightarrow \dfrac{\log 5}{\log 3} = \dfrac{(x+1)\log 3}{\log 3} \Rightarrow \dfrac{\log 5}{\log 3} = x+1$

$\Rightarrow x = \dfrac{\log 5}{\log 3} - 1 \approx 0.465$

25. $4.9^{x} - 87 = 0$

$4.9^{x} = 87$

$\log 4.9^{x} = \log 87$

$x\log 4.9 = \log 87$

$x = \dfrac{\log 87}{\log 4.9}$

$x \approx 2.810$

27. $19 = 2e^{4x}$

$\dfrac{19}{2} = \dfrac{2e^{4x}}{2}$

$\ln\left(\dfrac{19}{2}\right) = \ln e^{4x}$

$\ln\left(\dfrac{19}{2}\right) = 4x$

$\dfrac{\ln\left(\frac{19}{2}\right)}{4} = x$

$0.563 \approx x$

29. $7 + 3e^{5x} = 13$

$3e^{5x} = 6$

$e^{5x} = 2$

$\ln e^{5x} = \ln 2$

$5x = \ln 2$

$x = \dfrac{\ln 2}{5}$

$x \approx 0.139$

31. $\log_3 x = 4$

$x = 3^{4}$ Writing as an equivalent exponential equation

$x = 81$

33. $\log_2 x = -3$

$x = 2^{-3}$ Writing as an equivalent exponential equation

$x = \dfrac{1}{8}$

35. $\ln x = 5$

$x = e^{5}$ $\left(\ln x \text{ has base of } e\right)$

$x \approx 148.413$

37. $\ln 4x = 3$

$4x = e^{3}$

$x = \dfrac{e^{3}}{4}$

$x \approx 5.021$

39. $\log x = 2.5$

$x = 10^{2.5}$ $\left(\log x \text{ has base of } 10\right)$

$x \approx 316.228$

41. $\ln(2x+1) = 4$

$2x+1 = e^{4}$

$2x = e^{4} - 1$

$x = \dfrac{e^{4} - 1}{2}$

$x \approx 26.799$

43. $\ln x = 1$

$x = e^{1}$

$x \approx 2.718$

45. $5\ln x = -15$

$\ln x = -3$

$x = e^{-3}$

$x \approx 0.050$

47. $\log_2(8 - 6x) = 5$

$8 - 6x = 2^5$

$8 - 6x = 32$

$-6x = 24$

$x = -4$

The answer checks. The solution is -4.

49. $\log(x-9) + \log x = 1$

$\log_{10}\big[(x-9)(x)\big] = 1$

$x(x-9) = 10^1$

$x^2 - 9x = 10$

$x^2 - 9x - 10 = 0$

$(x+1)(x-10) = 0$

$x = -1 \ \text{ or } \ x = 10$

The number -1 does not check, because negative numbers do not have logarithms. The solution is 10.

51. $\log x - \log(x+3) = 1$

$\log_{10} \dfrac{x}{x+3} = 1$

$\dfrac{x}{x+3} = 10^1$

$x = 10(x+3)$

$x = 10x + 30$

$-9x = 30$

$x = -\dfrac{10}{3}$

The number $-\frac{10}{3}$ does not check. The equation has no solution.

53. $\log(2x+1) = \log(5) \Rightarrow 2x+1 = 5 \Rightarrow x = 2$

55. $\log_4(x+3) = 2 + \log_4(x-5)$

$\log_4(x+3) - \log_4(x-5) = 2$

$\log_4 \dfrac{x+3}{x-5} = 2$

$\dfrac{x+3}{x-5} = 4^2$

$\dfrac{x+3}{x-5} = 16$

$x + 3 = 16(x - 5)$

$x + 3 = 16x - 80$

$83 = 15x$

$\dfrac{83}{15} = x$

The number $\frac{83}{15}$ checks. It is the solution.

57. $\log_7(x+1) + \log_7(x+2) = \log_7 6$

$\log_7\big[(x+1)(x+2)\big] = \log_7 6$

$\log_7(x^2 + 3x + 2) = \log_7 6$

$x^2 + 3x + 2 = 6$

$x^2 + 3x - 4 = 0$

$(x+4)(x-1) = 0$

$x = -4 \ \text{ or } \ x = 1.$

Only the number 1 checks; it is the solution.

59. $\log_5(x+4) + \log_5(x-4) = \log_5 20$

$\log_5(x^2 - 16) = \log_5 20$

$x^2 - 16 = 20$

$x^2 = 36$

$x = \pm\sqrt{36} = \pm 6$

Only 6 checks; it is the solution.

61. $\ln(x+5) + \ln(x+1) = \ln 12$

$\ln(x^2 + 6x + 5) = \ln 12$

$x^2 + 6x + 5 = 12$

$x^2 + 6x - 7 = 0$

$(x+7)(x-1) = 0$

$x = -7 \ \text{ or } \ x = 1$

Only 1 checks; it is the solution.

63. $\log_2(x-3)+\log_2(x+3)=4$

$\log_2(x^2-9)=4$

$x^2-9=2^4$

$x^2-9=16$

$x^2=25$

$x=\pm\sqrt{25}=\pm 5$

Only 5 checks; it is the solution.

65. $\log_{12}(x+5)-\log_{12}(x-4)=\log_{12}3$

$\log_{12}\dfrac{x+5}{x-4}=\log_{12}3$

$\dfrac{x+5}{x-4}=3$

$x+5=3(x-4)$

$x+5=3x-12$

$17=2x$

$\dfrac{17}{2}=x$

The number $\frac{17}{2}$ checks and is the solution.

67. $\log_3(x-2)+\log_2 x=3$

$\log_2\big[(x-2)(x)\big]=3$

$x(x-2)=2^3$

$x^2-2x=8$

$x^2-2x-8=0$

$(x-4)(x+2)=0$

$x=4$ or $x=-2$

The number 4 checks, but -2 does not. The solution is 4.

69. $e^{0.5x}-7=2x+6$

Graph $y_1=e^{0.5x}-7$ and $y_2=2x+6$ in a window that shows the points of intersection of the graphs. One good choice is $[-10,10,-10,25]$, Yscl = 5. Use Intersect to find the first coordinates of the points of intersection. They are the solutions of the given equation. They are about -6.480 and 6.519.

71. $\ln(3x)=3x-8$

Graph $y_1=\ln(3x)$ and $y_2=3x-8$ in a window that shows the points of intersection of the graphs. One good choice is $[-5,5,-15,5]$. When we use Intersect in this window we can find only the coordinates of the right-hand point of intersection. They are about $(3.445,2.336)$, so one solution of the equation is about 3.445. To find the coordinates of the left-hand point of intersection we make the window smaller. One window that is appropriate is $[-1,1,-15,5]$. Using Intersect again we find that the other solution of the equation is about 0.0001. (The answer approximated to the nearest thousandth is 0.000, so we express it to the nearest ten-thousandth.)

73. Solve $\ln x=\log x$.

Graph $y_1=\ln x$ and $y_2=\log x$ in a window that shows the point of intersection of the graphs. One good choice is $[-5,5,5,5]$. Use Intersect to find the first coordinate of the point of intersection. It is solution of the given equation. It is 1.

75. *Thinking and Writing Exercise.*

77. Let x = the width of the rectangle, in feet. Then $x+6$ will equal the length, in feet. So:

$P=2l+2w\Rightarrow 2(x+6)+2x=26$

$\Rightarrow (x+6)+x=13\Rightarrow 2x+6=13$

$\Rightarrow 2x=7$

$\Rightarrow x=\dfrac{7}{2}$ ft = 3.5 ft

$\Rightarrow x+6=3.5+6=9.5$ ft

The width is 3.5 ft and the length is 9.5 ft.

79. Let x = the amount, in pounds, of the 25% sunflower seed brand, and y = the amount, in pounds, of the 40% sunflower seed brand. Then:

$x+y=50$ (1)

$0.25x+0.40y=0.33(50)$

or $25x+40y=33\cdot 50$

or $5x+8y=33\cdot 10=330$ (2)

Solve (1) for y:

$x + y = 50$ (1)

$\Rightarrow y = 50 - x$ (3)

Substitute $50 - x$ for y in (2) and solve for x:

$5x + 8y = 330$ (2)

$\Rightarrow 5x + 8(50 - x) = 330$

$\Rightarrow 5x + 400 - 8x = 330$

$\Rightarrow -3x = 330 - 400 = -70$

$\Rightarrow x = \dfrac{-70}{-3} = 23\dfrac{1}{3}$

Use (3) and the value of x to find y:

$y = 50 - x$ (3)

$\Rightarrow y = 50 - 23\dfrac{1}{3} = 49\dfrac{3}{3} - 23\dfrac{1}{3} = 26\dfrac{2}{3}$

Joanna should use $23\frac{1}{3}$ lbs of the 25% sunflower seed mix and $26\frac{2}{3}$ lbs of the 40% sunflower seed mix.

81. Max's rate is $\dfrac{1 \text{ score}}{2 \text{ hr}}$. Miles' rate is $\dfrac{1 \text{ score}}{3 \text{ hr}}$.

When working together their rate is::

$\dfrac{1 \text{ score}}{2 \text{ hr}} + \dfrac{1 \text{ score}}{3 \text{ hr}} = \dfrac{3 \text{ score}}{6 \text{ hr}} + \dfrac{2 \text{ score}}{6 \text{ hr}} = \dfrac{5 \text{ score}}{6 \text{ hr}}$

Therefore,

$\dfrac{5 \text{ score}}{6 \text{ hr}} \cdot t = 1 \text{ score}$

$\Rightarrow t = 1 \text{ score} \cdot \dfrac{6 \text{ hr}}{5 \text{ score}}$

$= \dfrac{6}{5}$ hr $= 1\dfrac{1}{5}$ hr $= 1$ hour, 12 min

83. *Thinking and Writing Exercise.*

85. $27^x = 81^{2x-3}$

$\left(3^3\right)^x = \left(3^4\right)^{2x-3}$

$3^{3x} = 3^{8x-12}$

$3x = 8x - 12$

$12 = 5x$

$\dfrac{12}{5} = x$

The solution is $\frac{12}{5}$.

87. $\log_x \left(\log_3 27\right) = 3$

$\log_3 27 = x^3$

$3 = x^3$ $\left(\log_3 27 = 3\right)$

$\sqrt[3]{3} = x$

The solution is $\sqrt[3]{3}$.

89. $x \cdot \log \dfrac{1}{8} = \log 8$

$x \cdot \log 8^{-1} = \log 8$

$x(-\log 8) = \log 8$ Using the power rule

$x = -1$

The solution is -1.

91. $2^{x^2+4x} = \dfrac{1}{8}$

$2^{x^2+4x} = \dfrac{1}{2^3}$

$2^{x^2+4x} = 2^{-3}$

$x^2 + 4x = -3$

$x^2 + 4x + 3 = 0$

$(x+3)(x+1) = 0$

$x = -3$ or $x = -1$

The solutions are -3 and -1.

93. $\log_5 |x| = 4$

$|x| = 5^4$

$|x| = 625$

$x = 625$ or $x = -625$

The solutions are 625 and -625.

95. $\log \sqrt{2x} = \sqrt{\log 2x}$

$\log (2x)^{1/2} = \sqrt{\log 2x}$

$\dfrac{1}{2} \log 2x = \sqrt{\log 2x}$

$\dfrac{1}{4}\left(\log 2x\right)^2 = \log 2x$ Squaring both sides

$\dfrac{1}{4}\left(\log 2x\right)^2 - \log 2x = 0$

Let $u = \log 2x$

$\dfrac{1}{4}u^2 - u = 0$

$u\left(\dfrac{1}{4}u - 1\right) = 0$

$u = 0$ or $\dfrac{1}{4}u - 1 = 0$

$u = 0$ or $\dfrac{1}{4}u = 1$

$u = 0$ or $u = 4$

$\log 2x = 0$ or $\log 2x = 4$ Replacing u with $\log 2x$

$2x = 10^0$ or $2x = 10^4$

$2x = 1$ or $2x = 10{,}000$

$x = \dfrac{1}{2}$ or $x = 5000$

Both numbers check. The solutions are $\frac{1}{2}$ and 5000.

97. $3^{x^2} \cdot 3^{4x} = \dfrac{1}{27}$

$3^{x^2 + 4x} = 3^{-3}$

$x^2 + 4x = -3$

$x^2 + 4x + 3 = 0$

$(x + 3)(x + 1) = 0$

$x = -3$ or $x = -1$

Both numbers check. The solutions are -3 and -1.

99. $\log x^{\log x} = 25$

$\log x(\log x) = 25$ Using the power rule

$(\log x)^2 = 25$

$\log x = \pm 5$

$x = 10^5$ or $x = 10^{-5}$

$x = 100{,}000$ or $x = \dfrac{1}{100{,}000}$

Both numbers check. The solutions are

$100{,}000$ and $\dfrac{1}{100{,}000}$.

101. $\left(81^{x-2}\right)\left(27^{x+1}\right) = 9^{2x-3}$

$\left[\left(3^4\right)^{x-2}\right]\left[\left(3^3\right)^{x+1}\right] = \left(3^2\right)^{2x-3}$

$\left(3^{4x-8}\right)\left(3^{3x+3}\right) = 3^{4x-6}$

$3^{7x-5} = 3^{4x-6}$

$7x - 5 = 4x - 6$

$3x = -1$

$x = -\dfrac{1}{3}$

The solution is $-\dfrac{1}{3}$.

103. $2^y = 16^{x-3}$ and $3^{y+2} = 27^x$

$2^y = \left(2^4\right)^{x-3}$ and $3^{y+2} = \left(3^3\right)^x$

$y = 4x - 12$ and $y + 2 = 3x$

$12 = 4x - y$ and $2 = 3x - y$

Solving this system of equations we get
$x = 10$ and $y = 28$. Then
$x + y = 10 + 28 = 38.$

Exercise Set 12.7

1. (a) $A(t) = 4000 \Rightarrow 77(1.283)^t = 4000$

$\Rightarrow 1.283^t = \dfrac{4000}{77}$

$\Rightarrow \log(1.283^t) = \log\left(\dfrac{4000}{77}\right)$

$\Rightarrow t\log(1.283) = \log\left(\dfrac{4000}{77}\right)$

$\Rightarrow t = \dfrac{\log\left(\dfrac{4000}{77}\right)}{\log(1.283)} \approx 15.85$

$1990 + 15.85 = 2005.85$. About 2006.

(b) $A(t) = 2 \cdot 77 = 77(1.283)^t \Rightarrow 1.283^t = 2$

$\Rightarrow t = \dfrac{\log(2)}{\log(1.283)} \approx 2.78$

The doubling time is approximately 2.8 years.

3. (a) $S(t) = 100 \Rightarrow 180(0.97)^t = 100$

$\Rightarrow 0.97^t = \dfrac{100}{180} \Rightarrow \log(0.97^t) = \log\left(\dfrac{100}{180}\right)$

$\Rightarrow t\log(0.97) = \log\left(\dfrac{100}{180}\right)$

$\Rightarrow t = \dfrac{\log\left(\dfrac{100}{180}\right)}{\log(0.97)} \approx 19.30$

1960 + 19.3 = 1979.3. About 1979.

(b) $S(t) = 25 \Rightarrow 180(0.97)^t = 25$

$\Rightarrow 0.97^t = \dfrac{25}{180} \Rightarrow \log(0.97^t) = \log\left(\dfrac{25}{180}\right)$

$\Rightarrow t\log(0.97) = \log\left(\dfrac{25}{180}\right)$

$\Rightarrow t = \dfrac{\log\left(\dfrac{25}{180}\right)}{\log(0.97)} \approx 64.81$

1960 + 64.81 = 2024.81. About 2025.

5. (a) $A(t) = 29{,}000(1.03)^t$

Let $A(t) = 35{,}000,$ and solve for t.

$35{,}000 = 29{,}000(1.03)^t$

$\dfrac{35{,}000}{29{,}000} = (1.03)^t$

$\log 1.2069 \approx \log(1.03)^t$

$\log 1.2069 \approx t\log 1.03$

$\dfrac{\log 1.2069}{\log 1.03} \approx t$

$6.4 \approx t$

The amount after about 6.4 years will reach $35,000.

(b) $2 \cdot P_0 = 2 \cdot 29{,}000 = 58{,}000$

Let $A(t) = 58{,}000,$ and solve for t.

$58{,}000 = 29{,}000(1.03)^t$

$\dfrac{58{,}000}{29{,}000} = (1.03)^t$

$\log 2 = \log 1.03^t$

$\log 2 = t\log 1.03$

$\dfrac{\log 2}{\log 1.03} = t$

$23.4 \approx t$

The doubling time is approximately 23.4 years.

7. (a) $m(t) = 48 \Rightarrow 1507(0.94)^t = 48$

$\Rightarrow 0.94^t = \dfrac{48}{1507}$

$\Rightarrow \log(0.94^t) = \log\left(\dfrac{48}{1507}\right)$

$\Rightarrow t\log(0.94) = \log\left(\dfrac{48}{1507}\right)$

$\Rightarrow t = \dfrac{\log\left(\dfrac{48}{1507}\right)}{\log(0.94)} \approx 55.70$

About $56°$ Fahrenheit.

(b) $m(t) = 15 \Rightarrow 1507(0.94)^t = 15$

$\Rightarrow 0.94^t = \dfrac{15}{1507}$

$\Rightarrow \log(0.94^t) = \log\left(\dfrac{15}{1507}\right)$

$\Rightarrow t\log(0.94) = \log\left(\dfrac{15}{1507}\right)$

$\Rightarrow t = \dfrac{\log\left(\dfrac{15}{1507}\right)}{\log(0.94)} \approx 74.50$

About $75°$ Fahrenheit.

9. (a) Note that 1 billion = 1000 million.

$S(t) = 1000 \Rightarrow 2.05(1.8)^t = 1000$

$\Rightarrow 1.8^t = \dfrac{1000}{2.05} \Rightarrow \log(1.8^t) = \log\left(\dfrac{1000}{2.05}\right)$

$\Rightarrow t\log(1.8) = \log\left(\dfrac{1000}{2.05}\right)$

$\Rightarrow t = \dfrac{\log\left(\dfrac{1000}{2.05}\right)}{\log(1.8)} \approx 10.53$

2002 + 10.53 = 2012.53. About 2013.

(b) $m(t) = 2 \cdot 2.05 \Rightarrow 2.05(1.8)^t \Rightarrow 1.8^t = 2$

$\Rightarrow t = \dfrac{\log(2)}{\log(1.8)} \approx 1.18$

The doubling time is approximately 1.2 years.

11. $\text{pH} = -\log\left[H^+\right]$

$= -\log\left[1.3 \times 10^{-5}\right]$

$\approx -(-4.886057)$

≈ 4.9

The pH of fresh-brewed coffee is about 4.9.

13. $\text{pH} = -\log\left[H^+\right]$

$7.0 = -\log\left[H^+\right]$

$-7.0 = \log\left[H^+\right]$

$10^{-7.0} = \left[H^+\right]$ Converting to an exponential equation

10^{-7} moles per liter

15. $L = 10 \cdot \log \dfrac{I}{I_0} = 10 \cdot \log \dfrac{10}{10^{-12}}$

$= 10 \cdot \log\left(10^{13}\right) = 10 \cdot 13 = 130$

130 decibels.

17. $L = 128.8 \Rightarrow 10 \cdot \log\left(\dfrac{I}{I_0}\right) = 128.8$

$\Rightarrow \log\left(\dfrac{I}{I_0}\right) = \dfrac{128.8}{10} = 12.88 \Rightarrow \dfrac{I}{I_0} = 10^{12.88}$

$\Rightarrow I = 10^{12.88} I_0 == 10^{12.88} 10^{-12} \text{ W/m}^2$

$= 10^{0.88} \text{ W/m}^2 \approx 7.6 \text{ W/m}^2$

19. $M = 7.5 \Rightarrow \log\left(\dfrac{v}{1.34}\right) = 7.5 \Rightarrow \dfrac{v}{1.34} = 10^{7.5}$

$\Rightarrow v = 1.34 \cdot 10^{7.5} \approx 4.24 \cdot 10^7$

Approximately 42.4 million emails a day.

21. $k = 2.5\%; \ 0.025$

(a) Using the exponential growth formula, we have $P(t) = P_0 e$

$P(t) = P_0 e^{0.025t}$

(b) $P_0 = \$5000$

After 1 year, $t = 1$.

$P(1) = 5000 e^{0.025 \cdot 1}$

$P(1) = 5000 e^{0.025}$

$\approx \$5126.58$

After 2 years, $t = 2$.

$P(2) = 5000 e^{0.025 \cdot 2}$

$= 5000 e^{0.05}$

$\approx \$5256.36$

(c) $2 \cdot 5,000 = 10,000$

$10,000 = 5000 e^{0.025t}$

$2 = e^{0.025t}$

$\ln 2 = \ln e^{0.025t}$

$\ln 2 = 0.025t$

$\dfrac{\ln 2}{0.025} = t$

$27.7 \approx t$

The investment will double in approximately 27.7 years.

23. (a) $P_0 = 310$ million, $k = 1.0\%$, or 0.01

$P(t) = P_0 e^{kt}$

$P(t) = 310 e^{0.01t}$, where $t =$ the number of years after 2010 and $P(t)$ is in millions.

(b) In 2016, $t = 2016 - 2010 = 6$

$P(6) = 310 e^{0.01 \cdot 6}$

$P(6) = 310 e^{0.06}$

$P(6) \approx 329$

In 2016, the U.S. population will be approximately 329 million.

(c)

$310 e^{0.01t} = 350 \Rightarrow e^{0.01t} = \dfrac{350}{310} = \dfrac{35}{31}$

$\Rightarrow 0.01t = \ln\left(\dfrac{35}{31}\right) \Rightarrow t = \dfrac{1}{0.01} \ln\left(\dfrac{35}{31}\right)$

$\Rightarrow t \approx 12.1$

The population will reach 350 million about 12 years after 2010, or about 2022.

25. $k = 12\%$, or 0.12 The doubling rate is when

$P(t) = 2P_0 \Rightarrow P_0 e^{0.12t} = 2P_0 \Rightarrow e^{0.12t} = 2$

$\Rightarrow 0.12t = \ln(2) \Rightarrow t = \dfrac{\ln(2)}{0.12} \approx 5.78$

The number of online college students doubles approximately every 5.8 years.

27. (a) $Y(10) = 87 \ln\left(\dfrac{10}{6.1}\right) \approx 43.00$

   $2000 + 43 = 2043$
   According to the model, the world population will reach 10 billion in about 2043.

   (b) $Y(12) = 87 \ln\left(\dfrac{12}{6.1}\right) \approx 58.87$

   $2000 + 58.87 = 2058.87$
   According to the model, the world population will reach 12 billion in late 2058.

   (c) Using the values we computed in parts (a) and (b) and any others we wish to calculate, we sketch the graph:

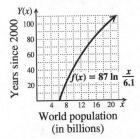

   World population
   (in billions)

29. (a) $S(0) = 68 - 20 \log(0+1)$

   $= 68 - 20 \log 1 = 68 - 20(0)$

   $= 68\%$

   (b) $S(4) = 68 - 20 \log(4+1)$

   $= 68 - 20 \log 5$

   $\approx 68 - 20(0.69897)$

   $\approx 54\%$

   $S(24) = 68 - 20 \log(24+1)$

   $= 68 - 20 \log 25$

   $\approx 68 - 20(1.39794)$

   $\approx 40\%$

   (c) Using the values we computed in parts (a) and (b) and any others we wish to calculate, we sketch the graph:

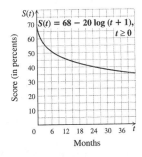

   Months

d. $50 = 68 - 20 \log(t+1)$

   $-18 = -20 \log(t+1)$

   $0.9 = \log(t+1)$

   $10^{0.9} = t+1$

   $7.9 \approx t+1$

   $6.9 \approx t$

   After about 6.9 months, the average score was 50.

31. (a) Let t = the year since 1990, and P be the power generated by wind in thousands of megawatts. Then:

   $P_0 = P(0) = 2$ and $P(19) = 35$ So:

   $P(t) = P_0 e^{kt} = 2e^{kt}$

   $\Rightarrow P(19) = 2e^{k19} = 35 \Rightarrow e^{19k} = \dfrac{35}{2} = 17.5$

   $\Rightarrow 19k = \ln(17.5)$

   $\Rightarrow k = \dfrac{\ln(17.5)}{19} \approx 0.150642$

   $\Rightarrow P(t) = 2e^{0.150642t}$

   (b) $P(t) = 50 \Rightarrow 2e^{0.150642t} = 50$

   $\Rightarrow e^{0.150642t} = 25 \Rightarrow 0.150642t = \ln(25)$

   $\Rightarrow t = \dfrac{\ln(25)}{0.150642} \approx 21.37$

   $1990 + 21.37 = 2011.37$
   According to the model, wind-power capacity will 50,000 MW of power in 2011.

33. (a) Let t = the year since 1997, and C be the cost per gigabit per second per mile in dollars. Then:

   $C_0 = C(0) = 8200$ and $C(10) = 500$ So:

   $C(t) = C_0 e^{kt} = 8200 e^{-kt}$

   $\Rightarrow C(10) = 8200 e^{-k10} = 500$

   $\Rightarrow e^{-10k} = \dfrac{500}{8200} = \dfrac{5}{82} \Rightarrow -10k = \ln\left(\dfrac{5}{82}\right)$

   $\Rightarrow k = -\dfrac{1}{10} \cdot \ln\left(\dfrac{5}{82}\right) \approx 0.2797$

   $\Rightarrow C(t) = 8200 e^{-0.2797t}$

   (b) $t = 2010 - 1997 = 13$

   $C(13) = 8200 e^{-0.2797 \cdot 13} \approx 216$

   According to the model, it cost approximately $216 per gigabit per second per mile to lay subsea cable in 2010.

(c) Use $C(t) = 1$.

$8200e^{-0.2797t} = 1$

$e^{-0.2797t} = \dfrac{1}{8200}$

$-0.2797t = \ln\left(\dfrac{1}{8200}\right)$

$t = \dfrac{\ln\left(\dfrac{1}{8200}\right)}{-0.2797}$

≈ 32.22

$1997 + 32 = 2029$

According to the model, it will cost $1 per gigabit per second per mile in about 2029.

35. We will use the function derived in Example 7:
$P(t) = P_0 e^{-0.00012t}$

If the seed lost 21% of its carbon-14 from an initial amount P_0, then

$(100\% - 21\%)P_0 = 79\%P_0 = 0.79P_0$ is the amount present. To find the age of the seed set $P(t) = 0.79P_0$ and solve for t.

$P(t) = 0.79P_0 \Rightarrow P_0 e^{-0.00012t} = 0.79P_0$

$\Rightarrow e^{-0.00012t} = 0.79 \Rightarrow -0.00012t = \ln(0.79)$

$\Rightarrow t = -\dfrac{\ln(0.79)}{0.00012} \approx 1964.35$

The seed was about 1964 years old.

37. The function $P(t) = P_0 e^{-kt}, k > 0$, can be used to model decay. For iodine-131, $k = 9.6\%$, or 0.096. To find the half-life we substitute 0.096 for k and $\frac{1}{2}P_0$ for $P(t)$, and solve for t.

$x = 2010 - 1980 = 30$

$\dfrac{1}{2}P_0 = P_0 e^{-0.096t}$, or $\dfrac{1}{2} = e^{-0.096t}$

$\ln\dfrac{1}{2} = \ln e^{-0.096t} = -0.096t$

$t = \dfrac{\ln 0.5}{-0.096} \approx \dfrac{-0.6931}{-0.096} \approx 7.2 \text{ days}$

39. (a) Use $P(t) = P_0 e^{-kt}, k > 0$ for exponential decay. To find k substitute $\dfrac{1}{2}P_0$ for $P(t)$ and 5 for t.

$P_0 e^{-k \cdot 5} = \dfrac{1}{2}P_0 \Rightarrow e^{-k \cdot 5} = \dfrac{1}{2}$

$\Rightarrow -5k = \ln\left(\dfrac{1}{2}\right) = -\ln(2)$

$\Rightarrow k = \dfrac{\ln(2)}{5} \approx 0.1386 \text{ or } 13.86\%$

$\Rightarrow P(t) = P_0 e^{-0.1386t}$

(b) If 95% has decayed then
$P(t) = (1 - 0.95)P_0 = 0.05P_0$

$P(t) = 0.05P_0 \Rightarrow P_0 e^{-0.1386t} = 0.05P_0$

$\Rightarrow e^{-0.1386t} = 0.05 \Rightarrow -0.1386t = \ln(0.05)$

$\Rightarrow t = -\dfrac{\ln(0.05)}{0.1386} \approx 21.6$

It takes approximately 21.6 hours for 95% of the caffeine to expelled.

41. (a) In 1990, $t = 0$ and $V_0 = 9$ In 2010, $t = 20$ and $V(20) = 104.3$ Substitute into the exponential growth formula and solve for k.

$V(t) = V_0 e^{kt} = 9e^{kt}$

$\Rightarrow V(20) = 9e^{20k} = 104.3$

$\Rightarrow e^{20k} = \dfrac{104.3}{9} \Rightarrow 20k = \ln\left(\dfrac{104.3}{9}\right)$

$\Rightarrow k = \dfrac{1}{20}\ln\left(\dfrac{104.3}{9}\right) \approx 0.1225$

$\Rightarrow V(t) = 9e^{0.1225t}$

(b) In 2020, $t = 2020 - 1990 = 30$
$V(30) = 9e^{0.1225 \cdot 30} = 9e^{3.675} \approx 355$

According to the model the sculpture will be worth approximately $355 million in 2020.

(c) $9e^{0.1225t} = 2 \cdot 9 \Rightarrow e^{0.1225t} = 2$

$\Rightarrow 0.1225t = \ln(2)$

$\Rightarrow t = \dfrac{\ln(2)}{0.1225} \approx 5.66$

The doubling time is about 5.7 years.

(d) $9e^{0.1225t} = 1000 \Rightarrow e^{0.1225t} = \dfrac{1000}{9}$

$\Rightarrow 0.1225t = \ln\left(\dfrac{1000}{9}\right)$

$\Rightarrow t = \dfrac{1}{0.1225} \cdot \ln\left(\dfrac{1000}{9}\right) \approx 38.45$

According to the model, the sculpture's value will exceed \$1 billion after about 38.3 years, or in 2028.

43. The number of transistors increases from 1974 to 2006 at a rate that makes it appear that an exponential function might fit the data.

45. The number of accidents increase then decrease at different rates. It does not appear that an exponential function would be a good model.

47. (a) Let t = the number of years after 1974, and n = the number of transistors per chip, in thousands. Then from the table, we have the points (0, 6), (4, 29), (8, 134), (15, 1200), (19, 3300), (25, 9500), and (32, 291,000). Enter the data on a graphing calculator and then use the exponential regression feature to get:

$n(t) = 7.8(1.3725)^{t}$

(b) $e^{k} = 1.3725$

$\Rightarrow k = \ln(1.3725) \approx 0.3166 = 31.66\%$

(c) For 2010, $t = 2010 - 1974 = 36$.

$n(36) = 7.8(1.3725)^{36} \approx 6.95888 \cdot 10^{5}$

Since n is in thousands, this corresponds to approximately 696 million transistors per chip.

49. (a) We enter the data from the table and then use the exponential regression feature. We have:

$f(x) = 20917152(0.87055)^{x}$

(b) $f(95) = 20917152(0.8705505633)^{95}$

≈ 4 hr.

51. *Thinking and Writing Exercise.*

53. $d^{2} = (x_{2} - x_{1})^{2} + (y_{2} - y_{1})^{2}$

$= (-3 - (-2))^{2} + (7 - 6)^{2} = 2$

$\Rightarrow d = \sqrt{2}$

55. $(x_{M}, y_{M}) = \left(\dfrac{x_{1} + x_{2}}{2}, \dfrac{y_{1} + y_{2}}{2}\right)$

$= \left(\dfrac{3 + 5}{2}, \dfrac{-8 + (-6)}{2}\right) = (4, -7)$

57. $x^{2} + 8x = 1 \Rightarrow x^{2} + 8x + \left(\dfrac{8}{2}\right)^{2} = 1 + \left(\dfrac{8}{2}\right)^{2}$

$\Rightarrow (x + 4)^{2} = 1 + 16 = 17$

$\Rightarrow x + 4 = \pm\sqrt{17} \Rightarrow x = -4 \pm \sqrt{17}$

59. $y = x^{2} - 5x - 6 = (x - 6)(x + 1)$

The function is quadratic with a positive leading coefficient. Therefore its graph is a parabola opening upward.
By the factored form we can see the x-intercepts are at $(-1, 0)$ and $(6, 0)$. Since the graph is symmetric along the axis of symmetry, the x-coordinate of the vertex must be halfway between -1 and 6, at $\dfrac{6 + (-1)}{2} = \dfrac{5}{2} = 2\dfrac{1}{2}$. To find the y-coordinate of the vertex find the value of the function at

$x = \dfrac{5}{2}$. $y\left(\dfrac{52}{2}\right) = \left(\dfrac{5}{2}\right)^{2} - 5\left(\dfrac{5}{2}\right) - 6$

$= \dfrac{25}{4} - \dfrac{25}{2} - 6 = \dfrac{25}{4} - \dfrac{50}{4} - \dfrac{24}{4}$

$= -\dfrac{49}{4} = -12\dfrac{1}{4}$

Since $y(0) = -6$, the y-intercept is at $(0, -6)$. Again, using the axis of symmetry and noting that $x = 0$ is $2\dfrac{1}{2}$ units to the left of the vertex, at $x = 5$, which is $2\dfrac{1}{2}$ units to the right of the vertex, y should also be -6.
Thus we have a parabola that opens upward, with vertex $\left(\dfrac{5}{2}, -\dfrac{49}{4}\right)$, x-intercepts of $(-1, 0)$ and $(6, 0)$, y-intercept of $(0, -6)$, and which passes through the point $(5, -6)$. This is

plenty of information to plot the graph:

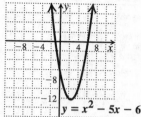

$y = x^2 - 5x - 6$

61. *Thinking and Writing Exercise.*

63. For continuous compounding, use the formula $P(t) = P_0 e^{kt}$, with $P_0 =$ the amount needed in 2008, in millions of dollars, to have \$20 million dollars in 2016, $k = 4\% = 0.04$, and $t = 2016 - 2008 = 8$ years.

$P(t) = 20 \Rightarrow P_0 e^{0.04 \cdot 8} = 20$

$\Rightarrow P_0 = \dfrac{20}{e^{0.32}} \approx 14.5$

If one could find an institution that offered continuous compounding for 8 years, \$14.5 million dollars would need to be invested to have \$20 million 8 years later.

65. $m(I) = -(19 + 2.5 \cdot \log I)$.

(a) Let $I = 1390 \, \text{W/m}^2$ and solve.

$m(I) = -(19 + 2.5 \log 1390)$

$\approx -(19 + 2.5 \cdot 3.1430)$

≈ -26.9

(b) Let $m(I) = 23$ and solve for I.

$23 = -(19 + 2.5 \cdot \log I)$

$23 = -19 - 2.5 \log I$

$42 = -2.5 \log I$

$\dfrac{42}{-2.5} = \log I$

$-16.8 = \log I \Rightarrow I = 10^{-16.8}$

$I \approx 1.58 \times 10^{-17}$

67. Since doubling time is the amount of time for $P(t)$ to equal $2 \cdot P_0$, substitute and solve for t.

$P(t) = P_0 e^{kt}$

$\dfrac{2 P_0}{P_0} = \dfrac{P_0 e^{kt}}{P_0}$

$2 = e^{kt}$

$\ln 2 = \ln e^{kt}$

$\ln 2 = kt$

$\dfrac{\ln 2}{k} = t.$

69. (a) Enter the data into a graphing calculator and choose the logistic option in the STAT CALC MENU to obtain:

$f(x) = \dfrac{62.2245}{1 + 2.2661 e^{-0.4893x}}$

(b) In 2010, $x = 2010 - 1997 = 13$

$f(13) = \dfrac{62.2245}{1 + 2.2661 e^{-0.4893(13)}}$

$\approx 0.62 = 62\%$

Chapter 12 Study Summary

1. $f(x) = 1 - 6x$, $g(x) = x^2 - 3$

$\Rightarrow (f \circ g)(x) = f(g(x)) = 1 - 6(g(x))$

$= 1 - 6(x^2 - 3) = 1 - 6x^2 + 18$

$= 19 - 6x^2$

2. By definition, *f(x)* is 1-1 if $f(x) = f(y) \Rightarrow x = y$.

For $f(x) = 5x - 7$,

$f(x) = f(y) \Rightarrow 5x - 7 = 5y - 7$

$\Rightarrow 5x = 5y \Rightarrow x = y \Rightarrow f$ is 1-1

3. First show that $f^{-1}(x)$ exist by showing that *f(x)* is one-to-one. For $f(x) = 5x + 1$,

$f(x) = f(y) \Rightarrow 5x + 1 = 5y + 1$

$\Rightarrow 5x = 5y \Rightarrow x = y \Rightarrow f$ is one-to-one.

Replace $f(x)$ by y: $y = 5x + 1$

Interchange x and y: $x = 5y + 1$

Solve for y: $y = \dfrac{x - 1}{5}$

Replace y by $f^{-1}(x)$: $f^{-1}(x) = \dfrac{x - 1}{5}$

4. Creating a table of values is convenient for graphing exponential functions.

x	$f(x) = 2^x$
-3	$2^{-3} = \dfrac{1}{2^3} = \dfrac{1}{8}$
-2	$2^{-2} = \dfrac{1}{2^2} = \dfrac{1}{4}$
-1	$2^{-1} = \dfrac{1}{2^1} = \dfrac{1}{2}$
0	$2^0 = 1$
1	$2^1 = 2$
2	$2^2 = 4$
3	$2^3 = 8$

From the table you can see that as x gets more negative 2^x gets smaller, but never becomes negative, or zero. (Note that a negative sign in an exponent does not make the expression negative, it just puts the base in the denominator when the sign of the exponent is changed.) This is equivalent to saying that as x goes to $-\infty$, 2^x approaches the line $y = 0$ (the x-axis) asymptotically.

As x gets more positive, however, 2^x grows without bound.

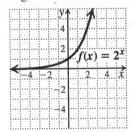

5. There are two ways to approach the graphs of log functions. One is to create a table of values, as in the previous exercise. The other is to recognize that $f(x) = \log_b(x)$ is the inverse function of the exponential function $g(x) = b^x$, and to remember that the graph of an inverse function is a reflection of the original function across the line $y = x$ (the line that goes through the points (x, x), such as $(-1, -1)$, $(0, 0)$, $(2, 2)$, and so on).

Therefore, since $f(x) = \log(x) = \log_{10}(x)$ is the inverse of $g(x) = 10^x$, we can plot the graph of 10^x, as the graph of 2^x was plotted in the previous

exercise, and then reflect the graph of 10^x across the line $y = x$.

As in exercise 4, create a table of values:

x	$g(x) = 10^x$
-3	$10^{-3} = \dfrac{1}{10^3} = \dfrac{1}{1000}$
-2	$10^{-2} = \dfrac{1}{10^2} = \dfrac{1}{100}$
-1	$10^{-1} = \dfrac{1}{10^1} = \dfrac{1}{10}$
0	$10^0 = 1$
1	$10^1 = 10$
2	$10^2 = 100$
3	$10^3 = 1000$

Again as x goes to $-\infty$, 10^x approaches the line $y = 0$ (the x-axis) asymptotically, and as x gets more positive 10^x grows without bound. Reflecting the resulting graph about the line $y = x$ gives the following graph for $f(x) = \log(x)$:

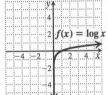

Note how the horizontal asymptote of $g(x) = 10^x$ has become a vertical asymptote for $f(x) = \log(x)$ upon reflection.

6. $5^4 = 625 \Rightarrow \log_5(5^4) = \log_5(625)$
    $\Rightarrow 4\log_5(5) = \log_5(625) \Rightarrow 4 \cdot 1 = \log_5(625)$
    $\Rightarrow \log_5(625) = 4$

7. $\log_9(xy) = \log_9(x) + \log_9(y)$

8. $\log_6\left(\dfrac{7}{10}\right) = \log_6(7) - \log_6(10)$

9. $\log(7^5) = 5\log(7)$

10. $\log_8(1) = 0$

11. $\log_7(7) = 1$

12. $\log_t(t^{12}) = 12\log_t(t) = 12 \cdot 1 = 12$

13. $\log(100) = \log_{10}(10^2) = 2\log_{10}(10) = 2$

14. $\ln(e) = \log_e(e) = 1$

15. $\log_2(5) = \dfrac{\log(5)}{\log(2)} \approx 2.3219$

16. $2^{3x} = 16 \Rightarrow 2^{3x} = 2^4 \Rightarrow 3x = 4 \Rightarrow x = \dfrac{4}{3}$

17. $e^{0.1x} = 10 \Rightarrow \ln(e^{0.1x}) = \ln(10)$

  $\Rightarrow 0.1x\ln(e) = \ln(10) \Rightarrow 0.1x = \ln(10)$

  $\Rightarrow x = \dfrac{\ln(10)}{0.1} \approx 23.02585$

18 (a) For exponential growth use the formula
  $P(t) = P_0 e^{kt}$ where k is the growth rate,
  expressed as a decimal, and P_0 is the
  initial population (the population at $t = 0$).
  Since we are given $P_0 = 15,000$ and
  $k = 2.3\% = 0.023,$ the function
  is: $P(t) = 15,000e^{0.023t}$

  (b) To find the doubling time, set
  $P(t) = 2P_0$ and solve for t.
  $P(t) = 2P_0 \Rightarrow 15,000e^{0.023t} = 2 \cdot 15,000$
  $\Rightarrow e^{0.023t} = 2 \Rightarrow \ln(e^{0.023t}) = \ln(2)$
  $\Rightarrow 0.023t = \ln(2) \Rightarrow t = \dfrac{\ln(2)}{0.023} \approx 30.1$
  The doubling time is approximately 30.1
  years.

19. For exponential decay use the formula
  $P(t) = P_0 e^{-kt}$ where k is the decay rate,
  expressed as a decimal, and P_0 is the initial
  amount (the amount at $t = 0$). Since we are
  given $k = 1.98\% = 0.0198,,$ the function
  is: $P(t) = P_0 e^{0.0198t}$
  Even though the initial amount is unknown,
  the half-life can still be determined by setting
  $P(t) = 0.5P_0$ and solving for t.

$P(t) = 0.5P_0 \Rightarrow P_0 e^{-0.0198t} = 0.5P_0$

$\Rightarrow e^{-0.0198t} = 0.5 \Rightarrow \ln(e^{-0.0198t}) = \ln(0.5)$

$\Rightarrow -0.0198t = \ln(0.5) \Rightarrow t = -\dfrac{\ln(0.5)}{0.0198} \approx 35.0$

The half-life is approximately 35 years.

Chapter 12 Review Exercises

 1. True

 2. True

 3. True

 4. False

 5. False

 6. True

 7. False

 8. False

 9. True

10. False

11. $f(x) = x^2 + 1, \qquad g(x) = 2x - 3$
  $(f \circ g)(x) = f(g(x)) = f(2x - 3)$
  $\qquad\qquad = (2x - 3)^2 + 1$
  $\qquad\qquad = 4x^2 - 12x + 9 + 1$
  $\qquad\qquad = 4x^2 - 12x + 10$
  $(g \circ f)(x) = g(f(x)) = g(x^2 + 1)$
  $\qquad\qquad = 2(x^2 + 1) - 3$
  $\qquad\qquad = 2x^2 + 2 - 3$
  $\qquad\qquad = 2x^2 - 1$

12. Possible answer: $h(x) = \sqrt{3 - x}$ is the square
  root of $(3 - x)$.
  $f(x) = \sqrt{x}; \; g(x) = 3 - x.$

13. The graph of $f(x) = 4 - x^2$ is a parabola
  which opens downward and does not pass the
  horizontal line test. The function is not 1-1.

14. $f(x) = x - 8$: $f(x)$ is a linear function and is 1:1.

Replace $f(x)$ by y: $y = x - 8$

Interchange x and y: $x = y - 8$

Solve for y: $x + 8 = y$

Replace y by $f^{-1}(x)$: $f^{-1}(x) = x + 8$

15. $g(x) = \dfrac{3x + 1}{2}$

$g(x)$ is one-to-one.

Replace $g(x)$ by y: $y = \dfrac{3x + 1}{2}$

Interchange x and y: $x = \dfrac{3y + 1}{2}$

Solve for y: $\dfrac{2x - 1}{3} = y$

Replace y by $g^{-1}(x)$: $g^{-1}(x) = \dfrac{2x - 1}{3}$

16. $f(x) = 27x^3$

$f(x)$ is one-to-one.

Replace $f(x)$ by y: $y = 27x^3$

Interchange x and y: $x = 27y^3$

Solve for y: $\sqrt[3]{\dfrac{x}{27}} = y$

Replace y by $f^{-1}(x)$: $f^{-1}(x) = \sqrt[3]{\dfrac{x}{27}}$ or

$f^{-1}(x) = \dfrac{\sqrt[3]{x}}{3}$

17. Graph: $f(x) = 3^x + 1$

Choose values of x and determine corresponding function values. Plot these points and connect them with a smooth curve.

$f(-2) = 3^{-2} + 1 = \dfrac{1}{9} + 1 = 1\dfrac{1}{9}$

$f(-1) = 3^{-1} + 1 = \dfrac{1}{3} + 1 = 1\dfrac{1}{3}$

$f(0) = 3^0 + 1 = 1 + 1 = 2$

$f(1) = 3^1 + 1 = 3 + 1 = 4$

$f(2) = 3^2 + 1 = 9 + 1 = 10$

x	$f(x)$, or $3^x + 1$
-2	$1\dfrac{1}{9}$
-1	$1\dfrac{1}{3}$
0	2
1	4
2	10

$f(x) = 3^x + 1$

18. Graph: $x = \left(\dfrac{1}{4}\right)^y$

Choose values of y and determine corresponding x values. Plot these points and connect them with a smooth curve.

For $y = -2$, $x = \left(\dfrac{1}{4}\right)^{-2} = 16$

For $y = -1$, $x = \left(\dfrac{1}{4}\right)^{-1} = 4$

For $y = 0$, $x = \left(\dfrac{1}{4}\right)^{0} = 1$

For $y = 1$, $x = \left(\dfrac{1}{4}\right)^{1} = \dfrac{1}{4}$

For $y = 2$, $x = \left(\dfrac{1}{4}\right)^{2} = \dfrac{1}{16}$

x, or $\left(\dfrac{1}{4}\right)^y$	y
16	-2
4	-1
1	0
$\dfrac{1}{4}$	1
$\dfrac{1}{16}$	2

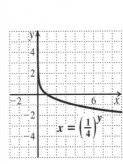

$x = \left(\dfrac{1}{4}\right)^y$

19. $y = \log_5 x$

The equation $y = \log_5 x$ is equivalent to $5^y = x$. Choose values of y and determine corresponding x values. Plot the points and connect them with a smooth curve.

For $y = -2$, $x = 5^{-2} = \dfrac{1}{25}$

For $y = -1$, $x = 5^{-1} = \dfrac{1}{5}$

For $y = 0$, $x = 5^{0} = 1$

For $y = 1$, $x = 5^{1} = 5$

For $y = 2$, $x = 5^{2} = 25$

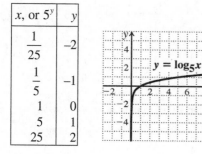

x, or 5^y	y
$\dfrac{1}{25}$	-2
$\dfrac{1}{5}$	-1
1	0
5	1
25	2

20. $\log_3 9$

Since $9 = 3^2$, $\log_3 3^2 = 2$

21. $\log_{10} \dfrac{1}{100}$

Since $\dfrac{1}{100} = \dfrac{1}{10^2} = 10^{-2}$, $\log_{10} 10^{-2} = -2$

22. $\log_5 5^7 = 7$ The logarithm is the exponent.

23. $\log_9 3$

Since $9^{\frac{1}{2}} = 3$, $\left[\left(3^2\right)^{\frac{1}{2}} = 3\right]$

$\log_9 3 = \dfrac{1}{2}$

24. $10^{-2} = \dfrac{1}{100} \Rightarrow \log_{10} \dfrac{1}{100} = -2$ The exponent is the logarithm

The base remains the same.

25. $25^{1/2} = 5 \Rightarrow \log_{25} 5 = \dfrac{1}{2}$ The exponent is the logarithm

The base remains the same.

26. $\log_4 16 = x \Rightarrow 16 = 4^x$ The exponent is the logarithm

The base remains the same

27. $\log_8 1 = 0 \Rightarrow 1 = 8^0$ The exponent is the logarithm

The base remains the same.

28. $\log_a x^4 y^2 z^3$

$= \log_a x^4 + \log_a y^2 + \log_a z^3$ Using the product rule

$= 4 \log_a x + 2 \log_a y + 3 \log_a z$ Using the power rule.

29. $\log_a \dfrac{x^5}{yz^2}$

$= \log_a x^5 - \log_a yz^2$ Using the quotient rule

$= \log_a x^5 - \left(\log_a y + \log_a z^2\right)$ Using the product rule

$= 5 \log_a x - \log_a y - 2 \log_a z$ Using the power rule

30. $\log \sqrt[4]{\dfrac{z^2}{x^3 y}} = \log \left(\dfrac{z^2}{x^3 y}\right)^{\frac{1}{4}}$

$= \dfrac{1}{4} \log \dfrac{z^2}{x^3 y}$ Using the power rule

$= \dfrac{1}{4}\left(\log z^2 - \log x^3 y\right)$ Using the quotient rule

$= \dfrac{1}{4}\left[\log z^2 - \left(\log x^3 + \log y\right)\right]$ Using the product rule

$= \dfrac{1}{4}\left(2 \log z - 3 \log x - \log y\right)$ Using the power rule

31. $\log_a 7 + \log_a 8 = \log_a (7 \cdot 8)$ Using the

$= \log_a 56$ product rule

32. $\log_a 72 - \log_a 12 = \log_a \dfrac{72}{12}$ Using the

$= \log_a 6$ quotient rule

33. $\dfrac{1}{2}\log a - \log b - 2\log c$

$= \log a^{\frac{1}{2}} - \log b - \log c^2$ Using the power rule

$= \log a^{\frac{1}{2}} - \left(\log b + \log c^2\right)$ Distributive law

$= \log a^{\frac{1}{2}} - \log bc^2$ Using the product rule

$= \log \dfrac{a^{\frac{1}{2}}}{bc^2}$ Using the quotient rule

34. $\dfrac{1}{3}\left[\log_a x - 2\log_a y\right]$

$= \dfrac{1}{3}\left[\log_a x - \log_a y^2\right]$ Using the power rule

$= \dfrac{1}{3}\log_a \dfrac{x}{y^2}$ Using the quotient rule

$= \log_a \left(\dfrac{x}{y^2}\right)^{\frac{1}{3}}$ Using the power rule

$= \log_a \sqrt[3]{\dfrac{x}{y^2}}$

35. $\log_m m = 1$, since $m^1 = m$

36. $\log_m 1 = 0$, since $m^0 = 1$

37. $\log_m m^{17} = 17$; the logarithm is the exponent.

38. $f(x) = 2[x-(-3)]^2 + 1$
    Vertex is $(-3, 1)$

$= \log_a 2 + \log_a 7$ (Product rule)

$= 1.8301 + 5.0999$

$= 6.93$

39. $\log_a \dfrac{2}{7} = \log_a 2 - \log_a 7$ (Quotient rule)

$= 1.8301 - 5.0999$

$= -3.2698$

40. $\log_a 28 = \log_a 2^2 \cdot 7$ (Prime factorization)

$= \log_a 2^2 + \log_a 7$ (Product rule)

$= 2\log_a 2 + \log_a 7$ (Power rule)

$= 2(1.8301) + 5.0999$

$= 8.7601$

41. $\log_a 3.5 = \log_a \dfrac{7}{2}$

$= \log_a 7 - \log_a 2$ (Quotient rule)

$= 5.0999 - 1.8301$

$= 3.2698$

42. $\log_a \sqrt{7} = \log_a 7^{\frac{1}{2}}$

$= \dfrac{1}{2}\log_a 7$ (Power rule)

$= \dfrac{1}{2} \cdot 5.0999$

$= 2.54995$

43. $\log_a \dfrac{1}{4} = \log_a 4^{-1}$

$= \log_a 2^{-2}$

$= -2\log_a 2$ (Power rule)

$= -2 \cdot 1.8301$

$= -3.6602$

44. $\log 75 \approx 1.8751$

45. $10^{1.789} \approx 61.5177$

46. $\ln 0.05 \approx -2.9957$

47. $e^{-0.98} \approx 0.3753$

48. Using common logarithms for the conversion, et $a = 10$, $b = 5$, and $m = 2$.

$\log_b m = \dfrac{\log_a m}{\log_a b}$

$\log_5 2 = \dfrac{\log_{10} 2}{\log_{10} 5}$

$\approx \dfrac{0.3010299957}{0.6989700043}$

≈ 0.4307

49. Using common logarithms for the conversion,
let $a = 10$, $b = 12$, and $m = 70$.

$$\log_b m = \frac{\log_a m}{\log_a b}$$

$$\log_{12} 70 = \frac{\log_{10} 70}{\log_{10} 12}$$

$$\approx \frac{1.84509804}{1.079181246}$$

$$\approx 1.7097$$

50. Graph: $f(x) = e^x - 1$

We find some function values, plot these
points, and draw the graph.

x	$e^x - 1$
-2	-0.9
-1	-0.6
0	0
1	1.7
2	6.4

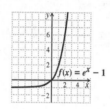

The domain is the set of real numbers and the
range is $(-1, \infty)$.

51. Graph: $g(x) = 0.6 \ln x$

We find some function values, plot these
points, and draw the graph.

x	$0.6 \ln x$
$0.$	-1.4
1	0
2	0.4
3	0.7
10	1.4

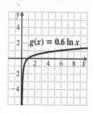

The domain is $(0, \infty)$. and the range is set of
real numbers.

52. $2^x = 32$

$2^x = 2^5$

$x = 5$

53. $3^{2x} = \dfrac{1}{9} = 9^{-1} = 3^{-2} \Rightarrow 2x = -2$

$\Rightarrow x = \dfrac{-2}{2} = -1$

54. $\log_3 x = -4 \Rightarrow x = 3^{-4} \quad x = 3^{-4}$

$$x = \frac{1}{81}$$

55. $\log_x 16 = 4 \Rightarrow 16 = x^4$

$2^4 = x^4$

$2 = x$

56. $\log x = -3 \Rightarrow x = 10^{-3}$

$$x = \frac{1}{1000}$$

57. $3 \ln x = -6$

$\ln x^3 = -6$

$x^3 = e^{-6}$

$\left(x^3\right)^{1/3} = \left(e^{-6}\right)^{1/3}$

$x = e^{-2}$

$x \approx 0.1353$

58. $4^{2x-5} = 19$

$$\log 4^{2x-5} = \log 19$$

$$(2x - 5)\log 4 = \log 19$$

$$2x \log 4 - 5 \log 4 = \log 19$$

$$2x \log 4 = \log 19 + 5 \log 4$$

$$\frac{2x \log 4}{2 \log 4} = \frac{\log 19 + 5 \log 4}{2 \log 4}$$

$$x = \frac{\log 19}{2 \log 4} + \frac{5}{2}$$

$$x \approx 3.5620$$

59. $2^x = 12 \Rightarrow \log\left(2^x\right) = \log(12)$

$\Rightarrow x \log(2) = \log(12)$

$$x = \frac{\log(12)}{\log(2)} \approx 3.585$$

60. $e^{-0.1t} = 0.03$

$\ln e^{-0.1t} = \ln 0.03$

$-0.1t = \ln 0.03$

$$t = \frac{\ln 0.03}{-0.1}$$

$$t \approx 35.0656$$

61. $2\ln(x) = -6 \Rightarrow \ln(x) = -3$

    $\Rightarrow x = e^{-3} \approx 0.049787$

62. $\log(2x - 5) = 1 \Rightarrow 2x - 5 = 10^1$

    $2x = 5$

    $x = \dfrac{5}{2}$

63. $\log_4 x - \log_4(x - 15) = 2$

    $\Rightarrow \log_4 \left[\dfrac{x}{x - 15} \right] = 2 \Rightarrow \dfrac{x}{x - 15} = 4^2 = 16$

    $\Rightarrow x = 16(x - 15) \Rightarrow x = 16x - 240$

    $\Rightarrow 15x = 240 \Rightarrow x = \dfrac{240}{15} = 16$

    The solution checks: $x = 16$.

64. $\log_3(x - 4) = 3 - \log_3(x + 4)$

    $\log_3(x - 4) + \log_3(x + 4) = 3$

    $\log_3\left(x^2 - 16\right) = 3$

    $x^2 - 16 = 3^3$

    $x^2 - 16 = 27$

    $x^2 = 43$

    $x = \pm\sqrt{43}$

    Only $\sqrt{43}$ checks and is the solution.

65. (a)

    $S(0) = 82 - 18\log(0 + 1)$

    $= 82 - 18\log 1$

    $= 82 - 18 \cdot 0 = 82$

    (b)

    $t = 6;\ 5(6) = 82 - 18\log(6 + 1)$

    $= 82 - 18\log 7 \approx 66.8$

    (c)

    Let $S(t) = 54$ and solve for t.

    $54 = 82 - 18\log(t + 1) \Rightarrow 18\log(t + 1) = 28$

    $\Rightarrow \log(t + 1) = \dfrac{28}{18} \Rightarrow t + 1 = 10^{\frac{14}{9}}$

    $\Rightarrow t + 1 \approx 35.938 \Rightarrow t \approx 34.938$

    Approximately 35 months.

66. (a) $V(t) = 900 \Rightarrow 1500(0.8)^t = 900$

    $\Rightarrow 0.8^t = \dfrac{900}{1500} = \dfrac{3}{5}$

    $\Rightarrow \ln\left(0.8^t\right) = \ln\left(\dfrac{3}{5}\right) = \ln(0.6)$

    $\Rightarrow t\ln(0.8) = \ln(0.6)$

    $\Rightarrow t = \dfrac{\ln(0.6)}{\ln(0.8)} \approx 2.29$

    The laptop will be worth \$900 in about 2.3 years.

    (b) $750 = 1500(0.8)^t$

    $\dfrac{750}{1500} = (0.8)^t$

    $\ln\dfrac{1}{2} = \ln 0.8^t$

    $\dfrac{\ln\dfrac{1}{2}}{\ln 0.8} = t$

    $3.11 \approx t$

    After about 3.1 years, the laptop will be worth half its original value.

67. For exponential growth use the formula:

    $A(t) = A_0 e^{kt}$

    (a) For 2007, $t = 0$, so $A_0 = 1.2$. To find k, use the value of A in 2012. In 2012, $t = 2012 - 2007 = 5$.

    Therefore:

    $A(5) = 2.1 \Rightarrow 1.2e^{5k} = 2.1$

    $\Rightarrow e^{5k} = \dfrac{2.1}{1.2} = 1.75 \Rightarrow \ln(e^{5k}) = \ln(1.75)$

    $\Rightarrow 5k = \ln(1.75) \Rightarrow k = \dfrac{\ln(1.75)}{5} \approx 0.112$

    $\Rightarrow A(t) = 1.2e^{0.112t}$

    (b) For 2015, $t = 2015 - 2007 = 8$.

    $A(8) = 1.2e^{0.112 \cdot 8} \approx 2.94$

    About \$2.94 billion will be spent in 2015.

(c) $A(t) = 4 \Rightarrow 1.2e^{0.112t} = 4 \Rightarrow e^{0.112t} = \dfrac{4}{1.2}$

$\Rightarrow \ln\left(e^{0.112t}\right) = \ln\left(\dfrac{4}{1.2}\right)$

$\Rightarrow 0.112t = \ln\left(\dfrac{4}{1.2}\right)$

$\Rightarrow t = \dfrac{\ln\left(\dfrac{4}{1.2}\right)}{0.112} \approx 10.75$

2007+10.75=2017.75
According to the model the amount spent on email marketing should reach $4 billion in late 2017.

(d) The doubling time only depends on k, not A_0.

$A(t) = 2A_0 \Rightarrow A_0 e^{0.112t} = 2A_0$

$\Rightarrow e^{0.112t} = 2 \Rightarrow \ln\left(e^{0.112t}\right) = \ln(2)$

$\Rightarrow 0.112t = \ln(2) \Rightarrow t = \dfrac{\ln(2)}{0.112} \approx 6.2$

According to the model the amount spent on e-mail marketing doubles approximately every 6.2 years.

68. For exponential decay use the formula:

$A(t) = A_0 e^{-kt}$

(a)The value of k is given as $13.7\% = 0.137$, and, since we are letting $t = 0$ in 2005, $A_0 = A$ in $2005 \Rightarrow A_0 = 3253$.

$\Rightarrow A(t) = 3253e^{-0.137t}$

(b) For 2012, $t = 2012 - 2005 = 7$.

$A(7) = 3253e^{-0.137 \cdot 7} \approx 1247$

In 2012, there will be 1247 spam messages per consumer.

(c) $A(t) = 100 \Rightarrow 3253e^{-0.137t} = 100$

$\Rightarrow e^{-0.137t} = \dfrac{100}{3253} \Rightarrow \ln\left(e^{-0.137t}\right) = \ln\left(\dfrac{100}{3253}\right)$

$\Rightarrow -0.137t = \ln\left(\dfrac{100}{3253}\right)$

$\Rightarrow t = -\dfrac{1}{0.137}\ln\left(\dfrac{100}{3253}\right) \approx 25.4$

2005+25.4=2030.4
According to the model the average number of spam messages per consumer should reach 100 in 2030.

69. Let $x =$ the number of years after 1995, and $f =$ the number of Hepatitis A cases in the U.S., in thousands. Then from the table, we have the points (0, 31.6), (5, 13.4), (8, 7.7), (9, 5.7), (10, 4.5), (11, 3.6), and (12, 3). Enter the data on a graphing calculator and then use the exponential regression feature to get:

$f(x) = 33.8684(0.8196)^x$

(b)
For 2010, $t = 2010 - 1995 = 15$.

$f(15) = 33.8684(0.8196)^{15} \approx 1.71$

According to the model, about 1.7 thousand cases, or 1700 cases, of Hepatitis A would occur in the U.S. in 2010.

(c)

$e^{-k} = 0.8196$

$\Rightarrow k = -\ln(0.8196) \approx 0.1989 = 19.89\%$

70. The value of the portfolio doubles in 6 years. Use this information to find k.

$2P_0 = P_0 e^{6k} \Rightarrow e^{6k} = 2$

$\Rightarrow \ln\left(e^{6k}\right) = \ln(2) \Rightarrow 6k = \ln(2)$

$\Rightarrow k = \dfrac{\ln(2)}{6} \approx 0.1155 = 11.55\%$

71. $2P_0 = 2 \cdot 7600;\ k = 4.2\% = .042,\ P_0 = 7600$.
Substitute into the exponential growth formula and solve for t.

$2 \cdot 7600 = 7600\, e^{0.042t}$

$2 = e^{0.042t}$

$\ln 2 = \ln e^{0.042t}$

$\ln 2 = 0.042t$

$\dfrac{\ln 2}{0.042} = t$

$16.5 \approx t$

$7600 will double to $13,200 in about 16.5 years.

72. If the skull has lost 34% of its carbon-14, $100\% - 34\% = 66\%$ remains. Substitute $66\% = 0.66$ into the decay formula and solve for t.

$$P(t) = P_0 e^{-0.00012t}$$

$$0.66 P_0 = P_0 e^{-0.00012t}$$

$$0.66 = e^{-0.00012t}$$

$$\ln 0.66 = \ln e^{-0.00012t}$$

$$\ln 0.66 = -0.00012t$$

$$\frac{\ln 0.66}{-0.00012} = t$$

$$3463 \approx t$$

The skull is about 3463 years-old.

73. $pH = -\log[H^+] = -\log[7.9 \cdot 10^{-6}] \approx 5.1$

74. Use $P(t) = P_0 e^{-kt}$, $k > 0$ for exponential

decay. To find k substitute $\frac{1}{2}P_0$ for $P(t)$

and 5 for t.

$$P_0 e^{-k \cdot 24,360} = \frac{1}{2}P_0 \Rightarrow e^{-24,360k} = \frac{1}{2}$$

$$\Rightarrow -24,360k = \ln\left(\frac{1}{2}\right) = -\ln(2)$$

$$\Rightarrow k = \frac{\ln(2)}{24,360} \approx 2.8454 \times 10^{-5}$$

$$\Rightarrow P(t) = P_0 e^{-2.8454 \times 10^{-5} t}$$

If 90% has decayed then

$$P(t) = (1 - 0.90)P_0 = 0.1P_0$$

$$P(t) = 0.1P_0 \Rightarrow P_0 e^{-2.8454 \times 10^{-5} \cdot t} = 0.1P_0$$

$$\Rightarrow e^{-2.8454 \times 10^{-5} \cdot t} = 0.1$$

$$\Rightarrow -2.8454 \times 10^{-5} \cdot t = \ln(0.1)$$

$$\Rightarrow t = -\frac{\ln(0.1)}{2.8454 \times 10^{-5}} \approx 80,923$$

It takes approximately 80.923 years for Plutonium-239 to loose 90% of its radioactivity.

75. Let $I = 2.5 \cdot 10^{-1} = 0.25$ W/m^2 and solve the given equation.

$$L = 10 \cdot \log \frac{0.25}{10^{-12}} \approx 114$$

This sound level is approximately 114 dB.

76. *Thinking and Writing Exercise.* Negative numbers do not have logarithms because logarithm bases are positive, and there is no exponent to which a positive number can be raised to yield a negative number.

77. *Thinking and Writing Exercise.* Taking the logarithm on each side of an equation produces an equivalent equation because the logarithm function is one-to-one. If two quantities are equal, their logarithms must be equal, and if the logarithms of two quantities are equal, the quantities must be the same.

78. $\ln(\ln x) = 3 \Rightarrow \ln x = e^3$

$\ln x = e^3 \Rightarrow x = e^{e^3}$

79. $2^{x^2 + 4x} = \frac{1}{8}$

$$2^{x^2 + 4x} = 2^{-3}$$

$$x^2 + 4x = -3$$

$$x^2 + 4x + 3 = 0$$

$$(x + 3)(x + 1) = 0$$

$$x = -3 \quad \text{or} \quad x = -1$$

Both numbers check and are the solutions.

80. $5^{x+y} = 25 \Rightarrow 5^{x+y} = 5^2$

$x + y = 2$

$2^{2x-y} = 64 \Rightarrow 2^{2x-y} = 2^6$

$2x - y = 6$

Solve the system: $x + y = 2$

$2x - y = 6$

Add the equations: $3x = 8$

$$x = \frac{8}{3}$$

$$\frac{8}{3} + y = 2$$

$$y = \frac{-2}{3}$$

The solution is $\left(\frac{8}{3}, \frac{-2}{3}\right)$.

81. $P_0 e^{5.32k} = 2P_0 \Rightarrow e^{5.32k} = 2$

$$\Rightarrow \ln\left(e^{5.32k}\right) = \ln(2) \Rightarrow 5.32k = \ln(2)$$

$$\Rightarrow k = \frac{\ln(2)}{5.32} \approx 0.1303 = 13.03\%$$

Chapter 12 Test

1. $f(x) = x + x^2, g(x) = 2x + 1$

$(f \circ g)(x) = f(g(x)) = f(2x+1)$
$= (2x+1) + (2x+1)^2$
$= 2x + 1 + 4x^2 + 4x + 1$
$= 4x^2 + 6x + 2$

$(g \circ f)(x) = g(f(x)) = g(x + x^2)$
$= 2(x + x^2) + 1$
$= 2x + 2x^2 + 1$
$= 2x^2 + 2x + 1$

2. Possible answer. $h(x)$ is the multiplicative inverse of $2x^2 + 1$.

$f(x) = \dfrac{1}{x}$ and $g(x) = 2x^2 + 1$

3. No. Example: $f(x) = 4$ when $x = 7$ and when $x = -1$. Also note, the graph of the function does not pass the horizontal line test.

4. $f(x) = 3x + 4$

$f(x)$ is one-to-one.

Replace $f(x)$ by y: $y = 3x + 4$

Interchange x and y: $x = 3y + 4$

Replace y by $f^{-1}(x)$: $f^{-1}(x) = \dfrac{x-4}{3}$

5. $g(x) = (x+1)^3$

$g(x)$ is one-to-one.

Replace $g(x)$ by y: $y = (x+1)^3$

Interchange x and y: $x = (y+1)^3$

Solve for y: $\sqrt[3]{x} = y + 1$

$\sqrt[3]{x} - 1 = y$

Replace y by $g^{-1}(x)$: $g^{-1}(x) = \sqrt[3]{x} - 1$

6. Graph: $f(x) = 2^x - 3$

Choose valves of x and determine the corresponding function values. Plot these points and connect them with a smooth

$f(-2) = 2^{-2} - 3 = \dfrac{1}{4} - 3 = -2\dfrac{3}{4}$

$f(-1) = 2^{-1} - 3 = \dfrac{1}{2} - 3 = -2\dfrac{1}{2}$

curve. $f(0) = 2^0 - 3 = 1 - 3 = -2$

$f(1) = 2^1 - 3 = 2 - 3 = -1$

$f(2) = 2^2 - 3 = 4 - 3 = 1$

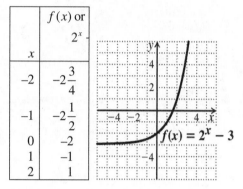

x	$f(x)$ or 2^x
-2	$-2\dfrac{3}{4}$
-1	$-2\dfrac{1}{2}$
0	-2
1	-1
2	1

7. Graph: $g(x) = \log_7 x$

The equation $y = \log_7 x$ is equivalent to $7^y = x$. Choose values of y and determine corresponding x values. Plot the points and connect them with a smooth curve.

For $y = -2$, $x = 7^{-2} = \dfrac{1}{49}$

For $y = -2$, $x = 7^{-1} = \dfrac{1}{7}$

For $y = 0$, $x = 7^0 = 1$

For $y = 1$, $x = 7^1 = 7$

For $y = 2$, $x = 7^2 = 49$

x, or 7^y	y
$\dfrac{1}{49}$	-2
$\dfrac{1}{7}$	-1
1	0
7	1
49	2

$g(x) = \log_7 x$

8. $\log_5 125$

Since $125 = 5^3$, $\log_5 5^3 = 3$

9. $\log_{100} 10$

Since $10 = 100^{\frac{1}{2}}$, $\log_{100} 100^{\frac{1}{2}} = \dfrac{1}{2}$

10. $3^{\log_3 18} = 18$

11. $\log_n (n) = 1$

12. $\log_c (1) = 0$

13. $\log_a (a^{19}) = 19 \log_a (a) = 19 \cdot 1 = 19$

14. $5^{-4} = \dfrac{1}{625} \Rightarrow \log_5 (5^{-4}) = \log_5 \left(\dfrac{1}{625} \right)$

$\Rightarrow (-4) \log_5 (5) = \log_5 \left(\dfrac{1}{625} \right)$

$\Rightarrow (-4) \cdot 1 = \log_5 \left(\dfrac{1}{625} \right)$

$\Rightarrow -4 = \log_5 \left(\dfrac{1}{625} \right)$

15. $m = \log_2 \left(\dfrac{1}{2} \right) \Rightarrow 2^m = 2^{\log_2 \left(\frac{1}{2} \right)} \Rightarrow 2^m = \dfrac{1}{2}$

16. $\log \dfrac{a^3 b^{1/2}}{c^2}$

$= \log (a^3 b^{1/2}) - \log c^2$ Using the quotient rule

$= \log a^3 + \log b^{1/2} - \log c^2$ Using the product rule

$= 3 \log a + \dfrac{1}{2} \log b - 2 \log c$ Using the power rule

17. $\dfrac{1}{3} \log_a x + 2 \log_a z$

$= \log_a x^{\frac{1}{3}} + \log_a z^2$ Using the power rule

$= \log_a x^{\frac{1}{3}} z^2$, or $\log_a z^2 \sqrt[3]{x}$ Using the product rule

18. $\log_a 14 = \log_a (2 \cdot 7)$

$= \log_a 2 + \log_2 7$ (Product Rule)

$= 0.301 + 0.845$

$= 1.146$

19. $\log_a 3 = \log_a \dfrac{6}{2}$

$= \log_a 6 - \log_a 2$ (Quotient Rule)

$= 0.778 - 0.301$

$= 0.477$

20. $\log_a 16 = \log_a 2^4$

$= 4 \log_a 2$ (Power Rule)

$= 4 \cdot 0.301$

$= 1.204$

21. $\log 12.3 \approx 1.0899$

22. $10^{-8} \approx 0.1585$

23. $\ln 0.4 \approx -0.9163$

24. $e^{4.8} \approx 121.5104$

25. Using common logarithms for the conversion, let $a = 10$, $b = 3$, and $m = 14$.

$\log_b m = \dfrac{\log_a m}{\log_a b}$

$\log_3 14 = \dfrac{\log_{10} 14}{\log_{10} 3}$

$\approx \dfrac{1.146128036}{0.4771212547}$

≈ 2.4022

26. Graph: $f(x) = e^x + 3$

We find some function values, plot these points, and draw the graph.

x	$f(x)$, or $e^x + 3$
-2	3.1
-1	3.4
0	4
1	5.7
2	10.4

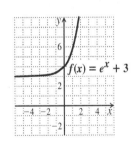

The domain is the set of real numbers and the range is $(3, \infty)$.

27. Graph: $g(x) = \ln(x-4)$

We find some function values, plot these points, and draw the graph

x	$f(x)$, or $\ln(x-4)$
4.1	−2.3
5	0
6	0.7
10	2.3

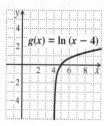

The domain is $(4, \infty)$ and the range is the set of real numbers.

28. $2^x = \dfrac{1}{32}$

$2^x = 2^{-5}$

$x = -5$

29. $\log_4 x = \dfrac{1}{2} \Rightarrow x = 4^{\frac{1}{2}} = (2^2)^{\frac{1}{2}} = 2$

30. $\log x = 4 \Rightarrow x = 10^4 = 10{,}000$

31. $5^{4-3x} = 87$

$\log 5^{4-3x} = \log 87$

$(4-3x)\log 5 = \log 87$

$4\log 5 - 3x\log 5 = \log 87$

$-3x\log 5 = \log 87 - 4\log 5$

$-\dfrac{1}{3}[-3x\log 5 = \log 87 - \log 5^4]$

$x\log 5 = -\dfrac{1}{3}(\log 87 - \log 5^4)$

$\dfrac{x\log 5}{\log 5} = \dfrac{-\dfrac{1}{3}(\log 87 - \log 5^4)}{\log 5}$

$x = -\dfrac{1}{3}\left(\dfrac{\log 87}{\log 5} - \dfrac{\log 5^4}{\log 5}\right)$

$x = -\dfrac{1}{3}\left(\dfrac{\log 87}{\log 5} - 4\right)$

$x \approx 0.4084$

32. $7^x = 1.2$

$\log 7^x = \log 1.2$

$x\log 7 = \log 1.2$

$x = \dfrac{\log 1.2}{\log 7}$

$x \approx 0.0937$

33. $\ln x = 3 \Rightarrow x = e^3 \approx 20.0855$

34. $\log(x-3) + \log(x+1) = \log 5$

$\log[(x-3)(x+1)] = \log 5$

$\log(x^2 - 2x - 3) = \log 5$

$x^2 - 2x - 3 = 5$

$x^2 - 2x - 8 = 0$

$(x-4)(x+2) = 0$

$x = 4 \ \text{ or } \ x = -2$

Only the number 4 checks, and it is the solution

35.   (a) Find R when $P = 383$

$R = 0.37\ln P + 0.05$

$= 0.37\ln(383) + 0.05 \approx 2.25$

The average walking speed is approximately 2.25 ft/sec.

(b) Find P when $R = 3$

$R = 0.37\ln P + 0.05$

$\Rightarrow 0.37\ln(P) = R - 0.05$

$\Rightarrow \ln(P) = \dfrac{R - 0.05}{0.37}$

$\Rightarrow P = e^{\frac{R-0.05}{0.37}} = e^{\frac{3-0.05}{0.37}} \approx 2901$

According to the model, the population of San Diego is approximately 2,901,000.

36.   (a) Let $t = 0$ in 2009. Then $P_0 = 8.2$ million.

The exponential growth rate is given as $k = 0.052\% = 0.00052$. Substitute this information into the formula for exponential growth.

$P(t) = P_0 e^{kt} = 8.2 e^{0.00052t}$

(b) In 2020, $t = 2020 - 2009 = 11$

$P(11) = 8.2 e^{0.00052 \cdot 11} \approx 8.247$

Population = 8,247,000.

In 2050, $t = 2050 - 2009 = 41$

$P(11) = 8.2 e^{0.00052 \cdot 41} \approx 8.3767$

Population = 8,376,700

(c) $P(t) = 9 \Rightarrow 8.2e^{0.00052t} = 9$

$\Rightarrow e^{0.00052t} = \dfrac{9}{8.2} \Rightarrow \ln\left(e^{0.00052t}\right) = \ln\left(\dfrac{9}{8.2}\right)$

$\Rightarrow 0.00052t = \ln\left(\dfrac{9}{8.2}\right)$

$\Rightarrow t = \dfrac{1}{0.00052}\ln\left(\dfrac{9}{8.2}\right) \approx 179$

2009+179=2188

According to the model, the population will reach 9 million in about 2188.

(d) $P(t) = 2P_0 \Rightarrow 8.2e^{0.00052t} = 2 \cdot 8.2$

$\Rightarrow e^{0.00052t} = 2 \Rightarrow \ln\left(e^{0.00052t}\right) = \ln(2)$

$\Rightarrow 0.00052t = \ln(2)$

$\Rightarrow t = \dfrac{\ln(2)}{0.00052} \approx 1333$

The doubling time is approximately 1333 years.

37. (a) Let $t = 0$ in 2001. Then $P_0 = 21,855$.

Therefore, $P(t) = P_0 e^{kt} = 21,855e^{kt}$. In 2010, $t = 2010 - 2001 = 9$. To find k, set $P(9) = 35,600$ and solve for k.

$P(9) = 35,600 \Rightarrow 21,855e^{9k} = 35,600$

$\Rightarrow e^{9k} = \dfrac{35,600}{21,855} = \dfrac{7120}{4371}$

$\Rightarrow \ln\left(e^{9k}\right) = \ln\left(\dfrac{7120}{4371}\right) \Rightarrow 9k = \ln\left(\dfrac{7120}{4371}\right)$

$\Rightarrow k = \dfrac{1}{9}\ln\left(\dfrac{7120}{4371}\right) \approx 0.0542$

$\Rightarrow P(t) = 21,855e^{0.0542t}$

(b) In 2015, $t = 2015 - 2001 = 14$

$P(14) = 21,855e^{0.0542 \cdot 14} \approx 46,676$

Tuition = \$46,676.

(c) $P(t) = 50,000 \Rightarrow 21,855e^{0.0542t} = 50,000$

$\Rightarrow e^{0.0542t} = \dfrac{50,000}{21,855} = \dfrac{10,000}{4371}$

$\Rightarrow \ln\left(e^{0.0542t}\right) = \ln\left(\dfrac{10,000}{4371}\right)$

$\Rightarrow 0.0542t = \ln\left(\dfrac{10,000}{4371}\right)$

$\Rightarrow t = \dfrac{1}{0.0542}\ln\left(\dfrac{10,000}{4371}\right) \approx 15.27$

2001+15.27=2016.27

According to the model, tuition will reach \$50,000 in 2016.

38. (a) Let x = the number of years after 2008 and $f(x)$ = the number of Kindle sales, in thousands. Then from the table, we have the points (0, 500), (1, 1027), and (2, 3533). Enter the data on a graphing calculator and then use the exponential regression feature to get:

$f(x) = 458.8188(2.6582)^x$

(b) For 2012, $x = 2012 - 2008 = 4$.

$f(4) = 458.8188(2.6582)^4 \approx 22,908$

According to the model, Kindle sales should reach approximately 22,908,000 in 2012.

39. P_0 is the initial investment, so $2P_0$ would be double that amount. $t = 15$. Use this information in the exponential growth formula to solve for k, the interest rate.

$2P_0 = P_0 \cdot e^{k \cdot 15}$

$2 = e^{15k}$

$\ln 2 = \ln e^{15k}$

$\ln 2 = 15k$

$\dfrac{\ln 2}{15} = k$

$0.046 \approx k$

$0.046 = 4.6\%$

The interest rate is 4.6%.

40. If 43% has been lost, $100\% - 43\%$, or 57% of the initial amount, P_0 remains.

$P(t) = P_0 e^{-0.00012t}$

$0.57P_0 = P_0 e^{-0.00012t}$

$0.57 = e^{-0.00012t}$

$\ln 0.57 = \ln e^{-0.00012t}$

$\ln 0.57 = -0.00012t$

$-\dfrac{\ln 0.57}{0.00012} = t$

$4684 \approx t$

The animal bone is approximately 4684 years-old.

41. $L = 10\log\left(\dfrac{I}{I_0}\right) = 10\log\left(\dfrac{I}{10^{-12}}\right)$

$L = 140 \Rightarrow 10\log\left(\dfrac{I}{10^{-12}}\right) = 140$

$\Rightarrow \log\left(\dfrac{I}{10^{-12}}\right) = \dfrac{140}{10} = 14$

$\dfrac{I}{10^{-12}} = 10^{14} \Rightarrow I = 10^{14} \cdot 10^{-12} = 10^{2}$

The intensity is 10^2 W/m^2.

42. Let $[H+] = 1.0 \times 10^{-7}$ and solve for pH.

$\text{pH} = -\log\left[H^+\right]$

$\text{pH} = -\log\left(1.0 \times 10^{-7}\right)$

$\text{pH} = -(-7)$

$\text{pH} = 7$

The pH is 7.

43. $\log_5 |2x - 7| = 4 \Rightarrow |2x - 7| = 5^4$

$\qquad |2x - 7| = 625$

$\qquad 2x - 7 = -625 \;\; \text{or} \;\; 2x - 7 = 625$

$\qquad\qquad 2x = -618 \;\; \text{or} \;\; 2x = 632$

$\qquad\qquad\quad x = -309 \;\; \text{or} \;\; x = 316$

Both numbers are the solutions.

44. Express $\log_a \dfrac{\sqrt[3]{x^2 z}}{\sqrt[3]{y^2 z - 1}}$ using

the individual logarithms of x, y, and z.

$\log_a \dfrac{\sqrt[3]{x^2 z}}{\sqrt[3]{y^2 z^{-1}}}$

$= \log_a \sqrt[3]{\dfrac{x^2 z}{y^2 z^{-1}}}$

$= \log_a \sqrt[3]{\dfrac{x^2 z^2}{y^2}}$

$= \log_a \left(\dfrac{x^2 z^2}{y^2}\right)^{1/3}$

$= \dfrac{1}{3}\log_a \left(\dfrac{x^2 z^2}{y^2}\right)$ Power Rule

$= \dfrac{1}{3}(\log_a x^2 z^2 - \log_a y^2)$ Quotient Rule

$= \dfrac{1}{3}(\log_a x^2 + \log_a z^2 - \log_a y^2)$ Product Rule

$= \dfrac{1}{3}(2\log_a x + 2\log_a z - 2\log_a y)$ Power Rule

Substituting the given information, we have

$= \dfrac{1}{3}[2(2) + 2(4) - 2(3)]$

$= \dfrac{1}{3}[4 + 8 - 6]$

$= \dfrac{1}{3} \cdot 6$

$= 2$

Chapters 1 – 12

Cumulative Review

1. $\dfrac{x^0 + y}{-z} = \dfrac{6^0 + 9}{-(-5)} = \dfrac{1+9}{5} = \dfrac{10}{5} = 2$

2. $(-2x^2 y^{-3})^{-4} = (-2)^{-4}(x^2)^{-4}(y^{-3})^{-4}$

   $\qquad = \dfrac{1}{16} \cdot x^{-8} \cdot y^{12}$

   $\qquad = \dfrac{y^{12}}{16x^8}$

3. $(-5x^4 y^{-3} z^2)(-4x^2 y^2)$

   $\qquad = (-5)(-4)x^{4+2} y^{-3+2} z^2$

   $\qquad = 20x^6 y^{-1} z^2$

   $\qquad = \dfrac{20x^6 z^2}{y}$

4. $\dfrac{3x^4 y^6 z^{-2}}{-9x^4 y^2 z^3} = \dfrac{3}{-9} x^{4-4} y^{6-2} z^{-2-3}$

   $\qquad = \dfrac{-1}{3} \cdot x^0 y^4 z^{-5} \qquad (x^0 = 1)$

   $\qquad = -\dfrac{y^4}{3z^5}$

5. $(1.5 \times 10^{-3})(4.2 \times 10^{-12})$

   $= (1.5 \cdot 4.2) \cdot (10^{-3} \times 10^{-12})$

   $= 6.3 \times 10^{-15}$

6. $3^3 + 2^2 - (32 \div 4 - 16 \div 8)$

   $\qquad = 27 + 4 - (8 - 2)$

   $\qquad = 27 + 4 - 6 = 25$

7. $3(2x - 3) = 9 - 5(2 - x)$

   $\quad 6x - 9 = 9 - 10 + 5x$

   $\quad 6x - 5x = -1 + 9$

   $\qquad x = 8$

8. $\quad (1)\ 4x - 3y = 15$

   $\quad (2)\ 3x + 5y = 4$

   Add: $5(1) + 3(2)$

   $20x - 15y = 75$

   $\underline{9x + 15y = 12}$

   $\qquad 29x = 87$

   $\qquad x = 3$

   By substitution,

   $4 \cdot 3 - 3y = 15$

   $\qquad -3y = 3$

   $\qquad y = -1$

   The solution is $(3, -1)$.

9. $\quad x + y - 3z = -1 \ (1)$

   $\quad 2x - y + z = 4 \ (2)$

   $\quad -x - y + z = 1 \ (3)$

   Add: $(1) + (2)$ to eliminate y.

   $\quad x\ +y\ -3z = -1$

   $\underline{\quad 2x\ -y\ +z = \ \ 4}$

   $\quad 3x \qquad -2z = \ \ 3$

   Add: $(1) + (3)$ to eliminate y.

   $\quad x\ +y\ -3z = -1$

   $\underline{\quad -x\ -y\ +z = \ \ 1}$

   $\qquad \quad -2z = \ \ 0$

   $\qquad \quad\ z = \ \ 0$

   Substitute to determine x.

   $3x - 2 \cdot 0 = 3$

   $\qquad 3x = 3$

   $\qquad x = 1$

   Substitute to determine y.

   $1 + y - 3 \cdot 0 = -1$

   $\qquad y = -2$

   The solution is $(1, -2, 0)$

10. $\qquad x(x - 3) = 70$

    $\qquad x^2 - 3x = 70$

    $\qquad x^2 - 3x - 70 = 0$

    $\qquad (x - 10)(x + 7) = 0$

    $\qquad x = 10 \text{ or } x = -7$

11. $\dfrac{7}{x^2-5x}-\dfrac{2}{x-5}=\dfrac{4}{x}$

[LCD is $x^2-5x = x(x-5)$]

Note: $x \neq 0,5$

$x(x-5)\cdot\left(\dfrac{7}{x^2-5x}-\dfrac{2}{x-5}\right)=x(x-5)\cdot\dfrac{4}{x}$

$7-2x = 4(x-5)$

$\Rightarrow 7-2x = 4x-20 \Rightarrow 27 = 6x$

$\Rightarrow x = \dfrac{27}{6} = \dfrac{9}{2}$

12. $\sqrt{4-5x} = 2x-1$

$\left(\sqrt{4-5x}\right)^2 = (2x-1)^2$

$4-5x = 4x^2-4x+1$

$0 = 4x^2+x-3$

$0 = (4x-3)(x+1)$

$x = \dfrac{3}{4} \ \text{ or } \ x = -1$

Only the number $\dfrac{3}{4}$ checks in the original

equation; $\dfrac{3}{4}$ is the solution.

13. $\sqrt[3]{2x} = 1$

$\left(\sqrt[3]{2x}\right)^3 = 1^3$

$2x = 1$

$x = \dfrac{1}{2}$

14. $3x^2+48 = 0$

$3(x^2+16) = 0$

$x^2+16 = 0$

$x^2 = -16$

$x = \pm\sqrt{-16}$

$x = \pm 4i$

15. $x^4-13x^2+36 = 0$

$(x^2-9)(x^2-4) = 0$

$(x-3)(x+3)(x-2)(x+2) = 0$

$x = 3 \text{ or } x = -3 \text{ or } x = 2 \text{ or } x = -2$

The solutions are $\pm 3, \pm 2$.

16. $\log_x 81 = 2$

$x^{\log_x 81} = x^2$

$81 = x^2$

$x = \pm 9$

However, we do not use negative numbers for base of logarithms. Therefore $x = 9$.

17. $3^{5x} = 7$

$\log 3^{5x} = \log 7$

$5x\log 3 = \log 7$

$\dfrac{5x\log 3}{5\log 3} = \dfrac{\log 7}{5\log 3}$

$x = \dfrac{\log 7}{5\log 3}$

≈ 0.3542

18. $\ln x - \ln(x-8) = 1$

$\ln\left(\dfrac{x}{x-8}\right) = 1$

$e^{\ln\left(\frac{x}{x-8}\right)} = e^1$

$\dfrac{x}{x-8} = e$

$(x-8)\cdot\dfrac{x}{x-8} = (x-8)\cdot e$

$x = ex-8e$

$8e = ex-x$

$8e = x(e-1)$

$x = \dfrac{8e}{e-1} \approx 12.6558$

19. $x^2+4x > 5$

$x^2+4x-5 > 0$

$(x+5)(x-1) > 0$

Solve the related equation.

$(x+5)(x-1) = 0$

$x = -5 \text{ or } x = 1$

Choose a number from each interval

$(-\infty,-5), (-5,1), \text{ and } (1,\infty)$ to determine

the solution.

The solution is $(-\infty,-5)\cup(1,\infty)$, or

$\{x\,|\,x<-5 \ or \ x>1\}$, since the numbers

chosen from those intervals make

the original inequality true.

20. $f(x) = x^2 + 6x;\ f(a) = 11$

$\quad f(a) = 11 \Rightarrow a^2 + 6a = 11 \Rightarrow x^2 + 6x - 11 = 0$

$\quad a = \dfrac{-B \pm \sqrt{B^2 - 4AC}}{2A}$

$\quad\quad = \dfrac{-6 \pm \sqrt{6^2 - 4\cdot 1\cdot(-11)}}{2\cdot 1}$

$\quad\quad = \dfrac{-6 \pm \sqrt{36 + 44}}{2}$

$\quad\quad = \dfrac{-6 \pm \sqrt{80}}{2} = \dfrac{-6 \pm 4\sqrt{5}}{2} = -3 \pm 2\sqrt{5}$

21. $f(x) = |2x - 3|;\ f(x) \geq 7$

$\quad f(x) \geq 7 \Rightarrow |2x - 3| \geq 7$

$\quad\quad 2x - 3 \leq -7 \quad \text{or} \quad 2x - 3 \geq 7$

$\quad\quad\quad 2x \leq -4 \quad \text{or} \quad\quad 2x \geq 10$

$\quad\quad\quad\quad x \leq -2 \quad \text{or} \quad\quad\quad x \geq 5$

$\quad \{x \mid x \leq -2 \text{ or } x \geq 5\}, \text{ or } (-\infty,\, -2] \cup [5,\, \infty)$

22. $D = \dfrac{ab}{b+a} \quad [LCD = b + a]$

$\quad (b+a)D = (b+a)\dfrac{ab}{b+a}$

$\quad Db + Da = ab$

$\quad\quad Db = ab - Da$

$\quad\quad Db = a(b - D)$

$\quad\quad \dfrac{Db}{b - D} = a$

23. $d = ax^2 + vx \Rightarrow ax^2 + vx - d = 0$

$\quad x = \dfrac{-B \pm \sqrt{B^2 - 4AC}}{2A}$

$\quad\quad = \dfrac{-v \pm \sqrt{v^2 - 4\cdot a\cdot(-d)}}{2\cdot a} = \dfrac{-v \pm \sqrt{v^2 + 4ad}}{2a}$

24. $f(x) = \dfrac{x+4}{3x^2 - 5x - 2}$

The domain in the set of real numbers, excluding values of x which make the denominator zero.

$\quad 3x^2 - 5x - 2 = 0$

$\quad (3x + 1)(x - 2) = 0$

$\quad\quad x = -\dfrac{1}{3} \quad \text{or} \quad x = 2$

$\{x \mid x \text{ is a real number}, x \neq -\tfrac{1}{3} \text{ and } x \neq 2\}$

25. $\quad (5p^2q^3 + 6pq - p^2 + p)$

$\quad\quad - (2p^2q^3 + p^2 - 5pq - 9)$

$\quad = 5p^2q^3 + 6pq - p^2 + p$

$\quad\quad - 2p^2q^3 - p^2 + 5pq + 9$

$\quad = 5p^2q^3 - 2p^2q^3 - p^2 - p^2$

$\quad\quad + 6pq + 5pq + p + 9$

$\quad = 3p^2q^3 - 2p^2 + 11pq + p + 9$

26. $(3x^2 - z^3)^2 = (3x^2)^2 - 2(3x^2)(z^3) + (z^3)^2$

$\quad\quad\quad\quad\quad = 9x^4 - 6x^2z^3 + z^6$

27. $\dfrac{1 + \frac{3}{x}}{x - 1 - \frac{12}{x}} \quad [LCD = x]$

$\quad \dfrac{x}{x} \cdot \dfrac{1 + \frac{3}{x}}{x - 1 - \frac{12}{x}} = \dfrac{x + 3}{x^2 - x - 12}$

$\quad\quad\quad = \dfrac{\cancel{x + 3}}{(x - 4)(\cancel{x + 3})}$

$\quad\quad\quad = \dfrac{1}{x - 4} \text{ for } x \neq \{-3, 0, 4\}$

28. $\dfrac{a^2 - a - 6}{a^3 - 27} \cdot \dfrac{a^2 + 3a + 9}{6}$

$\quad = \dfrac{(a - 3)(a + 2)}{(a - 3)(a^2 + 3a + 9)} \cdot \dfrac{a^2 + 3a + 9}{6}$

$\quad = \dfrac{(a - 3)(a + 2)(a^2 + 3a + 9)}{(a - 3)(a^2 + 3a + 9)\cdot 6}$

$\quad = \dfrac{(\cancel{a - 3})(a + 2)(\cancel{a^2 + 3a + 9})}{(\cancel{a - 3})(\cancel{a^2 + 3a + 9})\cdot 6}$

$\quad = \dfrac{a + 2}{6} \text{ for } a \neq 3$

29. $\dfrac{3}{x+6} - \dfrac{2}{x^2-36} + \dfrac{4}{x-6}$

[LCD is $x^2 - 36 = (x+6)(x-6)$]

$= \dfrac{x-6}{x-6} \cdot \dfrac{3}{x+6} - \dfrac{2}{(x+6)(x-6)}$

$+ \dfrac{x+6}{x+6} \cdot \dfrac{4}{x-6}$

$= \dfrac{3x-18-2+4x+24}{(x+6)(x-6)}$

$= \dfrac{7x+4}{(x+6)(x-6)}$

30. $\dfrac{\sqrt[3]{24xy^8}}{\sqrt[3]{3xy}} = \sqrt[3]{\dfrac{24xy^8}{3xy}} = \sqrt[3]{8y^7} = \sqrt[3]{2^3 y^6 \cdot y}$

$= 2y^2 \sqrt[3]{y}$

31. $\sqrt{x+5}\sqrt[5]{x+5} = (x+5)^{\frac{1}{2}}(x+5)^{\frac{1}{5}}$

$= (x+5)^{\frac{1}{2}+\frac{1}{5}} = (x+5)^{\frac{7}{10}}$, or $\sqrt[10]{(x+5)^7}$

32. $(2-i\sqrt{3})(6+i\sqrt{3})$

$= 12 + 2i\sqrt{3} - 6i\sqrt{3} - 3i^2$

$= 12 - 4i\sqrt{3} - 3(-1)$　$(i^2 = -1)$

$= 15 - 4i\sqrt{3}$

33. Since the divisor is of the form $x - c$. we can use synthetic division.

$$
\begin{array}{r|rrrrr}
3 & 1 & -8 & 15 & 1 & -3 \\
  &   & 3\cdot1 & 3\cdot(-5) & 3\cdot0 & 3\cdot1 \\
\hline
  & 1 & -5 & 0 & 1 & 0 \\
\end{array}
$$

$\Rightarrow (x^4 - 8x^3 + 15x^2 + x - 3) \div (x-3)$

$= x^3 - 5x^2 + 1$

34. $xy + 2xz - xw$

$= x \cdot y + x \cdot 2z - x \cdot w$

$= x(y + 2z - w)$

35. $6x^2 + 8xy - 8y^2$

$= 2(3x^2 + 4xy - 4y^2)$

$= 2(3x - 2y)(x + 2y)$

36. $x^4 - 4x^3 + 7x - 28$

$= (x^4 - 4x^3) + (7x - 28)$

$= x^3(x-4) + 7(x-4)$

$= (x^3 + 7)(x-4)$

37. $2m^2 + 12mn + 18n^2$

$= 2(m^2 + 6mn + 9n^2)$

$= 2(m + 3n)^2$

38. $x^4 - 16y^4 = (x^2)^2 - (4y^2)^2$

$= (x^2 + 4y^2)(x^2 - 4y^2)$

$= (x^2 + 4y^2)[(x)^2 - (2y)^2]$

$= (x^2 + 4y^2)(x + 2y)(x - 2y)$

39. $\dfrac{3 - \sqrt{y}}{2 - \sqrt{y}}$

[The conjugate of the denominator is $2 + \sqrt{y}$]

$\dfrac{3 - \sqrt{y}}{2 - \sqrt{y}} \cdot \dfrac{2 + \sqrt{y}}{2 + \sqrt{y}} = \dfrac{6 + \sqrt{y} - y}{4 - y}$

40. $f(x) = 9 - 2x$

$f(x)$ is one-to-one.

Replace $f(x)$ by y: $y = 9 - 2x$

Interchange x & y: $x = 9 - 2y$

Solve for y: $\dfrac{x-9}{-2} = y$,

or $\dfrac{9-x}{2} = y$

Replace y with $f^{-1}(x)$: $f^{-1}(x) = \dfrac{9-x}{2}$

41. Determine the slope of the line which contains $(0, -8)$ and $(-1, 2)$. Since $(0, -8)$ is the y-intercept use $y = mx + b$.

$m = \dfrac{y_2 - y_1}{x_2 - x_1} = \dfrac{2 - (-8)}{-1 - 0} = -10$

$y = -10x - 8$

42. The slope of $2x + y = 6$ is

$m = \dfrac{-A}{B} = \dfrac{-2}{1} = -2$. The slope of the line

perpendicular to $2x + y = 6$ is $\dfrac{1}{2}$, since

$l_1 \perp l_2 \Rightarrow m_1 \cdot m_2 = -1$. Since $b = 5$,

$mx + b = \dfrac{1}{2}x + 5$. The equation of the line is

$y = \dfrac{1}{2}x + 5$.

43. Graph: $5x = 15 + 3y$

Determine at least two points, plot these points, and graph the line.

x	y
0	–5
3	0

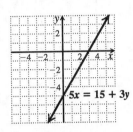

44. Graph: $y = \log_3 x \Rightarrow 3^y = x$

Choose values of y to determine corresponding values of x. Plot these points and connect with a smooth curve.

x_1 or 3^y	y
$3^{-2} = \dfrac{1}{9}$	–2
$3^{-1} = \dfrac{1}{3}$	–1
$3^0 = 1$	0
$3^1 = 3$	1
$3^2 = 9$	2

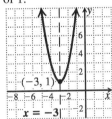

45. $-2x - 3y \le 12$

Graph the boundary line: $-2x - 3y = 12$ as a solid line. The point, $(0, 0)$ can be used to determine the correct half-plane to shade.
$-2(0) - 3(0) \le 12$?

$0 \le 12$ True

Shade the half-plane which includes $(0,0)$.

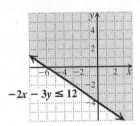

46. $f(x) = 2x^2 + 12x + 19$

(a) The vertex is at:

$-\dfrac{b}{2a} = -\dfrac{12}{2 \cdot 2} = -3,$

$f(-3) = 2(-3)^2 + 12(-3) + 19$

$= 2 \cdot 9 - 39 + 19 = 1$

Vertex: $\left(-\dfrac{b}{2a}, f\left(-\dfrac{b}{2a}\right)\right) = (-3, 1)$

(b) Axis of symmetry: $x = -3$

(c) The parabola open upward. Therefore the function takes on a minimum at its vertex of 1.

47. Graph: $f(x) = 2e^x$

Using values of x, determine ordered pairs of $f(x)$; plot these points and connect with a smooth curve.

x	$f(x)$, or $2e^x$
-2	0.3
-1	0.7
0	2
1	5.4
2	14.8

The domain is the set of real numbers and the range is $(0, \infty)$.

48. $3\log x - \dfrac{1}{2}\log y - 2\log z$

$= \log x^3 - \log y^{1/2} - \log z^2$ (Power Rule)

$= \log x^3 - (\log y^{1/2} + \log z^2)$

$= \log x^3 - \log y^{1/2} z^2$ (Product Rule)

$= \log\left(\dfrac{x^3}{y^{1/2} z^2}\right)$ (Quotient Rule)

49. c

50. b

51. a

52. d

53. Let x = the total volume of water carried by the Colorado River, in millions of acre-feet. Then the amount diverted for agricultural use is $0.90x$. Then:

$0.10x = 1.5 \Rightarrow x = \dfrac{1.5}{0.10} = 15$

$\Rightarrow 0.90x = 0.9 \cdot 15 = 13.5$

13.5 million acre-feet are diverted each year.

54. (a) For exponential growth use the function:

$D(t) = D_0 e^{kt} = 15e^{kt}$. Then:

$D(17) = 55 \Rightarrow 15e^{17k} = 55 \Rightarrow e^{17k} = \dfrac{55}{15} = \dfrac{11}{3}$

$\Rightarrow \ln\left(e^{17k}\right) = \ln\left(\dfrac{11}{3}\right) \Rightarrow 17k = \ln\left(\dfrac{11}{3}\right)$

$\Rightarrow k = \dfrac{1}{17}\ln\left(\dfrac{11}{3}\right) \approx 0.0764$

$\Rightarrow D(t) = 15e^{0.0764t}$

(b) In 2012, $t = 2012 - 1990 = 22$

$D(22) = 15e^{0.0764 \cdot 22} \approx 80.55$

According to the model 80.55 million m^3 of water will be able to be desalinated per day by 2012.

(c) $D(t) = 100 \Rightarrow 15e^{0.0764t} = 100$

$\Rightarrow e^{0.0764t} = \dfrac{100}{15} = \dfrac{20}{3}$

$\Rightarrow \ln\left(e^{0.0764t}\right) = \ln\left(\dfrac{20}{3}\right)$

$\Rightarrow 0.0764t = \ln\left(\dfrac{20}{3}\right)$

$\Rightarrow t = \dfrac{1}{0.0764}\ln\left(\dfrac{20}{3}\right) \approx 24.8$

1990 + 24.8 = 2014.8. Therefore, according to the model, the worldwide capacity will reach 100 million m^3 in late 2014.

55. Anne can do $\frac{1}{10}$ of the task in 1 minute. And clay can do $\frac{1}{12}$. Let t = time working together; use $r \cdot t = w$ to write their work equation. Solve this equation for t.

$$\frac{1}{10}t + \frac{1}{12}t = 1 \,[\text{LCD} = 60]$$

$$60\left(\frac{t}{10} + \frac{t}{12}\right) = 60 \cdot 1$$

$$6t + 5t = 60$$

$$11t = 60$$

$$t = \frac{60}{11}, \quad \text{or} \quad 5\frac{5}{11}$$

Working together, it will take $5\frac{5}{11}$ mins.

56. Let x = number of ounces of the 45% dressing and y = number of ounces of the 20% dressing.

Dressing	%	Number of ounces	Fat Calories
Thick & tasty	45%	x	$0.45\,x$
Light & lean	20%	y	$0.20\,x$
mix	30%	15	$0.30(15)$ $= 4.5$

Using the table, determine the equations
(1) $x + y = 15$ (total ounces) and
(2) $0.45x + 0.20 = 4.5$ (total fat calories).
Solve this system of equations using elimination

$$-x - y = -15$$
$$\underline{2.25x + y = 22.5}$$
$$1.25x = 7.5$$
$$x = 6$$
$$x + y = 15,$$
$$\text{so } 6 + y = 15$$
$$y = 9$$

To obtain the desired mix, use 6 oz of Thick and Tasty and 9 oz of Light and Lean.

57. Let x = the speed of the river in kph. The speed of the boat downstream is then $5 + x$, and the speed of the boat upstream is $5 - x$.

Use $d = rt \Rightarrow t = \dfrac{d}{r}$ to find x.

$$t_{\text{UPSTREAM}} = t_{\text{DOWNSTREAM}}$$

$$\frac{d_{\text{UPSTREAM}}}{r_{\text{UPSTREAM}}} = \frac{d_{\text{DOWNSTREAM}}}{r_{\text{DOWNSTREAM}}}$$

$$\frac{42}{5 + x} = \frac{12}{5 - x}$$

$$42(5 - x) = 12(5 + x)$$

$$210 - 42x = 60 + 12x$$

$$54x = 150$$

$$x = \frac{150}{54} = \frac{25}{9} = 2\frac{7}{9}\,\text{kph}$$

58. Since the function values increase at a steady rate, a linear function seems most appropriate.

59. Choose two ordered pairs and determine the slope, which signifies the rate of change.

$$m = \frac{y_z - y_1}{x_z - x_1}$$

We will use $(15, 18)$ and $(20, 23)$.

$$m = \frac{23 - 18}{20 - 15}$$
$$= \frac{5}{5}$$
$$= 1$$

The rate of change is \$1/min.

60. From Exercise 2, we have $m = 1$. Choose an ordered pair and substitute to determine the equation.

$$y_2 - y_1 = m(x_2 - x_1)$$
$$y - 18 = 1(x - 15)$$
$$y = x + 3$$

Express the equation as a function of x.
$$f(x) = x + 3$$

61. Let $x = 10$.
$$f(10) = 10 + 3 = 13$$
A 10-min. would cost \$13.

62. $m = 1$ signifies the cost per minute, and $b = 3$ signifies the fixed cost or "startup" cost of each massage.

63. From the graph enter the data points $(35, 14.85), (40, 18.53), (50, 33.05),$ and $(55, 47.25).$ Use the ExpReg option in the STATCALC MENU to obtain the function. $m(x) = 1.8937(1.0596)^x$

64. Let $x = 45,$ and solve for $m.$
$m(45) = 1.8937(1.0596)^{45}$
$m \approx 25.63$
The monthly premium would be about $\$25.63$ for a 45-year-old-male.

65. $\dfrac{5}{3x-3} + \dfrac{10}{3x+6} = \dfrac{5x}{x^2+x-2}$

$\dfrac{5}{3(x-1)} + \dfrac{10}{3(x+2)} = \dfrac{5x}{(x+2)(x-1)}$

[LCD is $3(x+2)(x-1)$]

Note: $x \ne -2, 1$

$3(x+2)(x-1)\left[\dfrac{5}{3(x-1)} + \dfrac{10}{3(x+2)} \right]$

$= 3(x+2)(x-1)\left[\dfrac{5x}{(x+2)(x-1)} \right]$

$5(x+2) + 10(x-1) = 3 \cdot 5x$

$5x + 10 + 10x - 10 = 15x$

$15x = 15x$

This is true for all values in the domain (reflexive property of equality) The solution is $\{x | x$ is a real number and $x \ne -2$ and $x \ne 1\}.$

66. $\log \sqrt{3x} = \sqrt{\log 3x}$

$\log(3x)^{1/2} = (\log 3x)^{1/2}$

$\dfrac{1}{2}\log 3x = (\log 3x)^{1/2}$

$\left(\dfrac{1}{2}\log 3x \right)^2 = ((\log 3x)^{1/2})^2$

$\dfrac{1}{4}(\log 3x)^2 = \log 3x$

$\dfrac{1}{4}(\log 3x)^2 - \log 3x = 0$

$\log 3x\left(\dfrac{1}{4}\log 3x - 1 \right) = 0$

$\log 3x = 0 \qquad$ or $\qquad \dfrac{1}{4}\log 3x - 1 = 0$

$3x = 10^0 \quad$ or $\qquad \dfrac{1}{4}\log 3x = 1$

$x = \dfrac{1}{3} \qquad$ or $\qquad \log 3x = 4$

$3x = 10^4$

$x = \dfrac{10,000}{3}$

Both numbers check and are the solutions.

67. Let x = the original speed

and $x + 5$ = the increased speed.

Using $T = \dfrac{D}{R}$, we determine the times.

Trips	Rate×	Time =	Distance
Original	x	$\dfrac{280}{x}$	280
Increased	$x + 5$	$\dfrac{280}{x+5}$	280

The time for the trip is 1 hour greater. We have the equation:

$$\frac{280}{x} = \frac{280}{x+5} + 1$$

$$[\text{LCD} = x(x+5)]$$

$$x(x+5) \cdot \frac{280}{x} = x(x+5)\left[\frac{280}{x+5} + 1\right]$$

$$280(x+5) = 280x + x^2 + 5x$$

$$280x + 1400 = 280x + x^2 + 5x$$

$$0 = x^2 + 5x - 1400$$

$$0 = (x+40)(x-35)$$

$$x = -40 \text{ or } x = 35$$

Since the speed cannot be negative, our solution is 35 mph.

Chapter 13
Conic Sections

Exercise Set 13.1

1. f

3. g

5. c

7. d

9.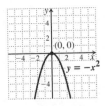

11.

13.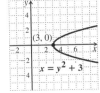

15.

17.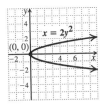

19.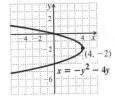

21.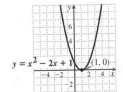

23.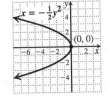

25.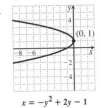

27.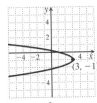

29. $(x-h)^2 + (y-k)^2 = r^2$ Standard form

$(x-0)^2 + (y-0)^2 = 6^2$ Substituting

$x^2 + y^2 = 36$ Simplifying

31. $(x-h)^2 + (y-k)^2 = r^2$ Standard form

$(x-7)^2 + (y-3)^2 = \left(\sqrt{5}\right)^2$ Substituting

$(x-7)^2 + (y-3)^2 = 5$

33. $(x-h)^2 + (y-k)^2 = r^2$

$[x-(-4)]^2 + (y-3)^2 = (4\sqrt{3})^2$

$(x+4)^2 + (y-3)^2 = 48$

35. $(x-h)^2 + (y-k)^2 = r^2$

$[x-(-7)]^2 + [y-(-2)]^2 = (5\sqrt{2})^2$

$(x+7)^2 + (y+2)^2 = 50$

37. Since the center is $(0,0)$, we have

$(x-0)^2 + (y-0)^2 = r^2$ or $x^2 + y^2 = r^2$. The

circle passes through $(-3,4)$. We find r^2 by

substituting -3 for x and 4 for y.

$(-3)^2 + 4^2 = r^2$

$9 + 16 = r^2$

$25 = r^2$

$x^2 + y^2 = 25$ is an equation of the circle.

39. Since the center is $(-4,1)$, we have

$[x-(-4)]^2 + (y-1)^2 = r^2$ or

$(x+4)^2 + (y-1)^2 = r^2$. The circle passes

through $(-2,5)$. We find r^2 by substituting

-2 for x and 5 for y.

$(-2+4)^2 + (5-1)^2 = r^2$

$4 + 16 = r^2$

$20 = r^2$

$(x+4)^2 + (y-1)^2 = 20$ is an equation of the

circle.

41. We write standard form

$(x-0)^2 + (y-0)^2 = 8^2$

Center: $(0,0)$; radius is 8.

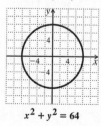

$x^2 + y^2 = 64$

43. We write standard form

$[x-(-1)]^2 + [y-(-3)]^2 = 6^2$

Center: $(-1,-3)$; radius is 6.

$(x+1)^2 + (y+3)^2 = 36$

45. We write standard form

$(x-4)^2 + [y-(-3)]^2 = (\sqrt{10})^2$

Center: $(4,-3)$; radius is $\sqrt{10}$.

$(x-4)^2 + (y+3)^2 = 10$

47. We write standard form

$(x-0)^2 + (y-0)^2 = (\sqrt{10})^2$

Center: $(0,0)$; radius is $\sqrt{10}$.

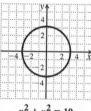

$x^2 + y^2 = 10$

49. We write standard form

$(x-5)^2 + (y-0)^2 = \left(\dfrac{1}{2}\right)^2$

Center: $(5,0)$; radius is $\dfrac{1}{2}$.

$(x-5)^2 + y^2 = \frac{1}{4}$

51. We write standard form
$$x^2 + 8x + y^2 - 6y = 15$$
$$\left(x^2 + 8x + 16\right) + \left(y^2 - 6y + 9\right) = 15 + 16 + 9$$
$$(x+4)^2 + (y-3)^2 = 40$$
$$\left[x - (-4)\right]^2 + (y-3)^2 = \left(\sqrt{40}\right)^2$$
Center: $(-4, 3)$; radius is $\sqrt{40}$, or $2\sqrt{10}$.

$x^2 + y^2 + 8x - 6y - 15 = 0$

53.
$$x^2 - 8x + y^2 + 2y = -13$$
$$\left(x^2 - 8x + 16\right) + \left(y^2 + 2y + 1\right) = -13 + 16 + 1$$
$$(x-4)^2 + (y+1)^2 = 4$$
$$(x-4)^2 + \left[y - (-1)\right]^2 = 2^2$$
Center: $(4, -1)$; radius is 2.

$x^2 + y^2 - 8x + 2y + 13 = 0$

55.
$$x^2 + y^2 + 10y = 75$$
$$x^2 + \left(y^2 + 10y + 25\right) = 75 + 25$$
$$(x-0)^2 + (y+5)^2 = 100$$
$$(x-0)^2 + \left[y - (-5)\right]^2 = 10^2$$
Center: $(0, -5)$; radius is 10.

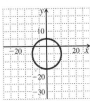

$x^2 + y^2 + 10y - 75 = 0$

57.
$$x^2 + 7x + y^2 - 3y = 10$$
$$\left(x^2 + 7x + \frac{49}{4}\right) + \left(y^2 - 3y + \frac{9}{4}\right) = 10 + \frac{49}{4} + \frac{9}{4}$$
$$\left(x + \frac{7}{2}\right)^2 + \left(y - \frac{3}{2}\right)^2 = \frac{98}{4}$$
$$\left[x - \left(-\frac{7}{2}\right)\right]^2 + \left(y - \frac{3}{2}\right)^2 = \left(\sqrt{\frac{98}{4}}\right)^2$$
Center: $\left(-\frac{7}{2}, \frac{3}{2}\right)$; radius is $\frac{7\sqrt{2}}{2}$.

$x^2 + y^2 + 7x - 3y - 10 = 0$

59.
$$36x^2 + 36y^2 = 1$$
$$\frac{1}{36} \cdot \left(36x^2 + 36y^2\right) = \frac{1}{36} \cdot 1$$
$$x^2 + y^2 = \left(\frac{1}{6}\right)^2$$
Center: $(0, 0)$; radius is $\frac{1}{6}$.

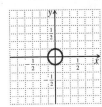

$36x^2 + 36y^2 = 1$

61. First we solve the equation for y.
$$x^2 + y^2 - 16 = 0$$
$$y^2 = 16 - x^2$$
$$y = \pm\sqrt{16 - x^2}$$
Then we graph $y_1 = \sqrt{16 - x^2}$ and $y_2 = -\sqrt{16 - x^2}$ on the same set of axes, choosing a squared window. We use $[-9, 9, -6, 6]$.

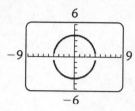

63. First we solve the equation for y. We can use the quadratic formula with $a = 1$, $b = -16$, and $c = x^2 + 14x + 54$ or we can complete the square on the y-terms and then proceed. We will complete the square.

$$x^2 + y^2 + 14x - 16y + 54 = 0$$

$$x^2 + 14x + y^2 - 16y + 64 - 64 + 54 = 0$$

$$x^2 + 14x + (y-8)^2 - 10 = 0$$

$$(y-8)^2 = 10 - x^2 - 14x$$

$$y = 8 \pm \sqrt{10 - x^2 - 14x}$$

Then we graph $y_1 = 8 + \sqrt{10 - x^2 - 14x}$ and $y_2 = 8 - \sqrt{10 - x^2 - 14x}$ on the same set of axes, choosing a squared window. We use $[-20, 7, -1, 17]$.

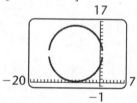

65. *Thinking and Writing Exercise.*

67. $\dfrac{y^2}{16} = 1$

$$y^2 = 16$$

$$y = \pm 4$$

69. $\dfrac{(x-1)^2}{25} = 1$

$$(x-1)^2 = 25$$

$$x - 1 = \pm 5$$

$$x = -4, 6$$

71. $\dfrac{1}{4} + \dfrac{(y+3)^2}{36} = 1$

$$\dfrac{(y+3)^2}{36} = \dfrac{3}{4}$$

$$(y+3)^2 = 27$$

$$y + 3 = \pm 3\sqrt{3}$$

$$y = -3 \pm 3\sqrt{3}$$

73. *Thinking and Writing Exercise.*

75. We make a drawing of the circle with center $(3, -5)$ and tangent to the y-axis.

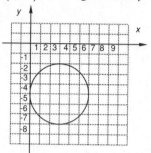

We see that the circle touches the y-axis at $(0, -5)$. Hence the radius is the distance between $(0, -5)$ and $(3, -5)$, or $\sqrt{(3-0)^2 + [-5-(-5)]^2}$, or 3. Now we write the equation of the circle.

$$(x-h)^2 + (y-k)^2 = r^2$$

$$(x-3)^2 + [y-(-5)]^2 = 3^2$$

$$(x-3)^2 + (y+5)^2 = 9$$

77. First we use the midpoint formula to find the center:

$$\left(\dfrac{7+(-1)}{2}, \dfrac{3+(-3)}{2} \right), \text{ or } \left(\dfrac{6}{2}, \dfrac{0}{2} \right), \text{ or } (3, 0)$$

The length of the radius is the distance between the center $(3, 0)$ and either endpoint of a diameter. We will use endpoint $(7, 3)$ in the distance formula:

$$r = \sqrt{(7-3)^2 + (3-0)^2} = \sqrt{25} = 5$$

Now we write the equation of the circle:

$$(x-h)^2 + (y-k)^2 = r^2$$
$$(x-3)^2 + (y-0)^2 = 5^2$$
$$(x-3)^2 + y^2 = 25$$

79. Let $(0, y)$ be the point on the y-axis that is equidistant from $(2,10)$ and $(6,2)$. Then the distance between $(2,10)$ and $(0, y)$ is the same as the distance between $(6,2)$ and $(0, y)$.

$$\sqrt{(0-2)^2 + (y-10)^2} = \sqrt{(0-6)^2 + (y-2)^2}$$
$$(-2)^2 + (y-10)^2 = (-6)^2 + (y-2)^2$$
$$4 + y^2 - 20y + 100 = 36 + y^2 - 4y + 4$$
$$64 = 16y$$
$$4 = y$$

The number checks. The point is $(0,4)$.

81. The outer circle has a radius of $\dfrac{9}{2}$ (from

$x^2 + y^2 = \dfrac{81}{4}$) and the inner edge of the red zone has a radius of 4 (from $x^2 + y^2 = 4^2$). The area of the red zone is the difference of their areas.

$$\pi \cdot \left(\frac{9}{2}\right)^2 - \pi \cdot (4^2) = \frac{81}{4}\pi - \frac{64}{4}\pi$$
$$= \frac{17}{4}\pi \approx 13.4 \text{ m}^2$$

83. The distance from the center of the circle at $(0, y)$ to the point at $(0, 23.5)$ is the same as the distance from the center to the point at $(580, 0)$. So, using the distance formula:

$$\sqrt{(0-0)^2 + (23.5-y)^2} = \sqrt{(0-580)^2 + (0-y)^2}$$
$$\sqrt{y^2 - 47y + 552.25} = \sqrt{336,400 + y^2}$$
$$y^2 - 47y + 552.25 = 336,400 + y^2$$
$$-47y = 335,847.75$$
$$y \approx -7145.7$$

The radius is the distance from the center to $(0, 23.5)$, or $7145.7 + 23.5 \approx 7169$ mm.

85. a) When the circle is positioned on a coordinate system as shown in the text, the center lies on the y-axis. To find the center, we will find the point on the y-axis that is equidistant from $(-4,0)$ and $(0,2)$. Let $(0, y)$ be this point.

$$\sqrt{[0-(-4)]^2 + (y-0)^2} =$$
$$\sqrt{(0-0)^2 + (y-2)^2}$$

Squaring both sides
$$16 + y^2 = y^2 - 4y + 4$$
$$12 = -4y$$
$$-3 = y$$

The center of the circle is $(0,-3)$.

b) We find the radius of the circle. Standard form:

$$(x-0)^2 + [y-(-3)]^2 = r^2$$
$$x^2 + (y+3)^2 = r^2$$

Substituting $(-4,0)$ for (x, y):

$$(-4)^2 + (0+3)^2 = r^2$$
$$16 + 9 = r^2$$
$$25 = r^2$$
$$5 = r$$

The radius is 5 ft.

87. We write the equation of a circle with center $(0,30.6)$ and radius 24.3:

$$x^2 + (y-30.6)^2 = 590.49$$

89.

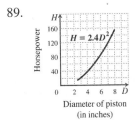

91. *Thinking and Writing Exercise.*

Exercise Set 13.2

1. True

3. False

5. True

7. True

9. $\dfrac{x^2}{1} + \dfrac{y^2}{9} = 1$

$\dfrac{x^2}{1^2} + \dfrac{y^2}{3^2} = 1$

The x-intercepts are $(1,0)$ and $(-1,0)$.

The y-intercepts are $(0,3)$ and $(0,-3)$.

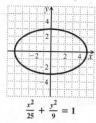

$\dfrac{x^2}{1} + \dfrac{y^2}{9} = 1$

11. $\dfrac{x^2}{25} + \dfrac{y^2}{9} = 1$

$\dfrac{x^2}{5^2} + \dfrac{y^2}{3^2} = 1$

The x-intercepts are $(5,0)$ and $(-5,0)$.

The y-intercepts are $(0,3)$ and $(0,-3)$.

$\dfrac{x^2}{25} + \dfrac{y^2}{9} = 1$

13. $4x^2 + 9y^2 = 36$

$\dfrac{1}{36} \cdot \left(4x^2 + 9y^2\right) = \dfrac{1}{36} \cdot 36$

$\dfrac{x^2}{9} + \dfrac{y^2}{4} = 1$

$\dfrac{x^2}{3^2} + \dfrac{y^2}{2^2} = 1$

The x-intercepts are $(3,0)$ and $(-3,0)$.

The y-intercepts are $(0,2)$ and $(0,-2)$.

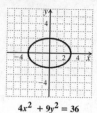

$4x^2 + 9y^2 = 36$

15. $16x^2 + 9y^2 = 144$

$\dfrac{1}{144} \cdot \left(16x^2 + 9y^2\right) = \dfrac{1}{144} \cdot 144$

$\dfrac{x^2}{9} + \dfrac{y^2}{16} = 1$

$\dfrac{x^2}{3^2} + \dfrac{y^2}{4^2} = 1$

The x-intercepts are $(3,0)$ and $(-3,0)$.

The y-intercepts are $(0,4)$ and $(0,-4)$.

$16x^2 + 9y^2 = 144$

17. $2x^2 + 3y^2 = 6$

$\dfrac{x^2}{3} + \dfrac{y^2}{2} = 1$

$\dfrac{x^2}{\left(\sqrt{3}\right)^2} + \dfrac{y^2}{\left(\sqrt{2}\right)^2} = 1$

The x-intercepts are $\left(\sqrt{3},0\right)$ and $\left(-\sqrt{3},0\right)$.

The y-intercepts are $\left(0,\sqrt{2}\right)$ and $\left(0,-\sqrt{2}\right)$.

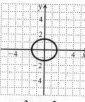

$2x^2 + 3y^2 = 6$

19. $5x^2 + 5y^2 = 125$

$x^2 + y^2 = 5^2$

This is the equation of a circle with center $(0,0)$ and radius 5.

$5x^2 + 5y^2 = 125$

21. $3x^2 + 7y^2 - 63 = 0$

$3x^2 + 7y^2 = 63$

$\dfrac{x^2}{21} + \dfrac{y^2}{9} = 1$

$\dfrac{x^2}{\left(\sqrt{21}\right)^2} + \dfrac{y^2}{3^2} = 1$

The x-intercepts are $\left(\sqrt{21},0\right)$ and $\left(-\sqrt{21},0\right)$.

The y-intercepts are $(0,3)$ and $(0,-3)$.

$3x^2 + 7y^2 - 63 = 0$

23. $16x^2 = 16 - y^2$

$16x^2 + y^2 = 16$

$\dfrac{x^2}{1} + \dfrac{y^2}{16} = 1$

$\dfrac{x^2}{(1)^2} + \dfrac{y^2}{(4)^2} = 1$

The x-intercepts are $(1,0)$ and $(-1,0)$.

The y-intercepts are $(0,4)$ and $(0,-4)$.

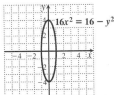

$16x^2 = 16 - y^2$

25. $16x^2 + 25y^2 = 1$

Note: $16 = \dfrac{1}{\frac{1}{16}}$ and $25 = \dfrac{1}{\frac{1}{25}}$

$\dfrac{x^2}{\frac{1}{16}} + \dfrac{y^2}{\frac{1}{25}} = 1$

$\dfrac{x^2}{\left(\frac{1}{4}\right)^2} + \dfrac{y^2}{\left(\frac{1}{5}\right)^2} = 1$

The x-intercepts are $\left(\dfrac{1}{4},0\right)$ and $\left(-\dfrac{1}{4},0\right)$.

The y-intercepts are $\left(0,\dfrac{1}{5}\right)$ and $\left(0,-\dfrac{1}{5}\right)$.

$16x^2 + 25y^2 = 1$

27. $\dfrac{(x-3)^2}{9} + \dfrac{(y-2)^2}{25} = 1$

$\dfrac{(x-3)^2}{3^2} + \dfrac{(y-2)^2}{5^2} = 1$

The center of the ellipse is $(3,2)$. Note that $a = 3$ and $b = 5$. We locate the center and then plot the points $(3+3,2)$, $(3-3,2)$, $(3,2+5)$ and $(3,2-5)$ or $(6,2)$, $(0,2)$, $(3,7)$ and $(3,-3)$.

$\dfrac{(x-3)^2}{9} + \dfrac{(y-2)^2}{25} = 1$

29. $\dfrac{(x+4)^2}{16} + \dfrac{(y-3)^2}{49} = 1$

$\dfrac{(x-(-4))^2}{4^2} + \dfrac{(y-3)^2}{7^2} = 1$

The center of the ellipse is $(-4,3)$. Note that $a = 4$ and $b = 7$. We locate the center and then plot the points $(-4+4,3)$, $(-4-4,3)$,

$(-4, 3+7)$ and $(-4, 3-7)$, or $(0,3)$, $(-8,3)$, $(-4,10)$ and $(-4,-4)$.

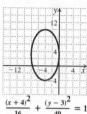

$$\frac{(x+4)^2}{16} + \frac{(y-3)^2}{49} = 1$$

31. $12(x-1)^2 + 3(y+4)^2 = 48$

$$\frac{(x-1)^2}{4} + \frac{(y+4)^2}{16} = 1$$

$$\frac{(x-1)^2}{2^2} + \frac{(y-(-4))^2}{4^2} = 1$$

The center of the ellipse is $(1,-4)$. Note that $a = 2$ and $b = 4$. We locate the center and then plot the points $(1+2,-4)$, $(1-2,-4)$, $(1,-4+4)$ and $(1,-4-4)$, or $(3,-4)$, $(-1,-4)$, $(1,0)$ and $(1,-8)$.

$12(x-1)^2 + 3(y+4)^2 = 48$

33. $4(x+3)^2 + 4(y+1)^2 - 10 = 90$

$$4(x+3)^2 + 4(y+1)^2 = 100$$

Observe that the x^2- and y^2-terms have the same coefficient. Dividing both sides by 4, we have

$$(x+3)^2 + (y+1)^2 = 25.$$

This is the equation of a circle with center $(-3,-1)$ and radius 5.

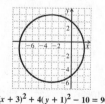

$4(x+3)^2 + 4(y+1)^2 - 10 = 90$

35. *Thinking and Writing Exercise.*

37. $x^2 - 5x + 3 = 0$

Use the quadratic formula:

$$\frac{-b \pm \sqrt{b^2 - 4ac}}{2a} = \frac{5 \pm \sqrt{(-5)^2 - 4(1)(3)}}{2(1)}$$

$$= \frac{5 \pm \sqrt{25 - 12}}{2}$$

$$= \frac{5 \pm \sqrt{13}}{2}$$

39. $\dfrac{4}{x+2} + \dfrac{3}{2x-1} = 2$

Note that $x+2$ is 0 when $x = -2$ and $2x-1$ is 0 when x is $\dfrac{1}{2}$, so -2 and $\dfrac{1}{2}$ cannot be solutions. We multiply by the LCD, $(x+2)(2x-1)$.

$$(x+2)(2x-1)\left(\frac{4}{x+2} + \frac{3}{2x-1}\right) =$$
$$(x+2)(2x-1)2$$

$$(2x-1) \cdot 4 + (x+2) \cdot 3 = 2(2x^2 + 3x - 2)$$

$$8x - 4 + 3x + 6 = 4x^2 + 6x - 4$$

$$11x + 2 = 4x^2 + 6x - 4$$

$$0 = 4x^2 - 5x - 6$$

$$0 = (4x+3)(x-2)$$

$$x = -\frac{3}{4}, 2$$

41. $x^2 = 11$

$x = \pm\sqrt{11}$

43. *Thinking and Writing Exercise.*

45. Plot the given points.

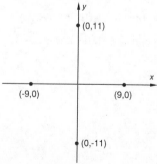

From the location of these points, we can see that the ellipse that contains them is centered at the origin with $a = 9$ and $b = 11$. We write the equation of the ellipse.

$$\frac{x^2}{9^2} + \frac{y^2}{11^2} = 1$$

$$\frac{x^2}{81} + \frac{y^2}{121} = 1$$

47. Plot the given points.

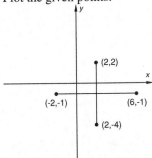

The vertical and horizontal segments intersect at $x = 2$ and $y = -1$. So the center of the ellipse is $(2, -1)$. Thus, the distance from $(-2, -1)$ to $(2, -1)$ is

$$\sqrt{\left[2 - (-2)\right]^2 + \left[-1 - (-1)\right]^2} = \sqrt{16} = 4 \text{, so}$$

$a = 4$. The distance from $(2, 2)$ to $(2, -1)$ is

$$\sqrt{(2 - 2)^2 + (-1 - 2)^2} = \sqrt{9} = 3 \text{, so } b = 3.$$

We write the equation of the ellipse.

$$\frac{(x - 2)^2}{4^2} + \frac{\left(y - (-1)\right)^2}{3^2} = 1$$

$$\frac{(x - 2)^2}{16} + \frac{(y + 1)^2}{9} = 1$$

49. The soloist is in the center of the ellipse, or $(0, 0)$. If the ellipse is 6 ft wide, we have the points $(-3, 0)$ and $(3, 0)$, and if it is 10 ft long, we have the points $(0, 5)$ and $(0, -5)$. Since $a = 3$ and $b = 5$, the equation is

$$\frac{x^2}{3^2} + \frac{y^2}{5^2} = 1 \text{, or } \frac{x^2}{9} + \frac{y^2}{25} = 1.$$

51. a) Let $F_1 = (-c, 0)$ and $F_2 = (c, 0)$. Then the sum of the distances from the foci to P is $2a$. By the distance formula,

$$\sqrt{(x + c)^2 + y^2} + \sqrt{(x - c)^2 + y^2} = 2a \text{, or}$$

$$\sqrt{(x + c)^2 + y^2} = 2a - \sqrt{(x - c)^2 + y^2}.$$

Squaring, we get

$$(x + c)^2 + y^2 =$$

$$4a^2 - 4a\sqrt{(x - c)^2 + y^2} + (x - c)^2 + y^2$$

or

$$x^2 + 2cx + c^2 + y^2$$

$$= 4a^2 - 4a\sqrt{(x - c)^2 + y^2}$$

$$+ x^2 - 2cx + c^2 + y^2$$

Thus

$$-4a^2 + 4cx = -4a\sqrt{(x - c)^2 + y^2}$$

$$a^2 - cx = a\sqrt{(x - c)^2 + y^2}$$

Squaring again, we get

$$a^4 - 2a^2cx + c^2x^2 =$$

$$a^2\left(x^2 - 2cx + c^2 + y^2\right)$$

$$a^4 - 2a^2cx + c^2x^2 =$$

$$a^2x^2 - 2a^2cx + a^2c^2 + a^2y^2,$$

$$a^4 - a^2c^2 = a^2x^2 - c^2x^2 + a^2y^2$$

$$a^2\left(a^2 - c^2\right) = x^2\left(a^2 - c^2\right) + a^2y^2$$

$$\frac{x^2}{a^2} + \frac{y^2}{a^2 - c^2} = 1.$$

   b) When P is at $(0, b)$, it follows that

$$b^2 = a^2 - c^2.$$

Substituting, we have

$$\frac{x^2}{a^2} + \frac{y^2}{b^2} = 1.$$

53. For the given ellipse, $a = 6/2$, or 3, and $b = 2/2$, or 1. The patient's mouth should be at a distance of $2c$ from the light source, where the coordinates of the foci of the ellipse are $(-c, 0)$ and $(c, 0)$. From Exercise 52, we know that $b^2 = a^2 - c^2$. We use this to find c.

$$b^2 = a^2 - c^2$$
$$1^2 = 3^2 - c^2 \quad \text{Substituting}$$
$$c^2 = 8$$
$$c = \sqrt{8}$$

Then $2c = 2\sqrt{8} \approx 5.66$. The patient's mouth should be about 5.66 ft from the light source.

55.
$$x^2 - 4x + 4y^2 + 8y - 8 = 0$$
$$x^2 - 4x + 4y^2 + 8y = 8$$
$$x^2 - 4x + 4(y^2 + 2y) = 8$$
$$(x^2 - 4x + 4) + 4(y^2 + 2y + 1) = 8 + 4 + 4 \cdot 1$$
$$(x - 2)^2 + 4(y + 1)^2 = 16$$
$$\frac{(x - 2)^2}{16} + \frac{(y + 1)^2}{4} = 1$$
$$\frac{(x - 2)^2}{4^2} + \frac{(y - (-1))^2}{2^2} = 1$$

The center of the ellipse is $(2, -1)$. Note that $a = 4$ and $b = 2$. We locate the center and then plot the points $(2 + 4, -1)$, $(2 - 4, -1)$, $(2, -1 + 2)$, $(2, -1 - 2)$, or $(6, -1)$, $(-2, -1)$, $(2, 1)$ and $(2, -3)$. Connect these points with an oval-shaped curve.

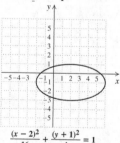

$$\frac{(x-2)^2}{16} + \frac{(y+1)^2}{4} = 1$$

Alternatively, we could write $x^2 - 4x + 4y^2 + 8y - 8 = 0$.

57. The sun is at the origin $(0, 0)$. Using the table of values, we see the maximum distance

of the earth from the sun is 152.1 million km., or 152,100,000 km.

Exercise Set 13.3

1. d

3. h

5. g

7. c

9. $\dfrac{y^2}{16} - \dfrac{x^2}{16} = 1$

$$\frac{y^2}{4^2} - \frac{x^2}{4^2} = 1$$

$a = 4$ and $b = 4$, so the asymptotes are $y = \dfrac{4}{4}x$ and $y = -\dfrac{4}{4}x$, or $y = x$ and $y = -x$. Replacing x with 0 and solving for y, we get $y = \pm 4$. The intercepts are $(0, 4)$ and $(0, -4)$.

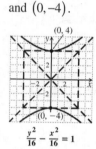

$$\frac{y^2}{16} - \frac{x^2}{16} = 1$$

11. $\dfrac{x^2}{4} - \dfrac{y^2}{25} = 1$

$$\frac{x^2}{2^2} - \frac{y^2}{5^2} = 1$$

$a = 2$ and $b = 5$, so the asymptotes are $y = \dfrac{5}{2}x$ and $y = -\dfrac{5}{2}x$. Replacing y with 0 and solving for x, we get $x = \pm 2$. The intercepts are $(2, 0)$ and $(-2, 0)$.

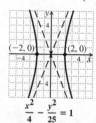

$$\frac{x^2}{4} - \frac{y^2}{25} = 1$$

13. $\dfrac{y^2}{36} - \dfrac{x^2}{9} = 1$

$\dfrac{y^2}{6^2} - \dfrac{x^2}{3^2} = 1$

$a = 3$ and $b = 6$, so the asymptotes are

$y = \dfrac{6}{3}x$ and $y = -\dfrac{6}{3}x$, or $y = 2x$ and

$y = -2x$. Replacing x with 0 and

solving for y, we get $y = \pm 6$. The

intercepts are $(0, 6)$ and $(0, -6)$.

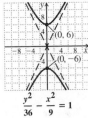

$\dfrac{y^2}{36} - \dfrac{x^2}{9} = 1$

15. $y^2 - x^2 = 25$

$\dfrac{y^2}{25} - \dfrac{x^2}{25} = 1$

$\dfrac{y^2}{5^2} - \dfrac{x^2}{5^2} = 1$

$a = 5$ and $b = 5$, so the asymptotes are

$y = \dfrac{5}{5}x$ and $y = -\dfrac{5}{5}x$, or $y = x$ and

$y = -x$. Replacing x with 0 and solving for

y, we get $y = \pm 5$. The intercepts are $(0, 5)$

and $(0, -5)$.

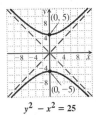

$y^2 - x^2 = 25$

17. $25x^2 - 16y^2 = 400$

$\dfrac{x^2}{16} - \dfrac{y^2}{25} = 1$

$\dfrac{x^2}{4^2} - \dfrac{y^2}{5^2} = 1$

$a = 4$ and $b = 5$, so the asymptotes are

$y = \dfrac{5}{4}x$ and $y = -\dfrac{5}{4}x$. Replacing y with 0

and solving for x, we get $x = \pm 4$. The

intercepts are $(4, 0)$ and $(-4, 0)$.

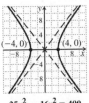

$25x^2 - 16y^2 = 400$

19. $xy = -6$

$y = -\dfrac{6}{x}$ Solving for y

We find some solutions, keeping the results in
a table.

x	y
$\dfrac{1}{2}$	-12
1	-6
6	-1
12	$-\dfrac{1}{2}$
$-\dfrac{1}{2}$	12
-1	6
-6	1
-12	$\dfrac{1}{2}$

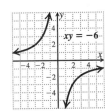

21. $xy = 4$

$y = \dfrac{4}{x}$ Solving for y

We find some solutions, keeping the results in a table.

x	y
$\dfrac{1}{2}$	8
1	4
4	1
8	$\dfrac{1}{2}$
$-\dfrac{1}{2}$	-8
-1	-4
-2	-2
-4	-1

23. $xy = -2$

$y = -\dfrac{2}{x}$ Solving for y

We find some solutions, keeping the results in a table.

x	y
$\dfrac{1}{2}$	-4
1	-2
2	-1
4	$-\dfrac{1}{2}$
$-\dfrac{1}{2}$	4
-1	2
-2	1
-4	$\dfrac{1}{2}$

25. $xy = 1$

$y = \dfrac{1}{x}$ Solving for y

We find some solutions, keeping the results in a table.

x	y
$\dfrac{1}{4}$	4
$\dfrac{1}{2}$	2
1	1
2	$\dfrac{1}{2}$
4	$\dfrac{1}{4}$
$-\dfrac{1}{4}$	-4
$-\dfrac{1}{2}$	-2
-1	-1
-2	$-\dfrac{1}{2}$
-4	$-\dfrac{1}{4}$

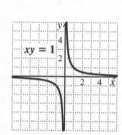

27. $x^2 + y^2 - 6x + 4y - 30 = 0$

$$\left(x^2 - 6x \quad\right) + \left(y^2 + 4y \quad\right) = 30$$

$$\left(x^2 - 6x + 9\right) + \left(y^2 + 4y + 4\right) = 30 + 9 + 4$$

$$(x - 3)^2 + (y + 2)^2 = 43$$

Both variables are squared, so the graph is not a parabola. The plus sign between x^2 and y^2 indicates that we have either a circle or an ellipse. Since the coefficients of x^2 and y^2 are the same, the graph is a circle.

29. $9x^2 + 4y^2 - 36 = 0$

$$9x^2 + 4y^2 = 36$$

$$\dfrac{x^2}{4} + \dfrac{y^2}{9} = 1$$

Both variables are squared, so the graph is not a parabola. The plus sign between x^2 and y^2 indicates that we have either a circle or an ellipse. Since the coefficients of x^2 and y^2 are different, the graph is an ellipse.

31. $4x^2 - 9y^2 - 72 = 0$

   $4x^2 - 9y^2 = 72$

   $\dfrac{x^2}{18} - \dfrac{y^2}{8} = 1$

   Both variables are squared, so the graph is not a parabola. The minus sign between x^2 and y^2 indicates that we have a hyperbola.

33. $y^2 = 20 - x^2$

   $x^2 + y^2 = 20$

   Both variables are squared, so the graph is not a parabola. The plus sign between x^2 and y^2 indicates that we have either a circle or an ellipse. Since the coefficients of x^2 and y^2 are the same, the graph is a circle.

35. $x - 10 = y^2 - 6y$

   $x = y^2 - 6y + 10$

   This equation has only one variable squared so we solve for the other variable. This is the equation for a parabola.

37. $x - \dfrac{8}{y} = 0$

   $x = \dfrac{8}{y}$

   $xy = 8$

   We have the product of x and y which indicates that we have a hyperbola.

39. $y + 6x = x^2 + 5$

   $y = x^2 - 6x + 5$

   This equation has only one variable squared so we solve for the other variable. This is the equation for a parabola.

41. $9y^2 = 36 + 4x^2$

   $9y^2 - 4x^2 = 36$

   $\dfrac{y^2}{4} - \dfrac{x^2}{9} = 1$

   Both variables are squared, so the graph is not a parabola. The minus sign between x^2 and y^2 indicates that we have a hyperbola.

43. $3x^2 + y^2 - x = 2x^2 - 9x + 10y + 40$

   $x^2 + y^2 + 8x - 10y = 40$

   Both variables are squared, so the graph is not a parabola. The plus sign between x^2 and y^2 indicates that we have either a circle or an ellipse. Since the coefficients of x^2 and y^2 are the same, the graph is a circle.

45. $16x^2 + 5y^2 - 12x^2 + 8y^2 - 3x + 4y = 568$

   $4x^2 + 13y^2 - 3x + 4y = 568$

   Both variables are squared, so the graph is not a parabola. The plus sign between x^2 and y^2 indicates that we have either a circle or an ellipse. Since the coefficients of x^2 and y^2 are different, the graph is an ellipse.

47. *Thinking and Writing Exercise.*

49. $5x + 2y = -3$

   $2x + 3y = 12$

   Multiply the first equation by 3 and the second by –2, then add.

   $15x + 6y = -9$

   $\underline{-4x - 6y = -24}$

   $11x = -33$

   $x = -3$

   Substitute into the first equation:

   $5(-3) + 2y = -3$

   $2y = 12$

   $y = 6$

   The solution is $(-3, 6)$.

51. $\dfrac{3}{4}x^2 + x^2 = 7$

   $\dfrac{7}{4}x^2 = 7$

   $x^2 = 4$

   $x = \pm 2$

53. $x^2 - 3x - 1 = 0$
Use the quadratic formula:

$$\frac{-b \pm \sqrt{b^2 - 4ac}}{2a} = \frac{3 \pm \sqrt{(-3)^2 - 4(1)(-1)}}{2(1)}$$

$$= \frac{3 \pm \sqrt{9 + 4}}{2}$$

$$= \frac{3 \pm \sqrt{13}}{2}$$

55. *Thinking and Writing Exercise.*

57. Since the intercepts are $(0,6)$ and $(0,-6)$, we know that the hyperbola is of the form
$\dfrac{y^2}{b^2} - \dfrac{x^2}{a^2} = 1$ and that $b = 6$. The equation of the asymptotes tell us that $b/a = 3$, so

$$\frac{6}{a} = 3$$

$$a = 2$$

The equation is $\dfrac{y^2}{6^2} - \dfrac{x^2}{2^2} = 1$, or $\dfrac{y^2}{36} - \dfrac{x^2}{4} = 1$.

59. $\dfrac{(x-5)^2}{36} - \dfrac{(y-2)^2}{25} = 1$

$$\frac{(x-5)^2}{6^2} - \frac{(y-2)^2}{5^2} = 1$$

$h = 5,\ k = 2,\ a = 6,\ b = 5$

Center: $(5,2)$

Vertices: $(5-6, 2)$ and $5 + 6, 2$ 9, or $(-1, 2)$ and $(11, 2)$

Asymptotes: $y - 2 = \dfrac{5}{6}(x - 5)$ and

$$y - 2 = -\frac{5}{6}(x - 5)$$

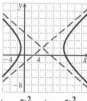

$\dfrac{(x-5)^2}{36} - \dfrac{(y-2)^2}{25} = 1$

61. $8(y+3)^2 - 2(x-4)^2 = 32$

$$\frac{(y+3)^2}{4} - \frac{(x-4)^2}{16} = 1$$

$$\frac{(y-(-3))^2}{2^2} - \frac{(x-4)^2}{4^2} = 1$$

$h = 4,\ k = -3,\ a = 4,\ b = 2$

Center: $(4,-3)$

Vertices: $(4, -3 + 2)$ and $(4, -3 - 2)$, or $(4, -1)$ and $(4, -5)$

Asymptotes: $y - (-3) = \dfrac{2}{4}(x - 4)$ and

$y - (-3) = -\dfrac{2}{4}(x-4)$, or $y + 3 = \dfrac{1}{2}(x-4)$

and $y + 3 = -\dfrac{1}{2}(x-4)$

$8(y + 3)^2 - 2(x - 4)^2 = 32$

63. $4x^2 - y^2 + 24x + 4y + 28 = 0$

$4(x^2 + 6x) - (y^2 - 4y) = -28$

$4(x^2 + 6x + 9 - 9) - (y^2 - 4y + 4 - 4) = -28$

$4(x^2 + 6x + 9) - (y^2 - 4y + 4) = -28 + 4 \cdot 9 - 4$

$4(x+3)^2 - (y-2)^2 = 4$

$$\frac{(x+3)^2}{1} - \frac{(y-2)^2}{4} = 1$$

$$\frac{(x-(-3))^2}{1^2} - \frac{(y-2)^2}{2^2} = 1$$

$h = -3,\ k = 2,\ a = 1,\ b = 2$

Center: $(-3, 2)$

Vertices: $(-3 - 1, 2)$ and $(-3 + 1, 2)$ or $(-4, 2)$ and $(-2, 2)$

Asymptotes: $y - 2 = \dfrac{2}{1}(x - (-3))$ and

$y - 2 = -\dfrac{2}{1}(x - (-3))$, or $y - 2 = 2(x + 3)$

and $y - 2 = -2(x + 3)$

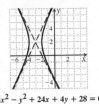

$4x^2 - y^2 + 24x + 4y + 28 = 0$

Mid-Chapter Review

Guided Solutions

1. $(x^2 - 4x) + (y^2 + 2y) = 6$

$(x^2 - 4x + 4) + (y^2 + 2y + 1) = 6 + 4 + 1$

$(x - 2)^2 + (y + 1)^2 = 11$

The center of the circle is $(2, -1)$.

The radius is $\sqrt{11}$.

2. $x^2 - \dfrac{y^2}{25} = 1$

Is there both an x^2-term and a y^2-term? Yes
Do both the x^2-term and the y^2-term have the same sign? No
The graph of the equation is a hyperbola.

Mixed Review

1. $y = 3(x - 4)^2 + 1$

The equation of the parabola is already in the correct form. The vertex of the parabola is (h, k), or $(4, 1)$.
Since the parabola has a vertical axis of symmetry, its equation is $x = h$, or $x = 4$.

2. $x = y^2 + 2y + 3$

$x = (y^2 + 2y + 1) - 1 + 3$

$x = (y + 1)^2 + 2$

The vertex of the parabola is (h, k), or $(2, -1)$.
Since the parabola has a horizontal axis of symmetry, it equation is $y = k$, or $y = -1$.

3. $(x - 3)^2 + (y - 2)^2 = 5$

Center: $(3, 2)$.

4. $x^2 + 6x + y^2 + 10y = 12$

$x^2 + 6x + 9 + y^2 + 10y + 25 = 12 + 9 + 25$

$(x + 3)^2 + (y + 5)^2 = 46$

Center: $(-3, -5)$.

5. $\dfrac{x^2}{144} + \dfrac{y^2}{81} = 1$

$\dfrac{x^2}{12^2} + \dfrac{y^2}{9^2} = 1$

The x-intercepts are $(12, 0)$ and $(-12, 0)$.

The y-intercepts are $(0, 9)$ and $(0, -9)$.

6. $\dfrac{x^2}{9} - \dfrac{y^2}{121} = 1$

$\dfrac{x^2}{3^2} - \dfrac{y^2}{11^2} = 1$

$a = 3$ and $b = 11$. Since the x term is positive, the vertices are $(3, 0)$ and $(-3, 0)$.

7. $4y^2 - x^2 = 4$

$\dfrac{4y^2}{4} - \dfrac{x^2}{4} = \dfrac{4}{4}$

$\dfrac{y^2}{1^2} - \dfrac{x^2}{2^2} = 1$

$a = 2$ and $b = 1$. Since the y term is positive, the vertices are $(0, 1)$ and $(0, -1)$.

8. $\dfrac{y^2}{9} - \dfrac{x^2}{4} = 1$

$\dfrac{y^2}{3^2} - \dfrac{x^2}{2^2} = 1$

$a = 2$ and $b = 3$, so the asymptotes are

$y = \dfrac{3}{2}x$ and $y = -\dfrac{3}{2}x$.

9. $x^2 + y^2 = 36$

$x^2 + y^2 = 6^2$

Both variables are squared, so the graph is not a parabola. The plus sign between x^2 and y^2 indicates that we have either a circle or an ellipse. Since the coefficients of x^2 and y^2 are the same, the graph is a circle. It has a center at $(0, 0)$ and a radius of 6.

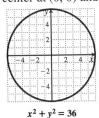

$x^2 + y^2 = 36$

10. $y = x^2 - 5$

This equation has only one variable squared. This is the equation for an upwards-opening parabola with a vertex at $(0, -5)$.

11. $\dfrac{x^2}{25} + \dfrac{y^2}{49} = 1$

$\dfrac{x^2}{5^2} + \dfrac{y^2}{7^2} = 1$

Both variables are squared, so the graph is not a parabola. The plus sign between x^2 and y^2 indicates that we have either a circle or an ellipse. Since the coefficients of x^2 and y^2 are different, the graph is an ellipse. Its intercepts are $(5, 0)$, $(-5, 0)$, $(0, 7)$, and $(0, -7)$.

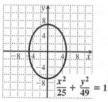

12. $\dfrac{x^2}{25} - \dfrac{y^2}{49} = 1$

$\dfrac{x^2}{5^2} - \dfrac{y^2}{7^2} = 1$

Both variables are squared, so the graph is not a parabola. The minus sign between x^2 and y^2 indicates that we have a hyperbola. Since the x-term is positive, it has a horizontal axis, with vertices at $(5, 0)$ and $(-5, 0)$ and asymptotes of $y = \pm\dfrac{7}{5}x$.

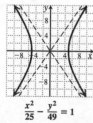

13. $x = (y + 3)^2 + 2$

This equation has only one variable squared. This is the equation for a rightwards-opening parabola with a vertex at $(2, -3)$.

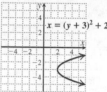

14. $4x^2 + 9y^2 = 36$

$\dfrac{4x^2}{36} + \dfrac{9y^2}{36} = 1$

$\dfrac{x^2}{3^2} + \dfrac{y^2}{2^2} = 1$

Both variables are squared, so the graph is not a parabola. The plus sign between x^2 and y^2 indicates that we have either a circle or an ellipse. Since the coefficients of x^2 and y^2 are different, the graph is an ellipse. Its intercepts are $(3, 0)$, $(-3, 0)$, $(0, 2)$, and $(0, -2)$.

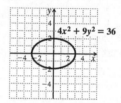

15. $xy = -4$

We have the product of x and y which indicates that we have a hyperbola. It passes through $(-4, 1)$, $(-2, 2)$, $(-1, 4)$, $(1, -4)$, $(2, -2)$, and $(4, -1)$.

16. $(x+2)^2 + (y-3)^2 = 1$

$(x+2)^2 + (y-3)^2 = 1^2$

Both variables are squared, so the graph is not a parabola. The plus sign between x^2 and y^2 indicates that we have either a circle or an ellipse. Since the coefficients of x^2 and y^2 are the same, the graph is a circle. It has a center at $(-2, 3)$ and a radius of 1.

17. $x^2 + y^2 - 8y - 20 = 0$

$x^2 + y^2 - 8y = 20$

$x^2 + y^2 - 8y + 16 = 20 + 16$

$x^2 + (y-4)^2 = 36$

$x^2 + (y-4)^2 = 6^2$

Both variables are squared, so the graph is not a parabola. The plus sign between x^2 and y^2 indicates that we have either a circle or an ellipse. Since the coefficients of x^2 and y^2 are the same, the graph is a circle. It has a center at $(0, 4)$ and a radius of 6.

18. $x = y^2 + 2y$

$x = y^2 + 2y + 1 - 1$

$x = (y+1)^2 - 1$

This equation has only one variable squared so we solve for the other variable. This is the equation for a rightwards-opening parabola with a vertex at $(-1, -1)$.

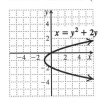

19. $16y^2 - x^2 = 16$

$\dfrac{16y^2}{16} - \dfrac{x^2}{16} = \dfrac{16}{16}$

$\dfrac{y^2}{1^2} - \dfrac{x^2}{4^2} = 1$

Both variables are squared, so the graph is not a parabola. The minus sign between x^2 and y^2 indicates that we have a hyperbola. Since the y-term is positive, it has a vertical axis, with vertices at $(0, 1)$ and $(0, -1)$ and asymptotes of $y = \pm\dfrac{1}{4}x$.

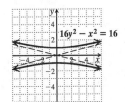

20. $x = \dfrac{9}{y}$

$xy = 9$

We have the product of x and y which indicates that we have a hyperbola. It passes through $(-9, -1)$, $(-3, -3)$, $(-1, -9)$, $(1, 9)$, $(3, 3)$, and $(9, 1)$.

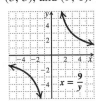

Exercise Set 13.4

1. True

3. False

5. True

7. $x^2 + y^2 = 25$ (1)

$y - x = 1$ (2)

First solve Eq. (2) for y.

$y = x + 1$ (3)

Then substitute $x + 1$ for y in Eq. (1) and solve for x.

$$x^2 + y^2 = 25$$
$$x^2 + (x+1)^2 = 25$$
$$x^2 + x^2 + 2x + 1 = 25$$
$$2x^2 + 2x + 1 = 25$$
$$2x^2 + 2x - 24 = 0$$
$$x^2 + x - 12 = 0$$
$$(x+4)(x-3) = 0$$
$$x+4 = 0 \ or \ x-3 = 0$$
$$x = -4 \ or \ x = 3$$

Now substitute these numbers in Eq. (3) and solve for y.
$$y = -4 + 1 = -3$$
$$y = 3 + 1 = 4$$
The pairs $(-4,-3)$ and $(3,4)$ check, so they are the solutions.

9. $4x^2 + 9y^2 = 36$ (1)
 $3y + 2x = 6$ (2)

First solve Eq. (2) for y.
$$y = \frac{6-2x}{3} \quad (3)$$

Then substitute $\dfrac{6-2x}{3}$ for y in Eq. (1) and solve for x.
$$4x^2 + 9y^2 = 36$$
$$4x^2 + 9\left(\frac{6-2x}{3}\right)^2 = 36$$
$$4x^2 + 9\left(\frac{36-24x+4x^2}{9}\right) = 36$$
$$4x^2 + 36 - 24x + 4x^2 = 36$$
$$8x^2 - 24x = 0$$
$$8x(x-3) = 0$$
$$8x = 0 \ or \ x-3 = 0$$
$$x = 0 \ or \ x = 3$$

Now substitute these numbers in Eq. (3) and solve for y.
$$y = \frac{6-2\cdot 0}{3} = 2$$
$$y = \frac{6-2\cdot 3}{3} = 0$$
The pairs $(0,2)$ and $(3,0)$ check, so they are the solutions.

11. $y^2 = x + 3$ (1)
 $2y = x + 4$ (2)

First solve Eq. (2) for x.
$$2y - 4 = x \quad (3)$$

Then substitute $2y-4$ for x in Eq. (1) and solve for y.
$$y^2 = x + 3$$
$$y^2 = (2y-4) + 3$$
$$y^2 = 2y - 1$$
$$y^2 - 2y + 1 = 0$$
$$(y-1)(y-1) = 0$$

$$y - 1 = 0 \ or \ y - 1 = 0$$
$$y = 1 \ or \ y = 1$$
Now substitute 1 for y in Eq. (3) and solve for x.
$$2 \cdot 1 - 4 = x$$
$$-2 = x$$
The pair $(-2,1)$ checks, so it is the solution.

13. $x^2 - xy + 3y^2 = 27$ (1)
 $x - y = 2$ (2)

First solve Eq. (2) for y.
$$x - 2 = y \quad (3)$$

Then substitute $x-2$ for y in Eq. (1) and solve for x.
$$x^2 - xy + 3y^2 = 27$$
$$x^2 - x(x-2) + 3(x-2)^2 = 27$$
$$x^2 - x^2 + 2x + 3x^2 - 12x + 12 = 27$$
$$3x^2 - 10x - 15 = 0$$
$$x = \frac{-(-10) \pm \sqrt{(-10)^2 - 4(3)(-15)}}{2 \cdot 3}$$
$$x = \frac{10 \pm \sqrt{100 + 180}}{6} = \frac{10 \pm \sqrt{280}}{6}$$
$$x = \frac{10 \pm 2\sqrt{70}}{6} = \frac{5 \pm \sqrt{70}}{3}$$

Now substitute these numbers in Eq. (3) and solve for y.

$$y = \frac{5+\sqrt{70}}{3} - 2 = \frac{-1+\sqrt{70}}{3}$$

$$y = \frac{5-\sqrt{70}}{3} - 2 = \frac{-1-\sqrt{70}}{3}$$

The pairs $\left(\dfrac{5+\sqrt{70}}{3}, \dfrac{-1+\sqrt{70}}{3}\right)$ and

$\left(\dfrac{5-\sqrt{70}}{3}, \dfrac{-1-\sqrt{70}}{3}\right)$ check, so they

are the solutions.

15. $x^2 + 4y^2 = 25$ (1)

   $x + 2y = 7$ (2)

First solve Eq. (2) for x.

$x = -2y + 7$ (3)

Then substitute $-2y + 7$ for x in Eq. (1) and

solve for y.

$$x^2 + 4y^2 = 25$$
$$(-2y+7)^2 + 4y^2 = 25$$
$$4y^2 - 28y + 49 + 4y^2 = 25$$
$$8y^2 - 28y + 24 = 0$$
$$2y^2 - 7y + 6 = 0$$
$$(2y-3)(y-2) = 0$$
$$2y - 3 = 0 \ \ or \ \ y - 2 = 0$$
$$y = \frac{3}{2} \ \ or \ \ y = 2$$

Now substitute these numbers in Eq. (3) and

solve for x.

$$x = -2 \cdot \frac{3}{2} + 7 = 4$$
$$x = -2 \cdot 2 + 7 = 3$$

The pairs $\left(4, \dfrac{3}{2}\right)$ and $(3, 2)$ check, so they

are the solutions.

17. $x^2 - xy + 3y^2 = 5$ (1)

   $x - y = 2$ (2)

First solve Eq. (2) for y.

$x - 2 = y$ (3)

Then substitute $x - 2$ for y in Eq. (1) and

solve for x.

$$x^2 - xy + 3y^2 = 5$$
$$x^2 - x(x-2) + 3(x-2)^2 = 5$$
$$x^2 - x^2 + 2x + 3x^2 - 12x + 12 = 5$$
$$3x^2 - 10x + 7 = 0$$
$$(3x-7)(x-1) = 0$$
$$3x - 7 = 0 \ \ or \ \ x - 1 = 0$$
$$x = \frac{7}{3} \ \ or \ \ x = 1$$

Now substitute these numbers in Eq. (3) and

solve for y.

$$y = \frac{7}{3} - 2 = \frac{1}{3}$$
$$y = 1 - 2 = -1$$

The pairs $\left(\dfrac{7}{3}, \dfrac{1}{3}\right)$ and $(1, -1)$ check, so they

are the solutions.

19. $3x + y = 7$ (1)

   $4x^2 + 5y = 24$ (2)

First solve Eq. (1) for y.

 $y = 7 - 3x$ (3)

Then substitute $7 - 3x$ for y in Eq. (2) and

solve for x.

$$4x^2 + 5y = 24$$
$$4x^2 + 5(7-3x) = 24$$
$$4x^2 + 35 - 15x = 24$$
$$4x^2 - 15x + 11 = 0$$
$$(4x-11)(x-1) = 0$$
$$4x - 11 = 0 \ \ or \ \ x - 1 = 0$$
$$x = \frac{11}{4} \ \ or \ \ x = 1$$

Now substitute these numbers in Eq. (3) and

solve for y.

$$y = 7 - 3 \cdot \frac{11}{4} = -\frac{5}{4}$$
$$y = 7 - 3 \cdot 1 = 4$$

The pairs $\left(\dfrac{11}{4}, -\dfrac{5}{4}\right)$ and $(1, 4)$ check, so they

are the solutions.

21. $a+b=6$ (1)

     $ab=8$ (2)

First solve Eq. (1) for a.

$a=6-b$ (3)

Then substitute $6-b$ for a in Eq. (2) and solve for b.

$(6-b)b=8$

$6b-b^2=8$

$b^2-6b+8=0$

$(b-4)(b-2)=0$

$\qquad\qquad b=2,4$

Now substitute these numbers in Eq. (3) and solve for a.

$a=6-b$

$a=6-2$

$a=4$

$a=6-b$

$a=6-4$

$a=2$

The pairs $(2,4)$ and $(4,2)$ check, so they are the solutions.

23. $2a+b=1$ (1)

    $b=4-a^2$ (2)

Equation (2) is already solved for b.

Substitute $4-a^2$ for b in Eq. (1) and solve for a.

$2a+4-a^2=1$

$\qquad 0=a^2-2a-3$

$\qquad\quad 0=(a-3)(a+1)$

$a-3=0$ or $a+1=0$

$\quad a=3$ or $a=-1$

Now substitute these numbers in Eq. (2) and solve for b.

$b=4-3^2=-5$

$b=4-(-1)^2=3$

The pairs $(3,-5)$ and $(-1,3)$ check, so they are the solutions.

25. $a^2+b^2=89$ (1)

    $a-b=3$ (2)

First solve Eq. (2) for a.

$a=b+3$ (3)

Then substitute $b+3$ for a in Eq. (1) and solve for b.

$(b+3)^2+b^2=89$

$b^2+6b+9+b^2=89$

$2b^2+6b-80=0$

$b^2+3b-40=0$

$(b+8)(b-5)=0$

$b+8=0$ or $b-5=0$

$\quad b=-8$ or $b=5$

Now substitute these numbers in Eq. (3) and solve for a.

$a=-8+3=-5$

$a=5+3=8$

The pairs $(-5,-8)$ and $(8,5)$ check, so they are the solutions.

27. $y=x^2$ (1)

    $x=y^2$ (2)

Eq. (1) is already solved for y. Substitute x^2 for y in Eq. (2) and solve for x.

$x=y^2$

$x=\left(x^2\right)^2$

$x=x^4$

$0=x^4-x$

$0=x\left(x^3-1\right)$

$0=x(x-1)\left(x^2+x+1\right)$

$x=0$ or $x=1$ or $x=\dfrac{-1\pm\sqrt{1^2-4\cdot1\cdot1}}{2}$

$x=0$ or $x=1$ or $x=-\dfrac{1}{2}\pm\dfrac{\sqrt{3}}{2}i$

Now substitute these numbers in Eq. (1) and solve for y.

$y=\left(-\dfrac{1}{2}+\dfrac{\sqrt{3}}{2}i\right)^2=-\dfrac{1}{2}-\dfrac{\sqrt{3}}{2}i$

$y=\left(-\dfrac{1}{2}-\dfrac{\sqrt{3}}{2}i\right)^2=-\dfrac{1}{2}+\dfrac{\sqrt{3}}{2}i$

The pairs $(0,0)$, $(1,1)$, and

$\left(-\dfrac{1}{2}+\dfrac{\sqrt{3}}{2}i,-\dfrac{1}{2}-\dfrac{\sqrt{3}}{2}i\right)$ and

$\left(-\dfrac{1}{2}-\dfrac{\sqrt{3}}{2}i,-\dfrac{1}{2}+\dfrac{\sqrt{3}}{2}i\right)$ check, so they

are the solutions.

29. $x^2+y^2=9$ (1)
     $x^2-y^2=9$ (2)

Here we use the elimination method.

$\begin{array}{rl} x^2+y^2=9 & (1) \\ x^2-y^2=9 & (2) \\ \hline 2x^2 \quad\;\; =18 & \text{Adding} \end{array}$

$x^2=9$

$x=\pm 3$

If $x=3$, $x^2=9$, and if $x=-3$, $x^2=9$, so substituting 3 or –3 in Eq. (1) gives us

$x^2+y^2=9$

$9+y^2=9$

$y^2=0$

$y=0.$

The pairs $(3,0)$ and $(-3,0)$ check. They are the solutions.

31. $x^2+y^2=25$ (1)
     $xy=12$ (2)

First we solve Eq. (2) for y.

$xy=12$

$y=\dfrac{12}{x}$

Then we substitute $\dfrac{12}{x}$ for y in Eq. (1) and

solve for x.

$x^2+y^2=25$

$x^2+\left(\dfrac{12}{x}\right)^2=25$

$x^2+\dfrac{144}{x^2}=25$

$x^4+144=25x^2$ Multiplying by x^2

$x^4-25x^2+144=0$

$u^2-25u+144=0$ Letting $u=x^2$

$(u-9)(u-16)=0$

$u=9$ or $u=16$

We now substitute x^2 for u and solve for x

$x^2=9$ or $x^2=16$

$x=\pm 3$ or $x=\pm 4$

Since $y=12/x$, if $x=3$, $y=4$; if $x=-3$,

$y=-4$; if $x=4$, $y=3$; and if $x=-4$,

$y=-3$. The pairs $(-4,-3)$, $(-3,-4)$, $(3,4)$

and $(4,3)$ check. They are the solutions.

33. $x^2+y^2=9$ (1)
     $25x^2+16y^2=400$ (2)

$\begin{array}{rl} -16x^2-16y^2=-144 & \text{Mult. (1) by }-16 \\ 25x^2+16y^2=\;\;\,400 & \\ \hline 9x^2 \qquad\quad\;\; =\;\;256 & \text{Adding} \end{array}$

$x=\pm\dfrac{16}{3}$

$\dfrac{256}{9}+y^2=9$ Substituting in (1)

$y=\pm\dfrac{5\sqrt{7}}{3}i$

The pairs $\left(\dfrac{16}{3},\dfrac{5\sqrt{7}}{3}i\right)$, $\left(\dfrac{16}{3},-\dfrac{5\sqrt{7}}{3}i\right)$,

$\left(-\dfrac{16}{3},\dfrac{5\sqrt{7}}{3}i\right)$, and $\left(-\dfrac{16}{3},-\dfrac{5\sqrt{7}}{3}i\right)$ check.

They are the solutions.

35. $x^2+y^2=14$ (1)
     $x^2-y^2=4$ (2)

$\begin{array}{rl} x^2+y^2=\;\;14 & \\ x^2-y^2=\quad 4 & \\ \hline 2x^2 \qquad =\;\;18 & \text{Adding} \end{array}$

$x^2=9$

$x=\pm 3$

$9+y^2=14$ Substituting in Eq. (1)

$y^2=5$

$y=\pm\sqrt{5}$

The pairs $\left(-3,-\sqrt{5}\right)$, $\left(-3,\sqrt{5}\right)$, $\left(3,-\sqrt{5}\right)$, and $\left(3,\sqrt{5}\right)$ check. They are the solutions.

37. $x^2 + y^2 = 20$ (1)
       $xy = 8$ (2)

First we solve Eq. (2) for y.

$$y = \frac{8}{x}$$

Then we substitute $\frac{8}{x}$ for y in Eq. (1) and solve for x.

$$x^2 + \left(\frac{8}{x}\right)^2 = 20$$

$$x^2 + \frac{64}{x^2} = 20$$

$$x^4 - 20x^2 + 64 = 0$$

$u^2 - 20u + 64 = 0$ Letting $u = x^2$

$(u-16)(u-4) = 0$

$u = 16$ or $u = 4$

$x^2 = 16$ or $x^2 = 4$

$x = \pm 4$ or $x = \pm 2$

$y = 8/x$, so if $x = 4$, $y = 2$; if $x = -4$, $y = -2$; if $x = 2$, $y = 4$; if $x = -2$, $y = -4$. The pairs $(4,2)$, $(-4,-2)$, $(2,4)$, and $(-2,-4)$ check. They are the solutions.

39. $x^2 + 4y^2 = 20$ (1)
       $xy = 4$ (2)

First we solve Eq. (2) for y.

$$y = \frac{4}{x}$$

Then we substitute $\frac{4}{x}$ for y in Eq. (1) and solve for x.

$$x^2 + 4y^2 = 20$$

$$x^2 + 4\left(\frac{4}{x}\right)^2 = 20$$

$$x^2 + \frac{64}{x^2} = 20$$

$$x^4 + 64 = 20x^2$$

$x^4 - 20x^2 + 64 = 0$

$u^2 - 20u + 64 = 0$ Letting $u = x^2$

$(u-16)(u-4) = 0$

$u = 16$ or $u = 4$

We now substitute x^2 for u and solve for x.

$x^2 = 16$ or $x^2 = 4$

$x = \pm 4$ or $x = \pm 2$

$y = 4/x$, so if $x = 4$, $y = 1$; if $x = -4$, $y = -1$; if $x = 2$, $y = 2$; and if $x = -2$, $y = -2$. The pairs $(4,1)$, $(-4,-1)$, $(2,2)$, and $(-2,-2)$ check. They are the solutions.

41. $2xy + 3y^2 = 7$ (1)
       $3xy - 2y^2 = 4$ (2)

$6xy + 9y^2 = 21$ Mult. (1) by 3
$\underline{-6xy + 4y^2 = -8}$ Mult. (2) by -2
$13y^2 = 13$

$y^2 = 1$

$y = \pm 1$

Substitute for y in Eq. (1) and solve for x.

When $y = 1$: $2 \cdot x \cdot 1 + 3 \cdot 1^2 = 7$

$2x = 4$

$x = 2$

When $y = -1$: $2 \cdot x \cdot (-1) + 3(-1)^2 = 7$

$-2x = 4$

$x = -2$

The pairs $(2,1)$ and $(-2,-1)$ check. They are the solutions.

43. $4a^2 - 25b^2 = 0$ (1)
       $2a^2 - 10b^2 = 3b + 4$ (2)

$4a^2 - 25b^2 = 0$

$\underline{-4a^2 + 20b^2 = -6b - 8}$ Mult. (2) by -2
$-5b^2 = -6b - 8$ Adding

$0 = 5b^2 - 6b - 8$

$0 = (5b + 4)(b - 2)$

Substitute for b in Eq. (1) and solve for y.

$$b = -\frac{4}{5} \text{ or } b = 2$$

Substitute for b in Eq. (1) and solve for a.

When $b = -\dfrac{4}{5}$: $4a^2 - 25\left(-\dfrac{4}{5}\right)^2 = 0$

$$4a^2 = 16$$
$$a^2 = 4$$
$$a = \pm 2$$

When $b = 2$: $4a^2 - 25(2)^2 = 0$

$$4a^2 = 100$$
$$a^2 = 25$$
$$a = \pm 5$$

The pairs $\left(2, -\dfrac{4}{5}\right)$, $\left(-2, -\dfrac{4}{5}\right)$, $(5,2)$ and

$(-5, 2)$ check. They are the solutions.

45. $ab - b^2 = -4$ (1)
$ab - 2b^2 = -6$ (2)

$$\begin{array}{l} ab - b^2 = -4 \\ \underline{-ab + 2b^2 = 6} \quad \text{Mult. } (2) \text{ by } -1 \\ b^2 = 2 \\ b = \pm\sqrt{2} \end{array}$$

Substitute for b in Eq. (1) and solve for a.

When $b = \sqrt{2}$:

$$a\left(\sqrt{2}\right) - \left(\sqrt{2}\right)^2 = -4$$
$$a\sqrt{2} = -2$$
$$a = -\dfrac{2}{\sqrt{2}} = -\sqrt{2}$$

When $b = -\sqrt{2}$:

$$a\left(-\sqrt{2}\right) - \left(-\sqrt{2}\right)^2 = -4$$
$$-a\sqrt{2} = -2$$
$$a = \dfrac{-2}{-\sqrt{2}} = \sqrt{2}$$

The pairs $\left(-\sqrt{2}, \sqrt{2}\right)$ and $\left(\sqrt{2}, -\sqrt{2}\right)$ check.

They are the solutions.

47. We let l and w represent the length and width, respectively.
The perimeter is 28 cm.
$2l + 2w = 28$, or $l + w = 14$
Using the Pythagorean theorem, we have another equation.
$l^2 + w^2 = 10^2$, or $l^2 + w^2 = 100$

We solve the system:
$l + w = 14$ (1)
$l^2 + w^2 = 100$ (2)
First solve Eq. 1 for w.
$w = 14 - l$ (3)
Then substitute $14 - l$ for w in Eq. (2) and solve for l.

$$l^2 + w^2 = 100$$
$$l^2 + (14 - l)^2 = 100$$
$$l^2 + 196 - 28l + l^2 = 100$$
$$2l^2 - 28l + 96 = 0$$
$$l^2 - 14l + 48 = 0$$
$$(l - 8)(l - 6) = 0$$

$l = 8$ or $l = 6$

If $l = 8$, then $w = 14 - 8$, or 6. If $l = 6$, then $w = 14 - 6$, or 8. Since the length is usually considered to be longer than the width, we have the solution $l = 8$ and $w = 6$, or $(8, 6)$. The length is 8 cm and the width is 6 cm.

49. Let l and w represent the length and width, respectively. We solve the system
$$lw = 2$$
$$2l + 2w = 6$$
We solve the first equation for l.
$$l = \dfrac{2}{w}$$

Then substitute the expression $\dfrac{2}{w}$ for l in the second equation.

$$2\left(\dfrac{2}{w}\right) + 2w = 6$$
$$4 + 2w^2 = 6w$$
$$2w^2 - 6w + 4 = 0$$
$$w^2 - 3w + 2 = 0$$
$$(w - 2)(w - 1) = 0$$
$$w = 2 \ \text{ or } \ w = 1$$

The solutions are $(1, 2)$ and $(2, 1)$. We choose the larger number to be the length, so the length is 2 yd and the width is 1 yd.

51. Let l equal the length and w equal the width of the cargo area.

The cargo area must be 60 ft^2, so we have one equation:

$lw = 60$

The Pythagorean equation gives us another equation:

$l^2 + w^2 = 13^2$, or $l^2 + w^2 = 169$

We solve the system of equations.

$$lw = 60 \quad (1)$$
$$l^2 + w^2 = 169 \quad (2)$$

First solve Eq. (1) for w:

$$lw = 60$$
$$w = \frac{60}{l} \quad (3)$$

Then substitute $60/l$ for w in Eq. (2) and solve for l.

$$l^2 + w^2 = 169$$
$$l^2 + \left(\frac{60}{l}\right)^2 = 169$$
$$l^2 + \frac{3600}{l^2} = 169$$
$$l^4 - 169l^2 + 3600 = 0$$

Let $u = l^2$ and $u^2 = l^4$ and substitute.

$$u^2 - 169u + 3600 = 0$$
$$(u - 144)(u - 25) = 0$$
$$u = 144 \ or \ u = 25$$
$$l^2 = 144 \ or \ l^2 = 25 \quad \text{Replacing } u \text{ with } l^2$$
$$l = \pm 12 \ or \ l = \pm 5$$

Since the length cannot be negative, we consider only 12 and 5. We substitute in Eq. (3) to find w. When $l = 12$, $w = 60/12 = 5$; when $l = 5$, $w = 60/5 = 12$. Since we usually consider length to be longer than width, length is 12 ft and width is 5 ft.

53. Let x and y represent the numbers. Solve the system:

$$xy = 90$$
$$x^2 + y^2 = 261$$

The solutions are $(6,15)$, $(-6,-15)$, $(15,6)$, and $(-15,-6)$. The numbers are 6 and 15 or −6 and −15.

55. Let x equal the length of a side of one bed and y equal the length of a side of the other bed. The area of the beds are x^2 and y^2, respectively.

Sum: $x^2 + y^2 = 832$

Difference: $x^2 - y^2 = 320$

Solve: Adding the two equations, we have:

$$2x^2 = 1152$$
$$x^2 = 576$$
$$x = \sqrt{576} = 24$$

Since the length cannot be negative, we use the positive value of x.

Substituting, we have

$$24^2 + y^2 = 832$$
$$y^2 = 256$$
$$y = 16 \ (16 \geq 0)$$

The length of the beds are 24 ft and 16 ft.

57. Let l equal the length of the rectangle and w equal the width.

Area: $lw = \sqrt{3} \quad (1)$

From the Pythagorean theorem:

$$l^2 + w^2 = 2^2 \quad (2)$$

We solve the system of equations.

We first solve Eq. (1) for w.

$$lw = \sqrt{3}$$
$$w = \frac{\sqrt{3}}{l}$$

Then we substitute $\dfrac{\sqrt{3}}{l}$ for w in Eq. (2) and solve for l.

$$l^2 + \left(\frac{\sqrt{3}}{l}\right)^2 = 4$$
$$l^2 + \frac{3}{l^2} = 4$$
$$l^4 + 3 = 4l^2$$
$$l^4 - 4l^2 + 3 = 0$$
$$u^2 - 4u + 3 = 0 \quad \text{Letting } u = l^2$$
$$(u - 3)(u - 1) = 0$$
$$u = 3 \ or \ u = 1$$

We now substitute l^2 for u and solve for l.

$$l^2 = 3 \ or \ l^2 = 1$$
$$l = \pm\sqrt{3} \ or \ l = \pm 1$$

Measurements cannot be negative, so we only need to consider $l = \sqrt{3}$ and $l = 1$. Since $w = \sqrt{3}/l$, if $l = \sqrt{3}$, $w = 1$ and if $l = 1$, $w = \sqrt{3}$. Length is usually considered to be longer than width, so we have the solution $l = \sqrt{3}$ and $w = 1$, or $\left(\sqrt{3}, 1\right)$. The length is $\sqrt{3}$ m and the width is 1 m.

59. *Thinking and Writing Exercise.*

61. $(-1)^9 (-3)^2 = -1 \cdot 9 = -9$

63. $\dfrac{(-1)^k}{k-6} = \dfrac{(-1)^7}{7-6} = \dfrac{-1}{1} = -1$

65. $\dfrac{n}{2}(3+n) = \dfrac{11}{2}(3+11) = \dfrac{11}{2} \cdot 14 = 77$

67. *Thinking and Writing Exercise.*

69. Let (h, k) be a point on the line $5x + 8y = -2$ which is the center of a circle that passes through the points $(-2, 3)$ and $(-4, 1)$. The distance between (h, k) and $(-2, 3)$ is the same as the distance between (h, k) and $(-4, 1)$. This gives us one equation:

$$\sqrt{\left[h-(-2)\right]^2 + (k-3)^2} =$$
$$\sqrt{\left[h-(-4)\right]^2 + (k-1)^2}$$
$$(h+2)^2 + (k-3)^2 = (h+4)^2 + (k-1)^2$$
$$h^2 + 4h + 4 + k^2 - 6k + 9 =$$
$$\qquad h^2 + 8h + 16 + k^2 - 2k + 1$$
$$4h - 6k + 13 = 8h - 2k + 17$$
$$-4h - 4k = 4$$
$$h + k = -1$$

We get a second equation by substituting (h, k) in $5x + 8y = -2$.

$$5h + 8k = -2$$

We now solve the following system:
$$h + k = -1$$
$$5h + 8k = -2$$

The solution, which is the center of the circle, is $(-2, 1)$.

Next, we find the length of the radius. We can find the distance between either $(-2, 3)$ or $(-4, 1)$ and the center $(-2, 1)$. We use $(-2, 3)$.

$$r = \sqrt{\left[-2-(-2)\right]^2 + (1-3)^2}$$
$$r = \sqrt{0^2 + (-2)^2}$$
$$r = \sqrt{4} = 2$$

We can write the equation of the circle with center $(-2, 1)$ and radius 2.

$$(x-h)^2 + (y-k)^2 = r^2$$
$$\left[x-(-2)\right]^2 + (y-1)^2 = 2^2$$
$$(x+2)^2 + (y-1)^2 = 4$$

71. $p^2 + q^2 = 13$ (1)

$\dfrac{1}{pq} = -\dfrac{1}{6}$ (2)

Solve Eq. (2) for p.

$$\dfrac{1}{q} = -\dfrac{p}{6}$$
$$-\dfrac{6}{q} = p$$

Substitute $-6/q$ for p in Eq. (1) and solve for q.

$$\left(-\dfrac{6}{q}\right)^2 + q^2 = 13$$
$$\dfrac{36}{q^2} + q^2 = 13$$
$$36 + q^4 = 13q^2$$
$$q^4 - 13q^2 + 36 = 0$$
$$u^2 - 13u + 36 = 0 \quad \text{Letting } u = q^2$$
$$(u-9)(u-4) = 0$$
$$u = 9 \text{ or } u = 4$$
$$q^2 = 9 \text{ or } q^2 = 4$$
$$q = \pm 3 \text{ or } q = \pm 2$$

Since $p = -6/q$, if $q = 3$, $p = -2$; if $q = -3$, $p = 2$; if $q = 2$, $p = -3$; and if $q = -2$, $p = 3$. The pairs $(-2, 3)$, $(2, -3)$, $(-3, 2)$, and $(3, -2)$ check. They are the solutions.

73. Let l equal the length of a side of the fence and w equal the length of the other side.

$l + w = 100$ (Length of fencing)

$lw = 2475$ (Area of rectangle)

Solve the first equation for w.

$l + w = 100$

$w = 100 - l$

Substitute $100 - l$ for w in the second equation to determine:

$(100 - l)l = 2475$

$0 = l^2 - 100l + 2475$

$0 = (l - 55)(l - 45)$

$l = 55 \ or \ l = 45$

Substituting:

$w = 100 - l = 100 - 55 = 45$

$w = 100 - l = 100 - 45 = 55$

We have the same pair of numbers for both solutions. We usually think of length as the greater of the two measures, so we have a rectangle with length of 55 ft and a width of 45 ft.

75. We let x and y represent the length and width of the base of the box, respectively. Make a drawing.

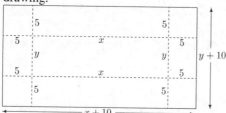

The dimensions of the metal sheet are $x + 10$ and $y + 10$. Solve the system:

$(x + 10)(y + 10) = 340$

$x \cdot y \cdot 5 = 350$

The solutions are $(10, 7)$ and $(7, 10)$.

Choosing the larger number as the length, we have the solution. The dimensions of the box are 10 in. by 7 in. by 5 in.

77. $\dfrac{\text{Length}}{\text{Height}} = \dfrac{16}{9}$

$l = \dfrac{16}{9} h$

Using the Pythagorean theorem, we have

$l^2 + h^2 = \text{Diagonal Measure}^2$

$l^2 + h^2 = 73^2$

$\left(\dfrac{16}{9} h\right)^2 + h^2 = 5329$

$\dfrac{337}{81} h^2 = 5329$

$h^2 \approx 1280.86$

$h \approx 35.8$

$l = \dfrac{16}{9} h$

$l = \dfrac{16}{9}(35.8)$

$l \approx 63.6$

The length is approximately 63.6 in., and the height is approximately 35.8 in.

79. $4xy - 7 = 0$

$x - 3y - 2 = 0$

Solve each equation for y.

$4xy - 7 = 0$

$4xy = 7$

$y = \dfrac{7}{4x}$

$x - 3y - 2 = 0$

$x - 2 = 3y$

$\dfrac{x - 2}{3} = y$

Using a graphing calculator, let

$y_1 = \dfrac{7}{4x}$ and $y_2 = \dfrac{x - 2}{3}$.

Use the INTERSECT feature to determine:

$(-1.50, -1.17)$ and $(3.50, 0.50)$.

These are the solutions.

Chapter 13 Study Summary

1. $x = y^2 + 6y + 7$

$x = y^2 + 6y + 9 - 9 + 7$

$x = (y + 3)^2 - 2$

This is a right-opening parabola, with a vertex at (−2, −3).

2. $x^2 + y^2 - 6x + 5 = 0$

   $x^2 - 6x + 9 + y^2 = -5 + 9$

   $(x - 3)^2 + y^2 = 2^2$

   The center is (3, 0) and the radius is 2

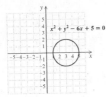

3. $\dfrac{x^2}{9} + y^2 = 1$

   The intercepts are (3, 0), (–3, 0), (0, 1), and

   (0, –1).

4. $\dfrac{y^2}{16} - \dfrac{x^2}{4} = 1$

   This is a hyperbola with a vertical axis, with vertices at (0, 4) and (0, –4) and asymptotes of $y = \pm 2x$.

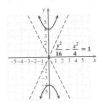

5. $x^2 + y^2 = 41$ (1)

   $y - x = 1$ (2)

   Solve (2) for y: $y = x + 1$. Then, substitute into (1):

   $x^2 + y^2 = 41$

   $x^2 + (x + 1)^2 = 41$

   $x^2 + x^2 + 2x + 1 = 41$

   $2x^2 + 2x - 40 = 0$

   $2(x + 5)(x - 4) = 0$

   $x = -5, 4$

   Substitute into (2):

$y - (-5) = 1$

$y = -4$

$y - 4 = 1$

$y = 5$

The solutions are (–5, –4) and (4, 5).

Chapter 13 Review Exercises

1. True

2. False

3. False

4. True

5. True

6. True

7. False

8. True

9. $(x + 3)^2 + (y - 2)^2 = 16$

   $[x - (-3)]^2 + (y - 2)^2 = (4)^2$

   Center: $(-3, 2)$; radius is 4.

10. $(x - 5)^2 + y^2 = 11$

    $(x - 5)^2 + (y - 0)^2 = (\sqrt{11})^2$

    Center: $(5, 0)$; radius is $\sqrt{11}$.

11. $x^2 - 6x + y^2 - 2y = -1$

    $(x^2 - 6x + 9) + (y^2 - 2y + 1) = -1 + 9 + 1$

    $(x - 3)^2 + (y - 1)^2 = 3^2$

    Center: $(3, 1)$; radius is 3.

12. $x^2 + 8x + y^2 - 6y = 20$

    $(x^2 + 8x + 16) + (y^2 - 6y + 9) = 20 + 16 + 9$

    $(x + 4)^2 + (y - 3)^2 = (\sqrt{45})^2$

    Center: $(-4, 3)$; radius is $3\sqrt{5}$.

13. $(x-h)^2 + (y-k)^2 = r^2$

$$[x-(-4)]^2 + (y-3)^2 = (4)^2$$

$$(x+4)^2 + (y-3)^2 = 16$$

14. $(x-h)^2 + (y-k)^2 = r^2$

$$(x-7)^2 + [y-(-2)]^2 = (2\sqrt{5})^2$$

$$(x-7)^2 + (y+2)^2 = 20$$

15. $5x^2 + 5y^2 = 80$

$$x^2 + y^2 = 16$$

Circle

$5x^2 + 5y^2 = 80$

16. $9x^2 + 2y^2 = 18$

$$\dfrac{x^2}{2} + \dfrac{y^2}{9} = 1$$

Ellipse

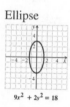

$9x^2 + 2y^2 = 18$

17. $y = -x^2 + 2x - 3$

Parabola

$y = -x^2 + 2x - 3$

18. $\dfrac{y^2}{9} - \dfrac{x^2}{4} = 1$

$$\dfrac{y^2}{3^2} - \dfrac{x^2}{2^2} = 1$$

Hyperbola

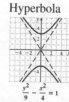

$\dfrac{y^2}{9} - \dfrac{x^2}{4} = 1$

19. $xy = 9$

$$y = \dfrac{9}{x}$$

Hyperbola

20. $x = y^2 + 2y - 2$

Parabola

$x = y^2 + 2y - 2$

21. $\dfrac{(x+1)^2}{3} + (y-3)^2 = 1$

Ellipse

$\dfrac{(x+1)^2}{3} + (y-3)^2 = 1$

22. $x^2 + y^2 + 6x - 8y - 39 = 0$

$$(x^2 + 6x + 9) + (y^2 - 8y + 16) = 39 + 9 + 16$$

$$(x+3)^2 + (y-4)^2 = 64$$

Circle

$x^2 + y^2 + 6x - 8y - 39 = 0$

23. $x^2 - y^2 = 21$ (1)

      $x + y = 3$ (2)

Solve for y using Eq. (2) and substitute into Eq. (1).

$$x + y = 3 \rightarrow y = 3 - x$$

$$x^2 - (3-x)^2 = 21$$

$$x^2 - 9 + 6x - x^2 = 21$$

$$6x = 30$$

$$x = 5$$

Substitute using $y = 3 - x$, $y = 3 - 5 = -2$.

This pair checks. Solution is $(5, -2)$.

24. $x^2 - 2x + 2y^2 = 8$ (1)

       $2x + y = 6$ (2)

Solve for y using Eq. (2) and substitute into

Eq. (1).

$2x + y = 6 \rightarrow y = 6 - 2x$

$x^2 - 2x + 2(6 - 2x)^2 = 8$

$x^2 - 2x + 72 - 48x + 8x^2 = 8$

$\qquad\qquad 9x^2 - 50x + 64 = 0$

$\qquad\qquad (9x - 32)(x - 2) = 0$

$x = \dfrac{32}{9}$ or $x = 2$

If $x = \dfrac{32}{9}$, $y = 6 - 2\left(\dfrac{32}{9}\right) = \dfrac{-10}{9}$.

If $x = 2$, $y = 6 - 2(2) = 2$.

We have $\left(\dfrac{32}{9}, \dfrac{-10}{9}\right)$ and $(2, 2)$. These pairs

check and are the solutions.

25. $x^2 - y = 5$ (1)

     $2x - y = 5$ (2)

Solve for y using Eq. (2) and substitute into

Eq. (1).

$2x - y = 5 \rightarrow y = 2x - 5$

$x^2 - (2x - 5) = 5$

$x^2 - 2x = 0$

$x(x - 2) = 0$

$x = 0$ or $x = 2$

If $x = 0$, $y = 2(0) - 5 = -5$.

If $x = 2$, $y = 2(2) - 5 = -1$.

We have $(0, -5)$ and $(2, -1)$. These pairs

check and are the solutions.

26. $x^2 + y^2 = 15$

     $x^2 - y^2 = 17$

Adding the two equations gives us:

$2x^2 = 32$

$\quad x^2 = 16$

$\quad\; x = \pm 4$

If $x = -4$, $(-4)^2 + y^2 = 15$

$\qquad\qquad\qquad y^2 = -1$

$\qquad\qquad\qquad y = \pm i$

If $x = 4$, $(4)^2 + y^2 = 15$

$\qquad\qquad\qquad y^2 = -1$

$\qquad\qquad\qquad y = \pm i$

We have $(-4, i)$, $(-4, -i)$, $(4, i)$, and

$(4, -i)$. These pairs check and are the

solutions.

27. $x^2 - y^2 = 3$

     $y = x^2 - 3$

Using substitution:

$x^2 - (x^2 - 3)^2 = 3$

$x^2 - (x^4 - 6x^2 + 9) = 3$

$0 = x^4 - 7x^2 + 12$

$0 = (x^2 - 4)(x^2 - 3)$

$x^2 - 4 = 0$ or $x^2 - 3 = 0$

$x = \pm 2$ or $x = \pm\sqrt{3}$

When $x = \pm 2$, $x^2 = 4$, so $y = 4 - 3 = 1$, and

we have the points $(2, 1)$ and $(-2, 1)$.

When $x = \pm\sqrt{3}$, $x^2 = 3$, so $3 - 3 = 0$, and we

have the points $(\sqrt{3}, 0)$ and $(-\sqrt{3}, 0)$.

These pairs check and are the solutions.

28. $x^2 + y^2 = 18$

     $2x + y = 3 \rightarrow y = 3 - 2x$

Using substitution:

$x^2 + (3 - 2x)^2 = 18$

$x^2 + 9 - 12x + 4x^2 = 18$

$5x^2 - 12x - 9 = 0$

$(5x + 3)(x - 3) = 0$

$x = -\dfrac{3}{5}$ or $x = 3$

When $x = -\dfrac{3}{5}$, $y = 3 - 2\left(-\dfrac{3}{5}\right) = \dfrac{21}{5}$.

When $x = 3$, $y = 3 - 2(3) = -3$.

The pairs $\left(-\dfrac{3}{5}, \dfrac{21}{5}\right)$ and $(3, -3)$ check and

are the solutions.

29. $x^2 + y^2 = 100 \rightarrow x^2 = 100 - y^2$

$2x^2 - 3y^2 = -120$

Using substitution:

$2(100 - y^2) - 3y^2 = -120$

$200 - 2y^2 - 3y^2 = -120$

$320 = 5y^2$

$64 = y^2$

$\pm 8 = y$

$x^2 = 100 - 64$

$x^2 = 36$

$x = \pm 6$

All the pairs check. The solutions are $(6, 8)$, $(6, -8)$, $(-6, 8)$, and $(-6, -8)$.

30. $x^2 + 2y^2 = 12$

$$xy = 4 \rightarrow x = \frac{4}{y}$$

Using substitution:

$$\left(\frac{4}{y}\right)^2 + 2y^2 = 12$$

$$\frac{16}{y^2} + 2y^2 = 12$$

$$16 + 2y^4 = 12y^2$$

$$2y^4 - 12y^2 + 16 = 0$$

$$y^4 - 6y^2 + 8 = 0$$

$$(y^2 - 4)(y^2 - 2) = 0$$

$$y = \pm 2 \ \ or \ \ y = \pm\sqrt{2}$$

When $y = -2$, $x = \dfrac{4}{-2} = -2$.

When $y = 2$, $x = \dfrac{4}{2} = 2$.

When $y = -\sqrt{2}$, $x = \dfrac{4}{-\sqrt{2}} = -2\sqrt{2}$.

When $y = \sqrt{2}$, $x = \dfrac{4}{\sqrt{2}} = 2\sqrt{2}$.

We have $(-2, -2)$, $(2, 2)$, $\left(-2\sqrt{2}, -\sqrt{2}\right)$, and $\left(2\sqrt{2}, \sqrt{2}\right)$. These pairs check and are the solutions.

31.   Using perimeter, we have $2l + 2w = 38$, or $l + w = 19$. Using area, we have $lw = 84$. Use substitution to solve.

$$l = 19 - w$$

$$(19 - w)w = 84$$

$$19w - w^2 = 84$$

$$0 = w^2 - 19w + 84$$

$$0 = (w - 12)(w - 7)$$

$$w = 12 \ \ or \ \ w = 7$$

$$l = 19 - 12 = 7 \ \ or \ \ l = 19 - 7 = 12$$

These are the same pair of numbers. We usually think of length as the greater measure, so length is 12 m and width is 7 m.

32.   Area: $lw = 108$

Using the Pythagorean theorem:

$l^2 + w^2 = 15^2$

Using substitution:

$$l = \frac{108}{w}$$

$$\left(\frac{108}{w}\right)^2 + w^2 = 15^2$$

$$11,664 + w^4 = 225w^2$$

$$w^4 - 225w^2 + 11,664 = 0$$

$$\left(w^2 - 144\right)\left(w^2 - 81\right) = 0$$

$$w = \pm 12 \ \ or \ \ w = \pm 9$$

Since measure cannot be negative, we disregard –12 and –9.

$$w = 12, \ l = \frac{108}{12} = 9$$

$$w = 9, \ l = \frac{108}{9} = 12$$

The length is 12 in. and the width is 9 in.

33.   Let x = length of side of lesser square and y = length of side of the greater. Using perimeter, we have

$4x + 12 = 4y$, or $x + 3 = y$.

Using area, we have

$x^2 + 39 = y^2$

Use substitution to solve the system of equations.

$$x^2 + 39 = (x + 3)^2$$

$$x^2 + 39 = x^2 + 6x + 9$$

$$30 = 6x$$

$$5 = x$$

$y = x + 3 = 5 + 3 = 8$

The perimeter of the lesser square is $4x = 4 \cdot 5$, or 20 cm, and the perimeter of the greater is $4y = 4 \cdot 8$, or 32 cm.

34. Let r_1 and r_2 represent the radii of the circles. $A = \pi r^2$, so we have:

$$\pi r_1^2 + \pi r_2^2 = 130\pi$$
$$r_1^2 + r_2^2 = 130$$

$C = 2\pi r_1$, so we have:

$$2\pi r_1 - 2\pi r_2 = 16\pi$$
$$2\pi(r_1 - r_2) = 16\pi$$
$$r_1 - r_2 = 8$$

Solve the system of equations:

$$r_1^2 + r_2^2 = 130$$
$$r_1 - r_2 = 8 \rightarrow r_1 = r_2 + 8$$
$$(r_2 + 8)^2 + r_2^2 = 130$$
$$r_2^2 + 16r_2 + 64 + r_2^2 = 130$$
$$2r_2^2 + 16r_2 - 66 = 0$$
$$r_2^2 + 8r_2 - 33 = 0$$
$$(r_2 + 11)(r_2 - 3) = 0$$
$$r_2 = -11 \ or \ r_2 = 3$$

The radius cannot have negative length, so $r_2 = 3$ and $r_1 = 3 + 8$, or 11. The radii have lengths of 3 ft and 11 ft.

35. *Thinking and Writing Exercise.* The graph of a parabola has one branch; whereas, the graph of a hyperbola has two. A hyperbola has asymptotes, but a parabola does not.

36. *Thinking and Writing Exercise.* Many of the relations discussed are not functions. Function notation could be used for vertical parabolas and for hyperbolas that have the axes as asymptotes.

37. $4x^2 - x - 3y^2 = 9$ (1)
    $-x^2 + x + y^2 = 2$ (2)

Multiply Eq. (2) by 3 and add to Eq. (1).

$$\begin{array}{r} 4x^2 - x - 3y^2 = 9 \\ -3x^2 + 3x + 3y^2 = 6 \quad 3 \times \text{Eq. (2)} \\ \hline x^2 + 2x \qquad = 15 \end{array}$$

$$x^2 + 2x - 15 = 0$$
$$(x + 5)(x - 3) = 0$$
$$x = -5 \ or \ x = 3$$

From Eq. (2), $y^2 = x^2 - x + 2$.

When $x = -5$, $y^2 = (-5)^2 - (-5) + 2$
$$y^2 = 25 + 5 + 2$$
$$y^2 = 32$$
$$y = \pm\sqrt{32} = \pm 4\sqrt{2}$$

When $x = 3$, $y^2 = 3^2 - 3 + 2$
$$y^2 = 9 - 3 + 2$$
$$y^2 = 8$$
$$y = \pm\sqrt{8} = \pm 2\sqrt{2}$$

The pairs check. The solutions are $\left(-5, -4\sqrt{2}\right)$, $\left(-5, 4\sqrt{2}\right)$, $\left(3, -2\sqrt{2}\right)$, and $\left(3, 2\sqrt{2}\right)$.

38. Let $(0, y)$ be the point on the y-axis that is equidistant from $(-8, 0)$ and $(8, 0)$. The distance is 10, so we have:

$$\sqrt{(-8 - 0)^2 + (y - 0)^2} = 10$$
$$\sqrt{64 + y^2} = 10$$
$$64 + y^2 = 100$$
$$y^2 = 36$$
$$y = \pm 6$$

The points are $(0, 6)$ and $(0, -6)$.

39. The three points are equidistant from the center of the circle, (h, k). Using each of the three points in the equation of a circle, we have:

$$\left[x - (-2)\right]^2 + \left[y - (-4)\right]^2 = r^2$$
$$x^2 + 4x + 4 + y^2 + 8y + 16 = r^2$$
$$x^2 + 4x + y^2 + 8y + 20 = r^2 \quad (1)$$
$$(x - 5)^2 + \left[y - (-5)\right]^2 = r^2$$
$$x^2 - 10x + 25 + y^2 + 10y + 25 = r^2$$
$$x^2 - 10x + y^2 + 10y + 50 = r^2 \quad (2)$$
$$(x - 6)^2 + (y - 2)^2 = r^2$$
$$x^2 - 12x + y^2 - 4y + 40 = r^2 \quad (3)$$

$(1) = (2):$

$$x^2 + 4x + y^2 + 8y + 20$$
$$= x^2 - 10x + y^2 + 10y + 50$$
$$14x - 2y = 30$$
$$7x - y = 15 \ (4)$$

$(1) = (3):$

$$x^2 + 4x + y^2 + 8y + 20$$
$$= x^2 - 12x + y^2 - 4y + 40$$
$$16x + 12y = 20$$
$$4x + 3y = 5 \ (5)$$

Using (4) and (5), solve the system of equations.

$$3(7x - y = 15) = 21x - 3y = 45$$
$$\underline{ 4x + 3y = 5}$$
$$25x = 50$$
$$x = 2$$
$$y = -1$$

The center of the circle is $(2, -1)$.

$$(x - 2)^2 + (y + 1)^2 = r^2$$

Choose any of the three points on the circle to determine r^2.

$$(6 - 2)^2 + (2 + 1)^2 = r^2$$
$$4^2 + 3^2 = r^2$$
$$25 = r^2$$

The equation of the circle is

$$(x - 2)^2 + (y + 1)^2 = 25$$

40. From the x-intercepts, $(-9, 0)$ and $(9, 0)$, we know $a = 9$; from the y-intercepts, $(0, -5)$ and $(0, 5)$, we know $b = 5$. So, we have the equation:

$$\frac{x^2}{9^2} + \frac{y^2}{5^2} = 1$$
$$\frac{x^2}{81} + \frac{y^2}{25} = 1$$

41. Let $(x, 0)$ be the point on the x-axis that is equidistant from $(-3, 4)$ and $(5, 6)$.

$$\sqrt{[x - (-3)]^2 + (0 - 4)^2} = \sqrt{(x - 5)^2 + (0 - 6)^2}$$
$$x^2 + 6x + 9 + 16 = x^2 - 10x + 25 + 36$$
$$16x = 36$$
$$x = \frac{36}{16} = \frac{9}{4}$$

The point is $\left(\frac{9}{4}, 0\right)$.

Chapter 13 Test

1. $(h, k) = (3, -4)$ and $r = 2\sqrt{3}$.
$$(x - h)^2 + (y - k)^2 = r^2$$
$$(x - 3)^2 + (y - (-4))^2 = (2\sqrt{3})^2$$
$$(x - 3)^2 + (y + 4)^2 = 12$$

2. $(x - 4)^2 + (y + 1)^2 = 5$
$$[x - (4)]^2 + (y - (-1))^2 = (\sqrt{5})^2$$
Center: $(4, -1)$; radius is $\sqrt{5}$.

3. $x^2 + y^2 + 4x - 6y + 4 = 0$
$$x^2 + 4x + y^2 - 6y = -4$$
$$x^2 + 4x + 4 + y^2 - 6y + 9 = -4 + 4 + 9$$
$$(x + 2)^2 + (y - 3)^2 = 3^2$$
Center: $(-2, 3)$; radius is 3.

4. $y = x^2 - 4x - 1$
Parabola

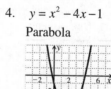

$$y = x^2 - 4x - 1$$

5. $x^2 + y^2 + 2x + 6y + 6 = 0$
Circle

$$x^2 + y^2 + 2x + 6y + 6 = 0$$

6. $\dfrac{x^2}{16} - \dfrac{y^2}{9} = 1$

Hyperbola

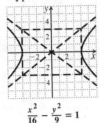

$\dfrac{x^2}{16} - \dfrac{y^2}{9} = 1$

7. $16x^2 + 4y^2 = 64$

Ellipse

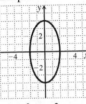

$16x^2 + 4y^2 = 64$

8. $xy = -5$

Hyperbola

$xy = -5$

9. $x = -y^2 + 4y$

Parabola

$x = -y^2 + 4y$

(4, 2)

10. $x^2 + y^2 = 36$ (1)

$3x + 4y = 24$ (2)

Solve Eq. (2) for x and substitute into

Eq. (1).

$3x + 4y = 24$

$x = \dfrac{24 - 4y}{3}$

$$x^2 + y^2 = 36$$

$$\left(\dfrac{24 - 4y}{3}\right)^2 + y^2 = 36$$

$$\dfrac{576 - 192y + 16y^2}{9} + y^2 = 36$$

$$576 - 192y + 25y^2 = 324$$

$$25y^2 - 192y + 252 = 0$$

$$(25y - 42)(y - 6) = 0$$

$$y = \dfrac{42}{25}, 6$$

If $y = \dfrac{42}{25}$, $x = \dfrac{24 - 4\left(\frac{42}{25}\right)}{3} = \dfrac{144}{25}$.

If $y = 6$, $x = \dfrac{24 - 4(6)}{3} = 0$.

The pairs check.

The solutions are $\left(\dfrac{144}{25}, \dfrac{42}{25}\right)$ and $(0, 6)$.

11. $x^2 - y = 3$ (1)

$2x + y = 5$ (2)

Add Eq. 2 and Eq. (1).

$x^2 - y = 3$

$\underline{2x + y = 5}$

$x^2 + 2x = 8$

$x^2 + 2x - 8 = 0$

$(x + 4)(x - 2) = 0$

$x = -4, 2$

$2x + y = 5$

$2(-4) + y = 5$

$y = 13$

$2(2) + y = 5$

$y = 1$

The pairs check. The solutions are $(-4, 13)$

and $(2, 1)$.

12. $x^2 - y^2 = 3$

$xy = 2$

Solve $xy = 2$ for x and substitute into the first

equation.

$$x = \frac{2}{y}$$

$$\left(\frac{2}{y}\right)^2 - y^2 = 3$$

$$\frac{4}{y^2} - y^2 = 3$$

$$4 - y^4 = 3y^2$$

$$0 = y^4 + 3y^2 - 4$$

$$0 = (y^2 + 4)(y^2 - 1)$$

$$y^2 + 4 = 0 \ \ or \ \ y^2 - 1 = 0$$

$$y = \pm\sqrt{-4} \ \ or \ \ y^2 = 1$$

$$y = \pm 2i \ \ or \ \ y = \pm 1$$

When $y = 2i$,

$$x = \frac{2}{2i} = \frac{1}{i} = \frac{i}{i^2} = -i.$$

When $y = -2i$,

$$x = \frac{2}{-2i} = \frac{1}{-i} = \frac{i}{-i^2} = i.$$

When $y = -1$, $x = \frac{2}{-1} = -2$.

When $y = 1$, $x = \frac{2}{1} = 2$.

The pairs check. The solutions are:
$(i, -2i)$, $(-i, 2i)$, $(2,1)$, and $(-2,-1)$.

13. $x^2 + y^2 = 10 \quad (1)$
     $x^2 = y^2 + 2 \quad (2)$

Using Eq. (2) substitute for x^2 into Eq. (1) and solve.

$$(y^2 + 2) + y^2 = 10$$

$$2y^2 = 8$$

$$y^2 = 4$$

$$y = \pm 2$$

When $y = -2$, $y^2 = 4$, and when $y = 2$, $y^2 = 4$.

$$x^2 = 4 + 2$$

$$x = \pm\sqrt{6}$$

The solutions are $\left(\sqrt{6}, 2\right)$, $\left(\sqrt{6}, -2\right)$, $\left(-\sqrt{6}, 2\right)$, and $\left(-\sqrt{6}, -2\right)$.

14. Area: $lw = 22$

Pythagorean Theorem: $l^2 + w^2 = \left(5\sqrt{5}\right)^2$.

Solving the first equation for l, we have:

$$l = \frac{22}{w}.$$

Using substitution:

$$\left(\frac{22}{w}\right)^2 + w^2 = 125$$

$$\frac{484}{w^2} + w^2 = 125$$

$$484 + w^4 = 125w^2$$

$$w^4 - 125w^2 + 484 = 0$$

$$\left(w^2 - 121\right)\left(w^2 - 4\right) = 0$$

$$w^2 - 121 = 0 \ \ or \ \ w^2 - 4 = 0$$

$$w = \pm 11 \ \ or \ \ w = \pm 2$$

Measure cannot be negative, so we use only $w = 11$ and $w = 2$.

If $w = 11$, then $l = \frac{22}{11} = 2$.

If $w = 2$, the $l = \frac{22}{2} = 11$.

We usually assign the greater measure to length, so length is 11 inches and width is 2 inches.

15. Let $x =$ the length of a side of the first square and $y =$ the length of a side of the second square. The areas are x^2 and y^2, respectively.

Sum: $x^2 + y^2 = 8$

Difference: $\underline{x^2 - y^2 = 2}$

$$2x^2 = 10 \quad \text{Add}$$

$$x^2 = 5$$

$$x = \pm\sqrt{5}$$

The length must be nonnegative.

If $x = \sqrt{5}$, $\quad x^2 + y^2 = 8$

$$\left(\sqrt{5}\right)^2 + y^2 = 8$$

$$y = \sqrt{3} \ \text{(Positive Value)}$$

The lengths of the sides of the squares are $\sqrt{5}$ m and $\sqrt{3}$ m.

16. Perimeter: $2l + 2w = 112$, or $l + w = 56$.
Pythagorean Theorem: $l^2 + w^2 = 40^2$.
From the first equation, $l = 56 - w$, using
substitution we have:

$$(56 - w)^2 + w^2 = 40^2$$

$$3136 - 112w + w^2 + w^2 = 1600$$

$$2w^2 - 112w + 1536 = 0$$

$$w^2 - 56w + 768 = 0$$

$$(w - 32)(w - 24) = 0$$

$w = 32 \ or \ w = 24$

Therefore, $l = 56 - 32$, or 24, and
$l = 56 - 24$, or 32.
We usually express length as the greater
measure. The length is 32 ft and the width is
24 ft.

17. $I = P \cdot R \cdot T$
For this problem, $T = 1$.
Brett invested P dollars at a rate of R.
$I = P \cdot R$

Erin invested $P + 240$ dollars at $\dfrac{5}{6}R$. They

both had interest of $72. We have two
equations:

$$72 = P \cdot R \qquad (1)$$

$$72 = (P + 240) \cdot \frac{5}{6}R \quad (2)$$

Solve Eq. (1) for R and substitute into Eq.
(2).

$$R = \frac{72}{P}$$

$$72 = (P + 240) \cdot \frac{5}{6} \cdot \frac{72}{P}$$

$$72 = (P + 240)\frac{60}{P}$$

$$72 = 60 + \frac{14,400}{P}$$

$$12 = \frac{14,400}{P}$$

$$P = \frac{14,400}{12}$$

$$P = 1200$$

$$R = \frac{72}{1200} = 0.06, \ or \ 6\%$$

Brett invested $1200 at 6%.

18. Plot the given points. From the location of
these points, we see that the ellipse which
contains them is centered at $(6, 3)$.

The distance from $(1, 3)$ to $(6, 3)$ is

$|6 - 1| = 5$, so $a = 5$. The distance from

$(6, 6)$ to $(6, 3)$ is $|6 - 3| = 3$, so $b = 3$.

We write the equation of the ellipse.

$$\frac{(x - 6)^2}{5^2} + \frac{(y - 3)^2}{3^2} = 1$$

$$\frac{(x - 6)^2}{25} + \frac{(y - 3)^2}{9} = 1$$

19. Let $(0, y)$ be the point on the y-axis which is
equidistant from $(-3, -5)$ and $(4, -7)$. We
equate their distances and solve

$$\sqrt{[0 - (-3)]^2 + [y - (-5)]^2}$$

$$= \sqrt{(0 - 4)^2 + [y - (-7)]^2}$$

$$\sqrt{9 + y^2 + 10y + 25} = \sqrt{16 + y^2 + 14y + 49}$$

$$y^2 + 10y + 34 = y^2 + 14y + 65$$

$$-4y = 31$$

$$y = \frac{-31}{4}$$

The point is $\left(0, \dfrac{-31}{4}\right)$.

20. Let x and y represent the two numbers.
  Sum: $x + y = 36$ (1)
 Product: $xy = 4$ (2)

Solve Eq. (2) for y and substitute into Eq.
(1).

$$y = \frac{4}{x}$$

$$x + \frac{4}{x} = 36$$

$$x^2 + 4 = 36x$$

$$x^2 - 36x + 4 = 0$$

We see that $a = 1$, $b = -36$, and $c = 4$.

$$x = \frac{-(-36) \pm \sqrt{(-36)^2 - 4 \cdot 1 \cdot 4}}{2 \cdot 1}$$

$$= \frac{36 \pm \sqrt{1280}}{2}$$

$$= \frac{36 \pm 16\sqrt{5}}{2}$$

$$= 18 \pm 8\sqrt{5}$$

When $x = 18 + 8\sqrt{5}$,

$$y = \frac{4}{18 + 8\sqrt{5}} \cdot \frac{18 - 8\sqrt{5}}{18 - 8\sqrt{5}}$$

$$= \frac{4\left(18 - 8\sqrt{5}\right)}{324 - 320} = 18 - 8\sqrt{5}.$$

When $x = 18 - 8\sqrt{5}$,

$$y = \frac{4}{18 - 8\sqrt{5}} = 18 + 8\sqrt{5}.$$

These are the same pairs of numbers.
We want the sum of their reciprocals:

$$\frac{1}{18 + 8\sqrt{5}} + \frac{1}{18 - 8\sqrt{5}}$$

$$= \frac{1}{18 + 8\sqrt{5}} \cdot \frac{18 - 8\sqrt{5}}{18 - 8\sqrt{5}} + \frac{1}{18 - 8\sqrt{5}} \cdot \frac{18 + 8\sqrt{5}}{18 + 8\sqrt{5}}$$

$$= \frac{18 - 8\sqrt{5}}{324 - 320} + \frac{18 + 8\sqrt{5}}{324 - 320} = \frac{36}{4} = 9$$

The sum of the reciprocals is 9.

21. Let the actor be in the center at $(0,0)$. Using

the information, we have $(-4,0)$ and $(4,0)$

and $(0,-7)$ and $(0,7)$. Thus, $a = 4$ and

$b = 7$. We write the equation of the ellipse.

$$\frac{x^2}{4^2} + \frac{y^2}{7^2} = 1$$

$$\frac{x^2}{16} + \frac{y^2}{49} = 1$$

Chapter 14

Sequences, Series, and the Binomial Theorem

1. f

3. d

5. c

7. $a_n = 2n - 3$
   $a_8 = 2 \cdot 8 - 3 = 13$

9. $a_n = (3n + 1)(2n - 5)$
   $a_9 = (3 \cdot 9 + 1)(2 \cdot 9 - 5) = 28 \cdot 13 = 364$

11. $a_n = (-1)^{n-1}(3.4n - 17.3)$
    $a_{12} = (-1)^{12-1}(3.4 \cdot 12 - 17.3) = -23.5$

13. $a_n = 3n^2(9n - 100)$
    $a_{11} = 3 \cdot 11^2(9 \cdot 11 - 100)$
    $\quad = 363 \cdot (-1)$
    $\quad = -363$

15. $a_n = \left(1 + \dfrac{1}{n}\right)^2$
    $a_{20} = \left(1 + \dfrac{1}{20}\right)^2 = \left(\dfrac{21}{20}\right)^2 = \dfrac{441}{400}$

17. $a_n = 2n + 3$
    $a_1 = 2 \cdot 1 + 3 = 5$
    $a_2 = 2 \cdot 2 + 3 = 7$
    $a_3 = 2 \cdot 3 + 3 = 9$
    $a_4 = 2 \cdot 4 + 3 = 11$
    $a_{10} = 2 \cdot 10 + 3 = 23$
    $a_{15} = 2 \cdot 15 + 3 = 33$

19. $a_n = n^2 + 2$
    $a_1 = 1^2 + 2 = 3$
    $a_2 = 2^2 + 2 = 6$
    $a_3 = 3^2 + 2 = 11$
    $a_4 = 4^2 + 2 = 18$
    $a_{10} = 10^2 + 2 = 102$
    $a_{15} = 15^2 + 2 = 227$

21. $a_n = \dfrac{n}{n+1}$
    $a_1 = \dfrac{1}{1+1} = \dfrac{1}{2},$
    $a_2 = \dfrac{2}{2+1} = \dfrac{2}{3},$
    $a_3 = \dfrac{3}{3+1} = \dfrac{3}{4},$
    $a_4 = \dfrac{4}{4+1} = \dfrac{4}{5};$
    $a_{10} = \dfrac{10}{10+1} = \dfrac{10}{11};$
    $a_{15} = \dfrac{15}{15+1} = \dfrac{15}{16}$

23. $a_n = \left(-\dfrac{1}{2}\right)^{n-1}$
    $a_1 = \left(-\dfrac{1}{2}\right)^{1-1} = 1$
    $a_2 = \left(-\dfrac{1}{2}\right)^{2-1} = -\dfrac{1}{2}$
    $a_3 = \left(-\dfrac{1}{2}\right)^{3-1} = \dfrac{1}{4}$
    $a_4 = \left(-\dfrac{1}{2}\right)^{4-1} = -\dfrac{1}{8}$
    $a_{10} = \left(-\dfrac{1}{2}\right)^{10-1} = -\dfrac{1}{512}$
    $a_{15} = \left(-\dfrac{1}{2}\right)^{15-1} = \dfrac{1}{16,384}$

25. $a_n = (-1)^n / n$

$a_1 = (-1)^1 / 1 = -1$

$a_2 = (-1)^2 / 2 = \dfrac{1}{2}$

$a_3 = (-1)^3 / 3 = -\dfrac{1}{3}$

$a_4 = (-1)^4 / 4 = \dfrac{1}{4}$

$a_{10} = (-1)^{10} / 10 = \dfrac{1}{10}$

$a_{15} = (-1)^{15} / 15 = -\dfrac{1}{15}$

27. $a_n = (-1)^n \left(n^3 - 1\right)$

$a_1 = (-1)^1 \left(1^3 - 1\right) = 0$

$a_2 = (-1)^2 \left(2^3 - 1\right) = 7$

$a_3 = (-1)^3 \left(3^3 - 1\right) = -26$

$a_4 = (-1)^4 \left(4^3 - 1\right) = 63$

$a_{10} = (-1)^{10} \left(10^3 - 1\right) = 999$

$a_{15} = (-1)^{15} \left(15^3 - 1\right) = -3374$

29. $-3, -1, 1, 3, 5$

31. $-1, -1, 1, 5, 11$

33. $\dfrac{1}{8}, \dfrac{4}{25}, \dfrac{1}{6}, \dfrac{8}{49}, \dfrac{5}{32}$

35. $2, 4, 6, 8, 10, \ldots$

These are even integers. $2n$

37. $1, -1, 1, -1, \ldots$

1 and -1 alternate, beginning with 1.

$(-1)^{n+1}$

39. $-1, 2, -3, 4, \ldots$

These are the first four natural numbers, but with alternating signs, beginning with a negative number. $(-1)^n \cdot n$

41. $3, 5, 7, 9, \ldots$

These are odd integers, beginning with 3.

$2n + 1$

43. $0, 3, 8, 15, 24, \ldots$

We can see a pattern if we write the sequence as

$0 \cdot 2, 1 \cdot 3, 2 \cdot 4, 3 \cdot 5, 4 \cdot 6$

$(n-1)(n+1)$, or $n^2 - 1$

45. $\dfrac{1}{2}, \dfrac{2}{3}, \dfrac{3}{4}, \dfrac{4}{5}, \dfrac{5}{6}, \ldots$

These are fractions in which the denominator is 1 greater than the numerator. Also, each numerator is 1 greater than the preceding numerator.

$\dfrac{n}{n+1}$

47. $5, 25, 125, 625, \ldots$

This sequence is powers of 5. 5^n

49. $-1, 4, -9, 16, \ldots$

This sequence is the squares of the first four natural numbers, but with alternating signs, beginning with a negative number.

$(-1)^n \cdot n^2$

51. $1, -2, 3, -4, 5, -6, \ldots; \ S_7$

$S_7 = 1 - 2 + 3 - 4 + 5 - 6 + 7 = 4$

53. $2, 4, 6, 8, \ldots; \ S_5$

$S_5 = 2 + 4 + 6 + 8 + 10 = 30$

55. $2, 3, \dfrac{11}{3}, \dfrac{25}{6}$

57. $-1, 3, -6, 10$

59. $\displaystyle\sum_{k=1}^{5} \dfrac{1}{2k} = \dfrac{1}{2 \cdot 1} + \dfrac{1}{2 \cdot 2} + \dfrac{1}{2 \cdot 3} + \dfrac{1}{2 \cdot 4} + \dfrac{1}{2 \cdot 5}$

$= \dfrac{1}{2} + \dfrac{1}{4} + \dfrac{1}{6} + \dfrac{1}{8} + \dfrac{1}{10}$

$= \dfrac{60}{120} + \dfrac{30}{120} + \dfrac{20}{120} + \dfrac{15}{120} + \dfrac{12}{120}$

$= \dfrac{137}{120}$

61. $\displaystyle\sum_{k=0}^{4} 10^k = 10^0 + 10^1 + 10^2 + 10^3 + 10^4$

$= 1 + 10 + 100 + 1000 + 10,000$

$= 11,111$

63. $\displaystyle\sum_{k=2}^{8} \frac{k}{k-1} = \frac{2}{2-1} + \frac{3}{3-1} + \frac{4}{4-1} + \frac{5}{5-1} + \frac{6}{6-1}$

$\qquad + \frac{7}{7-1} + \frac{8}{8-1}$

$= 2 + \frac{3}{2} + \frac{4}{3} + \frac{5}{4} + \frac{6}{5} + \frac{7}{6} + \frac{8}{7}$

$= \frac{840}{420} + \frac{630}{420} + \frac{560}{420} + \frac{525}{420}$

$\qquad + \frac{504}{420} + \frac{490}{420} + \frac{480}{42}$

$= \frac{4029}{420} = \frac{1343}{140}$

65. $\displaystyle\sum_{k=1}^{8} (-1)^{k+1} 2^k = (-1)^2 2^1 + (-1)^3 2^2 + (-1)^4 2^3$

$\qquad + (-1)^5 2^4 + (-1)^6 2^5 + (-1)^7 2^6$

$\qquad + (-1)^8 2^7 + (-1)^9 2^8$

$= 2 - 4 + 8 - 16 + 32 - 64$

$\qquad + 128 - 256$

$= -170$

67. $\displaystyle\sum_{k=0}^{5} (k^2 - 2k + 3)$

$= (0^2 - 2 \cdot 0 + 3) + (1^2 - 2 \cdot 1 + 3)$

$\qquad + (2^2 - 2 \cdot 2 + 3) + (3^2 - 2 \cdot 3 + 3)$

$\qquad + (4^2 - 2 \cdot 4 + 3) + (5^2 - 2 \cdot 5 + 3)$

$= 3 + 2 + 3 + 6 + 11 + 18$

$= 43$

69. $\displaystyle\sum_{k=3}^{5} \frac{(-1)^k}{k(k+1)} = \frac{(-1)^3}{3(3+1)} + \frac{(-1)^4}{4(4+1)} + \frac{(-1)^5}{5(5+1)}$

$= \frac{-1}{3 \cdot 4} + \frac{1}{4 \cdot 5} + \frac{-1}{5 \cdot 6}$

$= -\frac{1}{12} + \frac{1}{20} - \frac{1}{30}$

$= -\frac{4}{60} = -\frac{1}{15}$

71. $\dfrac{2}{3} + \dfrac{3}{4} + \dfrac{4}{5} + \dfrac{5}{6} + \dfrac{6}{7}$

This is a sum of fractions in which the denominator is one greater than the numerator. Also, each numerator is 1 greater than the preceding numerator.

$\displaystyle\sum_{k=1}^{5} \frac{k+1}{k+2}$.

73. $1 + 4 + 9 + 16 + 25 + 36$

This is the sum of the squares of the first six natural numbers.

$\displaystyle\sum_{k=1}^{6} k^2$.

75. $4 - 9 + 16 - 25 + \cdots + (-1)^n n^2$

This is a sum of terms of the form $(-1)^k k^2$, beginning with $k = 2$ and continuing through $k = n$.

$\displaystyle\sum_{k=2}^{n} (-1)^k k^2$.

77. $5 + 10 + 15 + 20 + 25 + \ldots$

This is a sum of multiples of 5, and it is an infinite series.

$\displaystyle\sum_{k=1}^{\infty} 5k$.

79. $\dfrac{1}{1 \cdot 2} + \dfrac{1}{2 \cdot 3} + \dfrac{1}{3 \cdot 4} + \dfrac{1}{4 \cdot 5} + \ldots$

This is a sum of fractions in which the numerator is 1 and the denominator is a product of two consecutive integers. The greater integer in each product is the lesser integer in the succeeding product. It is an infinite series.

$\displaystyle\sum_{k=1}^{\infty} \frac{1}{k(k+1)}$.

81. $u = 3n + 1$

Yscl = 5

83. $u = (-1)^\wedge n(n^2)$

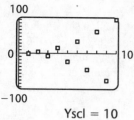

Yscl = 10

85. $u = 1/n$

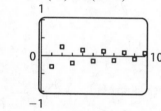

Yscl = 0.1

87. $u = (-1)^\wedge n / (n + 2)$

Yscl = 0.5

89. *Thinking and Writing Exercise.*

91. $\dfrac{7}{2}(a_1 + a_7) = \dfrac{7}{2}(8 + 20) = \dfrac{7}{2} \cdot 28 = 98$

93. $(a_1 + 3d) + d = a_1 + (3d + d)$
$\qquad\qquad\qquad = a_1 + 4d$

95. $(a_1 + a_n) + (a_1 + a_n) + (a_1 + a_n)$
$\quad = (a_1 + a_1 + a_1) + (a_n + a_n + a_n)$
$\quad = 3a_1 + 3a_n, \text{ or } 3(a_1 + a_n)$

97. *Thinking and Writing Exercise.*

99. $a_1 = 1, a_{n+1} = 5a_n - 2$
$a_1 = 1$
$a_2 = 5 \cdot 1 - 2 = 3$
$a_3 = 5 \cdot 3 - 2 = 13$
$a_4 = 5 \cdot 13 - 2 = 63$
$a_5 = 5 \cdot 63 - 2 = 313$
$a_6 = 5 \cdot 313 - 2 = 1563$

101. Find each term by multiplying the preceding term by 0.8:
$2500, $2000, $1600, $1280, $1024, $819.20, $655.36, $524.29, $419.43, $335.54

103. The sequence is –1, 1, –1, 1, –1, 1, …
Therefore, $S_2 = -1 + 1 = 0$,
$S_4 = -1 + 1 + (-1) + 1 = 0$,
$S_6 = -1 + 1 + (-1) + 1 + (-1) + 1 = 0$,
and so on until
$S_{100} = -1 + 1 + \ldots + (-1) + 1 = 0$.
$S_{101} = S_{100} + (-1) = 0 - 1 = -1$

105. $a_n = i^n$
$a_1 = i^1 = i$
$a_2 = i^2 = -1$
$a_3 = i^3 = i^2 \cdot i = -1 \cdot i = -i$
$a_4 = i^4 = \left(i^2\right)^2 = (-1)^2 = 1$
$a_5 = i^5 = \left(i^2\right)^2 \cdot i = (-1)^2 \cdot i = 1 \cdot i = i$
$S_5 = i - 1 - i + 1 + i = i$

107. Enter $y_1 = x^5 - 14x^4 + 6x^3 + 416x^2$
$-655x - 1050$. Then scroll through a table of values. We see that $y_1 = 6144$ when $x = 11$, so the 11th term of the sequence is 6144.

Exercise Set 14.2

1. True

3. False

5. True

7. False

9. $2, 6, 10, 14, \ldots$
$a_1 = 2 \quad d = 4$

11. $7, 3, -1, -5, \ldots$
$a_1 = 7 \quad d = -4$

13. $\dfrac{3}{2}, \dfrac{9}{4}, 3, \dfrac{15}{4}, \ldots$
$a_1 = \dfrac{3}{2} \quad d = \dfrac{3}{4}$

15. $\$5.12, \$5.24, \$5.36, \$5.48, \ldots$
$a_1 = \$5.12 \quad d = \$0.12$

17. $7, 10, 13, \ldots$
$a_1 = 7, \quad d = 3, \text{ and } n = 15$
$a_n = a_1 + (n-1)d$
$a_{15} = 7 + (15-1)3 = 7 + 14 \cdot 3 = 7 + 42 = 49$

19. $8, 2, -4, \ldots$
$a_1 = 8, d = -6, \text{ and } n = 18$
$a_n = a_1 + (n-1)d$
$a_{18} = 8 + (18-1)(-6) = 8 + 17(-6)$
$\phantom{a_{18}} = 8 - 102 = -94$

21. $\$1200, \$964.32, \$728.64, \ldots$
$a_1 = \$1200, d = -\$235.68, \text{ and } n = 13$
$a_n = a_1 + (n-1)d$
$a_{13} = \$1200 + (13-1)(-\$235.68)$
$\phantom{a_{13}} = \$1200 + 12(\$-\$235.68)$
$\phantom{a_{13}} = \$1200 - \$2828.16 = -\$1628.16$

23. $a_1 = 7, d = 3$
$a_n = 7 + (n-1)3$
Let $a_n = 82$
$82 = 7 + (n-1)3$
$82 = 7 + 3n - 3$
$78 = 3n$
$26 = n$
26th term

25. $a_1 = 8, d = -6$
$a_n = 8 + (n-1)(-6)$

Let $a_n = -328$
$-328 = 8 - 6n + 6$
$-342 = -6n$
$57 = n$
57th term

27. $a_n = a_1 + (n-1)d$
$a_{17} = 2 + (17-1)5$
$\phantom{a_{17}} = 2 + 16 \cdot 5 \qquad$ Substituting 17 for n,
$\phantom{a_{17}} = 2 + 80 \qquad\quad$ 2 for a_1, and 5 for d
$\phantom{a_{17}} = 82$

29. $a_n = a_1 + (n-1)d$
$33 = a_1 + (8-1)4$
$33 = a_1 + 28 \qquad$ Substituting 33 for a_8,
$ \quad$ 8 for n, and 4 for d
$5 = a_1$
(Note that this procedure is equivalent to subtracting d from a_8 seven times to get a_1: $33 - 7(4) = 33 - 28 = 5$)

31. $a_n = a_1 + (n-1)d$
$-76 = 5 + (n-1)(-3)$
$-76 = 5 - 3n + 3 \qquad$ Substituting
$-76 = 8 - 3n \qquad\quad$ -76 for a_n, 5 for a_1,
$-84 = -3n \qquad\qquad$ and -3 for d
$28 = n$

33. We know that $a_{17} = -40$ and $a_{28} = -73$. We would have to add d eleven times to get from a_{17} to a_{28}. That is,
$-40 + 11d = -73$
$11d = -33$
$d = -3.$
Since $a_{17} = -40$, we subtract d sixteen times to get to a_1.
$a_1 = -40 - 16(-3) = -40 + 48 = 8$
We write the first five terms of the sequence: $8, 5, 2, -1, -4$

35. $a_{13} = 13$ and $a_{54} = 54$
Observe that for this to be true, $a_1 = 1$ and $d = 1$.

37. $1+5+9+13+\ldots$

Note that $a_1 = 1, d = 4,$ and $n = 20.$ Before using the formula for S_n, we find a_{20}:

$a_{20} = 1 + (20-1)d$

$= 1 + 19 \cdot 4$ Substituting 4 into the formula for d

$= 77$

Then

$S_{20} = \dfrac{20}{2}(1+77)$

$= 10(78)$ Using the formula for S_n

$= 780.$

39. The sum is $1+2+3+\cdots+249+250.$ This is the sum of the arithmetic sequence for which $a_1 = 1, a_n = 250,$ and $n = 250.$ We use the formula for S_n.

$S_n = \dfrac{n}{2}(a_1 + a_n)$

$S_{300} = \dfrac{250}{2}(1+250) = 125(251) = 31,375$

41. The sum is $2+4+6+\cdots+98+100.$ This is the sum of the arithmetic sequence for which $a_1 = 2, a_n = 100,$ and $n = 50.$ Use the formula for S_n.

$S_n = \dfrac{n}{2}(a_1 + a_n)$

$S_{50} = \dfrac{50}{2}(2+100) = 2550$

43. The sum is $6+12+18+\cdots+96+102.$ This is the sum of the arithmetic sequence for which $a_1 = 6, a_n = 102,$ and $n = 17.$ We use the formula for S_n.

$S_n = \dfrac{n}{2}(a_1 + a_n)$

$S_{17} = \dfrac{17}{2}(6+102) = \dfrac{17}{2}(108) = 918$

45. Before using the formula for S_n, we find a_{20}:

$a_{20} = 4 + (20-1) \cdot 5$ Substituting into

$= 4 + 19 \cdot 5 = 99$ the formula for a_n

Then

$S_{20} = \dfrac{20}{2}(4+99)$ Using the formula for S_n

$= 10(103) = 1030.$

47. We want to find the fifteenth term and the sum of an arithmetic sequence with $a_1 = 7, d = 2,$ and $n = 15.$

$a_n = a_1 + (n-1)d$

$a_{15} = 7 + (15-1)2 = 7 + 14 \cdot 2 = 35$

$S_n = \dfrac{n}{2}(a_1 + a_2)$

$S_{15} = \dfrac{15}{2}(7+35) = \dfrac{15}{2} \cdot 42 = 315$

35 musicians in the last row, and a total of 315 musicians.

49. We have an arithmetic sequence $36, 32, 28, \ldots, 8, 4.$ $a_1 = 36,$ $d = -4,$ and $n = 9.$

$S_9 = \dfrac{9}{2}(36+4) = \dfrac{9}{2}(40) = 180$

There are 180 stones.

51. We have an arithmetic sequence $10, 20, 30, \ldots, 300, 310$

$a_1 = 10, d = 10,$ and $n = 31$

$S_{31} = \dfrac{31}{2}(310+10) = \dfrac{31}{2} \cdot 320 = 4960$

There will be 4960¢, or \$49.60.

53. We have the arithmetic sequence $20, 22, 24, \ldots a_{19}$

Determine $a_{19}.$

$a_1 = 20, d = 2, n = 19$

$a_{19} = 20 + (19-1)2 = 20 + 18 \cdot 2 = 20 + 36$

$= 56$

$S_{19} = \dfrac{19}{2}(20+56) = \dfrac{19}{2}(76) = 722$

There are 722 seats.

55. *Thinking and Writing Exercise.*

57. $m = \dfrac{1}{3}$ and $b = 10,$ so $y = \dfrac{1}{3}x + 10$

59. $2x + y = 8$ is the same as $y = -2x + 8$, so
$m = -2$.
$$y - y_1 = m(x - x_1)$$
$$y - 0 = -2(x - 5)$$
$$y = -2x + 10$$

61. $(x - h)^2 + (y - k)^2 = r^2$
$$(x - 0)^2 + (y - 0)^2 = 4^2$$
$$x^2 + y^2 = 16$$

63. *Thinking and Writing Exercise.*

65. The frog is at the bottom of a 100 foot well,
so $a_n = 100$. The frog climbs 4 ft. but slips
down 1 ft. with each jump, so $d = 4 - 1$, or 3,
and $a_1 = 3$
$$a_n = a_1 + (n - 1) \cdot 3$$
$$100 = 3 + (n - 1) \cdot 3$$
$$100 = 3 + 3n - 3$$
$$100 = 3n$$
$$33\tfrac{1}{3} = n$$
Since on the 33rd jump, the frog would be out
of the well, he would not slip; thus, we
disregard the $\tfrac{1}{3}$.
It takes the frog 33 jumps.

67. $a_1 = \$8760$
$a_2 = \$8760 + (-\$798.23) = \$7961.77$
$a_3 = \$8760 + 2(-\$798.23) = \$7163.54$
$a_4 = \$8760 + 3(-\$798.23) = \$6365.31$
$a_5 = \$8760 + 4(-\$798.23) = \$5567.08$
$a_6 = \$8760 + 5(-\$798.23) = \$4768.85$
$a_7 = \$8760 + 6(-\$798.23) = \$3970.62$
$a_8 = \$8760 + 7(-\$798.23) = \$3172.39$
$a_9 = \$8760 + 8(-\$798.23) = \$2374.16$
$a_{10} = \$8760 + 9(-\$798.23) = \$1575.93$

69. Let $d =$ the common difference. Since $p, m,$
and q form an arithmetic sequence,
$m = p + d$ and $q = p + 2d$. Then
$$\frac{p + q}{2} = \frac{p + (p + 2d)}{2} = p + d = m$$

71. Each integer from 501 through 750 is 500
more than the corresponding integer from 1
through 250. There are 250 integers from 501
through 750, so their sum is the sum of the
integers from 1 to 250 plus $250 \cdot 500$. From
Exercise 39, we know that the sum of the
integers from 1 through 250 is 31,375. Thus,
we have
$31,375 + 250 \cdot 500$, or $156,375$.

73. We graph the data points.

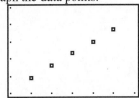

The points appear to be linear, so this could
be the graph of an arithmetic sequence. The
general term is $a_n = 0.7n + 68.3$, where $n = 1$
corresponds to a ball speed of 100 mph,
$n = 2$ corresponds to a ball speed of 101
mph, and so on.

75. We graph the data points, and they do not
appear to be linear. This is not the graph of an
arithmetic sequence.

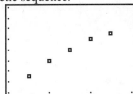

Exercise Set 14.3

1. Geometric sequence

3. Arithmetic sequence

5. Geometric series

7. Geometric series

9. $10, 20, 40, 80, \ldots$
$$\frac{20}{10} = 2, \frac{40}{20} = 2, \frac{80}{40} = 2$$
$$r = 2$$

11. $6, -0.6, 0.06, -0.006, \ldots$

$\dfrac{-0.6}{6} = -0.1, \dfrac{0.06}{-0.6} = -0.1, \dfrac{-0.006}{0.06} = -0.1$

$r = -0.1$

13. $\dfrac{1}{2}, -\dfrac{1}{4}, \dfrac{1}{8}, -\dfrac{1}{16}, \ldots$

$\dfrac{-\frac{1}{4}}{\frac{1}{2}} = -\dfrac{1}{4} \cdot \dfrac{2}{1} = -\dfrac{1}{2}$

$\dfrac{\frac{1}{8}}{-\frac{1}{4}} = \dfrac{1}{8} \cdot \dfrac{-4}{1} = -\dfrac{1}{2}$

$\dfrac{-\frac{1}{16}}{\frac{1}{8}} = -\dfrac{1}{16} \cdot \dfrac{8}{1} = -\dfrac{1}{2}$

$r = -\dfrac{1}{2}$

15. $75, 15, 3, \dfrac{3}{5}, \ldots$

$\dfrac{15}{75} = \dfrac{1}{5}, \dfrac{3}{15} = \dfrac{1}{5}, \dfrac{\frac{3}{5}}{3} = \dfrac{1}{5}$

$r = \dfrac{1}{5}$

17. $\dfrac{1}{m}, \dfrac{6}{m^2}, \dfrac{36}{m^3}, \dfrac{216}{m^4}$

$\dfrac{\frac{6}{m^2}}{\frac{1}{m}} = \dfrac{6}{m^2} \cdot \dfrac{m}{1} = \dfrac{6}{m}$

$\dfrac{\frac{36}{m^3}}{\frac{6}{m^2}} = \dfrac{36}{m^3} \cdot \dfrac{m^2}{6} = \dfrac{6}{m}$

$\dfrac{\frac{216}{m^4}}{\frac{36}{m^3}} = \dfrac{216}{m^4} \cdot \dfrac{m^3}{36} = \dfrac{6}{m}$

$r = \dfrac{6}{m}$

19. $3, 6, 12, \ldots$

$a_1 = 3, n = 7, \text{ and } r = \dfrac{6}{3} = 2$

We use the formula $a_n = a_1 r^{n-1}$.

$a_7 = 3 \cdot 2^{7-1} = 3 \cdot 2^6 = 3 \cdot 64 = 192$

21. $\sqrt{3}, 3, 3\sqrt{3}, \ldots$

$a_1 = \sqrt{3}, n = 10, r = \dfrac{3}{\sqrt{3}} = \sqrt{3}$

$a_n = a_1 r^{n-1}$

$a_{10} = \sqrt{3} \cdot \sqrt{3}^{10-1} = \sqrt{3} \cdot \sqrt{3}^9 = \sqrt{3} \cdot 3^4 \left(\sqrt{3}\right)$

$a_{10} = 3^5 = 243$

23. $-\dfrac{8}{243}, \dfrac{8}{81}, -\dfrac{8}{27}, \ldots$

$a_1 = -\dfrac{8}{243}, n = 14 \text{ and } r = \dfrac{\frac{8}{81}}{-\frac{8}{243}}$

$= \dfrac{8}{81}\left(-\dfrac{243}{8}\right) = -3$

$a_n = a_1 r^{n-1}$

$a_{14} = -\dfrac{8}{243} \cdot (-3)^{14-1}$

$= -\dfrac{8}{243} \cdot (-3)^{13}$

$= -\dfrac{8}{243} \cdot -1{,}594{,}323$

$= 52{,}488$

25. $a_1 = \$1000, n = 12, \text{ and } r = \dfrac{\$1080}{\$1000} = 1.08$

$a_n = a_1 r^{n-1}$

$a_{12} = \$1000(1.08)^{12-1}$

$\approx \$1000(2.331638997)$

$\approx \$2331.64$

27. $1, 5, 25, 125, \ldots$

$a_1 = 1; \; r = \dfrac{5}{1} = 5$

$a_n = a_1 r^{n-1}$

$a_n = 1 \cdot 5^{n-1} = 5^{n-1}$

29. $1, -1, 1, -1, \ldots$

$a_1 = 1; \; r = \dfrac{-1}{1} = -1$

$a_n = a_1 r^{n-1}$

$a_n = 1 \cdot (-1)^{n-1} = (-1)^{n-1}$

31. $\dfrac{1}{x}, \dfrac{1}{x^2}, \dfrac{1}{x^3}, \dots$

$a_1 = \dfrac{1}{x}; \; r = \dfrac{\frac{1}{x^2}}{\frac{1}{x}} = \dfrac{1}{x^2} \cdot \dfrac{x}{1} = \dfrac{1}{x}$

$a_n = a_1 r^{n-1}$

$a_n = \dfrac{1}{x} \cdot \left(\dfrac{1}{x}\right)^{n-1} = \left(\dfrac{1}{x}\right)^{1+n-1} = \left(\dfrac{1}{x}\right)^n$

$a_n = \left(\dfrac{1}{x}\right)^n, \text{ or } a_n = x^{-n}$

33. $6 + 12 + 24 + \dots$

$a_1 = 6, n = 9, r = \dfrac{12}{6} = 2$

$S_n = \dfrac{a_1\left(1 - r^n\right)}{1 - r}$

$S_9 = \dfrac{6\left(1 - 2^9\right)}{1 - 2} = \dfrac{6(-511)}{-1} = \dfrac{-3066}{-1}$

$S_9 = 3066$

35. $\dfrac{1}{18} - \dfrac{1}{6} + \dfrac{1}{2} - \dots$

$a_1 = \dfrac{1}{18}, n = 7, \text{ and } r = \dfrac{-\frac{1}{6}}{\frac{1}{18}} = -\dfrac{1}{6} \cdot \dfrac{18}{1} = -3$

$S_n = \dfrac{a_1\left(1 - r^n\right)}{1 - r}$

$S_7 = \dfrac{\frac{1}{18}\left[1 - (-3)^7\right]}{1 - (-3)} = \dfrac{\frac{1}{18}(1 + 2187)}{4} = \dfrac{\frac{1}{18}(2188)}{4}$

$= \dfrac{1}{18}(2188)\left(\dfrac{1}{4}\right) = \dfrac{547}{18}$

37. $1 + x + x^2 + x^3 + \dots$

$a_1 = 1, n = 8, \text{ and } r = \dfrac{x}{1}, \text{ or } x$

$S_n = \dfrac{a_1\left(1 - r^n\right)}{1 - r}$

$S_8 = \dfrac{1\left(1 - x^8\right)}{1 - x} = \dfrac{\left(1 + x^4\right)\left(1 - x^4\right)}{1 - x}$

$= \dfrac{\left(1 + x^4\right)\left(1 + x^2\right)\left(1 - x^2\right)}{1 - x}$

$= \dfrac{\left(1 + x^4\right)\left(1 + x^2\right)(1 + x)(1 - x)}{1 - x}$

$= \left(1 + x^4\right)\left(1 + x^2\right)(1 + x)$

39. $\$200, \$200(1.06), \$200(1.06)^2, \dots$

$a_1 = \$200, n = 16, \text{ and } r = \dfrac{\$200(1.06)}{\$200}$

$= 1.06$

$S_n = \dfrac{a_1\left(1 - r^n\right)}{1 - r}$

$S_{16} = \dfrac{\$200\left[1 - (1.06)^{16}\right]}{1 - 1.06}$

$\approx \dfrac{\$200(1 - 2.540351685)}{-0.06}$

$\approx \$5134.51$

41. $18 + 6 + 2 + \dots$

$|r| = \left|\dfrac{6}{18}\right| = \left|\dfrac{1}{3}\right| = \dfrac{1}{3}, \text{ and since } |r| < 1, \text{ the}$ series does have a limit.

$S_\infty = \dfrac{a_1}{1 - r} = \dfrac{18}{1 - \frac{1}{3}} = \dfrac{18}{\frac{2}{3}} = 18 \cdot \dfrac{3}{2} = 27$

43. $7 + 3 + \dfrac{9}{7} + \dots$

$|r| = \left|\dfrac{3}{7}\right| = \dfrac{3}{7}, \text{ and since } |r| < 1, \text{ the series}$ does have a limit.

$S_\infty = \dfrac{a_1}{1 - r} = \dfrac{7}{1 - \frac{3}{7}} = \dfrac{7}{\frac{4}{7}} = 7 \cdot \dfrac{7}{4} = \dfrac{49}{4}$

45. $3 + 15 + 75 + \dots$

$|r| = \left|\dfrac{15}{3}\right| = |5| = 5, \text{ and since } |r| \not< 1 \text{ the}$ series does not have a limit.

47. $4 - 6 + 9 - \dfrac{27}{2} + \ldots$

$|r| = \left|\dfrac{-6}{4}\right| = \left|-\dfrac{3}{2}\right| = \dfrac{3}{2}$, and since $|r| \not< 1$ the series does not have a limit.

49. $0.43 + 0.0043 + 0.000043 + \ldots$

$|r| = \left|\dfrac{0.0043}{0.43}\right| = |0.01| = 0.01$, and since $|r| < 1$, the series does have a limit.

$S_\infty = \dfrac{a_1}{1 - r} = \dfrac{0.43}{1 - 0.01} = \dfrac{0.43}{0.99} = \dfrac{43}{99}$

51. $\$500(1.02)^{-1} + \$500(1.02)^{-2}$

$\qquad + \$500(1.02)^{-3} + \ldots$

$|r| = \left|\dfrac{\$500(1.02)^{-2}}{\$500(1.02)^{-1}}\right| = |(1.02)^{-1}| = (1.02)^{-1}$,

or $\dfrac{1}{1.02}$, and since $|r| < 1$, the series does have a limit.

$S_\infty = \dfrac{a_1}{1 - r} = \dfrac{\$500(1.02)^{-1}}{1 - \left(\frac{1}{1.02}\right)} = \dfrac{\frac{\$500}{1.02}}{\frac{0.02}{1.02}}$

$\qquad = \dfrac{\$500}{1.02} \cdot \dfrac{1.02}{0.02} = \$25,000$

53. $0.7777\ldots = 0.7 + 0.07 + 0.007 + 0.0007 + \ldots$
This is an infinite geometric series with $a_1 = 0.7$.

$|r| = \left|\dfrac{0.07}{0.7}\right| = |0.1| = 0.1 < 1$, so the series has a sum.

$S_\infty = \dfrac{a_1}{1 - r} = \dfrac{0.7}{1 - 0.1} = \dfrac{0.7}{0.9} = \dfrac{7}{9}$

Fractional notation for $0.7777\ldots$ is $\dfrac{7}{9}$.

55. $8.3838\ldots = 8.3 + 0.083 + 0.00083 + \ldots$
This is an infinite geometric series with $a_1 = 8.3$.

$|r| = \left|\dfrac{0.083}{8.3}\right| = |0.01| = 0.01 < 1$, so the series has a sum.

$S_\infty = \dfrac{a_1}{1 - r} = \dfrac{8.3}{1 - 0.01} = \dfrac{8.3}{0.99} = \dfrac{830}{99}$

Fractional notation for $8.3838\ldots$ is $\dfrac{830}{99}$.

57. $0.15151515\ldots = 0.15 + 0.0015$
$\qquad\qquad\qquad + 0.000015 + \ldots$
This is an infinite geometric series with $a_1 = 0.15$.

$|r| = \left|\dfrac{0.0015}{0.15}\right| = |0.01| = 0.01 < 1$, so the series has a sum.

$S_\infty = \dfrac{a_1}{1 - r} = \dfrac{0.15}{1 - 0.01} = \dfrac{0.15}{0.99} = \dfrac{15}{99} = \dfrac{5}{33}$

Fractional notation for $0.15151515\ldots$ is $\dfrac{5}{33}$.

59. The rebound distances form a geometric sequence:

$\dfrac{1}{4} \times 20, \left(\dfrac{1}{4}\right)^2 \times 20, \left(\dfrac{1}{4}\right)^3 \times 20, \ldots$, or,

$5, \dfrac{1}{4} \times 5, \left(\dfrac{1}{4}\right)^2 \times 5, \ldots$

The height of the 6th rebound is the 6th term of the sequence.

$a_n = a_n r^{n-1}$, with

$a_1 = 5, \quad r = \dfrac{1}{4}, \quad n = 6:$

$a_6 = 5\left(\dfrac{1}{4}\right)^{6-1}$

$a_6 = 5\left(\dfrac{1}{4}\right)^5 = \dfrac{5}{1024}$

It will rebound $\dfrac{5}{1024}$ ft the 6th time.

61. In one year, the population will be $100,000 + 0.03(100,000)$, or $(1.03)100,000$. In two years, the population will be $(1.03)100,000 + 0.03(1.03)100,000$, or $(1.03)^2 100,000$.

Thus the populations form a geometric sequence:

$100,000, (1.03)100,000, (1.03)^2 100,000, \ldots$

The population in 15 years will be the 16th term of the sequence:

$$a_n = a_1 r^{n-1}$$

$$a_1 = 100,000, \quad r = 1.03, \quad n = 16:$$

$$a_{16} = 100,000(1.03)^{16-1}$$

$$a_{16} \approx 155,797.$$

In 15 years, the population will be about 155,797.

63.　We have a geometric sequence:

$$5000, 5000(0.96), 5000(0.96)^2, \ldots$$

The number of fruit flies remaining alive after 15 minutes is given by the 16th term of the sequence.

$$a_n = a_1 r^{n-1}$$

$$a_{16} = 5000 \cdot (0.96)^{16-1}$$

$$a_{16} \approx 2710$$

There will be approx. 2710 fruit flies remaining after 15 minutes.

65.　We have a geometric sequence.

$$17; 17(1.04); 17(1.04)^2 \ldots 17(1.04)^{n-1}$$

where n is the number of years after 2006.

$a_1 = 17$, $r = 1.04$, and $n = 9$ for 2015.

$$S_n = \frac{a_1(1 - r^n)}{1 - r}$$

$$S_9 = \frac{17(1 - 1.04^9)}{1 - 1.04}$$

$$S_9 \approx 179.9$$

Approximately 179.9 billion espresso-based coffees will be sold from 2007 through 2015.

67.　The lengths of the falls form a geometric sequence:

$$556, \left(\frac{3}{4}\right)556, \left(\frac{3}{4}\right)^2 556, \left(\frac{3}{4}\right)^3 556, \ldots$$

The total length of the first 6 falls is the sum of the first six terms of this sequence. The heights of the rebounds also form a geometric sequence:

$$\left(\frac{3}{4}\right)556, \left(\frac{3}{4}\right)^2 556, \left(\frac{3}{4}\right)^3 556, \ldots, \text{ or}$$

$$417, \left(\frac{3}{4}\right)417, \left(\frac{3}{4}\right)^2 417, \ldots$$

When the ball hits the ground for the 6th time, it will have rebounded 5 times. Thus the total length of the rebounds is the sum of the first five terms of this sequence.

We use the formula $S_n = \frac{a_1(1 - r^n)}{1 - r}$ twice, once with $a_1 = 556, r = \frac{3}{4}$, and $n = 6$ and a second time with $a_1 = 417, r = \frac{3}{4}$, and $n = 5$.

$D = $ Length of falls + length of rebounds

$$= \frac{556\left[1 - \left(\frac{3}{4}\right)^6\right]}{1 - \frac{3}{4}} + \frac{417\left[1 - \left(\frac{3}{4}\right)^5\right]}{1 - \frac{3}{4}}.$$

We use a calculator to obtain $D \approx 3100.35$. The ball will have traveled about 3100.35 ft.

69.　The heights of the stack form a geometric sequence:

$$0.02, 0.02(2), 0.02(2^2), \ldots$$

The height of the stack after it is doubled 10 times is given by the 11th term of this sequence.

$$a_1 = 0.02, \quad r = 2, \quad n = 11.$$

$$a_n = a_1 r^{n-1}.$$

$$a_{11} = 0.02(2^{11-1})$$

$$a_{11} = 0.02(1024) = 20.48$$

The final stack will be 20.48 in. high.

71.　The points lie on a straight line, so this is the graph of an arithmetic sequence.

73.　The points lie on the graph of an exponential function, so this is the graph of a geometric sequence.

75.　The points lie on the graph of an exponential function, so this is the graph of a geometric sequence.

77.　*Thinking and Writing Exercise.*

79.　$(x + y)^2 = (x + y)(x + y)$
$$= x^2 + xy + xy + y^2$$
$$= x^2 + 2xy + y^2$$

81.　$(x - y)^3 = (x - y)(x - y)(x - y)$
$$= (x^2 - 2xy + y^2)(x - y)$$
$$= x^3 - 2x^2 y + xy^2 - x^2 y + 2xy^2 - y^3$$
$$= x^3 - 3x^2 y + 3xy^2 - y^3$$

83. $(2x+y)^3 = (2x+y)(2x+y)(2x+y)$

$$= (4x^2 + 4xy + y^2)(2x+y)$$

$$= 8x^3 + 8x^2y + 2xy^2 + 4x^2y$$

$$+ 4xy^2 + y^3$$

$$= 8x^3 + 12x^2y + 6xy^2 + y^3$$

85. *Thinking and Writing Exercise*

87. $\displaystyle\sum_{k=1}^{\infty} 6(0.9)^k$

$a_1 = 6(0.9)^1 = 5.4$

$r = 0.9$

$S_\infty = \dfrac{a_1}{1-r} = \dfrac{5.4}{1-0.9} = \dfrac{5.4}{0.1} = 54$

89. $x^2 - x^3 + x^4 - x^5 + \ldots$

This is a geometric series with $a_1 = x^2$ and $r = -x$.

$$S_n = \frac{a_1\left(1-r^n\right)}{1-r} = \frac{x^2\left[1-(-x)^n\right]}{1-(-x)}$$

$$= \frac{x^2\left[1-(-x)^n\right]}{1+x}$$

91. The length of a side of the first square is 16 cm. The length of a side of the next square is the length of the hypotenuse of a right triangle with legs 8 cm and 8 cm, or $8\sqrt{2}$ cm. The length of a side of the next square is the length of the hypotenuse of a right triangle with legs $4\sqrt{2}$ cm and $4\sqrt{2}$ cm, or 8 cm. The areas of the squares form a sequence:

$(16)^2, \left(8\sqrt{2}\right)^2, (8)^2, \ldots$, or

$256, 128, 64, \ldots$

This is a geometric sequence with $a_1 = 256$ and $r = \frac{1}{2}$. We find the sum of the infinite geometric series $256 + 128 + 64 + \ldots$

$S_\infty = \dfrac{256}{1-\frac{1}{2}} = \dfrac{256}{\frac{1}{2}} = 512 \text{ cm}^2$

Mid-Chapter Review

Guided Solutions

1. $-6, -1, 4, 9, \ldots$

$a_n = a_1 + (n-1)d$

$n = 14, a_1 = -6, d = 5$

$a_{14} = -6 + (14-1)5$

$a_{14} = 59$

2. $\dfrac{1}{9}, -\dfrac{1}{3}, 1, -3, \ldots$

$a_n = a_1 r^{n-1}$

$n = 7, a_1 = \dfrac{1}{9}, r = -3$

$a_7 = \dfrac{1}{9} \cdot (-3)^{7-1}$

$a_7 = 81$

Mixed Review

1. $a_n = n^2 - 5n$

$a_{20} = 20^2 - 5 \cdot 20$

$= 400 - 100$

$= 300$

2. $\dfrac{1}{2}, \dfrac{1}{3}, \dfrac{1}{4}, \dfrac{1}{5}, \ldots$

These are fractions in which each denominator is 1 greater than the preceding denominator, and the numerator is always 1.

$\dfrac{1}{n+1}$

3. $1, 2, 3, 4, \ldots;\ S_{12}$

$S_{12} = 1+2+3+4+5+6+7+$

$8+9+10+11+12 = 78$

4. $\displaystyle\sum_{k=2}^{5} k^2 = 2^2 + 3^2 + 4^2 + 5^2$

$= 4+9+16+25$

$= 54$

5. $1-2+3-4+5-6$

This is a sum of terms of the form $(-1)^{k+1}k$, beginning with $k=1$ and continuing through $k=6$.

$$\sum_{k=1}^{6}(-1)^{k+1}k$$

6. $115,112,109,106,\ldots$

$112-115=-3,\ 109-112=-3,\ 106-109=-3$

$d=-3$

7. $10,15,20,25,\ldots$

$a_1=10,\ d=5,$ and $n=21$

$a_n=a_1+(n-1)d$

$a_{21}=10+(21-1)5=10+20\cdot5=110$

8. $a_1=10, d=0.2$

$a_n=10+(n-1)0.2$

Let $a_n=22$

$22=10+(n-1)0.2$

$22=10+0.2n-0.2$

$12.2=0.2n$

$61=n$

61st term

9. $a_n=a_1+(n-1)d$

$a_1=9,\ d=-2,$ and $n=25$

$a_{25}=9+(25-1)(-2)$

$a_{25}=9+(24)(-2)$

$a_{25}=9-48=-39$

10. $a_n=a_1+(n-1)d$

$a_5=65,\ d=11,$ and $n=5$

$65=a_1+(5-1)11$

$65=a_1+(4)11$

$21=a_1$

11. $a_n=a_1+(n-1)d$

$0=5+(n-1)\left(-\tfrac{1}{2}\right)$

$0=5-\tfrac{1}{2}n+\tfrac{1}{2}$

$-\tfrac{11}{2}=-\tfrac{1}{2}n$

$11=n$

Substituting 0 for a_n, 5 for a_1, and $-\tfrac{1}{2}$ for d

12. $2+12+22+32+\ldots$

$a_1=2, d=10,$ and $n=30$

$a_{30}=2+(30-1)(10)$

$=2+29(10)$

$=2+290=292$

$S_{30}=\dfrac{30}{2}[2+292]$

$=15(294)$

$=4410$

13. $\dfrac{1}{3},-\dfrac{1}{6},\dfrac{1}{12},-\dfrac{1}{24},\ldots$

$\dfrac{-\frac{1}{6}}{\frac{1}{3}}=-\dfrac{1}{2},\ \dfrac{\frac{1}{12}}{-\frac{1}{6}}=-\dfrac{1}{2},\ \dfrac{-\frac{1}{24}}{\frac{1}{12}}=-\dfrac{1}{2}$

$r=-\dfrac{1}{2}$

14. $5,10,20,40,\ldots$

$a_1=5, n=8,$ and $r=\dfrac{10}{5}=2$

We use the formula $a_n=a_1 r^{n-1}$.

$a_8=5\cdot2^{8-1}=5\cdot2^7=5\cdot128=640$

15. $2,-2,2,-2,\ldots$

$a_1=2;\ r=\dfrac{-2}{2}=-1$

$a_n=a_1 r^{n-1}$

$a_n=2(-1)^{n-1}$

16. $\$100+\$100(1.03)+\$100(1.03)^2+\ldots$

$a_1=\$100, n=10, r=\dfrac{\$100(1.03)}{\$100}=1.03$

$S_n=\dfrac{a_1\left(1-r^n\right)}{1-r}$

$S_{10}=\dfrac{100\left(1-(1.03)^{10}\right)}{1-1.03}$

$\approx\$1146.39$

17. $0.9+0.09+0.009+\ldots$

$|r|=\left|\dfrac{0.09}{0.9}\right|=|0.1|=0.1<1,$ so the series has a limit.

$$S_\infty = \frac{0.9}{1-0.1} = \frac{0.9}{0.9} = 1$$

18. $0.9 + 9 + 90 + \dots$

$|r| = \left|\dfrac{9}{0.9}\right| = |10| = 10,$ and since $|r| \not< 1$ the

series does not have a limit.

19.   We have the arithmetic sequence
$\$1, \$2, \$3, \$4 \dots a_{30}$

Determine a_{30}

$a_1 = 1, d = 1, n = 30$

$a_{30} = 1 + (30 - 1)1 = 1 + 29 = 30$

$S_{30} = \dfrac{30}{2}(1 + 30) = 15(31)$

     $= 465$

Renata earns $465 in June.

20.   We have a geometric sequence.
$\$1, \$2, \$4, \$8, \dots a_{30}$

$a_1 = 1, \ r = 2,$ and $n = 30$

$S_n = \dfrac{a_1\left(1 - r^n\right)}{1 - r}$

$S_{30} = \dfrac{1\left(1 - 2^{30}\right)}{1 - 2}$

$S_{30} = 1,073,741,823$

Dwight earns $1,073,741,823 in June.

Exercise Set 14.4

1. 2^5, or 32

3.   9

5. $\dbinom{8}{5}$, or $\dbinom{8}{3}$

7. $x^7 y^2$

9. $4! = 4 \cdot 3 \cdot 2 \cdot 1 = 24$

11. $11! = 11 \cdot 10 \cdot 9 \cdot 8 \cdot 7 \cdot 6 \cdot 5 \cdot 4 \cdot 3 \cdot 2 \cdot 1$
     $= 39,916,800$

13. $\dfrac{8!}{6!} = \dfrac{8 \cdot 7 \cdot 6!}{6!} = 8 \cdot 7 = 56$

15. $\dfrac{9!}{4!5!} = \dfrac{9 \cdot 8 \cdot 7 \cdot 6 \cdot 5!}{4 \cdot 3 \cdot 2 \cdot 5!} = \dfrac{9 \cdot 8 \cdot 7 \cdot 6}{4 \cdot 3 \cdot 2} = 126$

17. $\dbinom{7}{4} = \dfrac{7!}{3!4!} = \dfrac{7 \cdot 6 \cdot 5 \cdot 4!}{3!4!} = \dfrac{7 \cdot 6 \cdot 5}{3 \cdot 2} = 35$

19. $\dbinom{9}{9} = \dfrac{9!}{9!0!} = \dfrac{1}{0!} = 1$

21. $\dbinom{30}{2} = \dfrac{30!}{28!2!} = \dfrac{30 \cdot 29 \cdot 28!}{28!2!}$

     $= \dfrac{30 \cdot 29}{2}$

     $= 435$

23. $\dbinom{40}{38} = \dfrac{40!}{2!38!} = \dfrac{40 \cdot 39 \cdot 38!}{2!38!} = \dfrac{40 \cdot 39}{2}$

     $= 780$

25.   Expand $(a - b)^4$

Form 1: The expansion of $(a - b)^4$ has $4 + 1$, or 5 terms. The sum of the exponents in each term is 4. The exponents of a begin with 4 and decrease to 0. The exponents of $-b$ begin with 0 and increase to 4. We get the coefficients from the 5th row of Pascal's triangle.

1   4   6   4   1

$(a - b)^4 = 1 \cdot a^4 + 4a^3(-b) + 6a^2(-b)^2$

            $+ 4a(-b)^3 + (-b)^4$

     $= a^4 - 4a^3 b + 6a^2 b^2 - 4ab^3 + b^4$

Form 2: We have $a = a, b = -b, n = 4$

$(a - b)^4 = \dbinom{4}{0}a^4 + \dbinom{4}{1}a^3(-b) + \dbinom{4}{2}a^2(-b)^2$

            $+ \dbinom{4}{3}a(-b)^3 + \dbinom{4}{4}(-b)^4$

     $= \dfrac{4!}{4!0!}a^4 + \dfrac{4!}{3!1!}a^3(-b)$

        $+ \dfrac{4!}{2!2!}a^2(-b)^2 + \dfrac{4!}{1!3!}a(-b)^3$

        $+ \dfrac{4!}{0!4!}(-b)^4$

     $= a^4 - 4a^3 b + 6a^2 b^2 - 4ab^3 + b^4$

27. Expand $(p+q)^7$.

Form 1: We use the 8th row of Pascal's Triangle.

1   7   21   35   35   21   7   1

$(p+q)^7 = p^7 + 7p^6q + 21p^5q^2 + 35p^4q^3$
$+ 35p^3q^4 + 21p^2q^5 + 7pq^6 + q^7$

Form 2: We have $a=p, b=q, n=7$

$(p+q)^7 = \binom{7}{0}p^7 + \binom{7}{1}p^6q + \binom{7}{2}p^5q^2$

$+ \binom{7}{3}p^4q^3 + \binom{7}{4}p^3q^4$

$+ \binom{7}{5}p^2q^5 + \binom{7}{6}pq^6 + \binom{7}{7}q^7$

$= p^7 + 7p^6q + 21p^5q^2 + 35p^4q^3$
$+ 35p^3q^4 + 21p^2q^5 + 7pq^6 + q^7$

29. Expand $(3c-d)^7$.

Form 1: We use the eighth row of Pascal's triangle.

1   7   21   35   35   21   7   1

$(3c-d)^7 = 1 \cdot (3c)^7 + 7(3c)^6(-d)$
$+ 21(3c)^5(-d)^2 + 35(3c)^4(-d)^3$
$+ 35(3c)^3(-d)^4 + 21(3c)^2(-d)^5$
$+ 7(3c)(-d)^6 + 1 \cdot (-d)^7$
$= 2187c^7 - 5103c^6d + 5103c^5d^2$
$- 2835c^4d^3 + 945c^3d^4$
$- 189c^2d^5 + 21cd^6 - d^7$

Form 2: We have $a=3c, b=-d, n=7$

$(3c-d)^7 = \binom{7}{0}(3c)^7 + \binom{7}{1}(3c)^6(-d)$

$+ \binom{7}{2}(3c)^5(-d)^2$

$+ \binom{7}{3}(3c)^4(-d)^3$

$+ \binom{7}{4}(3c)^3(-d)^4$

$+ \binom{7}{5}(3c)^2(-d)^5$

$+ \binom{7}{6}(3c)(-d)^6 + \binom{7}{7}(-d)^7$

$= 2187c^7 - 5103c^6d + 5103c^5d^2$
$- 2835c^4d^3 + 945c^3d^4$
$- 189c^2d^5 + 21cd^6 - d^7$

31. Expand $(t^{-2}+2)^6$.

Form 1: We use the 7th row of Pascal's triangle.

1   6   15   20   15   6   1

$(t^{-2}+2)^6 = 1 \cdot (t^{-2})^6 + 6(t^{-2})^5(2)$

$+ 15(t^{-2})^4(2^2) + 20(t^{-2})^3(2^3)$

$+ 15(t^{-2})^2(2^4)$

$+ 6t^{-2}(2^5) + 1 \cdot 2^6$

$= t^{-12} + 12t^{-10} + 60t^{-8} + 160t^{-6}$
$+ 240t^{-4} + 192t^{-2} + 64$

Form 2: $a=t^{-2}, b=2, n=6$

$(t^{-2}+2)^6 = \binom{6}{0}(t^{-2})^6 + \binom{6}{1}(t^{-2})^5(2)$

$+ \binom{6}{2}(t^{-2})^4(2^2)$

$+ \binom{6}{3}(t^{-2})^3(2^3)$

$+ \binom{6}{4}(t^{-2})^2(2^4)$

$+ \binom{6}{5}(t^{-2})(2^5) + \binom{6}{6}2^6$

$= t^{-12} + 12t^{-10} + 60t^{-8} + 160t^{-6}$
$+ 240t^{-4} + 192t^{-2} + 64$

33. Expand $(x-y)^5$.

Form 1: We use the 6th row of Pascal's triangle.

1   5   10   10   5   1

$(x-y)^5 = 1 \cdot x^5 + 5x^4(-y) + 10x^3(-y)^2$
$+ 10x^2(-y)^3 + 5x(-y)^4 + 1 \cdot (-y)^5$
$= x^5 - 5x^4y + 10x^3y^2 - 10x^2y^3$
$+ 5xy^4 - y^5$

Form 2:

$$\left(x-y\right)^5 = \binom{5}{0}x^5 + \binom{5}{1}x^4\left(-y\right)$$

$$+ \binom{5}{2}x^3\left(-y\right)^2 + \binom{5}{3}x^2\left(-y\right)^3$$

$$+ \binom{5}{4}x\left(-y\right)^4 + \binom{5}{5}\left(-y\right)^5$$

$$= x^5 - 5x^4y + 10x^3y^2 - 10x^2y^3$$

$$+ 5xy^4 - y^5$$

35.   Expand $\left(3s + \dfrac{1}{t}\right)^9$.

Form 1: We use the tenth row of Pascal's triangle.

1 9 36 84 126 126 84 36 9 1

$$\left(3s + \frac{1}{t}\right)^9 = 1\cdot\left(3s\right)^9 + 9\left(3s\right)^8\left(\frac{1}{t}\right)$$

$$+ 36\left(3s\right)^7\left(\frac{1}{t}\right)^2 + 84\left(3s\right)^6\left(\frac{1}{t}\right)^3$$

$$+ 126\left(3s\right)^5\left(\frac{1}{t}\right)^4 + 126\left(3s\right)^4\left(\frac{1}{t}\right)^5$$

$$+ 84\left(3s\right)^3\left(\frac{1}{t}\right)^6 + 36\left(3s\right)^2\left(\frac{1}{t}\right)^7$$

$$+ 9\left(3s\right)\left(\frac{1}{t}\right)^8 + 1\cdot\left(\frac{1}{t}\right)^9$$

$$= 19{,}683s^9 + \frac{59{,}049s^8}{t}$$

$$+ \frac{78{,}732s^7}{t^2} + \frac{61{,}236s^6}{t^3}$$

$$+ \frac{30{,}618s^5}{t^4} + \frac{10{,}206s^4}{t^5}$$

$$+ \frac{2268s^3}{t^6} + \frac{324s^2}{t^7}$$

$$+ \frac{27s}{t^8} + \frac{1}{t^9}$$

Form 2:

$$\left(3s + \frac{1}{t}\right)^9 = \binom{9}{0}\left(3s\right)^9 + \binom{9}{1}\left(3s\right)^8\left(\frac{1}{t}\right)$$

$$+ \binom{9}{2}\left(3s\right)^7\left(\frac{1}{t}\right)^2 + \binom{9}{3}\left(3s\right)^6\left(\frac{1}{t}\right)^3$$

$$+ \binom{9}{4}\left(3s\right)^5\left(\frac{1}{t}\right)^4 + \binom{9}{5}\left(3s\right)^4\left(\frac{1}{t}\right)^5$$

$$+ \binom{9}{6}\left(3s\right)^3\left(\frac{1}{t}\right)^6 + \binom{9}{7}\left(3s\right)^2\left(\frac{1}{t}\right)^7$$

$$+ \binom{9}{8}\left(3s\right)\left(\frac{1}{t}\right)^8 + \binom{9}{9}\left(\frac{1}{t}\right)^9$$

$$= 19{,}683s^9 + \frac{59{,}049s^8}{t} + \frac{78{,}732s^7}{t^2}$$

$$+ \frac{61{,}236s^6}{t^3} + \frac{30{,}618s^5}{t^4}$$

$$+ \frac{10{,}206s^4}{t^5} + \frac{2268s^3}{t^6} + \frac{324s^2}{t^7}$$

$$+ \frac{27s}{t^8} + \frac{1}{t^9}$$

37.   Expand $\left(x^3 - 2y\right)^5$

Form 1: We use the sixth row of Pascal's triangle.

1 5 10 10 5 1

$$\left(x^3 - 2y\right)^5 = 1\cdot\left(x^3\right)^5 + 5\left(x^3\right)^4\left(-2y\right)$$

$$+ 10\left(x^3\right)^3\left(-2y\right)^2$$

$$+ 10\left(x^3\right)^2\left(-2y\right)^3 + 5\left(x^3\right)\left(-2y\right)^4$$

$$+ 1\cdot\left(-2y\right)^5$$

$$= x^{15} - 10x^{12}y + 40x^9y^2 - 80x^6y^3$$

$$+ 80x^3y^4 - 32y^5$$

Form 2:

$$\left(x^3-2y\right)^5=\binom{5}{0}\left(x^3\right)^5+\binom{5}{1}\left(x^3\right)^4(-2y)$$

$$+\binom{5}{2}\left(x^3\right)^3(-2y)^2$$

$$+\binom{5}{3}\left(x^3\right)^2(-2y)^3$$

$$+\binom{5}{4}\left(x^3\right)(-2y)^4+\binom{5}{5}(-2y)^5$$

$$=x^{15}-10x^{12}y+40x^9y^2-80x^6y^3$$

$$+80x^3y^4-32y^5$$

39. Expand $\left(\sqrt{5}+t\right)^6$.

Form 1: We use the 7th row of Pascal's triangle.

1 6 15 20 15 6 1

$$\left(\sqrt{5}+t\right)^6=1\cdot\left(\sqrt{5}\right)^6+6\left(\sqrt{5}\right)^5(t)$$

$$+15\left(\sqrt{5}\right)^4\left(t^2\right)+20\left(\sqrt{5}\right)^3\left(t^3\right)$$

$$+15\left(\sqrt{5}\right)^2\left(t^4\right)+6\left(\sqrt{5}\right)\left(t^5\right)+1\cdot t^6$$

$$=125+150\sqrt{5}t+375t^2+100\sqrt{5}t^3$$

$$+75t^4+6\sqrt{5}t^5+t^6$$

Form 2:

$$\left(\sqrt{5}+t\right)^6=\binom{6}{0}\left(\sqrt{5}\right)^6+\binom{6}{1}\left(\sqrt{5}\right)^5(t)$$

$$+\binom{6}{2}\left(\sqrt{5}\right)^4\left(t^2\right)+\binom{6}{3}\left(\sqrt{5}\right)^3\left(t^3\right)$$

$$+\binom{6}{4}\left(\sqrt{5}\right)^2\left(t^4\right)+\binom{6}{5}\left(\sqrt{5}\right)\left(t^5\right)$$

$$+\binom{6}{6}\left(t^6\right)$$

$$=125+150\sqrt{5}t+375t^2+100\sqrt{5}t^3$$

$$+75t^4+6\sqrt{5}t^5+t^6$$

41. Expand $\left(\dfrac{1}{\sqrt{x}}-\sqrt{x}\right)^6$.

Form 1: We use the 7th row of Pascal's triangle:

1 6 15 20 15 6 1

$$\left(\frac{1}{\sqrt{x}}-\sqrt{x}\right)^6=1\cdot\left(\frac{1}{\sqrt{x}}\right)^6+6\left(\frac{1}{\sqrt{x}}\right)^5\left(-\sqrt{x}\right)$$

$$+15\left(\frac{1}{\sqrt{x}}\right)^4\left(-\sqrt{x}\right)^2$$

$$+20\left(\frac{1}{\sqrt{x}}\right)^3\left(-\sqrt{x}\right)^3$$

$$+15\left(\frac{1}{\sqrt{x}}\right)^2\left(-\sqrt{x}\right)^4$$

$$+6\left(\frac{1}{\sqrt{x}}\right)\left(-\sqrt{x}\right)^5+1\cdot\left(-\sqrt{x}\right)^6$$

$$=x^{-3}-6x^{-2}+15x^{-1}-20$$

$$+15x-6x^2+x^3$$

Form 2:

$$\left(\frac{1}{\sqrt{x}}-\sqrt{x}\right)^6=\binom{6}{0}\left(\frac{1}{\sqrt{x}}\right)^6$$

$$+\binom{6}{1}\left(\frac{1}{\sqrt{x}}\right)^5\left(-\sqrt{x}\right)$$

$$+\binom{6}{2}\left(\frac{1}{\sqrt{x}}\right)^4\left(-\sqrt{x}\right)^2$$

$$+\binom{6}{3}\left(\frac{1}{\sqrt{x}}\right)^3\left(-\sqrt{x}\right)^3$$

$$+\binom{6}{4}\left(\frac{1}{\sqrt{x}}\right)^2\left(-\sqrt{x}\right)^4$$

$$+\binom{6}{5}\left(\frac{1}{\sqrt{x}}\right)\left(-\sqrt{x}\right)^5$$

$$+\binom{6}{6}\left(-\sqrt{x}\right)^6$$

$$=x^{-3}-6x^{-2}+15x^{-1}-20$$

$$+15x-6x^2+x^3$$

43. Find the 3rd term of $(a+b)^6$.

First, we note that $3=2+1$, $a=a$, $b=b$, and $n=6$. The 3rd term of the expansion is

$$\binom{6}{2}a^{6-2}b^2,\text{ or }\frac{6!}{4!2!}a^4b^2,\text{ or }15a^4b^2.$$

45. Find the 12th term of $(a-3)^{14}$.

First, we note that $12 = 11+1$, $a = a$, $b = -3$, and $n = 14$. The 12th term of the expansion is

$$\binom{14}{11} a^{14-11} \cdot (-3)^{11} = \frac{14!}{3!11!} a^3 (-177,147)$$

$$= 364 a^3 (-177,147)$$

$$= -64,481,508 a^3$$

47. Find the 5th term of $\left(2x^3 + \sqrt{y}\right)^8$.

First, we note that $5 = 4+1$, $a = 2x^3$, $b = \sqrt{y}$, and $n = 8$. The 5th term of the expansion is

$$\binom{8}{4}\left(2x^3\right)^{8-4}\left(\sqrt{y}\right)^4 = \frac{8!}{4!4!}\left(2x^3\right)^4\left(\sqrt{y}\right)^4$$

$$= 70\left(16x^{12}\right)\left(y^2\right)$$

$$= 1120 x^{12} y^2$$

49. The expansion of $\left(2u + 3v^2\right)^{10}$ has 11 terms so the 6th term is the middle term. Note that $6 = 5+1$, $a = 2u$, $b = 3v^2$, and $n = 10$. The 6th term of the expansion is

$$\binom{10}{5}\left(2u\right)^{10-5}\left(3v^2\right)^5 = \frac{10!}{5!5!}\left(2u\right)^5\left(3v^2\right)^5$$

$$= 252\left(32u^5\right)\left(243v^{10}\right)$$

$$= 1,959,552 u^5 v^{10}$$

51. The 9th term of $(x - y)^8$ is the last term, y^8.

53. *Thinking and Writing Exercise.*

55. $y = x^2 - 5$

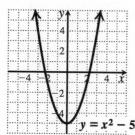

57. $y \geq x - 5$

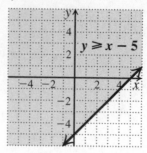

59. $f(x) = \log_5 x$

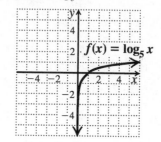

61. *Thinking and Writing Exercise.*

63. Consider a set of 5 elements, $\{a,b,c,d,e\}$.

List all the subsets of size 3:
$\{a,b,c\},\{a,b,d\},\{a,b,e\},\{a,c,d\},$
$\{a,c,e\},\{a,d,e\},\{b,c,d\},\{b,c,e\},$
$\{b,d,e\},\{c,d,e\}.$

There are exactly 10 subsets of size 3 and $\binom{5}{3} = 10$, so there are exactly $\binom{5}{3}$ ways of forming a subset of size 3 from a set of 5 elements.

65. $\binom{8}{5}\left(0.15\right)^3\left(0.85\right)^5 \approx 0.084$

67. Find and add the 7th through the 9th terms of $(0.15 + 0.85)^8$:

$$\binom{8}{6}\left(0.15\right)^2\left(0.85\right)^6 + \binom{8}{7}\left(0.15\right)\left(0.85\right)^7$$

$$+ \binom{8}{8}\left(0.85\right)^8 \approx 0.89$$

69. Prove: $\begin{pmatrix} n \\ r \end{pmatrix} = \begin{pmatrix} n \\ n-r \end{pmatrix}$

Evaluate the right expression:

$$\begin{pmatrix} n \\ n-r \end{pmatrix} = \frac{n!}{[n-(n-r)]!(n-r)!}$$

$$= \frac{n!}{r!(n-r)!} = \begin{pmatrix} n \\ r \end{pmatrix}$$

71.

$$\frac{\begin{pmatrix} 5 \\ 3 \end{pmatrix}(p^2)^2\left(-\frac{1}{2}p\sqrt[3]{q}\right)^3}{\begin{pmatrix} 5 \\ 2 \end{pmatrix}(p^2)^3\left(-\frac{1}{2}p\sqrt[3]{q}\right)^2} = \frac{-\frac{1}{8}p^7 q}{\frac{1}{4}p^8\sqrt[3]{q^2}}$$

$$= -\frac{\frac{1}{8}p^7 q}{\frac{1}{4}p^8 q^{2/3}}$$

$$= -\frac{1}{8} \cdot \frac{4}{1} \cdot p^{7-8} \cdot q^{1-2/3}$$

$$= -\frac{1}{2}p^{-1}q^{1/3} = -\frac{\sqrt[3]{q}}{2p}$$

73. $(x^2 + 2xy + y^2)(x^2 + 2xy + y^2)^2(x+y)$

$= (x+y)^2(x+y)^{2\cdot2}(x+y) = (x+y)^7$

We can find the given product by finding the binomial expansion of $(x+y)^7$. It is

$x^7 + 7x^6 y + 21x^5 y^2 + 35x^4 y^3$
$+ 35x^3 y^4 + 21x^2 y^5 + 7xy^6 + y^7$.

(See Exercise 27).

Chapter 14 Study Summary

1. $a_n = n^2 - 1$
$a_{12} = 12^2 - 1$
$= 144 - 1$
$= 143$

2. $-9, -8, -6, -3, 1, 6, 12$
$S_5 = -9 + (-8) + (-6) + (-3) + 1$
$= -25$

3. $\sum_{k=0}^{3} 5k = 5(0) + 5(1) + 5(2) + 5(3)$
$= 0 + 5 + 10 + 15$
$= 30$

4. $6, 6.5, 7, 7.5, \ldots$
$a_1 = 6, \quad d = 0.5, \text{ and } n = 20$
$a_n = a_1 + (n-1)d$
$a_{20} = 6 + (20-1)0.5 = 6 + 19 \cdot 0.5 = 15.5$

5. $6 + 6.5 + 7 + 7.5 + \ldots$
$a_1 = 6, d = 0.5, \text{ and } n = 20$
From Exercise 4, we know that $a_{20} = 15.5$.

$S_{20} = \frac{20}{2}[6 + (15.5)]$
$= 10(21.5)$
$= 215$

6. $-5, -10, -20, \ldots$
$a_1 = -5, n = 8, r = \frac{-10}{-5} = 2$
$a_n = a_1 r^{n-1}$
$a_8 = -5 \cdot 2^{8-1} = -5 \cdot 2^7 = -5 \cdot 128$
$a_8 = -640$

7. $-5 - 10 - 20 - \ldots$
$a_1 = -5, n = 12, r = \frac{-10}{-5} = 2$
$S_n = \frac{a_1(1 - r^n)}{1 - r}$
$S_{12} = \frac{-5(1 - 2^{12})}{1 - 2} = \frac{-5(-4095)}{-1}$
$S_{12} = -20,475$

8. $20 - 5 + \frac{5}{4} - \ldots$
$|r| = \left|\frac{-5}{20}\right| = \left|-\frac{1}{4}\right| = \frac{1}{4} < 1, \text{ so the series has a}$
limit.
$S_\infty = \frac{20}{1 - \left(-\frac{1}{4}\right)} = \frac{20}{\frac{5}{4}} = 20 \cdot \frac{4}{5} = 16$

9. $11! = 11 \cdot 10 \cdot 9 \cdot 8 \cdot 7 \cdot 6 \cdot 5 \cdot 4 \cdot 3 \cdot 2 \cdot 1$
$= 39,916,800$

10. $\begin{pmatrix} 9 \\ 3 \end{pmatrix} = \frac{9!}{6!3!} = \frac{9 \cdot 8 \cdot 7 \cdot 6!}{3 \cdot 2 \cdot 6!} = \frac{9 \cdot 8 \cdot 7}{3 \cdot 2} = 84$

11. Expand $\left(x^2 - 2\right)^5$

Form 1: We use the 6th Row of Pascal's triangle.

1   5   10   10   5   1

$$\left(x^2 - 2\right)^5 = 1\cdot\left(x^2\right)^5 + 5\cdot\left(x^2\right)^4\cdot(-2)$$
$$+ 10\cdot\left(x^2\right)^3\cdot(-2)^2$$
$$+ 10\cdot\left(x^2\right)^2\cdot(-2)^3$$
$$+ 5\cdot\left(x^2\right)\cdot(-2)^4 + 1\cdot(-2)^5$$
$$= x^{10} - 10x^8 + 40x^6 - 80x^4$$
$$+ 80x^2 - 32$$

Form 2: $a = x^2, b = -2, n = 5$

$$\left(x^2 - 2\right)^5 = \binom{5}{0}\left(x^2\right)^5 + \binom{5}{1}\left(x^2\right)^4(-2)$$
$$+ \binom{5}{2}\left(x^2\right)^3(-2)^2$$
$$+ \binom{5}{3}\left(x^2\right)^2(-2)^3$$
$$+ \binom{5}{4}x^2(-2)^4 + \binom{5}{5}(-2)^5$$
$$= x^{10} - 10x^8 + 40x^6 - 80x^4$$
$$+ 80x^2 - 32$$

12. Find the 4th term of $(t+3)^{10}$.

First, we note that $4 = 3+1$, $a = t$, $b = 3$, and $n = 10$. The 4th term of the expansion is

$$\binom{10}{3}(t)^{10-3}(3)^3 = \frac{10!}{3!7!}t^7\cdot 3^3$$
$$= 120\cdot 27t^7$$
$$= 3240t^7$$

Chapter 14 Review Exercises

1. False

2. True

3. True

4. False

5. False

6. True

7. False

8. False

9. $a_n = 4n - 3$
$a_1 = 4\cdot 1 - 3 = 1$
$a_2 = 4\cdot 2 - 3 = 5$
$a_3 = 4\cdot 3 - 3 = 9$
$a_4 = 4\cdot 4 - 3 = 13$
$a_8 = 4\cdot 8 - 3 = 29$
$a_{12} = 4\cdot 12 - 3 = 45$

10. $a_n = \dfrac{n-1}{n^2+1}$

$a_1 = \dfrac{1-1}{1^2+1} = 0$

$a_2 = \dfrac{2-1}{2^2+1} = \dfrac{1}{5}$

$a_3 = \dfrac{3-1}{3^2+1} = \dfrac{2}{10} = \dfrac{1}{5}$

$a_4 = \dfrac{4-1}{4^2+1} = \dfrac{3}{17}$

$a_8 = \dfrac{8-1}{8^2+1} = \dfrac{7}{65}$

$a_{12} = \dfrac{12-1}{12^2+1} = \dfrac{11}{145}$

11. $-5, -10, -15, -20, \ldots$
These are multiples of -5, where $a_1 = 1\cdot(-5)$, $a_2 = 2\cdot(-5)$, etc.
$a_n = -5n$

12. $-1, 3, -5, 7, -9, \ldots$
These could be odd counting/natural numbers with alternating signs, the first of which is negative.
$a_n = (-1)^n(2n-1)$

13. $\displaystyle\sum_{k=1}^{5}(-2)^k = (-2)^1 + (-2)^2 + (-2)^3$

$\qquad\qquad + (-2)^4 + (-2)^5$

$\qquad\quad = -2 + 4 + (-8) + 16 + (-32)$

$\qquad\quad = -22$

14. $\displaystyle\sum_{k=2}^{7}(1-2k) = (1-2\cdot2) + (1-2\cdot3) + (1-2\cdot4)$

$\qquad\qquad + (1-2\cdot5) + (1-2\cdot6)$

$\qquad\qquad + (1-2\cdot7)$

$\qquad\quad = -3 + (-5) + (-7) + (-9)$

$\qquad\qquad + (-11) + (-13)$

$\qquad\quad = -48$

15. $4 + 8 + 12 + 16 + 20$

$4\cdot1 + 4\cdot2 + 4\cdot3 + 4\cdot4 + 4\cdot5$

$\displaystyle\sum_{k=1}^{5} 4k$

16. $\dfrac{-1}{2} + \dfrac{1}{4} + \dfrac{-1}{8} + \dfrac{1}{16} + \dfrac{-1}{32}$

$\quad = -1\cdot\dfrac{1}{2^1} + \dfrac{1}{2^2} - 1\cdot\dfrac{1}{2^3} + \dfrac{1}{2^4} + -1\cdot\dfrac{1}{2^5}$

$\quad = \displaystyle\sum_{k=1}^{5}\dfrac{1}{(-2)^k}$

17. $-6,\, 1,\, 8,\, \ldots$

$\quad a_1 = -6,\, d = 7,\, n = 14$

$\quad a_{14} = -6 + (14-1)\cdot7$

$\qquad = -6 + 13\cdot7$

$\qquad = -6 + 91$

$\qquad = 85$

18. $a_n = a_1 + (n-1)d$

$\quad a_1 = 11,\, a_{13} = 43$

$\quad 43 = 11 + (13-1)d$

$\quad 32 = 12d$

$\quad \dfrac{32}{12} = d$

$\quad \dfrac{8}{3} = d$

19. $a_n = a_1 + (n-1)d$

$\quad 20 = a_1 + (8-1)d,\ 20 = a_1 + 7d$

$\quad 100 = a_1 + (24-1)d,\ 100 = a_1 + 23d$

Solve the system of equations:

$\qquad 20 = a_1 + 7d$

$\underline{(-)\,100 = a_1 + 23d}$

$\quad -80 = -16d$

$\qquad \dfrac{-80}{-16} = d$

$\qquad 5 = d$

$\quad 20 = a_1 + 7\cdot5$

$\quad 20 - 35 = a_1$

$\quad -15 = a_1$

20. $-8 + (-11) + (-14) + \cdots$

$\quad d = -3,\ a_1 = -8,\ n = 17$

$\quad a_n = a_1 + (n-1)d$

$\quad a_{17} = -8 + (17-1)(-3)$

$\quad a_{17} = -8 + 16(-3)$

$\quad a_{17} = -56$

$\quad S_n = \dfrac{n}{2}(a_1 + a_n)$

$\qquad = \dfrac{17}{2}(-8 + -56)$

$\qquad = \dfrac{17}{2}(-64)$

$\qquad = -544$

21. $5, 10, 15, \ldots, 500$

$\quad a_1 = 5,\, d = 5$

$\quad a_n = a_1 + (n-1)d$

$\quad 500 = 5 + (n-1)5$

$\quad 500 = 5 + 5n - 5$

$\quad 100 = n$

$\quad S_{100} = \dfrac{100}{2}(5 + 500)$

$\qquad = 50(505)$

$\qquad = 25,250$

22. $2, 2\sqrt{2}, 4, \cdots$

$a_1 = 2, r = \sqrt{2}, n = 20$

$a_{20} = 2 \cdot \sqrt{2}^{20-1}$

$\quad = 2\sqrt{2}^{19}$

$\quad = 2 \cdot 2^9 \sqrt{2}$

$\quad = 1024\sqrt{2}$

23. $40, 30, \dfrac{45}{2}, \cdots$

$\dfrac{30}{40} = \dfrac{3}{4}$

$\dfrac{\frac{45}{2}}{30} = \dfrac{45}{2} \cdot \dfrac{1}{30} = \dfrac{3}{4}$

$r = \dfrac{3}{4}$

24. $-2, 2, -2, \ldots$

These are negative and positive 2 alternating, beginning with -2.

$a_n = 2(-1)^n$

25. $3, \dfrac{3}{4}x, \dfrac{3}{16}x^2, \ldots$

$a_1 = 3$

$r = \dfrac{\frac{3x}{4}}{3} = \dfrac{x}{4}$

$a_n = 3\left(\dfrac{x}{4}\right)^{n-1}$

26. $3 + 12 + 48 + \cdots$

$a_1 = 3, n = 6, r = \dfrac{12}{3} = 4$

$S_n = \dfrac{a_1(1-r^n)}{1-r}$

$S_6 = \dfrac{3(1-4^6)}{1-4} = -(1-4^6)$

$\quad = 4095$

27. $3x - 6x + 12x - \ldots$

$a_1 = 3x, n = 12, r = \dfrac{-6x}{3x} = -2$

$S_n = \dfrac{a_1(1-r^n)}{1-r}$

$S_{12} = \dfrac{3x\left[1-(-2)^{12}\right]}{1-(-2)} = \dfrac{3x(1-4096)}{3}$

$\quad = -4095x$

28. $6 + 3 + 1.5 + 0.75 + \ldots$

$r = \dfrac{3}{6} = \dfrac{1}{2}; \left|\dfrac{1}{2}\right| < 1$

There is a limit.

$S_\infty = \dfrac{a_1}{1-r}$

$S_\infty = \dfrac{6}{1-\frac{1}{2}} = \dfrac{6}{\frac{1}{2}} = 6 \cdot 2 = 12$

29. $7 - 4 + \dfrac{16}{7} - \ldots$

$r = \dfrac{-4}{7}; \left|\dfrac{-4}{7}\right| < 1$

There is a limit.

$s_\infty = \dfrac{7}{1-\frac{-4}{7}} = \dfrac{7}{\frac{11}{7}} = 7 \cdot \dfrac{7}{11} = \dfrac{49}{11}$

30. $2 + (-2) + 2 + (-2) + \ldots$

$r = \dfrac{-2}{2} = -1; |-1| \nless 1$

No; there is not a limit.

31. $0.04 + 0.08 + 0.16 + 0.32 + \ldots$

$r = \dfrac{0.08}{0.04} = 2; |2| \nless 1$

No limit.

32. $\$2000 + \$1900 + \$1805 + \$1714.75 + \ldots$

$r = \dfrac{1900}{2000} = 0.95; |0.95| < 1$

There is a limit.

$S_\infty = \dfrac{2000}{1-0.95} = \dfrac{2000}{0.05} = \$40,000$

33. $0.555555\ldots = 0.5 + 0.05 + 0.005 + \ldots$
This is an infinite geometric series
with $a_1 = 0.5$.
$$|r| = \left|\frac{0.05}{0.5}\right| = 0.1$$
The series has a sum $(0.1 < 1)$.
$$S_\infty = \frac{0.5}{1 - 0.1} = \frac{0.5}{.9} = \frac{5}{9}$$
Fractional notation is $\frac{5}{9}$.

34. $1.39393939\ldots = 1 + 0.39393939\ldots$
$0.39393939\ldots$ is an infinite series
with $a_1 = 0.39$, $a_2 = 0.0039$, etc.
$$|r| = \left|\frac{0.0039}{0.39}\right| = 0.01$$
The series has a sum.
$$S_\infty = \frac{0.39}{1 - 0.01} = \frac{0.39}{0.99} = \frac{39}{99} = \frac{13}{33}$$
Since $1 = \frac{33}{33}$, $1.39393939\ldots = \frac{33}{33} + \frac{13}{33}$
Fractional notation is $\frac{46}{33}$.

35. Adams starting wage is $a_1 = 11.50$.
$d = 0.4$, the 40¢ raise; he received 4 raises per
year for 8 years, or 32 raises.
$n = 1 + 32 = 33$
$$a_{33} = 11.50 + (33 - 1)0.4$$
$$= 11.50 + 12.80$$
$$= \$24.30$$
At the end of 8 years, his hourly wage will be
$24.30.

36. We have an arithmetic sequence
$42, 41, 40, \ldots, 1$.
$a_1 = 42$, $a_{42} = 1$, $n = 42$
$$S_{40} = \frac{42}{2}(42 + 1) = 21 \cdot 43 = 903$$
There are 903 poles.

37. $\$12,000, 1.04(\$12,000), 1.04^2(\$12,000)$
$\ldots 1.04^n(\$12,000)$.
The amount to be repaid at the end of 7 years
is the amount owed at the beginning of the 8th
year.

$a_1 = 12,000$, $r = 1.04$, $n = 8$
$a_8 = 12,000 \cdot 1.04^{8-1}$
$\quad = 12,000 \cdot 1.04^7$
$\quad \approx \$15,791.18$

38. Since $r = \frac{1}{3}$ and $\left|\frac{1}{3}\right| < 1$, we determine the
sum.
$a_1 = 12$
$$S_\infty = \frac{12}{1 - \frac{1}{3}} = \frac{12}{\frac{2}{3}} = 12 \cdot \frac{3}{2} = 18$$
The total distance is 18, the fall distance is 12,
so the rebound distance is $18 - 12$, or 6.
The ball will rebound a total of 6 m.

39. $7! = 7 \cdot 6 \cdot 5 \cdot 4 \cdot 3 \cdot 2 \cdot 1 = 5040$

40. $\dbinom{8}{3} = \dfrac{8!}{5!3!} = \dfrac{8 \cdot 7 \cdot 6 \cdot 5!}{5!3!}$
$$= \frac{8 \cdot 7 \cdot 6}{3 \cdot 2} = 56$$

41. $(a + b)^{20}$
Note: $3 = 2 + 1$, $a = a$, $b = b$, and $n = 20$.
$$\binom{20}{2}a^{20-2}b^2 = 190a^{18}b^2$$

42. Expand: $(x - 2y)^4$
We use the 5th row of Pascal's triangle.
$$1 \quad 4 \quad 6 \quad 4 \quad 1$$
$(x - 2y)^4 = 1 \cdot x^4 + 4 \cdot x^3(-2y) + 6x^2(-2y)^2$
$\qquad + 4x(-2y)^3 + (-2y)^4$
$\qquad = x^4 - 8x^3y + 24x^2y^2 - 32xy^3 + 16y^4$

43. *Thinking and Writing Exercise.* For a
geometric series with $|r| < 1$, as n increases,
the absolute value of a_n decreases, since
$|r|^n$ decreases.

44. *Thinking and Writing Exercise*. The first form uses Pascal's triangle to determine coefficients; the second form uses factorial notation. The second form does not require finding the preceding rows of Pascal's triangle, and is generally easier when only one term is needed. When several terms of an expansion are needed and n is "not large" (say, $n = 8$), it is often easier to use Pascal's triangle.

45. $1 - x + x^2 - x^3 + \ldots$

$$a_1 = 1; r = \frac{-x}{1} = -x$$

$$S_n = \frac{a_1\left(1 - r^n\right)}{1 - r}$$

$$S_n = \frac{1\left[1 - \left(-x\right)^n\right]}{1 - \left(-x\right)} = \frac{1 - \left(-x\right)^n}{x + 1 \cdot}$$

46. Expand $\left(x^{-3} + x^3\right)^5$

Use the 6th row of Pascal's triangle.
1   5   10   10   5   1

$$a = x^{-3}, b = x^3, n = 5$$

$$\left(x^{-3} + x^3\right)^5 = 1 \cdot \left(x^{-3}\right)^5 + 5\left(x^{-3}\right)^4\left(x^3\right)$$

$$+ 10\left(x^{-3}\right)^3\left(x^3\right)^2 + 10\left(x^{-3}\right)^2 \cdot$$

$$\left(x^3\right)^3 + 5\left(x^{-3}\right)\left(x^3\right)^4 + \left(x^3\right)^5$$

$$= x^{-15} + 5x^{-12}x^3 + 10x^{-9}x^6$$

$$+ 10x^{-6}x^9 + 5x^{-3}x^{12} + x^{15}$$

$$= x^{-15} + 5x^{-9} + 10x^{-3} + 10x^3$$

$$+ 5x^9 + x^{15}$$

Chapter 14 Test

1. $a_n = \dfrac{1}{n^2 + 1}$

$$a_1 = \frac{1}{1^2 + 1} = \frac{1}{2}$$

$$a_2 = \frac{1}{2^2 + 1} = \frac{1}{5}$$

$$a_3 = \frac{1}{3^2 + 1} = \frac{1}{10}$$

$$a_4 = \frac{1}{4^2 + 1} = \frac{1}{17}$$

$$a_5 = \frac{1}{5^2 + 1} = \frac{1}{26}$$

$$a_{12} = \frac{1}{12^2 + 1} = \frac{1}{145}$$

2. $\dfrac{4}{3}, \dfrac{4}{9}, \dfrac{4}{27}, \ldots$

$$\frac{4}{3^1}, \frac{4}{3^2}, \frac{4}{3^3}, \ldots \frac{4}{3^n}$$

$$a_n = 4\left(\frac{1}{3}\right)^n$$

3. $\displaystyle\sum_{k=2}^{5}\left(1 - 2^k\right) = \left(1 - 2^2\right) + \left(1 - 2^3\right) + \left(1 - 2^4\right)$

$$+ \left(1 - 2^5\right)$$

$$= -3 + (-7) + (-15)$$

$$+ (-31)$$

$$= -56$$

4. $1 + (-8) + 27 + (-64) + 125$

$$1^3 + (-1) \cdot 2^3 + 3^3 + (-1) \cdot 4^3 + 5^3$$

$$\sum_{k=1}^{5}(-1)^{k+1} k^3$$

5. $\frac{1}{2}, 1, \frac{3}{2}, 2, \ldots$

$$a_1 = \frac{1}{2}, d = \frac{1}{2}, n = 13$$

$$a_n = a_1 + (n - 1)d$$

$$a_{13} = \frac{1}{2} + (13 - 1)\left(\frac{1}{2}\right)$$

$$= \frac{1}{2} + 12\left(\frac{1}{2}\right)$$

$$= \frac{13}{2}$$

6. $a_1 = 7, a_7 = -11$

Using substitution:

$a_n = a_1 + (n-1)d$

$a_7 = a_1 + (7-1)d$

$-11 = 7 + 6d$

$-18 = 6d$

$-3 = d$

7. Using substitution:

$16 = a_1 + (5-1)d, 16 = a_1 + 4d$

$-4 = a_1 + (10-1)d, -4 = a_1 + 9d$

Solve the system.

$16 = a_1 + 4d$

$(-) \dfrac{-4 = a_1 + 9d}{20 = \quad -5d}$

$-4 = d$

$-4 = a_1 + 9(-4)$

$-4 = a_1 - 36$

$32 = a_1$

8. 24, 36, 48, …, 240

$a_1 = 24, d = 12, n = 19, a_{19} = 240$

$S_n = \dfrac{n}{2}(a_1 + a_n)$

$S_{19} = \dfrac{19}{2}(24 + 240) = \dfrac{19}{2}(264)$

$= 2508$

9. $-3, 6, -12, \ldots$

$a_1 = -3, \quad n = 10, \quad r = \dfrac{6}{-3} = -2$

$a_n = a \cdot r^{n-1}$

$a_{10} = -3 \cdot (-2)^{10-1} = -3 \cdot (-2)^9 = -3 \cdot -512$

$a_{10} = 1536$

10. $22\dfrac{1}{2}, 15, 10, \ldots$

$\dfrac{15}{22\dfrac{1}{2}} = 15 \cdot \dfrac{2}{45} = \dfrac{2}{3}$

$\dfrac{10}{15} = \dfrac{2}{3}$

$r = \dfrac{2}{3}$

11. 3, 9, 27, ..

$3^1, 3^2, 3^3 \ldots 3^n$

The nth term is 3^n.

12. $11 + 22 + 44 + \ldots$

$a_1 = 11, n = 9, r = \dfrac{22}{11} = 2$

$S_n = \dfrac{a_1(1 - r^n)}{1 - r}$

$S_9 = \dfrac{(11)(1 - 2^9)}{1 - 2} = \dfrac{(11)(-511)}{-1}$

$S_9 = 5621$

13. $0.5 + 0.25 + 0.125 + \ldots$

$r = \dfrac{0.25}{0.5} = 0.5; \quad |0.5| < 1$

There is a limit.

$S_\infty = \dfrac{a_1}{1 - r}$

$S_\infty = \dfrac{0.5}{1 - 0.5} = \dfrac{0.5}{0.5} = 1$

14. $0.5 + 1 + 2 + 4 + \ldots$

$r = \dfrac{1}{0.5} = 2; |2| \not< 1$

No limit.

15. $\$1000 + \$80 + \$6.40 + \ldots$

$r = \dfrac{\$80}{\$1000} = 0.08; |0.08| < 1$

There is a limit.

$S_\infty = \dfrac{a_1}{1 - r}$

$S_\infty = \dfrac{1000}{1 - 0.08} = \dfrac{1000}{0.92} \approx \$1086.96$

16. $0.85858585\ldots$

$= 0.85 + 0.0085 + 0.000085 + \ldots$

$a_1 = 0.85$

$r = \dfrac{0.0085}{0.85} = 0.01$

$S_\infty = \dfrac{a_1}{1 - r}$

$S_\infty = \dfrac{0.85}{1 - 0.01} = \dfrac{0.85}{0.99} = \dfrac{85}{99}$

17. This is an arithmetic sequence.

$31, 33, 35, \ldots a_{18}$

$a_1 = 31, d = 2, n = 17$

$a_{17} = 31 + (17 - 1)2$

$\quad = 31 + 16 \cdot 2$

$\quad = 63$

There are 63 seats in Row 17.

18. This is an arithmetic sequence.

$100, 200, 300, \ldots, 1800$

$a_1 = 100, \ a_{18} = 1800, \ n = 18$

$S_n = \dfrac{n}{2}(a_1 + a_2)$

$S_{18} = \dfrac{18}{2}(100 + 1800) = 9 \cdot 1900$

$\quad = 17{,}100$

Her uncle gave her a total of $17,100.

19. This is a geometric sequence.

$10,000; 10,000 \cdot 0.95; 10,000 \cdot 0.95^2, \ldots$

$a_1 = 10{,}000, \ n = 11, \ r = 0.95$

$a_n = a_1 \cdot r^{n-1}$

$a_{11} = 10{,}000 \cdot (0.95)^{11-1}$

$a_{11} = 10{,}000 \cdot (0.95)^{10}$

$a_{11} \approx \$5987.37$

After 10 weeks, the boat will cost $5987.37

20. Since $r = \dfrac{2}{3}$ and $\left|\dfrac{2}{3}\right| < 1$, we determine the

sum.

$a_1 = 18$

$S_{\infty} = \dfrac{18}{1 - \frac{2}{3}} = \dfrac{18}{\frac{1}{3}} = 18 \cdot 3 = 54$

The total distance is 54 m, the fall distance is 18 m, so the rebound distance is $54 - 18$, or 36 m.

The rebound total distance is 36 m.

21. $\dbinom{12}{9} = \dfrac{12!}{3!9!} = \dfrac{12 \cdot 11 \cdot 10 \cdot 9!}{3!9!} = \dfrac{12 \cdot 11 \cdot 10}{3 \cdot 2}$

$\quad = 220$

22. Expand $(x - 3y)^5$

We will use the 6th row of Pascal's triangle.

$1 \quad 5 \quad 10 \quad 10 \quad 5 \quad 1$

$a = x, b = -3y, n = 5$

$(x - 3y)^5 = 1 \cdot (x)^5 + 5(x)^4(-3y)$

$\qquad + 10(x)^3(-3y)^2$

$\qquad + 10(x)^2(-3y)^3$

$\qquad + 5(x)(-3y)^4 + 1 \cdot (-3y)^5$

$= x^5 + 5x^4(-3y) + 10x^3 \cdot 9y^2$

$\quad + 10x^2(-27y^3)$

$\quad + 5x(81y^4) - 243y^5$

$= x^5 - 15x^4 y + 90x^3 y^2$

$\quad - 270x^2 y^3 + 405xy^4 - 243y^5$

23. $(a + x)^{12}$

Note: $4 = 3 + 1$

$a = a, b = x, n = 12$

$\dbinom{12}{3} a^{12-3} x^3 = 220a^9 x^3$

24. $2 + 4 + 6 + \ldots + 2n$

$a_1 = 2$

$S_n = \dfrac{n}{2}(2 + 2n)$

$S_n = \dfrac{n}{2} \cdot 2(1 + n)$

$S_n = n(n + 1)$

25. $1 + \dfrac{1}{x} + \dfrac{1}{x^2} + \dfrac{1}{x^3} + \ldots$

$a_1 = 1; \ r = \dfrac{\frac{1}{x}}{1} = \dfrac{1}{x}$

$S_n = \dfrac{1 \cdot \left(1 - \left(\frac{1}{x}\right)^n\right)}{1 - \frac{1}{x}} = \dfrac{1 - \left(\frac{1}{x}\right)^n}{1 - \frac{1}{x}}, \text{ or}$

$\dfrac{x^n - 1}{x^{n-1}(x - 1)}$

Chapters 1 – 14
Cumulative Review

1. $\left|-\dfrac{2}{3}+\dfrac{1}{5}\right| = \left|-\dfrac{10}{15}+\dfrac{3}{15}\right|$

   $\qquad = \left|-\dfrac{7}{15}\right|$

   $\qquad = \dfrac{7}{15}$

2. $y-[3-4(5-2y)-3y]$

   $y-[3-20+8y-3y]$

   $y-[-17+5y]$

   $-4y+17$

3. $(10\cdot 8-9\cdot 7)^2-54\div 9-3$

   $\qquad = (80-63)^2-6-3$

   $\qquad = 17^2-6-3$

   $\qquad = 289-6-3$

   $\qquad = 283-3$

   $\qquad = 280$

4. $(2.7\times 10^{-24})(3.1\times 10^9)$

   $\qquad = (2.7\times 3.1)(10^{-24}\times 10^9)$

   $\qquad = (8.37)(10^{-15})$

   $\qquad \approx 8.4\times 10^{-15}$

5. $\dfrac{ab-ac}{bc} = \dfrac{(-2)3-(-2)(-4)}{3(-4)}$

   $\qquad = \dfrac{-6-8}{-12}$

   $\qquad = \dfrac{-14}{-12} = \dfrac{7}{6}$

6. $(5a^2-3ab-7b^2)-(2a^2+5ab+8b^2)$

   $\qquad = (5-2)a^2+[(-3)-5]\,ab+[(-7)-8]b^2$

   $\qquad = 3a^2-8ab-15b^2$

7. $(2a-1)(2a+1) = 4a^2+2a-2a-1$

   $\qquad\qquad\qquad = 4a^2-1$

8. $(3a^2-5y)^2 = (3a^2)^2-2(3a^2)(5y)+(5y)^2$

   $\qquad\qquad\quad = 9a^4-30a^2y+25y^2$

9. $\dfrac{1}{x-2}-\dfrac{4}{x^2-4}+\dfrac{3}{x+2}$

   [LCD is $x^2-4=(x+2)(x-2)$]

   $= \dfrac{x+2}{x+2}\cdot\dfrac{1}{x-2}-\dfrac{4}{x^2-4}+\dfrac{x-2}{x-2}\cdot\dfrac{3}{x+2}$

   $= \dfrac{x+2-4+3(x-2)}{(x+2)(x-2)}$

   $= \dfrac{x+2-4+3x-6}{(x+2)(x-2)}$

   $= \dfrac{4x-8}{(x+2)(x-2)} = \dfrac{4(x-2)}{(x+2)(x-2)}$

   $= \dfrac{4}{x+2}$

10. $\dfrac{x^2-6x+8}{4x+12}\cdot\dfrac{x+3}{x^2-4}$

    $= \dfrac{(x-4)(x-2)}{4(x+3)}\cdot\dfrac{x+3}{(x-2)(x+2)}$

    $= \dfrac{(x-4)(x-2)(x+3)}{4(x+3)(x-2)(x+2)}$

    $= \dfrac{x-4}{4(x+2)}$

11. $\dfrac{3x+3y}{5x-5y}\div\dfrac{3x^2+3y^2}{5x^3-5y^3}$

    $= \dfrac{3x+3y}{5x-5y}\cdot\dfrac{5x^3-5y^3}{3x^2+3y^2}$

    $= \dfrac{3(x+y)}{5(x-y)}\cdot\dfrac{5(x-y)(x^2+xy+y^2)}{3(x^2+y^2)}$

    $= \dfrac{3(x+y)\cdot 5(x-y)(x^2+xy+y^2)}{5(x-y)\cdot 3(x^2+y^2)}$

    $= \dfrac{(x+y)(x^2+xy+y^2)}{x^2+y^2}$

12. $\dfrac{x - \frac{a^2}{x}}{1 + \frac{a}{x}}$ [LCD is x]

$$\dfrac{x}{x} \cdot \dfrac{x - \frac{a^2}{x}}{1 + \frac{a}{x}} = \dfrac{x^2 - a^2}{x + a}$$

$$= \dfrac{(\cancel{x+a})(x-a)}{\cancel{x+a}}$$

$$= x - a$$

13. $\sqrt{12a}\sqrt{12a^3 b} = \sqrt{12^2 \cdot a^4 \cdot b}$

$$= 12a^2 \sqrt{b}$$

14. $(-9x^2 y^5)(3x^8 y^{-7})$

$$= -9(3)x^{2+8} y^{5+(-7)}$$

$$= -27x^{10} y^{-2} = -\dfrac{27x^{10}}{y^2}$$

15. $(125x^6 y^{1/2})^{2/3} = (5^3)^{2/3}(x^6)^{2/3}(y^{1/2})^{2/3}$

$$= 5^2 x^4 y^{1/3}$$

$$= 25x^4 y^{1/3}$$

16. $\dfrac{\sqrt[3]{x^2 y^5}}{\sqrt[4]{xy^2}} = \dfrac{\sqrt[12]{x^8 y^{20}}}{\sqrt[12]{x^3 y^6}} = \sqrt[12]{x^5 y^{14}} = y\sqrt[12]{x^5 y^2}$

17. $(4 + 6i)(2 - i) = 4 \cdot 2 + 4(-i) + 6i \cdot 2 + 6i(-i)$

$$= 8 - 4i + 12i - 6i^2$$

$$= 8 + 8i + 6$$

$$= 14 + 8i$$

18. $4x^2 - 12x + 9 = (2x)^2 - 12x + 3^2$

$$= (2x - 3)^2$$

19. $27a^3 - 8 = (3a)^3 - (2)^3$

$$= (3a - 2)[(3a)^2 + 3a \cdot 2 + 2^2]$$

$$= (3a - 2)(9a^2 + 6a + 4)$$

20. $12s^4 - 48t^2 = 12(s^4 - 4t^2)$

$$= 12(s^2 - 2t)(s^2 + 2t)$$

21. $15y^4 + 33y^2 - 36 = 3(5y^4 + 11y^2 - 12)$

$$= 3(5y^2 - 4)(y^2 + 3)$$

22. $(7x^4 - 5x^3 + x^2 - 4) \div (x - 2)$

$$\underline{2}|7 - 5 \quad 1 \quad 0 \; -4$$

$$\underline{\quad\; 14 \; 18 \; 38 \; 76}$$

$$7 \; 9 \; 19 \; 38 \,|72$$

$$7x^3 + 9x^2 + 19x + 38 + \dfrac{72}{x - 2}$$

23. $f(x) = 3x^2 - 4x$

$$f(-2) = 3 \cdot (-2)^2 - 4(-2)$$

$$= 3 \cdot 4 + 8$$

$$= 20$$

24. $f(x) = \sqrt{2x - 8}$

The domain of $f(x)$ is the set of values for which $2x - 8 \geq 0$.

$$2x - 8 \geq 0$$

$$2x \geq 8$$

$$x \geq 4$$

Domain: $[4, \infty)$; or $\{x \,|\, x \geq 4\}$

25. $g(x) = \dfrac{x - 4}{x^2 - 10x + 25}$

The domain of $g(x)$ is the set of values for which $x^2 - 10x + 25 \neq 0$.

Solve: $x^2 - 10x + 25 = 0$, and exclude the value(s) from the domain.

$$x^2 - 10x + 25 = 0$$

$$(x - 5)^2 = 0$$

$$x = 5$$

Domain: $\{x \,|\, x \text{ is a real number and } x \neq 5\}$,

$$\text{or } (-\infty, 5) \cup (5, \infty)$$

26. $\dfrac{1 - \sqrt{x}}{1 + \sqrt{x}}$

[Conjugate of the denominator is $1 - \sqrt{x}$]

$$\dfrac{1 - \sqrt{x}}{1 + \sqrt{x}} \cdot \dfrac{1 - \sqrt{x}}{1 - \sqrt{x}} = \dfrac{1 - 2\sqrt{x} + x}{1 - x}$$

27. Parallel lines have equal slopes.
    $3x - y = 6$ has a slope of 3.
    $\left(m = \frac{-a}{b}, \frac{-3}{-1} = 3 \right)$, as does the line with
    y-intercept of $(0, -8)$.
    $y = mx + b$
    $y = 3x + (-8)$
    $y = 3x - 8$

28. $x = 5\sqrt{2}$ or $x = -5\sqrt{2}$
    $x - 5\sqrt{2} = 0$ or $x + 5\sqrt{2} = 0$
    $(x - 5\sqrt{2})(x + 5\sqrt{2}) = 0$
    $x^2 - (5\sqrt{2})^2 = 0$
    $x^2 - 50 = 0$

29. $x^2 + y^2 - 4x + 6y - 23 = 0$
    $x^2 - 4x + y^2 + 6y = 23$
    $(x^2 - 4x + 4) + (y^2 + 6y + 9) = 23 + 4 + 9$
    $(x - 2)^2 + (y + 3)^2 = 36$
    $(x - 2)^2 + [y - (-3)]^2 = 6^2$
    Center is $(2, -3)$
    Radius is 6.

30. $\frac{2}{3} \log_a x - \frac{1}{2} \log_a y + 5 \log_a z$
    $= \log_a x^{2/3} - \log_a y^{1/2} + \log_a z^5$
    $= \log_a \sqrt[3]{x^2} - \log_a \sqrt{y} + \log_a z^5$
    $= \log_a \frac{\sqrt[3]{x^2} \cdot z^5}{\sqrt{y}}$

31. $\log_a c = 5$
    $a^5 = c$

32. $\log 120 \approx 2.0792$

33. $\log_5 3 = \frac{\log 3}{\log 5}$
    ≈ 0.6826

34. $d = \sqrt{(x_2 - x_1)^2 + (y_2 - y_1)^2}$
    $d = \sqrt{[2 - (-1)]^2 + [-1 - (-5)]^2}$
    $d = \sqrt{3^2 + 4^2} = \sqrt{25} = 5$

35. $19, 12, 5, \ldots$
    $a_1 = 19, \quad d = -7, \quad n = 21$
    $a_{21} = a_1 + (n - 1)d$
    $a_{21} = 19 + (21 - 1) \cdot -7$
    $a_{21} = 19 + (-140)$
    $a_{21} = -121$

36. $-1 + 2 + 5 + \ldots$
    $a_1 = -1, \quad d = 3, \quad n = 25$
    $a_n = a_1 + (n - 1)d$
    $a_{25} = -1 + (25 - 1)3$
    $a_{25} = -1 + 72 = 71$
    $S_n = \frac{n}{2}(a_1 + a_n)$
    $S_{25} = \frac{25}{2}(-1 + 71)$
    $S_{25} = \frac{25}{2}(70) = 875$

37. $16, 4, 1, \ldots$
    $\frac{4}{16} = \frac{1}{4}, \frac{1}{4} = \frac{1}{4}$
    Thus, $r = \frac{1}{4}$; $a_1 = 16$
    $a_n = a_1 r^{n-1}$
    $a_n = 16 \left(\frac{1}{4} \right)^{n-1}$

38. $(a - 2b)^{10}; 7^{th}$ term
    $n = 10, \quad a = a, \quad b = -2b$
    $7 = 6 + 1, r = 6$
    $\binom{n}{r} a^{n-r} b^r$
    $\binom{10}{6} a^{10-6} (-2b)^6$
    $210 \cdot a^4 \cdot 64 b^6 = 13,440 a^4 b^6$

39. $4+6+9+\dots$

$$a_1 = 4, \quad r = \frac{6}{4} = 1.5, \quad n = 9$$

$$S_n = \frac{a_1(1-r^n)}{1-r}$$

$$S_9 = \frac{4(1-1.5^9)}{1-1.5}$$

$$S_9 \approx 299.546875$$

40. $8(x-1)-3(x-2)=1$

$$8x-8-3x+6=1$$
$$5x-2=1$$
$$5x=3$$
$$x=\frac{3}{5}$$

41. $\dfrac{6}{x}+\dfrac{6}{x+2}=\dfrac{5}{2}$

LCD is $2x(x+2)$

$$2x(x+2)\left[\frac{6}{x}+\frac{6}{x+2}\right]=2x(x+2)\cdot\frac{5}{2}$$

$$2x(x+2)\cdot\frac{6}{x}+2x(x+2)\cdot\frac{6}{x+2}$$
$$=2x(x+2)\cdot\frac{5}{2}$$

$$12(x+2)+12x=5x(x+2)$$
$$12x+24+12x=5x^2+10x$$
$$0=5x^2-14x-24$$
$$0=(5x+6)(x-4)$$

$5x+6=0 \text{ or } x-4=0$

$$x=-\frac{6}{5} \text{ or } x=4$$

42. $2x+1>5 \text{ or } x-7\le 3$

$$2x>4 \text{ or } x\le 10$$
$$x>2 \text{ or } x\le 10$$

All real numbers are either greater than 2, less than 10, or both.

$\mathbb{R}$, or $(-\infty, \infty)$

43. $5x+6y=-2 \quad (1)$

$3x+10y=2 \quad (2)$

$-3\cdot(1)+5\cdot(2)$

$$-15x-18y=6$$
$$\underline{15x+50y=10}$$
$$32y=16 \text{ (add)}$$

$$y=\frac{1}{2}$$

$$5x+6\left(\frac{1}{2}\right)=-2$$
$$5x=-5$$
$$x=-1$$

$$\left(-1,\frac{1}{2}\right)$$

44.

$$x+y-z=0 \,(1)$$
$$3x+y+z=6\,(2)$$
$$x-y+2z=5\,(3)$$

$(1)+(3) \quad x+y-z=0$

$$\underline{x-y+2z=5}$$
$$(4)\,2x \quad +z=5 \text{ (Add)}$$

$(2)+(3) \quad 3x+y+z=6$

$$\underline{x-y+2z=5}$$
$$(5)\,4x \quad +3z=11 \text{ (Add)}$$

Using (4) and (5) solve this system.

$$2x+z=5\,(4)$$
$$4x+3z=11\,(5)$$

$-2\cdot(4)+(5)$

$$-4x-2z=-10$$
$$\underline{4x+3z=11}$$
$$z=1$$

$$2x+1=5$$
$$2x=4$$
$$x=2$$

$$x+y-z=0$$
$$2+y-1=0$$
$$y=-1$$

$$(2,-1,1)$$

45. $3\sqrt{x-1} = 5-x$

$(3\sqrt{x-1})^2 = (5-x)^2$

$9(x-1) = 25-10x+x^2$

$9x-9 = x^2-10x+25$

$0 = x^2-19x+34$

$0 = (x-2)(x-17)$

$x-2 = 0$ or $x-17 = 0$

$x = 2$ or $x = 17$

Only the value 2 checks.
Solution is 2.

46. $x^4 - 29x^2 + 100 = 0$
Solve by Factoring.

$(x^2-25)(x^2-4) = 0$

$(x-5)(x+5)(x-2)(x+2) = 0$

$x-5 = 0$ or $x+5 = 0$ or $x-2 = 0$

or $x+2 = 0$

$x = 5$ or $x = -5$ or $x = 2$

or $x = -2$

Solutions are $\pm 2, \pm 5$.

47. $x^2 + y^2 = 8$

$\underline{x^2 - y^2 = 2}$

$2x^2 \quad\quad = 10$ (Add)

$x^2 = 5$

$x = \pm\sqrt{5}$

Substitute, to determine y.
Note:

$\left(\sqrt{5}\right)^2 = 5$

and

$\left(-\sqrt{5}\right)^2 = 5$

$\left(\pm\sqrt{5}\right)^2 + y^2 = 8$

$5 + y^2 = 8$

$y^2 = 3$

$y = \pm\sqrt{3}$

Solutions are $\left(\sqrt{5}, \sqrt{3}\right), \left(\sqrt{5}, -\sqrt{3}\right),$

$\left(-\sqrt{5}, \sqrt{3}\right)$ and $\left(-\sqrt{5}, -\sqrt{3}\right)$.

48. $4^x = 12$

$\log 4^x = \log 12$

$x\log 4 = \log 12$

$x = \dfrac{\log 12}{\log 4}$

$x \approx 1.7925$

49. $\log(x^2-25) - \log(x+5) = 3$

$\log\dfrac{x^2-25}{x+5} = 3$

$\log(x-5) = 3$

$10^3 = x-5$

$1005 = x$

50. $\log_5 x = -2$

$x = 5^{-2}$

$x = \dfrac{1}{25}$

51. $7^{2x+3} = 49$

$7^{2x+3} = 7^2$

$2x+3 = 2$

$2x = -1$

$x = -\dfrac{1}{2}$

52. $|2x-1| \le 5$

$-5 \le 2x-1 \le 5$

$-4 \le 2x \le 6$

$-2 \le x \le 3$

$\{x \mid -2 \le x \le 3\}$, or $[-2, 3]$

53. $15x^2 + 45 = 0$

$15x^2 = -45$

$x^2 = -3$

$x = \pm\sqrt{-3}$

$x = \pm i\sqrt{3}$

54. $x^2 + 4x = 3$

$x^2 + 4x + 4 = 3+4$

$(x+2)^2 = 7$

$x+2 = \pm\sqrt{7}$

$x = -2 \pm\sqrt{7}$

55. $y^2 + 3y > 10$

Solve the related equation

$$y^2 + 3y - 10 = 0$$
$$(y+5)(y-2) = 0$$
$$y + 5 = 0 \quad \text{or} \quad y - 2 = 0$$
$$y = -5 \quad \text{or} \qquad y = 2$$

These values give us the intervals $(-\infty, -5), (-5, 2)$, and $(2, \infty)$. Using a test point from each interval, we determine the solution. Since $>$, we do not include the zeros.

$\{y \mid y < -5 \text{ or } y > 2\}$, or

$$(-\infty, -5) \cup (2, \infty).$$

56. $f(x) = x^2 - 2x; f(a) = 80$

$$80 = x^2 - 2x$$
$$0 = x^2 - 2x - 80$$
$$0 = (x+8)(x-10)$$
$$x + 8 = 0 \quad \text{or} \quad x - 10 = 0$$
$$x = -8 \quad \text{or} \qquad x = 10$$

57. $f(x) = \sqrt{-x + 4} + 3; \; g(x) = \sqrt{x - 2} + 3$

Let $f(a) = g(a)$

$$\sqrt{-a + 4} + 3 = \sqrt{a - 2} + 3$$
$$\sqrt{-a + 4} = \sqrt{a - 2}$$
$$-a + 4 = a - 2$$
$$6 = 2a$$
$$3 = a$$

Since we squared both sides of the equation, we must check our solution in the original equation. 3 checks and is the solution.

58. $$V = P - Prt$$
$$V - P = -Prt$$
$$\frac{V - P}{-Pt} = r$$
$$\text{or } \frac{P - V}{Pt} = r$$

59. $$I = \frac{R}{R + r}$$
$$I(R + r) = R$$
$$IR + Ir = R$$
$$IR - R = -Ir$$
$$R(I - 1) = -Ir$$
$$R = \frac{-Ir}{I - 1}$$
$$\text{or } R = \frac{Ir}{1 - I}$$

60. a. Linear

b. $f(x) = 0$, when $x = 2$.

61. a. Logarithmic

b. $f(x) = 0$, when $x = -1$.

62. a. Quadratic

b. $f(x) = 0$, when $x = -1$ and when $x = 4$

63. a. Exponential

b. The graph is asymptotic to the x-axis; no real zeros.

Problems 64 – 71 are graphing exercises.

64. $3x - y = 7$

$$m = \frac{-a}{b} = \frac{-3}{-1} = 3$$
$$b = \frac{c}{b} = \frac{7}{-1} = -7$$

Using $m = 3$ and the y-intercept $(0, -7)$, we graph the line.

65. $x^2 + y^2 = 100$

$x^2 + y^2 = 10^2$

This graph is a circle whose center is $(0,0)$ and radius = 10.

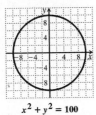

$x^2 + y^2 = 100$

66. $\dfrac{x^2}{36} - \dfrac{y^2}{9} = 1$

$\dfrac{x^2}{6^2} - \dfrac{y^2}{3^2} = 1$

$a = 6$ and $b = 3$. The asymptotes are

$y = \dfrac{3}{6}x$ and $y = -\dfrac{3}{6}x$, or

$y = \dfrac{1}{2}x$ and $y = -\dfrac{1}{2}x$

To help us sketch asymptotes and locate vertices, we use a and b, to form the pairs $(-6, 3), (6, 3), (-6, -3),$ and $(6, -3)$.

Plot these pairs and lightly sketch a rectangle. The asymptotes pass through the corners; since this is a horizontal hyperbola, the vertices are where the rectangle intersects the x-axis. Sketch the hyperbola.

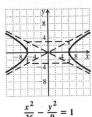

$\dfrac{x^2}{36} - \dfrac{y^2}{9} = 1$

67. $y = \log_2 x$

The graph contains the points $(1, 0)$ and $(2, 1)$, is increasing, and is asymptotic to the y-axis.

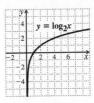

68. $f(x) = 2^x - 3$

$f(x) = 2^x$ contains the points $(0, 1)$ and $(1, 2)$, is increasing, and is asymptotic to the x-axis. We lightly sketch this function. To graph $f(x) = 2^x - 3$ we shift/translate the above graph 3 units downward.

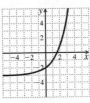

$f(x) = 2^x - 3$

69. $2x - 3y < -6$

Graph the related equation/line
$2x - 3y = -6$

You might use the intercepts, $(0, 2)$ and $(-3, 0)$. Use a dashed line, since $<$.

Choose a test point to determine the correct half-plane.

You might choose $(0, 0)$.

$2(0) - 3(0) < -6$

$0 < -6$

Since this is false, we shade the other half-plane.

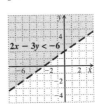

70. $f(x) = -2(x-3)^2 + 1$

These are to be labeled on the graph.

a. Vertex $(3, 1)$

b. Axis of symmetry is $x = 3$

c. Since $a = -2,$ the parabola opens downward. Thus, the maximum function value is 1, when $x = 3.$

$f(x) = -2(x - 3)^2 + 1$
Maximum: 1

71.

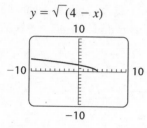

$$y = \sqrt{(4 - x)}$$

From the graph we determine

Domain: $(-\infty, 4]$

Range: $[0, \infty)$

72. Let $w =$ the width and $200 - 2w =$ the length, as shown in the diagram.

Area = Length × Width

$$A = (200 - 2w)w$$

$$A = 200w - 2w^2$$

The graph of this function is a parabola which opens downward, so a maximum exists.

$$A = -2(w^2 - 100w)$$

$$= -2(w^2 - 100w + 2500) - (-2)(2500)$$

$$= -2(w - 50)^2 + 5000$$

The maximum area is $5000 \text{ ft}^2.$

73. $P = 2l + 2w,$ or $34 = 2l + 2w,$ or $17 = l + w$

Using the Pythagorean theorem,

$$a^2 + b^2 = c^2$$

$$l^2 + w^2 = 13^2$$

Substitute for l using $l = 17 - w$ and solve.

$$(17 - w)^2 + w^2 = 13^2$$

$$289 - 34w + w^2 + w^2 = 169$$

$$2w^2 - 34w + 120 = 0$$

$$2(w - 12)(w - 5) = 0$$

$$w - 12 = 0 \text{ or } w - 5 = 0$$

$$w = 12 \text{ or } w = 5$$

We usually think of length as the greater of the two measures, so $w = 5$ and $l = 17 - 5$ or 12.

The dimensions are 5 ft by 12 ft.

74. Let $x =$ number of movies rented. Limited members pay $40 plus $2.45 per movie, or $40 + \$2.45x.$ Preferred members pay $60 plus $1.65 per movie, or $60 + \$1.65x.$ We want the values of x which satisfy

$$\$60 + \$1.65x < \$40 + \$2.45x$$

We solve for $x.$

$$60 + 1.65x < 40 + 2.45x$$

$$20 < 0.8x$$

$$25 < x$$

It will be less expensive to be a preferred member when more than 25 movies are rented.

75. Let $x =$ the number of ounces of the herbs costing $2.68/$oz.$ Let $y =$ the number of ounces of the herbs costing $4.60/$oz.$ When mixed, they cost $3.80/$oz,$ so

$$2.68x + 4.60y = 3.80(x + y).$$

We also know there are 24 oz in all, so $x + y = 24.$

We solve the system using substitution.

$$2.68x + 4.60y = 3.80(x + y)$$

$$x + y = 24, \ x = 24 - y$$

$$2.68(24 - y) + 4.60y = 3.80(24)$$

$$64.32 - 2.68y + 4.60y = 91.2$$

$$1.92y = 26.88$$

$$y = 14$$

$$x = 24 - y, \text{ or } 24 - 14 = 10$$

The mix should contain 10 oz of the herbs costing $2.68/oz and 14 oz of the herbs costing $4.60/oz.

76. Let r = speed of the plane in still air,
$r + 30$ = speed of the plane with the wind, and
$r - 30$ = speed of the plane against the wind.
Since the time for both is the same (equal), using
$t = \frac{d}{r}$, we have:

$$\frac{190}{r + 30} = \frac{160}{r - 30}$$

LCD is $(r + 30)(r - 30)$

$$(r + 30)(r - 30) \cdot \frac{190}{r + 30}$$
$$= (r + 30)(r - 30) \cdot \frac{160}{r - 30}$$
$$(r - 30) \cdot 190 = (r + 30) \cdot 160$$
$$190r - 5700 = 160r + 4800$$
$$30r = 10,500$$
$$r = 350$$

The plane can fly 350 mph in still air.

77. Jack can tap the trees in 21 hours, so he can do $\frac{1}{21}$ of the job in an hour. Delia can tap the trees in 14 hours, so she can do $\frac{1}{14}$ of the job in an hour. Let t = number of hours they work together. Work = rate × time
Jack's work + Delia's work = 1

$$\frac{1}{21}t + \frac{1}{14}t = 1, \text{ or } \frac{t}{21} + \frac{t}{14} = 1$$

LCD is 42.

$$42\left(\frac{t}{21} + \frac{t}{14}\right) = 42 \cdot 1$$
$$2t + 3t = 42$$
$$5t = 42$$
$$t = \frac{42}{5} = 8\frac{2}{5} \text{ hr}$$

Working together, they can tap the trees in $8\frac{2}{5}$ hr, or 8 hr 24 min.

78. F varies directly as v^2 (velocity squared) and inversely as r (radius), so

$$F = k \cdot \frac{v^2}{r}$$

Substitute the given value to determine the constant of variation k.

$$8 = k \cdot \frac{1^2}{10}$$
$$80 = k$$

We have the variation equation

$$F = 80 \cdot \frac{v^2}{r}$$

Substitute the known values and solve for F.

$$F = 80 \cdot \frac{2^2}{16}$$
$$F = 80 \cdot \frac{4}{16}$$
$$F = 20$$

79. a. Determine the slope using (0, 173) and (4, 181).

$$m = \frac{181 - 173}{4 - 0}$$
$$m = \frac{8}{4}$$
$$m = 2$$

The average rate of change is 2 million card holders per year.

b. $m = 2$ and $b = 173$
Using slope-intercept form, we have
$c = mt + b$
$c = 2t + 173$, where
c is in millions.

c. In 2015, $t = 9$
$c = 2(9) + 173$
$= 18 + 173$
$= 191$
There will be 191 million card holders in 2015.

d. Let $c = 250$:
$250 = 2t + 173$
$77 = 2t$
$38.5 = t$
2006 + 38.5 = 2044.5. There will be 250 million Americans with credit cards in 2044.

80. a. $P(t) = P_0 e^{kt}$
Using the given information,
$P_0 = 5$ billion. In 2006, $t = 27$ and
$P(27) = 60$.
Substitute to determine k.

$$60 = 5e^{k(27)}$$

$$\frac{60}{5} = e^{27k}$$

$$\ln 12 = 27k$$

$$\frac{\ln 12}{27} = k$$

$$0.092 \approx k$$

We have the exponential function $P(t) = 5e^{0.092t}$, where $P(t)$ is in billions of dollars.

b.   In 2012, $t = 33$

$$P(33) = 5e^{0.092(33)}$$

$$P(33) \approx 104$$

In 2012, there will be approximately $104 billion in electronic payments.

c. $$200 = 5e^{0.092t}$$

$$40 = e^{0.092t}$$

$$\ln 40 = 0.092t$$

$$\frac{\ln 40}{0.092} = t$$

$$40.1 \approx t$$

There will be $200 billion in electronic payments in $1979 + 40 = 2019$.

81.   This is a geometric sequence.
$2000, 2000 \cdot 1.05, 2000 \cdot 1.05^2, \ldots 2000 \cdot 1.05^n$
When Sarita is 62, $n = 40$.
$2000 \cdot 1.05^{40} \approx \$14,079.98$

82. $$\frac{9}{x} - \frac{9}{x+12} = \frac{108}{x^2 + 12x}$$
$x \neq 0; \quad x \neq -12$

LCD is $x^2 + 12x = x(x+12)$

$$x(x+12)\left(\frac{9}{x} - \frac{9}{x+12}\right) = (x^2 + 12x) \cdot \frac{108}{x^2 + 12x}$$

$$(x+12)9 - 9 \cdot x = 108$$

$$9x + 108 - 9x = 108$$

$$108 = 108$$

Since this is an identity (Reflexive Property of Equality), all values of x are solutions except 0 and –12.
$\{x \mid x \in \mathbb{R}, x \neq 0, x \neq -12\}$

83. $$\log_2 \left(\log_3 x\right) = 2$$

$$\log_3 x = 2^2$$

$$\log_3 x = 4$$

$$x = 3^4$$

$$x = 81$$

84. y varies directly as the cube of x, so $y = kx^3$.
If x is multiplied by 0.5,
$y = k\left(0.5x\right)^3$, or $y = k \cdot 0.125x^3$
y is divided by 8, since $0.125 = \frac{1}{8}$

85.   Let $x =$ the number of years Diaphantos lived, and $y =$ number of years his son lived.

Since Diaphantos spent $(1/6)x$ as a child, $(1/12)x$ as an adolescent, and $(1/7)x$ as a bachelor, we have

$$\frac{1}{6}x + \frac{1}{12}x + \frac{1}{7}x =$$

$$\frac{14}{84}x + \frac{7}{12}x + \frac{12}{84}x = \frac{33}{84}x = \frac{11}{28}x$$

Five years after he was married, or $\dfrac{33}{84}x + 5$,

his son was born. His son died 4 years before his father (Diaphantos), so Diaphantos' years can be represented as

$$\frac{33}{84}x + 5 + y + 4 \text{ or } x$$

which gives us the equation

$$\frac{33}{84}x + 5 + y + 4 = x$$

$$y + 9 = \frac{51}{84}x$$

We also know the son lived half as long as Diaphantos, so we also have the equation

$y = \dfrac{1}{2}x$. We solve this system using substitution.

$$y + 9 = \frac{51}{84}x$$

$$\left(\frac{1}{2}x\right) + 9 = \frac{51}{84}x$$

$$9 = \frac{9}{84}x$$

$$84 = x$$

Diaphantos lived 84 years.

Chapter R

Elementary Algebra Review

1. False

3. True

5. True

7. 4 is 4 units from 0 on the number line, so
$$|4| = 4$$

9. -1.3 is 1.3 units from 0 on the number line, so
$$|-1.3| = 1.3.$$

11. $(-13) + (-12) = -25$

13. $-\dfrac{1}{3} - \dfrac{2}{5} = \left(-\dfrac{1}{3}\right) + \left(-\dfrac{2}{5}\right)$ Rewrite as addition.

     $= \left(\dfrac{-5}{15}\right) + \left(\dfrac{-6}{15}\right)$ Add over common denominator.

     $= -\dfrac{11}{15}$

15. $4.2 - 10.7 = 4.2 + (-10.7)$ Rewrite as addition.

     $= -\left[|-10.7| - |4.2|\right]$ Subtract absolute values.

     $= -[6.5]$

     $= -6.5$

17. $-15 + 0 = -15$ (Identity Property of Zero)

19. $0 \div (-10) = 0 \cdot \left(-\dfrac{1}{10}\right) = 0$ (Multiplicative Property of Zero)

21. $\left(-\dfrac{3}{10}\right) + \left(-\dfrac{1}{5}\right) = \left(-\dfrac{3}{10}\right) + \left(-\dfrac{2}{10}\right)$

     $= -\dfrac{5}{10} = -\dfrac{1}{2}$

23. $-3.8 + 9.6 = 9.6 - 3.8 = 5.8$

25. $(-12) \div 4 = \dfrac{-12}{4} = -\dfrac{12}{4} = -3$

27. $32 - (-7) = 32 + 7 = 39$

29. $(-10)(-17.5) = (10)(17.5) = 175$

31. $(-68) + 36 = -(|-68| - |36|) = -32$

33. $2 + (-3) + 7 + 10 = 19 + (-3) = 16$

35. $3 \cdot (-2) \cdot (-1) \cdot (-1) = -(3 \cdot 2 \cdot 1 \cdot 1) = -6$

37. $(-1)^4 + 2^3 = 1 + 8 = 9$

39. $2 \cdot 6 - 3 \cdot 5 = 12 - 15 = 12 + (-15) = -3$

41. $3 - (2 \cdot 4 + 11) = 3 - (8 + 11) = 3 - 19 = -16$

43. $4 \cdot 5^2 = 4 \cdot 25 = 100$

45. $25 - 8 \cdot 3 + 1 = 25 - 24 + 1 = 1 + 1 = 2$

47. $2 - \left(3^3 + 16 \div (-2)^3\right) = 2 - \left(27 + 16 \div (-8)\right)$

     $= 2 - (27 + (-2))$

     $= 2 - (25)$

     $= -23$

49. $|6(-3)| + |(-2)(-9)| = |-18| + |18| = 18 + 18 = 36$

51. $\dfrac{7000 + (-10)^3}{10^2 \cdot (2 + 4)} = \dfrac{7000 + (-1000)}{100 \cdot (6)} = \dfrac{6000}{600} = 10$

53. $2 + 8 \div 2 \cdot 2 = 2 + 4 \cdot 2 = 2 + 8 = 10$

55. $y - x = 3 - 10 = -7$

57. $-3 - x^2 + 12x$

$$= -3 - (5)^2 + 12 \cdot 5$$
$$= -3 - 25 + 12 \cdot 5$$
$$= -3 - 25 + 60$$
$$= 32$$

59. $A = bh = (8)(3.5) = 28 \text{ cm}^2$

61. $4(2x + 7) = 4 \cdot 2x + 4 \cdot 7 = 8x + 28$

63. $-2(15 - 3x) = -2 \cdot 15 - (-2)(3x) = -30 + 6x$

65. $2(4a + 6b - 3c) = 2 \cdot 4a + 2 \cdot 6b - (2)(3c)$
$$= 8a + 12b - 6c$$

67. $-3(2x - y + z) = -3 \cdot 2x - (-3)y + (-3)z$
$$= -6x + 3y - 3z$$

69. $8x + 6y = 2 \cdot 4x + 2 \cdot 3y = 2(4x + 3y)$

71. $3 + 3w = 3 \cdot 1 + 3 \cdot w = 3(1 + w)$

73. $10x + 50y + 100 = 10 \cdot x + 10 \cdot 5y + 10 \cdot 10$
$$= 10(x + 5y + 10)$$

75. $3p - 2p = (3 - 2)p = 1p = p$

77. $4m + 10 - 5m + 12 = 4m - 5m + 10 + 12$
$$= -m + 22$$

79. $-6x + 7 + 9x = -6x + 9x + 7$
$$= 3x + 7$$

81. $2p - (7 - 4p) = 2p - 7 + 4p = 6p - 7$

83. $6x + 5y - 7(x - y) = 6x + 5y - 7x + 7y$
$$= -x + 12y$$

85. $6[2a + 4(a - 2b)] = 6[2a + 4a - 8b]$
$$= 6[6a - 8b]$$
$$= 36a - 48b$$

87. $3 - 2[5(x - 10y) - (3 + 2y)]$

$$= 3 - 2[5x - 50y - 3 - 2y]$$
$$= 3 - 2[5x - 52y - 3]$$
$$= 3 - 10x + 104y + 6$$
$$= -10x + 104y + 9$$

89. $3x - 2 = 10$ Write the equation.

$$\overline{3 \cdot 4 - 2} \mid 10 \quad \text{Substitute 4 for } x.$$
$$12 - 2 \mid 10$$
$$\overset{?}{10 = 10} \quad \text{TRUE}$$

Since $3 \cdot 4 - 2 = 10$ is true, 4 is a solution of $3x - 2 = 10$.

91. $4 - x = 1$ Write the equation.

$$\overline{4 - (-3)} \mid 1 \quad \text{Substitute } -3 \text{ for } x.$$
$$4 + 3 \mid 1$$
$$\overset{?}{7 = 1} \quad \text{FALSE}$$

Since $4 - (-3) = 1$ is false, -3 is not a solution of $4 - x = 1$.

93. $\dfrac{x}{2} = 2.3$ Write the equation.

$$\overline{\dfrac{4.6}{2}} \mid 2.3 \quad \text{Substitute 4.6 for } x.$$
$$\overset{?}{2.3 = 2.3} \quad \text{TRUE}$$

Since $\dfrac{4.6}{2} = 2.3$ is true, 4.6 is a solution of $\dfrac{x}{2} = 2.3$.

95. Let n represent the number; $3n = 348$

97. Let c represent the number of calories in a Taco Bell® Beef Burrito; $c + 69 = 500$

99. Let l represent the amount of water used to produce 1 lb of lettuce; $42 = 2l$

Exercise Set R.2

1. $-6 + x = 10$

$-6 + x + 6 = 10 + 6$

$x = 16$

Check:

$-6 + x = 10$

$\overline{-6 + 16 \mid 10}$

$\overset{?}{10 = 10}$ TRUE

The solution is 16.

3. $t + \dfrac{1}{3} = \dfrac{1}{4}$

$12\left(t + \dfrac{1}{3}\right) = 12\left(\dfrac{1}{4}\right)$

$12t + 4 = 3$

$12t + 4 - 4 = 3 - 4$

$12t = -1$

$\dfrac{12t}{12} = \dfrac{-1}{12}$

$t = -\dfrac{1}{12}$

Check:

$t + \dfrac{1}{3} = \dfrac{1}{4}$

$\begin{array}{c|c} -\dfrac{1}{12} + \dfrac{1}{3} & \dfrac{1}{4} \\ -\dfrac{1}{12} + \dfrac{1}{3} \cdot \dfrac{4}{4} & \dfrac{1}{4} \\ -\dfrac{1}{12} + \dfrac{4}{12} & \dfrac{1}{4} \\ \dfrac{3}{12} & \dfrac{1}{4} \end{array}$

$\dfrac{1}{4} \overset{?}{=} \dfrac{1}{4}$ TRUE

The solution is $-\dfrac{1}{12}$.

5. $-1.9 = x - 1.1$

$-1.9 + 1.1 = x - 1.1 + 1.1$

$-0.8 = x$

Check:

$-1.9 = x - 1.1$

$\overline{-1.9 \mid -0.8 - 1.1}$

$-1.9 \overset{?}{=} -1.9$ TRUE

The solution is -0.8.

7. $-x = \dfrac{5}{3}$

$(-1)(-x) = (-1)\left(\dfrac{5}{3}\right)$

$x = -\dfrac{5}{3}$

Check:

$-x = \dfrac{5}{3}$

$\begin{array}{c|c} -\left(-\dfrac{5}{3}\right) & \dfrac{5}{3} \\ \dfrac{5}{3} & \end{array}$

$\dfrac{5}{3} \overset{?}{=} \dfrac{5}{3}$ TRUE

The solution is $-\dfrac{5}{3}$.

9. $-\dfrac{2}{7}x = -12$

$(-7)\left(-\dfrac{2}{7}x\right) = (-7)(-12)$

$2x = 84$

$x = \dfrac{84}{2}$

$x = 42$

Check:

$-\dfrac{2}{7}x = -12$

$\begin{array}{c|c} \left(-\dfrac{2}{7}\right)(42) & -12 \\ \dfrac{-84}{7} & -12 \end{array}$

$-12 \overset{?}{=} -12$ TRUE

The solution is 42.

11. $\dfrac{-t}{5} = 1$

$5\left(\dfrac{-t}{5}\right) = 5 \cdot 1$

$-t = 5$

$\dfrac{-t}{-1} = \dfrac{5}{-1}$

$t = -5$

Check:

$\dfrac{-t}{5} = 1$

$\dfrac{-(-5)}{5} \Big| 1$

$\dfrac{5}{5} \Big| 1$

$\overset{?}{1 = 1}$ TRUE

The solution is –5.

13. $3x + 7 = 13$

$3x + 7 - 7 = 13 - 7$

$3x = 6$

$\dfrac{3x}{3} = \dfrac{6}{3}$

$x = 2$

Check:

$3x + 7 = 13$

$\dfrac{3 \cdot 2 + 7}{} \Big| 13$

$6 + 7 \Big| 13$

$\overset{?}{13 = 13}$ TRUE

The solution is 2.

15. $3y - 10 = 15$

$3y - 10 + 10 = 15 + 10$

$3y = 25$

$\dfrac{3y}{3} = \dfrac{25}{3}$

$y = \dfrac{25}{3}$

Check:

$3y - 10 = 15$

$\dfrac{3 \cdot \dfrac{25}{3} - 10}{} \Big| 15$

$25 - 10 \Big| 15$

$\overset{?}{15 = 15}$ TRUE

The solution is $\dfrac{25}{3}$.

17. $4x + 7 = 3 - 5x$

$4x + 7 - 7 = 3 - 7 - 5x$

$4x = -4 - 5x$

$4x + 5x = -4 - 5x + 5x$

$9x = -4$

$\dfrac{9x}{9} = \dfrac{-4}{9}$

$x = -\dfrac{4}{9}$

Check:

$4x + 7 = 3 - 5x$

$\dfrac{4 \cdot \left(-\dfrac{4}{9}\right) + 7}{} \Bigg| 3 - 5 \cdot \left(-\dfrac{4}{9}\right)$

$\dfrac{-16}{9} + 7 \Bigg| 3 + \dfrac{20}{9}$

$\dfrac{-16 + 7 \cdot 9}{9} \Bigg| \dfrac{3 \cdot 9 + 20}{9}$

$\dfrac{-16 + 63}{9} \Bigg| \dfrac{27 + 20}{9}$

$\dfrac{47}{9} \overset{?}{=} \dfrac{47}{9}$ TRUE

The solution is $-\dfrac{4}{9}$.

19. $2x - 7 = 5x + 1 - x$

$2x - 7 = 5x - x + 1$

$2x - 7 = 4x + 1$

$2x - 7 + 7 = 4x + 1 + 7$

$2x = 4x + 8$

$2x - 4x = 4x - 4x + 8$

$-2x = 8$

$\dfrac{-2x}{-2} = \dfrac{8}{-2}$

$x = -4$

21. $\dfrac{2}{5}+\dfrac{1}{3}t=5$

$15\left(\dfrac{2}{5}+\dfrac{1}{3}t\right)=15\cdot 5$

$6+5t=75$

$5t=75-6$

$5t=69$

$t=\dfrac{69}{5}$

23. $x+0.45=2.6x$

$0.45=2.6x-x$

$0.45=1.6x$

$\dfrac{0.45}{1.6}=x$

$x=\dfrac{45}{160}=\dfrac{9}{32}$

25. $8(3-m)+7=47$

$24-8m+7=47$

$31-8m=47$

$-8m=47-31$

$-8m=16$

$m=\dfrac{16}{-8}=-2$

27. $4-(6+x)=13$

$4-6-x=13$

$-2-x=13$

$-2-13=x$

$-15=x$

29. $2+3(4+c)=1-5(6-c)$

$2+12+3c=1-30+5c$

$14+3c=-29+5c$

$3c-5c=-29-14$

$-2c=-43$

$c=\dfrac{43}{2}$

31. $0.1(a-0.2)=1.2+2.4a$

$0.1a-0.02=1.2+2.4a$

$-0.02-1.2=2.4a-0.1a$

$-1.22=2.3a$

$\dfrac{-1.22}{2.3}=a$

$a=-\dfrac{122}{230}=-\dfrac{61}{115}$

33. $A=lw$

$\dfrac{A}{w}=\dfrac{lw}{w}$

$\dfrac{A}{w}=l$

35. $p=30q$

$\dfrac{p}{30}=\dfrac{30q}{30}$

$\dfrac{p}{30}=q$

37. $I=\dfrac{P}{V}$

$IV=\dfrac{P}{V}V$

$IV=P$

39. $q=\dfrac{p+r}{2}$

$2q=\dfrac{p+r}{2}\cdot 2$

$2q=p+r$

$2q-r=p$

41. $A=\pi r^2+\pi r^2 h$

$A=\pi\left(r^2+r^2 h\right)$

$\dfrac{A}{r^2+r^2 h}=\pi$

43. a. $x\le -5$

$5\le -5$ FALSE

The number 5 is not a solution of the inequality.

b. $x \le -5$

$-5 \le -5$ TRUE

The number -5 is a solution of the inequality.

c. $x \le -5$

$0 \le -5$ FALSE

The number 0 is not a solution of the inequality.

d. $x \le -5$

$-10 \le -5$ TRUE

The number -10 is a solution of the inequality.

45. $x + 3 \le 15$

$x \le 15 - 3$

$x \le 12$

$(-\infty, 12]$ or $\{x \mid x \le 12\}$

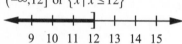

9   10   11   12   13   14   15

47. $m - 17 > -5$

$m > -5 + 17$

$m > 12$

$(12, \infty)$ or $\{m \mid m > 12\}$

9   10   11   12   13   14   15

49. $2x \ge -3$

$\dfrac{2x}{2} \ge \dfrac{-3}{2}$

$x \ge -\dfrac{3}{2}$

$\left[-\dfrac{3}{2}, \infty\right)$ or $\left\{x \mid x \ge -\dfrac{3}{2}\right\}$

-3 -2 -1   0   1   2   3

51. $-5t > 15$

$\dfrac{-5t}{-5} < \dfrac{15}{-5}$

$t < -3$

$(-\infty, -3)$ or $\{t \mid t < -3\}$

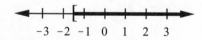

-6 -5 -4 -3 -2 -1   0

53. $2y - 7 > 13$

$2y > 20$

$\dfrac{2y}{2} > \dfrac{20}{2}$

$y > 10$

$(10, \infty)$ or $\{y \mid y > 10\}$

55. $6 - 5a \le a$

$6 \le a + 5a$

$6 \le 6a$

$\dfrac{6}{6} \le a$

$1 \le a$; or $a \ge 1$

$[1, \infty)$ or $\{a \mid a \ge 1\}$

57. $2(3 + 5x) \ge 7(10 - x)$

$6 + 10x \ge 70 - 7x$

$10x + 7x \ge 70 - 6$

$17x \ge 64$

$\dfrac{17x}{17} \ge \dfrac{64}{17}$

$x \ge \dfrac{64}{17}$

$\left[\dfrac{64}{17}, \infty\right)$ or $\left\{x \mid x \ge \dfrac{64}{17}\right\}$

59. $\dfrac{2}{3}(6 - x) < \dfrac{1}{4}(x + 3)$

$12 \cdot \dfrac{2}{3}(6 - x) < 12 \cdot \dfrac{1}{4}(x + 3)$

$8(6 - x) < 3(x + 3)$

$48 - 8x < 3x + 9$

$-8x - 3x < 9 - 48$

$-11x < -39$

$\dfrac{-11x}{-11} > \dfrac{-39}{-11}$

$x > \dfrac{39}{11}$

$\left(\dfrac{39}{11}, \infty\right)$ or $\left\{x \mid x > \dfrac{39}{11}\right\}$

61. $0.7(2+x) \geq 1.1x + 5.75$

$1.4 + 0.7x \geq 1.1x + 5.75$

$0.7x - 1.1x \geq 5.75 - 1.4$

$-0.4x \geq 4.35$

$\dfrac{-0.4x}{-0.4} \leq \dfrac{4.35}{-0.4}$

$x \leq -10.875$

$(-\infty, -10.875]$ or $\{x | x \leq -10.875\}$

63. ***Familiarize***. Let n represent the unknown number.

Translate.

$$\underbrace{\text{Three less than the sum of 2 and a number}}_{} \quad \underset{\downarrow}{\text{is}} \quad \underset{\downarrow}{6}$$

$$\underset{\downarrow}{} \qquad\qquad\qquad$$

$$n + 2 - 3 \qquad\qquad = \qquad 6$$

Carry out. Solve for n.

$n + 2 - 3 = 6$

$n - 1 = 6$

$n = 6 + 1$

$n = 7$

Check.

$$\begin{array}{c|c} n + 2 - 3 = 6 & \\ \hline 7 + 2 - 3 & 6 \\ 9 - 3 & 6 \\ & \overset{?}{} \\ 6 = 6 & \text{TRUE} \end{array}$$

State. The number is 7.

65. ***Familiarize***. Two consecutive even integers can be represented by n and $n + 2$.

Translate.

$$\underbrace{\text{The sum of two consecutive even integers}}_{} \quad \underset{\downarrow}{\text{is}} \quad \underset{\downarrow}{34}$$

$$n + (n + 2) \qquad = \qquad 34$$

Carry out. Solve for n.

$n + (n + 2) = 34$

$2n + 2 = 34$

$2n = 34 - 2$

$2n = 32$

$n = 16$

Check.

$$\begin{array}{c|c} n + (n + 2) = 34 & \\ \hline 16 + (16 + 2) & 34 \\ 16 + 18 & 34 \\ & \overset{?}{} \\ 34 = 34 & \text{TRUE} \end{array}$$

State. The numbers are $n = 16$ and $n + 2 = 18$.

67. ***Familiarize***. Let p represent the number of pages read so far. Then, the number of unread pages is $2p$ and the total number of pages is 500.

Translate.

$$\underbrace{\text{The sum of read and unread pages}}_{} \quad \underset{\downarrow}{\text{is}} \quad \underset{\downarrow}{500}$$

$$p + 2p \qquad = \qquad 500$$

Carry out. Solve for p.

$p + 2p = 500$

$3p = 500$

$p = \dfrac{500}{3}$

$p = 166\dfrac{2}{3}$

Check.

$$\begin{array}{c|c} p + 2p = 500 & \\ \hline \dfrac{500}{3} + 2 \cdot \dfrac{500}{3} & 500 \\[2mm] \dfrac{500}{3} + \dfrac{1000}{3} & 500 \\[2mm] \dfrac{1500}{3} & 500 \\ & \overset{?}{} \\ 500 = 500 & \text{TRUE} \end{array}$$

State. The number of read pages is $166\dfrac{2}{3}$.

69. *Familiarize*. Let l represent the length of the rectangle. The width is 5 cm less than the length: $w = l - 5$. The perimeter of a rectangle is twice the length plus twice the width.

Translate.

$$\underbrace{\text{The perimeter of a rectangle}} \overset{\text{is}}{\text{ }} \underbrace{\text{twice the length plus twice the width}}$$

$$\begin{array}{ccc} \downarrow & \downarrow & \downarrow \\ 28 & = & 2l + 2(l-5) \end{array}$$

Carry out. Solve for l.

$$28 = 2l + 2(l-5)$$
$$28 = 2l + 2l - 10$$
$$28 = 4l - 10$$
$$38 = 4l$$
$$\frac{38}{4} = l$$
$$\frac{19}{2} = l \quad \text{or} \quad l = 9\frac{1}{2} \text{ cm}$$

Check.

$$28 = 2l + 2(l-5)$$

$$\begin{array}{c|c} 28 & 2\left(\dfrac{19}{2}\right) + 2\left(\dfrac{19}{2} - 5\right) \\ 28 & 19 + 19 - 10 \end{array}$$

$$\overset{?}{28 = 28} \quad \text{TRUE}$$

State. The length of the rectangle is $9\frac{1}{2}$ cm and its width is $9\frac{1}{2} - 5 = 4\frac{1}{2}$ cm.

71. *Familiarize*. Let v represent the volume of water used in hundreds of cubic feet. Each bill is the sum of the service fee, \$9.70, plus the charge for the amount of water used, \$2.60v. The total bill is \$33.10.

Translate.

$$\underbrace{\text{The total bill}} \overset{\text{is}}{\text{ }} \underbrace{\begin{array}{c}\text{the service charge plus} \\ \text{the water usage charge}\end{array}}$$

$$\begin{array}{ccc} \downarrow & \downarrow & \downarrow \\ 33.10 & = & 9.70 + 2.60v \end{array}$$

Carry out. Solve for v.

$$33.10 = 9.70 + 2.60v$$
$$33.10 - 9.70 = 2.60v$$
$$23.40 = 2.60v$$
$$\frac{23.40}{2.60} = v$$
$$9 = v \quad \text{or} \quad 900 \text{ cubic feet}$$

Check.

$$33.10 = 9.70 + 2.60v$$

$$\begin{array}{c|c} 33.10 & 9.70 + 2.60 \cdot 9 \\ 33.10 & 9.70 + 23.40 \end{array}$$

$$\overset{?}{33.10 = 33.10} \quad \text{TRUE}$$

State. Nine hundred cubic feet of water were used.

73. *Familiarize*. Let p be the normal selling price. The sale price is 64¢, which is 20% off the normal price.

Translate.

$$\underbrace{\text{The sale price}} \overset{\text{is}}{\text{ }} \underbrace{\text{20\% off}} \underbrace{\text{the normal price}}$$

$$\begin{array}{cccc} \downarrow & \downarrow & \downarrow & \downarrow \\ 64 & = & (100\% - 20\%) & p \end{array}$$

Carry out. Solve for p.

$$64 = (100\% - 20\%)\,p$$
$$64 = 80\% \cdot p$$
$$64 = 0.8p$$
$$\frac{64}{0.8} = p$$
$$80 = p \quad 80¢$$

Check.

$$64 = 0.8p$$

$$\begin{array}{c|c} 64 & 0.8 \cdot 80 \end{array}$$

$$\overset{?}{64 = 64} \quad \text{TRUE}$$

State. The normal price is 80¢.

75. ***Familiarize***. Let t represent the time spent on the seventh day. The average time is the sum of the practice times divided by 7 and must be at least 15 minutes.

Translate.

The average time is at least 15 minutes

$$\downarrow \qquad \qquad \downarrow \qquad \downarrow$$

$$\frac{10+20+5+0+25+15+t}{7} \quad \geq \quad 15$$

Carry out. Solve for t.

$$\frac{10+20+5+0+25+15+t}{7} \geq 15$$

$$\frac{75+t}{7} \geq 15$$

$$75+t \geq 105$$

$$t \geq 105-75$$

$$t \geq 30 \quad \text{30 minutes or more}$$

Check. Let $t = 30$ minutes and compute the average:

$$\frac{75+t}{7} = \frac{75+30}{7} = \frac{105}{7} = 15$$

Thirty minutes of practice on the seventh day brings the average up to 15 minutes per day for the week. More than thirty minutes of practice on the seventh day will increase the average to more than 15 minutes per day.

State. Dierdre must practice at least 30 minutes on the seventh day to average 15 minutes per day for the week.

77. ***Familiarize***. Let h represent the number of hours that the room can be rented. The room rental consists of a cleaning fee of $75 plus $45 for each hour. The rental must be $200 or less.

Translate.

The room rental is no more than $200

$$\downarrow \qquad \quad \downarrow \qquad \downarrow$$

$$75+45h \qquad \leq \qquad 200$$

Carry out.

$$75+45h \leq 200$$

$$45h \leq 200-75$$

$$45h \leq 125$$

$$h \leq \frac{125}{45}$$

$$h \leq \frac{25}{9} \quad \text{or} \quad h \leq 2\frac{7}{9}$$

Check. Compute the rental amount for a meeting lasting $2\frac{7}{9} = \frac{25}{9}$ hours:

$$75+45h = 75+45\left(\frac{25}{9}\right)$$

$$= 75+5 \cdot 25$$

$$= 75+125 = 200$$

A meeting lasting $2\frac{7}{9}$ hours costs exactly $200. A meeting lasting longer than this will cost more than $200.

State. The room can be rented for $2\frac{7}{9}$ hours or less.

Exercise Set R.3

1.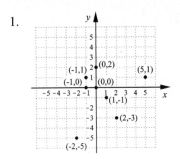

3. I

5. IV

7. I and IV

9. $y = 2x-5$

$$\begin{array}{c|c} 3 & 2 \cdot 1 - 5 \\ \hline 3 & 2-5 \end{array}$$

$$3 \stackrel{?}{=} -3 \quad \text{FALSE}$$

The ordered pair (1, 3) is not a solution of the equation.

11.

$$a - 5b = -3$$

$$\frac{2 - 5(1)\ \big|\ -3}{}$$

$$2 - 5\ \big|\ -3$$

$$\overset{?}{-3} = -3 \quad \text{TRUE}$$

The ordered pair (2, 1) is a solution of the equation.

13.

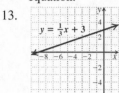

$y = \frac{1}{3}x + 3$

15.

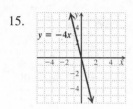

$y = -4x$

17.

$y = x^2 - 7$

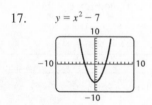

19.

$y = \text{abs}(x + 3)$

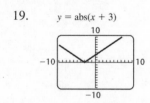

21.

$y = \sqrt{\ }(x^2 + 1)$

23. $m = \dfrac{y_2 - y_1}{x_2 - x_1}$

$$m = \frac{5 - 6}{2 - 3} = \frac{-1}{-1} = 1$$

25. $m = \dfrac{y_2 - y_1}{x_2 - x_1}$

$$m = \frac{-2 - 3}{1 - 0} = \frac{-5}{1} = -5$$

27. $m = \dfrac{y_2 - y_1}{x_2 - x_1}$

$$m = \frac{-\dfrac{1}{2} - \left(-\dfrac{1}{2}\right)}{5 - (-2)} = \frac{0}{7} = 0$$

29. $y = 2x - 5 = 2x + (-5)$

$$m = 2 \quad b = -5$$

Slope: 2; y-intercept: $(0, -5)$

31. $2x + 7y = 1$

$$7y = 1 - 2x$$

$$y = \frac{-2x + 1}{7} = -\frac{2}{7}x + \frac{1}{7}$$

$$m = -\frac{2}{7} \quad b = \frac{1}{7}$$

Slope: $-\dfrac{2}{7}$; y-intercept: $\left(0, \dfrac{1}{7}\right)$

33. $2x - y = 4$

To find the x-intercept, we let $y = 0$ and solve for x.

$$2x - 0 = 4$$

$$2x = 4$$

$$x = \frac{4}{2} = 2$$

The x-intercept is $(2, 0)$.

To find the y-intercept, we let $x = 0$ and solve for y.

$$2(0) - y = 4$$

$$0 - y = 4$$

$$y = -4$$

The y-intercept is $(0, -4)$.

We plot these points and finish by drawing a line through them.

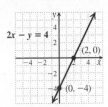

35. $y = x + 5$

To find the x-intercept, we let $y = 0$ and solve for x.
$$0 = x + 5$$
$$-5 = x$$

The x-intercept is $(-5, 0)$.

To find the y-intercept, we let $x = 0$ and solve for y.
$$y = 0 + 5$$
$$y = 5$$

The y-intercept is $(0, 5)$.

We plot these points and finish by drawing a line through them.

37. $3 - y = 2x$

To find the x-intercept, we let $y = 0$ and solve for x.
$$3 - 0 = 2x$$
$$\frac{3}{2} = x$$

The x-intercept is $\left(\frac{3}{2}, 0\right)$.

To find the y-intercept, we let $x = 0$ and solve for y.
$$3 - y = 2(0)$$
$$3 - y = 0$$
$$3 = y$$

The y-intercept is $(0, 3)$.

We plot these points and finish by drawing a line through them.

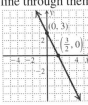

39. $y = 2x - 5$

The equation is in slope-intercept form. The y-intercept is $(0, -5)$. The slope is 2 which we can write as $\frac{2}{1}$. We plot the point $(0, -5)$ and move up 2 units and right 1 unit to obtain the point $(1, -3)$. We finish by drawing a line through these points.

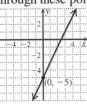

$y = 2x - 5$

41. $2y + 4x = 6$

We rewrite the equation in slope-intercept form.
$$2y + 4x = 6$$
$$y + 2x = 3$$
$$y = -2x + 3$$

The y-intercept is $(0, 3)$. The slope is -2 which we can write as $\frac{-2}{1}$. We plot the point $(0, 3)$ and move down 2 units and right 1 unit to obtain the point $(1, 1)$. We finish by drawing a line through these points.

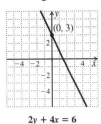

$2y + 4x = 6$

43. $y = 4$

We can write this equation as $y = 0 \cdot x + 4$. This is equation of a horizontal line with slope 0. Every point on the line is of the form $(x, 4)$.

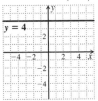

45. $x = 3$

This is the equation of a vertical line with x-intercept $(3,0)$. The slope of this line is undefined.

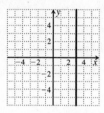

$$x = 3$$

47. $m = \frac{1}{3}; b = 1$

$y = mx + b$

$y = \frac{1}{3}x + 1$

49. y-intercept $(0,3) \rightarrow b = 3$

$m = \frac{4-3}{-1-0} = \frac{1}{-1} = -1$

$y = mx + b$

$y = -1 \cdot x + 3$

$y = -x + 3$

51. Find the slope of each line:

Equation 1   Equation 2

$x + y = 5$ $x - y = 1$

$y = -x + 5$ $-y = -x + 1$

$m_1 = -1$ $y = x - 1$

$m_2 = 1$

Since $m_1 m_2 = -1 \cdot 1 = -1$, the lines are perpendicular.

53. Find the slope of each line:

Equation 1   Equation 2

$2x + 3y = 1$ $2x - 3y = 5$

$3y = -2x + 1$ $-3y = -2x + 5$

$y = -\frac{2}{3}x + \frac{1}{3}$ $y = \frac{2}{3}x - \frac{5}{3}$

$m_1 = -\frac{2}{3}$ $m_2 = \frac{2}{3}$

Since $m_1 \neq m_2$, the lines are not parallel.

Since $m_1 m_2 = \left(-\frac{2}{3}\right)\left(\frac{2}{3}\right) = -\frac{4}{9} \neq -1$,

the lines are not perpendicular.

The lines are neither parallel nor perpendicular.

55. a. $g(-3) = \frac{1}{3}(-3) + 7 = -1 + 7 = 6$

b. $g(4) = \frac{1}{3}(4) + 7 = 1\frac{1}{3} + 7 = 8\frac{1}{3}$

c. $g(a+3) = \frac{1}{3}(a+3) + 7$

$= \frac{1}{3}a + \frac{3}{3} + 7$

$= \frac{1}{3}a + 1 + 7$

$= \frac{1}{3}a + 8$

57. a. $f(-1) = 5(-1)^2 + 2(-1) + 3 = 5 - 2 + 3 = 6$

b. $f(0) = 5(0)^2 + 2(0) + 3 = 3$

c. $f(2a) = 5(2a)^2 + 2(2a) + 3$

$= 5 \cdot 4a^2 + 4a + 3$

$= 20a^2 + 4a + 3$

59. A vertical line can be drawn so that it crosses the graph more than once. The y-axis is such a line. Since the graph fails the Vertical-Line Test, it is not the graph of a function.

61. A vertical line crosses the graph in at most one point. Since the graph passes the Vertical-Line Test, it is a graph of a function.

63. $f(x) = \dfrac{x}{x-3}$

The function is undefined when the denominator is equal to zero. The denominator is zero for $x = 3$ and non-zero for all other real values of x. So the domain is $\{x | x \text{ is a real number } and\ x \neq 3\}$.

65. $g(x) = 2x + 3$

The function is defined for all real values of x. The domain is all real numbers $\mathbb{R}$.

67. $f(x) = \sqrt{\dfrac{1}{2}x + 3}$

The function is defined for all values of x such that $\dfrac{1}{2}x + 3 \geq 0$. Solve for x:

$\dfrac{1}{2}x + 3 \geq 0$

$\dfrac{1}{2}x \geq -3$

$x \geq -6$

The domain is all real values of x greater than or equal to –6:

$\{x | x \geq -6\}$, or $[-6, \infty)$

Exercise Set R.4

1. $x + y = 7$
$\quad x - y = 1$

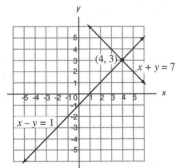

The solution is $(4, 3)$.

3. $\quad y = -2x + 5$
$\quad x + y = 4$

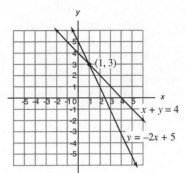

The solution is $(1, 3)$.

5. $\quad y = x - 3$
$\quad y = -2x + 3$

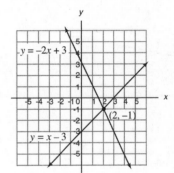

The solution is $(2, -1)$.

7. $\quad 4x - 20 = 5y$
$\quad 8x - 10y = 12$

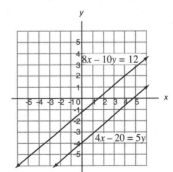

The system has no solution.

9. $x = 6$

$y = -1$

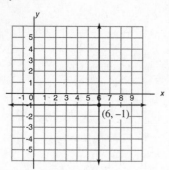

The solution is $(6, -1)$.

11. $y = \frac{1}{5}x + 4$

$2y = \frac{2}{5}x + 8$

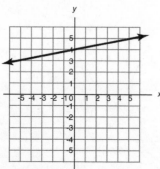

The solution is $\left\{(x, y)\middle| y = \frac{1}{5}x + 4\right\}$.

13. $y = 5 - 4x$ (1)

$2x + y = 1$ (2)

Substitute (1) into (2) and solve for x:

$2x + (5 - 4x) = 1$

$2x + 5 - 4x = 1$

$-2x + 5 = 1$

$-2x = 1 - 5$

$-2x = -4$

$x = \dfrac{-4}{-2} = 2$

Substitute $x = 2$ into (1):

$y = 5 - 4 \cdot 2 = 5 - 8 = -3$

The solution is $(2, -3)$.

15. $3x + 5y = 3$ (1)

$x = 8 - 4y$ (2)

Substitute (2) into (1) and solve for y:

$3(8 - 4y) + 5y = 3$

$24 - 12y + 5y = 3$

$-7y + 24 = 3$

$-7y = 3 - 24$

$-7y = -21$

$y = \dfrac{-21}{-7} = 3$

Substitute $y = 3$ into (2):

$x = 8 - 4(3) = 8 - 12 = -4$

The solution is $(-4, 3)$.

17. $3s - 4t = 14$ (1)

$t = 8 - 5s$ (2)

Substitute (2) into (1):

$3s - 4(8 - 5s) = 14$

$3s - 32 + 20s = 14$

$23s - 32 = 14$

$23s = 14 + 32$

$23s = 46$

$s = \dfrac{46}{23} = 2$

Substitute $s = 2$ into (2):

$t = 8 - 5 \cdot 2 = 8 - 10 = -2$

The solution is $(2, -2)$.

19. $4x - 2y = 6$ (1)

$2x - 3 = y$ (2)

Substitute (2) into (1):

$4x - 2(2x - 3) = 6$

$4x - 4x + 6 = 6$

$6 = 6$

The system is *consistent* and has infinitely many solutions.
Using (2), the solutions are:

$\left\{(x, y)\middle| 2x - 3 = y\right\}$

21.
$$x - 4y = 3 \quad (1)$$
$$5x + 3y = 4 \quad (2)$$

Solve (1) for x:

$$x = 3 + 4y \quad (3)$$

Substitute (3) into (2):

$$5(3 + 4y) + 3y = 4$$
$$15 + 20y + 3y = 4$$
$$23y + 15 = 4$$
$$23y = -11$$
$$y = -\frac{11}{23}$$

Substitute $y = -\dfrac{11}{23}$ into (3):

$$x = 3 + 4\left(-\frac{11}{23}\right) = \frac{69}{23} - \frac{44}{23} = \frac{25}{23}$$

The solution is $\left(\dfrac{25}{23}, -\dfrac{11}{23}\right)$.

23.
$$2x - 3 = y \quad (1)$$
$$y - 2x = 1 \quad (2)$$

Substitute (1) into (2):

$$(2x - 3) - 2x = 1$$
$$-3 = 1$$

The system is *inconsistent*.
There are no solutions.

25.
$$x + 3y = 7 \quad (1)$$
$$-x + 4y = 7 \quad (2)$$

Add equations (1) and (2):

$$\begin{array}{r} x + 3y = 7 \\ + \ -x + 4y = 7 \\ \hline 7y = 14 \end{array}$$

$$y = \frac{14}{7} = 2$$

Substitute into (1) and solve:

$$x + 3 \cdot 2 = 7$$
$$x + 6 = 7$$
$$x = 1$$

The solution is $(1, 2)$.

27.
$$2x - y = -3 \quad (1)$$
$$x + y = 9 \quad (2)$$

Add equations (1) and (2):

$$\begin{array}{r} 2x - y = -3 \\ + \ x + y = \ 9 \\ \hline 3x = 6 \end{array}$$

$$x = \frac{6}{3} = 2$$

Substitute into (2) and solve:

$$2 + y = 9$$
$$y = 7$$

The solution is $(2, 7)$.

29.
$$3x + y = -1 \quad (1)$$
$$2x - 3y = -8 \quad (2)$$

To eliminate y, multiply (1) by 3 and add:

$$\begin{array}{r} 9x + 3y = -3 \\ + \ 2x - 3y = -8 \\ \hline 11x = -11 \end{array}$$

$$x = -1$$

Substitute into (1) and solve:

$$3(-1) + y = -1$$
$$-3 + y = -1$$
$$y = 2$$

The solution is $(-1, 2)$.

31. $4x + 3y = 6$ (1)
     $2x - 2y = 1$ (2)

To eliminate x, multiply (2)

by -2 and add :

$$4x + 3y = 6$$
$$+\ -4x + 4y = -2$$
$$\overline{7y = 4}$$

$$y = \frac{4}{7}$$

Substitute into (1) and solve:

$$4x + 3\left(\frac{4}{7}\right) = 6$$

$$4x + \frac{12}{7} = 6$$

$$4x = \frac{42}{7} - \frac{12}{7}$$

$$x = \frac{30}{7} \cdot \frac{1}{4}$$

$$x = \frac{15}{14}$$

The solution is $\left(\dfrac{15}{14}, \dfrac{4}{7}\right)$.

33. $5r - 3s = 24$ (1)
     $3r + 5s = 28$ (2)

To eliminate r, multiply (1)

by 3 and multiply (2) by -5, then add :

$$15r - 9s = 72$$
$$+\ -15r - 25s = -140$$
$$\overline{-34s = -68}$$

$$s = \frac{-68}{-34} = 2$$

Substitute into (1) and solve:

$$5r - 3 \cdot 2 = 24$$
$$5r - 6 = 24$$
$$5r = 30$$
$$r = 6$$

The solution is $(6, 2)$.

35. $6s + 9t = 12$ (1)
     $4s + 6t = 5$ (2)

To eliminate s, multiply (1) by 2 and

multiply (2) by -3, then add :

$$12s + 18t = 24$$
$$+\ -12s - 18t = -15$$
$$\overline{0 = 9}$$

The system is *inconsistent*.

There are no solutions.

37. $12x - 6y = -15$ (1)
     $-4x + 2y = 5$ (2)

To eliminate x, multiply (2)

by 3 and add :

$$12x - 6y = -15$$
$$+\ -12x + 6y = 15$$
$$\overline{0 = 0}$$

The system is *consistent*

and has infinitely many

solutions. Using (2), the

solutions are:

$$\{(x, y) | -4x + 2y = 5\}$$

39. *Familiarize*. Let x and y represent the
     unknown numbers.
     Translate. The sum of the two numbers is 89:
     $x + y = 89$ (1)
     One number is 3 more than the other:
     $y = x + 3$ (2)
     Carry out. Solve by substituting (2) into (1).
     $$x + (x + 3) = 89$$
     $$2x + 3 = 89$$
     $$2x = 86$$
     $$x = 43$$
     Substitute $x = 43$ into (2).
     $$y = x + 3 = 43 + 3 = 46$$
     Check. Substitute $x = 43$ and $y = 46$ into (1).
     $$x + y = 89$$
     $$\overline{43 + 46 \,|\, 89}$$
     $$89 \overset{?}{=} 89 \quad \text{TRUE}$$
     Substitute $x = 43$ and $y = 46$ into (2).

$y = x + 3$

$$\overline{46 \,\big|\, 43 + 3}$$

$$\overset{?}{46 = 46} \quad \text{TRUE}$$

State. The numbers are 43 and 46.

41. *Familiarize*. Let x and y represent the measures of the two angles.

Translate. The two angles are supplementary:
$x + y = 180$ (1)

One angle is 33° more than twice the other:
$x = 2y + 33$ (2)

Carry out. Solve the system by substituting (2) into (1):

$(2y + 33) + y = 180$

$3y + 33 = 180$

$3y = 147$

$y = 49$

Substitute $y = 49$ into (2):

$x = 2 \cdot 49 + 33$

$x = 98 + 33 = 131$

Check. Substitute $x = 131$ and $y = 49$ into (1).

$x + y = 180$

$$\overline{131 + 49 \,\big|\, 180}$$

$$\overset{?}{180 = 180} \quad \text{TRUE}$$

Substitute $x = 131$ and $y = 49$ into (2).

$x = 2y + 33$

$$\overline{131 \,\big|\, 2 \cdot 49 + 33}$$

$$131 \,\big|\, \quad 98 + 33$$

$$\overset{?}{131 = 131} \quad \text{TRUE}$$

State. The angle measures are 131° and 49°.

43. *Familiarize*. Let x and y represent the numbers of two-point and three-point shots, respectively.

Translate. The total number of made shots is 36:

$x + y = 36$ (1)

The total points scored from the made shots is 76:

$2x + 3y = 76$ (2)

Carry out. To eliminate x, multiply (1) by –2 and add:

$$-2x - 2y = -72$$

$$\underline{+ \quad 2x + 3y = \quad 76}$$

$$y = \quad 4$$

Substitute $y = 4$ into (1) and solve:

$x + 4 = 36$

$x = 32$

Check. Substitute $x = 32$ and $y = 4$ into (1).

$x + y = 36$

$$\overline{32 + 4 \,\big|\, 36}$$

$$\overset{?}{36 = 36} \quad \text{TRUE}$$

Substitute $x = 32$ and $y = 4$ into (2).

$2x + 3y = 76$

$$\overline{2 \cdot 32 + 3 \cdot 4 \,\big|\, 76}$$

$$64 + 12 \,\big|\, 76$$

$$\overset{?}{76 = 76} \quad \text{TRUE}$$

State. The Spurs scored 32 2-point shots and 4 3-point shots.

45. *Familiarize*. Let g and p represent the numbers of group and private lessons, respectively.

Translate. There were 14 students:

$g + p = 14$ (1)

Group lessons are \$12 and private lessons are \$20, and Jean collected \$216:

$12g + 20p = 216$ (2)

Carry out. To eliminate g, multiply (1) by –12 and add:

$$-12g - 12p = -168$$

$$\underline{+ \quad 12g + 20p = \quad 216}$$

$$8p = \quad 48$$

$$p = 6$$

Substitute $p = 6$ into (1) and solve:

$g + 6 = 14$

$g = 8$

Check. Substitute $g = 8$ and $p = 6$ into (1).

$g + p = 14$

$$\overline{8 + 6 \,\big|\, 14}$$

$$\overset{?}{14 = 14} \quad \text{TRUE}$$

Substitute $g = 8$ and $p = 6$ into (2).

$$12g + 20p = 216$$

$$\frac{12\cdot 8 + 20\cdot 6 \,|\, 216}{96 + 120 \,|\, 216}$$

$$\overset{?}{216} = 216 \quad \text{TRUE}$$

State. There were 6 private lessons and 8 group lessons.

47. **Familiarize**. Let p and b represent the numbers of pounds of peanuts and Brazil nuts, respectively.
Translate. The total number of pounds is 480:

$$p + b = 480 \quad (1)$$

Peanuts cost $3.20 per pound and Brazil nuts cost $6.40 per pound, and the total value of both nuts, once mixed, is ($5.50/lb)(480 lb) = $2640.00:

$$3.20p + 6.40b = 2640 \quad (2)$$

Carry out. Solve (1) for p, $p = 480 - b$, then substitute into (2) and solve:

$$3.20(480 - b) + 6.40b = 2640$$

$$1536 - 3.20b + 6.40b = 2640$$

$$3.20b = 1104$$

$$b = \frac{1104}{3.20} = 345$$

Substitute $b = 345$ into $p = 480 - b$ to get
$p = 480 - 345 = 135$
Check. Substitute $p = 135$ and $b = 345$ into (1).

$$p + b = 480$$

$$\overline{135 + 345 \,|\, 480}$$

$$\overset{?}{480} = 480 \quad \text{TRUE}$$

Substitute $p = 135$ and $b = 345$ into (2).

$$3.20p + 6.40b = 2640.00$$

$$\frac{3.20\cdot 135 + 6.40\cdot 345 \,|\, 2640.00}{432 + 2208 \,|\, 2640.00}$$

$$\overset{?}{2640.00} = 2640.00 \quad \text{TRUE}$$

State. The mix should contain 135 lb of peanuts and 345 lb of Brazil nuts.

49. **Familiarize**. Let x and y represent the numbers of ounces of Streakfree and Sunstream, respectively, in the final solution.
Translate. The total number of ounces of mixed solution is 90:

$$x + y = 90 \quad (1)$$

The final solution will have (20%)(90 oz) = (0.2)(90) = 18 oz of alcohol. Of this total, $0.12x$ oz will come from the Streakfree (12% alcohol) and $0.30y$ oz will come from the Sunstream (30% alcohol):

$$0.12x + 0.30y = 18 \quad (2)$$

These data are summarized below:

Type of Solution	Streakfree	Sunstream	Mixture
Amount of Solution	x	y	90
Percent Alcohol	12%	30%	20%
Amount of Alcohol in Solution	$0.12x$	$0.3y$	0.2(90) =18

Carry out. Solve (1) for x, $x = 90 - y$, then substitute into (2) and solve:

$$0.12(90 - y) + 0.30y = 18$$

$$10.8 - 0.12y + 0.30y = 18$$

$$10.8 + 0.18y = 18$$

$$0.18y = 7.2$$

$$y = 40$$

Substitute $y = 40$ into $x = 90 - y$ to get
$x = 90 - 40 = 50$
Check. Substitute $x = 50$ and $y = 40$ into (1).

$$x + y = 90$$

$$\overline{50 + 40 \,|\, 90}$$

$$\overset{?}{90} = 90 \quad \text{TRUE}$$

Substitute $x = 50$ and $y = 40$ into (2).

$$0.12x + 0.30y = 18$$

$$\frac{0.12\cdot 50 + 0.30\cdot 40 \,|\, 18}{6 + 12 \,|\, 18}$$

$$\overset{?}{18} = 18 \quad \text{TRUE}$$

State. The mixture should be mixed using 50 ounces of Streakfree and 40 ounces of Sunstream.

51. **Familiarize**. Let t represent the travel time and let d represent the combined distance traveled. In t hours, the combined distance is $75t + 88t$. This distance must total 489 km.
Translate.
$d = 75t + 88t$ (1)
$d = 489$ (2)
Carry out. Substitute (2) into (1) and solve.
$489 = 75t + 88t$ (3)
$489 = 163t$
$\frac{489}{163} = t$ or $t = 3$
Check. Substitute $t = 3$ into (3).
$489 = 75t + 88t$
$$\begin{array}{c|c} 489 & 75 \cdot 3 + 88 \cdot 3 \\ \hline 489 & 225 + 264 \end{array}$$
$489 \stackrel{?}{=} 489$ TRUE
State. The cars will be 489 km apart at 3 hours.

53. **Familiarize**. Let a represent the airspeed of the plane and let w represent the speed of the headwind. The total distance is 2900 mi. Against the headwind w, the speed of the plane is $a - w$ and the travel time is 5 hours. Against the headwind $w/2$, the travel time is 4 hr 50 min $= \frac{29}{6}$ hr and the speed of the plane relative to the ground is $a - w/2$. For either trip, $d = r \cdot t$.
Translate.
With a headwind of w:
$2900 = (a - w) \cdot 5$ (1)
With a headwind of $w/2$:
$2900 = \left(a - \frac{w}{2}\right) \cdot \frac{29}{6}$ (2)
Carry out. Solve (2) for a.
$2900 \cdot \frac{6}{29} = a - \frac{w}{2}$
$600 = a - \frac{w}{2}$
$600 + \frac{w}{2} = a$ (3)
Substitute (3) into (1) and solve.

$2900 = \left(\left(600 + \frac{w}{2}\right) - w\right) \cdot 5$
$2900 = \left(600 - \frac{w}{2}\right) 5$
$580 = 600 - \frac{w}{2}$
$-20 = -\frac{w}{2}$
$40 = w$
Substitute $w = 40$ into (3) and solve.
$a = 600 + \frac{40}{2} = 600 + 20 = 620$
Check. With the headwind of speed w:
$2900 = (a - w) \cdot 5$
$$\begin{array}{c|c} 2900 & (620 - 40) \cdot 5 \\ \hline 2900 & 580 \cdot 5 \end{array}$$
$2900 \stackrel{?}{=} 2900$ TRUE
With the headwind of speed $w/2$:
$2900 = \left(a - \frac{w}{2}\right) \cdot \frac{29}{6}$
$$\begin{array}{c|c} 2900 & \left(620 - \frac{40}{2}\right) \cdot \frac{29}{6} \\ \hline 2900 & 600 \cdot \frac{29}{6} \end{array}$$
$2900 \stackrel{?}{=} 2900$ TRUE
State. The speed of the plane in still air is $a = 620$ mph and the speed of the headwind is $w = 40$ mph.

55. Set $f(x) = 0$ and solve for x.
$f(x) = 7 - x = 0$
$7 = x$

57. Set $f(x) = 0$ and solve for x.
$f(x) = 2x + 9 = 0$
$2x = -9$
$x = -\frac{9}{2}$

59. $2x - 5 = 3x + 1$

We use the Zero method, first getting 0 on one side of the equation.

$2x - 5 = 3x + 1$

$-x - 5 = 1$ Subtracting $3x$ from both sides

$-x - 6 = 0$ Subtracting 1 from both sides

We let $f(x) = -x - 6$, and find the zero of the function by locating the x-intercept of its graph.

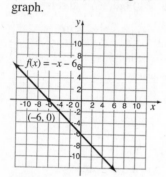

Since $f(-6) = 0$, the zero of the function is –6. We check –6 in the original equation.

$2x - 5 = 3x + 1$

$$
\begin{array}{c|c}
2(-6) - 5 & 3(-6) + 1 \\
-12 - 5 & -18 + 1 \\
\end{array}
$$

$$\overset{?}{-17 = -17} \qquad \text{TRUE}$$

The solution is –6.

61. $3x = x + 6$

We use the Zero method, first getting 0 on one side of the equation.

$3x = x + 6$

$2x = 6$ Subtracting x from both sides

$2x - 6 = 0$ Subtracting 6 from both sides

We let $f(x) = 2x - 6$, and find the zero of the function by locating the x-intercept of its graph.

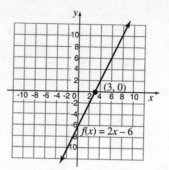

Since $f(3) = 0$, the zero of the function is 3. We check 3 in the original equation.

$3x = x + 6$

$$\overline{3(3) \mid 3 + 6}$$

$$\overset{?}{9 = 9} \qquad \text{TRUE}$$

The solution is 3.

63. $5x + 1 = 11x - 3$

We graph two equations, $y_1 = 5x + 1$ and $y_2 = 11x - 3$. Using the INTERSECT feature, we see that the point of intersection is $(0.666667, 4.333333)$, or $\left(\frac{2}{3}, 4\frac{1}{3}\right)$. The solution is 0.666667, or $\frac{2}{3}$.

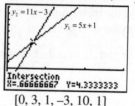

$[0, 3, 1, -3, 10, 1]$

Exercise Set R.5

1. For all real $a \neq 0$, $a^0 = 1$. So, $(-25)^0 = 1$.

3. $4^0 - 4^1 = 1 - 4 = -3$

5. $8^{-2} = \dfrac{1}{8^2} = \dfrac{1}{64}$

7. $10x^{-5} = \dfrac{10}{x^5}$

9. $(ab)^{-2} = \dfrac{1}{(ab)^2}$

11. $\dfrac{1}{y^{-10}} = \dfrac{1}{\left(\dfrac{1}{y^{10}}\right)} = 1 \cdot \dfrac{y^{10}}{1} = y^{10}$

13. $\dfrac{1}{y^4} = y^{-4}$

15. $\dfrac{1}{x^t} = x^{-t}$

17. $x^5 \cdot x^8 = x^{5+8} = x^{13}$

19. $\dfrac{a}{a^{-5}} = a^{1-(-5)} = a^{1+5} = a^6$

21. $\dfrac{(4x)^{10}}{(4x)^2} = (4x)^{10-2} = (4x)^8$

23. $\left(7^8\right)^5 = 7^{8\cdot5} = 7^{40}$

25. $\left(x^{-2}y^{-3}\right)^{-4} = \left(x^{-2}\right)^{-4} \cdot \left(y^{-3}\right)^{-4}$
$= x^{(-2)\cdot(-4)} \, y^{(-3)\cdot(-4)}$
$= x^8 y^{12}$

27. $\left(\dfrac{y^2}{4}\right)^3 = \dfrac{\left(y^2\right)^3}{4^3} = \dfrac{y^{2\cdot3}}{64} = \dfrac{y^6}{64}$

29. $\left(\dfrac{2p^3}{3q^4}\right)^{-2} = \dfrac{\left(2p^3\right)^{-2}}{\left(3q^4\right)^{-2}}$
$= \dfrac{\left(3q^4\right)^2}{\left(2p^3\right)^2}$
$= \dfrac{3^2 q^{4\cdot2}}{2^2 p^{3\cdot2}}$
$= \dfrac{9q^8}{4p^6}$

31. $8x^3, -6x^2, x, -7$

33. $18x^3 + 36x^9 - 7x + 3$
Coefficients (for terms from left to right): 18, 36, –7, 3
Degrees (for terms from left to right): 3, 9, 1, 0
Degree of polynomial: 9

35. $-x^2y + 4y^3 - 2xy$
Coefficients (for terms from left to right): –1, 4, –2
Degrees (for terms from left to right): 3, 3, 2
Degree of polynomial: 3

37. Leading term: $8p^4$; leading coefficient: 8

39. $3x^3 - x^2 + x^4 + x^2 = x^4 + 3x^3 - x^2 + x^2$
$= x^4 + 3x^3$

41. $3 - 2t^2 + 8t - 3t - 5t^2 + 7$
$= -2t^2 - 5t^2 + 8t - 3t + 3 + 7$
$= -7t^2 + 5t + 10$

43. $3(-2)^2 - 7(-2) + 10 = 3 \cdot 4 + 14 + 10 = 36$

45. $a^2b^3 + 2b^2 - 6a = (2)^2(-1)^3 + 2(-1)^2 - 6(2)$
$= 4(-1) + 2(1) - 12$
$= -14$

47. $s = 16(3)^2 = 144$ feet

49. $f(3) = 3^2 - 3 + 5$
$= 9 - 3 + 5$
$= 6 + 5 = 11$

51. $p(-1) = -(-1)^3 - (-1)^2 + 4(-1)$
$= -(-1) - (1) - 4$
$= 1 - 1 - 4 = -4$

53. $\left(3x^3 + 2x^2 + 8x\right) + \left(x^3 - 5x^2 + 7\right)$
$= 3x^3 + x^3 + 2x^2 - 5x^2 + 8x + 7$
$= 4x^3 - 3x^2 + 8x + 7$

55. $\left(8y^2-2y-3\right)-\left(9y^2-7y-1\right)$

$\quad = 8y^2-9y^2-2y+7y-3+1$

$\quad = -y^2+5y-2$

57. $\left(-x^2y+2y^2+y\right)-\left(3y^2+2x^2y-7y\right)$

$\quad = -x^2y-2x^2y+2y^2-3y^2+y+7y$

$\quad = -3x^2y-y^2-8y$

59. $4x^2\left(3x^3-7x+7\right)$

$\quad = 4x^2\left(3x^3\right)+4x^2\left(-7x\right)+4x^2\left(7\right)$

$\quad = 12x^5-28x^3+28x^2$

61. $\left(2a+y\right)\left(4a+b\right)$

$\quad = \left(2a\right)\left(4a\right)+\left(2a\right)b+y\left(4a\right)+yb$

$\quad = 8a^2+2ab+4ay+by$

63. $\left(x+7\right)\left(x^2-3x+1\right)$

$\quad = x\left(x^2-3x+1\right)+7\left(x^2-3x+1\right)$

$\quad = x^3-3x^2+x+7x^2-21x+7$

$\quad = x^3+4x^2-20x+7$

65. $\left(x+7\right)\left(x-7\right)=x^2-7^2$

$\quad$ (Difference of squares)

$\quad = x^2-49$

67. $\left(x+y\right)^2=x^2+2xy+y^2$

$\quad$ (Square of a binomial)

69. $\left(2x^2+7\right)\left(3x^2-2\right)$

$\quad = \left(2x^2\right)\left(3x^2-2\right)+7\left(3x^2-2\right)$

$\quad = 6x^4-4x^2+21x^2-14$

$\quad = 6x^4+17x^2-14$

71. $\left(a-3b\right)^2=\left(a\right)^2-2\left(a\right)\left(3b\right)+\left(3b\right)^2$

$\quad$ (Square of a binomial)

$\quad = a^2-6ab+9b^2$

73. $\left(6a-5y\right)\left(7a+3y\right)$

$\quad = 6a\left(7a+3y\right)-5y\left(7a+3y\right)$

$\quad = 42a^2+18ay-35ay-15y^2$

$\quad = 42a^2-17ay-15y^2$

75. $\dfrac{3t^5+9t^3-6t^2+15t}{-3t}$

$\quad = \dfrac{3t^5}{-3t}+\dfrac{9t^3}{-3t}+\dfrac{-6t^2}{-3t}+\dfrac{15t}{-3t}$

$\quad = -t^4-3t^2+2t-5$

Check:

$\left(-t^4-3t^2+2t-5\right)\left(-3t\right)$

$\quad = 3t^5+9t^3-6t^2+15t$

77.
$$\begin{array}{r}
5x+3 \\
3x-5\overline{\smash{)}15x^2-16x-15} \\
\underline{15x^2-25x} \\
9x-15 \\
\underline{9x-15} \\
0
\end{array}$$

Quotient: $5x+3$

Check:

$\left(3x-5\right)\left(5x+3\right)=3x\left(5x+3\right)-5\left(5x+3\right)$

$\quad = 15x^2+9x-25x-15$

$\quad = 15x^2-16x-15$

79.
$$\begin{array}{r}
2x^2-3x+3 \\
x+1\overline{\smash{)}2x^3-x^2+0x+1} \\
\underline{2x^3+2x^2} \\
-3x^2+0x \\
\underline{-3x^2-3x} \\
3x+1 \\
\underline{3x+3} \\
-2
\end{array}$$

Quotient: $2x^2-3x+3+\dfrac{-2}{x+1}$

Check:

$$\left(x+1\right)\left(2x^2-3x+3+\frac{-2}{x+1}\right)$$

$$=\left(x+1\right)\left(2x^2\right)+\left(x+1\right)\left(-3x\right)$$

$$\qquad+\left(x+1\right)(3)+\left(x+1\right)\left(\frac{-2}{x+1}\right)$$

$$=2x^3+2x^2-3x^2-3x+3x+3-2$$

$$=2x^3-x^2+1$$

81.

$$\begin{array}{r} 5x + 3 \\ x^2+0x-1\overline{)5x^3 + 3x^2 - 5x + 0} \\ \underline{5x^3 + 0x^2 - 5x} \\ 3x^2 + 0x + 0 \\ \underline{3x^2 - 0x - 3} \\ 3 \end{array}$$

Quotient: $5x+3+\dfrac{3}{x^2-1}$

Check:

$$\left(x^2-1\right)\left(5x+3+\frac{3}{x^2-1}\right)$$

$$=\left(x^2-1\right)(5x)+\left(x^2-1\right)(3)$$

$$\qquad+\left(x^2-1\right)\left(\frac{3}{x^2-1}\right)$$

$$=5x^3-5x+3x^2-3+3$$

$$=5x^3+3x^2-5x$$

Exercise Set R.6

1. $18t^5-12t^4+6t^3=6t^3\left(3t^2-2t+1\right)$

3. $y^2-6y+9=(y-3)(y-3)$
    $$=(y-3)^2$$

5. $2p^3\left(p+2\right)+\left(p+2\right)=\left(p+2\right)\left(2p^3+1\right)$

7. x^2+100
    Prime

9. $8t^3+27=\left(2t\right)^3+3^3=\left(2t+3\right)\left(4t^2-6t+9\right)$

11. $m^2+13m+42=\left(m+6\right)\left(m+7\right)$

13. $x^4-81=\left(x^2+9\right)\left(x^2-9\right)$
    $$=\left(x^2+9\right)(x+3)(x-3)$$

15. $8x^2+22x+15=\left(2x+3\right)\left(4x+5\right)$

17. $x^3+2x^2-x-2=x^3+2x^2-\left(x+2\right)$
    $$=x^2\left(x+2\right)-\left(x+2\right)$$
    $$=\left(x+2\right)\left(x^2-1\right)$$
    $$=\left(x+2\right)(x+1)(x-1)$$

19. $0.001t^6-0.008=\left(0.1t^2\right)^3-\left(0.2\right)^3$
    $$=\left(0.1t^2-0.2\right)\left[\left(0.1t^2\right)^2+\left(0.1t^2\right)(0.2)+\left(0.2\right)^2\right]$$
    $$=\left(0.1t^2-0.2\right)\left(0.01t^4+0.02t^2+0.04\right)$$

21. $-\frac{1}{16}+x^4=x^4-\frac{1}{16}=\left(x^2+\frac{1}{4}\right)\left(x^2-\frac{1}{4}\right)$
    $$=\left(x^2+\frac{1}{4}\right)\left(x+\frac{1}{2}\right)\left(x-\frac{1}{2}\right)$$

23. $mn-2m+3n-6=m\left(n-2\right)+3\left(n-2\right)$
    $$=\left(m+3\right)\left(n-2\right)$$

25. $5mn+m^2-150n^2=m^2+5mn-150n^2$
    $$=\left(m+15n\right)\left(m-10n\right)$$

27. $24x^2y-6y-10xy=2y\left(12x^2-3-5x\right)$
    $$=2y\left(12x^2-5x-3\right)$$
    $$=2y\left(3x+1\right)\left(4x-3\right)$$

29. $y^2-121=\left(y+11\right)\left(y-11\right)$

31. $\left(x-2\right)\left(x+7\right)=0$
    $x-2=0$ *or* $x+7=0$
    $x=2$ *or* $x=-7$
    The solutions are –7 and 2.

33. $8x\left(11-x\right)=0$
    $8x=0$ *or* $11-x=0$
    $x=0$ *or* $11=x$
    The solutions are 0 and 11.

35. $x^2 = 100$

$x^2 - 100 = 0$

$(x+10)(x-10) = 0$

$x+10 = 0 \ or \ x-10 = 0$

$x = -10 \ or \ x = 10$

The solutions are –10 and 10.

37. $4x^2 - 18x = 70$

$4x^2 - 18x - 70 = 0$

$2x^2 - 9x - 35 = 0$

$(2x+5)(x-7) = 0$

$2x+5 = 0 \ or \ x-7 = 0$

$2x = -5 \ or \ x = 7$

$x = -\frac{5}{2} \ or \ x = 7$

The solutions are $-\frac{5}{2}$ and 7.

39. $2x^3 - 10x^2 = 0$

$x^3 - 5x^2 = 0$

$x^2(x-5) = 0$

$x = 0 \ or \ x-5 = 0$

$x = 0 \ or \ x = 5$

The solutions are 0 and 5.

41. $(a+1)(a-5) = 7$

$a^2 - 5a + a - 5 = 7$

$a^2 - 4a - 5 = 7$

$a^2 - 4a - 5 - 7 = 0$

$a^2 - 4a - 12 = 0$

$(a+2)(a-6) = 0$

$a+2 = 0 \ or \ a-6 = 0$

$a = -2 \ or \ a = 6$

The solutions are –2 and 6.

43. $x^2 + 6x - 55 = 0$

$(x+11)(x-5) = 0$

$x+11 = 0 \ or \ x-5 = 0$

$x = -11 \ or \ x = 5$

The solutions are –11 and 5.

45. $\frac{1}{2}x^2 + 5x + \frac{25}{2} = 0$

$x^2 + 10x + 25 = 0$

$(x+5)(x+5) = 0$

$x+5 = 0 \ or \ x+5 = 0$

$x = -5 \ or \ x = -5$

The solution is –5.

47. ***Familiarize.*** Let h = the height of the triangular fountain, in feet. Since the base of the triangle is 2 ft longer than the height, the base is $h + 2$ ft. Recall that the area of a triangle is given by the formula $A = \frac{1}{2} \cdot base \cdot height$.

Translate. $\frac{1}{2}(h+2)h = 24$

Solve.

$\frac{1}{2}(h+2)h = 24$

$(h+2)h = 48$

$h^2 + 2h = 48$

$h^2 + 2h - 48 = 0$

$(h+8)(h-6) = 0$

$h+8 = 0 \ or \ h-6 = 0$

$h = -8 \ or \ h = 6$

Since length cannot be negative, we discard the –8 and retain the 6 as the height. If the height is 6 ft, then the base is 6 + 2, or 8 ft.
Check. The area of a triangle with height 6 and base 8 is $\frac{1}{2}(8)(6) = 24$ ft^2. This checks.
State. The base of the rectangular garden is 8 ft and the height is 6 ft.

49. ***Familiarize.*** Let x = the length of one leg and $2x - 1$ = the length of the other leg. Note that the Pythagorean Theorem states that $a^2 + b^2 = c^2$, where a and b represent the lengths of the legs of a right triangle and c is the length of the hypotenuse. We make a sketch.

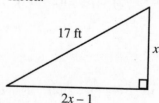

Translate.

$x^2 + (2x-1)^2 = 17^2$

Solve.

$x^2 + (2x-1)^2 = 17^2$

$x^2 + 4x^2 - 4x + 1 = 289$

$5x^2 - 4x + 1 = 289$

$5x^2 - 4x - 288 = 0$

$(x-8)(5x+36) = 0$

$x - 8 = 0 \ or \ 5x + 36 = 0$

$x - 8 = 0 \ or \ 5x = -36$

$x = 8 \ or \ x = -\frac{36}{5}$

Since length cannot be negative, we discard the $-\frac{36}{5}$ and retain 8 as the solution. If one leg is 8 ft, the other leg is $2 \cdot 8 - 1$, or 15 ft.

Check. Since the hypotenuse is 17 ft, we can use the Pythagorean Theorem to check the result.

$8^2 + 15^2 = 64 + 225 = 289$

$17^2 = 289$

This result checks.

State. The length of the legs are 8 ft and 15 ft.